Properties of Real Numbers

Associative properties:
of addition $(a + b) + c = a + (b + c)$
of multiplication $(ab)c = a(bc)$

Commutative properties:
of addition $a + b = b + a$
of multiplication $ab = ba$

Distributive Property:
$a(b + c) = ab + ac$

Properties of Radicals

If all radicals are real numbers and there are no divisions by 0, then

$$\sqrt[n]{ab} = \sqrt[n]{a}\,\sqrt[n]{b}$$

$$\sqrt[n]{\frac{a}{b}} = \frac{\sqrt[n]{a}}{\sqrt[n]{b}}$$

$$\sqrt[m]{\sqrt[n]{a}} = \sqrt[n]{\sqrt[m]{a}} = \sqrt[mn]{a}$$

Properties of Equality

If $a = b$ and c is a number, then

$$a + c = b + c \quad \text{and} \quad a - c = b - c$$

$$ac = bc \quad \text{and} \quad \frac{a}{c} = \frac{b}{c} \quad (c \neq 0)$$

Quadratic Formula

$$x = \frac{-b \pm \sqrt{b^2 - 4ac}}{2a} \quad (a \neq 0)$$

Rules of Exponents

If there are no divisions by 0,

$$x^m x^n = x^{m+n} \qquad (x^m)^n = x^{mn}$$

$$(xy)^n = x^n y^n \qquad \left(\frac{x}{y}\right)^n = \frac{x^n}{y^n}$$

$$x^0 = 1(x \neq 0) \qquad x^{-n} = \frac{1}{x^n}$$

$$\frac{x^m}{x^n} = x^{m-n} \qquad \left(\frac{x}{y}\right)^{-n} = \left(\frac{y}{x}\right)^n$$

Factoring Formulas

Factoring the difference of two squares:
$$x^2 - y^2 = (x + y)(x - y)$$

Factoring trinomial squares:
$$x^2 + 2xy + y^2 = (x + y)^2$$
$$x^2 - 2xy + y^2 = (x - y)^2$$

Factoring the sum and difference of two cubes:
$$x^3 + y^3 = (x + y)(x^2 - xy + y^2)$$
$$x^3 - y^3 = (x - y)(x^2 + xy + y^2)$$

The Distance Formula

The distance d between points (x_1, y_1) and (x_2, y_2) is given by
$$= \quad x_2 - x_1)^2 + (y_2 - y_1)^2$$

Midpoint Formula

The midpoint of the line segment joining (x_1, y_1) and (x_2, y_2) is the point M with coordinates
$$M = \left(\frac{x_1 + x_2}{2}, \frac{y_1 + y_2}{2}\right)$$

Properties of Inequalities

If a, b, and c are real numbers:

If $a < b$, then $a + c < b + c$ and $a - c < b - c$.

If $a < b$ and $c > 0$, then $ac < bc$ and $\dfrac{a}{c} < \dfrac{b}{c}$.

If $a < b$ and $c < 0$, then $ac > bc$ and $\dfrac{a}{c} > \dfrac{b}{c}$.

Trichotomy Property:
$a < b$, $a = b$, or $a > b$

Transitive Property:
If $a < b$ and $b < c$, then $a < c$.

Slope

The slope of the nonvertical line passing through points $P(x_1, y_1)$ and $Q(x_2, y_2)$ is

$$m = \frac{\text{change in } y}{\text{change in } x} = \frac{y_2 - y_1}{x_2 - x_1} \quad (x_2 \neq x_1)$$

Equations of Lines

Point-slope form:
An equation of the line passing through $P(x_1, y_1)$ and with slope m is
$$y - y_1 = m(x - x_1)$$

Slope-intercept form:
An equation of the line with slope m and y-intercept $(0, b)$ is
$$y = mx + b$$

Standard form of an equation of a line:
$$Ax + By = C$$

General form of an equation of a line:
$$Ax + By + C = 0$$

Vertical and Horizontal Lines

Equation of a vertical line through (a, b):
$$x = a$$

Equation of a horizontal line through (a, b):
$$y = b$$

Slopes of horizontal and vertical lines:
- The slope of a horizontal line (a line with an equation of the form $y = b$) is 0.
- The slope of a vertical line (a line with an equation of the form $x = a$) is not defined.

Definition of Absolute Value

$$|x| = \begin{cases} x \text{ when } x \geq 0 \\ -x \text{ when } x < 0 \end{cases}$$

Absolute Value Equations
- If $k \geq 0$, then $|x| = k$ is equivalent to $x = k$ or $x = -k$.
- If a and b are algebraic expressions, $|a| = |b|$ is equivalent to $a = b$ or $a = -b$.

Absolute Value Inequalities
- If $k > 0$, then $|x| < k$ is equivalent to $-k < x < k$.
- If $k > 0$, then $|x| > k$ is equivalent to $x > k$ or $x < -k$.

These two properties hold for $\leq$ and $\geq$ also.

Circles

The standard form of an equation of a circle with center (h, k) and radius r:
$$(x - h)^2 + (y - k)^2 = r^2$$

The standard form of an equation of a circle with center $(0, 0)$ and radius r:
$$x^2 + y^2 = r^2$$

Quadratic Functions
A **quadratic function** is a second-degree polynomial function in one variable of the form

$$f(x) = ax^2 + bx + c \text{ or } y = ax^2 + bx + c,$$

where a, b, and c are real numbers and $a \neq 0$.

The graph of a quadratic function of the form

$$f(x) = ax^2 + bx + c \quad (a \neq 0)$$

is a **parabola** with vertex at

$$\left(-\frac{b}{2a}, c - \frac{b^2}{4a} \right)$$

- If $a > 0$, the parabola **opens upward**.
- If $a < 0$, the parabola **opens downward**.

The **standard form of an equation of a quadratic function** is

$$y = f(x) = a(x - h)^2 + k \quad (a \neq 0)$$

The vertex is at (h, k).
- The parabola opens upward when $a > 0$ and downward when $a < 0$.
- The axis of symmetry of the parabola is the vertical line graph of the equation $x = h$.

Tests for Symmetry

Test for x-axis symmetry:
To test for x-axis symmetry, replace y with $-y$. If the resulting equation is equivalent to the original one, the graph is symmetric about the x-axis.

Test for y-axis symmetry:
To test for y-axis symmetry, replace x with $-x$. If the resulting equation is equivalent to the original one, the graph is symmetric about the y-axis.

Test for origin symmetry:
To test for symmetry about the origin, replace x with $-x$ and y with $-y$. If the resulting equation is equivalent to the original one, the graph is symmetric about the origin.

Graphs of Common Functions

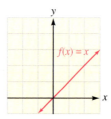
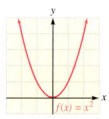
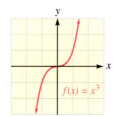
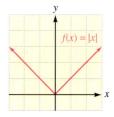
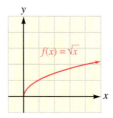
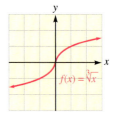

Greatest-Integer Function

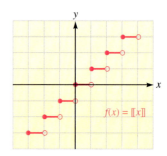

Interest Formulas

Compound interest formula:
If P dollars are deposited in an account earning interest at an annual rate r, compounded n times each year, the amount A in the account after t years is given by

$$A = P\left(1 + \frac{r}{n} \right)^{nt}$$

Continuous compound interest formula:
If P dollars are deposited in an account earning interest at an annual rate r, compounded continuously, the amount A after t years is given by the formula

$$A = Pe^{rt}$$

College Algebra

Twelfth Edition

R. David Gustafson
Rock Valley College

Jeffrey D. Hughes
Hinds Community College

CENGAGE
Learning·

Australia · Brazil · Mexico · Singapore · United Kingdom · United States

College Algebra, Twelfth Edition
R. David Gustafson, Jeffrey D. Hughes

Product Director: Terry Boyle

Product Manager: Gary Whalen

Content Developer: Stacy Green

Associate Content Developer: Samantha Lugtu

Product Assistant: Katharine Werring

Media Developer: Lynh Pham

Marketing Manager: Mark Linton

Content Project Manager: Jennifer Risden

Art Director: Vernon Boes

Manufacturing Planner: Becky Cross

Production Service: MPS Limited

Photo Researcher: Lumina Datamatics

Text Researcher: Lumina Datamatics

Copy Editor: Patricia Daley

Illustrator: Lori Heckelman; MPS Limited

Text Designer: Diane Beasley

Cover Designer: Terri Wright

Cover Image: ©Seth deRoulet

Compositor: MPS Limited

For product information and technology assistance, contact us at **Cengage Learning Customer & Sales Support, 1-800-354-9706.**

For permission to use material from this text or product, submit all requests online at **www.cengage.com/permissions**. Further permissions questions can be e-mailed to **permissionrequest@cengage.com**.

Library of Congress Control Number: 2015935653

Student Edition
ISBN: 978-1-305-65223-1

Loose-leaf Edition
ISBN: 978-1-305-94504-3

Cengage Learning
20 Channel Center Street
Boston, MA 02210
USA

Cengage Learning is a leading provider of customized learning solutions with employees residing in nearly 40 different countries and sales in more than 125 countries around the world. Find your local representative at **www.cengage.com**.

Cengage Learning products are represented in Canada by Nelson Education, Ltd.

To learn more about Cengage Learning Solutions, visit **www.cengage.com**.

Purchase any of our products at your local college store or at our preferred online store **www.cengagebrain.com**.

Printed in the United States of America
Print Number: 01   Print Year: 2015

Contents

Chapter 7

©Shots Studio/Shutterstock.com

Conic Sections and Quadratic Systems   697

Chapter 8

©Andrey_Popov/Shutterstock.com

Sequences, Series, and Probability   769

Preface

To the Instructor

It is with great delight that we present the twelfth edition of *College Algebra*. This revised edition maintains the same philosophy of the highly successful previous editions but is enhanced to meet the current expectations of today's students and instructors. An instructional experience for students and instructors is provided to increase problem-solving skills while at the same time preparing students for success in trigonometry, calculus, statistics, and other disciplines of study. The textbook has been revised for greater clarity of design and instruction.

Our goal is to write a textbook that

- presents solid mathematics written in way that is easy to understand for students with diverse abilities, backgrounds, and future career goals;
- provides a strong instructional experience that will help students master algebra and that emphasizes the important concept of a function;
- motivates learning by using popular culture and real-life applications;
- improves critical-thinking abilities in all students;
- develops algebra skills needed for future success in mathematics courses.

> **Tip**
>
> As you begin to learn the graphs of common functions, often referred to as parent or basic functions, you should place these in your library, that is, your library of functions. It's simply a collection of important functions. As you continue to learn mathematics, this collection will grow.

©Angela Waye/Shutterstock.com

> **Take Note**
>
> Some equations represent functions and some do not. If for some input x in an equation there corresponds more than one output y, the equation will not represent a function.

> **Critical Thinking**
>
> *Determine if the statement is true or false. If the statement is false, then correct it and make it true.*
>
> **89.** All functions have inverses.
>
> **90.** If $f(x) = x^3 + 7$, then $f^{-1}(x) = \dfrac{1}{x^3 + 7}$.

We believe that we have accomplished this goal through a successful blending of content and pedagogy. We present thorough coverage of classic college algebra topics, and we incorporate these topics into a contemporary framework of tested teaching strategies. We provide features that are written to appeal to both students and instructors. This book emphasizes conceptual understanding, problem solving, and the appropriate use of technology.

New Features

- **Tip Boxes to Provide Helpful Strategies for Studying and Learning Content** Tip boxes appear throughout the textbook, and many of these boxes provide a visual representation of the tip to help students remember the tip.

- **Take Note Boxes to Provide Additional Insights into Specific Content** These noteworthy statements highlight important information in the book. The statements often provide clarification on a specific step or concept in an example and also emphasize theory, which is sometimes overlooked.

- **Critical Thinking Exercises to Encourage Algebraic Thinking and to Assess Conceptual Understanding** Exercise sets now include Critical Thinking exercises. These exercises are primarily true–false statements. If a statement is false, students are then asked to make the necessary change to produce a true statement. Matching exercises are also used.

Continued and Updated Features

We have kept and updated the pedagogical features that made the previous editions of the book so successful.

- **Student-Friendly Writing Style to Alleviate Student Anxiety about Reading and Understanding a Mathematics Textbook** The exposition is clear, concise, and reader friendly. The writing level is informal yet accurate. Students and instructors alike should find the reading both interesting and inviting.

Careers and Mathematics: **Cryptanalyst**

Cryptanalysts are essential to the implementation of protection systems employed by corporations and private citizens to keep hackers out of important data systems. They also play an important part in the safety of cities, states, and countries by analyzing and deciphering secret coding systems used by terrorists and other enemies. With increased reliance on the Internet to conduct business, store data, and pass information between groups, the job of a cryptanalyst is becoming increasingly more important to ensure both private and governmental security of data.

4.1 Quadratic Functions

In this section, we will learn to

1. Recognize the characteristics of a quadratic function.
2. Find the vertex of a parabola whose equation is in standard form.
3. Graph a quadratic function.
4. Find the vertex of a parabola whose equation is in general form.
5. Use a quadratic function to solve maximum and minimum problems.

Quadratic functions are important because we can use them to model many real-life problems. For example, the path of a basketball jump shot by LeBron James and the path of a guided missile can be modeled with quadratic functions. Businesses like Coca Cola and Best Buy can use quadratic functions to help maximize the profit and revenue for the products they produce and sell.

1. Recognize the Characteristics of a Quadratic Function

The linear function $f(x) = mx + b$ $(m \neq 0)$ is a first-degree polynomial function, because its right side is a first-degree polynomial in the variable x. A function defined by a polynomial of second degree is called a **quadratic function**.

EXAMPLE 7 · **Evaluating and Graphing a Piecewise-Defined Function**

Use the piecewise-defined function shown to find each of the following:

$$f(x) = \begin{cases} -2 & \text{if } x \leq 0 \\ x + 1 & \text{if } x > 0 \end{cases}$$

a. $f(-3)$ **b.** $f(0)$ **c.** $f(1)$ **d.** $f(2)$ **e.** the graph of $f(x)$

SOLUTION To evaluate the piecewise-defined function, we use the top part of the definition if $x \leq 0$ and the corresponding value of $f(x)$ is -2. We will use the bottom part of the definition if $x > 0$ and the corresponding value of $f(x)$ is $x + 1$. We will then consdier both pieces of the function and sketch its graph.

a. $f(-3) = -2$ -3 is less than or equal to 0. The function's value is -2.
b. $f(0) = -2$ 0 is the less than or equal to 0. The function's value is -2.
c. $f(1) = 1 + 1 = 2$ 1 is greater than 0. Evaluate $f(x) = x + 1$ at 1.
d. $f(2) = 2 + 1 = 3$ 2 is greater than 0. Evaluate $f(x) = x + 1$ at 2.
e. In the interval $(-\infty, 0]$, the function value is always -2 and the graph is the horizontal line $y = -2$. In the interval $(0, \infty)$, the graph of the function is ↓

- **Careers in Mathematics Chapter Openers to Encourage Students to Explore Careers That Use Mathematics and to Make a Connection between Math and Real Life** Each chapter opens with "Careers in Mathematics." New, exciting careers are featured in this edition. These snapshots include information on how professionals use math in their work and who employs them. Most information is taken from the *Occupational Outlook Handbook*. A web address is provided so students can learn more about the career discussed in each of these chapter opener features.

- **Section Openers to Peak Interest and Motivate Students to Read the Material** Each section begins with a contemporary photo and a real-life application.

- **Numbered Objectives to Keep Students Focused** Numbered learning objectives are given at the beginning of each section and appear as subheadings throughout each section.

- **Titled Examples to Clearly Identify the Purpose of Each Example and an Example Structure to Help Students Gain a Deeper Understanding of How to Solve Each Problem** Descriptive titles have been added to example identifiers. Solutions begin with a stated approach. The examples are engaging, and step-by-step solutions with annotations are provided.

- **Self Checks That Actively Reinforce Student Understanding of Concepts and Example Solutions** Each example is followed immediately by a Self Check Exercise. The answers for students are offered at the end of each section. To assist instructors, the answers to Self Checks appear next to the problem in the AIE (Annotated Instructor's Edition), printed in blue.

- **Now Try Exercises to Provide Students an Additional Opportunity to Assess Their Understanding of the Concept Related to Each Worked Example** A reference to an exercise follows all Examples and Self Check problems. These references also show students a correspondence between the examples in the book and the exercise sets.

- **Application Examples to Answer the Student Question: When Will I Ever Use This Math?** Applications from a wide range of disciplines demonstrate how mathematics is used to solve real problems. These applications motivate the material and help students become better problem solvers.

- **Accent on Technology Features to Encourage Students to Become Intelligent Users of Technology and to Present Concepts Graphically**   Accent on Technology boxes appear throughout the textbook. These illustrate and guide the use of a TI-84 graphing calculator for specific problems. Although graphing calculators are incorporated into the book, their use is not required. All graphing topics are fully discussed in traditional ways.

Strategy for Graphing a Quadratic Function	To graph a quadratic function
	1. Determine whether the parabola opens upward or downward.
	2. Find the vertex of the parabola.
	3. Find the x-intercept(s).
	4. Find the y-intercept.
	5. Identify one additional point on the graph.
	6. Draw a smooth curve through the points found in Steps 2–5.

- **Strategy Boxes to Enable Students to Build on Their Mathematical Reasoning and Approach Problems with Confidence**   Strategy boxes offer problem-solving techniques and steps at appropriate points in the material.

- **Caution Boxes to Alert Students to Common Errors and Misunderstandings**   Caution boxes appear throughout the text and reinforce correct mathematics.

Caution

A common mistake students make is to forget to insert 0 if a term is missing. Just as a missing tooth would need to be replaced, a missing term needs to be replaced. We use 0 as a placeholder when using synthetic division.

©BlueSkyImage/Shutterstock.com

- **Getting Ready Exercises to Test Student Understanding of Concepts and Proper Use of Mathematical Vocabulary**   Each problem set begins with "Getting Ready" exercises. Students should be able to answer these fill-in-the blank questions before moving on to the Practice exercises.

- **Comprehensive Exercise Sets to Improve Mathematical Skills and Cement Understanding, and Interesting Applications to Emphasize Problem Solving**   The exercise sets progress from routine to more challenging. The mathematics in each exercise set is sound, but not so rigorous that it will confuse students. All exercise sets include Getting Ready, Practice, Application, Discovery and Writing, and Critical Thinking problems. The book contains more than 4000 exercises, many of them new. New application exercises have been added and others updated. All application problems have titles.

- **Chapter Reviews and Chapter Tests to Give Students the Best Opportunity for Study and Exam Preparation**   Each chapter closes with a Chapter Review and Chapter Test. Chapter Reviews are comprehensive and consist of three parts: definitions and concepts, examples, and review exercises. Chapter Tests cover all the important topics and yet are brief enough to emulate a "real-time" test, so students can practice not only the math but their test-taking aptitude. As an additional quick reference, endpapers offer the important formulas and graphs developed in the book.

- **Cumulative Reviews to Reinforce Student Learning and Improve Students Retention of Concepts**   Cumulative Review Exercises appear after every two chapters. These comprehensive reviews revisit all the essential topics covered in prior chapters.

Content and Organizational Changes

Each chapter and section in the text has been edited to fine-tune the presentation of topics for better flow of concepts and for clarity. There are many new exercises and applications. Key changes made to specific chapters include the following:

Chapter 0: A Review of Basic Algebra
- Section 0.1—The empty set or null set is defined and illustrated.
- Section 0.2—A Tip Box for remembering exponent rules is included.
- Section 0.3—A table showing commonly occurring fourth and fifth roots is added. A Tip Box showing differences in rationalizing square roots and cube roots is provided.
- Section 0.4—A table of special products, showing the formulas and an example of each are provided.
- Section 0.5—Two factoring Tip Boxes are provided to help students remember the signs when factoring.

Chapter 1: Equations and Inequalities
- Section 1.3—This section is titled Complex Numbers; in the previous edition complex numbers were discussed in Section 1.5. Section 1.3 does not include complex solutions to quadratic equations; these are covered in Section 1.4.
- Section 1.4—This section now includes all content about solving quadratic equations; this section also discusses the discriminant. Three new examples—solving a quadratic equation by using the Square Root Property, solving a quadratic

equation by Completing the Square, and solving a quadratic equation using the Quadratic Formula—have been added. These examples have solutions involving *i*. Many exercises previously found in Section 1.5 have been moved to this section.

- Section 1.5—This was Section 1.3 in the previous edition.
- Section 1.6—This section is now titled Other Types of Equations. A new objective, Solving Equations Quadratic in Form, has been added. A new example of this type is included in the section, and the Substitution Method is introduced.

Chapter 2: Function and Graphs

- Section 2.1—This section is titled Functions and Function Notation. It was Section 3.1 in the previous edition. Functions are now introduced at the beginning of Chapter 2. Graphing of nonlinear functions remains in Chapter 3. The concept of a relation is introduced in this section, and four new examples are presented.
- Section 2.2—This was Section 2.1 in the previous edition and has been retitled as The Rectangular Coordinate System and Graphing Lines. A linear function definition box is included.
- Section 2.3—This was Section 2.2 in the previous edition and has been retitled as Linear Functions and Slope. The average rate of change of a function on a close interval is introduced.
- Section 2.4—This was Section 2.3 in the previous edition. A new method for deriving $y = mx + b$ is shown. A linear regression example is included.
- Section 2.5—This was Section 2.4 in the previous edition.
- Section 2.6—This was Section 2.5 in the previous edition.
- Chapter 2 Review and Test—The definition of a relation is now included.

Chapter 3: Functions

- Section 3.1—This section is now titled Graphs of Functions. A new section opener is provided. Determining function values graphically is a section objective. Two new examples have been added to the section: determining function values graphically and finding the domain and the range given the graph of a function. The terminology of *library of functions* or *parent functions* is used in this section.
- Section 3.2—This was Section 3.4 in the previous edition. A new example, applying transformations of graphs, has been added.
- Section 3.3—This section now includes more information on functions; the discussion of polynomial functions has been moved to Chapter 4. A more through explanation of symmetry is given. Five new examples have been added: using a graph to determine if a function is even, odd, or neither; determining algebraically whether a function is even, odd, or neither; determining the open intervals where a function is increasing, decreasing, or constant; using a graph to identify local extrema; and evaluating and graphing piecewise-defined functions.
- Section 3.4—This was Section 3.6 in the previous edition. A new example is provided: evaluating the sum, difference, product, and quotient of two functions.
- Section 3.5—This was Section 3.7 in the previous edition.

Chapter 4: Polynomial and Rational Functions

- Section 4.1—This was Section 3.2 in the previous edition. A new Accent on Technology feature titled Quadratic Regression is included. Students are asked to find the vertex and intercepts when graphing a parabola.
- Section 4.2—This was Section 3.3 in the previous edition and has been rewritten. The Leading Coefficient Test and end behavior are now discussed in this section. There are three new examples: determining whether functions are polynomial functions; using factoring to find the zeros of a polynomial function; and using the Leading Coefficient Test to determine end behavior.
- Section 4.3—This was previously Section 5.1. Material from Chapter 5 of the previous edition has been moved to Chapter 4. The Division Algorithm is formally stated. There is also more emphasis on using zero terminology instead of root.
- Section 4.4—This was previously Section 5.2.
- Section 4.5—This was previously Section 5.3 and is now titled Zeros of Polynomial Functions. The emphasis is on finding zeros of polynomial functions instead of roots of polynomial equations.
- Section 4.6—This was Section 3.5 in the previous edition.
- Chapter 4 Review and Test—A summary of the Leading Coefficient Test and the four cases are now included, as well as the Intermediate Value Theorem.

Chapter 5: Exponential and Logarithmic Functions

Note: Chapter 5 was Chapter 4 in the previous edition.

- Section 5.2—Photos have been added to complement each application example. Two exponential regression exercises have been added.

Chapter 6: Linear Systems
- Section 6.1—The Addition Method is now primarily referred to as the Elimination Method.
- Section 6.2—A Tip Box on how to place a matrix in row-echelon form is provided.
- Section 6.3—There are two new examples in this section: performing scalar multiplication and solving matrix equations.
- Section 6.4—An alternative method for finding the inverse of a 2×2 matrix is shown in a Tip Box.
- Section 6.5—A Tip Box to help students remember the sign pattern for cofactors is provided.

Chapter 8: Sequences, Series, and Probability
- Section 8.2—Two new examples are presented: writing the terms of a sequence and finding a special term in sequence.
- Chapter 8 Review and Test—The nth term or general term or a sequence notation is included.

Organization and Coverage

This text can be used in a variety of ways. To maintain optimum flexibility, many chapters are sufficiently independent to allow instructors to pick and choose topics that are relevant to their students.

Ancillaries for the Instructor

MindTap for Mathematics
Experience matters when you want to improve student success. With MindTap for Mathematics, instructors can:
- Personalize the Learning Path to match the course syllabus by rearranging content or appending original material to the online content
- Improve the learning experience and outcomes by streamlining the student workflow
- Customize online assessments and assignments
- Connect a Learning Management System portal to the online course
- Track student engagement, progress, and comprehension
- Promote student success through interactivity, multimedia, and exercises

Instructors who use a Learning Management System (such as Blackboard, Canvas, or Moodle) for tracking course content, assignments, and grading can seamlessly access the MindTap suite of content and assessments for this course.

Learn more at **www.cengage.com/mindtap**.

Instructor Companion Site
Everything you need for your course in one place! This collection of book-specific lecture and class tools is available online via **www.cengage.com/login**. Access and download PowerPoint presentations, images, instructor's manual, and more.

Cengage Learning Testing Powered by Cognero (978-1-305-88231-7)
CLT is a flexible online system that allows you to author, edit, and manage test bank content; create multiple test versions in an instant; and deliver tests from your LMS, your classroom or wherever you want. This is available online via **www.cengage.com/login**.

Complete Solutions Manual
This manual contains solutions to all exercises from the text, including Chapter Review Exercises, Chapter Tests, and Cumulative Review Exercises. Located on the instructor companion website.

Test Bank
The test bank includes six tests per chapter as well as three final exams. The tests are made up of a combination of multiple-choice, free-response, true/false, and fill-in-the-blank questions. Located on the instructor companion website.

Ancillaries for the Student

MindTap for Mathematics
MindTap for Mathematics is a digital-learning solution that places learning at the center of the experience. In addition to algorithmically generated problems, immediate feedback, and a powerful answer evaluation and grading system, MindTap for Mathematics gives you a personalized path of dynamic assignments, a focused improvement plan, and just-in-time, integrated remediation that turns cookie cutter into cutting edge, apathy into engagement, and memorizers into higher-level thinkers.

Learn more at **www.cengage.com/mindtap**.

CengageBrain.com
To access additional course materials, please visit **www.cengagebrain.com**. At the CengageBrain.com home page, search for the ISBN of your title (from the back cover of your book) using the search box at the top of the page. This will take you to the product page where these resources can be found.

Student Solutions Manual (978-1-305-87874-7)
Go beyond the answers: Students can see what it takes to understand the concepts and improve their grades! This manual provides worked-out, step-by-step solutions to the odd-numbered problems in the text, thus giving students the information they need to understand how these problems are solved.

Text-Specific DVDs (978-1-305-87877-8)
These text-specific instructional videos provide students with visual reinforcement of concepts and explanations in easy-to-understand terms with detailed examples and sample problems. A flexible format offers versatility for quickly accessing topics or catering lectures to self-paced, online, or hybrid courses. Closed captioning is provided for the hearing impaired.

To the Student

Congratulations!! You now own a state-of-art textbook that has been written especially for you. Cowabunga! Are you ready to catch an exciting algebra wave and do some algebraic surfing? An adventure in mathematics truly awaits you. Just as a surfer would need a surfboard to surf and guidance from a patient and knowledgeable surfing instructor, we provide the same for you with the content in this textbook. Easy-to-follow examples are provided, and Tips, Take Notes, and Cautions are shared to help you conquer whatever algebra "waves" come your way. Have no fear, because Definition boxes, Property and Theorem boxes, and Strategy boxes are color coded and stand out easily in the book to help you grasp the algebra vocabulary and concepts and avoid a wipeout. The mathematics is also supported graphically with technology. So what are you waiting for? Are you ready to hit the ocean water and hang ten? If so, grab your favorite pencil, some paper, and your textbook and dive in. You'll soon experience the thrill of riding this college algebra wave successfully. We wish you the very best on your mathematics journey. Have fun learning, stay positive, and keep focused on your career goal.

Acknowledgments

We are grateful to the following people, who reviewed previous editions of the text or the current manuscript in its various stages. All of them provided valuable suggestions that have been incorporated into this book.

Catherine Aguilar-Morgan, New Mexico State University–Alamogordo

Ebrahim Ahmadizadeh, Northampton Community College

Ricardo Alfaro, University of Michigan–Flint

Sue Allen, Michigan State University

Richard Andrews, University of Wisconsin

James Arnold, University of Wisconsin

Ronald Atkinson, Tennessee State University

Wilson Banks, Illinois State University

Chad Bemis, Riverside Community College

Anjan Biswas, Tennessee State College

Jerry Bloomberg, Essex Community College

Elaine Bouldin, Middle Tennessee State University

Dale Boye, Schoolcraft College

Eddy Joe Brackin, University of North Alabama

Susan Williams Brown, Gadsden State Community College

Jana Bryant, Manatee Community College

Lee R. Clancy, Golden West College

Krista Blevins Cohlmia, Odessa College

Dayna Coker, Southwestern Oklahoma State University—Sayre campus

Jan Collins, Embry Riddle College

Cecilia Cooper, William & Harper College

John S. Cross, University of Northern Iowa

Charles D. Cunningham, Jr., James Madison University

M. Hilary Davies, University of Alaska–Anchorage

Elias Deeba, University of Houston–Downtown

Grace DeVelbiss, Sinclair Community College

Lena Dexter, Faulkner State Junior College

Emily Dickinson, University of Arkansas

Mickey P. Dunlap, University of Tennessee, Martin

Gerard G. East, Southwestern Oklahoma State University

Eric Ellis, Essex Community College

Eunice F. Everett, Seminole Community College

Dale Ewen, Parkland College

Harold Farmer, Wallace Community College–Hanceville

Ronald J. Fischer, Evergreen Valley College

Mary Jane Gates, University of Arkansas at Little Rock

Lee R. Gibson, University of Louisville

Marvin Goodman, Monmouth College

Edna Greenwood, Tarrant County College

Jerry Gustafson, Beloit College

Jerome Hahn, Bradley University

Douglas Hall, Michigan State University

Robert Hall, University of Wisconsin

David Hansen, Monterey Peninsula College

Shari Harris, John Wood Community College

Sheyleah Harris-Plant, South Plains College

Kevin Hastings, University of Delaware

William Hinrichs, Rock Valley College

Arthur M. Hobbs, Texas A&M University

Jack E. Hofer, California Polytechnic State University

Ingrid Holzner, University of Wisconsin

Wayne Humphrey, Cisco College

Warren Jaech, Tacoma Community College

Joy St. John Johnson, Alabama A&M University

Nancy Johnson, Broward Community College

Patricia H. Jones, Methodist College

William B. Jones, University of Colorado

Barbara Juister, Elgin Community College

David Kinsey, University of Southern Indiana

Helen Kriegsman, Pittsburg State University

Marjorie O. Labhart, University of Southern Indiana

Betty J. Larson, South Dakota State University

Paul Lauritsen, Brown College

Jaclyn LeFebvre, Illinois Central College

Susan Loveland, University of Alaska–Anchorage

James Mark, Eastern Arizona College

Marcel Maupin, Oklahoma State University, Oklahoma City

Robert O. McCoy, University of Alaska–Anchorage

Judy McKinney, California Polytechnic Institute at Pomona

Sandra McLaurin, University of North Carolina

Marcus McWaters, University of Southern Florida

Donna Menard, University of Massachusetts, Dartmouth

James W. Mettler, Pennsylvania State University

Eldon L. Miller, University of Mississippi

Stuart E. Mills, Louisiana State University–Shreveport

Mila Mogilevskaya, Wichita State University

Gilbert W. Nelson, North Dakota State

Marie Neuberth, Catonsville City College

Nam Nguyen, University of Texas–PanAm

Charles Odion, Houston Community College

C. Altay Özgener, State College of Florida

Anthony Peressini, University of Illinois

David L. Phillips, University of Southern Colorado

William H. Price, Middle Tennessee State University

Ronald Putthoff, University of Southern Mississippi

Brooke P. Quinlan, Hillsborough Community College

Leela Rakesh, Carnegie Mellon University

Janet P. Ray, Seattle Central Community College

Robert K. Rhea, J. Sargeant Reynolds Community College

Barbara Riggs, Tennessee Technological University

Minnie Riley, Hinds Community College

Renee Roames, Purdue University

Paul Schaefer, SUNY, Geneseo

Vincent P. Schielack, Jr., Texas A&M University

Robert Sharpton, Miami Dade Community College

L. Thomas Shiflett, Southwest Missouri State University

Richard Slinkman, Bemidji State University

Merreline Smith, California Polytechnic Institute at Pomona

John Snyder, Sinclair Community College

Sandra L. Spain, Thomas Nelson Community College

Warren Strickland, Del Mar College

Paul K. Swets, Angelo State College

Ray Tebbetts, San Antonio College

Faye Thames, Lamar State University

Douglas Tharp, University of Houston–Downtown

Carolyn A. Wailes, University of Alabama, Birmingham

Carol M. Walker, Hinds Community College

William Waller, University of Houston–Downtown

Richard H. Weil, Brown College

Carroll G. Wells, Western Kentucky University

William H. White, University of South Carolina at Spartanburg

Clifton Whyburn, University of Houston

Charles R. Williams, Midwestern State University

Harry Wolff, University of Wisconsin

Roger Zarnowski, Angelo State University

Albert Zechmann, University of Nebraska

We wish to thank the staff at Cengage Learning, especially Gary Whalen and Jennifer Risden, for their support in the production process. Stacy Green, our development editor, provided constant encouragement and valuable input. We appreciate her presence by our side during this entire journey. Special thanks to Lynn Lustberg at MPS North America for her patience throughout the production process, and thanks to everyone at MPS North America for their excellent copyediting and typesetting skills. We appreciate the important work of our accuracy reviewer Carol Walker.

R. David Gustafson

Jeffrey Hughes

A Review of Basic Algebra

In this chapter, we review many concepts and skills learned in previous algebra courses. Be sure to master this material now, because it is the basis for the rest of this course.

Careers and Mathematics: Pharmacist

©Tyler Olson/Shutterstock.com

Pharmacists distribute prescription drugs to individuals. They also advise patients, physicians, and other healthcare workers on the selection, dosages, interactions, and side effects of medications. They also monitor patients to ensure that they are using their medications safely and effectively. Some pharmacists specialize in oncology, nuclear pharmacy, geriatric pharmacy, or psychiatric pharmacy.

Education and Mathematics Required

- Pharmacists are required to possess a Pharm.D. degree from an accredited college or school of pharmacy. This degree generally takes four years to complete. To be admitted to a Pharm.D. program, at least two years of college must be completed, which includes courses in the natural sciences, mathematics, humanities, and the social sciences. A series of examinations must also be passed to obtain a license to practice pharmacy.
- College Algebra, Trigonometry, Statistics, and Calculus I are courses required for admission to a Pharm.D. program.

How Pharmacists Use Math and Who Employs Them

- Pharmacists use math throughout their work to calculate dosages of various drugs. These dosages are based on weight and whether the medication is given in pill form, by infusion, or intravenously.
- Most pharmacists work in a community setting, such as a retail drugstore, or in a healthcare facility, such as a hospital.

Career Outlook and Salary

- Employment of pharmacists is expected to grow by 14% between 2012 and 2022, about as fast as the average for all occupations.
- Median annual wages of wage and salary pharmacists is approximately $116,670.

For more information see: www.bls.gov/oco

0.1 Sets of Real Numbers

In this section, we will learn to

1. Identify sets of real numbers.
2. Identify properties of real numbers.
3. Graph subsets of real numbers on the number line.
4. Graph intervals on the number line.
5. Define absolute value.
6. Find distances on the number line.

Sudoku, a game that involves number placement, is very popular. The objective is to fill a 9 by 9 grid so that each column, each row, and each of the 3 by 3 blocks contains the numbers from 1 to 9. A man playing Sudoku on his tablet is shown in the margin.

To solve Sudoku puzzles, logic and the set of numbers, {1, 2, 3, 4, 5, 6, 7, 8, 9}, are used. Sets of numbers are important in mathematics, and we begin our study of algebra with this topic.

A **set** is a collection of objects, such as a set of dishes or a set of golf clubs. The set of vowels in the English language can be denoted as {a, e, i, o, u}, where the braces { } are read as "the set of."

If every member of one set B is also a member of another set A, we say that B is a **subset** of A. We can denote this by writing $B \subset A$, where the symbol $\subset$ is read as "is a subset of." (See Figure 0-1 below.) If set B equals set A, we can write $B \subseteq A$.

If A and B are two sets, we can form a new set consisting of all members that are in set A or set B or both. This set is called the **union** of A and B. We can denote this set by writing $A \cup B$, where the symbol $\cup$ is read as "union." (See Figure 0-1 below.)

We can also form the set consisting of all members that are in both set A and set B. This set is called the **intersection** of A and B. We can denote this set by writing $A \cap B$, where the symbol $\cap$ is read as "intersection." (See Figure 0-1 below.)

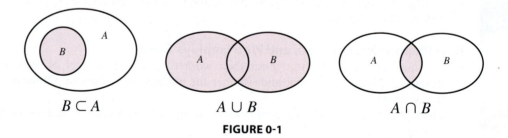

$$B \subset A \qquad\qquad A \cup B \qquad\qquad A \cap B$$

FIGURE 0-1

Understanding Subsets and Finding the Union and Intersection of Two Sets

EXAMPLE 1

Let $A = \{a, e, i\}$, $B = \{c, d, e\}$, and $V = \{a, e, i, o, u\}$.

a. Is $A \subset V$?   **b.** Find $A \cup B$.   **c.** Find $A \cap B$.

SOLUTION

a. Since each member of set A is also a member of set V, $A \subset V$.

b. The union of set A and set B contains the members of set A, set B, or both. Thus, $A \cup B = \{a, c, d, e, i\}$.

c. The intersection of set A and set B contains the members that are in both set A and set B. Thus, $A \cap B = \{e\}$.

Self Check 1   **a.** Is $B \subset V$?   **b.** Find $B \cup V$
c. Find $A \cap V$

Now Try Exercise 33.

If a set has no elements, it is called the **empty set**, or the **null set**, and is represented by either { } or $\varnothing$. Here is an example of how the empty set can result when finding the intersection of two sets.

$$\{5, 10, 15, 20\} \cap \{25, 30, 35, 40\} = \varnothing$$

Because the sets have no common elements, their intersection has no elements and is the empty set.

1. Identify Sets of Real Numbers

There are several sets of numbers that we use in everyday life.

Natural Numbers, Whole Numbers, and Integers	**Natural numbers** The numbers that we use for counting: $\{1, 2, 3, 4, 5, 6, \ldots\}$ **Whole numbers** The set of natural numbers including 0: $\{0, 1, 2, 3, 4, 5, 6, \ldots\}$ **Integers** The set of whole numbers and their negatives: $\{\ldots, -5, -4, -3, -2, -1, 0, 1, 2, 3, 4, 5, \ldots\}$

In the definitions above, each group of three dots (called an *ellipsis*) indicates that the numbers continue forever in the indicated direction.

Two important subsets of the natural numbers are the *prime* and *composite* numbers. A **prime number** is a natural number greater than 1 that is divisible only by itself and 1. A **composite number** is a natural number greater than 1 that is not prime.

- **The set of prime numbers:** $\{2, 3, 5, 7, 11, 13, 17, 19, 23, 29, 31, \ldots\}$
- **The set of composite numbers:** $\{4, 6, 8, 9, 10, 12, 14, 15, 16, 18, 20, 21, \ldots\}$

Two important subsets of the set of integers are the *even* and *odd integers*. The **even integers** are the integers that are exactly divisible by 2. The **odd integers** are the integers that are not exactly divisible by 2.

- **The set of even integers:** $\{\ldots, -10, -8, -6, -4, -2, 0, 2, 4, 6, 8, 10, \ldots\}$
- **The set of odd integers:** $\{\ldots, -9, -7, -5, -3, -1, 1, 3, 5, 7, 9, \ldots\}$

So far, we have listed numbers inside braces to specify sets. This method is called the **roster method**. When we give a rule to determine which numbers are in a set, we are using **set-builder notation**. To use set-builder notation to denote the set of prime numbers, we write

$$\{x \mid x \text{ is a prime number}\}$$

Read as "the set of all numbers x such that x is a prime number." Recall that when a letter stands for a number, it is called a variable.

variable   such that   rule that determines membership in the set

Take Note
Remember that the denominator of a fraction can **never** be 0.

The fractions of arithmetic are called *rational numbers*.

Rational Numbers   **Rational numbers** are fractions that have an integer numerator and a nonzero integer denominator. Using set-builder notation, the rational numbers are

$$\left\{ \frac{a}{b} \mid a \text{ is an integer and } b \text{ is a nonzero integer} \right\}$$

Rational numbers can be written as fractions or decimals. Some examples of rational numbers are

$$5 = \frac{5}{1}, \qquad \frac{3}{4} = 0.75, \qquad -\frac{1}{3} = -0.333\ldots, \quad \text{and} \quad -\frac{5}{11} = -0.454545\ldots$$

The = sign indicates that two quantities are equal.

These examples suggest that the decimal forms of all rational numbers are either *terminating decimals* or *repeating decimals*.

Determining whether the Decimal Form of a Fraction Terminates or Repeats

EXAMPLE 2   Determine whether the decimal form of each fraction terminates or repeats:

   a. $\dfrac{7}{16}$ **b.** $\dfrac{65}{99}$

SOLUTION   In each case, we perform a long division and write the quotient as a decimal.

   a. To change $\frac{7}{16}$ to a decimal, we perform a long division to get $\frac{7}{16} = 0.4375$. Since 0.4375 terminates, we can write $\frac{7}{16}$ as a terminating decimal.

   b. To change $\frac{65}{99}$ to a decimal, we perform a long division to get $\frac{65}{99} = 0.656565\ldots$. Since $0.656565\ldots$ repeats, we can write $\frac{65}{99}$ as a repeating decimal.

   We can write repeating decimals in compact form by using an overbar. For example, $0.656565\ldots = 0.\overline{65}$.

Self Check 2   Determine whether the decimal form of each fraction terminates or repeats:

   a. $\dfrac{38}{99}$ **b.** $\dfrac{7}{8}$

Now Try Exercise 35.

Some numbers have decimal forms that neither terminate nor repeat. These nonterminating, nonrepeating decimals are called **irrational numbers**. Three examples of irrational numbers are

$$1.010010001000010\ldots, \qquad \sqrt{2} = 1.414213562\ldots, \qquad \text{and} \qquad \pi = 3.141592654\ldots$$

The union of the set of rational numbers (the terminating and repeating decimals) and the set of irrational numbers (the nonterminating, nonrepeating decimals) is the set of *real numbers* (the set of all decimals).

Real Numbers   A **real number** is any number that is rational or irrational. Using set-builder notation, the set of real numbers is

$$\{x \mid x \text{ is a rational or an irrational number}\}$$

Classifying Real Numbers

EXAMPLE 3   In the set $\{-3, -2, 0, \frac{1}{2}, 1, \sqrt{5}, 2, 4, 5, 6\}$, list all

   a. even integers   **b.** prime numbers   **c.** rational numbers

SOLUTION   We will check to see whether each number is a member of the set of even integers, the set of prime numbers, and the set of rational numbers.

a. even integers: $-2, 0, 2, 4, 6$

b. prime numbers: $2, 5$

c. rational numbers: $-3, -2, 0, \frac{1}{2}, 1, 2, 4, 5, 6$

Self Check 3   In the set in Example 3, list all   **a.** odd integers   **b.** composite numbers   **c.** irrational numbers.

Now Try Exercise 43.

Figure 0-2 shows how the previous sets of numbers are related.

> **Tip**
> Learning several sets of numbers at one time can initially cause your head to spin. To help, examine closely the figure shown. You will easily grasp connections between the sets of numbers in algebra and not be overwhelmed.

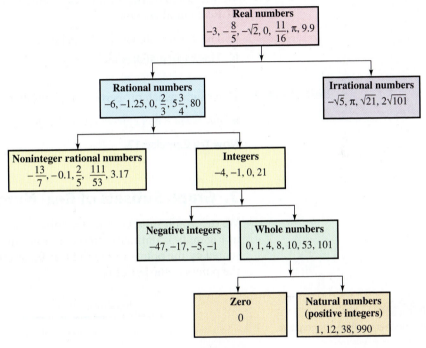

FIGURE 0-2

2. Identify Properties of Real Numbers

When we work with real numbers, we will use the following properties.

Properties of Real Numbers   If a, b, and c are real numbers,

The Commutative Properties for Addition and Multiplication

$$a + b = b + a \qquad\qquad ab = ba$$

The Associative Properties for Addition and Multiplication

$$(a + b) + c = a + (b + c) \qquad (ab)c = a(bc)$$

The Distributive Property of Multiplication over Addition or Subtraction

$$a(b + c) = ab + ac \qquad \text{or} \qquad a(b - c) = ab - ac$$

The Double Negative Rule

$$-(-a) = a$$

The Distributive Property also applies when more than two terms are within parentheses.

EXAMPLE 4

Identifying Properties of Real Numbers

Determine which property of real numbers justifies each statement.

a. $(9 + 2) + 3 = 9 + (2 + 3)$ **b.** $3(x + y + 2) = 3x + 3y + 3 \cdot 2$

SOLUTION   We will compare the form of each statement to the forms listed in the Properties of Real Numbers box.

a. This form matches the Associative Property of Addition.

b. This form matches the Distributive Property.

Self Check 4   Determine which property of real numbers justifies each statement:

a. $mn = nm$ **b.** $(xy)z = x(yz)$ **c.** $p + q = q + p$

Now Try Exercise 17.

3. Graph Subsets of Real Numbers on the Number Line

We can graph subsets of real numbers on the **number line**. The number line shown in Figure 0-3 continues forever in both directions. The **positive numbers** are represented by the points to the right of 0, and the **negative numbers** are represented by the points to the left of 0.

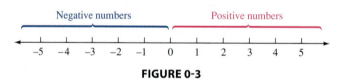

FIGURE 0-3

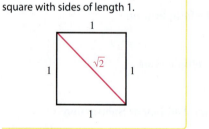

Figure 0-4(a) shows the graph of the natural numbers from 1 to 5. The point associated with each number is called the *graph* of the number, and the number is called the *coordinate* of its point.

Figure 0-4(b) shows the graph of the prime numbers that are less than 10.
Figure 0-4(c) shows the graph of the integers from -4 to 3.
Figure 0-4(d) shows the graph of the real numbers $-\frac{7}{3}$, $-\frac{3}{4}$, $0.\overline{3}$, and $\sqrt{2}$.

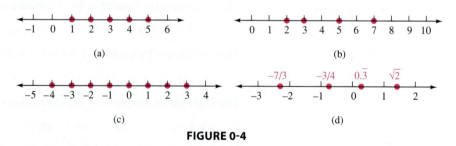

FIGURE 0-4

The graphs in Figure 0-4 suggest that there is a **one-to-one correspondence** between the set of real numbers and the points on a number line. This means that to each real number there corresponds exactly one point on the number line, and to each point on the number line there corresponds exactly one real-number coordinate.

EXAMPLE 5   **Graphing a Set of Numbers on a Number Line**

Graph the set $\left\{-3, -\frac{4}{3}, 0, \sqrt{5}\right\}$.

SOLUTION   We will mark (plot) each number on the number line. To the nearest tenth, $\sqrt{5} = 2.2$.

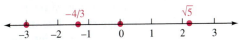

Self Check 5   Graph the set $\left\{-2, \frac{3}{4}, \sqrt{3}\right\}$. (*Hint:* To the nearest tenth, $\sqrt{3} = 1.7$.)

Now Try Exercise 51.

4. Graph Intervals on the Number Line

To show that two quantities are not equal, we can use an **inequality symbol**.

Symbol	Read as	Examples		
$\neq$	"is not equal to"	$5 \neq 8$	and	$0.25 \neq \frac{1}{3}$
$<$	"is less than"	$12 < 20$	and	$0.17 < 1.1$
$>$	"is greater than"	$15 > 9$	and	$\frac{1}{2} > 0.2$
$\leq$	"is less than or equal to"	$25 \leq 25$	and	$1.7 \leq 2.3$
$\geq$	"is greater than or equal to"	$19 \geq 19$	and	$15.2 \geq 13.7$
$\approx$	"is approximately equal to"	$\sqrt{2} \approx 1.414$	and	$\sqrt{3} \approx 1.732$

It is possible to write an inequality with the inequality symbol pointing in the opposite direction. For example,

- $12 < 20$ is equivalent to $20 > 12$
- $2.3 \geq -1.7$ is equivalent to $-1.7 \leq 2.3$

In Figure 0-3, the coordinates of points get larger as we move from left to right on a number line. Thus, if a and b are the coordinates of two points, the one to the right is the greater. This suggests the following facts:

- If $a > b$, point a lies to the right of point b on a number line.
- If $a < b$, point a lies to the left of point b on a number line.

Figure 0-5(a) shows the graph of the *inequality* $x > -2$ (or $-2 < x$). This graph includes all real numbers x that are greater than -2. The parenthesis at -2 indicates that -2 is not included in the graph. Figure 0-5(b) shows the graph of $x \leq 5$ (or $5 \geq x$). The bracket at 5 indicates that 5 is included in the graph.

> **Take Note**
>
> Parentheses indicate that endpoints are not included in an interval. Square brackets indicate that endpoints are included in an interval.

       (a)   (b)

FIGURE 0-5

Sometimes two inequalities can be written as a single expression called a **compound inequality**. For example, the compound inequality

$$5 < x < 12$$

is a combination of the inequalities $5 < x$ and $x < 12$. It is read as "5 is less than x, and x is less than 12," and it means that x is between 5 and 12. Its graph is shown in Figure 0-6.

FIGURE 0-6

The graphs shown in Figures 0-5 and 0-6 are portions of a number line called **intervals**. The interval shown in Figure 0-7(a) is denoted by the inequality $-2 < x < 4$, or in **interval notation** as $(-2, 4)$. The parentheses indicate that the endpoints are not included. The interval shown in Figure 0-7(b) is denoted by the inequality $x > 1$, or as $(1, \infty)$ in interval notation. The symbol ∞ (infinity) is not a real number. It is used to indicate that the graph in Figure 0-7(b) extends infinitely far to the right.

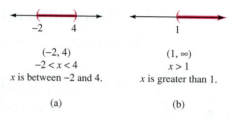

$(-2, 4)$ $\qquad$ $(1, \infty)$
$-2 < x < 4$ $\qquad$ $x > 1$
x is between -2 and 4. $\qquad$ x is greater than 1.

(a) $\qquad\qquad$ (b)

FIGURE 0-7

A compound inequality such as $-2 < x < 4$ can be written as two separate inequalities:

$$x > -2 \qquad \text{and} \qquad x < 4$$

This expression represents the **intersection** of two intervals. In interval notation, this expression can be written as

$(-2, \infty) \cap (-\infty, 4)$ $\qquad$ Read the symbol $\cap$ as "intersection."

Since the graph of $-2 < x < 4$ will include all points whose coordinates satisfy both $x > -2$ and $x < 4$ at the same time, its graph will include all points that are larger than -2 but less than 4. This is the interval $(-2, 4)$, whose graph is shown in Figure 0-7(a).

> **Tip**
> Parentheses are always used with ∞ or $-\infty$.

Writing an Inequality in Interval Notation and Graphing the Inequality

EXAMPLE 6 Write the inequality $-3 < x < 5$ in interval notation and graph it.

SOLUTION This is the interval $(-3, 5)$. Its graph includes all real numbers between -3 and 5, as shown in Figure 0-8.

FIGURE 0-8

Self Check 6 Write the inequality $-2 < x \leq 5$ in interval notation and graph it.

Now Try Exercise 63.

If an interval extends forever in one direction, it is called an **unbounded interval**.

Unbounded Intervals

Interval	Inequality	Graph
(a, ∞)	$x > a$	
$(2, \infty)$	$x > 2$	
$[a, \infty)$	$x \geq a$	
$[2, \infty)$	$x \geq 2$	
$(-\infty, a)$	$x < a$	
$(-\infty, 2)$	$x < 2$	
$(-\infty, a]$	$x \leq a$	
$(-\infty, 2]$	$x \leq 2$	
$(-\infty, \infty)$	$-\infty < x < \infty$	

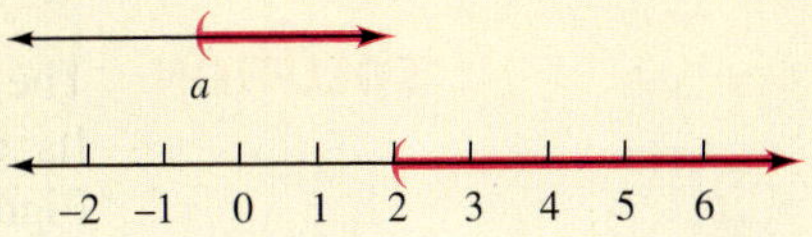

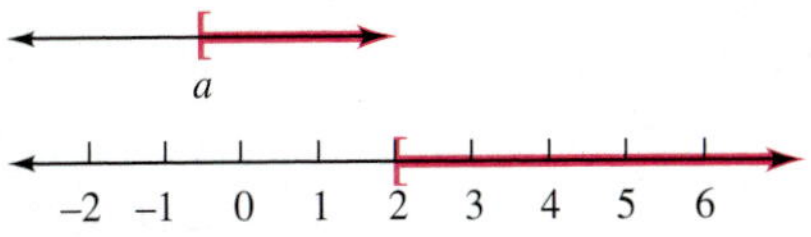

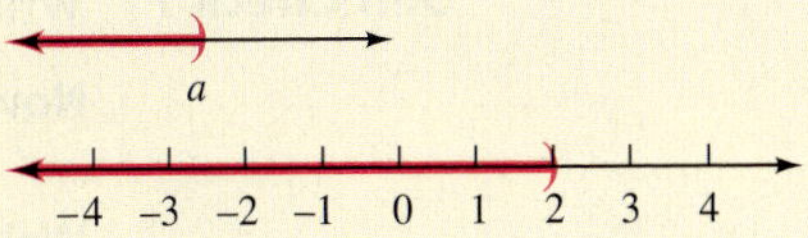

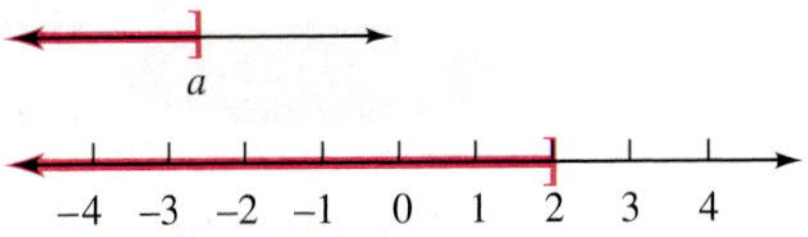

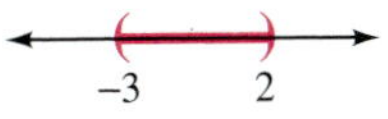

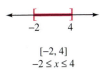

$(-3, 2)$
$-3 < x < 2$

(a)

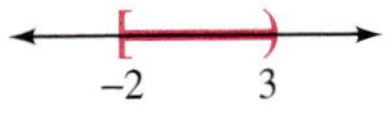

$[-2, 3)$
$-2 \leq x < 3$

(b)

FIGURE 0-9

A bounded interval with no endpoints is called an **open interval**. Figure 0-9(a) shows the open interval between -3 and 2. A bounded interval with one endpoint is called a **half-open interval**. Figure 0-9(b) shows the half-open interval between -2 and 3, including -2.

Intervals that include two endpoints are called **closed intervals**. Figure 0-10 shows the graph of a closed interval from -2 to 4.

$[-2, 4]$
$-2 \leq x \leq 4$

FIGURE 0-10

Open Intervals, Half-Open Intervals, and Closed Intervals

Interval	Inequality	Graph
Open		
(a, b)	$a < x < b$	
$(-2, 3)$	$-2 < x < 3$	
Half-Open		
$[a, b)$	$a \leq x < b$	
$[-2, 3)$	$-2 \leq x < 3$	
Half-Open		
$(a, b]$	$a < x \leq b$	
$(-2, 3]$	$-2 < x \leq 3$	
Closed		
$[a, b]$	$a \leq x \leq b$	
$[-2, 3]$	$-2 \leq x \leq 3$	

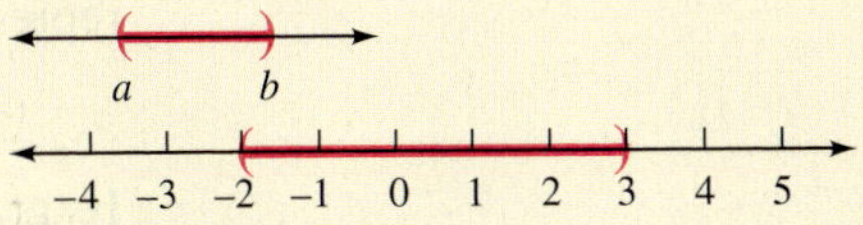

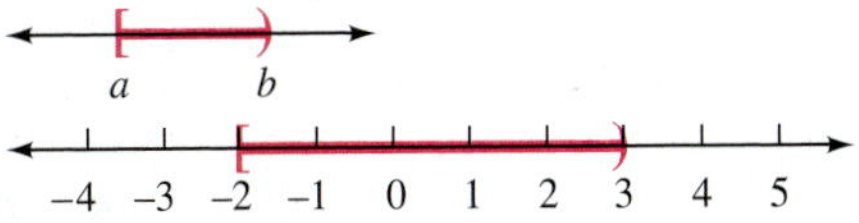

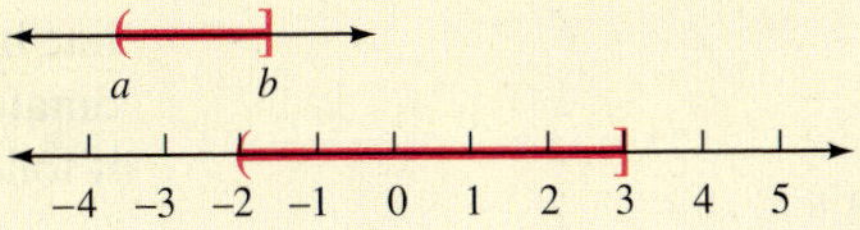

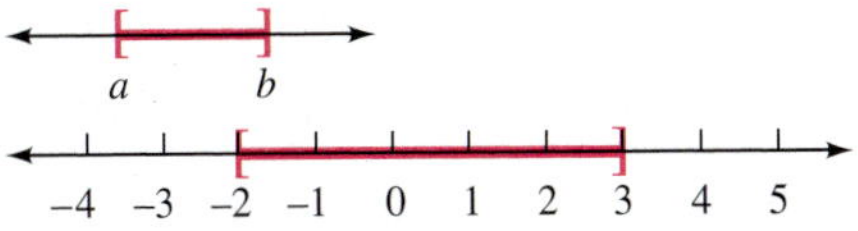

EXAMPLE 7

Writing an Inequality in Interval Notation and Graphing the Inequality

Write the inequality $3 \leq x$ in interval notation and graph it.

SOLUTION

The inequality $3 \leq x$ can be written in the form $x \geq 3$. This is the interval $[3, \infty)$. Its graph includes all real numbers greater than or equal to 3, as shown in Figure 0-11.

3

FIGURE 0-11

Self Check 7   Write the inequality $5 > x$ in interval notation and graph it.

Now Try Exercise 65.

EXAMPLE 8

Writing an Inequality in Interval Notation and Graphing the Inequality

Write the inequality $5 \geq x \geq -1$ in interval notation and graph it.

SOLUTION

The inequality $5 \geq x \geq -1$ can be written in the form

$$-1 \leq x \leq 5$$

This is the interval $[-1, 5]$. Its graph includes all real numbers from -1 to 5. The graph is shown in Figure 0-12.

−1   5

FIGURE 0-12

Self Check 8   Write the inequality $0 \leq x \leq 3$ in interval notation and graph it.

Now Try Exercise 69.

The expression

$$x < -2 \text{ or } x \geq 3 \qquad \text{Read as "x is less than -2 or x is greater than or equal to 3."}$$

represents the *union* of two intervals. In interval notation, it is written as

$$(-\infty, -2) \cup [3, \infty) \qquad \text{Read the symbol } \cup \text{ as "union."}$$

Its graph is shown in Figure 0-13.

−2   3

FIGURE 0-13

5. Define Absolute Value

The **absolute value** of a real number x (denoted as $|x|$) is the distance on a number line between 0 and the point with a coordinate of x. For example, points with coordinates of 4 and -4 both lie four units from 0, as shown in Figure 0-14. Therefore, it follows that

$$|-4| = |4| = 4$$

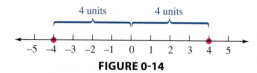

FIGURE 0-14

In general, for any real number x,

$$|-x| = |x|$$

We can define absolute value algebraically as follows.

Absolute Value   If x is a real number, then

$$|x| = x \qquad \text{when } x \geq 0$$

$$|x| = -x \qquad \text{when } x < 0$$

Take Note

Remember that x is **not** always positive and $-x$ is **not** always negative.

This definition indicates that when x is positive or 0, then x is its own absolute value. However, when x is negative, then $-x$ (which is positive) is its absolute value. Thus, $|x|$ is always nonnegative.

$$|x| \geq 0 \qquad \text{for all real numbers } x$$

EXAMPLE 9   **Using the Definition of Absolute Value**

Write each number without using absolute value symbols:

a. $|3|$ **b.** $|-4|$ **c.** $|0|$ **d.** $-|-7|$

SOLUTION   In each case, we will use the definition of absolute value.

a. $|3| = 3$ **b.** $|-4| = 4$ **c.** $|0| = 0$ **d.** $-|-7| = -(7) = -7$

Self Check 9   Write each number without using absolute value symbols:

a. $|-10|$ **b.** $|12|$ **c.** $-|6|$

Now Try Exercise 85.

In Example 10, we must determine whether the number inside the absolute value is positive or negative.

EXAMPLE 10   **Simplifying an Expression with Absolute Value Symbols**

Write each number without using absolute value symbols:

a. $|\pi - 1|$ **b.** $|2 - \pi|$ **c.** $|2 - x|$ if $x \geq 5$

SOLUTION   **a.** Since $\pi \approx 3.1416$, $\pi - 1$ is positive, and $\pi - 1$ is its own absolute value.

$$|\pi - 1| = \pi - 1$$

b. Since $2 - \pi$ is negative, its absolute value is $-(2 - \pi)$.

$$|2 - \pi| = -(2 - \pi) = -2 - (-\pi) = -2 + \pi = \pi - 2$$

c. Since $x \geq 5$, the expression $2 - x$ is negative, and its absolute value is $-(2 - x)$.

$$|2 - x| = -(2 - x) = -2 + x = x - 2 \text{ provided } x \geq 5$$

Self Check 10   Write each number without using absolute value symbols. (*Hint:* $\sqrt{5} \approx 2.236$.)

a. $|2 - \sqrt{5}|$ **b.** $|2 - x|$ if $x \leq 1$

Now Try Exercise 89.

6. Find Distances on the Number Line

On the number line shown in Figure 0-15, the distance between the points with coordinates of 1 and 4 is $4 - 1$, or 3 units. However, if the subtraction were done in the other order, the result would be $1 - 4$, or -3 units. To guarantee that the distance between two points is always positive, we can use absolute value symbols. Thus, the distance d between two points with coordinates of 1 and 4 is

$$d = |4 - 1| = |1 - 4| = 3$$

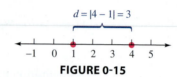

FIGURE 0-15

In general, we have the following definition for the distance between two points on the number line.

Distance between Two Points   If a and b are the coordinates of two points on the number line, the distance d between the points is given by the formula

$$d = |b - a|$$

EXAMPLE 11   **Finding the Distance between Two Points on a Number Line**

Find the distance on a number line between the points with the given coordinates.
a. 3 and 5   **b.** -2 and 3   **c.** -5 and -1

SOLUTION   We will use the formula for finding the distance between two points.

a. $d = |5 - 3| = |2| = 2$

b. $d = |3 - (-2)| = |3 + 2| = |5| = 5$

c. $d = |-1 - (-5)| = |-1 + 5| = |4| = 4$

Self Check 11   Find the distance on a number line between the points with the given coordinates.
a. 4 and 10   **b.** -2 and -7

Now Try Exercise 99.

Self Check Answers

1. a. no   **b.** $\{a, c, d, e, i, o, u\}$ **c.** $\{a, e, i\}$ **2. a.** repeats
b. terminates   **3. a.** $-3, 1, 5$ **b.** 4, 6   **c.** $\sqrt{5}$ **4. a.** Commutative
Property of Multiplication   **b.** Associative Property of Multiplication
c. Commutative Property of Addition   **5.**

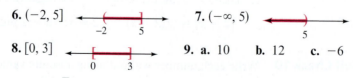

6. $(-2, 5]$ **7.** $(-\infty, 5)$

8. $[0, 3]$ **9. a.** 10   **b.** 12   **c.** -6

10. a. $\sqrt{5} - 2$ **b.** $2 - x$ **11. a.** 6   **b.** 5

Exercises 0.1

Getting Ready
You should be able to complete these vocabulary and concept statements before you proceed to the practice exercises.

Fill in the blanks.

1. A _____ is a collection of objects.

2. If every member of one set *B* is also a member of a second set *A*, then *B* is called a _____ of *A*.

3. If *A* and *B* are two sets, the set that contains all members that are in sets *A* and *B* or both is called the _____ of *A* and *B*.

4. If *A* and *B* are two sets, the set that contains all members that are in both sets is called the _____ of *A* and *B*.

5. A real number is any number that can be expressed as a _____.

6. A _____ is a letter that is used to represent a number.

7. The smallest prime number is ___ .

8. All integers that are exactly divisible by 2 are called _____ integers.

9. Natural numbers greater than 1 that are not prime are called _____ numbers.

10. Fractions such as $\frac{2}{3}, \frac{8}{2}$, and $-\frac{7}{9}$ are called _____ numbers.

11. Irrational numbers are _____ that don't terminate and don't repeat.

12. The symbol _____ is read as "is less than or equal to."

13. On a number line, the _____ numbers are to the left of 0.

14. The only integer that is neither positive nor negative is ___ .

15. The Associative Property of Addition states that $(x + y) + z =$ _____ .

16. The Commutative Property of Multiplication states that $xy =$ ____ .

17. Use the Distributive Property to complete the statement: $5(m + 2) =$ _____ .

18. The statement $(m + n)p = p(m + n)$ illustrates the _____ Property of _____ .

19. The graph of an _____ is a portion of a number line.

20. The graph of an open interval has ____ endpoints.

21. The graph of a closed interval has _____ endpoints.

22. The graph of a _____ interval has one endpoint.

23. Except for 0, the absolute value of every number is _____ .

24. The _____ between two distinct points on a number line is always positive.

Let

$\mathbf{N}$ = *the set of natural numbers*

$\mathbf{W}$ = *the set of whole numbers*

$\mathbf{Z}$ = *the set of integers*

$\mathbf{Q}$ = *the set of rational numbers*

$\mathbf{R}$ = *the set of real numbers*

Determine whether each statement is true or false. Read the symbol $\subset$ as "is a subset of."

25. $\mathbf{N} \subset \mathbf{W}$

26. $\mathbf{Q} \subset \mathbf{R}$

27. $\mathbf{Q} \subset \mathbf{N}$

28. $\mathbf{Z} \subset \mathbf{Q}$

29. $\mathbf{W} \subset \mathbf{Z}$

30. $\mathbf{R} \subset \mathbf{Z}$

Practice
Let $A = \{a, b, c, d, e\}$, $B = \{d, e, f, g\}$, and $C = \{a, c, e, f\}$. Find each set.

31. $A \cup B$

32. $A \cap B$

33. $A \cap C$

34. $B \cup C$

Determine whether the decimal form of each fraction terminates or repeats.

35. $\dfrac{9}{16}$

36. $\dfrac{3}{8}$

37. $\dfrac{3}{11}$

38. $\dfrac{5}{12}$

Consider the following set:
$\{-5, -4, -\frac{2}{3}, 0, 1, \sqrt{2}, 2, 2.75, 6, 7\}$

39. Which numbers are natural numbers?

40. Which numbers are whole numbers?

41. Which numbers are integers?

42. Which numbers are rational numbers?

43. Which numbers are irrational numbers?

44. Which numbers are prime numbers?

45. Which numbers are composite numbers?

46. Which numbers are even integers?

47. Which numbers are odd integers?

48. Which numbers are negative numbers?

Graph each subset of the real numbers on a number line.

49. The natural numbers between 1 and 5

50. The composite numbers less than 10

51. The prime numbers between 10 and 20

52. The integers from -2 to 4

53. The integers between -5 and 0

54. The even integers between -9 and -1

55. The odd integers between -6 and 4

56. -0.7, 1.75, and $3\frac{7}{8}$

Write each inequality in interval notation and graph the interval.

57. $x > 2$

58. $x < 4$

59. $0 < x < 5$

60. $-2 < x < 3$

61. $x > -4$

62. $x < 3$

63. $-2 \le x < 2$

64. $-4 < x \le 1$

65. $x \le 5$

66. $x \ge -1$

67. $-5 < x \le 0$

68. $-3 \le x < 4$

69. $-2 \le x \le 3$

70. $-4 \le x \le 4$

71. $6 \ge x \ge 2$

72. $3 \ge x \ge -2$

Write each pair of inequalities as the intersection of two intervals and graph the result.

73. $x > -5$ and $x < 4$

74. $x \ge -3$ and $x < 6$

75. $x \ge -8$ and $x \le -3$

76. $x > 1$ and $x \le 7$

Write each inequality as the union of two intervals and graph the result.

77. $x < -2$ or $x > 2$

78. $x \le -5$ or $x > 0$

79. $x \le -1$ or $x \ge 3$

80. $x < -3$ or $x \ge 2$

Write each expression without using absolute value symbols.

81. $|13|$ **82.** $|-17|$

83. $|0|$ **84.** $-|63|$

85. $-|-8|$ **86.** $|-25|$

87. $-|32|$ **88.** $-|-6|$

89. $|\pi - 5|$ **90.** $|8 - \pi|$

91. $|\pi - \pi|$ **92.** $|2\pi|$

93. $|x + 1|$ and $x \ge 2$ **94.** $|x + 1|$ and $x \le -2$

95. $|x - 4|$ and $x < 0$ **96.** $|x - 7|$ and $x > 10$

Find the distance between each pair of points on the number line.

97. 3 and 8   **98.** -5 and 12

99. -8 and -3 **100.** 6 and -20

Applications

101. What subset of the real numbers would you use to describe the populations of Memphis and Miami?

102. What subset of the real numbers would you use to describe the subdivisions of an inch on a ruler?

103. What subset of the real numbers would you use to report temperatures in London and Lisbon?

104. What subset of the real numbers would you use to describe the prices of hoodies at Aéropostale?

105. **Temperature** The average low temperature in International Falls, Minnesota, in January is $-7°F$. The average high temperature is $15°F$. Determine the degrees difference between the average high and the average low.

106. Temperature Harbin, China, is one of the world's coldest cities and known for its ice and snow festivals. In February, the average nightly low temperature is $-20°C$ and the average daily high temperature is $-7°C$. What is the temperature drop from day to night?

©The Curious Travelers/Shutterstock.com

Discovery and Writing

107. Explain why $-x$ could be positive.

108. Explain why every integer is a rational number.

109. Is the statement $|ab| = |a| \cdot |b|$ always true? Explain.

110. Is the statement $\left|\dfrac{a}{b}\right| = \dfrac{|a|}{|b|}$ $(b \neq 0)$ always true? Explain.

111. Is the statement $|a + b| = |a| + |b|$ always true? Explain.

112. Explain why it is incorrect to write $a < b > c$ if $a < b$ and $b > c$.

Critical Thinking

Determine if the statement is true or false. If the statement is false, then correct it and make it true.

113. There are six integers between -3 and 3.

114. $\dfrac{725}{0}$ is a rational number because 725 and 0 are integers.

115. ∞ is a real number.

116. $|a - b| = |b - a|$

117. $\varnothing \subset \left\{5, \pi, \sqrt{3}, \dfrac{13}{4}\right\}$

118. $\varnothing \subset \varnothing$

119. There are six subsets of $\{11, 22, 33\}$.

120. A set is always a subset of itself.

0.2 Integer Exponents and Scientific Notation

In this section, we will learn to

1. Define natural-number exponents.

2. Apply the rules of exponents.

3. Apply the rules for order of operations to evaluate expressions.

4. Express numbers in scientific notation.

5. Use scientific notation to simplify computations.

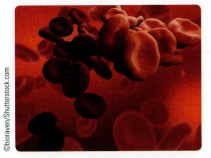

©bioraven/Shutterstock.com

The number of cells in the human body is approximated to be one hundred trillion or 100,000,000,000,000. One hundred trillion is $(10)(10)(10)\cdots(10)$, where ten occurs fourteen times. Fourteen factors of ten can be written as 10^{14}.

In this section, we will use integer exponents to represent repeated multiplication of numbers.

1. Define Natural-Number Exponents

When two or more quantities are multiplied together, each quantity is called a *factor* of the product. The exponential expression x^4 indicates that x is to be used as a factor four times.

$$x^4 = \overbrace{x \cdot x \cdot x \cdot x}^{4 \text{ factors of } x}$$

In general, the following is true.

Natural-Number Exponents For any natural number n,

$$x^n = \overbrace{x \cdot x \cdot x \cdot \cdots \cdot x}^{n \text{ factors of } x}$$

In the **exponential expression** x^n, x is called the **base**, and n is called the **exponent** or the **power** to which the base is raised. The expression x^n is called a **power of x**. From the definition, we see that a natural-number exponent indicates how many times the base of an exponential expression is to be used as a factor in a product. If an exponent is 1, the 1 is usually not written:

$$x^1 = x$$

EXAMPLE 1

Using the Definition of Natural-Number Exponents

Write each expression without using exponents:

a. 4^2 **b.** $(-4)^2$ **c.** -5^3 **d.** $(-5)^3$ **e.** $3x^4$ **f.** $(3x)^4$

SOLUTION In each case, we apply the definition of natural-number exponents.

a. $4^2 = 4 \cdot 4 = 16$ Read 4^2 as "four squared."

b. $(-4)^2 = (-4)(-4) = 16$ Read $(-4)^2$ as "negative four squared."

c. $-5^3 = -5(5)(5) = -125$ Read -5^3 as "the negative of five cubed."

d. $(-5)^3 = (-5)(-5)(-5) = -125$ Read $(-5)^3$ as "negative five cubed."

e. $3x^4 = 3 \cdot x \cdot x \cdot x \cdot x$ Read $3x^4$ as "3 times x to the fourth power."

f. $(3x)^4 = (3x)(3x)(3x)(3x) = 81 \cdot x \cdot x \cdot x \cdot x$ Read $(3x)^4$ as "$3x$ to the fourth power."

Self Check 1 Write each expression without using exponents:

a. 7^3 **b.** $(-3)^2$ **c.** $5a^3$ **d.** $(5a)^4$

Now Try Exercise 19.

Take Note

Note the distinction between ax^n and $(ax)^n$:

$$ax^n = a \cdot \overbrace{x \cdot x \cdot x \cdot \cdots \cdot x}^{n \text{ factors of } x} \qquad\qquad (ax)^n = \overbrace{(ax)(ax)(ax) \cdot \cdots \cdot (ax)}^{n \text{ factors of } ax}$$

Also note the distinction between $-x^n$ and $(-x)^n$:

$$-x^n = -(\overbrace{x \cdot x \cdot x \cdot \cdots \cdot x}^{n \text{ factors of } x}) \qquad\qquad (-x)^n = \overbrace{(-x)(-x)(-x) \cdot \cdots \cdot (-x)}^{n \text{ factors of } -x}$$

Accent on Technology

Using a Calculator to Find Powers

We can use a graphing calculator to find powers of numbers. For example, consider 2.35^3.

- Input 2.35 and press the [^] key.
- Input 3 and press [ENTER].

Take Note

To find powers on a scientific calculator use the $\boxed{y^x}$ key.

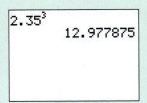

2.35^3

12.977875

FIGURE 0-16

We see that $2.35^3 = 12.977875$, as Figure 0-16 shows.

2. Apply the Rules of Exponents

We begin the review of the rules of exponents by considering the product $x^m x^n$. Since x^m indicates that x is to be used as a factor m times, and since x^n indicates that x is to be used as a factor n times, there are $m + n$ factors of x in the product $x^m x^n$.

$$x^m x^n = \overbrace{\underbrace{x \cdot x \cdot x \cdot \cdots \cdot x}_{m \text{ factors of } x} \cdot \underbrace{x \cdot x \cdot x \cdot \cdots \cdot x}_{n \text{ factors of } x}}^{m + n \text{ factors of } x} = x^{m+n}$$

This suggests that to multiply exponential expressions with the same base, we *keep the base and add the exponents.*

Product Rule for Exponents   If m and n are natural numbers, then

$$x^m x^n = x^{m+n}$$

Take Note

The Product Rule applies to exponential expressions with the same base. A product of two powers with different bases, such as $x^4 y^3$, cannot be simplified.

To find another property of exponents, we consider the exponential expression $(x^m)^n$. In this expression, the exponent n indicates that x^m is to be used as a factor n times. This implies that x is to be used as a factor mn times.

$$(x^m)^n = \overbrace{\underbrace{(x^m)(x^m)(x^m) \cdot \cdots \cdot (x^m)}_{n \text{ factors of } x^m}}^{mn \text{ factors of } x} = x^{mn}$$

This suggests that to raise an exponential expression to a power, we *keep the base and multiply the exponents.*

To raise a product to a power, we raise each factor to that power.

$$(xy)^n = \overbrace{(xy)(xy)(xy) \cdot \cdots \cdot (xy)}^{n \text{ factors of } xy} = \underbrace{(x \cdot x \cdot x \cdot \cdots \cdot x)}_{n \text{ factors of } x}\underbrace{(y \cdot y \cdot y \cdot \cdots \cdot y)}_{n \text{ factors of } y} = x^n y^n$$

To raise a fraction to a power, we raise both the numerator and the denominator to that power. If $y \neq 0$, then

$$\left(\frac{x}{y}\right)^n = \overbrace{\left(\frac{x}{y}\right)\left(\frac{x}{y}\right)\left(\frac{x}{y}\right) \cdot \cdots \cdot \left(\frac{x}{y}\right)}^{n \text{ factors of } \frac{x}{y}}$$

$$= \frac{\overbrace{xxx \cdot \cdots \cdot x}^{n \text{ factors of } x}}{\underbrace{yyy \cdot \cdots \cdot y}_{n \text{ factors of } y}}$$

$$= \frac{x^n}{y^n}$$

The previous three results are called the *Power Rules of Exponents.*

Power Rules of Exponents If m and n are natural numbers, then

$$(x^m)^n = x^{mn} \qquad (xy)^n = x^n y^n \qquad \left(\frac{x}{y}\right)^n = \frac{x^n}{y^n} \qquad (y \neq 0)$$

EXAMPLE 2

Using Exponent Rules to Simplify Expressions with Natural-Number Exponents

Simplify: **a.** $x^5 x^7$ **b.** $x^2 y^3 x^5 y$ **c.** $(x^4)^9$ **d.** $(x^2 x^5)^3$

e. $\left(\dfrac{x}{y^2}\right)^5$ **f.** $\left(\dfrac{5x^2 y}{z^3}\right)^2$

SOLUTION In each case, we will apply the appropriate rule of exponents.

a. $x^5 x^7 = x^{5+7} = x^{12}$ **b.** $x^2 y^3 x^5 y = x^{2+5} y^{3+1} = x^7 y^4$

c. $(x^4)^9 = x^{4 \cdot 9} = x^{36}$ **d.** $(x^2 x^5)^3 = (x^7)^3 = x^{21}$

e. $\left(\dfrac{x}{y^2}\right)^5 = \dfrac{x^5}{(y^2)^5} = \dfrac{x^5}{y^{10}} \quad (y \neq 0)$

f. $\left(\dfrac{5x^2 y}{z^3}\right)^2 = \dfrac{5^2 (x^2)^2 y^2}{(z^3)^2} = \dfrac{25 x^4 y^2}{z^6} \quad (z \neq 0)$

Self Check 2 Simplify:

a. $(y^3)^2$ **b.** $(a^2 a^4)^3$ **c.** $(x^2)^3 (x^3)^2$ **d.** $\left(\dfrac{3a^3 b^2}{c^3}\right)^3 \quad (c \neq 0)$

Now Try Exercise 49.

If we assume that the rules for natural-number exponents hold for exponents of 0, we can write

$$x^0 x^n = x^{0+n} = x^n = \mathbf{1} x^n$$

Since $x^0 x^n = \mathbf{1} x^n$, it follows that if $x \neq 0$, then $x^0 = 1$.

Zero Exponent $x^0 = 1 \quad (x \neq 0)$

Take Note

0 raised to the power of 0 is not defined.

If we assume that the rules for natural-number exponents hold for exponents that are negative integers, we can write

$$x^{-n} x^n = x^{-n+n} = x^0 = 1 \quad (x \neq 0)$$

However, we know that

$$\frac{1}{x^n} \cdot x^n = 1 \quad (x \neq 0) \qquad \frac{1}{x^n} \cdot x^n = \frac{x^n}{x^n}, \text{ and any nonzero number divided by itself is 1.}$$

Since $x^{-n} x^n = \dfrac{1}{x^n} \cdot x^n$, it follows that $x^{-n} = \dfrac{1}{x^n} \ (x \neq 0)$.

Negative Exponents If n is an integer and $x \neq 0$, then

$$x^{-n} = \frac{1}{x^n} \qquad \text{and} \qquad \frac{1}{x^{-n}} = x^n$$

Because of the previous definitions, all of the rules for natural-number exponents will hold for integer exponents.

Simplifying Expressions with Integer Exponents

EXAMPLE 3

Simplify and write all answers without using negative exponents:

a. $(3x)^0$ **b.** $3(x^0)$ **c.** x^{-4} **d.** $\dfrac{1}{x^{-6}}$ **e.** $x^{-3}x$ **f.** $(x^{-4}x^8)^{-5}$

SOLUTION

We will use the definitions of zero exponent and negative exponents to simplify each expression.

a. $(3x)^0 = 1$ **b.** $3(x^0) = 3(1) = 3$ **c.** $x^{-4} = \dfrac{1}{x^4}$ **d.** $\dfrac{1}{x^{-6}} = x^6$

e. $x^{-3}x = x^{-3+1}$ **f.** $(x^{-4}x^8)^{-5} = (x^4)^{-5}$

$\qquad = x^{-2}$ $\qquad\qquad = x^{-20}$

$\qquad = \dfrac{1}{x^2}$ $\qquad\qquad = \dfrac{1}{x^{20}}$

> **Caution**
>
> A common mistake is to give an answer of 0 when a number is raised to the power of 0. This is incorrect! Avoid making this error.
>
> $17^0 \neq 0,\ \left(-\dfrac{3}{5}\right)^0 \neq 0,$ and $(8x)^0 \neq 0$
>
> These expressions equal 1 and not 0.

Self Check 3

Simplify and write all answers without using negative exponents:

a. $7a^0$ **b.** $3a^{-2}$ **c.** $a^{-4}a^2$ **d.** $(a^3a^{-7})^3$

Now Try Exercise 59.

To develop the Quotient Rule for Exponents, we proceed as follows:

$$\frac{x^m}{x^n} = x^m\left(\frac{1}{x^n}\right) = x^m x^{-n} = x^{m+(-n)} = x^{m-n} \quad (x \neq 0)$$

This suggests that to divide two exponential expressions with the same nonzero base, we *keep the base and subtract the exponent in the denominator from the exponent in the numerator.*

Quotient Rule for Exponents

If m and n are integers, then

$$\frac{x^m}{x^n} = x^{m-n} \quad (x \neq 0)$$

Simplifying Expressions with Integer Exponents

EXAMPLE 4

Simplify and write all answers without using negative exponents:

a. $\dfrac{x^8}{x^5}$ **b.** $\dfrac{x^2x^4}{x^{-5}}$

SOLUTION

We will apply the Product and Quotient Rules of Exponents.

a. $\dfrac{x^8}{x^5} = x^{8-5}$ **b.** $\dfrac{x^2x^4}{x^{-5}} = \dfrac{x^6}{x^{-5}}$

$\qquad = x^3$ $\qquad\qquad = x^{6-(-5)}$

$\qquad\qquad\qquad\qquad = x^{11}$

Self Check 4

Simplify and write all answers without using negative exponents:

a. $\dfrac{x^{-6}}{x^2}$ **b.** $\dfrac{x^4x^{-3}}{x^2}$

Now Try Exercise 69.

EXAMPLE 5   **Simplifying Expressions with Integer Exponents**

Simplify and write all answers without using negative exponents:

a. $\left(\dfrac{x^3y^{-2}}{x^{-2}y^3}\right)^{-2}$ **b.** $\left(\dfrac{x}{y}\right)^{-n}$

SOLUTION   We will apply the appropriate rules of exponents.

a. $\left(\dfrac{x^3y^{-2}}{x^{-2}y^3}\right)^{-2} = (x^{3-(-2)}y^{-2-3})^{-2}$

$= (x^5y^{-5})^{-2}$

$= x^{-10}y^{10}$

$= \dfrac{y^{10}}{x^{10}}$

b. $\left(\dfrac{x}{y}\right)^{-n} = \dfrac{x^{-n}}{y^{-n}}$

$= \dfrac{x^{-n}x^ny^n}{y^{-n}x^ny^n}$ Multiply numerator and denominator by 1 in the form $\dfrac{x^ny^n}{x^ny^n}$

$= \dfrac{x^0y^n}{y^0x^n}$ $x^{-n}x^n = x^0$ and $y^{-n}y^n = y^0$

$= \dfrac{y^n}{x^n}$ $x^0 = 1$ and $y^0 = 1$

$= \left(\dfrac{y}{x}\right)^n$

Self Check 5   Simplify and write all answers without using negative exponents:

a. $\left(\dfrac{x^4y^{-3}}{x^{-3}y^2}\right)^2$ **b.** $\left(\dfrac{2a}{3b}\right)^{-3}$

Now Try Exercise 75.

Part b of Example 5 establishes the following rule.

A Fraction to a Negative Power   If n is a natural number, then

$$\left(\dfrac{x}{y}\right)^{-n} = \left(\dfrac{y}{x}\right)^n \qquad (x \neq 0 \quad \text{and} \quad y \neq 0)$$

Tip

Learning and applying several exponent rules at one time can often be confusing. Some students admit to being intimidated by them because the formal statements of the rules involve x's, y's, m's, and n's.

It can be very helpful to rewrite the rules in your own words. Here are some examples.

- When I multiply exponential expressions with the same bases, then I keep the common base and add the exponents.
- When I raise a power to a power, then I multiply the exponents.
- When I raise a product to a power, then I raise each factor to the power.
- When I raise a quotient to a power, then I raise both the numerator and the denominator to the power.
- When I divide exponential expressions with the same bases, then I keep the common base and subtract the exponents.
- When I raise an exponential expression to a negative power, I can invert the exponential expression and raise it to a positive power.

3. Apply the Rules for Order of Operations to Evaluate Expressions

When several operations occur in an expression, we must perform the operations in the following order to get the correct result.

Strategy for Evaluating Expressions Using Order of Operations

If an expression does not contain grouping symbols such as parentheses or brackets, follow these steps:

1. Find the values of any exponential expressions.
2. Perform all multiplications and/or divisions, working from left to right.
3. Perform all additions and/or subtractions, working from left to right.

- If an expression contains grouping symbols such as parentheses, brackets, or braces, use the rules above to perform the calculations within each pair of grouping symbols, working from the innermost pair to the outermost pair.
- In a fraction, simplify the numerator and the denominator of the fraction separately. Then simplify the fraction, if possible.

Take Note

Many students remember the Order of Operations Rule with the acronym **PEMDAS:**

- **P**arentheses
- **E**xponents
- **M**ultiplication
- **D**ivision
- **A**ddition
- **S**ubtraction

For example, to simplify $\dfrac{3[4 - (6 + 10)]}{2^2 - (6 + 7)}$, we proceed as follows:

$$\frac{3[4 - (6 + 10)]}{2^2 - (6 + 7)} = \frac{3(4 - 16)}{2^2 - (6 + 7)} \qquad \text{Simplify within the inner parentheses: } 6 + 10 = 16.$$

$$= \frac{3(-12)}{2^2 - 13} \qquad \text{Simplify within each parentheses: } 4 - 16 = -12 \text{ and } 6 + 7 = 13.$$

$$= \frac{3(-12)}{4 - 13} \qquad \text{Evaluate the power: } 2^2 = 4.$$

$$= \frac{-36}{-9} \qquad 3(-12) = -36; \, 4 - 13 = -9$$

$$= 4$$

EXAMPLE 6

Evaluating Algebraic Expressions

If $x = -2$, $y = 3$, and $z = -4$, evaluate

a. $-x^2 + y^2 z$ **b.** $\dfrac{2z^3 - 3y^2}{5x^2}$

SOLUTION In each part, we will substitute the numbers for the variables, apply the rules of order of operations, and simplify.

a. $-x^2 + y^2 z = -(-2)^2 + 3^2(-4)$

$\qquad\qquad = -(4) + 9(-4)$ Evaluate the powers.

$\qquad\qquad = -4 + (-36)$ Do the multiplication.

$\qquad\qquad = -40$ Do the addition.

b. $\dfrac{2z^3 - 3y^2}{5x^2} = \dfrac{2(-4)^3 - 3(3)^2}{5(-2)^2}$

$\qquad\qquad = \dfrac{2(-64) - 3(9)}{5(4)}$ Evaluate the powers.

$\qquad\qquad = \dfrac{-128 - 27}{20}$ Do the multiplications.

$\qquad\qquad = \dfrac{-155}{20}$ Do the subtraction.

$\qquad\qquad = -\dfrac{31}{4}$ Simplify the fraction.

Self Check 6 If $x = 3$ and $y = -2$, evaluate $\dfrac{2x^2 - 3y^2}{x - y}$.

Now Try Exercise 91.

4. Express Numbers in Scientific Notation

Scientists often work with numbers that are very large or very small. These numbers can be written compactly by expressing them in *scientific notation*.

Scientific Notation A number is written in *scientific notation* when it is written in the form

$\qquad N \times 10^n$

where $1 \le |N| < 10$ and n is an integer.

©Frtommy/Shutterstock.com

Light travels 29,980,000,000 centimeters per second. To express this number in scientific notation, we must write it as the product of a number between 1 and 10 and some integer power of 10. The number 2.998 lies between 1 and 10. To get 29,980,000,000, the decimal point in 2.998 must be moved ten places to the right. This is accomplished by multiplying 2.998 by 10^{10}.

Standard notation ⟶ $29{,}980{,}000{,}000 = 2.998 \times 10^{10}$ ⟵ Scientific notation

One meter is approximately 0.0006214 mile. To express this number in scientific notation, we must write it as the product of a number between 1 and 10 and some integer power of 10. The number 6.214 lies between 1 and 10. To get 0.0006214, the decimal point in 6.214 must be moved four places to the left. This is accomplished by multiplying 6.214 by $\frac{1}{10^4}$ or by multiplying 6.214 by 10^{-4}.

Standard notation ⟶ $0.0006214 = 6.214 \times 10^{-4}$ ⟵ Scientific notation

To write each of the following numbers in scientific notation, we start to the right of the first nonzero digit and count to the decimal point. The exponent gives the number of places the decimal point moves, and the sign of the exponent indicates the direction in which it moves.

a. $3\,7\,2\,0\,0\,0 = 3.72 \times 10^5$ 5 places to the right

b. $0.\,0\,0\,0\,5\,3\,7 = 5.37 \times 10^{-4}$ 4 places to the left

c. $7.36 = 7.36 \times 10^0$ No movement of the decimal point

Writing Numbers in Scientific Notation

EXAMPLE 7

Write each number in scientific notation: **a.** 62,000   **b.** −0.0027

SOLUTION   **a.** We must express 62,000 as a product of a number between 1 and 10 and some integer power of 10. This is accomplished by multiplying 6.2 by 10^4.

$$62,000 = 6.2 \times 10^4$$

b. We must express −0.0027 as a product of a number whose absolute value is between 1 and 10 and some integer power of 10. This is accomplished by multiplying −2.7 by 10^{-3}.

$$-0.0027 = -2.7 \times 10^{-3}$$

Self Check 7   Write each number in scientific notation:

a. −93,000,000   **b.** 0.0000087

Now Try Exercise 103.

Writing Numbers in Standard Notation

EXAMPLE 8

Write each number in standard notation:

a. 7.35×10^2 **b.** 3.27×10^{-5}

SOLUTION   **a.** The factor of 10^2 indicates that 7.35 must be multiplied by 2 factors of 10. Because each multiplication by 10 moves the decimal point one place to the right, we have

$$7.35 \times 10^2 = 735$$

b. The factor of 10^{-5} indicates that 3.27 must be divided by 5 factors of 10. Because each division by 10 moves the decimal point one place to the left, we have

$$3.27 \times 10^{-5} = 0.0000327$$

Self Check 8   Write each number in standard notation: **a.** 6.3×10^3 **b.** 9.1×10^{-4}

Now Try Exercise 111.

5. Use Scientific Notation to Simplify Computations

Another advantage of scientific notation becomes evident when we multiply and divide very large and very small numbers.

Using Scientific Notation to Simplify Computations

EXAMPLE 9

Use scientific notation to calculate $\dfrac{(3,400,000)(0.00002)}{170,000,000}$.

SOLUTION After changing each number to scientific notation, we can do the arithmetic on the numbers and the exponential expressions separately.

$$\frac{(3{,}400{,}000)(0.00002)}{170{,}000{,}000} = \frac{(3.4 \times 10^6)(2.0 \times 10^{-5})}{1.7 \times 10^8}$$

$$= \frac{6.8}{1.7} \times 10^{6+(-5)-8}$$

$$= 4.0 \times 10^{-7}$$

$$= 0.0000004$$

Self Check 9 Use scientific notation to simplify $\dfrac{(192{,}000)(0.0015)}{(0.0032)(4500)}$.

Now Try Exercise 119.

Accent on Technology

Scientific Notation

Graphing calculators will often give an answer in scientific notation. For example, consider 21^8.

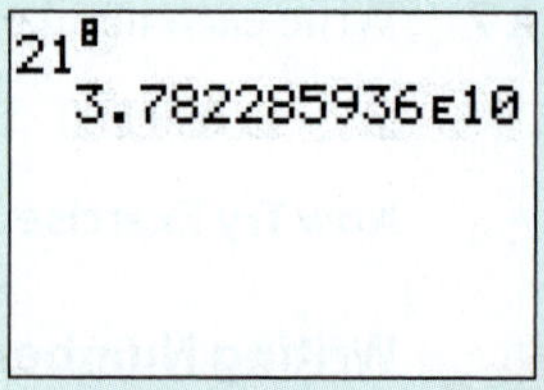

```
21^8
     3.782285936E10
```

FIGURE 0-17

We see in Figure 0-17 that the answer given is 3.782285936E10, which means $3.782285936 \times 10^{10}$.

Take Note

To calculate an expression like $\dfrac{21^8}{0.000000000061}$ on a scientific calculator, it is necessary to convert the denominator to scientific notation because the number has too many digits to fit on the screen.

- For scientific notation, we enter these numbers and press these keys: 6.1 [EXP] 11 [+/−].

- To evaluate the expression above, we enter these numbers and press these keys:

  21 [y^x] 8 [=] [÷] 6.1 [EXP] 11 [+/−] [=]

The display will read [6.200468748²⁰]. In standard notation, the answer is approximately 620,046,874,800,000,000,000.

Self Check Answers

1. **a.** $7 \cdot 7 \cdot 7 = 343$ **b.** $(-3)(-3) = 9$ **c.** $5 \cdot a \cdot a \cdot a$
d. $(5a)(5a)(5a)(5a) = 625 \cdot a \cdot a \cdot a \cdot a$ **2. a.** y^6 **b.** a^{18} **c.** x^{12}
d. $\dfrac{27a^9b^6}{c^9}$ **3. a.** 7 **b.** $\dfrac{3}{a^2}$ **c.** $\dfrac{1}{a^2}$ **d.** $\dfrac{1}{a^{12}}$ **4. a.** $\dfrac{1}{x^8}$ **b.** $\dfrac{1}{x}$
5. a. $\dfrac{x^{14}}{y^{10}}$ **b.** $\dfrac{27b^3}{8a^3}$ **6.** $\dfrac{6}{5}$ **7. a.** -9.3×10^7 **b.** 8.7×10^{-6}
8. a. 6300 **b.** 0.00091 **9.** 20

Exercises 0.2

Getting Ready

You should be able to complete these vocabulary and concept statements before you proceed to the practice exercises.

Fill in the blanks.

1. Each quantity in a product is called a _____ of the product.

2. A _____ number exponent tells how many times a base is used as a factor.

3. In the expression $(2x)^3$, ____ is the exponent and ____ is the base.

4. The expression x^n is called an _____ expression.

5. A number is in _____ notation when it is written in the form $N \times 10^n$, where $1 \le |N| < 10$ and n is an _____.

6. Unless _____ indicate otherwise, _____ are performed before additions.

Complete each exponent rule. Assume $x \ne 0$.

7. $x^m x^n = $ ____

8. $(x^m)^n = $ ____

9. $(xy)^n = $ ____

10. $\dfrac{x^m}{x^n} = $ ____

11. $x^0 = $ __

12. $x^{-n} = $ ____

Practice

Write each number or expression without using exponents.

13. 13^2

14. 10^3

15. -5^2

16. $(-5)^2$

17. $4x^3$

18. $(4x)^3$

19. $(-5x)^4$

20. $-6x^2$

21. $-8x^4$

22. $(-8x)^4$

Write each expression using exponents.

23. $7xxx$

24. $-8yyyy$

25. $(-x)(-x)$

26. $(2a)(2a)(2a)$

27. $(3t)(3t)(-3t)$

28. $-(2b)(2b)(2b)(2b)$

29. $xxxyy$

30. $aaabbbb$

Use a calculator to simplify each expression.

31. 2.2^3

32. 7.1^4

33. -0.5^4

34. $(-0.2)^4$

Simplify each expression. Write all answers without using negative exponents. Assume that all variables are restricted to those numbers for which the expression is defined.

35. $x^2 x^3$

36. $y^3 y^4$

37. $(z^2)^3$

38. $(t^6)^7$

39. $(y^5 y^2)^3$

40. $(a^3 a^6)a^4$

41. $(z^2)^3 (z^4)^5$

42. $(t^3)^4 (t^5)^2$

43. $(a^2)^3 (a^4)^2$

44. $(a^2)^4 (a^3)^3$

45. $(3x)^3$

46. $(-2y)^4$

47. $(x^2 y)^3$

48. $(x^3 z^4)^6$

49. $\left(\dfrac{a^2}{b}\right)^3$

50. $\left(\dfrac{x}{y^3}\right)^4$

51. $(-x)^0$

52. $4x^0$

53. $(4x)^0$

54. $-2x^0$

55. z^{-4}

56. $\dfrac{1}{t^{-2}}$

57. $y^{-2} y^{-3}$

58. $-m^{-2} m^3$

59. $(x^3 x^{-4})^{-2}$

60. $(y^{-2} y^3)^{-4}$

61. $\dfrac{x^7}{x^3}$

62. $\dfrac{r^5}{r^2}$

63. $\dfrac{a^{21}}{a^{17}}$

64. $\dfrac{t^{13}}{t^4}$

65. $\dfrac{(x^2)^2}{x^2 x}$

66. $\dfrac{s^9 s^3}{(s^2)^2}$

67. $\left(\dfrac{m^3}{n^2}\right)^3$

68. $\left(\dfrac{t^4}{t^3}\right)^3$

69. $\dfrac{(a^3)^{-2}}{aa^2}$

70. $\dfrac{r^9 r^{-3}}{(r^{-2})^3}$

71. $\left(\dfrac{a^{-3}}{b^{-1}}\right)^{-4}$

72. $\left(\dfrac{t^{-4}}{t^{-3}}\right)^{-2}$

73. $\left(\dfrac{r^4 r^{-6}}{r^3 r^{-3}}\right)^2$

74. $\dfrac{(x^{-3} x^2)^2}{(x^2 x^{-5})^{-3}}$

75. $\left(\dfrac{x^5 y^{-2}}{x^{-3} y^2}\right)^4$

76. $\left(\dfrac{x^{-7} y^5}{x^7 y^{-4}}\right)^3$

77. $\left(\dfrac{5x^{-3} y^{-2}}{3x^2 y^{-3}}\right)^{-2}$

78. $\left(\dfrac{3x^2 y^{-5}}{2x^{-2} y^{-6}}\right)^{-3}$

79. $\left(\dfrac{3x^5 y^{-3}}{6x^{-5} y^3}\right)^{-2}$

80. $\left(\dfrac{12x^{-4} y^3 z^{-5}}{4x^4 y^{-3} z^5}\right)^3$

81. $\dfrac{(8^{-2} z^{-3} y)^{-1}}{(5y^2 z^{-2})^3 (5yz^{-2})^{-1}}$

82. $\dfrac{(m^{-2} n^3 p^4)^{-2} (mn^{-2} p^3)^4}{(mn^{-2} p^3)^{-4} (mn^2 p)^{-1}}$

Simplify each expression.

83. $-\dfrac{5[6^2 + (9 - 5)]}{4(2 - 3)^2}$

84. $\dfrac{6[3 - (4 - 7)^2]}{-5(2 - 4^2)}$

Let x = −2, y = 0, and z = 3 and evaluate each expression.

85. x^2

86. $-x^2$

87. x^3

88. $-x^3$

89. $(-xz)^3$

90. $-xz^3$

91. $\dfrac{-(x^2z^3)}{z^2 - y^2}$

92. $\dfrac{z^2(x^2 - y^2)}{x^3z}$

93. $5x^2 - 3y^3z$

94. $3(x - z)^2 + 2(y - z)^3$

95. $\dfrac{-3x^{-3}z^{-2}}{6x^2z^{-3}}$

96. $\dfrac{(-5x^2z^{-3})^2}{5xz^{-2}}$

Express each number in scientific notation.

97. 372,000

98. 89,500

99. −177,000,000

100. −23,470,000,000

101. 0.007

102. 0.00052

103. −0.000000693

104. −0.000000089

105. one trillion

106. one millionth

Express each number in standard notation.

107. 9.37×10^5

108. 4.26×10^9

109. 2.21×10^{-5}

110. 2.774×10^{-2}

111. 0.00032×10^4

112. $9,300.0 \times 10^{-4}$

113. -3.2×10^{-3}

114. -7.25×10^3

Use the method of Example 9 to do each calculation. Write all answers in scientific notation.

115. $\dfrac{(65,000)(45,000)}{250,000}$

116. $\dfrac{(0.000000045)(0.00000012)}{45,000,000}$

117. $\dfrac{(0.00000035)(170,000)}{0.00000085}$

118. $\dfrac{(0.0000000144)(12,000)}{600,000}$

119. $\dfrac{(45,000,000,000)(212,000)}{0.00018}$

120. $\dfrac{(0.00000000275)(4750)}{500,000,000,000}$

Applications

Use scientific notation to compute each answer. Write all answers in scientific notation.

121. Speed of sound The speed of sound in air is 3.31×10^4 centimeters per second. Compute the speed of sound in meters per minute.

122. Volume of a box Calculate the volume of a box that has dimensions of 6000 by 9700 by 4700 millimeters.

123. Mass of a proton The mass of one proton is 0.0000000000000000000000000167248 gram. Find the mass of one billion protons.

124. Speed of light The speed of light in a vacuum is approximately 30,000,000,000 centimeters per second. Find the speed of light in miles per hour. (160,934.4 cm = 1 mile.)

125. License plates License plates come in various forms. The number of different license plates of the form three digits followed by three letters, is $10 \cdot 10 \cdot 10 \cdot 26 \cdot 26 \cdot 26$. Write this expression using exponents. Then evaluate it and express the result in scientific notation.

©Leonard Zhukovsky/Shutterstock.com

126. Astronomy The distance d, in miles, of the nth planet from the sun is given by the formula

$$d = 9,275,200[3(2^{n-2}) + 4]$$

To the nearest million miles, find the distance of Earth and the distance of Mars from the sun. Give each answer in scientific notation.

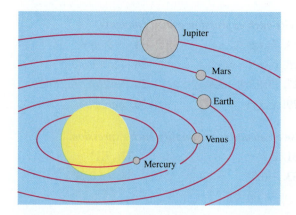

127. New way to the center of the Earth The spectacular "blue marble" image is the most detailed true-color image of the entire Earth to date. A new NASA-developed technique estimates Earth's center of mass within 1 millimeter (0.04 inch) a year by using a combination of four space-based techniques.

The distance from the Earth's center to the North Pole (the **polar radius**) measures approximately 6356.750 km, and the distance from the center to the equator (the **equatorial radius**) measures approximately 6378.135 km. Express each distance using scientific notation.

128. Refer to Exercise 127. Given that 1 km is approximately equal to 0.62 miles, use scientific notation to express each distance in miles.

Discovery and Writing

Write each expression with a single base.

129. $x^n x^2$

130. $\dfrac{x^m}{x^3}$

131. $\dfrac{x^m x^2}{x^3}$

132. $\dfrac{x^{3m+5}}{x^2}$

133. $x^{m+1} x^3$

134. $a^{n-3} a^3$

135. Explain why $-x^4$ and $(-x)^4$ represent different numbers.

136. Explain why $-x^{55}$ and $(-x)^{55}$ represent equal numbers.

137. Explain how to write a number in scientific notation.

138. Explain why 32×10^2 is not in scientific notation.

139. Explain why $x^{11} \cdot x^{11} \neq x^{121}$.

140. Explain why $11^2 \cdot 11^3 \neq 121^5$.

141. Explain why $\dfrac{y^{50}}{y^{10}} \neq y^5$.

142. Explain why $(6xyz)^6 \neq 6x^6 y^6 z^6$.

Critical Thinking

In Exercises 143–148, determine if the statement is true or false. If the statement is false, then correct it and make it true.

143. $0^0 = 1$

144. $\pi^0 = 1$

145. $x^{-n} = -\dfrac{1}{x^n}$

146. $(x + y)^{-n} = \dfrac{1}{x^n} + \dfrac{1}{y^n}$

147. $2^{-1} > 2^{-2}$

148. $(-2)^{-1} < (-2)^{-2}$

149. Young adults between the ages of 18 and 24 send an average of 110 text messages per day. If there are approximately 31.5 million young adults in America in this age group, how many text messages are sent in one year? Write the answer using scientific notation.

150. Health authorities recommend that we drink eight 8-ounce glasses of water each day. How many glasses of water would you drink over a lifetime of 80 years? Write the answer using scientific notation.

0.3 Rational Exponents and Radicals

In this section, we will learn to

1. Define rational exponents whose numerators are 1.
2. Define rational exponents whose numerators are not 1.
3. Define radical expressions.
4. Simplify and combine radicals.
5. Rationalize denominators and numerators.

"Dead Man's Curve" is a 1964 hit song by the rock and roll duo Jan Berry and Dean Torrence. The song details a teenage drag race that ends in an accident. Today, dead man's curve is a commonly used expression given to dangerous curves on our roads. Every curve has a "critical speed." If we exceed this speed, regardless of how skilled a driver we are, we will lose control of the vehicle.

The radical expression $3.9\sqrt{r}$ gives the critical speed in miles per hour when we travel a curved road with a radius of r feet. A knowledge of square roots and radicals is important and used in the construction of safe highways and roads. We will study the topic of radicals in this section.

1. Define Rational Exponents Whose Numerators Are 1

If we apply the rule $(x^m)^n = x^{mn}$ to $(25^{1/2})^2$, we obtain

$$(25^{1/2})^2 = 25^{(1/2)2} \qquad \text{Keep the base and multiply the exponents.}$$
$$= 25^1 \qquad \tfrac{1}{2} \cdot 2 = 1$$
$$= 25$$

Thus, $25^{1/2}$ is a real number whose square is 25. Although both $5^2 = 25$ and $(-5)^2 = 25$, we define $25^{1/2}$ to be the positive real number whose square is 25:

$$25^{1/2} = 5 \qquad \text{Read } 25^{1/2} \text{ as "the square root of 25."}$$

In general, we have the following definition.

> **Take Note**
>
> In the expression $a^{1/n}$, there is **no** real-number nth root of a when n is even and $a < 0$. For example, $(-64)^{1/2}$ is **not** a real number, because the square of **no** real number is -64.

Rational Exponents   If $a \geq 0$ and n is a natural number, then $a^{1/n}$ (read as "the nth root of a") is the nonnegative real number b such that

$$b^n = a$$

Since $b = a^{1/n}$, we have $b^n = (a^{1/n})^n = a$.

EXAMPLE 1   **Simplifying Expressions with Rational Exponents**

In each case, we will apply the definition of rational exponents.

a. $16^{1/2} = 4$ Because $4^2 = 16$. Read $16^{1/2}$ as "the square root of 16."

b. $27^{1/3} = 3$ Because $3^3 = 27$. Read $27^{1/3}$ as "the cube root of 27."

c. $\left(\dfrac{1}{81}\right)^{1/4} = \dfrac{1}{3}$ Because $\left(\dfrac{1}{3}\right)^4 = \dfrac{1}{81}$. Read $\left(\dfrac{1}{81}\right)^{1/4}$ as "the fourth root of $\dfrac{1}{81}$."

    d. $\;-32^{1/5} = -(\mathbf{32^{1/5}})$ Read $32^{1/5}$ as "the fifth root of 32."

                 $= -(\mathbf{2})$ Because $2^5 = 32$.

                 $= -2$

Self Check 1   Simplify: **a.** $100^{1/2}$ **b.** $243^{1/5}$

Now Try Exercise 15.

    If n is even in the expression $a^{1/n}$ and the base contains variables, we often use absolute value symbols to guarantee that an even root is nonnegative.

    $(49x^2)^{1/2} = 7|x|$ Because $(7|x|)^2 = 49x^2$. Since x could be negative, absolute value symbols are necessary to guarantee that the square root is nonnegative.

    $(16x^4)^{1/4} = 2|x|$ Because $(2|x|)^4 = 16x^4$. Since x could be negative, absolute value symbols are necessary to guarantee that the fourth root is nonnegative.

    $(729x^{12})^{1/6} = 3x^2$ Because $(3x^2)^6 = 729x^{12}$. Since x^2 is always nonnegative, no absolute value symbols are necessary. Read $(729x^{12})^{1/6}$ as "the sixth root of $729x^{12}$."

    If n is an odd number in the expression $a^{1/n}$, the base a can be negative.

Simplifying Expressions with Rational Exponents

EXAMPLE 2

Simplify by using the definition of rational exponent.

    a. $(-8)^{1/3} = -2$ Because $(-2)^3 = -8$.

    b. $(-3125)^{1/5} = -5$ Because $(-5)^5 = -3125$.

    c. $\left(-\dfrac{1}{1000}\right)^{1/3} = -\dfrac{1}{10}$ Because $\left(-\dfrac{1}{10}\right)^3 = -\dfrac{1}{1000}$.

Self Check 2   Simplify: **a.** $(-125)^{1/3}$ **b.** $(-100{,}000)^{1/5}$

Now Try Exercise 19.

    If n is odd in the expression $a^{1/n}$, and the base contains variables, we do not use absolute value symbols, because odd roots can be negative.

    $(-27x^3)^{1/3} = -3x$ Because $(-3x)^3 = -27x^3$.

    $(-128a^7)^{1/7} = -2a$ Because $(-2a)^7 = -128a^7$.

    We summarize the definitions concerning $a^{1/n}$ as follows.

Summary of Definitions of $a^{1/n}$

If n is a natural number and a is a real number in the expression $a^{1/n}$, then

If $a \geq 0$, then $a^{1/n}$ is the nonnegative real number b such that $b^n = a$.

If $a < 0$ $\begin{cases} \text{and } n \text{ is odd, then } a^{1/n} \text{ is the real number } b \text{ such that } b^n = a. \\ \text{and } n \text{ is even, then } a^{1/n} \text{ is not a real number.} \end{cases}$

    The following chart also shows the possibilities that can occur when simplifying $a^{1/n}$.

Strategy for Simplifying Expressions of the Form $a^{1/n}$

a	n	$a^{1/n}$	Examples
$a = 0$	n is a natural number.	$0^{1/n}$ is the real number 0 because $0^n = 0$.	$0^{1/2} = 0$ because $0^2 = 0$. $0^{1/5} = 0$ because $0^5 = 0$.
$a > 0$	n is a natural number.	$a^{1/n}$ is the non-negative real number such that $(a^{1/n})^n = a$.	$16^{1/2} = 4$ because $4^2 = 16$. $27^{1/3} = 3$ because $3^3 = 27$.
$a < 0$	n is an odd natural number.	$a^{1/n}$ is the real number such that $(a^{1/n})^n = a$.	$(-32)^{1/5} = -2$ because $(-2)^5 = -32$. $(-125)^{1/3} = -5$ because $(-5)^3 = -125$.
$a < 0$	n is an even natural number.	$a^{1/n}$ is not a real number.	$(-9)^{1/2}$ is not a real number. $(-81)^{1/4}$ is not a real number.

2. Define Rational Exponents Whose Numerators Are Not 1

The definition of $a^{1/n}$ can be extended to include rational exponents whose numerators are not 1. For example, $4^{3/2}$ can be written as either

$$(4^{1/2})^3 \text{ or } (4^3)^{1/2} \qquad \text{Because of the Power Rule, } (x^m)^n = x^{mn}.$$

This suggests the following rule.

Rule for Rational Exponents   If m and n are positive integers, the fraction $\frac{m}{n}$ is in lowest terms, and $a^{1/n}$ is a real number, then

$$a^{m/n} = (a^{1/n})^m = (a^m)^{1/n}$$

Tip

Keep a positive attitude when approaching problems with fractions as exponents. This can help you master them.

©karen roach/Shutterstock.com

In the previous rule, we can view the expression $a^{m/n}$ in two ways:

1. $(a^{1/n})^m$: the mth power of the nth root of a
2. $(a^m)^{1/n}$: the nth root of the mth power of a

For example, $(-27)^{2/3}$ can be simplified in two ways:

$$(-27)^{2/3} = [(-27)^{1/3}]^2 \qquad \text{or} \qquad (-27)^{2/3} = [(-27)^2]^{1/3}$$
$$= (-3)^2 \qquad\qquad\qquad\qquad\qquad = (729)^{1/3}$$
$$= 9 \qquad\qquad\qquad\qquad\qquad\qquad = 9$$

As this example suggests, it is usually easier to take the root of the base first to avoid large numbers.

Tip

It is helpful to think of the phrase *power over root* when we see a rational exponent. The numerator of the fraction represents the *power* and the denominator represents the *root*. Begin with the *root* when simplifying to avoid large numbers.

Negative Rational Exponents   If m and n are positive integers, the fraction $\frac{m}{n}$ is in lowest terms and $a^{1/n}$ is a real number, then

$$a^{-m/n} = \frac{1}{a^{m/n}} \quad \text{and} \quad \frac{1}{a^{-m/n}} = a^{m/n} \quad (a \neq 0)$$

EXAMPLE 3

Simplifying Expressions with Rational Exponents

We will apply the rules for rational exponents.

a. $25^{3/2} = (25^{1/2})^3$

$= 5^3$

$= 125$

b. $\left(-\dfrac{x^6}{1000}\right)^{2/3} = \left[\left(-\dfrac{x^6}{1000}\right)^{1/3}\right]^2$

$= \left(-\dfrac{x^2}{10}\right)^2$

$= \dfrac{x^4}{100}$

c. $32^{-2/5} = \dfrac{1}{32^{2/5}}$

$= \dfrac{1}{(32^{1/5})^2}$

$= \dfrac{1}{2^2}$

$= \dfrac{1}{4}$

d. $\dfrac{1}{81^{-3/4}} = 81^{3/4}$

$= (81^{1/4})^3$

$= 3^3$

$= 27$

Self Check 3   Simplify: **a.** $49^{3/2}$ **b.** $16^{-3/4}$ **c.** $\dfrac{1}{(27x^3)^{-2/3}}$

Now Try Exercise 43.

Because of the definition, rational exponents follow the same rules as integer exponents.

EXAMPLE 4

Using Exponent Rules to Simplify Expressions with Rational Exponents

Simplify each expression. Assume that all variables represent positive numbers, and write answers without using negative exponents.

a. $(36x)^{1/2} = 36^{1/2}x^{1/2}$

$= 6x^{1/2}$

b. $\dfrac{(a^{1/3}b^{2/3})^6}{(y^3)^2} = \dfrac{a^{6/3}b^{12/3}}{y^6}$

$= \dfrac{a^2b^4}{y^6}$

c. $\dfrac{a^{x/2}a^{x/4}}{a^{x/6}} = a^{x/2 + x/4 - x/6}$

$= a^{6x/12 + 3x/12 - 2x/12}$

$= a^{7x/12}$

d. $\left[\dfrac{-c^{-2/5}}{c^{4/5}}\right]^{5/3} = (-c^{-2/5 - 4/5})^{5/3}$

$= [(-1)(c^{-6/5})]^{5/3}$

$= (-1)^{5/3}(c^{-6/5})^{5/3}$

$= -1c^{-30/15}$

$= -c^{-2}$

$= -\dfrac{1}{c^2}$

Self Check 4   Use the directions for Example 4:

a. $\left(\dfrac{y^2}{49}\right)^{1/2}$ b. $\dfrac{b^{3/7}b^{2/7}}{b^{4/7}}$ c. $\dfrac{(9r^2s)^{1/2}}{rs^{-3/2}}$

Now Try Exercise 59.

3. Define Radical Expressions

Radical signs can also be used to express roots of numbers.

Definition of $\sqrt[n]{a}$ If n is a natural number greater than 1 and if $a^{1/n}$ is a real number, then

$$\sqrt[n]{a} = a^{1/n}$$

In the **radical expression** $\sqrt[n]{a}$, the symbol $\sqrt{}$ is the **radical sign**, a is the **radicand**, and n is the **index** (or the **order**) of the radical expression. If the order is 2, the expression is a **square root**, and we do not write the index.

$$\sqrt{a} = \sqrt[2]{a}$$

If the index of a radical is 3, we call the radical a **cube root**.

---

nth Root of a Nonnegative Number   If n is a natural number greater than 1 and $a \geq 0$, then $\sqrt[n]{a}$ is the nonnegative number whose nth power is a.

$$\left(\sqrt[n]{a}\right)^n = a$$

> **Take Note**
>
> In the expression $\sqrt[n]{a}$, there is **no** real-number nth root of a when n is even and $a < 0$. For example, $\sqrt{-64}$ is **not** a real number, because the square of **no** real number is -64.

If 2 is substituted for n in the equation $\left(\sqrt[n]{a}\right)^n = a$, we have

$$\left(\sqrt[2]{a}\right)^2 = \left(\sqrt{a}\right)^2 = \sqrt{a}\sqrt{a} = a \text{ for } a \geq 0$$

This shows that if a number a can be factored into two equal factors, either of those factors is a square root of a. Furthermore, if a can be factored into n equal factors, any one of those factors is an nth root of a.

If n is an odd number greater than 1 in the expression $\sqrt[n]{a}$, the radicand can be negative.

Finding nth Roots of Real Numbers

EXAMPLE 5

We apply the definitions of cube root and fifth root.

a. $\sqrt[3]{-27} = -3$ Because $(-3)^3 = -27$.

b. $\sqrt[3]{-8} = -2$ Because $(-2)^3 = -8$.

c. $\sqrt[3]{-\dfrac{27}{1000}} = -\dfrac{3}{10}$ Because $\left(-\dfrac{3}{10}\right)^3 = -\dfrac{27}{1000}$.

d. $-\sqrt[5]{-243} = -\left(\sqrt[5]{-243}\right)$

$$= -(-3)$$

$$= 3$$

Self Check 5   Find each root: **a.** $\sqrt[3]{216}$ **b.** $\sqrt[5]{-\dfrac{1}{32}}$

Now Try Exercise 69.

We summarize the definitions concerning $\sqrt[n]{a}$ as follows.

Summary of Definitions of $\sqrt[n]{a}$	If n is a natural number greater than 1 and a is a real number, then
	If $a \geq 0$, then $\sqrt[n]{a}$ is the nonnegative real number such that $\left(\sqrt[n]{a}\right)^n = a$.
	If $a < 0$ $\begin{cases}\text{and } n \text{ is odd, then } \sqrt[n]{a} \text{ is the real number such that } \left(\sqrt[n]{a}\right)^n = a. \\ \text{and } n \text{ is even, then } \sqrt[n]{a} \text{ is not a real number.}\end{cases}$

The following chart also shows the possibilities that can occur when simplifying $\sqrt[n]{a}$.

Strategy for Simplifying Expressions of the Form $\sqrt[n]{a}$

a	n	$\sqrt[n]{a}$	Examples
$a = 0$	n is a natural number greater than 1.	$\sqrt[n]{0}$ is the real number 0 because $0^n = 0$.	$\sqrt[3]{0} = 0$ because $0^3 = 0$. $\sqrt[5]{0} = 0$ because $0^5 = 0$.
$a > 0$	n is a natural number greater than 1.	$\sqrt[n]{a}$ is the nonnegative real number such that $\left(\sqrt[n]{a}\right)^n = a$.	$\sqrt{16} = 4$ because $4^2 = 16$. $\sqrt[3]{27} = 3$ because $3^3 = 27$.
$a < 0$	n is an odd natural number greater than 1.	$\sqrt[n]{a}$ is the real number such that $\left(\sqrt[n]{a}\right)^n = a$.	$\sqrt[5]{-32} = -2$ because $(-2)^5 = -32$. $\sqrt[3]{-125} = -5$ because $(-5)^3 = -125$.
$a < 0$	n is an even natural number.	$\sqrt[n]{a}$ is not a real number.	$\sqrt{-9}$ is not a real number. $\sqrt[4]{-81}$ is not a real number.

We have seen that if $a^{1/n}$ is real, then $a^{m/n} = (a^{1/n})^m = (a^m)^{1/n}$. This same fact can be stated in radical notation.

$$a^{m/n} = \left(\sqrt[n]{a}\right)^m = \sqrt[n]{a^m}$$

Thus, *the mth power of the nth root of a is the same as the nth root of the mth power of a*. For example, to find $\sqrt[3]{27^2}$, we can proceed in either of two ways:

$$\sqrt[3]{27^2} = \left(\sqrt[3]{27}\right)^2 = 3^2 = 9 \text{ or } \sqrt[3]{27^2} = \sqrt[3]{729} = 9$$

By definition, $\sqrt{a^2}$ represents a nonnegative number. If a could be negative, we must use absolute value symbols to guarantee that $\sqrt{a^2}$ will be nonnegative. Thus, if a is unrestricted,

$$\sqrt{a^2} = |a|$$

A similar argument holds when the index is any even natural number. The symbol $\sqrt[4]{a^4}$, for example, means the *positive* fourth root of a^4. Thus, if a is unrestricted,

$$\sqrt[4]{a^4} = |a|$$

EXAMPLE 6

Simplifying Radical Expressions

If x is unrestricted, simplify **a.** $\sqrt[6]{64x^6}$ **b.** $\sqrt[3]{x^3}$ **c.** $\sqrt{9x^8}$

SOLUTION We apply the definitions of sixth roots, cube roots, and square roots.

a. $\sqrt[6]{64x^6} = 2|x|$ Use absolute value symbols to guarantee that the result will be nonnegative.

b. $\sqrt[3]{x^3} = x$ Because the index is odd, no absolute value symbols are needed.

c. $\sqrt{9x^8} = 3x^4$ Because $3x^4$ is always nonnegative, no absolute value symbols are needed.

Self Check 6 Use the directions for Example 6:

a. $\sqrt[4]{16x^4}$ **b.** $\sqrt[3]{27y^3}$ **c.** $\sqrt[4]{x^8}$

Now Try Exercise 73.

> **Tip**
>
> If you do not know whether or not to include absolute value in your answers when simplifying expressions with rational exponents or radical expressions, these two tips should help.
>
> 1. If n is **even** in the expression $a^{m/n}$ and your answer involves a variable raised to the **odd** power, then include the variable part of your answer in absolute value.
>
> 2. If the index of a radical is **even** and your answer involves a variable raised to the **odd** power, then include the variable part of your answer in absolute value.

4. Simplify and Combine Radicals

Many properties of exponents have counterparts in radical notation. For example, since $a^{1/n}b^{1/n} = (ab)^{1/n}$ and $\frac{a^{1/n}}{b^{1/n}} = \left(\frac{a}{b}\right)^{1/n}$ and $(b \neq 0)$, we have the following.

Multiplication and Division Properties of Radicals	If all expressions represent real numbers,
	$\sqrt[n]{a}\sqrt[n]{b} = \sqrt[n]{ab}$ $\dfrac{\sqrt[n]{a}}{\sqrt[n]{b}} = \sqrt[n]{\dfrac{a}{b}}$ $(b \neq 0)$

In words, we say

The product of two nth roots is equal to the nth root of their product.

The quotient of two nth roots is equal to the nth root of their quotient.

These properties involve the nth root of the product of two numbers or the nth root of the quotient of two numbers. There is no such property for sums or differences.

> **Caution**
>
> Some students mistakenly think there are sum and difference properties for radicals and use them to simplify radical expressions. Avoid this common error. For example,
>
> $$\sqrt{9 + 4} \neq \sqrt{9} + \sqrt{4}, \text{ because}$$
>
> $$\sqrt{9 + 4} = \sqrt{13} \qquad \text{but} \qquad \sqrt{9} + \sqrt{4} = 3 + 2 = 5$$
>
> and $\sqrt{13} \neq 5$. In general,
>
> $$\sqrt{a + b} \neq \sqrt{a} + \sqrt{b} \qquad \text{and} \qquad \sqrt{a - b} \neq \sqrt{a} - \sqrt{b}$$

Numbers that are squares of positive integers, such as

1, 4, 9, 16, 25, and 36

are called **perfect squares**. Expressions such as $4x^2$ and $\frac{1}{9}x^6$ are also perfect squares, because each one is the square of another expression with integer exponents and rational coefficients.

$$4x^2 = (2x)^2 \text{ and } \frac{1}{9}x^6 = \left(\frac{1}{3}x^3\right)^2$$

Numbers that are cubes of positive integers, such as

1, 8, 27, 64, 125, and 216

are called **perfect cubes**. Expressions such as $64x^3$ and $\frac{1}{27}x^9$ are also perfect cubes, because each one is the cube of another expression with integer exponents and rational coefficients.

$$64x^3 = (4x)^3 \text{ and } \frac{1}{27}x^9 = \left(\frac{1}{3}x^3\right)^3$$

There are also perfect fourth powers, perfect fifth powers, and so on.

We can use perfect powers and the Multiplication Property of Radicals to simplify many radical expressions. For example, to simplify $\sqrt{12x^5}$, we factor $12x^5$ so that one factor is the largest perfect square that divides $12x^5$. In this case, it is $4x^4$. We then rewrite $12x^5$ as $4x^4 \cdot 3x$ and simplify.

$$\sqrt{12x^5} = \sqrt{4x^4 \cdot 3x} \qquad \text{Factor } 12x^5 \text{ as } 4x^4 \cdot 3x.$$
$$= \sqrt{4x^4}\sqrt{3x} \qquad \text{Use the Multiplication Property of Radicals: } \sqrt{ab} = \sqrt{a}\sqrt{b}.$$
$$= 2x^2\sqrt{3x} \qquad \sqrt{4x^4} = 2x^2$$

To simplify $\sqrt[3]{432x^9y}$, we find the largest perfect-cube factor of $432x^9y$ (which is $216x^9$) and proceed as follows:

$$\sqrt[3]{432x^9y} = \sqrt[3]{216x^9 \cdot 2y} \qquad \text{Factor } 432x^9y \text{ as } 216x^9 \cdot 2y.$$
$$= \sqrt[3]{216x^9}\sqrt[3]{2y} \qquad \text{Use the Multiplication Property of Radicals: } \sqrt[3]{ab} = \sqrt[3]{a}\sqrt[3]{b}.$$
$$= 6x^3\sqrt[3]{2y} \qquad \sqrt[3]{216x^9} = 6x^3$$

Radical expressions with the same index and the same radicand are called **like** or **similar radicals**. We can combine the like radicals in $3\sqrt{2} + 2\sqrt{2}$ by using the Distributive Property.

$$3\sqrt{2} + 2\sqrt{2} = (3 + 2)\sqrt{2}$$
$$= 5\sqrt{2}$$

This example suggests that to combine like radicals, we *add their numerical coefficients and keep the same radical.*

When radicals have the same index but different radicands, we can often change them to equivalent forms having the same radicand. We can then combine them. For example, to simplify $\sqrt{27} - \sqrt{12}$, we simplify both radicals and combine like radicals.

$$\sqrt{27} - \sqrt{12} = \sqrt{9 \cdot 3} - \sqrt{4 \cdot 3} \qquad \text{Factor 27 and 12.}$$
$$= \sqrt{9}\sqrt{3} - \sqrt{4}\sqrt{3} \qquad \sqrt{ab} = \sqrt{a}\sqrt{b}$$
$$= 3\sqrt{3} - 2\sqrt{3} \qquad \sqrt{9} = 3 \text{ and } \sqrt{4} = 2.$$
$$= \sqrt{3} \qquad \text{Combine like radicals.}$$

Tip

There are commonly occurring fourth and fifth roots. Memorizing the ones listed in the following table can be very helpful when simplifying radicals. Assume that the variables represent positive numbers.

Fourth Roots	Fifth Roots
$\sqrt[4]{1} = 1$	$\sqrt[5]{1} = 1$
$\sqrt[4]{16} = 2$	$\sqrt[5]{32} = 2$
$\sqrt[4]{81} = 3$	$\sqrt[5]{243} = 3$
$\sqrt[4]{256} = 4$	$\sqrt[5]{1024} = 4$
$\sqrt[4]{625} = 5$	$\sqrt[5]{x^5} = 1$
$\sqrt[4]{x^4} = x$	$\sqrt[5]{x^{10}} = x^2$
$\sqrt[4]{x^8} = x^2$	$\sqrt[5]{x^{15}} = x^3$
$\sqrt[4]{x^{12}} = x^3$	$\sqrt[5]{x^{20}} = x^4$

Adding and Subtracting Radical Expressions

EXAMPLE 7

Simplify: **a.** $\sqrt{50} + \sqrt{200}$ **b.** $3z\sqrt[5]{64z} - 2\sqrt[5]{2z^6}$

SOLUTION   We will simplify each radical expression and then combine like radicals.

$$\textbf{a.} \quad \sqrt{50} + \sqrt{200} = \sqrt{25 \cdot 2} + \sqrt{100 \cdot 2}$$
$$= \sqrt{25}\sqrt{2} + \sqrt{100}\sqrt{2}$$
$$= 5\sqrt{2} + 10\sqrt{2}$$
$$= 15\sqrt{2}$$

$$\textbf{b.} \quad 3z\sqrt[5]{64z} - 2\sqrt[5]{2z^6} = 3z\sqrt[5]{32 \cdot 2z} - 2\sqrt[5]{z^5 \cdot 2z}$$
$$= 3z\sqrt[5]{32}\,\sqrt[5]{2z} - 2\sqrt[5]{z^5}\sqrt[5]{2z}$$
$$= 3z(2)\sqrt[5]{2z} - 2z\sqrt[5]{2z}$$
$$= 6z\sqrt[5]{2z} - 2z\sqrt[5]{2z}$$
$$= 4z\sqrt[5]{2z}$$

Self Check 7   Simplify: **a.** $\sqrt{18} - \sqrt{8}$ **b.** $2\sqrt[3]{81a^4} + a\sqrt[3]{24a}$

Now Try Exercise 85.

5. Rationalize Denominators and Numerators

By **rationalizing the denominator**, we can write a fraction such as

$$\frac{\sqrt{5}}{\sqrt{3}}$$

as a fraction with a rational number in the denominator. All that we must do is multiply both the numerator and the denominator by $\sqrt{3}$. (Note that $\sqrt{3}\sqrt{3}$ is the rational number 3.)

$$\frac{\sqrt{5}}{\sqrt{3}} = \frac{\sqrt{5}\sqrt{3}}{\sqrt{3}\sqrt{3}} = \frac{\sqrt{15}}{3}$$

Take Note

When we rationalize the denominator, a radical will no longer appear in the denominator. When we rationalize the numerator, a radical will no longer appear in the numerator.

To rationalize the numerator, we multiply both the numerator and the denominator by $\sqrt{5}$. (Note that $\sqrt{5}\sqrt{5}$ is the rational number 5.)

$$\frac{\sqrt{5}}{\sqrt{3}} = \frac{\sqrt{5}\sqrt{5}}{\sqrt{3}\sqrt{5}} = \frac{5}{\sqrt{15}}$$

Rationalizing the Denominator of a Radical Expression

EXAMPLE 8

Rationalize each denominator and simplify. Assume that all variables represent positive numbers.

a. $\dfrac{1}{\sqrt{7}}$ **b.** $\sqrt[3]{\dfrac{3}{4}}$ **c.** $\sqrt{\dfrac{3}{x}}$ **d.** $\sqrt{\dfrac{3a^3}{5x^5}}$

SOLUTION   We will multiply both the numerator and the denominator by a radical that will make the denominator a rational number.

a. $\dfrac{1}{\sqrt{7}} = \dfrac{1\sqrt{7}}{\sqrt{7}\sqrt{7}}$

$= \dfrac{\sqrt{7}}{7}$

b. $\sqrt[3]{\dfrac{3}{4}} = \dfrac{\sqrt[3]{3}}{\sqrt[3]{4}}$

$= \dfrac{\sqrt[3]{3}\sqrt[3]{2}}{\sqrt[3]{4}\sqrt[3]{2}}$ Multiply numerator and denominator by $\sqrt[3]{2}$, because $\sqrt[3]{4}\sqrt[3]{2} = \sqrt[3]{8} = 2$.

$= \dfrac{\sqrt[3]{6}}{\sqrt[3]{8}}$

$= \dfrac{\sqrt[3]{6}}{2}$

c. $\sqrt{\dfrac{3}{x}} = \dfrac{\sqrt{3}}{\sqrt{x}}$

$= \dfrac{\sqrt{3}\sqrt{x}}{\sqrt{x}\sqrt{x}}$

$= \dfrac{\sqrt{3x}}{x}$

d. $\sqrt{\dfrac{3a^3}{5x^5}} = \dfrac{\sqrt{3a^3}}{\sqrt{5x^5}}$

$= \dfrac{\sqrt{3a^3}\sqrt{5x}}{\sqrt{5x^5}\sqrt{5x}}$ Multiply numerator and denominator by $\sqrt{5x}$, because $\sqrt{5x^5}\sqrt{5x} = \sqrt{25x^6} = 5x^3$.

$= \dfrac{\sqrt{15a^3x}}{\sqrt{25x^6}}$

$= \dfrac{\sqrt{a^2}\sqrt{15ax}}{5x^3}$

$= \dfrac{a\sqrt{15ax}}{5x^3}$

Self Check 8   Use the directions for Example 8:

a. $\dfrac{6}{\sqrt{6}}$ **b.** $\sqrt[3]{\dfrac{2}{5x}}$

Now Try Exercise 101.

Tip

Expressions with square roots and cube roots are rationalized differently.

- To rationalize the denominator of $\dfrac{7}{\sqrt{5}}$, we multiply by $\dfrac{\sqrt{5}}{\sqrt{5}}$ and obtain $\dfrac{7\sqrt{5}}{\sqrt{25}} = \dfrac{7\sqrt{5}}{5}$.

- To rationalize the denominator of $\dfrac{7}{\sqrt[3]{5}}$, we do **not** multiply by $\dfrac{\sqrt[3]{5}}{\sqrt[3]{5}}$. If we do, we obtain $\dfrac{7\sqrt[3]{5}}{\sqrt[3]{25}}$ and the denominator isn't rationalized.

It is easy to memorize how to rationalize a square root. It requires a higher level of thinking to rationalize a cube root. Consider $\sqrt[3]{5}$. Since *cube root* means "3rd root," we need to multiply by a number so that we obtain 5^3 or 125 inside the cube root. Multiplying $\sqrt[3]{5}$ by $\sqrt[3]{25}$ does that for us.

$$\frac{7\sqrt[3]{25}}{\sqrt[3]{5}\sqrt[3]{25}} = \frac{7\sqrt[3]{25}}{\sqrt[3]{125}} = \frac{7\sqrt[3]{25}}{5}$$

Rationalizing roots of higher order will require a higher level of thinking.

Rationalizing the Numerator of a Radical Expression

EXAMPLE 9   Rationalize each numerator and simplify. Assume that all variables represent positive numbers: **a.** $\dfrac{\sqrt{x}}{7}$ **b.** $\dfrac{2\sqrt[3]{9x}}{3}$

SOLUTION   We will multiply both the numerator and the denominator by a radical that will make the numerator a rational number.

a. $\dfrac{\sqrt{x}}{7} = \dfrac{\sqrt{x} \cdot \sqrt{x}}{7\sqrt{x}}$

$= \dfrac{x}{7\sqrt{x}}$

b. $\dfrac{2\sqrt[3]{9x}}{3} = \dfrac{2\sqrt[3]{9x} \cdot \sqrt[3]{3x^2}}{3\sqrt[3]{3x^2}}$

$= \dfrac{2\sqrt[3]{27x^3}}{3\sqrt[3]{3x^2}}$

$= \dfrac{2(3x)}{3\sqrt[3]{3x^2}}$

$= \dfrac{2x}{\sqrt[3]{3x^2}}$ *Divide out the 3's.*

Self Check 9   Use the directions for Example 9:

a. $\dfrac{\sqrt{2x}}{5}$ b. $\dfrac{3\sqrt[3]{2y^2}}{6}$

Now Try Exercise 111.

After rationalizing denominators, we often can simplify an expression.

EXAMPLE 10   **Rationalizing Denominators and Simplifying**

Simplify: $\sqrt{\dfrac{1}{2}} + \sqrt{\dfrac{1}{8}}$.

SOLUTION   We will rationalize the denominators of each radical and then combine like radicals.

$\sqrt{\dfrac{1}{2}} + \sqrt{\dfrac{1}{8}} = \dfrac{1}{\sqrt{2}} + \dfrac{1}{\sqrt{8}}$ $\sqrt{\dfrac{1}{2}} = \dfrac{\sqrt{1}}{\sqrt{2}} = \dfrac{1}{\sqrt{2}}; \ \sqrt{\dfrac{1}{8}} = \dfrac{\sqrt{1}}{\sqrt{8}} = \dfrac{1}{\sqrt{8}}$

$= \dfrac{1\sqrt{2}}{\sqrt{2}\sqrt{2}} + \dfrac{1\sqrt{2}}{\sqrt{8}\sqrt{2}}$

$= \dfrac{\sqrt{2}}{2} + \dfrac{\sqrt{2}}{\sqrt{16}}$

$= \dfrac{\sqrt{2}}{2} + \dfrac{\sqrt{2}}{4}$

$= \dfrac{3\sqrt{2}}{4}$

Self Check 10   Simplify: $\sqrt[3]{\dfrac{x}{2}} - \sqrt[3]{\dfrac{x}{16}}$.

Now Try Exercise 115.

Another property of radicals can be derived from the properties of exponents. If all of the expressions represent real numbers,

$$\sqrt[n]{\sqrt[m]{x}} = \sqrt[n]{x^{1/m}} = (x^{1/m})^{1/n} = x^{1/(mn)} = \sqrt[mn]{x}$$

$$\sqrt[m]{\sqrt[n]{x}} = \sqrt[m]{x^{1/n}} = (x^{1/n})^{1/m} = x^{1/(nm)} = \sqrt[mn]{x}$$

These results are summarized in the following *theorem* (a fact that can be proved).

Theorem   If all of the expressions involved represent real numbers, then

$$\sqrt[m]{\sqrt[n]{x}} = \sqrt[n]{\sqrt[m]{x}} = \sqrt[mn]{x}$$

We can use the previous theorem to simplify many radicals. For example,

$$\sqrt[3]{\sqrt{8}} = \sqrt{\sqrt[3]{8}} = \sqrt{2}$$

Rational exponents can be used to simplify many radical expressions, as shown in the following example.

EXAMPLE 11   **Simplifying Radicals Using the Previously Stated Theorem**

Simplify. Assume that x and y are positive numbers.

a. $\sqrt[6]{4}$ **b.** $\sqrt[12]{x^3}$ **c.** $\sqrt[9]{8y^3}$

SOLUTION   In each case, we will write the radical as an exponential expression, simplify the resulting expression, and write the final result as a radical.

a. $\sqrt[6]{4} = 4^{1/6} = (2^2)^{1/6} = 2^{2/6} = 2^{1/3} = \sqrt[3]{2}$

b. $\sqrt[12]{x^3} = x^{3/12} = x^{1/4} = \sqrt[4]{x}$

c. $\sqrt[9]{8y^3} = (2^3y^3)^{1/9} = (2y)^{3/9} = (2y)^{1/3} = \sqrt[3]{2y}$

Self Check 11   Simplify: **a.** $\sqrt[4]{4}$ **b.** $\sqrt[9]{27x^3}$

Now Try Exercise 117.

Self Check Answers

1. a. 10   **b.** 3   **2. a.** -5 **b.** -10 **3. a.** 343   **b.** $\dfrac{1}{8}$

c. $9x^2$ **4. a.** $\dfrac{y}{7}$ **b.** $b^{1/7}$ **c.** $3s^2$ **5. a.** 6   **b.** $-\dfrac{1}{2}$ **6. a.** $2|x|$

b. $3y$ **c.** x^2 **7. a.** $\sqrt{2}$ **b.** $8a\sqrt[3]{3a}$ **8. a.** $\sqrt{6}$ **b.** $\dfrac{\sqrt[3]{50x^2}}{5x}$

9. a. $\dfrac{2x}{5\sqrt{2x}}$ **b.** $\dfrac{y}{\sqrt[3]{4y}}$ **10.** $\dfrac{\sqrt[3]{4x}}{4}$ **11. a.** $\sqrt{2}$ **b.** $\sqrt[3]{3x}$

Exercises 0.3

Getting Ready

You should be able to complete these vocabulary and concept statements before you proceed to the practice exercises.

Fill in the blanks.

1. If $a = 0$ and n is a natural number, then $a^{1/n} = $ ____ .
2. If $a > 0$ and n is a natural number, then $a^{1/n}$ is a _____ number.
3. If $a < 0$ and n is an even number, then $a^{1/n}$ is ____ a real number.

4. $6^{2/3}$ can be written as _____ or _____ .
5. $\sqrt[n]{a} = $ ____
6. $\sqrt{a^2} = $ ____
7. $\sqrt[n]{a}\sqrt[n]{b} = $ ____
8. $\sqrt[n]{\dfrac{a}{b}} = $ ____
9. $\sqrt{x + y}$ ____ $\sqrt{x} + \sqrt{y}$
10. $\sqrt[m]{\sqrt[n]{x}}$ or $\sqrt[n]{\sqrt[m]{x}}$ can be written as ____ .

Practice

Simplify each expression.

11. $9^{1/2}$

12. $8^{1/3}$

13. $\left(\dfrac{1}{25}\right)^{1/2}$

14. $\left(\dfrac{16}{625}\right)^{1/4}$

15. $-81^{1/4}$

16. $-\left(\dfrac{8}{27}\right)^{1/3}$

17. $(10{,}000)^{1/4}$

18. $1024^{1/5}$

19. $\left(-\dfrac{27}{8}\right)^{1/3}$

20. $-64^{1/3}$

21. $(-64)^{1/2}$

22. $(-125)^{1/3}$

Simplify each expression. Use absolute value symbols when necessary.

23. $(16a^2)^{1/2}$

24. $(25a^4)^{1/2}$

25. $(16a^4)^{1/4}$

26. $(-64a^3)^{1/3}$

27. $(-32a^5)^{1/5}$

28. $(64a^6)^{1/6}$

29. $(-216b^6)^{1/3}$

30. $(256t^8)^{1/4}$

31. $\left(\dfrac{16a^4}{25b^2}\right)^{1/2}$

32. $\left(-\dfrac{a^5}{32b^{10}}\right)^{1/5}$

33. $\left(-\dfrac{1000x^6}{27y^3}\right)^{1/3}$

34. $\left(\dfrac{49t^2}{100z^4}\right)^{1/2}$

Simplify each expression. Write all answers without using negative exponents.

35. $4^{3/2}$

36. $8^{2/3}$

37. $-16^{3/2}$

38. $(-8)^{2/3}$

39. $-1000^{2/3}$

40. $100^{3/2}$

41. $64^{-1/2}$

42. $25^{-1/2}$

43. $64^{-3/2}$

44. $49^{-3/2}$

45. $-9^{-3/2}$

46. $(-27)^{-2/3}$

47. $\left(\dfrac{4}{9}\right)^{5/2}$

48. $\left(\dfrac{25}{81}\right)^{3/2}$

49. $\left(-\dfrac{27}{64}\right)^{-2/3}$

50. $\left(\dfrac{125}{8}\right)^{-4/3}$

Simplify each expression. Assume that all variables represent positive numbers. Write all answers without using negative exponents.

51. $(100s^4)^{1/2}$

52. $(64u^6v^3)^{1/3}$

53. $(32y^{10}z^5)^{-1/5}$

54. $(625a^4b^8)^{-1/4}$

55. $(x^{10}y^5)^{3/5}$

56. $(64a^6b^{12})^{5/6}$

57. $(r^8s^{16})^{-3/4}$

58. $(-8x^9y^{12})^{-2/3}$

59. $\left(-\dfrac{8a^6}{125b^9}\right)^{2/3}$

60. $\left(\dfrac{16x^4}{625y^8}\right)^{3/4}$

61. $\left(\dfrac{27r^6}{1000s^{12}}\right)^{-2/3}$

62. $\left(-\dfrac{32m^{10}}{243n^{15}}\right)^{-2/5}$

63. $\dfrac{a^{2/5}a^{4/5}}{a^{1/5}}$

64. $\dfrac{x^{6/7}x^{3/7}}{x^{2/7}x^{5/7}}$

Simplify each radical expression.

65. $\sqrt{49}$

66. $\sqrt{81}$

67. $\sqrt[3]{125}$

68. $\sqrt[3]{-64}$

69. $\sqrt[3]{-125}$

70. $\sqrt[5]{-243}$

71. $\sqrt[5]{-\dfrac{32}{100{,}000}}$

72. $\sqrt[4]{\dfrac{256}{625}}$

Simplify each expression, using absolute value symbols when necessary. Write answers without using negative exponents.

73. $\sqrt{36x^2}$

74. $-\sqrt{25y^2}$

75. $\sqrt{9y^4}$

76. $\sqrt{a^4b^8}$

77. $\sqrt[3]{8y^3}$

78. $\sqrt[3]{-27z^9}$

79. $\sqrt[4]{\dfrac{x^4y^8}{z^{12}}}$

80. $\sqrt[5]{\dfrac{a^{10}b^5}{c^{15}}}$

Simplify each expression. Assume that all variables represent positive numbers so that no absolute value symbols are needed.

81. $\sqrt{8}-\sqrt{2}$

82. $\sqrt{75}-2\sqrt{27}$

83. $\sqrt{200x^2}+\sqrt{98x^2}$

84. $\sqrt{128a^3}-a\sqrt{162a}$

85. $2\sqrt{48y^5}-3y\sqrt{12y^3}$

86. $y\sqrt{112y}+4\sqrt{175y^3}$

87. $2\sqrt[3]{81}+3\sqrt[3]{24}$

88. $3\sqrt[4]{32}-2\sqrt[4]{162}$

89. $\sqrt[4]{768z^5}+\sqrt[4]{48z^5}$

90. $-2\sqrt[5]{64y^2}+3\sqrt[5]{486y^2}$

91. $\sqrt{8x^2y}-x\sqrt{2y}+\sqrt{50x^2y}$

92. $3x\sqrt{18x}+2\sqrt{2x^3}-\sqrt{72x^3}$

93. $\sqrt[3]{16xy^4}+y\sqrt[3]{2xy}-\sqrt[3]{54xy^4}$

94. $\sqrt[4]{512x^5}-\sqrt[4]{32x^5}+\sqrt[4]{1250x^5}$

Rationalize each denominator and simplify. Assume that all variables represent positive numbers.

95. $\dfrac{3}{\sqrt{3}}$

96. $\dfrac{6}{\sqrt{5}}$

97. $\dfrac{2}{\sqrt{x}}$

98. $\dfrac{8}{\sqrt{y}}$

99. $\dfrac{2}{\sqrt[3]{2}}$

100. $\dfrac{4d}{\sqrt[3]{9}}$

101. $\dfrac{5a}{\sqrt[3]{25a}}$

102. $\dfrac{7}{\sqrt[3]{36c}}$

103. $\dfrac{2b}{\sqrt[4]{3a^2}}$

104. $\sqrt{\dfrac{x}{2y}}$

105. $\sqrt[3]{\dfrac{2u^4}{9v}}$

106. $\sqrt[3]{-\dfrac{3s^5}{4r^2}}$

Rationalize each numerator and simplify. Assume that all variables are positive numbers.

107. $\dfrac{\sqrt{5}}{10}$

108. $\dfrac{\sqrt{y}}{3}$

109. $\dfrac{\sqrt[3]{9}}{3}$

110. $\dfrac{\sqrt[3]{16b^2}}{16}$

111. $\dfrac{\sqrt[5]{16b^3}}{64a}$

112. $\sqrt{\dfrac{3x}{57}}$

Rationalize each denominator and simplify.

113. $\sqrt{\dfrac{1}{3}} - \sqrt{\dfrac{1}{27}}$

114. $\sqrt[3]{\dfrac{1}{2}} + \sqrt[3]{\dfrac{1}{16}}$

115. $\sqrt{\dfrac{x}{8}} - \sqrt{\dfrac{x}{2}} + \sqrt{\dfrac{x}{32}}$

116. $\sqrt[3]{\dfrac{y}{4}} + \sqrt[3]{\dfrac{y}{32}} - \sqrt[3]{\dfrac{y}{500}}$

Simplify each radical expression.

117. $\sqrt[4]{9}$

118. $\sqrt[6]{27}$

119. $\sqrt[10]{16x^6}$

120. $\sqrt[6]{27x^9}$

Applications

121. Collage A square-shaped collage of photos has an area of 120 square inches. What is the length of each of its sides?

©nito/Shutterstock.com

122. Volume The volume of a cube-shaped box is 2000 square inches. What is the length of each of its sides?

Discovery and Writing

We often can multiply and divide radicals with different indices. For example, to multiply $\sqrt{3}$ by $\sqrt[3]{5}$, we first write each radical as a sixth root

$\sqrt{3} = 3^{1/2} = 3^{3/6} = \sqrt[6]{3^3} = \sqrt[6]{27}$

$\sqrt[3]{5} = 5^{1/3} = 5^{2/6} = \sqrt[6]{5^2} = \sqrt[6]{25}$

and then multiply the sixth roots.

$\sqrt{3}\sqrt[3]{5} = \sqrt[6]{27}\sqrt[6]{25} = \sqrt[6]{(27)(25)} = \sqrt[6]{675}$

Division is similar.

Use this idea to write each of the following expressions as a single radical.

123. $\sqrt{2}\sqrt[3]{2}$

124. $\sqrt{3}\sqrt[3]{5}$

125. $\dfrac{\sqrt[4]{3}}{\sqrt{2}}$

126. $\dfrac{\sqrt[3]{2}}{\sqrt{5}}$

127. Explain why $a^{1/n}$ is undefined if n is even and a represents a negative number.

128. For what values of x does $\sqrt[4]{x^4} = x$? Explain.

129. Explain what is meant by rationalizing the denominator of a radical expression.

130. If all of the radicals involved represent real numbers and $y \neq 0$, explain why

$$\sqrt[n]{\dfrac{x}{y}} = \dfrac{\sqrt[n]{x}}{\sqrt[n]{y}}$$

131. If all of the radicals involved represent real numbers and there is no division by 0, explain why

$$\left(\dfrac{x}{y}\right)^{-m/n} = \sqrt[n]{\dfrac{y^m}{x^m}}$$

132. The definition of $x^{m/n}$ requires that $\sqrt[n]{x}$ be a real number. Explain why this is important. (*Hint:* Consider what happens when n is even, m is odd, and x is negative.)

Critical Thinking

In Exercises 133–140, match each expression on the left with an equivalent expression on the right. Assume all variables represent positive numbers.

133. $(-16)^{1/4}$

a. -2

134. $(1024)^{1/10}$

b. $x\sqrt[87]{x}$

135. $0^{111/19}$

c. 2

136. $(-1)^{-12/19}$

d. 0

137. $\sqrt[87]{-1}$

e. $\dfrac{\sqrt[87]{x^{86}}}{x}$

138. $\sqrt[87]{x^{88}}$

f. 1

139. $\dfrac{1}{\sqrt[87]{x}}$

g. undefined

140. $\sqrt[3]{\sqrt[3]{-512}}$

h. -1

0.4 Polynomials

In this section, we will learn to

1. Define polynomials.

2. Add and subtract polynomials.

3. Multiply polynomials.

4. Rationalize denominators.

5. Divide polynomials.

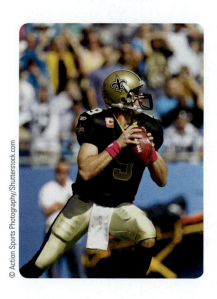

Football is one of the most popular sports in the United States. Drew Brees, one of the National Football League's (NFL's) most talented quarterbacks, plays for the New Orleans Saints. He led the Saints to their first Super Bowl victory and was named Most Valuable Player (MVP) of Super Bowl XLIV. Since joining the Saints, Brees has led all NFL quarterbacks in touchdowns, passing yards, and 300-yard games. *Sports Illustrated* named Brees its Sportsman of the Year a few years ago.

An algebraic expression can be used to model the trajectory or path of a football when passed by Drew Brees. Suppose the height in feet, t seconds after the football leaves Drew's hand, is given by the algebraic expression

$$-0.1t^2 + t + 5.5$$

At a time of $t = 3$ seconds, we see that the height of the football is

$$-0.1(\mathbf{3})^2 + \mathbf{3} + 5.5 = 7.6 \text{ ft.}$$

Algebraic expressions like $-0.1t^2 + t + 5.5$ are called **polynomials**, and we will study them in this section.

1. Define Polynomials

A **monomial** is a real number or the product of a real number and one or more variables with whole-number exponents. The number is called the **coefficient** of the variables. Some examples of monomials are

$$3x, \qquad 7ab^2, \qquad -5ab^2c^4, \qquad x^3, \qquad \text{and} \qquad -12$$

with coefficients of 3, 7, -5, 1, and -12, respectively.

The **degree** of a monomial is the sum of the exponents of its variables. All nonzero constants (except 0) have a degree of 0.

- The degree of $3x$ is 1.
- The degree of $-5ab^2c^4$ is 7.
- The degree of -12 is 0 (since $-12 = -12x^0$).

- The degree of $7ab^2$ is 3.
- The degree of x^3 is 3.
- 0 has no defined degree.

A monomial or a sum of monomials is called a **polynomial**. Each monomial in that sum is called a **term** of the polynomial. A polynomial with two terms is called a **binomial**, and a polynomial with three terms is called a **trinomial**.

Monomials	Binomials	Trinomials
$3x^2$	$2a + 3b$	$x^2 + 7x - 4$
$-25xy$	$4x^3 - 3x^2$	$4y^4 - 2y + 12$
a^2b^3c	$-2x^3 - 4y^2$	$12x^3y^2 - 8xy - 24$

The **degree of a polynomial** is the degree of the term in the polynomial with highest degree. The only polynomial with no defined degree is 0, which is called the **zero polynomial**. Here are some examples.

- $3x^2y^3 + 5xy^2 + 7$ is a trinomial of 5th degree, because its term with highest degree (the first term) is 5.
- $3ab + 5a^2b$ is a binomial of degree 3.
- $5x + 3y^2 + \sqrt[4]{3}z^4 - \sqrt{7}$ is a polynomial, because its variables have whole-number exponents. It is of degree 4.
- $-7y^{1/2} + 3y^2 + \sqrt[5]{3}z$ is not a polynomial, because one of its variables (y in the first term) does not have a whole-number exponent.

If two terms of a polynomial have the same variables with the same exponents, they are **like** or **similar** terms. To combine the like terms in the sum $3x^2y + 5x^2y$ or the difference $7xy^2 - 2xy^2$, we use the Distributive Property:

Take Note

To combine like terms, we add (or subtract) their coefficients and keep the same variables and the same exponents.

$$3x^2y + 5x^2y = (3 + 5)x^2y \qquad\qquad 7xy^2 - 2xy^2 = (7 - 2)xy^2$$
$$= 8x^2y \qquad\qquad\qquad\qquad\qquad = 5xy^2$$

This illustrates that *to combine like terms, we add (or subtract) their coefficients and keep the same variables and the same exponents.*

2. Add and Subtract Polynomials

Recall that we can use the Distributive Property to remove parentheses enclosing the terms of a polynomial. When the sign preceding the parentheses is $+$, we simply drop the parentheses:

$$+(a + b - c) = +1(a + b - c)$$
$$= 1a + 1b - 1c$$
$$= a + b - c$$

When the sign preceding the parentheses is $-$, we drop the parentheses and the $-$ sign and change the sign of each term within the parentheses.

$$-(a + b - c) = -1(a + b - c)$$
$$= -1a + (-1)b - (-1)c$$
$$= -a - b + c$$

Take Note

To add or subtract polynomials, we remove parentheses and combine like terms.

We can use these facts to add and subtract polynomials. *To add (or subtract) polynomials, we remove parentheses (if necessary) and combine like terms.*

Adding Polynomials

EXAMPLE 1   Add: $(3x^3y + 5x^2 - 2y) + (2x^3y - 5x^2 + 3x)$.

SOLUTION   To add the polynomials, we remove parentheses and combine like terms.

$$(3x^3y + 5x^2 - 2y) + (2x^3y - 5x^2 + 3x)$$
$$= 3x^3y + 5x^2 - 2y + 2x^3y - 5x^2 + 3x$$
$$= 3x^3y + 2x^3y + 5x^2 - 5x^2 - 2y + 3x \qquad \text{Use the Commutative Property to rearrange terms.}$$
$$= 5x^3y - 2y + 3x \qquad\qquad\qquad\qquad \text{Combine like terms.}$$

Alternate strategy: We can add the polynomials in a vertical format by writing like terms in a column and adding the like terms, column by column.

$$\begin{aligned} 3x^3y + 5x^2 - 2y \\ \underline{2x^3y - 5x^2 \qquad + 3x} \\ 5x^3y \qquad - 2y + 3x \end{aligned}$$

Self Check 1   Add: $(4x^2 + 3x - 5) + (3x^2 - 5x + 7)$.

Now Try Exercise 21.

EXAMPLE 2   **Subtracting Polynomials**

Subtract: $(2x^2 + 3y^2) - (x^2 - 2y^2 + 7)$.

SOLUTION   To subtract the polynomials, we remove parentheses and combine like terms.

$$
\begin{aligned}
(2x^2 + 3y^2) &- (x^2 - 2y^2 + 7) \\
&= 2x^2 + 3y^2 - x^2 + 2y^2 - 7 \\
&= 2x^2 - x^2 + 3y^2 + 2y^2 - 7 &&\text{Use the Commutative Property to rearrange terms.} \\
&= x^2 + 5y^2 - 7 &&\text{Combine like terms.}
\end{aligned}
$$

Alternate strategy: We can subtract the polynomials in a vertical format by writing like terms in a column and subtracting the like terms, column by column.

$$
\begin{array}{l}
2x^2 + 3y^2 \\
\underline{-(x^2 - 2y^2 + 7)} \\
x^2 + 5y^2 - 7
\end{array}
\qquad
\begin{array}{l}
2x^2 - x^2 = (2 - 1)x^2 \\
3y^2 - (-2)y^2 = 3y^2 + 2y^2 = (3 + 2)y^2 \\
0 - 7 = -7
\end{array}
$$

Self Check 2   Subtract: $(4x^2 + 3x - 5) - (3x^2 - 5x + 7)$.

Now Try Exercise 23.

Caution

A common error students make when subtracting polynomials is to distribute the negative sign only to the first term of the second polynomial. Always distribute the negative sign through all the terms in the second polynomial.

We can also use the Distributive Property to remove parentheses enclosing several terms that are multiplied by a constant. For example,

$$4(3x^2 - 2x + 6) = 4(3x^2) - 4(2x) + 4(6)$$
$$= 12x^2 - 8x + 24$$

This example suggests that *to add multiples of one polynomial to another, or to subtract multiples of one polynomial from another, we remove parentheses and combine like terms.*

Using the Distributive Property and Combining Like Terms

EXAMPLE 3   Simplify: $7x(2y^2 + 13x^2) - 5(xy^2 - 13x^3)$.

SOLUTION $7x(2y^2 + 13x^2) - 5(xy^2 - 13x^3)$

$= 14xy^2 + 91x^3 - 5xy^2 + 65x^3$ Use the Distributive Property to remove parentheses.

$= 14xy^2 - 5xy^2 + 91x^3 + 65x^3$ Use the Commutative Property to rearrange terms.

$= 9xy^2 + 156x^3$ Combine like terms.

Self Check 3   Simplify: $3(2b^2 - 3a^2b) + 2b(b + a^2)$.

Now Try Exercise 27.

3. Multiply Polynomials

To find the product of $3x^2y^3z$ and $5xyz^2$, we proceed as follows:

$(3x^2y^3z)(5xyz^2) = 3 \cdot x^2 \cdot y^3 \cdot z \cdot 5 \cdot x \cdot y \cdot z^2$

$= 3 \cdot 5 \cdot x^2 \cdot x \cdot y^3 \cdot y \cdot z \cdot z^2$ Use the Commutative Property to rearrange terms.

$= 15x^3y^4z^3$

This illustrates that *to multiply two monomials, we multiply the coefficients and then multiply the variables.*

To find the product of a monomial and a polynomial, we use the Distributive Property.

> **Take Note**
>
> 1. To multiply two monomials, we multiply the coefficients and then multiply the variables.
> 2. To multiply a polynomial by a monomial, we multiply each term of the polynomial by the monomial.

$3xy^2(2xy + x^2 - 7yz) = 3xy^2(2xy) + (3xy^2)(x^2) - (3xy^2)(7yz)$

$= 6x^2y^3 + 3x^3y^2 - 21xy^3z$

This illustrates that *to multiply a polynomial by a monomial, we multiply each term of the polynomial by the monomial.*

To multiply one binomial by another, we use the Distributive Property twice.

Multiplying Binomials

EXAMPLE 4   Multiply: **a.** $(x + y)(x + y)$ **b.** $(x - y)(x - y)$ **c.** $(x + y)(x - y)$

SOLUTION   **a.** $(x + y)(x + y) = (x + y)x + (x + y)y$

$= x^2 + xy + xy + y^2$

$= x^2 + 2xy + y^2$

b. $(x - y)(x - y) = (x - y)x - (x - y)y$

$= x^2 - xy - xy + y^2$

$= x^2 - 2xy + y^2$

c. $(x + y)(x - y) = (x + y)x - (x + y)y$

$= x^2 + xy - xy - y^2$

$= x^2 - y^2$

Self Check 4   Multiply: **a.** $(x + 2)(x + 2)$ **b.** $(x - 3)(x - 3)$
c. $(x + 4)(x - 4)$

Now Try Exercise 45.

The products in Example 4 are called **special products**. Because they occur so often, it is worthwhile to learn their forms.

Special Product Formulas

$(x + y)^2 = (x + y)(x + y) = x^2 + 2xy + y^2$

$(x - y)^2 = (x - y)(x - y) = x^2 - 2xy + y^2$

$(x + y)(x - y) = x^2 - y^2$

Using the Special Product Formulas is easy and can save you time. Three examples are shown in the table.

Special Product	Formula	Example
Squaring a Binomial Sum	$(x + y)^2 =$ $x^2 + 2xy + y^2$	$(4x + 7y)^2 = (4x)^2 + 2 \cdot 4x \cdot 7y + (7y)^2$ $= 16x^2 + 56xy + 49y^2$
Squaring a Binomial Difference	$(x - y)^2 =$ $x^2 - 2xy + y^2$	$(5x - 9y)^2 = (5x)^2 - 2 \cdot 5x \cdot 9y + (9y)^2$ $= 25x^2 - 90xy + 81y^2$
Sum and Difference of Two Terms	$(x + y)(x - y) =$ $x^2 - y^2$	$(11x + 3y)(11x - 3y) = (11x)^2 - (3y)^2$ $= 121x^2 - 9y^2$

Caution

An error students sometimes make is to square a binomial and get a binomial. Always remember that $(x + y)^2$ and $(x - y)^2$ have trinomials for their products and that

$(x + y)^2 \neq x^2 + y^2$ and $(x - y)^2 \neq x^2 - y^2$

For example,

$(3 + 5)^2 \neq 3^2 + 5^2$ and $(3 - 5)^2 \neq 3^2 - 5^2$

$8^2 \neq 9 + 25$ $(-2)^2 \neq 9 - 25$

$64 \neq 34$ $4 \neq -16$

We can use the **FOIL method** to multiply one binomial by another. FOIL is an acronym for First terms, Outer terms, Inner terms, and Last terms.

F	**O**	**I**	**L**
Stands for First	Stands for Outer	Stands For Inner	Stands for Last

To use this method to multiply $3x - 4$ by $2x + 5$, we write

First terms   Last terms

$(3x - 4)(2x + 5) = 3x(2x) + 3x(5) - 4(2x) - 4(5)$

Inner terms $= 6x^2 + 15x - 8x - 20$

Outer terms $= 6x^2 + 7x - 20$

In this example,

- the product of the first terms is $6x^2$,
- the product of the outer terms is $15x$,
- the product of the inner terms is $-8x$, and
- the product of the last terms is -20.

The resulting like terms of the product are then combined.

EXAMPLE 5

Using the FOIL Method to Multiply Binomials

Use the FOIL method to multiply: $(\sqrt{3} + x)(2 - \sqrt{3}x)$.

SOLUTION

$$(\sqrt{3} + x)(2 - \sqrt{3}x) = 2\sqrt{3} - \sqrt{3}\sqrt{3}x + 2x - x\sqrt{3}x$$
$$= 2\sqrt{3} - 3x + 2x - \sqrt{3}x^2$$
$$= 2\sqrt{3} - x - \sqrt{3}x^2$$

Self Check 5   Multiply: $(2x + \sqrt{3})(x - \sqrt{3})$.

Now Try Exercise 63.

To multiply a polynomial with more than two terms by another polynomial, we multiply each term of one polynomial by each term of the other polynomial and combine like terms whenever possible.

EXAMPLE 6

Multiplying Polynomials

Multiply: **a.** $(x + y)(x^2 - xy + y^2)$ **b.** $(x + 3)^3$

SOLUTION

a. $(x + y)(x^2 - xy + y^2) = x^3 - x^2y + xy^2 + yx^2 - xy^2 + y^3$
$$= x^3 + y^3$$

b. $(x + 3)^3 = (x + 3)(x + 3)^2$
$$= (x + 3)(x^2 + 6x + 9)$$
$$= x^3 + 6x^2 + 9x + 3x^2 + 18x + 27$$
$$= x^3 + 9x^2 + 27x + 27$$

> **Take Note**
>
> When multiplying polynomials, when neither is a monomial, simply multiply each term of one polynomial by each term of the other polynomial. Then combine like terms. See Example 6a.

Self Check 6   Multiply: $(x + 2)(2x^2 + 3x - 1)$.

Now Try Exercise 67.

> **Tip**
>
> Two formats, one horizontal and the other vertical, were shown for adding, subtracting, and multiplying polynomials. Practice both formats and choose the method that is easier for you.

Alternate strategy: We can use a vertical format to multiply two polynomials, such as the polynomials given in Self Check 6. We first write the polynomials as follows and draw a line beneath them. We then multiply each term of the upper polynomial by each term of the lower polynomial and write the results so that like terms appear in each column. Finally, we combine like terms column by column.

$$
\begin{array}{r}
2x^2 + 3x - 1 \\
x + 2 \\
\hline
4x^2 + 6x - 2 \\
2x^3 + 3x^2 - x \\
\hline
2x^3 + 7x^2 + 5x - 2
\end{array}
$$

Multiply $2x^2 + 3x - 1$ by 2.

Multiply $2x^2 + 3x - 1$ by x.

In each column, combine like terms.

If n is a whole number, the expressions $a^n + 1$ and $2a^n - 3$ are polynomials and we can multiply them as follows:

$$(a^n + 1)(2a^n - 3) = 2a^{2n} - 3a^n + 2a^n - 3$$

$$= 2a^{2n} - a^n - 3 \qquad \text{Combine like terms.}$$

We can also use the methods previously discussed to multiply expressions that are not polynomials, such as $x^{-2} + y$ and $x^2 - y^{-1}$.

$$(x^{-2} + y)(x^2 - y^{-1}) = x^{-2+2} - x^{-2}y^{-1} + x^2y - y^{1-1}$$

$$= x^0 - \frac{1}{x^2y} + x^2y - y^0$$

$$= 1 - \frac{1}{x^2y} + x^2y - 1 \qquad x^0 = 1 \text{ and } y^0 = 1.$$

$$= x^2y - \frac{1}{x^2y}$$

4. Rationalize Denominators

If the denominator of a fraction is a binomial containing square roots, we can use the product formula $(x + y)(x - y)$ to rationalize the denominator. For example, to rationalize the denominator of

$$\frac{6}{\sqrt{7} + 2} \qquad \text{To rationalize a denominator means to change the denominator into a rational number.}$$

we multiply the numerator and denominator by $\sqrt{7} - 2$ and simplify.

$$\frac{6}{\sqrt{7} + 2} = \frac{6(\sqrt{7} - 2)}{(\sqrt{7} + 2)(\sqrt{7} - 2)} \qquad \frac{\sqrt{7} - 2}{\sqrt{7} - 2} = 1$$

$$= \frac{6(\sqrt{7} - 2)}{7 - 4}$$

$$= \frac{6(\sqrt{7} - 2)}{3} \qquad \text{Here the denominator is a rational number.}$$

$$= 2(\sqrt{7} - 2)$$

In this example, we multiplied both the numerator and the denominator of the given fraction by $\sqrt{7} - 2$. This binomial is the same as the denominator of the given fraction $\sqrt{7} + 2$, except for the sign between the terms. Such binomials are called **conjugate binomials** or **radical conjugates**.

| Conjugate Binomials | **Conjugate binomials** are binomials that are the same except for the sign between their terms. The conjugate of $a + b$ is $a - b$, and the conjugate of $a - b$ is $a + b$. |

Rationalizing the Denominator of a Radical Expression

EXAMPLE 7 Rationalize the denominator: $\dfrac{\sqrt{3x} - \sqrt{2}}{\sqrt{3x} + \sqrt{2}}$ ($x > 0$).

SOLUTION We multiply the numerator and the denominator by $\sqrt{3x} - \sqrt{2}$ (the conjugate of $\sqrt{3x} + \sqrt{2}$) and simplify.

$$\frac{\sqrt{3x} - \sqrt{2}}{\sqrt{3x} + \sqrt{2}} = \frac{\left(\sqrt{3x} - \sqrt{2}\right)\left(\sqrt{3x} - \sqrt{2}\right)}{\left(\sqrt{3x} + \sqrt{2}\right)\left(\sqrt{3x} - \sqrt{2}\right)} \qquad \frac{\sqrt{3x} - \sqrt{2}}{\sqrt{3x} - \sqrt{2}} = 1$$

$$= \frac{\sqrt{3x}\sqrt{3x} - \sqrt{3x}\sqrt{2} - \sqrt{2}\sqrt{3x} + \sqrt{2}\sqrt{2}}{\left(\sqrt{3x}\right)^2 - \left(\sqrt{2}\right)^2}$$

$$= \frac{3x - \sqrt{6x} - \sqrt{6x} + 2}{3x - 2}$$

$$= \frac{3x - 2\sqrt{6x} + 2}{3x - 2}$$

Self Check 7 Rationalize the denominator: $\dfrac{\sqrt{x} + 2}{\sqrt{x} - 2}$.

Now Try Exercise 89.

In calculus, we often rationalize a numerator.

Rationalizing the Numerator of a Radical Expression

EXAMPLE 8 Rationalize the numerator: $\dfrac{\sqrt{x + h} - \sqrt{x}}{h}$.

SOLUTION To rid the numerator of radicals, we multiply the numerator and the denominator by the conjugate of the numerator and simplify.

$$\frac{\sqrt{x + h} - \sqrt{x}}{h} = \frac{\left(\sqrt{x + h} - \sqrt{x}\right)\left(\sqrt{x + h} + \sqrt{x}\right)}{h\left(\sqrt{x + h} + \sqrt{x}\right)} \qquad \frac{\sqrt{x + h} + \sqrt{x}}{\sqrt{x + h} + \sqrt{x}} = 1$$

$$= \frac{x + h - x}{h\left(\sqrt{x + h} + \sqrt{x}\right)} \qquad \text{Here the numerator has no radicals.}$$

$$= \frac{h}{h\left(\sqrt{x + h} + \sqrt{x}\right)}$$

$$= \frac{1}{\sqrt{x + h} + \sqrt{x}} \qquad \text{Divide out the common factor of } h.$$

Self Check 8 Rationalize the numerator: $\dfrac{\sqrt{4 + h} - 2}{h}$.

Now Try Exercise 99.

5. Divide Polynomials

To divide monomials, we write the quotient as a fraction and simplify by using the rules of exponents. For example,

Take Note

1. To divide monomials, we write the quotient as a fraction and simplify.
2. To divide a polynomial by a monomial, we write the fraction as a sum of separate fractions, and we simplify each.
3. To divide two polynomials when neither is a monomial, we use long division.

$$\frac{6x^2y^3}{-2x^3y} = -3x^{2-3}y^{3-1}$$

$$= -3x^{-1}y^2$$

$$= -\frac{3y^2}{x}$$

To divide a polynomial by a monomial, we write the quotient as a fraction, write the fraction as a sum of separate fractions, and simplify each one. For example, to divide $8x^5y^4 + 12x^2y^5 - 16x^2y^3$ by $4x^3y^4$, we proceed as follows:

$$\frac{8x^5y^4 + 12x^2y^5 - 16x^2y^3}{4x^3y^4} = \frac{8x^5y^4}{4x^3y^4} + \frac{12x^2y^5}{4x^3y^4} + \frac{-16x^2y^3}{4x^3y^4}$$

$$= 2x^2 + \frac{3y}{x} - \frac{4}{xy}$$

To divide two polynomials, we can use long division. To illustrate, we consider the division

$$\frac{2x^2 + 11x - 30}{x + 7}$$

which can be written in long division form as

$$x + 7\overline{)2x^2 + 11x - 30}$$

Tip

Spend time learning vocabulary in algebra. The terms *divisor, dividend,* and *quotient* are important to know in mathematics.

The binomial $x + 7$ is called the **divisor**, and the trinomial $2x^2 + 11x - 30$ is called the **dividend**. The final answer, called the **quotient**, will appear above the long division symbol.

We begin the division by asking "What monomial, when multiplied by x, gives $2x^2$?" Because $x \cdot 2x = 2x^2$, the answer is $2x$. We place $2x$ in the quotient, multiply each term of the divisor by $2x$, subtract, and bring down the -30.

$$
\begin{array}{r}
2x \\
x + 7\overline{)2x^2 + 11x - 30} \\
\underline{2x^2 + 14x} \\
-\ 3x - 30
\end{array}
$$

We continue the division by asking "What monomial, when multiplied by x, gives $-3x$?" We place the answer, -3, in the quotient, multiply each term of the divisor by -3, and subtract. This time, there is no number to bring down.

$$
\begin{array}{r}
2x -\ 3 \\
x + 7\overline{)2x^2 + 11x - 30} \\
\underline{2x^2 + 14x} \\
-\ 3x - 30 \\
\underline{-\ 3x - 21} \\
-9
\end{array}
$$

Because the degree of the remainder, -9, is less than the degree of the divisor, the division process stops, and we can express the result in the form

$$\text{quotient} + \frac{\text{remainder}}{\text{divisor}}$$

Thus,

$$\frac{2x^2 + 11x - 30}{x + 7} = 2x - 3 + \frac{-9}{x + 7}$$

Using Long Division to Divide Polynomials

EXAMPLE 9

Divide $6x^3 - 11$ by $2x + 2$.

SOLUTION

We set up the division, leaving spaces for the missing powers of x in the dividend.

$$2x + 2 \overline{)6x^3 \qquad\qquad - 11}$$

The division process continues as usual, with the following results:

$$
\begin{array}{r}
3x^2 - 3x + 3 \\
2x + 2 \overline{)6x^3 \qquad\qquad\; - 11} \\
\underline{6x^3 + 6x^2} \\
-6x^2 \\
\underline{-6x^2 - 6x} \\
+6x - 11 \\
\underline{+6x + 6} \\
-17
\end{array}
$$

> **Take Note**
>
> In Example 9, we could write the missing powers of x using coefficients of 0.
>
> $$2x + 2 \overline{)6x^3 + 0x^2 + 0x - 11}$$

Thus, $\dfrac{6x^3 - 11}{2x + 2} = 3x^2 - 3x + 3 + \dfrac{-17}{2x + 2}$.

Self Check 9   Divide: $3x + 1 \overline{)9x^2 - 1}$.

Now Try Exercise 113.

Using Long Division to Divide Polynomials

EXAMPLE 10

Divide $-3x^3 - 3 + x^5 + 4x^2 - x^4$ by $x^2 - 3$.

SOLUTION

The division process works best when the terms in the divisor and dividend are written with their exponents in descending order.

> **Tip**
>
> The subtraction steps in a long division problem challenge many students. Keep in mind that after the multiplication step, you simply draw a horizontal line, change the signs, and add.

$$
\begin{array}{r}
x^3 - x^2 + 1 \\
x^2 - 3 \overline{)x^5 - x^4 - 3x^3 + 4x^2 \quad - 3} \\
\underline{x^5 \qquad\quad - 3x^3} \\
-x^4 \qquad\quad + 4x^2 \\
\underline{-x^4 \qquad\quad + 3x^2} \\
x^2 - 3 \\
\underline{x^2 - 3} \\
0
\end{array}
$$

Thus, $\dfrac{-3x^3 - 3 + x^5 + 4x^2 - x^4}{x^2 - 3} = x^3 - x^2 + 1$.

Self Check 10   Divide: $x^2 + 1 \overline{)3x^2 - x + 1 - 2x^3 + 3x^4}$.

Now Try Exercise 115.

Self Check Answers

1. $7x^2 - 2x + 2$ **2.** $x^2 + 8x - 12$ **3.** $8b^2 - 7a^2b$

4. a. $x^2 + 4x + 4$ **b.** $x^2 - 6x + 9$ **c.** $x^2 - 16$ **5.** $2x^2 - x\sqrt{3} - 3$

6. $2x^3 + 7x^2 + 5x - 2$ **7.** $\dfrac{x + 4\sqrt{x} + 4}{x - 4}$ **8.** $\dfrac{1}{\sqrt{4 + h} + 2}$

9. $3x - 1$ **10.** $3x^2 - 2x + \dfrac{x + 1}{x^2 + 1}$

Exercises 0.4

Getting Ready

You should be able to complete these vocabulary and concept statements before you proceed to the practice exercises.

Fill in the blanks.

1. A _____ is a real number or the product of a real number and one or more _____.
2. The _____ of a monomial is the sum of the exponents of its _____.
3. A _____ is a polynomial with three terms.
4. A _____ is a polynomial with two terms.
5. A monomial is a polynomial with _____ term.
6. The constant 0 is called the _____ polynomial.
7. Terms with the same variables with the same exponents are called _____ terms.
8. The _____ of a polynomial is the same as the degree of its term of highest degree.
9. To combine like terms, we add their _____ and keep the same _____ and the same exponents.
10. The conjugate of $3\sqrt{x} + 2$ is _____.

Determine whether the given expression is a polynomial. If so, tell whether it is a monomial, a binomial, or a trinomial, and give its degree.

11. $x^2 + 3x + 4$
12. $5xy - x^3$
13. $x^3 + y^{1/2}$
14. $x^{-3} - 5y^{-2}$
15. $4x^2 - \sqrt{5}x^3$
16. x^2y^3
17. $\sqrt{15}$
18. $\dfrac{5}{x} + \dfrac{x}{5} + 5$
19. 0
20. $3y^3 - 4y^2 + 2y$

Practice

Perform the operations and simplify.

21. $(x^3 - 3x^2) + (5x^3 - 8x)$
22. $(2x^4 - 5x^3) + (7x^3 - x^4 + 2x)$
23. $(y^5 + 2y^3 + 7) - (y^5 - 2y^3 - 7)$
24. $(3t^7 - 7t^3 + 3) - (7t^7 - 3t^3 + 7)$
25. $2(x^2 + 3x - 1) - 3(x^2 + 2x - 4) + 4$
26. $5(x^3 - 8x + 3) + 2(3x^2 + 5x) - 7$

27. $8(t^2 - 2t + 5) + 4(t^2 - 3t + 2) - 6(2t^2 - 8)$
28. $-3(x^3 - x) + 2(x^2 + x) + 3(x^3 - 2x)$
29. $y(y^2 - 1) - y^2(y + 2) - y(2y - 2)$
30. $-4a^2(a + 1) + 3a(a^2 - 4) - a^2(a + 2)$
31. $xy(x - 4y) - y(x^2 + 3xy) + xy(2x + 3y)$
32. $3mn(m + 2n) - 6m(3mn + 1) - 2n(4mn - 1)$
33. $2x^2y^3(4xy^4)$
34. $-15a^3b(-2a^2b^3)$
35. $-3m^2n(2mn^2)\left(-\dfrac{mn}{12}\right)$
36. $-\dfrac{3r^2s^3}{5}\left(\dfrac{2r^2s}{3}\right)\left(\dfrac{15rs^2}{2}\right)$
37. $-4rs(r^2 + s^2)$
38. $6u^2v(2uv^2 - y)$
39. $6ab^2c(2ac + 3bc^2 - 4ab^2c)$
40. $-\dfrac{mn^2}{2}(4mn - 6m^2 - 8)$
41. $(a + 2)(a + 2)$
42. $(y - 5)(y - 5)$
43. $(a - 6)^2$
44. $(t + 9)^2$
45. $(x + 4)(x - 4)$
46. $(z + 7)(z - 7)$
47. $(x - 3)(x + 5)$
48. $(z + 4)(z - 6)$
49. $(u + 2)(3u - 2)$
50. $(4x + 1)(2x - 3)$
51. $(5x - 1)(2x + 3)$
52. $(4x - 1)(2x - 7)$
53. $(3a - 2b)^2$
54. $(4a + 5b)(4a - 5b)$
55. $(3m + 4n)(3m - 4n)$
56. $(4r + 3s)^2$
57. $(2y - 4x)(3y - 2x)$
58. $(-2x + 3y)(3x + y)$
59. $(9x - y)(x^2 - 3y)$
60. $(8a^2 + b)(a + 2b)$
61. $(5z + 2t)(z^2 - t)$
62. $(y - 2x^2)(x^2 + 3y)$
63. $(\sqrt{5} + 3x)(2 - \sqrt{5}x)$
64. $(\sqrt{2} + x)(3 + \sqrt{2}x)$
65. $(3x - 1)^3$
66. $(2x - 3)^3$

67. $(3x + 1)(2x^2 + 4x - 3)$

68. $(2x - 5)(x^2 - 3x + 2)$

69. $(3x + 2y)(2x^2 - 3xy + 4y^2)$

70. $(4r - 3s)(2r^2 + 4rs - 2s^2)$

Multiply the expressions as you would multiply polynomials.

71. $2y^n(3y^n + y^{-n})$

72. $3a^{-n}(2a^n + 3a^{n-1})$

73. $-5x^{2n}y^n(2x^{2n}y^{-n} + 3x^{-2n}y^n)$

74. $-2a^{3n}b^{2n}(5a^{-3n}b - ab^{-2n})$

75. $(x^n + 3)(x^n - 4)$

76. $(a^n - 5)(a^n - 3)$

77. $(2r^n - 7)(3r^n - 2)$

78. $(4z^n + 3)(3z^n + 1)$

79. $x^{1/2}(x^{1/2}y + xy^{1/2})$

80. $ab^{1/2}(a^{1/2}b^{1/2} + b^{1/2})$

81. $(a^{1/2} + b^{1/2})(a^{1/2} - b^{1/2})$

82. $(x^{3/2} + y^{1/2})^2$

Rationalize each denominator.

83. $\dfrac{2}{\sqrt{3} - 1}$

84. $\dfrac{1}{\sqrt{5} + 2}$

85. $\dfrac{3x}{\sqrt{7} + 2}$

86. $\dfrac{14y}{\sqrt{2} - 3}$

87. $\dfrac{x}{x - \sqrt{3}}$

88. $\dfrac{y}{2y + \sqrt{7}}$

89. $\dfrac{y + \sqrt{2}}{y - \sqrt{2}}$

90. $\dfrac{x - \sqrt{3}}{x + \sqrt{3}}$

91. $\dfrac{\sqrt{2} - \sqrt{3}}{1 - \sqrt{3}}$

92. $\dfrac{\sqrt{3} - \sqrt{2}}{1 + \sqrt{2}}$

93. $\dfrac{\sqrt{x} - \sqrt{y}}{\sqrt{x} + \sqrt{y}}$

94. $\dfrac{\sqrt{2x} + y}{\sqrt{2x} - y}$

Rationalize each numerator.

95. $\dfrac{\sqrt{2} + 1}{2}$

96. $\dfrac{\sqrt{x} - 3}{3}$

97. $\dfrac{y - \sqrt{3}}{y + \sqrt{3}}$

98. $\dfrac{\sqrt{a} - \sqrt{b}}{\sqrt{a} + \sqrt{b}}$

99. $\dfrac{\sqrt{x + 3} - \sqrt{x}}{3}$

100. $\dfrac{\sqrt{2 + h} - \sqrt{2}}{h}$

Perform each division and write all answers without using negative exponents.

101. $\dfrac{36a^2b^3}{18ab^6}$

102. $\dfrac{-45r^2s^5t^3}{27r^6s^2t^8}$

103. $\dfrac{16x^6y^4z^9}{-24x^9y^6z^0}$

104. $\dfrac{32m^6n^4p^2}{26m^6n^7p^2}$

105. $\dfrac{5x^3y^2 + 15x^3y^4}{10x^2y^3}$

106. $\dfrac{9m^4n^9 - 6m^3n^4}{12m^3n^3}$

107. $\dfrac{24x^5y^7 - 36x^2y^5 + 12xy}{60x^5y^4}$

108. $\dfrac{9a^3b^4 + 27a^2b^4 - 18a^2b^3}{18a^2b^7}$

Perform each division. If there is a nonzero remainder, write the answer in quotient $+ \frac{\text{remainder}}{\text{divisor}}$ form.

109. $x + 3\overline{)3x^2 + 11x + 6}$

110. $3x + 2\overline{)3x^2 + 11x + 6}$

111. $2x - 5\overline{)2x^2 - 19x + 37}$

112. $x - 7\overline{)2x^2 - 19x + 35}$

113. $\dfrac{2x^3 + 1}{x - 1}$

114. $\dfrac{2x^3 - 9x^2 + 13x - 20}{2x - 7}$

115. $x^2 + x - 1\overline{)x^3 - 2x^2 - 4x + 3}$

116. $x^2 - 3\overline{)x^3 - 2x^2 - 4x + 5}$

117. $\dfrac{x^5 - 2x^3 - 3x^2 + 9}{x^3 - 2}$

118. $\dfrac{x^5 - 2x^3 - 3x^2 + 9}{x^3 - 3}$

119. $\dfrac{x^5 - 32}{x - 2}$

120. $\dfrac{x^4 - 1}{x + 1}$

121. $11x - 10 + 6x^2\overline{)36x^4 - 121x^2 + 120 + 72x^3 - 142x}$

122. $x + 6x^2 - 12\overline{)-121x^2 + 72x^3 - 142x + 120 + 36x^4}$

Applications

123. Geometry Find an expression that represents the area of the brick wall.

$(x - 2)$ ft

$(x + 5)$ ft

124. Geometry The area of the triangle shown in the illustration is represented as $(x^2 + 3x - 40)$ square feet. Find an expression that represents its height.

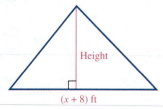

Height

$(x + 8)$ ft

125. Gift boxes The corners of a 12 in.-by-12 in. piece of cardboard are folded inward and glued to make a box. Write a polynomial that represents the volume of the resulting box.

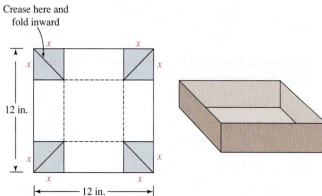

Crease here and fold inward

x

x

x

x

12 in.

x

x

x

x

12 in.

126. Travel Complete the following table, which shows the rate (mph), time traveled (hr), and distance traveled (mi) by a family on vacation.

r	·	t	=	d
$3x + 4$				$3x^2 + 19x + 20$

Discovery and Writing

127. Show that a trinomial can be squared by using the formula
$$(a + b + c)^2 = a^2 + b^2 + c^2 + 2ab + 2bc + 2ac.$$

128. Show that $(a + b + c + d)^2 = a^2 + b^2 + c^2 + d^2 + 2ab + 2ac + 2ad + 2bc + 2bd + 2cd$.

129. Explain the FOIL method.

130. Explain how to rationalize the numerator of $\frac{\sqrt{x} + 2}{x}$.

131. Explain why $(a + b)^2 \neq a^2 + b^2$.

132. Explain why $\sqrt{a^2 + b^2} \neq \sqrt{a^2} + \sqrt{b^2}$.

Critical Thinking

In Exercises 133–138, determine if the statement is true or false. If the statement is false, then correct it and make it true.

133. All polynomials are trinomials.

134. All binomials are polynomials.

135. $(12x - 5y)^2 = 144x^2 - 120xy + 25y^2$

136. $(6x + y)^2 = 36x^2 + y^2$

137. $(x^{1/3} - 6)(4x^{1/3} + 7) = 4x^{1/9} - 17x^{1/3} - 42$

138. $(x^{-3} + 5)(x^{-3} - 5) = x^9 - 25$

In Exercises 139 and 140, the revenue associated with selling x units of a product is $x^2 + 200x$ dollars, and the cost associated with producing x units of the product is $-200x + 500$ dollars.

139. Determine the polynomial that represents the profit in dollars of making x units of the product.

140. If 100 units are produced and sold, would the profit exceed $50,000?

0.5 Factoring Polynomials

In this section, we will learn to

1. Factor out the greatest common factor.
2. Factor by grouping.
3. Factor the difference of two squares.
4. Factor trinomials.
5. Factor trinomials by grouping.
6. Factor the sum and difference of two cubes.
7. Factor miscellaneous polynomials.

Television shows involving crime, murder, and criminal investigation are some of the most-watched in America. *CSI: Crime Scene Investigation* and its spinoffs,

CSI: NY, Miami, and *Cyber,* have been watched by millions of people. Also, *NCIS: Naval Criminal Investigative Service* and its spinoffs, *NCIS: Los Angeles* and *New Orleans,* have been top rated. Fans of these shows enjoy watching a team of forensic scientists uncover the circumstances that led to an unusual death or crime. These mysteries are captivating because many viewers are interested in the field of forensic science.

In this section, we will investigate a mathematics mystery. We will be given a polynomial and asked to unveil or uncover the two or more polynomials that were multiplied together to obtain the given polynomial. The process that we will use to solve this mystery is called *factoring.*

When two or more polynomials are multiplied together, each one is called a **factor** of the resulting product. For example, the factors of $7(x + 2)(x + 3)$ are

$$7, \qquad x + 2, \qquad \text{and} \qquad x + 3$$

The process of writing a polynomial as the product of several factors is called **factoring**.

In this section, we will discuss factoring where the coefficients of the polynomial and the polynomial factors are integers. If a polynomial cannot be factored by using integers only, we call it a **prime polynomial**.

1. Factor Out the Greatest Common Factor

The simplest type of factoring occurs when we see a common monomial factor in each term of the polynomial. In this case, our strategy is to use the Distributive Property and factor out the greatest common factor, abbreviated GCF.

EXAMPLE 1

Factoring by Removing the GCF

Factor: $3xy^2 + 6x$.

SOLUTION   We note that each term contains a greatest common factor of $3x$:

$$3xy^2 + 6x = \mathbf{3x}(y^2) + \mathbf{3x}(2)$$

We can then use the Distributive Property to factor out the common factor of $3x$:

$$3xy^2 + 6x = \mathbf{3x}(y^2 + 2) \qquad \text{We can check by multiplying: } 3x(y^2 + 2) = 3xy^2 + 6x.$$

Self Check 1   Factor: $4a^2 - 8ab$.

Now Try Exercise 11.

EXAMPLE 2

Factoring by Removing the GCF

Factor: $x^2y^2z^2 - xyz$.

SOLUTION   We factor out the greatest common factor of xyz:

$$x^2y^2z^2 - xyz = \mathbf{xyz}(xyz) - \mathbf{xyz}(1)$$
$$= \mathbf{xyz}(xyz - 1) \qquad \text{We can check by multiplying.}$$

The last term in the expression $x^2y^2z^2 - xyz$ has an understood coefficient of 1. When the xyz is factored out, the 1 must be written.

> **Take Note**
> The smallest power or exponent that appears on the variable in both terms is used for the GCF.

Self Check 2   Factor: $a^2b^2c^2 + a^3b^3c^3$.

Now Try Exercise 13.

2. Factor by Grouping

A strategy that is often helpful when factoring a polynomial with four or more terms is called *grouping*. Terms with common factors are grouped together and then their greatest common factors are factored out using the Distributive Property.

EXAMPLE 3

Factoring by Grouping

Factor: $ax + bx + a + b$.

SOLUTION Although there is no factor common to all four terms, we can factor x out of the first two terms and write the expression as

$$ax + bx + a + b = x(a + b) + (a + b)$$

We can now factor out the common factor of $a + b$.

$$\begin{aligned} ax + bx + a + b &= x(a + b) + (a + b) \\ &= x(a + b) + 1(a + b) \\ &= (a + b)(x + 1) \qquad \text{We can check by multiplying.} \end{aligned}$$

Self Check 3 Factor: $x^2 + xy + 2x + 2y$.

Now Try Exercise 17.

3. Factor the Difference of Two Squares

A binomial that is the difference of the squares of two quantities factors easily. The strategy used is to write the polynomial as the product of two factors. The first factor is the sum of the quantities and the other factor is the difference of the quantities.

EXAMPLE 4

Factoring the Difference of Two Squares

Factor: $49x^2 - 4$.

SOLUTION We observe that each term is a perfect square:

$$49x^2 - 4 = (7x)^2 - 2^2$$

The difference of the squares of two quantities is the product of two factors. One is the sum of the quantities, and the other is the difference of the quantities. Thus, $49x^2 - 4$ factors as

$$\begin{aligned} 49x^2 - 4 &= (7x)^2 - 2^2 \\ &= (7x + 2)(7x - 2) \qquad \text{We can check by multiplying.} \end{aligned}$$

> **Caution**
>
> The sum of two squares (with no common factor) cannot be factored using real numbers. If Example 4 had been Factor: $49x^2 + 4$, we would state *prime* as our answer. An error commonly made is to factor the sum of two squares as $(7x + 2)(7x + 2)$. Avoid making this mistake!

Self Check 4 Factor: $9a^2 - 16b^2$.

Now Try Exercise 21.

Example 4 suggests a formula for factoring the difference of two squares.

Factoring the Difference of Two Squares $x^2 - y^2 = (x + y)(x - y)$

EXAMPLE 5   **Factoring the Difference of Two Squares Twice in One Problem**

Factor: $16m^4 - n^4$.

SOLUTION   The binomial $16m^4 - n^4$ can be factored as the difference of two squares:

$$16m^4 - n^4 = (4m^2)^2 - (n^2)^2$$
$$= (4m^2 + n^2)(4m^2 - n^2)$$

The first factor is the sum of two squares and is prime. The second factor is a difference of two squares and can be factored:

$$16m^4 - n^4 = (4m^2 + n^2)[(2m)^2 - n^2]$$
$$= (4m^2 + n^2)(2m + n)(2m - n) \qquad \text{We can check by multiplying.}$$

Self Check 5   Factor: $a^4 - 81b^4$.

Now Try Exercise 23.

EXAMPLE 6   **Removing a GCF and Factoring the Difference of Two Squares**

Factor: $18t^2 - 32$.

SOLUTION   We begin by factoring out the common monomial factor of 2.

$$18t^2 - 32 = 2(9t^2 - 16)$$

Since $9t^2 - 16$ is the difference of two squares, it can be factored.

$$18t^2 - 32 = 2(9t^2 - 16)$$
$$= 2(3t + 4)(3t - 4) \qquad \text{We can check by multiplying.}$$

Self Check 6   Factor: $-3x^2 + 12$.

Now Try Exercise 65.

4. Factor Trinomials

Trinomials that are squares of binomials can be factored by using the following formulas.

Factoring Trinomial Squares

(1) $x^2 + 2xy + y^2 = (x + y)(x + y) = (x + y)^2$

(2) $x^2 - 2xy + y^2 = (x - y)(x - y) = (x - y)^2$

For example, to factor $a^2 - 6a + 9$, we note that it can be written in the form

$$a^2 - 2(3a) + 3^2 \qquad x = a \text{ and } y = 3$$

which matches the left side of Equation 2 above. Thus,

$$a^2 - 6a + 9 = a^2 - 2(3a) + 3^2$$
$$= (a - 3)(a - 3)$$
$$= (a - 3)^2 \qquad \text{We can check by multiplying.}$$

Factoring trinomials that are not squares of binomials often requires some guesswork. If a trinomial with no common factors is factorable, it will factor into the product of two binomials.

EXAMPLE 7

Factoring a Trinomial with Leading Coefficient of 1

Factor: $x^2 + 3x - 10$.

SOLUTION

To factor $x^2 + 3x - 10$, we must find two binomials $x + a$ and $x + b$ such that

$$x^2 + 3x - 10 = (x + a)(x + b)$$

where the product of a and b is –10 and the sum of a and b is 3.

$$ab = -10 \quad \text{and} \quad a + b = 3$$

To find such numbers, we list the possible factorizations of –10:

$$10(-1) \qquad 5(-2) \qquad -10(1) \qquad -5(2)$$

Only in the factorization $5(-2)$ do the factors have a sum of 3. Thus, $a = 5$ and $b = -2$, and

$$x^2 + 3x - 10 = (x + a)(x + b)$$

(3) $\quad x^2 + 3x - 10 = (x + 5)(x - 2)$ We can check by multiplying.

Because of the Commutative Property of Multiplication, the order of the factors in Equation 3 is not important. Equation 3 can also be written as

$$x^2 + 3x - 10 = (x - 2)(x + 5)$$

Self Check 7   Factor: $p^2 - 5p - 6$.

Now Try Exercise 29.

EXAMPLE 8

Factoring a Trinomial with Leading Coefficient Not 1

Factor: $2x^2 - x - 6$.

SOLUTION

Since the first term is $2x^2$, the first terms of the binomial factors must be $2x$ and x:

$$2x^2 - x - 6 = (2x \qquad)(x \qquad)$$

The product of the last terms must be -6, and the sum of the products of the outer terms and the inner terms must be $-x$. Since the only factorization of -6 that will cause this to happen is $3(-2)$, we have

$$2x^2 - x - 6 = (2x + 3)(x - 2)$$ We can check by multiplying.

Self Check 8   Factor: $6x^2 - x - 2$.

Now Try Exercise 33.

It is not easy to give specific rules for factoring trinomials, because some guesswork is often necessary. However, the following hints are helpful.

Strategy for Factoring a General Trinomial with Integer Coefficients

1. Write the trinomial in descending powers of one variable.
2. Factor out any greatest common factor, including -1 if that is necessary to make the coefficient of the first term positive.
3. When the sign of the first term of a trinomial is $+$ and the sign of the third term is $+$, the sign between the terms of each binomial factor is the same as the sign of the middle term of the trinomial.

When the sign of the first term is $+$ and the sign of the third term is $-$, one of the signs between the terms of the binomial factors is $+$ and the other is $-$.

4. Try various combinations of first terms and last terms until you find one that works. If no possibilities work, the trinomial is prime.
5. Check the factorization by multiplication.

Factoring a Trinomial Completely

EXAMPLE 9

Factor: $10xy + 24y^2 - 6x^2$.

SOLUTION   We write the trinomial in descending powers of x and then factor out the common factor of -2.

$$10xy + 24y^2 - 6x^2 = -6x^2 + 10xy + 24y^2$$
$$= -2(3x^2 - 5xy - 12y^2)$$

Since the sign of the third term of $3x^2 - 5xy - 12y^2$ is $-$, the signs between the binomial factors will be opposite. Since the first term is $3x^2$, the first terms of the binomial factors must be $3x$ and x:

$$-2(3x^2 - 5xy - 12y^2) = -2(3x \qquad)(x \qquad)$$

The product of the last terms must be $-12y^2$, and the sum of the outer terms and the inner terms must be $-5xy$. Of the many factorizations of $-12y^2$, only $4y(-3y)$ leads to a middle term of $-5xy$. So we have

$$10xy + 24y^2 - 6x^2 = -6x^2 + 10xy + 24y^2$$
$$= -2(\mathbf{3x^2 - 5xy - 12y^2})$$
$$= -2(\mathbf{3x + 4y})(\mathbf{x - 3y}) \qquad \text{We can check by multiplying.}$$

Self Check 9   Factor: $-6x^2 - 15xy - 6y^2$.

Now Try Exercise 69.

5. Factor Trinomials by Grouping

Another way of factoring trinomials involves factoring by grouping. This method can be used to factor trinomials of the form $ax^2 + bx + c$. For example, to factor $6x^2 + 5x - 6$, we note that $a = 6$, $b = 5$, and $c = -6$, and proceed as follows:

1. Find the product ac: $6(-6) = -36$. This number is called the **key number**.
2. Find two factors of the key number (-36) whose sum is $b = 5$. Two such numbers are 9 and -4.

$$9(-4) = -36 \qquad \text{and} \qquad 9 + (-4) = 5$$

3. Use the factors 9 and -4 as coefficients of two terms to be placed between $6x^2$ and -6.

$$6x^2 + \mathbf{5x} - 6 = 6x^2 + \mathbf{9x - 4x} - 6$$

4. Factor by grouping:

$$6x^2 + 9x - 4x - 6 = 3x(\mathbf{2x + 3}) - 2(\mathbf{2x + 3})$$
$$= (\mathbf{2x + 3})(3x - 2) \qquad \text{Factor out } 2x + 3.$$

Factoring a Trinomial by Grouping

EXAMPLE 10

Factor: $15x^2 + x - 2$.

SOLUTION Since $a = 15$ and $c = -2$ in the trinomial, $ac = -30$. We now find factors of -30 whose sum is $b = 1$. Such factors are 6 and -5. We use these factors as coefficients of two terms to be placed between $15x^2$ and -2.

$$15x^2 + 6x - 5x - 2$$

Finally, we factor by grouping.

$$3x(5x + 2) - 1(5x + 2) = (5x + 2)(3x - 1)$$

Self Check 10 Factor: $15a^2 + 17a - 4$.

Now Try Exercise 39.

We can often factor polynomials with variable exponents. For example, if n is a natural number,

$$a^{2n} - 5a^n - 6 = (a^n + 1)(a^n - 6)$$

because

$$(a^n + 1)(a^n - 6) = a^{2n} - 6a^n + a^n - 6$$
$$= a^{2n} - 5a^n - 6 \qquad \text{Combine like terms.}$$

6. Factor the Sum and Difference of Two Cubes

Two other types of factoring involve binomials that are the sum or the difference of two cubes. Like the difference of two squares, they can be factored by using a formula.

| **Factoring the Sum and Difference of Two Cubes** | $x^3 + y^3 = (x + y)(x^2 - xy + y^2)$ |
| | $x^3 - y^3 = (x - y)(x^2 + xy + y^2)$ |

Tip

SOAP can be an easy way to remember the signs of the binomial and trinomial terms on the right side of the factoring formulas. **S** represents same, **O** represents opposite, and **AP** represents always positive.

| $x^3 - y^3 = (x - y)(x^2 + xy + y^2)$ | $x^3 + y^3 = (x + y)(x^2 - xy + y^2)$ |
| S   O   AP | S   O   AP |

Steven van Soldt/iStockphoto.com

Factoring the Sum of Two Cubes

EXAMPLE 11

Factor: $27x^6 + 64y^3$.

SOLUTION We can write this expression as the sum of two cubes and factor it as follows:

$$27x^6 + 64y^3 = (\mathbf{3x^2})^3 + (\mathbf{4y})^3$$
$$= (\mathbf{3x^2} + \mathbf{4y})[(\mathbf{3x^2})^2 - (\mathbf{3x^2})(\mathbf{4y}) + (\mathbf{4y})^2]$$
$$= (3x^2 + 4y)(9x^4 - 12x^2y + 16y^2) \qquad \text{We can check by multiplying.}$$

Self Check 11   Factor: $8a^3 + 1000b^6$.

Now Try Exercise 43.

Factoring the Difference of Two Cubes

EXAMPLE 12   Factor: $125x^3 - 8$.

SOLUTION   This binomial can be written as $(5x)^3 - 2^3$, which is the difference of two cubes. Substituting into the formula for the difference of two cubes gives

$$125x^3 - 8 = (5x)^3 - 2^3$$
$$= (5x - 2)[(5x)^2 + (5x)(2) + 2^2]$$
$$= (5x - 2)(25x^2 + 10x + 4) \qquad \text{We can check by multiplying.}$$

Self Check 12   Factor: $p^3 - 64$.

Now Try Exercise 45.

Tip

The factoring formulas for the sum and difference of two cubes are easier to remember if we use F for the FIRST quantity cubed and L for the LAST quantity cubed.

Factoring the Sum of Two Cubes	Factoring the Difference of Two Cubes
$F^3 + L^3 = (F + L)(F^2 - FL + L^2)$	$F^3 - L^3 = (F - L)(F^2 + FL + L^2)$
To factor the sum of the FIRST cubed and the LAST cubed, we simply multiply the FIRST plus the LAST times the FIRST squared minus the FIRST times the LAST plus the LAST squared.	To factor the difference of the FIRST cubed and the LAST cubed, we simply multiply the FIRST minus the LAST times the FIRST squared plus the FIRST times the LAST plus the LAST squared.

7. Factor Miscellaneous Polynomials

Factoring a Miscellaneous Polynomial

EXAMPLE 13   Factor: $x^2 - y^2 + 6x + 9$.

SOLUTION   Here we will factor a trinomial and a difference of two squares.

$$x^2 - y^2 + 6x + 9 = x^2 + 6x + 9 - y^2 \qquad \text{Use the Commutative Property to rearrange the terms.}$$
$$= (x + 3)^2 - y^2 \qquad \text{Factor } x^2 + 6x + 9.$$
$$= (x + 3 + y)(x + 3 - y) \qquad \text{Factor the difference of two squares.}$$

We could try to factor this expression in another way.

$$x^2 - y^2 + 6x + 9 = (x + y)(x - y) + 3(2x + 3) \qquad \text{Factor } x^2 - y^2 \text{ and } 6x + 9.$$

However, we are unable to finish the factorization. If grouping in one way doesn't work, try various other ways.

Self Check 13   Factor: $a^2 + 8a - b^2 + 16$.

Now Try Exercise 101.

Factoring a Miscellaneous Trinomial

EXAMPLE 14   Factor: $z^4 - 3z^2 + 1$.

SOLUTION   This trinomial cannot be factored as the product of two binomials, because no combination will give a middle term of $-3z^2$. However, if the middle term were $-2z^2$, the trinomial would be a perfect square, and the factorization would be easy:

$$z^4 - 2z^2 + 1 = (z^2 - 1)(z^2 - 1)$$
$$= (z^2 - 1)^2$$

We can change the middle term in $z^4 - 3z^2 + 1$ to $-2z^2$ by adding z^2 to it. However, to make sure that adding z^2 does not change the value of the trinomial, we must also subtract z^2. We can then proceed as follows.

$$z^4 - 3z^2 + 1 = z^4 - 3z^2 + z^2 + 1 - z^2 \qquad \text{Add and subtract } z^2.$$
$$= z^4 - 2z^2 + 1 - z^2 \qquad \text{Combine } -3z^2 \text{ and } z^2.$$
$$= (z^2 - 1)^2 - z^2 \qquad \text{Factor } z^4 - 2z^2 + 1.$$
$$= (z^2 - 1 + z)(z^2 - 1 - z) \qquad \text{Factor the difference of two squares.}$$

In this type of problem, we will always try to add and subtract a perfect square in hopes of making a perfect-square trinomial that will lead to factoring a difference of two squares.

Self Check 14   Factor: $x^4 + 3x^2 + 4$.

Now Try Exercise 107.

It is helpful to identify the problem type when we must factor polynomials that are given in random order.

Factoring Strategy

1. Factor out all common monomial factors.
2. If an expression has two terms, check whether the problem type is
   a. The difference of two squares:
   $$x^2 - y^2 = (x + y)(x - y)$$
   b. The sum of two cubes:
   $$x^3 + y^3 = (x + y)(x^2 - xy + y^2)$$
   c. The difference of two cubes:
   $$x^3 - y^3 = (x - y)(x^2 + xy + y^2)$$
3. If an expression has three terms, try to factor it as a *trinomial*.
4. If an expression has four or more terms, try factoring by *grouping*.
5. Continue until each individual factor is prime.
6. Check the results by multiplying.

Self Check Answers

1. $4a(a - 2b)$ 2. $a^2b^2c^2(1 + abc)$ 3. $(x + y)(x + 2)$
4. $(3a + 4b)(3a - 4b)$ 5. $(a^2 + 9b^2)(a + 3b)(a - 3b)$
6. $-3(x + 2)(x - 2)$ 7. $(p - 6)(p + 1)$ 8. $(3x - 2)(2x + 1)$
9. $-3(x + 2y)(2x + y)$ 10. $(3a + 4)(5a - 1)$
11. $8(a + 5b^2)(a^2 - 5ab^2 + 25b^4)$ 12. $(p - 4)(p^2 + 4p + 16)$
13. $(a + 4 + b)(a + 4 - b)$ 14. $(x^2 + 2 + x)(x^2 + 2 - x)$

Exercises 0.5

Getting Ready

You should be able to complete these vocabulary and concept statements before you proceed to the practice exercises.

Fill in the blanks.

1. When polynomials are multiplied together, each polynomial is a _____ of the product.
2. If a polynomial cannot be factored using _____ coefficients, it is called a _____ polynomial.

Complete each factoring formula.

3. $ax + bx =$ _____
4. $x^2 - y^2 =$ _____
5. $x^2 + 2xy + y^2 =$ _____
6. $x^2 - 2xy + y^2 =$ _____
7. $x^3 + y^3 =$ _____
8. $x^3 - y^3 =$ _____

Practice

In each expression, factor out the greatest common monomial.

9. $3x - 6$
10. $5y - 15$
11. $8x^2 + 4x^3$
12. $9y^3 + 6y^2$
13. $7x^2y^2 + 14x^3y^2$
14. $25y^2z - 15yz^2$

In each expression, factor by grouping.

15. $a(x + y) + b(x + y)$
16. $b(x - y) + a(x - y)$
17. $4a + b - 12a^2 - 3ab$
18. $x^2 + 4x + xy + 4y$

In each expression, factor the difference of two squares.

19. $4x^2 - 9$
20. $36z^2 - 49$
21. $4 - 9r^2$
22. $16 - 49x^2$
23. $81x^4 - 1$
24. $81 - x^4$
25. $(x + z)^2 - 25$
26. $(x - y)^2 - 9$

In each expression, factor the trinomial.

27. $x^2 + 8x + 16$
28. $a^2 - 12a + 36$
29. $b^2 - 10b + 25$
30. $y^2 + 14y + 49$

31. $m^2 + 4mn + 4n^2$
32. $r^2 - 8rs + 16s^2$
33. $12x^2 - xy - 6y^2$
34. $8x^2 - 10xy - 3y^2$

In each expression, factor the trinomial by grouping.

35. $x^2 + 10x + 21$
36. $x^2 + 7x + 10$
37. $x^2 - 4x - 12$
38. $x^2 - 2x - 63$
39. $6p^2 + 7p - 3$
40. $4q^2 - 19q + 12$

In each expression, factor the sum of two cubes.

41. $t^3 + 343$
42. $r^3 + 8s^3$
43. $125y^3 + 216z^3$
44. $27y^3 + 1000z^3$

In each expression, factor the difference of two cubes.

45. $8z^3 - 27$
46. $125a^3 - 64$
47. $343y^3 - z^3$
48. $27y^3 - 512z^3$

Factor each expression completely. If an expression is prime, so indicate.

49. $3a^2bc + 6ab^2c + 9abc^2$
50. $5x^3y^3z^3 + 25x^2y^2z^2 - 125xyz$
51. $3x^3 + 3x^2 - x - 1$
52. $4x + 6xy - 9y - 6$
53. $2txy + 2ctx - 3ty - 3ct$
54. $2ax + 4ay - bx - 2by$
55. $ax + bx + ay + by + az + bz$
56. $6x^2y^3 + 18xy + 3x^2y^2 + 9x$
57. $x^2 - (y - z)^2$
58. $z^2 - (y + 3)^2$
59. $(x - y)^2 - (x + y)^2$
60. $(2a + 3)^2 - (2a - 3)^2$
61. $x^4 - y^4$
62. $z^4 - 81$
63. $3x^2 - 12$
64. $3x^3y - 3xy$
65. $18xy^2 - 8x$
66. $27x^2 - 12$

67. $x^2 - 2x + 15$

68. $x^2 + x + 2$

69. $-15 + 2a + 24a^2$

70. $-32 - 68x + 9x^2$

71. $6x^2 + 29xy + 35y^2$

72. $10x^2 - 17xy + 6y^2$

73. $12p^2 - 58pq - 70q^2$

74. $3x^2 - 6xy - 9y^2$

75. $-6m^2 + 47mn - 35n^2$

76. $-14r^2 - 11rs + 15s^2$

77. $-6x^3 + 23x^2 + 35x$

78. $-y^3 - y^2 + 90y$

79. $6x^4 - 11x^3 - 35x^2$

80. $12x + 17x^2 - 7x^3$

81. $x^4 + 2x^2 - 15$

82. $x^4 - x^2 - 6$

83. $a^{2n} - 2a^n - 3$

84. $a^{2n} + 6a^n + 8$

85. $6x^{2n} - 7x^n + 2$

86. $9x^{2n} + 9x^n + 2$

87. $4x^{2n} - 9y^{2n}$

88. $8x^{2n} - 2x^n - 3$

89. $10y^{2n} - 11y^n - 6$

90. $16y^{4n} - 25y^{2n}$

91. $2x^3 + 2000$

92. $3y^3 + 648$

93. $(x + y)^3 - 64$

94. $(x - y)^3 + 27$

95. $64a^6 - y^6$

96. $a^6 + b^6$

97. $a^3 - b^3 + a - b$

98. $(a^2 - y^2) - 5(a + y)$

99. $64x^6 + y^6$

100. $z^2 + 6z + 9 - 225y^2$

101. $x^2 - 6x + 9 - 144y^2$

102. $x^2 + 2x - 9y^2 + 1$

103. $(a + b)^2 - 3(a + b) - 10$

104. $2(a + b)^2 - 5(a + b) - 3$

105. $x^6 + 7x^3 - 8$

106. $x^6 - 13x^4 + 36x^2$

107. $x^4 + x^2 + 1$

108. $x^4 + 3x^2 + 4$

109. $x^4 + 7x^2 + 16$

110. $y^4 + 2y^2 + 9$

111. $4a^4 + 1 + 3a^2$

112. $x^4 + 25 + 6x^2$

Applications

113. Candy To find the amount of chocolate used in the outer coating of one of the malted-milk balls shown, we can find the volume V of the chocolate shell using the formula $V = \frac{4}{3}\pi r_1^3 - \frac{4}{3}\pi r_2^3$. Factor the expression on the right side of the formula.

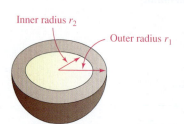

Inner radius r_2

Outer radius r_1

114. Movie stunts The formula that gives the distances a stuntwoman is above the ground t seconds after she falls over the side of a 144-foot tall building is $s = 144 - 16t^2$. Factor the right side.

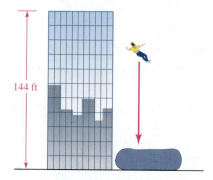

144 ft

Discovery and Writing

115. Explain how to factor the difference of two squares.

116. Explain how to factor the difference of two cubes.

117. Explain how to factor by grouping.

118. Explain what is meant by *factor completely*.

Factor the indicated monomial from the given expression.

119. $3x + 2$; 2

120. $5x - 3$; 5

121. $x^2 + 2x + 4$; 2

122. $3x^2 - 2x - 5$; 3

123. $a + b; a$

124. $a - b; b$

125. $x + x^{1/2}; x^{1/2}$

126. $x^{3/2} - x^{1/2}; x^{1/2}$

127. $2x + \sqrt{2}y; \sqrt{2}$

128. $\sqrt{3}a - 3b; \sqrt{3}$

129. $ab^{3/2} - a^{3/2}b; ab$

130. $ab^2 + b; b^{-1}$

Factor each expression by grouping three terms and two terms.

131. $x^2 + x - 6 + xy - 2y$

132. $2x^2 + 5x + 2 - xy - 2y$

133. $a^4 + 2a^3 + a^2 + a + 1$

134. $a^4 + a^3 - 2a^2 + a - 1$

Critical Thinking

In Exercises, 135–140, determine if the statement is true or false. If the statement is false, then correct it and make it true.

135. The GCF of $22x^{22}y^{44} - 44x^{44}y^{22}$ is $22x^{22}y^{22}$.

136. $25x^{200}z^{200} + 36$ factors completely as $(5x^{100}z^{100} + 6)^2$.

137. $p^3q^3r^3 + 64$ factors completely as $(pqr + 4)^3$.

138. $9x^2 + 15xy + 25y^2$ is a factor of $27x^3 - 125y^3$.

139. $(2x + 5y)^2 - (7z - 9w)^2$ factors completely as $(2x + 5y + 7z - 9w)(2x + 5y - 7z + 9w)$.

140. The polynomial $x^2 + kx + 12$ can be factored for integer values $k = 7, 8,$ and 13 only.

0.6   Rational Expressions

In this section, we will learn to

1. Define rational expressions.

2. Simplify rational expressions.

3. Multiply and divide rational expressions.

4. Add and subtract rational expressions.

5. Simplify complex fractions.

Tim Boyle/Getty Images

Abercrombie and Fitch (A&F) is a very successful American clothing company founded in 1892 by David Abercrombie and Ezra Fitch. Today the stores are popular shopping destinations for university students wanting to keep up with the latest styles and trends.

Suppose that a clothing manufacturer finds that the cost in dollars of producing x fleece vintage shirts is given by the algebraic expression $13x + 1000$. The average cost of producing each shirt could be obtained by dividing the production cost, $13x + 1000$, by the number of shirts produced, x. The algebraic fraction

$$\frac{13x + 1000}{x}$$

represents the average cost per shirt. We see that the average cost of producing 200 shirts would be $18.

$$\frac{13(\textcolor{red}{200}) + 1000}{\textcolor{red}{200}} = 18$$

An understanding of algebraic fractions is important in solving many real-life problems.

1. Define Rational Expressions

If x and y are real numbers, the quotient $\frac{x}{y}$ ($y \neq 0$) is called a **fraction**. The number x is called the **numerator**, and the number y is called the **denominator**.

Algebraic fractions are quotients of algebraic expressions. If the expressions are polynomials, the fraction is called a **rational expression**. The first two of the following algebraic fractions are rational expressions. The third is not, because the numerator and denominator are not polynomials.

$$\frac{5y^2 + 2y}{y^2 - 3y - 7} \qquad \frac{8ab^2 - 16c^3}{2x + 3} \qquad \frac{x^{1/2} + 4x}{x^{3/2} - x^{1/2}}$$

We summarize some of the properties of fractions as follows:

Properties of Fractions   If a, b, c, and d are real numbers and no denominators are 0, then

Equality of Fractions

$$\frac{a}{b} = \frac{c}{d} \qquad \text{if and only if} \qquad ad = bc$$

Fundamental Property of Fractions

$$\frac{ax}{bx} = \frac{a}{b}$$

Multiplication and Division of Fractions

$$\frac{a}{b} \cdot \frac{c}{d} = \frac{ac}{bd} \qquad \text{and} \qquad \frac{a}{b} \div \frac{c}{d} = \frac{a}{b} \cdot \frac{d}{c} = \frac{ad}{bc}$$

Addition and Subtraction of Fractions

$$\frac{a}{b} + \frac{c}{b} = \frac{a + c}{b} \qquad \text{and} \qquad \frac{a}{b} - \frac{c}{b} = \frac{a - c}{b}$$

The first two examples illustrate each of the previous properties of fractions.

EXAMPLE 1   **Illustrating the Properties of Fractions**

Assume that no denominators are 0.

a. $\dfrac{2a}{3} = \dfrac{4a}{6}$ Because $2a(6) = 3(4a)$

b. $\dfrac{6xy}{10xy} = \dfrac{3(2xy)}{5(2xy)}$ Factor the numerator and denominator and divide out the common factors.

$= \dfrac{3}{5}$

Self Check 1   a. Is $\dfrac{3y}{5} = \dfrac{15z}{25}$?   b. Simplify: $\dfrac{15a^2b}{25ab^2}$.

Now Try Exercise 9.

Illustrating the Properties of Fractions

EXAMPLE 2

Assume that no denominators are 0.

a. $\dfrac{2r}{7s} \cdot \dfrac{3r}{5s} = \dfrac{2r \cdot 3r}{7s \cdot 5s}$

$= \dfrac{6r^2}{35s^2}$

b. $\dfrac{3mn}{4pq} \div \dfrac{2pq}{7mn} = \dfrac{3mn}{4pq} \cdot \dfrac{7mn}{2pq}$

$= \dfrac{21m^2n^2}{8p^2q^2}$

c. $\dfrac{2ab}{5xy} + \dfrac{ab}{5xy} = \dfrac{2ab + ab}{5xy}$

$= \dfrac{3ab}{5xy}$

d. $\dfrac{6uv^2}{7w^2} - \dfrac{3uv^2}{7w^2} = \dfrac{6uv^2 - 3uv^2}{7w^2}$

$= \dfrac{3uv^2}{7w^2}$

Self Check 2   Perform each operation:

a. $\dfrac{3a}{5b} \cdot \dfrac{2a}{7b}$ **b.** $\dfrac{2ab}{3rs} \div \dfrac{2rs}{4ab}$ **c.** $\dfrac{5pq}{3t} + \dfrac{3pq}{3t}$ **d.** $\dfrac{5mn^2}{3w} - \dfrac{mn^2}{3w}$

Now Try Exercise 19.

To add or subtract rational expressions with unlike denominators, we write each expression as an equivalent expression with a common denominator. We can then add or subtract the expressions. For example,

$\dfrac{3x}{5} + \dfrac{2x}{7} = \dfrac{3x(7)}{5(7)} + \dfrac{2x(5)}{7(5)}$

$= \dfrac{21x}{35} + \dfrac{10x}{35}$

$= \dfrac{21x + 10x}{35}$

$= \dfrac{31x}{35}$

$\dfrac{4a^2}{15} - \dfrac{3a^2}{10} = \dfrac{4a^2(2)}{15(2)} - \dfrac{3a^2(3)}{10(3)}$

$= \dfrac{8a^2}{30} - \dfrac{9a^2}{30}$

$= \dfrac{8a^2 - 9a^2}{30}$

$= \dfrac{-a^2}{30}$

$= -\dfrac{a^2}{30}$

A rational expression is **in lowest terms** if all factors common to the numerator and the denominator have been removed. To **simplify a rational expression** means to write it in lowest terms.

2. Simplify Rational Expressions

To simplify rational expressions, we use the Fundamental Property of Fractions. This enables us to divide out all factors that are common to the numerator and the denominator.

Simplifying a Rational Expression

EXAMPLE 3

Simplify: $\dfrac{x^2 - 9}{x^2 - 3x}$ $(x \neq 0, 3)$.

SOLUTION   We factor the difference of two squares in the numerator, factor out x in the denominator, and divide out the common factor of $x - 3$.

$\dfrac{x^2 - 9}{x^2 - 3x} = \dfrac{(x + 3)(x - 3)}{x(x - 3)}$ $\dfrac{x - 3}{x - 3} = 1$

$= \dfrac{x + 3}{x}$

Self Check 3   Simplify: $\dfrac{a^2 - 4a}{a^2 - a - 12}$ $(a \neq 4, -3)$.

Now Try Exercise 23.

We will encounter the following properties of fractions in the next examples.

Properties of Fractions   If a and b represent real numbers and there are no divisions by 0, then

- $\dfrac{a}{1} = a$

- $\dfrac{a}{b} = \dfrac{-a}{-b} = -\dfrac{a}{-b} = -\dfrac{-a}{b}$

- $\dfrac{a}{a} = 1$

- $-\dfrac{a}{b} = \dfrac{a}{-b} = \dfrac{-a}{b} = \dfrac{-a}{-b}$

EXAMPLE 4   **Simplifying a Rational Expression**

Simplify: $\dfrac{x^2 - 2xy + y^2}{y - x}$ $(x \neq y)$.

SOLUTION   We factor the trinomial in the numerator, factor -1 from the denominator, and divide out the common factor of $x - y$.

$$\frac{x^2 - 2xy + y^2}{y - x} = \frac{(x - y)(x - y)}{-1(x - y)} \qquad \frac{x - y}{x - y} = 1$$

$$= \frac{x - y}{-1}$$

$$= -\frac{x - y}{1}$$

$$= -(x - y)$$

> **Tip**
>
> Expressions like $x - y$ and $y - x$ are like twins. They look alike but are very different. When we divide them, we do not get 1. We always get -1.
>
>

Self Check 4   Simplify: $\dfrac{a^2 - ab - 2b^2}{2b - a}$ $(2b - a \neq 0)$.

Now Try Exercise 25.

EXAMPLE 5   **Simplifying a Rational Expression**

Simplify: $\dfrac{x^2 - 3x + 2}{x^2 - x - 2}$ $(x \neq 2, -1)$.

SOLUTION   We factor the numerator and denominator and divide out the common factor of $x - 2$.

$$\frac{x^2 - 3x + 2}{x^2 - x - 2} = \frac{(x - 1)(x - 2)}{(x + 1)(x - 2)} \qquad \frac{x - 2}{x - 2} = 1$$

$$= \frac{x - 1}{x + 1}$$

Self Check 5   Simplify: $\dfrac{a^2 + 3a - 4}{a^2 + 2a - 3}$ $(a \neq 1, -3)$.

Now Try Exercise 27.

3. Multiply and Divide Rational Expressions

EXAMPLE 6

Multiplying Rational Expressions

Multiply: $\dfrac{x^2 - x - 2}{x^2 - 1} \cdot \dfrac{x^2 + 2x - 3}{x - 2}$ $(x \neq 1, -1, 2)$.

SOLUTION   To multiply the rational expressions, we multiply the numerators, multiply the denominators, and divide out the common factors.

$$\frac{x^2 - x - 2}{x^2 - 1} \cdot \frac{x^2 + 2x - 3}{x - 2} = \frac{(x^2 - x - 2)(x^2 + 2x - 3)}{(x^2 - 1)(x - 2)}$$

$$= \frac{(x - 2)(x + 1)(x - 1)(x + 3)}{(x + 1)(x - 1)(x - 2)} \qquad \frac{x-2}{x-2} = 1, \frac{x+1}{x+1} = 1, \frac{x-1}{x-1} = 1$$

$$= x + 3$$

Self Check 6   Simplify: $\dfrac{x^2 - 9}{x^2 - x} \cdot \dfrac{x - 1}{x^2 - 3x}$ $(x \neq 0, 1, 3)$.

Now Try Exercise 33.

EXAMPLE 7

Dividing Rational Expressions

Divide: $\dfrac{x^2 - 2x - 3}{x^2 - 4} \div \dfrac{x^2 + 2x - 15}{x^2 + 3x - 10}$ $(x \neq 2, -2, 3, -5)$.

SOLUTION   To divide the rational expressions, we multiply by the reciprocal of the second rational expression. We then simplify by factoring the numerator and denominator and dividing out the common factors.

$$\frac{x^2 - 2x - 3}{x^2 - 4} \div \frac{x^2 + 2x - 15}{x^2 + 3x - 10}$$

$$= \frac{x^2 - 2x - 3}{x^2 - 4} \cdot \frac{x^2 + 3x - 10}{x^2 + 2x - 15}$$

$$= \frac{(x^2 - 2x - 3)(x^2 + 3x - 10)}{(x^2 - 4)(x^2 + 2x - 15)}$$

$$= \frac{(x - 3)(x + 1)(x - 2)(x + 5)}{(x + 2)(x - 2)(x + 5)(x - 3)} \qquad \frac{x-3}{x-3} = 1, \frac{x-2}{x-2} = 1, \frac{x+5}{x+5} = 1$$

$$= \frac{x + 1}{x + 2}$$

> **Tip**
>
> Dividing rational expressions is easy if you remember the following. Never worry! Never cry! Invert the second rational expression! Multiply!
>
>
>
> ©Anson0618/Shutterstock.com

Self Check 7   Simplify: $\dfrac{a^2 - a}{a + 2} \div \dfrac{a^2 - 2a}{a^2 - 4}$ $(a \neq 0, 2, -2)$.

Now Try Exercise 39.

EXAMPLE 8

Using Multiplication and Division to Simplify a Rational Expression

Simplify: $\dfrac{2x^2 - 5x - 3}{3x - 1} \cdot \dfrac{3x^2 + 2x - 1}{x^2 - 2x - 3} \div \dfrac{2x^2 + x}{3x}$ $\left(x \neq \dfrac{1}{3}, -1, 3, 0, -\dfrac{1}{2}\right)$.

SOLUTION   We can change the division to a multiplication, factor, and simplify.

$$\frac{2x^2 - 5x - 3}{3x - 1} \cdot \frac{3x^2 + 2x - 1}{x^2 - 2x - 3} \div \frac{2x^2 + x}{3x}$$

$$= \frac{2x^2 - 5x - 3}{3x - 1} \cdot \frac{3x^2 + 2x - 1}{x^2 - 2x - 3} \cdot \frac{3x}{2x^2 + x}$$

$$= \frac{(2x^2 - 5x - 3)(3x^2 + 2x - 1)(3x)}{(3x - 1)(x^2 - 2x - 3)(2x^2 + x)}$$

$$= \frac{(x - 3)(2x + 1)(3x - 1)(x + 1)3x}{(3x - 1)(x + 1)(x - 3)x(2x + 1)} \qquad \frac{x - 3}{x - 3} = 1, \ \frac{2x + 1}{2x + 1} = 1, \ \frac{3x - 1}{3x - 1} = 1, \ \frac{x + 1}{x + 1} = 1, \ \frac{x}{x} = 1$$

$$= 3$$

Self Check 8   Simplify: $\dfrac{x^2 - 25}{x - 2} \div \dfrac{x^2 - 5x}{x^2 - 2x} \cdot \dfrac{x^2 + 2x}{x^2 + 5x}$ $(x \neq 0, 2, 5, -5)$.

Now Try Exercise 45.

4. Add and Subtract Rational Expressions

To add (or subtract) rational expressions with like denominators, we add (or subtract) the numerators and keep the common denominator.

EXAMPLE 9   **Adding Rational Expression with Like Denominators**

Add: $\dfrac{2x + 5}{x + 5} + \dfrac{3x + 20}{x + 5}$ $(x \neq -5)$.

SOLUTION
$$\frac{2x + 5}{x + 5} + \frac{3x + 20}{x + 5} = \frac{5x + 25}{x + 5} \qquad \text{Add the numerators and keep the common denominator.}$$

$$= \frac{5(x + 5)}{(x + 5)} \qquad \text{Factor out 5 and divide out the common factor of } x + 5.$$

$$= 5$$

Self Check 9   Add: $\dfrac{3x - 2}{x - 2} + \dfrac{x - 6}{x - 2}$ $(x \neq 2)$.

Now Try Exercise 49.

To add (or subtract) rational expressions with unlike denominators, we must find a common denominator, called the **least** (or lowest) **common denominator (LCD)**. Suppose the unlike denominators of three rational expressions are 12, 20, and 35. To find the LCD, we first find the prime factorization of each number.

$$12 = 4 \cdot 3 \qquad\qquad 20 = 4 \cdot 5 \qquad\qquad 35 = 5 \cdot 7$$
$$\quad = 2^2 \cdot 3 \qquad\qquad\quad = 2^2 \cdot 5$$

Because the LCD is the smallest number that can be divided by 12, 20, and 35, it must contain factors of 2^2, 3, 5, and 7. Thus, the

$$\text{LCD} = 2^2 \cdot 3 \cdot 5 \cdot 7 = 420$$

That is, 420 is the smallest number that can be divided without remainder by 12, 20, and 35.

When finding an LCD, we always factor each denominator and then create the LCD by using each factor the greatest number of times that it appears in any one denominator. The product of these factors is the LCD.

Take Note

Remember: To find the *least* common denominator, use each factor the *greatest* number of times that it occurs.

This rule also applies if the unlike denominators of the rational expressions contain variables. Suppose the unlike denominators are $x^2(x-5)$ and $x(x-5)^3$. To find the LCD, we use x^2 and $(x-5)^3$. Thus, the LCD is the product $x^2(x-5)^3$.

Adding Rational Expressions with Unlike Denominators

EXAMPLE 10

Add: $\dfrac{1}{x^2-4} + \dfrac{2}{x^2-4x+4}$ $(x \ne 2, -2)$.

SOLUTION   We factor each denominator and find the LCD.

$$x^2 - 4 = (x+2)(x-2)$$

$$x^2 - 4x + 4 = (x-2)(x-2) = (x-2)^2$$

The LCD is $(x+2)(x-2)^2$. We then write each rational expression with its denominator in factored form, convert each rational expression into an equivalent expression with a denominator of $(x+2)(x-2)^2$, add the expressions, and simplify.

$$\frac{1}{x^2-4} + \frac{2}{x^2-4x+4} = \frac{1}{(x+2)(x-2)} + \frac{2}{(x-2)(x-2)}$$

$$= \frac{1(x-2)}{(x+2)(x-2)(x-2)} + \frac{2(x+2)}{(x-2)(x-2)(x+2)} \qquad \frac{x-2}{x-2}=1, \frac{x+2}{x+2}=1$$

$$= \frac{1(x-2) + 2(x+2)}{(x+2)(x-2)(x-2)}$$

$$= \frac{x-2 + 2x + 4}{(x+2)(x-2)(x-2)}$$

$$= \frac{3x+2}{(x+2)(x-2)^2}$$

Take Note
Always attempt to simplify the final result. In this case, the final fraction is already in lowest terms.

Self Check 10   Add: $\dfrac{3}{x^2-6x+9} + \dfrac{1}{x^2-9}$ $(x \ne 3, -3)$.

Now Try Exercise 59.

Combining and Simplifying Rational Expressions with Unlike Denominators

EXAMPLE 11

Simplify: $\dfrac{x-2}{x^2-1} - \dfrac{x+3}{x^2+3x+2} + \dfrac{3}{x^2+x-2}$ $(x \ne 1, -1, -2)$.

SOLUTION   We factor the denominators to find the LCD.

$$x^2 - 1 = (x+1)(x-1)$$

$$x^2 + 3x + 2 = (x+2)(x+1)$$

$$x^2 + x - 2 = (x+2)(x-1)$$

The LCD is $(x+1)(x-1)(x+2)$. We now write each rational expression as an equivalent expression with this LCD, and proceed as follows:

$$\frac{x-2}{x^2-1} - \frac{x+3}{x^2+3x+2} + \frac{3}{x^2+x-2}$$

$$= \frac{x-2}{(x+1)(x-1)} - \frac{x+3}{(x+1)(x+2)} + \frac{3}{(x-1)(x+2)}$$

$$= \frac{(x-2)(x+2)}{(x+1)(x-1)(x+2)} - \frac{(x+3)(x-1)}{(x+1)(x+2)(x-1)} + \frac{3(x+1)}{(x-1)(x+2)(x+1)} \qquad \frac{x+2}{x+2}=1, \frac{x-1}{x-1}=1, \frac{x+1}{x+1}=1$$

$$= \frac{(x^2-4) - (x^2+2x-3) + (3x+3)}{(x+1)(x-1)(x+2)}$$

$$= \frac{x^2-4-x^2-2x+3+3x+3}{(x+1)(x-1)(x+2)}$$

$$= \frac{\cancel{x+2}}{(x+1)(x-1)\cancel{(x+2)}}$$

$$= \frac{1}{(x+1)(x-1)} \qquad \text{Divide out the common factor of } x+2. \ \frac{x+2}{x+2}=1$$

Self Check 11   Simplify: $\dfrac{4y}{y^2-1} - \dfrac{2}{y+1} + 2 \ (y \neq 1, -1)$.

Now Try Exercise 61.

5. Simplify Complex Fractions

A **complex fraction** is a fraction that has a fraction in its numerator or a fraction in its denominator. There are two methods generally used to simplify complex fractions. These are stated here for you.

Strategies for Simplifying Complex Fractions

Method 1: Multiply the Complex Fraction by 1

- Determine the LCD of all fractions in the complex fraction.
- Multiply both numerator and denominator of the complex fraction by the LCD. Note that when we multiply by $\frac{LCD}{LCD}$ we are multiplying by 1.

Method 2: Simplify the Numerator and Denominator and then Divide

- Simplify the numerator and denominator so that both are single fractions.
- Perform the division by multiplying the numerator by the reciprocal of the denominator.

EXAMPLE 12

Simplifying Complex Fractions

Simplify: $\dfrac{\dfrac{1}{x} + \dfrac{1}{y}}{\dfrac{x}{y}} \ (x, y \neq 0)$.

Method 1: We note that the LCD of the three fractions in the complex fraction is xy. So we multiply the numerator and denominator of the complex fraction by xy and simplify:

$$\frac{\dfrac{1}{x}+\dfrac{1}{y}}{\dfrac{x}{y}} = \frac{xy\left(\dfrac{1}{x}+\dfrac{1}{y}\right)}{xy\left(\dfrac{x}{y}\right)} = \frac{\dfrac{xy}{x}+\dfrac{xy}{y}}{\dfrac{xyx}{y}} = \frac{y+x}{x^2}$$

Tip

In general, Method 1 is easier. If you find complex fractions difficult to master, then concentrate your study time on Method 1.

Method 2: We combine the fractions in the numerator of the complex fraction to obtain a single fraction over a single fraction.

$$\frac{\dfrac{1}{x}+\dfrac{1}{y}}{\dfrac{x}{y}}=\frac{\dfrac{1(y)}{x(y)}+\dfrac{1(x)}{y(x)}}{\dfrac{x}{y}}=\frac{\dfrac{y+x}{xy}}{\dfrac{x}{y}}$$

Then we use the fact that any fraction indicates a division:

$$\frac{\dfrac{y+x}{xy}}{\dfrac{x}{y}}=\frac{y+x}{xy}\div\frac{x}{y}=\frac{y+x}{xy}\cdot\frac{y}{x}=\frac{(y+x)\overset{}{y}}{xy\,x}=\frac{y+x}{x^2}$$

Self Check 12　Simplify: $\dfrac{\dfrac{1}{x}-\dfrac{1}{y}}{\dfrac{1}{x}+\dfrac{1}{y}}$ $(x, y \neq 0)$.

Now Try Exercise 81.

Self Check Answers

1. a. no　**b.** $\dfrac{3a}{5b}$　**2. a.** $\dfrac{6a^2}{35b^2}$　**b.** $\dfrac{4a^2b^2}{3r^2s^2}$　**c.** $\dfrac{8pq}{3t}$　**d.** $\dfrac{4mn^2}{3w}$

3. $\dfrac{a}{a+3}$　**4.** $-(a+b)$　**5.** $\dfrac{a+4}{a+3}$　**6.** $\dfrac{x+3}{x^2}$　**7.** $a-1$

8. $x+2$　**9.** 4　**10.** $\dfrac{2(2x+3)}{(x+3)(x-3)^2}$　**11.** $\dfrac{2y}{y-1}$　**12.** $\dfrac{y-x}{y+x}$

Exercises 0.6

Getting Ready

You should be able to complete these vocabulary and concept statements before you proceed to the practice exercises.

Fill in the blanks.

1. In the fraction $\frac{a}{b}$, a is called the _____.
2. In the fraction $\frac{a}{b}$, b is called the _____.
3. $\frac{a}{b} = \frac{c}{d}$ if and only if _____.
4. The denominator of a fraction can never be _____.

Complete each formula.

5. $\dfrac{a}{b} \cdot \dfrac{c}{d} =$ _____

6. $\dfrac{a}{b} \div \dfrac{c}{d} =$ _____

7. $\dfrac{a}{b} + \dfrac{c}{b} =$ _____

8. $\dfrac{a}{b} - \dfrac{c}{b} =$ _____

Determine whether the fractions are equal. Assume that no denominators are 0.

9. $\dfrac{8x}{3y}, \dfrac{16x}{6y}$

10. $\dfrac{3x^2}{4y^2}, \dfrac{12y^2}{16x^2}$

11. $\dfrac{25xyz}{12ab^2c}, \dfrac{50a^2bc}{24xyz}$

12. $\dfrac{15rs^2}{4rs^2}, \dfrac{37.5a^3}{10a^3}$

Practice

Simplify each rational expression. Assume that no denominators are 0.

13. $\dfrac{7a^2b}{21ab^2}$

14. $\dfrac{35p^3q^2}{49p^4q}$

Perform the operations and simplify, whenever possible. Assume that no denominators are 0.

15. $\dfrac{4x}{7} \cdot \dfrac{2}{5a}$

16. $-\dfrac{5y}{2z} \cdot \dfrac{4}{y^2}$

17. $\dfrac{8m}{5n} \div \dfrac{3m}{10n}$

18. $\dfrac{15p}{8q} \div \dfrac{-5p}{16q^2}$

19. $\dfrac{3z}{5c} + \dfrac{2z}{5c}$

20. $\dfrac{7a}{4b} - \dfrac{3a}{4b}$

21. $\dfrac{15x^2y}{7a^2b^3} - \dfrac{x^2y}{7a^2b^3}$

22. $\dfrac{8rst^2}{15m^4t^2} + \dfrac{7rst^2}{15m^4t^2}$

Simplify each fraction. Assume that no denominators are 0.

23. $\dfrac{2x - 4}{x^2 - 4}$

24. $\dfrac{x^2 - 16}{x^2 - 8x + 16}$

25. $\dfrac{4 - x^2}{x^2 - 5x + 6}$

26. $\dfrac{25 - x^2}{x^2 + 10x + 25}$

27. $\dfrac{6x^3 + x^2 - 12x}{4x^3 + 4x^2 - 3x}$

28. $\dfrac{6x^4 - 5x^3 - 6x^2}{2x^3 - 7x^2 - 15x}$

29. $\dfrac{x^3 - 8}{x^2 + ax - 2x - 2a}$

30. $\dfrac{xy + 2x + 3y + 6}{x^3 + 27}$

Perform the operations and simplify, whenever possible. Assume that no denominators are 0.

31. $\dfrac{x^2 - 1}{x} \cdot \dfrac{x^2}{x^2 + 2x + 1}$

32. $\dfrac{y^2 - 2y + 1}{y} \cdot \dfrac{y + 2}{y^2 + y - 2}$

33. $\dfrac{3x^2 + 7x + 2}{x^2 + 2x} \cdot \dfrac{x^2 - x}{3x^2 + x}$

34. $\dfrac{x^2 + x}{2x^2 + 3x} \cdot \dfrac{2x^2 + x - 3}{x^2 - 1}$

35. $\dfrac{x^2 + x}{x - 1} \cdot \dfrac{x^2 - 1}{x + 2}$

36. $\dfrac{x^2 + 5x + 6}{x^2 + 6x + 9} \cdot \dfrac{x + 2}{x^2 - 4}$

37. $\dfrac{2x^2 + 32}{8} \div \dfrac{x^2 + 16}{2}$

38. $\dfrac{x^2 + x - 6}{x^2 - 6x + 9} \div \dfrac{x^2 - 4}{x^2 - 9}$

39. $\dfrac{z^2 + z - 20}{z^2 - 4} \div \dfrac{z^2 - 25}{z - 5}$

40. $\dfrac{ax + bx + a + b}{a^2 + 2ab + b^2} \div \dfrac{x^2 - 1}{x^2 - 2x + 1}$

41. $\dfrac{3x^2 + 5x - 2}{x^3 + 2x^2} \div \dfrac{6x^2 + 13x - 5}{2x^3 + 5x^2}$

42. $\dfrac{x^2 + 13x + 12}{8x^2 - 6x - 5} \div \dfrac{2x^2 - x - 3}{8x^2 - 14x + 5}$

43. $\dfrac{x^2 + 7x + 12}{x^3 - x^2 - 6x} \cdot \dfrac{x^2 - 3x - 10}{x^2 + 2x - 3} \cdot \dfrac{x^3 - 4x^2 + 3x}{x^2 - x - 20}$

44. $\dfrac{x(x - 2) - 3}{x(x + 7) - 3(x - 1)} \cdot \dfrac{x(x + 1) - 2}{x(x - 7) + 3(x + 1)}$

45. $\dfrac{x^2 - 2x - 3}{21x^2 - 50x - 16} \cdot \dfrac{3x - 8}{x - 3} \div \dfrac{x^2 + 6x + 5}{7x^2 - 33x - 10}$

46. $\dfrac{x^3 + 27}{x^2 - 4} \div \left(\dfrac{x^2 + 4x + 3}{x^2 + 2x} \div \dfrac{x^2 + x - 6}{x^2 - 3x + 9} \right)$

47. $\dfrac{3}{x + 3} + \dfrac{x + 2}{x + 3}$

48. $\dfrac{3}{x + 1} + \dfrac{x + 2}{x + 1}$

49. $\dfrac{4x}{x - 1} - \dfrac{4}{x - 1}$

50. $\dfrac{6x}{x - 2} - \dfrac{3}{x - 2}$

51. $\dfrac{2}{5 - x} + \dfrac{1}{x - 5}$

52. $\dfrac{3}{x - 6} - \dfrac{2}{6 - x}$

53. $\dfrac{3}{x + 1} + \dfrac{2}{x - 1}$

54. $\dfrac{3}{x + 4} + \dfrac{x}{x - 4}$

55. $\dfrac{a + 3}{a^2 + 7a + 12} + \dfrac{a}{a^2 - 16}$

56. $\dfrac{a}{a^2 + a - 2} + \dfrac{2}{a^2 - 5a + 4}$

57. $\dfrac{x}{x^2 - 4} - \dfrac{1}{x + 2}$

58. $\dfrac{b^2}{b^2 - 4} - \dfrac{4}{b^2 + 2b}$

59. $\dfrac{3x - 2}{x^2 + 2x + 1} - \dfrac{x}{x^2 - 1}$

60. $\dfrac{2t}{t^2 - 25} - \dfrac{t + 1}{t^2 + 5t}$

61. $\dfrac{2}{y^2 - 1} + 3 + \dfrac{1}{y + 1}$

62. $2 + \dfrac{4}{t^2 - 4} - \dfrac{1}{t - 2}$

63. $\dfrac{1}{x-2} + \dfrac{3}{x+2} - \dfrac{3x-2}{x^2-4}$

64. $\dfrac{x}{x-3} - \dfrac{5}{x+3} + \dfrac{3(3x-1)}{x^2-9}$

65. $\left(\dfrac{1}{x-2} + \dfrac{1}{x-3}\right) \cdot \dfrac{x-3}{2x}$

66. $\left(\dfrac{1}{x+1} - \dfrac{1}{x-2}\right) \div \dfrac{1}{x-2}$

67. $\dfrac{3x}{x-4} - \dfrac{x}{x+4} - \dfrac{3x+1}{16-x^2}$

68. $\dfrac{7x}{x-5} + \dfrac{3x}{5-x} + \dfrac{3x-1}{x^2-25}$

69. $\dfrac{1}{x^2+3x+2} - \dfrac{2}{x^2+4x+3} + \dfrac{1}{x^2+5x+6}$

70. $\dfrac{-2}{x-y} + \dfrac{2}{x-z} - \dfrac{2z-2y}{(y-x)(z-x)}$

71. $\dfrac{3x-2}{x^2+x-20} - \dfrac{4x^2+2}{x^2-25} + \dfrac{3x^2-25}{x^2-16}$

72. $\dfrac{3x+2}{8x^2-10x-3} + \dfrac{x+4}{6x^2-11x+3} - \dfrac{1}{4x+1}$

Simplify each complex fraction. Assume that no denominators are 0.

73. $\dfrac{\dfrac{3a}{b}}{\dfrac{6ac}{b^2}}$

74. $\dfrac{\dfrac{3t^2}{9x}}{\dfrac{t}{18x}}$

75. $\dfrac{\dfrac{3a^2b}{ab}}{27}$

76. $\dfrac{\dfrac{3u^2v}{4t}}{\dfrac{3uv}{}}$

77. $\dfrac{\dfrac{x-y}{ab}}{\dfrac{y-x}{ab}}$

78. $\dfrac{\dfrac{x^2-5x+6}{2x^2y}}{\dfrac{x^2-9}{2x^2y}}$

79. $\dfrac{\dfrac{1}{x} + \dfrac{1}{y}}{xy}$

80. $\dfrac{xy}{\dfrac{11}{x} + \dfrac{11}{y}}$

81. $\dfrac{\dfrac{1}{x} + \dfrac{1}{y}}{\dfrac{1}{x} - \dfrac{1}{y}}$

82. $\dfrac{\dfrac{1}{x} - \dfrac{1}{y}}{\dfrac{1}{x} + \dfrac{1}{y}}$

83. $\dfrac{\dfrac{3a}{b} - \dfrac{4a^2}{x}}{\dfrac{1}{b} + \dfrac{1}{ax}}$

84. $\dfrac{1 - \dfrac{x}{y}}{\dfrac{x^2}{y^2} - 1}$

85. $\dfrac{x + 1 - \dfrac{6}{x}}{x + 5 + \dfrac{6}{x}}$

86. $\dfrac{2z}{1 - \dfrac{3}{z}}$

87. $\dfrac{3xy}{1 - \dfrac{1}{xy}}$

88. $\dfrac{x - 3 + \dfrac{1}{x}}{-\dfrac{1}{x} - x + 3}$

89. $\dfrac{3x}{x + \dfrac{1}{x}}$

90. $\dfrac{2x^2 + 4}{2 + \dfrac{4x}{5}}$

91. $\dfrac{\dfrac{x}{x+2} - \dfrac{2}{x-1}}{\dfrac{3}{x+2} + \dfrac{x}{x-1}}$

92. $\dfrac{\dfrac{2x}{x-3} + \dfrac{1}{x-2}}{\dfrac{3}{x-3} - \dfrac{x}{x-2}}$

Write each expression without using negative exponents, and simplify the resulting complex fraction. Assume that no denominators are 0.

93. $\dfrac{1}{1 + x^{-1}}$

94. $\dfrac{y^{-1}}{x^{-1} + y^{-1}}$

95. $\dfrac{3(x+2)^{-1} + 2(x-1)^{-1}}{(x+2)^{-1}}$

96. $\dfrac{2x(x-3)^{-1} - 3(x+2)^{-1}}{(x-3)^{-1}(x+2)^{-1}}$

Applications

97. Engineering   The stiffness k of the shaft shown in the illustration is given by the following formula where k_1 and k_2 are the individual stiffnesses of each section. Simplify the complex fraction on the right side of the formula.

$$k = \dfrac{1}{\dfrac{1}{k_1} + \dfrac{1}{k_2}}$$

Section 1   Section 2

98. Electronics   The combined resistance R of three resistors with resistances of R_1, R_2, and R_3 is given by the following formula. Simplify the complex fraction on the right side of the formula.

$$R = \dfrac{1}{\dfrac{1}{R_1} + \dfrac{1}{R_2} + \dfrac{1}{R_3}}$$

99. Explain what a rational expression is.

100. Describe how to simplify a rational expression.

101. Explain why the denominator of a rational expression cannot be 0.

102. Describe how to add or subtract rational expressions.

103. Explain how to multiply rational expressions.

104. Explain how to divide rational expressions.

105. Explain why the formula $\dfrac{a}{b} + \dfrac{c}{d} = \dfrac{ad + bc}{bd}$ is valid.

106. Explain why the formula $\dfrac{a}{b} \div \dfrac{c}{d} = \dfrac{a}{b} \cdot \dfrac{d}{c}$ is valid.

Discovery and Writing

Simplify each complex fraction. Assume that no denominators are 0.

107. $\dfrac{x}{1 + \dfrac{1}{3x^{-1}}}$

108. $\dfrac{ab}{2 + \dfrac{3}{2a^{-1}}}$

109. $\dfrac{1}{1 + \dfrac{1}{1 + \dfrac{1}{x}}}$

110. $\dfrac{y}{2 + \dfrac{2}{2 + \dfrac{2}{y}}}$

Critical Thinking

In Exercises 111–114, determine if the statement is true or false. If the statement is false, then correct it and make it true.

111. The numerator of a rational expression can never be 0.

112. The denominator of a rational expression can never be 0.

113. $\dfrac{x + 7}{x + 7} = 1$ for all values of x.

114. $\dfrac{x - 7}{7 - x} = -1$ for all values of x.

In Exercises 115–118, determine if the statement is true or false. If the statement is false, then correct it and make it true. Assume no denominators are 0.

115. $-\dfrac{(x - y)^3}{(y - x)^3} = 1$

116. $\dfrac{25 + x}{25} = 1 + x$

117. $\dfrac{5}{x} + \dfrac{5}{y} = \dfrac{10}{x + y}$

118. $10 - \dfrac{1}{x} = \dfrac{9}{x}$

119. Domain The set of all real numbers for which an algebraic expression is defined is called the domain. What is the domain of the rational expression $\dfrac{3x + 8}{x - 6}$?

120. Domain What is the domain of the rational expression $\dfrac{3}{x^2 + 3}$? Refer to Exercise 119.

CHAPTER REVIEW

0.1 Sets of Real Numbers

Definitions and Concepts	Examples
Natural numbers: The numbers that we count with.	$1, 2, 3, 4, 5, 6, 7, 8, 9, 10, \ldots$
Whole numbers: The natural numbers and 0.	$0, 1, 2, 3, 4, 5, 6, 7, 8, 9, 10, \ldots$
Integers: The whole numbers and the negatives of the natural numbers.	$\ldots, -5, -4, -3, -2, -1, 0, 1, 2, 3, 4, 5, \ldots$
Rational numbers: $\{x \mid x$ can be written in the form $\frac{a}{b}$ $(b \neq 0)$, where a and b are integers.$\}$ All decimals that either terminate or repeat.	$2 = \dfrac{2}{1}, -5 = -\dfrac{5}{1}, \dfrac{2}{3}, -\dfrac{7}{3}, 0.25 = \dfrac{1}{4}$ $\dfrac{3}{4} = 0.25, \dfrac{13}{5} = 2.6, \dfrac{2}{3} = 0.666\ldots = 0.\overline{6}, \dfrac{7}{11} = 0.\overline{63}$
Irrational numbers: Nonrational real numbers. All decimals that neither terminate nor repeat.	$\sqrt{2}, -\sqrt{26}, \pi, 0.232232223\ldots$

Definitions and Concepts	Examples
Real numbers: Any number that can be expressed as a decimal.	$-6, -\dfrac{9}{13}, 0, \pi, \sqrt{31}, 10.73$
Prime numbers: A natural number greater than 1 that is divisible only by itself and 1.	$2, 3, 5, 7, 11, 13, 17, \ldots$
Composite numbers: A natural number greater than 1 that is not prime.	$4, 6, 8, 9, 10, 12, 14, 15, \ldots$
Even integers: The integers that are exactly divisible by 2.	$\ldots, -6, -4, -2, 0, 2, 4, 6, \ldots$
Odd integers: The integers that are not exactly divisible by 2.	$\ldots, -5, -3, -1, 1, 3, 5, \ldots$
Associative Properties: of addition $(a + b) + c = a + (b + c)$ of multiplication $(ab)c = a(bc)$	$5 + (4 + 7) = (5 + 4) + 7$ $(5 \cdot 4) \cdot 7 = 5(4 \cdot 7)$
Commutative Properties: of addition $a + b = b + a$ of multiplication $ab = ba$	$4 + 7 = 7 + 4 \qquad 1.7 + 2.5 = 2.5 + 1.7$ $4 \cdot 7 = 7 \cdot 4 \qquad 1.7(2.5) = 2.5(1.7)$
Distributive Property: $a(b + c) = ab + ac$	$3(x + 6) = 3x + 3 \cdot 6 \qquad 0.2(y - 10) = 0.2y - 0.2(10)$
Double Negative Rule: $-(-a) = a$	$-(-5) = 5 \qquad -(-a) = a$
Open intervals have no endpoints. **Closed intervals** have two endpoints. **Half-open intervals** have one endpoint.	$(-3, 2)$ $[-3, 2]$ $(-3, 2]$
Absolute value: If $x \geq 0$, then $\lvert x \rvert = x$. If $x < 0$, then $\lvert x \rvert = -x$.	$\lvert 7 \rvert = 7 \qquad \lvert 3.5 \rvert = 3.5 \qquad \left\lvert \dfrac{7}{2} \right\rvert = \dfrac{7}{2} \qquad \lvert 0 \rvert = 0$ $\lvert -7 \rvert = 7 \qquad \lvert -3.5 \rvert = 3.5 \qquad \left\lvert -\dfrac{7}{2} \right\rvert = \dfrac{7}{2}$
Distance: The **distance** d between points a and b on a number line is $d = \lvert b - a \rvert$.	The distance on the number line between points with coordinates of -3 and 2 is $d = \lvert 2 - (-3) \rvert = \lvert 5 \rvert = 5$.

Exercises

Consider the set $\{-6, -3, 0, \frac{1}{2}, 3, \pi, \sqrt{5}, 6, 8\}$. List the numbers in this set that are

1. natural numbers.
2. whole numbers.
3. integers.
4. rational numbers.
5. irrational numbers.
6. real numbers.

Consider the set $\{-6, -3, 0, \frac{1}{2}, 3, \pi, \sqrt{5}, 6, 8\}$. List the numbers in this set that are

7. prime numbers.
8. composite numbers.
9. even integers.
10. odd integers.

Determine which property of real numbers justifies each statement.

11. $(a + b) + 2 = a + (b + 2)$

12. $a + 7 = 7 + a$

13. $4(2x) = (4 \cdot 2)x$

14. $3(a + b) = 3a + 3b$

15. $(5a)7 = 7(5a)$

16. $(2x + y) + z = (y + 2x) + z$

17. $-(-6) = 6$

Graph each subset of the real numbers:

18. the prime numbers between 10 and 20

19. the even integers from 6 to 14

Graph each interval on the number line.

20. $-3 < x \leq 5$

21. $x \geq 0$ or $x < -1$

22. $(-2, 4]$

23. $(-\infty, 2) \cap (-5, \infty)$

24. $(-\infty, -4) \cup [6, \infty)$

Write each expression without absolute value symbols.

25. $|6|$ **26.** $|-25|$

27. $|1 - \sqrt{2}|$ **28.** $|\sqrt{3} - 1|$

29. On a number line, find the distance between points with coordinates of -5 and 7.

0.2 Integer Exponents and Scientific Notation

Definitions and Concepts	Examples
Natural-number exponents: $x^n = \overbrace{x \cdot x \cdot x \cdot \cdots \cdot x}^{n \text{ factors of } x}$	$x^5 = x \cdot x \cdot x \cdot x \cdot x$
Rules of exponents: If there are no divisions by 0, • $x^m x^n = x^{m+n}$ • $(x^m)^n = x^{mn}$ • $(xy)^n = x^n y^n$ • $\left(\dfrac{x}{y}\right)^n = \dfrac{x^n}{y^n}$ • $x^0 = 1 \ (x \neq 0)$ • $x^{-n} = \dfrac{1}{x^n}$ • $\dfrac{x^m}{x^n} = x^{m-n}$ • $\left(\dfrac{x}{y}\right)^{-n} = \left(\dfrac{y}{x}\right)^n$	$x^2 x^5 = x^{2+5} = x^7$ $(x^2)^7 = x^{2\cdot7} = x^{14}$ $(xy)^5 = x^5 y^5$ $\left(\dfrac{x}{y}\right)^5 = \dfrac{x^5}{y^5}$ $6^0 = 1$ $6^{-2} = \dfrac{1}{6^2} = \dfrac{1}{36}$ $\dfrac{x^6}{x^4} = x^{6-4} = x^2$ $\left(\dfrac{x}{6}\right)^{-3} = \left(\dfrac{6}{x}\right)^3 = \dfrac{6^3}{x^3} = \dfrac{216}{x^3}$
Scientific notation: A number is written in **scientific notation** when it is written in the form $N \times 10^n$, where $1 \leq |N| < 10$ and n is an integer.	Write each number in scientific notation. $386{,}000 = 3.86 \times 10^5$ $0.0025 = 2.5 \times 10^{-3}$ Write each number in standard form. $7.3 \times 10^3 = 7300$ $5.25 \times 10^{-4} = 0.000525$

Exercises

Write each expression without using exponents.

30. $-5a^3$

31. $(-5a)^2$

Write each expression using exponents.

32. $3 \cdot t \cdot t \cdot t$

33. $(-2b)(3b)$

Simplify each expression.

34. $n^2 n^4$

35. $(p^3)^2$

36. $(x^3 y^2)^4$

37. $\left(\dfrac{a^4}{b^2}\right)^3$

38. $(m^{-3} n^0)^2$

39. $\left(\dfrac{p^{-2} q^2}{2}\right)^3$

40. $\dfrac{a^5}{a^8}$

41. $\left(\dfrac{a^2}{b^3}\right)^{-2}$

42. $\left(\dfrac{3x^2 y^{-2}}{x^2 y^2}\right)^{-2}$

43. $\left(\dfrac{a^{-3} b^2}{ab^{-3}}\right)^{-2}$

44. $\left(\dfrac{-3x^3 y}{xy^3}\right)^{-2}$

45. $\left(-\dfrac{2m^{-2} n^0}{4m^2 n^{-1}}\right)^{-3}$

46. If $x = -3$ and $y = 3$, evaluate $-x^2 - xy^2$.

Write each number in scientific notation.

47. 6750

48. 0.00023

Write each number in standard notation.

49. 4.8×10^2

50. 0.25×10^{-3}

51. Use scientific notation to simplify $\dfrac{(45{,}000)(350{,}000)}{0.000105}$.

0.3   Rational Exponents and Radicals

Definitions and Concepts	Examples
Summary of $a^{1/n}$ definitions: • If $a \geq 0$, then $a^{1/n}$ is the nonnegative number b such that $b^n = a$. • If $a < 0$ and n is odd, then $a^{1/n}$ is the real number b such that $b^n = a$. • If $a < 0$ and n is even, then $a^{1/n}$ is not a real number.	$16^{1/2} = 4$ because $4^2 = 16$. $(-27)^{1/3} = -3$ because $(-3)^3 = -27$. $(-16)^{1/2}$ is not a real number because no real number squared is -16.
Rule for rational exponents: If m and n are positive integers, $\frac{m}{n}$ is in lowest terms, and $a^{1/n}$ is a real number, then $$a^{m/n} = (a^{1/n})^m = (a^m)^{1/n}$$	$8^{2/3} = (8^{1/3})^2 = 2^2 = 4$ or $(8^2)^{1/3} = 64^{1/3} = 4$
Definition of $\sqrt[n]{a}$: $$\sqrt[n]{a} = a^{1/n}$$	$\sqrt[3]{125} = 125^{1/3} = 5$
Properties of radicals: If all radicals are real numbers and there are no divisions by 0, then • $\sqrt[n]{ab} = \sqrt[n]{a}\,\sqrt[n]{b}$ • $\sqrt[n]{\dfrac{a}{b}} = \dfrac{\sqrt[n]{a}}{\sqrt[n]{b}}$ • $\sqrt[m]{\sqrt[n]{a}} = \sqrt[n]{\sqrt[m]{a}} = \sqrt[mn]{a}$	$\sqrt[5]{32x^{10}} = \sqrt[5]{32}\,\sqrt[5]{x^{10}} = 2x^2$ $\sqrt[4]{\dfrac{x^{12}}{625}} = \dfrac{\sqrt[4]{x^{12}}}{\sqrt[4]{625}} = \dfrac{x^{12/4}}{625^{1/4}} = \dfrac{x^3}{5}$ $\sqrt[3]{\sqrt{64}} = \sqrt[3]{8},\quad \sqrt{\sqrt[3]{64}} = \sqrt{4},\quad \sqrt[2 \cdot 3]{64} = \sqrt[6]{64}$ $\qquad = 2 \qquad\qquad = 2 \qquad\qquad = 2$

Exercises

Simplify each expression, if possible.

52. $121^{1/2}$

53. $\left(\dfrac{27}{125}\right)^{1/3}$

54. $(32x^5)^{1/5}$

55. $(81a^4)^{1/4}$

56. $(-1000x^6)^{1/3}$

57. $(-25x^2)^{1/2}$

58. $(x^{12}y^2)^{1/2}$

59. $\left(\dfrac{x^{12}}{y^4}\right)^{-1/2}$

60. $\left(\dfrac{-c^{2/3}c^{5/3}}{c^{-2/3}}\right)^{1/3}$

61. $\left(\dfrac{a^{-1/4}a^{3/4}}{a^{9/2}}\right)^{-1/2}$

Simplify each expression.

62. $64^{2/3}$

63. $32^{-3/5}$

64. $\left(\dfrac{16}{81}\right)^{3/4}$

65. $\left(\dfrac{32}{243}\right)^{2/5}$

66. $\left(\dfrac{8}{27}\right)^{-2/3}$

67. $\left(\dfrac{16}{625}\right)^{-3/4}$

68. $(-216x^3)^{2/3}$

69. $\dfrac{p^{a/2}p^{a/3}}{p^{a/6}}$

Simplify each expression.

70. $\sqrt{36}$

71. $-\sqrt{49}$

72. $\sqrt{\dfrac{9}{25}}$

73. $\sqrt[3]{\dfrac{27}{125}}$

74. $\sqrt{x^2y^4}$

75. $\sqrt[3]{x^3}$

76. $\sqrt[4]{\dfrac{m^8n^4}{p^{16}}}$

77. $\sqrt[5]{\dfrac{a^{15}b^{10}}{c^5}}$

Simplify and combine terms.

78. $\sqrt{50}+\sqrt{8}$

79. $\sqrt{12}+\sqrt{3}-\sqrt{27}$

80. $\sqrt[3]{24x^4}-\sqrt[3]{3x^4}$

Rationalize each denominator.

81. $\dfrac{\sqrt{7}}{\sqrt{5}}$

82. $\dfrac{8}{\sqrt{8}}$

83. $\dfrac{1}{\sqrt[3]{2}}$

84. $\dfrac{2}{\sqrt[3]{25}}$

Rationalize each numerator.

85. $\dfrac{\sqrt{2}}{5}$

86. $\dfrac{\sqrt{5}}{5}$

87. $\dfrac{\sqrt{2x}}{3}$

88. $\dfrac{3\sqrt[3]{7x}}{2}$

0.4 Polynomials

Definitions and Concepts	Examples
Monomial: A polynomial with one term.	2, $3x$, $4x^2y$, $-x^3y^2z$
Binomial: A polynomial with two terms.	$2t+3$, $3r^2-6r$, $4m+5n$
Trinomial: A polynomial with three terms.	$3p^2-7p+8$, $3m^2+2n-p$
The **degree of a monomial** is the sum of the exponents on its variables.	The degree of $4x^2y^3$ is $2+3=5$.
The **degree of a polynomial** is the degree of the term in the polynomial with highest degree.	The degree of the first term of $3p^2q^3-6p^3q^4+9pq^2$ is $2+3=5$, the degree of the second term is $3+4=7$, and the degree of the third term is $1+2=3$. The degree of the polynomial is the largest of these. It is 7.
Multiplying a monomial times a polynomial: $\quad a(b+c+d+\cdots)=ab+ac+ad+\cdots$	$3(x+2)=3x+6$ $\\ 2x(3x^2-2y+3)=6x^3-4xy+6x$

Definitions and Concepts	Examples
Addition and subtraction of polynomials: To add or subtract polynomials, remove parentheses and combine like terms.	Add: $(3x^2 + 5x) + (2x^2 - 2x)$. $$(3x^2 + 5x) + (2x^2 - 2x) = 3x^2 + 5x + 2x^2 - 2x$$ $$= 3x^2 + 2x^2 + 5x - 2x$$ $$= 5x^2 + 3x$$ Subtract: $(4a^2 - 5b) - (3a^2 - 7b)$. $$(4a^2 - 5b) - (3a^2 - 7b) = 4a^2 - 5b - 3a^2 + 7b$$ $$= 4a^2 - 3a^2 - 5b + 7b$$ $$= a^2 + 2b$$
Special products: $$(x + y)^2 = x^2 + 2xy + y^2$$ $$(x - y)^2 = x^2 - 2xy + y^2$$ $$(x + y)(x - y) = x^2 - y^2$$	$$(2m + 3)^2 = 4m^2 + 12m + 9$$ $$(4t - 3s)^2 = 16t^2 - 24ts + 9s^2$$ $$(2m + n)(2m - n) = 4m^2 - n^2$$
Multiplying a binomial times a binomial: Use the FOIL method.	$$(2a + b)(a - b) = 2a(a) + 2a(-b) + ba + b(-b)$$ $$= 2a^2 - 2ab + ab - b^2$$ $$= 2a^2 - ab - b^2$$
The **conjugate** of $a + b$ is $a - b$. To rationalize the denominator of a radical expression, multiply both the numerator and denominator of the rational expression by the conjugate of the denominator.	Rationalize the denominator: $\dfrac{x}{\sqrt{x} + 2}$. $$\frac{x}{\sqrt{x} + 2} = \frac{x(\sqrt{x} - 2)}{(\sqrt{x} + 2)(\sqrt{x} - 2)} \quad \text{Multiply the numerator and denominator by the conjugate of } \sqrt{x} + 2.$$ $$= \frac{x\sqrt{x} - 2x}{x - 4} \quad \text{Simplify.}$$
Division of polynomials: To divide polynomials, use long division.	Divide: $2x + 3\overline{)6x^3 + 7x^2 - x + 3}$. $$\begin{array}{r} 3x^2 - x + 1 \\ 2x + 3\overline{)6x^3 + 7x^2 - x + 3} \\ \underline{6x^3 + 9x^2} \\ -2x^2 - x \\ \underline{-2x^2 - 3x} \\ +2x + 3 \\ \underline{+2x + 3} \\ 0 \end{array}$$

Exercises

Give the degree of each polynomial and tell whether the polynomial is a monomial, a binomial, or a trinomial.

89. $x^3 - 8$

90. $8x - 8x^2 - 8$

91. $\sqrt{3x^2}$

92. $4x^4 - 12x^2 + 1$

Perform the operations and simplify.

93. $2(x + 3) + 3(x - 4)$

94. $3x^2(x - 1) - 2x(x + 3) - x^2(x + 2)$

95. $(3x + 2)(3x + 2)$

96. $(3x + y)(2x - 3y)$

97. $(4a + 2b)(2a - 3b)$

98. $(z + 3)(3z^2 + z - 1)$

99. $(a^n + 2)(a^n - 1)$

100. $\left(\sqrt{2} + x\right)^2$

101. $\left(\sqrt{2} + 1\right)\left(\sqrt{3} + 1\right)$

102. $\left(\sqrt[3]{3} - 2\right)\left(\sqrt[3]{9} + 2\sqrt[3]{3} + 4\right)$

Rationalize each denominator.

103. $\dfrac{2}{\sqrt{3} - 1}$

104. $\dfrac{-2}{\sqrt{3} - \sqrt{2}}$

105. $\dfrac{2x}{\sqrt{x} - 2}$

106. $\dfrac{\sqrt{x} - \sqrt{y}}{\sqrt{x} + \sqrt{y}}$

Rationalize each numerator.

107. $\dfrac{\sqrt{x} + 2}{5}$

108. $\dfrac{1 - \sqrt{a}}{a}$

Perform each division.

109. $\dfrac{3x^2 y^2}{6x^3 y}$

110. $\dfrac{4a^2 b^3 + 6ab^4}{2b^2}$

111. $2x + 3 \overline{)2x^3 + 7x^2 + 8x + 3}$

112. $x^2 - 1 \overline{)x^5 + x^3 - 2x - 3x^2 - 3}$

0.5 Factoring Polynomials

Definitions and Concepts	Examples
Factoring out the greatest common factor:	
$ab + ac = a(b + c)$	$3p^3 - 6p^2 q + 9p = 3p(p^2 - 2pq + 3)$
Factoring the difference of two squares:	
$x^2 - y^2 = (x + y)(x - y)$	$4x^2 - 9 = (2x + 3)(2x - 3)$
Factoring trinomials:	
• Trinomial squares	
$x^2 + 2xy + y^2 = (x + y)^2$	$9a^2 + 12ab + 4b^2 = (3a + 2b)(3a + 2b) = (3a + 2b)^2$
$x^2 - 2xy + y^2 = (x - y)^2$	$r^2 - 4rs + 4s^2 = (r - 2s)(r - 2s) = (r - 2s)^2$
• To factor general trinomials use trial and error or grouping.	$6x^2 - 5x - 6 = (2x - 3)(3x + 2)$
Factoring the sum and difference of two cubes:	
$x^3 + y^3 = (x + y)(x^2 - xy + y^2)$	$r^3 + 8 = (r + 2)(r^2 - 2r + 4)$
$x^3 - y^3 = (x - y)(x^2 + xy + y^2)$	$27a^3 - 8b^3 = (3a - 2b)(9a^2 + 6ab + 4b^2)$

Exercises

Factor each expression completely, if possible.

113. $3t^3 - 3t$

114. $5r^3 - 5$

115. $6x^2 + 7x - 24$

116. $3a^2 + ax - 3a - x$

117. $8x^3 - 125$

118. $6x^2 - 20x - 16$

119. $x^2 + 6x + 9 - t^2$

120. $3x^2 - 1 + 5x$

121. $8z^3 + 343$

122. $1 + 14b + 49b^2$

123. $121z^2 + 4 - 44z$

124. $64y^3 - 1000$

125. $2xy - 4zx - wy + 2zw$

126. $x^8 + x^4 + 1$

0.6 Algebraic Fractions

Definitions and Concepts	Examples
Properties of fractions: If there are no divisions by 0, then	

Properties of fractions:
If there are no divisions by 0, then

- $\dfrac{a}{b} = \dfrac{c}{d}$ if and only if $ad = bc$.

- $\dfrac{a}{b} = \dfrac{ax}{bx}$

- $\dfrac{a}{b} \cdot \dfrac{c}{d} = \dfrac{ac}{bd}$

- $\dfrac{a}{b} \div \dfrac{c}{d} = \dfrac{ad}{bc}$

- $\dfrac{a}{b} + \dfrac{c}{b} = \dfrac{a+c}{b}$ $\dfrac{a}{b} - \dfrac{c}{b} = \dfrac{a-c}{b}$

- $a \cdot 1 = a$ $\dfrac{a}{1} = a$ $\dfrac{a}{a} = 1$

- $\dfrac{a}{b} = \dfrac{-a}{-b} = -\dfrac{a}{-b} = -\dfrac{-a}{b}$

- $-\dfrac{a}{b} = \dfrac{a}{-b} = \dfrac{-a}{b} = -\dfrac{-a}{-b}$

Examples

$\dfrac{3x}{4} = \dfrac{6x}{8}$ because $(3x)8$ and $4(6x)$ both equal $24x$.

$\dfrac{6x^2}{8x^3} = \dfrac{3 \cdot 2 \cdot \not x \cdot \not x}{2 \cdot 4 \cdot \not x \cdot \not x \cdot x} = \dfrac{3}{4x}$

$\dfrac{3p}{2q} \cdot \dfrac{2p}{6q} = \dfrac{3p \cdot 2p}{2q \cdot 6q} = \dfrac{3p \cdot 2p}{2q \cdot 3 \cdot 2q} = \dfrac{p^2}{2q^2}$

$\dfrac{2t}{3s} \div \dfrac{2t}{6s} = \dfrac{2t}{3s} \cdot \dfrac{6s}{2t} = \dfrac{2t \cdot 6s}{3s \cdot 2t} = \dfrac{2t \cdot 2 \cdot 3\not s}{3\not s 2\not t} = 2$

$\dfrac{x}{4} + \dfrac{y}{4} = \dfrac{x+y}{4}$ $\dfrac{3p}{2q} - \dfrac{p}{2q} = \dfrac{3p-p}{2q} = \dfrac{2p}{2q} = \dfrac{p}{q}$

$7 \cdot 1 = 7$ $\dfrac{7}{1} = 7$ $\dfrac{7}{7} = 1$

$\dfrac{7}{2} = \dfrac{-7}{-2} = -\dfrac{7}{-2} = -\dfrac{-7}{2}$

$-\dfrac{7}{2} = \dfrac{7}{-2} = \dfrac{-7}{2} = -\dfrac{-7}{-2}$

Simplifying rational expressions:
To simplify a rational expression, factor the numerator and denominator, if possible, and divide out factors that are common to the numerator and denominator.

To simplify $\dfrac{2-x}{2x-4}$, factor -1 from the numerator and 2 from the denominator to get

$$\frac{2-x}{2x-4} = \frac{-1(-2+x)}{2(x-2)} = \frac{-1(x-2)}{2(x-2)} = \frac{-1}{2} = -\frac{1}{2}$$

Adding or subtracting rational expressions:
To add or subtract rational expressions with unlike denominators, find the LCD of the expressions, write each expression with a denominator that is the LCD, add or subtract the expressions, and simplify the result, if possible.

$$\frac{2x}{x+2} - \frac{2x}{x-3} = \frac{2x(x-3)}{(x+2)(x-3)} - \frac{2x(x+2)}{(x-3)(x+2)}$$
$$= \frac{2x(x-3) - 2x(x+2)}{(x+2)(x-3)}$$
$$= \frac{2x^2 - 6x - 2x^2 - 4x}{(x+2)(x-3)}$$
$$= \frac{-10x}{(x+2)(x-3)}$$

Simplify complex fractions:
To simplify complex fractions, multiply the numerator and denominator of the complex fraction by the LCD of all the fractions.

$$\frac{1 - \dfrac{y}{2}}{\dfrac{1}{y} + \dfrac{1}{2}} = \frac{2y\left(1 - \dfrac{y}{2}\right)}{2y\left(\dfrac{1}{y} + \dfrac{1}{2}\right)} = \frac{2y - y^2}{2 + y} = \frac{y(2-y)}{2+y}$$

Exercises

Simplify each rational expression.

127. $\dfrac{2-x}{x^2-4x+4}$

128. $\dfrac{a^2-9}{a^2-6a+9}$

Perform each operation and simplify. Assume that no denominators are 0.

129. $\dfrac{x^2-4x+4}{x+2} \cdot \dfrac{x^2+5x+6}{x-2}$

130. $\dfrac{2y^2-11y+15}{y^2-6y+8} \cdot \dfrac{y^2-2y-8}{y^2-y-6}$

131. $\dfrac{2t^2+t-3}{3t^2-7t+4} \div \dfrac{10t+15}{3t^2-t-4}$

132. $\dfrac{p^2+7p+12}{p^3+8p^2+4p} \div \dfrac{p^2-9}{p^2}$

133. $\dfrac{x^2+x-6}{x^2-x-6} \cdot \dfrac{x^2-x-6}{x^2+x-2} \div \dfrac{x^2-4}{x^2-5x+6}$

134. $\left(\dfrac{2x+6}{x+5} \div \dfrac{2x^2-2x-4}{x^2-25}\right)\dfrac{x^2-x-2}{x^2-2x-15}$

135. $\dfrac{2}{x-4} + \dfrac{3x}{x+5}$

136. $\dfrac{5x}{x-2} - \dfrac{3x+7}{x+2} + \dfrac{2x+1}{x+2}$

137. $\dfrac{x}{x-1} + \dfrac{x}{x-2} + \dfrac{x}{x-3}$

138. $\dfrac{x}{x+1} - \dfrac{3x+7}{x+2} + \dfrac{2x+1}{x+2}$

139. $\dfrac{3(x+1)}{x} - \dfrac{5(x^2+3)}{x^2} + \dfrac{x}{x+1}$

140. $\dfrac{3x}{x+1} + \dfrac{x^2+4x+3}{x^2+3x+2} - \dfrac{x^2+x-6}{x^2-4}$

Simplify each complex fraction. Assume that no denominators are 0.

141. $\dfrac{\dfrac{5x}{2}}{\dfrac{3x^2}{8}}$

142. $\dfrac{\dfrac{3x}{y}}{\dfrac{6x}{y^2}}$

143. $\dfrac{\dfrac{1}{x}+\dfrac{1}{y}}{x-y}$

144. $\dfrac{x^{-1}+y^{-1}}{y^{-1}-x^{-1}}$

CHAPTER TEST

Consider the set $\{-7, -\frac{2}{3}, 0, 1, 3, \sqrt{10}, 4\}$.

1. List the numbers in the set that are odd integers.

2. List the numbers in the set that are prime numbers.

Determine which property justifies each statement.

3. $(a+b)+c = (b+a)+c$

4. $a(b+c) = ab+ac$

Graph each interval on a number line.

5. $-4 < x \le 2$

6. $(-\infty, -3) \cup [6, \infty)$

Write each expression without using absolute value symbols.

7. $|-17|$

8. $|x-7|$, when $x < 0$

Find the distance on a number line between points with the following coordinates.

9. -4 and 12

10. -20 and -12

Simplify each expression. Assume that all variables represent positive numbers, and write all answers without using negative exponents.

11. $x^4 x^5 x^2$

12. $\dfrac{r^2 r^3 s}{r^4 s^2}$

13. $\dfrac{(a^{-1}a^2)^{-2}}{a^{-3}}$

14. $\left(\dfrac{x^0 x^2}{x^{-2}}\right)^6$

Write each number in scientific notation.

15. $450{,}000$

16. 0.000345

Write each number in standard notation.

17. 3.7×10^3

18. 1.2×10^{-3}

Simplify each expression. Assume that all variables represent positive numbers, and write all answers without using negative exponents.

19. $(25a^4)^{1/2}$

20. $\left(\dfrac{36}{81}\right)^{3/2}$

21. $\left(\dfrac{8t^6}{27s^9}\right)^{-2/3}$

22. $\sqrt[3]{27a^6}$

23. $\sqrt{12} + \sqrt{27}$

24. $2\sqrt[3]{3x^4} - 3x\sqrt[3]{24x}$

25. Rationalize the denominator: $\dfrac{x}{\sqrt{x} - 2}$.

26. Rationalize the numerator: $\dfrac{\sqrt{x} - \sqrt{y}}{\sqrt{x} + \sqrt{y}}$.

Perform each operation.

27. $(a^2 + 3) - (2a^2 - 4)$

28. $(3a^3b^2)(-2a^3b^4)$

29. $(3x - 4)(2x + 7)$

30. $(a^n + 2)(a^n - 3)$

31. $(x^2 + 4)(x^2 - 4)$

32. $(x^2 - x + 2)(2x - 3)$

33. $x - 3 \overline{)6x^2 + x - 23}$

34. $2x - 1 \overline{)2x^3 + 3x^2 - 1}$

Factor each polynomial completely.

35. $3x + 6y$

36. $x^2 - 100$

37. $45x^2 - 20y^2$

38. $10t^2 - 19tw + 6w^2$

39. $64m^3 + 125n^3$

40. $3a^3 - 648$

41. $x^4 - x^2 - 12$

42. $6x^4 + 11x^2 - 10$

Simplify each rational expression. Assume no denominators are 0.

43. $\dfrac{44p^3q^6}{33p^4q^2}$

44. $\dfrac{49 - x^2}{x^2 + 14x + 49}$

Perform each operation and simplify if possible. Assume that no denominators are 0.

45. $\dfrac{x}{x + 2} + \dfrac{2}{x + 2}$

46. $\dfrac{x}{x + 1} - \dfrac{x}{x - 1}$

47. $\dfrac{x^2 + x - 20}{x^2 - 16} \cdot \dfrac{x^2 - 25}{x - 5}$

48. $\dfrac{x + 2}{x^2 + 2x + 1} \div \dfrac{x^2 - 4}{x + 1}$

Simplify each complex fraction. Assume that no denominators are 0.

49. $\dfrac{\dfrac{1}{a} + \dfrac{1}{b}}{\dfrac{1}{b}}$

50. $\dfrac{x^{-1}}{x^{-1} + y^{-1}}$

Equations and Inequalities

1

The main topic of this chapter is equations—one of the most important concepts in algebra. Equations are used in almost every academic discipline and especially in chemistry, physics, medicine, computer science, and business.

Careers and Mathematics: Marketing

People who work in the field of marketing coordinate their companies' market research, marketing strategy, sales, advertising, promotion, pricing, product development, and public relations activities. Advertising managers direct a firm's promotional campaign while marketing managers promote products and services. Promotions managers work on incentives to increase sales while public relations managers plan and direct the public image for the employer. Sales managers plan and direct the distribution of the product to the customer.

Education and Mathematics Required
- For marketing, sales, and promotions management positions, employers often prefer a bachelor's or master's degree in business administration with an emphasis on marketing. For advertising management positions, some employers prefer a bachelor's degree in advertising or journalism. For public relations management positions, some employers prefer a bachelor's or master's degree in public relations or journalism.
- College Algebra, Business Calculus I and II, and Economic Statistics are required.

How Marketing Managers Use Math and Who Employs Them
- Statistics is used for predicting sales and effectiveness of advertising campaigns.
- These managers are found in virtually every industry. They are employed in wholesale trade, retail trade, manufacturing, the finance and insurance industries, as well as professional, scientific, and technical services, public and private educational services, and healthcare.

Career Outlook and Salary
- Overall employment of advertising, marketing, promotions, public relations, and sales managers is expected to increase by 12% through 2022.
- The median annual wages in May 2012 were $88,590 for advertising and promotions managers and $119,480 for marketing managers.

For more information see: www.bls.gov/oco

1.1   Linear Equations and Rational Equations

In this section, we will learn to

1. Use vocabulary related to solving equations and identify types of equations.
2. Solve linear equations.
3. Solve rational equations.
4. Solve formulas for a specific variable.

It's been said that "weddings today are as beautiful as they are expensive." Suppose a couple budgets $5000 for their wedding reception at a historic home. If $800 is charged for renting the home and there is a $42-per-person fee for food and beverages, how many guests can the couple accommodate at their reception?

If we let the **variable** x represent the number of guests, the expression $800 + 42x$ represents the cost for the reception. That is, $800 for the home rental plus $42 times the number of guests, x. We want to know the value of x that makes the expression equal $5000.

We can write the statement $800 + 42x = 5000$ to indicate that the two quantities are equal. A statement indicating that two quantities are equal is called an **equation.** In this section, we will learn how to solve equations of this type.

If $x = 100$, the equation is true because when we substitute 100 for x, we obtain a true statement.

$$800 + 42(\mathbf{100}) = 5000$$

$$800 + 4200 = 5000$$

$$5000 = 5000$$

The couple can accommodate 100 guests at their wedding reception.

1.   Use Vocabulary Related to Solving Equations and Identify Types of Equations

An equation can be either true or false. For example, the equation $2 + 2 = 4$ is true, and the equation $2 + 3 = 6$ is false. An equation such as $3x - 2 = 10$ can be true or false depending on the value of x.

If $x = 4$, the equation is true, because 4 satisfies the equation.

$$3\mathbf{x} - 2 = 10$$

$$3(\mathbf{4}) - 2 \overset{?}{=} 10 \qquad \text{Substitute 4 for } x. \text{ Read } \overset{?}{=} \text{ as "is possibly equal to."}$$

$$12 - 2 \overset{?}{=} 10$$

$$10 = 10$$

This equation is false for all other values of x.

Any number that satisfies an equation is called a **solution** or **root** of the equation. The set of all solutions of an equation is called its **solution set.** We have seen that the solution set of $3x - 2 = 10$ is $\{4\}$. To **solve** an equation means to find its solution set.

There can be restrictions on the values of a variable. For example, in the fraction

$$\frac{x^2 + 4}{x - 2}$$

we cannot replace x with 2, because that would make the denominator equal to 0.

EXAMPLE 1   **Finding the Restrictions on the Values of a Variable**

Find the restrictions on the values of b in the equation: $\dfrac{7b}{b+6} = \dfrac{2}{b-1}$.

SOLUTION   For $\frac{7b}{b+6}$ to be a real number, b cannot be -6 because that would make the denominator 0. For $\frac{2}{b-1}$ to be a real number, b cannot be 1 because that would make the denominator 0. Thus, the values of b are restricted to the set of all real numbers except 1 or -6.

Self Check 1   Find the restrictions on a: $\dfrac{5}{a+5} = \dfrac{3a}{a-2}$.

Now Try Exercise 13.

There are three types of equations: identities, contradictions, and conditional equations. These are defined and illustrated in the following table.

Type of Equation	Definition	Example
Identity	Every acceptable real number replacement for the variable is a solution.	$x^2 - 9 = (x+3)(x-3)$ Every real number x is a solution.
Contradiction	No real number is a solution.	$x = x + 1$ The equation has no solution. No real number can be 1 greater than itself.
Conditional Equation	Solution set contains some but not all real numbers.	$3x - 2 = 10$ The equation has one solution, the number 4.

Two equations with the same solution set are called **equivalent equations**.

There are certain properties of equality that we can use to transform equations into equivalent but less complicated equations. If we use these properties, the resulting equations will be equivalent and will have the same solution set.

Properties of Equality

The Addition and Subtraction Properties
If a, b, and c are real numbers and $a = b$, then

$$a + c = b + c \qquad \text{and} \qquad a - c = b - c$$

The Multiplication and Division Properties
If a, b, and c are real numbers and $a = b$, then

$$ac = bc \qquad \text{and} \qquad \frac{a}{c} = \frac{b}{c} \qquad (c \neq 0)$$

The Substitution Property
In an equation, a quantity may be substituted for its equal without changing the truth of the equation.

Tip

1. The Addition and Subtraction Properties allow the same number to be added to or subtracted from each side of an equation without changing the solution set.
2. The Multiplication and Division Properties allow each side of an equation to be multiplied by or divided by the same nonzero number without changing the solution set.

2. Solve Linear Equations

The easiest equations to solve are the **first-degree** or **linear equations**. Since these equations involve first-degree polynomials, they are also called **first-degree polynomial equations**.

Linear Equations | A **linear equation in one variable** (say, x) is any equation that can be written in the form

$$ax + b = 0 \qquad (a \text{ and } b \text{ are real numbers and } a \neq 0)$$

To solve the linear equation $2x + 3 = 0$, we subtract 3 from both sides of the equation and divide both sides by 2.

$$2x + 3 = 0$$

$$2x + 3 - 3 = 0 - 3 \qquad \text{To undo the addition of 3, subtract 3 from both sides.}$$

$$2x = -3$$

$$\frac{2x}{2} = -\frac{3}{2} \qquad \text{To undo the multiplication by 2, divide both sides by 2.}$$

$$x = -\frac{3}{2}$$

To verify that $-\frac{3}{2}$ satisfies the equation, we substitute $-\frac{3}{2}$ for x and simplify:

$$2x + 3 = 0$$

$$2\left(-\frac{3}{2}\right) + 3 \stackrel{?}{=} 0 \qquad \text{Substitute } -\frac{3}{2} \text{ for } x.$$

$$-3 + 3 \stackrel{?}{=} 0 \qquad 2\left(-\frac{3}{2}\right) = -3$$

$$0 = 0$$

Since both sides of the equation are equal, the solution checks. $2x + 3 = 0$ is a conditional equation.

In Exercise 74, you will be asked to solve the general linear equation $ax + b = 0$ for x, thereby showing that every conditional linear equation has exactly one solution.

EXAMPLE 2 | **Solving a Linear Equation**

Find the solution set: $3(x + 2) = 5x + 2$.

SOLUTION | We proceed as follows:

$$3(x + 2) = 5x + 2$$

$$3x + 6 = 5x + 2 \qquad \text{Use the Distributive Property and remove parentheses.}$$

$$3x - 3x + 6 = 5x - 3x + 2 \qquad \text{Subtract } 3x \text{ from both sides.}$$

$$6 = 2x + 2 \qquad \text{Combine like terms.}$$

$$6 - 2 = 2x + 2 - 2 \qquad \text{Subtract 2 from both sides.}$$

$$4 = 2x \qquad \text{Simplify.}$$

$$\frac{4}{2} = \frac{2x}{2} \qquad \text{Divide both sides by 2.}$$

$$2 = x \qquad \text{Simplify.}$$

Since all of the above equations are equivalent, the solution set of the original equation is $\{2\}$. The equation is a conditional equation. Verify that 2 satisfies the equation.

Self Check 2   Find the solution set: $4(x - 3) = 7x - 3$.

Now Try Exercise 35.

Checking Solutions to Linear Equations

Using the table feature on a graphing calculator, we can easily check the solution of a linear equation. We enter each side of the equation into the graph editor; go to the table, and then enter the value of x that we found as the solution. The value of both entries in the table should be the same. These steps are shown in Figure 1-1 below for Example 2. We see that both Y1 and Y2 equal 12 when $x = 2$. This verifies the solution.

Enter each side of the equation.   **Go to the table and enter the value $x = 2$.**

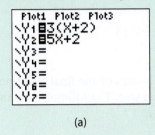

(a)   (b)

FIGURE 1-1

EXAMPLE 3   **Solving a Linear Equation with Fractions**

Find the solution set: $\dfrac{3}{2}y - \dfrac{2}{3} = \dfrac{1}{5}y$.

SOLUTION   To clear the equation of fractions, we multiply both sides by the least common denominator (LCD) of the three fractions and proceed as follows:

$$\frac{3}{2}y - \frac{2}{3} = \frac{1}{5}y$$

> **Take Note**
>
> Equations without fractions are always easier to solve. We can eliminate the fractions by multiplying by the LCD of the fractions.

$$30\left(\frac{3}{2}y - \frac{2}{3}\right) = 30\left(\frac{1}{5}y\right) \qquad \text{Multiply both sides by 30, the LCD of } \frac{3}{2}, \frac{2}{3}, \text{ and } \frac{1}{5}.$$

$$45y - 20 = 6y \qquad \text{Remove parentheses and simplify.}$$

$$45y - 20 + 20 = 6y + 20 \qquad \text{Add 20 to both sides.}$$

$$45y = 6y + 20 \qquad \text{Simplify.}$$

$$45y - 6y = 6y - 6y + 20 \qquad \text{Subtract } 6y \text{ from both sides.}$$

$$39y = 20 \qquad \text{Combine like terms.}$$

$$\frac{39y}{39} = \frac{20}{39} \qquad \text{Divide both sides by 39.}$$

$$y = \frac{20}{39} \qquad \text{Simplify.}$$

The solution set is $\left\{\dfrac{20}{39}\right\}$. The equation is a conditional equation. Verify that $\dfrac{20}{39}$ satisfies the equation.

Self Check 3   Find the solution set: $\dfrac{2}{3}p - 3 = \dfrac{p}{6}$.

Now Try Exercise 43.

EXAMPLE 4

Classifying Linear Equations

Solve and classify as an identity, a conditional equation, or a contradiction:
a. $3(x + 5) = 3(1 + x)$ **b.** $5 + 5(x + 2) - 2x = 3x + 15$

SOLUTION **a.**

$$3(x + 5) = 3(1 + x)$$

$$3x + 15 = 3 + 3x \qquad \text{Remove parentheses.}$$

$$3x - 3x + 15 = 3 + 3x - 3x \qquad \text{Subtract } 3x \text{ from both sides.}$$

$$15 = 3 \qquad \text{Combine like terms.}$$

> **Take Note**
> 1. If solving leads to a false statement, the equation is a contradiction.
> 2. If solving leads to a true statement such as $0 = 0$, the equation is an identity.

Since $15 = 3$ is false, the equation has no roots. Its solution set is the empty set, which is denoted as $\varnothing$. This equation is a contradiction.

b.

$$5 + 5(x + 2) - 2x = 3x + 15$$

$$5 + 5x + 10 - 2x = 3x + 15 \qquad \text{Remove parentheses.}$$

$$3x + 15 = 3x + 15 \qquad \text{Simplify.}$$

$$0 = 0 \qquad \text{Subtract } 3x + 15 \text{ from both sides.}$$

Because both sides of the final equation are identical, every value of x will make the equation true. The solution set is the set of all real numbers. This equation is an identity.

Self Check 4 Solve and classify as an identity, a conditional equation, or a contradiction:
a. $-2(x - 4) + 6x = 4(x + 1)$
b. $2(x + 1) + 4 = 2(x + 3)$

Now Try Exercise 29.

3. Solve Rational Equations

Rational equations are equations that contain rational expressions. Some examples of rational equations are

$$\frac{2}{x - 3} = 7, \qquad \frac{x + 1}{x - 2} = \frac{3}{x - 2}, \qquad \text{and} \qquad \frac{x + 2}{x + 3} + \frac{1}{x^2 + 2x - 3} = 1$$

When solving these equations, we will multiply both sides by a quantity containing a variable. When we do this, we could inadvertently multiply both sides of an equation by 0 and obtain a solution that makes the denominator of a fraction 0. In this case, we have found a false solution, called an **extraneous solution**. These solutions do not satisfy the equation and must be discarded.

The following equation has an extraneous solution.

> **Caution**
> A common error made when solving rational equations with variables in the denominator is to forget to check the proposed solution. Always "trash" or reject any proposed solution that makes any denominator 0.

$$\frac{x + 1}{x - 2} = \frac{3}{x - 2}$$

$$(x - 2)\left(\frac{x + 1}{x - 2}\right) = (x - 2)\left(\frac{3}{x - 2}\right) \qquad \text{Multiply both sides by } x - 2.$$

$$x + 1 = 3 \qquad \frac{x - 2}{x - 2} = 1$$

$$x = 2 \qquad \text{Subtract 1 from both sides.}$$

If we check by substituting 2 for x, we obtain 0's in the denominator. Thus, 2 is not a root. The solution set is $\varnothing$ and the equation is a contradiction.

By checking the proposed solution $x = 2$, we knew to reject 2 as a solution because we obtained 0's in the denominators. Another approach is to initially note

any restrictions on the variables. In this case, we note that x cannot be 2. Thus, if 2 surfaces as our answer for x when we solve the equation, we "trash" or reject it.

Solving a Rational Equation

EXAMPLE 5

Solve: $\dfrac{x+2}{x+3} + \dfrac{1}{x^2+2x-3} = 1$.

SOLUTION   Note that x cannot be -3, because that would cause the denominator of the first fraction to be 0. To find other restrictions, we factor the trinomial in the denominator of the second fraction.

$$x^2 + 2x - 3 = (x+3)(x-1)$$

Since the denominator will be 0 when $x = -3$ or $x = 1$, x cannot be -3 or 1.

$$\frac{x+2}{x+3} + \frac{1}{x^2+2x-3} = 1$$

$$\frac{x+2}{x+3} + \frac{1}{(x+3)(x-1)} = 1 \qquad \text{Factor } x^2 + 2x - 3.$$

$$(x+3)(x-1)\left[\frac{x+2}{x+3} + \frac{1}{(x+3)(x-1)}\right] = (x+3)(x-1)1 \qquad \text{Multiply both sides by } (x+3)(x-1).$$

$$(x+3)(x-1)\left(\frac{x+2}{x+3}\right) + (x+3)(x-1)\frac{1}{(x+3)(x-1)} = (x+3)(x-1)1 \qquad \text{Remove brackets.}$$

$$(x-1)(x+2) + 1 = (x+3)(x-1) \qquad \text{Simplify.}$$

$$x^2 + x - 2 + 1 = x^2 + 2x - 3 \qquad \text{Multiply the binomials.}$$

$$x - 1 = 2x - 3 \qquad \text{Subtract } x^2 \text{ from both sides and combine like terms.}$$

$$2 = x \qquad \text{Add 3 and subtract } x \text{ from both sides.}$$

Because 2 is a meaningful replacement for x, it is a root. However, it is a good idea to check it.

$$\frac{x+2}{x+3} + \frac{1}{x^2+2x-3} = 1$$

$$\frac{2+2}{2+3} + \frac{1}{2^2+2(2)-3} \stackrel{?}{=} 1 \qquad \text{Substitute 2 for } x.$$

$$\frac{4}{5} + \frac{1}{5} \stackrel{?}{=} 1$$

$$1 = 1$$

Since 2 satisfies the equation, it is a root. The equation is a conditional equation. ■

Self Check 5   Solve: $\dfrac{3}{5} + \dfrac{7}{x+2} = 2$.

Now Try Exercise 61.

4. Solve Formulas for a Specific Variable

Many equations, called **formulas**, contain several variables. For example, the formula that converts degrees Celsius to degrees Fahrenheit is $F = \frac{9}{5}C + 32$. If we want to change a large number of Fahrenheit readings to degrees Celsius, it would be tedious to substitute each value of F into the formula and then repeatedly solve it for C. It is better to solve the formula for C, substitute the values for F, and evaluate C directly.

EXAMPLE 6

Solving a Formula for a Specific Variable

Solve $F = \dfrac{9}{5}C + 32$ for C.

SOLUTION   We use the same methods as for solving linear equations.

$$F = \frac{9}{5}C + 32$$

$$F - 32 = \frac{9}{5}C \qquad \text{Subtract 32 from both sides.}$$

$$\frac{5}{9}(F - 32) = \frac{5}{9}\left(\frac{9}{5}C\right) \qquad \text{Multiply both sides by } \frac{5}{9}.$$

$$\frac{5}{9}(F - 32) = C \qquad \text{Simplify.}$$

This result can also be written in the form $C = \dfrac{5F - 160}{9}$.

Self Check 6   Solve $C = \dfrac{5}{9}(F - 32)$ for F.

Now Try Exercise 79.

EXAMPLE 7

Solving a Formula for a Specific Variable

The formula $A = P + Prt$ is used to find the amount of money in a savings account at the end of a specified time. A represents the amount, P represents the principal (the original deposit), r represents the rate of simple interest per unit of time, and t represents the number of units of time. Solve this formula for P.

SOLUTION   We factor P from both terms on the right side of the equation and proceed as follows:

> **Tip**
>
> Placing a circle around the variable you are solving for in a formula can help you focus on the specific variable you are isolating.

$$A = P + Prt$$

$$A = P(1 + rt) \qquad \text{Factor out } P.$$

$$\frac{A}{1 + rt} = P \qquad \text{Divide both sides by } 1 + rt.$$

$$P = \frac{A}{1 + rt}$$

Self Check 7   Solve $pq = fq + fp$ for f.

Now Try Exercise 85.

Self Check Answers

1. all real numbers except 2 or -5 2. $\{-3\}$ 3. $\{6\}$

4. **a.** no solution; contradiction   **b.** all real numbers; identity

5. 3   6. $F = \dfrac{9}{5}C + 32$ 7. $f = \dfrac{pq}{q + p}$

Exercises 1.1

Getting Ready

You should be able to complete these vocabulary and concept statements before you proceed to the practice exercises.

Fill in the blanks.

1. If a number satisfies an equation, it is called a _____ or a _____ of the equation.
2. If an equation is true for all values of its variable, it is called an _____.
3. A contradiction is an equation that is true for _____ values of its variable.
4. A _____ equation is true for some values of its variable and is not true for others.
5. An equation of the form $ax + b = 0$ is called a _____ equation.
6. If an equation contains rational expressions, it is called a _____ equation.
7. A conditional linear equation has _____ root.
8. The _____ of a fraction can never be 0.

Practice

Each quantity represents a real number. Find any restrictions on x.

9. $2x + 5 = -17$

10. $\dfrac{1}{2}x - 7 = 14$

11. $\dfrac{1}{x} = 12$

12. $\dfrac{3}{x - 2} = 9x$

13. $\dfrac{8}{x - 6} = \dfrac{5}{x + 2}$

14. $\dfrac{x}{x - 3} = -\dfrac{4}{x + 4}$

15. $\dfrac{1}{x - 3} = \dfrac{5x}{x^2 - 16}$

16. $\dfrac{1}{x^2 - 3x - 4} = \dfrac{5}{x} + 2$

Solve each equation, if possible. Classify each one as an identity, a conditional equation, or a contradiction.

17. $2x + 5 = 15$
18. $3x + 2 = x + 8$
19. $2(n + 2) - 5 = 2n$
20. $3(m + 2) = 2(m + 3) + m$
21. $\dfrac{x + 7}{2} = 7$
22. $\dfrac{x}{2} - 7 = 14$
23. $2(a + 1) = 3(a - 2) - a$

24. $x^2 = (x + 4)(x - 4) + 16$

25. $3(x - 3) = \dfrac{6x - 18}{2}$

26. $x(x + 2) = (x + 1)^2$

27. $\dfrac{3}{b - 3} = 1$

28. $x^2 - 8x + 15 = (x - 3)(x + 5)$

29. $2x^2 + 5x - 3 = (2x - 1)(x + 3)$

30. $2x^2 + 5x - 3 = 2x\left(x + \dfrac{19}{2}\right)$

Solve each equation. If an equation has no solution, so indicate.

31. $2x + 7 = 10 - x$ 32. $9a - 3 = 15 + 3a$

33. $5(x - 2) = 2(x + 4)$ 34. $5(r - 4) = -5(r - 4)$

35. $7(2x + 5) - 6(x + 8) = 7$

36. $6(x - 5) - 4(x + 2) = -1$

37. $\dfrac{5}{3}z - 8 = 7$ 38. $\dfrac{4}{3}y + 12 = -4$

39. $\dfrac{z}{5} + 2 = 4$ 40. $\dfrac{3p}{7} - p = -4$

41. $\dfrac{3x - 2}{3} = 2x + \dfrac{7}{3}$ 42. $\dfrac{7}{2}x + 5 = x + \dfrac{15}{2}$

43. $\dfrac{3x + 1}{20} = \dfrac{1}{2}$ 44. $2x - \dfrac{7}{6} + \dfrac{x}{6} = \dfrac{4x + 3}{6}$

45. $\dfrac{3 + x}{3} + \dfrac{x + 7}{2} = 4x + 1$

46. $2(2x + 1) - \dfrac{3x}{2} = \dfrac{-3(4 + x)}{2}$

47. $\dfrac{3}{2}(3x - 2) - 10x - 4 = 0$

48. $\dfrac{a(a - 3) + 5}{7} = \dfrac{(a - 1)^2}{7}$

49. $\dfrac{(y + 2)^2}{3} = y + 2 + \dfrac{y^2}{3}$

50. $(t + 1)(t - 1) = (t + 2)(t - 3) + 4$

51. $x(x + 2) = (x + 1)^2 - 1$

52. $(x - 2)(x - 3) = (x + 3)(x + 4)$

Solve each rational equation. Check for false or extraneous solutions.

53. $\dfrac{4}{x} = \dfrac{2}{5} = \dfrac{6}{x}$

54. $\dfrac{3}{x} + \dfrac{1}{2} = \dfrac{4}{x}$

55. $\dfrac{2}{x+1} + \dfrac{1}{3} = \dfrac{1}{x+1}$

56. $\dfrac{3}{x-2} + \dfrac{1}{x} = \dfrac{3}{x-2}$

57. $\dfrac{9t+6}{t(t+3)} = \dfrac{7}{t+3}$

58. $x + \dfrac{2(-2x+1)}{3x+5} = \dfrac{3x^2}{3x+5}$

59. $\dfrac{2}{(a-7)(a+2)} = \dfrac{4}{(a+3)(a+2)}$

60. $\dfrac{2}{n-2} + \dfrac{1}{n+1} = \dfrac{1}{n^2-n-2}$

61. $\dfrac{2x+3}{x^2+5x+6} + \dfrac{3x-2}{x^2+x-6} = \dfrac{5x-2}{x^2-4}$

62. $\dfrac{3x}{x^2+x} - \dfrac{2x}{x^2+5x} = \dfrac{x+2}{x^2+6x+5}$

63. $\dfrac{3x+5}{x^3+8} + \dfrac{3}{x^2-4} = \dfrac{2(3x-2)}{(x-2)(x^2-2x+4)}$

64. $\dfrac{1}{n+8} - \dfrac{3n-4}{5n^2+42n+16} = \dfrac{1}{5n+2}$

65. $\dfrac{1}{11-n} - \dfrac{2(3n-1)}{-7n^2+74n+33} = \dfrac{1}{7n+3}$

66. $\dfrac{4}{a^2-13a-48} - \dfrac{2}{a^2-18a+32} = \dfrac{1}{a^2+a-6}$

67. $\dfrac{5}{y+4} + \dfrac{2}{y+2} = \dfrac{6}{y+2} - \dfrac{1}{y^2+6y+8}$

68. $\dfrac{6}{2a-6} - \dfrac{3}{3-3a} = \dfrac{1}{a^2-4a+3}$

69. $\dfrac{3y}{6-3y} + \dfrac{2y}{2y+4} = \dfrac{8}{4-y^2}$

70. $\dfrac{3+2a}{a^2+6+5a} - \dfrac{2-3a}{a^2-6+a} = \dfrac{5a-2}{a^2-4}$

71. $\dfrac{a}{a+2} - 1 = -\dfrac{3a+2}{a^2+4a+4}$

72. $\dfrac{x-1}{x+3} + \dfrac{x-2}{x-3} = \dfrac{1-2x}{3-x}$

Solve each formula for the specified variable.

73. $f = ma;\ m$

74. $ax + b = 0;\ x$

75. $P = 2l + 2w;\ w$

76. $V = \dfrac{1}{3}\pi r^2 h;\ h$

77. $V = \dfrac{1}{3}\pi r^2 h;\ r^2$

78. $z = \dfrac{x-\mu}{\sigma};\ \mu$

79. $P_n = L + \dfrac{si}{f};\ s$

80. $P_n = L + \dfrac{si}{f};\ f$

81. $F = \dfrac{mMg}{r^2};\ m$

82. $\dfrac{1}{f} = \dfrac{1}{p} + \dfrac{1}{q};\ f$

83. $\dfrac{x}{a} + \dfrac{y}{b} = 1;\ y$

84. $\dfrac{x}{a} - \dfrac{y}{b} = 1;\ a$

85. $\dfrac{1}{r} = \dfrac{1}{r_1} + \dfrac{1}{r_2};\ r$

86. $\dfrac{1}{r} = \dfrac{1}{r_1} + \dfrac{1}{r_2};\ r_1$

87. $l = a + (n-1)d;\ n$

88. $l = a + (n-1)d;\ d$

89. $a = (n-2)\dfrac{180}{n};\ n$

90. $S = \dfrac{a-lr}{1-r};\ a$

91. $R = \dfrac{1}{\dfrac{1}{r_1} + \dfrac{1}{r_2} + \dfrac{1}{r_3}};\ r_1$

92. $R = \dfrac{1}{\dfrac{1}{r_1} + \dfrac{1}{r_2} + \dfrac{1}{r_3}};\ r_3$

Discovery and Writing

93. Explain the difference between an identity and a contradiction. Give examples of each.

94. Share a strategy that can be used to identify the restrictions on a variable in a rational equation.

95. Explain why a conditional linear equation always has exactly one root.

96. Define an extraneous solution and explain how such a solution occurs.

Critical Thinking

Determine if the statement is true or false. If the statement is false, then correct it and make it true.

97. The equation $4x + 5(x-3) = 9x - 15$ is a contradiction.

98. The equation $4x + 5(x - 3) = 9x + 15$ is an identity.

99. $\sqrt{7}, -4.5$, and π would be included in the solution set for $-2x - 8 = -(2x + 8)$.

100. The equation $x + 188,424 = x + 188,425$ has an infinite number of solutions.

101. The solution set of
$$\frac{1}{\left(\dfrac{1}{x-3}\right)} + \frac{1}{\left(\dfrac{1}{x-4}\right)} = 7 \text{ is } \{7\}.$$

102. If $y_1 = \dfrac{1}{x - \pi}$ and $y_2 = 1$, then $y_1 = y_2$ when $x = 2\pi$.

1.2 Applications of Linear Equations

In this section, we will learn to

1. Solve number problems.

2. Solve geometric problems.

3. Solve investment and percent problems.

4. Solve break-point analysis problems.

5. Solve shared-work problems.

6. Solve mixture problems.

7. Solve uniform motion problems.

©Debby Wong/Shutterstock.com

Luke Bryan is one of the most successful country music singers in the business today. He signed with Capitol Records in 2007 and since then has had several number one singles, including "Crash My Party" and "That's My Kind of Night." He was the recipient of the Academy of Country Music Awards' Entertainer of the Year award in 2013.

Suppose you hear that for one day only Luke Bryan concert tickets can be purchased for 30% off the original price. Knowing this is a bargain, you immediately purchase a ticket for $77, the selling price. Later that day, this question comes to mind: What was the original price of the concert ticket?

A linear equation can be used to model this problem. We can let x represent the original price of the concert ticket and subtract 30% of x (the discount) and we will get the selling price $77. The linear equation is $x - 0.3x = 77$. We can solve this equation to determine the original price of the concert ticket.

$$x - 0.3x = 77$$
$$0.7x = 77 \qquad \text{Combine like terms.}$$
$$x = \frac{77}{0.7} \qquad \text{Divide both sides by 0.7.}$$
$$x = 110$$

The original price of the concert ticket is $110.

In this section, we will use the equation-solving techniques discussed in the previous section to solve applied problems (often called *word problems*). To solve these problems, we must translate the verbal description of the problem into an equation. The process of finding the equation that describes the words of the problem is called **mathematical modeling**. The equation itself is often called a **mathematical model** of the situation described in the word problem.

The following list of steps provides a strategy to follow when we try to find the equation that models an applied problem.

Strategy for Modeling Equations	
	1. Analyze the problem to see what you are to find. Often, drawing a diagram or making a table will help you visualize the facts.
	2. Pick a variable to represent the quantity that is to be found, and write a sentence telling what that variable represents. Express all other quantities mentioned in the problem as expressions involving this single variable.
	3. Find a way to express a quantity in two different ways. This might involve a formula from geometry, finance, or physics.
	4. Form an equation indicating that the two quantities found in Step 3 are equal.
	5. Solve the equation.
	6. Answer the questions asked in the problem.
	7. Check the answers in the words of the problem.

This list does not apply to all situations, but it can be used for a wide range of problems with only slight modifications.

1. Solve Number Problems

Solving a Number Problem

EXAMPLE 1

A student has scores of 74, 78, and 70 on three music theory exams. What score is needed on a fourth exam for the student to earn an average grade of 80?

SOLUTION   To find an equation that models the problem, we can let x represent the required grade on the fourth exam. The average grade will be one-fourth of the sum of the four grades. We know this average is to be 80.

The average of the four grades	equals	the required average grade.
$\dfrac{74 + 78 + 70 + x}{4}$	$=$	80

We can solve this equation for x.

$$\frac{222 + x}{4} = 80 \qquad \text{\color{red}{74 + 78 + 70 = 222}}$$
$$222 + x = 320 \qquad \text{\color{red}{Multiply both sides by 4.}}$$
$$x = 98 \qquad \text{\color{red}{Subtract 222 from both sides.}}$$

To earn an average of 80, the student must score 98 on the fourth exam.

Self Check 1   A student scores 82, 96, 91, and 92 on four college algebra exams. What score is needed on a fifth exam for the student to earn an average grade of 90?

Now Try Exercise 11.

2. Solve Geometric Problems

Solving a Geometric Problem

EXAMPLE 2

A city ordinance requires a man to install a fence around the swimming pool shown in Figure 1-2. He wants the border around the pool to be of uniform width. If he has 154 feet of fencing, find the width of the border.

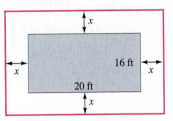

FIGURE 1-2

SOLUTION

We can let x represent the width of the border. The distance around the large rectangle, called its **perimeter**, P is given by the formula $P = 2l + 2w$, where l is the length, $20 + 2x$, and w is the width, $16 + 2x$. Since the man has 154 feet of fencing, the perimeter will be 154 feet. To find an equation that models the problem, we substitute these values into the formula for perimeter.

$P = 2l + 2w$	The formula for the perimeter of a rectangle.
$154 = 2(20 + 2x) + 2(16 + 2x)$	Substitute 154 for P, $20 + 2x$ for l, and $16 + 2x$ for w.
$154 = 40 + 4x + 32 + 4x$	Use the Distributive Property to remove parentheses.
$154 = 72 + 8x$	Combine like terms.
$82 = 8x$	Subtract 72 from both sides.
$10\dfrac{1}{4} = x$	Divide both sides by 8.

This border will be $10\frac{1}{4}$ feet wide.

Self Check 2

In Example 2, if 168 feet of fencing is available, find the border's width.

Now Try Exercise 19.

3. Solve Investment and Percent Problems

Solving an Investment Problem

EXAMPLE 3

Laurie invested \$10,000, part at 9% and the rest at 14%. If the annual income from these investments is \$1275, how much did she invest at each rate?

SOLUTION

We can let x represent the amount invested at 9%. Then $10,000 - x$ represents the amount invested at 14%. Since the annual income from any investment is the product of the interest rate and the amount invested, we have the following information.

Type of Investment	Rate	Amount Invested	Interest Earned
9% investment	0.09	x	$0.09x$
14% investment	0.14	$10,000 - x$	$0.14(10,000 - x)$

The total income from these two investments can be expressed in two ways: as \$1275 and as the sum of the incomes from the two investments.

The income from the 9% investment	plus	the income from the 14% investment	equals	the total income.
$0.09x$	$+$	$0.14(10,000 - x)$	$=$	1275

We can solve this equation for x.

$$0.09x + 0.14(10,000 - x) = 1275$$

$9x + 14(10,000 - x) = 127,500$ To eliminate the decimal points, multiply both sides by 100.

$9x + 140,000 - 14x = 127,500$ Use the Distributive Property to remove parentheses.

$-5x + 140,000 = 127,500$ Combine like terms.

$-5x = -12,500$ Subtract 140,000 from both sides.

$x = 2500$ Divide both sides by -5.

The amount invested at 9% was $2500, and the amount invested at 14% was $7500 ($10,000 − $2500). These amounts are correct, because 9% of $2500 is $225, 14% of $7500 is $1050, and the sum of these amounts is $1275.

Self Check 3   A man invests $12,000, part at 7% and the rest at 9%. If the annual income from these investments is $965, how much was invested at each rate?

Now Try Exercise 25.

4. Solve Break-Point Analysis Problems

Running a machine involves two costs—**setup costs** and **unit costs**. Setup costs include the cost of installing a machine and preparing it to do a job. Unit cost is the cost to manufacture one item, which includes the costs of material and labor. If there are two machines to do a job, the **break point** is the number of units manufactured at which the cost on each machine is the same.

Solving a Break-Point Analysis Problem

EXAMPLE 4

Suppose that one machine has a setup cost of $400 and a unit cost of $1.50, and a second machine has a setup cost of $500 and a unit cost of $1.25. Find the break point.

SOLUTION   We can let x represent the number of items to be manufactured. The cost C_1 of using machine 1 is

$$C_1 = 400 + 1.5x$$

and the cost C_2 of using machine 2 is

$$C_2 = 500 + 1.25x$$

The break point occurs when these two costs are equal.

The cost of using machine 1	equals	the cost of using machine 2.
$400 + 1.5x$	$=$	$500 + 1.25x$

We can solve this equation for x.

$$400 + 1.5x = 500 + 1.25x$$

$$1.5x = 100 + 1.25x \qquad \text{Subtract 400 from both sides.}$$

$$0.25x = 100 \qquad \text{Subtract } 1.25x \text{ from both sides.}$$

$$x = 400 \qquad \text{Divide both sides by 0.25.}$$

The break point is 400 units. This result is correct, because it will cost the same amount to manufacture 400 units with either machine.

$$C_1 = \$400 + \$1.5(400) = \$1000 \quad \text{and} \quad C_2 = \$500 + \$1.25(400) = \$1000$$

Self Check 4   An ATM has a setup cost of $3000 and operating costs averaging $1 per transaction. Another ATM has a setup cost of $3500 and an operating cost of $0.50 per transaction. Find the number of transactions at which the costs for each ATM is the same.

Now Try Exercise 37.

5. Solve Shared-Work Problems

Solving a Shared-Work Problem

EXAMPLE 5

The Toll Way Authority needs to pave 100 miles of interstate highway before freezing temperatures come in about 60 days. Sjostrom and Sons has estimated that it can do the job in 110 days. Scandroli and Sons has estimated that it can do the job in 140 days. If the authority hires both contractors, will the job get done in time?

SOLUTION   Since Sjostrom can do the job in 110 days, it can do $\frac{1}{110}$ of the job in one day. Since Scandroli can do the job in 140 days, it can do $\frac{1}{140}$ of the job in one day. If we let n represent the number of days it will take to pave the highway if both contractors work together, they can do $\frac{1}{n}$ of the job in one day. The work that they can do together in one day is the sum of what each can do in one day.

Take Note

The times stated in the problem will appear in the denominator of the equation.

The part Sjostrom can pave in one day	plus	the part Scandroli can pave in one day	equals	the part they can pave together in one day.
$\dfrac{1}{110}$	$+$	$\dfrac{1}{140}$	$=$	$\dfrac{1}{n}$

We can solve this equation for n.

$$\frac{1}{110} + \frac{1}{140} = \frac{1}{n}$$

$$(110)(140)n\left(\frac{1}{110} + \frac{1}{140}\right) = (110)(140)n\left(\frac{1}{n}\right) \qquad \text{Multiply both sides by } (110)(140)n \text{ to eliminate the fractions.}$$

$$\frac{(110)(140)n}{110} + \frac{(110)(140)n}{140} = \frac{(110)(140)n}{n} \qquad \text{Use the Distributive Property to remove parentheses.}$$

$$140n + 110n = 15{,}400 \qquad \frac{110}{110} = 1, \frac{140}{140} = 1, \text{ and } \frac{n}{n} = 1.$$

$$250n = 15{,}400 \qquad \text{Combine like terms.}$$

$$n = 61.6 \qquad \text{Divide both sides by 250.}$$

It will take the contractors about 62 days to pave the highway. With any luck, the job will be done in time.

Self Check 5 John and Eric both work at Firestone Auto Care. John can install a new set of tires in 45 minutes. Eric is faster and can install a set in 30 minutes. If John and Eric work together, how long will it take them to install one set of tires?

Now Try Exercise 41.

6. Solve Mixture Problems

Solving a Mixture Problem

EXAMPLE 6

A container is partially filled with 20 liters of whole milk containing 4% butter-fat. How much 1% milk must be added to obtain a mixture that is 2% butterfat?

SOLUTION Since the first container shown in Figure 1-3 contains 20 liters of 4% milk, it contains 0.04(20) liters of butterfat. To this amount, we will add the contents of the second container, which holds 0.01(l) liters of butterfat.

The sum of these two amounts will equal the number of liters of butterfat in the third container, which is 0.02(20 + l) liters of butterfat. This information is presented in table form in Figure 1-4.

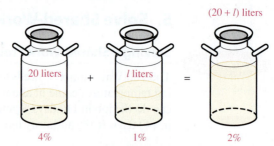

FIGURE 1-3

The butterfat in the 4% milk	plus	the butterfat in the 1% milk	equals	the butterfat in the 2% milk.
4% of 20 liters	+	1% of l liters	=	2% of (20 + l) liters

	Percentage of Butterfat	·	Amount of Milk	=	Amount of Butterfat
4% milk	0.04		20		0.04 (20)
1% milk	0.01		l		0.01 (l)
2% milk	0.02		20 + l		0.02(20 + l)

FIGURE 1-4

We can solve this equation for l.

$$0.04(20) + 0.01(l) = 0.02(20 + l)$$

$$4(20) + l = 2(20 + l) \qquad \text{Multiply both sides by 100.}$$

$$80 + l = 40 + 2l \qquad \text{Remove parentheses.}$$

$$40 = l \qquad \text{Subtract 40 and l from both sides.}$$

To dilute the 20 liters of 4% milk to a 2% mixture, 40 liters of 1% milk must be added. To check, we note that the final mixture contains 0.02(60) = 1.2 liters of pure butterfat, and that this is equal to the amount of pure butterfat in the 4% milk and the 1% milk; 0.04(20) + 0.01(40) = 1.2 liters.

Self Check 6 Milk containing 4% butterfat is mixed with 8 gallons of milk containing 1% butterfat to make a low-fat cottage cheese mixture containing 2% butterfat. How many gallons of the richer milk is used?

Now Try Exercise 47.

7. Solve Uniform Motion Problems

Solving a Uniform Motion Problem

EXAMPLE 7

A man leaves home driving his Ford F-150 truck at the rate of 50 mph. When his daughter discovers than he has forgotten his iPhone, she drives her Ford Mustang after him at a rate of 65 mph. How long will it take her to catch her dad if he had a 15-minute head start?

SOLUTION

Uniform motion problems are based on the formula $d = rt$, where d is the distance, r is the rate, and t is the time. We can draw a diagram and organize the information given in the problem in a chart as shown in Figures 1-5 and 1-6. In the chart, t represents the number of hours the daughter must drive to overtake her father. Because the father has a 15-minute, or $\frac{1}{4}$ hour, head start, he has been on the road for $\left(t + \frac{1}{4}\right)$ hours.

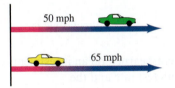

FIGURE 1-5

	d	$=$	r	$\cdot$	t
Man	$50\left(t + \frac{1}{4}\right)$		50		$t + \frac{1}{4}$
Daughter	$65t$		65		t

FIGURE 1-6

We can set up the following equation and solve it for t.

The distance the man drives	equals	the distance the daughter drives.
$50\left(t + \dfrac{1}{4}\right)$	$=$	$65t$

We can solve this equation for t.

$$50\left(t + \frac{1}{4}\right) = 65t$$

$$50t + \frac{25}{2} = 65t \qquad \text{Remove parentheses.}$$

$$\frac{25}{2} = 15t \qquad \text{Subtract } 50t \text{ from both sides.}$$

$$\frac{5}{6} = t \qquad \text{Divide both sides by 15 and simplify.}$$

It will take the daughter $\frac{5}{6}$ hour, or 50 minutes, to overtake her dad.

Self Check 7 On a reality television show, an officer traveling in a police cruiser at 90 mph pursues Jennifer who has a 3-minute head start. If the officer overtakes Jennifer in 12 minutes, how fast is Jennifer traveling?

Now Try Exercise 63.

Self Check Answers

1. 89   **2.** 12 feet   **3.** $5750 at 7%; $6250 at 9%   **4.** 1000
5. 18 minutes   **6.** 4   **7.** 72 mph

Exercises 1.2

Getting Ready
You should be able to complete these vocabulary and concept statements before you proceed to the practice exercises.

Fill in the blanks.

1. To average n scores, _____ the scores and divide by n.

2. The formula for the _____ of a rectangle is $P = 2l + 2w$.

3. The simple annual interest earned on an investment is the product of the interest rate and the _____ invested.

4. The number of units manufactured at which the cost on two machines is equal is called the _____.

5. Distance traveled is the product of the _____ and the _____.

6. 5% of 30 liters is _____ liters.

Practice
Solve each problem.

7. **Algebra scores** Dylan has completed all assignments in college algebra except for taking his comprehensive final exam. His current test scores are shown in the table.

Test 1	Test 2	Test 3	Test 4	Online Homework	Final Exam
60	78	80	90	88	?

If his online homework average for the semester is weighted as a test grade and his final exam grade is weighted as two test grades, what must he score on the final exam to have an 80 average in the course?

8. **Psychology scores** Mandy has completed all assignments in psychology except for taking her comprehensive final exam. Her current scores are shown in the table.

Test 1	Test 2	Test 3	Test 4	Test 5	Final Exam
70	88	93	85	88	?

If the final exam score is weighted as a test grade and also replaces her lowest test score, what must she make on the final exam to have a 90 average in the course?

9. **Test scores** Tate scored 5 points higher on his midterm and 13 points higher on his final than he did on his first exam. If his mean (average) score was 90, what was his score on the first exam?

10. **Test scores** Courtney took four tests in chemistry class. On each successive test, her score improved by 3 points. If her mean score was 69.5, what did she score on the first test?

11. **Teacher certification** On the Illinois certification test for teachers specializing in learning disabilities, a teacher earned the scores shown in the accompanying table. What was the teacher's score in program development?

Human development with special needs	82
Assessment	90
Program development and instruction	?
Professional knowledge and legal issues	78
AVERAGE SCORE	86

12. **Golfing** Par on a golf course is 72. If a golfer shot rounds of 76, 68, and 70 in a tournament, what will she need to shoot on the final round to average par?

13. **Replacing locks** A locksmith at Pop-A-Lock charges $40 plus $28 for each lock installed. How many locks can be replaced for $236?

14. Delivering ads A University of Florida student earns $20 per day delivering advertising brochures door-to-door, plus $1.50 for each person he interviews. How many people did he interview on a day when he earned $56?

15. Electronic LED billboard An electronic LED billboard in Times Square is 26 feet taller than it is wide. If its perimeter is 92 feet, find the dimensions of the billboard.

16. Hockey rink A National Hockey League rink is 115 feet longer than it is wide. If the perimeter of the rink is 570 feet, find the dimensions of the rink?

17. Width of a picture frame The picture frame with the dimensions shown in the illustration was built with 14 feet of framing material. Find x its width.

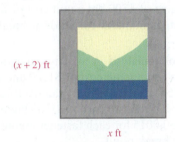

$(x + 2)$ ft

x ft

18. Fencing a garden If a gardener fences in the total rectangular area shown in the illustration instead of just the square area, he will need twice as much fencing to enclose the garden. How much fencing will he need?

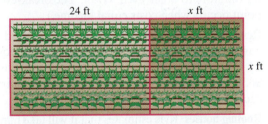

24 ft x ft

x ft

19. Swimming pool A rectangular swimming pool measures 12 meters by 6 meters and is surrounded by a 116 meter rectangular wooden fence. If the fence forms a border around the pool of uniform width, determine the width of the border.

20. Aquarium A rectangular glass aquarium has a length of 15 yards, a width of 10 yards, and is placed inside a rectangular room for viewing at a museum. If the room has a perimeter of 146 yards and forms a walkway of uniform width that surrounds the aquarium, determine the width of the walkway.

21. Wading pool dimensions The area of the triangular swimming pool shown in the illustration is doubled by adding a rectangular wading pool. Find the dimensions of the wading pool. (*Hint:* The area of

a triangle $= \frac{1}{2}bh$, and the area of a rectangle $= lw$.)

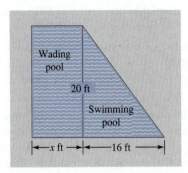

Wading pool

20 ft

Swimming pool

x ft   16 ft

22. House construction A builder wants to install a triangular window with the angles shown in the illustration. What angles will he have to cut to make the window fit? (*Hint:* The sum of the angles in a triangle equals 180°.)

$x°$

$(x + 30)°$ $(x + 30)°$

23. Length of a living room If a carpenter adds a porch with dimensions shown in the illustration to the living room, the living area will be increased by 50%. Find the length of the living room.

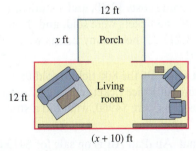

12 ft

x ft   Porch

12 ft   Living room

$(x + 10)$ ft

24. Depth of water in a trough The trough in the illustration has a cross-sectional area of 54 square inches. Find the depth, d, of the trough. (*Hint:* Area of a trapezoid $= \frac{1}{2}h(b_1 + b_2)$.)

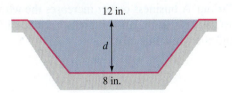

12 in.

d

8 in.

25. Investment Jeffrey invested $16,000 in two accounts paying 4% and 6% annual interest. If the total interest earned in one year was $815, how much did he invest at each rate?

26. Investment An executive invests $22,000, some at 7% and the rest at 6% annual interest. If he receives an annual return of $1420, how much is invested at each rate?

27. Financial planning After inheriting some money, a woman wants to invest enough to have an annual income of $5000. If she can invest $20,000 at 9% annual interest, how much more will she have to invest at 7% to achieve her goal? (See the table.)

Type	Rate	Amount	Income
9% investment	0.09	20,000	.09(20,000)
7% investment	0.07	x	.07x

28. Investment A woman invests $37,000, part at 8% and the rest at $9\frac{1}{2}$% annual interest. If the $9\frac{1}{2}$% investment provides $452.50 more income than the 8% investment, how much is invested at each rate?

29. Investment Equal amounts are invested at 6%, 7%, and 8% annual interest. If the three investments yield a total of $2037 annual interest, find the total investment.

30. Investment Aaron invested equal amounts of money at 5%, 7%, 9%, and 11% annual interest. If the four investments yielded a total of $1440 annual interest, find the total amount invested.

31. Ticket sales A full-price ticket for a college basketball game costs $2.50, and a student ticket costs $1.75. If 585 tickets were sold, and the total receipts were $1,217.25, how many tickets were student tickets?

32. Ticket sales Of the 800 tickets sold to a movie, 480 were full-price tickets costing $7 each. If the gate receipts were $4960, what did a student ticket cost?

33. Discount An iPad Air is on sale for $413.08. What was the original price of the iPad if it was discounted 8%?

34. Discount After being discounted 20%, a weather radio sells for $63.96. Find the original price.

35. Markup A business owner increases the wholesale cost of a kayak by 70% and sells it for $365.50. Find the wholesale cost.

©Kalmatsuy/Shutterstock.com

36. Markup A merchant increases the wholesale cost of a surfboard by 30% to determine the selling price. If the surfboard sells for $588.90, find the wholesale cost.

37. Break-point analysis A machine to mill a brass plate has a setup cost of $600 and a unit cost of $3 for each plate manufactured. A bigger machine has a setup cost of $800 but a unit cost of only $2 for each plate manufactured. Find the break point.

38. Break-point analysis A machine to manufacture fasteners has a setup cost of $1200 and a unit cost of $0.005 for each fastener manufactured. A newer machine has a setup cost of $1500 but a unit cost of only $0.0015 for each fastener manufactured. Find the break point.

39. Computer sales A computer store has fixed costs of $8925 per month and a unit cost of $850 for every computer it sells. If the store can sell all the computers it can get for $1275 each, how many must be sold for the store to break even? (*Hint:* The break-even point occurs when costs equal income.)

40. Restaurant management A restaurant has fixed costs of $137.50 per day and an average unit cost of $4.75 for each meal served. If a typical meal costs $6, how many customers must eat at the restaurant each day for the owner to break even?

41. Roofing houses Kyle estimates that it will take him 7 days to roof his house. A professional roofer estimates that it will take him 4 days to roof the same house. How long will it take if they work together?

42. Sealing asphalt One crew can seal a parking lot in 8 hours and another in 10 hours. How long will it take to seal the parking lot if the two crews work together?

43. Mowing lawns Julie can mow a lawn with a lawn tractor in 2 hours, and her husband can mow the same lawn with a push mower in 4 hours. How long will it take to mow the lawn if they work together?

44. Filling swimming pools A garden hose can fill a swimming pool in 3 days, and a larger hose can fill the pool in 2 days. How long will it take to fill the pool if both hoses are used?

45. Filling swimming pools An empty swimming pool can be filled in 10 hours. When full, the pool can be drained in 19 hours. How long will it take to fill the empty pool if the drain is left open?

46. Preparing seafood Kevin stuffs shrimp in his job as a seafood chef. He can stuff 1000 shrimp in 6 hours. When his sister helps him, they can stuff 1000 shrimp in 4 hours. If Kevin gets sick, how long will it take his sister to stuff 500 shrimp?

47. Diluting solutions How much water should be added to 20 ounces of a 15% solution of alcohol to dilute it to a 10% solution?

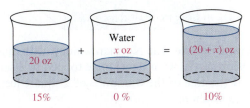

48. Increasing concentrations The beaker shown below contains a 2% saltwater solution.
   a. How much water must be boiled away to increase the concentration of the salt solution from 2% to 3%?
   b. Where on the beaker would the new water level be?

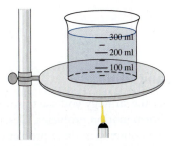

49. Winterizing cars A car radiator has a 6-liter capacity. If the liquid in the radiator is 40% antifreeze, how much liquid must be replaced with pure antifreeze to bring the mixture up to a 50% solution?

50. Mixing milk If a bottle holding 3 liters of milk contains $3\frac{1}{2}$% butterfat, how much skimmed milk must be added to dilute the milk to 2% butterfat?

51. Preparing solutions A nurse has 1 liter of a solution that is 20% alcohol. How much pure alcohol must she add to bring the solution up to a 25% concentration?

52. Diluting solutions If there are 400 cubic centimeters of a chemical in 1 liter of solution, how many cubic centimeters of water must be added to dilute it to a 25% solution? (*Hint:* 1000 cc = 1 liter.)

53. Cleaning swimming pools A swimming pool contains 15,000 gallons of water. How many gallons of chlorine must be added to "shock the pool" and bring the water to a $\frac{3}{100}$% solution?

54. Mixing fuels An automobile engine can run on a mixture of gasoline and a substitute fuel. If gas costs $3.50 per gallon and the substitute fuel costs $2 per gallon, what percent of a mixture must be substitute fuel to bring the cost down to $2.75 per gallon?

55. Evaporation How many liters of water must evaporate to turn 12 liters of a 24% salt solution into a 36% solution?

56. Increasing concentrations A beaker contains 320 ml of a 5% saltwater solution. How much water should be boiled away to increase the concentration to 6%?

57. Lowering fat How many pounds of extra-lean hamburger that is 7% fat must be mixed with 30 pounds of hamburger that is 15% fat to obtain a mixture that is 10% fat?

58. Dairy foods How many gallons of cream that is 22% butterfat must be mixed with milk that is 2% butterfat to get 20 gallons of milk containing 4% butterfat?

59. Mixing solutions How many gallons of a 5% alcohol solution must be mixed with 90 gallons of 1% solution to obtain a 2% solution?

60. Preparing medicines A doctor prescribes an ointment that is 2% hydrocortisone. A pharmacist has 1% and 5% concentrations in stock. How much of each should the pharmacist use to make a 1-ounce tube?

61. Feeding cattle A cattleman wants to mix 2400 pounds of cattle feed that is to be 14% protein. Barley (11.7% protein) will make up 25% of the mixture. The remaining 75% will be made up of oats (11.8% protein) and soybean meal (44.5% protein). How many pounds of each will he use?

62. Feeding cattle If the cattleman in Exercise 61 wants only 20% of the mixture to be barley, how many pounds of each should he use?

63. Driving rates John drove to Daytona Beach, Florida, in 5 hours. When he returned, there was less traffic, and the trip took only 3 hours. If John averaged 26 mph faster on the return trip, how fast did he drive each way?

64. Distance problem Allison drove home at 60 mph, but her brother Austin, who left at the same time, could drive at only 48 mph. When Allison arrived, Austin still had 60 miles to go. How far did Allison drive?

65. Distance problem Two cars leave Hinds Community College traveling in opposite directions. One car travels at 60 mph and the other at 64 mph. In how many hours will they be 310 miles apart?

66. Bank robbery Some bank robbers leave town, speeding at 70 mph. Ten minutes later, the police give chase, traveling at 78 mph. How long, after the robbery, will it take the police to overtake the robbers?

67. Jogging problem Two Michigan State University cross-country runners are 440 yards apart and are running toward each other, one at 8 mph and the other at 10 mph. In how many seconds will they meet?

68. Driving rates One morning, Justin drove 5 hours before stopping to eat lunch at Chick-fil-A. After lunch, he increased his speed by 10 mph. If he completed a 430-mile trip in 8 hours of driving time, how fast did he drive in the morning?

©Paul Brennan/Shutterstock.com

69. Boating problem A Johnson motorboat goes 5 miles upstream in the same time it requires to go 7 miles downstream. If the river flows at 2 mph, find the speed of the boat in still water.

70. Wind velocity A plane can fly 340 mph in still air. If it can fly 200 miles downwind in the same amount of time it can fly 140 miles upwind, find the velocity of the wind.

Use a calculator to help solve each problem.

71. Machine tool design 712.51 cubic millimeters of material was removed by drilling the blind hole as shown in the illustration. Find the depth of the hole. (*Hint:* The volume of a cylinder is given by $V = \pi r^2 h$.)

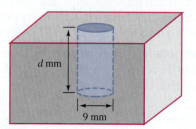

72. Architecture The Norman window with dimensions as shown is a rectangle topped by a semicircle. If the area of the window is 68.2 square feet, find its height h.

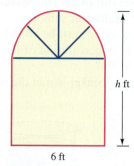

Discovery and Writing

73. Consider the strategy you use to solve investment and uniform motion problems. Describe any similarities you observe in these problem types.

74. Which type of application was hardest for you to solve? Why? What strategy or approach works best for you when approaching solving this problem?

75. Explain why the solution to an application problem should be checked in the original wording of the problem and not in the equation obtained from the words.

1.3   Complex Numbers

In this section, we will learn to

1. Simplify imaginary numbers.
2. Perform operations on complex numbers.
3. Find powers of i.
4. Determine the absolute value of a complex number.
5. Factor the sum of two squares.

©Sybille Yates/Shutterstock.com

Fractals are beautiful art designs that are computer generated from numbers called *complex numbers*. We will study these numbers in this section. One general class of fractal sets, called the Mandelbrot set, is named after the mathematician Benoit Mandelbrot, who studied and popularized fractals. One Mandelbrot set fractal is shown here. Fractals exhibit a repeating pattern that displays at every scale.

Disney's animated feature film *Frozen* was a huge blockbuster success. The movie's most popular song, "Let It Go," won the Academy Award for best song and included the word *fractals*. In the movie, Queen Elsa sings, "My soul is spiraling in frozen *fractals* all around."

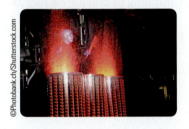
©Joe Seer/Shutterstock.com

Members of The Blue Man Group, a highly successful group of creative performers, cover themselves in blue grease paint, wear latex bald caps, and dress in black. They combine rock music, comedy, multimedia theatrics, and sophisticated lighting to entertain their audiences. The group has used fractals as an art theme in its shows.

We have explored the set of real numbers to great extent. In this section, we will extend this set and explore numbers called *complex numbers*.

1.   Simplify Imaginary Numbers

For many years, mathematicians considered numbers such as $\sqrt{-1}$, $\sqrt{-5}$, and $\sqrt{-9}$ to make no sense. Sir Isaac Newton (1642–1727) called them "impossible numbers." In the 17th century, these symbols were called **imaginary numbers** by René Descartes. Today, they have important uses, such as describing the behavior of alternating current in electronics. The imaginary numbers are based on the imaginary unit, and the imaginary unit is denoted by the letter i.

©Photobank.ch/Shutterstock.com

Imaginary Unit i   The **imaginary unit** is denoted by the letter i and is defined as

$$i = \sqrt{-1}$$

From the definition it follows that $i^2 = -1$.

Because imaginary numbers follow the rules for exponents, we have

$$(3i)^2 = 3^2 i^2 = 9(-1) = -9$$

Since $(3i)^2 = -9$, $3i$ is the square root of -9, and we can write

$$\sqrt{-9} = \sqrt{(-1)(9)}$$

$$= \sqrt{-1}\sqrt{9} \qquad \sqrt{ab} = \sqrt{a}\sqrt{b}$$

$$= 3i \qquad \sqrt{-1} = i$$

The Multiplication Property of Radicals can be used to simplify imaginary numbers. Four examples are shown in the table below.

> **Imaginary Numbers Written in Terms of *i***
>
> $\sqrt{-25} = \sqrt{(-1)(25)} = \sqrt{-1}\sqrt{25} = (i)(5) = 5i$
>
> $-\sqrt{-169} = -\sqrt{(-1)(169)} = -\sqrt{-1}\sqrt{169} = -(i)(13) = -13i$
>
> $\sqrt{-24} = \sqrt{(-1)(24)} = \sqrt{-1}\sqrt{24} = (i)(2\sqrt{6}) = 2\sqrt{6}i$ or $2i\sqrt{6}$
>
> $\sqrt{-\dfrac{8}{49}} = \sqrt{(-1)\left(\dfrac{8}{49}\right)} = \sqrt{-1}\sqrt{\dfrac{8}{49}} = (i)\left(\dfrac{\sqrt{8}}{\sqrt{49}}\right) = (i)\left(\dfrac{2\sqrt{2}}{7}\right) = \dfrac{2\sqrt{2}}{7}i$ or $\dfrac{2i\sqrt{2}}{7}$

Take Note

When $\sqrt{b}\,i$ is handwritten, we write *i* first so that it is clear that *i* is not under the radical symbol.

Caution

If *a* and *b* both are negative, then $\sqrt{ab} \neq \sqrt{a}\sqrt{b}$. For example, the correct simplification of $\sqrt{-16}\sqrt{-4}$ is

$\sqrt{-16}\sqrt{-4} = (4i)(2i) = 8i^2 = 8(-1) = -8$

The following simplification is incorrect. Since both numbers are negative, the Multiplication Property of Radicals does not apply.

$\sqrt{-16}\sqrt{-4} = \sqrt{(-16)(-4)} = \sqrt{64} = 8$

Accent on Technology

Simplifying Imaginary Numbers

If we place our graphing calculator in $a + bi$ mode, it can be used to simplify many imaginary numbers. Please note that on a graphing calculator *i* is located above the decimal point key.

In Figure 1-7, we see the graphing calculator results for $\sqrt{-81}$, $\sqrt{-\frac{4}{25}}$, and $\sqrt{-108}$.

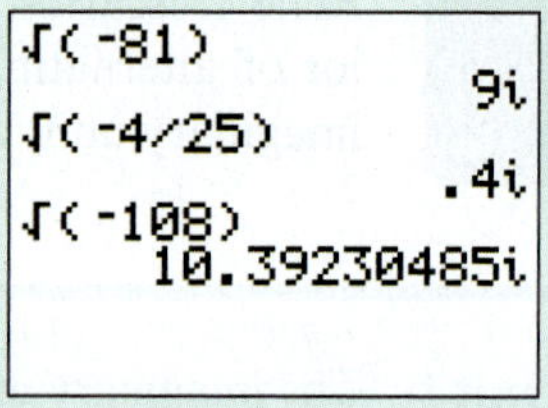

```
√(-81)
                    9i
√(-4/25)
                   .4i
√(-108)
          10.39230485i
```

FIGURE 1-7

2. Perform Operations on Complex Numbers

Numbers that are the sum or difference of a real number and an imaginary number, such as $3 + 4i$, $-5 + 7i$, and $-1 - 9i$, are called *complex numbers*.

Complex Numbers

A **complex number** is a number that can be written in the form $a + bi$, where *a* and *b* are real numbers and $i = \sqrt{-1}$.

The number *a* is called the **real part**, and *b* is called the **imaginary part**.

If $b = 0$, the complex number $a + bi$ is the real number a. If $a = 0$ and $b \neq 0$, the complex number $a + bi$ is the imaginary number bi. It follows that the set of real numbers and the set of imaginary numbers are subsets of the set of complex numbers. Figure 1-8 illustrates how the various sets of numbers are related.

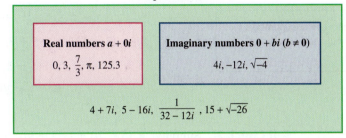

FIGURE 1-8

To determine whether two complex numbers are equal, we can use the following definition.

Equality of Complex Numbers

Two complex numbers are **equal** if their real parts are equal and their imaginary parts are equal. If $a + bi$ and $c + di$ are two complex numbers, then

$$a + bi = c + di \quad \text{if and only if} \quad a = c \quad \text{and} \quad b = d$$

EXAMPLE 1   **Using the Definition of Equality of Complex Numbers**

For what numbers x and y is $3x + 4i = (2y + x) + xi$?

SOLUTION   Since the numbers are equal, their imaginary parts must be equal: $x = 4$. Since their real parts are equal, $3x = 2y + x$. We can solve the system

$$\begin{cases} x = 4 \\ 3x = 2y + x \end{cases}$$

by substituting 4 for x in the second equation and solving for y. We find that $y = 4$. The solution is $x = 4$ and $y = 4$.

Self Check 1   Find x: $a + (x + 3)i = a - (2x - 1)i$.

Now Try Exercise 21.

Complex numbers can be added and subtracted as if they were binomials.

Addition and Subtraction of Complex Numbers

Two complex numbers such as $a + bi$ and $c + di$ are **added** and **subtracted** as if they were binomials:

$$(a + bi) + (c + di) = (a + c) + (b + d)i$$
$$(a + bi) - (c + di) = (a - c) + (b - d)i$$

Because of the preceding definition, the sum or difference of two complex numbers is another complex number.

EXAMPLE 2

Adding and Subtracting Complex Numbers

Simplify: **a.** $(3 + 4i) + (2 + 7i)$ **b.** $(-5 + 8i) - (2 - 12i)$

SOLUTION

a. $(3 + 4i) + (2 + 7i) = 3 + 4i + 2 + 7i$

$$= 3 + 2 + 4i + 7i$$

$$= 5 + 11i$$

> **Tip**
> Add or subtract complex numbers just as you would add or subtract polynomials.

b. $(-5 + 8i) - (2 - 12i) = -5 + 8i - 2 + 12i$

$$= -5 - 2 + 8i + 12i$$

$$= -7 + 20i$$

Self Check 2   Simplify: **a.** $(5 - 2i) + (-3 + 9i)$ **b.** $(2 + 5i) - (6 + 7i)$

Now Try Exercise 25.

Complex numbers can also be multiplied as if they were binomials.

Multiplication of Complex Numbers

Two complex numbers $a + bi$ and $c + di$ are **multiplied** as if they were binomials, with $i^2 = -1$:

$$(a + bi)(c + di) = (ac - bd) + (ad + bc)i$$

Because of this definition, the product of two complex numbers is another complex number.

EXAMPLE 3

Multiplying Complex Numbers

Multiply: **a.** $(3 + 4i)(2 + 7i)$ **b.** $(5 - 7i)(1 + 3i)$

SOLUTION

a. $(3 + 4i)(2 + 7i) = 6 + 21i + 8i + 28i^2$

$$= 6 + 21i + 8i + 28(-1) \qquad i^2 = -1$$

$$= 6 - 28 + 29i$$

$$= -22 + 29i$$

b. $(5 - 7i)(1 + 3i) = 5 + 15i - 7i - 21i^2$

$$= 5 + 15i - 7i - 21(-1) \qquad i^2 = -1$$

$$= 5 + 21 + 8i$$

$$= 26 + 8i$$

Self Check 3   Multiply: $(2 - 5i)(3 + 2i)$.

Now Try Exercise 39.

To avoid errors in determining the sign of the result, always express numbers in $a + bi$ form before attempting any algebraic manipulations.

EXAMPLE 4

Multiplying Complex Numbers

Multiply: $\left(-2 + \sqrt{-16}\right)\left(4 - \sqrt{-9}\right)$.

SOLUTION   We change each number to $a + bi$ form:

$$-2 + \sqrt{-16} = -2 + \sqrt{16}\sqrt{-1} = -2 + 4i$$
$$4 - \sqrt{-9} = 4 - \sqrt{9}\sqrt{-1} = 4 - 3i$$

and then find the product.

$$(-2 + 4i)(4 - 3i) = -8 + 6i + 16i - 12i^2$$
$$= -8 + 6i + 16i - 12(-1) \qquad i^2 = -1$$
$$= -8 + 12 + 22i$$
$$= 4 + 22i$$

Self Check 4   Multiply: $\left(3 + \sqrt{-25}\right)\left(2 - \sqrt{-9}\right)$.

Now Try Exercise 43.

Before we discuss the division of complex numbers, we introduce the concept of a complex conjugate.

Complex Conjugates   The complex numbers $a + bi$ and $a - bi$ are called **complex conjugates** of each other.

For example,

- $2 + 5i$ and $2 - 5i$ are complex conjugates.
- $-\frac{1}{2} + 4i$ and $-\frac{1}{2} - 4i$ are complex conjugates.
- $13i$ and $-13i$ are complex conjugates. That is, the complex conjugate of $0 + 13i$ is $0 - 13i$.

What makes this concept important is the fact that the product of two complex conjugates is always a real number. For example,

$$(2 + 5i)(2 - 5i) = 4 - 10i + 10i - 25i^2$$
$$= 4 - 25(-1) \qquad i^2 = -1$$
$$= 4 + 25$$
$$= 29$$

In general, we have

$$(a + bi)(a - bi) = a^2 - abi + abi - b^2 i^2$$
$$= a^2 - b^2(-1) \qquad i^2 = -1$$
$$= a^2 + b^2$$

To divide complex numbers, we use the concept of complex conjugates to rationalize the denominator.

Take Note

When we multiplied the complex conjugates $2 + 5i$ and $2 - 5i$, our result was the real number 29. To arrive at 29 quickly, we can simply square 2 and get 4, square 5 and get 25, then add 4 and 25 to get 29. How cool is that!

©vita khorzhevska/Shutterstock.com

EXAMPLE 5 **Dividing Complex Numbers**

Divide and write the result in $a + bi$ form: $\dfrac{3}{2 + i}$.

SOLUTION To divide, we rationalize the denominator and simplify.

$$\frac{3}{2 + i} = \frac{3(\mathbf{2 - i})}{(2 + i)(\mathbf{2 - i})}$$

To make the denominator a real number, multiply the numerator and denominator by the complex conjugate of $2 + i$, which is $2 - i$.

$$= \frac{6 - 3i}{4 - 2i + 2i - i^2}$$

Multiply.

$$= \frac{6 - 3i}{4 + 1}$$

Simplify the denominator.

$$= \frac{6 - 3i}{5}$$

$$= \frac{6}{5} - \frac{3}{5}i$$

It is common to accept $\frac{6}{5} - \frac{3}{5}i$ as a substitute for $\frac{6}{5} + \left(-\frac{3}{5}\right)i$.

Self Check 5 Divide and write the result in $a + bi$ form: $\dfrac{3}{3 - i}$.

Now Try Exercise 51.

EXAMPLE 6 **Dividing Complex Numbers**

Divide and write the result in $a + bi$ form: $\dfrac{2 - \sqrt{-16}}{3 + \sqrt{-1}}$.

SOLUTION $\dfrac{2 - \sqrt{-16}}{3 + \sqrt{-1}} = \dfrac{2 - 4i}{3 + i}$

Change each number to $a + bi$ form.

$$= \frac{(2 - 4i)(\mathbf{3 - i})}{(3 + i)(\mathbf{3 - i})}$$

To make the denominator a real number, multiply the numerator and denominator by $3 - i$.

$$= \frac{6 - 2i - 12i + 4i^2}{9 - 3i + 3i - i^2}$$

Remove parentheses.

$$= \frac{2 - 14i}{9 + 1}$$

Combine like terms; $i^2 = -1$.

$$= \frac{2}{10} - \frac{14i}{10}$$

$$= \frac{1}{5} - \frac{7}{5}i$$

Self Check 6 Divide and write the result in $a + bi$ form: $\dfrac{3 + \sqrt{-25}}{2 - \sqrt{-1}}$.

Now Try Exercise 59.

Examples 5 and 6 illustrate that the quotient of two complex numbers is another complex number.

3. Find Powers of i

The powers of i with natural number exponents produce an interesting pattern as shown in the table.

Powers of i

$i^1 = \sqrt{-1} = i$	$i^5 = i^4 i = 1i = i$
$i^2 = \left(\sqrt{-1}\right)^2 = -1$	$i^6 = i^4 i^2 = 1(-1) = -1$
$i^3 = i^2 i = -1i = -i$	$i^7 = i^4 i^3 = 1(-i) = -i$
$i^4 = i^2 i^2 = (-1)(-1) = 1$	$i^8 = i^4 i^4 = 1(1) = 1$

The pattern continues: $i, -1, -i, 1, \ldots$.

Take Note

i^0 is 1 because any number except 0 raised to the 0th power is 1.

Simplifying Powers of i

EXAMPLE 7

Simplify: i^{365}.

SOLUTION   Since $i^4 = 1$, each occurrence of i^4 is a factor of 1. To determine how many factors of i^4 are in i^{365}, we divide 365 by 4. The quotient is 91, and the remainder is 1.

$$i^{365} = (i^4)^{91} \cdot i^1$$

$$= 1^{91} \cdot i^1 \qquad i^4 = 1$$

$$= i \qquad\qquad 1^{91} = 1 \text{ and } 1 \cdot i = i.$$

Self Check 7   Simplify: i^{1999}.

Now Try Exercise 69.

The result of Example 7 illustrates the following theorem.

Powers of i   If n is a natural number that has a remainder of r when divided by 4, then

$$i^n = i^r$$

When n is divisible by 4, the remainder r is 0 and $i^0 = 1$.

We can also simplify powers of i that involve negative integer exponents.

$$i^{-1} = \frac{1}{i} = \frac{1 \cdot i}{i \cdot i} = \frac{i}{-1} = -i \qquad\qquad i^{-2} = \frac{1}{i^2} = \frac{1}{-1} = -1$$

$$i^{-3} = \frac{1}{i^3} = \frac{1 \cdot i}{i^3 \cdot i} = \frac{i}{i^4} = \frac{i}{1} = i \qquad\qquad i^{-4} = \frac{1}{i^4} = \frac{1}{1} = 1$$

Accent on Technology

Performing Operations on Complex Numbers

If we place our graphing calculator in $a + bi$ mode we can use it to perform operations on complex numbers. Addition, subtraction, and multiplication of $7 - 2i$ and $2 + i$ are shown in Figure 1-9(a). Division of $7 - 2i$ and $2 + i$ is shown in

Figure 1-9(b). We can also find powers of i using our graphing calculator. i^{24} and i^{25} are shown in Figure 1-9(c). Since 4E^-13i can be approximated as 0, $i^{24} = 1$. Since -5E^-13 can be approximated as 0, $i^{25} = i$.

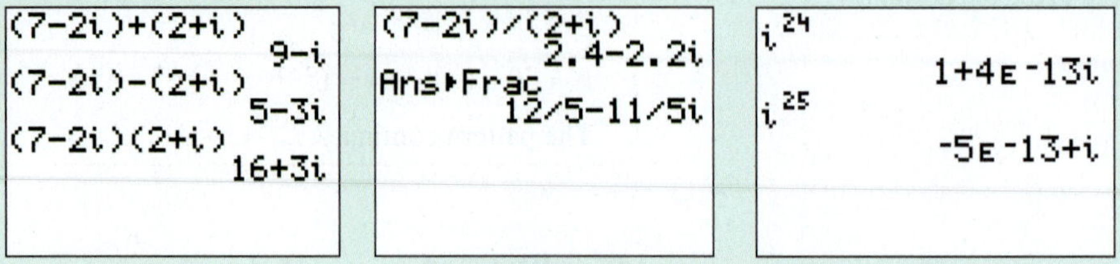

(7-2i)+(2+i) ... 9-i (7-2i)-(2+i) ... 5-3i (7-2i)(2+i) ... 16+3i	(7-2i)/(2+i) ... 2.4-2.2i Ans▶Frac ... 12/5-11/5i	i²⁴ ... 1+4E-13i i²⁵ ... -5E-13+i
(a)	(b)	(c)

FIGURE 1-9

4. Determine the Absolute Value of a Complex Number

Absolute Value of a Complex Number

If $a + bi$ is a complex number, then

$$|a + bi| = \sqrt{a^2 + b^2}$$

Because of the previous definition, the absolute value of a complex number is a real number. For this reason, i does not appear in the result.

EXAMPLE 8 **Determine the Absolute Value of a Complex Number**

Write without absolute value symbols: **a.** $|3 + 4i|$ **b.** $|4 - 6i|$

SOLUTION In each case, we apply the definition of absolute value of a complex number.

a. $|3 + 4i| = \sqrt{3^2 + 4^2}$

$= \sqrt{9 + 16}$

$= \sqrt{25}$

$= 5$

b. $|4 - 6i| = \sqrt{4^2 + (-6)^2}$

$= \sqrt{16 + 36}$

$= \sqrt{52}$

$= \sqrt{4 \cdot 13}$

$= \sqrt{4}\sqrt{13}$

$= 2\sqrt{13}$

Self Check 8 Write without absolute value symbols: $|2 - 5i|$.

Now Try Exercise 81.

EXAMPLE 9 **Determine the Absolute Value of a Complex Number**

Write without absolute value symbols: **a.** $\left| \dfrac{2i}{3 + i} \right|$ **b.** $|a + 0i|$

SOLUTION **a.** We first write $\frac{2i}{3 + i}$ in $a + bi$ form:

$$\frac{2i}{3 + i} = \frac{2i(3 - i)}{(3 + i)(3 - i)} = \frac{6i - 2i^2}{9 - i^2} = \frac{6i + 2}{10} = \frac{1}{5} + \frac{3}{5}i$$

and then find the absolute value of $\frac{1}{5} + \frac{3}{5}i$.

$$\left| \frac{2i}{3 + i} \right| = \left| \frac{1}{5} + \frac{3}{5}i \right| = \sqrt{\left(\frac{1}{5}\right)^2 + \left(\frac{3}{5}\right)^2} = \sqrt{\frac{10}{25}} = \frac{\sqrt{10}}{5}$$

b. $|a + 0i| = \sqrt{a^2 + 0^2} = \sqrt{a^2} = |a|$

From part b, we see that $|a| = \sqrt{a^2}$.

Self Check 9   Write without absolute value symbols: $\left| \dfrac{3i}{2 - i} \right|$.

Now Try Exercise 89.

5. Factor the Sum of Two Squares

We have seen that the sum of two squares cannot be factored over the set of integers. However, it is possible to factor the sum of two squares over the set of complex numbers. For example, to factor $9x^2 + 16y^2$ we proceed as follows:

$$9x^2 + 16y^2 = 9x^2 - (\mathbf{-1})16y^2$$

$$= 9x^2 - \mathbf{i^2}(16y^2) \qquad i^2 = -1$$

$$= 9x^2 - 16y^2i^2$$

$$= (3x + 4yi)(3x - 4yi) \qquad \text{Factor the difference of two squares.}$$

EXAMPLE 10   **Factoring the Sum of Two Squares**

Factor: $100x^2 + 144y^2$.

SOLUTION   We will rewrite $100x^2 + 144y^2$ as the difference of two squares and then factor.

$$100x^2 + 144y^2 = 100x^2 - (\mathbf{-1})144y^2$$

$$= 100x^2 - \mathbf{i^2}144y^2 \qquad i^2 = -1$$

$$= 100x^2 - 144y^2i^2$$

$$= (10x + 12yi)(10x - 12yi) \qquad \text{Factor the difference of two squares.}$$

Self Check 10   Factor $x^2 + 225y^2$.

Now Try Exercise 97.

Self Check Answers

1. $-\dfrac{2}{3}$ **2. a.** $2 + 7i$ **b.** $-4 - 2i$ **3.** $16 - 11i$ **4.** $21 + i$

5. $\dfrac{9}{10} + \dfrac{3}{10}i$ **6.** $\dfrac{1}{5} + \dfrac{13}{5}i$ **7.** $-i$ **8.** $\sqrt{29}$ **9.** $\dfrac{3\sqrt{5}}{5}$

10. $(x + 15yi)(x - 15yi)$

Exercises 1.3

Getting Ready

You should be able to complete these vocabulary and concept statements before you proceed to the practice exercises.

Fill in the blanks.

1. $\sqrt{-3}$, $\sqrt{-9}$, and $\sqrt{-12}$ are examples of _____ numbers.

2. In the complex number $a + bi$, a is the _____ part, and b is the _____ part.

3. If $a = 0$ and $b \neq 0$ in the complex number $a + bi$, the number is an _____ number.

4. If $b = 0$ in the complex number $a + bi$, the number is a _____ number.

5. The complex conjugate of $2 + 5i$ is _____.

6. By definition, $|a + bi| = $ _____.

7. The absolute value of a complex number is a _____ number.

8. The product of two complex conjugates is a _____ number.

Practice

Simplify the imaginary numbers.

9. $\sqrt{-144}$

10. $-\sqrt{-225}$

11. $-\sqrt{-128}$

12. $\sqrt{-108}$

13. $-2\sqrt{-24}$

14. $7\sqrt{-48}$

15. $\sqrt{-\dfrac{50}{9}}$

16. $-\sqrt{-\dfrac{72}{25}}$

17. $-7\sqrt{-\dfrac{3}{8}}$

18. $5\sqrt{-\dfrac{5}{27}}$

Find the values of x and y.

19. $x + (x + y)i = 3 + 8i$

20. $x + 5i = y - yi$

21. $3x - 2yi = 2 + (x + y)i$

22. $\begin{cases} 2 + (x + y)i = 2 - i \\ x + 3i = 2 + 3i \end{cases}$

Perform all operations. Give all answers in a + bi form.

23. $(2 - 7i) + (3 + i)$

24. $(-7 + 2i) + (2 - 8i)$

25. $(5 - 6i) - (7 + 4i)$

26. $(11 + 2i) - (13 - 5i)$

27. $(14i + 2) + (2 - \sqrt{-16})$

28. $(5 + \sqrt{-64}) - (23i - 32)$

29. $(3 + \sqrt{-4}) - (2 + \sqrt{-9})$

30. $(7 - \sqrt{-25}) + (-8 + \sqrt{-1})$

31. $(4 + 7i) + (8 - 2i) - (5 + 4i)$

32. $(5 - 7i) - (4 - 2i) + (8 + i)$

33. $(3 + \sqrt{-16}) - (4 - \sqrt{-36}) + (5 - \sqrt{-144})$

34. $(-1 + \sqrt{-1}) - (2 + \sqrt{-81}) + (8 - \sqrt{-121})$

35. $-5(3 + 5i)$

36. $5(2 - i)$

37. $7i(4 - 8i)$

38. $-2i(3 - 7i)$

39. $(2 + 3i)(3 + 5i)$

40. $(5 - 7i)(2 + i)$

41. $(2 + 3i)^2$

42. $(3 - 4i)^2$

43. $(11 + \sqrt{-25})(2 - \sqrt{-36})$

44. $(6 + \sqrt{-49})(6 - \sqrt{-49})$

45. $(\sqrt{-16} + 3)(2 + \sqrt{-9})$

46. $(12 - \sqrt{-4})(-7 + \sqrt{-25})$

47. $\dfrac{1}{-i}$

48. $\dfrac{3}{i}$

49. $\dfrac{-4}{3i}$

50. $\dfrac{10}{7i}$

51. $\dfrac{1}{2 + i}$

52. $\dfrac{-2}{3 - i}$

53. $\dfrac{2i}{7 + i}$

54. $\dfrac{-3i}{2 + 5i}$

55. $\dfrac{2 + i}{3 - i}$

56. $\dfrac{3 - i}{1 + i}$

57. $\dfrac{4 - 5i}{2 + 3i}$

58. $\dfrac{34 + 2i}{2 - 4i}$

59. $\dfrac{5 - \sqrt{-16}}{-8 + \sqrt{-4}}$

60. $\dfrac{3 - \sqrt{-9}}{2 - \sqrt{-1}}$

61. $\dfrac{2 + i\sqrt{3}}{3 + i}$

62. $\dfrac{3 + i}{4 - i\sqrt{2}}$

Simplify each expression.

63. i^9

64. i^{26}

65. i^{38}

66. i^{99}

67. i^{87}

68. i^{44}

69. i^{100}

70. i^{201}

71. i^{-6}

72. i^0

73. i^{-10}

74. i^{-31}

75. $\dfrac{1}{i^3}$

76. $\dfrac{3}{i^5}$

77. $\dfrac{-4}{i^{10}}$

78. $\dfrac{-10}{i^{24}}$

Write without absolute value symbols.

79. $|3 + 4i|$

80. $|5 + 12i|$

81. $|2 + 3i|$

82. $|5 - i|$

83. $|-7 + \sqrt{-49}|$

84. $|-2 - \sqrt{-16}|$

85. $\left|\dfrac{1}{2} + \dfrac{1}{2}i\right|$

86. $\left|\dfrac{1}{2} - \dfrac{1}{4}i\right|$

87. $|-6i|$

88. $|5i|$

89. $\left|\dfrac{2}{1 + i}\right|$

90. $\left|\dfrac{3}{3 + i}\right|$

91. $\left|\dfrac{-3i}{2 + i}\right|$

92. $\left|\dfrac{5i}{i - 2}\right|$

93. $\left|\dfrac{i + 2}{i - 2}\right|$

94. $\left|\dfrac{2 + i}{2 - i}\right|$

Factor each expression over the set of complex numbers.

95. $x^2 + 4$

96. $16a^2 + 9$

97. $25p^2 + 36q^2$

98. $100r^2 + 49s^2$

99. $2y^2 + 8z^2$

100. $12b^2 + 75c^2$

101. $50m^2 + 2n^2$

102. $64a^4 + 4b^2$

Applications

In electronics, the formula $V = IR$ is called Ohm's Law. It gives the relationship in a circuit between the voltage V (in volts), the current I (in amperes), and the resistance R (in ohms).

103. Electronics Find V when $I = 3 - 2i$ amperes and $R = 3 + 6i$ ohms.

104. Electronics Find R when $I = 2 - 3i$ amperes and $V = 21 + i$ volts.

105. Electronics The impedance Z in an AC (alternating current) circuit is a measure of how much the circuit impedes (hinders) the flow of current through it. The impedance is related to the voltage V and the current I by the following formula.

$$V = IZ$$

If a circuit has a current of $(0.5 + 2.0i)$ amps and an impedance of $(0.4 - 3.0i)$ ohms, find the voltage.

106. Fractals Complex numbers are fundamental in the creation of the intricate geometric shape shown below, called a *fractal*. The process of creating this image is based on the following sequence of steps, which begins by picking any complex number, which we will call z.

1. Square z, and then add that result to z.
2. Square the result from step 1, and then add it to z.
3. Square the result from step 2, and then add it to z.

If we begin with the complex number i, what is the result after performing steps 1, 2, and 3?

Discovery and Writing

107. Show that the addition of two complex numbers is commutative by adding the complex numbers $a + bi$ and $c + di$ in both orders and observing that the sums are equal.

108. Show that the multiplication of two complex numbers is commutative by multiplying the complex numbers $a + bi$ and $c + di$ in both orders and observing that the products are equal.

109. Show that the addition of complex numbers is associative.

110. Explain how to determine whether two complex numbers are equal.

111. Define the complex conjugate of a complex number.

112. Explain how to divide two complex numbers.

Critical Thinking

Determine if the statement is true or false. If the statement is false, then correct it and make it true.

113. $\sqrt{-300} = -10\sqrt{3}$

114. $\sqrt[3]{-125} = 5i$

115. $\dfrac{\pi}{i} = -\pi i$

116. $(2 + 3i)^3 = 8 - 27i$

117. $4444i^{4444} = 4444$

118. $\sqrt{-10}\sqrt{-7} = \sqrt{70}$

119. $(5 - 6i)(5 + 6i)(2 - i)(2 + i)$ is a real number.

120. $81x^2 + 100y^2$ can be factored.

1.4   Quadratic Equations

In this section, we will learn to

1. Solve quadratic equations using factoring and the Square Root Property.
2. Solve quadratic equations using completing the square.
3. Solve quadratic equations using the Quadratic Formula.
4. Determine the easiest strategy to use to solve a quadratic equation.
5. Solve formulas for a variable that is squared.
6. Define and use the discriminant.
7. Write rational equations in quadratic form and solve the equations.

Fenway Park, America's most beloved ballpark, is home to the Boston Red Sox baseball club. The park opened in 1912 and is the oldest major league baseball stadium.

In baseball, the distance between home plate and first base is 90 feet and the distance between first base and second base is 90 feet. To find the distance between home plate and second base, we can use the *Pythagorean Theorem*, which states that

The sum of the squares of the two legs of a right triangle is equal to the square of its hypotenuse.

Because home plate, first base, and second base form a right triangle, we can let x represent the distance between home plate and second base (the **hypotenuse** of the right triangle) and 90 feet represent the length of each leg. We can then apply the Pythagorean Theorem and write the equation $90^2 + 90^2 = x^2$. To find the distance between home plate and second base, we must solve this equation.

$$90^2 + 90^2 = x^2$$

$$8100 + 8100 = x^2 \qquad \text{Square 90 two times.}$$

$$16{,}200 = x^2 \qquad \text{Simplify.}$$

To find x, we must determine what positive number squared gives 16,200. From Chapter 0, we know that this number is the square root of 16,200.

$$\sqrt{16{,}200} \approx 127.3 \qquad \text{Use a calculator and round to the nearest tenth.}$$

To the nearest tenth, the distance between home plate and second base is 127.3 feet.

Since this equation contains the term x^2, it is an example of a new type of equation, called a *quadratic equation*. In this section, we will learn several strategies for solving these equations.

1.   Solve Quadratic Equations Using Factoring and the Square Root Property

Polynomial equations such as $2x^2 + 11x - 21 = 0$ and $3x^2 - x + 2 = 0$ are called *quadratic* or *second-degree* equations.

Quadratic Equation   A **quadratic equation** is an equation that can be written in the form $ax^2 + bx + c = 0$, where a, b, and c are real numbers and $a \neq 0$. The given form is called **standard form**.

To solve quadratic equations by factoring, we can use the following theorem.

Zero-Factor Theorem   If a and b are real numbers, and if $ab = 0$, then

$$a = 0 \qquad \text{or} \qquad b = 0$$

PROOF   Suppose that $ab = 0$. If $a = 0$, we are finished, because at least one of a or b is 0.

If $a \neq 0$, then a has a reciprocal $\frac{1}{a}$, and we can multiply both sides of the equation $ab = 0$ by $\frac{1}{a}$ to obtain

$$ab = 0$$

$$\frac{1}{a}(ab) = \frac{1}{a}(0) \qquad \text{Multiply both sides by } \frac{1}{a}.$$

$$\left(\frac{1}{a} \cdot a\right)b = 0 \qquad \text{Use the Associative Property to group } \frac{1}{a} \text{ and } a \text{ together.}$$

$$1b = 0 \qquad \frac{1}{a} \cdot a = 1$$

$$b = 0$$

Thus, if $a \neq 0$, then b must be 0, and the theorem is proved.

Solving a Quadratic Equation by Factoring

EXAMPLE 1

Solve: $2x^2 - 9x = 35$.

SOLUTION   First we write the quadratic equation in standard form: $2x^2 - 9x - 35 = 0$. The left side can be factored and written as

$$(2x + 5)(x - 7) = 0$$

This product can be 0 if and only if one of the factors is 0. So we can use the Zero-Factor Theorem and set each factor equal to 0. We can then solve each equation for x.

$$2x + 5 = 0 \qquad \text{or} \qquad x - 7 = 0$$
$$2x = -5 \qquad\qquad\qquad x = 7$$
$$x = -\frac{5}{2}$$

Because $(2x + 5)(x - 7) = 0$ only if one of its factors is zero, $-\frac{5}{2}$ and 7 are the only solutions of the equation.

Verify that each one satisfies the equation.

> **Take Note**
>
> The Zero-Factor Theorem can be used only when there is a constant term of 0 on one side of the equation.

Self Check 1   Solve: $6x^2 - 3 = -7x$.

Now Try Exercise 17.

In many quadratic equations, the quadratic expression does not factor over the set of integers. For example, the left side of $x^2 - 5x + 3 = 0$ is a prime polynomial and cannot be factored over the set of integers.

To develop a method to solve these equations, we consider the equation $x^2 = c$. If c is positive, it has two real roots that can be found by adding $-c$ to both sides, factoring $x^2 - c$ over the set of real numbers, setting each factor equal to 0, and solving for x.

$$x^2 = c$$

$$x^2 - c = 0 \qquad \text{Subtract } c \text{ from both sides.}$$

$$x^2 - \left(\sqrt{c}\right)^2 = 0 \qquad \left(\sqrt{c}\right)^2 = c$$

$$\left(x - \sqrt{c}\right)\left(x + \sqrt{c}\right) = 0 \qquad \text{Factor the difference of two squares.}$$

$$x - \sqrt{c} = 0 \quad \text{or} \quad x + \sqrt{c} = 0 \qquad \text{Set each factor equal to 0.}$$

$$x = \sqrt{c} \qquad\qquad x = -\sqrt{c}$$

The roots of $x^2 = c$ are $x = \sqrt{c}$ and $x = -\sqrt{c}$. This fact is summarized in the **Square Root Property**.

Square Root Property   If $x^2 = c$, then

$$x = \sqrt{c} \quad \text{or} \quad x = -\sqrt{c}$$

The roots of the quadratic equation can be written in a more compact form using **double-sign notation**. We often refer to this form as "plus or minus" form.

$$x = \pm\sqrt{c} \qquad \text{This is read informally as "} x \text{ equals plus or minus square root of } c.\text{"}$$

EXAMPLE 2

Solving a Quadratic Equation by Using the Square Root Property

Solve: $x^2 - 8 = 0$.

SOLUTION   We solve for x^2 and use the Square Root Property.

$$x^2 - 8 = 0$$

$$x^2 = 8$$

$$x = \sqrt{8} \quad \text{or} \quad x = -\sqrt{8}$$

$$x = 2\sqrt{2} \qquad\quad x = -2\sqrt{2} \qquad \sqrt{8} = \sqrt{4}\sqrt{2} = 2\sqrt{2}$$

> **Caution**
>
> A common error made is to list only the positive root. There will be both a positive and negative root.

These solutions can be written using double-sign notation as $x = \pm 2\sqrt{2}$. Verify that each root satisfies the equation.

Self Check 2   Solve: $x^2 - 12 = 0$.

Now Try Exercise 21.

EXAMPLE 3

Solving a Quadratic Equation Using the Square Root Property

Solve: $(x + 4)^2 = -27$

SOLUTION   Again we will use the Square Root Property.

$$(x + 4)^2 = -27$$

$$x + 4 = \sqrt{-27} \qquad \text{or} \qquad x + 4 = -\sqrt{-27}$$

$$x + 4 = 3i\sqrt{3} \qquad\qquad x + 4 = -3i\sqrt{3} \qquad \sqrt{-27} = \sqrt{-1}\sqrt{9}\sqrt{3} = 3i\sqrt{3}$$

$$x = -4 + 3i\sqrt{3} \qquad\qquad x = -4 - 3i\sqrt{3}$$

> **Caution**
>
> Students are often tempted to square the binomial on the left side of the equation $(x + 4)^2 = -27$ and not use the Square Root Property. This approach should be avoided because it can produce a more challenging equation, one where additional knowledge is needed to solve it.

We can write the roots using "plus or minus" notation as $x = -4 \pm 3i\sqrt{3}$. Verify that each root satisfies the equation.

Self Check 3   Solve: $(2x + 5)^2 = -45$.

Now Try Exercise 45.

2. Solve Quadratic Equations Using Completing the Square

Another way to solve quadratic equations is called **completing the square**. This method is based on the following products:

$$x^2 + 2ax + a^2 = (x + a)^2 \quad \text{and} \quad x^2 - 2ax + a^2 = (x - a)^2$$

The trinomials $x^2 + 2ax + a^2$ and $x^2 - 2ax + a^2$ are perfect-square trinomials, because each one factors as the square of a binomial. In each case, the coefficient of the first term is 1. If we take one-half of the coefficient of x in the middle term and square it, we obtain the third term.

$$\left[\frac{1}{2}(2a)\right]^2 = a^2 \quad \text{and} \quad \left[\frac{1}{2}(-2a)\right]^2 = (-a)^2 = a^2$$

This suggests that to make $x^2 + bx$ a perfect-square trinomial, we find one-half of b, square it, and add the result to the binomial. For example, to make $x^2 + 10x$ a perfect-square trinomial, we find one-half of 10 to get 5, square 5 to get 25, and add 25 to $x^2 + 10x$.

$$x^2 + \mathbf{10}x + \left[\frac{1}{2}(\mathbf{10})\right]^2 = x^2 + 10x + (5)^2$$

$$= x^2 + 10x + 25 \qquad \text{Note that } x^2 + 10x + 25 = (x + 5)^2.$$

To make $x^2 - 11x$ a perfect-square trinomial, we find one-half of -11 to get $-\frac{11}{2}$, square $-\frac{11}{2}$ to get $\frac{121}{4}$, and add $\frac{121}{4}$ to $x^2 - 11x$.

$$x^2 - \mathbf{11}x + \left[\frac{1}{2}(-\mathbf{11})\right]^2 = x^2 - 11x + \left(-\frac{11}{2}\right)^2$$

$$= x^2 - 11x + \frac{121}{4} \qquad \text{Note that } x^2 - 11x + \frac{121}{4} = \left(x - \frac{11}{2}\right)^2.$$

To solve a quadratic equation in x by completing the square, we follow these steps.

Strategy for Completing the Square

1. If the coefficient of x^2 is not 1, make it 1 by dividing both sides of the equation by the coefficient of x^2.

2. If necessary, add a number to both sides of the equation to get the constant on the right side of the equation.

3. Complete the square on x:

   a. Identify the coefficient of x, take one-half of it, and square the result.

   b. Add the number found in part **a** to both sides of the equation.

4. Factor the perfect-square trinomial and combine like terms on the other side.

5. Solve the resulting quadratic equation by using the Square Root Property.

To use completing the square to solve $x^2 - 10x + 24 = 0$, we note that the coefficient of x^2 is 1. We move on to Step 2 and subtract 24 from both sides to get the constant term on the right side of the equal sign.

$$x^2 - 10x = -24$$

Tip

Be sure to apply the five steps on several exercises so that the process becomes automatic. Work problems where the coefficient of x^2 is 1 and not 1. Work problems where the coefficient of x is even as well as odd.

We can then complete the square by adding $\left[\frac{1}{2}(-10)\right]^2 = 25$ to both sides.

$$x^2 - 10x + \mathbf{25} = -24 + \mathbf{25}$$

$$x^2 - 10x + 25 = 1 \qquad \text{Simplify on the right side.}$$

We then factor the perfect square trinomial on the left side.

$$(x - 5)^2 = 1$$

Finally, we use the Square Root Property to solve this equation.

$$x - 5 = 1 \qquad \text{or} \qquad x - 5 = -1$$
$$x = 6 \qquad\qquad\qquad x = 4$$

EXAMPLE 4

Solving a Quadratic Equation by Completing the Square

Use completing the square to solve $x^2 + 4x - 6 = 0$.

SOLUTION Here the coefficient of x^2 is already 1. We move to Step 2 and add 6 to both sides to isolate the binomial $x^2 + 4x$.

$$x^2 + 4x = 6$$

We then find the number to add to both sides by completing the square. Since one-half of 4 (the coefficient of x) is 2 and $2^2 = 4$, we add 4 to both sides.

$$x^2 + 4x + \mathbf{4} = 6 + \mathbf{4} \qquad\qquad \text{Add 4 to both sides.}$$

$$x^2 + 4x + 4 = 10$$

$$(x + 2)^2 = 10 \qquad\qquad \text{Factor } x^2 + 4x + 4.$$

$$x + 2 = \sqrt{10} \qquad \text{or} \qquad x + 2 = -\sqrt{10} \qquad \text{Use the Square Root Property.}$$

$$x = -2 + \sqrt{10} \qquad\qquad x = -2 - \sqrt{10}$$

The solutions can be written compactly as $x = -2 \pm \sqrt{10}$. Verify that each root satisfies the original equation.

Self Check 4 Solve: $x^2 - 2x - 9 = 0$.

Now Try Exercise 59.

EXAMPLE 5

Solving a Quadratic Equation by Completing the Square

Use completing the square to solve $x(x + 3) = 9$.

SOLUTION We remove parentheses to get

$$x^2 + 3x = 9$$

Since the coefficient of x^2 is 1 and the constant is on the right side, we move to Step 3 and find the number to be added to both sides to complete the square. Since one-half of 3 (the coefficient of x) is $\frac{3}{2}$ and the square of $\frac{3}{2}$ is $\frac{9}{4}$, we add $\frac{9}{4}$ to both sides.

$$x^2 + 3x + \frac{\mathbf{9}}{\mathbf{4}} = 9 + \frac{\mathbf{9}}{\mathbf{4}} \qquad\qquad \text{Add } \frac{9}{4} \text{ to both sides.}$$

$$\left(x + \frac{3}{2}\right)^2 = \frac{45}{4} \qquad\qquad \text{Factor } x^2 + 3x + \frac{9}{4} \text{ and combine terms on right.}$$

$$x + \frac{3}{2} = \frac{\sqrt{45}}{2} \qquad \text{or} \qquad x + \frac{3}{2} = -\frac{\sqrt{45}}{2} \qquad \text{Use the Square Root Property.}$$

$$x + \frac{3}{2} = \frac{3\sqrt{5}}{2} \qquad \text{or} \qquad x + \frac{3}{2} = -\frac{3\sqrt{5}}{2} \qquad \sqrt{45} = \sqrt{9}\sqrt{5} = 3\sqrt{5}$$

$$x = \frac{-3 + 3\sqrt{5}}{2} \qquad\qquad x = \frac{-3 - 3\sqrt{5}}{2} \qquad \text{Subtract } \frac{3}{2} \text{ from both sides.} \\ \text{and combine.}$$

The solutions can be written as $\dfrac{-3 \pm 3\sqrt{5}}{2}$. Verify that each root satisfies the original equation.

Self Check 5   Solve: $x(x + 5) = 1$.

Now Try Exercise 63.

EXAMPLE 6

Solving a Quadratic Equation Using Completing the Square

Use completing the square to solve $2x^2 - 5x + 5 = 0$.

SOLUTION   We begin by dividing both sides of the equation by 2 to make the coefficient of x^2 equal to 1. Then we proceed as follows:

$$2x^2 - 5x + 5 = 0$$

$$x^2 - \frac{5}{2}x + \frac{5}{2} = 0 \qquad\qquad \text{Divide both sides by 2.}$$

$$x^2 - \frac{5}{2}x = -\frac{5}{2} \qquad\qquad \text{Subtract } \tfrac{5}{2} \text{ from both sides.}$$

$$x^2 - \frac{5}{2}x + \frac{25}{16} = -\frac{5}{2} + \frac{25}{16} \qquad \text{Add } \left[\tfrac{1}{2} \cdot \left(-\tfrac{5}{2}\right)\right]^2 \text{ or } \tfrac{25}{16} \text{ to both sides.}$$

$$\left(x - \frac{5}{4}\right)^2 = -\frac{15}{16} \qquad\qquad \text{Factor } x^2 - \tfrac{5}{2}x + \tfrac{25}{16} \text{ and combine} \\ \text{terms on the right side of the equation.}$$

$$x - \frac{5}{4} = \sqrt{-\frac{15}{16}} \qquad \text{or} \qquad x - \frac{5}{4} = -\sqrt{-\frac{15}{16}} \qquad \text{Use the Square Root Property.}$$

$$x - \frac{5}{4} = \frac{\sqrt{15}}{4}i \qquad\qquad x - \frac{5}{4} = -\frac{\sqrt{15}}{4}i \qquad \sqrt{-\frac{15}{16}} = \frac{\sqrt{15}}{4}i$$

$$x = \frac{5}{4} + \frac{\sqrt{15}}{4}i \qquad\qquad x = \frac{5}{4} + \frac{\sqrt{15}}{4}i \qquad \text{Add } \tfrac{5}{4} \text{ to both sides.}$$

We can write the two solutions compactly as $x = \dfrac{5}{4} \pm \dfrac{\sqrt{15}}{4}i$. Verify that each root satisfies the original equation.

Self Check 6   Solve: $3x^2 - x + 1 = 0$.

Now Try Exercise 65.

3. Solve Quadratic Equations Using the Quadratic Formula

We can solve the equation $ax^2 + bx + c = 0$ $(a \neq 0)$ by completing the square. The result will be a formula that we can use to solve quadratic equations.

$$ax^2 + bx + c = 0$$

$$\frac{ax^2}{a} + \frac{b}{a}x + \frac{c}{a} = \frac{0}{a} \qquad\qquad \text{Divide both sides by } a.$$

> **Take Note**
>
> The end result of solving $ax^2 + bx + c = 0$ by completing the square is a very important formula that we can use to solve all quadratic equations. It is the most "famous" formula in college algebra. It is the Quadratic Formula.

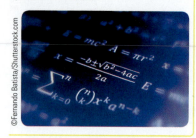

$$x^2 + \frac{b}{a}x = -\frac{c}{a}$$

Simplify and subtract $\frac{c}{a}$ from both sides.

$$x^2 + \frac{b}{a}x + \frac{b^2}{4a^2} = \frac{b^2}{4a^2} - \frac{4ac}{4aa}$$

Add $\frac{b^2}{4a^2}$ to both sides and multiply the numerator and denominator of $\frac{c}{a}$ by $4a$.

$$\left(x + \frac{b}{2a}\right)^2 = \frac{b^2 - 4ac}{4a^2}$$

Factor the left side and add the fractions on the right side.

We can now use the Square Root Property.

$$x + \frac{b}{2a} = \sqrt{\frac{b^2 - 4ac}{4a^2}} \qquad \text{or} \qquad x + \frac{b}{2a} = -\sqrt{\frac{b^2 - 4ac}{4a^2}}$$

$$x = -\frac{b}{2a} + \frac{\sqrt{b^2 - 4ac}}{2a} \qquad\qquad x = -\frac{b}{2a} - \frac{\sqrt{b^2 - 4ac}}{2a}$$

$$x = \frac{-b + \sqrt{b^2 - 4ac}}{2a} \qquad\qquad x = \frac{-b - \sqrt{b^2 - 4ac}}{2a}$$

These values of x are the two roots of the equation $ax^2 + bx + c = 0$. They are usually combined into a single expression, called the **Quadratic Formula**.

Quadratic Formula The solutions of the general quadratic equation, $ax^2 + bx + c = 0$, are

$$x = \frac{-b \pm \sqrt{b^2 - 4ac}}{2a} \qquad (a \neq 0)$$

The Quadratic Formula should be read twice, once using the $+$ sign and once using the $-$ sign. The Quadratic Formula implies that

$$x = \frac{-b + \sqrt{b^2 - 4ac}}{2a} \qquad \text{or} \qquad x = \frac{-b - \sqrt{b^2 - 4ac}}{2a}$$

> **Caution**
>
> Be sure to write the Quadratic Formula correctly. Do **not** write the Quadratic Formula as
>
> $$x = -b \pm \frac{\sqrt{b^2 - 4ac}}{2a}$$

Solving a Quadratic Equation Using the Quadratic Formula

EXAMPLE 7

Use the Quadratic Formula to solve $x^2 - 4x + 5 = 0$.

SOLUTION In this equation, $a = 1$, $b = -4$, and $c = 5$.

> **Tip**
>
> 1. Use parentheses and substitute carefully to avoid mistakes.
> 2. After you simplify the radical, factor the numerator and then divide.

$$x = \frac{-b \pm \sqrt{b^2 - 4ac}}{2a}$$

$$x = \frac{-(-4) \pm \sqrt{(-4)^2 - 4(1)(5)}}{2(1)}$$

Substitute 1 for a, -4 for b, and 5 for c.

$$x = \frac{4 \pm \sqrt{16 - 20}}{2}$$

$$x = \frac{4 \pm \sqrt{-4}}{2}$$

$$x = \frac{4 \pm 2i}{2}$$

$$\sqrt{-4} = \sqrt{4}\sqrt{-1} = 2i$$

$$x = \frac{2(2 \pm i)}{2}$$

Factor $4 + 2i$ as $2(2 + i)$.

$$x = 2 \pm i$$

Simplify.

The roots $x = 2 + i$ and $x = 2 - i$ both satisfy the equation. Note that the roots are complex conjugates.

Self Check 7   Solve: $x^2 + 3x + 4 = 0$.

Now Try Exercise 85.

Solving a Quadratic Equation Using the Quadratic Formula

EXAMPLE 8

Use the Quadratic Formula to solve $2x^2 + 8x - 7 = 0$.

SOLUTION   In this equation, $a = 2$, $b = 8$, and $c = -7$. We will substitute these values into the Quadratic Formula.

$$x = \frac{-b \pm \sqrt{b^2 - 4ac}}{2a}$$

This is the Quadratic Formula.

$$x = \frac{-8 \pm \sqrt{8^2 - 4(2)(-7)}}{2(2)}$$

Substitute 2 for a, 8 for b, and -7 for c.

$$x = \frac{-8 \pm \sqrt{120}}{4}$$

$8^2 - 4(2)(-7) = 64 + 56 = 120$

$$x = \frac{-8 \pm 2\sqrt{30}}{4}$$

$\sqrt{120} = \sqrt{4 \cdot 30} = 2\sqrt{30}$

$$x = \frac{2(-4 \pm \sqrt{30})}{4}$$

Factor out 2 in the numerator.

$$x = \frac{-4 \pm \sqrt{30}}{2}$$

Simplify.

Both values satisfy the original equation.

Self Check 8   Solve: $4x^2 + 16x - 13 = 0$.

Now Try Exercise 81.

4. Determine the Easiest Strategy to Use To Solve a Quadratic Equation

So far, we have solved quadratic equations by *factoring*, by the *square root method*, by *completing the square*, and by the *Quadratic Formula*. With so many methods available, it is useful to think about which one will be the easiest way to solve a specific quadratic equation. Although we have used completing the square to develop the Quadratic Formula, it is usually the most complicated way to solve a quadratic equation. Therefore, unless specified, we will usually not use this method. However, we will complete the square again later in the book to write certain equations in specific forms.

The following chart summarizes the different types of quadratic equations that can occur, a suggested method for solving them, and an example.

Type of Quadratic Equation	Easiest Strategy to Solve It	Example
Equations of the form $ax^2 + bx + c = 0$ where the left side factors easily.	Use factoring and the Zero-Factor Theorem.	Solve: $6x^2 - 11x + 3 = 0$ $(2x - 3)(3x - 1) = 0$ $2x - 3 = 0$ or $3x - 1 = 0$ $x = \dfrac{3}{2}$ $\quad$ $x = \dfrac{1}{3}$
Equations of the form $ax^2 + bx = 0$ where the constant term is missing and the left side factors easily.	Use factoring and the Zero-Factor Theorem.	Solve: $9x^2 + 6x = 0$ $3x(3x + 2) = 0$ $3x = 0$ or $3x + 2 = 0$ $x = 0$ $\quad$ $x = -\dfrac{2}{3}$
Equations of the form $ax^2 - c = 0$ where the term involving x is missing and the left side factors easily.	Use factoring and the Zero-Factor Theorem.	Solve: $4x^2 - 9 = 0$ $(2x + 3)(2x - 3) = 0$ $2x + 3 = 0$ or $2x - 3 = 0$ $x = -\dfrac{3}{2}$ $\quad$ $x = \dfrac{3}{2}$
Equations of the form $ax^2 - c = 0$ or $x^2 = k$ where k is a constant.	Use the Square Root Property.	Solve: $2x^2 - 5 = 0$ $x^2 = \dfrac{5}{2}$ $x = \pm\sqrt{\dfrac{5}{2}}$ $x = \pm\dfrac{\sqrt{10}}{2}$
Equations of the form $x^2 + bx + c = 0$ where b is an even number.	Use completing the square.	Solve: $x^2 - 4x + 6 = 0$ $x^2 - 4x = -6$ $x^2 - 4x + 4 = -6 + 4$ $(x - 2)^2 = -2$ $x - 2 = \pm\sqrt{-2}$ $x = 2 \pm i\sqrt{2}$
Equations of the form $ax^2 + bx + c = 0$ where the left side cannot be factored easily or cannot be factored at all.	Use the Quadratic Formula.	Solve: $3x^2 - x - 5 = 0$. $x = \dfrac{-b \pm \sqrt{b^2 - 4ac}}{2a}$ $x = \dfrac{-(-1) \pm \sqrt{(-1)^2 - 4(3)(-5)}}{2(3)}$ $\quad a = 3$ $\quad b = -1$ $\quad c = -5$ $x = \dfrac{1 \pm \sqrt{1 + 60}}{6}$ $x = \dfrac{1 \pm \sqrt{61}}{6}$

5. Solve Formulas for a Variable That Is Squared

Many formulas involve quadratic equations. For example, if an object is fired from ground level, straight up into the air with an initial velocity of 88 feet per second, its height is given by the formula $h = 88t - 16t^2$, where h represents its height (in feet) and t represents the elapsed time (in seconds) since it was fired.

To solve this formula for t, we use the Quadratic Formula.

$$h = 88t - 16t^2$$

$$16t^2 - 88t + h = 0 \qquad \text{Add } 16t^2 \text{ and } -88t \text{ to both sides.}$$

$$t = \dfrac{-(-88) \pm \sqrt{(-88)^2 - 4(16)(h)}}{2(16)} \qquad \text{Substitute into the Quadratic Formula.}$$

$$t = \frac{88 \pm \sqrt{7744 - 64h}}{32} \qquad \text{Simplify.}$$

6. Define and Use the Discriminant

We can predict the number and type of solutions or roots a quadratic equation will have before we solve it. Suppose that the coefficients a, b, and c in the equation $ax^2 + bx + c = 0$ $(a \neq 0)$ are real numbers. Then the two roots of the equation are given by the Quadratic Formula

$$x = \frac{-b \pm \sqrt{b^2 - 4ac}}{2a} \qquad (a \neq 0)$$

The value of $b^2 - 4ac$, called the **discriminant**, determines the number and nature of the roots. The possibilities are summarized in the table as follows.

Discriminant	Number and Type of Roots
0	One repeated rational number
Positive and a perfect square	Two different rational numbers
Positive and not a perfect square	Two different irrational numbers
Negative	Two different nonreal complex numbers

Take Note

The discriminant is $b^2 - 4ac$ and not $\sqrt{b^2 - 4ac}$.

Using the Discriminant to Determine the Number and Type of Roots of a Quadratic Equation

EXAMPLE 9 Determine the number and type of roots of $3x^2 + 4x + 1 = 0$.

SOLUTION We calculate the discriminant $b^2 - 4ac$.

$$b^2 - 4ac = 4^2 - 4(3)(1) \qquad \text{Substitute 4 for } b, \text{ 3 for } a, \text{ and 1 for } c.$$
$$= 16 - 12$$
$$= 4$$

Since a, b, and c are real numbers and the discriminant is positive and a perfect square, then the roots will be two different rational numbers.

Self Check 9 Determine the number and type of the roots of $4x^2 - 3x - 2 = 0$.

Now Try Exercise 107.

Using the Discriminant to Find the Constant k

EXAMPLE 10 If k is a constant, many quadratic equations are represented by the equation

$$(k - 2)x^2 + (k + 1)x + 4 = 0$$

Find the values of k that will give an equation with roots that are equal or repeated rational numbers.

SOLUTION We calculate the discriminant $b^2 - 4ac$ and set it equal to 0.

$$b^2 - 4ac = (k + 1)^2 - 4(k - 2)(4)$$
$$0 = k^2 + 2k + 1 - 16k + 32$$
$$0 = k^2 - 14k + 33$$
$$0 = (k - 3)(k - 11)$$

$$k - 3 = 0 \quad \text{or} \quad k - 11 = 0$$
$$k = 3 \qquad\qquad k = 11$$

When $k = 3$ or $k = 11$, the equation will have equal roots. As a check, we let $k = 3$ and note that the equation $(k - 2)x^2 + (k + 1)x + 4 = 0$ becomes

$$(\mathbf{3} - 2)x^2 + (\mathbf{3} + 1)x + 4 = 0$$
$$x^2 + 4x + 4 = 0$$

The roots of this equation are equal rational numbers, as expected:

$$x^2 + 4x + 4 = 0$$
$$(x + 2)(x + 2) = 0$$
$$x + 2 = 0 \quad \text{or} \quad x + 2 = 0$$
$$x = -2 \qquad\qquad x = -2$$

Similarly, $k = 11$ will give an equation with equal rational roots.

Self Check 10   Find k such that $(k - 2)x^2 - (k + 3)x + 9 = 0$ will have equal roots.

Now Try Exercise 111.

7. Write Rational Equations in Quadratic Form and Solve the Equations

If an equation can be written in quadratic form, it can be solved with the techniques used for solving quadratic equations.

EXAMPLE 11   **Solving a Rational Equation**

Solve: $\dfrac{1}{x - 1} + \dfrac{3}{x + 1} = 2$.

SOLUTION   Since neither denominator can be zero, $x \neq 1$ and $x \neq -1$. If either number appears as a root, it must be discarded.

$$\frac{1}{x - 1} + \frac{3}{x + 1} = 2$$

$$(x - 1)(x + 1)\left[\frac{1}{x - 1} + \frac{3}{x + 1}\right] = (x - 1)(x + 1)2 \qquad \text{Multiply both sides by } (x - 1)(x + 1).$$

$$(x + 1) + 3(x - 1) = 2(x^2 - 1) \qquad \text{Remove brackets and simplify.}$$

$$4x - 2 = 2x^2 - 2 \qquad \text{Remove parentheses and simplify.}$$

$$0 = 2x^2 - 4x \qquad \text{Add } 2 - 4x \text{ to both sides.}$$

The resulting equation is a quadratic equation that we can solve by factoring.

$$2x^2 - 4x = 0$$

$$2x(x - 2) = 0 \qquad \text{Factor } 2x^2 - 4x.$$

$$2x = 0 \quad \text{or} \quad x - 2 = 0$$
$$x = 0 \qquad\qquad x = 2$$

Verify these results by checking each root in the original equation.

Self Check 11   Solve: $\dfrac{1}{x-1} + \dfrac{2}{x+1} = 1$.

Now Try Exercise 121.

Self Check Answers

1. $\dfrac{1}{3}, -\dfrac{3}{2}$

2. $\pm 2\sqrt{3}$

3. $-\dfrac{5}{2} \pm \dfrac{3\sqrt{5}}{2}i$

4. $1 \pm \sqrt{10}$

5. $\dfrac{-5 \pm \sqrt{29}}{2}$

6. $\dfrac{1}{6} \pm \dfrac{\sqrt{11}}{6}i$

7. $-\dfrac{3}{2} \pm \dfrac{\sqrt{7}}{2}i$

8. $\dfrac{-4 \pm \sqrt{29}}{2}$

9. two different irrational numbers

10. $3, 27$

11. $0, 3$

Exercises 1.4

Getting Ready

You should be able to complete these vocabulary and concept statements before you proceed to the practice exercises.

Fill in the blanks.

1. A quadratic equation is an equation that can be written in the form _____, where $a \neq 0$.

2. If a and b are real numbers and _____, then $a = 0$ or $b = 0$.

3. The equation $x^2 = c$ has two roots. They are $x =$ _____ and $x =$ _____.

4. The Quadratic Formula is _____ $(a \neq 0)$.

5. If a, b, and c are real numbers and if $b^2 - 4ac = 0$, the two roots of the quadratic equation are repeated _____.

6. If a, b, and c are real numbers and $b^2 - 4ac < 0$, the two roots of the quadratic equation are

_____.

Practice

Solve each equation by factoring.

7. $x^2 - x - 6 = 0$

8. $x^2 + 8x + 15 = 0$

9. $x^2 - 144 = 0$

10. $x^2 + 4x = 0$

11. $2x^2 + x - 10 = 0$

12. $3x^2 + 4x - 4 = 0$

13. $5x^2 - 13x + 6 = 0$

14. $2x^2 + 5x - 12 = 0$

15. $15x^2 + 16x = 15$

16. $6x^2 - 25x = -25$

17. $12x^2 + 9 = 24x$

18. $24x^2 + 6 = 24x$

Use the Square Root Property to solve each equation.

19. $x^2 = 9$

20. $x^2 = 64$

21. $x^2 = -169$

22. $x^2 = -81$

23. $y^2 - 50 = 0$

24. $x^2 - 75 = 0$

25. $y^2 + 54 = 0$

26. $x^2 + 125 = 0$

27. $2x^2 = 40$

28. $5x^2 = 400$

29. $2x^2 = -90$

30. $5x^2 = -200$

31. $4x^2 = 7$

32. $16x^2 = 11$

33. $9x^2 = -7$

34. $25x^2 = -11$

35. $2x^2 - 13 = 0$

36. $-3x^2 = -11$

37. $2x^2 + 15 = 0$

38. $-5x^2 = 11$

39. $(x + 1)^2 - 8 = 0$

40. $(y + 2)^2 - 98 = 0$

41. $(x + 1)^2 + 12 = 0$

42. $(y + 2)^2 + 120 = 0$

43. $(2x + 1)^2 = 27$

44. $(5y + 2)^2 - 48 = 0$

45. $(5x + 1)^2 = -8$

46. $(7y + 2)^2 + 48 = 0$

Complete the square to make each a perfect-square trinomial.

47. $x^2 + 6x$

48. $x^2 + 8x$

49. $x^2 - 4x$

50. $x^2 - 12x$

51. $a^2 + 5a$

52. $t^2 + 9t$

53. $r^2 - 11r$

54. $s^2 - 7s$

55. $y^2 + \dfrac{3}{4}y$

56. $p^2 + \dfrac{3}{2}p$

57. $q^2 - \dfrac{1}{5}q$

58. $m^2 - \dfrac{2}{3}m$

Solve each equation by completing the square.

59. $x^2 + 12x = -8$

60. $x^2 - 6x = -1$

61. $x^2 - 10x + 37 = 0$

62. $a^2 + 16a + 82 = 0$

63. $x^2 + 5 = -5x$

64. $x^2 + 1 = -4x$

65. $y^2 + 11y = -49$

66. $x^2 - 5x = -22$

67. $2x^2 - 20x = -49$

68. $4x^2 + 8x = 7$

69. $3x^2 = 1 - 4x$

70. $3x^2 + 4x = 5$

71. $2x^2 = 3x + 1$

72. $2x^2 + 5x = 14$

Use the Quadratic Formula to solve each equation.

73. $9x^2 = 18x - 14$

74. $7z^2 = -14z - 13$

75. $2x^2 = 14x - 30i$

76. $5x^2 + x = -5$

77. $3x^2 = -5x - 1$

78. $2x^2 = 5x + 11$

79. $x^2 + 1 = -7x$

80. $13x^2 + 1 = -10x$

81. $3x^2 + 6x = -1$

82. $2x(x + 3) = -1$

83. $7x^2 = 2x + 2$

84. $5x\left(x + \dfrac{1}{5}\right) = 3$

85. $x^2 + 2x + 2 = 0$

86. $a^2 + 4a + 8 = 0$

87. $y^2 + 4y + 5 = 0$

88. $x^2 + 2x + 5 = 0$

89. $x^2 - 2x = -5$

90. $z^2 - 3z = -8$

91. $x^2 - \dfrac{2}{3}x = -\dfrac{2}{9}$

92. $x^2 + \dfrac{5}{4} = x$

Solve each formula for the indicated variable.

93. $h = \dfrac{1}{2}gt^2$; t

94. $x^2 + y^2 = r^2$; x

95. $h = 64t - 16t^2$; t

96. $y = 16x^2 - 4$; x

97. $\dfrac{x^2}{a^2} + \dfrac{y^2}{b^2} = 1$; y

98. $\dfrac{x^2}{a^2} - \dfrac{y^2}{b^2} = 1$; x

99. $\dfrac{x^2}{a^2} - \dfrac{y^2}{b^2} = 1$; a

100. $\dfrac{x^2}{a^2} - \dfrac{y^2}{b^2} = 1$; b

101. $x^2 + xy - y^2 = 0$; x

102. $x^2 - 3xy + y^2 = 0$; y

Use the discriminant to determine the number and type of roots. Do not solve the equation.

103. $x^2 + 6x + 9 = 0$

104. $-3x^2 + 2x = 21$

105. $3x^2 - 2x + 5 = 0$

106. $9x^2 + 42x + 49 = 0$

107. $10x^2 + 29x = 21$

108. $10x^2 + x = 21$

109. $x^2 - 5x + 2 = 0$

110. $-8x^2 - 2x = 13$

111. Find two values of k such that $x^2 + kx + 3k - 5 = 0$ will have two roots that are equal.

112. For what value(s) of b will the solutions of $x^2 - 2bx + b^2 = 0$ be equal?

Change each rational equation to quadratic form and solve it by the most efficient method.

113. $x + 1 = \dfrac{12}{x}$

114. $x - 2 = \dfrac{15}{x}$

115. $8x - \dfrac{3}{x} = 10$

116. $15x - \dfrac{4}{x} = 4$

117. $\dfrac{5}{x} = \dfrac{4}{x^2} - 6$

118. $\dfrac{6}{x^2} + \dfrac{1}{x} = 12$

119. $x\left(30 - \dfrac{13}{x}\right) = \dfrac{10}{x}$

120. $x\left(20 - \dfrac{17}{x}\right) = \dfrac{10}{x}$

121. $\dfrac{1}{x} + \dfrac{3}{x + 2} = 2$

122. $\dfrac{1}{x - 1} + \dfrac{1}{x - 4} = \dfrac{5}{4}$

123. $\dfrac{1}{x + 1} + \dfrac{5}{2x - 4} = 1$

124. $\dfrac{x(2x + 1)}{x - 2} = \dfrac{10}{x - 2}$

125. $x + 1 + \dfrac{x + 2}{x - 1} = \dfrac{3}{x - 1}$

126. $\dfrac{1}{4 - y} = \dfrac{1}{4} + \dfrac{1}{y + 2}$

127. $\dfrac{4 + a}{2a} = \dfrac{a - 2}{3}$

128. $\dfrac{(a - 2)(a + 4)}{10} = \dfrac{a(a - 3)}{5}$

129. $x + \dfrac{36}{x} = 0$

130. $x + \dfrac{5}{x} = 2$

Discovery and Writing

131. Explain why the Zero-Factor Theorem is true.

132. Explain how to complete the square on $x^2 - 17x$.

133. If r_1 and r_2 are the roots of $ax^2 + bx + c = 0$, show that $r_1 + r_2 = -\dfrac{b}{a}$.

134. If r_1 and r_2 are the roots of $ax^2 + bx + c = 0$, show that $r_1 r_2 = \dfrac{c}{a}$.

In Exercises 135 and 136, a stone is thrown straight upward, higher than the top of a tree. The stone is even with the top of the tree at time t_1 on the way up and at time t_2 on the way down. If the height of the tree is h feet, both t_1 and t_2 are solutions of $h = v_0 t - 16t^2$.

135. Show that the tree is $16t_1 t_2$ feet tall.

136. Show that v_0 is $16(t_1 + t_2)$ feet per second.

Critical Thinking

In Exercises 137–140, match each quadratic equation on the left with the easiest strategy to use to solve it on the right.

137. $6x^2 + 76 = 0$ **a.** Factoring

138. $6x^2 + 35x - 6 = 0$ **b.** Square Root Property

139. $x^2 - 6x = 6$ **c.** Completing the Square

140. $6x^2 = 6x + 1$ **d.** Quadratic Formula

Determine if the statement is true or false. If the statement is false, then correct it and make it true.

141. $1492x^2 + 1984x - 1776 = 0$ has real number solutions.

142. $2004x^2 + 10x + 1994 = 0$ has real number solutions.

1.5 Applications of Quadratic Equations

In this section, we will learn to

1. Solve geometric problems.
2. Solve uniform motion problems.
3. Solve falling body problems.
4. Solve business problems.
5. Solve shared-work problems.

Bennie Thornton/Alamy

The Grand Canyon Skywalk is a tourist attraction located along the Colorado River in the state of Arizona. The glass walkway is shaped like a horseshoe and is 4000 feet above the floor of the canyon. The Grand Canyon, known for its overwhelming size and beautiful landscape, is awe-inspiring and one of our nation's most astounding natural wonders.

If a Clif energy bar is accidentally dropped over the side of the skywalk, how long will it take it to hit the canyon floor?

If t represents the time in seconds, the quadratic equation $-16t^2 + 4000 = 0$ models the time it takes the energy bar to fall to the canyon floor. We can solve this equation by using the Square Root Property.

$$-16t^2 + 4000 = 0$$

$-16t^2 = -4000$	Subtract 4000 from both sides.
$t^2 = 250$	Divide both sides by -16.
$t = \pm\sqrt{250}$	Use the Square Root Property.
$t \approx \pm 15.8$	Round to the nearest tenth.

Because time cannot be negative, we disregard the negative answer. The time it will take the energy bar to reach the canyon floor is about 15.8 seconds.

As this example illustrates, the solutions of many applied problems involve quadratic equations.

> **Tip**
>
> Here are suggestions on how to best approach application problems in this section.
> 1. Understand the problem.
> 2. Write a quadratic equation that can be used to solve the problem.
> 3. Solve the quadratic equation.
> 4. Check your solution.

1. Solve Geometric Problems

Solving an Area Problem

EXAMPLE 1

The length of a rectangle exceeds its width by 3 feet. If its area is 40 square feet, find its dimensions.

SOLUTION

To find an equation that models the problem, we can let w represent the width of the rectangle. Then, $w + 3$ will represent its length (see Figure 1-10). Since the formula for the area of a rectangle is $A = lw$ (area = length × width), the area of the rectangle is $(w + 3)w$, which is equal to 40.

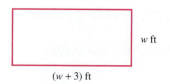

w ft

$(w + 3)$ ft

FIGURE 1-10

The length of the rectangle	times	the width of the rectangle	equals	the area of the rectangle.
$(w + 3)$	$\cdot$	w	$=$	40

We can solve this quadratic equation for w.

$(w + 3)w = 40$	
$w^2 + 3w = 40$	
$w^2 + 3w - 40 = 0$	Subtract 40 from both sides.
$(w - 5)(w + 8) = 0$	Factor.
$w - 5 = 0 \quad$ or $\quad w + 8 = 0$	
$w = 5 \qquad\qquad w = -8$	

> **Take Note**
>
> -8 is discarded because dimensions cannot be negative.

When $w = 5$, the length is $w + 3 = 8$. The solution -8 must be discarded, because a rectangle cannot have a negative width.

We can verify that this solution is correct by observing that a rectangle with dimensions of 5 feet by 8 feet has an area of 40 square feet.

Self Check 1   The length of a rectangle exceeds its width by 10 feet. If its area is 375 square feet, find its dimensions.

Now Try Exercise 5.

Solving a Right Triangle Problem

EXAMPLE 2

On a college campus, a sidewalk 85 meters long (represented by the red lines in Figure 1-11) joins a dormitory building D with the student center C. However, the students prefer to walk directly from D to C. If segment DC is 65 meters long, how long is each piece of the existing sidewalk?

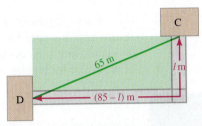

FIGURE 1-11

SOLUTION   We note that the triangle shown in the figure is a right triangle, with a **hypotenuse** that is 65 meters long. If we let the shorter leg of the triangle be l meters long, the length of the longer leg will be $(85 - l)$ meters. By the **Pythagorean Theorem**, we know that the sum of the squares of the two legs of a right triangle is equal to the square of the hypotenuse. Thus, we can form the equation

$$l^2 + (85 - l)^2 = 65^2 \qquad \text{In a right triangle, } a^2 + b^2 = c^2.$$

which we can solve as follows.

$$l^2 + 7225 - 170l + l^2 = 4225 \qquad \text{Expand } (85 - l)^2.$$

$$2l^2 - 170l + 3000 = 0 \qquad \text{Combine like terms and subtract 4225 from both sides.}$$

$$l^2 - 85l + 1500 = 0 \qquad \text{Divide both sides by 2.}$$

Since the left side is difficult to factor, we will solve this equation using the Quadratic Formula.

$$l = \frac{-b \pm \sqrt{b^2 - 4ac}}{2a}$$

$$l = \frac{-(-85) \pm \sqrt{(-85)^2 - 4(1)(1500)}}{2(1)}$$

$$l = \frac{85 \pm \sqrt{1225}}{2}$$

$$l = \frac{85 \pm 35}{2}$$

$$l = \frac{85 + 35}{2} \quad \text{or} \quad l = \frac{85 - 35}{2}$$

$$= 60 \qquad\qquad\qquad = 25$$

The length of the shorter leg is 25 meters. The length of the longer leg is $(85 - 25)$ meters, or 60 meters.

Self Check 2 The length of a video screen is 21 feet shorter than its height. If the diagonal of the screen is 39 feet, find the height of the screen.

Now Try Exercise 15.

2. Solve Uniform Motion Problems

Solving a Uniform Motion Problem

EXAMPLE 3

A man drives 600 miles to attend a technology conference in Tampa, Florida. On the return trip, he is able to increase his speed by 10 mph and save 2 hours of driving time. How fast did he drive in each direction?

SOLUTION We can let s represent the car's speed (in mph) driving to the conference. On the return trip, his speed was $s + 10$ mph. Recall that the distance traveled by an object moving at a constant rate for a certain time is given by the formula $d = rt$. If we divide both sides of this formula by r, we will have a formula for time.

$$t = \frac{d}{r}$$

We can organize the information given in this problem as shown in the following table.

©Jon Bilous/Shutterstock.com

	d	=	r	$\cdot$	t
Outbound trip	600		s		$\dfrac{600}{s}$
Return trip	600		$s + 10$		$\dfrac{600}{s + 10}$

Although neither the outbound nor the return travel time is given, we know the difference of those times.

The longer time of the outbound trip	minus	the shorter time of the return trip	equals	the difference in travel times.
$\dfrac{600}{s}$	$-$	$\dfrac{600}{s + 10}$	$=$	2

We can solve this equation for s.

$$\frac{600}{s} - \frac{600}{s + 10} = 2$$

$$s(s + 10)\left(\frac{600}{s} - \frac{600}{s + 10}\right) = s(s + 10)(2) \qquad \text{Multiply both sides by } s(s + 10) \text{ to clear the equation of fractions.}$$

$$600(s + 10) - 600s = 2s(s + 10) \qquad \text{Simplify.}$$

$$600s + 6000 - 600s = 2s^2 + 20s \qquad \text{Remove parentheses.}$$

$$6000 = 2s^2 + 20s \qquad \text{Combine like terms.}$$

$$0 = 2s^2 + 20s - 6000 \qquad \text{Subtract 6000 from both sides.}$$

$$0 = s^2 + 10s - 3000 \qquad \text{Divide both sides by 2.}$$

$$0 = (s - 50)(s + 60) \qquad \text{Factor.}$$

$$s - 50 = 0 \quad \text{or} \quad s + 60 = 0 \qquad \text{Set each factor equal to 0.}$$

$$s = 50 \qquad \qquad s = -60$$

Take Note

The solution -60 is discarded because time cannot be negative.

The solution $s = -60$ must be discarded. The man drove 50 mph to the conference and $50 + 10$, or 60, mph on the return trip.

These answers are correct, because a 600-mile trip at 50 mph would take $\frac{600}{50}$, or 12 hours. At 60 mph, the same trip would take only 10 hours, which is 2 hours less time.

Self Check 3   Terrence drives his motorcycle 600 miles to Key West, Florida. On the return trip, he is able to increase his speed by 10 mph and save 3 hours of driving time. How fast did he drive in each direction?

Now Try Exercise 19.

3. Solve Falling Body Problems

Solving a Falling Body Problem

EXAMPLE 4   If a water balloon launcher propels a water balloon straight up into the air with an initial velocity of 144 feet per second, its height is given by the formula $h = 144t - 16t^2$, where h represents its height (in feet) and t represents the time (in seconds) since it was launched. How long will it take for the water balloon to return to the point from which it was launched?

SOLUTION   When the water balloon returns to its starting point, its height is again 0. Thus, we can set h equal to 0 and solve for t.

Martin Tanner/iStockphoto.com

$$\boldsymbol{h} = 144t - 16t^2$$

$$\boldsymbol{0} = 144t - 16t^2 \qquad \text{Let } h = 0.$$

$$0 = 16t(9 - t) \qquad \text{Factor.}$$

$$16t = 0 \quad \text{or} \quad 9 - t = 0 \qquad \text{Set each factor equal to 0.}$$

$$t = 0 \qquad \qquad t = 9$$

At $t = 0$, the water balloon's height is 0, because it was just released. When $t = 9$, the height is again 0, and the water balloon has returned to its starting point.

Self Check 4   How long does it take the balloon in Example 4 to reach a height of 324 feet?

Now Try Exercise 23.

4. Solve Business Problems

Solving a Business Problem

EXAMPLE 5   A bus company shuttles 1120 passengers daily between Rockford, Illinois, and O'Hare Airport. The current one-way fare is $10. For each 25¢ increase in the fare, the company predicts that it will lose 48 passengers. What increase in fare will produce daily revenue of $10,208?

SOLUTION   Let q represent the number of quarters the fare will be increased. Then the new fare will be $(10 + 0.25q)$. Since the company will lose 48 passengers for each 25¢ increase, $48q$ passengers will be lost when the rate increases by q quarters. The passenger load will then be $(1120 - 48q)$ passengers.

Since the daily revenue of $10,208 will be the product of the rate and the number of passengers, we have

©Mechanik/Shutterstock.com

$$(10 + 0.25q)(1120 - 48q) = 10{,}208$$

$$11{,}200 - 480q + 280q - 12q^2 = 10{,}208 \qquad \text{Remove parentheses.}$$

$$-12q^2 - 200q + 992 = 0 \qquad \text{Combine like terms and subtract 10,208 from both sides.}$$

$$3q^2 + 50q - 248 = 0 \qquad \text{Divide both sides by } -4.$$

Since the left side is difficult to factor, we will solve this equation with the Quadratic Formula.

$$q = \frac{-b \pm \sqrt{b^2 - 4ac}}{2a}$$

$$q = \frac{-50 \pm \sqrt{50^2 - 4(3)(-248)}}{2(3)} \qquad \text{Substitute 3 for } a \text{, 50 for } b \text{, and } -248 \text{ for } c.$$

$$q = \frac{-50 \pm \sqrt{2500 + 2976}}{6}$$

$$q = \frac{-50 \pm \sqrt{5476}}{6}$$

$$q = \frac{-50 \pm 74}{6}$$

$$q = \frac{-50 + 74}{6} \qquad \text{or} \qquad q = \frac{-50 - 74}{6}$$

$$= \frac{24}{6} \qquad\qquad\qquad = \frac{-124}{6}$$

$$= 4 \qquad\qquad\qquad = -\frac{62}{3}$$

Take Note

$-\frac{62}{3}$ is discarded. The number of riders cannot be negative.

Since the number of riders cannot be negative, the result of $-\frac{62}{3}$ must be discarded. To generate \$10,208 in daily revenues, the company should raise the fare by 4 quarters, or \$1, to \$11.

Self Check 5   A rock band has been drawing average crowds of 400 people. It is projected that for every \$1 increase in the \$10 ticket price, the average attendance will decrease by 20. At what ticket price will nightly receipts be \$4500?

Now Try Exercise 31.

5. Solve Shared-Work Problems

Solving a Shared-Work Problem

EXAMPLE 6

One environmental company can clean up an oil spill in 2 days less time than its competitor. Working together they were able to clean up the spill in 10 days. How long would it have taken the first company to clean up the spill if it worked alone?

SOLUTION   Suppose the first company can clean up the spill in x days. Then the first company can do $\frac{1}{x}$ of the job each day. Because the first company can do the work in 2 days less than its competitor, it will take the competitor $(x + 2)$ days to clean up the spill. The competitor can do $\frac{1}{x + 2}$ of the job each day.

Working together, they can clean up the spill in 10 days. So together they can do $\frac{1}{10}$ of the job each day. The sum of the work each can do in one day is equal to the work that they can do together in one day.

©Nightman1965/Shutterstock.com

The part the first company can clean up in one day	plus	the part the second company can clean up in one day	equals	the part they can clean up together in one day.
$\dfrac{1}{x}$	$+$	$\dfrac{1}{x+2}$	$=$	$\dfrac{1}{10}$

We can solve this equation for x.

$$\frac{1}{x} + \frac{1}{x+2} = \frac{1}{10}$$

$$\mathbf{10x(x+2)}\left(\frac{1}{x} + \frac{1}{x+2}\right) = \mathbf{10x(x+2)}\left(\frac{1}{10}\right) \qquad \text{Multiply both sides by } 10x(x+2) \text{ to eliminate the fractions.}$$

$$\frac{10x(x+2)}{x} + \frac{10x(x+2)}{x+2} = \frac{10x(x+2)}{10} \qquad \text{Distribute the multiplication by } 10x(x+2).$$

$$10(x+2) + 10x = x(x+2) \qquad \frac{x}{x} = 1, \frac{x+2}{x+2} = 1, \text{ and } \frac{10}{10} = 1.$$

$$10x + 20 + 10x = x^2 + 2x \qquad \text{Use the Distributive Property to remove parentheses.}$$

$$0 = x^2 - 18x - 20 \qquad \text{Subtract } 20x \text{ and } 20 \text{ from both sides.}$$

Since the right side cannot be factored over the integers, we will solve the equation with the Quadratic Formula.

$$x = \frac{-b \pm \sqrt{b^2 - 4ac}}{2a}$$

$$x = \frac{-(\mathbf{-18}) \pm \sqrt{(\mathbf{-18})^2 - 4(\mathbf{1})(\mathbf{-20})}}{2(\mathbf{1})} \qquad \text{Substitute 1 for } a, -18 \text{ for } b, \text{ and } -20 \text{ for } c.$$

$$x = \frac{18 \pm \sqrt{324 + 80}}{2}$$

$$x = \frac{18 \pm \sqrt{404}}{2}$$

$$x = \frac{18 \pm 20.09975124}{2}$$

$$x = \frac{18 + 20.09975124}{2} \quad \text{or} \quad x = \frac{18 - 20.09975124}{2}$$

$$\approx 19.05 \qquad\qquad\qquad \approx -1.05$$

Since the work cannot be completed in a negative number of days, we discard the solution of −1.05. Thus, the first company can complete the job working alone in a little over 19 days.

Self Check 6 A hose can fill a swimming pool in 7 hours. Another hose needs 2 more hours to fill the pool than the two hoses combined. How long would it take the second hose to fill the pool?

Now Try Exercise 39.

Self Check Answers

1. 15 feet by 25 feet **2.** 36 feet **3.** 40 mph to Key West and 50 mph returning
4. 4.5 seconds **5.** $15 **6.** ≈4.9 hours

Exercises 1.5

Getting Ready

You should be able to complete these vocabulary and concept statements before you proceed to the practice exercises.

Fill in the blanks.

1. The formula for the area of a rectangle is _____.
2. The formula that relates distance, rate, and time is _____.

Practice

Solve each problem.

3. **Geometric problem** A rectangle is 4 feet longer than it is wide. If its area is 32 square feet, find its dimensions.

4. **Geometric problem** A rectangle is 5 times as long as it is wide. If the area is 125 square feet, find its perimeter.

5. **Dallas Cowboys video screen** The Dallas Cowboys stadium has the world's largest video screen. The rectangular screen's length is 88 feet more than its width. If the video screen has an area of 11,520 square feet, find the dimensions of the screen.

Ken Durden/Shutterstock.com

6. **IMAX screen** A large movie screen is in the Panasonic IMAX theater at Darling Harbor, Sydney, Australia. The rectangular screen has an area of 11,349 square feet. Find the dimensions of the screen if it is 20 feet longer than it is wide.

7. **Geometric problem** The side of a square is 4 centimeters shorter than the side of a second square. If the sum of their areas is 106 square centimeters, find the length of one side of the larger square.

8. **Geometric problem** If two opposite sides of a square are increased by 10 meters and the other sides are decreased by 8 meters, the area of the rectangle that is formed is 63 square meters. Find the area of the original square.

9. **Geometric problem** Find the dimensions of a rectangle whose area is 180 cm2 and whose perimeter is 54 cm.

10. **Flags** In 1912, an order by President Taft fixed the width and length of the U.S. flag in the ratio of 1 to 1.9. If 100 square feet of cloth are to be used to make a U.S. flag, estimate its dimensions to the nearest $\frac{1}{4}$ foot.

11. **Metal fabrication** A piece of tin, 12 inches on a side, is to have four equal squares cut from its corners, as in the illustration. If the edges are then to be folded up to make a box with a floor area of 64 square inches, find the depth of the box.

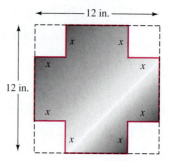

12. **Making gutters** A piece of sheet metal, 18 inches wide, is bent to form the gutter shown in the illustration. If the cross-sectional area is 36 square inches, find the depth of the gutter.

13. **Geometric problem** The base of a triangle is one-third as long as its height. If the area of the triangle is 24 square meters, how long is its base?

14. **Geometric problem** The base of a triangle is one-half as long as its height. If the area of the triangle is 100 square yards, find its height.

15. **Right triangle** If one leg of a right triangle is 14 meters shorter than the other leg, and the hypotenuse is 26 meters, find the length of the two legs.

16. **Right triangle** If one leg of a right triangle is five times the other leg, and the hypotenuse is $10\sqrt{26}$ centimeters, find the length of the two legs.

17. **Manufacturing** A manufacturer of television sets for a news studio received an order for sets with a 46-inch screen (measured along the diagonal). If the televisions are $17\frac{1}{2}$ inches wider than they are high,

find the dimensions of the screen to the nearest tenth of an inch.

18. Finding dimensions An oriental rug is 2 feet longer than it is wide. If the diagonal of the rug is 12 feet, to the nearest tenth of a foot, find its dimensions.

19. Cycling rates A cyclist rides from DeKalb to Rockford, a distance of 40 miles. His return trip takes 2 hours longer, because his speed decreases by 10 mph. How fast does he ride each way?

20. Travel times Jake drives a tractor from one town to another, a distance of 120 kilometers. He drives 10 kilometers per hour faster on the return trip, cutting 1 hour off the time. How fast does he drive each way?

21. Uniform motion problem If the speed were increased by 10 mph, a 420-mile trip would take 1 hour less time. How long will the trip take at the slower speed?

22. Uniform motion problem By increasing her usual speed by 25 kilometers per hour, a bus driver decreases the time on a 25-kilometer trip by 10 minutes. Find the usual speed.

23. Ballistics The height of a projectile fired upward with an initial velocity of 400 feet per second is given by the formula $h = -16t^2 + 400t$, where h is the height in feet and t is the time in seconds. Find the time required for the projectile to return to earth.

24. Ballistics The height of an object tossed upward with an initial velocity of 104 feet per second is given by the formula $h = -16t^2 + 104t$, where h is the height in feet and t is the time in seconds. Find the time required for the object to return to its point of departure.

25. Falling coins An object will fall s feet in t seconds, where $s = 16t^2$. How long will it take for a penny to hit the ground if it is dropped from the top of the Sears Tower in Chicago? (*Hint:* The tower is 1454 feet tall.)

26. Movie stunts According to the *Guinness Book of World Records, 1998,* stuntman Dan Koko fell a distance of 312 feet into an airbag after jumping from the Vegas World Hotel and Casino. The distance d in feet traveled by a free-falling object in t seconds is given by the formula $d = 16t^2$. To the nearest tenth of a second, how long did the fall last?

27. Accidents The height h (in feet) of an object that is dropped from a height of s feet is given by the formula $h = s - 16t^2$, where t is the time the object has been falling. A 5-foot-tall woman on a sidewalk looks directly overhead and sees a window washer drop a bottle from 4 stories up. How long does she have to get out of the way? Round to the nearest tenth. (A story is 12 feet.)

28. Ballistics The height of an object thrown upward with an initial velocity of 32 feet per second is given by the formula $h = -16t^2 + 32t$, where t is the time in seconds. How long will it take the object to reach a height of 16 feet?

29. Setting fares A bus company has 3000 passengers daily, paying a 25¢ fare. For each nickel increase in fare, the company projects that it will lose 80 passengers. What fare increase will produce $994 in daily revenue?

30. Jazz concerts A jazz group on tour has been drawing average crowds of 500 persons. It is projected that for every $1 increase in the $12 ticket price, the average attendance will decrease by 50. At what ticket price will nightly receipts be $5600?

31. Concert receipts Tickets for the annual symphony orchestra pops concert cost $15, and the average attendance at the concerts has been 1200 persons. Management projects that for each 50¢ decrease in ticket price, 40 more patrons will attend. How many people attended the concert if the receipts were $17,280?

32. Projecting demand The *Vilas County News* earns a profit of $20 per year for each of its 3000 subscribers. Management projects that the profit per subscriber would increase by 1¢ for each additional subscriber over the current 3000. How many subscribers are needed to bring a total profit of $120,000?

33. Investment problems Morgan and Chloe each have a bank CD. Morgan's is $1000 larger than Chloe's, but the interest rate is 1% less. Last year Morgan received interest of $280, and Chloe received $240. Find the rate of interest for each CD.

34. Investment problem Scott and Laura have both invested some money. Scott invested $3000 more than Laura and at a 2% higher interest rate. If Scott received $800 annual interest and Laura received $400, how much did Scott invest?

35. Buying microwave ovens Some mathematics professors would like to purchase a $150 microwave oven for the department workroom. If four of the professors don't contribute, everyone's share will increase by $10. How many professors are in the department?

36. Digital cameras A merchant could sell one model of digital cameras at list price for $180. If he had three more cameras, he could sell each one for $10 less and still receive $180. Find the list price of each camera.

37. Filling storage tanks Two pipes are used to fill a water storage tank. The first pipe can fill the tank in 4 hours, and the two pipes together can fill the tank in 2 hours less time than the second pipe alone. How long would it take for the second pipe to fill the tank?

38. Filling swimming pools A hose can fill a swimming pool in 6 hours. Another hose needs 3 more hours to fill the pool than the two hoses combined. How long would it take the second hose to fill the pool?

39. Mowing lawns Kristy can mow a lawn in 1 hour less time than her brother Steven. Together they can finish the job in 5 hours. How long would it take Kristy if she worked alone?

40. Cleaning the garage Working together, Sarah and Heidi can clean the garage in 2 hours. If they work alone, it takes Heidi 3 hours longer than it takes Sarah. How long would it take Heidi to clean the garage alone?

41. Planting windscreens A farmer intends to construct a windscreen by planting trees in a quarter-mile row. His daughter points out that 44 fewer trees will be needed if they are planted 1 foot farther apart. If her dad takes her advice, how many trees will be needed? A row starts and ends with a tree. (*Hint:* 1 mile = 5280 feet.)

42. Angle between spokes If a wagon wheel had 10 more spokes, the angle between spokes would decrease by 6°. How many spokes does the wheel have?

43. Architecture A **golden rectangle** is one of the most visually appealing of all geometric forms. The front of the Parthenon, built in Athens in the 5th century B.C. and shown in the illustration, is a golden rectangle. In a golden rectangle, the length l and the height h of the rectangle must satisfy the following equation. If a rectangular billboard is to have a height of 15 feet, how long should it be if it is to form a golden rectangle? Round to the nearest tenth of a foot.

$$\frac{l}{h} = \frac{h}{l - h}$$

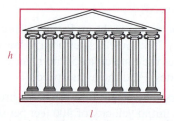

44. Golden ratio Rectangle $ABCD$, shown here, will be a **golden rectangle** if $\frac{AB}{AD} = \frac{BC}{BE}$ where $AE = AD$. Let $AE = 1$ and find the ratio of AB to AD.

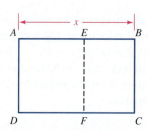

45. Automobile engines As the piston shown moves upward, it pushes a cylinder of a gasoline/air mixture that is ignited by the spark plug. The formula that gives the volume of a cylinder is $V = \pi r^2 h$, where r is the radius and h the height. Find the radius of the piston (to the nearest hundredth of an inch) if it displaces 47.75 cubic inches of gasoline/air mixture as it moves from its lowest to its highest point.

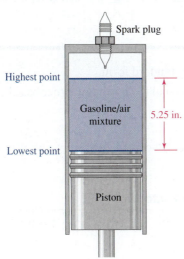

Spark plug

Highest point

Gasoline/air mixture

5.25 in.

Lowest point

Piston

46. History One of the important cities of the ancient world was Babylon. Greek historians wrote that the city was square-shaped. Its area numerically exceeded its perimeter by about 124. Find its dimensions in miles. (Round to the nearest tenth.)

Discovery and Writing

47. Summarize the general strategy used to solve application problems in this section.

48. Describe why it is important to check your solutions to an application problem.

49. Which of the preceding application problems did you find the hardest? Why?

50. Is it possible for a rectangle to have a width that is 3 units shorter than its diagonal and a length that is 4 units longer than its diagonal?

1.6 Other Types of Equations

In this section, we will learn to

1. Solve polynomial equations by factoring.

2. Solve other equations by factoring.

3. Solve equations quadratic in form.

4. Solve radical equations.

5. Solve applications of radical equations.

Pike's Peak is in the Rocky Mountain range, near Colorado Springs, Colorado. Standing at its summit, a person can see for miles.

The distance a person can see from the peak of the mountain is called the **horizon distance**. If this distance, d, is measured in miles and the height of the observer, h, is measured in feet, d and h are related by the formula $d = \sqrt{1.5h}$.

Since the height of Pike's Peak is approximately 14,000 feet, we can substitute 14,000 for h into the formula and simplify.

$$d = \sqrt{1.5h}$$

$$d = \sqrt{1.5(\mathbf{14{,}000})} \qquad \text{Substitute 14,000 for } h.$$

$$= \sqrt{21{,}000}$$

$$\approx 144.9137675$$

From the top of Pike's Peak, a person can see about 145 miles.

Since this equation contains a radical, it is called a *radical equation,* one of the topics of the section.

1. Solve Polynomial Equations by Factoring

The equation $ax^2 + bx + c = 0$ is a polynomial equation of second degree, because its left side contains a second-degree polynomial. Many polynomial equations of higher degree can be solved by factoring. A strategy for solving a polynomial equation is shown here.

Strategy for Solving Polynomial Equations	**1.** Write the polynomial equation in standard form. • Arrange the terms of the polynomial in descending order based on their degrees. • Set the polynomial equal to 0. **2.** Use the Zero-Factor Theorem and solve by factoring.

EXAMPLE 1

Solving Polynomial Equations by Factoring

Solve by factoring: **a.** $6x^3 - x^2 - 2x = 0$ **b.** $x^4 - 5x^2 + 4 = 0$

SOLUTION   We will solve each equation by factoring.

a.
$$6x^3 - x^2 - 2x = 0$$
$$x(6x^2 - x - 2) = 0 \qquad \text{Factor out } x.$$
$$x(3x - 2)(2x + 1) = 0 \qquad \text{Factor } 6x^2 - x - 2.$$

We set each factor equal to 0.

$$x = 0 \quad \text{or} \quad 3x - 2 = 0 \quad \text{or} \quad 2x + 1 = 0$$
$$x = \frac{2}{3} \qquad\qquad x = -\frac{1}{2}$$

Verify that each solution satisfies the original equation.

b.
$$x^4 - 5x^2 + 4 = 0$$
$$(x^2 - 4)(x^2 - 1) = 0 \qquad \text{Factor } x^4 - 5x^2 + 4.$$
$$(x + 2)(x - 2)(x + 1)(x - 1) = 0 \qquad \text{Factor each difference of two squares.}$$

We set each factor equal to 0.

$$x + 2 = 0 \quad \text{or} \quad x - 2 = 0 \quad \text{or} \quad x + 1 = 0 \quad \text{or} \quad x - 1 = 0$$
$$x = -2 \qquad\quad x = 2 \qquad\quad x = -1 \qquad\quad x = 1$$

Verify that each solution satisfies the original equation.

Tip

If factoring is challenging for you, review Section 0.5, where you will find guidelines that will help you with factoring.

Take Note

The polynomial equation of degree 3 had three solutions. The polynomial equation of degree 4 had four solutions.

Self Check 1   Solve by factoring: $2x^3 + 3x^2 - 2x = 0$.

Now Try Exercise 9.

2. Solve Other Equations by Factoring

To solve another type of equation by factoring, we use a property that states that equal powers of equal numbers are equal.

Power Property of Real Numbers	If a and b are numbers, n is an integer, and $a = b$, then $$a^n = b^n$$

When we raise both sides of an equation to the same power, the resulting equation might not be equivalent to the original one. For example, if we raise both sides of

(1) $x = 4$ with a solution set of $\{4\}$

to the second power, we obtain

(2) $x^2 = 16$ with a solution set of $\{4, -4\}$

Equations 1 and 2 have different solution sets, and the solution -4 of Equation 2 does not satisfy Equation 1. Because raising both sides of an equation to the same power often introduces **extraneous solutions** (false solutions that don't satisfy the original equation), we must check all suspected roots to be certain that they satisfy the original equation.

The following equation has an extraneous solution.

$$x - x^{1/2} - 6 = 0$$
$$(x^{1/2} - 3)(x^{1/2} + 2) = 0 \qquad \textcolor{red}{\text{Factor } x - x^{1/2} - 6.}$$
$$x^{1/2} - 3 = 0 \quad \text{or} \quad x^{1/2} + 2 = 0 \qquad \textcolor{red}{\text{Set each factor equal to 0.}}$$
$$x^{1/2} = 3 \qquad\qquad x^{1/2} = -2$$

Because equal powers of equal numbers are equal, we can square both sides of the previous equations to get

$$(x^{1/2})^2 = (3)^2 \quad \text{or} \quad (x^{1/2})^2 = (-2)^2$$
$$x = 9 \qquad\qquad x = 4$$

The number 9 satisfies the equation $x - x^{1/2} - 6 = 0$ but 4 does not, as the following check shows:

If $x = 9$

$$x - x^{1/2} - 6 = 0$$
$$9 - 9^{1/2} - 6 \overset{?}{=} 0$$
$$9 - 3 - 6 \overset{?}{=} 0$$
$$0 = 0$$

If $x = 4$

$$x - x^{1/2} - 6 = 0$$
$$4 - 4^{1/2} - 6 \overset{?}{=} 0$$
$$4 - 2 - 6 \overset{?}{=} 0$$
$$-4 \neq 0$$

The number 9 is the only root.

Tip

1. Raising both sides of an equation to an even power does not always produce an equivalent equation, and proposed solutions must be checked.

2. Raising both sides of an equation to an odd power does produce an equivalent equation, and proposed solutions do not have to be checked.

Regardless of the type of power used, it is still wise to check all proposed solutions.

Take Note

Raising both sides of the equation to the second power produced an extraneous root of $x = 4$, which must be discarded.

Solving Other Types of Equations by Factoring

EXAMPLE 2   Solve by factoring: **a.** $2x^{2/5} - 5x^{1/5} - 3 = 0$ **b.** $3(3x - 2x^{1/2}) = -1$

SOLUTION   We will solve each equation by factoring.

a.
$$2x^{2/5} - 5x^{1/5} - 3 = 0$$
$$(2x^{1/5} + 1)(x^{1/5} - 3) = 0 \qquad \textcolor{red}{\text{Factor } 2x^{2/5} - 5x^{1/5} - 3.}$$
$$2x^{1/5} + 1 = 0 \quad \text{or} \quad x^{1/5} - 3 = 0 \qquad \textcolor{red}{\text{Set each factor equal to 0.}}$$
$$2x^{1/5} = -1 \qquad\qquad x^{1/5} = 3$$
$$x^{1/5} = -\frac{1}{2}$$

Tip

After factoring, to solve the equation with the rational exponent, apply these steps.

1. Isolate the expression with the rational exponent.

2. Raise both sides to the reciprocal of the exponent.

3. Check each proposed solution.

We can raise both sides of each of the previous equations to the fifth power to obtain

$$(x^{1/5})^5 = \left(-\frac{1}{2}\right)^5 \qquad \text{or} \qquad (x^{1/5})^5 = (3)^5$$

$$x = -\frac{1}{32} \qquad\qquad\qquad x = 243$$

Verify that each solution satisfies the original equation.

b. $\qquad 3(3x - 2x^{1/2}) = -1$

$\qquad\qquad 9x - 6x^{1/2} + 1 = 0 \qquad\qquad$ Remove parentheses and add 1 to both sides.

$\qquad (3x^{1/2} - 1)(3x^{1/2} - 1) = 0 \qquad\qquad$ Factor.

$\qquad 3x^{1/2} - 1 = 0 \quad \text{or} \quad 3x^{1/2} - 1 = 0 \qquad$ Set each factor equal to 0.

$\qquad\qquad x^{1/2} = \frac{1}{3} \qquad\qquad x^{1/2} = \frac{1}{3} \qquad$ Solve each equation.

We can square both sides of each of the previous equations to obtain

$$(x^{1/2})^2 = \left(\frac{1}{3}\right)^2 \qquad \text{or} \qquad (x^{1/2})^2 = \left(\frac{1}{3}\right)^2$$

$$x = \frac{1}{9} \qquad\qquad\qquad x = \frac{1}{9}$$

Here, the solutions are the same. Verify that $x = \frac{1}{9}$ satisfies the equation.

Self Check 2 Solve by factoring: $x^{2/5} - x^{1/5} - 2 = 0$.

Now Try Exercise 19.

3. Solve Equations Quadratic in Form

The equations we solved in Example 2 were not quadratic equations in x. However, these two equations are described as being **quadratic in form**. This means we can write the two equations in the form

$$au^2 + bu + c = 0$$

where $a \neq 0$ and u is an algebraic expression involving x. We can do this by making an appropriate substitution.

Equation	Substitution	$au^2 + bu + c = 0$ Form
$2x^{2/5} - 5x^{1/5} - 3 = 0$	$u = x^{1/5}$	$2u^2 - 5u - 3 = 0$
$9x - 6x^{1/2} + 1 = 0$	$u = x^{1/2}$	$9u^2 - 6u + 1 = 0$

Some students find solving equations quadratic in form easier to solve by making a substitution. Let's consider this approach now.

Solving an Equation Quadratic in Form by Making an Appropriate Substitution

EXAMPLE 3

Solve by substitution: $2x^{2/5} - 5x^{1/5} - 3 = 0$.

SOLUTION We will let $u = x^{1/5}$ and then let $u^2 = (x^{1/5})^2 = x^{2/5}$.

Tip

Learning to use u and making a substitution is important. Substitution is used often in calculus.

$$2u^2 - 5u - 3 = 0$$

$$(2u + 1)(u - 3) = 0 \qquad\qquad \text{Factor } 2u^2 - 5u - 3.$$

$$2u + 1 = 0 \qquad \text{or} \qquad u - 3 = 0 \qquad \text{Set each factor equal to 0.}$$

$$u = -\frac{1}{2} \qquad\qquad\qquad u = 3 \qquad \text{Solve for } u.$$

$$x^{1/5} = -\frac{1}{2} \qquad\qquad\qquad x^{1/5} = 3 \qquad \text{Replace } u \text{ with } x^{1/5}.$$

$$\left(x^{1/5}\right)^5 = \left(-\frac{1}{2}\right)^5 \qquad\qquad \left(x^{1/5}\right)^5 = (3)^5 \qquad \text{Raise both sides to the fifth power.}$$

$$x = -\frac{1}{32} \qquad\qquad\qquad x = 243$$

Verify that each solution satisfies the original equation.

Self Check 3   Solve $9x - 6x^{1/2} + 1 = 0$ by making an appropriate substitution.

Now Try Exercises 25.

4. Solve Radical Equations

Radical equations are equations containing radicals with variables in the radicand. To solve such equations, we use the Power Property of Real Numbers. A strategy for solving radical equations with square roots or cube roots is stated here.

Strategy for Solving Radical Equations with Square Roots or Cube Roots

1. Isolate one radical on the left side of the equation.
2. Raise both sides of the equation to the same power (the index of the radical).
   - If the radical is a square root, square both sides.
   - If the radical is a cube root, cube both sides.
3. If a radical remains, repeat Steps 1 and 2.
4. Solve the resulting equation and check your answers.

We can extend the strategy stated above to solve radical equations containing fourth roots, fifth roots, . . . , nth roots.

Solving a Radical Equation with One Square Root

EXAMPLE 4   Solve: $\sqrt{x + 3} - 4 = 7$.

SOLUTION   We will isolate the radical on the left side and then square both sides.

$$\sqrt{x + 3} - 4 = 7$$

$$\sqrt{x + 3} = 11 \qquad \text{Add 4 to both sides to isolate the radical.}$$

$$\left(\sqrt{x + 3}\right)^2 = (11)^2 \qquad \text{Square both sides.}$$

$$x + 3 = 121 \qquad \text{Simplify.}$$

$$x = 118 \qquad \text{Subtract 3 from both sides.}$$

Since squaring both sides might introduce extraneous roots, we must check the result of 118.

$$\sqrt{x + 3} - 4 = 7$$

$$\sqrt{118 + 3} - 4 \stackrel{?}{=} 7 \qquad \text{Substitute 118 for } x.$$

$$\sqrt{121} - 4 \stackrel{?}{=} 7$$

$$11 - 4 \stackrel{?}{=} 7$$

$$7 = 7$$

> **Take Note**
>
> Remember to check all roots when solving radical equations, because raising both sides of an equation to a power can introduce extraneous roots.

Because it checks, 118 is a root of the equation.

Self Check 4   Solve: $\sqrt{x - 3} + 4 = 7$.

Now Try Exercise 31.

Solving a Radical Equation with One Square Root

EXAMPLE 5

Solve: $\sqrt{x + 3} = 3x - 1$.

SOLUTION   We will square both sides of the equation to eliminate the radical and solve the resulting equations by factoring.

$$\sqrt{x + 3} = 3x - 1$$

$$\left(\sqrt{x + 3}\right)^2 = (3x - 1)^2 \qquad \text{Square both sides.}$$

$$x + 3 = 9x^2 - 6x + 1 \qquad \text{Remove parentheses.}$$

$$0 = 9x^2 - 7x - 2 \qquad \text{Add } -x - 3 \text{ to both sides.}$$

$$0 = (9x + 2)(x - 1) \qquad \text{Factor } 9x^2 - 7x - 2.$$

$$9x + 2 = 0 \quad \text{or} \quad x - 1 = 0 \qquad \text{Set each factor equal to 0.}$$

$$x = -\frac{2}{9} \quad \bigg| \quad x = 1$$

Since squaring both sides can introduce extraneous roots, we must check each result.

$$\sqrt{x + 3} = 3x - 1 \qquad \text{or} \qquad \sqrt{x + 3} = 3x - 1$$

$$\sqrt{-\frac{2}{9} + 3} \stackrel{?}{=} 3\left(-\frac{2}{9}\right) - 1 \qquad\qquad \sqrt{1 + 3} \stackrel{?}{=} 3(1) - 1$$

$$\sqrt{\frac{25}{9}} \stackrel{?}{=} -\frac{2}{3} - 1 \qquad\qquad\qquad \sqrt{4} \stackrel{?}{=} 3 - 1$$

$$\frac{5}{3} \neq -\frac{5}{3} \qquad\qquad\qquad\qquad\qquad 2 = 2$$

Since $-\frac{2}{9}$ does not satisfy the equation, it is extraneous. Since 1 checks, it is the only solution.

> **Caution**
>
> If we *forget* to check solution(s) to a radical equation, we risk having an *incorrect* solution set. *Failure* to check both $x = -\frac{2}{9}$ and $x = 1$ in Example 5 would have resulted in an *error*.

Self Check 5   Solve: $\sqrt{x - 2} = 2x - 10$.

Now Try Exercise 39.

Solving a Radical Equation with One Cube Root

EXAMPLE 6   Solve: $\sqrt[3]{x^3 + 56} = x + 2$.

SOLUTION   To eliminate the radical, we cube both sides of the equation.

$$\sqrt[3]{x^3 + 56} = x + 2$$

$$\left(\sqrt[3]{x^3 + 56}\right)^3 = (x + 2)^3 \qquad \text{Cube both sides.}$$

$$x^3 + 56 = x^3 + 6x^2 + 12x + 8 \qquad \text{Remove parentheses.}$$

$$0 = 6x^2 + 12x - 48 \qquad \text{Simplify.}$$

$$0 = x^2 + 2x - 8 \qquad \text{Divide both sides by 6.}$$

$$0 = (x + 4)(x - 2) \qquad \text{Factor } x^2 + 2x - 8.$$

$$x + 4 = 0 \qquad \text{or} \qquad x - 2 = 0 \qquad \text{Set each factor equal to 0.}$$

$$x = -4 \qquad\qquad\qquad x = 2$$

We check each suspected solution to see whether either is extraneous.

For $x = -4$	*For $x = 2$*
$\sqrt[3]{x^3 + 56} = x + 2$	$\sqrt[3]{x^3 + 56} = x + 2$
$\sqrt[3]{(-4)^3 + 56} \overset{?}{=} -4 + 2$	$\sqrt[3]{2^3 + 56} \overset{?}{=} 2 + 2$
$\sqrt[3]{-64 + 56} \overset{?}{=} -2$	$\sqrt[3]{8 + 56} \overset{?}{=} 4$
$\sqrt[3]{-8} \overset{?}{=} -2$	$\sqrt[3]{64} \overset{?}{=} 4$
$-2 = -2$	$4 = 4$

Since both values satisfy the equation, -4 and 2 are roots.

Self Check 6   Solve: $\sqrt[3]{x^3 + 7} = x + 1$.

Now Try Exercise 65.

Solving a Radical Equation with Two Square Roots

EXAMPLE 7   Solve: $\sqrt{2x + 3} + \sqrt{x - 2} = 4$.

SOLUTION   We can write the equation in the form

$$\sqrt{2x + 3} = 4 - \sqrt{x - 2} \qquad \text{Subtract } \sqrt{x - 2} \text{ from both sides.}$$

so that the left side contains one radical. We then square both sides to get

$$\left(\sqrt{2x + 3}\right)^2 = \left(4 - \sqrt{x - 2}\right)^2$$

$$2x + 3 = 16 - 8\sqrt{x - 2} + x - 2$$

$$2x + 3 = 14 - 8\sqrt{x - 2} + x \qquad \text{Combine like terms.}$$

$$x - 11 = -8\sqrt{x - 2} \qquad \text{Subtract 14 and } x \text{ from both sides.}$$

We then square both sides again to eliminate the radical.

$$(x - 11)^2 = \left(-8\sqrt{x - 2}\right)^2$$

$$x^2 - 22x + 121 = 64(x - 2)$$

$$x^2 - 22x + 121 = 64x - 128$$

Tip

Always isolate one radical on a side before squaring. If not, the equation will become more complicated.

Caution

When squaring both sides, make sure you do not square each term individually. It is incorrect to do that.

$$x^2 - 86x + 249 = 0$$

$$(x - 3)(x - 83) = 0$$

$$x - 3 = 0 \quad \text{or} \quad x - 83 = 0 \qquad \text{Set each factor equal to 0.}$$

$$x = 3 \qquad \qquad x = 83$$

Substituting these results into the equation will show that 83 doesn't check; it is extraneous. However, 3 does satisfy the equation and is a root.

Self Check 7 Solve: $\sqrt{2x + 1} + \sqrt{x + 5} = 6$.

Now Try Exercise 53.

5. Solve Applications of Radical Equations

Many applications can be solved using radical equations.

Solving an Application Problem

EXAMPLE 8

A highway curve banked at 8° will accommodate traffic traveling s mph if the radius of the curve is r feet, according to the formula $s = 1.45\sqrt{r}$. Find what radius is necessary to accommodate 70-mph traffic. (See Figure 1-12.)

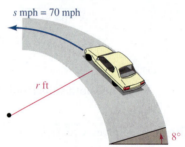

FIGURE 1-12

SOLUTION We can substitute 70 for s in the formula and solve for r.

$$s = 1.45\sqrt{r}$$

$$70 = 1.45\sqrt{r} \qquad \text{Substitute 70 for } s.$$

$$\frac{70}{1.45} = \sqrt{r} \qquad \text{Divide both sides by 1.45.}$$

$$\left(\frac{70}{1.45}\right)^2 = r \qquad \text{Square both sides.}$$

$$2330.558859 = r$$

We can use a calculator to find r. The radius of the curve is approximately 2331 feet.

Self Check 8 Find the radius necessary to accommodate 65-mph traffic.

Now Try Exercise 71.

Self Check Answers

1. $0, \dfrac{1}{2}, -2$ **2.** $-1, 32$ **3.** $\dfrac{1}{9}$ **4.** 12 **5.** 6 **6.** $1, -2$ **7.** 4
8. approximately 2010 feet

Exercises 1.6

Getting Ready

You should be able to complete these vocabulary and concept statements before you proceed to the practice exercises.

Fill in the blanks.

1. Equal powers of equal real numbers are _____.
2. If a and b are real numbers and $a = b$ then $a^2 =$ ___.
3. False solutions that don't satisfy the equation are called _____ solutions.
4. Radical equations contain radicals with variables in their _____.

Practice

Use factoring to solve each equation.

5. $x^3 + 9x^2 + 20x = 0$

6. $x^3 + 4x^2 - 21x = 0$

7. $6a^3 - 5a^2 - 4a = 0$

8. $8b^3 - 10b^2 + 3b = 0$

9. $y^4 - 26y^2 + 25 = 0$

10. $y^4 - 13y^2 + 36 = 0$

11. $2y^4 - 46y^2 = -180$

12. $2x^4 - 102x^2 = -196$

13. $x^4 = 8x^2 + 9$

14. $x^4 - 12x^2 = 64$

15. $4y^4 + 7y^2 - 36 = 0$

16. $9y^4 + 56y^2 - 225 = 0$

Solve each equation by factoring or by making an appropriate substitution.

17. $x^4 - 37x^2 + 36 = 0$

18. $x^4 - 50x^2 + 49 = 0$

19. $2m^{2/3} + 3m^{1/3} - 2 = 0$

20. $6t^{2/5} + 11t^{1/5} + 3 = 0$

21. $x - 13x^{1/2} + 12 = 0$

22. $p + p^{1/2} - 20 = 0$

23. $6p + p^{1/2} = 1$

24. $3r - r^{1/2} = 2$

25. $2t^{1/3} + 3t^{1/6} - 2 = 0$

26. $z^3 - 7z^{3/2} - 8 = 0$

27. $x^{-2} - 10x^{-1} + 16 = 0$

28. $2y^{-2} + 9y^{-1} - 5 = 0$

29. $z^{3/2} - z^{1/2} = 0$

30. $r^{5/2} - r^{3/2} = 0$

Find all real solutions of each equation.

31. $\sqrt{x - 2} - 3 = 2$

32. $\sqrt{a - 3} - 5 = 0$

33. $3\sqrt{x + 1} = \sqrt{6}$

34. $\sqrt{x + 3} = 2\sqrt{x}$

35. $\sqrt{5a - 2} = \sqrt{a + 6}$

36. $\sqrt{16x + 4} = \sqrt{x + 4}$

37. $2\sqrt{x^2 + 3} = \sqrt{-16x - 3}$

38. $\sqrt{x^2 + 1} = \dfrac{\sqrt{-7x + 11}}{\sqrt{6}}$

39. $\sqrt{x^2 + 21} = x + 3$

40. $\sqrt{5 - x^2} = -(x + 1)$

41. $\sqrt{x + 37} = x - 5$

42. $\sqrt{10 + x} - x - 4 = 0$

43. $\sqrt{3z + 1} = z - 1$

44. $\sqrt{y + 2} = 4 - y$

45. $x - \sqrt{7x - 12} = 0$

46. $x - \sqrt{4x - 4} = 0$

47. $x + 4 = \sqrt{\dfrac{6x + 6}{5}} + 3$

48. $\sqrt{\dfrac{8x + 43}{3}} - 1 = x$

49. $\sqrt{\dfrac{x^2 - 1}{x - 2}} = 2\sqrt{2}$

50. $\dfrac{\sqrt{x^2 - 1}}{\sqrt{3x - 5}} = \sqrt{2}$

51. $\sqrt{2p + 1} - 1 = \sqrt{p}$

52. $\sqrt{r} + \sqrt{r + 2} = 2$

53. $\sqrt{x + 3} = \sqrt{2x + 8} - 1$

54. $\sqrt{x + 2} + 1 = \sqrt{2x + 5}$

55. $\sqrt{y + 8} - \sqrt{y - 4} = -2$

56. $\sqrt{z + 5} - 2 = \sqrt{z - 3}$

57. $\sqrt{2b + 3} - \sqrt{b + 1} = \sqrt{b - 2}$

58. $\sqrt{a + 1} + \sqrt{3a} = \sqrt{5a + 1}$

59. $\sqrt{\sqrt{b} + \sqrt{b + 8}} = 2$

60. $\sqrt{\sqrt{x + 19} - \sqrt{x - 2}} = \sqrt{3}$

61. $\sqrt[3]{7x + 1} = 4$

62. $\sqrt[3]{11a - 40} = 5$

63. $\sqrt[3]{x^3 + 7} = x + 1$

64. $\sqrt[3]{x^3 - 7} + 1 = x$

65. $\sqrt[3]{8x^3 + 61} = 2x + 1$

66. $\sqrt[3]{8x^3 - 37} = 2x - 1$

67. $\sqrt[4]{30t + 25} = 5$

68. $\sqrt[4]{3z + 1} = 2$

69. $\sqrt[5]{2x - 11} = \sqrt[5]{14}$

70. $\sqrt[5]{x^2 - 24} = 1$

Applications

71. Height of a bridge The distance d (in feet) that an object will fall in t seconds is given by the following formula. To find the height of a bridge above a river, a man drops a stone into the water. (See the illustration.) If it takes the stone 5 seconds to hit the water, how high is the bridge?

$$t = \sqrt{\frac{d}{16}}$$

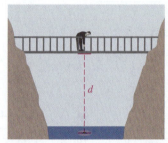

72. Horizon distance The higher a lookout tower, the farther an observer can see. (See the illustration.) The distance d (called the **horizon distance**, measured in miles) is related to the height h of the observer (measured in feet) by the following formula.

$$d = \sqrt{1.5h}$$

How tall must a tower be for the observer to see 30 miles?

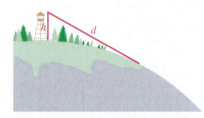

73. Carpentry During construction, carpenters often brace walls, as shown in the illustration. The appropriate length of the brace is given by the following formula.

$$l = \sqrt{f^2 + h^2}$$

If a carpenter nails a 10-foot brace to the wall 6 feet above the floor, how far from the base of the wall should he nail the brace to the floor?

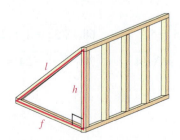

74. Windmills The power generated by a windmill is related to the velocity of the wind by the following formula where P is the power (in watts) and v is the velocity of the wind (in mph).

$$v = \sqrt[3]{\frac{P}{0.02}}$$

To the nearest 10 watts, find the power generated when the velocity of the wind is 31 mph.

©Wdg Photo/Shutterstock.com

75. Diamonds The *effective rate of interest r* earned by an investment is given by the following formula where P is the initial investment that grows to value A after n years.

$$r = \sqrt[n]{\frac{A}{P}} - 1$$

If a diamond buyer got $4000 for a 1.03-carat diamond that he had purchased 4 years earlier and earned an annual rate of return of 6.5% on the investment, what did he originally pay for the diamond?

76. Theater productions The ropes, pulleys, and sandbags shown in the illustration are part of a mechanical system used to raise and lower scenery for a stage play. For the scenery to be in the proper position, the following formula must apply:

$$w_2 = \sqrt{w_1^2 + w_3^2}$$

If $w_2 = 12.5$ lb and $w_3 = 7.5$ lb, find w_1.

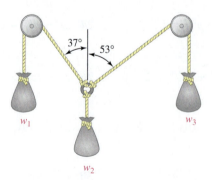

Discovery and Writing

77. Explain the Power Property of Real Numbers.

78. Describe what it means for an equation to be quadratic in form.

79. Identify two methods that can be used to solve the equation $x^4 - 6x^2 - 7 = 0$. Compare and contrast the two methods.

80. Outline a strategy that can be used to solve radical equations.

81. Explain why squaring both sides of an equation might introduce extraneous roots.

82. Can cubing both sides of an equation introduce extraneous roots? Explain.

Critical Thinking

Determine if the statement is true or false. If the statement is false, then correct it and make it true.

83. Factoring can be used to solve $x^4 + 6x^3 + 5 = 0$.

84. The first step used to solve the equation $4x^4 = 2x^2$ is to divide both sides of the equation by $2x^2$.

85. To solve the equation $5y^{\frac{2}{3}} - 4y^{\frac{1}{3}} - 1 = 0$, we can make the substitution $u = y^{\frac{1}{3}}$.

86. The equation $x^{\frac{1}{8}} + 7x^{\frac{1}{4}} + 12 = 0$ is quadratic in form.

87. If $\sqrt{\sqrt{\sqrt{x}}} = 2$ then $x = 16$

88. To solve the radical equation $\sqrt{z} - \sqrt{z+2} = 2$, we square each term individually.

89. To solve the radical equation $\sqrt{x+1} - \sqrt{2x+3} = -1$, the first step is to square both sides.

90. If $\sqrt[999]{x^2} = -1$, then $x = \pm i$.

1.7   Inequalities

In this section, we will learn to

1. Use the properties of inequalities.

2. Solve linear inequalities and applications.

3. Solve compound inequalities.

4. Solve quadratic inequalities.

5. Solve rational inequalities.

Studying abroad is a wonderful opportunity for students. Suppose you read about a program to study Spanish in Costa Rica for several weeks during the summer. The cost of the program includes $900 for round-trip airfare plus $350 per week, which covers tuition, meals per day, and living accommodations. If you have $3000, how many weeks can you afford to spend in Costa Rica?

If x represents the number of weeks you can spend in Costa Rica, we can write an inequality that represents the cost.

Airfare	plus	cost of $350 per week for x weeks	is less than or equal to	$3000.
900	+	350x	≤	3000

To solve this inequality, we can proceed as follows:

$$900 + 350x \leq 3000$$

$$350x \leq 2100 \qquad \text{Subtract 900 from both sides.}$$

$$x \leq 6 \qquad \text{Divide both sides by 350.}$$

There is enough money to spend up to 6 weeks in Costa Rica.

In this section, we will review the inequality symbols, learn inequality properties, and learn strategies that can be applied to solve several types of inequalities. We previously introduced the following symbols.

Symbol	Read as	Examples
$\neq$	"is not equal to"	$8 \neq 10$ and $25 \neq 12$
$<$	"is less than"	$8 < 10$ and $12 < 25$
$>$	"is greater than"	$30 > 10$ and $100 > -5$
$\leq$	"is less than or equal to"	$-6 \leq 12$ and $-8 \leq -8$
$\geq$	"is greater than or equal to"	$12 \geq -5$ and $9 \geq 9$
$\approx$	"is approximately equal to"	$7.49 \approx 7.5$ and $\frac{1}{3} \approx 0.33$

Since the coordinates of points get larger as we move from left to right on the number line,

$a > b$ if point a lies to the right of point b on a number line.

$a < b$ if point a lies to the left of point b on a number line.

Using Inequality Symbols

EXAMPLE 1

a. $5 > -3$, because 5 lies to the right of -3 on the number line.

b. $-7 < -2$, because -7 lies to the left of -2 on the number line.

c. $3 \leq 3$, because $3 = 3$.

d. $2 \leq 3$, because $2 < 3$.

e. $x + 1 > x$, because $x + 1$ lies one unit to the right of x on the number line.

Self Check 1   Write an inequality symbol to make the true statement:

a. $25 \,\square\, 12$ **b.** $-5 \,\square\, -5$ **c.** $-12 \,\square\, -20$

Now Try Exercise 1.

1. Use the Properties of Inequalities

The Trichotomy Property   For any real numbers a and b, one of the following statements is true.

$$a < b, \qquad a = b, \qquad \text{or} \qquad a > b$$

The Trichotomy Property indicates that one of the following statements is true about two real numbers. Either the first is less than the second, or the first is equal to the second, or the first is greater than the second.

The Transitive Property   If a, b, and c are real numbers, then

if $a < b$ and $b < c$, then $a < c$.

if $a > b$ and $b > c$, then $a > c$.

The first part of the Transitive Property indicates that if a first number is less than a second and the second number is less than a third, then the first number is less than the third.

The second part of the Transitive Property is similar, with the words "is greater than" substituted for "is less than."

Addition and Subtraction Properties of Inequality   Let a, b, and c represent real numbers.

If $a < b$, then $a + c < b + c$.

If $a < b$, then $a - c < b - c$.

Similar properties exist for $>$, $\leq$, and $\geq$.

This property states that *any real number can be added to (or subtracted from) both sides of an inequality to obtain another inequality with the same order (direction).* For example, if we add 4 to (or subtract 4 from) both sides of $8 < 12$, we get

$$8 < 12 \qquad\qquad\qquad 8 < 12$$
$$8 + 4 < 12 + 4 \qquad\qquad 8 - 4 < 12 - 4$$
$$12 < 16 \qquad\qquad\qquad 4 < 8$$

and the $<$ symbol is unchanged.

Multiplication and Division Properties of Inequality   Let a, b, and c represent real numbers.

Part 1: If $a < b$ and $c > 0$, then $ca < cb$.

If $a < b$ and $c > 0$, then $\dfrac{a}{c} < \dfrac{b}{c}$.

Part 2: If $a < b$ and $c < 0$, then $ca > cb$.

If $a < b$ and $c < 0$, then $\dfrac{a}{c} > \dfrac{b}{c}$.

Similar properties exist for $>$, $\leq$, and $\geq$.

This property has two parts.

- Part 1 states that *both sides of an inequality can be multiplied (or divided) by the same positive number to obtain another inequality with the same order.*

For example, if we multiply (or divide) both sides of $8 < 12$ by 4, we get

$$8 < 12 \qquad\qquad\qquad 8 < 12$$
$$8(4) < 12(4) \qquad\qquad \dfrac{8}{4} < \dfrac{12}{4}$$
$$32 < 48 \qquad\qquad\qquad 2 < 3$$

and the $<$ symbol is unchanged.

- Part 2 states that *both sides of an inequality can be multiplied (or divided) by the same negative number to obtain another inequality with the opposite order.*
For example, if we multiply (or divide) both sides of $8 < 12$ by -4, we get

<table>
<tr><td>$8 < 12$</td><td>$8 < 12$</td></tr>
<tr><td>$8(-4) > 12(-4)$</td><td>$\dfrac{8}{-4} > \dfrac{12}{-4}$</td></tr>
<tr><td>$-32 > -48$</td><td>$-2 > -3$</td></tr>
</table>

and the $<$ symbol is changed to a $>$ symbol.

> **Take Note**
>
> The properties of inequalities are the same as the properties of equality, unless we are multiplying or dividing by a negative number.

2. Solve Linear Inequalities and Applications

Linear inequalities are inequalities such as $ax + c < 0$ or $ax - c \geq 0$, where $a \neq 0$. Numbers that make an inequality true when substituted for the variable are solutions of the inequality. Inequalities with the same solution set are called **equivalent inequalities**. Because of the previous properties, we can solve inequalities as we do equations. However, we must always remember to change the order of an inequality when multiplying (or dividing) both sides by a negative number.

Solving a Linear Inequality

EXAMPLE 2   Solve: $3(x + 2) < 8$.

SOLUTION   We proceed as with equations.

$$3(x + 2) < 8$$
$$3x + 6 < 8 \qquad \text{Remove parentheses.}$$
$$3x < 2 \qquad \text{Subtract 6 from both sides.}$$
$$x < \frac{2}{3} \qquad \text{Divide both sides by 3.}$$

All numbers that are less than $\frac{2}{3}$ are solutions of the inequality. The solution set can be expressed in interval notation as $\left(-\infty, \frac{2}{3}\right)$ and be graphed as in Figure 1-13.

FIGURE 1-13

Self Check 2   Solve: $5(p - 4) > 25$.

Now Try Exercise 15.

Solving a Linear Inequality

EXAMPLE 3   Solve: $-5(x - 2) \leq 20 + x$.

SOLUTION   We proceed as with equations.

$$-5(x - 2) \leq 20 + x$$
$$-5x + 10 \leq 20 + x \qquad \text{Remove parentheses.}$$
$$-6x + 10 \leq 20 \qquad \text{Subtract } x \text{ from both sides.}$$
$$-6x \leq 10 \qquad \text{Subtract 10 from both sides.}$$

We now divide both sides of the inequality by -6, which changes the order of the inequality.

$$x \geq \frac{10}{-6} \qquad \text{Divide both sides by } -6.$$

$$x \geq -\frac{5}{3} \qquad \text{Simplify the fraction.}$$

The graph of the solution set is shown in Figure 1-14. It is the interval $\left[-\frac{5}{3}, \infty\right)$.

FIGURE 1-14

Self Check 3   Solve: $-4(x + 3) \geq 16$.

Now Try Exercise 19.

> **Caution**
>
> The most common error made when solving inequalities is forgetting to reverse the inequality symbol when multiplying or dividing both sides by a negative number. In Example 3, both sides were divided by -6. If the order of the inequality isn't reversed, the solution will be incorrect.

Solving an Application of a Linear Inequality

EXAMPLE 4   An empty truck with driver weighs 4350 pounds. It is loaded with Georgia peaches weighing 31 pounds per bushel. Between farm and market is a bridge with a 10,000-pound load limit. How many bushels can the truck legally carry?

©Hurst Photo/Shutterstock.com

SOLUTION   The empty truck with driver weighs 4350 pounds, and peaches weighs 31 pounds per bushel. If we let b represent the number of bushels in a legal load, the weight of the peaches will be $31b$ pounds. Since the combined weight of the truck, driver, and cargo cannot exceed 10,000 pounds, we can form the following inequality.

The weight of the empty truck with driver	plus	the weight of the peaches	must be less than or equal to	10,000 pounds.
4350	+	$31b$	$\leq$	10,000

We can solve the inequality as follows.

$$4350 + 31b \leq 10,000$$

$$31b \leq 5650 \qquad \text{Subtract 4350 from each side.}$$

$$b \leq 182.2580645 \qquad \text{Divide both sides by 31.}$$

The truck can legally carry approximately $182\frac{1}{4}$ bushels or less.

Self Check 4   If the empty truck and driver in Example 4 weigh 3800 pounds, how many bushels can the truck legally carry?

Now Try Exercise 85.

> **Take Note**
>
> Remember that $2 < x < 5$ means that $x > 2$ and $x < 5$. The word *and* indicates that both inequalities must be true at the same time.

> **Caution**
>
> It is incorrect to write $x \geq 5$ or $x \leq 2$ as $2 \geq x \geq 5$ because this would mean that $2 \geq 5$, which is false.

3. Solve Compound Inequalities

The statement that x is between 2 and 5 implies two inequalities,

$$x > 2 \text{ and } x < 5$$

It is customary to write both inequalities as one **compound inequality**:

$$2 < x < 5 \quad \text{Read as "2 is less than } x \text{ and } x \text{ is less than 5."}$$

To express that x is not between 2 and 5, we must convey the idea that either x is greater than or equal to 5, or that x is less than or equal to 2. This is equivalent to the statement

$$x \geq 5 \quad \text{or} \quad x \leq 2$$

This inequality is satisfied by all numbers x that satisfy one or both of its parts.

EXAMPLE 5

Solving a Compound Inequality by Isolating x between the Inequality Symbols

Solve: $5 < 3x - 7 \leq 8$.

SOLUTION We can isolate x between the inequality symbols by adding 7 to each part of the inequality to get

$$5 + 7 < 3x - 7 + 7 \leq 8 + 7 \qquad \text{Add 7 to each part.}$$
$$12 < 3x \leq 15 \qquad \text{Do the additions.}$$

and dividing all parts by 3 to get

$$4 < x \leq 5$$

The solution set is the interval $(4, 5]$, whose graph appears in Figure 1-15.

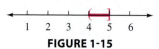

FIGURE 1-15

Self Check 5 Solve: $-5 \leq 2x + 1 < 9$.

Now Try Exercise 33.

EXAMPLE 6

Solving a Compound Inequality by Solving Each Inequality Separately

Solve: $3 + x \leq 3x + 1 < 7x - 2$.

SOLUTION Because it is impossible to isolate x between the inequality symbols, we must solve each inequality separately.

> **Take Note**
>
> At first it seems impossible to solve the inequality because we cannot isolate x between the inequality symbols. An alternate approach, separating the inequalities, makes it possible to solve the problem.

$$
\begin{array}{ll}
3 + x \leq 3x + 1 & \text{and} \quad 3x + 1 < 7x - 2 \\
3 \leq 2x + 1 & \qquad\quad 1 < 4x - 2 \\
2 \leq 2x & \qquad\quad 3 < 4x \\
1 \leq x & \qquad\quad \dfrac{3}{4} < x \\
x \geq 1 & \qquad\quad x > \dfrac{3}{4}
\end{array}
$$

Since the connective in this inequality is *and*, the solution set is the intersection (or overlap) of the intervals $[1, \infty)$ and $\left(\frac{3}{4}, \infty\right)$, which is $[1, \infty)$. The graph is shown in Figure 1-16.

FIGURE 1-16

Self Check 6   Solve: $x + 1 < 2x - 3 \le 3x - 5$.

Now Try Exercise 47.

Take Note

It is possible for an inequality to be true for all values of its variable. It is also possible for an inequality to have no solutions.

For example,

- $x < x + 1$ is true for all numbers x.

- $x > x + 1$ is true for no numbers x.

---

4. Solve Quadratic Inequalities

If $a \ne 0$, inequalities like $ax^2 + bx + c < 0$ and $ax^2 + bx + c \ge 0$ are called **quadratic inequalities**. We will begin by giving two methods for solving quadratic inequalities.

Strategy for Solving Quadratic Inequalities

Method 1: Constructing a Table and Testing Numbers

- Solve the quadratic equation and use the roots of the equation to establish intervals on a number line.
- Construct a table. To do so, write down each interval, select a number to test from each interval, test the selected value to determine if it satisfies the inequality, and then write the result.
- Use the results from the table and write the solution of the quadratic inequality.

Method 2: Constructing a Sign Graph

- Solve the quadratic equation and use the roots of the equation to establish intervals on the number line.
- Construct a sign graph by determining the sign of each factor on each interval. Write these signs on the number line.
- Use these results to write the solution to the quadratic inequality.

The roots of the quadratic equation *will be included* in the solution if the quadratic inequality involves $<$ or $\ge$. The roots *will not be included* in the solution if the quadratic inequality involves $<$ or $>$.

In our first example, we will solve the quadratic inequality using both of the strategies or methods outlined earlier.

Solving a Quadratic Inequality

EXAMPLE 7

Solve: $x^2 - x - 6 > 0$.

SOLUTION   **Method 1:**   First we solve the equation $x^2 - x - 6 = 0$.

$$x^2 - x - 6 = 0$$

$$(x + 2)(x - 3) = 0$$

$$x + 2 = 0 \quad \text{or} \quad x - 3 = 0$$

$$x = -2 \quad \big| \quad x = 3$$

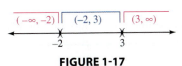

FIGURE 1-17

The graphs of these solutions establish the three intervals shown in Figure 1-17. To determine which intervals are solutions, we construct a table and test a number in each interval and see whether it satisfies the inequality.

Interval	Test Value	Inequality $x^2 - x - 6 > 0$	Result
$(-\infty, -2)$	-6	$(-6)^2 - (-6) - 6 \overset{?}{>} 0$ $36 > 0$ True	The numbers in this interval are solutions.
$(-2, 3)$	0	$0^2 - 0 - 6 \overset{?}{>} 0$ $-6 > 0$ False	The numbers in this interval are not solutions.
$(3, \infty)$	5	$5^2 - 5 - 6 \overset{?}{>} 0$ $14 > 0$ True	The numbers in this interval are solutions.

The solutions are in the intervals $(-\infty, -2)$ or $(3, \infty)$ as shown in Figure 1-18. Note that we write a parenthesis next to -2 and 3 because they do not satisfy the quadratic inequality. The quadratic inequality given contains $>$.

FIGURE 1-18

> **Tip**
>
> Method 1 is easier to apply, but the concepts presented in Method 2 are useful in understanding calculus. It is recommended that you become familiar with both methods.

Method 2: A second method relies on the number line and a notation that keeps track of the signs of the factors of $x^2 - x - 6$, which are $(x - 3)(x + 2)$.

- First, we consider the factor $x - 3$.

  If $x = 3$, then $x - 3 = 0$.

  If $x < 3$, then $x - 3$ is negative.

  If $x > 3$, then $x - 3$ is positive.

- Next, we consider the factor $x + 2$.

  If $x = -2$, then $x + 2 = 0$.

  If $x < -2$, then $x + 2$ is negative.

  If $x > -2$, then $x + 2$ is positive.

We construct a **sign graph** as shown in Figure 1-19 by using $+$ and $-$ signs and writing these above the number line.

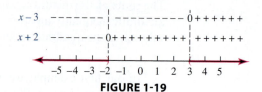

FIGURE 1-19

Only to the left of -2 and to the right of 3 do the signs of both factors agree. Only there is the product positive. The solutions are in the intervals $(-\infty, -2)$ or $(3, \infty)$.

Self Check 7   Solve: $x(x + 1) - 6 \geq 0$.

Now Try Exercise 59.

EXAMPLE 8   **Solving a Quadratic Inequality**

Solve: $x(x + 3) \leq -2$.

SOLUTION   We remove parentheses and add 2 to both sides to make the right side of the equation equal to 0 and solve $x^2 + 3x + 2 \leq 0$.

Tip

Method 1: Think "determine" true or false.

Method 2: Think "determine" positive or negative.

Method 1:   First we solve the equation $x^2 + 3x + 2 = 0$.

$$x^2 + 3x + 2 = 0$$

$$(x + 2)(x + 1) = 0$$

$$x + 2 = 0 \quad \text{or} \quad x + 1 = 0$$

$$x = -2 \quad\quad\quad x = -1$$

These solutions establish the intervals shown in Figure 1-20.

FIGURE 1-20

The solutions of $x^2 + 3x + 2 \leq 0$ will be the numbers in one or more of these intervals. To determine which intervals are solutions, we test a number in each interval to see whether it satisfies the inequality.

Interval	Test Value	Inequality $x^2 + 3x + 2 \leq 0$		Result
$(-\infty, -2)$	-7	$(-7)^2 + 3(-7) + 2 \overset{?}{\leq} 0$ $30 \leq 0$	False	The numbers in this interval are not solutions.
$(-2, -1)$	$-\dfrac{3}{2}$	$\left(-\dfrac{3}{2}\right)^2 + 3\left(-\dfrac{3}{2}\right) + 2 \overset{?}{\leq} 0$ $-\dfrac{1}{4} \leq 0$	True	The numbers in this interval are solutions.
$(-1, \infty)$	0	$0^2 - 0 + 2 \overset{?}{\leq} 0$ $2 \leq 0$	False	The numbers in this interval are not solutions.

The solution is the interval $[-2, -1]$ whose graph appears in Figure 1-21. Note that we write a bracket next to -2 and -1 because they satisfy the quadratic inequality. The quadratic inequality given contains $\leq$.

FIGURE 1-21

Method 2:   We construct a sign graph.

- First, we consider the factor $x + 1$.

  If $x = -1$, then $x + 1 = 0$

  If $x < -1$, then $x + 1$ is negative.

  If $x > -1$, then $x + 1$ is positive.

- Next, we consider the factor $x + 2$.

  If $x = -2$ then $x + 2 = 0$.

  If $x < -2$ then $x + 2$ is negative.

  If $x > -2$ then $x + 2$ is positive.

  The sign graph is shown in Figure 1-22.

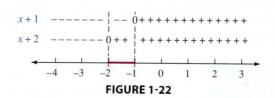

FIGURE 1-22

Only between -2 and -1 do the factors have opposite signs. Here, the product is negative. The solution is the interval $[-2, -1]$.

Self Check 8 Solve: $x^2 - 5x - 6 \leq 0$.

Now Try Exercise 61.

EXAMPLE 9

Solving a Quadratic Inequality

Solve: $x^2 - 5 \geq 0$.

SOLUTION This equation will not factor using only integers. However, we can solve it by adding 5 to both sides and using the SquareRoot Property. We will then use Method 1 and solve the inequality.

$$x^2 - 5 = 0$$

$$x^2 - 5 + 5 = 0 + 5 \qquad \text{Add 5 to both sides.}$$

$$x^2 = 5$$

$$x = \sqrt{5} \quad \text{or} \quad x = -\sqrt{5}$$

These solutions establish the intervals shown in the table and Figure 1-23. To decide which ones are solutions, we test a number in each interval to see whether it is a solution.

Interval	Test Value	Inequality $x^2 - 5 \geq 0$	Result
$(-\infty, -\sqrt{5})$	-3	$(-3)^2 - 5 \overset{?}{\geq} 0$ $4 \geq 0$ True	The numbers in this interval are solutions.
$(-\sqrt{5}, \sqrt{5})$	0	$0^2 - 5 \overset{?}{\geq} 0$ $-5 \geq 0$ False	The numbers in this interval are not solutions.
$(\sqrt{5}, \infty)$	3	$3^2 - 5 \overset{?}{\geq} 0$ $4 \geq 0$ True	The numbers in this interval are solutions.

As shown in Figure 1-23, the solution is the union of two intervals:

$$\left(-\infty, -\sqrt{5}\,\right] \cup \left[\sqrt{5}, \infty\right).$$

FIGURE 1-23

Self Check 9 Solve $x^2 - 7 \leq 0$.

Now Try Exercise 67.

4. Solve Rational Inequalities

Inequalities that contain fractions with polynomial numerators and denominators are called **rational inequalities**. To solve them, we can use the same strategies that we use to solve quadratic inequalities.

EXAMPLE 10

Solving a Rational Inequality

Solve the rational inequality: $\dfrac{x^2 - x - 2}{x^2 - 4x + 3} \le 0$.

SOLUTION   **Method 1:** The intervals are found by solving $x^2 - x - 2 = 0$ and $x^2 - 4x + 3 = 0$. The solutions of the first equation are –1 and 2, and the solutions of the second equation are 1 and 3. These solutions establish the five intervals shown in Figure 1-24.

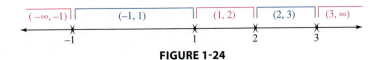

FIGURE 1-24

The solutions of $\frac{x^2 - x - 2}{x^2 - 4x + 3} \le 0$ will be the numbers in one or more of these intervals. To determine which intervals are solutions, we test a number in each interval to see whether it satisfies the inequality.

Interval	Test Value	Inequality $\dfrac{x^2 - x - 2}{x^2 - 4x + 3} \le 0$	Result
$(-\infty, -1)$	-2	$\dfrac{(-2)^2 - (-2) - 2}{(-2)^2 - 4(-2) + 3} \le 0$ $\dfrac{4}{15} \le 0$ False	The numbers in this interval are not solutions.
$(-1, 1)$	0	$\dfrac{0^2 - 0 - 2}{0^2 - 4(0) + 3} \le 0$ $-\dfrac{2}{3} \le 0$ True	The numbers in this interval are solutions.
$(1, 2)$	$\dfrac{3}{2}$	$\dfrac{\left(\frac{3}{2}\right)^2 - \left(\frac{3}{2}\right) - 2}{\left(\frac{3}{2}\right)^2 - 4\left(\frac{3}{2}\right) + 3} \le 0$ $\dfrac{5}{3} \le 0$ False	The numbers in this interval are not solutions.
$(2, 3)$	$\dfrac{5}{2}$	$\dfrac{\left(\frac{5}{2}\right)^2 - \left(\frac{5}{2}\right) - 2}{\left(\frac{5}{2}\right)^2 - 4\left(\frac{5}{2}\right) + 3} \le 0$ $-\dfrac{7}{3} \le 0$ True	The numbers in this interval are solutions.
$(3, \infty)$	4	$\dfrac{4^2 - 4 - 2}{4^2 - 4(4) + 3} \le 0$ $\dfrac{10}{3} \le 0$ False	The numbers in this interval are not solutions.

The numbers in the intervals $(-1, 1)$ and $(2, 3)$ satisfy the inequality, but the numbers in the intervals $(-\infty, -1)$, $(1, 2)$, and $(3, \infty)$ do not. The solution set is $[-1, 1) \cup [2, 3)$. The graph of the solution set is shown in Figure 1-25.

FIGURE 1-25

Because $x = -1$ and $x = 2$ make the numerator 0, they satisfy the inequality. Thus, their graphs are drawn with brackets to show that -1 and 2 are included. Because 1 and 3 give 0's in the denominator, the parentheses at $x = 1$ and $x = 3$ show that 1 and 3 are not in the solution set.

Method 2: We factor each trinomial and write the inequality in the form

$$\frac{(x - 2)(x + 1)}{(x - 3)(x - 1)} \le 0$$

We then construct the sign graph shown in Figure 1-26. The value of the fraction will be 0 when $x = 2$ and $x = -1$. The value will be negative when there is an odd number of negative factors. This happens between -1 and 1 and between 2 and 3.

The graph of the solution set also appears in Figure 1-26. The brackets at -1 and 2 show that these numbers are in the solution set. The parentheses at 1 and 3 show that these numbers are not in the solution set.

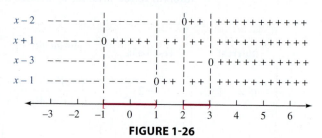

FIGURE 1-26

Self Check 10   Solve: $\dfrac{x^2 + 2x - 3}{x^2 + 4x + 3} > 0$.

Now Try Exercise 73.

EXAMPLE 11   **Solving a Rational Inequality**

Solve: $\dfrac{6}{x} > 2$.

SOLUTION   We will first rewrite the rational inequality and get a 0 on the right side. Then we can use either method to solve the inequality.

To get a 0 on the right side, we subtract 2 from both sides. We then combine like terms on the left side.

$$\frac{6}{x} > 2$$

$$\frac{6}{x} - 2 > 0 \qquad \textcolor{red}{\text{Subtract 2 from both sides.}}$$

$$\frac{6}{x} - \frac{2x}{x} > 0 \qquad \textcolor{red}{\frac{x}{x} = 1}$$

$$\frac{6 - 2x}{x} > 0 \qquad \textcolor{red}{\text{Add the numerators and keep the common denominator.}}$$

The inequality now has the form of a rational inequality. We can use either method to solve the inequality. The intervals are found by solving $6 - 2x = 0$ and $x = 0$. The solution of the first equation is 3, and the solution of the second equation is 0. This determines the intervals $(-\infty, 0)$, $(0, 3)$, and $(3, \infty)$. Because only the numbers in the interval $(0, 3)$ satisfy the original inequality, the solution set is $(0, 3)$. The graph is shown in Figure 1-27.

FIGURE 1-27

We could construct a sign graph as in Figure 1-28 and obtain the same solution set.

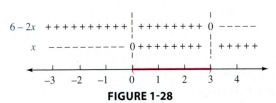

FIGURE 1-28

Self Check 11   Solve: $\dfrac{2}{x} < 4$.

Now Try Exercise 77.

> **Caution**
>
> It is tempting to solve Example 11 by multiplying both sides by x and solving the inequality $6 > 2x$. However, multiplying both sides by x gives $6 > 2x$ only when x is positive. If x is negative, multiplying both sides by x will reverse the direction of the $>$ symbol, and the inequality $\frac{6}{x} > 2$ will be equivalent to $6 < 2x$. If you fail to consider both cases, you will get a wrong answer.

Self Check Answers

1. **a.** $>$ or $\geq$ **b.** $\leq$ or $\geq$ **c.** $>$ or $\geq$ 2. $(9, \infty)$

3. $(-\infty, -7]$ 4. 200 bushels or less   5. $[-3, 4)$

6. $(4, \infty)$ 7. $(-\infty, -3] \cup [2, \infty)$ 8. $[-1, 6]$

9. $\left[-\sqrt{7}, \sqrt{7}\right]$ 10. $(-\infty, -3) \cup (-3, -1) \cup (1, \infty)$

11. $(-\infty, 0) \cup \left(\dfrac{1}{2}, \infty\right)$

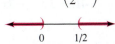

Exercises 1.7

Fill in the blanks.

1. If $x > y$, then x lies to the _____ of y on a number line.

2. $a < b$, _____, or $a > b$.

3. If $a < b$ and $b < c$, then _____.

4. If $a < b$ then $a + c <$ _____.

5. If $a < b$ then $a - c <$ _____.

6. If $a < b$ and $c > 0$, then ac _____ bc.

7. If $a < b$ and $c < 0$, then ac _____ bc.

8. If $a < b$ and $c < 0$, then $\dfrac{a}{c}$ _____ $\dfrac{b}{c}$.

9. $3x - 5 < 12$ and $ax + c > 0$ $(a \neq 0)$ are examples of _____ inequalities.

10. $ax^2 + bx - c \geq 0$ $a \neq 0$ and $3x^2 - 6x < 0$ are examples of _____ inequalities.

11. If two inequalities have the same solution set, they are called _____ inequalities.

12. An inequality that contains a fraction with a polynomial numerator and denominator is called a _____ inequality.

Practice

Solve each inequality, graph the solution set, and write the answer in interval notation. Do not worry about drawing your graphs exactly to scale.

13. $3x + 2 < 5$

14. $-2x + 4 < 6$

15. $3x + 2 \geq 5$

16. $-2x + 4 \geq 6$

17. $-5x + 3 > -2$

18. $4x - 3 > -4$

19. $-5x + 3 \leq -2$

20. $4x - 3 \leq -4$

21. $2(x - 3) \leq -2(x - 3)$

22. $3(x + 2) \leq 2(x + 5)$

23. $\dfrac{3}{5}x + 4 > 2$

24. $\dfrac{1}{4}x - 3 > 5$

25. $\dfrac{x + 3}{4} < \dfrac{2x - 4}{3}$

26. $\dfrac{x + 2}{5} > \dfrac{x - 1}{2}$

27. $\dfrac{6(x - 4)}{5} \geq \dfrac{3(x + 2)}{4}$

28. $\dfrac{3(x + 3)}{2} < \dfrac{2(x + 7)}{3}$

29. $\dfrac{5}{9}(a + 3) - a \geq \dfrac{4}{3}(a - 3) - 1$

30. $\dfrac{2}{3}y - y \leq -\dfrac{3}{2}(y - 5)$

31. $\dfrac{2}{3}a - \dfrac{3}{4}a < \dfrac{3}{5}\left(a + \dfrac{2}{3}\right) + \dfrac{1}{3}$

32. $\dfrac{1}{4}b + \dfrac{2}{3}b - \dfrac{1}{2} > \dfrac{1}{2}(b + 1) + b$

33. $4 < 2x - 8 \leq 10$

34. $3 \leq 2x + 2 < 6$

35. $9 \geq \dfrac{x - 4}{2} > 2$

36. $5 < \dfrac{x - 2}{6} < 6$

37. $0 \leq \dfrac{4 - x}{3} \leq 5$

38. $0 \geq \dfrac{5 - x}{2} \geq -10$

39. $-2 \geq \dfrac{1 - x}{2} \geq -10$

40. $-2 \leq \dfrac{1 - x}{2} < 10$

41. $-3x > -2x > -x$

42. $-3x < -2x < -x$

43. $x < 2x < 3x$

44. $x > 2x > 3x$

45. $2x + 1 < 3x - 2 < 12$

46. $2 - x < 3x + 5 < 18$

47. $2 + x < 3x - 2 < 5x + 2$

48. $x > 2x + 3 > 4x - 7$

49. $3 + x > 7x - 2 > 5x - 10$

50. $2 - x < 3x + 1 < 10x$

51. $x \le x + 1 \le 2x + 3$

52. $-x \ge -2x + 1 \ge -3x + 1$

53. $x^2 + 7x + 12 < 0$

54. $x^2 - 13x + 12 \le 0$

55. $x^2 - 5x + 6 \ge 0$

56. $6x^2 + 5x - 6 > 0$

57. $x^2 + 5x + 6 < 0$

58. $x^2 + 9x + 20 \ge 0$

59. $6x^2 + 5x + 1 \ge 0$

60. $x^2 + 9x + 20 < 0$

61. $6x^2 - 5x < -1$

62. $9x^2 + 24x > -16$

63. $2x^2 \ge 3 - x$

64. $9x^2 \le 24x - 16$

65. $x^2 - 3 \ge 0$

66. $x^2 - 7 \le 0$

67. $x^2 - 11 < 0$

68. $x^2 - 20 > 0$

69. $\dfrac{x + 3}{x - 2} < 0$

70. $\dfrac{x + 3}{x - 2} > 0$

71. $\dfrac{x^2 + x}{x^2 - 1} > 0$

72. $\dfrac{x^2 - 4}{x^2 - 9} < 0$

73. $\dfrac{x^2 + 5x + 6}{x^2 + x - 6} \ge 0$

74. $\dfrac{x^2 + 10x + 25}{x^2 - x - 12} \le 0$

75. $\dfrac{6x^2 - x - 1}{x^2 + 4x + 4} > 0$

76. $\dfrac{6x^2 - 3x - 3}{x^2 - 2x - 8} < 0$

77. $\dfrac{3}{x} > 2$

78. $\dfrac{3}{x} < 2$

79. $\dfrac{6}{x} < 4$

80. $\dfrac{6}{x} > 4$

81. $\dfrac{3}{x - 2} \le 5$

82. $\dfrac{3}{x + 2} \le 4$

83. $\dfrac{6}{x^2 - 1} < 1$

84. $\dfrac{6}{x^2 - 1} > 1$

Applications

Solve each problem.

85. Golfing lessons Macy decides to take golfing lessons. If her new set of golf clubs cost $250 and private lessons are $60 per hour lesson, what is the maximum number of lessons she can take if the total spent for lessons and purchasing clubs is at most $970?

86. **Surfing lessons** Dylan and Dusty plan to take weekly surfing lessons together. If the 2-hour lessons are $40 per person and they plan to spend $200 each on new surfboards, what is the maximum number of lessons the two can take if the total amount spent for lessons and surfboards is at most $960?

87. **Long distance** A long-distance telephone call costs 40¢ for the first three minutes and 10¢ for each additional minute. At most how many minutes can a person talk and not exceed $2?

88. **Buying a computer** A student who can afford to spend up to $2000 sees the ad shown in the illustration. If she buys a touch-screen laptop, how many games can she buy?

Big Sale!!!!

◀ ‖ $1695.95

Games
$19.95

89. **Musical items** Andy can spend up to $275 on a guitar and some music books. If he can buy a guitar for $150 and music books for $9.75, what is the greatest number of music books that he can buy?

90. **Buying DVDs** Audrey wants to spend less than $600 for a DVD recorder and some DVDs. If the recorder of her choice costs $425 and DVDs cost $7.50 each, how many DVDs can she buy?

91. **Buying a refrigerator** Madeline, who has $1200 to spend, wants to buy a refrigerator. Refer to the following table and write an inequality that shows how much she can pay p for the refrigerator.

State sales tax	6.5%
City sales tax	0.25%

92. **Renting a rototiller** The cost of renting a rototiller is $17.50 for the first hour and $8.95 for each additional hour. How long, to the nearest hour, can a person have the rototiller if the cost must be less than $75?

93. **Profit** Profit occurs when revenue exceeds cost. If the revenue R in dollars from producing and selling x Hugo Boss polo shirts is $R = 26x$ dollars and the cost C is $C = 6x + 3660$ dollars, what production level produces a profit?

94. **Profit** The revenue R in dollars of producing and selling x Yankee candles is $R = 19x$ and the cost C is $C = 3x + 2800$. At what production level will revenue exceed cost and the company obtain a profit?

95. **Real estate taxes** A city council has proposed the following two methods of taxing real estate:

Method 1	$2200 + 4% of assessed value
Method 2	$1200 + 6% of assessed value

For what range of assessments a would the first method benefit the taxpayer?

96. **Medical plans** A college provides its employees with a choice of the two medical plans shown in the following table. For what size hospital bills is Plan 2 better for the employee than Plan 1? (*Hint:* The cost to the employee includes both the deductible payment and the employee's coinsurance payment.)

Plan 1	Plan 2
Employee pays $100	Employee pays $200
Plan pays 70% of the rest	Plan pays 80% of the rest

97. **Medical plans** To save costs, the college in Exercise 96 raised the employee deductible, as shown in the following table. For what size hospital bills is Plan 2 better for the employee than Plan 1? (*Hint:* The cost to the employee includes both the deductible payment and the employee's coinsurance payment.)

Plan 1	Plan 2
Employee pays $200	Employee pays $400
Plan pays 70% of the rest	Plan pays 80% of the rest

98. **Geometry** The perimeter of a rectangle is to be between 180 inches and 200 inches. Find the range of values for its length l when its width is 40 inches.

99. **Geometry** The perimeter of an equilateral triangle is to be between 50 centimeters and 60 centimeters. Find the range of lengths of one side s.

100. **Geometry** The perimeter of a square is to be from 25 meters to 60 meters. Find the range of values for its area A.

101. Projectile height If a Nerf sports bash ball is projected from ground level with an initial velocity of 160 feet per second, its height s in feet t seconds after being projected is given by the equation $s = -16t^2 + 160t$. When will the height of the bash ball exceed 144 feet?

102. Projectile height If a jumbo hypercharged pop sky ball is projected from ground level with an initial velocity of 192 feet per second, its height s in feet t seconds after being projected is given by the equation $s = -16t^2 + 240t$. When will the height of the pop sky ball exceed 576 feet?

Discovery and Writing

103. The techniques used for solving linear equations and linear inequalities are similar, yet different. Explain.

104. When graphing the solution set of an inequality, what does a bracket indicate on the number line? What does a parenthesis indicate on the number line?

105. What is a quadratic inequality? Give two examples.

106. What is a rational inequality? Give two examples.

Critical Thinking

Determine if the statement is true or false. If the statement is false, then correct it and make it true.

107. The solution set of the inequality $x^2 - 100 \le 0$ is $(-\infty, 10]$.

108. The solution set of $x^2 > 0$ is all real numbers.

109. To solve the rational inequality $\dfrac{1}{x - 10} < 2$, the first step is multiply both sides by $x - 10$ to clear the inequality of fractions.

110. The solution set of the inequality $\dfrac{100}{x} \ge 10$ is $[0, 10]$.

1.8 Absolute Value

In this section, we will learn to

1. Define and use absolute value.
2. Solve equations of the form $|x| = k$.
3. Solve equations with two absolute values.
4. Solve inequalities of the forms $|x| < k$ and $|x| \le k$.
5. Solve inequalities of the forms $|x| > k$ and $|x| \ge k$.
6. Solve compound inequalities with absolute value.
7. Solve inequalities with two absolute values.

Cancun, Mexico, is a popular spring-break destination for college students. The beautiful beaches, exciting nightlife, and tropical weather make it a pleasurable experience.

The average annual temperature in Cancun is 78°F with fluctuations of approximately 7 degrees. We can represent this temperature range using absolute value notation. If we let x represent the temperature at a given time, the absolute value of the difference between x and 78 is less than or equal to 7. We can write this as $|x - 78| \le 7$.

In this section, we will review absolute value and examine its consequences in greater detail.

1. Define and Use Absolute Value

Absolute Value   The **absolute value** of the real number x, denoted by $|x|$, is defined as follows:

If $x \ge 0$, then $|x| = x$.

If $x < 0$, then $|x| = -x$.

This definition provides a way to associate a nonnegative real number with any real number.

- If $x \geq 0$, then x (which is positive or 0) is its own absolute value.
- If $x < 0$, then $-x$ (which is positive) is the absolute value.

Either way, $|x|$ is positive or 0:

$$|x| \geq 0 \qquad \text{for all real numbers } x$$

Using the Definition of Absolute Value

EXAMPLE 1   Write each expression without using absolute value symbols:

a. $|7|$ **b.** $|-3|$ **c.** $-|-7|$ **d.** $|x - 2|$

SOLUTION   In each case, we will apply the definition of absolute value.

a. Because 7 is positive, $|7| = 7$.

b. Because -3 is negative, $|-3| = -(-3) = 3$.

c. The expression $-|-7|$ means "the negative of the absolute value of -7." Thus, $-|-7| = -(7) = -7$.

d. To denote the absolute value of a variable quantity, we must give a conditional answer.

If $x - 2 \geq 0$, then $|x - 2| = x - 2$.

If $x - 2 < 0$, then $|x - 2| = -(x - 2) = -x + 2$.

Self Check 1   Write each expression without using absolute value symbols:

a. $|0|$ **b.** $|-17|$ **c.** $|x + 5|$

Now Try Exercise 13.

2. Solve Equations of the Form $|x| = k$

In the equation $|x| = 8$, x can be either 8 or -8, because $|8| = 8$ and $|-8| = 8$. In general, the following is true.

Absolute Value Equations	If $k \geq 0$, then $\qquad |x| = k \quad$ is equivalent to $\quad x = k \quad$ or $\quad x = -k$

The absolute value of a number represents the distance on the number line from a point to the origin. The solutions of $|x|$ are the coordinates of the two points that lie exactly k units from the origin. (See Figure 1-29.)

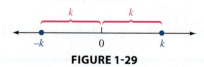

FIGURE 1-29

The equation $|x - 3| = 7$ indicates that a point on the number line with a coordinate of $x - 3$ is 7 units from the origin. Thus, $x - 3$ can be 7 or -7.

$$x - 3 = 7 \qquad \text{or} \qquad x - 3 = -7$$
$$x = 10 \qquad\qquad\qquad x = -4$$

The solutions of 10 and −4 are shown in Figure 1-30. Both of these numbers satisfy the equation.

FIGURE 1-30

$$
\begin{array}{c|c}
|x - 3| = 7 \quad \text{and} & |x - 3| = 7 \\
|\mathbf{10} - 3| = 7 & |\mathbf{-4} - 3| = 7 \\
|7| = 7 & |-7| = 7 \\
7 = 7 & 7 = 7
\end{array}
$$

> **Take Note**
>
> In general, absolute value equations will have two solutions. Consider these exceptions:
>
> - Note that if $k < 0$, then $|x| = k$ has no solution.
>
>   For example, the equation $|x| = -5$ has no solution.
>
> - If $k = 0$, then $|x| = k$ has only one solution.
>
>   For example, the equation $|x| = 0$ has one solution, $x = 0$.

Solving an Absolute Value Equation

EXAMPLE 2

Solve: $|3x - 5| + 3 = 10$.

SOLUTION

We will first write the equation in the form $|x| = k$. To do so, we will subtract 3 from both sides of the equation.

$$
\begin{aligned}
|3x - 5| + 3 &= 10 \\
|3x - 5| + 3 - 3 &= 10 - 3 \\
|3x - 5| &= 7
\end{aligned}
$$

The equation $|3x - 5| = 7$ is equivalent to two equations

$$3x - 5 = 7 \quad \text{or} \quad 3x - 5 = -7$$

which can be solved separately:

$$
\begin{array}{c|c}
3x - 5 = 7 \quad \text{or} & 3x - 5 = -7 \\
3x = 12 & 3x = -2 \\
x = 4 & x = -\dfrac{2}{3}
\end{array}
$$

The solution set consists of the points shown in Figure 1-31.

> **Caution**
>
> Be sure to isolate the absolute value expression on one side before applying the property. If you do not do this, your solution set will be only partially correct.

> **Tip**
>
> it is easy to forget to do the second half of the problem. In this example, don't forget to set $3x - 5$ equal to -7.

FIGURE 1-31

Self Check 2   Solve: $|2x + 3| = 7$.

Now Try Exercise 23.

3. Solve Equations with Two Absolute Values

The equation $|a| = |b|$ is true when $a = b$ or when $a = -b$. For example,

$$
\begin{array}{c|c}
|3| = |3| \quad \text{or} & |3| = |-3| \\
3 = 3 & 3 = 3
\end{array}
$$

In general, the following is true.

Equations with Two Absolute Values	If a and b represent algebraic expressions, the equation $|a| = |b|$ is equivalent to $$a = b \quad \text{or} \quad a = -b$$

Solving an Equation with Two Absolute Values

EXAMPLE 3

Solve: $|2x| = |x - 3|$.

SOLUTION   The equation $|2x| = |x - 3|$ will be true when $2x$ and $x - 3$ are equal or when they are negatives. This gives two equations, which can be solved separately:

$$2x = x - 3 \quad \text{or} \quad 2x = -(x - 3)$$
$$x = -3 \qquad\qquad 2x = -x + 3$$
$$3x = 3$$
$$x = 1$$

> **Caution**
>
> A common error made in solving the second equation is to set $2x$ equal to $x + 3$. This is incorrect.

Verify that -3 and 1 satisfy the equation.

Self Check 3   Solve: $|3x + 1| = |5x - 3|$.

Now Try Exercise 39.

We will now turn our attention to inequalities and solve inequalities containing absolute values.

4. Solve Inequalities of the Forms $|x| < k$ and $|x| \le k$

The inequality $|x| < 5$ indicates that a point with coordinate x is less than 5 units from the origin. (See Figure 1-32.) Thus, x is between -5 and 5, and

$$|x| < 5 \quad \text{is equivalent to} \quad -5 < x < 5$$

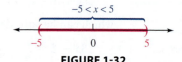

FIGURE 1-32

In general, the inequality $|x| < k \ (k > 0)$ indicates that a point with coordinate x is less than k units from the origin. (See Figure 1-33.)

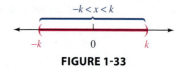

FIGURE 1-33

Similarly, the inequality $|x| \le k \ (k > 0)$ indicates that a point with coordinate x is less than or equal to k units from the origin.

Inequalities of the Forms $|x| < k$ and $|x| \le k$	If $k > 0$, then • $|x| < k$ is equivalent to $-k < x < k$. • $|x| \le k$ is equivalent to $-k \le x \le k$.

EXAMPLE 4

Solving an Absolute Value Inequality of the Form $|x| < k$

Solve: $|x - 2| < 7$.

SOLUTION

The inequality $|x - 2| < 7$ is equivalent to

$$-7 < x - 2 < 7$$

We can add 2 to each part of this inequality to get

$$-5 < x < 9$$

The solution set is the interval $(-5, 9)$, shown in Figure 1-34.

FIGURE 1-34

Self Check 4

Solve: $|x + 3| < 9$.

Now Try Exercise 47.

5. Solve Inequalities of the Forms $|x| > k$ and $|x| \geq k$

The inequality $|x| > 5$ indicates that a point with coordinate x is more than 5 units from the origin. (See Figure 1-35.) Thus, $x < -5$ or $x > 5$.

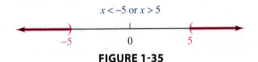

FIGURE 1-35

In general, the inequality $|x| > k$ $(k > 0)$ indicates that a point with coordinate x is more than k units from the origin. (See Figure 1-36.)

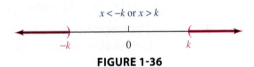

FIGURE 1-36

Similarly, the inequality $|x| \geq k$ $(k > 0)$ indicates that a point with coordinate x is k or more units from the origin and thus, $x \leq -5$ or $x \geq 5$.

Inequalities of the Forms $|x| > k$ **and** $|x| \geq k$

If $k > 0$, then

- $|x| > k$ is equivalent to $x < -k$ or $x > k$.
- $|x| \geq k$ is equivalent to $x \leq -k$ or $x \geq k$.

EXAMPLE 5

Solving an Absolute Value Inequality of the Form $|x| \geq k$

Solve: $\left| \dfrac{2x + 3}{2} \right| + 7 \geq 12$.

SOLUTION

We begin by subtracting 7 from both sides of the inequality to isolate the absolute value on the left side.

$$\left| \frac{2x + 3}{2} \right| \geq 5$$

This result is equivalent to two inequalities that can be solved separately.

$$\frac{2x + 3}{2} \le -5 \quad \text{or} \quad \frac{2x + 3}{2} \ge 5$$

$$2x + 3 \le -10 \qquad\qquad 2x + 3 \ge 10$$

$$2x \le -13 \qquad\qquad 2x \ge 7$$

$$x \le -\frac{13}{2} \qquad\qquad x \ge \frac{7}{2}$$

The solution set is the union of the intervals $\left(-\infty, -\frac{13}{2}\right]$ and $\left[\frac{7}{2}, \infty\right)$. Its graph appears in Figure 1-37.

FIGURE 1-37

Self Check 5 Solve: $\left|\dfrac{3x - 6}{3}\right| + 2 \ge 12$.

Now Try Exercise 59.

6. Solve Compound Inequalities with Absolute Value

Solving a Compound Inequality with Absolute Value

EXAMPLE 6

Solve: $0 < |x - 5| \le 3$.

SOLUTION   The inequality $0 < |x - 5| \le 3$ consists of two inequalities that can be solved separately. The solution will be the intersection of the inequalities

$$0 < |x - 5| \quad \text{and} \quad |x - 5| \le 3$$

The inequality $0 < |x - 5|$ is true for all x except 5. The inequality $|x - 5| \le 3$ is equivalent to the inequality

$$-3 \le x - 5 \le 3$$

$$2 \le x \le 8 \qquad \text{Add 5 to each part.}$$

The solution set is the intersection of these two solutions, which is the interval $[2, 8]$, except 5. This is the union of the intervals $[2, 5)$ and $(5, 8]$, as shown in Figure 1-38.

FIGURE 1-38

Self Check 6 Solve: $0 < |x + 2| \le 5$.

Now Try Exercise 63.

7. Solve Inequalities with Two Absolute Values

In Example 9b of Section 1.3, we saw that $|a|$ could be defined as

$$|a| = \sqrt{a^2}$$

We will use this fact in the next example.

EXAMPLE 7

Solving an Inequality with Two Absolute Values

Solve $|x + 2| > |x + 1|$ and give the result in interval notation.

SOLUTION

$$|x + 2| > |x + 1|$$

$$\sqrt{(x + 2)^2} > \sqrt{(x + 1)^2} \qquad \text{Use } |a| = \sqrt{a^2}.$$

$$(x + 2)^2 > (x + 1)^2 \qquad \text{Square both sides.}$$

$$x^2 + 4x + 4 > x^2 + 2x + 1 \qquad \text{Expand each binomial.}$$

$$4x > 2x - 3 \qquad \text{Subtract } x^2 \text{ and 4 from both sides.}$$

$$2x > -3 \qquad \text{Subtract } 2x \text{ from both sides.}$$

$$x > -\frac{3}{2} \qquad \text{Divide both sides by 2.}$$

The solution set is the interval $\left(-\frac{3}{2}, \infty\right)$. Check several numbers in this interval to verify that this interval is the solution.

Self Check 7   Solve $|x - 3| \leq |x + 2|$ and give the result in interval notation.

Now Try Exercise 77.

Three other properties of absolute value are sometimes useful.

Properties of Absolute Value

If a and b are real numbers, then

1. $|ab| = |a||b|$ **2.** $\left|\dfrac{a}{b}\right| = \dfrac{|a|}{|b|} \ (b \neq 0)$ **3.** $|a + b| \leq |a| + |b|$

- Properties 1 and 2 above indicate that the absolute value of a product (or a quotient) is the product (or the quotient) of the absolute values.
- Property 3 indicates that the absolute value of a sum is either equal to or less than the sum of the absolute values.

Self Check Answers

1. a. 0   **b.** 17   **c.** if $x + 5 \geq 0$, $|x + 5| = x + 5$; if $x + 5 < 0$,

$|x + 5| = -x - 5$ **2.** $-5, 2$ **3.** $2, \dfrac{1}{4}$

4. $(-12, 6)$ **5.** $(-\infty, -8] \cup [12, \infty)$ **6.** $[-7, -2) \cup (-2, 3]$ **7.** $\left[\dfrac{1}{2}, \infty\right)$

Exercises 1.8

Getting Ready

You should be able to complete these vocabulary and concept statements before you proceed to the practice exercises.

Fill in the blanks.

1. If $x \geq 0$, then $|x| = $ ___.

2. If $x < 0$, then $|x| = $ _____.

3. $|x| = k$ is equivalent to _____.

4. $|a| = |b|$ is equivalent to $a = b$ or _____.

5. $|x| < k$ is equivalent to _____.

6. $|x| > k$ is equivalent to _____.

7. $|x| \geq k$ is equivalent to _____.

8. $\sqrt{a^2} = $ _____.

Practice

Write each expression without absolute value symbols.

9. $|7|$

10. $|-9|$

11. $|0|$

12. $|3 - 5|$

13. $|5| - |-3|$

14. $|-3| + |5|$

15. $|\pi - 2|$

16. $|\pi - 4|$

17. $|x - 5|$ and $x \geq 5$

18. $|x - 5|$ and $x \leq 5$

19. $|x^3|$

20. $|2x|$

Solve each absolute value equation for x.

21. $|x + 2| = 2$

22. $|2x + 5| = 3$

23. $|3x - 1| - 7 = -2$

24. $|7x - 5| + 5 = 8$

25. $\left|\dfrac{3x - 4}{2}\right| = 5$

26. $\left|\dfrac{10x + 1}{2}\right| = \dfrac{9}{2}$

27. $\left|\dfrac{2x - 4}{5}\right| + 6 = 8$

28. $\left|\dfrac{3x + 11}{7}\right| - 15 = -14$

29. $\left|\dfrac{x - 3}{4}\right| = -2$

30. $\left|\dfrac{x + 5}{2}\right| + 3 = 2$

31. $\left|\dfrac{x - 5}{3}\right| = 0$

32. $\left|\dfrac{x + 7}{9}\right| = 0$

33. $\left|\dfrac{4x - 2}{x}\right| = 3$

34. $\left|\dfrac{2(x - 3)}{3x}\right| = 6$

35. $|x| = x$

36. $|x| + x = 2$

37. $|x + 3| = |x|$

38. $|x + 5| = |5 - x|$

39. $|x - 3| = |2x + 3|$

40. $|x - 2| = |3x + 8|$

41. $|x + 2| = |x - 2|$

42. $|2x - 3| = |3x - 5|$

43. $\left|\dfrac{x + 3}{2}\right| = |2x - 3|$

44. $\left|\dfrac{x - 2}{3}\right| = |6 - x|$

45. $\left|\dfrac{3x - 1}{2}\right| = \left|\dfrac{2x + 3}{3}\right|$

46. $\left|\dfrac{5x + 2}{3}\right| = \left|\dfrac{x - 1}{4}\right|$

Solve each absolute value inequality. Express the solution set in interval notation, and graph it.

47. $|x - 3| < 6$

48. $|x - 2| \geq 4$

49. $|x + 3| > 6$

50. $|x + 2| \leq 4$

51. $|2x + 4| \geq 10$

52. $|5x - 2| < 7$

53. $|3x + 5| + 1 \leq 9$

54. $|2x - 7| - 3 > 2$

55. $|x + 3| > 0$

56. $|x - 3| \leq 0$

57. $\left|\dfrac{5x + 2}{3}\right| < 1$

58. $\left|\dfrac{3x + 2}{4}\right| > 2$

59. $3\left|\dfrac{3x - 1}{2}\right| > 5$

60. $2\left|\dfrac{8x + 2}{5}\right| \leq 1$

61. $\dfrac{|x - 1|}{-2} > -3$

62. $\dfrac{|2x - 3|}{-3} < -1$

Solve each compound inequality with absolute value. Express the solution set in interval notation, and graph it.

63. $0 < |2x + 1| < 3$

64. $0 < |2x - 3| < 1$

65. $8 > |3x - 1| > 3$

66. $8 > |4x - 1| > 5$

67. $2 < \left|\dfrac{x - 5}{3}\right| < 4$

68. $3 < \left|\dfrac{x - 3}{2}\right| < 5$

69. $10 > \left|\dfrac{x - 2}{2}\right| > 4$

70. $5 \geq \left|\dfrac{x + 2}{3}\right| > 1$

71. $2 \le \left| \dfrac{x+1}{3} \right| < 3$

72. $8 > \left| \dfrac{3x+1}{2} \right| > 2$

Solve each inequality and express the solution using interval notation.

73. $|x+1| \ge |x|$

74. $|x+1| < |x+2|$

75. $|2x+1| < |2x-1|$

76. $|3x-2| \ge |3x+1|$

77. $|x+1| < |x|$

78. $|x+2| \le |x+1|$

79. $|2x+1| \ge |2x-1|$

80. $|3x-2| < |3x+1|$

Applications

81. Finding temperature ranges The temperatures on a summer day satisfy the inequality $|t - 78°| \le 8°$, where t is the temperature in degrees Fahrenheit. Express this range without using absolute value symbols.

82. Finding operating temperatures A car CD player has an operating temperature of $|t - 40°| < 80°$, where t is the temperature in degrees Fahrenheit. Express this range without using absolute value symbols.

83. Range of camber angles The specifications for a certain car state that the camber angle c of its wheels should be $0.6° \pm 0.5°$. Express this range with an inequality containing an absolute value.

84. Tolerance of a sheet of steel A sheet of steel is to be 0.25 inch thick, with a tolerance of 0.015 inch. Express this specification with an inequality containing an absolute value.

85. Humidity level A Steinway piano should be placed in an environment where the relative humidity h is between 38% and 72%. Express this range with an inequality containing an absolute value.

iLexx/iStockphoto.com

86. Light bulbs A light bulb is expected to last h hours, where $|h - 1500| \le 200$. Express this range without using absolute value symbols.

87. Error analysis In a lab, students measured the percent of copper p in a sample of copper sulfate. The students know that copper sulfate is actually 25.46% copper by mass. They are to compare their results to the actual value and find the amount of *experimental error.*

   a. Which measurements shown in the illustration satisfy the absolute value inequality $|p - 25.46| \le 1.00$?

   b. What can be said about the amount of error for each of the trials listed in part a?

Lab 4	Section A
Title: "Percent copper (CU) in copper sulfate (CuSO$_4$·5H$_2$O)"	

Results

	% Copper
Trial #1:	22.91%
Trial #2:	26.45%
Trial #3:	26.49%
Trial #4:	24.76%

88. Error analysis See Exercise 87.

   a. Which measurements satisfy the absolute value inequality $|p - 25.46| > 1.00$?

   b. What can be said about the amount of error for each of the trials listed in part a?

Discovery and Writing

89. Explain how to find the absolute value of a number.

90. Explain why the equation $|x| + 9 = 0$ has no solution.

91. If $k > 0$, explain the differences between the solution sets of $|x| < k$ and $|x| > k$.

92. If $k < 0$, explain why the solution set of $|x| < k$ has no solution.

93. If $k < 0$, explain why the solution set of of $|x| > k$ is all real numbers.

94. Explain how to solve an inequality with two absolute values.

Critical Thinking

Determine if the statement is true or false. If the statement is false, then correct it and make it true.

95. Absolute value equations always have two solutions.

96. $|-x| = x$

97. The solution set of $|x| \geq 5$ is $[5, \infty)$.

98. $|a + b| \geq |a| + |b|$

99. $|x| + 555 < 554$ has no solution.

100. The solution set of $|x| + 555 > 554$ is all real numbers.

CHAPTER REVIEW

1.1 Linear Equations and Rational Equations

Definitions and Concepts	Examples
An **equation** is a statement indicating that two quantities are equal. There can be restrictions on the variable in an equation.	**Equations:** $2x - 5 = 10$, $\quad \dfrac{2(x - 2)}{x - 3} = \dfrac{7x + 3}{x + 2}$ In the equation $2x - 5 = 10$, x can be any real number. In the equation $\dfrac{2(x - 2)}{x - 3} = \dfrac{7x + 3}{x + 2}$, x cannot be 3 or -2, because this would give a 0 in the denominator.
Properties of equality: If $a = b$ and c is a number, then $\quad a + c = b + c \quad$ and $\quad a - c = b - c$ $\quad ac = bc \qquad$ and $\quad \dfrac{a}{c} = \dfrac{b}{c} \quad (c \neq 0)$	If $a = b$, then $\quad a + 7 = b + 7 \quad$ and $\quad a - 7 = b - 7$ $\quad 7a = 7b \qquad$ and $\quad \dfrac{a}{7} = \dfrac{b}{7}$
A **linear equation** is an equation that can be written in the form $ax + b = 0$ $(a \neq 0)$. To solve a linear equation, use the properties of equality to isolate x on one side of the equation.	Solve $3x - 5 = 4$. $\qquad 3x - 5 = 4$ $\quad 3x - 5 + 5 = 4 + 5 \qquad$ Add 5 to both sides. $\qquad\qquad 3x = 9 \qquad$ Combine like terms. $\qquad\qquad \dfrac{3x}{3} = \dfrac{9}{3} \qquad$ Divide both sides by 3. $\qquad\qquad x = 3$
An **identity** is an equation that is true for all acceptable replacements for its variable. A **contradiction** is an equation that is false for all acceptable replacements for its variable.	**Identities:** $x + x = 2x$, $\quad 2(x + 1) = 2x + 2$ **Contradictions:** $x + 1 = x$, $\quad 2(x + 1) = 2x + 3$
Rational equations are equations that contain rational expressions. To solve rational equations, multiply both sides of the equation by an expression that will remove the denominators and solve the resulting equation. Be sure to check the answers to identify any **extraneous solutions**.	Solve $\dfrac{2x}{x - 3} = \dfrac{6}{x - 3}$. $\qquad \dfrac{2x}{x - 3} = \dfrac{6}{x - 3}$ $\quad (x - 3)\left(\dfrac{2x}{x - 3}\right) = (x - 3)\left(\dfrac{6}{x - 3}\right) \qquad$ Multiply both sides by $x - 3$. $\qquad\qquad 2x = 6 \qquad$ Simplify. $\qquad\qquad x = 3 \qquad$ Divide both sides by 2 and simplify. The result of 3 is extraneous because when we substitute 3 into the original equation, we get a denominator of 0.

Definitions and Concepts	Examples
Formulas can be solved for a specific variable.	Solve $A = \dfrac{1}{2}bh$ for h.

$$A = \frac{1}{2}bh$$

$$2A = bh \qquad \text{Multiply both sides by 2.}$$

$$\frac{2A}{b} = \frac{bh}{b} \qquad \text{Divide both sides by } b.$$

$$\frac{2A}{b} = h \qquad \text{Simplify.}$$

Exercises

Find the restrictions on x, if any.

1. $3x + 7 = 4$

2. $x + \dfrac{1}{x} = 2$

3. $\dfrac{1}{x-1} = 4$

4. $\dfrac{1}{x-2} = \dfrac{2}{x-3}$

Solve each equation and classify it as an identity, a conditional equation, or a contradiction.

5. $3(9x + 4) = 28$

6. $\dfrac{3}{2}a = 7(a + 11)$

7. $8(3x - 5) - 4(x + 3) = 12$

8. $\dfrac{x+3}{x+4} + \dfrac{x+3}{x+2} = 2$

9. $\dfrac{3}{x-1} = \dfrac{1}{2}$

10. $\dfrac{8x^2 + 72x}{9 + x} = 8x$

11. $\dfrac{3x}{x-1} - \dfrac{5}{x+3} = 3$

12. $x + \dfrac{1}{2x-3} = \dfrac{2x^2}{2x-3}$

13. $\dfrac{4}{x^2 - 13x - 48} - \dfrac{1}{x^2 + x - 6} = \dfrac{2}{x^2 - 18x + 32}$

14. $\dfrac{a-1}{a+3} + \dfrac{2a-1}{3-a} = \dfrac{2-a}{a-3}$

Solve each formula for the indicated variable.

15. $C = \dfrac{5}{9}(F - 32)$; F

16. $P_n = l + \dfrac{si}{f}$; f

17. $\dfrac{1}{f} = \dfrac{1}{f_1} + \dfrac{1}{f_2}$; f_1

18. $S = \dfrac{a - lr}{1 - r}$; l

1.2 Application of Linear Equations

Definitions and Concepts	Examples
Use the following steps to solve an application problem:	Two students leave their dorm in two cars traveling in opposite directions. If one student drives at a rate of 55 mph and the other at a rate of 50 mph, how long will it take for them to be 210 miles apart?

Use the following steps to solve an application problem:

1. Analyze the problem.
2. Pick a variable to represent the quantity to be found.
3. Form an equation.
4. Solve the equation.
5. Check the solution in the words of the problem.

Analyze the problem and pick a variable. We can organize the facts of the problem in the following chart. Since each student drives the same amount of time, let t represent that time.

	d	$=$	r	$\cdot$	t
Student 1	$55t$		55		t
Student 2	$50t$		50		t

Definitions and Concepts	Examples
	Form and solve an equation. Since the students are driving in opposite directions, the distance they are apart in t hours is the sum of the distances they drive, a total of 210 miles. We can form and solve the following equation: $$55t + 50t = 210$$ $$105t = 210 \qquad \text{Combine like terms.}$$ $$t = 2 \qquad \text{Divide both sides by 105.}$$ **Check.** In 2 hours, student 1 drives 55(2) miles and student 2 drives 50(2) miles, or 110 miles plus 100 miles. At this time, they will be 210 miles apart.

Exercises

19. Test scores Carlos took four tests in an English class. On each successive test, his score improved by 4 points. If his mean score was 66%, what did he score on the first test?

20. Fencing a garden A homeowner has 100 ft of fencing to enclose a rectangular garden. If the garden is to be 5 ft longer than it is wide, find its dimensions.

21. Travel Two women leave a shopping center by car traveling in opposite directions. If one car averages 45 mph and the other 50 mph, how long will it take for the cars to be 285 miles apart?

22. Travel Two taxis leave an airport and travel in the same direction. If the average speed of one taxi is 40 mph and the average speed of the other taxi is 46 mph, how long will it take before the cars are 3 miles apart?

23. Preparing a solution A liter of fluid is 50% alcohol. How much water must be added to dilute it to a 20% solution?

24. Washing windows Scott can wash 37 windows in 3 hours, and Bill can wash 27 windows in 2 hours. How long will it take the two of them to wash 100 windows?

25. Filling a tank A tank can be filled in 9 hours by one pipe and in 12 hours by another. How long will it take both pipes to fill the empty tank?

26. Producing brass How many ounces of pure zinc must be alloyed with 20 ounces of brass that is 30% zinc and 70% copper to produce brass that is 40% zinc?

27. Lending money A bank lends $10,000, part of it at 11% annual interest and the rest at 14%. If the annual income is $1265, how much was loaned at each rate?

28. Producing oriental rugs An oriental rug manufacturer can use one loom with a setup cost of $750 that can weave a rug for $115. Another loom, with a setup cost of $950, can produce a rug for $95. How many rugs are produced if the costs are the same on each loom?

1.3   Complex Numbers

Definitions and Concepts	Examples
Complex numbers: Numbers that can be written in the form $a + bi$, where a and b are real numbers and $i = \sqrt{-1}$, are **complex numbers**.	$7 - 2i$, $9 + 5i$, and $2 + \sqrt{7}i$ are complex numbers. $\sqrt{-36} = \sqrt{(-1)(36)} = \sqrt{-1}\sqrt{36} = (i)(6) = 6i$
Equality of complex numbers: $a + bi = c + di$ if and only if $a = c$ and $b = d$	$3 + \sqrt{4}i = \frac{6}{2} + 2i$ because $3 = \frac{6}{2}$ and $\sqrt{4} = 2$.

Definitions and Concepts	Examples
Adding, subtracting, and multiplying complex numbers:	
• $(a + bi) + (c + di) = (a + c) + (b + d)i$	$(-3 + 4i) + (2 + 7i) = (-3 + 2) + (4 + 7)i = -1 + 11i$
• $(a + bi) - (c + di) = (a - c) + (b - d)i$	$(-3 + 4i) - (2 + 7i) = (-3 - 2) + (4 - 7)i = -5 - 3i$
• $(a + bi)(c + di) = (ac - bd) + (ad + bc)i$ or multiply them as if they were binomials.	$(-3 + 4i)(2 + 7i) = -3(2) - 3(7i) + 4i(2) + 4i(7i)$ $= -6 - 21i + 8i + 28i^2$ $= -6 - 21i + 8i - 28$ $= -34 - 13i$
The **complex conjugate** of $a + bi$ is $a - bi$.	$3 + 4i$ and $3 - 4i$ are complex conjugates.
Division of complex numbers: To divide complex numbers, rationalize the denominator.	Divide $2 + i$ by $2 - i$. $\dfrac{2 + i}{2 - i} = \dfrac{(2 + i)(2 + i)}{(2 - i)(2 + i)}$ $\dfrac{2 + i}{2 + i} = 1$ $= \dfrac{4 + 2i + 2i + i^2}{4 + 2i - 2i - i^2}$ $= \dfrac{4 + 4i - 1}{4 - (-1)}$ $= \dfrac{3 + 4i}{5}$ $= \dfrac{3}{5} + \dfrac{4}{5}i$
Powers of i: If n is a natural number that has a remainder of r when divided by 4, then $i^n = i^r$.	$i^2 = -1, i^3 = -i, i^4 = 1, i^5 = i, i^6 = -i, i^7 = -i,$ $i^8 = 1, \ldots$
Absolute value of a complex number: $\lvert a + bi \rvert = \sqrt{a^2 + b^2}$	$\lvert 5 - 7i \rvert = \sqrt{5^2 + (-7)^2} = \sqrt{25 + 49} = \sqrt{74}$

Exercises

Perform all operations and express all answers in $a + bi$ form.

29. $3\sqrt{-300}$

30. $-\sqrt{\dfrac{-45}{4}}$

31. $(2 - 3i) + (-4 + 2i)$

32. $(3 - \sqrt{-36}) + (\sqrt{-16} + 2)$

33. $(2 - 3i) - (4 + 2i)$

34. $(5 - 11i)(5 + 11i)$

35. $(8 - 3i)^2$

36. $(3 + \sqrt{-9})(2 - \sqrt{-25})$

37. $\dfrac{3}{i}$

38. $-\dfrac{5}{6i}$

39. $\dfrac{3}{1 + i}$

40. $\dfrac{2i}{2 - i}$

41. $\dfrac{3 + i}{3 - i}$

42. $\dfrac{3 - 2i}{1 + i}$

43. Simplify: i^{53}.

44. Simplify: i^{103}.

45. $-\dfrac{2}{i^3}$

46. $\lvert 3 - i \rvert$

47. $\left\lvert \dfrac{1 + i}{1 - i} \right\rvert$

48. Factor $64r^2 + 9s^2$ over the set of complex numbers.

1.4 Quadratic Equations

Definitions and Concepts	Examples
A **quadratic equation** is an equation that can be written in the form $ax^2 + bx + c = 0$, where a, b, and c are real numbers and $a \neq 0$.	$3x^2 - 5x - 7 = 0, \qquad 5x^2 - 25 = 0, \qquad 7x^2 + 14x = 0$
Zero-Factor Theorem: If $ab = 0$, then $a = 0$ or $b = 0$.	Solve $x^2 - x - 6 = 0$ using the Zero-Factor Theorem. $$x^2 - x - 6 = 0$$ $$(x + 2)(x - 3) = 0 \qquad \text{Factor } x^2 - x - 6.$$ $$x + 2 = 0 \quad \text{or} \quad x - 3 = 0$$ $$x = -2 \quad \mid \quad x = 3$$
Square Root Property: If $x^2 = c$, then $\quad x = \sqrt{c} \quad \text{or} \quad x = -\sqrt{c}$	If $x^2 = -32$, then $$x = \sqrt{-32} \qquad \text{or} \qquad x = -\sqrt{-32}$$ $$= \sqrt{-16 \cdot 2} \qquad\qquad = -\sqrt{-16 \cdot 2}$$ $$= 4i\sqrt{2} \qquad\qquad\quad = -4i\sqrt{2}$$
Steps to complete the square: 1. Make the coefficient of x^2 equal to 1. 2. Get the constant on the right side of the equation. 3. Complete the square on x. Take one-half the coefficient of x, square it, and add it to both sides of the equation. 4. Factor the resulting perfect-square trinomial and combine like terms. 5. Solve the resulting quadratic equation by using the Square Root Property.	Solve $x^2 - x - 6 = 0$ by completing the square: 1. Since the coefficient of x^2 is 1, we go to Step 2. 2. Add 6 to both sides to get the constant on the right side: $x^2 - x = 6$. 3. $x^2 - x + \left(-\dfrac{1}{2}\right)^2 = 6 + \left(-\dfrac{1}{2}\right)^2$ 4. $\left(x - \dfrac{1}{2}\right)^2 = \dfrac{25}{4}$ 5. $x - \dfrac{1}{2} = \pm\sqrt{\dfrac{25}{4}}$ $\qquad x = \dfrac{1}{2} \pm \dfrac{5}{2}$ $x = \dfrac{6}{2} = 3 \quad \text{or} \quad x = -\dfrac{4}{2} = -2$
Quadratic Formula: $x = \dfrac{-b \pm \sqrt{b^2 - 4ac}}{2a} \quad (a \neq 0)$	If $3x^2 - 5x + 1 = 0$, then $a = 3$, $b = -5$, and $c = 1$. So $$x = \frac{-b \pm \sqrt{b^2 - 4ac}}{2a} = \frac{-(-5) \pm \sqrt{(-5)^2 - 4(3)(1)}}{2(3)}$$ $$= \frac{5 \pm \sqrt{25 - 12}}{6} = \frac{5 \pm \sqrt{13}}{6}$$
Discriminant: The value $b^2 - 4ac$ is the **discriminant**. • If $b^2 - 4ac = 0$, there is one root of $ax^2 + bx + c = 0$ and it is a repeated rational number. • If $b^2 - 4ac > 0$ and a perfect square, the roots of $ax^2 + bx + c = 0$ are two different rational numbers.	In $4x^2 - 12x + 9 = 0$, the discriminant is $b^2 - 4ac = (-12)^2 - 4(4)(9) = 0$. So there is one root and it is a repeated rational number. In $x^2 - x - 6 = 0$, the discriminant is $b^2 - 4ac = (-1)^2 - 4(1)(-6) = 25$. Since 25 is positive and a perfect square, the roots are two different rational numbers.

Definitions and Concepts	Examples
• If $b^2 - 4ac > 0$ and not a perfect square, the roots of $ax^2 + bx + c = 0$ are two different irrational numbers.	In $x^2 - x - 5 = 0$, the discriminant is $b^2 - 4ac = (-1)^2 - 4(1)(-5) = 21$. Since 21 is positive and not a perfect square, the roots are two different irrational numbers.
• If $b^2 - 4ac < 0$, the roots of $ax^2 + bx + c = 0$ are two different nonreal complex numbers.	In $3x^2 - 2x + 1 = 0$, the discriminant is $b^2 - 4ac = (-2)^2 - 4(3)(1) = -8 < 0$. So the roots are two different nonreal complex numbers.

Exercises

Solve each equation by factoring.

49. $2x^2 - x - 6 = 0$

50. $12x^2 + 13x = 4$

51. $5x^2 - 8x = 0$

52. $27x^2 = 30x - 8$

Solve each equation by using the Square Root Property.

53. $2x^2 = 16$

54. $12x^2 = -60$

55. $(4z - 5)^2 = 32$

56. $(5x - 7)^2 = -45$

Solve each equation by completing the square.

57. $x^2 - 8x + 15 = 0$

58. $3x^2 + 18x = -24$

59. $5x^2 - x - 1 = 0$

60. $5x^2 - x = 0$

61. Solve: $3x^2 - 2x + 1 = 0$.

Use the Quadratic Formula to solve each equation.

62. $x^2 + 5x - 14 = 0$

63. $3x^2 - 25x = 18$

64. $5x^2 = 1 - x$

65. $5 = a^2 + 2a$

66. Solve: $3x^2 + 4 = 2x$.

67. Calculate the discriminant associated with the equation $6x^2 + 5x + 1 = 0$.

68. Determine the number and nature of the roots of the equation in Exercise 67.

69. Find the value of k that will make the roots of $kx^2 + 4x + 12 = 0$ equal.

70. Find the values of k that will make the roots of $4y^2 + (k + 2)y = 1 - k$ equal.

71. $\dfrac{3x}{2} - \dfrac{2x}{x - 1} = x - 3$

72. Solve: $\dfrac{4}{a - 4} + \dfrac{4}{a - 1} = 5$.

1.5 Applications of Quadratic Equations

Definitions and Concepts	Examples
Many real-life problems are modeled by quadratic equations.	If a missile is launched straight up into the air with an initial velocity of 128 feet per second, its height will be given by the formula $h = -16t^2 + 128t$, where h represents its height (in feet) and t represents the time (in seconds) since it was launched. How long will it take the missile to return to its starting point?
	When the missile returns to its starting point, its height will again be 0. So we let $h = 0$ and solve for t.
	$$0 = -16t^2 + 128t$$ $$0 = -16t^2(t - 8)$$ $$-16t = 0 \quad \text{or} \quad t - 8 = 0$$ $$t = 0 \qquad \qquad t = 8$$
	The missile will leave its starting point at 0 seconds and return at 8 seconds.

Exercises

73. Fencing a field A farmer wishes to enclose a rectangular garden with 300 yards of fencing. A river runs along one side of the garden, so no fencing is needed there. Find the dimensions of the rectangle if the area is 10,450 square yards.

74. Flying rates A jet plane, flying 120 mph faster than a propeller-driven plane, travels 3520 miles in 3 hours less time than the propeller plane requires to fly the same distance. How fast does each plane fly?

75. Flight of a ball A ball thrown into the air reaches a height h (in feet) according to the formula $h = -16t^2 + 64t$, where t is the time elapsed since the ball was thrown. Find the shortest time it will take the ball to reach a height of 48 feet.

76. Width of a walk A man built a walk of uniform width around a rectangular pool. If the area of the walk is 117 square feet and the dimensions of the pool are 16 feet by 20 feet, how wide is the walk?

1.6 Other Types of Equations

Definitions and Concepts	Examples
Polynomial equations: Many polynomial equations of higher degree can be solved by factoring.	Solve $x^3 - 5x^2 + 6x = 0$. $x^3 - 5x^2 + 6x = 0$ $x(x^2 - 5x + 6) = 0$ Factor out x. $x(x - 2)(x - 3) = 0$ Factor $x^2 - 5x + 6$. $x = 0$　or　$x - 2 = 0$　or　$x - 3 = 0$ 　　　　　　　　$x = 2$　　　　$x = 3$ The solution set is $\{0, 2, 3\}$.
Power property of real numbers: If $a = b$, then $a^2 = b^2$.	If $x = 5$, then $x^2 = 5^2$ or $x^2 = 25$.
Factoring can be used to solve certain nonpolynomial equations. These equations are quadratic in form and can also be solved by making an appropriate substitution. Check all solutions because extraneous roots can be introduced.	Solve $2x - 5x^{1/2} + 3 = 0$. $2x - 5x^{1/2} + 3 = 0$ $(2x^{1/2} - 3)(x^{1/2} - 1) = 0$ $2x^{1/2} - 3 = 0$　or　$x^{1/2} - 1 = 0$ $x^{1/2} = \dfrac{3}{2}$　　　　$x^{1/2} = 1$ $x = \dfrac{9}{4}$　　　　　$x = 1$ Both roots check.
Radical equations: To solve radical equations, use the Power Property of Real Numbers.	Solve $\sqrt{2x - 3} = x - 1$. $\sqrt{2x - 3} = x - 1$ $\left(\sqrt{2x - 3}\right)^2 = (x - 1)^2$ Square both sides. $2x - 3 = x^2 - 2x + 1$ $0 = x^2 - 4x + 4$ $0 = (x - 2)(x - 2)$ $x - 2 = 0$　or　$x - 2 = 0$ $x = 2$　　　　$x = 2$ Since 2 checks, it is a root.

Exercises

Solve each equation.

77. $x^3 + 4x^2 - 12x = 0$

78. $3x^3 + 4x^2 - 4x = 0$

79. $x^4 - 2x^2 + 1 = 0$

80. $x^4 - 36 = -35x^2$

81. $a - a^{1/2} - 6 = 0$

82. $x^{2/3} + x^{1/3} - 6 = 0$

83 $6y^{-2} + 13y^{-1} - 5 = 0$

84. $\sqrt{5x - 11} - 5 = -3$

85. $\sqrt{x - 1} + x = 7$

86. $\sqrt{a + 9} - \sqrt{a} = 3$

87. $\sqrt{5 - x} + \sqrt{5 + x} = 4$

88. $\sqrt{y + 5} + \sqrt{y} = 1$

89. $\sqrt[3]{4x - 9} + 3 = 2$

90. $\sqrt[4]{x - 2} + 3 = 5$

1.7 Inequalities

Definitions and Concepts	Examples
Addition, Subtraction, Multiplication, and Division Properties of Inequalities:	
If a, b, and c are real numbers:	
If $a < b$, $a + c < b + c$ and $a - c < b - c$.	If $x < 10$, then $x + 6 < 10 + 6$ and $x - 6 < 10 - 6$.
If $a < b$ and $c > 0$, $ac < bc$ and $\dfrac{a}{c} < \dfrac{b}{c}$.	If $x < 12$, then $3x < 3(12)$ and $\dfrac{x}{3} < \dfrac{12}{3}$.
If $a < b$ and $c < 0$, $ac > bc$ and $\dfrac{a}{c} > \dfrac{b}{c}$.	If $x \le 12$, then $-3x \ge -3(12)$ and $\dfrac{x}{-3} \ge \dfrac{12}{-3}$.
Trichotomy Property:	
$\quad a < b$, $a = b$, or $a > b$	Either $x < 3$, $x = 3$, or $x > 3$.
Transitive Property:	
$\quad$ If $a < b$ and $b < c$, then $a < c$.	If $x < 4$ and $4 < y$, then $x < y$.
Types of inequalities:	
• Linear inequality	$-4(x - 3) \ge 7$
• Compound inequality	$-5 \le 2x + 1 < 3$
• Quadratic inequality	$x^2 - 2x + 3 \le 0$
• Rational inequality	$\dfrac{x + 1}{x - 2} > 0$
Solving a linear inequality:	Solve the *linear* inequality: $-4(x - 3) \ge 7$.
Use the same steps to solve a linear inequality as you would use to solve a linear equation. However, remember to reverse the order of the inequality when you multiply (or divide) both sides of an inequality by a negative number.	$-4(x - 3) \ge 7$ $-4x + 12 \ge 7$ Remove parentheses. $-4x \ge -5$ Subtract 12 from both sides. $x \le \dfrac{5}{4}$ Divide both sides by -4. $\dfrac{5}{4}$ In interval notation, the solution is $\left(-\infty, \dfrac{5}{4}\right]$.

Definitions and Concepts	Examples
Solving a compound inequality: For a compound inequality, isolate x in the middle, if possible. If not, solve each inequality separately. The intersection of the intervals is the solution.	Solve the *compound* inequality: $-5 \le 2x + 1 < 3$. $-5 \le 2x + 1 < 3$ $-6 \le 2x < 2$ Subtract 1 from all three parts. $-3 \le x < 1$ Divide each part by 2. In interval notation, the solution is $[-3, 1)$.

Solving quadratic and rational inequalities:

Quadratic and rational inequalities can be solved by constructing a table and testing values or by constructing a sign graph.

Solve the *rational* inequality: $\frac{x+1}{x-2} > 0$.

First note that the solution of $x + 1 = 0$ ($x = -1$) and the solution of $x - 2 = 0$ ($x = 2$) form three intervals: $(-\infty, -1)$, $(-1, 2)$, and $(2, \infty)$. To determine which intervals are solutions, we test a number in each interval to see whether it satisfies the inequality.

		Inequality $\dfrac{x+1}{x-2} > 0$	
Interval	Test Value		Result
$(-\infty, -1)$	-2	$\dfrac{-2+1}{-2-2} = \dfrac{1}{4} > 0$ True	The numbers in this interval are solutions.
$(-1, 2)$	0	$\dfrac{0+1}{0-2} = -\dfrac{1}{2} > 0$ False	The numbers in this interval are not solutions.
$(2, \infty)$	3	$\dfrac{3+1}{3-2} = 4 > 0$ True	The numbers in this interval are solutions.

The solution is the union of two intervals:

$(-\infty, -1) \cup (2, \infty)$

Exercises

Solve each inequality; graph the solution set and write the answer in interval notation.

91. $2x - 9 < 5$

92. $5x + 3 \ge 2$

93. $\dfrac{5(x-1)}{2} < x$

94. $\dfrac{1}{4}x + \dfrac{2}{3}x - x > \dfrac{1}{2} + \dfrac{1}{2}(x + 1)$

95. $0 \le \dfrac{3+x}{2} < 4$

96. $2 + a < 3a - 2 \le 5a + 2$

97. $(x + 2)(x - 4) > 0$

98. $(x - 1)(x + 4) < 0$

99. $x^2 - 2x - 3 \le 0$

100. $2x^2 + x > 3$

101. $\dfrac{x+2}{x-3} \ge 0$

102. $\dfrac{x-1}{x+4} \le 0$

103. $\dfrac{x^2 + x - 2}{x - 3} \ge 0$

104. $\dfrac{5}{x} < 2$

1.8 Absolute Value

Definitions and Concepts	Examples
Definition of absolute value of x: $$\|x\| = \begin{cases} x \text{ when } x \ge 0 \\ -x \text{ when } x < 0 \end{cases}$$	$\|5\| = 5 \qquad \|-7\| = 7 \qquad -\|-10\| = -10$
Absolute value equations: • If $k \ge 0$, then $\|x\| = k$ is equivalent to $x = k$ or $x = -k$. • If a and b are algebraic expressions, $\|a\| = \|b\|$ is equivalent to $a = b$ or $a = -b$.	Solve: $\|x - 2\| = 6$. $x - 2 = 6 \quad$ or $\quad x - 2 = -6$ $\qquad x = 8 \qquad\qquad\quad x = -4$ Solve: $\|3x\| = \|x - 2\|$. $3x = x - 2 \quad$ or $\quad 3x = -(x - 2)$ $2x = -2 \qquad\qquad\quad 3x = -x + 2$ $\quad x = -1 \qquad\qquad\quad 4x = 2$ $\qquad\qquad\qquad\qquad\qquad x = \dfrac{1}{2}$ The solutions to both examples check.
Absolute value inequality properties: • If $k > 0$, then $\|x\| < k$ is equivalent to $-k < x < k$. • If $k > 0$, then $\|x\| > k$ is equivalent to $x > k$ or $x < -k$. These two properties also hold for $\le$ and $\ge$.	Solve: $\|x - 2\| < 6$. $-6 < x - 2 < 6$ $-4 < x < 8 \qquad$ Add 2 to all three parts. The solution in interval notation is $(-4, 8)$. Solve: $\|x - 2\| > 6$. $x - 2 > 6 \quad$ or $\quad x - 2 < -6$ $\quad x > 8 \qquad\qquad\quad x < -4 \qquad$ Add 2 to both parts. Thus, $x < -4$ or $x > 8$ The solution in interval notation is $(-\infty, -4) \cup (8, \infty)$.
Properties of absolute value: **1.** $\|ab\| = \|a\|\|b\|$ **2.** $\left\|\dfrac{a}{b}\right\| = \dfrac{\|a\|}{\|b\|} \quad (b \ne 0)$ **3.** $\|a + b\| \le \|a\| + \|b\|$	$\|-3x\| = \|-3\|\|x\| = 3\|x\|$ $\left\|\dfrac{-3}{x}\right\| = \dfrac{\|-3\|}{\|x\|} = \dfrac{3}{\|x\|} \quad (x \ne 0)$ $\|x + 3\| \le \|x\| + \|3\|$

Exercises

Solve each equation or inequality.

105. $|x + 1| = 6$

106. $\left|\dfrac{3x + 11}{7}\right| - 1 = 0$

112. $\left|\dfrac{x + 2}{3}\right| + 5 < 6$

113. $\left|\dfrac{x - 3}{4}\right| > 8$

107. $\left|\dfrac{2a - 6}{3a}\right| - 6 = 0$

108. $|2x - 1| = |2x + 1|$

114. $1 < |2x + 3| < 4$

115. $0 < |3x - 4| < 7$

109. $|3x - 11| + 16 = 5$

110. $|x + 3| < 3$

111. $|3x - 7| \geq 1$

CHAPTER TEST

Find all restrictions on x.

1. $\dfrac{x}{x(x - 1)} = 2$

2. $\dfrac{4}{3x - 2} + 3 = 7$

Solve each equation.

3. $7(2a + 5) - 7 = 6(a + 8)$

4. $\dfrac{1}{a - 2} - \dfrac{1}{5a} = \dfrac{3}{2a}$

5. Solve for x: $z = \dfrac{x - \mu}{\sigma}$.

6. Solve for a: $\dfrac{1}{a} = \dfrac{1}{b} + \dfrac{1}{c}$.

7. Test scores A student's average on three tests is 75. If the final is to count as two one-hour tests, what grade must the student make to bring the average up to 80?

8. Investment A woman invested part of $20,000 at 6% interest and the rest at 7%. If her annual interest is $1260, how much did she invest at 6%?

Simplify the imaginary numbers.

9. $3\sqrt{-96}$

10. $\sqrt{-\dfrac{18}{5}}$

Perform each operation and write all answers in a + bi form.

11. $(4 - 5i) - (-3 + 7i)$

12. $(4 - 5i)(3 - 7i)$

13. $\dfrac{2}{2 - i}$

14. $\dfrac{1 + i}{1 - i}$

Simplify each expression.

15. i^{13}

16. $7i^4$

Solve each equation.

17. $4x^2 - 8x + 3 = 0$

18. $2b^2 - 12 = -5b$

19. $5x^2 = -135$

20. Use completing the square to solve $x^2 - 14x = 23$.

21. Use the Quadratic Formula to solve
$3x^2 - 5x - 9 = 0$.

22. $\dfrac{3}{x^2 - 5x - 14} = \dfrac{4}{x^2 + 5x + 6}$

23. Find k such that $x^2 + (k + 1)x + k + 4 = 0$ will have two equal roots.

24. Height of a projectile The height h (in feet) of a projectile shot up into the air, at time t (in seconds), is given by the formula $h = -16t^2 + 128t$. Find the time t required for the projectile to return to its starting point.

Find each absolute value.

25. $|5 - 12i|$

26. $\left|\dfrac{1}{3 + i}\right|$

Solve each equation.

27. $z^4 - 13z^2 + 36 = 0$

28. $2p^{2/5} - p^{1/5} - 1 = 0$

29. $\sqrt{x + 5} = 12$

30. $\sqrt{2z + 3} = 1 - \sqrt{z + 1}$

Solve each inequality; graph the solution set and write the answer using interval notation.

31. $5x - 3 \leq 7$

32. $\dfrac{x+3}{4} > \dfrac{2x-4}{3}$

33. $5 \leq 2x - 1 < 7$

34. $1 + x < 3x - 3 < 4x - 2$

35. $x^2 - 7x - 8 \geq 0$

36. $\dfrac{x+2}{x-1} \leq 0$

Solve each equation.

37. $\left| \dfrac{3x+2}{2} \right| = 4$

38. $|x + 3| = |x - 3|$

Solve each inequality; graph the solution set and write the answer using interval notation.

39. $|2x - 5| > 2$

40. $\left| \dfrac{2x+3}{3} \right| \leq 5$

CUMULATIVE REVIEW EXERCISES

Consider the set $\{-5, -3, -2, 0, 1, \sqrt{2}, 2, \frac{5}{2}, 5, 6, 11\}$.

1. Which numbers are even integers?

2. Which numbers are prime numbers?

Write each inequality as an interval and graph it.

3. $-4 \leq x < 7$

4. $x \geq 2$ or $x < 0$

Determine which property of the real numbers justifies each expression.

5. $(a + b) + c = c + (a + b)$

6. If $x < 3$ and $3 < y$, then $x < y$.

Simplify each expression. Assume that all variables represent positive numbers. Give all answers with positive exponents.

7. $(81a^4)^{1/2}$

8. $81(a^4)^{1/2}$

9. $(a^{-3}b^{-2})^{-2}$

10. $\left(\dfrac{4x^4}{12x^2y} \right)^{-2}$

11. $\left(\dfrac{4x^0y^2}{x^2y} \right)^{-2}$

12. $\left(\dfrac{4x^{-5}y^2}{6x^{-2}y^{-3}} \right)^2$

13. $(a^{1/2}b)^2(ab^{1/2})^2$

14. $(a^{1/2}b^{1/2}c)^2$

Rationalize each denominator and simplify.

15. $\dfrac{3}{\sqrt{3}}$

16. $\dfrac{2}{\sqrt[3]{4x}}$

17. $\dfrac{3}{y - \sqrt{3}}$

18. $\dfrac{3x}{\sqrt{x} - 1}$

Simplify each expression and combine like terms.

19. $\sqrt{75} - 3\sqrt{5}$

20. $\sqrt{18} + \sqrt{8} - 2\sqrt{2}$

21. $\left(\sqrt{2} - \sqrt{3} \right)^2$

22. $\left(3 - \sqrt{5} \right)\left(3 + \sqrt{5} \right)$

Perform the operations and simplify when necessary.

23. $(3x^2 - 2x + 5) - 3(x^2 + 2x - 1)$

24. $5x^2(2x^2 - x) + x(x^2 - x^3)$

25. $(3x - 5)(2x + 7)$

26. $(z + 2)(z^2 - z + 2)$

27. $3x + 2\overline{)6x^3 + x^2 + x + 2}$

28. $x^2 + 2\overline{)3x^4 + 7x^2 - x + 2}$

Factor each polynomial.

29. $3t^2 - 6t$

30. $3x^2 - 10x - 8$

31. $x^8 - 2x^4 + 1$

32. $x^6 - 1$

Perform the operations and simplify.

33. $\dfrac{x^2 - 4}{x^2 + 5x + 6} \cdot \dfrac{x^2 - 2x - 15}{x^2 + 3x - 10}$

34. $\dfrac{6x^3 + x^2 - x}{x + 2} \div \dfrac{3x^2 - x}{x^2 + 4x + 4}$

35. $\dfrac{2}{x + 3} + \dfrac{5x}{x - 3}$

36. $\dfrac{x - 2}{x + 3}\left(\dfrac{x + 3}{x^2 - 4} - 1\right)$

37. $\dfrac{\dfrac{1}{a} + \dfrac{1}{b}}{\dfrac{1}{ab}}$

38. $\dfrac{x^{-1} - y^{-1}}{x - y}$

Solve each equation.

39. $\dfrac{3x}{x + 5} = \dfrac{x}{x - 5}$

40. $8(2x - 3) - 3(5x + 2) = 4$

Solve each formula for the indicated variable.

41. $\dfrac{1}{R} = \dfrac{1}{R_1} + \dfrac{1}{R_2}$; R

42. $S = \dfrac{a - lr}{1 - r}$; r

43. Gardening A gardener wishes to enclose her rectangular raspberry patch with 40 feet of fencing. The raspberry bushes are planted along the garage, so no fencing is needed on that side. Find the dimensions if the total area is to be 192 square feet.

44. Financial planning A college student invested part of a $25,000 inheritance at 7% interest and the rest at 6%. If his annual interest is $1670, how much did he invest at 6%?

Perform the operations. If the result is not real, express the answer in a + bi form.

45. $\dfrac{2 + i}{2 - i}$

46. $\dfrac{i(3 - i)}{(1 + i)(1 + i)}$

47. $|3 + 4i|$

48. $\dfrac{5}{i^7} + 5i$

Solve each equation.

49. $15x^2 - 16x - 7 = 0$

50. $(7x - 4)^2 = -8$

51. $\dfrac{x + 3}{x - 1} - \dfrac{6}{x} = 1$

52. $x^4 + 36 = 13x^2$

53. $\sqrt{y + 2} + \sqrt{11 - y} = 5$

54. $z^{2/3} - 13z^{1/3} + 36 = 0$

Solve each inequality; graph the solution set and write the answer using interval notation.

55. $5x - 7 \le 4$

56. $x^2 - 8x + 15 > 0$

57. $\dfrac{x^2 + 4x + 3}{x - 2} \ge 0$

58. $\dfrac{9}{x} > x$

59. $|2x - 3| \ge 5$

60. $\left|\dfrac{3x - 5}{2}\right| < 2$

Functions and Graphs

Careers and Mathematics: **Cartographer**

Mark & Audrey Gibson/Stock Connection Blue/Alamy

Cartographers collect, analyze, interpret, and map geographic information about the Earth's surface. Their work involves geographical research and compiling data to produce maps. They analyze latitude and longitude, elevation and distance, as well as population density, land-use patterns, precipitation levels, and demographic characteristics. Their maps are prepared in either digital or graphic form, using information provided by geodetic surveys, aerial cameras, satellites, light-imaging detection, and Geographic Information Systems (GIS).

Education and Mathematics Required
- Cartographers usually have a bachelor's degree in cartography, geography, surveying, engineering, forestry, computer science, or a physical science.
- College Algebra, Trigonometry, Calculus I and II, Elementary Statistics, and Spatial Statistics are usually required.

How Cartographers Use Math and Who Employs Them
- Math helps cartographers with map scale, coordinate systems, and map projection. Map scale is the relationship between distances on a map and the corresponding distances on the Earth's surface. Coordinate systems are numeric methods of representing locations on the Earth's surface. Map projection is a function or transformation that relates coordinates of points on a curved surface to coordinates of points on a plane.
- Most cartographers work with engineering, architectural, and surveying firms. Government agencies hire cartographers to work in highway departments and in areas such as land management, natural resources planning, and national defense.

Career Outlook and Salary
- Employment of cartographers is expected to grow 20% from 2012 to 2022.
- The median annual wage in May 2012 was $57,440.

For more information see: www.bls.gov/oco

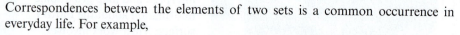

2.1   Functions and Function Notation

In this section, we will learn to

1. Define relation, domain, and range.
2. Understand the concept of a function.
3. Determine whether an equation represents a function.
4. Find the domain of a function.
5. Evaluate a function.
6. Evaluate the difference quotient for a function.
7. Solve applications involving functions.

©Twin Design/Shutterstock.com

Correspondences between the elements of two sets is a common occurrence in everyday life. For example,

- To every Apple iPhone, there corresponds exactly one phone number.
- To every Honda Civic car, there corresponds exactly one vehicle identification number.
- To every NFL football team, there corresponds exactly one team city.
- To every item's barcode at the Target store, there corresponds exactly one price.

This table shows the four highest-grossing movies of all time in the United States and the year each movie was released.

Movie	Year
Avatar	2009
Titanic	1997
The Avengers	2012
The Dark Knight	2008

The information shown in the table sets up a correspondence between a movie and the year it was released. Note that for each of the movies, there corresponds exactly one year in which it was released.

For each of the correspondences mentioned already, for each input there corresponded exactly one output. These correspondences represent functions, the topic of the section. To fully grasp the concept of a function we will first consider relations.

1.   Define Relation, Domain, and Range

The movie data in the table can be displayed as a set of ordered pairs, where the **first component** is the movie and the **second component** is the year it was released.

{(*Avatar*, 2009), (*Titanic*, 1997), (*The Avengers*, 2012), (*The Dark Knight*, 2008)}

Sets of ordered pairs like this are called **relations**. A **relation** is a correspondence between a set of input values, called the **domain**, and a set of output values, called the **range**. For example, a golf course has 18 holes, and each hole has a length in yards. The domain of the relation would be the numbers 1 through 18, and the range would be the list of the yardages. In mathematics, relations can be represented in a variety of ways. We can use a set of ordered pairs, a table of values, an equation, or a graph to represent the relation.

Relation   A **relation** is a set of ordered pairs. The set created by the first components in the ordered pairs is called the **domain** (the inputs). The set created by the second components in the ordered pairs is the **range** (the outputs).

Finding the Domain and Range of a Relation

EXAMPLE 1   Find the domain and the range of the relation: $\{(2, 7), (-3, 13), (5, -11), (-10, 13)\}$.

SOLUTION   The set of first components is the domain and the set of second components is the range. The domain is $\{2, -3, 5, -10\}$. The range is $\{7, 13, -11\}$. Note that if a value is repeated, we list it only once.

Self Check 1   Find the domain and the range of the relation: $\{(-2, -7), (3, -13), (-5, 11), (10, -13)\}$.

Now Try Exercise 11.

Writing the Ordered Pairs in a Relation and Identifying the Domain and Range

EXAMPLE 2   The scorecard for a golf course is shown in Figure 2-1 and represents a relation. Write the set of ordered pairs (hole, yards) represented by the relation and state the domain and range created by this set of ordered pairs.

Hole	1	2	3	4	5	6	7	8	9	10	11	12	13	14	15	16	17	18
Yards	269	315	140	390	365	120	451	358	375	465	298	165	315	495	135	330	358	420

FIGURE 2-1

SOLUTION   The set of ordered pairs is $\{(1, 269), (2, 315), (3, 140), (4, 390), (5, 365), (6, 120), (7, 451), (8, 358), (9, 375), (10, 465), (11, 298), (12, 165), (13, 315), (14, 495), (15, 135), (16, 330), (17, 358), (18, 420)\}$.

The domain is the set created by the first components in the ordered pairs. The domain is

$$\{1, 2, 3, 4, 5, 6, 7, 8, 9, 10, 11, 12, 13, 14, 15, 16, 17, 18\}$$

The range is the set created by the second components in the ordered pairs. The range is $\{269, 315, 140, 390, 365, 120, 451, 358, 375, 465, 298, 165, 315, 495, 135, 330, 358, 420\}$.

Self Check 2   Five Ford cars and the starting manufacturer retail selling price (MSRP) for each are shown in the table.

Ford Car	MSRP
Fiesta	$13,965
Focus	$17,170
Fusion	$22,010
Mustang	$23,800
Taurus	$27,055

Write the set of ordered pairs (car, price) represented by the data for the relation. State the domain and range created by this set of ordered pairs.

Now Try Exercise 15.

Relations can also be defined using an **arrow** or **mapping diagram**. Our next example will illustrate this.

EXAMPLE 3

Determining the Ordered Pairs in a Relation and Identifying the Domain and Range

Use Figure 2-2 to list the ordered pairs in each relation illustrated in the mapping. The relations are R1 and R2. State the domain and range of each relation.

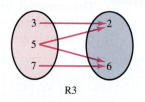

(a)

(b)

R1   R2

FIGURE 2-2

SOLUTION   **(a)** The relation R1 is $\{(a, w), (b, x), (c, y)\}$. The domain is $\{a, b, c\}$. The range is $\{w, x, y\}$.

(b) The relation R2 is $\{(4, 3), (5, 3), (6, 3)\}$. The domain is $\{4, 5, 6\}$. The range is $\{3\}$.

Self Check 3   Use Figure 2-3 to list the ordered pairs in the relation R3 illustrated. State the domain and range of the relation.

R3

FIGURE 2-3

Now Try Exercise 13.

Now let's look at a special and very important type of relation called a function.

2. Understanding the Concept of a Function

Any relation that assigns exactly one output value for each input value is called a **function**. The solar-powered aircraft *Pathfinder* recorded the following data relating altitude to temperature. This table, shown in Figure 2-4, is an example of a function. For each altitude, there is exactly one temperature.

Altitude (1000 ft)	Temperature (°C)
15	4.5
20	−5.9
25	−16.1
30	−27.9
35	−39.8
40	−50.2
45	−62.9

FIGURE 2-4

In Example 3, the relations established in parts **(a)** and **(b)** are also functions, because to every value in the domain, there is assigned only one value in the range. However, the relation in Self Check 3 is not a function because the value 5 in the domain is assigned two values in the range, 2 and 6. In a relation that is also a function, a number in the range (the second component in the ordered pair) can appear in more than one ordered pair, as in part (**b**) in Example 3. However, a number in the domain can be in only one ordered pair. This leads us to an important fact: Every function is a relation, but not every relation is a function.

Equations are relations and are frequently used in mathematics. Some equations represent functions because exactly one quantity corresponds to (or depends on) another quantity according to some specific rule. For example, the equation $y = x^2 - 1$ sets up a correspondence between two infinite sets of real numbers, x and y, according to the rule *square x and subtract 1*.

Equation: $y = x^2 - 1$

Correspondence: Each real number x determines exactly one real number y.

Rule: Square x and subtract 1.

Since the value of y depends on the number x, we call y the **dependent variable** and x the **independent variable**.

The equation $y = x^2 - 1$ determines what **output value** y will result from each **input value** x. This idea of inputs and outputs is shown in Figure 2-5(a). In the equation $y = x^2 - 1$, if the input x is 2, the output y is

$$y = 2^2 - 1 = 3$$

This is illustrated in Figure 2-5(b).

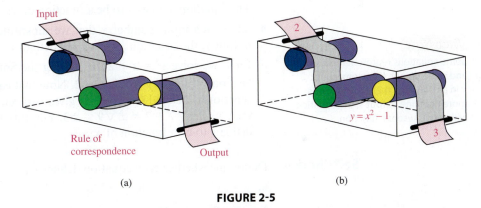

(a)   (b)

FIGURE 2-5

If we also input values of -2, -1, 0, and 1 for x into the equation and determine the corresponding values for y, we can write our inputs x and outputs y in table form as shown in Figure 2-6. We see that for each number we input for x there corresponds exactly one output for y. The equation represents y as a function of x.

$y = x^2 - 1$		
x	y	(x, y)
-2	3	$(-2, 3)$
-1	0	$(-1, 0)$
0	-1	$(0, -1)$
1	0	$(1, 0)$
2	3	$(2, 3)$

FIGURE 2-6

Any correspondence that assigns exactly one value of y to each number x is called a **function**. We refer to the set of inputs as the **domain** of the function and the set of outputs as the **range** of the function.

Function	A **function** f is a correspondence between a set of input values x and a set of output values y, where to each x-value there corresponds exactly one y-value.
Domain	The set of input values x is called the **domain** of the function.
Range	The set of output values y is called the **range** of the function.

3. Determine Whether an Equation Represents a Function

EXAMPLE 4

Determining Whether an Equation Represents a Function

Determine whether the following equations define y to be a function of x.

a. $7x + y = 5$ **b.** $y = |x| + 1$ **c.** $y^2 = x + 2$

SOLUTION In each case, we will solve each equation for y, if necessary, and determine whether each input value x determines exactly one output value y.

a. First, we solve $7x + y = 5$ for y to get

$$y = 5 - 7x$$

From this equation, we see that for each input value x there corresponds exactly one output value y. For example, if $x = 3$, then $y = 5 - 7(3) = -16$. The equation defines y to be a function of x.

b. Since each input number x that we substitute for x determines exactly one output y, the equation $y = |x| + 1$ defines y to be a function of x.

c. First, we solve $y^2 = x + 2$ for y using the Square Root Property covered in Section 1.4 to get $y = \pm\sqrt{x + 2}$. Note that each input value x for which the function is defined (except -2) gives two outputs y. For example, if $x = 7$, then $y = \pm\sqrt{7 + 2} = \pm\sqrt{9} = \pm 3$. For this reason, the equation does not define y to be a function of x.

> **Take Note**
>
> Some equations represent functions and some do not. If for some input x in an equation there corresponds more than one output y, the equation will not represent a function.

Self Check 4 Determine whether each equation defines y to be a function of x.

a. $y = |x|$ **b.** $y = \sqrt{x}$ **c.** $y^2 = 2x$

Now Try Exercise 33.

4. Find the Domain of a Function

The **domain** of a function is the set of all real numbers x for which the function is defined. Thus, to find the domain of a function we must find the set of numbers that are permissible inputs for x. It is often helpful to determine any restrictions on the input values for x. For example, consider the polynomial, radical, and rational functions shown in the table. These types of functions are important.

Function	Restriction	Domain
$y = x^2 + x - 5$	There are no restrictions on x. The right side of the equation is a polynomial.	all real numbers, $(-\infty, \infty)$
$y = \sqrt{x}$	The restriction that x cannot be a negative real number is placed on x because the square root of a negative number is an imaginary number.	$x \geq 0$, the interval $[0, \infty)$
$y = \dfrac{1}{x}$	The restriction that x cannot be 0 is placed on x because division by 0 isn't defined.	$x \neq 0$, $(-\infty, 0) \cup (0, \infty)$

Finding the Domain of a Function

EXAMPLE 5   Find the domain of the function defined by each equation:

a. $y = 2x - 5$ **b.** $y = \sqrt{3x - 2}$ **c.** $y = \dfrac{3x}{x + 2}$

SOLUTION   We must determine what numbers are permissible inputs for x. This set of numbers is the domain.

a. Any real number that we input for x can be multiplied by 2 and then 5 can be subtracted from the result. Thus, the domain is the interval $(-\infty, \infty)$.

b. Since the radicand must be nonnegative, we have

$$3x - 2 \geq 0$$

$$3x \geq 2 \qquad \text{Add 2 to both sides.}$$

$$x \geq \frac{2}{3} \qquad \text{Divide both sides by 3.}$$

Thus, the domain is the interval $\left[\frac{2}{3}, \infty\right)$.

c. Since the fraction $\frac{3x}{x + 2}$ is undefined when $x = -2$, -2 is not a permissible input value. Since all other values of x are permissible inputs, the domain is $(-\infty, -2) \cup (-2, \infty)$.

> **Tip**
> Notice that to determine the domain of the functions like $y = \dfrac{3x}{x + 2}$, we do not have to inspect the numerator. It can be any value, including 0.

Self Check 5   Find the domain of each function: **a.** $y = |x| + 2$ **b.** $y = \sqrt[3]{x + 2}$

Now Try Exercise 39.

Finding the Domain of a Function

EXAMPLE 6   Find the domain of the function defined by the equation $y = \dfrac{1}{x^2 - 5x - 6}$.

SOLUTION   We can factor the denominator to see what values of x will give 0 in the denominator. These values are not in the domain.

$$x^2 - 5x - 6 = 0$$

$$(x - 6)(x + 1) = 0$$

$$x - 6 = 0 \quad \text{or} \quad x + 1 = 0$$

$$x = 6 \qquad \qquad x = -1$$

The domain is $(-\infty, -1) \cup (-1, 6) \cup (6, \infty)$.

Self Check 6   Find the domain of the function defined by the equation $y = \dfrac{2}{x^2 - 16}$.

Now Try Exercise 53.

5. Evaluate a Function

To indicate that y is a function of x, we often use **function notation** and write

$$y = f(x) \qquad \text{Read as "y is a function of x."}$$

> **Caution**
> The symbol $f(x)$ denotes a function and does not mean f times x.

The notation $y = f(x)$ provides a way of denoting the value of y (the dependent variable) that corresponds to some input number x (the independent variable). For example, if $y = f(x)$, the value of y that is determined when $x = 2$ is denoted by $f(2)$, read as "f of 2." If $f(x) = 5 - 7x$, we can evaluate $f(2)$ by substituting 2 for x.

$$f(x) = 5 - 7x$$

$$f(2) = 5 - 7(2) \qquad \text{Substitute the input 2 for } x.$$

$$= -9$$

If $x = 2$, then $y = f(2) = -9$.

To evaluate $f(-5)$, we substitute -5 for x.

$$f(x) = 5 - 7x$$

$$f(-5) = 5 - 7(-5) \qquad \text{Substitute the input } -5 \text{ for } x.$$

$$= 40$$

If $x = -5$, then $y = f(-5) = 40$.

Take Note

To see why function notation is helpful, consider the following sentences. Note that the second sentence is much more concise.

1. In the function $y = 3x^2 + x - 4$, find the value of y when $x = -3$.
2. In the function $f(x) = 3x^2 + x - 4$, find $f(-3)$.

In this context, the notations y and $f(x)$ both represent the output of a function and can be used interchangeably, but function notation is more concise.

Sometimes functions are denoted by letters other than f. The notations $y = g(x)$ and $y = h(x)$ also denote functions involving the independent variable x.

Evaluating a Function

EXAMPLE 7

Let $g(x) = 3x^2 + x - 4$. Find **a.** $g(-3)$ **b.** $g(k)$ **c.** $g(-t^3)$ **d.** $g(k + 1)$

SOLUTION In each case, we will substitute the input value into the function and simplify.

a. $g(x) = 3x^2 + x - 4$ **b.** $g(x) = 3x^2 + x - 4$

$g(-3) = 3(-3)^2 + (-3) - 4$ $g(k) = 3k^2 + k - 4$

$= 3(9) - 3 - 4$

$= 20$

c. $g(x) = 3x^2 + x - 4$ **d.** $g(x) = 3x^2 + x - 4$

$g(-t^3) = 3(-t^3)^2 + (-t^3) - 4$ $g(k + 1) = 3(k + 1)^2 + (k + 1) - 4$

$= 3t^6 - t^3 - 4$ $= 3(k^2 + 2k + 1) + k + 1 - 4$

$= 3k^2 + 6k + 3 + k + 1 - 4$

$= 3k^2 + 7k$

Self Check 7 Evaluate: **a.** $g(0)$ **b.** $g(2)$ **c.** $g(k - 1)$

Now Try Exercise 63.

Accent on Technology

Evaluating a Function

Functions can be easily evaluated on a graphing calculator. There are several ways to do this, but one of the easiest is to use the graph editor and the table. Press [Y=] and input $3x^2 + x - 4$ in the graph editor. Next, press [2nd] [WINDOW] and use the table setup shown in Figure 2-7. Then press [2nd] [GRAPH] and enter values for x. The function values will appear in the table.

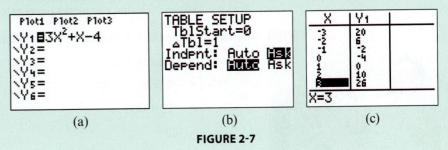

(a)   (b)   (c)

FIGURE 2-7

6. Evaluate the Difference Quotient for a Function

The fraction $\frac{f(x+h)-f(x)}{h}$ is called the **difference quotient** and is important in calculus. The difference quotient can be used to find quantities such as the velocity of a guided missile or the rate of change of a company's profit.

Finding the Difference Quotient

EXAMPLE 8

If $f(x) = x^2 - 2x - 5$, evaluate $\dfrac{f(x+h)-f(x)}{h}$.

SOLUTION   We will evaluate the difference quotient in three steps. Find $f(x+h)$. Then subtract $f(x)$. Then divide by h.

Step 1: Find $f(x+h)$.

$$f(x) = x^2 - 2x - 5$$

$$f(x+h) = (x+h)^2 - 2(x+h) - 5 \qquad \text{Substitute } x+h \text{ for } x.$$

$$= x^2 + 2xh + h^2 - 2x - 2h - 5$$

Step 2: Find $f(x+h) - f(x)$. We can use the result from Step 1.

$$f(x+h) - f(x) = x^2 + 2xh + h^2 - 2x - 2h - 5 - (x^2 - 2x - 5) \qquad \text{Subtract } f(x).$$

$$= x^2 + 2xh + h^2 - 2x - 2h - 5 - x^2 + 2x + 5 \qquad \text{Remove parentheses.}$$

$$= 2xh + h^2 - 2h \qquad \text{Combine like terms.}$$

> **Tip**
>
> After the completion of Step 2, in a polynomial function, when $f(x+h) - f(x)$ is simplified, each term will always include an h. Use that fact to check your work as you progress through the problem.

Step 3: Find the difference quotient $\frac{f(x+h)-f(x)}{h}$. We can use the result from Step 2.

$$\frac{f(x+h)-f(x)}{h} = \frac{2xh + h^2 - 2h}{h} \qquad \text{Divide both sides by } h.$$

$$= \frac{h(2x + h - 2)}{h} \qquad \text{In the numerator, factor out } h.$$

$$= 2x + h - 2 \qquad \text{Divide out } h \text{: } \frac{h}{h} = 1.$$

Self Check 8   If $f(x) = x^2 + 2$, evaluate $\dfrac{f(x+h)-f(x)}{h}$.

Now Try Exercise 85.

7. Solve Applications Involving Functions

Functions are used in many real-life situations.

EXAMPLE 9

Solving an Application Problem

The target heart rate, $f(x)$, in beats per minute, at which a person should train to get an effective workout is a function of the person's age x in years. If $f(x) = -0.6x + 132$, find the target heart rate for a 20-year-old college student.

SOLUTION

We can evaluate the function at $x = 20$ to determine the target heart rate.

$$f(x) = -0.6x + 132$$
$$f(20) = -0.6(20) + 132 \qquad \text{Substitute 20 in for } x.$$
$$= -12 + 132$$
$$= 120$$

The target heart rate would be 120 beats per minute.

Self Check 9 Find the target heart rate for a 30-year-old person.

Now Try Exercise 95.

EXAMPLE 10

Using a Function to Model an Application

Cost of a Fraternity Dance The cost associated with a fraternity dance is $300 for Country Club rental and $24 for each couple that attends.

a. Write the cost C of the dance in terms of the number of couples x attending.

b. Find the cost if 45 couples attend the dance.

SOLUTION Use the information stated in the problem to write a function that models the application.

a. The cost C for the dance is $24 per couple plus the $300 rental fee. If x couples attend, the cost is $24x$ plus 300. Therefore, the cost function is

$$C(x) = 24x + 300$$

b. To find the cost when 45 couples attend, we find $C(45)$.

$$C(x) = 24x + 300$$
$$C(45) = 24(45) + 300 \qquad \text{Substitute 45 for } x.$$
$$= 1080 + 300$$
$$= 1380$$

If 45 couples attend, the cost of the dance is $1380.

Self Check 10 Find the cost when 60 couples attend.

Now Try Exercise 103.

Self Check Answers

1. The domain is $\{-2, 3, -5, 10\}$. The range is $\{-7, -13, 11\}$.   **2.** The set of ordered pairs is {(Fiesta, $13,965), (Focus, $17,170), (Fusion, $22,010), (Mustang, $23,800), (Taurus, $27,055)}. The domain is {Fiesta, Focus, Fusion, Mustang, Taurus}. The range is {$13,965, $17,170, $22,010, $23,800, $27,055}.   **3.** The relation R3 is {(3, 2), (5, 2), (5, 6), (7, 6)}. The domain is $\{3, 5, 7\}$. The range is $\{2, 6\}$.   **4. a.** a function   **b.** a function   **c.** not a function   **5. a.** $(-\infty, \infty)$ **b.** $(-\infty, \infty)$ **6.** $(-\infty, -4) \cup (-4, 4) \cup (4, \infty)$ **7. a.** -4 **b.** 10   **c.** $3k^2 - 5k - 2$ **8.** $2x + h$ **9.** 114 beats per minute   **10.** $1740

Exercises 2.1

Getting Ready

You should be able to complete these vocabulary and concept statements before you proceed to the practice exercises.

Fill in the blanks.

1. A correspondence that assigns exactly one value of y to any number x is called a _____.
2. A set of ordered pairs is called a _____.
3. The set of input numbers x in a function is called the _____ of the function.
4. The set of all output values y in a function is called the _____ of the function.
5. The statement "y is a function of x" can be written as the equation _____.
6. In the function of Exercise 5, ___ is called the independent variable.
7. In the function of Exercise 5, y is called the _____ variable.
8. The expression $\frac{f(x+h)-f(x)}{h}$, $h \neq 0$, is called the _____.

Practice

A relation is given. (a) State the domain and range. (b) Determine if the relation is a function.

9. $\{(2, 3), (3, 4), (4, 5), (5, 6)\}$ 10. $\{(5, 4), (6, 4), (7, 4), (8, 4)\}$

11. $\{(1, 3), (1, 4), (2, 5), (-5, 2)\}$

12. $\{(-1, 2), (2, -1), (0, 1), (0, 3)\}$

In Exercises 13–16, a relation is given. (a) Write the ordered pairs of the relation. (b) State the domain and the range. (c) Determine whether the relation is a function.

13. University   Mascot

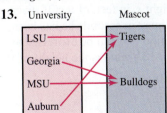

14.   City   State

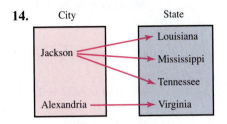

15.

Golf Score	Date
76	September 9
76	October 12
78	May 10
80	June 1

16.

Occupation	Median Salary
Architect	$73,090
Dentist	$149,310
Microbiologist	$66,260
Actuary	$93,680

Assume that all variables represent real numbers. Determine whether each equation determines y to be a function of x.

17. $y = x$ 18. $y - 2x = 0$
19. $y^2 = x$ 20. $y^2 - 4x = 1$
21. $y = x^2$ 22. $y + 1 = 5x^3$
23. $|y| = x$ 24. $2|y| = x - 4$
25. $|x - 2| = y$ 26. $y - |x| = 3$
27. $|x| = |y|$ 28. $|y| = |x - 2|$
29. $y = 7$ 30. $x = 7$
31. $y - 7 = \sqrt{x}$ 32. $y - \sqrt[3]{x} = 8$
33. $x^2 + y^2 = 25$
34. $(x - 1)^2 + y^2 = 16$

Let the function f be defined by the equation $y = f(x)$, where x and $f(x)$ are real numbers. Find the domain of each function.

35. $f(x) = 3x + 5$ 36. $f(x) = -5x + 2$

37. $f(x) = x^2 - x + 1$ **38.** $f(x) = x^3 - 3x + 2$

39. $f(x) = \sqrt{x - 2}$ **40.** $f(x) = \sqrt{2x + 3}$

41. $f(x) = \sqrt{4 - x}$ **42.** $f(x) = 3\sqrt{2 - x}$

43. $f(x) = \sqrt{x^2 - 1}$ **44.** $f(x) = \sqrt{x^2 - 2x - 3}$

45. $f(x) = \sqrt[3]{x + 1}$ **46.** $f(x) = \sqrt[3]{5 - x}$

47. $f(x) = \dfrac{3}{x + 1}$ **48.** $f(x) = \dfrac{-7}{x + 3}$

49. $f(x) = \dfrac{x}{x - 3}$ **50.** $f(x) = \dfrac{x + 2}{x - 1}$

51. $f(x) = \dfrac{x}{x^2 - 4}$

52. $f(x) = \dfrac{2x}{x^2 - 9}$

53. $f(x) = \dfrac{1}{x^2 - 4x - 5}$

54. $f(x) = \dfrac{x}{2x^2 - 16x + 30}$

55. $f(x) = |x| + 3$

56. $f(x) = 2|x - 1|$

Let the function f be defined by $y = f(x)$, where x and $f(x)$ are real numbers. Find $f(2)$, $f(-3)$, $f(k)$, and $f(k^2 - 1)$.

57. $f(x) = 3x - 2$

58. $f(x) = 5x + 7$

59. $f(x) = \dfrac{1}{2}x + 3$

60. $f(x) = \dfrac{2}{3}x + 5$

61. $f(x) = x^2$

62. $f(x) = 3 - x^2$

63. $f(x) = x^2 + 3x - 1$

64. $f(x) = -x^2 - 2x + 1$

65. $f(x) = x^3 - 2$

66. $f(x) = -x^3$

67. $f(x) = |x^2 + 1|$

68. $f(x) = |x^2 + x + 4|$

69. $f(x) = \dfrac{2}{x + 4}$

70. $f(x) = \dfrac{3}{x - 5}$

71. $f(x) = \dfrac{1}{x^2 - 1}$

72. $f(x) = \dfrac{3}{x^2 + 3}$

73. $f(x) = \sqrt{x^2 + 1}$
74. $f(x) = \sqrt{x^2 - 1}$

75. $f(x) = \sqrt[3]{x} - 1$
76. $f(x) = \sqrt[3]{x} + 1$

Evaluate the difference quotient for each function $f(x)$.

77. $f(x) = 3x + 1$ **78.** $f(x) = 5x - 1$

79. $f(x) = -7x + 8$ **80.** $f(x) = -8x - 1$

81. $f(x) = x^2 + 1$ **82.** $f(x) = x^2 - 3$

83. $f(x) = 4x^2 - 6$ **84.** $f(x) = 5x^2 + 3$

85. $f(x) = x^2 + 3x - 7$ **86.** $f(x) = x^2 - 5x + 1$

87. $f(x) = 2x^2 - 4x + 2$ **88.** $f(x) = 3x^2 + 2x - 3$

89. $f(x) = -x^2 + x - 3$
90. $f(x) = -3x^2 + 5x - 1$
91. $f(x) = x^3$
92. $f(x) = -x^3$
93. $f(x) = \dfrac{1}{x}$

94. $f(x) = \sqrt{x}$

Applications

95. Target heart rate The target heart rate $f(x)$, in beats per minute, at which a person should train to get an effective workout is a function of the person's age x in years. If $f(x) = -0.6x + 132$, find the target heart rate for a 25-year-old college student.

96. Temperature conversion The Fahrenheit temperature reading F is a function of the Celsius reading C. The function can be written as $F(C) = \frac{9}{5}C + 32$. Find the Fahrenheit temperature for the Celsius temperatures: $C = 0$; $C = -40$; $C = 10$.

97. Free-falling objects The velocity v of a falling object is a function of the time t it has been falling. If v as a function of t can be expressed as $v(t) = -32t + 15$, where v is in feet per second and t is in seconds, when will the velocity be 0?

98. Cliff divers The height s, in feet, of a cliff diver is a function of the time t in seconds he has been falling. If s as a function of t can be expressed as $s(t) = -16t^2 + 10t + 300$, what is the height of the diver at 3 seconds?

©lassedesignen/Shutterstock.com

99. Go green The typical American family uses about 300 gallons of water a day. If the number of gallons g used expressed in terms of days d is $g(d) = 300d$, find the number of gallons used in one year.

100. Volume of a basketball The volume V of a sphere can be expressed in terms of its radius r according to the function $V(r) = \frac{4}{3}\pi r^3$. Find the volume of a men's NCAA basketball if the diameter of the ball is 29.5 centimeters. Round to the nearest square centimeter.

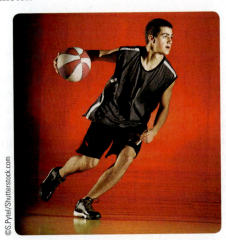

©S.Pytel/Shutterstock.com

101. Formulas The area A of a rectangle is determined by the length and width. If the length of a rectangle is x inches and the width is 5 inches more than the length, express the area as a function of the length.

102. Formulas The volume V of a rectangular box is determined by the length, the width, and the height. For a particular set of boxes, the height is 4 feet, the length is given as x feet, and the width is $3x$ feet. Express the volume as a function of x.

103. Cost of t-shirts A chapter of Phi Theta Kappa, an honors society for two-year college students, is purchasing t-shirts for each of its members. A local company has agreed to make the shirts for $8 each plus a graphic arts fee of $75.

   a. Write a function that describes the cost C for the shirts in terms of x, the number of t-shirts ordered.

   b. Find the total cost of 85 t-shirts.

104. Service projects The Circle "K" Club is planning a service project for children at a local children's home. They plan to rent a "Dora the Explorer Moonwalk" for the event. The cost of the moonwalk will include a $60 delivery fee and $45 for each hour it is used. Write a function that describes the cost C for renting the moonwalk in terms of x, the number of hours used.

105. Cell phone plans A grandmother agrees to purchase a cell phone for emergency use only. AT&T now offers

such a plan for $9.99 per month and $0.07 for each minute x the phone is used.

   a. Write a function that describes the monthly cost C in terms of the time in minutes x the phone is used.

   b. If the grandmother uses her phone for 20 minutes during the first month, what was her bill?

106. Concessions A concessionaire at a football game pays a vendor $40 per game for selling hot dogs at $2.50 each.

   a. Write a function that describes the income I the vendor earns for the concessionaire during the game if the vendor sells x hot dogs.

   b. Find the income if the vendor sells 175 hot dogs.

Discovery and Writing

107. Using words, state three real-life correspondences that represent relations.

108. Using words, state three real-life correspondences that represent functions.

109. Explain why some equations represent y as a function of x, and some do not.

110. Explain why all functions are relations, but not all relations are functions.

111. Describe what is meant by the domain of a function.

112. Are the domains of the functions $f(x) = \sqrt{x-4}$ and $g(x) = \sqrt{4-x}$ the same or different?

113. Describe how you would find the domain of $f(x) = \sqrt{x^2 - 16}$.

114. Describe how you would find the domain of $f(x) = \dfrac{\sqrt{x+3}}{x^2 - 5x + 6}$.

115. Are the functions $f(x) = x + 3$ and $g(x) = \dfrac{x^2 - 9}{x - 3}$ the same? Explain why or why not.

116. Explain why the difference quotient for $f(x) = 5$ is 0.

Critical Thinking

In Exercises 117–124, match the function with its range. Some answers can be repeated.

117. $f(x) = 5x$ **a.** $(-\infty, 0) \cup (0, \infty)$

118. $f(x) = x^2 + 1$ **b.** $(-\infty, 0]$

119. $f(x) = -x^2$ **c.** $[-1, \infty)$

120. $f(x) = x^3$ **d.** $[1, \infty)$

121. $f(x) = |x|$ **e.** $(-\infty, \infty)$

122. $f(x) = \sqrt{x} - 1$ **f.** $[0, \infty)$

123. $f(x) = \sqrt[3]{x}$

124. $f(x) = \dfrac{1}{x}$

2.2 The Rectangular Coordinate System and Graphing Lines

In this section, we will learn to

1. Plot points in the rectangular coordinate system.
2. Graph linear equations.
3. Graph vertical and horizontal lines.
4. Solve applications using linear functions.
5. Find the distance between two points.
6. Find the midpoint of a line segment.

We often say that a picture is worth a thousand words. In fact, pictures and graphs are an effective way to present information. For this reason, they appear frequently in newspapers and magazines. The graph in Figure 2-8 shows the average retail price per gallon for gasoline in the United States during a two-year period. Ordered pairs of the form (date, price per gallon) are plotted and a smooth curve is then drawn through them. Note that for each day shown, there corresponds exactly one price per gallon. The graph represents a function. We can see that the price per gallon was approximately $2.36 on December 22, 2014.

24 Month Average Retail Price Chart

FIGURE 2-8
Source: www.gasbuddy.com. Reprinted with permission from GasBuddy/OpenStore LLC

In mathematics, graphs are also an effective way to present information. In this chapter, we will draw graphs of equations containing two variables and then discuss the information that we can derive from graphs.

The solutions of an equation with variables x and y such as $y = -\frac{1}{2}x + 4$ are ordered pairs of real numbers (x, y) that satisfy the equation. To find some ordered pairs that satisfy the equation, we substitute **input values** of x into the equation and find the corresponding **output values** of y. For example, if we substitute 2 for x, we obtain

$$y = -\frac{1}{2}x + 4$$

$$y = -\frac{1}{2}(2) + 4 \qquad \text{Substitute 2 for } x.$$

$$= -1 + 4$$

$$= 3$$

Since $y = 3$ when $x = 2$, the ordered pair $(2, 3)$ is a solution of the equation. The first coordinate, 2, of the ordered pair is usually called the **x-coordinate**. The second coordinate, 3, is usually called the **y-coordinate**. The solution $(2, 3)$ and several other solutions are listed in the table of values shown in Figure 2-9.

> **Tip**
>
> To complete the table of solutions, we first pick values for x. Next we compute each y-value. Then we write each solution as an ordered pair. Note that we choose x-values that are multiples of the denominator, 2. This makes the computations easier when multiplying the x-value by $-\frac{1}{2}$ to find the corresponding y-value.

$y = -\frac{1}{2}x + 4$		
x	y	(x, y)
-4	6	$(-4, 6)$
-2	5	$(-2, 5)$
0	4	$(0, 4)$
2	3	$(2, 3)$
4	2	$(4, 2)$

Pick values for x. ——— Write each solution as an ordered pair.

Compute each y-value.

FIGURE 2-9

Accent on Technology

Courtesy of Texas Instruments

Generating Table Values

If an equation in x and y is solved for y, we can use a graphing calculator to generate a table of solutions. The instructions in this discussion are for a TI-84 Plus graphing calculator. For details about other brands, please consult the owner's manual.

To construct a table of solutions for $x + 2y = 8$, we first solve the equation for y.

$$x + 2y = 8$$

$$2y = -x + 8 \qquad \text{Subtract } x \text{ from both sides.}$$

$$y = -\frac{1}{2}x + 4 \qquad \text{Divide both sides by 2 and simplify.}$$

- To construct a table of values for $y = -\frac{1}{2}x + 4$, we first enter the equation. We press [Y=] and enter $-(1/2)x + 4$, as shown in Figure 2-10(a).
- Next, we press [2nd] [WINDOW] and enter one value for x on the line labeled TblStart=. In Figure 2-10(b), -4 has been entered on this line. Other values for x that will appear in the table are determined by setting an **increment value** on the line labeled ΔTbl=. In Figure 2-10(b), an increment value of 2 has been entered. This means that each x-value in the table will be 2 units larger than the previous one.
- Finally, we press [2nd] [GRAPH] to obtain the table of values shown in Figure 2-10(c). This table contains all of the solutions listed in Figure 2-9, plus the two additional solutions $(6, 1)$ and $(8, 0)$.

To see other values, we simply scroll up and down the screen by pressing the up and down arrow keys.

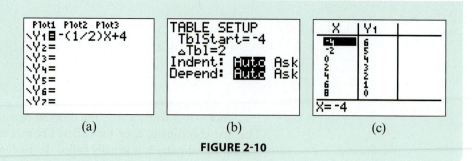

FIGURE 2-10

Before we can present the table of solutions shown in Figure 2-9 in graphical form, we need to discuss the rectangular coordinate system.

1. Plot Points in the Rectangular Coordinate System

The **rectangular coordinate system** consists of two perpendicular number lines that divide the plane into four **quadrants**, numbered as shown in Figure 2-11. The horizontal number line is called the **x-axis**, and the vertical number line is called the **y-axis**. These axes intersect at a point called the **origin**, which is 0 on each axis. The positive direction on the x-axis is to the right, the positive direction on the y-axis is upward, and the same unit distance is used on both axes, unless otherwise indicated.

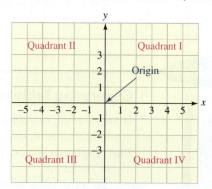

FIGURE 2-11

To plot (or graph) the point associated with the pair $x = 2$ and $y = 3$, denoted as $(2, 3)$, we start at the origin, count 2 units to the right, and then count 3 units up. (See Figure 2-12.) Point P (which lies in the first quadrant) is the graph of the ordered pair $(2, 3)$. The ordered pair $(2, 3)$ gives the **coordinates** of point P.

To plot point Q with coordinates $(-4, 6)$, we start at the origin, count 4 units to the left, and then count 6 units up. Point Q lies in the second quadrant. Point R with coordinates $(6, -4)$ lies in the fourth quadrant.

> **Take Note**
>
> The ordered pairs $(-4, 6)$ and $(6, -4)$ represent **different** points. $(-4, 6)$ is in the second quadrant and $(6, -4)$ is in the fourth quadrant.

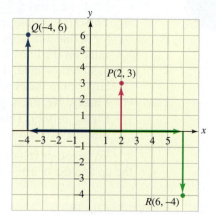

FIGURE 2-12

2. Graph Linear Equations

The **graph of the equation** $y = -\frac{1}{2}x + 4$ is the graph of all points (x, y) on the rectangular coordinate system whose coordinates satisfy the equation. To graph $y = -\frac{1}{2}x + 4$, we plot the pairs listed in the table of solutions shown in Figure 2-13. These points lie on the line shown in the figure. This line is the graph of the equation.

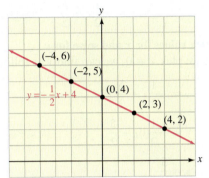

$y = -\frac{1}{2}x + 4$		
x	y	(x, y)
-4	6	$(-4, 6)$
-2	5	$(-2, 5)$
0	4	$(0, 4)$
2	3	$(2, 3)$
4	2	$(4, 2)$

FIGURE 2-13

> **Take Note**
>
> The equation $y = -\frac{1}{2}x + 4$ represents y as a function of x and therefore can be written as $f(x) = -\frac{1}{2}x + 4$.

Note that for each value for x we input, there corresponds exactly one output value for y, and the equation is a function. Specifically, we refer to it as a linear function. The domain and range are all real numbers.

When we say that the graph of an equation is a line, we imply two things:

1. Every point with coordinates that satisfy the equation will lie on the line.
2. Every point on the line will have coordinates that satisfy the equation.

When the graph of an equation is a line, we call the equation a **linear equation**. These equations are often written in **standard form** as $Ax + By = C$, where A, B, and C are specific numbers (called **constants**) and x and y are variables. Either A or B can be 0, but A and B cannot both be 0.

Standard Form of an Equation of a Line

The **standard form of an equation of a line** is

$$Ax + By = C$$

where A, B, and C are real numbers and A and B are not both 0.

Here are four examples of linear equations written in standard form.

Linear Equation	Values of A, B, and C
$3x + 2y = 6$	$A = 3, B = 2, C = 6$
$5x - 2y = -10$	$A = 5, B = -2, C = -10$
$2y = 7$	$A = 0, B = 2, C = 7$
$x = -4$	$A = 1, B = 0, C = -4$

EXAMPLE 1

Graphing a Linear Equation

Graph: $x + 2y = 5$.

SOLUTION   We will solve the equation for y and form a table of solutions by picking values for x, substituting them into the equation, and solving for the other variable y. We will plot the points represented in the table of solutions and draw a line through the points.

Solve the equation for y.

$$x + 2y = 5$$

$$x - x + 2y = 5 - x \qquad \text{Subtract } x \text{ from both sides.}$$

$$2y = -x + 5 \qquad \text{Simplify.}$$

$$y = -\frac{1}{2}x + \frac{5}{2} \qquad \text{Divide both sides by 2.}$$

Pick values for x and solve for y.

If we pick $x = 0$, we can find y as follows:

$$y = -\frac{1}{2}x + \frac{5}{2}$$

$$y = -\frac{1}{2}(0) + \frac{5}{2} \qquad \text{Substitute 0 in for } x.$$

$$y = \frac{5}{2} \qquad \text{Simplify.}$$

The ordered pair $\left(0, \dfrac{5}{2}\right)$ satisfies the equation.

To find another ordered pair, we pick $x = 1$ and find y.

$$y = -\frac{1}{2}x + \frac{5}{2}$$

$$y = -\frac{1}{2}(1) + \frac{5}{2} \qquad \text{Substitute 1 in for } x.$$

$$y = 2 \qquad \text{Simplify.}$$

The ordered pair $(1, 2)$ satisfies the equation.

These pairs and others that satisfy the equation are shown in Figure 2-14. We plot the points and join them with a line to get the graph of the equation. Note that the equation represents a linear function.

Tip

To graph a line, we only need to plot two points because two points determine a line. However, it is a good idea to find a third point as a check.

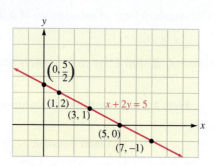

$x + 2y = 5$		
x	y	(x, y)
0	$\dfrac{5}{2}$	$\left(0, \dfrac{5}{2}\right)$
1	2	$(1, 2)$
3	1	$(3, 1)$
5	0	$(5, 0)$
7	-1	$(7, -1)$

FIGURE 2-14

Self Check 1   Graph: $3x - 2y = 6$.

Now Try Exercise 35.

EXAMPLE 2   **Graphing a Linear Equation**

Graph: $3(y + 2) = 2x - 3$.

SOLUTION   We will solve the equation for y and find ordered pairs (x, y) that satisfy the equation. Then, we will plot the points and graph the line.

Solve the equation for y.

$$3(y + 2) = 2x - 3$$

$$3y + 6 = 2x - 3 \qquad \text{Use the Distributive Property to remove parentheses.}$$

$$3y = 2x - 9 \qquad \text{Subtract 6 from both sides.}$$

$$y = \frac{2}{3}x - 3 \qquad \text{Divide both sides by 3.}$$

Pick values for x, solve for y.

We now substitute numbers for x to find the corresponding values of y. If we let $x = 0$ and find y, we get

$$y = \frac{2}{3}x - 3$$

$$y = \frac{2}{3}(0) - 3 \qquad \text{Substitute 0 for } x.$$

$$y = -3 \qquad \text{Simplify.}$$

The point $(0, -3)$ lies on the graph.

If we let $x = 3$, we get

$$y = \frac{2}{3}x - 3$$

$$y = \frac{2}{3}(3) - 3 \qquad \text{Substitute 3 for } x.$$

$$y = 2 - 3 \qquad \text{Simplify.}$$

$$y = -1$$

The point $(3, -1)$ lies on the graph.

We plot these points and others, as in Figure 2-15, and draw the line that passes through the points. Note that the equation represents a linear function. The domain and range are all real numbers.

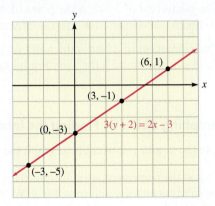

$3(y + 2) = 2x - 3$		
x	y	(x, y)
-3	-5	$(-3, -5)$
0	-3	$(0, -3)$
3	-1	$(3, -1)$
6	1	$(6, 1)$

FIGURE 2-15

Self Check 2   Graph: $2(x - 1) = 6 - 8y$.

Now Try Exercise 39.

Accent on Technology

Graphing Equations

We can graph equations with a graphing calculator. To see a graph, we must choose the minimum and maximum values of the x- and y-coordinates that will appear on the calculator's window. A window with standard settings of

$$\text{Xmin} = -10 \qquad \text{Xmax} = 10 \qquad \text{Ymin} = -10 \qquad \text{Ymax} = 10$$

will produce a graph where the value of x is in the interval $[-10, 10]$, and the value of y is in the interval $[-10, 10]$.

- To use a graphing calculator to graph $3x + 2y = 12$, we first solve the equation for y.

$$2y = -3x + 12 \qquad \text{Subtract } 3x \text{ from both sides.}$$

$$y = -\frac{3}{2}x + 6 \qquad \text{Divide both sides by 2.}$$

- Next, we press [Y=] and enter the right side of the equation. The screen is shown in Figure 2-16(a).
- We then press [GRAPH] to obtain the graph shown in Figure 2-16(b).

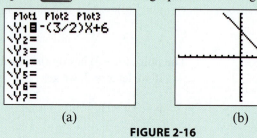

(a) (b)

FIGURE 2-16

In Figure 2-16, the graph intersects the y-axis at the point $(0, 6)$, which is called the **y-intercept**. It intersects the x-axis at the point $(4, 0)$, which is called the **x-intercept**.

Intercepts of a Line	The **y-intercept** of a line is the point $(0, b)$, where the line intersects the y-axis. To find b, substitute 0 for x in the equation of the line and solve for y.
	The **x-intercept** of a line is the point $(a, 0)$, where the line intersects the x-axis. To find a, substitute 0 for y in the equation of the line and solve for x.

Graphing a Line by Finding the Intercepts

EXAMPLE 3

Use the x- and y-intercepts to graph the equation $3x + 2y = 12$.

SOLUTION To find the y-intercept, we substitute 0 for x and solve for y. To find the x-intercept, we substitute 0 for y and solve for x. We will also find a third point as a check and then plot the points and draw the graph.

Find the y-intercept.
To find the y-intercept, we substitute 0 for x and solve for y.

$$3x + 2y = 12$$

$$3(0) + 2y = 12 \qquad \text{Substitute 0 for } x.$$

$$2y = 12 \qquad \text{Simplify.}$$

$$y = 6 \qquad \text{Divide both sides by 2.}$$

The y-intercept is the point $(0, 6)$.

Find the *x*-intercept.

To find the *x*-intercept, we substitute 0 for *y* and solve for *x*.

$$3x + 2y = 12$$

$$3x + 2(\textbf{0}) = 12 \qquad \text{Substitute 0 for } y.$$

$$3x = 12 \qquad \text{Simplify.}$$

$$x = 4 \qquad \text{Divide both sides by 3.}$$

The *x*-intercept is the point $(4, 0)$.

Find a third point as a check.

If we let $x = 2$, we will find that $y = 3$.

$$3x + 2y = 12$$

$$3(\textbf{2}) + 2y = 12 \qquad \text{Substitute 2 for } x.$$

$$6 + 2y = 12 \qquad \text{Simplify.}$$

$$2y = 6 \qquad \text{Subtract 6 from both sides.}$$

$$y = 3 \qquad \text{Divide both sides by 2.}$$

The point $(2, 3)$ satisfies the equation.

We plot each pair (as in Figure 2-17) and join them with a line to get the graph of the equation. Note that the equation represents a linear function. The domain and range are all real numbers.

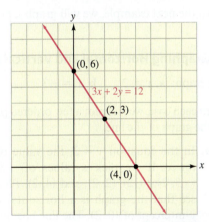

$3x + 2y = 12$		
x	y	(x, y)
0	6	$(0, 6)$
2	3	$(2, 3)$
4	0	$(4, 0)$

FIGURE 2-17

Self Check 3   Graph: $2x - 3y = 12$.

Now Try Exercise 47.

Accent on Technology

Finding Intercepts Using Zoom and Trace

We can use the trace feature on a graphing calculator to find the approximate coordinates of any point on a graph. When we press TRACE, a flashing cursor will appear on the screen. The coordinates of the cursor will also appear at the bottom of the screen.

- To find the *y*-intercept of the graph of $2y = -5x - 7$ (or $y = -\frac{5}{2}x - \frac{7}{2}$), we graph the equation, using $[-10, 10]$ for *x* and $[-10, 10]$ for *y*, and press TRACE to get Figure 2-18(a). We see from the figure that the *y*-intercept is $(0, -3.5)$.

- We can approximate the x-intercept by using the left arrow key and moving the cursor toward the x-intercept until we arrive at a point with the coordinates shown in Figure 2-18(b). We see from the graph that the x-coordinate of the x-intercept is approximately -1.489362.

To get better results, we can zoom in by pressing [ZOOM] to get a magnified picture, trace again, and move the cursor to the point with coordinates shown in Figure 2-18(c). Since the y-coordinate is almost 0, we now have a good approximation for the x-intercept. The x-coordinate of the x-intercept is approximately -1.382979. We can achieve better results with repeated zooms.

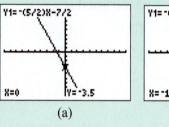
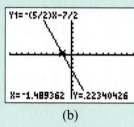
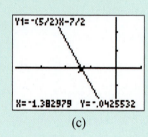

(a)   (b)   (c)

FIGURE 2-18

3. Graph Horizontal and Vertical Lines

In the next example, we will graph a horizontal and a vertical line.

Graphing Horizontal and Vertical Lines

EXAMPLE 4

Graph: **a.** $y = 2$ **b.** $x = -3$

SOLUTION   In each case, we will plot a few ordered pairs that satisfy the equation and then draw the graph of the line.

a. In the equation $y = 2$, the value of y is always 2. Any value can be used for x. If we pick x-values of $-3, 0, 2,$ and 4, we get the ordered pairs: $(-3, 2), (0, 2), (2, 2),$ and $(4, 2)$. Plotting the pairs shown in Figure 2-19, we see that the graph is a horizontal line, parallel to the x-axis and having a y-intercept of $(0, 2)$. The line has no x-intercept.

b. In the equation $x = -3$, the value of x is always -3. Any value can be used for y. If we pick y-values of $-2, 0, 2,$ and 3, we get the ordered pairs: $(-3, -2), (-3, 0), (-3, 2),$ and $(-3, 3)$. After plotting the pairs shown in Figure 2-19, we see that the graph is a vertical line, parallel to the y-axis and having an x-intercept of $(-3, 0)$. The line has no y-intercept.

Take Note

1. Horizontal lines are functions. Consider $y = 2$. For each x-value we input, there corresponds exactly one output value for y, 2.
2. Vertical lines are not functions. Consider $x = -3$. There is more than one y-value paired with the x-value -3.

$y = 2$		
x	y	(x, y)
-3	2	$(-3, 2)$
0	2	$(0, 2)$
2	2	$(2, 2)$
4	2	$(4, 2)$

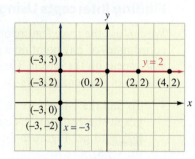

$x = -3$		
x	y	(x, y)
-3	-2	$(-3, -2)$
-3	0	$(-3, 0)$
-3	2	$(-3, 2)$
-3	3	$(-3, 3)$

FIGURE 2-19

Self Check 4   Graph: **a.** $x = 2$ **b.** $y = -3$

Now Try Exercise 51.

Example 4 suggests the following facts.

Equations of Vertical and Horizontal Lines

If a and b are real numbers, then

- The graph of the equation $x = a$ is a vertical line with x-intercept of $(a, 0)$. If $a = 0$, the line $x = 0$ is the y-axis.
- The graph of the equation $y = b$ is a horizontal line with y-intercept of $(0, b)$. If $b = 0$, the line $y = 0$ is the x-axis.

---

4. Solve Applications Using Linear Functions

As we have seen in this section, all equations of lines are functions except for the equations of vertical lines.

Linear Function

A **linear function** is a function determined by the form

$$f(x) = ax + b$$

for all real numbers a and b. If $a \neq 0$, both the domain and range are $(-\infty, \infty)$.

Linear functions can be used to solve a variety of application problems.

EXAMPLE 5

Solving an Application Problem

A high-definition curved-screen smart television purchased for $2750 is expected to depreciate according to the formula $f(x) = -550x + \$2750$, where $f(x)$ is the value of the television after x years. When will the television be worth nothing?

©Jordache/Shutterstock.com

SOLUTION   The television will have no value when its value $f(x)$ is 0. To find x when $f(x) = 0$, we substitute 0 for $f(x)$ and solve for x.

$$\begin{aligned} f(x) &= -550x + 2750 \\ 0 &= -550x + 2750 \\ -2750 &= -550x \qquad \text{Subtract 2750 from both sides.} \\ 5 &= x \qquad \text{Divide both sides by } -550. \end{aligned}$$

The television will have no value in 5 years.

Self Check 5   When will the value of the television be $1650?

Now Try Exercise 97.

5. Find the Distance between Two Points

To derive the formula used to find the distance between two points on a rectangular coordinate system, we use **subscript notation** and denote the points as

$P(x_1, y_1)$ Read as "point P with coordinates of x sub 1 and y sub 1."

$Q(x_2, y_2)$ Read as "point Q with coordinates of x sub 2 and y sub 2."

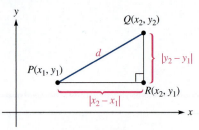

FIGURE 2-20

If $P(x_1, y_1)$ and $Q(x_2, y_2)$ are two points in Figure 2-20 and point R has coordinates (x_2, y_1), triangle PQR is a right triangle. By the Pythagorean Theorem, the square of the hypotenuse of right triangle PQR is equal to the sum of the squares of the two legs. Because leg RQ is vertical, the square of its length is $(y_2 - y_1)^2$. Since leg PR is horizontal, the square of its length is $(x_2 - x_1)^2$. Thus, we have

(1) $d^2 = (x_2 - x_1)^2 + (y_2 - y_1)^2$

Because equal positive numbers have equal positive square roots, we can take the positive square root of both sides of Equation 1 to obtain the **Distance Formula**.

The Distance Formula The distance d between points $P(x_1, y_1)$ and $Q(x_2, y_2)$ is given by

$$d = \sqrt{(x_2 - x_1)^2 + (y_2 - y_1)^2}$$

EXAMPLE 6

Finding the Distance between Two Points

Find the distance between $P(-1, -2)$ and $Q(-7, 8)$.

SOLUTION We use the Distance Formula, $d = \sqrt{(x_2 - x_1)^2 + (y_2 - y_1)^2}$, to find the distance between $P(-1, -2)$ and $Q(-7, 8)$.

If we let $P(-1, -2) = P(x_1, y_1)$ and $Q(-7, 8) = Q(x_2, y_2)$, we can substitute -1 for x_1, -2 for y_1, -7 for x_2, and 8 for y_2 into the formula and simplify.

$$d(PQ) = \sqrt{(\mathbf{x_2} - \mathbf{x_1})^2 + (\mathbf{y_2} - \mathbf{y_1})^2} \qquad \text{Read } d(PQ) \text{ as "the length of segment } PQ.\text{"}$$

$$d(PQ) = \sqrt{[\mathbf{-7} - (\mathbf{-1})]^2 + [\mathbf{8} - (\mathbf{-2})]^2}$$

$$= \sqrt{(-6)^2 + (10)^2}$$

$$= \sqrt{36 + 100}$$

$$= \sqrt{136}$$

$$= \sqrt{4 \cdot 34}$$

$$= 2\sqrt{34} \qquad \sqrt{4 \cdot 34} = \sqrt{4}\sqrt{34} = 2\sqrt{34}$$

Self Check 6 Find the distance between $P(-2\ -5)$ and $Q(3, 7)$.

Now Try Exercise 73.

6. Find the Midpoint of a Line Segment

If point M in Figure 2-21 lies midway between points $P(x_1, y_1)$ and $Q(x_2, y_2)$, point M is called the **midpoint** of segment PQ. To find the coordinates of M, we find the average of the x-coordinates and the average of the y-coordinates of P and Q.

The Midpoint Formula The midpoint of the line segment with endpoints at $P(x_1, y_1)$ and $Q(x_2, y_2)$ is the point M with coordinates of

$$M = \left(\frac{x_1 + x_2}{2}, \frac{y_1 + y_2}{2} \right)$$

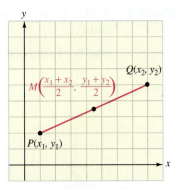

FIGURE 2-21

You will be asked to prove this formula in Exercise 111 by using the Distance Formula to show that $d(PM) + d(MQ) = d(PQ)$.

Finding the Midpoint of a Line Segment

EXAMPLE 7   Find the midpoint of the segment joining $P(-7, 2)$ and $Q(1, -4)$.

SOLUTION   We use the Midpoint Formula, $M = \left(\dfrac{x_1 + x_2}{2}, \dfrac{y_1 + y_2}{2}\right)$ to find the midpoint of the line segment joining $P(-7, 2)$ and $Q(1, -4)$. To do so, we substitute $P(-7, 2)$ for $P(x_1, y_1)$ and $Q(1, -4)$ for $Q(x_2, y_2)$ into the Midpoint Formula to get

$$x_M = \frac{x_1 + x_2}{2} \quad \text{and} \quad y_M = \frac{y_1 + y_2}{2}$$
$$= \frac{-7 + 1}{2} \qquad\qquad = \frac{2 + (-4)}{2}$$
$$= \frac{-6}{2} \qquad\qquad = \frac{-2}{2}$$
$$= -3 \qquad\qquad = -1$$

The midpoint is $M(-3, -1)$.

Self Check 7   Find the midpoint of the segment joining $P(-7, -8)$ and $Q(-2, 10)$.

Now Try Exercise 83.

Using Midpoint Formula to Find Coordinates

EXAMPLE 8   The midpoint of the segment joining $P(-3, 2)$ and $Q(x_2, y_2)$ is $M(1, 4)$. Find the coordinates of Q.

SOLUTION   We can let $P(x_1, y_1) = P(-3, 2)$ and $M(x_M, y_M) = M(1, 4)$, and then find the coordinates x_2 and y_2 of point $Q(x_2, y_2)$.

$$x_M = \frac{x_1 + x_2}{2} \quad \text{and} \quad y_M = \frac{y_1 + y_2}{2}$$
$$1 = \frac{-3 + x_2}{2} \qquad\qquad 4 = \frac{2 + y_2}{2}$$
$$2 = -3 + x_2 \qquad\qquad 8 = 2 + y_2 \qquad \text{Multiply both sides by 2.}$$
$$5 = x_2 \qquad\qquad 6 = y_2$$

The coordinates of point Q are $(5, 6)$.

Self Check 8 If the midpoint of a segment PQ is $M(2, -5)$ and one endpoint is $Q(6, 9)$, find P.

Now Try Exercise 87.

Self Check Answers

1.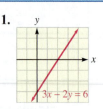

2.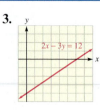

3.

4.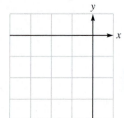

5. 2 years   6. 13   7. $M\left(-\dfrac{9}{2}, 1\right)$ 8. $(-2, -19)$

Exercises 2.2

Getting Ready

You should be able to complete these vocabulary and concept statements before you proceed to the practice exercises.

Fill in the blanks.

1. The coordinate axes divide the plane into four _____.

2. The coordinate axes intersect at the _____.

3. The positive direction on the x-axis is _____.

4. The positive direction on the y-axis is _____.

5. The x-coordinate is the _____ coordinate in an ordered pair.

6. The y-coordinate is the _____ coordinate in an ordered pair.

7. A _____ equation is an equation whose graph is a line.

8. The point where a line intersects the _____ is called the y-intercept.

9. The point where a line intersects the x-axis is called the _____.

10. The graph of the equation $x = a$ will be a _____ line.

11. The graph of the equation $y = b$ will be a _____ line.

12. Complete the Distance Formula:

    $d = $ _____.

13. If a point divides a segment into two equal segments, the point is called the _____ of the segment.

14. The midpoint of the segment joining $P(x_1, y_1)$ and

    $Q(x_2, y_2)$ is $M = $ _____.

Practice

Refer to the illustration and determine the coordinates of each point.

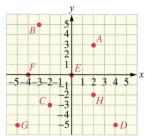

15. A	**16.** B
17. C	**18.** D
19. E	**20.** F
21. G	**22.** H

Graph each point. Indicate the quadrant in which the point lies, or the axis on which it lies.

23. $(2, 5)$	**24.** $(-3, 4)$
25. $(-4, -5)$	**26.** $(6, 2)$
27. $(5, 2)$	**28.** $(3, -4)$
29. $(4, 0)$	**30.** $(0, 2)$

Solve each equation for y and graph the equation. Then check your graph with a graphing calculator.

31. $y - 2x = 7$ **32.** $y + 3 = -4x$

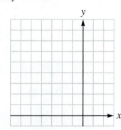

33. $y + 5x = 5$

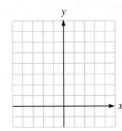

34. $y - 3x = 6$

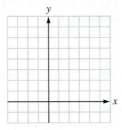

43. $2x - y = 4$

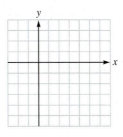

44. $3x + y = 9$

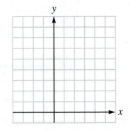

35. $6x - 3y = 10$

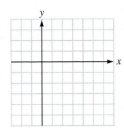

36. $4x + 8y - 1 = 0$

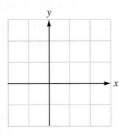

45. $3x + 2y = 6$

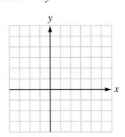

46. $2x - 3y = 6$

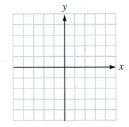

37. $3x = 6y - 1$

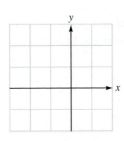

38. $2x + 1 = 4y$

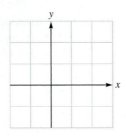

47. $4x - 5y = 20$

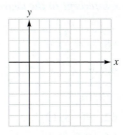

48. $3x - 5y = 15$

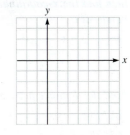

39. $2(x + y + 1) = x + 2$

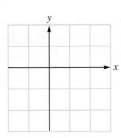

40. $5(x + 2) = 3y - x$

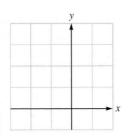

Graph each equation.

49. $y = 3$

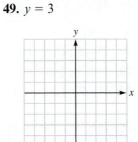

50. $x = -4$

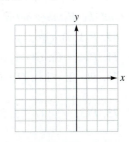

Find the x- and y-intercepts and use them to graph each equation.

41. $x + y = 5$

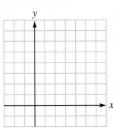

42. $x - y = 3$

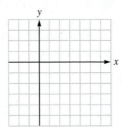

51. $3x + 5 = -1$

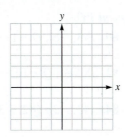

52. $7y - 1 = 6$

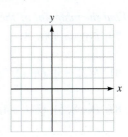

53. $3(y + 2) = y$

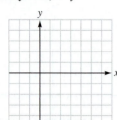

54. $4 + 3y = 3(x + y)$

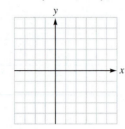

55. $3(y + 2x) = 6x + y$

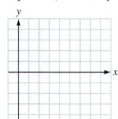

56. $5(y - x) = x + 5y$

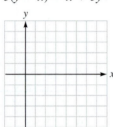

Use a graphing calculator to graph each equation and then find the x-coordinate of the x-intercept to the nearest hundredth.

57. $y = 3.7x - 4.5$

58. $y = \dfrac{3}{5}x + \dfrac{5}{4}$

59. $1.5x - 3y = 7$

60. $0.3x + y = 7.5$

Find the distance between P and O(0, 0).

61. $P(4, -3)$

62. $P(-5, 12)$

63. $P(-3, 2)$

64. $P(5, 0)$

65. $P(1, 1)$

66. $P(6, -8)$

67. $P(\sqrt{3}, 1)$

68. $P(\sqrt{7}, \sqrt{2})$

Find the distance between P and Q.

69. $P(3, 7)$; $Q(6, 3)$

70. $P(4, 9)$; $Q(9, 21)$

71. $P(4, -6)$; $Q(-1, 6)$

72. $P(0, 5)$; $Q(6, -3)$

73. $P(-2, -15)$; $Q(-6, -21)$

74. $P(-7, 11)$; $Q(-11, 7)$

75. $P(3, -3)$; $Q(-5, 5)$

76. $P(6, -3)$; $Q(-3, 2)$

77. $P(\pi, -2)$; $Q(\pi, 5)$

78. $P(\sqrt{5}, 0)$; $Q(0, 2)$

Find the midpoint of the line segment PQ.

79. $P(2, 4)$; $Q(6, 8)$

80. $P(3, -6)$; $Q(-1, -6)$

81. $P(2, -5)$; $Q(-2, 7)$

82. $P(0, 3)$; $Q(-10, -13)$

83. $P(-8, 5)$; $Q(6, -4)$

84. $P(3, -2)$; $Q(2, -3)$

85. $P(0, 0)$; $Q(\sqrt{5}, \sqrt{5})$

86. $P(\sqrt{3}, 0)$; $Q(0, -\sqrt{5})$

One endpoint P and the midpoint M of line segment PQ are given. Find the coordinates of the other endpoint, Q.

87. $P(1, 4)$; $M(3, 5)$

88. $P(2, -7)$; $M(-5, 6)$

89. $P(5, -5)$; $M(5, 5)$

90. $P(-7, 3)$; $M(0, 0)$

91. Show that a triangle with vertices at $(13, -2)$, $(9, -8)$, and $(5, -2)$ is isosceles.

92. Show that a triangle with vertices at $(-1, 2)$, $(3, 1)$, and $(4, 5)$ is isosceles.

93. In the illustration, points M and N are the midpoints of AC and BC, respectively. Find the length of MN.

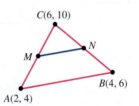

94. In the illustration, points M and N are the midpoints of AC and BC, respectively. Show that $d(MN) = \frac{1}{2}[d(AB)]$.

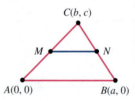

95. In the illustration, point M is the midpoint of the hypotenuse of right triangle AOB. Show that the area of rectangle $OLMN$ is one-half of the area of triangle AOB.

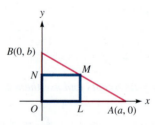

96. Rectangle $ABCD$ in the illustration is twice as long as it is wide, and its sides are parallel to the coordinate axes. If the perimeter is 42, find the coordinates of point C.

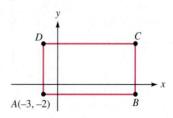

Applications

97. House appreciation A lake house purchased for $325,000 is expected to appreciate according to the formula $f(x) = 17,500x + 325,000$, where $f(x)$ is the value of the lake house after x years. Find the value of the house 5 years later.

©StefanoT/Shutterstock.com

98. Car depreciation A car purchased for $24,000 is expected to depreciate according to the formula $f(x) = -1920x + 24,000$, where $f(x)$ is the value of the car after x years. When will the car be worthless?

©G-stockstudio/Shutterstock.com

99. Demand equations The number of photo scanners that consumers buy depends on price. The higher the price, the fewer photo scanners people will buy. The equation that relates price to the number of photo scanners sold at that price is called a **demand equation**. If the demand for a photo scanner is $p = -\frac{1}{10}q + 170$, where p is the price and q is the number of photo scanners sold at that price, how many photo scanners will be sold at a price of $150?

100. Supply equations The number of television sets that manufacturers produce depends on price. The higher the price, the more TVs manufacturers will produce. The equation that relates price to the number of TVs produced at that price is called a **supply equation**. If the supply equation for a 25-inch TV is $p = \frac{1}{10}q + 130$, where p is the price and q is the number of TVs produced for sale at that price, how many TVs will be produced if the price is $150?

101. Meshing gears The rotational speed V of a large gear (with N teeth) is related to the speed v of the smaller gear (with n teeth) by the equation $V = \frac{nv}{N}$. If the larger gear in the illustration is making 60 revolutions per minute, how fast is the smaller gear spinning?

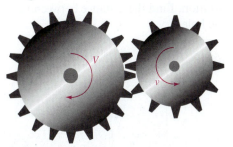

102. Crime prevention The number $f(x)$ of incidents of family violence requiring police response appears to be related to x, the money spent on crisis intervention, by the equation $f(x) = 430 - 0.005x$. What expenditure would reduce the number of incidents to 350?

103. Football Suppose Tony Romo, quarterback for the Dallas Cowboys, throws a football to his wide receiver. If Tony's location on the football field is represented by $Q(30, 25)$, 30 yards from the end zone and 25 yards from the sideline, and his wide receiver location is represented by $R(0, 10)$, on the end zone line and 10 yards from the sideline, find the actual distance the football was thrown.

104. Baseball If home plate on a baseball field is represented by the origin and second base is represented by the $P(90, 90)$, find the actual distance between home plate and second base. The units are in feet.

105. Football Use the information stated in Exercise 103. If Tony Romo's pass is intercepted at the midpoint between the wide receiver and Tony, find the point of interception.

106. Baseball Use the information stated in Exercise 104 to identify the midpoint between home plate and second base.

107. Navigation See the illustration. An ocean liner is located 23 miles east and 72 miles north of Pigeon Cove Lighthouse, and its home port is 47 miles west and 84 miles south of the lighthouse. How far is the ship from port?

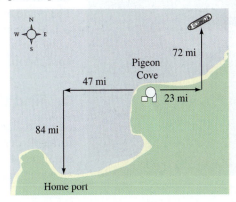

108. Engineering Two holes are to be drilled at locations specified by the engineering drawing shown in the illustration. Find the distance between the centers of the holes.

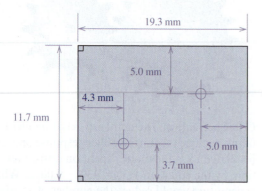

Discovery and Writing

109. Explain how to determine the quadrant in which the point $P(a, b)$ lies.

110. Explain how to graph a line using the intercept method.

111. In Figure 2-21, show that $d(PM) + d(MQ) = d(PQ)$.

112. Use the result of Exercise 111 to explain why point M is the midpoint of segment PQ.

Critical Thinking

Determine if the statement is true or false. If the statement is false, then correct it and make it true.

113. The domain of $y = 8$ is $(-\infty, \infty)$.

114. The range of $y = 8$ is $\{8\}$.

115. All linear equations are functions.

116. Three points are always required to graph a line.

117. A line can have at most one y-intercept.

118. A line must have at least one x-intercept.

119. The distance between the origin and $P(x, y)$ is $\sqrt{x^2 + y^2}$.

120. The midpoint between $P(x, y)$ and $Q(-x, -y)$ is the origin.

2.3 Linear Functions and Slope

In this section, we will learn to

1. Find the slope of a line.

2. Interpret slope as a rate of change.

3. Find slopes of horizontal and vertical lines.

4. Find slopes of parallel and perpendicular lines.

Dorling Kindersley ltd/Alamy

The world's steepest passenger railway is the Lookout Mountain Incline Railway in Chattanooga, Tennessee. Passengers experience breathtaking views of the city and surrounding mountains as the trolley-style railcars travel up Lookout Mountain. The grade or steepness of the track is 72.7% near the top.

Mathematicians use the term **slope** to represent the measure of the steepness of a line. We will explore the topic of slope in this section because it has many real-life applications.

1. Find the Slope of a Line

Suppose that a college student rents a room for $300 per month, plus a $200 nonrefundable deposit. The table shown in Figure 2-22(b) gives the cost (y) for different numbers of months (x). If we construct a graph from these data, we get the line shown in Figure 2-22(a).

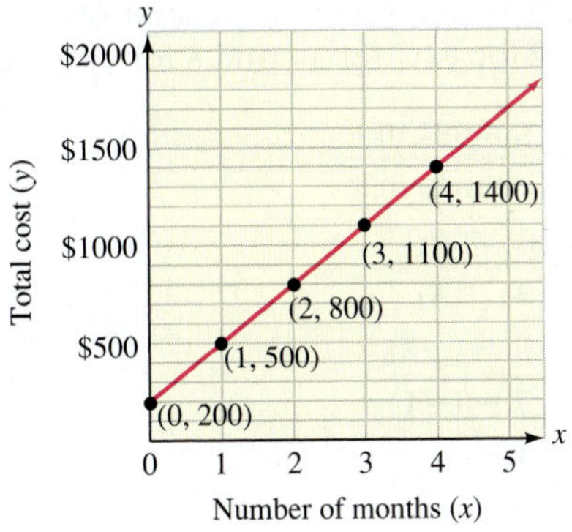

Time in Months	Total Cost
x	y
0	200
1	500
2	800
3	1100
4	1400

(b)

FIGURE 2-22

From the graph, we can see that if x changes from 0 to 1, y changes from 200 to 500. As x changes from 1 to 2, y changes from 500 to 800, and so on. The ratio of the change in y divided by the change in x is the constant 300.

$$\frac{\text{Change in } y}{\text{Change in } x} = \frac{500 - 200}{1 - 0} = \frac{800 - 500}{2 - 1} = \frac{1100 - 800}{3 - 2} = \frac{1400 - 1100}{4 - 3} = \frac{300}{1} = 300$$

The ratio of the change in y divided by the change in x between any two points on any line is always a constant. This constant rate of change is called the **slope** of the line.

The Slope of a Nonvertical Line

The **slope of the nonvertical line** (see Figure 2-23) passing through points $P(x_1, y_1)$ and $Q(x_2, y_2)$ is

$$m = \frac{\text{change in } y}{\text{change in } x} = \frac{y_2 - y_1}{x_2 - x_1} \qquad (x_2 \neq x_1)$$

Tip

You can use the coordinates of any two points on a line to compute the slope of the line.

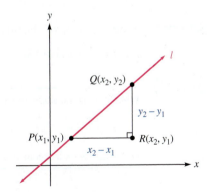

FIGURE 2-23

Finding the Slope of a Line Given Two Points

EXAMPLE 1

Find the slope of the line passing through $P(-1, -2)$ and $Q(7, 8)$.

SOLUTION   We will substitute the points $P(-1, -2)$ and $Q(7, 8)$ into the slope formula, $m = \dfrac{\text{change in } y}{\text{change in } x} = \dfrac{y_2 - y_1}{x_2 - x_1}$, to find the slope of the line.

Let $P(x_1, y_1) = P(-1, -2)$ and $Q(x_2, y_2) = Q(7, 8)$. Then we substitute -1 for x_1, -2 for y_1, 7 for x_2, and 8 for y_2 to get

$$m = \frac{\text{change in } y}{\text{change in } x}$$

$$m = \frac{y_2 - y_1}{x_2 - x_1}$$

$$= \frac{8 - (-2)}{7 - (-1)}$$

$$= \frac{10}{8}$$

$$= \frac{5}{4}$$

The slope of the line is $\frac{5}{4}$. See Figure 2-24.

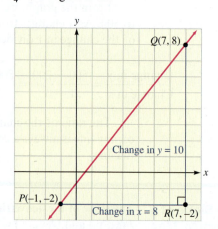

FIGURE 2-24

> **Take Note**
>
> If the slope of the line is a positive number, then its graph will rise from left to right. The linear function is described as increasing on its domain.

We would have obtained the same result if we had let $P(x_1, y_1) = P(7, 8)$ and $Q(x_2, y_2) = Q(-1, -2)$.

Self Check 1   Find the slope of the line passing through $P(-3, -4)$ and $Q(5, 9)$.

Now Try Exercise 13.

> **Caution**
>
> When calculating slope, always subtract the y-values and the x-values in the same order.
>
> $$m = \frac{y_2 - y_1}{x_2 - x_1} \quad \text{or} \quad m = \frac{y_1 - y_2}{x_1 - x_2}$$
>
> Otherwise, we will obtain an *incorrect* result.

A slope can be a positive real number, 0, or a negative real number. If the denominator of the slope formula is 0, slope is not defined.

The change in y (often denoted as Δy) is the **rise** of the line between points P and Q. The change in x (often denoted as Δx) is the **run**. Using this terminology, we can define slope to be the ratio of the rise to the run:

$$m = \frac{y_2 - y_1}{x_2 - x_1} = \frac{\Delta y}{\Delta x} = \frac{\text{rise}}{\text{run}} \qquad (\Delta x \neq 0)$$

EXAMPLE 2

Finding the Slope of a Line Given Its Equation in Standard Form

Find the slope of the line determined by $5x + 2y = 10$. (See Figure 2-25.)

SOLUTION   We will find the coordinates of the x- and y-intercepts and substitute into slope formula, $m = \dfrac{\text{change in } y}{\text{change in } x} = \dfrac{y_2 - y_1}{x_2 - x_1}$, to find the slope of the line.

- If $y = 0$, then $x = 2$, and the point $(2, 0)$ lies on the line.
- If $x = 0$, then $y = 5$, and the point $(0, 5)$ lies on the line.

We then find the slope of the line between $P(2, 0)$ and $Q(0, 5)$.

$$m = \frac{\text{change in } y}{\text{change in } x}$$

$$m = \frac{y_2 - y_1}{x_2 - x_1}$$

$$= \frac{5 - 0}{0 - 2}$$

$$= -\frac{5}{2}$$

The slope is $-\frac{5}{2}$.

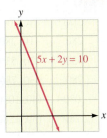

FIGURE 2-25

> **Take Note**
>
> If the slope of the line is a negative number, then its graph will fall from left to right. The linear function is described as decreasing on its domain.

Self Check 2   Find the slope of the line determined by $3x - 2y = 9$.

Now Try Exercise 31.

2. Interpreting Slope as a Rate of Change

By definition, slope is defined as the ratio of the change in y divided by the change in x. This number indicates how fast y is changing with respect to x. Specifically, slope indicates the **rate of change** of the dependent variable per unit change in the independent variable. For a linear function, the rate of change is constant. If the graph of our function is not a line, we can use slope formula to calculate an **average rate of change** of the function on a given interval.

Functional notation can be used as another way of writing slope formula. For a function f defined on an interval $[a, b]$ the **average rate of change from a to b** is

$$\frac{f(b) - f(a)}{b - a}$$

EXAMPLE 3

Interpreting Slope as a Rate of Change

If carpet for a hotel room costs \$25 per square yard plus a delivery charge of \$30, the total cost $C(x)$ of x square yards is given by the linear function

Total cost	equals	cost per square yard	times	the number of square yards purchased	plus	the delivery charge.
$C(x)$	$=$	25	$\cdot$	x	$+$	30

Graph the function $C(x) = 25x + 30$. Find and interpret the slope of the line.

SOLUTION   We will complete a table of solutions and graph the equation on a coordinate system with a vertical $C(x)$-axis and a horizontal x-axis. Figure 2-26 shows a table of ordered pairs and the graph.

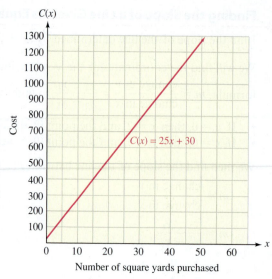

$C(x) = 25x + 30$		
x	$C(x)$	$(x, C(x))$
10	280	(10, 280)
20	530	(20, 530)
30	780	(30, 780)
40	1030	(40, 1030)
50	1280	(50, 1280)

FIGURE 2-26

If we pick the points (30, 780) and (50, 1280) to find the slope, we have

$$m = \frac{\Delta C}{\Delta x}$$

$$= \frac{C_2 - C_1}{x_2 - x_1}$$

$$= \frac{1280 - 780}{50 - 30} \qquad \text{Substitute 1280 for } C_2, \text{ 780 for } C_1, \text{ 50 for } x_2, \text{ and 30 for } x_1.$$

$$= \frac{500}{20}$$

$$= 25$$

The rate of change of the cost of the carpet is $25 when the square footage changes by one yard. The slope of 25 (in dollars/square yard) is the cost per square yard of the carpet.

Self Check 3 If the cost of the carpet in Example 3 increases to $35 per square yard, find the slope of the line.

Now Try Exercise 87.

EXAMPLE 4 **Interpreting Slope as an Average Rate of Change**

It takes a skier 25 minutes to complete the course shown in Figure 2-27. Find his average rate of descent in feet per minute.

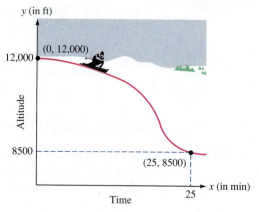

FIGURE 2-27

SOLUTION   To find the average rate of descent, we will find the ratio of the change in altitude to the change in time. To find this ratio, we will calculate the slope of the line passing through the points $(0, 12{,}000)$ and $(25, 8500)$.

$$\begin{aligned}\text{Average rate} \atop \text{of descent} &= \frac{8500 - 12{,}000}{25 - 0}\\[4pt] &= \frac{-3500}{25}\\[4pt] &= -140\end{aligned}$$

The average rate of descent is 140 ft/min.

Self Check 4   If it takes the skier 20 minutes to complete the course, find his average rate of descent in feet per minute.

Now Try Exercise 89.

3. Find Slopes of Horizontal and Vertical Lines

If $P(x_1, y_1)$ and $Q(x_2, y_2)$ are points on the horizontal line shown in Figure 2-28(a), then $y_1 = y_2$, and the numerator of the fraction is 0.

$$\frac{y_2 - y_1}{x_2 - x_1} \qquad \text{On a horizontal line, } x_2 \neq x_1.$$

Thus, the value of the fraction is 0, and the slope of the horizontal line is 0.

If $P(x_1, y_1)$ and $Q(x_2, y_2)$ are points on the vertical line shown in Figure 2-28(b), then $x_1 = x_2$, and the denominator of the fraction is 0.

$$\frac{y_2 - y_1}{x_2 - x_1} \qquad \text{On a vertical line, } y_2 \neq y_1.$$

Since the denominator of a fraction cannot be 0, the slope of a vertical line is not defined.

Tip

An easy way to remember that the slope of a vertical line is undefined is to think of a rock climber who is climbing a vertical mountain. If the climber isn't very careful, he/she could be undefined.

©Fongfong/Shutterstock.com

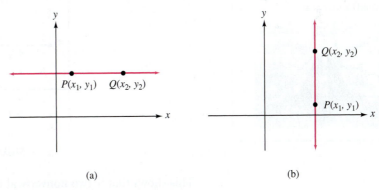

(a)   (b)

FIGURE 2-28

Slopes of Horizontal and Vertical Lines   The slope of a horizontal line (a line with an equation of the form $y = b$ or $f(x) = b$) is 0.

The slope of a vertical line (a line with an equation of the form $x = a$) is not defined.

Here are a few facts about slope. See Figure 2-29.

- If a line rises as we follow it from left to right, as in Figure 2-29(a), its slope is positive.

- If a line drops as we follow it from left to right, as in Figure 2-29(b), its slope is negative.
- If a line is horizontal, as in Figure 2-29(c), its slope is 0.
- If a line is vertical, as in Figure 2-29(d), it has no defined slope.

Slope Concepts

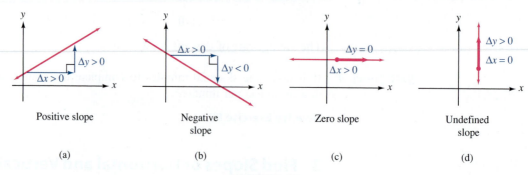

Positive slope   Negative slope   Zero slope   Undefined slope

(a)   (b)   (c)   (d)

FIGURE 2-29

4. Find Slopes of Parallel and Perpendicular Lines

To see a relationship between parallel lines and their slopes, we refer to the parallel lines l_1 and l_2 shown in Figure 2-30, with slopes of m_1 and m_2, respectively. Because right triangles ABC and DEF are similar, it follows that

$$m_1 = \frac{\Delta y \text{ of } l_1}{\Delta x \text{ of } l_1} = \frac{\Delta y \text{ of } l_2}{\Delta x \text{ of } l_2} = m_2$$

> **Tip**
>
> Consider escalators in a department store. The escalators are parallel. Their slopes would be identical.
>
>
>
> ©MarchCattle/Shutterstock.com

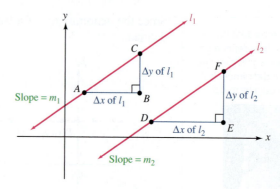

FIGURE 2-30

This shows that if two nonvertical lines are parallel, they have the same slope. It is also true that when two lines have the same slope, they are parallel.

Slopes of Parellel Lines	Nonvertical parallel lines have the same slope, and lines having the same slope are parallel.
	Since vertical lines are parallel, lines with undefined slopes are parallel.

EXAMPLE 5   **Solving a Slope Problem Involving Parallel Lines**

The lines in Figure 2-31 are parallel. Find y.

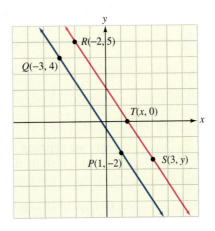

FIGURE 2-31

SOLUTION   Since the lines are parallel, their slopes are equal. To find y, we will find the slope of each line, set them equal, and solve the resulting equation.

Slope of PQ = Slope of RS

$$\frac{-2-4}{1-(-3)} = \frac{y-5}{3-(-2)}$$

$$\frac{-6}{4} = \frac{y-5}{5} \qquad \text{Simplify.}$$

$$-30 = 4(y-5) \qquad \text{Multiply both sides by 20.}$$

$$-30 = 4y - 20 \qquad \text{Remove parentheses and simplify.}$$

$$-10 = 4y \qquad \text{Add 20 to both sides.}$$

$$-\frac{5}{2} = y \qquad \text{Divide both sides by 4 and simplify.}$$

Thus, $y = -\frac{5}{2}$.

Self Check 5   Find x in Figure 2-31.

Now Try Exercise 65.

The following theorem relates perpendicular lines and their slopes.

Slopes of Perpendicular Lines   If two nonvertical lines are perpendicular, the product of their slopes is -1.

If the product of the slopes of two lines is -1, the lines are perpendicular.

PROOF   Suppose l_1 and l_2 are lines with slopes of m_1 and m_2 that intersect at some point. See Figure 2-32. Then superimpose a coordinate system over the lines so that the intersection point is the origin. Let $P(a, b)$ be a point on l_1, and let $Q(c, d)$ be a point on l_2. Neither point P nor point Q can be the origin.

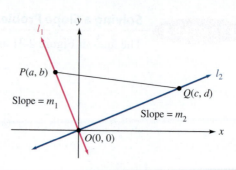

FIGURE 2-32

First, we suppose that l_1 and l_2 are perpendicular. Then triangle POQ is a right triangle with its right angle at O. By the Pythagorean Theorem,

$$d(OP)^2 + d(OQ)^2 = d(PQ)^2$$
$$(a - 0)^2 + (b - 0)^2 + (c - 0)^2 + (d - 0)^2 = (a - c)^2 + (b - d)^2$$
$$a^2 + b^2 + c^2 + d^2 = a^2 - 2ac + c^2 + b^2 - 2bd + d^2$$
$$0 = -2ac - 2bd$$
$$bd = -ac$$

(1)
$$\frac{b}{a} \cdot \frac{d}{c} = -1 \qquad \text{Divide both sides by } ac.$$

The coordinates of P are (a, b), and the coordinates of O are $(0, 0)$. Using the definition of slope, we have

$$m_1 = \frac{b - 0}{a - 0} = \frac{b}{a}$$

Similarly, we have

$$m_2 = \frac{d}{c}$$

Take Note

If the product of two numbers is -1, the numbers are called **negative reciprocals**.

We substitute m_1 for $\frac{b}{a}$ and m_2 for $\frac{d}{c}$ in Equation 1 to obtain

$$m_1 m_2 = -1$$

Hence, if lines l_1 and l_2 are perpendicular, the product of their slopes is -1.

Conversely, we suppose that the product of the slopes of lines l_1 and l_2 is -1. Because the steps in the previous discussion are reversible, we have $d(OP)^2 + d(OQ)^2 = d(PQ)^2$. By the Pythagorean Theorem, triangle POQ is a right triangle. Thus, l_1 and l_2 are perpendicular.

EXAMPLE 6   **Solving a Slope Problem Involving Perpendicular Lines**

Are the lines shown in Figure 2-33 perpendicular?

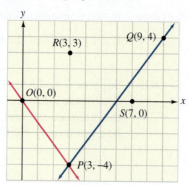

FIGURE 2-33

SOLUTION   We will determine the slopes of the lines and see whether their product is –1.

$$\text{Slope of } OP = \frac{\Delta y}{\Delta x}$$

$$= \frac{y_2 - y_1}{x_2 - x_1}$$

$$= \frac{-4 - 0}{3 - 0}$$

$$= -\frac{4}{3}$$

$$\text{Slope of } PQ = \frac{\Delta y}{\Delta x}$$

$$= \frac{y_2 - y_1}{x_2 - x_1}$$

$$= \frac{4 - (-4)}{9 - 3}$$

$$= \frac{8}{6}$$

$$= \frac{4}{3}$$

Since the product of the slopes is $-\frac{16}{9}$ and not -1, the lines are not perpendicular.

Self Check 6   Is either line in Figure 2-33 perpendicular to the line passing through R and S?

Now Try Exercise 61.

Self Check Answers

1. $\frac{13}{8}$ 2. $\frac{3}{2}$ 3. 35   4. 175 ft/min   5. $\frac{4}{3}$ 6. yes

Exercises 2.3

Getting Ready
You should be able to complete these vocabulary and concept statements before you proceed to the practice exercises.

Fill in the blanks.

1. The slope of a nonvertical line is defined to be the change in y _____ by the change in x.

2. The change in ___ is often called the rise.

3. The change in x is often called the ____.

4. When computing the slope from the coordinates of two points, always subtract the y-values and the x-values in the _____.

5. The symbol Δy means _____ y.

6. The slope of a _____ line is 0.

7. The slope of a _____ line is undefined.

8. If the slopes of two lines are equal, the lines are _____.

9. If the product of the slopes of two lines is -1, the lines are _____.

10. If two lines are perpendicular, the product of their slopes is ____.

Practice
Find the slope of the line passing through each pair of points, if possible.

11. $P(2, 2)$; $Q(-1, -1)$

12. $P(3, -1)$; $Q(5, 3)$

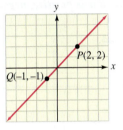

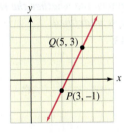

13. $P(-6, 3)$; $Q(6, -2)$

14. $P(2, 5)$; $Q(3, 10)$

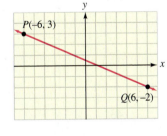

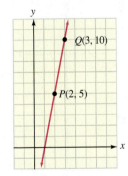

15. $P(3, -2)$; $Q(-1, 5)$

16. $P(3, 7)$; $Q(6, 16)$

17. $P(8, -7)$; $Q(4, 1)$

18. $P(5, 17)$; $Q(17, 17)$

19. $P(-7, -14)$; $Q(2, -14)$

20. $P(-4, 3)$; $Q(-4, -3)$

21. $P(-5, 3)$; $Q(-5, -2)$

22. $P(2, \sqrt{7})$; $Q(\sqrt{7}, 2)$

23. $P\left(\dfrac{3}{2}, \dfrac{2}{3}\right)$; $Q\left(\dfrac{5}{2}, \dfrac{7}{3}\right)$

24. $P\left(-\dfrac{2}{5}, \dfrac{1}{3}\right)$; $Q\left(\dfrac{3}{5}, -\dfrac{5}{3}\right)$

25. $P(a + b, c)$; $Q(b + c, a)$ assume $c \neq a$

26. $P(b, 0)$; $Q(a + b, a)$ assume $a \neq 0$

Find two points on the line and use slope formula to find the slope of the line.

27. $y = 3x + 2$

28. $f(x) = 5x - 8$

29. $f(x) = 4x - 6$

30. $f(x) = -\dfrac{1}{3}x + 5$

31. $5x - 10y = 3$

32. $8y + 2x = 5$

33. $3(y + 2) = 2x - 3$

34. $4(x - 2) = 3y + 2$

35. $3(y + x) = 3(x - 1)$

36. $2x + 5 = 2(y + x)$

Find the slope of the line, if possible.

37. $y = 7$

38. $2y = 5$

39. $f(x) = \dfrac{1}{4}$

40. $f(x) = \pi$

41. $x = -\dfrac{1}{2}$

42. $x - 7 = 0$

Determine whether the slope of the line is positive, negative, 0, or undefined.

43.

44.

45.

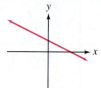

46.

47.

48.

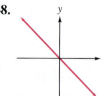

Determine whether the lines with the given slopes are parallel, perpendicular, or neither.

49. $m_1 = 3$; $m_2 = -\dfrac{1}{3}$

50. $m_1 = \dfrac{2}{3}$; $m_2 = \dfrac{3}{2}$

51. $m_1 = \sqrt{8}$; $m_2 = 2\sqrt{2}$

52. $m_1 = 1$; $m_2 = -1$

53. $m_1 = -\sqrt{2}$; $m_2 = \dfrac{\sqrt{2}}{2}$

54. $m_1 = 2\sqrt{7}$; $m_2 = \sqrt{28}$

55. $m_1 = -0.125$; $m_2 = 8$

56. $m_1 = 0.125$; $m_2 = \dfrac{1}{8}$

57. $m_1 = ab^{-1}$; $m_2 = -a^{-1}b$ $(a \neq 0, b \neq 0)$

58. $m_1 = \left(\dfrac{a}{b}\right)^{-1}$; $m_2 = -\dfrac{b}{a}$ $(a \neq 0, b \neq 0, a \neq b)$

Determine whether the line through the given points and the line through $R(-3, 5)$ and $S(2, 7)$ are parallel, perpendicular, or neither.

59. $P(2, 4)$; $Q(7, 6)$

60. $P(-3, 8)$; $Q(-13, 4)$

61. $P(-4, 6)$; $Q(-2, 1)$

62. $P(0, -9)$; $Q(4, 1)$

63. $P(a, a)$; $Q(3a, 6a)$ $(a \neq 0)$

64. $P(b, b)$; $Q(-b, 6b)$ $(b \neq 0)$

Lines PQ and RS are either parallel or perpendicular. Find x or y.

65. Parallel: $P(-3, 7)$; $Q(2, 9)$; $R(10, -4)$; $S(x, -6)$

66. Parallel: $P(2, -3)$; $Q(5, 7)$; $R(3, -1)$; $S(6, y)$

67. Perpendicular: $P(2, -7)$; $Q(1, 0)$; $R(-9, 5)$; $S(-2, y)$

68. Perpendicular: $P(1, -2)$; $Q(3, 4)$; $R(x, 6)$; $S(6, 5)$

Find the slopes of lines PQ and PR, and determine whether points P, Q, and R lie on the same line.

69. $P(-2, 8)$; $Q(-6, 9)$; $R(2, 5)$

70. $P(1, -1)$; $Q(3, -2)$; $R(-3, 0)$

71. $P(-a, a)$; $Q(0, 0)$; $R(a, -a)$

72. $P(a, a + b)$; $Q(a + b, b)$; $R(a - b, a)$

Determine which, if any, of the three lines PQ, PR, and QR are perpendicular.

73. $P(5, 4)$; $Q(2, -5)$; $R(8, -3)$

74. $P(8, -2)$; $Q(4, 6)$; $R(6, 7)$

75. $P(1, 3)$; $Q(1, 9)$; $R(7, 3)$

76. $P(2, -3)$; $Q(-3, 2)$; $R(3, 8)$

77. $P(0, 0)$; $Q(a, b)$; $R(-b, a)$

78. $P(a, b)$; $Q(-b, a)$; $R(a - b, a + b)$

79. **Right triangles** Show that the points $A(-1, -1)$, $B(-3, 4)$, and $C(4, 1)$ are the vertices of a right triangle.

80. **Right triangles** Show that the points $D(0, 1)$, $E(-1, 3)$, and $F(3, 5)$ are the vertices of a right triangle.

81. Squares Show that the points $A(1, -1)$, $B(3, 0)$, $C(2, 2)$, and $D(0, 1)$ are the vertices of a square.

82. Squares Show that the points $E(-1, -1)$, $F(3, 0)$, $G(2, 4)$, and $H(-2, 3)$ are the vertices of a square.

83. Parallelograms Show that the points $A(-2, -2)$, $B(3, 3)$, $C(2, 6)$, and $D(-3, 1)$ are the vertices of a parallelogram. (Show that both pairs of opposite sides are parallel.)

84. Trapezoids Show that points $E(1, -2)$, $F(5, 1)$, $G(3, 4)$, and $H(-3, 4)$ are the vertices of a trapezoid. (Show that only one pair of opposite sides is parallel.)

85. Geometry In the illustration, points M and N are midpoints of CB and BA, respectively. Show that MN is parallel to AC.

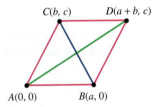

86. Geometry In the illustration, $d(AB) = d(AC)$. Show that AD is perpendicular to BC.

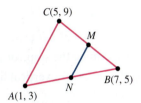

Applications

87. Rate of growth When a college started an aviation program, the administration agreed to predict enrollments using a straight-line method. If the enrollment during the first year was 14, and the enrollment during the fifth year was 42, find the average rate of growth per year (the slope of the line). See the illustration.

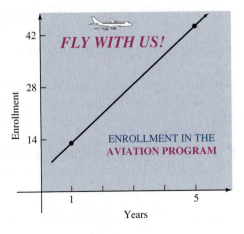

88. Rate of growth A small business predicts sales according to a straight-line method. If sales were $50,000 in the first year and $110,000 in the third year, find the average rate of growth in dollars per year (the slope of the line).

89. Rate of decrease The price of computers has been dropping steadily for the past ten years. If a desktop PC cost $6700 ten years ago, and the same computing power cost $2200 three years ago, find the average rate of decrease per year. (Assume a straight-line model.)

90. Hospital costs The table shows the changing mean daily cost for a hospital room. For the ten-year period, find the rate of change per year of the portion of the room cost that is absorbed by the hospital.

Year	Total Cost to the Hospital	Amount Passed on to Patient
2000	$459	$212
2005	$670	$295
2010	$812	$307

91. Charting temperature changes The following Fahrenheit temperature readings were recorded over a four-hour period.

Time	12:00	1:00	2:00	3:00	4:00
Temperature	47°	53°	59°	65°	71°

Let t represent the time (in hours), with 12:00 corresponding to $t = 0$. Let T represent the temperature.

Plot the points (t, T), and draw the line through those points. Explain the meaning of $\frac{\Delta T}{\Delta t}$.

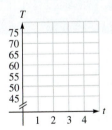

92. Tracking the Dow The Dow Jones Industrial Averages at the close of trade on three consecutive days were as follows:

Day	Monday	Tuesday	Wednesday
Close	12,981	12,964	12,947

Let d represent the day, with $d = 0$ corresponding to Monday, and let D represent the Dow Jones average. Plot the points (d, D), and draw the graph. Explain the meaning of $\frac{\Delta D}{\Delta d}$.

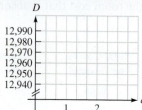

93. Speed of an airplane A pilot files a flight plan indicating her intention to fly at a constant speed of 590 mph. Write an equation that expresses the distance traveled in terms of the flying time. Then graph the equation and interpret the slope of the line. (*Hint: d = rt.*)

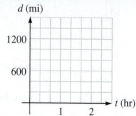

94. Growth of savings A student deposits $25 each month in a Holiday Club account at her bank. The account pays no interest. Write an equation that expresses the amount A in her account in terms of the number of deposits n. Then graph the line, and interpret the slope of the line.

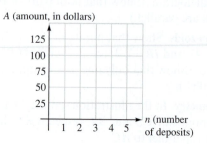

Discovery and Writing

95. Explain what the slope of a line is.

96. How do you determine the slope of a line?

97. Explain why the slope of a vertical line is undefined.

98. Explain how to determine whether two lines are parallel, perpendicular, or neither.

Critical Thinking

Determine if the statement is true or false. If the statement is false, then correct it and make it true.

99. Slope formula is $m = \dfrac{x_2 - x_1}{y_2 - y_1}$.

100. The slope of a linear function is never undefined.

101. The slope of the line passing through $\left(\sqrt{5}, \dfrac{152}{99}\right)$ and $\left(-\sqrt{5}, \dfrac{152}{99}\right)$ is 0.

102. The slope of the line passing through $\left(\dfrac{152}{99}, \sqrt{5}\right)$ and $\left(\dfrac{152}{99}, -\sqrt{5}\right)$ is undefined.

103. The slope of the line parallel to $f(x) = \pi$ is undefined.

104. The slope of the line perpendicular to $f(x) = \pi$ is 0.

105. If the price of a movie ticket increased from $6.95 in 2008 to $10.25 in 2014, then the average rate of change in ticket price during this time period was approximately $0.55/year.

106. If the cost of college tuition in 2010 was $6015 and in 2014 was $9139, then the average rate of change in tuition during this time period was $781/year.

2.4 Writing and Graphing Equations of Lines

In this section, we will learn to

1. Use slope-intercept form to write an equation of a line.
2. Graph linear equations using the slope and *y*-intercept.
3. Determine whether linear equations represent lines that are parallel, perpendicular, or neither.
4. Use point-slope form to write an equation of a line.
5. Write equations of parallel and perpendicular lines.
6. Write an equation of a line that models a real-life problem.
7. Use linear curve fitting to solve problems.

Suppose we purchase a Kawasaki motorcycle for $10,000 and know that it depreciates $1200 in value each year.

We can use the facts given to write a linear equation that represents the value *y* of the motorcycle *x* years after it was purchased. Because the motorcycle's value decreases $1200 each year, the slope of the line's graph is -1200. Because the purchase price is $10,000, we know that when we let $x = 0$, the value of *y* will equal 10,000. A linear equation that satisfies these two conditions is $y = -1200x + 10,000$. This equation represents the straight-line depreciation of the motorcycle.

In this section, we will write equations of lines given specific characteristics or features of the line.

1. Use Slope-Intercept Form to Write an Equation of a Line

To derive one form of the equation of a line, consider the line shown in Figure 2-34 with *y*-intercept of $P(0, b)$, and let $Q(x, y)$ be another point on the line. If we let $(x_1, y_1) = (0, b)$ and $(x_2, y_2) = (x, y)$ the slope of the line is given by

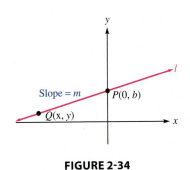

FIGURE 2-34

$$m = \frac{y - b}{x - 0}$$
Substitute $(0, b)$ and (x, y) into slope formula.

$$m = \frac{y - b}{x}$$
Multiply both sides by x to clear the equation of fractions.

$$mx = y - b$$

$$y = mx + b$$
Write the equation with y on the left side.

Because the equation displays the slope m and the y-coordinate b of the y-intercept, it is called the **slope-intercept form** of the equation of the line.

Slope-Intercept Form of an Equation of a line
An equation of the line with slope m and y-intercept $(0, b)$ is $$y = mx + b$$

Three examples of linear equations written in slope-intercept form are shown below.

Take Note

Note that when a line is written in slope-intercept form the coefficient of x is the slope and the constant term is the y-coordinate of the y-intercept.

Example	Slope	*y*-Intercept
$y = 2x + 7$	$m = 2$	$b = 7; (0, 7)$
$y = \frac{2}{3}x - 5$	$m = \frac{2}{3}$	$b = -5; (0, -5)$
$y = -4x + \frac{1}{5}$	$m = -4$	$b = \frac{1}{5}; \left(0, \frac{1}{5}\right)$

EXAMPLE 1

Using Slope-Intercept Form to Write an Equation of a Line

Use slope-intercept form to write an equation of the line with slope 4 that passes through $P(5, 9)$.

SOLUTION   Since we know that $m = 4$ and that the ordered pair $(5, 9)$ satisfies the equation, we substitute 4 for m, 5 for x, and 9 for y in the equation $y = mx + b$ and solve for b.

$$y = mx + b \qquad \text{This is the slope-intercept form.}$$
$$9 = 4(5) + b \qquad \text{Substitute 4 for } m, \text{ 5 for } x, \text{ and 9 for } y.$$
$$9 = 20 + b \qquad \text{Simplify.}$$
$$-11 = b \qquad \text{Subtract 20 from both sides.}$$

Because $m = 4$ and $b = -11$, an equation is $y = 4x - 11$.

> **Tip**
>
> If a point lies on the line, then its coordinates will satisfy the equation. Note that $(5, 9)$ satisfies $y = 4x - 11$. This is an easy way to check your answer.

Self Check 1   Use slope-intercept form to write an equation of the line with slope $\frac{7}{3}$ and passing through $(3, 1)$.

Now Try Exercise 17.

2. Graph Linear Equations Using the Slope and *y*-Intercept

It is easy to graph a linear equation when it is written in slope-intercept form. For example, to graph $y = \frac{4}{3}x - 2$ we note that $b = -2$ and that the y-intercept is $(0, b) = (0, -2)$. (See Figure 2-35.)

Because the slope is $\frac{\Delta y}{\Delta x} = \frac{4}{3}$, we can locate another point Q on the line by starting at point P and counting 3 units to the right and 4 units up. The change in x from point P to point Q is $\Delta x = 3$, and the corresponding change in y is $\Delta y = 4$. The line joining points P and Q is the graph of the equation.

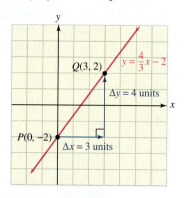

FIGURE 2-35

EXAMPLE 2

Finding the Slope and *y*-Intercept of a Line and Graphing the Line

Find the slope and the y-intercept of the line with equation $3(y + 2) = 6x - 1$ and graph it.

SOLUTION   We will write the equation in the form $y = mx + b$ to find the slope m and the y-intercept $(0, b)$. Then we will use m and b to graph the line.

$$3(y + 2) = 6x - 1$$
$$3y + 6 = 6x - 1 \qquad \text{Remove parentheses.}$$
$$3y = 6x - 7 \qquad \text{Subtract 6 from both sides.}$$
$$y = 2x - \frac{7}{3} \qquad \text{Divide both sides by 3.}$$

> **Take Note**
>
> $y = 2x - \dfrac{7}{3}$ is a linear function and can be written as $f(x) = 2x - \dfrac{7}{3}$.

The slope of the graph is 2, and the y-intercept is $\left(0, -\frac{7}{3}\right)$. We plot the y-intercept. Then we find a second point on the line by moving 1 unit to the right and 2 units up to the point $\left(1, -\frac{1}{3}\right)$. To get the graph, we draw a line through the two points, as shown in Figure 2-36.

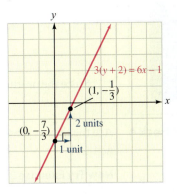

FIGURE 2-36

Self Check 2 Find the slope and the y-intercept of the line with equation $2(x - 3) = -3(y + 5)$. Then graph it.

Now Try Exercise 31.

3. Determine Whether Linear Equations Represent Lines that Are Parallel, Perpendicular, or Neither

EXAMPLE 3 **Determining Whether Lines Are Parallel, Perpendicular, or Neither**

Determine whether the lines represented by $4x + 8y = 10$ and $2x = 12 - 4y$ are parallel, perpendicular, or neither.

SOLUTION We will find the slope of each line and compare their slopes. If their slopes are the same, the lines are parallel. If the product of their slopes is -1, the lines are perpendicular. Otherwise, they are neither parallel nor perpendicular.

We solve each equation for y and write each equation in slope-intercept form.

$$4x + 8y = 10$$
$$8y = -4x + 10$$
$$y = \frac{-4x}{8} + \frac{10}{8}$$
$$y = -\frac{1}{2}x + \frac{5}{4}$$

$$2x = 12 - 4y$$
$$4y = -2x + 12$$
$$y = \frac{-2x}{4} + \frac{12}{4}$$
$$y = -\frac{1}{2}x + 3$$

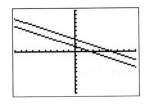

FIGURE 2-37

Since the values of b are different, the lines are distinct. Since each slope is $-\frac{1}{2}$, the lines are parallel. See Figure 2-37. The graphs of the parallel lines are shown.

Self Check 3 Are the lines represented by $y = 3x + 2$ and $6x - 2y = 5$ parallel, perpendicular, or neither?

Now Try Exercise 37.

EXAMPLE 4 **Determining Whether Lines Are Parallel, Perpendicular, or Neither**

Determine whether the lines represented by $4x + 8y = 10$ and $4x - 2y = 21$ are parallel, perpendicular, or neither.

SOLUTION We will find the slope of each line and compare their slopes. If their slopes are the same, the lines are parallel. If the product of their slopes is -1, the lines are perpendicular. Otherwise, they are neither parallel nor perpendicular.

Tip

Perpendicular means that when the two lines are graphed they would intersect forming a right angle.

We solve each equation for y and write each equation in slope-intercept form.

$$4x + 8y = 10 \qquad\qquad 4x - 2y = 21$$

$$8y = -4x + 10 \qquad\qquad -2y = -4x + 21$$

$$y = \frac{-4x}{8} + \frac{10}{8} \qquad\qquad y = \frac{-4x}{-2} + \frac{21}{-2}$$

$$y = -\frac{1}{2}x + \frac{5}{4} \qquad\qquad y = 2x - \frac{21}{2}$$

Since the product of their slopes ($-\frac{1}{2}$ and 2) is -1, the lines are perpendicular. ∎

Self Check 4 Are the lines represented by $3x + 2y = 7$ and $y = \frac{2}{3}x + 3$ parallel, perpendicular, or neither?

Now Try Exercise 35.

4. Use Point-Slope Form to Write an Equation of a Line

Suppose that line l in Figure 2-38 has a slope of m and passes through the point $P(x_1, y_1)$. If $Q(x, y)$ is any other point on line l, we have

$$m = \frac{y - y_1}{x - x_1}$$

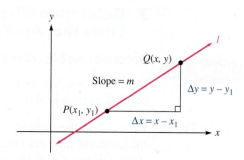

FIGURE 2-38

Take Note

$y - y_1 = m(x - x_1)$ is often referred to as **point-slope formula** because we substitute a point and slope into it to write an equation of a line.

If we multiply both sides by $x - x_1$, we have

(1) $\qquad y - y_1 = m(x - x_1)$

Since Equation 1 displays the coordinates of the point (x_1, y_1) on the line and the slope m of the line, it is called the **point-slope form** of the equation of a line.

Point-Slope Form of an Equation of a Line

The equation of a line passing through $P(x_1, y_1)$ and with slope m is

$$y - y_1 = m(x - x_1).$$

Finding an Equation of the Line with a Given Slope Passing through a Given Point

EXAMPLE 5

Find an equation of the line with slope $-\frac{5}{3}$ and passing through $P(3, -1)$. Write the equation in standard form.

SOLUTION We will substitute $-\frac{5}{3}$ for m, 3 for x_1, and -1 for y_1 in the point-slope form and simplify.

$$y - y_1 = m(x - x_1) \qquad \text{This is the point-slope form.}$$

$$y - (-1) = -\frac{5}{3}(x - 3) \qquad \text{Substitute } -\frac{5}{3} \text{ for } m, 3 \text{ for } x_1, \text{ and } -1 \text{ for } y_1.$$

$$y + 1 = -\frac{5}{3}x + 5 \qquad \text{Remove parentheses.}$$

$$y = -\frac{5}{3}x + 4 \qquad \text{Subtract 1 from both sides.}$$

An equation of the line in slope-intercept form is $y = -\frac{5}{3}x + 4$.

We now write the equation in the standard form $Ax + By = C$.

$$\frac{5}{3}x + y = 4 \qquad \text{Add } \frac{5}{3}x \text{ to both sides.}$$

$$5x + 3y = 12 \qquad \text{Multiply both sides by 3.}$$

An equation of the line in standard form is $5x + 3y = 12$.

Self Check 5   Find an equation of the line with slope $-\frac{2}{3}$ and passing through $P(-4, 5)$. Write the equation in standard form.

Now Try Exercise 49.

Finding an Equation of the Line that Passes through Two Given Points

EXAMPLE 6

Find an equation of the line passing through $P(3, 7)$ and $Q(-5, 3)$. Write the equation in the standard form $Ax + By = C$.

SOLUTION   We will find the slope of the line and then choose either point P or point Q and substitute both the slope and coordinates of the point into the point-slope form.

First we find the slope of the line.

$$m = \frac{y_2 - y_1}{x_2 - x_1} \qquad \text{This is the slope formula.}$$

$$= \frac{3 - 7}{-5 - 3} \qquad \text{Substitute 3 for } y_2, 7 \text{ for } y_1, -5 \text{ for } x_2, \text{ and 3 for } x_1.$$

$$= \frac{-4}{-8}$$

$$= \frac{1}{2}$$

We can choose either point P or point Q and substitute its coordinates into the point-slope form. If we choose $P(3, 7)$, we substitute $\frac{1}{2}$ for m, 3 for x_1, and 7 for y_1.

$$y - y_1 = m(x - x_1) \qquad \text{This is the point-slope form.}$$

$$y - 7 = \frac{1}{2}(x - 3) \qquad \text{Substitute } \frac{1}{2} \text{ for } m, 3 \text{ for } x_1, \text{ and 7 for } y_1.$$

$$y = \frac{1}{2}x - \frac{3}{2} + 7 \qquad \text{Remove parentheses and add 7 to both sides.}$$

$$y = \frac{1}{2}x + \frac{11}{2} \qquad -\frac{3}{2} + 7 = -\frac{3}{2} + \frac{14}{2} = \frac{11}{2}$$

An equation of the line in slope-intercept form is $y = \frac{1}{2}x + \frac{11}{2}$.

We now write the equation in the standard form $Ax + By = C$.

$$-\frac{1}{2}x + y = \frac{11}{2} \qquad \text{Add } -\frac{1}{2}x \text{ to both sides.}$$

$$x - 2y = -11 \qquad \text{Multiply both sides by } -2.$$

An equation of the line in standard form is $x - 2y = -11$. See Figure 2-39. The line passes through $P(3, 7)$ and $Q(-5, 3)$.

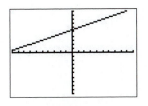

FIGURE 2-39

Self Check 6 Find an equation in standard form of the line passing through $P(-5, 4)$ and $Q(8, -6)$.

Now Try Exercise 61.

5. Write Equations of Parallel and Perpendicular Lines

EXAMPLE 7 **Finding an Equation of a Parallel Line**

Find an equation of the line passing through $P(-2, 5)$ and parallel to the line $y = 8x - 3$. Write the equation in slope-intercept form.

SOLUTION We will substitute the coordinates of $P(-2, 5)$ and the slope of the line parallel to $y = 8x - 3$ into point-slope form and simplify the results to write an equation of the parallel line.

The slope of the line given by $y = 8x - 3$ is 8, the coefficient of x. Since the graph of the desired equation is to be parallel to the graph of $y = 8x - 3$, its slope must also be 8.

We will substitute -2 for x_1, 5 for y_1, and 8 for m into the point-slope form and simplify.

$$y - y_1 = m(x - x_1)$$
$$y - 5 = 8[x - (-2)] \qquad \text{Substitute 5 for } y_1, 8 \text{ for } m, \text{ and } -2 \text{ for } x_1.$$
$$y - 5 = 8(x + 2) \qquad -(-2) = 2$$
$$y - 5 = 8x + 16 \qquad \text{Use the Distributive Property to remove parentheses.}$$
$$y = 8x + 21 \qquad \text{Add 5 to both sides.}$$

An equation of the desired line in slope-intercept form is $y = 8x + 21$.

Self Check 7 Find an equation of the line passing through $Q(1, 2)$ and parallel to the line $y = 8x - 3$. Write the equation in slope-intercept form.

Now Try Exercise 67.

EXAMPLE 8 **Finding an Equation of a Perpendicular Line**

Find an equation of the line passing through $P(-2, 5)$ and perpendicular to the line $y = 8x - 3$. Write the equation in slope-intercept form.

SOLUTION We will substitute the coordinates of $P(-2, 5)$ and the slope of the line perpendicular to $y = 8x - 3$ into point-slope form and simplify to write an equation of the perpendicular line.

Because the slope of the given line is 8, the slope of the desired perpendicular line must be $-\frac{1}{8}$.

We substitute -2 for x_1, 5 for y_1, and $-\frac{1}{8}$ for m into the point-slope form and simplify.

$$y - y_1 = m(x - x_1)$$
$$y - 5 = -\frac{1}{8}[x - (-2)] \qquad \text{Substitute 5 for } y_1, -\frac{1}{8} \text{ for } m, \text{ and } -2 \text{ for } x_1.$$
$$y - 5 = -\frac{1}{8}(x + 2) \qquad -(-2) = 2$$
$$y = -\frac{1}{8}x - \frac{1}{4} + 5 \qquad \text{Remove parentheses and add 5 to both sides.}$$
$$y = -\frac{1}{8}x + \frac{19}{4} \qquad -\frac{1}{4} + 5 = -\frac{1}{4} + \frac{20}{4} = \frac{19}{4}$$

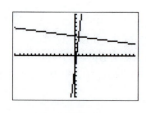

FIGURE 2-40

An equation of the line in slope-intercept form is $y = -\frac{1}{8}x + \frac{19}{4}$. See Figure 2-40, which shows the graphs of the two perpendicular lines.

Self Check 8   Find an equation of the line passing through $Q(1, 2)$ and perpendicular to $y = 8x - 3$. Write the equation in slope-intercept form.

Now Try Exercise 73.

We summarize the various forms of an equation of a line as follows.

> **Tip**
>
> It is very important that you become familiar with the various forms of an equation of a line. Read the directions to a problem very carefully so that you will write the equation in the correct form.
>
>
> ©Barang/Shutterstock.com

Summary of Forms of an Equation of a Line	
Standard form	$Ax + By = C$ A and B cannot both be 0.
Slope-intercept form	$y = mx + b$ The slope is m, and the y-intercept is $(0, b)$.
Point-slope form	$y - y_1 = m(x - x_1)$ The slope is m, and the line passes through (x_1, y_1).
A horizontal line	$y = b$ The slope is 0, and the y-intercept is $(0, b)$.
A vertical line	$x = a$ There is no defined slope, and the x-intercept is $(a, 0)$.

> **Take Note**
>
> When writing equations in $Ax + By = C$ form, we usually clear the equation of fractions and make A positive. For example, the equation $-x + \frac{5}{2}y = 2$ can be changed to $2x - 5y = -4$ by multiplying both sides by -2. We will also divide out any common integer factors of A, B, and C. For example, we would write $4x + 8y = 12$ as $x + 2y = 3$.

6. Write an Equation of the Line that Models a Real-Life Problem

For tax purposes, many businesses use *straight-line depreciation* to find the declining value of aging equipment.

Solving an Application Problem

EXAMPLE 9

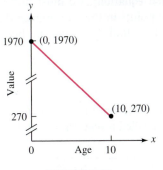
©Fotoslaz/Shutterstock.com

A business purchases a digital multimedia projector for $1970 and expects it to last for ten years. It can then be sold as scrap for a *salvage value* of $270.

If y is the value of the projector after x years of use, and y and x are related by the equation of a line,

a. Find an equation of the line.
b. Find the value of the projector after $2\frac{1}{2}$ years.
c. Find the economic meaning of the y-intercept of the line.
d. Find the economic meaning of the slope of the line.

SOLUTION   **a.** We will find the slope and use the point-slope form to write an equation of the line. (See Figure 2-41.)

When the projector is new, its age x is 0, and its value y is $1970. When the projector is 10 years old, $x = 10$ and $y = $270. Since the line passes through the points $(0, 1970)$ and $(10, 270)$ the slope of the line is

FIGURE 2-41

$$m = \frac{y_2 - y_1}{x_2 - x_1} \qquad \text{This is the slope formula.}$$

$$= \frac{270 - 1970}{10 - 0} \qquad \text{Substitute 270 for } y_2, \text{ 1970 for } y_1, \text{ 10 for } x_2, \text{ and 0 for } x_1.$$

$$= \frac{-1700}{10}$$

$$= -170$$

To find an equation of the line, we substitute -170 for m, 0 for x_1, and 1970 for y_1 in the point-slope form and simplify.

$$y - y_1 = m(x - x_1)$$

$$y - 1970 = -170(x - 0)$$

(3) $$y = -170x + 1970$$

The value y of the projector is related to its age x by the equation $y = -170x + 1970$.

b. To find the value after $2\frac{1}{2}$ years, we will substitute 2.5 for x in Equation 3 and solve for y.

$$y = -170x + 1970$$

$$= -170(2.5) + 1970 \qquad \text{Substitute 2.5 for } x.$$

$$= -425 + 1970$$

$$= 1545$$

In $2\frac{1}{2}$ years, the projector will be worth \$1545.

c. The y-intercept of the graph is $(0, b)$, where b is the value of y when $x = 0$.

$$y = -170x + 1970$$

$$y = -170(0) + 1970 \qquad \text{Substitute 0 for } x.$$

$$y = 1970$$

The y-coordinate b of the y-intercept is the value of a 0-year-old projector, which is the projector's original cost, \$1970.

d. Each year, the value decreases by \$170, because the slope of the line is -170. The slope of the depreciation line is called the *annual depreciation rate*.

Problems that have an annual appreciation rate can be worked similarly.

Self Check 9 A business purchases a Canon copier for \$2700 and expects it to last for ten years. It can then be sold for \$300. Write a straight-line depreciation line for the copier.

Now Try Exercise 87.

7. Use Linear Curve Fitting to Solve Problems

In statistics, the process of using one variable to predict another is called **regression**. For example, if we know a woman's height, we can make a good prediction about her weight, because taller women usually weigh more than shorter women. To write a **prediction equation** (sometimes called a **regression equation**), we must find the equation of the line that comes closest to all of the points in the scattergram than any other possible line. There are statistical methods to find this equation, but we can only approximate the linear equation here.

Using Linear Curve Fitting

EXAMPLE 10

Figure 2-42 shows the result of sampling ten women and finding their heights and weights. The graph of the ordered pairs (h, w) is called a **scattergram** and is also shown.

a. Write an approximation of the regression equation.

b. Use the regression equation to approximate the weight of a woman who is 66 inches tall.

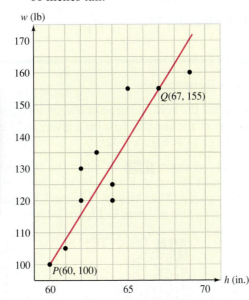

Woman	Height (h) in Inches	Weight (w) in Pounds
1	60	100
2	61	105
3	62	120
4	62	130
5	63	135
6	64	120
7	64	125
8	65	155
9	67	155
10	69	160

FIGURE 2-42

SOLUTION **a.** We will place a straightedge on the scattergram as shown in Figure 2-42 and draw the line joining two points that seems to best fit all the points. In the figure, line PQ is drawn, where point P has coordinates of $(60, 100)$ and point Q has coordinates of $(67, 155)$.

Our approximation of the regression equation will be the equation of the line passing through points P and Q. To find the equation of this line, we first find its slope.

$$m = \frac{y_2 - y_1}{x_2 - x_1} \qquad \text{This is the slope formula.}$$

$$= \frac{155 - 100}{67 - 60} \qquad \text{Substitute 155 for } y_2, 100 \text{ for } y_1, 67 \text{ for } x_2, \text{ and } 60 \text{ for } x_1.$$

$$= \frac{55}{7}$$

We can then use point-slope form to find an equation of the line.

$$y - y_1 = m(x - x_1) \qquad \text{This is the point-slope form.}$$

$$y - 100 = \frac{55}{7}(x - 60) \qquad \text{Choose } (60, 100) \text{ for } (x_1, y_1).$$

$$y = \frac{55}{7}x - \frac{3300}{7} + 100 \qquad \text{Remove parentheses and add 100 to both sides.}$$

$$y = \frac{55}{7}x - \frac{2600}{7} \qquad \text{Simplify.}$$

Our approximation of the regression equation is $y = \frac{55}{7}x - \frac{2600}{7}$.

b. To predict the weight of a woman who is 66 inches tall, for example, we substitute 66 for x in the equation and simplify.

$$y = \frac{55}{7}x - \frac{2600}{7}$$

$$y = \frac{55}{7}(66) - \frac{2600}{7}$$

$$y \approx 147.1428571$$

We would predict that a 66-inch-tall woman chosen at random will weigh about 147 pounds.

Self Check 10 Use the regression equation found in Example 10 to predict the weight of a 67-inch-tall woman.

Now Try Exercise 105.

Linear Regression

We can use the linear regression feature on a graphing calculator to find an equation of the line that best fits a given set of data points. We will do that for the data given in Figure 2-42.

- First, we press `STAT` to enter the statistics menu on the calculator. This screen is shown in Figure 2-43(a). Next we press `ENTER` to input our data. We can input our heights into the L1 column and our weights into the L2 column. This is shown in Figure 2-43(b).

 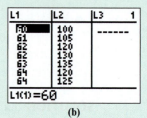

(a)   (b)

FIGURE 2-43

- To obtain an equation of the regression line, we press `STAT` and then the right-arrow key once to access the calculate menu. This screen is shown in Figure 2-44(a). To calculate a linear regression equation, we press `4` to select LinReg ($ax + b$) and then `ENTER` to obtain an equation. The screen is shown in Figure 2-44(b).

(a)   (b)

FIGURE 2-44

Note that the regression line is of the form $y = ax + b$, where a is the slope and b is the y-coordinate of the y-intercept. If we substitute the values shown in Figure 2-44(b) for a and b and round to hundredths, we can write an equation of the line that best fits the data. The regression line is $y = 6.78x - 301.18$.

Self Check Answers

1. $y = \dfrac{7}{3}x - 6$ **2.** $-\dfrac{2}{3}, (0, -3)$ **3.** parallel   **4.** perpendicular

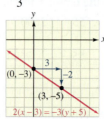

5. $2x + 3y = 7$ **6.** $10x + 13y = 2$ **7.** $y = 8x - 6$ **8.** $y = -\dfrac{1}{8}x + \dfrac{17}{8}$

9. $y = -240x + 2700$ **10.** 155 pounds

Exercises 2.4

Getting Ready

You should be able to complete these vocabulary and concept statements before you proceed to the practice exercises.

Fill in the blanks.

1. The equation $y = mx + b$ is called the _____ form of the equation of a line.

2. In the equation $y = mx + b$, ___ is the slope of the graph of the line.

3. In the equation $y = mx + b$, $(0, b)$ is the _____.

4. The formula for the point-slope form of a line is _____.

5. The standard form of an equation of a line is _____.

6. In statistics a _____ line is a linear equation that best fits given data.

Practice

Use slope-intercept form to write an equation of the line with the given properties.

7. $m = 3; b = -2$

8. $m = -\dfrac{1}{3}; b = \dfrac{2}{3}$

9. $m = 5; b = -\dfrac{1}{5}$

10. $m = \sqrt{2}; b = \sqrt{2}$

11. $m = a; b = \dfrac{1}{a}$

12. $m = a; b = 2a$

13. $m = a; b = a$

14. $m = \dfrac{1}{a}; b = a$

Use slope-intercept form to write an equation of a line passing through the given point and having the given slope. Express the answer in standard form.

15. $P(0, 0); m = \dfrac{3}{2}$

16. $P(-3, -7); m = -\dfrac{2}{3}$

17. $P(-3, 5); m = -3$

18. $P(-5, 1); m = 1$

19. $P\left(0, \sqrt{2}\right); m = \sqrt{2}$

20. $P\left(-\sqrt{3}, 0\right); m = 2\sqrt{3}$

Find the slope and the y-intercept of the lines determined by the given equations.

21. $3x - 2y = 8$

22. $-2x + 4y = 12$

23. $-2(x + 3y) = 5$

24. $5(2x - 3y) = 4$

25. $x = \dfrac{2y - 4}{7}$

26. $3x + 4 = -\dfrac{2(y - 3)}{5}$

Write each equation in slope-intercept form to determine the slope and y-intercept. Then use the slope and y-intercept to graph the line.

27. $x - y = 1$

28. $x + y = 2$

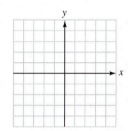

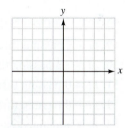

29. $x = \dfrac{3}{2}y - 3$

30. $x = -\dfrac{4}{5}y + 2$

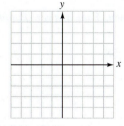

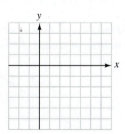

31. $3(y - 4) = -2(x - 3)$

32. $-4(2x + 3) = 3(3y + 8)$

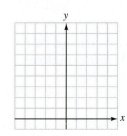

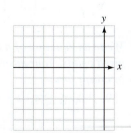

Determine whether the graphs of each pair of equations are parallel, perpendicular, or neither.

33. $y = 3x + 4,\ y = 3x - 7$

34. $y = 4x - 13,\ y = \dfrac{1}{4}x + 13$

35. $x + y = 2,\ y = x + 5$

36. $x = y + 2,\ y = x + 3$

37. $y = 3x + 7,\ 2y = 6x - 9$

38. $2x + 3y = 9,\ 3x - 2y = 5$

39. $3x + 6y = 1,\ y = \dfrac{1}{2}x$

40. $x = 3y + 4,\ y = -3x + 7$

41. $y = 3,\ x = 4$

42. $y = -3,\ y = -7$

43. $x = \dfrac{y - 2}{3},\ 3(y - 3) + x = 0$

44. $2y = 8,\ 3(2 + x) = 3(y + 2)$

Write an equation of the line with the given properties. Your answer should be written in standard form.

45. $m = 2$ passing through $P(2, 4)$

46. $m = -3$ passing through $P(3, 5)$

47. $m = 2$ passing through $P\left(-\dfrac{3}{2}, \dfrac{1}{2}\right)$

48. $m = -6$ passing through $P\left(\dfrac{1}{4}, -2\right)$

49. $m = \dfrac{2}{5}$ passing through $P(-1, 1)$

50. $m = -\dfrac{1}{5}$ passing through $P(-2, -3)$

51. $m = 0$ passing through $P(-6, -3)$

52. $m = 0$ passing through $P(7, 5)$

53. m is undefined passing through $P(-6, -3)$

54. m is undefined passing through $P(6, -1)$

55. $m = \pi$ passing through $P(\pi, 0)$

56. $m = \pi$ passing through $P(0, \pi)$

Find an equation of each line shown. Your answer should be written in standard form.

57.

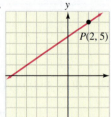

58.

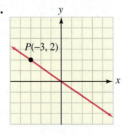

Write an equation of a line that passes through the two given points. Your answer should be written in slope-intercept form.

59. $P(0, 0),\ Q(4, 4)$

60. $P(-5, -5),\ Q(0, 0)$

61. $P(3, 4),\ Q(0, -3)$

62. $P(4, 0),\ Q(6, -8)$

Write an equation in slope-intercept form of each line shown.

63.

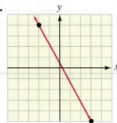

64.

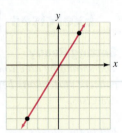

Write an equation of the line that passes through the given point and is parallel to the given line. Your answer should be written in slope-intercept form.

65. $P(0, 0),\ y = 4x - 7$

66. $P(0, 0),\ x = -3y - 12$

67. $P(2, 5),\ 4x - y = 7$

68. $P(-6, 3),\ y + 3x = -12$

69. $P(4, -2),\ x = \dfrac{5}{4}y - 2$

70. $P(1, -5),\ x = -\dfrac{3}{4}y + 5$

Write an equation of the line that passes through the given point and is perpendicular to the given line. Your answer should be written in slope-intercept form.

71. $P(0, 0),\ y = 4x - 7$

72. $P(0, 0),\ x = -3y - 12$

73. $P(2, 5),\ 4x - y = 7$

74. $P(-6, 3),\ y + 3x = -12$

75. $P(4, -2),\ x = \dfrac{5}{4}y - 2$

76. $P(1, -5),\ x = -\dfrac{3}{4}y + 5$

77. Find an equation of the line perpendicular to the line $y = 3$ and passing through the midpoint of the segment joining $(2, 4)$ and $(-6, 10)$.

78. Find an equation of the line parallel to the line $y = -8$ and passing through the midpoint of the segment joining $(-4, 2)$ and $(-2, 8)$.

79. Find an equation of the line parallel to the line $x = 3$ and passing through the midpoint of the segment joining $(2, -4)$ and $(8, 12)$.

80. Find an equation of the line perpendicular to the line $x = 3$ and passing through the midpoint of the segment joining $(-2, 2)$ and $(4, -8)$.

Applications

In Exercises 81–91, assume straight-line depreciation or straight-line appreciation.

81. Depreciation A surfing van was purchased for $24,300. Its salvage value at the end of its 7-year useful life is expected to be $1900. Find a depreciation equation.

82. Depreciation A small business purchases the laptop computer shown. It will be depreciated over a 4-year period, when its salvage value will be $300. Find a depreciation equation.

$2700

83. Appreciation A condominium in San Diego was purchased for $475,000. The owners expect the condominium to double in value in 10 years. Find an appreciation equation.

84. Appreciation A house purchased for $112,000 is expected to double in value in 12 years. Find an appreciation equation.

85. Depreciation Find a depreciation equation for the TV in the following want ad.

> *For Sale*: 3-year-old 54-inch TV, $1900 new. Asking $1190.
> Call 875-5555. Ask for Mike.

86. Depreciation A Bose Wave Radio cost $555 when new and is expected to be worth $80 after 5 years. What will it be worth after 3 years?

87. Salvage value A copier cost $1050 when new and will be depreciated at the rate of $120 per year. If the useful life of the copier is 8 years, find its salvage value.

88. Rate of depreciation A jet ski that cost $13,800 when new will have no salvage value after 6 years. Find its annual rate of depreciation.

89. Value of an antique An antique table is expected to appreciate $40 each year. If the table will be worth $450 in 2 years, what will it be worth in 13 years?

90. Value of an antique An antique clock is expected to be worth $350 after 2 years and $530 after 5 years. What will the clock be worth after 7 years?

91. Purchase price of real estate A cottage that was purchased 3 years ago is now appraised at $47,700. If the property has been appreciating $3500 per year, find its original purchase price.

92. Computer repair A computer repair company charges a fixed amount, plus an hourly rate, for a service call. Use the information in the illustration to find the hourly rate

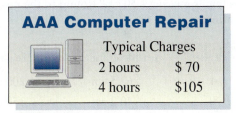

AAA Computer Repair

Typical Charges	
2 hours	$ 70
4 hours	$105

93. Automobile repair An auto repair shop charges an hourly rate, plus the cost of parts. If the cost of labor for a $1\frac{1}{2}$-hour radiator repair is $69, find the cost of labor for a 5-hour transmission overhaul.

94. Printer charges A printer charges a fixed setup cost, plus $1 for every 100 copies. If 700 copies cost $52, how much will it cost to print 1000 copies?

95. Predicting fires A local fire department recognizes that city growth and the number of reported fires are related by a linear equation. City records show that 300 fires were reported in a year when the local population was 57,000 persons, and 325 fires were reported in a year when the population was 59,000 persons. How many fires can be expected in the year when the population reaches 100,000 persons?

96. Estimating the cost of rain gutter A neighbor tells you that an installer of rain gutter charges $60, plus a dollar amount per foot. If the neighbor paid $435 for the installation of 250 feet of gutter, how much will it cost you to have 300 feet installed?

97. Converting temperatures Water freezes at 32° F, or 0° C. Water boils at 212° F, or 100° C. Find a formula for converting a temperature from degrees Fahrenheit to degrees Celsius.

98. Converting units A speed of 1 mile per hour is equal to 88 feet per minute, and of course, 0 miles per hour is 0 feet per minute. Find an equation for converting a speed x, in miles per hour, to the corresponding speed y, in feet per minute.

99. Smoking The percent y of 18- to 25-year-old smokers in the United States has been declining at a constant rate since 1974. If about 47% of this group smoked in 1974 and about 29% smoked in 1994, find a linear equation that models this decline. If this trend continues, estimate what percent will smoke in 2024.

100. Forensic science Scientists believe there is a linear relationship between the height h (in centimeters) of a male and the length f (in centimeters) of his femur bone. Use the data in the table to find a linear equation that expresses the height h in terms of f. Round all constants to the nearest thousandth. How tall would you expect a man to be if his femur measures 50 cm? Round to the nearest centimeter.

Person	Length of Femur (f)	Height (h)
A	62.5 cm	200 cm
B	40.2 cm	150 cm

101. Predicting stock prices The value of the stock of ABC Corporation has been increasing by the same fixed dollar amount each year. The pattern is expected to continue. Let 2010 be the base year corresponding to $x = 0$ with $x = 1, 2, 3, \ldots$ corresponding to later years. ABC stock was selling at $37\frac{1}{2}$ in 2010 and at $45 in 2012. If y represents the price of ABC stock, find the equation $y = mx + b$ that relates x and y, and predict the price in the year 2020.

102. Estimating inventory Inventory of unsold goods showed a surplus of 375 units in January and 264 in April. Assume that the relationship between inventory and time is given by the equation of a line, and estimate the expected inventory in March. Because March lies between January and April, this estimation is called **interpolation**.

103. Oil depletion When a Petroland oil well was first brought on line, it produced 1900 barrels of crude oil per day. In each later year, owners expect its daily production to drop by 70 barrels. Find the daily production after $3\frac{1}{2}$ years.

104. Waste management The corrosive waste in industrial sewage limits the useful life of the piping in a waste processing plant to 12 years. The piping system was originally worth $137,000, and it will cost the company $33,000 to remove it at the end of its 12-year useful life. Find a depreciation equation.

105. Crickets The table shows the approximate chirping rate at various temperatures for one type of cricket.

©EncikAri/Shutterstock.com

Temperature (°F)	Chirps per Minute
50	20
60	80
70	115
80	150
100	250

a. Construct a scattergram below.

Chirps/min

(graph with vertical axis marked 25, 50, 75, 100, 125, 150, 175, 200, 225, 250 and horizontal axis Temp (°F) marked 50 60 70 80 90 100)

b. Assume a linear relationship and write a regression equation.

c. Estimate the chirping rate at a temperature of 90°F.

106. Fishing The table shows the lengths and weights of seven muskies captured by the Department of Natural Resources in Catfish Lake in Eagle River, Wisconsin.

Musky	Length (in.)	Weight (lb)
1	26	5
2	27	8
3	29	9
4	33	12
5	35	14
6	36	14
7	38	19

a. Construct a scattergram for the data.

b. Assume a linear relationship and write a regression equation.

c. Estimate the weight of a musky that is 32 inches long.

 107. Use the linear regression feature on a graphing calculator to determine an equation of the line that best fits the data given in Exercise 105. Round to the hundredths.

 108. Use the linear regression feature on a graphing calculator to determine an equation of the line that best fits the data given in Exercise 106. Round to the hundredths.

Discovery and Writing

109. Explain how to find an equation of a line passing through two given points.

110. Explain how to find an equation of a line that passes through a given point and is parallel to a given line.

111. Describe how to find an equation of a line that passes through a given point and is perpendicular to a given line.

112. In straight-line depreciation, explain why the slope of the line is called the *rate of depreciation*.

113. Prove that an equation of a line with x-intercept of $(a, 0)$ and y-intercept of $(0, b)$ can be written in the form

$$\frac{x}{a} + \frac{y}{b} = 1$$

114. Find the x- and y-intercepts of the line $bx + ay = ab$.

Investigate the properties of slope and the y-intercept by experimenting with the following problems.

115. Graph $y = mx + 2$ for several positive values of m. What do you notice?

116. Graph $y = mx + 2$ for several negative values of m. What do you notice?

117. Graph $y = 2x + b$ for several increasing positive values of b. What do you notice?

118. Graph $y = 2x + b$ for several decreasing negative values of b. What do you notice?

Critical Thinking

Determine if the statement is true or false. If the statement is false, then correct it and make it true.

119. The slope of the graph of $Ax + By = C$ ($A \neq 0$ and $B \neq 0$) is $-\dfrac{B}{A}$.

120. The y-intercept of the graph of $Ax + By = C$ ($A \neq 0, B \neq 0$) is $\left(0, \dfrac{B}{C}\right)$.

121. $y = \pi$ and $y = -\pi$ are parallel lines.

122. $x = \dfrac{\sqrt{11}}{11}$ and $y = -\dfrac{\sqrt{11}}{11}$ are perpendicular lines.

123. The equation of the line passing through $(-99, 99)$ and parallel to $x = 99$ is $y = -99$.

124. The equation of the line passing through $(99, -99)$ and perpendicular to $y = 99$ is $x = 99$.

125. The equations $\sqrt{5}x + \sqrt{10}y = \sqrt{15}$ and $x + \sqrt{2}y = \sqrt{3}$ describe the same line.

126. To determine whether lines are parallel, perpendicular, or neither, we always graph the equations and inspect them.

2.5 Graphs of Equations and Circles

In this section, we will learn to

1. Find the x- and y-intercepts of a graph.

2. Use symmetry to help graph equations.

3. Identify the center and radius of a circle.

4. Write equations of circles.

5. Graph circles.

6. Solve equations by using a graphing calculator.

The London Eye is one of the tallest and most beautiful observation wheels in the world.

　　It stands 443 feet high and was the vision of the architects David Marks and Julia Barfield. Its circular design was used as a metaphor for the turning of the century. The Eye has been described as a breathtaking feat of design and engineering.

　　The graphs of many equations are curves and circles. In this section, we will plot several points (x, y) that satisfy such an equation and join them with a smooth curve. Usually the shape of the graph will become evident.

1. Find the x- and y-Intercepts of a Graph

In Figure 2-45(a), the **x-intercepts** of the graph are $(a, 0)$ and $(b, 0)$, the points where the graph intersects the x-axis. In Figure 2-45(b), the **y-intercept** is $(0, c)$, the point where the graph intersects the y-axis.

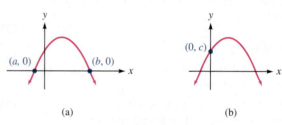

(a)　　　　　　　　　　　　　　　(b)

FIGURE 2-45

To graph the equation $y = x^2 - 4$, we first find the x- and y-intercepts. To find the x-intercepts, we let $y = 0$ and solve for x.

$$y = x^2 - 4$$

$$0 = x^2 - 4 \qquad \text{Substitute 0 for } y.$$

$$0 = (x + 2)(x - 2) \qquad \text{Factor } x^2 - 4.$$

$$x + 2 = 0 \quad \text{or} \quad x - 2 = 0 \qquad \text{Set each factor equal to 0.}$$

$$x = -2 \quad | \quad \quad x = 2$$

Since $y = 0$ when $x = -2$ and $x = 2$, the x-intercepts are $(-2, 0)$ and $(2, 0)$. (See Figure 2-46.)

　　To find the y-intercept, we let $x = 0$ and solve for y.

$$y = x^2 - 4$$

$$y = 0^2 - 4 \qquad \text{Substitute 0 for } x.$$

$$y = -4$$

Since $y = -4$ when $x = 0$, the y-intercept is $(0, -4)$.

We can find other pairs (x, y) that satisfy the equation by substituting numbers for x and finding the corresponding values of y. For example, if $x = -3$, then $y = (-3)^2 - 4 = 5$ and the point $(-3, 5)$ lies on the graph.

The coordinates of the intercepts and other points appear in Figure 2-46. If we plot the points and draw a curve through them, we obtain the graph of the equation.

> **Take Note**
>
> For each input value x there corresponds exactly one output value y. Therefore, the equation $y = x^2 - 4$ represents y as a function of x. The equation can be written using functional notation as $f(x) = x^2 - 4$.

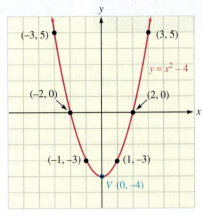

$y = x^2 - 4$		
x	y	(x, y)
-3	5	$(-3, 5)$
-2	0	$(-2, 0)$
-1	-3	$(-1, -3)$
0	-4	$(0, -4)$
1	-3	$(1, -3)$
2	0	$(2, 0)$
3	5	$(3, 5)$

FIGURE 2-46

This graph is called a **parabola**. Its lowest point, $V(0, -4)$, is called the **vertex**. Because the y-axis divides the parabola into two congruent halves, it is called an **axis of symmetry**. We say that the parabola is **symmetric about the y-axis**.

Accent on Technology

Graphing an Equation

A graphing calculator can be used to draw the graph of the equation $y = x^2 - 4$. When using a graphing calculator, we must enter the equation and then set the window.

- To enter the equation, press [Y=], which opens the window shown in Figure 2-47(a) and enter the right side of the equation.
- To set the window, press [WINDOW] as shown in Figure 2-47(b). We can experiment with different windows, or we can use the [ZOOM] menu to select some default windows.
- Press [GRAPH], and the equation is graphed as shown in Figure 2-47(c).

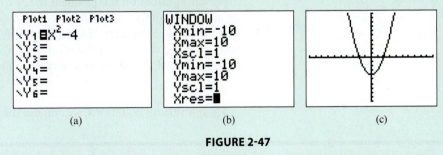

(a)   (b)   (c)

FIGURE 2-47

2. Use Symmetry to Help Graph Equations

Symmetry is a tool that we can use to help graph equations. There are many examples of symmetry in the real world.

There are several ways in which a graph can have symmetry.

1. If the point $(x, -y)$ lies on the graph whenever the point (x, y) does, the graph is **symmetric about the x-axis**. See Figure 2-48(a). This implies that a graph is symmetric about the x-axis if we get the same x-coordinate when we evaluate its equation at y or at $-y$.

2. If the point $(-x, y)$ lies on a graph whenever the point (x, y) does, the graph is **symmetric about the y-axis**. See Figure 2-48(b). This implies that a graph is symmetric about the y-axis if we get the same y-coordinate when we evaluate its equation at x or at $-x$.

3. If the point $(-x, -y)$ lies on the graph whenever the point (x, y) does, the graph is **symmetric about the origin**. See Figure 2-48(c). This implies that the graph is symmetric about the origin if we get opposite values of y when we evaluate its equation at x or at $-x$.

> **Take Note**
>
> If a graph is symmetric with respect to the x-axis and y-axis, then it is symmetric about the origin.

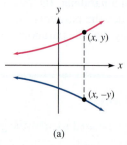

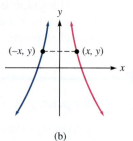

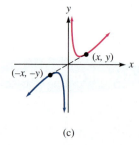

(a)　　　　　　　　　　(b)　　　　　　　　　　(c)

FIGURE 2-48

Symmetry Tests

- **Test for x-Axis Symmetry** To test for x-axis symmetry, replace y with $-y$. If the resulting equation is equivalent to the original one, the graph is symmetric about the x-axis.

- **Test for y-Axis Symmetry** To test for y-axis symmetry, replace x with $-x$. If the resulting equation is equivalent to the original one, the graph is symmetric about the y-axis.

- **Test for Origin Symmetry** To test for symmetry about the origin, replace x with $-x$ and y with $-y$. If the resulting equation is equivalent to the original one, the graph is symmetric about the origin.

To graph various equations, we will use the following strategy.

Strategy for Graphing Equations To graph various equations, we will use the following three steps:

1. Find the x- and y-intercepts.

2. Test for symmetries.

3. Graph the equation by plotting points, making use of symmetry, and joining the points with a smooth curve.

Graphing an Equation Using Intercepts and Symmetry

EXAMPLE 1

Graph: $y = |x|$.

SOLUTION To graph $y = |x|$, we will find the x- and y-intercepts, test for symmetries, plot points, and join the points with a smooth curve.

Step 1: Find the x- and y-intercepts.   To find the x-intercepts, we let $y = 0$ and solve for x.

$$y = |x|$$
$$0 = |x| \qquad \text{Substitute 0 for } y.$$
$$|x| = 0$$
$$x = 0$$

The x-intercept is $(0, 0)$. (See Figure 2-49.)

To find the y-intercepts, we let $x = 0$ and solve for y.

$$y = |x|$$
$$y = |0| \qquad \text{Substitute 0 for } x.$$
$$y = 0$$

The y-intercept is $(0, 0)$.

Step 2: Test for symmetries.   To test for x-axis symmetry, we replace y with $-y$.

(1) $y = |x|$ This is the original equation.

(2) $-y = |x|$ Replace y with $-y$.

Since Equations 1 and 2 are different, the graph is not symmetric about the x-axis.

To test for y-axis symmetry, we replace x with $-x$.

(1) $y = |x|$ This is the original equation.

$y = |-x|$ Replace x with $-x$.

(3) $y = |x|$ $|-x| = |x|$

Since Equations 1 and 3 are the same, the graph is symmetric about the y-axis.

To test for symmetry about the origin, we replace x with $-x$ and y with $-y$.

(1) $y = |x|$ This is the original equation.

$-y = |-x|$ Replace x with $-x$ and y with $-y$.

(4) $-y = |x|$ $|-x| = |x|$

Since Equations 1 and 4 are different, the graph is not symmetric about the origin.

Step 3: Graph the equation. To graph the equation, we plot the x- and y-intercepts and several other pairs (x, y) with positive values of x. We can use the y-axis symmetry to draw the graph for negative values of x. See Figure 2-49.

Tip

Keep in mind that to draw the graph of an equation, intercepts and symmetry are not required. A table of values and plotting of points are required. Intercepts and symmetry are simply helpful tools for graphing.

©Mania-room/Shutterstock.com

Take Note

For each input value x, there corresponds exactly one output value y. Therefore, the equation $y = |x|$ is a function. We can write it using functional notation as $f(x) = |x|$.

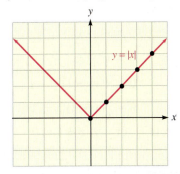

\multicolumn{3}{c}{$y = |x|$}		
x	y	(x, y)
0	0	(0, 0)
1	1	(1, 1)
2	2	(2, 2)
3	3	(3, 3)
4	4	(4, 4)

FIGURE 2-49

Self Check 1   Graph: $y = -|x|$.

Now Try Exercise 53.

Accent on Technology

Graphing an Absolute Value Equation

A graphing calculator can be used to draw the graph of the equation $y = |0.5x|$.

- To enter the equation, press [Y=], which opens the window shown in Figure 2-50(a). To enter the absolute value equation into Y_1, we need to access the absolute value function on our calculator. It is #1 in the [MATH] NUM menu as shown in Figure 2-50(b). Press [ENTER] and input $0.5x$, and the equation will appear in the graph window. Figure 2-50(c).
- To set the window, press [ZOOM], scroll down to 6: [see Figure 2-50(d)], and then press [ENTER].
- The equation is graphed as shown in Figure 2-50(e).

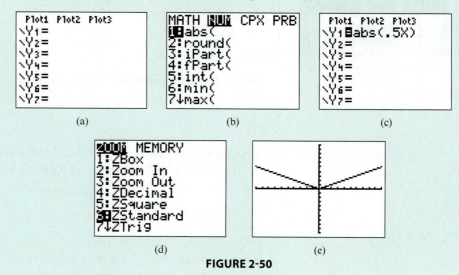

(a)   (b)   (c)

(d)   (e)

FIGURE 2-50

EXAMPLE 2

Graphing an Equation Using Intercepts and Symmetry

Graph: $y = x^3 - x$.

SOLUTION   To graph $y = x^3 - x$, we will find the x- and y-intercepts, test for symmetries, plot points, and join the points with a smooth curve.

Step 1: Find the x- and y-intercepts.   To find the x-intercepts, we let $y = 0$ and solve for x.

$$y = x^3 - x$$

$$0 = x^3 - x \qquad \text{Substitute 0 for } y.$$

$$0 = x(x^2 - 1) \qquad \text{Factor out } x.$$

$$0 = x(x + 1)(x - 1) \qquad \text{Factor } x^2 - 1.$$

$$x = 0 \quad \text{or} \quad x + 1 = 0 \quad \text{or} \quad x - 1 = 0 \qquad \text{Set each factor equal to 0.}$$

$$x = -1 \qquad\qquad x = 1$$

The x-intercepts are $(0, 0)$, $(-1, 0)$, and $(1, 0)$.

To find the y-intercepts, we let $x = 0$ and solve for y.

$$y = x^3 - x$$

$$y = 0^3 - 0 \qquad \text{Substitute 0 for } x.$$

$$y = 0$$

The y-intercept is $(0, 0)$.

Step 2: Test for symmetries.   We test for symmetry about the x-axis by replacing y with $-y$.

(1) $y = x^3 - x$ This is the original equation.

        $-y = x^3 - x$ Replace y with $-y$.

(2) $y = -x^3 + x$ Multiply both sides by -1.

Since Equations 1 and 2 are different, the graph is not symmetric about the x-axis.

To test for y-axis symmetry, we replace x with $-x$.

(1) $y = x^3 - x$ This is the original equation.

        $y = (-x)^3 - (-x)$ Replace x with $-x$.

(3) $y = -x^3 + x$ Simplify.

Since Equations 1 and 3 are different, the graph is not symmetric about the y-axis.

To test for symmetry about the origin, we replace x with $-x$ and y with $-y$.

(1) $y = x^3 - x$ This is the original equation.

        $-y = (-x)^3 - (-x)$ Replace x with $-x$ and y with $-y$.

        $-y = -x^3 + x$ Simplify.

(4) $y = x^3 - x$ Multiply both sides by -1.

Since Equations 1 and 4 are the same, the graph is symmetric about the origin.

Step 3: Graph the equation.   To graph the equation, we plot the x- and y-intercepts and several other pairs (x, y) with positive values of x. We can use the property of symmetry about the origin to draw the graph for negative values of x. See Figure 2-51.

> **Take Note**
>
> For each input value x, there corresponds exactly one output value y. The equation $y = x^3 - x$ is a function. We can write it using functional notation as $f(x) = x^3 - x$.

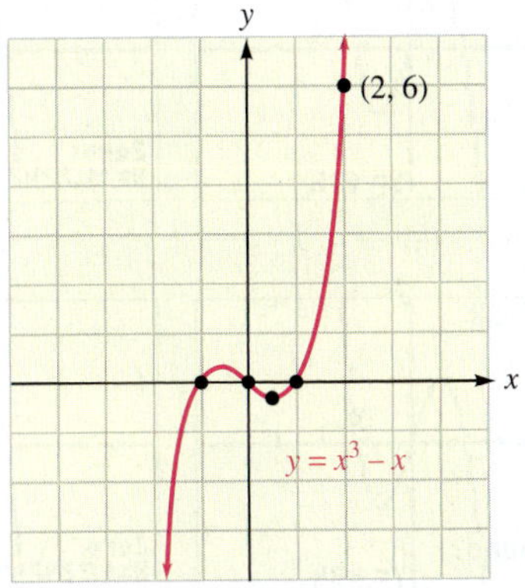

$y = x^3 - x$		
x	y	(x, y)
-1	0	$(-1, 0)$
0	0	$(0, 0)$
$\dfrac{1}{2}$	$-\dfrac{3}{8}$	$\left(\dfrac{1}{2}, -\dfrac{3}{8}\right)$
1	0	$(1, 0)$
2	6	$(2, 6)$

FIGURE 2-51

Self Check 2   Graph: $y = x^3 - 9x$.

Now Try Exercise 49.

Finding the Intercepts of a Graph Using the Zero Feature

A graphing calculator can find the intercepts of an equation, such as $y = 2x^3 - 3x$. The easiest way to do this is to use the graph of the equation. The x-coordinate of the x-intercept is also called a **zero** of the equation.

- Enter the function into Y1. See Figure 2-52(a).
- Set the window. Press ZOOM, scroll down to 4:, and then then press ENTER for this graph. See Figures 2-52(a), (b), and (c).
- To find the zeros, we need to access the CALC menu, which is found by pressing 2nd TRACE. Select 2:ZERO. See Figure 2-52(d).
- Move the cursor until it is to the left of the zero we are trying to find and press ENTER. See Figure 2-52(e).
- Repeat the previous step by moving the cursor to the right of the zero. Press ENTER **two times**. Do not stop when the screen shows GUESS. See Figures 2-52 (f), (g), and (h).
- The zero is $x = -1.224745$.
- Repeat the steps to find the zero $x = 1.2247449$. See Figures 2-52(i), (j), and (k).
- We can see that another zero is $x = 0$.

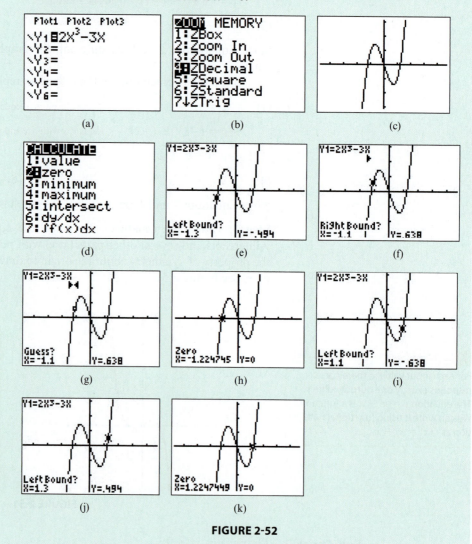

FIGURE 2-52

The x-intercepts are $(-1.224745, 0)$, $(1.2247449, 0)$, and $(0, 0)$.

Graphing an Equation Using Intercepts and Symmetry

EXAMPLE 3   Graph: $y = \sqrt{x}$.

SOLUTION   To graph $y = \sqrt{x}$, we will find the x- and y-intercepts, test for symmetries, plot points, and join the points with a smooth curve.

Step 1: Find the x- and y-intercepts.   We can see that the x- and y-intercepts of $y = \sqrt{x}$ are both $(0, 0)$. The graph passes through the origin.

Step 2: Test for symmetries.   We test for symmetry about the x-axis by replacing y with $-y$.

(1) $y = \sqrt{x}$ This is the original equation.

        $-y = \sqrt{x}$ Replace y with $-y$.

(2) $y = -\sqrt{x}$ Multiply both sides by -1.

Since Equations 1 and 2 are different, the graph is not symmetric about the x-axis.

To test for y-axis symmetry, we replace x with $-x$.

(1) $y = \sqrt{x}$ This is the original equation.

(3) $y = \sqrt{-x}$ Replace x with $-x$.

Since Equations 1 and 3 are different, the graph is not symmetric about the y-axis.

To test for symmetry about the origin, we replace x with $-x$ and y with $-y$.

(1) $y = \sqrt{x}$ This is the original equation.

        $-y = \sqrt{-x}$ Replace x with $-x$ and y with $-y$.

(4) $y = -\sqrt{-x}$ Multiply both sides by -1.

Since Equations 1 and 4 are different, the graph is not symmetric about the origin.

The graph of this equation has no symmetries.

Step 3: Graph the equation.   We plot several points to obtain the graph in Figure 2-53.

> **Take Note**
>
> For each input value x, there corresponds exactly one output value y. The equation $y = \sqrt{x}$ represents a function. We can use functional notation and write it as $f(x) = \sqrt{x}$.

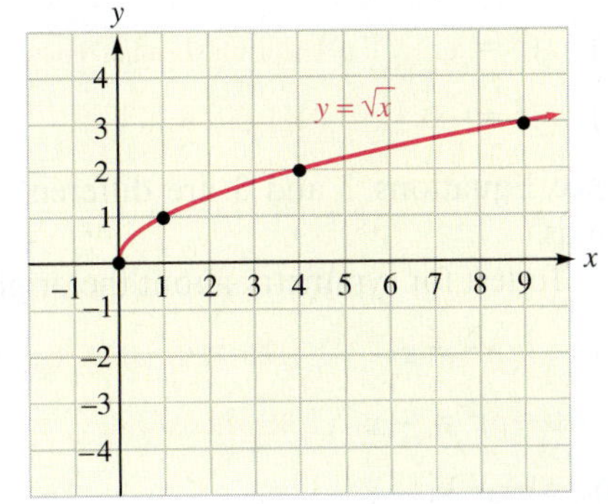

$y = \sqrt{x}$		
x	y	(x, y)
0	0	$(0, 0)$
1	1	$(1, 1)$
4	2	$(4, 2)$
9	3	$(9, 3)$

FIGURE 2-53

Self Check 3   Graph: $y = -\sqrt{x}$.

Now Try Exercise 59.

Graphing a Radical Equation.

To draw the graph of $y = \sqrt{3.5x}$ using a graphing calculator, enter the equation into the graph menu. See Figure 2-54(a); set the window using the values shown in Figure 2-54(b), and press $\boxed{\text{GRAPH}}$ to graph the equation. See Figure 2-54(c).

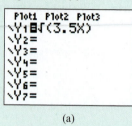
(a)

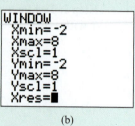
(b)

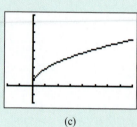
(c)

FIGURE 2-54

EXAMPLE 4

Graphing an Equation Using Intercepts and Symmetry

Graph: $y^2 = x$.

SOLUTION To graph $y^2 = x$, we will find the x- and y-intercepts, test for symmetries, plot points, and join the points with a smooth curve.

Step 1: Find the x- and y-intercepts. We can see that the x- and y-intercepts of $y^2 = x$ are both $(0, 0)$. The graph passes through the origin.

Step 2: Test for symmetries. We test for symmetry about the x-axis by replacing y and $-y$.

(1) $y^2 = x$ This is the original equation.

        $(-y)^2 = x$ Replace y with $-y$.

(2) $y^2 = x$ Simplify.

Since Equations 1 and 2 are the same, the graph is symmetric about the x-axis.
    We test for symmetry about the y-axis by replacing x with $-x$.

(1) $y^2 = x$ This is the original equation.

(3) $y^2 = -x$ Replace x with $-x$.

Since Equations 1 and 3 are different, the graph is not symmetric about the y-axis.
    To test for symmetry about the origin, we replace x with $-x$ and y with $-y$.

(1) $y^2 = x$ This is the original equation.

        $(-y)^2 = -x$ Replace x with $-x$ and y with $-y$.

(4) $y^2 = -x$ Simplify.

Since Equations 1 and 4 are different, the graph is not symmetric about the origin.

Step 3: Graph the equation. To graph the equation, we plot the x- and y-intercepts and several other pairs (x, y) with positive values of y. We can use the property of x-axis symmetry to draw the graph for negative values of y. See Figure 2-55.

Take Note

The equation $y^2 = x$ does not represent y as a function of x. Consider an input x-value of 1. There would correspond two output values 1 and -1 for y, making it not a function.

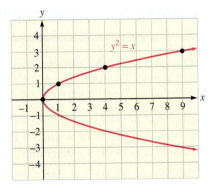

$y^2 = x$		
x	y	(x, y)
0	0	$(0, 0)$
1	1	$(1, 1)$
4	2	$(4, 2)$
9	3	$(9, 3)$

FIGURE 2-55

Self Check 4   Graph: $y^2 = -x$.

Now Try Exercise 57.

The graphs in Examples 3 and 4 are related. We have solved the equation $y^2 = x$ for y, and two equations resulted.

$$y = \sqrt{x} \quad \text{and} \quad y = -\sqrt{x}$$

The equation, $y = \sqrt{x}$ was graphed in Example 3. It is the top half of the parabola shown in Example 4. The equation, $y = -\sqrt{x}$ is the bottom half.

Accent on Technology

Graphing an Equation with a y^2 Term

A graphing calculator is programmed to graph equations with y defined in terms of x. To graph the equation $y^2 = 2x$, it is necessary to solve for y and then enter the resulting equations to see the graph.

$$y^2 = 2x$$
$$y = \pm\sqrt{2x}$$
$$y = \sqrt{2x} \quad \text{or} \quad y = -\sqrt{2x}$$

- We enter the two equations as shown in Figure 2-56(a).
- Next we set the WINDOW as shown in Figure 2-56(b).

  Then we GRAPH the equation(s) as shown in Figure 2-56(c).

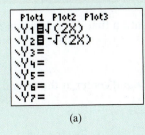

(a)

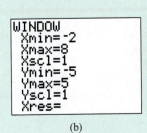

(b)

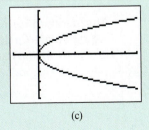

(c)

FIGURE 2-56

3. Identify the Center and Radius of a Circle

Circle	A **circle** is the set of all points in a plane that are a fixed distance from a point called its **center**. The fixed distance is the **radius** of the circle.

To find an equation of a circle with radius r and center at $C(h, k)$, we must find all points $P(x, y)$ in the xy-plane such that the length of line segment PC is r. (See Figure 2-57.)

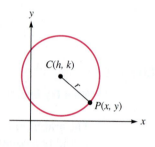

FIGURE 2-57

We can use the Distance Formula to find the length of CP, which is r:

$$r = \sqrt{(x - h)^2 + (y - k)^2}$$

After squaring both sides, we get

$$r^2 = (x - h)^2 + (y - k)^2$$

This equation is called the **standard form of an equation of a circle**.

The Standard Form of an Equation of a Circle with Center at (h, k) and Radius r	The graph of any equation that can be written in the standard form $$(x - h)^2 + (y - k)^2 = r^2$$ is a circle with radius r and center at point (h, k).

If $r = 0$, the circle is a single point called a **point circle**. If the center of a circle is the origin, then $(h, k) = (0, 0)$ and we have the following result.

The Standard Form of an Equation of a Circle with Center at $(0, 0)$ and Radius r	The graph of any equation that can be written in the standard form $$x^2 + y^2 = r^2$$ is a circle with radius r and center at the origin.

The equations of four circles written in standard form and their graphs are shown in Figure 2-58.

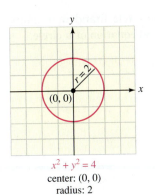

$x^2 + y^2 = 4$
center: (0, 0)
radius: 2
(a)

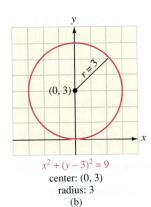

$x^2 + (y - 3)^2 = 9$
center: (0, 3)
radius: 3
(b)

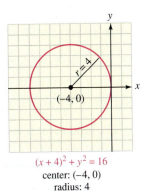

$(x + 4)^2 + y^2 = 16$
center: (−4, 0)
radius: 4
(c)

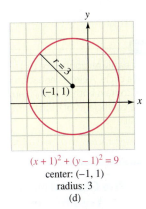

$(x + 1)^2 + (y - 1)^2 = 9$
center: (−1, 1)
radius: 3
(d)

FIGURE 2-58

If we are given an equation of a circle in standard form, we can easily determine the coordinates of its center and the length of its radius.

EXAMPLE 5

Finding the Center and Radius of a Circle in Standard Form

Find the center and radius of the circle with the equation $(x - 3)^2 + (y + 2)^2 = 36$.

SOLUTION   Since the equation is written in standard form, we can identify h, k, and r by comparing the equation to the equation $(x - h)^2 + (y - k)^2 = r^2$.

Standard Form: $(x - \mathbf{h})^2 + (y - \mathbf{k})^2 = \mathbf{r}^2$

Given Form: $(x - \mathbf{3})^2 + [y - (\mathbf{-2})]^2 = \mathbf{6}^2$

We see that $h = 3$, $k = -2$, and $r = 6$. The center (h, k) of the circle is at $(3, -2)$ and the radius is 6.

Self Check 5   Identify the center and radius of the circle with equation of $(x - 4)^2 + y^2 = 49$.

Now Try Exercise 71.

4. Write Equations of Circles

EXAMPLE 6

Writing an Equation of a Circle Given the Center and Radius

Write an equation of the circle with center at $(-2, 4)$ and radius 1.

SOLUTION   We will substitute the coordinates of the center and the radius into the standard form $(x - h)^2 + (y - k)^2 = r^2$.

$(x - \mathbf{h})^2 + (y - \mathbf{k})^2 = \mathbf{r}^2$

$[x - (\mathbf{-2})]^2 + (y - \mathbf{4})^2 = \mathbf{1}^2$ Substitute −2 for h, 4 for k, and 1 for r.

$(x + 2)^2 + (y - 4)^2 = 1$ Simplify.

An equation of the circle with center at $(-2, 4)$ and radius 1 is $(x + 2)^2 + (y - 4)^2 = 1$.

Self Check 6   Write an equation in standard form of the circle with center $(2, -4)$ and radius $\sqrt{5}$.

Now Try Exercise 81.

If we square the binomials in the equation of the circle found in Example 6, we get another form of the equation, called the **general equation of a circle**.

$$(x + 2)^2 + (y - 4)^2 = 1 \qquad \text{This is the equation of Example 6.}$$
$$x^2 + 4x + 4 + y^2 - 8y + 16 = 1 \qquad \text{Remove parentheses.}$$
$$x^2 + y^2 + 4x - 8y + 19 = 0 \qquad \text{Subtract 1 from both sides and simplify.}$$

The general form is $x^2 + y^2 + 4x - 8y + 19 = 0$.

> **Tip**
>
> We can easily recognize the equation of a circle. A circle's equation will always contain both x^2 and y^2 terms, and the coefficients of both terms will be equal.

The General Form of an Equation of a Circle

The **general form** of an equation of a circle is

$$x^2 + y^2 + cx + dy + e = 0$$

where c, d, and e are real numbers.

Finding the General Form of an Equation of a Circle Given the Center and Radius

EXAMPLE 7

Find the general form of an equation of the circle with radius 5 and center at $(3, 2)$.

SOLUTION   We will substitute 5 for r, 3 for h, and 2 for k in the standard form of the equation of a circle and simplify.

$$(x - \mathbf{h})^2 + (y - \mathbf{k})^2 = \mathbf{r}^2 \qquad \text{This is the standard equation.}$$
$$(x - \mathbf{3})^2 + (y - \mathbf{2})^2 = \mathbf{5}^2 \qquad \text{Substitute.}$$
$$x^2 - 6x + 9 + y^2 - 4y + 4 = 25 \qquad \text{Remove parentheses.}$$
$$x^2 + y^2 - 6x - 4y - 12 = 0 \qquad \text{Subtract 25 from both sides and simplify.}$$

The general form is $x^2 + y^2 - 6x - 4y - 12 = 0$.

Self Check 7   Find the general form of an equation of a circle with radius 6 and center at $(-2, 5)$.

Now Try Exercise 85.

Finding the General Form of an Equation of a Circle Given the Endpoints of Its Diameter

EXAMPLE 8

Find the general form of an equation of the circle with endpoints of its diameter at $(8, -3)$ and $(-4, 13)$.

SOLUTION   We will find the center and the radius of the circle, substitute into the standard equation, and simplify. To determine the center, we will find the midpoint of the diameter. To find the radius, we will find the distance from the center to one endpoint of the diameter.

Step 1: Find the center of the circle.   We will find the midpoint of its diameter. Since $(x_1, y_1) = (8, -3)$ and $(x_2, y_2) = (-4, 13)$, we know that $x_1 = 8$, $x_2 = -4$, $y_1 = -3$, and $y_2 = 13$.

$$h = \frac{x_1 + x_2}{2} \qquad\qquad k = \frac{y_1 + y_2}{2} \qquad \text{Use the Midpoint Formula.}$$

$$h = \frac{\mathbf{8 + (-4)}}{2} \qquad\qquad k = \frac{\mathbf{-3 + 13}}{2}$$

$$= \frac{4}{2} \qquad\qquad = \frac{10}{2}$$

$$= 2 \qquad\qquad = 5$$

The center of the circle is at $(h, k) = (2, 5)$.

Step 2: Find the radius of the circle.   To find the radius, we find the distance between the center and one endpoint of the diameter. The center is at $(2, 5)$ and one endpoint is $(8, -3)$.

$$r = \sqrt{(\boldsymbol{x_2} - \boldsymbol{x_1})^2 + (\boldsymbol{y_2} - \boldsymbol{y_1})^2} \qquad \text{Use the Distance Formula.}$$

$$r = \sqrt{(\boldsymbol{8} - \boldsymbol{2})^2 + (\boldsymbol{-3} - \boldsymbol{5})^2} \qquad \text{Substitute 8 for } x_2, \text{ 2 for } x_1, -3 \text{ for } y_2, \text{ and 5 for } y_1.$$

$$= \sqrt{6^2 + (-8)^2}$$

$$= \sqrt{36 + 64}$$

$$= 10 \qquad\qquad \sqrt{36 + 64} = \sqrt{100} = 10$$

The radius of the circle is 10 units.

Step 3: Substitute and simplify.   To find an equation of a circle with center at $(2, 5)$ and radius 10, we substitute 2 for h, 5 for k, and 10 for r in the standard equation of the circle and simplify:

$$(x - \boldsymbol{h})^2 + (y - \boldsymbol{k})^2 = \boldsymbol{r}^2 \qquad \text{This is the standard equation.}$$

$$(x - \boldsymbol{2})^2 + (y - \boldsymbol{5})^2 = \boldsymbol{10}^2$$

$$x^2 - 4x + 4 + y^2 - 10y + 25 = 100 \qquad \text{Remove parentheses.}$$

$$x^2 + y^2 - 4x - 10y - 71 = 0 \qquad \text{Subtract 100 from both sides and simplify.}$$

The general form of the equation is $x^2 + y^2 - 4x - 10y - 71 = 0$.

Self Check 8   Find an equation of a circle with endpoints of its diameter at $(-2, 2)$ and $(6, 8)$.

Now Try Exercise 89.

5. Graph Circles

We can convert the general form of the equation of a circle into standard form by completing the square on x and y.

Graphing a Circle Whose Equation Is in General Form

EXAMPLE 9   Graph the circle whose equation is $2x^2 + 2y^2 - 8x + 4y - 40 = 0$.

SOLUTION   We will convert the general equation of the circle into standard form by completing the square on x and y. We will then use the coordinates of the center of the circle and its radius to draw its graph.

Step 1: Complete the square on x and y.   First, we divide both sides of the equation by 2 to make the coefficients of x^2 and y^2 equal to 1.

$$2x^2 + 2y^2 - 8x + 4y - 40 = 0$$

$$x^2 + y^2 - 4x + 2y - 20 = 0$$

To find the coordinates of the center and the radius, we add 20 to both sides and write the equation in standard form by completing the square on both x and y:

$$x^2 + y^2 - 4x + 2y = 20$$

$$x^2 - 4x + y^2 + 2y = 20$$

$$x^2 - 4x \textcolor{red}{+ 4} + y^2 + 2y \textcolor{red}{+ 1} = 20 \textcolor{red}{+ 4 + 1} \qquad \text{Add 4 and 1 to both sides to complete the square.}$$

$$(x - 2)^2 + (y + 1)^2 = 25 \qquad \text{Factor } x^2 - 4x + 4 \text{ and } y^2 + 2y + 1.$$

$$(x - 2)^2 + [y - (-1)]^2 = 5^2$$

Step 2: Graph the circle. From the standard form of the equation of the circle, we see that its radius is 5 and that the coordinates of its center are $h = 2$ and $k = -1$. Thus, the center of the circle is at is $(2, -1)$.

To graph the circle, we plot its center $(2, -1)$ and locate points on the circle that are 5 units from its center. The graph of the circle is shown in Figure 2-59.

Because the radius of the circle is 5, the easiest points to locate on the circle to draw the graph are the points that are 5 units to the right, 5 units to the left, 5 units above, and 5 units below the center.

Those points are $(7, -1), (-3, -1), (2, 4),$ and $(2, -6)$.

> **Take Note**
>
> Circles aren't functions. See Figure 2-59 and note that the points $(2, 4)$ and $(2, -6)$ lie on the circle. Since the x-value 2 is paired with two y-values, 4 and -6, the equation doesn't represent a function.

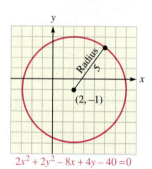

$$2x^2 + 2y^2 - 8x + 4y - 40 = 0$$

FIGURE 2-59

Self Check 9 Graph: $2x^2 + 2y^2 + 4x - 8y + 2 = 0$.

Now Try Exercise 109.

Accent on Technology

Graphing a Circle

When we graphed $y^2 = 2x$ using a graphing calculator, we saw that it is necessary to solve the equation for y in terms of x. The same is true when we graph a circle on a graphing calculator.

- To graph $(x - 2)^2 + (y + 1)^2 = 25$, we must solve the equation for y:

$$(x - 2)^2 + (y + 1)^2 = 25$$

$$(y + 1)^2 = 25 - (x - 2)^2$$

$$y + 1 = \pm\sqrt{25 - (x - 2)^2}$$

$$y = -1 \pm \sqrt{25 - (x - 2)^2}$$

This last expression represents two equations: $y = -1 + \sqrt{25 - (x - 2)^2}$ and $y = -1 - \sqrt{25 - (x - 2)^2}$.

- We graph both of these equations separately on the same coordinate axes by entering the first equation as Y1 and the second as Y2, as shown in Figure 2-60(a). Depending on the window setting of the maximum and minimum values of x and y, the graph may not appear to be a circle. However, if we use the ZOOM 5: ZSquare window, the graph will be circular. See Figures 2-60(b) and (c). If it appears that there are gaps in the graph, that is due to the way the calculator draws graphs by darkening pixels.

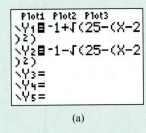
(a)

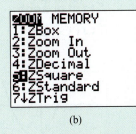
(b)

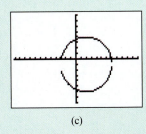
(c)

FIGURE 2-60

Note that in this example, it is easier to graph the circle by hand than with a calculator.

6. Solve Equations by Using a Graphing Calculator

We can solve many equations using the graphing concepts discussed in this chapter and a graphing calculator. For example, the solutions of $x^2 - x - 3 = 0$ will be the numbers x that will make $y = 0$ in the equation $y = x^2 - x - 3$. These numbers will be the x-coordinates of the x-intercepts of the graph of $y = x^2 - x - 3$.

Accent on Technology

Solving Equations

To use a graphing calculator to solve $x^2 - x - 3 = 0$, we graph the equation $y = x^2 - x - 3$, and find the x-coordinates of the x-intercepts. These will be the solutions of the equation. The graph of $y = x^2 - x - 3$ is shown in Figure 2-61(c).

- Use the ZERO command shown earlier in this section to solve the equation. Notice the screen in Figures 2-61(h) and (i) looks different than in Figures 2-61(e) and (f). On some calculators we can just enter the value we want as the left and right bound rather than using the cursor to get to a bound. We see that the two solutions to the equation are $x \approx -1.302776$ and $x \approx 2.3027756$.
- Every equation does not have a solution over the real numbers. For example, to try and solve the equation $x^2 + 2x + 2 = 0$, graph the equation $y = x^2 + 2x + 2$. We can see that it does not intersect the x-axis, and there is no solution to the equation. See Figure 2-61(k).

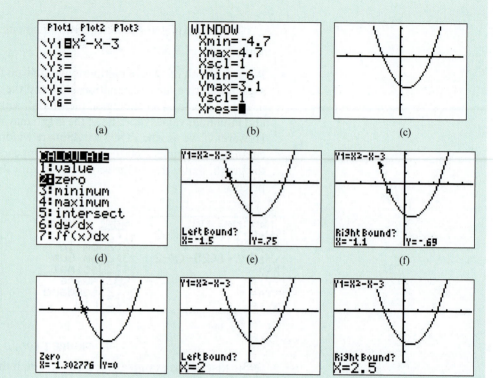

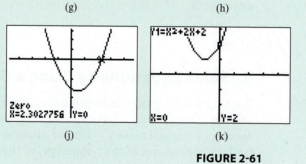

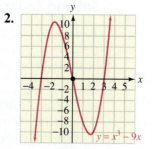

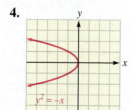

FIGURE 2-61

Self Check Answers

1.

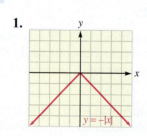

2.

$y = x^3 - 9x$

3.

$y = -\sqrt{x}$

4.

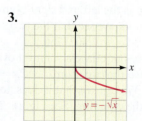

$y^2 = -x$

5. $(4, 0); 7$ **6.** $(x - 2)^2 + (y + 4)^2 = 5$ **7.** $x^2 + y^2 + 4x - 10y - 7 = 0$

8. $x^2 + y^2 - 4x - 10y + 4 = 0$ **9.**

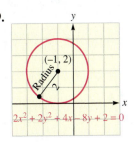

$2x^2 + 2y^2 + 4x - 8y + 2 = 0$

Exercises 2.5

Getting Ready

You should be able to complete these vocabulary and concept statements before you proceed to the practice exercises.

Fill in the blanks.

1. The point where a graph intersects the x-axis is called the _____.

2. The y-intercept is the point where a graph intersects the _____.

3. If a line divides a graph into two congruent halves, we call the line an _____.

4. If the point $(-x, y)$ lies on a graph whenever (x, y) does, the graph is symmetric about the _____.

5. If the point $(x, -y)$ lies on a graph whenever (x, y) does, the graph is symmetric about the _____.

6. If the point $(-x, -y)$ lies on a graph whenever (x, y) does, the graph is symmetric about the _____.

7. A _____ is the set of all points in a plane that are fixed distance from a point called its _____.

8. A _____ is the distance from the center of a circle to a point on the circle.

9. The standard form of an equation of a circle with center at the origin and radius r is _____.

10. The standard form of an equation of a circle with center at (h, k) and radius r is _____.

Practice

Find the x- and y-intercepts of each graph. Do not graph the equation.

11. $y = x^2 - 4$ **12.** $y = x^2 - 9$

13. $y = 4x^2 - 2x$ **14.** $y = 2x - 4x^2$

15. $y = x^2 - 4x - 5$ **16.** $y = x^2 - 10x + 21$

17. $y = x^2 + x - 2$ **18.** $y = x^2 + 2x - 3$

19. $y = x^3 - 9x$ **20.** $y = x^3 + x$

21. $y = x^4 - 1$ **22.** $y = x^4 - 25x^2$

Graph each equation. Check your graph with a graphing calculator.

23. $y = x^2$

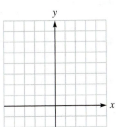

24. $y = -x^2$

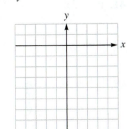

25. $y = -x^2 + 2$

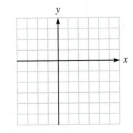

26. $y = x^2 - 1$

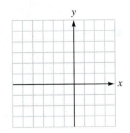

27. $y = x^2 - 4x$

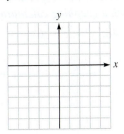

28. $y = x^2 + 2x$

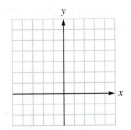

29. $y = \frac{1}{2}x^2 - 2x$

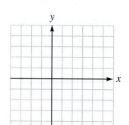

30. $y = \frac{1}{2}x^2 + 3$

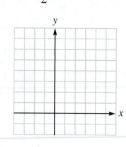

51. $y = |x - 2|$

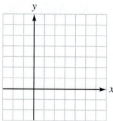

52. $y = |x| - 2$

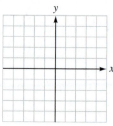

Find the symmetries, if any, of the graph of each equation. Do not graph the equation.

31. $y = x^2 + 2$

32. $y = 3x + 2$

33. $y^2 + 1 = x$

34. $y^2 + y = x$

35. $y^2 = x^2$

36. $y = 3x + 7$

37. $y = 3x^2 + 7$

38. $x^2 + y^2 = 1$

39. $y = 3x^3 + 7$

40. $y = 3x^3 + 7x$

41. $y^2 = 3x$

42. $y = 3x^4 + 7$

43. $y = |x|$

44. $y = |x + 1|$

45. $|y| = x$

46. $|y| = |x|$

53. $y = -|x| + 3$

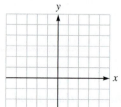

54. $y = 3|x|$

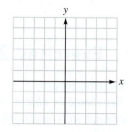

55. $y^2 = -x$

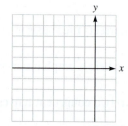

56. $y^2 = 4x$

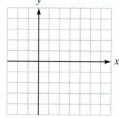

57. $y^2 = 9x$

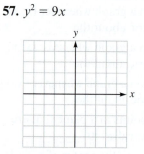

58. $y^2 = -4x$

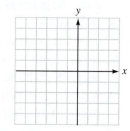

Graph each equation. Be sure to find any intercepts and symmetries. Check your graph with a graphing calculator.

47. $y = x^2 + 4x$

48. $y = x^2 - 6x$

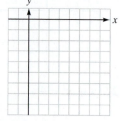

49. $y = x^3$

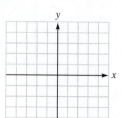

50. $y = x^3 + x$

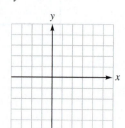

59. $y = \sqrt{x} - 1$

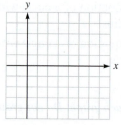

60. $y = 1 - \sqrt{x}$

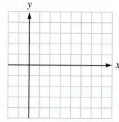

61. $xy = 4$

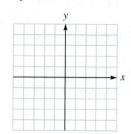

62. $xy = -9$

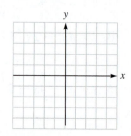

63. $y = \sqrt[3]{x}$

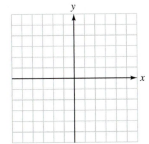

64. $y = -\sqrt[3]{x}$

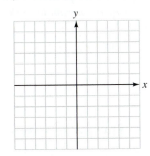

Identify the center and radius of each circle written in standard form.

65. $x^2 + y^2 = 100$

66. $x^2 + y^2 = 81$

67. $x^2 + (y - 5)^2 = 49$

68. $x^2 + (y + 3)^2 = 8$

69. $(x + 6)^2 + y^2 = \dfrac{1}{4}$

70. $(x - 5)^2 + y^2 = \dfrac{16}{25}$

71. $(x - 4)^2 + (y - 1)^2 = 9$

72. $(x + 11)^2 + (x + 7)^2 = 121$

73. $\left(x - \dfrac{1}{4}\right)^2 + (y + 2)^2 = 45$

74. $\left(x + \sqrt{5}\right)^2 + (y - 3)^2 = 1$

Write an equation in standard form of the circle with the given properties.

75. Center at the origin; $r = 5$

76. Center at the origin; $r = \sqrt{3}$

77. Center at $(0, -6)$; $r = 6$

78. Center at $(0, 7)$; $r = 9$

79. Center at $(8, 0)$; $r = \dfrac{1}{5}$

80. Center at $(-10, 0)$; $r = \sqrt{11}$

81. Center at $(-2, 12)$, $r = 13$

82. Center at $\left(\dfrac{2}{7}, -5\right)$; $r = 7$

Write an equation in general form of the circle with the given properties.

83. Center at the origin; $r = 1$

84. Center at the origin; $r = 4$

85. Center at $(6, 8)$; $r = 4$

86. Center at $(5, 3)$; $r = 2$

87. Center at $(3, -4)$; $r = \sqrt{2}$

88. Center at $(-9, 8)$; $r = 2\sqrt{3}$

89. Ends of diameter at $(3, -2)$ and $(3, 8)$

90. Ends of diameter at $(5, 9)$ and $(-5, -9)$

91. Center at $(-3, 4)$ and passing through the origin

92. Center at $(-2, 6)$ and passing through the origin

Convert the general form of each circle given into standard form.

93. $x^2 + y^2 - 6x + 4y + 4 = 0$

94. $x^2 + y^2 + 4x - 8y - 5 = 0$

95. $x^2 + y^2 - 10x - 12y + 57 = 0$

96. $x^2 + y^2 + 2x + 18y + 57 = 0$

97. $2x^2 + 2y^2 - 8x - 16y + 22 = 0$

98. $3x^2 + 3y^2 + 6x - 30y + 3 = 0$

Graph each circle.

99. $x^2 + y^2 - 25 = 0$

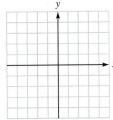

100. $x^2 + y^2 - 8 = 0$

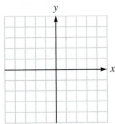

101. $(x - 1)^2 + (y + 2)^2 = 4$

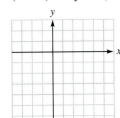

102. $(x + 1)^2 + (y - 2)^2 = 9$

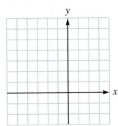

103. $x^2 + y^2 + 2x - 24 = 0$ **104.** $x^2 + y^2 - 4y = 12$ **110.** $9x^2 + 9y^2 - 6x + 18y + 1 = 0$

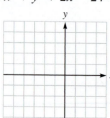

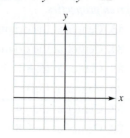

105. $x^2 + y^2 + 4x + 2y - 11 = 0$

 Use a graphing calculator to graph each equation. Then find the coordinates of the vertex of the parabola to the nearest hundredth.

111. $y = 2x^2 - x + 1$ **112.** $y = x^2 + 5x - 6$

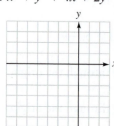

113. $y = 7 + x - x^2$ **114.** $y = 2x^2 - 3x + 2$

106. $x^2 + y^2 - 6x + 2y + 1 = 0$

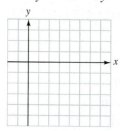

 Use a graphing calculator to solve each equation. Round to the nearest hundredth.

115. $x^2 - 7 = 0$ **116.** $x^2 - 3x + 2 = 0$

107. $9x^2 + 9y^2 - 12y = 5$ **108.** $4x^2 + 4y^2 + 4y = 15$

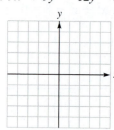

117. $x^3 - 3 = 0$ **118.** $3x^3 - x^2 - x = 0$

Applications

119. Golfing Phil Mickelson's tee shot follows a path given by $y = 64t - 16t^2$, where y is the height of the ball (in feet) after t seconds of flight. How long will it take for the ball to strike the ground?

120. Golfing Halfway through its flight, the golf ball of Exercise 119 reaches the highest point of its trajectory. How high is that?

109. $4x^2 + 4y^2 - 4x + 8y + 1 = 0$

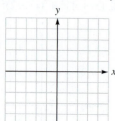

121. Stopping distances The stopping distance D (in feet) for a Ford Fusion car moving V miles per hour is given by $D = 0.08V^2 + 0.9V$. Graph the equation for velocities between 0 and 60 mph.

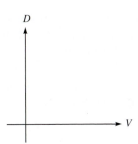

122. Stopping distances See Exercise 121. How much farther does it take to stop at 60 mph than at 30 mph?

123. Basketball court The center circle of the Kansas Jayhawks basketball court is a circle with a 12-ft diameter. If the center of the circle is located at the origin, find an equation in standard form that models the circle.

124. Oil spill Oil spills from a tanker in the Gulf of Mexico and surfaces continuously at coordinates (0, 0). If oil spreads in a circular pattern for ten hours and the circle's radius increases at a rate of 2 inches per hour, write an equation of the circle that models the range of the spill's effect.

125. Super Loop The Fire Ball Super Loop is a rollercoaster ride that is shaped like a circle. Find an equation of the loop in standard form if it is positioned 5 feet off of the ground, has a diameter of 60 feet, and its center is at coordinates (0, 35).

126. Hurricane As a hurricane strengthens, an eye begins to form at the center of the storm. At a wind speed of 80 mph the eye of a hurricane is circular when viewed from above and is 30 miles in diameter. If the eye is located at map coordinates

(5, 10), find an equation, in standard form, of the circle that models the eye of the hurricane.

127. CB radios The CB radio of a trucker covers the circular area shown in the illustration. Find an equation of that circle, in general form.

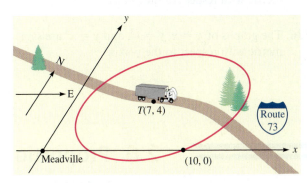

128. Firestone tires Two 24-inch-diameter Firestone tires stand against a wall, as shown in the illustration. Find equations in general form of the circular boundaries of the tires.

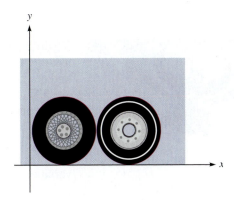

Discovery and Writing

129. Draw three graphs: one that is symmetric to the x-axis, one that is symmetric to the y-axis, and one that is symmetric to the origin.

130. Explain how you test for symmetry with respect to the x-axis, y-axis, and origin.

131. How do you recognize the equation of a circle?

132. Describe the process of converting a circle in general form into standard form.

When converting a circle's equation from general to standard form, it is possible to obtain a constant term on the right side that is zero or negative. If the constant term is zero, the graph is a single point. If the constant term is negative, the graph is nonexistent. Determine whether the graph of the equation is a single point or nonexistent.

133. $x^2 - 4x + y^2 - 6y + 13 = 0$

134. $x^2 - 12x + y^2 + 4y + 43 = 0$

Critical Thinking

Determine if the statement is true or false. If the statement is false, then correct it and make it true.

135. The graphs of $y = x^2$, $y = x^4$, and $y = x^6$ are symmetric with respect to the x-axis.

136. The graphs of $y = x$, $y = x^3$, and $y = x^5$ are symmetric with respect to the y-axis.

137. If the graph of an equation is symmetric with respect to the x-axis and y-axis, then it is symmetric with respect to the origin.

138. If the graph of an equation is symmetric with respect to the origin, then it is symmetric with respect to the x-axis and y-axis.

139. The center of the circle
$$\left(x + \frac{3}{8}\right)^2 + \left(y - \frac{\sqrt[3]{11}}{\pi}\right)^2 = \sqrt[4]{2} \text{ is } \left(-\frac{3}{8}, \frac{\sqrt[3]{11}}{\pi}\right).$$

140. The radius of the circle $\left(x - \frac{2}{9}\right)^2 + \left(y + \frac{\sqrt[3]{6}}{5\pi}\right)^2 = \sqrt[4]{2}$ is $\sqrt[8]{2}$.

141. The graph of the equation $(x - 4)^2 + \left(y + \frac{1}{7}\right)^2 = 0$ is the single point $\left(-4, \frac{1}{7}\right)$.

142. The graph of the equation $x^2 + y^2 = -9$ is nonexistent.

2.6 Proportion and Variation

In this section, we will learn to

1. Solve proportions.
2. Use direct variation to solve problems.
3. Use inverse variation to solve problems.
4. Use joint variation to solve problems.
5. Use combined variation to solve problems.

Skydiving is an adventurous sport. Jumping out of an aircraft at a height of 14,000 feet and free falling delivers a rush of adrenaline that is an exhilarating experience and one that skydivers never forget.

The unique experience allows skydivers to fall approximately 1300 feet every five seconds and reach speeds between 120 and 150 miles per hour.

For a free-falling object, we can calculate the distance fallen given the amount of time. In fact, the distance of the fall is proportional to the square of the time. In this section we will consider how variables are related. These ways include direct variation, inverse variation, joint variation and combinations of these.

1. Solve Proportions

The quotient of two numbers is often called a **ratio**. For example, the fraction $\frac{3}{2}$ (or the expression 3:2) can be read as "the ratio of 3 to 2." Some examples are

$$\frac{3}{5}, \qquad 7:9, \qquad \frac{x+1}{9}, \qquad \frac{a}{b}, \qquad \text{and} \qquad \frac{x^2 - 4}{x + 5}$$

An equation indicating that two ratios are equal is called a **proportion**. Some examples of proportions are

$$\frac{2}{3} = \frac{4}{6}, \qquad \frac{x}{y} = \frac{3}{5}, \qquad \text{and} \qquad \frac{x^2 + 8}{2(x + 3)} = \frac{17(x + 3)}{2}$$

In the proportion $\frac{a}{b} = \frac{c}{d}$, the numbers a and d are called the **extremes**, and the numbers b and c are called the **means**.

To develop an important property of proportions, we suppose that

$$\frac{a}{b} = \frac{c}{d}$$

and multiply both sides by bd to get

$$bd\left(\frac{a}{b}\right) = bd\left(\frac{c}{d}\right)$$

$$\frac{bda}{b} = \frac{bdc}{d}$$

$$da = bc$$

Thus, if $\frac{a}{b} = \frac{c}{d}$ then $ad = bc$.

The same products ad and bc can be found by multiplying diagonally in the proportion $\frac{a}{b} = \frac{c}{d}$. We call ad and bc **cross products**.

Property of Proportions   In any proportion, the product of the extremes is equal to the product of the means.

We can use this property to solve proportions.

EXAMPLE 1   **Solving a Proportion**

Solve the proportion: $\dfrac{x}{5} = \dfrac{2}{x + 3}$.

SOLUTION   We will use the property of proportions to solve the proportion.

$$\frac{x}{5} = \frac{2}{x + 3}$$

$$x(x + 3) = 5 \cdot 2 \qquad \text{The product of the extremes equals the product of the means.}$$

$$x^2 + 3x = 10 \qquad \text{Remove parentheses and simplify.}$$

$$x^2 + 3x - 10 = 0 \qquad \text{Subtract 10 from both sides.}$$

$$(x - 2)(x + 5) = 0 \qquad \text{Factor the trinomial.}$$

$$x - 2 = 0 \quad \text{or} \quad x + 5 = 0 \qquad \text{Set each factor equal to 0.}$$

$$x = 2 \qquad \qquad x = -5$$

Thus, $x = 2$ or $x = -5$. Verify each solution.

Self Check 1   Solve: $\dfrac{2}{5} = \dfrac{3}{x - 4}$.

Now Try Exercise 13.

EXAMPLE 2

Solving an Application Problem Involving a Proportion

Gasoline and oil for a Nissan outboard boat motor are to be mixed in a 50:1 ratio. How many ounces of oil should be mixed with 6 gallons of gasoline?

SOLUTION

We will first express 6 gallons in terms of ounces.

$$6 \text{ gallons} = 6 \cdot 128 \text{ ounces} = 768 \text{ ounces}$$

Then, we let x represent the number of ounces of oil needed, set up the proportion, and solve it.

$$\frac{50}{1} = \frac{768}{x}$$

$$50x = 768 \qquad \text{The product of the extremes equals the product of the means.}$$

$$x = \frac{768}{50} \qquad \text{Divide both sides by 50.}$$

$$x = 15.36$$

Approximately 15 ounces of oil should be added to 6 gallons of gasoline.

©Brenda Carson/Shutterstock.com

Self Check 2

How many ounces of oil should be mixed with 6 gallons of gas if the ratio is to be 40 parts of gas to 1 part of oil?

Now Try Exercise 15.

2. Use Direct Variation to Solve Problems

Two variables are said to **vary directly** or be **directly proportional** if their ratio is a constant. The variables x and y vary directly when

$$\frac{y}{x} = k \quad \text{or, equivalently,} \quad y = kx \quad (k \text{ is a constant})$$

Direct Variation

The words "**y varies directly with x,**" or "**y is directly proportional to x,**" mean that $y = kx$ for some real-number constant k.

The number k is called the **constant of proportionality**.

EXAMPLE 3

Solving a Direct Variation Problem

Distance traveled in a given time varies directly with the speed. If an eagle travels 70 miles at 30 mph, how far will it travel in the same time at 45 mph?

SOLUTION

We will use direct variation to solve the problem.

The phrase *distance varies directly with speed* translates into the formula $d = ks$, where d represents the distance traveled and s represents the speed. The constant of proportionality k can be found by substituting 70 for d and 30 for s in the equation $d = ks$.

$$d = ks$$

$$70 = k(30)$$

$$k = \frac{7}{3}$$

©Igor Kovalenko/Shutterstock.com

To evaluate the distance d traveled at 45 mph, we substitute $\frac{7}{3}$ for k and 45 for s into the formula $d = ks$.

$$d = \frac{7}{3}s$$

$$= \frac{7}{3}(45)$$

$$= 105$$

In the time it takes to go 70 miles at 30 mph, the eagle could travel 105 miles at 45 mph.

Self Check 3   How far will the eagle travel in the same time if its speed is 60 mph?

Now Try Exercise 39.

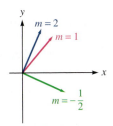

FIGURE 2-62

The statement *y varies directly with x* is equivalent to the equation $y = mx$, where m is the constant of proportionality. Because the equation is in the form $y = mx + b$ with $b = 0$, its graph is a line with slope m and y-intercept at $(0, 0)$. The graphs of $y = mx$ for several values of m are shown in Figure 2-62.

The graph of the relationship of direct variation is always a line that passes through the origin.

3. Use Inverse Variation to Solve Problems

Two variables are said to **vary inversely** or be **inversely proportional** if their product is a constant.

$$xy = k \quad \text{or, equivalently,} \quad y = \frac{k}{x} \quad (k \text{ is a constant})$$

Inverse Variation   The words **"y varies inversely with x,"** or **"y is inversely proportional to x,"** mean that $y = \frac{k}{x}$ for some real-number constant k.

EXAMPLE 4

Solving an Inverse Variation Problem

Intensity of illumination from a light source varies inversely with the square of the distance from the source. If the intensity of a light source is 100 lumens at a distance of 20 feet, find the intensity at 30 feet.

SOLUTION   We will use inverse variation to solve the problem.

If I is the intensity and d is the distance from the light source, the phrase *intensity varies inversely with the square of the distance* translates into the formula

$$I = \frac{k}{d^2}$$

We can evaluate k by substituting 100 for I and 20 for d in the formula and solving for k.

$$I = \frac{k}{d^2}$$

$$100 = \frac{k}{20^2}$$

$$k = 40{,}000$$

To find the intensity at a distance of 30 feet, we substitute 40,000 for k and 30 for d in the formula

$$I = \frac{k}{d^2}$$

$$I = \frac{40{,}000}{30^2}$$

$$= \frac{400}{9}$$

At 30 feet, the intensity of light would be $\frac{400}{9}$ lumens per square centimeter.

Self Check 4 Find the intensity at 50 feet.

Now Try Exercise 43.

The statement *y is inversely proportional to x* is equivalent to the equation $y = \frac{k}{x}$, where k is a constant. Figure 2-63 shows the graphs of $y = \frac{k}{x}$ $(x > 0)$ for three values of k. In each case, the equation determines one branch of a curve called a **hyperbola**. Verify these graphs with a graphing calculator.

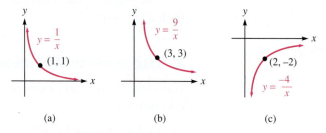

(a)   (b)   (c)

FIGURE 2-63

4. Use Joint Variation to Solve Problems

Joint Variation   The words **"*y* varies jointly with *w* and *x*"** mean that $y = kwx$ for some real-number constant k.

EXAMPLE 5

Solving a Joint Variation Problem

Kinetic energy of an object varies jointly with its mass and the square of its velocity. A 25-gram mass moving at the rate of 30 centimeters per second has a kinetic energy of 11,250 dyne-centimeters. Find the kinetic energy of a 10-gram mass that is moving at 40 centimeters per second.

SOLUTION   We will use joint variation to solve the problem. If we let E, m, and v represent the kinetic energy, mass, and velocity, respectively, the phrase *energy varies jointly with the mass and the square of its velocity* translates into the formula

$$E = kmv^2$$

The constant k can be evaluated by substituting 11,250 for E, 25 for m, and 30 for v in the formula.

$$E = kmv^2$$

$$11{,}250 = k(25)(30)^2$$

Tip

The constant of variation is usually positive because most real-life applications involve only positive quantities. However, the definitions of direct, inverse, joint, and combined variation allow for a negative constant of variation.

$$11{,}250 = 22{,}500k$$

$$k = \frac{1}{2}$$

We can now substitute $\frac{1}{2}$ for k, 10 for m, and 40 for v in the formula and evaluate E.

$$E = \textcolor{red}{kmv^2}$$

$$= \frac{1}{2}(10)(40)^2$$

$$= 8000$$

A 10-gram mass that is moving at 40 centimeters per second has a kinetic energy of 8000 dyne-centimeters.

Self Check 5   Find the kinetic energy of a 25-gram mass that is moving at 100 centimeters per second.

Now Try Exercise 45.

5. Use Combined Variation to Solve Problems

The preceding terminology can be used in various combinations. In each of the statements shown in the following table, the formula on the left translates into the words on the right.

Formula	Words
$y = \dfrac{kx}{z}$	y varies directly with x and inversely with z.
$y = kx^2\sqrt[3]{z}$	y varies jointly with the square of x and the cube root of z.
$y = \dfrac{kx\sqrt{z}}{\sqrt[3]{t}}$	y varies jointly with x and the square root of z and inversely with the cube root of t.
$y = \dfrac{k}{xz}$	y varies inversely with the product of x and z.

EXAMPLE 6

Using Combined Variation to Solve a Problem

The time it takes to build a highway varies directly with the length of the road but inversely with the number of workers. If it takes 100 workers 4 weeks to build 2 miles of the Pacific Coast Highway, how long will it take 80 workers to build 10 miles of the highway?

SOLUTION   We will use combined variation to solve the problem.

We can let t represent the time in weeks, l represent the length in miles, and w represent the number of workers. Because time varies directly with the length of the highway but inversely with the number of workers, the relationship between these variables can be expressed by the equation

$$t = \frac{kl}{w}$$

©Doug Meek/Shutterstock.com

We substitute 4 for t, 100 for w, and 2 for l to find k:

$$4 = \frac{k(2)}{100}$$

$400 = 2k$ Multiply both sides by 100.

$200 = k$ Divide both sides by 2.

We now substitute 80 for w, 10 for l, and 200 for k in the equation $t = \frac{kl}{w}$ and simplify:

$$t = \frac{kl}{w}$$

$$t = \frac{200(10)}{80}$$

$$= 25$$

It will take 25 weeks for 80 workers to build 10 miles of highway.

Self Check 6   How long will it take 50 workers to build 5 miles of highway?

Now Try Exercise 47.

Self Check Answers

1. $\dfrac{23}{2}$ 2. 19.2 oz   3. 140 mi   4. 16 lumens per cm^2

5. 125,000 dyne-centimeters   6. 20 weeks

Exercises 2.6

Getting Ready
You should be able to complete these vocabulary and concept statements before you proceed to the practice exercises.

Fill in the blanks.

1. A ratio is the _____ of two numbers.

2. A proportion is a statement that two _____ are equal.

3. In the proportion $\frac{a}{b} = \frac{c}{d}$, b and c are called the _____.

4. In the proportion $\frac{a}{b} = \frac{c}{d}$, a and d are called the _____.

5. In a proportion, the product of the _____ is equal to the product of the _____.

6. The equation $y = kx$ indicates _____ variation.

7. The equation $y = \frac{k}{x}$ indicates _____ variation.

8. In the equation $y = kx$, k is called the _____ of proportionality.

9. The equation $y = kxz$ represents _____ variation.

10. In the equation $y = \frac{kx^2}{z}$, y varies directly with ___ and inversely with ___.

Practice
Solve each proportion.

11. $\dfrac{4}{x} = \dfrac{2}{7}$

12. $\dfrac{5}{2} = \dfrac{x}{6}$

13. $\dfrac{x}{2} = \dfrac{3}{x + 1}$

14. $\dfrac{x + 5}{6} = \dfrac{7}{8 - x}$

Set up and solve a proportion to answer each question.

15. The ratio of women to men in a mathematics class is 3:5. How many women are in the class if there are 30 men?

16. The ratio of lime to sand in mortar is 3:7. How much lime must be mixed with 21 bags of sand to make mortar?

Find the constant of proportionality.

17. y is directly proportional to x. If $x = 30$, then $y = 15$.

18. z is directly proportional to t. If $t = 7$, then $z = 21$.

19. I is inversely proportional to R. If $R = 20$, then $I = 50$.

20. R is inversely proportional to the square of I. If $I = 25$, then $R = 100$.

21. E varies jointly with I and R. If $R = 25$ and $I = 5$, then $E = 125$.

22. z is directly proportional to the sum of x and y. If $x = 2$ and $y = 5$, then $z = 28$.

Solve each problem.

23. y is directly proportional to x. If $y = 15$ when $x = 4$, find y when $x = \frac{7}{5}$.

24. w is directly proportional to z. If $w = -6$ when $z = 2$, find w when $z = -3$.

25. w is inversely proportional to z. If $w = 10$ when $z = 3$, find w when $z = 5$.

26. y is inversely proportional to x. If $y = 100$ when $x = 2$, find y when $x = 50$.

27. P varies jointly with r and s. If $P = 16$ when $r = 5$ and $s = -8$, find P when $r = 2$ and $s = 10$.

28. m varies jointly with the square of n and the square root of q. If $m = 24$ when $n = 2$ and $q = 4$, find m when $n = 5$ and $q = 9$.

Determine whether the graph could represent direct variation, inverse variation, or neither.

29.

30.

31.

32.

Applications

Set up and solve the required proportion.

33. **Cellphones** A country has 221 mobile cellular telephones per 250 inhabitants. If the country's population is about 280,000, how many mobile cellular telephones does the country have?

34. **Caffeine** Many convenience stores sell supersize 44-ounce soft drinks in refillable cups. For each of the products listed in the table, find the amount of caffeine contained in one of the supersize cups. Round to the nearest milligram.

©Luciano Mortula/Shutterstock.com

Soft Drink 12 oz	Caffeine (mg)
Mountain Dew	55
Coca-Cola Classic	47
Pepsi	37

Based on data from the *Los Angeles Times*

35. **Wallpapering** Read the instructions on the label of wallpaper adhesive. Estimate the amount of adhesive needed to paper 500 square feet of kitchen walls if a heavy wallpaper will be used.

> COVERAGE: One-half gallon will hang approximately 4 single rolls (140 square feet), depending on the weight of the wall covering and the condition of the wall.

36. **Recommended dosages** The recommended child's dose of the sedative hydroxine is 0.006 gram per kilogram of body mass. Find the dosage for a 30-kilogram child in milligrams.

37. **Gas laws** The volume of a gas varies directly with the temperature and inversely with the pressure. When the temperature of a certain gas is 330°C, the pressure is 40 pounds per square inch and the volume is 20 cubic feet. Find the volume when the pressure increases 10 pounds per square inch and the temperature decreases to 300°C.

38. **Hooke's Law** The force f required to stretch a spring a distance d is directly proportional to d. A force of 5 newtons stretches a spring 0.2 meter. What force will stretch the spring 0.35 meter?

39. **Free-falling objects** The distance that an object will fall in t seconds varies directly with the square of t. An object falls 16 feet in 1 second. How long will it take the object to fall 144 feet?

40. Heat dissipation The power, in watts, dissipated as heat in a resistor varies directly with the square of the voltage and inversely with the resistance. If 20 volts are placed across a 20-ohm resistor, it will dissipate 20 watts. What voltage across a 10-ohm resistor will dissipate 40 watts?

41. Period of a pendulum The time required for one complete swing of a pendulum is called the **period** of the pendulum. The period varies directly with the square of its length. If a 1-meter pendulum has a period of 1 second, find the length of a pendulum with a period of 2 seconds.

42. Frequency of vibration The **pitch**, or **frequency**, of a vibrating string varies directly with the square root of the tension. If a string vibrates at a frequency of 144 hertz due to a tension of 2 pounds, find the frequency when the tension is 18 pounds.

43. Illumination Intensity of illumination from a light source varies inversely with the square of the distance from the source. If the intensity of a light source is 60 lumens at a distance of 10 feet, find the intensity at 20 feet.

44. Illumination Intensity of illumination from a light source varies inversely with the square of the distance from the source. If the intensity of a light source is 100 lumens at a distance of 15 feet, find the intensity at 25 feet.

45. Kinetic energy The kinetic energy of an object varies jointly with its mass and the square of its velocity. What happens to the energy when the mass is doubled and the velocity is tripled?

46. Heat dissipation The power, in watts, dissipated as heat in a resistor varies jointly with the resistance, in ohms, and the square of the current, in amperes. A 10-ohm resistor carrying a current of 1 ampere dissipates 10 watts. How much power is dissipated in a 5-ohm resistor carrying a current of 3 amperes?

47. Gravitational attraction The gravitational attraction between two massive objects varies jointly with their masses and inversely with the square of the distance between them. What happens to this force if each mass is tripled and the distance between them is doubled?

48. Gravitational attraction In Problem 47, what happens to the force if one mass is doubled and the other tripled and the distance between them is halved?

49. Plane geometry The area of an equilateral triangle varies directly with the square of the length of a side. Find the constant of proportionality.

50. Solid geometry The diagonal of a cube varies directly with the length of a side. Find the constant of proportionality.

Discovery and Writing

51. Explain the terms *extremes* and *means*.

52. Distinguish between a *ratio* and a *proportion*.

53. What is k in a variation problem?

54. Describe a strategy to solve a variation problem.

55. Explain why $\frac{y}{x} = k$ indicates that y varies directly with x.

56. Explain why $xy = k$ indicates that y varies inversely with x.

57. Explain the term *joint variation* and give an example.

58. As temperature increases on the Fahrenheit scale, it also increases on the Celsius scale. Is this direct variation? Explain.

Critical Thinking

In Exercises 59–62, match each variation sentence on the left with a variation equation on the right.

59. B varies inversely as the cube of r. **a.** $B = \dfrac{kst^2}{r^3}$

60. B varies jointly as the cube of r and the square of t. **b.** $B = \dfrac{kt^2}{r^3}$

61. B varies directly as the square of t and inversely as the cube of r. **c.** $B = kr^3t^2$

62. B varies inversely as the cube of r and jointly as s and the square of t. **d.** $B = \dfrac{k}{r^3}$

In Exercises 63–66, match each variation equation on the left with a variation sentence on the right.

63. $M = \dfrac{kn}{p^4}$

64. $M = knq^2p^4$

65. $M = \dfrac{kp^4}{q^2}$

66. $M = \dfrac{kq^2p^4}{n}$

a. M varies directly as the fourth power of p and inversely as the square of q.

b. M varies jointly as the square of q, the fourth power of p, and inversely as n.

c. M varies directly as n and inversely as the fourth power of p.

d. M varies jointly as n, the square of q, and the fourth power of p.

CHAPTER REVIEW

2.1 Functions and Function Notation

Definitions and Concepts	Examples
A **relation** is a set of ordered pairs. The set created by the first components in the ordered pairs is called the **domain** (the inputs). The set created by the second components in the ordered pairs is the **range** (the outputs).	Find the domain and the range of the relation: $$\{(3, 8), (-2, 12), (6, -10), (-9, 12)\}$$ The set of first components is the domain and the set of second components is the range. The domain is $\{3, -2, 6, -9\}$. The range is $\{8, 12, -10\}$.
A **function** f is a correspondence between a set of input values x and a set of output values y, where to each x-value there corresponds exactly one y-value.	Determine whether the equation $y^2 = 4x$ defines y to be a function of x. First we solve for y. $$y = \pm\sqrt{4x}$$ $$y = \pm 2\sqrt{x}$$ Since for each real number input for x (except 0) there corresponds an output of two y-values, y is not a function of x.
The set of input values x is called the **domain** of a function. The set of output values y is called the **range** of the function.	Find the domain of $g(x) = \sqrt{x - 4}$. Since $x - 4$ must be non-negative $$x - 4 \geq 0$$ $$x \geq 4$$ The domain is $[4, \infty)$.
To evaluate a function $f(x)$ at a given input value x, we substitute the input value for x.	Let $f(x) = \frac{6}{x - 6}$. Find $f(-3)$. $$f(-3) = \frac{6}{-3 - 6} = \frac{6}{-9} = -\frac{2}{3}$$
The fraction $\frac{f(x + h) - f(x)}{h}$ is called the **difference quotient** and is important in calculus.	See Example 8 in Section 2.1.

Exercises

A relation is given. (a) State the domain and the range. (b) Determine if the relation is a function.

1. $\{(3, 4), (4, 5), (5, 6), (6, 7)\}$

2. $\{(2, 4), (2, 5), (3, 6), (-4, 3)\}$

Determine whether each equation defines y to be a function of x. Assume that all variables represent real numbers.

3. $y = 3$

4. $y + 5x^2 = 2$

5. $y^2 - x = 5$

6. $y = |x| + x$

Find the domain of each function. Write each answer using interval notation.

7. $f(x) = 3x^2 - 5$

8. $f(x) = \frac{3x}{x - 5}$

9. $f(x) = \frac{3x}{4x^2 - 16}$

10. $f(x) = \sqrt{x - 1}$

11. $f(x) = \sqrt{5 - x}$

12. $f(x) = \sqrt{x^2 + 1}$

Find $f(2)$, $f(-3)$, and $f(k)$.

13. $f(x) = 5x - 2$

14. $f(x) = \dfrac{6}{x - 5}$

15. $f(x) = |x - 2|$

16. $f(x) = \dfrac{x^2 - 3}{x^2 + 3}$

Evaluate the difference quotient for each function $f(x)$.

17. $f(x) = 5x - 6$

18. $f(x) = 2x^2 - 7x + 3$

19. Target heart rate The target heart rate $f(x)$, in beats per minute, at which a person should train to get an effective workout is a function of the person's age x in years. If $f(x) = -0.6x + 132$, find the target heart rate for a 45-year-old college professor.

20. Concessions A concessionaire at a basketball game pays a vendor \$50 per game for selling hamburgers at \$3.50 each.

   a. Write a function that describes the income I the vendor earns for the concessionaire during the game if the vendor sells x hamburgers.

   b. Find the income if the vendor sells 200 hamburgers.

2.2 The Rectangular Coordinate System and Graphing Lines

Definitions and Concepts	**Examples**
The rectangular coordinate system divides the plane into four quadrants.	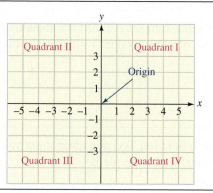

The **graph** of an equation in x and y is the set of all points (x, y) that satisfy the equation.	Use the x- and y-intercepts to graph the equation $6x + 4y = 24$.
The **y-intercept** of a line is the point $(0, b)$, where the line intersects the y-axis. To find b, substitute 0 for x in the equation of the line and solve for y.	**Find the y-intercept.** To find the y-intercept, we substitute 0 for x and solve for y.

$$6x + 4y = 24$$

$$6(0) + 4y = 24 \qquad \text{Substitute 0 in for } x.$$

$$4y = 24 \qquad \text{Simplify.}$$

$$y = 6 \qquad \text{Divide both sides by 4.}$$

The y-intercept is the point $(0, 6)$.

The **x-intercept** of a line is the point $(a, 0)$, where the line intersects the x-axis. To find a, substitute 0 for y in the equation of the line and solve for x.

Find the x-intercept. To find the x-intercept, we substitute 0 for y and solve for x.

$$6x + 4y = 24$$

$$6x + 4(0) = 24 \qquad \text{Substitute 0 for } y.$$

$$6x = 24 \qquad \text{Simplify.}$$

$$x = 4 \qquad \text{Divide both sides by 6.}$$

The x-intercept is the point $(4, 0)$.

Definitions and Concepts	**Examples**
	Find a third point as a check. If we let $x = 2$, we will find that $y = 3$.

$$6x + 4y = 24$$

$$6(2) + 4y = 24 \qquad \text{Substitute 2 for } x.$$

$$12 + 4y = 24 \qquad \text{Simplify.}$$

$$4y = 12 \qquad \text{Subtract 12 from both sides.}$$

$$y = 3 \qquad \text{Divide both sides by 4.}$$

The point $(2, 3)$ satisfies the equation.

We plot each pair and join them with a line to get the graph of the equation.

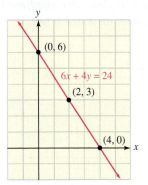

Equation of a vertical line through (a, b):

$$x = a$$

Equation of a horizontal line through (a, b):

$$y = b$$

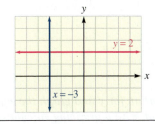

The Distance Formula:

The distance d between points (x_1, y_1) and (x_2, y_2) is given by

$$d = \sqrt{(x_2 - x_1)^2 + (y_2 - y_1)^2}$$

Find the distance between $P(-5, 2)$ and $Q(-3, 4)$. We can use the Distance Formula

$$d = \sqrt{(x_2 - x_1)^2 + (y_2 - y_1)^2}$$

to find the distance between $P(-5, 2)$ and $Q(-3, 4)$. If we let $P(-5, 2) = P(x_1, y_1)$ and $Q(-3, 4) = Q(x_2, y_2)$, we can substitute -5 for x_1, 2 for y_1, -3 for x_2, and 4 for y_2 into the formula and simplify.

$$d(PQ) = \sqrt{(x_2 - x_1)^2 + (y_2 - y_1)^2}$$

$$d(PQ) = \sqrt{[-3 - (-5)]^2 + [4 - 2]^2}$$

$$= \sqrt{(2)^2 + (2)^2}$$

$$= \sqrt{4 + 4} = \sqrt{8} = \sqrt{4 \cdot 2} = 2\sqrt{2}$$

The distance between the two points is $2\sqrt{2}$.

The Midpoint Formula:

The midpoint of the line segment joining (x_1, y_1) and (x_2, y_2) is the point M with coordinates

$$M = \left(\frac{x_1 + x_2}{2}, \frac{y_1 + y_2}{2} \right)$$

To find the midpoint of the segment with endpoints at $(-4, 5)$ and $(6, 7)$, average the x-coordinates and average the y-coordinates:

The midpoint is $\left(\frac{-4 + 6}{2}, \frac{5 + 7}{2} \right) = \left(\frac{2}{2}, \frac{12}{2} \right) = (1, 6)$.

Exercises

Refer to the illustration and find the coordinates of each point.

21. *A*   **22.** *B*

23. *C*   **24.** *D*

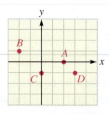

Graph each point. Indicate the quadrant in which the point lies or the axis on which it lies.

25. $(-3, 5)$ **26.** $(5, -3)$

27. $(0, -7)$ **28.** $\left(-\dfrac{1}{2}, 0\right)$

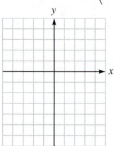

Solve each equation for y and graph the equation. Then check your graph with a graphing calculator.

29. $2x - y = 6$ **30.** $2x + 5y = -10$

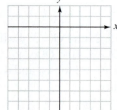

Use the x- and the y-intercepts to graph each equation.

31. $3x - 5y = 15$ **32.** $x + y = 7$

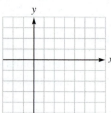 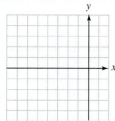

33. $x + y = -7$ **34.** $x - 5y = 5$

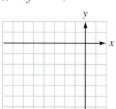 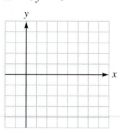

Graph each equation.

35. $y = 4$ **36.** $x = -2$

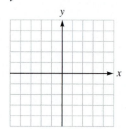 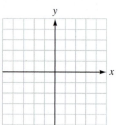

37. Depreciation A Ford Mustang purchased for $18,750 is expected to depreciate according to the formula $f(x) = -2200x + 18,750$, where $f(x)$ is the value of the Mustang after x years. Find its value after 3 years.

38. House appreciation A house purchased for $250,000 is expected to appreciate according to the formula $f(x) = 16,500x + 250,000$, where $f(x)$ is the value of the house after x years. Find the value of the house 5 years later.

Find the length of the segment PQ.

39. $P(-3, 7);\ Q(3, -1)$ **40.** $P(-8, 6);\ Q(-12, 10)$

41. $P\left(\sqrt{3}, 9\right);\ Q\left(\sqrt{3}, 7\right)$ **42.** $P(a, -a);\ Q(-a, a)$

Find the midpoint of the segment PQ.

43. $P(-3, 7);\ Q(3, -1)$ **44.** $P(0, 5);\ Q(-12, 10)$

45. $P\left(\sqrt{3}, 9\right);\ Q\left(\sqrt{3}, 7\right)$ **46.** $P(a, -a);\ Q(-a, a)$

2.3 Linear Functions and Slope

Definitions and Concepts	Examples
The slope of a nonvertical line passing through points $P(x_1, y_1)$ and $Q(x_2, y_2)$ is $$m = \frac{\text{change in } y}{\text{change in } x} = \frac{y_2 - y_1}{x_2 - x_1} \quad (x_2 \neq x_1)$$	Find the slope of the line passing through $P(-1, -3)$ and $Q(7, 9)$. We will substitute the points $P(-1, -3)$ and $Q(7, 9)$ into the slope formula, $$m = \frac{\text{change in } y}{\text{change in } x} = \frac{y_2 - y_1}{x_2 - x_1}$$ to find the slope of the line. Let $P(x_1, y_1) = P(-1, -3)$ and $Q(x_2, y_2) = Q(7, 9)$. Then we substitute -1 for x_1, -3 for y_1, 7 for x_2, and 9 for y_2 to get $$m = \frac{\text{change in } y}{\text{change in } x}$$ $$m = \frac{y_2 - y_1}{x_2 - x_1}$$ $$= \frac{9 - (-3)}{7 - (-1)} = \frac{12}{8} = \frac{3}{2}$$ The slope of the line is $\frac{3}{2}$.
Slopes of horizontal and vertical lines: The slope of a horizontal line (a line with an equation of the form $y = b$) is 0. The slope of a vertical line (a line with an equation of the form $x = a$) is not defined.	The slope of the graph of $y = 7$ is 0. The slope of the line $x = 6$ is not defined.
Slopes of parallel lines: Nonvertical parallel lines have the same slope. **Slopes of perpendicular lines:** The product of the slopes of two perpendicular lines is -1, provided neither line is vertical.	Determine whether the lines with the given slopes are parallel, perpendicular, or neither. $$m_1 = -6; m_2 = \frac{1}{6}$$ The product of the slopes $m_1 = -6$ and $m_2 = \frac{1}{6}$ is -1. The lines are perpendicular.

Exercises

Find the slope of the line PQ, if possible.

47. $P(3, -5)$; $Q(1, 7)$ **48.** $P(2,7)$; $Q(-5,-7)$

49. $P(5, -8)$; $Q(5, \frac{1}{2})$ **50.** $P(\frac{2}{3}, -8)$; $Q(-1, -8)$

51. $P(b, a)$; $Q(a, b)$ **52.** $P(a + b, b)$; $Q(b, b - a)$

Find two points on the line and find the slope of the line.

53. $y = 3x + 6$ **54.** $y = -\frac{1}{5}x - 6$

Determine whether the slope of each line is 0 or undefined.

55.   **56.**

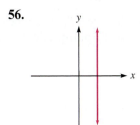

Determine whether the slope of each line is positive or negative.

57.

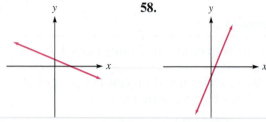

58.

61. A line passes through $(-2, 5)$ and $(6, 10)$. A line parallel to it passes through $(2, 2)$ and $(10, y)$. Find y.

62. A line passes through $(-2, 5)$ and $(6, 10)$. A line perpendicular to it passes through $(-2, 5)$ and $(x, -3)$. Find x.

63. Rate of descent If an airplane descends 3000 feet in 15 minutes, what is the average rate of descent in feet per minute?

64. Rate of growth A small business predicts sales according to a straight-line method. If sales were $50,000 in the first year and $147,500 in the third year, find the rate of growth in dollars per year (the slope of the line).

Determine whether the lines with the given slopes are parallel, perpendicular, or neither.

59. $m_1 = 5; m_2 = -\dfrac{1}{5}$ **60.** $m_1 = \dfrac{2}{7}; m_2 = \dfrac{7}{2}$

2.4   Writing and Graphing Equations of Lines

Definitions and Concepts	**Examples**
Point-slope form: An equation of the line passing through $P(x_1, y_1)$ and with slope m is $y - y_1 = m(x - x_1)$	Write an equation of the line with slope $-\frac{4}{3}$ and passing through $P(3, -2)$. 　　We will substitute $-\frac{4}{3}$ for m, 3 for x_1, and -2 for y_1 in the point-slope form $y - y_1 = m(x - x_1)$ and simplify. $y - \boldsymbol{y_1} = \boldsymbol{m}(x - \boldsymbol{x_1})$ $y - (\boldsymbol{-2}) = -\dfrac{4}{3}(x - 3)$　　Substitute $-\dfrac{4}{3}$ for m, 3 for x_1, and -2 for y_1. $y + 2 = -\dfrac{4}{3}x + 4$　　Remove parentheses. $y = -\dfrac{4}{3}x + 2$　　Subtract 2 from both sides.
Slope-intercept form: An equation of the line with slope m and y-intercept $(0, b)$ is $y = mx + b$. **Standard form of an equation of a line:** 　$Ax + By = C$ Slope-intercept form can be used to find the slope and the y-intercept from the equation of a line.	The equation of the line $y = -\frac{5}{3}x + 4$ is written in slope-intercept form. The above equation written in standard form $Ax + By = C$ is $5x + 3y = 12$. Find the slope and the y-intercept of the line with equation $2x + 5y = -10$. 　　We will write the equation in the form $y = mx + b$ to find the slope m and the y-intercept $(0, b)$. $2x + 5y = -10$ $5y = -2x - 10$　　Subtract $2x$ from both sides. $y = -\dfrac{2}{5}x - 2$　　Divide both sides by 5. The slope of the graph is $-\frac{2}{5}$, and the y-intercept is $(0, -2)$.

Definitions and Concepts	Examples
Horizontal line: $y = b$ The slope is 0, and the y-intercept is $(0, b)$.	The equation of the horizontal line with slope 0 and y-intercept $(0, 7)$ is $y = 7$.
Vertical line: $x = a$ There is no defined slope, and the x-intercept is $(a, 0)$.	The equation of the vertical line with no defined slope and the x-intercept $(-6, 0)$ is $x = -6$.

Exercises

Use slope-intercept form to write an equation of each line.

65. The line has a slope of $\frac{2}{3}$ and a y-intercept of 3.

66. The slope is $-\frac{3}{2}$ and the line passes through $(0, -5)$.

Find the slope and the y-intercept of the graph of each line.

67. $3x - 2y = 10$

68. $2x + 4y = -8$

69. $-2y = -3x + 10$

70. $2x = -4y - 8$

71. $5x + 2y = 7$

72. $3x - 4y = 14$

Use slope-intercept form to graph each equation.

73. $y = \dfrac{3}{5}x - 2$

74. $y = -\dfrac{4}{3}x + 3$

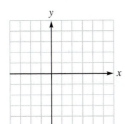
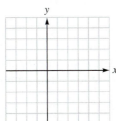

Determine whether the graphs of each pair of equations are parallel, perpendicular, or neither.

75. $y = 3x + 8$, $2y = 6x - 19$

76. $2x + 3y = 6$, $3x - 2y = 15$

Use point-slope form to write an equation of each line. Write the answer in standard form.

77. The line passes through the origin and the point $(-5, 7)$.

78. The line passes through $(-2, 1)$ and has a slope of -4.

79. The line passes through $(2, -1)$ and has a slope of $-\frac{1}{5}$.

80. The line passes through $(7, -5)$ and $(4, 1)$.

Write an equation of each line.

81. The line has a slope of 0 and passes through $(-5, 17)$.

82. The line has no defined slope and passes through $(-5, 17)$.

Write an equation of each line. Write the answer in slope-intercept form.

83. The line is parallel to $3x - 4y = 7$ and passes through $(2, 0)$.

84. The line passes through $(7, -2)$ and is parallel to the line segment joining $(2, 4)$ and $(4, -10)$.

85. The line passes through $(0, 5)$ and is perpendicular to the line $x + 3y = 4$.

86. The line passes through $(7, -2)$ and is perpendicular to the line segment joining $(2, 4)$ and $(4, -10)$.

87. **Billing for services** Angie's Painting and Decorating Service charges a fixed amount for accepting a wallpapering job and adds a fixed dollar amount for each roll hung. If the company bills a customer $177 to hang 11 rolls and $294 to hang 20 rolls, find the cost to hang 27 rolls.

88. **Paying for college** Rolf must earn $5040 for next semester's tuition. Assume he works x hours tutoring algebra at $14 per hour and y hours tutoring Spanish at $18 per hour and makes his goal. Write an equation expressing the relationship between x and y, and graph the equation. If Rolf tutors algebra for 180 hours, how long must he tutor Spanish?

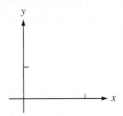

2.5 Graphs of Equations and Circles

Definitions and Concepts	**Examples**

To graph an equation:

1. Find the x- and y-intercepts.
2. Find the symmetries of the graph.
3. Plot some additional points, if necessary, and draw the graph.

Graph: $y = x^3 - 4x$.

To graph $y = x^3 - 4x$, we will find the x- and y-intercepts, test for symmetries, plot points, and join the points with a smooth curve.

Intercepts of a graph:
To find the x-intercepts, let $y = 0$ and solve for x.
To find the y-intercepts, let $x = 0$ and solve for y.

Step 1: Find the x- and y-intercepts. To find the x-intercepts, we let $y = 0$ and solve for x.

$y = x^3 - 4x$

$0 = x^3 - 4x$ Substitute 0 for y.

$0 = x(x^2 - 4)$ Factor out x.

$0 = x(x + 2)(x - 2)$ Factor $x^2 - 4$.

$x = 0$ or $x + 2 = 0$ or $x - 2 = 0$ Set each factor equal to 0.

$x = -2$ $x = 2$

The x-intercepts are $(0, 0)$, $(-2, 0)$, and $(2, 0)$.
To find the y-intercepts, we let $x = 0$ and solve for y.

$y = x^3 - 4x$

$y = 0^3 - 4(0)$ Substitute 0 in for x.

$y = 0$

The y-intercept is $(0, 0)$.

Test for x-axis symmetry:
To test for x-axis symmetry, replace y with $-y$. If the resulting equation is equivalent to the original one, the graph is symmetric about the x-axis.

Step 2: Test for symmetries. We test for symmetry about the x-axis by replacing y with $-y$.

(1) $y = x^3 - 4x$ This is the original equation.

    $-y = x^3 - 4x$ Replace y with $-y$.

(2) $y = -x^3 + 4x$ Multiply both sides by -1.

Since Equations 1 and 2 are different, the graph is not symmetric about the x-axis.

Test for y-axis symmetry:
To test for y-axis symmetry, replace x with $-x$. If the resulting equation is equivalent to the original one, the graph is symmetric about the y-axis.

To test for y-axis symmetry, we replace x with $-x$.

(1) $y = x^3 - 4x$ This is the original equation.

    $y = (-x)^3 - 4(-x)$ Replace x with $-x$.

(3) $y = -x^3 + 4x$ Simplify.

Since Equations 1 and 3 are different, the graph is not symmetric about the y-axis.

Test for origin symmetry:
To test for symmetry about the origin, replace x with $-x$ and y with $-y$. If the resulting equation is equivalent to the original one, the graph is symmetric about the origin.

To test for symmetry about the origin, we replace x with $-x$ and y with $-y$.

(1) $y = x^3 - 4x$ This is the original equation.

    $-y = (-x)^3 - 4(-x)$ Replace x with $-x$ and y with $-y$.

    $-y = -x^3 + 4x$ Simplify.

(4) $y = x^3 - 4x$ Multiply both sides by -1.

Since Equations 1 and 4 are the same, the graph is symmetric about the origin.

Definitions and Concepts	**Examples**
Plot some additional points and draw the graph.	**Step 3: Graph the equation.** To graph the equation, we plot the x- and y-intercepts and several other pairs (x, y) with positive values of x. We can use the property of symmetry about the origin to draw the graph for negative values of x.

$y = x^3 - 4x$		
x	y	(x, y)
-2	0	$(-2, 0)$
0	0	$(0, 0)$
1	-3	$(1, -3)$
2	0	$(2, 0)$
3	15	$(3, 15)$

Circles:

A **circle** is the set of all points in a plane that are a fixed distance from a point called its **center**. The fixed distance is the **radius** of the circle.

The standard equation of a circle with center (h, k) and radius r:

The graph of any equation that can be written in the form

$$(x - h)^2 + (y - k)^2 = r^2$$

is a circle with radius r and center at point (h, k).

The standard equation of a circle with center $(0, 0)$ and radius r:

The graph of any equation that can be written in the form

$$x^2 + y^2 = r^2$$

is a circle with radius r and center at the origin.

Find the center and radius of the circle with the equation $(x - 6)^2 + (y + 5)^2 = 4$.

Standard Form: $(x - \mathbf{h})^2 + (y - \mathbf{k})^2 = \mathbf{r}^2$

Given Form: $(x - \mathbf{6})^2 + [y - (\mathbf{-5})]^2 = \mathbf{2}^2$

We see that $h = 6$, $k = -5$, and $r = 2$. The center (h, k) of the circle is at $(6, -5)$ and the radius is 2.

The general form of an equation of a circle:

The general form of an equation of a circle is

$$x^2 + y^2 + cx + dy + e = 0$$

where c, d, and e are real numbers.

To convert the general form of the equation of the circle $x^2 + y^2 + 4x - 2y - 20 = 0$ into standard form, we must complete the square on both x and y:

$$x^2 + y^2 + 4x - 2y - 20 = 0$$
$$x^2 + y^2 + 4x - 2y = 20$$
$$x^2 + 4x + y^2 - 2y = 20$$
$$x^2 + 4x + \mathbf{4} + y^2 - 2y + \mathbf{1} = 20 + \mathbf{4} + \mathbf{1}$$
    Add 4 and 1 to both sides to complete the square.

$$(x + 2)^2 + (y - 1)^2 = 25$$
    Factor $x^2 - 4x + 4$ and $y^2 - 2y + 1$.

Exercises

Find the x- and y-intercepts of each graph. Do not graph the equation.

89. $y = 4x - 8x^2$

90. $y = x^2 - 10x - 24$

Find the symmetries, if any, of the graph of each equation. Do not graph the equation.

91. $y^2 = 8x$

92. $y = 3y^4 + 6$

93. $y = -2|x|$

94. $y = |x + 2|$

Graph each equation. Find all intercepts and symmetries.

95. $y = x^2 + 2$

96. $y = -x^2 + 9$

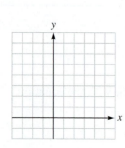
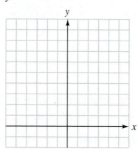

97. $y = x^3 - 2$

98. $y = \sqrt{x} + 2$

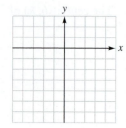
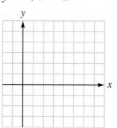

99. $y = -\sqrt{x - 4}$

100. $y = \frac{1}{2}|x|$

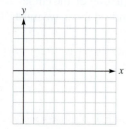
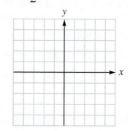

101. $y = |x + 1| + 2$

102. $y = \sqrt[3]{x} - 1$

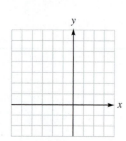
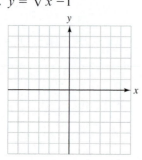

Use a graphing calculator to graph each equation.

103. $y = |x - 4| + 2$

104. $y = -\sqrt{x + 2} + 3$

105. $y = x + 2|x|$

106. $y^2 = x - 3$

Identify the center and radius of each circle given in standard form.

107. $x^2 + y^2 = 64$

108. $x^2 + (y - 6)^2 = 100$

109. $(x + 7)^2 + y^2 = \frac{1}{4}$

110. $(x - 5)^2 + (y + 1)^2 = 9$

Write an equation of each circle in standard form.

111. Center at $(0, 0)$; $r = 7$

112. Center at $(3, 0)$; $r = \frac{1}{5}$

113. Center at $(-2, 12)$, $r = 5$

114. Center at $\left(\frac{2}{7}, 5\right)$; $r = 9$

Write an equation of each circle in standard form and general form.

115. Center at $(-3, 4)$; radius 12

116. Ends of diameter at $(-6, -3)$ and $(5, 8)$

Convert the general form of each circle given into standard form.

117. $x^2 + y^2 + 6x - 4y + 4 = 0$

118. $2x^2 + 2y^2 - 8x - 16y - 10 = 0$

Graph each circle.

119. $x^2 + y^2 - 16 = 0$

120. $x^2 + y^2 - 4x = 5$

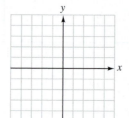
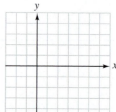

121. $x^2 + y^2 - 2y = 15$

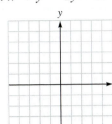

122. $x^2 + y^2 - 4x + 2y = 4$

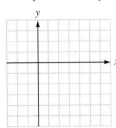

Use a graphing calculator to solve each equation. If an answer is not exact, round to the nearest hundredth.

123. $x^2 - 11 = 0$

124. $x^3 - x = 0$

125. $|x^2 - 2| - 1 = 0$

126. $x^2 - 3x = 5$

2.6 Proportion and Variation

Definitions and Concepts	Examples
Proportion: An equation indicating that two ratios are equal is called a **proportion**. In the proportion $\frac{a}{b} = \frac{c}{d}$, the numbers a and d are called the **extremes**, and the numbers b and c are called the **means**.	In the proportion $\frac{3}{5} = \frac{9}{15}$, 5 and 9 are called the means, and 3 and 15 are called the extremes.
Property of proportions: In any proportion, the product of the extremes is equal to the product of the means. Use the property of proportions to solve a proportion for a variable.	In the proportion on the previous page, the product of the means is equal to the product of the extremes: $$5(9) = 3(15) = 45$$ Solve the proportion: $\dfrac{x}{6} = \dfrac{2}{x + 11}$ for x. We will use the property of proportions to solve the proportion.

$$\frac{x}{6} = \frac{2}{x + 11}$$

$x(x + 11) = 6 \cdot 2$ The product of the extremes equals the product of the means.

$x^2 + 11x = 12$ Remove parentheses and simplify.

$x^2 + 11x - 12 = 0$ Subtract 12 from both sides.

$(x + 12)(x - 1) = 0$ Factor the trinomial.

$x + 12 = 0$ or $x - 1 = 0$ Set each factor equal to 0.

$x = -12$ | $x = 1$

Thus, $x = -12$ or $x = 1$.

Definitions and Concepts	Examples
Direct variation: The words **"y varies directly with x,"** or **"y is directly proportional to x,"** mean that $y = kx$ for some real-number constant k. The number k is called the **constant of proportionality**.	In the direct variation formula, $y = 3x$, find y when $x = 5$. $$y = 3x = 3(5) = 15$$
Inverse variation: The words **"y varies inversely with x,"** or **"y is inversely proportional to x,"** mean that $y = \frac{k}{x}$ for some real-number constant k.	In the inverse variation formula $y = \frac{k}{x}$, find the constant of variation if $y = 5$ when $x = 20$. $$y = \frac{k}{x}$$ $$5 = \frac{k}{20} \qquad \text{Substitute 5 for } y \text{ and 20 for } x.$$ $$k = 100$$
Joint variation: The words **"y varies jointly with w and x"** mean that $y = kwx$ for some real-number constant k.	Kinetic energy of an object varies jointly with its mass and the square of its velocity. A 50-gram mass moving at the rate of 20 centimeters per second has a kinetic energy of 40,000 dyne-centimeters. Find the kinetic energy of a 10-gram mass that is moving at 60 centimeters per second. We will use joint variation to solve the problem. If we let E, m, and v represent the kinetic energy, mass, and velocity, respectively, the phrase *energy varies jointly with the mass and the square of its velocity* translates into the formula $E = kmv^2$. The constant k can be evaluated by substituting 40,000 for E, 50 for m, and 20 for v in the formula $$E = kmv^2$$ $$40{,}000 = k(50)(20)^2$$ $$40{,}000 = 20{,}000k$$ $$k = 2$$ We can now substitute 2 for k, 10 for m, and 60 for v in the formula and evaluate E. $$E = kmv^2$$ $$= 2(10)(60)^2$$ $$= 72{,}000$$ A 10-gram mass that is moving at 60 centimeters per second has a kinetic energy of 72,000 dyne-centimeters.

Exercises

Solve each proportion.

127. $\dfrac{x+3}{10} = \dfrac{x-1}{x}$

128. $\dfrac{x-1}{2} = \dfrac{12}{x+1}$

129. **Recommended dosages** The recommended dosage of a medication is 250 mg per 110 pounds of body weight. Find the dosage for a person weighing 176 pounds.

130. **Hooke's Law** The force required to stretch a spring is directly proportional to the amount of stretch. If a 3-pound force stretches a spring 5 inches, what force would stretch the spring 3 inches?

131. **Kinetic energy** A moving body has a kinetic energy directly proportional to the square of its velocity. By what factor does the kinetic energy of an automobile increase if its speed increases from 30 mph to 50 mph?

132. **Gas laws** The volume of gas in a balloon varies directly as the temperature and inversely as the pressure. If the volume is 400 cubic centimeters when the temperature is 300 K and the pressure is 25 dynes per square centimeter, find the volume when the temperature is 200 K and the pressure is 20 dynes per square centimeter.

133. **Area** The area of a rectangle varies jointly with its length and width. Find the constant of proportionality.

134. **Electrical resistance** The resistance of a wire varies directly as the length of the wire and inversely as the square of its diameter. A 1000-foot length of wire, 0.05 inch in diameter, has a resistance of 200 ohms. What would be the resistance of a 1500-foot length of wire that is 0.08 inch in diameter?

CHAPTER TEST

Find the domain of each function. Write each answer using interval notation.

1. $f(x) = \dfrac{3}{2x - 5}$

2. $f(x) = \sqrt{x + 3}$

Find $f(-1)$ and $f(2)$.

3. $f(x) = \dfrac{x}{x - 1}$

4. $f(x) = \sqrt{x + 7}$

Find the difference quotient.

5. $f(x) = x^2 - x + 5$

Indicate the quadrant in which the point lies or the axis on which it lies.

6. $(-3, \pi)$

7. $(0, -8)$

Find the x- and y-intercepts and use them to graph the equation.

8. $x + 3y = 6$

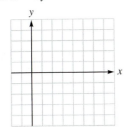

9. $2x - 5y = 10$

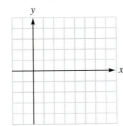

Graph each equation.

10. $2(x + y) = 3x + 5$

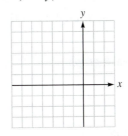

11. $3x - 5y = 3(x - 5)$

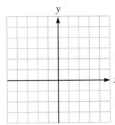

12. $\dfrac{1}{2}(x - 2y) = y - 1$

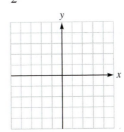

13. $\dfrac{x + y - 5}{7} = 3x$

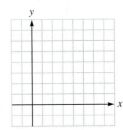

Find the distance between points P and Q.

14. $P(1, -1); Q(-3, 4)$

15. $P(0, \pi); Q(-\pi, 0)$

Find the midpoint of the line segment PQ.

16. $P(3, -7); Q(-3, 7)$

17. $P(0, \sqrt{2}); Q(\sqrt{8}, \sqrt{18})$

Find the slope of the line PQ.

18. $P(3, -9)$; $Q(-5, 1)$

19. $P(\sqrt{3}, 3)$; $Q(-\sqrt{12}, 0)$

Determine whether the two lines are parallel, perpendicular, or neither.

20. $y = 3x - 2$; $y = 2x - 3$

21. $2x - 3y = 5$; $3x + 2y = 7$

Write an equation of the line with the given properties. Your answers should be written in slope-intercept form, if possible.

22. Passing through $(3, -5)$; $m = 2$

23. $m = 3$; $b = \dfrac{1}{2}$

24. Parallel to $2x - y = 3$; $b = 5$

25. Perpendicular to $2x - y = 3$; $b = 5$

26. Passing through $\left(2, -\dfrac{3}{2}\right)$ and $\left(3, \dfrac{1}{2}\right)$

27. Parallel to the y-axis and passing through $(3, -4)$

Find the x- and y-intercepts of each graph.

28. $y = x^3 - 16x$

29. $y = |x - 4|$

Find the symmetries of each graph.

30. $y^2 = x - 1$ **31.** $y = x^4 + 1$

Graph each equation. Find all intercepts and symmetries.

32. $y = x^2 - 9$ **33.** $x = |y|$

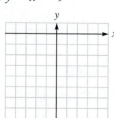   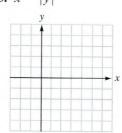

34. $y = 2\sqrt{x}$ **35.** $x = y^3$

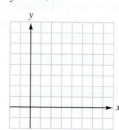   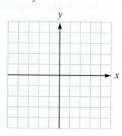

Write an equation of each circle in standard form.

36. Center at $(5, 7)$; radius of 8

37. Center at $(2, 4)$; passing through $(6, 8)$

Graph each equation.

38. $x^2 + y^2 = 9$ **39.** $x^2 - 4x + y^2 + 3 = 0$

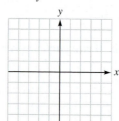   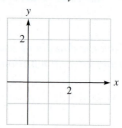

Write each statement as an equation.

40. y varies directly as the square of z.

41. w varies jointly with r and the square of s.

42. P varies directly with Q. $P = 7$ when $Q = 2$. Find P when $Q = 5$.

43. y is directly proportional to x and inversely proportional to the square of z, and $y = 16$ when $x = 3$ and $z = 2$. Find x when $y = 2$ and $z = 3$.

Use a graphing calculator to find the positive root of each equation. Round to two decimal places.

44. $x^2 - 7 = 0$ **45.** $x^2 - 5x - 5 = 0$

Functions

CAREERS AND MATHEMATICS: Computer and Information Research Scientist

Computer and information research scientists are highly trained innovative workers who design and invent new technology. They solve complex business, scientific, and general computing problems. Computer and information research scientists conduct research on a variety of topics, including computer hardware, virtual reality, and robotics.

Education and Mathematics Required

- Most are required to possess a Ph.D. in computer science, computer engineering, or a closely related discipline. An aptitude for math is important.
- College Algebra, Trigonometry, Calculus, Linear Algebra, Ordinary Differential Equations, Theory of Analysis, Abstract Algebra, Graph Theory, Numerical Methods, and Combinatorics are math courses required.

How Computer and Information Research Scientists Use Math and Who Employs Them

- Computer and information research scientists use mathematics as they span a range of topics from theoretical studies of algorithms to the computation of implementing computing systems in hardware and software.
- Many computer and information research scientists are employed by Internet service providers; Web search portals; and data processing, hosting, and related services firms. Others work for government, manufacturers of computer and electronic products, insurance companies, financial institutions, and universities.

Career Outlook and Earnings

- Employment of computer and information research scientists is expected to grow by 15% through 2022, which is much faster than the average for all occupations.
- The median annual wages of computer and information research scientists is approximately $102,190. Some earn more than $150,000 a year.

For more information see: www.bls.gov/oco

3.1 Graphs of Functions

In this section, we will learn to

1. Graph a function by plotting points.

2. Use the Vertical Line Test to identify functions.

3. Determine function values graphically.

4. Determine the domain and range of a function from its graph.

5. Recognize the graphs of common functions.

The Super Bowl is the annual championship game of the National Football League. The Super Bowl has frequently been the most-watched, American television broadcast of the year and in recent years has averaged over 110 million viewers.

The television broadcast includes many high-profile and funny commercials. A number of major brands, such as Doritos, have been well known for making repeated appearances during the Super Bowl. However, the prominence of airing a commercial during the Super Bowl has also carried an increasingly high price.

The graph in Figure 3-1 shows the cost of a 30-second commercial in dollars for Super Bowls 1–49. Super Bowl 1 was played in 1967 and Super Bowl 49 was played in 2015. From the graph we can note many interesting facts.

- The cost of a 30-second commercial is increasing.
- The cost of a 30-second commercial increased by approximately a million dollars from Super Bowl 30 to Super Bowl 35.
- The cost of a 30-second commercial was approximately $4.5 million for Super Bowl 49.

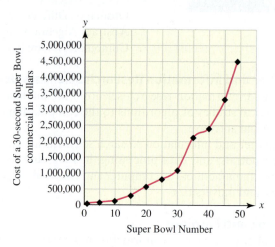

FIGURE 3-1

The graph shown in Figure 3-1 is the graph of a function. Ordered pairs of the form (Super Bowl Number, Cost) are plotted and a smooth curve is then drawn through them that produces the graph of the function. Note that for each Super Bowl played, there corresponds exactly one cost for a commercial. The definition of a function is satisfied.

Graphs of functions convey very important information. In this section we will continue our study of functions by learning to identify the graph of a function and exploring properties of the graphs of functions.

1. Graph a Function by Plotting Points

In Chapter 2, we graphed equations by making a table of solutions and plotting the points given in the table. Then we connected the points by drawing a smooth curve through them. We will use the same strategy to graph functions.

| **Graph of a Function** | The **graph** of a function f in the xy-plane is the set of all points (x, y) where x is in the domain of f, y is in the range of f, and $y = f(x)$. |

EXAMPLE 1

Graphing a Function by Plotting Points

Graph the functions. **a.** $f(x) = -2|x| + 3$ **b.** $f(x) = \sqrt{x - 2}$

SOLUTION In each case, we will make a table of solutions and plot the points given by the table. Then we will connect the points by drawing a smooth curve through them and obtain the graph of the function. See Figures 3-2 and 3-3.

a. $f(x) = -2|x| + 3$

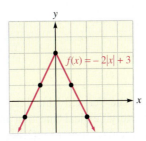

$f(x) = -2|x| + 3$		
x	$f(x)$	$(x, f(x))$
-2	-1	$(-2, -1)$
-1	1	$(-1, 1)$
0	3	$(0, 3)$
1	1	$(1, 1)$
2	-1	$(2, -1)$

FIGURE 3-2

b. $f(x) = \sqrt{x - 2}$

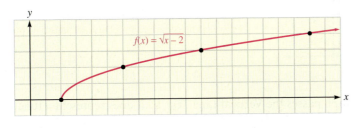

$f(x) = \sqrt{x - 2}$		
x	$f(x)$	$(x, f(x))$
2	0	$(2, 0)$
6	2	$(6, 2)$
11	3	$(11, 3)$
18	4	$(18, 4)$

FIGURE 3-3

Self Check 1 Graph: $f(x) = |x + 3|$.

Now Try Exercise 27.

2. Use the Vertical Line Test to Identify Functions

We can use a **Vertical Line Test** to determine whether a graph represents a function.

| **Vertical Line Test** | • If every vertical line that can be drawn intersects the graph in no more than one point, the graph represents a function. See Figure 3-4(a). |
| | • If a vertical line can be drawn that intersects the graph at more than one point, the graph does not represent a function. See Figure 3-4(b). |

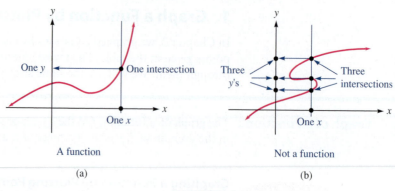

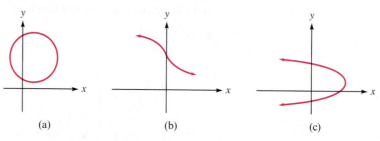

A function

(a)

Not a function

(b)

FIGURE 3-4

Using the Vertical Line Test to Identify Functions

EXAMPLE 2

Determine which of the following graphs represent functions.

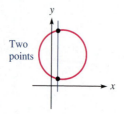

(a)   (b)   (c)

SOLUTION We will use the Vertical Line Test by drawing several vertical lines through each graph. If every vertical line that intersects the graph does so exactly once, the graph represents a function. Otherwise, the graph does not represent a function.

Take Note

Some graphs represent functions and some do not. Graphs that pass the Vertical Line Test are functions.

a. This graph fails the Vertical Line Test, so it does not represent a function.

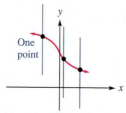

The vertical line drawn crosses the graph at two points.

b. This graph passes the Vertical Line Test, so it does represent a function.

Three vertical lines are drawn. A vertical line will never cross the graph at more than one point.

c. This graph fails the Vertical Line Test, so it does not represent a function.

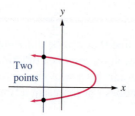

The vertical line drawn crosses the graph at two points.

Self Check 2   Which graphs represent functions?

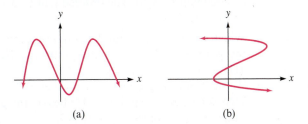

(a)   (b)

Now Try Exercise 49.

Not all equations define functions. For example, the equation $x = |y|$ does not define a function, because two values of y can correspond to one number x. For example, if $x = 2$, then y can be either 2 or -2. The graph of the equation is shown in Figure 3-5. Since the graph does not pass the Vertical Line Test, it does not represent a function.

Take Note

The graph fails the Vertical Line Test and is not a function. We cannot write the equation using functional notation.

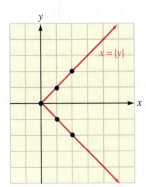

$x = |y|$		
x	y	(x, y)
2	-2	$(2, -2)$
1	-1	$(1, -1)$
0	0	$(0, 0)$
1	1	$(1, 1)$
2	2	$(2, 2)$

FIGURE 3-5

3. Determine Function Values Graphically

In many applications that involve functions, we may not be given the equation of the function but will instead have only data. It is important that we learn how to read the value of a function from the graph of that function. Remember, the value of the function is y. It is not a point; it is a y-value.

Reading the Value of a Function from a Graph

EXAMPLE 3

Refer to the graph of function $f(x)$ shown in Figure 3-6.
a. Find $f(-3)$.   **b.** Find $f(1)$.   **c.** Find the value of x for which $f(x) = 17$.

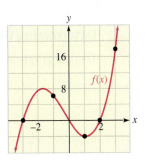

FIGURE 3-6

SOLUTION   **a.** When $x = -3$, we see from the graph that the value of y is 0. This is because the point $(-3, 0)$ lies on the graph of the function. Therefore, we write $f(-3) = 0$.

It is important to note that the value of the function at -3 is 0 and not the ordered pair $(-3, 0)$. The ordered pair $(-3, 0)$ is simply a point on the graph of the function.

b. When $x = 1$, we see from the graph that the value of y is -4. This is because the point $(1, -4)$ lies on the graph of the function. Therefore, we write $f(1) = -4$.

c. We see from the graph that a y-value of 18 corresponds to an x-value of 3. That is because the point $(3, 18)$ lies on the graph of the function. There the value of x for which $f(x) = 18$ is 3.

Self Check 3   Use Figure 3-6 and find $f(2)$.

Now Try Exercise 53.

4. Determine the Domain and Range of a Function from Its Graph

> **Tip**
> 1. To find the domain, project the graph onto the x-axis.
> 2. To find the range, project the graph onto the y-axis.
>   The projection is the "shadow" the graph makes with each axis.

Both the **domain** and the **range** of a function can be identified by viewing the graph of the function. The inputs or x-values that correspond to points on the graph of the function can be identified on the x-axis and used to state the domain of the function. The outputs or $f(x)$ values that correspond to points on the graph of the function can be identified on the y-axis and used to state the range of the function. (See Figure 3-7.)

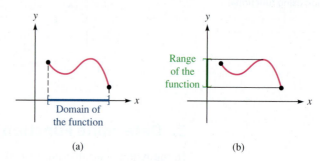

(a)   (b)

FIGURE 3-7

From Figure 3-2, we can see that the domain of the function $f(x) = -2|x| + 3$ is the set of all real numbers, and that the range is the set of real numbers that are less than or equal to 3.

From Figure 3-3, we can see that the domain of the function $f(x) = \sqrt{x - 2}$ is the set of all real numbers greater than or equal to 2, and that the range is the set of all real numbers greater than or equal to 0.

EXAMPLE 4   **Determining the Domain and Range of a Function from the Graph**

Use the graph of each function to determine its domain and range.

a. 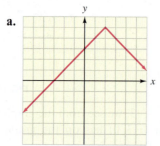   **b.**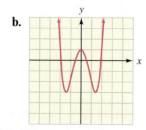

c.

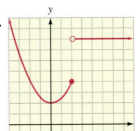

d.

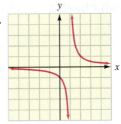

SOLUTION   For the graph of each function, we will highlight the domain in blue on the *x*-axis and highlight the range in green on the *y*-axis.

a. The domain is $(-\infty, \infty)$ and is highlighted in blue on the *x*-axis. The range is $(-\infty, 5]$ and is highlighted in green on the *y*-axis.

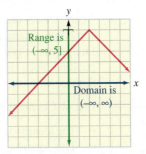

b. The domain is $(-\infty, \infty)$ and is highlighted in blue on the *x*-axis. The range is $[-3, \infty)$ and is highlighted in green on the *y*-axis.

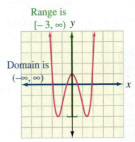

c. The domain is $(-\infty, \infty)$ and is highlighted in blue on the *x*-axis. The range is $[2, \infty)$ and is highlighted in green on the *y*-axis.

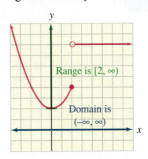

d. The domain is $(-\infty, 1) \cup (1, \infty)$ and is highlighted in blue on the *x*-axis. The range is $(-\infty, 0) \cup (0, \infty)$ and is highlighted in green on the *y*-axis.

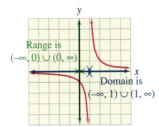

Self Check 4 Use the graph of the function to determine its domain and range.

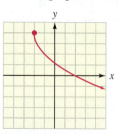

Now Try Exercise 83.

5. Recognize the Graphs of Common Functions

There are six common functions that occur frequently in algebra. These graphs are sometimes referred to as **basic functions** or **parent functions** and can be sketched by plotting points or generated using a graphing calculator. These common functions are the building blocks for graphing more complicated functions in algebra. Once you become familiar with these graphs, you should be able to sketch them quickly by hand.

List of Common Functions

- The **identify function** $f(x) = x$ pairs each real number with itself. See Figure 3-8(a).
- The **squaring function** $f(x) = x^2$ pairs each real number with its square. See Figure 3-8(b).
- The **cubing function** $f(x) = x^3$ pairs each real number with its cube. See Figure 3-8(c).
- The **absolute value function** $f(x) = |x|$ pairs each real number with its absolute value. See Figure 3-8(d).
- The **square root function** $f(x) = \sqrt{x}$ pairs each real number with its principal square root. See Figure 3-8(e).
- The **cube root function** $f(x) = \sqrt[3]{x}$ pairs each real number with its cube root. See Figure 3-8(f).

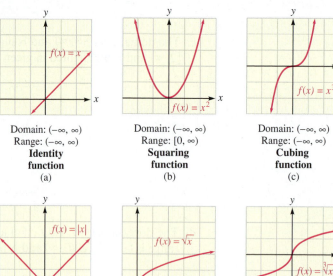

Domain: $(-\infty, \infty)$
Range: $(-\infty, \infty)$
Identity function
(a)

Domain: $(-\infty, \infty)$
Range: $[0, \infty)$
Squaring function
(b)

Domain: $(-\infty, \infty)$
Range: $(-\infty, \infty)$
Cubing function
(c)

Domain: $(-\infty, \infty)$
Range: $[0, \infty)$
Absolute value function
(d)

Domain: $[0, \infty)$
Range: $[0, \infty)$
Square root function
(e)

Domain: $(-\infty, \infty)$
Range: $(-\infty, \infty)$
Cube root function
(f)

FIGURE 3-8

The graphs of the common functions can easily be generated using a graphing calculator. The graphs of the absolute value function and the cube root function are shown in the Accent on Technology.

Accent on Technology

Graphing the Absolute Value Function

To graph the absolute value function on a graphing calculator, we must call up the function abs(into the graphing window. We press MATH, then scroll right to NUM as shown below in Figure 3-9.

Press Y=. Go to MATH NUM and press 1. Input x and GRAPH. Use ZOOM 4 as the window.

  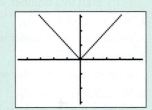 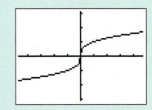

FIGURE 3-9

Accent on Technology

Graphing the Cube Root Function

We can graph this function on a graphing calculator using the cube root key by selecting MATH 4 from the graph window as shown in Figure 3-10. Input x and GRAPH. Use ZOOM 4 as the window.

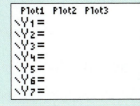   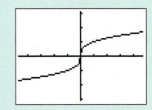

FIGURE 3-10

Self Check Answers

1.
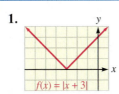

2. (a)   **3.** 0   **4.** Domain is $[-2, \infty)$. Range is $(-\infty, 4]$.

Exercises 3.1

Getting Ready

You should be able to complete these vocabulary and concept statements before you proceed to the practice exercises.

Fill in the blanks.

1. The graph of a function $y = f(x)$ in the xy-plane is the set of all points _____ that satisfy the equation, where x is in the _____ of f and y is in the _____ of f.

2. If every _____ line that intersects a graph does so _____, the graph represents a function.

3. We call $f(x) = x$ the _____ function because it pairs each real number with itself.

4. We call $f(x) = x^2$ the _____ function because it pairs each real number with its square.

5. We call $f(x) = x^3$ the _____ function because it pairs each real number with its cube.

6. We call $f(x) = |x|$ the _____ function because it pairs each real number with its absolute value.

7. We call $f(x) = \sqrt{x}$ the _____ function because it pairs each real number with its principal square root.

8. We call $f(x) = \sqrt[3]{x}$ the _____ function because it pairs each real number with its cube root.

Practice

Graph each function. Use the graph to identify the domain and range of each function.

9. $f(x) = 2x + 3$

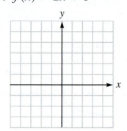

10. $f(x) = -2x - 4$

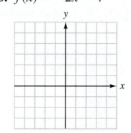

11. $f(x) = -\dfrac{3}{4}x + 4$

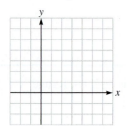

12. $f(x) = \dfrac{1}{2}x - 3$

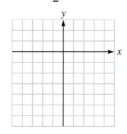

13. $f(x) = x^2 - 4$

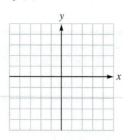

14. $f(x) = -x^2 + 3$

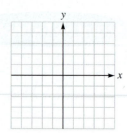

15. $f(x) = -\dfrac{1}{2}x^2 + 5$

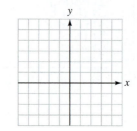

16. $f(x) = \dfrac{1}{3}x^2 - 2$

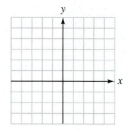

17. $f(x) = 3(x + 2)^2$

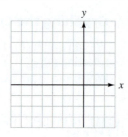

18. $f(x) = -x^2 + 2x - 1$

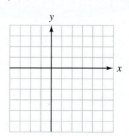

19. $f(x) = x^3 - 2$

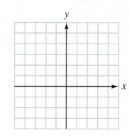

20. $f(x) = -x^3 + 2$

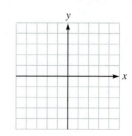

21. $f(x) = -x^3 + 1$

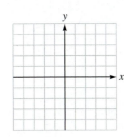

22. $f(x) = \dfrac{1}{4}x^3 + 1$

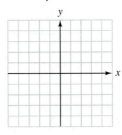

29. $f(x) = \left| \dfrac{1}{2}x + 3 \right|$

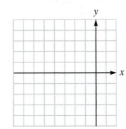

30. $f(x) = -\left| \dfrac{1}{2}x + 3 \right|$

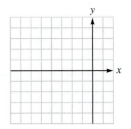

23. $f(x) = -\dfrac{1}{2}x^3 - 4$

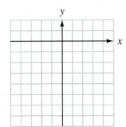

24. $f(x) = (x - 1)^3$

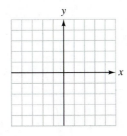

31. $f(x) = 4|x| + 1$

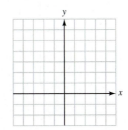

32. $f(x) = \dfrac{1}{4}|x| - 2$

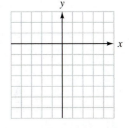

25. $f(x) = -|x|$

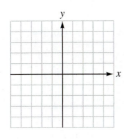

26. $f(x) = -|x| - 3$

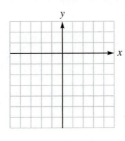

33. $f(x) = \sqrt{x} + 2$

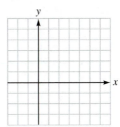

34. $f(x) = -\sqrt{x} + 1$

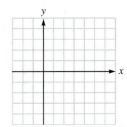

27. $f(x) = |x - 2|$

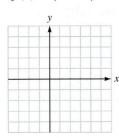

28. $f(x) = -|x - 2|$

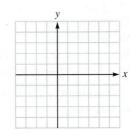

35. $f(x) = 2\sqrt{x} - 3$

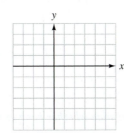

36. $f(x) = -\dfrac{1}{2}\sqrt{x} - 4$

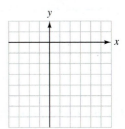

37. $f(x) = \sqrt{2x - 4}$

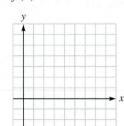

38. $f(x) = -\sqrt{2x - 4}$

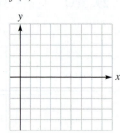

39. $f(x) = \sqrt[3]{x} + 2$

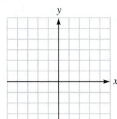

40. $f(x) = -\sqrt[3]{x} + 1$

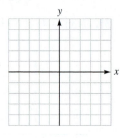

41. $f(x) = 3\sqrt[3]{x}$

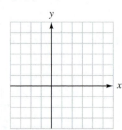

42. $f(x) = -2\sqrt[3]{x} + 5$

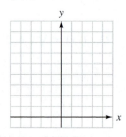

43. $f(x) = -2\sqrt[3]{x} + 1$

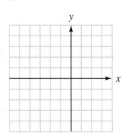

44. $f(x) = \sqrt[3]{x - 1} + 7$

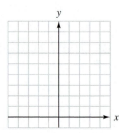

Use the Vertical Line Test to determine whether each graph represents a function.

45.

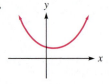

46.

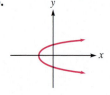

47.

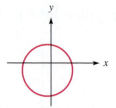

48.

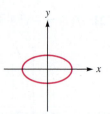

49.

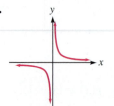

50.

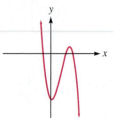

51.

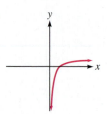

52.

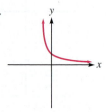

Use the graph of the function f shown to determine each of the following.

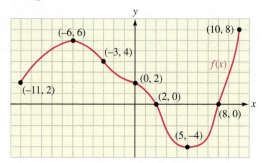

53. $f(-11)$　　　　　　　**54.** $f(-3)$
55. $f(2)$　　　　　　　　**56.** $f(10)$
57. x-intercepts
58. y-intercept
59. an x-value for which $f(x) = 6$
60. the x-value for which $f(x) = -4$

Use the graph of the function f shown to determine each of the following.

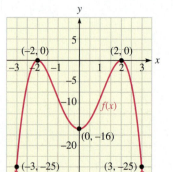

61. $f(-3)$

62. $f(3)$

63. x-intercepts

64. y-intercept

65. an x-value for which $f(x) = -16$

66. the x-values for which $f(x) = -25$

Use the graph of the function f shown to determine each of the following.

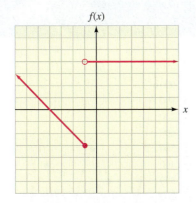

67. $f(-6)$ **68.** $f(-2)$

69. $f(-1)$ **70.** $f\left(\dfrac{10}{3}\right)$

71. x-intercept

72. y-intercept

73. the x-value for which $f(x) = 1$

74. the x-values for which $f(x) = 4$

Use the graph to determine each function's domain and range.

75.
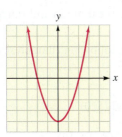

76.
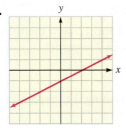

77.
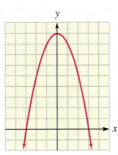

78.
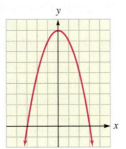

79.
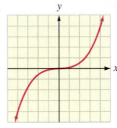

80.
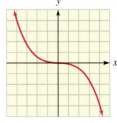

81.
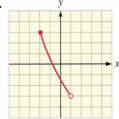

82.
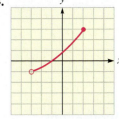

83.
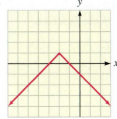

84.
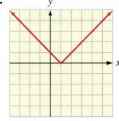

85.
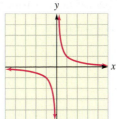

86.
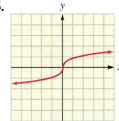

87.
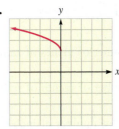

88.
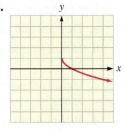

89.

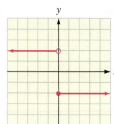

90.

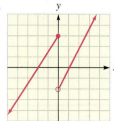

Use a graphing calculator to graph each function. Then determine the domain and range of the function.

91. $f(x) = |3x + 2|$

92. $f(x) = \sqrt{2x - 5}$

93. $f(x) = \sqrt[3]{5x - 1}$

94. $f(x) = -\sqrt[3]{3x + 2}$

Applications

95. Rain in Dallas, Texas The graph shows the average number of inches of rain, per month, in Dallas, Texas, for the months of May through October.

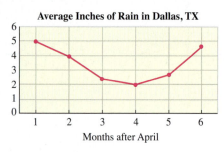

Use the graph to approximate the following.

a. Domain and range

b. Identify the average number of inches of rain in May.

c. Identify the average number of inches of rain in June.

d. What month is the average number of inches of rain 2?

96. Height of terrain The graph of the function displays the height in yards of a terrain traveled by a motorcyclist at specific mile markers along

a highway. Use the graph to find each of the following.

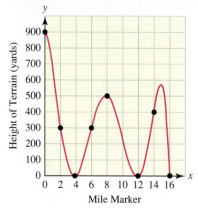

a. Determine the domain and range.

b. Determine the height at mile marker 2.

c. Determine the height at mile marker 8.

d. At what mile markers is the terrain flat?

e. Determine the y-intercept.

f. Between mile markers 4 and 12, how many times is the height of the terrain 200 yards?

Discovery and Writing

97. Describe a strategy for sketching the graph of a function.

98. Describe how to use the Vertical Line Test to determine whether a graph represents a function.

99. Explain how to determine the domain and range of a function's graph.

100. Draw a graph that has the following characteristics:
- Domain: $[-5, 10]$
- Range: $[-10, 10]$
- x-intercepts: $(-5, 0)$ and $(5, 0)$
- y-intercept: $(0, -10)$
- passes through $(-2, 3)$ and $(10, 10)$

101. Use a graphing calculator to graph the function $f(x) = \sqrt{x}$, and use $\boxed{\text{TRACE}}$ and $\boxed{\text{ZOOM}}$ to find $\sqrt{5}$ to three decimal places.

102. Use a graphing calculator to graph the function $f(x) = \sqrt[3]{x}$ and use the TRACE and ZOOM features to find $\sqrt[3]{2}$ to three decimal places.

Critical Thinking

Determine if the statement is true or false. If the statement is false, then correct it and make it true.

103. Functions always pass the Horizontal Line Test.

104. A function can have at most one y-intercept.

105. If the domain of a function is all real numbers, then the range is all real numbers.

106. The domain of the square root function and cube root function is all real numbers.

Match the graphs of the functions on the left with the graph of the function on the right so that both have identical domains and ranges.

107.

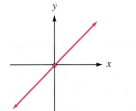

a.

108.

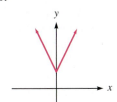

b.

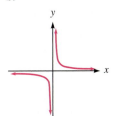

109.

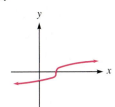

c.

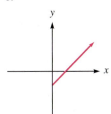

110.

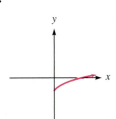

d.

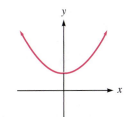

3.2 **Transformations of the Graphs of Functions**

In this section, we will learn to

1. Use vertical translations to graph functions.
2. Use horizontal translations to graph functions.
3. Graph functions using two translations.
4. Use reflections about the x- and y-axes to graph functions.
5. Use vertical stretching and shrinking to graph functions.
6. Use horizontal stretching and shrinking to graph functions.
7. Graph functions using a combination of transformations.

We can often transform the graph of a function into the graph of another function by shifting the graph vertically or horizontally. Also, we can reflect a graph about the x- or y-axis, and stretch or shrink a graph horizontally or vertically to transform the graph of a function into the graph of another function. In this section, we will graph new functions from known ones using these methods.

Consider a white water rafting trip on the Ocoee River in Tennessee. Suppose one company charges a group of students $20 for each hour on the river, plus $100 for a guide and equipment. The cost of the rafting trip can be represented by the function

$$C_1(t) = 20t + 100$$

where $C_1(t)$ represents the cost in dollars to raft t hours on the river.

If the company increases its charge for the guide and equipment to $150, the new cost function can be represented by

$$C_2(t) = 20t + 150$$

The graphs of the two cost functions are shown in Figure 3-11.

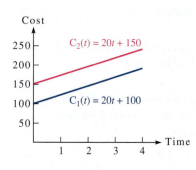

FIGURE 3-11

Tip

You can use the term *shift* instead of *translate* if you prefer.

Note that if we shift the graph of $C_1(t)$ 50 units vertically, we obtain the graph of $C_2(t)$. This shift is called a *translation*.

As we continue our study of translations, it will be helpful to review the graphs of the basic functions that are shown in Figure 3-8 in Section 3.1. In this section, the graphs of $f(x) = x^2$, $f(x) = x^3$, $f(x) = |x|$, $f(x) = \sqrt{x}$, and $f(x) = \sqrt[3]{x}$ will be translated and stretched in various ways.

1. Use Vertical Translations to Graph Functions

The graphs of functions can be identical except for their position in the xy-plane. For example, Figure 3-12 shows the graph of $y = x^2 + k$ for three values of k. If $k = 0$, we have the graph of $y = x^2$. The graph of $y = x^2 + 2$ is identical to the graph of $y = x^2$, except that it is shifted 2 units upward. The graph of $y = x^2 - 3$ is identical to the graph of $y = x^2$, except that it is shifted 3 units downward. These shifts are called **vertical translations**.

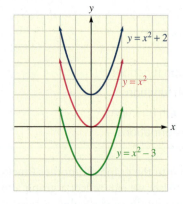

$y = x^2$			$y = x^2 + 2$			$y = x^2 - 3$		
x	y	(x, y)	x	y	(x, y)	x	y	(x, y)
-2	4	$(-2, 4)$	-2	6	$(-2, 6)$	-2	1	$(-2, 1)$
-1	1	$(-1, 1)$	-1	3	$(-1, 3)$	-1	-2	$(-1, -2)$
0	0	$(0, 0)$	0	2	$(0, 2)$	0	-3	$(0, -3)$
1	1	$(1, 1)$	1	3	$(1, 3)$	1	-2	$(1, -2)$
2	4	$(2, 4)$	2	6	$(2, 6)$	2	1	$(2, 1)$

FIGURE 3-12

In general, we can make the following observations.

Vertical Translations   If f is a function and k is a positive number, then

- The graph of $y = f(x) + k$ is identical to the graph of $y = f(x)$ except that it is translated k units upward.
- The graph of $y = f(x) - k$ is identical to the graph of $y = f(x)$ except that it is translated k units downward.

EXAMPLE 1   **Using Vertical Translations to Graph Functions**

Graph each function: **a.** $g(x) = |x| - 2$ **b.** $h(x) = |x| + 3$

SOLUTION   We will use vertical translations of $f(x) = |x|$ to graph each function.

a. The graph of $g(x) = |x| - 2$ is identical to the graph of $f(x) = |x|$, except that it is translated 2 units downward. It is translated downward because 2 is subtracted from $|x|$. The graph of $g(x)$ is shown in Figure 3-13(a).

b. The graph of $h(x) = |x| + 3$ is identical to the graph of $f(x) = |x|$, except that it is translated 3 units upward. It is translated upward because 3 is added to $|x|$. The graph of $h(x)$ is shown in Figure 3-13(b).

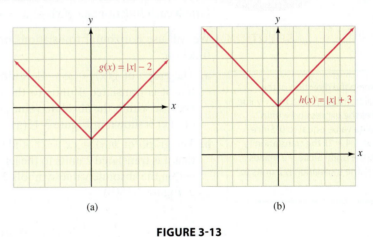

(a)   (b)

FIGURE 3-13

Self Check 1   *Fill in the blanks:* The graph of $g(x) = x^2 + 3$ is identical to the graph of $f(x) = x^2$, except that it is translated ___ units _____. The graph of $h(x) = x^2 - 4$ is identical to the graph of $f(x) = x^2$, except that it is translated ___ units _____.

Now Try Exercise 27.

2. Use Horizontal Translations to Graph Functions

Figure 3-14 shows the graph of $y = (x + k)^2$ for three values of k. If $k = 0$, we have the graph of $y = x^2$. The graph of $y = (x - 2)^2$ is identical to the graph of $y = x^2$, except that it is shifted 2 units to the right. The graph of $y = (x + 3)^2$ is identical to the graph of $y = x^2$, except that it is shifted 3 units to the left. These shifts are called **horizontal translations**.

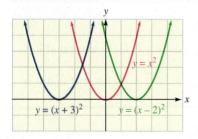

$y = x^2$		
x	y	(x, y)
-2	4	$(-2, 4)$
-1	1	$(-1, 1)$
0	0	$(0, 0)$
1	1	$(1, 1)$
2	4	$(2, 4)$

$y = (x - 2)^2$		
x	y	(x, y)
0	4	$(0, 4)$
1	1	$(1, 1)$
2	0	$(2, 0)$
3	1	$(3, 1)$
4	4	$(4, 4)$

$y = (x + 3)^2$		
x	y	(x, y)
-5	4	$(-5, 4)$
-4	1	$(-4, 1)$
-3	0	$(-3, 0)$
-2	1	$(-2, 1)$
-1	4	$(-1, 4)$

FIGURE 3-14

In general, we can make the following observations.

Horizontal Translations   If f is a function and k is a positive number, then

- The graph of $y = f(x - k)$ is identical to the graph of $y = f(x)$ except that it is translated k units to the right.
- The graph of $y = f(x + k)$ is identical to the graph of $y = f(x)$ except that it is translated k units to the left.

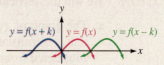

Using Horizontal Translations to Graph Functions

EXAMPLE 2

Graph each function: **a.** $g(x) = |x - 4|$ **b.** $h(x) = |x + 2|$

SOLUTION   We will use horizontal translations of the graph of $f(x) = |x|$ to graph each function.

> **Tip**
>
> To determine the direction of the horizontal translation, find the x-value that makes the expression within the absolute value equal to 0. Since 4 makes $x - 4 = 0$, the translation is 4 units to the right. Since -2 makes $x + 2 = 0$, the translation is 2 units to the left.

a. The graph of $g(x) = |x - 4|$ is identical to the graph of $f(x) = |x|$, except that it is translated 4 units to the right. It is translated 4 units to the right because within the absolute value symbols 4 is subtracted from x. The graph of $g(x)$ is shown in Figure 3-15(a).

b. The graph of $h(x) = |x + 2|$ is identical to the graph of $f(x) = |x|$, except that it is translated 2 units to the left. It is translated 2 units to the left because within the absolute value symbols 2 is added to x. The graph of $h(x)$ is shown in Figure 3-15(b).

(a)

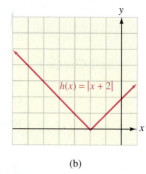

(b)

FIGURE 3-15

Self Check 2   *Fill in the blanks:* The graph of $g(x) = (x - 3)^2$ is identical to the graph of $f(x) = x^2$ except that it is translated ___ units to the _____. The graph of $h(x) = (x + 2)^2$ is identical to the graph of $f(x) = x^2$ except that it is translated ___ units to the ____.

Now Try Exercise 29.

Caution

When using horizontal translations to graph a function, it is easy to shift the function in the *wrong* direction.

- If we see a positive constant subtracted from x, we have the tendency to shift the graph left. This is **incorrect**.
- If we see a positive constant added to x, we have the tendency to shift the graph right. This too is **incorrect**.

We should avoid making these common errors.

3. Graph Functions Using Two Translations

Sometimes we can obtain a graph by using both a horizontal and a vertical translation.

EXAMPLE 3

Using a Horizontal and a Vertical Translation to Graph a Function

Graph each function:

a. $g(x) = (x - 5)^3 + 4$ **b.** $h(x) = (x + 2)^2 - 2$

SOLUTION

By inspection, we see that the function in part **a** involves two translations of $f(x) = x^3$ and the function in part **b** involves two translations of $f(x) = x^2$. We will perform the horizontal translation first, followed by a vertical translation to obtain the graph of each function.

a. The graph of $g(x) = (x - 5)^3 + 4$ is identical to the graph of $f(x) = x^3$, except that it is translated 5 units to the right and 4 units upward, as shown in Figure 3-16(a).

b. The graph of $h(x) = (x + 2)^2 - 2$ is identical to the graph of $f(x) = x^2$, except that it is translated 2 units to the left and 2 units downward, as shown in Figure 3-16(b).

Take Note

We would obtain the same graph if we performed the vertical translation first followed by the horizontal translation.

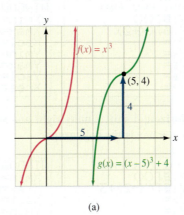

(a)

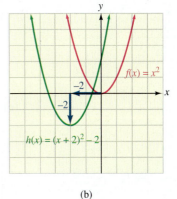

(b)

FIGURE 3-16

Self Check 3

Fill in the blanks: The graph of $g(x) = |x - 4| + 5$ is identical to the graph of $f(x) = |x|$, except that it is translated ___ units to the _____ and ___ units _____.

Now Try Exercise 31.

Translations of Functions

We can use a graphing calculator to help understand translations of functions. Compare the graph of each function listed to the graph of $f(x) = \sqrt{x}$.

    a. $g(x) = \sqrt{x} - 4$ **b.** $h(x) = \sqrt{x + 2}$ **c.** $k(x) = \sqrt{x - 2} + 3$

a. Graph the functions $f(x) = \sqrt{x}$ and $g(x) = \sqrt{x} - 4$ on the same screen using the window shown in Figure 3-17(a). From these graphs, we can see that the graph of g is obtained by shifting the graph of f downward 4 units.

b. Graph the functions $f(x) = \sqrt{x}$ and $h(x) = \sqrt{x + 2}$ on the same screen using the window shown in Figure 3-17(b). From these graphs, we can see that the graph of h is obtained by shifting the graph of f to the left 2 units.

c. Graph the functions $f(x) = \sqrt{x}$ and $k(x) = \sqrt{x - 2} + 3$ on the same screen using the window shown in Figure 3-17(c). From these graphs, we can see that the graph of k is obtained by shifting the graph of f right 2 units and upward 3 units.

(a) Vertical shift downward 4 units. The graph of $f(x) = \sqrt{x}$ is the bold graph.

> **Take Note**
> To make the graph of $f(x) = \sqrt{x}$ appear bold, scroll to the left of Y_1 and press [ENTER].

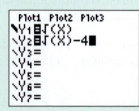  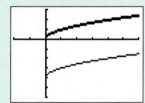

(b) Horizontal shift left 2 units. The graph of $f(x) = \sqrt{x}$ is the bold graph.

 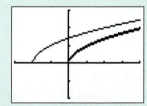

(c) Horizontal shift 2 units right and vertical shift 3 units upward. The graph of $f(x) = \sqrt{x}$ is the bold graph.

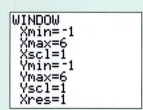 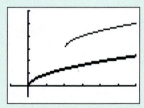

FIGURE 3-17

4. Use Reflections about the x- and y- Axes to Graph Functions

Figure 3-18(a) shows that the graph of $y = -\sqrt{x}$ is identical to the graph of $y = \sqrt{x}$ except that it is reflected about the x-axis. Figure 3-18(b) shows that the graph of $y = \sqrt{-x}$ is identical to the graph of $y = \sqrt{x}$ except that it is reflected about the y-axis.

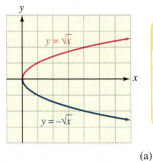

$y = -\sqrt{x}$		
x	y	(x, y)
0	0	$(0, 0)$
1	-1	$(1, -1)$
4	-2	$(4, -2)$

(a)

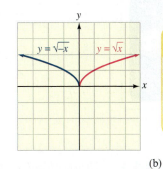

$y = \sqrt{-x}$		
x	y	(x, y)
0	0	$(0, 0)$
-1	1	$(-1, 1)$
-4	2	$(-4, 2)$

(b)

FIGURE 3-18

In general, we can make the following observations.

Reflections   If f is a function, then

- The graph of $y = -f(x)$ is identical to the graph of $y = f(x)$ except that it is reflected about the x-axis.

- The graph of $y = f(-x)$ is identical to the graph of $y = f(x)$ except that it is reflected about the y-axis.

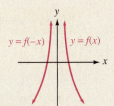

EXAMPLE 4

Using Reflections to Graph Functions

Graph each function: **a.** $g(x) = -|x + 1|$ **b.** $h(x) = |-x + 1|$

SOLUTION   By inspection, we see that the function given in part **a** involves a reflection about the x-axis and the function given in part **b** involves a reflection about the y-axis. We will use reflections to draw the graph of each function.

a. The graph of $g(x) = -|x + 1|$ is identical to the graph of $f(x) = |x + 1|$, except that it is reflected about the x-axis. This is because $g(x) = -f(x)$. The graphs of both functions are shown in Figure 3-19(a).

b. The graph of $h(x) = |-x + 1|$ is identical to the graph of $f(x) = |x + 1|$, except that it is reflected about the y-axis. This is because $f(-x) = h(x)$. The graphs of both functions are shown in Figure 3-19(b).

Tip

It is often helpful to think of a reflection as a mirror image of the graph about the x- or y-axis.

©Maria Mylnikova/Shutterstock.com

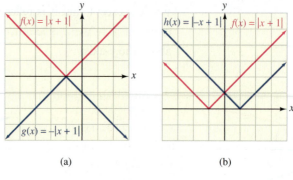

(a)　　　　(b)

FIGURE 3-19

Self Check 4 *Fill in the blanks:* The graph of $g(x) = -\sqrt[3]{x}$ is identical to the graph of $f(x) = \sqrt[3]{x}$, except that it is reflected about the ____ axis. The graph of $h(x) = \sqrt{-x - 4}$ is identical to the graph of $f(x) = \sqrt{x - 4}$, except that it is reflected about the ____ axis.

Now Try Exercise 51.

5. Use Vertical Stretching and Shrinking to Graph Functions

Figure 3-20 shows the graphs of $y = x^2$, $y = 3x^2$, and $y = \frac{1}{3}x^2$.

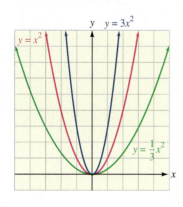

	$y = x^2$	
x	y	(x, y)
-2	4	$(-2, 4)$
-1	1	$(-1, 1)$
0	0	$(0, 0)$
1	1	$(1, 1)$
2	4	$(2, 4)$

	$y = 3x^2$	
x	y	(x, y)
-2	12	$(-2, 12)$
-1	3	$(-1, 3)$
0	0	$(0, 0)$
1	3	$(1, 3)$
2	12	$(2, 12)$

	$y = \frac{1}{3}x^2$	
x	y	(x, y)
-2	$\frac{4}{3}$	$\left(-2, \frac{4}{3}\right)$
-1	$\frac{1}{3}$	$\left(-1, \frac{1}{3}\right)$
0	0	$(0, 0)$
1	$\frac{1}{3}$	$\left(1, \frac{1}{3}\right)$
2	$\frac{4}{3}$	$\left(2, \frac{4}{3}\right)$

FIGURE 3-20

Because each value of $y = 3x^2$ is 3 times greater than the corresponding value of $y = x^2$, its graph is stretched vertically by a factor of 3. Because each value of $y = \frac{1}{3}x^2$ is 3 times smaller than the corresponding value of $y = x^2$, its graph shrinks vertically by a factor of $\frac{1}{3}$.

In general, we can make the following observations.

Vertical Stretching and Shrinking

If f is a function and $k > 1$, then

• The graph of $y = kf(x)$ can be obtained by stretching the graph of $y = f(x)$ vertically by multiplying each value of $f(x)$ by k.

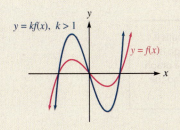

If f is a function and $0 < k < 1$, then

• The graph of $y = kf(x)$ can be obtained by shrinking the graph of $y = f(x)$ vertically by multiplying each value of $f(x)$ by k.

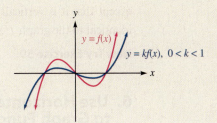

EXAMPLE 5

Graphing a Function by Vertical Stretching or Shrinking

Graph each function: **a.** $g(x) = 2|x|$ **b.** $h(x) = \dfrac{1}{2}|x|$

SOLUTION We will vertically stretch or vertically shrink the basic function $f(x) = |x|$ to graph each of the given functions.

a. The graph of $g(x) = 2|x|$ is identical to the graph of $f(x) = |x|$ except that it is vertically stretched by a factor of 2. This is because each value of $|x|$ is multiplied by 2. The graphs of both functions are shown in Figure 3-21.

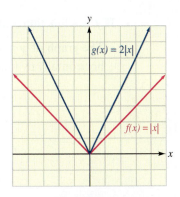

FIGURE 3-21

b. The graph of $g(x) = \frac{1}{2}|x|$ is identical to the graph of $f(x) = |x|$ except that it is vertically shrunk by a factor of $\frac{1}{2}$. This is because each value of $|x|$ is multiplied by $\frac{1}{2}$. The graphs of both functions are shown in Figure 3-22.

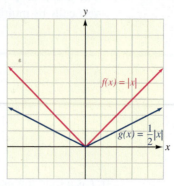

FIGURE 3-22

Self Check 5 *Fill in the blanks:* The graph of $g(x) = 5x^3$ is identical to the graph of $f(x) = x^3$, except that it is vertically _____ by a factor of ___. The graph of $h(x) = \frac{1}{5}x^3$ is identical to the graph $f(x) = x^3$, except that it is vertically _____ by a factor of ___.

Now Try Exercise 59.

6. Use Horizontal Stretching and Shrinking to Graph Functions

Functions can also be graphed by using horizontal stretchings and shrinkings.

Horizontal Stretching and Shrinking   If f is a function and $k > 1$, then

- The graph of $y = f(kx)$ can be obtained by shrinking the graph of $y = f(x)$ horizontally by multiplying each x-value of $f(x)$ by $\frac{1}{k}$.

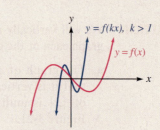

If f is a function and $0 < k < 1$, then

- The graph of $y = f(kx)$ can be obtained by stretching the graph of $y = f(x)$ horizontally by multiplying each x-value of $f(x)$ by $\frac{1}{k}$.

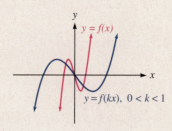

Because we multiply each x-value of $f(x)$ by k to obtain the graph of $y = f(kx)$, the graphs are horizontally stretched or shrunk by a factor of $\frac{1}{k}$.

Graphing a Function by Horizontally Shrinking or Stretching

EXAMPLE 6

Graph $y = (3x)^2 - 1$ using the graph of the function $y = x^2 - 1$.

SOLUTION

Since $3 > 1$, the graph of $y = (3x)^2 - 1$ can be obtained by shrinking the graph of $y = x^2 - 1$ horizontally by multiplying each x-coordinate of $y = x^2 - 1$ by $\frac{1}{3}$.

First, we complete the table of solutions for $y = x^2 - 1$, shown in Figure 3-23. Next, we multiply each x-value in the table by $\frac{1}{3}$ to obtain the table of solutions for $y = (3x)^2 - 1$, also shown in Figure 3-23.

We now draw each graph as shown in Figure 3-23.

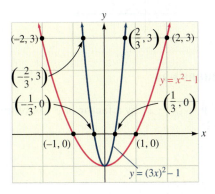

$y = x^2 - 1$		
x	y	(x, y)
-2	3	$(-2, 3)$
-1	0	$(-1, 0)$
0	-1	$(0, -1)$
1	0	$(1, 0)$
2	3	$(2, 3)$

$y = (3x)^2 - 1$		
x	y	(x, y)
$-\frac{2}{3}$	3	$\left(-\frac{2}{3}, 3\right)$
$-\frac{1}{3}$	0	$\left(-\frac{1}{3}, 0\right)$
0	-1	$(0, -1)$
$\frac{1}{3}$	0	$\left(\frac{1}{3}, 0\right)$
$\frac{2}{3}$	3	$\left(\frac{2}{3}, 3\right)$

FIGURE 3-23

Self Check 6   *Fill in the blank:* The graph of $g(x) = \left(\frac{1}{3}x\right)^2 - 1$ is identical to the graph of $f(x) = x^2 - 1$, except that it is horizontally _____ by a factor of __.

Now Try Exercise 69.

7. Graph Functions Using a Combination of Transformations

To graph functions involving a combination of transformations, we must apply each translation or stretching to the function.

Strategy for Graphing Using a Sequence of Transformations

To graph a function using a combination of transformations, perform the transformations in the following order:

1. horizontal translation
2. stretching or shrinking
3. reflection
4. vertical translation

EXAMPLE 7

Graphing a Function by Using a Combination of Transformations

Use the graph of the function $f(x) = |x|$ to graph $g(x) = 3|x - 2| + 4$.

SOLUTION We will graph $g(x) = 3|x - 2| + 4$ by applying three transformations to the graph of the basic function $f(x) = |x|$:

Step 1: Translate the graph of $f(x) = |x|$ horizontally 2 units to the right.

Step 2: Vertically stretch the graph by a factor of 3.

Step 3: Translate the graph vertically 4 units upward.

Step 1: Translate the graph of $f(x) = |x|$ horizontally 2 units to the right to obtain the graph of $y = |x - 2|$. See Figure 3-24.

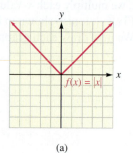
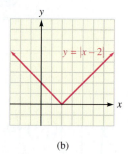

(a)   (b)

FIGURE 3-24

Step 2: Vertically stretch the graph of $y = |x - 2|$ by a factor of 3 to obtain the graph of $y = 3|x - 2|$. See Figure 3-25.

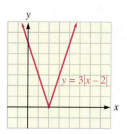

FIGURE 3-25

Step 3: Translate the graph of $y = 3|x - 2|$ vertically 4 units upward to obtain the graph of $g(x) = 3|x - 2| + 4$. See Figure 3-26.

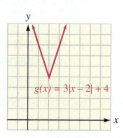

FIGURE 3-26

Self Check 7 Use the graph of the function $f(x) = x^3$ to graph $g(x) = \frac{1}{3}(x + 1)^3 - 2$.

Now Try Exercise 75.

There are many examples of transformations in real life. For example, many people collect Transformers toys made by Hasbro. One moment the most highly desired toy might be a high-performance vehicle racing down the highway, the next moment it might be a towering robot charging into battle. We are also familiar the science fiction action films *Transformers,* which are based on the Transformers toy. Nature also provides examples of transformation. Consider a butterfly. The four-stage development process of a butterfly is called metamorphosis, a Greek term meaning "transformation" or "change of shape."

We can summarize the transformation ideas in this section as follows.

Summary of Transformations

If f is a function and k represents a positive number, then

The graph of	*can be obtained by graphing $y = f(x)$ and*
$y = f(x) + k$	translating the graph k units upward.
$y = f(x) - k$	translating the graph k units downward.
$y = f(x + k)$	translating the graph k units to the left.
$y = f(x - k)$	translating the graph k units to the right.
$y = -f(x)$	reflecting the graph about the x-axis.
$y = f(-x)$	reflecting the graph about the y-axis.
$y = kf(x) \quad k > 1$	stretching the graph vertically by multiplying each value of $f(x)$ by k.
$y = kf(x) \quad 0 < k < 1$	shrinking the graph vertically by multiplying each value $f(x)$ by k.
$y = f(kx) \quad k > 1$	shrinking the graph horizontally by multiplying each x-value of $f(x)$ by $\frac{1}{k}$.
$y = f(kx) \quad 0 < k < 1$	stretching the graph horizontally by multiplying each x-value of $f(x)$ by $\frac{1}{k}$.

EXAMPLE 8

Applying Transformations of Graphs

Figure 3-27 shows the graph of $y = f(x)$. Use this graph and a translation to sketch the graph of **a.** $y = f(x) + 2$ **b.** $y = f(x - 2)$ **c.** $y = 2f(x)$

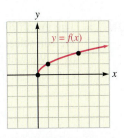

FIGURE 3-27

SOLUTION   We will use the summary of transformations in the section given earlier to graph each function.

a. The graph of $y = f(x) + 2$ is identical to the graph of $y = f(x)$, except that it is translated 2 units upward. See Figure 3-28(a).

b. The graph of $y = f(x - 2)$ is identical to the graph of $y = f(x)$, except that it is translated 2 units to the right. See Figure 3-28(b).

c. The graph of $y = 2f(x)$ is identical to the graph of $y = f(x)$, except that it is stretched vertically by multiplying each y-value of $f(x)$ by 2. See Figure 3-28(c).

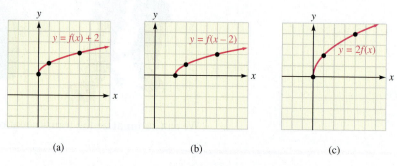

(a)   (b)   (c)

FIGURE 3-28

Self Check 8   Use Figure 3-27 and a reflection to sketch the graph of **a.** $y = -f(x)$
b. $y = f(-x)$

Now Try Exercise 85.

Applying Transformations of Graphs

EXAMPLE 9

The graph of $y = f(x)$ is shown in Figure 3-29. Use this graph to sketch the graph of the indicated functions.

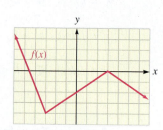

FIGURE 3-29

a. $y = \frac{1}{3}f(x)$ **b.** $y = f(3x)$ **c.** $y = -2f(x + 1) - 2$

SOLUTION   **a.** The graph of $y = \frac{1}{3}f(x)$ is obtained by vertical shrinking the graph of $f(x)$. We will select five points on the graph of $f(x)$ and multiply each y-coordinate by $\frac{1}{3}$. We will then plot the new points and draw the graph of $y = \frac{1}{3}f(x)$. See Figure 3-30. The graphs of $f(x)$ and $y = \frac{1}{3}f(x)$ are indicated.

$$(-5, \mathbf{1}) \to \left(-5, \tfrac{1}{3}\right)$$

$$(-3, \mathbf{-4}) \to \left(-3, -\tfrac{4}{3}\right)$$

$$(0, \mathbf{-2}) \to \left(0, -\tfrac{2}{3}\right)$$

$$(3, \mathbf{0}) \to (3, \mathbf{0})$$

$$(6, \mathbf{-2}) \to \left(6, -\tfrac{2}{3}\right)$$

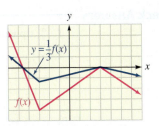

FIGURE 3-30

b. The graph of $y = f(3x)$ is obtained by horizontally shrinking the graph of $f(x)$. We will select five points on the graph of $f(x)$ and multiply each x-coordinate by $\tfrac{1}{3}$. We will then plot the new points and draw the graph of $y = f(3x)$. See Figure 3-31. The graphs of $f(x)$ and $y = f(3x)$ are indicated.

$$(\mathbf{-5}, 1) \to \left(-\tfrac{5}{3}, 1\right)$$

$$(\mathbf{-3}, -4) \to (\mathbf{-1}, -4)$$

$$(\mathbf{0}, -2) \to (\mathbf{0}, -2)$$

$$(\mathbf{3}, 0) \to (\mathbf{1}, 0)$$

$$(\mathbf{6}, -2) \to (\mathbf{2}, -2)$$

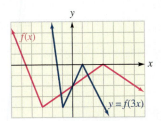

FIGURE 3-31

c. The graph of $y = -2f(x + 1) - 2$ is a transformation of the graph of $f(x)$. We will select five points on the graph of $f(x)$ and apply a sequence of four transformations. The order is important. First, we shift the graph to the left 1 unit; then we will stretch the graph by a factor of 2. Next we apply an x-axis reflection. Finally, we shift the graph down vertically 2 units. We will then plot the new points and draw the graph of $y = -2f(x + 1) - 2$. See Figure 3-32. The graphs of both $f(x)$ and $y = -2f(x + 1) - 2$ are indicated.

The four steps of the transformation are shown in the table.

Original Points:	$(-5, 1)$	$(-3, -4)$	$(0, -2)$	$(3, 0)$	$(6, -2)$
Step 1: Left 1 (subtract 1 from each x-value)	$(\mathbf{-6}, 1)$	$(\mathbf{-4}, -4)$	$(\mathbf{-1}, -2)$	$(\mathbf{2}, 0)$	$(\mathbf{5}, -2)$
Step 2: Stretch y by 2 (multiply y by 2)	$(-6, \mathbf{2})$	$(-4, \mathbf{-8})$	$(-1, \mathbf{-4})$	$(2, 0)$	$(5, \mathbf{-4})$
Step 3: x-axis reflection (multiply y by -1)	$(-6, \mathbf{-2})$	$(-4, \mathbf{8})$	$(-1, \mathbf{4})$	$(2, \mathbf{0})$	$(5, \mathbf{4})$
Step 4: Down 2 (subtract 2 from each y-value)	$(-6, \mathbf{-4})$	$(-4, \mathbf{6})$	$(-1, \mathbf{2})$	$(2, \mathbf{-2})$	$(5, \mathbf{2})$

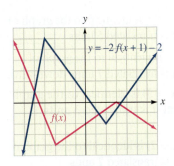

FIGURE 3-32

Self Check 9   Use the graph of $f(x)$ shown in Figure 3-29 to sketch the graphs of the indicated functions. **a.** $y = 3f(x)$ **b.** $y = f\left(\tfrac{1}{3}x\right)$

Now Try Exercise 103.

Self Check Answers

1. 3, upward; 4, downward   2. 3, right; 2, left   3. 4, right; 5, upward

4. x-; y-   5. stretched, 5; shrunk, $\dfrac{1}{5}$ 6. stretched, 3

7.
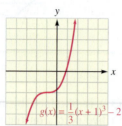

8. a.
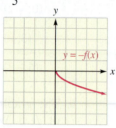

b.
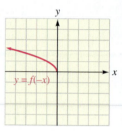

9. a.
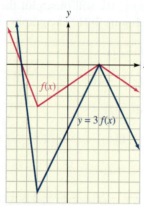

b.
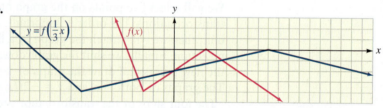

Exercises 3.2

Getting Ready

You should be able to complete these vocabulary and concept statements before you proceed to the practice exercises.

Fill in the blanks.

1. The graph of $y = f(x) + 5$ is identical to the graph of $y = f(x)$ except that it is translated 5 units _____.

2. The graph of _____ is identical to the graph of $y = f(x)$ except that it is translated 7 units downward.

3. The graph of $y = f(x - 3)$ is identical to the graph of $y = f(x)$ except that it is translated 3 units _____.

4. The graph of $y = f(x + 2)$ is identical to the graph of $y = f(x)$ except that it is translated 2 units _____.

5. To draw the graph of $y = (x + 2)^2 - 3$, translate the graph of $y = x^2$ ___ units to the left and 3 units _____.

6. To draw the graph of $y = (x - 3)^3 + 1$, translate the graph of $y = x^3$ 3 units to the _____ and 1 unit _____.

7. The graph of $y = f(-x)$ is a reflection of the graph of $y = f(x)$ about the _____.

8. The graph of _____ is a reflection of the graph of $y = f(x)$ about the x-axis.

9. The graph of $y = f(4x)$ shrinks the graph of $y = f(x)$ _____ by multiplying each x-value of $f(x)$ by $\frac{1}{4}$.

10. The graph of $y = 8f(x)$ stretches the graph of $y = f(x)$ _____ by a factor of 8.

Practice

The graph of each function is a translation of the graph of $f(x) = x^2$. Graph each function.

11. $g(x) = x^2 - 2$

12. $g(x) = (x - 2)^2$

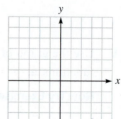

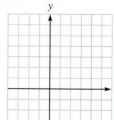

13. $g(x) = (x + 3)^2$

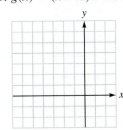

14. $g(x) = x^2 + 3$

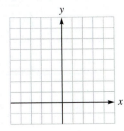

23. $h(x) = (x - 2)^3 - 3$

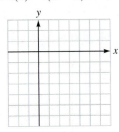

24. $h(x) = (x + 1)^3 + 4$

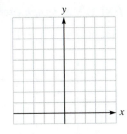

15. $h(x) = (x + 1)^2 + 2$

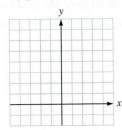

16. $h(x) = (x - 3)^2 - 1$

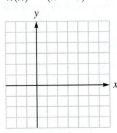

25. $y + 2 = x^3$

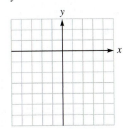

26. $y - 7 = (x - 5)^3$

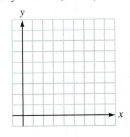

17. $h(x) = \left(x + \dfrac{1}{2}\right)^2 - \dfrac{1}{2}$

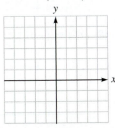

18. $h(x) = \left(x - \dfrac{3}{2}\right)^2 + \dfrac{5}{2}$

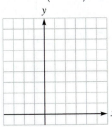

The graph of each function is a translation of the graph of $f(x) = |x|$. *Graph each function.*

27. $g(x) = |x| + 2$

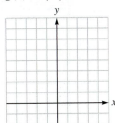

28. $g(x) = |x| - 2$

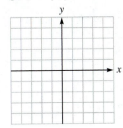

The graph of each function is a translation of the graph of $f(x) = x^3$. *Graph each function.*

19. $g(x) = x^3 + 1$

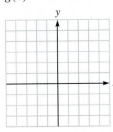

20. $g(x) = x^3 - 3$

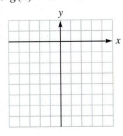

29. $g(x) = |x - 5|$

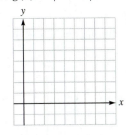

30. $g(x) = |x + 4|$

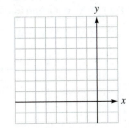

21. $g(x) = (x - 2)^3$

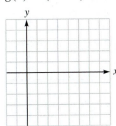

22. $g(x) = (x + 3)^3$

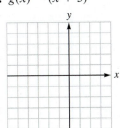

31. $f(x) = |x + 2| - 1$

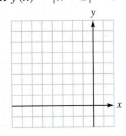

32. $h(x) = |x - 3| + 3$

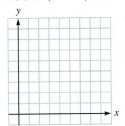

The graph of each function is a translation of the graph of $f(x) = \sqrt{x}$*. Graph each function.*

33. $g(x) = \sqrt{x} + 1$

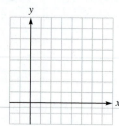

34. $g(x) = \sqrt{x} - 3$

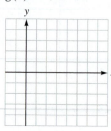

35. $g(x) = \sqrt{x + 2}$

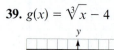
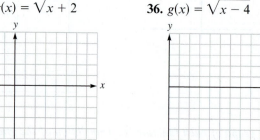

36. $g(x) = \sqrt{x - 4}$

37. $h(x) = \sqrt{x - 2} - 1$

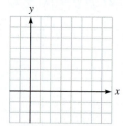

38. $h(x) = \sqrt{x + 2} + 3$

The graph of each function is a translation of the graph of $f(x) = \sqrt[3]{x}$*. Graph each function.*

39. $g(x) = \sqrt[3]{x} - 4$

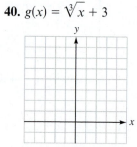

40. $g(x) = \sqrt[3]{x} + 3$

41. $g(x) = \sqrt[3]{x - 2}$

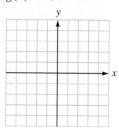
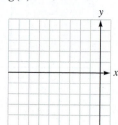

42. $g(x) = \sqrt[3]{x + 5}$

43. $h(x) = \sqrt[3]{x + 1} - 1$

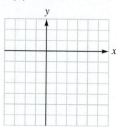

44. $h(x) = \sqrt[3]{x - 1} - 1$

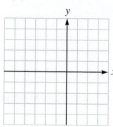

The graph of each function is a reflection of the graph of $y = x^2$*,* $y = x^3$*,* $y = |x|$*,* $y = \sqrt{x}$*, or* $y = \sqrt[3]{x}$*. Graph each function.*

45. $f(x) = -x^2$

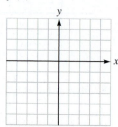

46. $g(x) = (-x)^2$

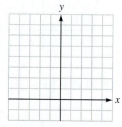

47. $h(x) = -x^3$

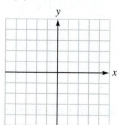

48. $g(x) = (-x)^3$

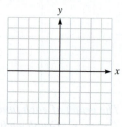

49. $f(x) = -|x|$

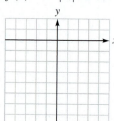

50. $f(x) = |-x|$

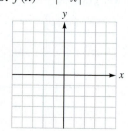

51. $f(x) = -\sqrt{x}$

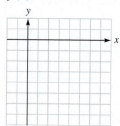

52. $f(x) = \sqrt{-x}$

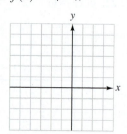

53. $f(x) = -\sqrt[3]{x}$

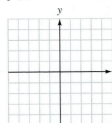

54. $g(x) = \sqrt[3]{-x}$

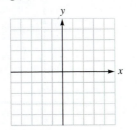

63. $f(x) = 3\sqrt{x}$

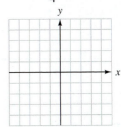

64. $f(x) = \frac{1}{4}\sqrt{x}$

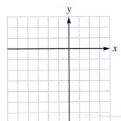

The graph of each function is a vertical stretching or shrinking of the graph of $y = x^2$, $y = x^3$, $y = |x|$, $y = \sqrt{x}$, or $y = \sqrt[3]{x}$. Graph each function.

55. $f(x) = 2x^2$

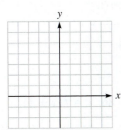

56. $g(x) = \frac{1}{2}x^2$

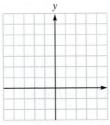

65. $f(x) = \frac{1}{2}\sqrt[3]{x}$

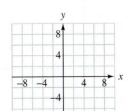

66. $f(x) = 4\sqrt[3]{x}$

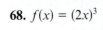

The graph of each function is a horizontal stretching or shrinking of the graph of $y = x^2$ or $y = x^3$. Graph each function.

57. $h(x) = -3x^2$

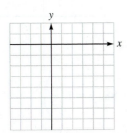

58. $f(x) = -\frac{1}{3}x^2$

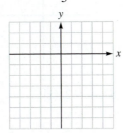

67. $f(x) = \left(\frac{1}{2}x\right)^3$

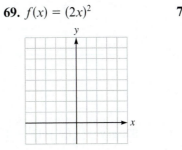

68. $f(x) = (2x)^3$

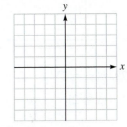

59. $f(x) = \frac{1}{2}x^3$

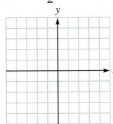

60. $g(x) = 2x^3$

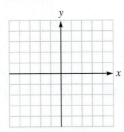

69. $f(x) = (2x)^2$

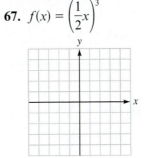

70. $f(x) = (-2x)^3$

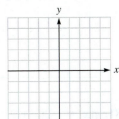

61. $h(x) = -3|x|$

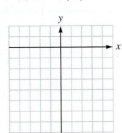

62. $f(x) = \frac{1}{3}|x|$

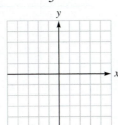

Graph each function using a combination of transformations applied to the graph of a basic function.

71. $g(x) = 3(x + 2)^2 - 1$

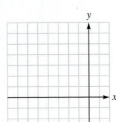

72. $g(x) = -\frac{1}{3}(x + 1)^2 + 1$

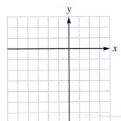

73. $h(x) = -2|x| + 3$

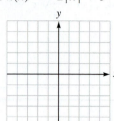

74. $f(x) = -2|x + 3|$

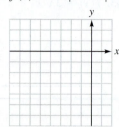

83. $f(x) = 2\sqrt[3]{x} + 4$

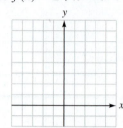

84. $f(x) = -2\sqrt[3]{x + 1}$

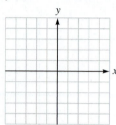

75. $f(x) = 2|x - 2| + 1$

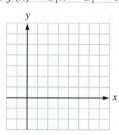

76. $f(x) = -3|x + 5| - 2$

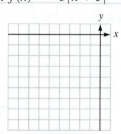

Use the following graph and transformations to graph $g(x)$. Sketch the graph of each function.

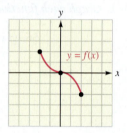

77. $f(x) = 2\sqrt{x} + 3$
$(x \geq 0)$

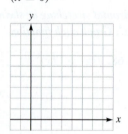

78. $g(x) = 2\sqrt{x + 3}$
$(x \geq -3)$

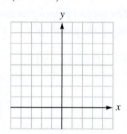

85. $y = f(x) + 1$

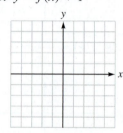

86. $y = f(x + 1)$

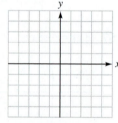

79. $h(x) = 2\sqrt{x - 2} + 1$
$(x \geq 2)$

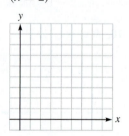

80. $h(x) = \frac{1}{2}\sqrt{x + 5} - 2$
$(x \geq -5)$

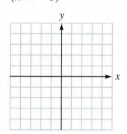

87. $y = 2f(x)$

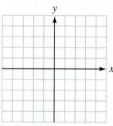

88. $y = f\left(\frac{x}{2}\right)$

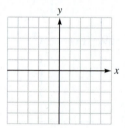

81. $g(x) = -2(x + 2)^3 - 1$

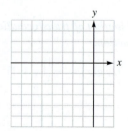

82. $g(x) = \frac{1}{3}(x + 1)^3 - 1$

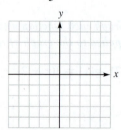

89. $y = f(x - 2) + 1$

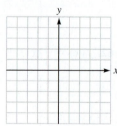

90. $y = -f(x)$

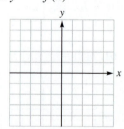

91. $y = 2f(-x)$

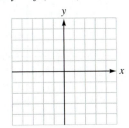

92. $y = f(x + 1) - 2$

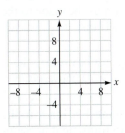

99. $y = f(-x)$

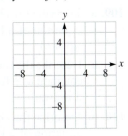

100. $y = -f(x)$

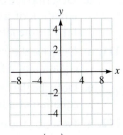

The figure shows the graph of $f(x)$. Use the given graph and transformations to graph each function.

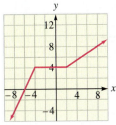

101. $y = 4f(x)$

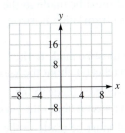

102. $y = \frac{1}{4}f(x)$

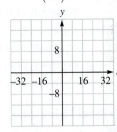

103. $y = f(4x)$

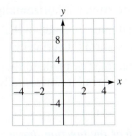

104. $y = f\left(\frac{1}{4}x\right)$

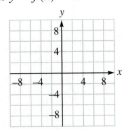

93. $y = f(x) + 2$

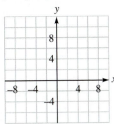

94. $y = f(x) - 2$

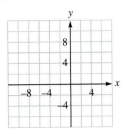

95. $y = f(x + 2)$

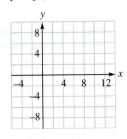

96. $y = f(x - 2)$

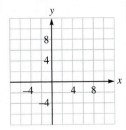

97. $y = f(x - 4) - 2$

98. $y = f(x + 4) + 2$

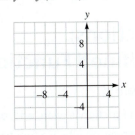

Discovery and Writing

Use a graphing calculator to perform each experiment. Write a brief paragraph describing your findings.

105. Investigate the translations of the graph of a function by graphing the parabola $y = (x - k)^2 + k$ for several values of k. What do you observe about successive positions of the vertex?

106. Investigate the translations of the graph of a function by graphing the parabola $y = (x - k)^2 + k^2$ for several values of k. What do you observe about successive positions of the vertex?

107. Investigate the horizontal stretching of the graph of a function by graphing $y = \sqrt{ax}$ for several values of a. What do you observe?

108. Investigate the vertical stretching of the graph of a function by graphing $y = b\sqrt{x}$ for several values of b. What do you observe? Are these graphs different from the graphs in Exercise 107?

Write a paragraph using your own words.

109. a. Describe the change that must be made to the equation of a function to translate it vertically upward.

    b. Describe the change that must be made to the equation of a function to translate it vertically downward.

110. a. Describe the change that must be made to the equation of a function to translate it horizontally to the right.

    b. Describe the change that must be made to the equation of a function to translate it horizontally to the left.

111. a. Describe the change that must be made to the equation of a function to reflect it about the *x*-axis.

    b. Describe the change that must be made to the equation of a function to reflect it about the *y*-axis.

112. a. Describe the change that must be made to the equation of a function to stretch it vertically.

    b. Describe the change that must be made to the equation of a function to stretch it horizontally.

113. a. Describe the change that must be made to the equation of a function to shrink it vertically.

    b. Describe the change that must be made to the equation of a function to shrink it horizontally.

Critical Thinking

Write the equation for the graph of the function shown.

114.
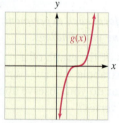

115.
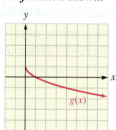

116.
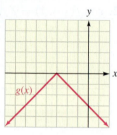

117.
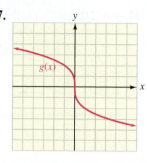

118.
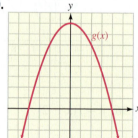

119.

120.
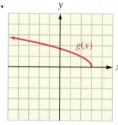

3.3 More on Functions; Piecewise-Defined Functions

In this section, we will learn to

1. Determine whether a function is even, odd, or neither.

2. Identify the open intervals on which a function is increasing, decreasing, or constant.

3. Identify local maxima and local minima on the graph of a function.

4. Evaluate and graph piecewise-defined functions.

5. Evaluate and graph the greatest-integer function.

the line $f(x) = x + 1$. This is a linear function with slope $m = 1$ and y-intercept $(0, b) = (0, 1)$. The graph of this piecewise-defined function is shown in Figure 3-43.

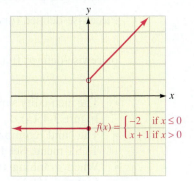

$$f(x) = \begin{cases} -2 & \text{if } x \le 0 \\ x + 1 & \text{if } x > 0 \end{cases}$$

FIGURE 3-43

Self Check 7   Graph: $f(x) = \begin{cases} 2x & \text{if } x \le 0 \\ x - 1 & \text{if } x > 0 \end{cases}$.

Now Try Exercise 55.

Graphing a Piecewise-Defined Function

EXAMPLE 8

Graph the function: $f(x) = \begin{cases} -x & \text{if } x < 0 \\ x^2 & \text{if } 0 \le x \le 1. \\ 1 & \text{if } x > 1 \end{cases}$

SOLUTION   This piecewise-defined function is defined in three parts. We will graph each part of the function. That is, we will graph $f(x) = -x$ in the interval $(-\infty, 0)$, $f(x) = x^2$ in the interval $[0, 1]$, and $f(x) = 1$ in the interval $(1, \infty)$.

If $x < 0$, the value of $f(x)$ is determined by the equation $f(x) = -x$. We graph the line with slope $m = -1$ and y-intercept $(0, b) = (0, 0)$ in the interval $(-\infty, 0)$. In the interval $(-\infty, 0)$, the function is decreasing.

If $0 \le x \le 1$, the value of $f(x)$ is x^2. We graph the squaring function in the interval $[0, 1]$. In the open interval $(0, 1)$ the function is increasing.

If $x > 1$, the value of $f(x)$ is 1. In the open interval $(1, \infty)$, the function is constant and its graph is the same as the graph of $y = 1$.

The graph of the piecewise function appears in Figure 3-44.

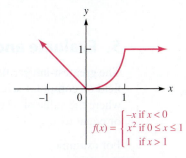

$$f(x) = \begin{cases} -x & \text{if } x < 0 \\ x^2 & \text{if } 0 \le x \le 1 \\ 1 & \text{if } x > 1 \end{cases}$$

FIGURE 3-44

Self Check 8   Graph the function: $f(x) = \begin{cases} -x^2 & \text{if } x \le -1 \\ -1 & \text{if } -1 < x < 1 \\ x + 2 & \text{if } x \ge 1 \end{cases}$

Now Try Exercise 61.

Piecewise-Defined Functions

Piecewise-defined functions can be graphed on a calculator. We will graph the

piecewise-defined function $f(x) = \begin{cases} -x & \text{if } x < 0 \\ x^2 & \text{if } 0 \leq x \leq 1 \\ 1 & \text{if } x > 1 \end{cases}$ given in Example 8.

- The inequality symbols are found by pressing [2nd] [MATH] (TEST) shown in Figure 3-45(a).
- The "and" command is found in the LOGIC menu which is accessed via the TEST menu. See Figure 3-45(b).
- When graphing these functions, place the calculator in DOT mode. See Figure 3-45(c).
- Enter the function as shown in Figure 3-45(d). Note that the use of parentheses is essential. A viewing window of $[-5, 5]$ for x and $[-3, 3]$ for y is used here.

The graph of the function is shown in Figure 3-45(e).

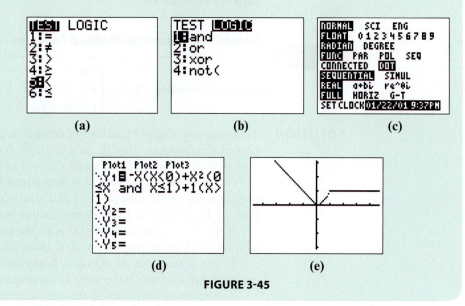

FIGURE 3-45

5. Evaluate and Graph the Greatest-Integer Function

The **greatest-integer function** is important in many business applications and in the field of computer science. This function is determined by the equation $f(x) = [\![x]\!]$, where the value of $f(x)$ that corresponds to x is the greatest integer that is less than or equal to x.

For example,

$$f(2.71) = [\![2.71]\!] = 2$$

$$f(23.5) = [\![23.5]\!] = 23$$

$$f(10) = [\![10]\!] = 10$$

$$f(\pi) = [\![\pi]\!] = 3$$

$$f(-2.5) = [\![-2.5]\!] = -3$$

EXAMPLE 9

Graphing the Greatest-Integer Function

Graph: $f(x) = [\![x]\!]$.

SOLUTION   We will list several intervals and determine the corresponding values of the greatest-integer function. Then we will use these values to graph the function.

$[0, 1)$ $f(x) = [\![x]\!] = 0$ For numbers from 0 to 1 (not including 1), the greatest integer in the interval is 0.

$[1, 2)$ $f(x) = [\![x]\!] = 1$ For numbers from 1 to 2 (not including 2), the greatest integer in the interval is 1.

$[2, 3)$ $f(x) = [\![x]\!] = 2$ For numbers from 2 to 3 (not including 3), the greatest integer in the interval is 2.

Within each interval, the values of y are constant, but they jump by 1 at integer values of x. The graph is shown in Figure 3-46. From the graph, we can see that the domain of the greatest integer function is the interval $(-\infty, \infty)$. The range is the set of integers.

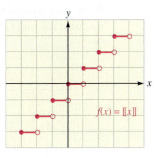

$f(x) = [\![x]\!]$

FIGURE 3-46

Self Check 9   Find **a.** $[\![7.61]\!]$ and **b.** $[\![-3.75]\!]$.

Now Try Exercise 67.

Since the greatest-integer function is made up of a series of horizontal line segments, it is an example of a group of functions called **step functions**.

EXAMPLE 10

Graphing a Step Function Occurring in an Application

To print business forms, a printing company charges customers $10 for the order, plus $20 for each box containing 200 forms. The printing company counts any portion of a box as a full box. Graph this step function.

SOLUTION   To graph the step function, we will determine the cost for printing various amounts of boxes of forms. Then we will graph our result.

If we order the forms and then change our minds before the forms are printed, the cost will be $10. Thus, the ordered pair $(0, 10)$ will be on the graph.

If we purchase up to one full box, the cost will be $10 for the order and $20 for the printing, for a total of $30. Thus, the ordered pair $(1, 30)$ will be on the graph.

The cost for $1\frac{1}{2}$ boxes will be the same as the cost for 2 full boxes, or $50. Thus, the ordered pairs $(1.5, 50)$ and $(2, 50)$ are on the graph.

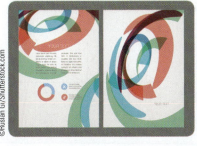

The complete graph is shown in Figure 3-47.

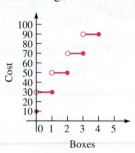

FIGURE 3-47

Self Check 10 Find the cost of $4\frac{1}{2}$ boxes.

Now Try Exercise 73.

Self Check Answers

1. even **2. a.** odd **b.** even **3.** odd **4.** no intervals
5. **a.** increasing on $(-\infty, -4) \cup (8, 14)$ **b.** decreasing on $(-4, 1)$
c. constant on $(1, 8)$ **6.** local maxima is 0; local minima is -1
7. 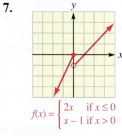 8. 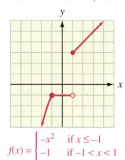 **9. a.** 7 **b.** -4

10. $110

Exercises 3.3

Getting Ready
You should be able to complete these vocabulary and concept statements before you proceed to the practice exercises.

Fill in the blanks.

1. If the graph of a function is symmetric about the _____, it is called an even function.

2. If the graph of a function is symmetric about the origin, it is called an ____ function.

3. If a function is even, then $f(-x) =$ ____.

4. If a function is odd, then $f(-x) =$ _____.

5. If the values of $f(x)$ get larger as x increases on an interval, we say that the function is _____ on the open interval.

6. If the values of $f(x)$ get smaller as x increases on an interval, we say that the function is _____ on the open interval.

7. If the values of $f(x)$ do not change as x increases on an interval, we say that the function is _____ on the open interval.

8. A local _____ occurs where a function changes from increasing to decreasing.

9. A local _____ occurs where a function changes from decreasing to increasing.

10. _____ functions are defined by different equations for different intervals in their domains.

Practice
Determine whether each function is even, odd, or neither.

11. 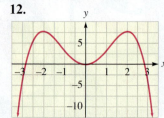 12.

13.
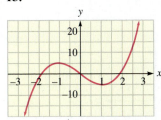

14.
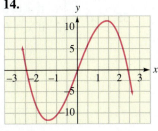

37.
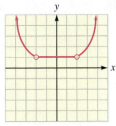

38.
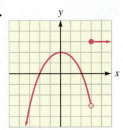

15.
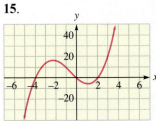

16.
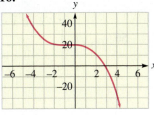

Use the graph to identify any local maxima and local minima.

39.

40.
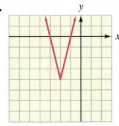

Determine algebraically whether each function is even, odd, or neither.

17. $f(x) = x^4 + x^2$

18. $f(x) = x^3 - 2x$

19. $f(x) = x^3 + x^2$

20. $f(x) = x^6 - x^2$

21. $f(x) = x^5 + x^3$

22. $f(x) = x^3 - x^2$

23. $f(x) = 2x^3 - 3x$

24. $f(x) = 4x^2 - 5$

25. $f(x) = \dfrac{x}{x^2 - 1}$

26. $f(x) = \dfrac{2x}{x^2 - 9}$

27. $f(x) = \dfrac{1}{x^4}$

28. $f(x) = -\dfrac{2}{x^2}$

29. $f(x) = \sqrt{x + 1}$

30. $f(x) = \sqrt{2x - 5}$

31. $f(x) = \dfrac{|x|}{x}$

32. $f(x) = 2x - |x|$

41.
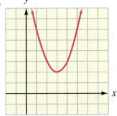

42.
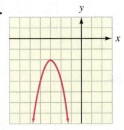

State the open intervals where each function is increasing, decreasing, or constant.

33.
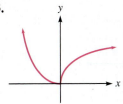

34.
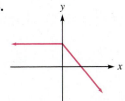

43.
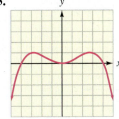

44.
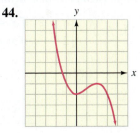

35.
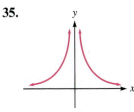

36.
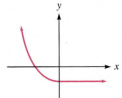

45.
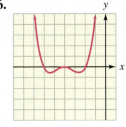

46.

47.

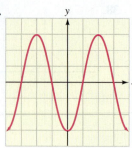

48.

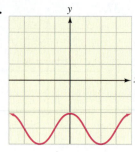

55. $f(x) = \begin{cases} x & \text{if } x \leq 0 \\ 2 & \text{if } x > 0 \end{cases}$

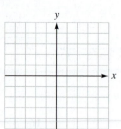

Evaluate each piecewise-defined function.

49. $f(x) = \begin{cases} 2x + 2 & \text{if } x < 0 \\ 3 & \text{if } x \geq 0 \end{cases}$

    a. $f(-2)$ **b.** $f(0)$

50. $f(x) = \begin{cases} x - 2 & \text{if } x < 1 \\ x^2 & \text{if } x \geq 1 \end{cases}$

    a. $f(-1)$ **b.** $f(5)$

51. $f(x) = \begin{cases} 2 & \text{if } x < 0 \\ 2 - x & \text{if } 0 \leq x < 2 \\ x + 1 & \text{if } x \geq 2 \end{cases}$

    a. $f(-1)$ **b.** $f(1)$ **c.** $f(2)$

52. $f(x) = \begin{cases} 2x & \text{if } x < 0 \\ 3 - x & \text{if } 0 \leq x < 2 \\ |x| & \text{if } x \geq 2 \end{cases}$

    a. $f(-0.5)$ **b.** $f(0)$ **c.** $f(2)$

56. $f(x) = \begin{cases} -x & \text{if } x < 0 \\ \dfrac{1}{2}x & \text{if } x > 0 \end{cases}$

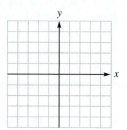

Graph each piecewise-defined function.

53. $f(x) = \begin{cases} x + 2 & \text{if } x < 0 \\ 2 & \text{if } x \geq 0 \end{cases}$

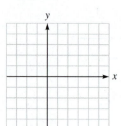

54. $f(x) = \begin{cases} 2x & \text{if } x < 0 \\ -2x & \text{if } x \geq 0 \end{cases}$

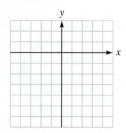

57. $f(x) = \begin{cases} -4 - x & \text{if } x < 1 \\ 3 & \text{if } x \geq 1 \end{cases}$

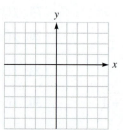

58. $f(x) = \begin{cases} -5 - x & \text{if } x < 1 \\ -3 & \text{if } x \geq 1 \end{cases}$

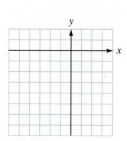

59. $f(x) = \begin{cases} -x & \text{if } x < 0 \\ x^2 & \text{if } x \geq 0 \end{cases}$

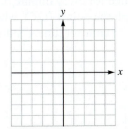

60. $f(x) = \begin{cases} |x| & \text{if } x < 0 \\ \sqrt{x} & \text{if } x \geq 0 \end{cases}$

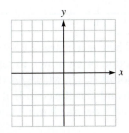

61. $f(x) = \begin{cases} 0 & \text{if } x < 0 \\ x^2 & \text{if } 0 \leq x \leq 2 \\ 4 - 2x & \text{if } x > 2 \end{cases}$

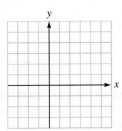

62. $f(x) = \begin{cases} 2 & \text{if } x < 0 \\ 2 - x & \text{if } 0 \leq x < 2 \\ x & \text{if } x \geq 2 \end{cases}$

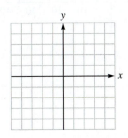

Evaluate each function at the indicated x-values.

63. $f(x) = [\![x]\!]$ **a.** $f(3)$ **b.** $f(-4)$ **c.** $f(-2.3)$

64. $f(x) = [\![3x]\!]$ **a.** $f(4)$ **b.** $f(-2)$ **c.** $f(-1.2)$

65. $f(x) = [\![x + 3]\!]$ **a.** $f(-1)$ **b.** $f\left(\dfrac{2}{3}\right)$ **c.** $f(1.3)$

66. $f(x) = [\![4x]\!] - 1$ **a.** $f(-3)$ **b.** $f(0)$ **c.** $f(\pi)$

Graph each function.

67. $y = [\![2x]\!]$

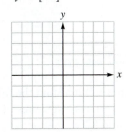

68. $y = \left[\!\left[\dfrac{1}{3}x + 3\right]\!\right]$

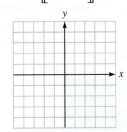

69. $y = [\![x]\!] - 1$

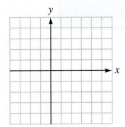

70. $y = [\![x + 2]\!]$

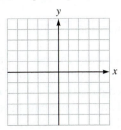

Applications

71. Grading scales A mathematics instructor assigns letter grades according to the following scale.

From	Up to but Less Than	Grade
60%	70%	D
70%	80%	C
80%	90%	B
90%	100% (including 100%)	A

Graph the ordered pairs (p, g), where p represents the percent and g represents the grade. Find the final semester grade of a student who has test scores of 67%, 73%, 84%, 87%, and 93%.

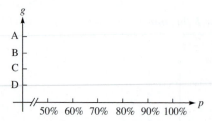

72. Calculating grades See Exercise 71 and find the final semester grade of a student who has test scores of 53%, 65%, 64%, 73%, 89%, and 82%.

73. Renting a jeep A rental company charges $20 to rent a Jeep for one day, plus $4 for every 100 miles (or portion of 100 miles) that it is driven. Graph the ordered pairs (m, C), where m represents the miles driven and C represents the cost. Find the cost if the Jeep is driven 275 miles in one day.

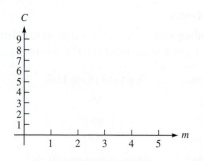

©Bikeriderlondon/Shutterstock.com

74. Riding in a taxi A taxicab company charges $3 for a trip up to 1 mile, and $2 for every extra mile (or portion of a mile). Graph the ordered pairs (m, C), where m represents the miles traveled and C represents the cost. Find the cost to ride $10\frac{1}{4}$ miles.

75. Computer communications An on-line information service charges for connect time at a rate of $12 per hour, computed for every minute or fraction of a minute. Graph the points (t, C), where C is the cost of t minutes of connect time. Find the cost of $7\frac{1}{2}$ minutes.

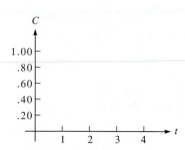

76. iPad repair There is a charge of $30, plus $40 per hour (or fraction of an hour), to repair an iPad. Graph the points (t, C), where t is the time it takes to do the job and C is the cost. If it takes 4 hours to repair the iPad, how much did it cost?

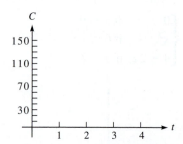

77. Rounding numbers Measurements are rarely exact; they are often *rounded* to an appropriate precision. Graph the points (x, y), where y is the result of rounding the number x to the nearest ten.

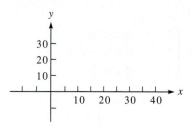

78. Signum function Computer programmers often use the following function, denoted by $y = \text{sgn } x$. Graph this function and find its domain and range.

$$y = \begin{cases} -1 & \text{if } x < 0 \\ 0 & \text{if } x = 0 \\ 1 & \text{if } x > 0 \end{cases}$$

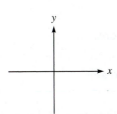

79. Graph the function defined by $y = \dfrac{|x|}{x}$ and compare it to the graph in Exercise 78. Are the graphs the same?

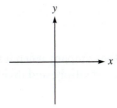

80. Graph: $y = x + |x|$.

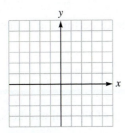

Discovery and Writing

81. If you are given a function's graph, how do you determine whether the function is even, odd, or neither?

82. If you are given a function's equation, how do you determine whether the function is even, odd, or neither?

83. What does it mean for a function to be increasing on an interval? Give two examples of real-life functions that are increasing.

84. What does it mean for a function to be decreasing on an interval? Give two examples of real-life functions that are decreasing.

85. Describe what happens at the point where the graph of a function changes from increasing to decreasing.

86. Describe what happens at the point where the graph of a function changes from decreasing to increasing.

87. In this section, we discussed maximum and minimum values; state two real-life situations where a maximum or minimum value is important.

88. Use postal rates for a first-class postage stamp to create a step function for calculating costs of mailing a first-class letter.

89. Six Flags White Water in Atlanta, Georgia, charges the following prices for daily admission to the park: general admission, $39.99; children under 48 inches, $34.99. Describe how this information can be represented by a piecewise-defined function.

90. Construct a piecewise-defined function that occurs in everyday life.

Critical Thinking

Determine if the statement is true or false. If the statement is false, then correct it and make it true.

91. The function $f(x) = x^{100} + x^{50}$ is an even function.

92. The function $g(x) = x^{101} + x^{51}$ is an odd function.

93. The function $f(x) = \sqrt[7]{x}$ is an odd function.

94. The function $f(x) = \sqrt[8]{x}$ is an even function.

95. The quotient of two odd functions is an even function.

96. All functions have a local maximum value and a local minimum value.

97. Local maximum and minimum values can be the same.

98. If function f decreases on the interval $(-\infty, x_1)$ and increases on the interval (x_1, ∞), then $f(x_1)$ is a local maximum value.

99. If function f increases on the interval $(-\infty, x_1)$ and decreases on the interval (x_1, ∞), then $f(x_1)$ is a local minimum value.

100. $\pi - \lfloor -\pi \rfloor = \pi + 4$

3.4   Operations on Functions

In this section, we will learn to

1. Add, subtract, multiply, and divide functions, specifying domains.
2. Write functions as sums, differences, products, or quotients of other functions.
3. Evaluate composite functions.
4. Determine domains for composite functions.
5. Write functions as compositions.
6. Use operations on functions to solve problems.

©EpicStockMedia/Shutterstock.com

Functions can be combined by addition, subtraction, multiplication, and division. In this section, we will explore these operations on functions and give careful attention to their domains and ranges.

Suppose that the functions $R(x) = 140x$ and $C(x) = 120{,}000 + 40x$ model a company's yearly revenue and cost for producing and selling surfboards. By subtracting the functions, $R(x) - C(x)$, we would arrive at a new function represented by

$$(R - C)(x) = 140x - (120{,}000 + 40x)$$

$$= 140x - 120{,}000 - 40x \qquad \text{Remove parentheses.}$$

$$= 100x - 120{,}000$$

This function represents the profit made by the company when it sells x surfboards. We will now discuss how to add, subtract, multiply, and divide functions.

1. Add, Subtract, Multiply, and Divide Functions, Specifying Domains

With the following definitions, it is possible to perform arithmetic operations on algebraic functions.

Adding, Subtracting, Multiplying, and Dividing Functions

If the ranges of functions f and g are subsets of the real numbers, then

1. The **sum** of f and g, denoted as $f + g$, is defined by

   $$(f + g)(x) = f(x) + g(x)$$

2. The **difference** of f and g, denoted as $f - g$, is defined by

   $$(f - g)(x) = f(x) - g(x)$$

3. The **product** of f and g, denoted as $f \cdot g$, is defined by

   $$(f \cdot g)(x) = f(x)g(x)$$

4. The **quotient** of f and g, denoted as f/g, is defined by

   $$(f/g)(x) = \frac{f(x)}{g(x)} \quad (g(x) \neq 0)$$

The **domain** of each function, unless otherwise restricted, is the set of real numbers x that are in the domains of both f and g. In the case of the quotient f/g, there is the restriction that $g(x) \neq 0$.

Finding the Sum and Difference of Two Functions and Specifying Domains

EXAMPLE 1

Let $f(x) = 3x^2 - 5x + 1$ and $g(x) = 2x^2 + x - 3$. Find each function and its domain: **a.** $f + g$ **b.** $f - g$

SOLUTION   We will find the sum of f and g by using the definition

$$(f + g)(x) = f(x) + g(x)$$

We will find the difference of f and g by using the definition

$$(f - g)(x) = f(x) - g(x).$$

To find the domain of each result, we will consider the domains of both f and g.

a. $(f + g)(x) = f(x) + g(x)$

$$= (3x^2 - 5x + 1) + (2x^2 + x - 3)$$

$$= 5x^2 - 4x - 2$$

Since the domain of both f and g is the set of real numbers, the domain of $f + g$ is the interval $(-\infty, \infty)$.

b. $(f - g)(x) = f(x) - g(x)$

$$= (3x^2 - 5x + 1) - (2x^2 + x - 3)$$

$$= 3x^2 - 5x + 1 - 2x^2 - x + 3 \qquad \text{Remove parentheses.}$$

$$= x^2 - 6x + 4$$

Since the domain of both f and g is the set of real numbers, the domain of $f - g$ is the interval $(-\infty, \infty)$.

Self Check 1   Find $g - f$.

Now Try Exercise 15.

Finding the Product and Quotient of Two Functions and Specifying Domains

EXAMPLE 2

Let $f(x) = 3x + 1$ and $g(x) = 2x - 3$. Find each function and its domain: **a.** $f \cdot g$ **b.** f/g

SOLUTION   We will multiply f and g by using the definition

$$(f \cdot g)(x) = f(x)g(x)$$

We will divide f by g by using the definition

$$(f/g)(x) = \frac{f(x)}{g(x)} \qquad (g(x) \neq 0)$$

We will find the domains of each result by considering the domains of both f and g.

a. $(f \cdot g)(x) = f(x) \cdot g(x)$

$$= (3x + 1)(2x - 3)$$

$$= 6x^2 - 7x - 3$$

Since the domain of both f and g is the set of real numbers, the domain of $f \cdot g$ is the interval $(-\infty, \infty)$.

b. $(f/g)(x) = \dfrac{f(x)}{g(x)}$ $(g(x) \neq 0)$

$$= \dfrac{3x + 1}{2x - 3} \quad (2x - 3 \neq 0)$$

Since $\frac{3}{2}$ will make $2x - 3$ equal to 0, the domain of f/g is the set of all real numbers except $\frac{3}{2}$. This is $\left(-\infty, \frac{3}{2}\right) \cup \left(\frac{3}{2}, \infty\right)$.

Self Check 2   Find g/f and its domain.

Now Try Exercise 17.

EXAMPLE 3   **Using Operations on Functions and Specifying Domains**

Let $f(x) = x^2 - 4$ and $g(x) = \sqrt{x}$. Find each function and its domain: **a.** $f + g$ **b.** $f \cdot g$ **c.** f/g **d.** g/f

SOLUTION   To determine each result, we use the definitions of the sum, product, and quotient of two functions. We will determine the domain of each result by considering the domains of both f and g.

First, we find the domains of f and g. Because 4 can be subtracted from any real number squared, the domain of f is the interval $(-\infty, \infty)$. Because $\sqrt{x}$ is to be a real number, the domain of g is the interval $[0, \infty)$.

a. $(f + g)(x) = f(x) + g(x)$

$$= x^2 - 4 + \sqrt{x}$$

The domain of $f + g$ consists of the numbers x that are in the domain of both f and g. This is $(-\infty, \infty) \cap [0, \infty)$, which is $[0, \infty)$. The domain of $f + g$ is $[0, \infty)$.

b. $(f \cdot g)(x) = f(x)g(x)$

$$= (x^2 - 4)\sqrt{x}$$

$$= x^2\sqrt{x} - 4\sqrt{x} \qquad \textcolor{red}{\text{Distribute the multiplication of } \sqrt{x}.}$$

The domain of $f \cdot g$ consists of the numbers x that are in the domain of both f and g. The domain of $f \cdot g$ is $[0, \infty)$.

c. $(f/g)(x) = \dfrac{f(x)}{g(x)}$ $(g(x) \neq 0)$

$$= \dfrac{x^2 - 4}{\sqrt{x}}$$

The domain of f/g consists of the numbers x that are in the domain of both f and g, except 0 (because division by 0 is undefined). The domain of f/g is $(0, \infty)$.

To write $(f/g)(x)$ in a different form, we can rationalize the denominator of $\frac{x^2 - 4}{\sqrt{x}}$ and simplify.

$$(f/g)(x) = \dfrac{x^2 - 4}{\sqrt{x}}$$

$$= \dfrac{(x^2 - 4)\sqrt{x}}{\sqrt{x} \cdot \sqrt{x}} \qquad \textcolor{red}{\text{Multiply numerator and denominator by } \sqrt{x}.}$$

$$= \dfrac{x^2\sqrt{x} - 4\sqrt{x}}{x}$$

d. $(g/f)(x) = \dfrac{g(x)}{f(x)}$ $(f(x) \neq 0)$

$\qquad\quad = \dfrac{\sqrt{x}}{x^2 - 4}$

The domain of g/f consists of the numbers x that are in $[0, \infty)$, the domain of both f and g, except 2 (because division by 0 is undefined). The domain of g/f is $[0, 2) \cup (2, \infty)$.

Self Check 3   Find $g - f$ and its domain.

Now Try Exercise 23.

We can perform these operations using a graphing calculator.

Operations on Functions

A graphing calculator can be used to graph operations on function. Consider Example 3.

- We will use a bold setting to indicate the function $f + g$. In Figure 3-48, we see in the left graph that the graph of $f + g$ appears to be the same as f. However, if we zoom in to a small part of the graph as shown in the right graph, we can see that the functions are different. To input Y_1 and Y_2, press $\boxed{\text{VARS}}$, scroll right to Y-VARS, press $\boxed{\text{ENTER}}$, and select the desired function.

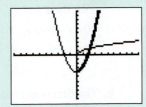

FIGURE 3-48

- We use the bold setting to indicate the function $f \cdot g$. Again, by zooming in to just part of the window, we can see the graphs are different. See Figure 3-49.

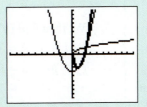
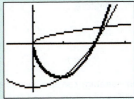

FIGURE 3-49

- We demonstrate the graph of f/g using a graphing calculator. See Figure 3-50. f/g appears bold.

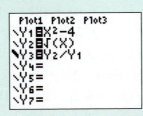
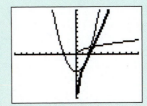

FIGURE 3-50

- We demonstrate the graph of g/f using a graphing calculator. See Figure 3-51. g/f appears bold.

FIGURE 3-51

Evaluating the Sum, Difference, Product, and Quotient of Two Functions

EXAMPLE 4

Let $f(x) = x^2 + 1$ and $g(x) = 2x + 1$. Find each of the following:

a. $(f + g)(3)$ **b.** $(f - g)(-5)$ **c.** $(f \cdot g)(-1)$ **d.** $(f/g)(-2)$

SOLUTION **a.** The first step is to find $(f + g)(x)$.

$$(f + g)(x) = f(x) + g(x)$$
$$= x^2 + 1 + 2x + 1$$
$$= x^2 + 2x + 2$$

We then find $(f + g)(3)$.
$$(f + g)(x) = x^2 + 2(x) + 2$$
$$(f + g)(3) = 3^2 + 2(3) + 2 \qquad \text{Substitute 3 in for } x.$$
$$= 17$$

b. The first step is to find $(f - g)(x)$.
$$(f - g)(x) = f(x) - g(x)$$
$$= x^2 + 1 - (2x + 1)$$
$$= x^2 + 1 - 2x - 1 \qquad \text{Remove parentheses}$$
$$= x^2 - 2x$$

We then find $(f - g)(-5)$.
$$(f - g)(x) = x^2 - 2x$$
$$(f - g)(-5) = (-5)^2 - 2(-5) \qquad \text{Substitute } -5 \text{ in for } x.$$
$$= 25 + 10$$
$$= 35$$

c. The first step is to find $(f \cdot g)(x)$.
$$(f \cdot g)(x) = f(x) \cdot g(x)$$
$$= (x^2 + 1)(2x + 1)$$
$$= 2x^3 + x^2 + 2x + 1 \qquad \text{Use the Distributive Property.}$$

We then find $(f \cdot g)(-1)$.
$$(f \cdot g)(x) = 2x^3 + x^2 + 2x + 1$$
$$(f \cdot g)(-1) = 2(-1)^3 + (-1)^2 + 2(-1) + 1 \qquad \text{Substitute } -1 \text{ in for } x.$$
$$= -2 + 1 - 2 + 1$$
$$= -2$$

> **Tip**
>
> In Example 4, we found $(f + g)(x)$, $(f - g)(x)$, $(f \cdot g)(x)$, and $(f/g)(x)$ first and then evaluated each at the given x-value. An easier approach for some students is to evaluate f and g first and then determine the combined function value. Consider $(f + g)(3)$. Note that $f(3) = 10$ and $g(3) = 7$. Therefore,
>
> $$(f + g)(3) = f(3) + g(3)$$
> $$= 10 + 7$$
> $$= 17$$
>
> This same strategy can also be applied to the difference, product, and quotient functions.

d. The first step is to find $(f/g)(x)$.

$$(f/g)(x) = \frac{f(x)}{g(x)} \quad (g(x) \neq 0)$$

$$= \frac{x^2 + 1}{2x + 1}$$

We then find $(f/g)(-2)$.

$$(f/g)(x) = \frac{x^2 + 1}{2x + 1}$$

$$(f/g)(-2) = \frac{(-2)^2 + 1}{2(-2) + 1} \qquad \text{Substitute } -2 \text{ in for } x.$$

$$= \frac{5}{-3}$$

$$= -\frac{5}{3}$$

Self Check 4   Find $(f \cdot g)(-2)$.

Now Try Exercise 29.

2. Write Functions as Sums, Differences, Products, or Quotients of Other Functions

EXAMPLE 5   **Writing Functions as Combinations of Other Functions**

Let $h(x) = x^2 + 3x + 2$. Find two functions f and g such that **a.** $f + g = h$
b. $f \cdot g = h$

SOLUTION   For part **a**, we must find two functions f and g whose sum is h. For part **b**, we must find two functions f and g whose product is h.

a. There are many possibilities. One is $f(x) = x^2$ and $g(x) = 3x + 2$, then

$$(f + g)(x) = f(x) + g(x)$$

$$= (x^2) + (3x + 2)$$

$$= x^2 + 3x + 2$$

$$= h(x)$$

Another possibility is $f(x) = x^2 + 2x$ and $g(x) = x + 2$.

b. Again, there are many possibilities. One is suggested by factoring $x^2 + 3x + 2$.

$$x^2 + 3x + 2 = (x + 1)(x + 2)$$

If we let $f(x) = x + 1$ and $g(x) = x + 2$, then

$$(f \cdot g)(x) = f(x) \cdot g(x)$$

$$= (x + 1)(x + 2)$$

$$= x^2 + 3x + 2$$

$$= h(x)$$

Another possibility is $f(x) = 3$ and $g(x) = \frac{x^2}{3} + x + \frac{2}{3}$.

Self Check 5 Find two functions f and g such that $f - g = h$.

Now Try Exercise 47.

3. Evaluate Composite Functions

Often one quantity is a function of a second quantity that depends, in turn, on a third quantity. For example, the cost of a car trip is a function of the gasoline consumed. The amount of gasoline consumed, in turn, is a function of the number of miles driven. Such chains of dependence are analyzed mathematically as *composition of functions*.

Suppose that $y = f(x)$ and $y = g(x)$ define two functions. Any number x in the domain of g will produce a corresponding value $g(x)$ in the range of g. If $g(x)$ is in the domain of function f, then $g(x)$ can be substituted into f, and a corresponding value $f(g(x))$ will be determined. This two-step process defines a new function, called a **composite function**, denoted by $f \circ g$. (See Figure 3-52.)

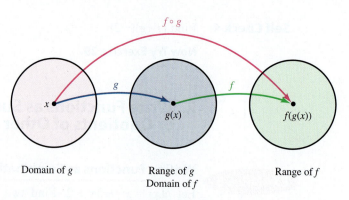

FIGURE 3-52

Composite Function	The **composite function** $f \circ g$ is defined by

$$(f \circ g)(x) = f(g(x))$$

The **domain** of $f \circ g$ consists of all those numbers in the domain of g for which $g(x)$ is in the domain of f.

To illustrate the previous definition, we consider the functions $f(x) = 5x + 1$ and $g(x) = 4x - 3$ and find $(f \circ g)(x)$ and $(g \circ f)(x)$.

$(f \circ g)(x) = f(g(x))$	$(g \circ f)(x) = g(f(x))$
$= f(4x - 3)$	$= g(5x + 1)$
$= 5(4x - 3) + 1$	$= 4(5x + 1) - 3$
$= 20x - 14$	$= 20x + 1$

> **Take Note**
>
> Note that for this example $(f \circ g)(x) \neq (g \circ f)(x)$.

Since we get different results, the composition of functions is not commutative.

We have seen that a function can be represented by a machine. If we put a number from the domain into the machine (the input), a number from the range comes out (the output). For example, if we put 2 into the machine shown in Figure 3-53(a), the number $f(2) = 5(2) - 2 = 8$ comes out. In general, if we put x into the machine shown in Figure 3-53(b), the value $f(x)$ comes out.

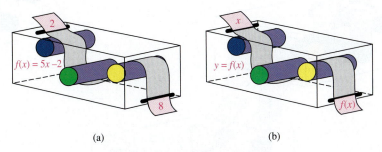

(a)   (b)

FIGURE 3-53

The function machines shown in Figure 3-54 illustrate the composition $f \circ g$. When we put a number x into the function g, the value $g(x)$ comes out. The value $g(x)$ then goes into function f, and $f(g(x))$ comes out.

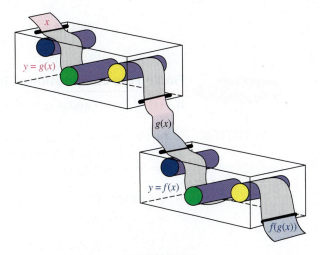

FIGURE 3-54

To further illustrate these ideas, we let $f(x) = 2x + 1$ and $g(x) = x - 4$.

- $(f \circ g)(9)$ means $f(g(9))$. In Figure 3-55(a), function g receives the number 9 and subtracts 4, and the number $g(x) = 5$ comes out. The 5 goes into the f function, which doubles it and adds 1. The final result, 11, is the output of the composite function $f \circ g$:

$$(f \circ g)(9) = f(g(9)) = f(5) = 2(5) + 1 = 11$$

- $(f \circ g)(x)$ means $f(g(x))$. In Figure 3-55(a), function g receives the number x and subtracts 4, and the number $x - 4$ comes out. The $x - 4$ goes into the f function, which doubles it and adds 1. The final result, $2x - 7$, is the output of the composite function $f \circ g$.

$$(f \circ g)(x) = f(g(x)) = f(x - 4) = 2(x - 4) + 1 = 2x - 7$$

- $(g \circ f)(-2)$ means $g(f(-2))$. In Figure 3-55(b), function f receives the number -2, doubles it and adds 1, and releases -3 into the g function. Function g subtracts 4 from -3 and releases a final output of -7. Thus,

$$(g \circ f)(-2) = g(f(-2)) = g(-3) = -3 - 4 = -7$$

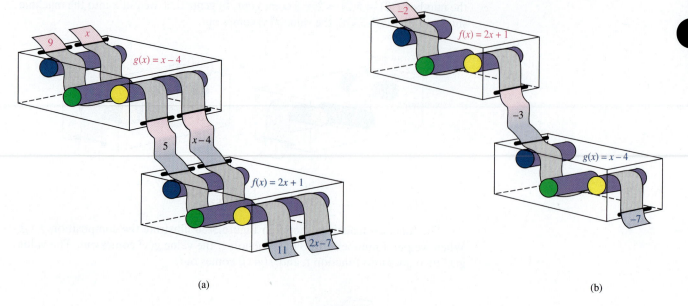

(a)   (b)

FIGURE 3-55

Finding the Composition of Two Functions

EXAMPLE 6

If $f(x) = 2x + 7$ and $g(x) = x^2 - 1$, find **a.** $(f \circ g)(x)$ **b.** $(g \circ f)(x)$

SOLUTION In part **a**, because $(f \circ g)(x)$ means $f(g(x))$, we will replace x in $f(x) = 2x + 7$ with $g(x)$. In part **b**, because $(g \circ f)(x)$ means $g(f(x))$ we will replace x in $g(x) = x^2 - 1$ with $f(x)$.

a. $(f \circ g)(x) = f(g(x))$

$\qquad\qquad = f(x^2 - 1)$ Substitute $x^2 - 1$ for $g(x)$.

$\qquad\qquad = 2(x^2 - 1) + 7$ Evaluate $f(x^2 - 1)$.

$\qquad\qquad = 2x^2 - 2 + 7$ Remove parentheses.

$\qquad\qquad = 2x^2 + 5$

b. $(g \circ f)(x) = g(f(x))$

$\qquad\qquad = g(2x + 7)$ Substitute $2x + 7$ for $f(x)$.

$\qquad\qquad = (2x + 7)^2 - 1$ Evaluate $g(2x + 7)$.

$\qquad\qquad = 4x^2 + 28x + 49 - 1$ Square the binomial.

$\qquad\qquad = 4x^2 + 28x + 48$

> **Tip**
>
> The function to the right of the composition symbol will be substituted into the function to the left of the composition symbol. To find $f \circ g$, we substitute g into f. To find $g \circ f$, we substitute f into g.

Self Check 6 If $f(x) = 2x + 7$ and $h(x) = x + 1$, find $(f \circ h)(x)$.

Now Try Exercise 59.

Evaluating the Composition of Two Functions

EXAMPLE 7

If $f(x) = 3x - 2$ and $g(x) = 3x^2 + 6x - 5$, find $(f \circ g)(-2)$.

SOLUTION Because $(f \circ g)(-2)$ means $f(g(-2))$, we first find $g(-2)$. We then find $f(g(-2))$.

Take Note

We can also determine $(f \circ g)(-2)$ by first finding $(f \circ g)(x)$ and then evaluating it at -2.

$$g(x) = 3x^2 + 6x - 5$$

$$g(-2) = 3(-2)^2 + 6(-2) - 5$$

$$= 3(4) - 12 - 5$$

$$= -5$$

$$(f \circ g)(-2) = f(g(-2))$$

$$= f(-5)$$

$$= 3(-5) - 2$$

$$= -17$$

Self Check 7   Find $(g \circ f)(-1)$.

Now Try Exercise 81.

4. Determine Domains of Composite Functions

To be in the domain of the composite function $f \circ g$, a number x has to be in the domain of g, and the output of g must be in the domain of f. Thus, the domain of $f \circ g$ consists of those inputs x that are in the domain of g and for which $g(x)$ is in the domain of f.

Strategy to Determine the Domain of $f \circ g$

To determine the domain of $(f \circ g)(x) = f(g(x))$, apply the following restrictions to the composition:

1. If x is not in the domain of g, it will not be in the domain of $f \circ g$.

2. Any x that has an output $g(x)$ that is not in the domain of f will not be in the domain of $f \circ g$.

EXAMPLE 8

Finding the Domain of Composite Functions

Let $f(x) = \sqrt{x}$ and $g(x) = x - 3$. Find the domain of   **a.** $f \circ g$ **b.** $g \circ f$

SOLUTION   We will first find the domains of $f(x)$ and $g(x)$. Then, we will find the domain of $f \circ g$ and $g \circ f$ by applying the restrictions stated above.

For $\sqrt{x}$ to be a real number, x must be a nonnegative real number. Thus, the domain of f is the interval $[0, \infty)$. Since any real number x can be an input into g, the domain of g is the interval $(-\infty, \infty)$.

a. The domain of $f \circ g$ is the set of real numbers x such that x is in the domain of g and $g(x)$ is in the domain of f. We have seen that all values of x are in the domain of g. However, $g(x)$ must be nonnegative, because $g(x)$ must be in the domain of f. So we must find the values of x such that $g(x)$ is greater than or equal to 0.

$$g(x) \geq 0 \qquad \text{\small\textit{g(x) must be nonnegative.}}$$

$$x - 3 \geq 0 \qquad \text{\small\textit{Substitute } x - 3 \text{ for } g(x).}$$

$$x \geq 3 \qquad \text{\small\textit{Add 3 to both sides.}}$$

Since $x \geq 3$, the domain of $f \circ g$ is the interval $[3, \infty)$.

b. The domain of $g \circ f$ is the set of real numbers x such that x is in the domain of f and $f(x)$ is in the domain of g. We have seen that only nonnegative values of x are in the domain of f. Because all values of $f(x)$ are in the domain of g, the domain of $g \circ f$ is the domain of f, which is the interval $[0, \infty)$.

Self Check 8   Find the domain of $f \circ f$.

Now Try Exercise 63.

Accent on Technology

Composition of Functions and Domain

Consider the two functions given in Example 8(a). We can use a graphing calculator to graph the composite function $f \circ g$. To enter the composition, we enter $y_1(y_2)$ using **VARS**. See Figure 3-56. From the graph, we can determine the domain of $f \circ g$. We see that the domain is $[3, \infty)$.

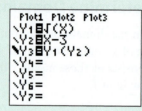
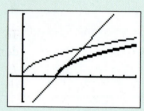

FIGURE 3-56

EXAMPLE 9   **Finding the Domain and Evaluating Composite Functions**

Let $f(x) = \dfrac{x + 3}{x - 2}$ and $g(x) = \dfrac{1}{x}$.

a. Find the domain of $f \circ g$.   **b.** Find $f \circ g$.

SOLUTION   We will first find the domains of $f(x)$ and $g(x)$. Then, we will find the domain of $f \circ g$ and $g \circ f$ by applying the restrictions stated earlier.

Since any real number except 2 can be an input into f, the domain of f is $(-\infty, 2) \cup (2, \infty)$.

For $\frac{1}{x}$ to be a real number, x cannot be 0. Thus, the domain of g is $(-\infty, 0) \cup (0, \infty)$.

a. The domain of $f \circ g$ is the set of real numbers x such that x is in the domain of g and $g(x)$ is in the domain of f. We have seen that all values of x but 0 are in the domain of g and that all values of $g(x)$ but 2 are in the domain of f. So we must exclude 0 from the domain of $f \circ g$ and all values of x where $g(x) = 2$. To find the excluded values, we proceed as follows:

$$g(x) = 2$$

$$\frac{1}{x} = 2 \qquad \text{Substitute } \frac{1}{x} \text{ for } g(x).$$

$$1 = 2x \qquad \text{Since } x \neq 0, \text{ we can multiply both sides by } x.$$

$$x = \frac{1}{2} \qquad \text{Divide both sides by 2.}$$

The domain of $f \circ g$ is the set of all real numbers except 0 and $\frac{1}{2}$, which is $(-\infty, 0) \cup \left(0, \frac{1}{2}\right) \cup \left(\frac{1}{2}, \infty\right)$.

b. To find $f \circ g$, we proceed as follows.

$$(f \circ g)(x) = f(g(x))$$

$$= f\left(\frac{1}{x}\right) \qquad \text{Substitute } \frac{1}{x} \text{ for } g(x).$$

$$= \frac{\dfrac{1}{x} + 3}{\dfrac{1}{x} - 2} \qquad \text{Substitute } \frac{1}{x} \text{ for } x \text{ in } f.$$

$$= \frac{1 + 3x}{1 - 2x} \qquad \text{Multiply numerator and denominator by } x.$$

Thus, $(f \circ g)(x) = \frac{1 + 3x}{1 - 2x}$.

Self Check 9   Let $f(x) = \dfrac{x}{x - 1}$ and $g(x) = \dfrac{1}{x}$. **a.** Find the domain of $f \circ g$ **b.** Find $f \circ g$

Now Try Exercise 71.

> **Caution**
>
> Example 9 illustrates that the domain of the composite function $f \circ g$ **cannot** always be found by finding $f \circ g$ and analyzing it. In the example, we found that the domain is all real numbers, except for 0 and $\frac{1}{2}$. If we only analyze the form $f \circ g$, we would state the domain incorrectly as all real numbers except $\frac{1}{2}$.

5. Write Functions as Compositions

When we form a composite function $f \circ g$, we obtain a function h. It is possible to reverse this composition process and begin with a function h and express it as a composition of two functions. This process is called **decomposition**.

For example, consider $h(x) = (2x^2 - 6x + 3)^4$. The function h takes $2x^2 - 6x + 3$ and raises it to the fourth power. To write the function h as a composition of functions f and g, we can let

$$f(x) = x^4 \quad \text{and} \quad g(x) = 2x^2 - 6x + 3$$

Then

$$(f \circ g)(x) = f(g(x))$$

$$= f(2x^2 - 6x + 3)$$

$$= (2x^2 - 6x + 3)^4$$

$$= h(x)$$

> **Take Note**
>
> The result of a decomposition is not unique. There are several possibilities. In the example to the right, another possibility is $f(x) = x^2$ and $g(x) = (2x^2 - 6x + 3)^2$.

Writing a Function as a Composition of Two Functions

EXAMPLE 10   Let $h(x) = \sqrt{x + 1}$. Find two functions f and g such that $f \circ g = h$.

Tip

To write a function as a composition of two functions, you must think backward. Be sure and check your choices for f and g because you may have them reversed.

SOLUTION   Because h takes the square root of the algebraic function $x + 1$, we let $f(x) = \sqrt{x}$ and $g(x) = x + 1$.

$$f(x) = \sqrt{x} \quad \text{and} \quad g(x) = x + 1$$

We can check the composition $f \circ g$ to see that it gives the original function h.

$$(f \circ g)(x) = f(g(x))$$
$$= f(x + 1)$$
$$= \sqrt{x + 1}$$
$$= h(x)$$

Self Check 10   Let $h(x) = \sqrt[3]{x^2 - 5}$. Find functions f and g such that $f \circ g = h$.

Now Try Exercise 99.

6. Use Operations on Functions to Solve Problems

EXAMPLE 11

Solving an Application Using Composition of Functions

A laboratory sample is removed from a cooler at a temperature of 15°F. Technicians then warm the sample at the rate of 3°F per hour. Express the sample's temperature in degrees Celsius as a function of the time t (in hours) since it was removed from the cooler.

SOLUTION   We first write a Fahrenheit temperature function that represents the warming of the sample. We then use composition of functions to write degrees Celsius as a function of degrees Fahrenheit.

The temperature of the sample is 15°F when $t = 0$. Because the sample warms at 3°F per hour, it warms $3t°$ after t hours. Thus, the Fahrenheit temperature after t hours is given by the function

$$F(t) = 3t + 15 \qquad \text{\textcolor{red}{$F(t)$ is the Fahrenheit temperature and t represents the time in hours.}}$$

The Celsius temperature is a function of the Fahrenheit temperature $F(t)$, given by the formula

$$C(F(t)) = \frac{5}{9}(F(t) - 32)$$

To express the sample's Celsius temperature as a function of time, we find the composition function $C \circ F$.

$$(C \circ F)(t) = C(F(t))$$
$$= C(3t + 15) \qquad \qquad \text{\textcolor{red}{Substitute for $F(t)$.}}$$
$$= \frac{5}{9}[(3t + 15) - 32] \qquad \text{\textcolor{red}{Substitute $3t + 15$ for $F(t)$ in $C(F(t))$.}}$$
$$= \frac{5}{9}(3t - 17) \qquad \qquad \text{\textcolor{red}{Simplify.}}$$
$$= \frac{15}{9}t - \frac{85}{9}$$
$$= \frac{5}{3}t - \frac{85}{9}$$

Self Check 11   Find the $(C \circ F)(10)$.

Now Try Exercise 119.

Exercises 3.4

Getting Ready
You should be able to complete these vocabulary and concept statements before you proceed to the practice exercises.

Fill in the blanks.

1. $(f + g)(x) =$ _____
2. $(f - g)(x) =$ _____
3. $(f \cdot g)(x) =$ _____
4. $(f/g)(x) =$ _____, where $g(x) \neq 0$
5. The domain of $f + g$ is the _____ of the domains of f and g.
6. $(f \circ g)(x) =$ _____
7. $(g \circ f)(x) =$ _____
8. To determine $(f \circ g)(-5)$, first find _____.
9. Composition of functions is not _____.
10. To be in the domain of the composite function $f \circ g$, a number x has to be in the _____ of g, and the output of g must be in the _____ of f.

Practice
Let $f(x) = 2x + 1$ and $g(x) = 3x - 2$. Find each function and its domain.

11. $f + g$ 12. $f - g$

13. $f \cdot g$ 14. f/g

Let $f(x) = x^2 + x$ and $g(x) = x^2 - 1$. Find each function and its domain.

15. $f - g$ 16. $f + g$

17. f/g 18. $f \cdot g$

Let $f(x) = x^2 - 7x + 3$ and $g(x) = x^2 - 5x + 6$. Find each function and its domain.

19. $f + g$
20. $f - g$
21. $f \cdot g$
22. f/g

Let $f(x) = x^2 - 7$ and $g(x) = \sqrt{x}$. Find each function and its domain.

23. $f + g$ 24. $f - g$

25. f/g 26. $f \cdot g$

Let $f(x) = x^2 - 1$ and $g(x) = 3x - 2$. Find each value, if possible.

27. $(f + g)(2)$

28. $(f + g)(-3)$

29. $(f - g)(0)$

30. $(f - g)(-5)$

31. $(f \cdot g)(2)$

32. $(f \cdot g)(-1)$

33. $(f/g)\left(\dfrac{2}{3}\right)$

34. $(f/g)(0)$

Let $f(x) = 2x - 5$ and $g(x) = \sqrt[3]{x}$. Find each value.

35. $(f + g)(8)$

36. $(f + g)(-8)$

37. $(f - g)(-27)$

38. $(f - g)(8)$

39. $(f \cdot g)(-1)$

40. $(f \cdot g)(1)$

41. $(f/g)\left(\dfrac{1}{8}\right)$

42. $(f/g)\left(-\dfrac{1}{8}\right)$

Find two functions f and g such that $h(x)$ can be expressed as the function indicated. Several answers are possible.

43. $h(x) = 3x^2 + 2x; f + g$

44. $h(x) = 3x^2; f \cdot g$

45. $h(x) = \dfrac{3x^2}{x^2 - 1}; f/g$

46. $h(x) = 5x + x^2; f - g$

47. $h(x) = x(3x^2 + 1); f - g$

48. $h(x) = (3x - 2)(3x + 2); f + g$

49. $h(x) = x^2 + 7x - 18; f \cdot g$

50. $h(x) = 5x^5; f/g$

Let $f(x) = 3x$ and $g(x) = x + 1$. Determine the domain of each composite function and then find the composite function.

51. $f \circ g$

52. $g \circ f$

53. $f \circ f$

54. $g \circ g$

Let $f(x) = x^2$ and $g(x) = 2x$. Determine the domain of each composite function and then find the composite function.

55. $g \circ f$

56. $f \circ g$

57. $g \circ g$

58. $f \circ f$

Let $f(x) = 2x^2 - 3x + 7$ and $g(x) = 4x - 1$. Determine the domain of the composite function and then find the composite function.

59. $f \circ g$

60. $g \circ f$

61. $f \circ f$

62. $g \circ g$

Let $f(x) = \sqrt{x}$ and $g(x) = x + 1$. Determine the domain of each composite function and then find the composite function.

63. $f \circ g$

64. $g \circ f$

65. $f \circ f$

66. $g \circ g$

Let $f(x) = \sqrt{x + 1}$ and $g(x) = x^2 - 1$. Determine the domain of each composite function and then find the composite function.

67. $g \circ f$

68. $f \circ g$

69. $g \circ g$

70. $f \circ f$

Let $f(x) = \frac{1}{x - 1}$ and $g(x) = \frac{1}{x - 2}$. Determine the domain of each composite function and then find the composite function.

71. $f \circ g$

72. $g \circ f$

73. $f \circ f$

74. $g \circ g$

Let $f(x) = 2x - 5$ and $g(x) = 5x - 2$. Find each value.

75. $(f \circ g)(2)$

76. $(f \circ g)(-2)$

77. $(g \circ f)(-3)$

78. $(g \circ f)(3)$

79. $(f \circ f)\left(-\dfrac{1}{2}\right)$

80. $(g \circ g)\left(\dfrac{3}{5}\right)$

Let $f(x) = 3x^2 - 2$ and $g(x) = 4x + 4$. Find each value.

81. $(f \circ g)(-3)$

82. $(f \circ g)\left(\dfrac{1}{4}\right)$

83. $(g \circ f)(3)$

84. $(g \circ f)\left(\dfrac{1}{3}\right)$

85. $(f \circ f)\left(\sqrt{3}\right)$

86. $(g \circ g)(-4)$

Let $f(x) = \dfrac{2}{x}$ and $g(x) = \sqrt{x}$. Find each value.

87. $(f \circ g)(100)$

88. $(f \circ g)(8)$

89. $(g \circ f)\left(\dfrac{1}{32}\right)$

90. $(g \circ f)(8)$

91. $(g \circ g)\left(\dfrac{81}{256}\right)$

92. $(f \circ f)\left(-\dfrac{3}{5}\right)$

Find two functions f and g such that the composition
f ∘ g = h expresses the given correspondence. Several
answers are possible.

93. $h(x) = 3x - 2$

94. $h(x) = 7x - 5$

95. $h(x) = x^2 - 2$

96. $h(x) = x^3 - 3$

97. $h(x) = (x - 2)^2$

98. $h(x) = (x - 3)^3$

99. $h(x) = \sqrt{x + 2}$

100. $h(x) = \dfrac{1}{x - 5}$

101. $h(x) = \sqrt{x} + 2$

102. $h(x) = \dfrac{1}{x} - 5$

103. $h(x) = x$

104. $f(x) = 3$

Use the graphs of functions f and g to answer each problem.

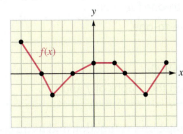

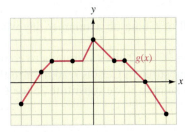

105. $(f + g)(-4)$

106. $(f - g)(1)$

107. $(f \cdot g)(5)$

108. $(f/g)(-1)$

109. $(f \circ g)(3)$

110. $(g \circ f)(2)$

111. $(f \circ f)(-2)$

112. $(g \circ g)(-5)$

Use the tables of values of f and g to answer each problem.

x	$f(x)$
2	4
4	9
6	13
13	17

x	$g(x)$
0	0
2	4
3	9
4	16

113. $(f + g)(2)$

114. $(f/g)(4)$

115. $(f \circ g)(2)$

116. $(g \circ f)(2)$

Applications

117. DVD camcorder Suppose that the functions $R(x) = 300x$ and $C(x) = 60,000 + 40x$ model a company's monthly revenue and cost for producing and selling DVD camcorders.

  a. Find $(R - C)(x)$, the function that models the monthly profit, $P(x)$.

  b. Find the company's profit if 500 camcorders are produced and sold in one month.

©Volt Collection/Shutterstock.com

118. TV screen The height of the television screen shown is 13 inches.

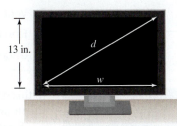

  a. Write a formula to find the area of the viewing screen.

  b. Use the Pythagorean Theorem to write a formula to find the width w of the screen.

  c. Write a formula to find the area of the screen as a function of the diagonal d.

119. Area of an oil spill Suppose an oil spill from a tanker is spreading in the shape of a circular ripple. If the function $d(t) = 3t$ represents the diameter of the spill in inches at time t minutes, express the area, A, of the oil spill as a function of time. Find the area of the oil spill after 2 hours. Round to one decimal place.

120. **Area of a square** Write a formula for the area A of a square in terms of its perimeter P.

121. **Perimeter of a square** Write a formula for the perimeter P of a square in terms of its area A.

122. **Ceramics** When the temperature of a pot in a kiln is 1200°F, an artist turns off the heat and leaves the pot to cool at a controlled rate of 81°F per hour. Express the temperature of the pot in degrees Celsius as a function of the time t (in hours) since the kiln was turned off.

Discovery and Writing

123. Describe how to determine the composition of two functions.

124. Explain how to determine the domain of the composition of two functions.

125. Let $f(x) = 3x$. Show that $(f + f)(x) = f(x + x)$.

126. Let $g(x) = x^2$. Show that $(g + g)(x) \neq g(x + x)$.

127. Let $f(x) = \frac{x-1}{x+1}$. Find $(f \circ f)(x)$.

128. Let $g(x) = \frac{x}{x-1}$. Find $(g \circ g)(x)$.

Let $f(x) = x^2 - x$, $g(x) = x - 3$, *and* $h(x) = 3x$. *Use a graphing calculator to graph both functions on the same axes. Write a brief paragraph summarizing your observations.*

129. f and $f \circ g$

130. f and $g \circ f$

131. f and $f \circ h$

132. f and $h \circ f$

Critical Thinking

Determine if the statement is true or false. If the statement is false, then correct it and make it true.

133. $(f - g)(x) = (g - f)(x)$

134. $(f \cdot g)(x) = (g \cdot f)(x)$

135. $(f \circ g)(x)$ sometimes equals $(g \circ f)(x)$

136. If $f(x) = x^2$, then $(f \circ f \circ f)(x) = x^6$

137. If $g(x) = -x^3$, then $(g \circ g \circ g)(x) = -x^9$

138. If $f(x) = \frac{5}{x}$, then $(f \circ f \circ f \circ f)(x) = x$

139. If $f(x) = x^{975}$ and $g(x) = x^{864}$ then $(f \circ g)(-1) = 1$

140. If $f(x) = \sqrt[99]{x}$ and $g(x) = \sqrt[77]{x}$ then $(f \circ g)(-1) = 1$

3.5 Inverse Functions

In this section, we will learn to

1. Understand the definition of a one-to-one function.

2. Determine whether a function is one-to-one.

3. Verify inverse functions.

4. Find the inverse of a one-to-one function.

5. Understand the relationship between the graphs of f and f^{-1}.

©eXpose/Shutterstock.com

In this section, we will discuss inverse functions. A function and its inverse do opposite things.

Suppose we climb the Great Wall of China on a summer day when the temperature reaches a high of 35°C.

The linear function defined by $F = \frac{9}{5}C + 32$ gives a formula to convert degrees Celsius to degrees Fahrenheit. If we substitute a Celsius reading into the formula, a Fahrenheit reading comes out. For example, if we substitute 35 for C, we obtain a Fahrenheit reading of 95°:

$$F = \frac{9}{5}C + 32$$

$$= \frac{9}{5}(35) + 32$$

$$= 63 + 32$$

$$= 95$$

If we want to find a Celsius reading from a Fahrenheit reading, we need a formula into which we can substitute a Fahrenheit reading and have a Celsius reading come out. Such a formula is $C = \frac{5}{9}(F - 32)$, which takes the Fahrenheit reading of 95° and turns it back into a Celsius reading of 35°.

$$C = \frac{5}{9}(F - 32)$$

$$= \frac{5}{9}(95 - 32)$$

$$= \frac{5}{9}(63)$$

$$= 35$$

The functions defined by these two formulas do opposite things. The first turns 35°C into 95°F, and the second turns 95°F back into 35°C. Such functions are called *inverse functions*.

Some functions have inverses that are functions and some do not. To guarantee that the inverse of a function will also be a function, we must know that the function is *one-to-one*.

1. Understand the Definition of a One-to-One Function

In this section, we will find inverses of functions that are one-to-one. *One-to-one functions* are functions whose inverses are also functions.

We now examine what it means for a function to be one-to-one. Consider the following two functions:

Function 1: To each student, there corresponds exactly one eye color

Function 2: To each student, there corresponds exactly one college identification number

Function 1 **is not a one-to-one function** because two different students can have the same eye color.

Function 2 **is a one-to-one function** because two different students will always have two different ID numbers.

Recall that each element x in the domain of a function has a single output y. For some functions, different numbers x in the domain can have the same output. See Figure 3-57(a). For other functions, called **one-to-one functions,** different numbers x have different outputs. See Figure 3-57(b).

Take Note

All functions are not one-to-one functions. Each student has exactly one eye color, which represents a function. Because two people can have the same eye color, the function is not a one-to-one function. Note that both students shown have green eyes.

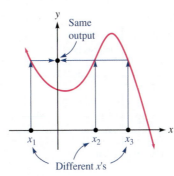

Not a one-to-one function

(a)

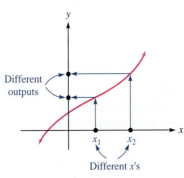

A one-to-one function

(b)

FIGURE 3-57

One-to-One Functions	A function f from a set **X** to a set **Y** is called a **one-to-one function** if and only if different numbers in the domain of f have different outputs in the range of f.

The previous definition implies that if x_1 and x_2 are two numbers in the domain of f and $x_1 \neq x_2$, then $f(x_1) \neq f(x_2)$.

2. Determine Whether a Function Is One-to-One

Determining Whether a Function Is One-to-One

EXAMPLE 1

Determine whether each function is one-to-one.
a. $f(x) = x^4 + x^2$ **b.** $f(x) = x^3$

SOLUTION

We will examine the functions and determine whether the definition of a one-to-one function applies. If different x-values always produce different y-values, the function is one-to-one.

a. The function $f(x) = x^4 + x^2$ is not one-to-one, because different numbers in the domain have the same output. For example, 2 and -2 have the same output: $f(2) = f(-2) = 20$.

b. The function $f(x) = x^3$ is one-to-one, because different numbers x produce different outputs $f(x)$. This is because different numbers have different cubes.

Self Check 1   Determine whether $f(x) = \sqrt{x}$ is one-to-one.

Now Try Exercise 15.

A **Horizontal Line Test** can be used to determine whether the graph of a function represents a one-to-one function. If every horizontal line that intersects the graph of a function does so exactly once, the function passes the Horizontal Line Test and is one-to-one. See Figure 3-58(a). If any horizontal line intersects the graph of a function more than once, the function fails the Horizontal Line Test and is not one-to-one. See Figure 3-58(b).

> **Take Note**
>
> A one-to-one function satisfies both the Horizontal and Vertical Line Tests.

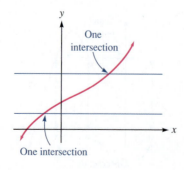

A one-to-one function

(a)

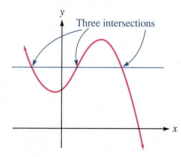

Not a one-to-one function

(b)

FIGURE 3-58

Using the Horizontal Line Test

EXAMPLE 2

Use the Horizontal Line Test to determine whether each graph represents a one-to-one function.

a.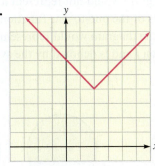

b.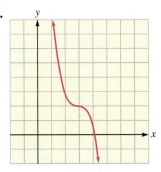

SOLUTION   We will use the Horizontal Line Test and draw many horizontal lines. If every horizontal line that intersects the graph does so exactly once, the function is one-to-one. If any horizontal line intersects the graph more than once, the function is not one-to-one.

a. Because the horizontal line drawn in Figure 3-59 intersects the graph in two places, we know that the function fails the Horizontal Line Test and is not a one-to-one function.

b. Several horizontal lines are drawn in Figure 3-60, and each one intersects the graph exactly once. We conclude that the graph passes the Horizontal Line Test and represents a one-to-one function.

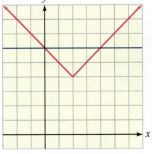

FIGURE 3-59

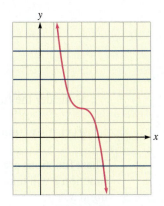

FIGURE 3-60

Self Check 2   Determine whether the graph below represents a one-to-one function.

Now Try Exercise 25.

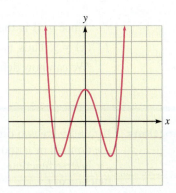

3. Verify Inverse Functions

Figure 3-61(a) illustrates a function f from set **X** to set **Y**. Since three arrows point to a single y, the function f is not one-to-one. If the arrows in Figure 3-61(a) were reversed, the diagram would not represent a function.

If the arrows of the one-to-one function f in Figure 3-61(b) were reversed, as in Figure 3-61(c), the diagram would represent a function. This function is called the **inverse of function f** and is denoted by the symbol f^{-1}.

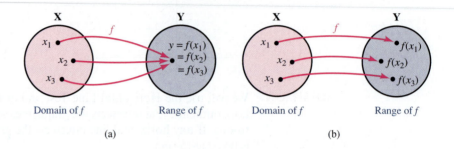

(a)   (b)

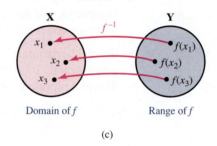

(c)

FIGURE 3-61

Consider the functions:

$$f(x) = 4x \quad \text{and} \quad g(x) = \frac{x}{4}$$

These functions are inverses of each other because f multiplies any input x by 4 and function g will take the result and divide it by 4. The final result will be the original input x.

We can show that the composition of these functions (in either order) is the identity function.

$$(f \circ g)(x) = f(g(x)) = f\left(\frac{x}{4}\right) = 4\left(\frac{x}{4}\right) = x \quad \text{and} \quad (g \circ f)(x) = g(f(x)) = g(4x) = \frac{4x}{4} = x$$

Since $g(x)$ is the inverse of $f(x)$, we can write $g(x)$ using inverse notation as $f^{-1}(x) = \frac{x}{4}$. Thus, $(f \circ f^{-1}) = x$ and $(f^{-1} \circ f)(x) = x$.

We can now define inverse functions.

Inverse Functions If f and g are two one-to-one functions such that $(f \circ g)(x) = x$ for every x in the domain of g and $(g \circ f)(x) = x$ for every x in the domain of f, then f and g are **inverse functions**. Function g can be denoted as f^{-1} and is called the **inverse function of f**.

We can also list two important properties of one-to-one functions.

**Properties of a
One-to-One Function**

Property 1: If f is a one-to-one function, there is a one-to-one function $f^{-1}(x)$ such that

$$(f^{-1} \circ f)(x) = x \quad \text{and} \quad (f \circ f^{-1})(x) = x.$$

Property 2: The domain of f is the range of f^{-1} and the range of f is the domain of f^{-1}.

Figure 3-62 shows a one-to-one function f and its inverse f^{-1}. To the number x in the domain of f, there corresponds an output $f(x)$ in the range of f. Since $f(x)$ is in the domain of f^{-1}, the output for $f(x)$ under the function f^{-1} is $f^{-1}(f(x)) = x$. Thus, $(f^{-1} \circ f)(x) = f^{-1}(f(x)) = x$.

> **Take Note**
>
> To show that one function is the inverse of another, we must show that their compositions are the **identity function**, x.

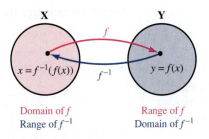

Domain of f
Range of f^{-1}

Range of f
Domain of f^{-1}

FIGURE 3-62

Verifying That Two Functions Are Inverses

EXAMPLE 3

Verify that $f(x) = x^3$ and $g(x) = \sqrt[3]{x}$ are inverse functions.

SOLUTION   To show that f and g are inverse functions, we must show that $f \circ g$ and $g \circ f$ are x, the identity function.

$$(f \circ g)(x) = f(g(x)) = f(\sqrt[3]{x}) = (\sqrt[3]{x})^3 = x$$

$$(g \circ f)(x) = g(f(x)) = g(x^3) = \sqrt[3]{x^3} = x$$

Because g is the inverse of f, we can use inverse notation and write $g(x) = \sqrt[3]{x}$ as $f^{-1}(x) = \sqrt[3]{x}$. Because f is the inverse of g, we can use inverse notation and write $f(x) = x^3$ as $g^{-1}(x) = x^3$.

Self Check 3   If $x \geq 0$, are $f(x) = x^2$ and $g(x) = \sqrt{x}$ inverse functions?

Now Try Exercise 31.

4. Find the Inverse of a One-to-One Function

If f is the one-to-one function $y = f(x)$, then f^{-1} reverses the correspondence of f. That is, if $f(a) = b$, then $f^{-1}(b) = a$. To determine f^{-1}, we follow these steps.

**Strategy for Finding f^{-1}
from a Given Function $f(x)$**

Step 1: Replace $f(x)$ with y.

Step 2: Interchange the variables x and y.

Step 3: Solve the resulting equation for y.

Step 4: Replace y with $f^{-1}(x)$.

Once $f^{-1}(x)$ is determined, it should be verified by showing that $(f \circ f^{-1})(x) = x$ and $(f^{-1} \circ f)(x) = x$.

EXAMPLE 4 ### Finding the Inverse of a One-to-One Function

Find the inverse of $f(x) = \dfrac{3}{2}x + 2$ and verify the result.

SOLUTION We will use the strategy given above to find f^{-1}. We then will verify the result by showing that $(f \circ f^{-1})(x) = x$ and $(f^{-1} \circ f)(x) = x$.

To find f^{-1}, we use the following steps.

Step 1: Replace $f(x)$ with y.

$$f(x) = \frac{3}{2}x + 2$$

$$y = \frac{3}{2}x + 2$$

Step 2: Interchange the variables x and y.

$$x = \frac{3}{2}y + 2$$

Step 3: Solve the resulting equation for y.

$$x = \frac{3}{2}y + 2$$

$$2x = 3y + 4 \qquad \text{Multiply both sides by 2.}$$

$$2x - 4 = 3y \qquad \text{Subtract 4 from both sides.}$$

$$y = \frac{2x - 4}{3} \qquad \text{Divide both sides by 3.}$$

Step 4: Replace y with $f^{-1}(x)$.

$$y = \frac{2x - 4}{3}$$

$$f^{-1}(x) = \frac{2x - 4}{3}$$

The inverse of $f(x) = \dfrac{3}{2}x + 2$ is $f^{-1}(x) = \dfrac{2x - 4}{3}$.

To verify the result, we will use $f(x) = \dfrac{3}{2}x + 2$ and $f^{-1}(x) = \dfrac{2x - 4}{3}$ and show that $(f \circ f^{-1})(x) = x$ and $(f^{-1} \circ f)(x) = x$.

$$(f \circ f^{-1})(x) = f(f^{-1}(x))$$

$$= f\left(\frac{2x - 4}{3}\right)$$

$$= \frac{3}{2}\left(\frac{2x - 4}{3}\right) + 2$$

$$= x - 2 + 2$$

$$= x$$

$$(f^{-1} \circ f)(x) = f^{-1}(f(x))$$

$$= f^{-1}\left(\frac{3}{2}x + 2\right)$$

$$= \frac{2\left(\frac{3}{2}x + 2\right) - 4}{3}$$

$$= \frac{3x + 4 - 4}{3}$$

$$= x$$

Take Note

After completing Step 3, if y does not represent a function of x, the process ends and f does not have an inverse.

Tip

You might be tempted to determine the inverse of a one-to-one function "mentally." That works for a basic function like $f(x) = x^3$. Its inverse is the cube root function, $f^{-1}(x) = \sqrt[3]{x}$. For more complicated functions, always apply the four steps.

Self Check 4   Find $f(2)$. Then find $f^{-1}(5)$. Explain the significance of the results.

Now Try Exercise 39.

5. Understand the Relationship between the Graphs of f and f^{-1}

Because we interchange the positions of x and y to find the inverse of a function, the point (b, a) lies on the graph of $y = f^{-1}(x)$ whenever the point (a, b) lies on the graph of $y = f(x)$. Thus, the graph of a function and its inverse are reflections of each other about the line $y = x$.

Finding f^{-1} and Graphing Both f and f^{-1}

EXAMPLE 5

Find the inverse of $f(x) = x^3 + 3$. Graph the function and its inverse on the same set of coordinate axes.

SOLUTION   We will find the inverse of the function $f(x)$ using the strategy given in the section. We will use translations to graph both f and f^{-1}.

We first find f^{-1} and proceed as follows:

Step 1: Replace $f(x)$ with y.

$$f(x) = x^3 + 3$$
$$y = x^3 + 3$$

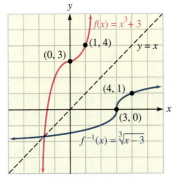

FIGURE 3-63

Step 2: Interchange the variables x and y.

$$x = y^3 + 3$$

Step 3: Solve the resulting equation for y.

$$x - 3 = y^3$$
$$y = \sqrt[3]{x - 3}$$

Step 4: Replace y with $f^{-1}(x)$.

$$f^{-1}(x) = \sqrt[3]{x - 3}$$

We now graph f and f^{-1}. To graph $f(x) = x^3 + 3$, we translate the graph of $y = x^3$ vertically upward 3 units. To graph $f^{-1}(x) = \sqrt[3]{x - 3}$ we translate the graph of $y = \sqrt[3]{x}$ horizontally 3 units to the right. The graphs of f and f^{-1} are shown in Figure 3-63. In the graph that appears in Figure 3-63, the line $y = x$ is the axis of symmetry.

Take Note

We can also graph f and f^{-1} by completing a table of solutions. The x and y columns of f can be reversed to obtain the table of solutions for f^{-1}.

Self Check 5   Find $f(2)$. Then find $f^{-1}(11)$. Explain the significance of the result.

Now Try Exercise 57.

In the next example, we will consider a function that is not one-to-one but becomes so when we restrict its domain. By restricting the domain of the function and making it one-to-one, we are able to find its inverse and examine the function and its inverse graphically.

EXAMPLE 6

Restricting the Domain of f to Make It One-to-One; Finding f^{-1}; Graphing f and f^{-1}; Stating Domain and Range

The function $y = f(x) = x^2 + 3$ is not one-to-one. However, it becomes one-to-one when we restrict its domain to the interval $(-\infty, 0]$. Under this restriction,

a. Find the inverse of f.

b. Graph each function and state each one's domain and range.

SOLUTION   We will find f^{-1} by using the four-step strategy given in the section. We will then graph f and f^{-1} by using translations and then identify the domain and range from the graphs of each.

a. We first find f^{-1} and follow these steps:

Step 1: Replace $f(x)$ with y.

$$f(x) = x^2 + 3 \quad (x \le 0)$$
$$y = x^2 + 3$$

Step 2: Interchange the variables x and y.

$$x = y^2 + 3 \quad (y \le 0) \qquad \textcolor{red}{\text{Interchange } x \text{ and } y.}$$

Step 3: Solve the resulting equation for y.

$$x - 3 = y^2$$

To solve this equation for y, we take the square root of both sides. Because $y \le 0$, we have $-\sqrt{x - 3} = y \quad (y \le 0)$

Step 4: Replace y with $f^{-1}(x)$.

The inverse of f is defined by $f^{-1}(x) = -\sqrt{x - 3}$.

b. We graph the function $f(x) = x^2 + 3$ with domain $(-\infty, 0]$ by translating the graph of the parabola $y = x^2$ with domain $(-\infty, 0]$ vertically upward 3 units. From the graph, we see that the y coordinates are 3 and above and thus the range is the interval $[3, \infty)$. (See Figure 3-64.)

We graph the function $f^{-1}(x) = -\sqrt{x - 3}$ by translating the graph of $y = \sqrt{x}$ horizontally to the right 3 units and then reflecting the graph about the x-axis. It has domain $[3, \infty)$ and range $(-\infty, 0]$. (See Figure 3-64.) Note that the line of symmetry is shown and is $y = x$.

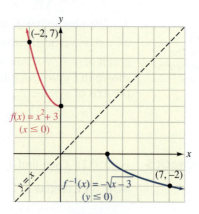

FIGURE 3-64

Domain of f and **Range** of f^{-1}: $(-\infty, 0]$

Range of f and **Domain** of f^{-1}: $[3, \infty)$

Self Check 6   Find the inverse of f when its domain is restricted to the interval $[0, \infty)$.

Now Try Exercise 69.

If a function is defined by the equation $y = f(x)$, we can often find the domain of f by inspection. Finding the range can be more difficult. One way to find the range of f is to find the domain of f^{-1}.

EXAMPLE 7

Using the Domain of $f^{-1}(x)$ to Find the Range of $f(x)$

Find the domain and range of $f(x) = \frac{2}{x} + 3$. Find its range by finding the domain of $f^{-1}(x)$.

SOLUTION   We will find the domain of $f(x) = \frac{2}{x} + 3$ by identifying the values of x that make the function undefined. We will then find $f^{-1}(x)$ and find its domain. The domain of $f^{-1}(x)$ will be the range of $f(x)$.

Because x cannot be 0, the domain of f is $(-\infty, 0) \cup (0, \infty)$. Next, we find $f^{-1}(x)$.

Step 1: Replace $f(x)$ with y.

$$f(x) = \frac{2}{x} + 3$$

$$y = \frac{2}{x} + 3$$

Step 2: Interchange the variables x and y.

$$x = \frac{2}{y} + 3 \qquad \text{Interchange } x \text{ and } y.$$

Step 3: Solve the resulting equation for y.

$$xy = 2 + 3y \qquad \text{Multiply both sides by } y.$$
$$xy - 3y = 2 \qquad \text{Subtract } 3y \text{ from both sides.}$$
$$y(x - 3) = 2 \qquad \text{Factor out } y.$$
$$y = \frac{2}{x - 3} \qquad \text{Divide both sides by } x - 3.$$

Step 4: Replace y with $f^{-1}(x)$.

$$f^{-1}(x) = \frac{2}{x - 3}$$

The domain of $f^{-1}(x) = \frac{2}{x-3}$ is $(-\infty, 3) \cup (3, \infty)$ because x cannot be 3. Because the range of f is the domain of f^{-1}, the range of f is $(-\infty, 3) \cup (3, \infty)$.

Self Check 7   Find the range of $f(x) = \frac{3}{x} - 1$.

Now Try Exercise 77.

Self Check Answers

1. yes   **2.** no   **3.** yes   **4.** 5; 2   **5.** 11; 2   **6.** $f^{-1}(x) = \sqrt{x - 3}$
7. $(-\infty, -1) \cup (-1, \infty)$

Exercises 3.5

Getting Ready

You should be able to complete these vocabulary and concept statements before you proceed to the practice exercises.

Fill in the blanks.

1. If different numbers in the domain of a function have different outputs, the function is called a _____ function.

2. If every _____ line intersects the graph of a function only once, the function is one-to-one.

3. Two functions f and g are inverses if their composition in either order is the _____ function.

4. The graph of a function and its inverse are reflections of each other about the line _____.

Practice

Determine whether each function is one-to-one.

5. $f(x) = 5$

6. $f(x) = -5$

7. $f(x) = 3x$

8. $f(x) = \frac{1}{2}x$

9. $f(x) = x^2 + 3$

10. $f(x) = x^4 - x^2$

11. $f(x) = x^3 + 5$

12. $f(x) = (x - 1)^3$

13. $f(x) = x^3 - x$

14. $f(x) = x^2 - x$

15. $f(x) = |x|$

16. $f(x) = |x - 3|$

17. $\sqrt{x}$

18. $f(x) = \sqrt{x - 5}$

19. $f(x) = \sqrt[3]{x}$

20. $f(x) = \sqrt[3]{x} - 2$

21. $f(x) = (x - 2)^2; \; x \geq 2$

22. $f(x) = \frac{1}{x}$

Use the Horizontal Line Test to determine whether each graph represents a one-to-one function.

23.

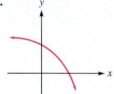

24.

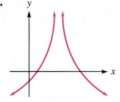

25.

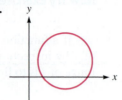

26.

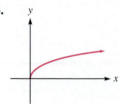

27.

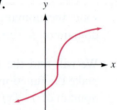

28.

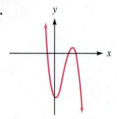

Verify that the functions are inverses by showing that $f \circ g$ and $g \circ f$ are the identity function.

29. $f(x) = 5x$ and $g(x) = \frac{1}{5}x$

30. $f(x) = 4x + 5$ and $g(x) = \frac{x - 5}{4}$

31. $f(x) = x^3 + 8$ and $g(x) = \sqrt[3]{x - 8}$

32. $f(x) = 8x^3$ and $g(x) = \frac{\sqrt[3]{x}}{2}$

33. $f(x) = \sqrt[5]{x} - 1$ and $g(x) = (x + 1)^5$

34. $f(x) = x^5 - 2$ and $g(x) = \sqrt[5]{x + 2}$

35. $f(x) = \frac{x + 1}{x}$ and $g(x) = \frac{1}{x - 1}$

36. $f(x) = \frac{x + 1}{x - 1}$ and $g(x) = \frac{x + 1}{x - 1}$

Each equation defines a one-to-one function f. Determine f^{-1} and verify that $f \circ f^{-1}$ and $f^{-1} \circ f$ are both the identity function.

37. $f(x) = 3x$

38. $f(x) = \frac{1}{3}x$

39. $f(x) = 3x + 2$

40. $f(x) = 2x - 5$

41. $f(x) = x^3 + 2$

42. $f(x) = (x + 2)^3$

43. $f(x) = \sqrt[5]{x}$

44. $f(x) = \sqrt[5]{x} + 4$

45. $f(x) = \frac{1}{x + 3}$

46. $f(x) = \frac{1}{x - 2}$

47. $f(x) = \dfrac{1}{2x}$ **48.** $f(x) = \dfrac{1}{x^3}$

59. $f(x) = (x - 6)^3$ **60.** $f(x) = -x^3 + 4$

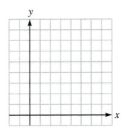 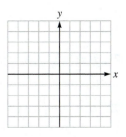

Find the inverse of each one-to-one function and graph both the function and its inverse on the same set of coordinate axes.

49. $y = 5x$ **50.** $y = \dfrac{3}{2}x$

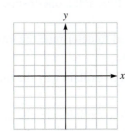 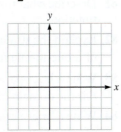

61. $f(x) = \dfrac{1}{2x}$

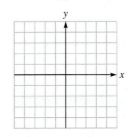

51. $y = 2x - 4$ **52.** $y = \dfrac{3}{2}x - 2$

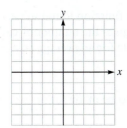 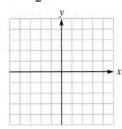

62. $f(x) = \dfrac{1}{x - 3}$

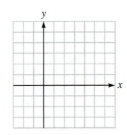

53. $x - y = 2$ **54.** $x + y = 0$

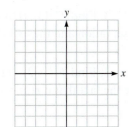 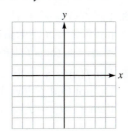

63. $f(x) = \dfrac{x + 1}{x - 1}$ **64.** $f(x) = \dfrac{x - 1}{x}$

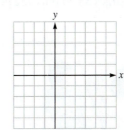 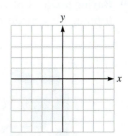

55. $2x + y = 4$ **56.** $3x + 2y = 6$

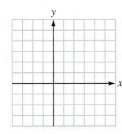 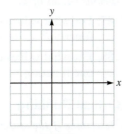

The function f defined by the given equation is one-to-one on the given domain. Find $f^{-1}(x)$.

65. $f(x) = x^2 + 5$ $(x \geq 0)$

66. $f(x) = x^2 - 5$ $(x \geq 0)$

57. $f(x) = \sqrt[3]{x - 4}$ **58.** $f(x) = \sqrt[3]{x + 3}$

67. $f(x) = 4x^2$ $(x \geq 0)$

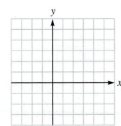 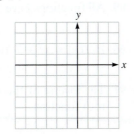

68. $f(x) = -4x^2$ $(x \geq 0)$

69. $f(x) = x^2 - 3$ $(x \leq 0)$

70. $f(x) = \dfrac{1}{x^2}$ $(x > 0)$

71. $f(x) = x^4 - 8$ $(x \geq 0)$

72. $f(x) = \dfrac{-1}{x^4}$ $(x < 0)$

73. $f(x) = \sqrt{4 - x^2}$ $(0 \leq x \leq 2)$

74. $f(x) = \sqrt{x^2 - 1}$ $(x \leq -1)$

Find the domain and the range of f. Find the range by finding the domain of f^{-1}.

75. $f(x) = \dfrac{x}{x - 2}$

76. $f(x) = \dfrac{x - 2}{x + 3}$

77. $f(x) = \dfrac{1}{x} - 2$

78. $f(x) = \dfrac{3}{x} - \dfrac{1}{2}$

Applications

79. Buying pizza A pizzeria charges \$8.50 plus 75¢ per topping for a medium pizza.

©Syda Productions/Shutterstock.com

  a. Find a linear function that expresses the cost $f(x)$ of a medium pizza in terms of the number of toppings x.

  b. Find the cost of a pizza that has four toppings.

  c. Find the inverse of the function found in part (a) to find a formula that gives the number of toppings $f^{-1}(x)$ in terms of the cost x.

  d. If Josh has \$10, how many toppings can he afford?

80. Cell phone bills A phone company charges \$11 per month plus a nickel per call.

  a. Find a rational function that expresses the average cost $f(x)$ of a call in a month when x calls were made.

  b. To the nearest tenth of a cent, find the average cost of a call in a month when 68 calls were made.

  c. Find the inverse of the function found in part (a) to find a formula that gives the number of calls $f^{-1}(x)$ that can be made for an average cost x.

  d. How many calls need to be made for an average cost of 15¢ per call?

Discovery and Writing

81. Describe what makes a function a one-to-one function.

82. Explain the strategy used to determine the inverse of a one-to-one function.

83. Write a brief paragraph to explain why the range of f is the domain of f^{-1}.

84. Write a brief paragraph to explain why the graphs of a function and its inverse are reflections about the line $y = x$.

85. Let $f(x) = x^5 + x^3 + x + 3$. Find $f^{-1}(3)$. (Hint: Do not find $f^{-1}(x)$. Use observation and the fact that if $f(a) = b$, then $f^{-1}(b) = a$.)

86. Let $f(x) = x^5 + x^3 + x - 3$. Find $f^{-1}(-3)$. (Hint: Do not find $f^{-1}(x)$. Use the fact that if $f(a) = b$, then $f^{-1}(b) = a$.)

Use a graphing calculator to graph each function for various values of a.

87. For what values of a is $f(x) = x^3 + ax$ a one-to-one function?

88. For what values of a is $f(x) = x^3 + ax^2$ a one-to-one function?

Critical Thinking

Determine if the statement is true or false. If the statement is false, then correct it and make it true.

89. All functions have inverses.

90. If $f(x) = x^3 + 7$, then $f^{-1}(x) = \dfrac{1}{x^3 + 7}$.

91. The inverse of the squaring function is the cube root function.

92. The inverse of the cube root function is the cubing function.

93. If $f(x) = x^{123}$ then $f^{-1}(x) = \sqrt[123]{x}$.

94. $f(x) = x^{888}$ is not a one-to-one function.

95. The graph of a function and its inverse are symmetric about the y-axis.

96. Functions that are either increasing or decreasing on their domains have inverses.

CHAPTER REVIEW

3.1 Graphs of Functions

Definitions and Concepts	Examples

The **graph** of a function f in the xy-plane is the set of all points (x, y) where x is in the domain of f, y is in the range of f, and $y = f(x)$.

Graph the function $f(x) = -2|x| + 5$ and determine the domain and range of the function.

To graph the function, we make a table of values and plot the points by drawing a smooth curve through them.

$$f(x) = -2|x| + 5$$

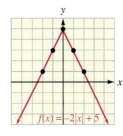

$f(x) = -2|x| + 5$		
x	$f(x)$	$(x, f(x))$
-2	1	$(-2, 1)$
-1	3	$(-1, 3)$
0	5	$(0, 5)$
1	3	$(1, 3)$
2	1	$(2, 1)$

The domain and the range of a function can be identified by viewing the graph of the function. The inputs or x-values that correspond to points on the graph of the function can be identified on the x-axis and used to state the domain of the function. The outputs or $f(x)$ values that correspond to points on the graph of the function can be identified on the y-axis and used to state the range of the function.

The domain of $f(x) = -2|x| + 5$ is $(-\infty, \infty)$. The range is $(-\infty, 5]$. Note that the graph of the function passes the Vertical Line Test.

Vertical Line Test:
If every vertical line that intersects a graph does so exactly once, every number x determines exactly one value of y, and the graph represents a function.

Exercises

Use the graph to determine the function's domain and range.

1.

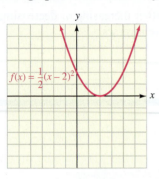

$f(x) = \frac{1}{2}(x-2)^2$

Graph each function. Use the graph to identify the domain and range of each function.

2. $f(x) = -x^2 + 4$

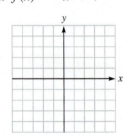

3. $f(x) = 3|x - 2|$

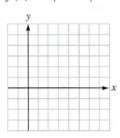

4. $f(x) = -\frac{1}{2}|x| + 3$

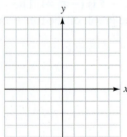

5. $f(x) = 2x^3 + 2$

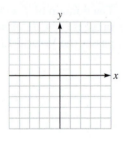

6. $f(x) = -(x - 4)^3$

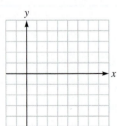

7. $f(x) = \sqrt{x + 5} + 1$

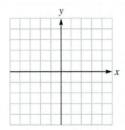

8. $f(x) = -\sqrt{x} - 4$

9. $f(x) = 2\sqrt[3]{x}$

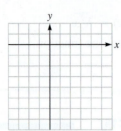

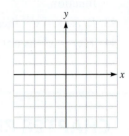

10. $f(x) = -\sqrt[3]{x - 1}$

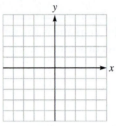

Use the Vertical Line Test to determine whether each graph represents a function.

11.

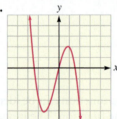

12.

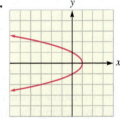

Use the graph of the function f shown to determine each of the following.

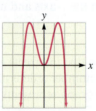

13. Domain and range

14. $f(2)$

15. $f(-1)$

16. the x-values for which $f(x) = 0$

Use the graph of the function f shown to determine each of the following.

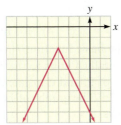

17. Domain and range

18. $f(-1)$

19. $f(-4)$

20. the x-values for which $f(x) = -8$

21. Target heart rate The target heart rate $f(x)$, in beats per minute, at which a person should train to get an effective workout is a function of his age x in years. If $f(x) = -0.6x + 132$, find the target heart rate for a 19-year-old college student. Round to the nearest whole number.

22. Cliff divers The height s, in feet, of a cliff diver is a function of the time t in seconds she has been falling. If s as a function of t can be expressed as $s(t) = -16t^2 + 10t + 300$, what is the height of the diver at 2.5 seconds?

3.2 Transformations of the Graphs of Functions

Definitions and Concepts

Vertical translations:

If $k > 0$, the graph of $\begin{cases} y = f(x) + k \\ y = f(x) - k \end{cases}$

is identical to the graph of $y = f(x)$, except that it is

translated k units $\begin{cases} \text{upward} \\ \text{downward} \end{cases}$.

Horizontal translations:

If $k > 0$, the graph of $\begin{cases} y = f(x - k) \\ y = f(x + k) \end{cases}$

is identical to the graph of $y = f(x)$, except that it is

translated k units to the $\begin{cases} \text{right} \\ \text{left} \end{cases}$.

Vertical stretchings:

If f is a function and $k > 1$, then

* The graph of $y = kf(x)$ can be obtained by stretching the graph of $y = f(x)$ vertically by multiplying each value of $f(x)$ by k.

If f is a function and $0 < k < 1$, then

* The graph of $y = kf(x)$ can be obtained by shrinking the graph of $y = f(x)$ vertically by multiplying each value of $f(x)$ by k.

Horizontal stretchings:

If f is a function and $k > 1$, then

* The graph of $y = f(kx)$ can be obtained by shrinking the graph of $y = f(x)$ horizontally by multiplying each x-value of $f(x)$ by $\frac{1}{k}$.

If f is a function and $0 < k < 1$,

* The graph of $y = f(kx)$ can be obtained by stretching the graph of $y = f(x)$ horizontally by multiplying each x-value of $f(x)$ by $\frac{1}{k}$.

Examples

The function $g(x) = \sqrt{x + 3} - 2$ is a translation of the graph of $f(x) = \sqrt{x}$. Graph both on one set of coordinate axes.

By inspection, we see that the function $g(x) = \sqrt{x + 3} - 2$ involves two translations of $f(x) = \sqrt{x}$. The graph of $g(x) = \sqrt{x + 3} - 2$ is identical to the graph of $f(x) = \sqrt{x}$ except it is translated 3 units to the left and 2 units downward as shown in the figure.

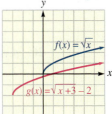

Graph: $g(x) = -\frac{1}{3}|x|$.

The graph of $g(x) = -\frac{1}{3}|x|$ is identical to the graph of $f(x) = |x|$ except that it is vertically shrunk by a factor of $\frac{1}{3}$ and reflected about the x-axis. This is because each value of $|x|$ is multiplied by $-\frac{1}{3}$. The graphs of both functions are shown in the figure.

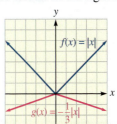

Definitions and Concepts	Examples
Reflections: If f is a function, then • The graph of $y = -f(x)$ is identical to the graph of $y = f(x)$ except that it is reflected about the x-axis. • The graph of $y = f(-x)$ is identical to the graph of $y = f(x)$ except that it is reflected about the y-axis. To graph functions involving a combination of transformations, we must apply each transformation to the basic function. We will apply these transformations in the following order: **1.** Horizontal translation **2.** Stretching or shrinking **3.** Reflection **4.** Vertical translation	Graph $g(x) = 2(x - 4)^3 + 1$. We will graph $g(x) = 2(x - 4)^3 + 1$ by applying three translations to the basic function $f(x) = x^3$: translate $f(x) = x^3$ horizontally 4 units to the right, stretch the graph by a factor of 2, and translate the graph vertically 1 unit upward. The graphs of both functions are shown in the figure. 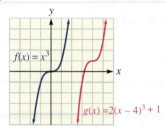

Exercises

Each function is a translation of a basic function.
Graph both on one set of coordinate axes.

23. $g(x) = x^2 + 5$

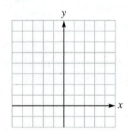

24. $g(x) = (x - 7)^3$

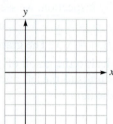

25. $g(x) = \sqrt{x + 2} + 3$

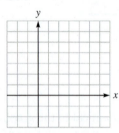

26. $g(x) = |x - 4| + 2$

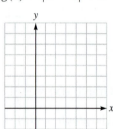

Each function is a stretching of $f(x) = x^3$. Graph both on one set of coordinate axes.

27. $g(x) = \dfrac{1}{3}x^3$

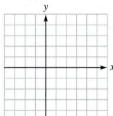

28. $g(x) = (-5x)^3$

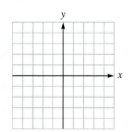

Graph each function using a combination of translations, stretchings, and reflections.

29. $f(x) = 2(x - 6)^2 - 8$

30. $f(x) = \dfrac{1}{2}(x + 2)^2 + 6$

31. $g(x) = -|x - 4| + 3$

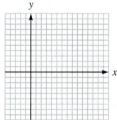

32. $g(x) = \dfrac{1}{4}|x - 4| + 1$

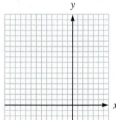

33. $g(x) = 3\sqrt{x + 3} + 2$

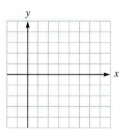

34. $g(x) = \dfrac{1}{3}(x + 3)^3 + 2$

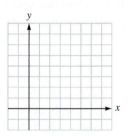

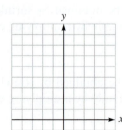

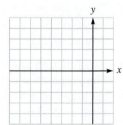

35. $f(x) = \sqrt{-x} + 3$

36. $g(x) = 2\sqrt[3]{x} - 5$

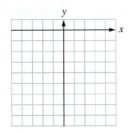

3.3 More on Functions; Piecewise-Defined Functions

Definitions and Concepts	**Examples**
If $f(-x) = f(x)$ for all x in the domain of f, the graph of the function is symmetric about the y-axis, and the function is called an **even function**.	The graph of $f(x) = -2|x| + 5$ is symmetric about the y-axis and is an example of an even function. 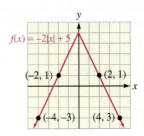 Note that $f(-x) = f(x)$. $$f(-x) = -2|-x| + 5$$ $$= -2x + 5$$ $$= f(x)$$
If $f(-x) = -f(x)$ for all x in the domain of f, the function is symmetric about the origin, and the function is called an **odd function**.	The graph of $f(x) = 3\sqrt[3]{x}$ is symmetric with respect to the origin and is an example of an odd function. 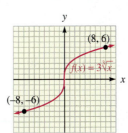 Note that $f(-x) = -f(x)$. $$f(-x) = 3\sqrt[3]{-x}$$ $$= -3\sqrt[3]{x}$$ $$= -f(x)$$

Definitions and Concepts	**Examples**
If we trace the graph of a function from left to right and the values $f(x)$ increase, we say that the function is **increasing on the interval** (a, b). If the values $f(x)$ decrease, we say that the function is **decreasing on the interval** (a, b). If the values $f(x)$ remain unchanged as x increases, we say that the function is **constant on the interval** (a, b).	Consider the graph of the function shown. 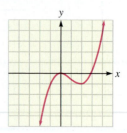
$f(c)$ is a **local maximum** if there exists an interval (a, b) with $a < c < b$ such that $f(x) \leq f(c)$ for all x in (a, b).	
$f(c)$ is a **local minimum** if there exists an interval (a, b) with $a < c < b$ such that $f(x) \geq f(c)$ for all x in (a, b).	Note: 1. The function increases on $(-\infty, 0) \cup (2, \infty)$. 2. The function decreases on $(0, 2)$. 3. $f(0) = 0$ is a local maximum. 4. $f(2) = -1$ is a local minimum.
Some functions, called **piecewise-defined functions,** are defined by using different equations for different intervals in their domains.	$f(x) = \begin{cases} x - 2 & \text{if } x < 3 \\ x^2 & \text{if } x \geq 3 \end{cases}$ • $f(2) = 2 - 2 = 0$ • $f(4) = 4^2 = 16$
The **greatest-integer function** is important in many business applications and in the field of computer science. This function is determined by the equation $f(x) = [x]$, where the value of $f(x)$ that corresponds to x is the greatest integer that is less than or equal to x.	Let $f(x) = [x - 3]$. Find $f(2.2)$. $f(x) = [x - 3]$ $f(2.2) = [2.2 - 3] = [-0.8] = -1$ -1 is the greatest integer less than or equal to -0.8.

Exercises

Determine whether each function is even, odd, or neither.

37.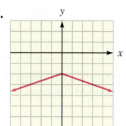

38.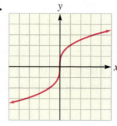

39.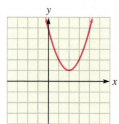

40.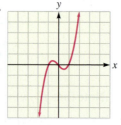

41. $y = x^3 - x$

42. $y = x^2 - 4x$

43. $y = x^3 - x^2$

44. $y = 1 - x^4$

Determine the open intervals on which the graph of the function is increasing, decreasing, or constant.

45.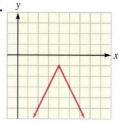

46.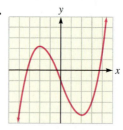

State the values of any local maxima or minima.

47.

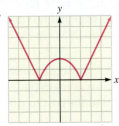

48.

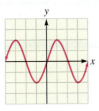

Evaluate each piecewise-defined function.

49. $f(x) = \begin{cases} x - 2 & \text{if } x < 3 \\ x^2 & \text{if } x \geq 3 \end{cases}$

    a. $f(-2)$ **b.** $f(3)$

50. $f(x) = \begin{cases} 2 & \text{if } x < 0 \\ 2 - x & \text{if } 0 \leq x < 2 \\ x + 1 & \text{if } x \geq 2 \end{cases}$

    a. $f\left(\dfrac{3}{2}\right)$ **b.** $f(2)$

Graph each piecewise-defined function and determine the open intervals on which it is increasing, decreasing, or constant.

51. $y = f(x) = \begin{cases} x + 3 & \text{if } x \leq 0 \\ 3 & \text{if } x > 0 \end{cases}$

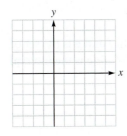

52. $y = f(x) = \begin{cases} x + 5 & \text{if } x \leq 0 \\ 5 - x & \text{if } x > 0 \end{cases}$

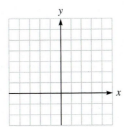

53. $f(x) = \begin{cases} -3x + 1 & \text{if } x < 0 \\ \dfrac{1}{3}x^2 - 4 & \text{if } x \geq 0 \end{cases}$

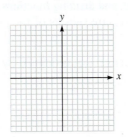

54. $f(x) = \begin{cases} \dfrac{1}{2}(x + 1)^2 + 2 & \text{if } x \leq -1 \\ -2 & \text{if } -1 < x < 1 \\ 2x + \dfrac{1}{2} & \text{if } x \geq 1 \end{cases}$

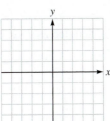

Evaluate each function at the indicated x-values.

55. $f(x) = \llbracket 2x \rrbracket$ Find $f(1.7)$.

56. $f(x) = \llbracket x - 5 \rrbracket$ Find $f(4.99)$.

Graph each function.

57. $f(x) = [\![x]\!] + 2$

58. $f(x) = [\![x - 1]\!]$

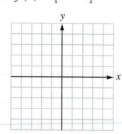

59. Renting a Jeep A rental company charges $20 to rent a Jeep Wrangler for one day, plus $8 for every 100 miles (or portion of 100 miles) that it is driven. Find the cost if the Jeep is driven 295 miles in one day.

60. Riding in a taxi A taxicab company charges $4 for a trip up to 1 mile, and $2 for every extra mile (or portion of a mile). Find the cost to ride $11\frac{1}{2}$ miles.

3.4 Operations on Functions

Definitions and Concepts

Examples

Adding, subtracting, multiplying, and dividing functions:
If the ranges of functions f and g are subsets of the real numbers, then

Let $f(x) = 3x + 5$ and $g(x) = 4x - 7$. Find each function and its domain: **a.** $f + g$ **b.** $f - g$ **c.** $f \cdot g$ **d.** f/g

1. The **sum of f and g**, denoted as $f + g$, is defined by $(f + g)(x) = f(x) + g(x)$.

a. $(f + g)(x) = f(x) + g(x)$
$$= (3x + 5) + (4x - 7)$$
$$= 7x - 2$$

Since the domain of both f and g is the set of real numbers, the domain of $f + g$ is the interval $(-\infty, \infty)$.

2. The **difference of f and g**, denoted as $f - g$, is defined by $(f - g)(x) = f(x) - g(x)$.

b. $(f - g)(x) = f(x) - g(x)$
$$= (3x + 5) - (4x - 7)$$
$$= 3x + 5 - 4x + 7$$
$$= -x + 12$$

Since the domain of both f and g is the set of real numbers, the domain of $f - g$ is the interval $(-\infty, \infty)$.

3. The **product of f and g**, denoted as $f \cdot g$, is defined by $(f \cdot g)(x) = f(x) \cdot g(x)$.

c. $(f \cdot g)(x) = f(x) \cdot g(x)$
$$= (3x + 5)(4x - 7)$$
$$= 12x^2 - x - 35$$

Since the domain of both f and g is the set of real numbers, the domain of $f \cdot g$ is the interval $(-\infty, \infty)$.

4. The **quotient of f and g**, denoted as f/g, is defined by $(f/g)(x) = \frac{f(x)}{g(x)}, g(x) \neq 0$.

The domain of each function, unless otherwise restricted, is the set of real numbers x that are in the domains of both f and g. In the case of the quotient f/g, there is the restriction that $g(x) \neq 0$.

d. $(f/g)(x) = \dfrac{f(x)}{g(x)}$
$$= \frac{3x + 5}{4x - 7} \quad (4x - 7 \neq 0)$$

Since $\frac{7}{4}$ will make $4x - 7$ equal to 0, the domain of f/g is the set of all real numbers except $\frac{7}{4}$. This is $\left(-\infty, \frac{7}{4}\right) \cup \left(\frac{7}{4}, \infty\right)$.

Definitions and Concepts	Examples
The **composite function** $f \circ g$ is defined by $(f \circ g)(x) = f(g(x))$. The domain of $f \circ g$ consists of all those numbers in the domain of g for which $g(x)$ is in the domain of f.	If $f(x) = 2x + 7$ and $g(x) = x^2 + 1$, find $(f \circ g)(x)$ and its domain. Because $(f \circ g)(x)$ means $f(g(x))$, we will replace x in $f(x) = 2x + 7$ with $g(x)$. $$(f \circ g)(x) = f(g(x))$$ $$= f(x^2 + 1)$$ $$= 2(x^2 + 1) + 7$$ $$= 2x^2 + 9$$ The domain of $(f \circ g)(x)$ is the interval $(-\infty, \infty)$ because the domain of both f and g consists of all real numbers.

Exercises

Let $f(x) = x^2 - 1$ and $g(x) = 2x + 1$. Find each function and its domain.

61. $f + g$ **62.** $f \cdot g$

63. $f - g$ **64.** f/g

Let $f(x) = 2x^2 - 1$ and $g(x) = 2x - 1$. Find each value, if possible.

65. $(f + g)(-3)$ **66.** $(f - g)(-5)$

67. $(f \cdot g)(2)$ **68.** $(f/g)\left(\dfrac{1}{2}\right)$

Let $f(x) = x^2 - 1$ and $g(x) = 2x + 1$. Find each function and its domain.

69. $f \circ g$ **70.** $g \circ f$

Let $f(x) = x^2 - 5$ and $g(x) = 3x + 1$. Find each value.

71. $(f \circ g)(-2)$ **72.** $(g \circ f)(-2)$

Find two functions f and g such that the composition $f \circ g = h$ expresses the given correspondence. Several answers are possible.

73. $h(x) = \sqrt{x - 5}$ **74.** $h(x) = (x + 6)^3$

3.5 Inverse Functions

Definitions and Concepts	Examples
A function f from a set **X** to a set **Y** is called a **one-to-one function** if and only if different numbers in the domain of f have different outputs in the range of f. A **Horizontal Line Test** can be used to determine whether the graph of a function represents a one-to-one function. If every horizontal line that intersects the graph of a function does so exactly once, the function passes the Horizontal Line Test and is one-to-one.	Determine whether the function $f(x) = x^4 - 2x^2$ is one-to-one. The function $f(x) = x^4 - 2x^2$ is not one-to-one, because different numbers in the domain have the same output. For example, 2 and -2 have the same output: $f(2) = f(-2) = 8$.

Definitions and Concepts	Examples
Inverse functions: If f and g are two one-to-one functions such that $(f \circ g)(x) = x$ for every x in the domain of g and $(g \circ f)(x) = x$ for every x in the domain of f, then f and g are **inverse functions**. Function g is denoted as f^{-1} and is called the **inverse function of f**.	Verify that $f(x) = x^5$ and $g(x) = \sqrt[5]{x}$ are inverse functions. To show that f and g are inverse functions, we must show that $f \circ g$ and $g \circ f$ are x, the identity function. $$(f \circ g)(x) = f(g(x)) = f\left(\sqrt[5]{x}\right) = \left(\sqrt[5]{x}\right)^5 = x$$ $$(g \circ f)(x) = g(f(x)) = g(x^5) = \sqrt[5]{x^5} = x$$
Properties of a one-to-one function: **Property 1:** If f is a one-to-one function, there is a one-to-one function $f^{-1}(x)$ such that $(f^{-1} \circ f)(x) = x$ and $(f \circ f^{-1})(x) = x$. **Property 2:** The domain of f is the range of f^{-1}, and the range of f is the domain of f^{-1}.	Because g is the inverse of f, we can use inverse notation and write $f(x) = x^5$ and $f^{-1}(x) = \sqrt[5]{x}$. Because f is the inverse of g, we can use inverse notation and write $g(x) = \sqrt[5]{x}$ and $g^{-1}(x) = x^5$.
Strategy for finding f^{-1}: **Step 1:** Replace $f(x)$ with y. **Step 2:** Interchange the variables x and y. **Step 3:** Solve the resulting equation for y. **Step 4:** Replace y with $f^{-1}(x)$. The graph of a function and its inverse are reflections of each other about the line $y = x$.	Find the inverse of $f(x) = x^3 + 5$. We will find the inverse of the function using the strategy given in the section. **Step 1:** Replace $f(x)$ with y. $$f(x) = x^3 + 5$$ $$y = x^3 + 5$$ **Step 2:** Interchange the variables x and y. $$x = y^3 + 5$$ **Step 3:** Solve the resulting equation for y. $$x - 5 = y^3$$ $$y = \sqrt[3]{x - 5}$$ **Step 4:** Replace y with $f^{-1}(x)$. $$f^{-1}(x) = \sqrt[3]{x - 5}$$

Exercises

Determine whether each function is one-to-one.

75. $f(x) = x^2 + 7$ **76.** $f(x) = x^3$

Use the Horizontal Line Test to determine whether each graph represents a one-to-one function.

77.

78.
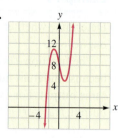

Verify that the functions are inverses by showing that $f \circ g$ and $g \circ f$ are the identity function.

79. $f(x) = 8x - 3$ **80.** $f(x) = \dfrac{1}{2 - x}$

Each equation defines a one-to-one function. Find f^{-1} and verify that $f \circ f^{-1}$ and $f^{-1} \circ f$ are the identity function.

81. $y = 7x - 1$ **82.** $f(x) = 5x - 8$

83. $f(x) = x^3 - 10$ **84.** $f(x) = \sqrt[3]{x + 5}$

85. $y = \dfrac{5}{x}$ **86.** $y = \dfrac{1}{2 - x}$

87. $y = \dfrac{x}{1-x}$

88. $y = \dfrac{3}{x^3}$

90. Find the range of $y = \frac{2x+3}{5x-10}$ by finding the domain of f^{-1}.

89. Find the inverse of the one-to-one function $f(x) = 2x - 5$ and graph both the function and its inverse on the same set of coordinate axes.

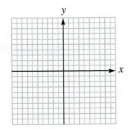

CHAPTER TEST

Graph each function by plotting points.

1. $f(x) = 2|x+1| + 2$

2. $f(x) = -2x^3 - 4$

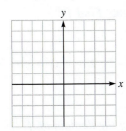

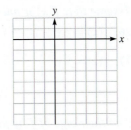

Use the graph of the function shown to determine the following.

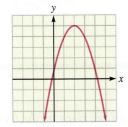

3. domain and range

4. $f(1)$

Use transformations to graph each function.

5. $f(x) = (x-3)^2 + 1$

6. $f(x) = \sqrt{x-1} + 5$

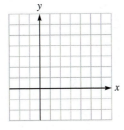

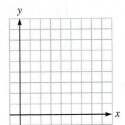

7. $f(x) = -(x-1)^3 + 3$

8. $f(x) = -\dfrac{1}{2}|x+5| - 2$

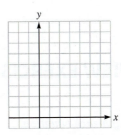

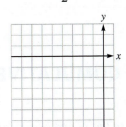

9. $f(x) = 2\sqrt[3]{x-6} - 1$

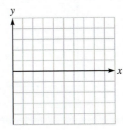

Determine whether the functions are even, odd, or neither.

10.

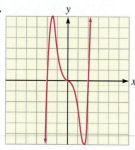

11. $f(x) = 2x^4 - 3x^2 - 7$

Use the graph to determine any local maxima or minima.

12.

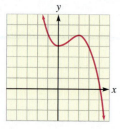

Use the piecewise-defined function shown to find each value.

$$f(x) = \begin{cases} 2x & \text{if } x < 0 \\ 3 - x & \text{if } 0 \le x < 2 \\ |x| & \text{if } x \ge 2 \end{cases}$$

13. $f\left(\dfrac{3}{2}\right)$

14. $f(5)$

15. Graph $f(x) = \begin{cases} -x - 1 & \text{if } x < 1 \\ 4 & \text{if } x \ge 1 \end{cases}$.

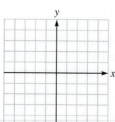

Let $f(x) = 3x$ and $g(x) = x^2 + 2$. Find each function.

16. $f + g$

17. f/g

18. $g \circ f$

19. $f \circ g$

Let $f(x) = 2x^2 - 5x + 1$ and $g(x) = 5x + 1$. Find each function value.

20. $(f + g)(-2)$ **21.** $(f - g)(2)$

22. $(f \cdot g)(-1)$ **23.** $(f/g)(0)$

24. $(f \circ g)(-1)$ **25.** $(g \circ f)(-3)$

Assume that $f(x)$ is one-to-one. Find f^{-1}.

26. $f(x) = 5x - 2$

27. $f(x) = \dfrac{x + 1}{x - 1}$ **28.** $f(x) = x^3 - 3$

Find the range of f by finding the domain of f^{-1}.

29. $y = \dfrac{3}{x} - 2$ **30.** $y = \dfrac{3x - 1}{x - 3}$

CUMULATIVE REVIEW EXERCISES

Use the x- and y-intercepts to graph each equation.

1. $5x - 3y = 15$ **2.** $3x + 2y = 12$

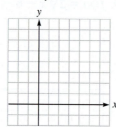

Find the length, the midpoint, and the slope of the line segment PQ.

3. $P\left(-2, \dfrac{7}{2}\right)$; $Q\left(3, -\dfrac{1}{2}\right)$

4. $P(3, 7)$; $Q(-7, 3)$

Find the slope of the line passing through the two given points.

5. $P(-1, 9)$ and $Q(-4, -6)$

6. $P\left(2, -\dfrac{1}{3}\right)$ and $Q\left(5, -\dfrac{1}{3}\right)$

Write the equation of the line with the given properties. Give the answer in slope-intercept form.

7. The line passes through $(-3, 5)$ and $(3, -7)$.

8. The line passes through $\left(\frac{3}{2}, \frac{5}{2}\right)$ and has a slope of $\frac{7}{2}$.

9. The line is parallel to $3x - 5y = 7$ and passes through $(-5, 3)$.

10. The line is perpendicular to $x - 4y = 12$ and passes through the origin.

Graph each equation. Make use of intercepts and symmetries.

11. $x^2 = y - 2$

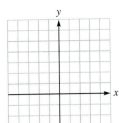

12. $y^2 = x - 2$

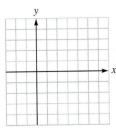

Identify the center and radius of the circles.

13. $x^2 + (y - 7)^2 = \dfrac{1}{4}$

14. $(x - 5)^2 + (y + 4)^2 = 144$

Graph each circle.

15. $x^2 + y^2 = 100$

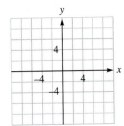

16. $x^2 - 2x + y^2 = 8$

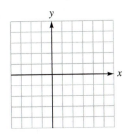

Solve each proportion.

17. $\dfrac{x - 2}{x} = \dfrac{x - 6}{5}$

18. $\dfrac{x + 2}{x - 6} = \dfrac{3x + 1}{2x - 11}$

19. Dental billing The billing schedule for dental X-rays specifies a fixed amount for the office visit plus a fixed amount for each X-ray exposure. If 2 X-rays cost $37 and 4 cost $54, find the cost of 5 exposures.

20. Automobile collisions The energy dissipated in an automobile collision varies directly with the square of the speed. By what factor does the energy increase in a 50-mph collision compared with a 20-mph collision?

Graph each function.

21. $f(x) = -2|x - 2| - 1$

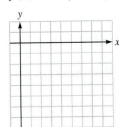

22. $f(x) = x^2 - 4$

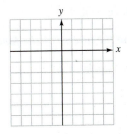

23. $f(x) = -x^2 + 4$

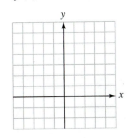

24. $f(x) = -x^3 - 5$

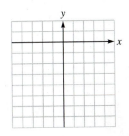

25. $f(x) = 2\sqrt{x + 4} - 1$

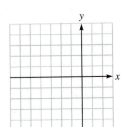

26. $f(x) = \sqrt[3]{x - 1} - 3$

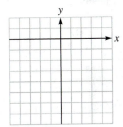

Find the domain and range of the function.

27.

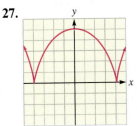

Let $f(x) = 3x - 4$ and $g(x) = x^2 + 1$. Find each function and its domain.

28. $(f - g)(x)$

29. $(f \cdot g)(x)$

30. $(f/g)(x)$

Let $f(x) = 3x - 4$ and $g(x) = x^2 + 1$. Find each value.

31. $(f \circ g)(2)$ **32.** $(g \circ f)(2)$

33. $(f \circ g)(x)$

34. $(g \circ f)(x)$

Find the inverse of the function defined by each equation.

35. $y = 3x + 2$ **36.** $y = \dfrac{1}{x - 3}$

37. $y = x^2 + 5 \ (x \geq 0)$ **38.** $3x - y = 1$

Write each sentence as an equation.

39. y varies directly with the product of w and z.

40. y varies directly with x and inversely with the square of t.

Polynomial and Rational Functions

Careers and Mathematics: Applied Mathematician

Applied mathematicians use theories and techniques, such as computational methods and mathematical modeling, to formulate and solve practical problems that arise in engineering, business, science, government, as well as in the social, life, and physical sciences. Applications are diverse and range from analyzing the effects and safety of new drugs to determining the most efficient way to schedule airline routes between cities.

Education and Mathematics Required

- In general, a Ph.D. in mathematics is required for jobs other than those in the federal government.
- College Algebra; Trigonometry; Calculus I, II, III; Linear Algebra; Ordinary Differential Equations; Real Analysis; Abstract Algebra; Theory of Analysis; and Complex Analysis form a basic list. Most mathematicians will study additional topics in their area of specialty.

How Applied Mathematicians Use Math and Who Employs Them

- Mathematical theories, algorithms, computational techniques, and computer technology can be used to solve economic, scientific, engineering, and business problems.
- A large number of mathematicians work for the federal government, with about 80% of those working for the U.S. Department of Defense. There are also positions in NASA. Many mathematicians are employed by universities as faculty members and divide their time between teaching and research. Mathematicians are also employed by research-and-development laboratories as part of technical teams.

Career Outlook and Earnings

- Much faster than average employment growth is expected for mathematicians. Ph.D. holders with a strong background in mathematics and a related field, such as computer science or engineering, will have excellent employment opportunities. Employment of mathematicians is expected to increase by 23% between 2012 and 2022.
- The median annual salary for mathematicians is $101,360, with the top 10% of earners having a salary of $142,460.

For more information see: www.bls.gov/oco

4.1 Quadratic Functions

In this section, we will learn to

1. Recognize the characteristics of a quadratic function.
2. Find the vertex of a parabola whose equation is in standard form.
3. Graph a quadratic function.
4. Find the vertex of a parabola whose equation is in general form.
5. Use a quadratic function to solve maximum and minimum problems.

Quadratic functions are important because we can use them to model many real-life problems. For example, the path of a basketball jump shot by LeBron James and the path of a guided missile can be modeled with quadratic functions. Businesses like Coca Cola and Best Buy can use quadratic functions to help maximize the profit and revenue for the products they produce and sell.

1. Recognize the Characteristics of a Quadratic Function

The linear function $f(x) = mx + b$ ($m \neq 0$) is a first-degree polynomial function, because its right side is a first-degree polynomial in the variable x. A function defined by a polynomial of second degree is called a **quadratic function**.

Quadratic Function	A **quadratic function** is a second-degree polynomial function in one variable of the form

$$f(x) = ax^2 + bx + c \quad \text{or} \quad y = ax^2 + bx + c,$$

where a, b, and c are real numbers and $a \neq 0$.

Some examples of quadratic functions are

$$f(x) = x^2 - 2x - 3 \quad \text{and} \quad f(x) = -2x^2 - 8x - 3.$$

Quadratic functions can be graphed by plotting points. For example, to graph the function $f(x) = x^2 - 2x - 3$, we plot several points with coordinates that satisfy the equation. We then join them with a smooth curve to obtain the graph shown in Figure 4-1.

Take Note

The graph of the quadratic function decreases on the open interval $(-\infty, 1)$, increases on the open interval $(1, \infty)$, and has a local minimum value of $f(1) = -4$.

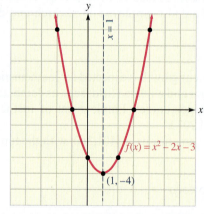

$f(x) = x^2 - 2x - 3$		
x	$f(x)$	$(x, f(x))$
-2	5	$(-2, 5)$
-1	0	$(-1, 0)$
0	-3	$(0, -3)$
1	-4	$(1, -4)$
2	-3	$(2, -3)$
3	0	$(3, 0)$
4	5	$(4, 5)$

Domain: $(-\infty, \infty)$, **Range:** $[-4, \infty)$

FIGURE 4-1

A table of values and the graph of $f(x) = -2x^2 - 8x - 3$ are shown in Figure 4-2.

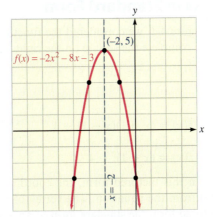

$f(x) = -2x^2 - 8x - 3$		
x	$f(x)$	$(x, f(x))$
-4	-3	$(-4, -3)$
-3	3	$(-3, 3)$
-2	5	$(-2, 5)$
-1	3	$(-1, 3)$
0	-3	$(0, -3)$

Domain: $(-\infty, \infty)$, **Range:** $(-\infty, 5]$

FIGURE 4-2

The graph of a quadratic function is called a **parabola**, a cup-shaped curve that either opens upward $\cup$ or downward $\cap$. The graphs in Figures 4-1 and 4-2 suggest that the graph of a quadratic function has the following characteristics.

Characteristics of Quadratic Functions

Characteristics	Examples	
Equation of a quadratic function $f(x) = ax^2 + bx + c$	$f(x) = x^2 - 2x - 3$ (See Figure 4-1.)	$f(x) = -2x^2 - 8x - 3$ (See Figure 4-2.)
If $a > 0$, the parabola **opens up**. If $a < 0$, the parabola **opens down**.	$a = 1$, opens up	$a = -2$, opens down
The **vertex** is the turning point of the parabola.	Vertex is $(1, -4)$	Vertex is $(-2, 5)$
The **minimum** or **maximum** **point** occurs at the vertex.	$(1, -4)$ is the minimum or lowest point on the graph.	$(-2, 5)$ is the maximum or highest point on the graph.
The **axis of symmetry** is the vertical line that intersects the parabola at the vertex. The parabola is symmetric about this vertical line.	The graph of $x = 1$ is the axis of symmetry.	The graph of $x = -2$ is the axis of symmetry.

Accent on Technology

Graphing Quadratic Functions

We can use a graphing calculator to graph quadratic functions. If we use window settings of $[-10, 10]$ for x and $[-10, 10]$ for y, the graph of $f(x) = x^2 - 2x - 3$ will look like Figure 4-3(a). The graph of $f(x) = -2x^2 - 8x - 3$ will look like Figure 4-3(b).

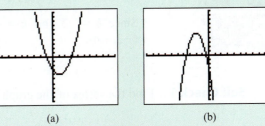

(a)   (b)

FIGURE 4-3

2. Find the Vertex of a Parabola Whose Equation Is in Standard Form

In Figure 4-1, we considered the function $f(x) = x^2 - 2x - 3$ with vertex $(1, -4)$. If we complete the square on the right side of the equation we will obtain

$$f(x) = (x^2 - 2x + 1) - 3 - 1 \qquad \text{One-half of } -2 \text{ is } -1, \text{ and } (-1)^2 \text{ is } 1.$$
$$\text{Add 1 and subtract 1 on the right side of the equation.}$$

$$f(x) = (x - 1)(x - 1) - 4 \qquad \text{Factor } x^2 - 2x + 1.$$

$$f(x) = (x - 1)^2 - 4$$

In this factored form, the coordinates of the vertex of the parabola can be read from the equation. The vertex of $f(x) = (x - 1)^2 - 4$ is $(1, -4)$. We call this factored form the **standard form** of an equation of a quadratic function.

Standard Form of an Equation of a Quadratic Function	The graph of a quadratic function $$f(x) = a(x - h)^2 + k \ (a \neq 0)$$ is a parabola with vertex at (h, k). The parabola opens upward when $a > 0$ and downward when $a < 0$. The axis of symmetry of the parabola is the vertical line graph of the equation $x = h$.

EXAMPLE 1 — Finding the Vertex of a Parabola in Standard Form

Find the vertex of the graph of each quadratic function:
a. $f(x) = 2(x - 3)^2 + 5$ **b.** $f(x) = -3(x + 2)^2 - 4$

SOLUTION In each case, the equation of the quadratic function is given in standard form. From the equations, we can identify h and k, because the vertex is the point with coordinates (h, k).

a. We identify the values of h and k.

Standard Form: $f(x) = a(x - h)^2 + k$

Given Function: $f(x) = 2(x - 3)^2 + 5$ $h = 3$ and $k = 5$.

Since $h = 3$ and $k = 5$, the vertex is the point with coordinates of $(3, 5)$.

b. We identify the values of h and k.

Standard Form: $f(x) = a(x - h)^2 + k$

Given Function: $f(x) = -3(x + 2)^2 - 4$

$$f(x) = -3[x - (-2)]^2 + (-4) \qquad h = -2 \text{ and } k = -4.$$

Since $h = -2$ and $k = -4$, the vertex is the point with coordinates of $(-2, -4)$.

Take Note

The vertex of each of the functions is exactly what we would expect based on our previous study of transformations of functions.

1. For $f(x) = 2(x - 3)^2 + 5$, the squaring function is shifted 3 units right and 5 units upward. Therefore, the vertex is $(3, 5)$.
2. For $f(x) = -3(x + 2)^2 - 4$, the squaring function is shifted 2 units left and 4 units downward. Therefore, the vertex is $(-2, -4)$.

©iQoncept/Shutterstock.com

Self Check 1 Find the vertex of the graph of the quadratic function $f(x) = 2(x + 5)^2 - 4$.

Now Try Exercise 19.

3. Graph a Quadratic Function

The easiest way to graph a quadratic function is to follow these steps.

> **Strategy for Graphing a Quadratic Function**
>
> To graph a quadratic function
>
> 1. Determine whether the parabola opens upward or downward.
> 2. Find the vertex of the parabola.
> 3. Find the x-intercept(s).
> 4. Find the y-intercept.
> 5. Identify one additional point on the graph.
> 6. Draw a smooth curve through the points found in Steps 2–5.

EXAMPLE 2

Graphing a Quadratic Function Written in Standard Form

Graph the quadratic function $f(x) = 2(x + 1)^2 - 8$.

SOLUTION We first determine whether the parabola opens upward or downward. Then we will find the vertex and the x- and y-intercepts. Finally, we will find one additional point and draw a smooth curve through the plotted points.

Step 1: Determine whether the parabola opens upward or downward.

Standard Form: $f(x) = a(x - h)^2 + k$

Given Form: $f(x) = 2(x + 1)^2 - 8$

Since $a = 2$ and 2 is positive, the parabola opens upward.

Step 2: Find the vertex of the parabola.

Standard Form: $f(x) = a(x - h)^2 + k$

Given Form: $f(x) = 2(x + 1)^2 - 8$

$$f(x) = 2[x - (-1)]^2 + (-8)$$

Since $h = -1$ and $k = -8$ the vertex is the point with coordinates of $(-1, -8)$.

Step 3: Find the x-intercept(s).

> **Tip**
>
> 1. The graph of a quadratic function can have one, two, or no x-intercepts.
> 2. The graph of a quadratic function will always have one y-intercept.

To find the x-intercepts, we substitute 0 for $f(x)$ and solve for x.

$f(x) = 2(x + 1)^2 - 8$	
$0 = 2(x + 1)^2 - 8$	Substitute 0 for $f(x)$.
$8 = 2(x + 1)^2$	Add 8 to both sides of the equation.
$4 = (x + 1)^2$	Divide both sides by 2.
$x + 1 = \pm 2$	Write $(x + 1)^2$ on the left side and use the Square Root Property.
$x = -1 \pm 2$	Subtract 1 from both sides.
$x = 1$ or $x = -3$	

The x-intercepts are the points with coordinates of $(1, 0)$ and $(-3, 0)$.

Step 4: Find the *y*-intercept.

To find the *y*-intercept, we substitute 0 in for *x* and solve for *y*.

$$f(x) = 2(\textcolor{red}{x} + 1)^2 - 8$$

$$y = 2(\textcolor{red}{x} + 1)^2 - 8 \qquad \text{Substitute } y \text{ for } f(x).$$

$$y = 2(\textcolor{red}{0} + 1)^2 - 8 \qquad \text{Substitute } 0 \text{ in for } x.$$

$$y = 2(1)^2 - 8$$

$$y = 2 - 8$$

$$y = -6$$

The *y*-intercept is the point with coordinates of $(0, -6)$.

Step 5: Identify one additional point on the graph.

Because of symmetry, the point $(-2, -6)$ is on the graph.

Step 6: Draw a smooth curve through the points found in Steps 2–5.

We can now draw the graph of the function as shown in Figure 4-4.

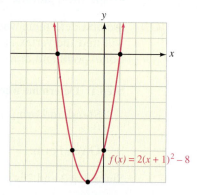

$$f(x) = 2(x + 1)^2 - 8$$

FIGURE 4-4

> **Tip**
>
> Sometimes the vertex of the graph of a quadratic function occurs at the origin, $(0, 0)$. In this case, both $h = 0$ and $k = 0$, and the equation of the parabola is of the form $f(x) = a(x - 0)^2 + 0$ or $f(x) = ax^2$ and $a \neq 0$. Some examples are: $f(x) = 2x^2$, $f(x) = -4x^2$, and $f(x) = \frac{2}{3}x^2$.

Self Check 2   Graph the function: $f(x) = -(x - 2)^2 + 4$.

Now Try Exercise 41.

4. Find the Vertex of a Parabola Whose Equation Is in General Form

To graph a quadratic function given in the form $f(x) = ax^2 + bx + c \; (a \neq 0)$, called **general form**, we must find the coordinates of the vertex of the parabola. To do so, we can complete the square on $ax^2 + bx$ to change the equation into standard form $f(x) = a(x - h)^2 + k$. As we have seen, we can read the coordinates (h, k) of the vertex from this form. Once the quadratic function is in standard form, we can graph the function following the steps stated earlier.

EXAMPLE 3   **Finding the Vertex of a Parabola Written in General Form**

Find the vertex of the parabola whose equation is $f(x) = -2x^2 + 12x - 16$.

SOLUTION   We will complete the square on *x*, write the equation in standard form, and identify *h* and *k*, the coordinates of the vertex.

We begin by completing the square on $-2x^2 + 12x$.

$f(x) = -2x^2 + 12x - 16$	Identify a: $a = -2$.
$f(x) = -2(x^2 - 6x) - 16$	Factor $a = -2$ from $-2x^2 + 12x$.
$f(x) = -2(x^2 - 6x + 9 - 9) - 16$	One-half of -6 is -3 and $(-3)^2 = 9$. Add and subtract 9 within the parentheses.
$f(x) = -2(x^2 - 6x + 9) - 2(-9) - 16$	Distribute the multiplication by -2.
$f(x) = -2(x - 3)^2 + 18 - 16$	Factor $x^2 - 6x + 9$ and multiply.
$f(x) = -2(x - 3)^2 + 2$	Simplify.

The equation is now in standard form with $h = 3$ and $k = 2$. Therefore, the vertex is the point with coordinates $(h, k) = (3, 2)$.

Self Check 3   Find the vertex of the graph of $f(x) = 4x^2 - 16x + 19$.

Now Try Exercise 27.

To find formulas for the coordinates of the vertex of a parabola defined by $f(x) = ax^2 + bx + c$ $(a \neq 0)$, we can complete the square on x to write the equation in standard form $(f(x) = a(x - h)^2 + k)$:

$f(x) = ax^2 + bx + c$	
$f(x) = a\left(x^2 + \dfrac{b}{a}x\right) + c$	Factor a from $ax^2 + bx$.
$f(x) = a\left(x^2 + \dfrac{b}{a}x + \dfrac{b^2}{4a^2} - \dfrac{b^2}{4a^2}\right) + c$	Add and subtract $\dfrac{b^2}{4a^2}$ within the parentheses.
$f(x) = a\left(x^2 + \dfrac{b}{a}x + \dfrac{b^2}{4a^2}\right) - a\left(\dfrac{b^2}{4a^2}\right) + c$	Distribute the multiplication of a.
$f(x) = a\left(x + \dfrac{b}{2a}\right)^2 + c - \dfrac{b^2}{4a}$	Factor $x^2 + \dfrac{b}{a}x + \dfrac{b^2}{4a^2}$ and simplify $a\left(\dfrac{b^2}{4a^2}\right)$.
$f(x) = a\left[x - \left(-\dfrac{b}{2a}\right)\right]^2 + c - \dfrac{b^2}{4a}$	$-\left(-\dfrac{b}{2a}\right) = \dfrac{b}{2a}$

Tip

You don't need to memorize the formula for the y-coordinate of the vertex of a parabola. It is usually convenient to find the y-coordinate by substituting $-\dfrac{b}{2a}$ for x in the function and solving for y.

If we compare the last equation to the form $f(x) = a(x - h)^2 + k$, we see that $h = -\dfrac{b}{2a}$ and $k = c - \dfrac{b^2}{4a}$. This result gives the following fact.

Vertex of a Parabola   The graph of the function $f(x) = ax^2 + bx + c$ $(a \neq 0)$ is a parabola with vertex at $\left(-\dfrac{b}{2a}, c - \dfrac{b^2}{4a}\right)$ or $\left(-\dfrac{b}{2a}, f\left(-\dfrac{b}{2a}\right)\right)$.

Graphing a Quadratic Function Written in General Form

EXAMPLE 4

Graph the function: $f(x) = -2x^2 - 5x + 3$.

SOLUTION   We begin by determining whether the parabola opens upward or downward. Then we find the vertex by using the formula $h = -\dfrac{b}{2a}$. Next we find the x- and y-intercepts and one additional point and then draw a smooth curve through the plotted points.

Step 1: Determine whether the parabola opens up or downward.

The equation has the form $f(x) = ax^2 + bx + c$, where $a = -2, b = -5,$ and $c = 3$. Since $a < 0$, the parabola opens downward.

Step 2: Find the vertex.

To find the x-coordinate of the vertex, we substitute the values of a and b into the formula $x = -\frac{b}{2a}$.

$$x = -\frac{b}{2a} = -\frac{-5}{2(-2)} = -\frac{5}{4}$$

The x-coordinate of the vertex is $-\frac{5}{4}$. To find the y-coordinate, we substitute $-\frac{5}{4}$ for x in the equation and solve for y.

$$y = -2x^2 - 5x + 3$$

$$y = -2\left(-\frac{5}{4}\right)^2 - 5\left(-\frac{5}{4}\right) + 3 \qquad \text{Substitute } -\frac{5}{4} \text{ for } x.$$

$$= -2\left(\frac{25}{16}\right) + \frac{25}{4} + 3$$

$$= -\frac{25}{8} + \frac{50}{8} + \frac{24}{8}$$

$$= \frac{49}{8}$$

Since the vertex is the point $\left(-\frac{5}{4}, \frac{49}{8}\right)$, we can plot it on the coordinate system in Figure 4-5(a) and draw the axis of symmetry.

Step 3: Find the x-intercept(s).

To find the x-intercepts, we substitute 0 for $f(x)$, and solve for x.

$$y = -2x^2 - 5x + 3$$

$$0 = -2x^2 - 5x + 3 \qquad \text{Substitute 0 for } y.$$

$$0 = 2x^2 + 5x - 3 \qquad \text{Divide both sides by } -1 \text{ to make the leading coefficient positive.}$$

$$0 = (2x - 1)(x + 3) \qquad \text{Factor the trinomial.}$$

$$2x - 1 = 0 \quad \text{or} \quad x + 3 = 0 \qquad \text{Set each factor equal to 0.}$$

$$x = \frac{1}{2} \quad \bigg| \quad x = -3 \qquad \text{Solve each linear equation.}$$

The x-intercepts are $\left(\frac{1}{2}, 0\right)$ and $(-3, 0)$. We plot these intercepts as shown in Figure 4-5(a).

Step 4: Find the y-intercept.

To find the y-intercept, we let $x = 0$, and solve for $f(x)$.

$$y = -2x^2 - 5x + 3$$

$$= -2(0)^2 - 5(0) + 3 \qquad \text{Substitute 0 for } x.$$

$$= 0 - 0 + 3$$

$$= 3$$

The y-intercept is $(0, 3)$. We plot the intercept as shown in Figure 4-5(a).

Step 5: Plot one additional point.

Because of symmetry, we know that the point $\left(-2\frac{1}{2}, 3\right)$ is on the graph. We plot this point on the coordinate system in Figure 4-5(a).

> **Take Note**
>
> The y-intercept of a parabola written in the general form $f(x) = ax^2 + bx + c\ (a \neq 0)$ is the point $(0, c)$. This is because when we substitute 0 for x, y is always c.

Step 6: We can now draw the graph of the function, as shown in Figure 4-5(b).

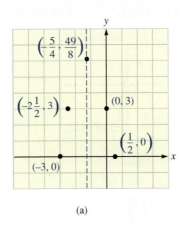

(a)

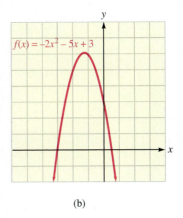

(b)

FIGURE 4-5

Self Check 4   Graph the function: $f(x) = -3x^2 + 7x - 2$.

Now Try Exercise 57.

Accent on Technology

Finding the Maximum Point or Minimum Point (Vertex) of a Parabola

We can use a graphing calculator to find the maximum point or minimum point (vertex) of a parabola. Consider the $f(x) = -2x^2 - 5x + 3$ given in Example 4. We will apply the following steps to determine the maximum point or vertex of the parabola.

1. Enter the function.

2. Set a window that shows the function.

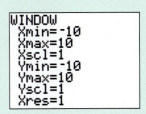

3. Graph the function.

4. Go to the CALC menu by pressing [2nd] [TRACE] and select maximum.

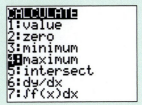

5. Use the [TRACE] key to get a left bound.

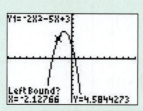

6. Use the [TRACE] key to get a right bound.

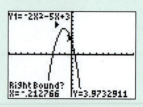

7. Use the [TRACE] key to make a guess.

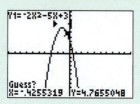

8. Press [ENTER] and find the maximum point.

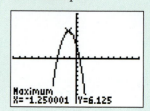

FIGURE 4-6

The process shown in Figure 4-6 leads us to we see that the maximum point or the vertex is $(-1.250001, 6.125)$.

5. Use a Quadratic Function to Solve Maximum and Minimum Problems

EXAMPLE 5

Using a Quadratic Function to Solve a Maximum Area Problem

The Montana Dude Rancher's Association has 400 feet of fencing to enclose a rectangular corral. To save money and fencing, the association intends to use the bank of a river as one boundary of the corral, as in Figure 4-7. Find the dimensions that will enclose the largest area.

SOLUTION   We will represent the fenced area with a quadratic function. Since its parabolic graph opens downward, the largest or maximum area will occur at the vertex. We can use the vertex formula to find the vertex.

Step 1: Represent the area with a quadratic function.
Let x represent the width of the fenced area. Then $400 - 2x$ represents the length. Because the area A of a rectangle is the product of the length and the width, we have

$$A = (400 - 2x)x \quad \text{or} \quad A(x) = -2x^2 + 400x$$

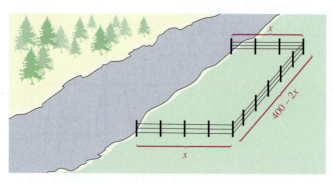

FIGURE 4-7

The graph of this area function is a parabola. Since the coefficient of x^2 is negative, the parabola opens downward and its vertex is its highest point. The A-coordinate of the vertex (x, A) represents the maximum area, and the x-coordinate represents the width of the corral that will give the maximum area.

Step 2: Find the vertex of the parabola.
We compare the equations

$$A(x) = -2x^2 + 400x \quad \text{and} \quad f(x) = ax^2 + bx + c$$

to see that $a = -2$, $b = 400$, and $c = 0$. Using the vertex formula, the vertex of the parabola is the point with coordinates

$$\left(-\frac{b}{2a}, c - \frac{b^2}{4a} \right) = \left(-\frac{400}{2(-2)}, 0 - \frac{400^2}{4(-2)} \right) = (100, 20{,}000)$$

Take Note

Note that we could have determined the y-coordinate of 20,000 by finding $A(100)$.

$$A(x) = -2x^2 + 400x$$

$$A(100) = -2(100)^2 + 400(100) \qquad \text{Substitute 100 for } x.$$

$$= -2(10{,}000) + 40{,}000$$

$$= -20{,}000 + 40{,}000$$

$$= 20{,}000$$

If the fence runs 100 feet out from the river, 200 feet parallel to the river, and 100 feet back to the river, it will enclose the largest possible area, which is 20,000 square feet.

Self Check 5   Find the largest area possible if the association has 1200 feet of fencing available.

Now Try Exercise 63.

Using a Quadratic Function to Solve a Minimum Cost Problem

EXAMPLE 6

A company that makes and sells stand-up paddleboards has found that the total weekly cost $C(x)$, in dollars, of producing x paddleboards is given by the function $C(x) = 0.5x^2 - 210x + 26{,}250$. Find the production level that minimizes the weekly cost and find that weekly minimum cost.

SOLUTION   The weekly cost function $C(x)$ is a quadratic function whose graph is a parabola that opens upward. The minimum value of $C(x)$ occurs at the vertex of the parabola. We will use the vertex formula to find the vertex of the parabola.

Since the coefficient of x^2 is 0.5 (a positive real number), the x-coordinate of the vertex is the production level that will minimize the cost, and the y-coordinate is that minimum cost. We compare the equations

$$C(x) = 0.5x^2 - 210x + 26{,}250 \quad \text{and} \quad f(x) = ax^2 + bx + c$$

to see that $a = 0.5$, $b = -210$, and $c = 26{,}250$. Using the vertex formula, we see that the vertex of the parabola is the point with coordinates

$$\left(-\frac{b}{2a}, c - \frac{b^2}{4a} \right) = \left(-\frac{-210}{2(0.5)}, 26{,}250 - \frac{(-210)^2}{4(0.5)} \right) = (210, 4200)$$

> **Take Note**
>
> We can also determine the y-coordinate of 4200 by finding $C(210)$.
>
> $$C(x) = 0.5x^2 - 210x + 26{,}250$$
>
> $$C(\mathbf{210}) = 0.5(\mathbf{210})^2 - 210(\mathbf{210}) + 26{,}250 \qquad \text{Substitute 210 for } x.$$
>
> $$= 0.5(44{,}100) - 44{,}100 + 26{,}250$$
>
> $$= 22{,}050 - 17{,}850$$
>
> $$= 4200$$

If the company makes 210 paddleboards each week, it will minimize its production cost. The minimum weekly cost will be $4200.

Self Check 6 A company that makes and sells baseball caps has found that the total monthly cost C in dollars of producing x caps is given by the function $C(x) = 0.2x^2 - 80x + 9000$. Find the production level that will minimize the monthly cost and find the minimum cost.

Now Try Exercise 77.

Accent on Technology

Quadratic Regression

Quadratic regression is a process used to find the equation of a quadratic function that is a "best fit" for a set of data. These models have many uses. In order to solve problems of this type, we use a graphing calculator with built-in functions that analyze the entered data.

The distance a vehicle will travel while braking depends on the speed the car is traveling at the time the brakes are applied. This distance is known as the **braking** or **stopping distance**. The table shows stopping distances in feet for a vehicle traveling at various speeds in miles per hour.

Speed in mph	10	20	30	40	50	60	70
Stopping Distance	27	63	109	164	229	304	388

©lightpoet/Shutterstock.com

We will use a graphing calculator to find the quadratic function that best fits this data.

1. Enter the data into LIST in a calculator. Press 2nd STAT to access the list window and then enter the values. Enter the mph values into L1 and the Stopping Distance values into L2 as shown in Figure 4-8.

L1	L2	L3	1
10	27	------	
20	63		
30	109		
40	164		
50	229		
60	304		
70	388		

L1 = {10, 20, 30, 40…

FIGURE 4-8

2. Set up a window for graphing. Use the values in the table shown in Figure 4-9.

FIGURE 4-9

3. Create a scatterplot and graph as shown in Figure 4-10.

 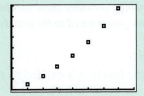

FIGURE 4-10

4. Press STAT and go to the CALC menu. Scroll to 5: QuadReg and press ENTER. Then press ENTER once again. This gives the equation $y = 0.0482x^2 + 2.1607x + 0.5714$ as shown in Figure 4-11. Note that we rounded to four decimals.

 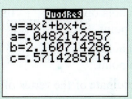

FIGURE 4-11

5. Graph the quadratic function as shown in Figure 4-12.

 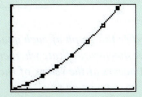

FIGURE 4-12

Our regression equation graph fits our data extremely well. We can use the regression equation we found to answer many questions. For example, what would be the stopping distance for an automobile traveling at a speed of 55 mph? We substitute 55 into our regression equation for x and determine y.

$$y = 0.0482x^2 + 2.1607x + 0.5714$$

$$y = 0.0482(55)^2 + 2.1607(55) + 0.5714$$

$$y = 265.2149$$

The stopping distance is approximately 265 feet.

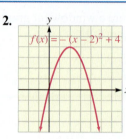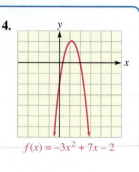
Exercises 4.1

Getting Ready
You should be able to complete these vocabulary and concept statements before you proceed to the practice exercises.

Fill in the blanks.

1. A quadratic function is defined by the equation
   _____ $(a \neq 0)$.
2. The standard form for the equation of a parabola is
   _____ $(a \neq 0)$.
3. The vertex of the parabolic graph of the equation
   $y = 2(x - 3)^2 + 5$ will be at _____.
4. The vertical line that intersects the parabola at its
   vertex is the _____.
5. If the parabola opens _____ the vertex will be a
   minimum point.
6. If the parabola opens _____ the vertex will be a
   maximum point.
7. The x-coordinate of the vertex of the parabolic graph
   of $f(x) = ax^2 + bx + c$ is ____.
8. The y-coordinate of the vertex of the parabolic graph
   of $f(x) = ax^2 + bx + c$ is _____.

Practice
Determine whether the graph of each quadratic function opens upward or downward. State whether a maximum or minimum point occurs at the vertex of the parabola.

9. $f(x) = \dfrac{1}{2}x^2 + 3$ 10. $f(x) = 2x^2 - 3x$

11. $f(x) = -3(x + 1)^2 + 2$ 12. $f(x) = -5(x - 1)^2 - 1$

13. $f(x) = -2x^2 + 5x - 1$ 14. $f(x) = 2x^2 - 3x + 1$

Find the vertex of each parabola.

15. $f(x) = x^2 - 1$ 16. $f(x) = -x^2 + 2$
17. $f(x) = (x - 3)^2 + 5$ 18. $f(x) = -2(x - 3)^2 + 4$

19. $f(x) = -2(x + 6)^2 - 4$ 20. $f(x) = \dfrac{1}{3}(x + 1)^2 - 5$

21. $f(x) = \dfrac{2}{3}(x - 3)^2$ 22. $f(x) = 7(x + 2)^2 + 8$

23. $f(x) = x^2 - 4x + 4$ 24. $f(x) = x^2 - 10x + 25$

25. $f(x) = x^2 + 6x - 3$ 26. $f(x) = -x^2 + 9x - 2$

27. $f(x) = -2x^2 + 12x - 17$ 28. $f(x) = 2x^2 + 16x + 33$

29. $f(x) = 3x^2 - 4x + 5$ 30. $f(x) = -4x^2 + 3x + 4$

31. $f(x) = \dfrac{1}{2}x^2 + 4x - 3$ 32. $f(x) = -\dfrac{2}{3}x^2 + 3x - 5$

Graph each quadratic function given in standard form. Identify the vertex, intercepts, and axis of symmetry.

33. $f(x) = x^2 - 4$ 34. $f(x) = x^2 + 1$

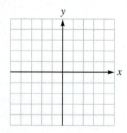   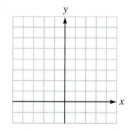

35. $f(x) = -3x^2 + 6$

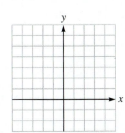

36. $f(x) = -4x^2 + 4$

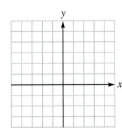

41. $f(x) = 2(x + 1)^2 - 2$

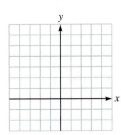

42. $f(x) = -\dfrac{3}{4}(x - 2)^2$

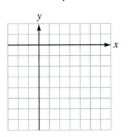

37. $f(x) = -\dfrac{1}{2}x^2 + 8$

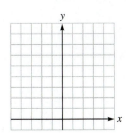

38. $f(x) = \dfrac{1}{2}x^2 - 2$

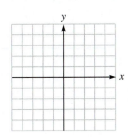

43. $f(x) = -(x + 4)^2 + 1$

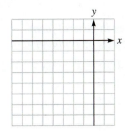

44. $f(x) = -3(x - 4)^2 + 3$

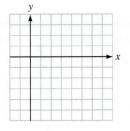

45. $f(x) = -3(x - 2)^2 + 6$

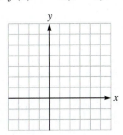

46. $f(x) = 2(x - 3)^2 - 4$

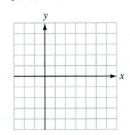

39. $f(x) = (x - 3)^2 - 1$

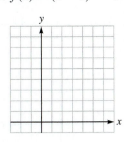

40. $f(x) = (x + 3)^2 - 1$

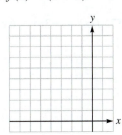

47. $f(x) = \dfrac{1}{3}(x - 1)^2 - 3$ **48.** $f(x) = -\dfrac{1}{2}(x + 1)^2 + 8$

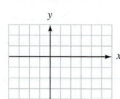   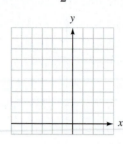

53. $f(x) = -x^2 - 4x + 1$ **54.** $f(x) = -x^2 - x + 6$

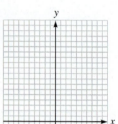

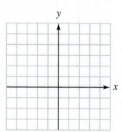

Graph each quadratic function given in general form. Identify the vertex, intercepts, and axis of symmetry.

49. $f(x) = x^2 + 2x$ **50.** $f(x) = x^2 - 6x$

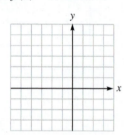   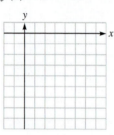

55. $f(x) = 2x^2 - 12x + 10$ **56.** $f(x) = -3x^2 - 3x + 18$

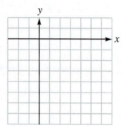

51. $f(x) = x^2 - 6x - 7$ **52.** $f(x) = x^2 - 4x + 1$

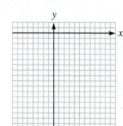   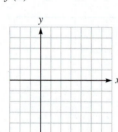

57. $f(x) = -3x^2 - 6x - 9$ **58.** $f(x) = -4x^2 - 4x + 3$

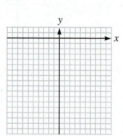

59. $f(x) = \dfrac{1}{2}x^2 - 2x - \dfrac{5}{2}$　　**60.** $f(x) = -\dfrac{1}{2}x^2 - x + 4$

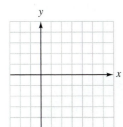

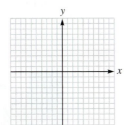

partition for the remaining two sides. What dimensions will maximize the area?

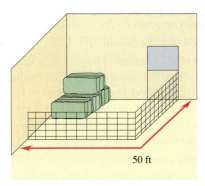

50 ft

66. Maximizing grazing area A rancher wishes to enclose a rectangular partitioned corral with 1800 feet of fencing. (See the illustration.) What dimensions of the corral would enclose the largest possible area? Find the maximum area.

Applications

61. Police investigations A police officer seals off the scene of an accident using a roll of yellow tape that is 300 feet long. What dimensions should be used to seal off the maximum rectangular area around the collision? Find the maximum area.

62. Maximizing area A rectangular flower bed has a width of x feet and a perimeter of 100 feet. Find x such that the area of the rectangle is maximized.

©Forewer/Shutterstock.com

63. Maximizing land area Jake has 800 feet of fencing to enclose a rectangular plot of land that borders a river. If Jake doesn't need a fence along the side of the river, find the length and width of the plot that will maximize the area. What is the largest area that can be enclosed?

64. Maximizing parking lot area A rectangular parking lot is being constructed for your college football stadium. If the parking lot is bordered on one side by a street and there are 750 yards of fencing available for the other three sides, find the length and width of the lot that will maximize the area. What is the largest area that can be enclosed?

65. Maximizing storage area A farmer wants to partition a rectangular feed storage area in a corner of his barn, as shown in the illustration. The barn walls form two sides of the stall, and the farmer has 50 feet of

67. Sheet metal fabrication A 24-inch-wide sheet of metal is to be bent into a rectangular trough with the cross section shown in the illustration. Find the dimensions that will maximize the amount of water the trough can hold. That is, find the dimensions that will maximize the cross-sectional area.

Depth

24 in.

Width

68. Maximizing cross-sectional area A 90-foot-wide sheet of metal is to be bent to form a rectangular trough from which your animals will drink water. Find the dimensions that will maximize the amount of water the trough can hold. That is, find the cross-sectional area of the trough.

69. Architecture A parabolic arch has an equation of $x^2 + 20y - 400 = 0$, where x is measured in feet. Find the maximum height of the arch.

70. Path of a guided missile A guided missile is propelled from the origin of a coordinate system with the x-axis along the ground and the y-axis vertical. Its path, or **trajectory**, is given by the equation $y = 400x - 16x^2$. Find the object's maximum height.

71. Height of a basketball The path of a basketball thrown from the free throw line can be modeled by the quadratic function $f(x) = -0.06x^2 + 1.5x + 6$, where x is the horizontal distance (in feet) from the free throw line and $f(x)$ is the height (in feet) of the ball. Find the maximum height of the basketball.

72. Projectile motion Devin throws a ball up a hill that makes an angle of 45° with the horizontal. The ball lands 100 feet up the hill. Its trajectory is a parabola with equation $y = -x^2 + ax$ for some number a. Find a.

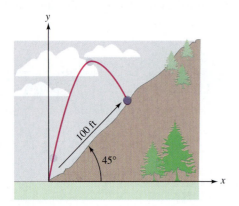

73. Height of a football A football is thrown by a quarterback from the 10-yard line and caught by the wide receiver on the 50-yard line. The football's path on this interval can be modeled by the quadratic function $f(x) = -\frac{1}{20}x^2 + 3x - 19$, where x is the horizontal distance in yards from the goal line and $f(x)$ is the height of the football in feet. Find the maximum height reached by the football.

74. Maximizing height A ball is thrown straight up from the top of a building 144 ft. tall with an initial velocity of 64 ft per second. The height $s(t)$ (in feet) of the ball from the ground, at time t (in seconds), is given by $s(t) = 144 + 64t - 16t^2$. Find the maximum height attained by the ball.

75. Flat-screen television sets A wholesaler of appliances finds that she can sell $(1200 - x)$ flat-screen television sets each week when the price is x dollars. What price will maximize revenue?

76. Maximizing revenue A seller of contemporary desks finds that he can sell $(820 - x)$ desks each month when the price is x dollars. What price will maximize revenue?

77. Minimizing cost A company that produces and sells digital cameras has determined that the total weekly cost $C(x)$, in dollars, of producing x digital cameras is given by the function $C(x) = 1.5x^2 - 144x + 5856$. Determine the production level that minimizes the weekly cost for producing the digital cameras and find that weekly minimum cost.

78. Maximizing profit A company that produces and sells chandeliers has determined that the total monthly profit $P(x)$ in dollars of producing and selling x chandeliers is given by the function $P(x) = -1.5x^2 + 153x + 7215$. Determine the production level that maximizes the monthly profit, and find that maximum profit.

79. Finding mass transit fares The Municipal Transit Authority serves 150,000 commuters daily when the fare is $1.80. Market research has determined that every penny decrease in the fare will result in 1000 new riders. What fare will maximize revenue?

80. Selling concert tickets Tickets for a concert are cheaper when purchased in quantity. The first 100 tickets are priced at $10 each, but each additional block of 100 tickets purchased decreases the cost of each ticket by 50¢. How many blocks of tickets should be sold to maximize the revenue?

81. Finding hotel rates A 300-room hotel is two-thirds filled when the nightly room rate is $90. Experience has shown that each $5 increase in cost results in 10 fewer occupied rooms. Find the nightly rate that will maximize income.

82. Finding Hilton rates A 500-room Hilton hotel is 80% filled when the nightly room rate is $160. Experience has shown that each $5 increase in the rate results in 10 fewer occupied rooms. Find the nightly rate that will maximize the nightly revenue.

An object is tossed vertically upward from ground level. Its height s(t), in feet, at time t seconds is given by the position function $s(t) = -16t^2 + 80t$. Use the position function for Exercises 83–86.

83. In how many seconds does the object reach its maximum height?

84. In how many seconds does the object return to the point from which it was thrown?

85. What is the maximum height reached by the object?

86. Show that it takes the same amount of time for the object to reach its maximum height as it does to return from that height to the point from which it was thrown.

 Use a graphing calculator to determine the coordinates of the vertex of each parabola. You will have to select appropriate viewing windows.

87. $y = 2x^2 + 9x - 56$

88. $y = 14x - \dfrac{x^2}{5}$

89. $y = (x - 7)(5x + 2)$

90. $y = -x(0.2 + 0.1x)$

 Use a graphing calculator and quadratic regression to find the quadratic function that best fits the given set of data.

91. $\{(-1, 6), (0, -1), (1, -3), (2, -1.5), (3, 5), (4, 10)\}$
Round to three decimal places.

92. $\{(-3, 4), (-1, 5), (0, 7), (2, 9), (5, 7), (6, 5)\}$
Round to three decimal places.

93. Alligators The length (in inches) and weight (in pounds) of 25 alligators is shown in the table. Find the quadratic function that best fits the data. Round a, b, and c to six decimal places. Use the regression function to estimate the weight of an alligator that is 130 inches long. Round the weight to the nearest pound.

©Raffaella Calzoni/Shutterstock.com

Length	Weight	Length	Weight	Length	Weight
94	130	72	38	90	106
74	51	128	366	89	84
147	640	85	84	68	39
58	28	82	80	76	42
86	80	86	83	114	197
94	110	88	70	90	102
63	33	72	61	78	57
86	90	74	54		
69	36	61	44		

94. Alligators Refer to Exercise 94. If an alligator weighs 125 pounds, what is its approximate length? Round to the nearest inch.

Discovery and Writing Exercises

95. What is a quadratic function?

96. Describe two ways of finding the vertex of a parabola given in general form.

97. What is an axis of symmetry of a parabola?

98. Share the strategy you would use to solve a maximum or minimum application problem.

99. Find the dimensions of the largest rectangle that can be inscribed in the right triangle ABC shown in the illustration.

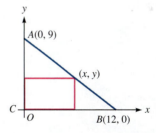

100. Point P lies in the first quadrant and on the line $x + y = 1$ in such a position that the area of triangle OPA is maximum. Find the coordinates of P. (See the illustration.)

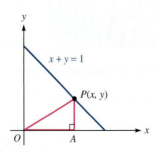

101. The sum of two numbers is 6, and the sum of the squares of those two numbers is as small as possible. What are the numbers?

102. What number most exceeds its square?

Critical Thinking

Determine if the statement is true or false. If the statement is false, then correct it and make it true.

103. If the sum of a number and its square is a minimum, then the two numbers are $-\dfrac{1}{2}$ and $\dfrac{1}{4}$.

104. The graphs of some quadratic functions have no x-intercepts.

105. The graphs of some quadratic functions have no y-intercepts.

106. The graph of a quadratic function is never constant.

107. If $f(x) = a(x - h)^2 + k$, and $a > 0$, then the range is $(-\infty, k]$.

108. If $f(x) = a(x - h)^2 + k$, and $a < 0$, then the graph of $f(x)$ is increasing on $(-\infty, h)$.

109. If $g(x) = ax^2 + bx + c$, and $a > 0$, then the graph of the function is increasing on $\left[-\dfrac{b}{2a}, \infty \right)$.

110. The axis of symmetry of the parabola $f(x) = 444x^2 - 888x + 222$ is $x = -1$.

4.2 Polynomial Functions

In this section, we will learn to

1. Recognize polynomial functions.

2. Understand characteristics of the graphs of polynomial functions.

3. Find zeros of polynomial functions by factoring.

4. Determine end behavior.

5. Graph polynomial functions.

6. Use the Intermediate Value Theorem.

So far, we have discussed two types of polynomial functions—first-degree (or linear) functions, and second-degree (or quadratic) functions. In this section, we will discuss polynomial functions of higher degree.

Polynomial functions can be used to model the path of a roller coaster or to model the fluctuation of gasoline prices over the past few months. The Hollywood Rip Ride Rockit is a roller coaster located at Universal Studios in Orlando, Florida. With a length of 3800 feet, a height of 167 feet, and a top speed of 65 miles per hour, it is one of the largest roller coasters ever built. Riders are video recorded for the entire ride and can choose from one of thirty songs to listen to during the experience. Portions of Rip Ride Rockit's tracks can be modeled with a polynomial function.

©Allen.G/Shutterstock.com

1. Recognize Polynomial Functions

Polynomial Function

A **polynomial function in one variable (say, x)** is a function of the form

$$f(x) = a_n x^n + a_{n-1} x^{n-1} + \cdots + a_1 x + a_0$$

where $a_n, a_{n-1}, \ldots, a_1$, and a_0 are real numbers and n is a whole number.

The **degree of a polynomial function** is the largest power of x that appears in the polynomial. The real number a_n is referred to as the **leading coefficient**.

It is important that we know the definition of a polynomial function and can identify polynomial functions. We have already covered basic polynomial functions. These were constant, linear, and quadratic functions. In this section, we will focus on polynomial functions of degree 3 or higher. The table shows examples of polynomial functions and examples of functions that are not polynomials.

Polynomial Function	Not a Polynomial Function
$f(x) = 5$ is constant function of degree 0.	$g(x) = \sqrt{x - 2} + 4$
$f(x) = \dfrac{3}{5}x - 7$ is a linear function of degree 1.	$g(x) = \sqrt[3]{x} - 7$
$f(x) = -2x^2 + 4x - 5$ is a quadratic function of degree 2.	$g(x) = 2\lvert x + 4 \rvert - 3$
$f(x) = \dfrac{1}{2}x^3 - 5x^2 + 3x - 11$ is a polynomial function of degree 3.	$g(x) = \dfrac{2x}{x^2 - 4}$
$f(x) = -5x^4 - 3x^2 + \sqrt{2}x - 8$ is a polynomial function of degree 4.	$g(x) = 3x^{-2} + 2x^{-1} - 7$
$f(x) = 2x^5 - 3x^4 - x^3 + 2x - 6$ is a polynomial function of degree 5.	$g(x) = x^{2/3} - 5x + 9$

EXAMPLE 1

Identifying Polynomial Functions

Determine whether or not the functions are a polynomial. For those that are, state the degree.

a. $f(x) = 7x^6 - 4x^3 - 5x + \sqrt{2}$ **b.** $g(x) = \sqrt[3]{2}x^8 - 9x^4 + \dfrac{10}{x}$

c. $h(x) = 2x^{-4} - 5x^{-2} + 4$

SOLUTION We use the definition of a polynomial function. The largest power of x will be the degree.

a. $f(x) = 7x^6 - 4x^3 - 5x + \sqrt{2}$ is a polynomial function of degree 6. Note that each exponent on the variable x is a whole number and that 6 is the largest power of x.

b. $g(x) = \sqrt[3]{2}x^8 - 9x^4 + \dfrac{10}{x}$ is not a polynomial function. Note that even though each exponent on the variable x is a whole number, x appears in the denominator. For that reason, the definition of a polynomial function is not satisfied.

c. $h(x) = 2x^{-4} - 7x^{-2} + 6$ is not a polynomial function. Note that the exponents -4 and -2 are not whole numbers.

Self Check 1 Determine whether or not the functions are polynomial functions. For those that are, state the degree. **a.** $f(x) = \sqrt[3]{2}x^4 - 6x^2 + 1$

b. $g(x) = \dfrac{1}{x^2} + \dfrac{5}{x} - 5$

Now Try Exercise 15.

2. Understand the Characteristics of the Graphs of Polynomial Functions

There are several characteristics of the graphs of polynomial functions that should be noted. We will list five of them. Consider the graphs of the two polynomial functions shown in Figures 4-13 and 4-14.

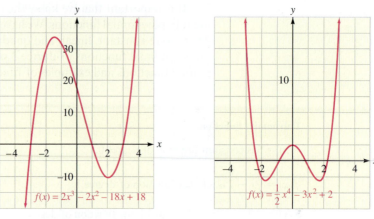

FIGURE 4-13   FIGURE 4-14

Characteristic 1: The graph of a polynomial function is a smooth and continuous curve.

Like the graphs of linear and quadratic functions, the graphs of higher-degree polynomial functions are smooth continuous curves. See Figures 4-13 and 4-14. Because their graphs are smooth, they have no cusps or corners. Because they are continuous, their graphs have no breaks or holes. Polynomial functions can always be drawn without lifting the pencil from the paper. See Figures 4-15 and 4-16.

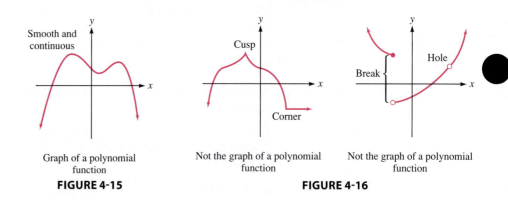

Graph of a polynomial function

FIGURE 4-15

Not the graph of a polynomial function

Not the graph of a polynomial function

FIGURE 4-16

Characteristic 2: The graph of a polynomial function has a specific end behavior and will either rise or fall on the far left and far right.

By end behavior, we mean what happens on the far left and far right on the graph. A polynomial function's graph will either rise or fall on its ends. Note that the graph of the polynomial function in Figure 4-13 falls on the left and rises on the right. Note that the polynomial function in Figure 4-14 rises on both the left and right.

Characteristic 3: The graph of a polynomial function often has x-intercepts.

We see in Figure 4-13 that the graph of the polynomial function has three x-intercepts. We see in Figure 4-14 that the graph of the polynomial function has four x-intercepts. The x-values of the x-intercepts are known as **zeros** of the polynomial function. A zero of a polynomial function is a value of x for which $f(x) = 0$. It is interesting to note that the graph of a polynomial function will either touch and turn at the zero of the polynomial function or cross the x-axis there. We will examine this concept in greater detail in this section.

Characteristic 4: The graph of a polynomial function often has turning points.

We see in Figure 4-13 that the graph of the polynomial function turns two times. We see in Figure 4-14 that the graph of the polynomial function turns three times.

These turning points on the graph are where maxima and minima occur. Finding the exact turning points often requires calculus techniques.

Characteristic 5: The graph of a polynomial function can be symmetric about the y-axis or the origin.

We see in Figure 4-14 that the graph of the polynomial function is symmetric with respect to the y-axis. Recall that a function's graph is symmetric about the y-axis if $f(-x) = f(x)$. A function's graph is symmetric about the origin if $f(-x) = -f(x)$. Symmetry can be very helpful when graphing a polynomial function. See Figure 4-17.

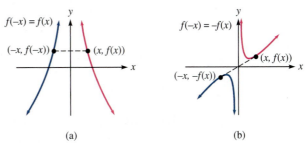

(a)   (b)

FIGURE 4-17

Now that we have discussed several characteristics of the graphs of polynomial functions, we will the continue learning the concepts necessary to graph them.

3. Find Zeros of Polynomial Functions by Factoring

In Chapter 1, we learned how to determine roots or solutions of several types of equations, including polynomial equations. We used factoring to solve the polynomial equations. If $f(x)$ represents a polynomial function, then the solutions or roots of the polynomial equation $f(x) = 0$ are known as **zeros** of the polynomial function. In this section, we will find zeros of polynomial functions which can be factored. However, not all polynomials can be factored using the factoring techniques we have learned; we will study those types of polynomials in subsequent sections. Each zero of the polynomial function appears as an x-intercept on the graph of the polynomial function.

EXAMPLE 2

Using Factoring to Find the Zeros of a Polynomial Function

Find the zeros of each polynomial function.
a. $f(x) = x^3 + 4x^2 - x - 4$ **b.** $f(x) = x^5 - 9x^3$

SOLUTION   We will set the polynomial function $f(x)$ equal to 0 and solve the polynomial equation to determine the zeros of each polynomial function.

a. We set $f(x) = 0$ and factor by grouping to find the zeros.

$$x^3 + 4x^2 - x - 4 = 0$$

$$x^2(x + 4) - 1(x + 4) = 0 \qquad \text{Factor out } x^2 \text{ from the first two terms and } -1 \text{ from the last two terms.}$$

$$(x + 4)(x^2 - 1) = 0 \qquad \text{Factor out the common factor } x + 4.$$

$$(x + 4)(x + 1)(x - 1) = 0 \qquad \text{Factor } x^2 - 1.$$

$$x + 4 = 0 \quad \text{or} \quad x + 1 = 0 \quad \text{or} \quad x - 1 = 0 \qquad \text{Set each factor equal to 0 and solve.}$$

$$x = -4 \qquad \qquad x = -1 \qquad \qquad x = 1$$

The zeros of the polynomial function are $x = -4$, $x = -1$, and $x = 1$. The graph of the polynomial function has x-intercepts $(-4, 0)$, $(-1, 0)$, and $(1, 0)$ as shown in Figure 4-18.

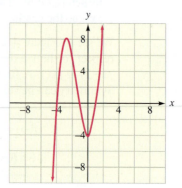

FIGURE 4-18

b. We set $f(x) = 0$ and use factoring to find the zeros.

$$x^5 - 9x^3 = 0$$

$$x^3(x^2 - 9) = 0 \qquad \text{Factor out } x^3.$$

$$x^3(x + 3)(x - 3) = 0 \qquad \text{Factor } x^2 - 9.$$

$$x^3 = 0 \quad \text{or} \quad x + 3 = 0 \quad \text{or} \quad x - 3 = 0 \qquad \text{Set each factor equal to 0 and solve.}$$

$$x = 0 \qquad\qquad x = -3 \qquad\qquad x = 3$$

The zeros of the polynomial function are $x = 0$, $x = -3$, and $x = 3$. The graph of the polynomial function has x-intercepts $(0, 0)$, $(-3, 0)$, and $(3, 0)$ as shown in Figure 4-19.

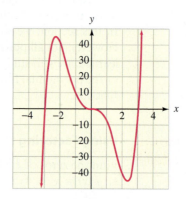

FIGURE 4-19

Self Check 2 Find the zeros of each polynomial function.

a. $f(x) = x^3 - x^2 - 2x$ **b.** $f(x) = -2x^4 + 2x^2$

Now Try Exercise 29.

4. Determine End Behavior

A basic polynomial function, called a **power function**, is a polynomial function of the form $f(x) = ax^n$, where a is a real number, $a \neq 0$, and n is a nonnegative integer. Power functions of the form $f(x) = x^n$ have simple graphs. Look at the graphs shown in Figure 4-20.

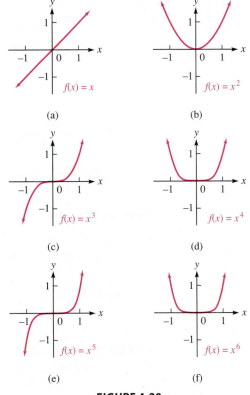

FIGURE 4-20

Notice that when n is even, the function is an even function, and the graph is symmetric with respect to the y-axis. The graph has the same general shape as $f(x) = x^2$. When n is odd, the function is an odd function, and the graph is symmetric with respect to the origin. When n is odd and greater than 1, the graph has the same general shape as $f(x) = x^3$. Note that the graphs are flatter at the origin and the functions increase more rapidly as n increases.

Now let's turn our attention to other polynomials and explore characteristics of their graphs. The end behavior of the graph of a polynomial function is the same as the graph of its term with highest degree. The leading coefficient, a_n, and the degree, n, of a polynomial will be our starting point. These will determine the end behavior. This is summarized in the **Leading Coefficient Test**.

Leading Coefficient Test and End Behavior

For $f(x) = a_n x^n + a_{n-1} x^{n-1} + \cdots + a_1 x + a_0$, $a_n \neq 0$, as x increases without bound ($x \to \infty$) or decreases without bound ($x \to -\infty$), the function will eventually increase without bound ($f(x) \to \infty$) or decrease without bound ($f(x) \to -\infty$) as shown below.

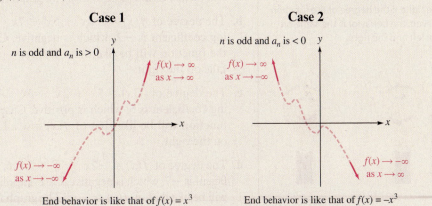

Case 3

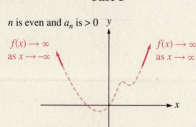

n is even and a_n is > 0

$f(x) \to \infty$ as $x \to -\infty$

$f(x) \to \infty$ as $x \to \infty$

End behavior is like that of $f(x) = x^2$

Case 4

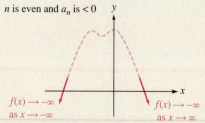

n is even and a_n is < 0

$f(x) \to -\infty$ as $x \to -\infty$

$f(x) \to -\infty$ as $x \to \infty$

End behavior is like that of $f(x) = -x^2$

Summary of Cases

- **Case 1:** If the degree of the polynomial is odd and the leading coefficient is positive, then the graph of the polynomial function falls on the left and rises on the right.

- **Case 2:** If the degree of the polynomial is odd and the leading coefficient is negative, then the graph of the polynomial function rises on the left and falls on the right.

- **Case 3:** If the degree of the polynomial is even and the leading coefficient is positive, then the graph of the polynomial function rises on the left and rises on the right.

- **Case 4:** If the degree of the polynomial is even and the leading coefficient is negative, then the graph of the polynomial function falls on the left and falls on the right.

 EXAMPLE 3

Using the Leading Coefficient Test to Determine End Behavior

Use the Leading Coefficient Test to describe the end behavior of each function.

a. $f(x) = 4x^3 - 2x + 1$ **b.** $f(x) = -2x^5 + 5x^4 - 7x - 3$

c. $f(x) = 5x^4 - 2x^3 + 3x + 4$ **d.** $f(x) = -7x^6 - x^3 + 2$

SOLUTION

We will identify the degree of the polynomial and the sign of the leading coefficient and then apply the Leading Coefficient Test to determine the end behavior of each function.

a. The degree of $f(x) = 4x^3 - 2x + 1$ is **3**, which is odd, and the leading coefficient is **4**, which is positive. Case 1 applies. The end behavior of the function will be like that of $f(x) = x^3$. The graph falls on the left and rises on the right.

b. The degree of $f(x) = -2x^5 + 5x^4 - 7x - 3$ is **5**, which is odd, and the leading coefficient is **−2**, which is negative. Case 2 applies. The end behavior of the function will be like that of $f(x) = -x^3$. The graph rises on the left and falls on the right.

c. The degree of $f(x) = 5x^4 - 2x^3 + 3x + 4$ is **4**, which is even, and the leading coefficient is **5**, which is positive. Case 3 applies. The end behavior of the function will be like that of $f(x) = x^2$. The graph rises on the left and rises on the right.

d. The degree of $f(x) = -7x^6 - x^3 + 2$ is **6**, which is even, and the leading coefficient is **−7**, which is negative. Case 4 applies. The end behavior of the function will be like that of $f(x) = -x^2$. The graph falls on the left and falls on the right.

Tip

End behavior is often referred to as right-and left-hand behavior. If the degree of the function is odd, the behavior is opposite on the left and the right. If the degree of the function is even, the behavior is the same on the left and the right.

Self Check 3   Describe the end behavior of each polynomial.

a. $f(x) = 2x^7 - 3x^4$

b. $f(x) = -\dfrac{2}{3}x^8 + 5x^3 - 3$

Now Try Exercise 43.

5. Graph Polynomial Functions

Now that we can find zeros of polynomial functions by factoring and know how to determine the end behavior of a polynomial function, we are almost prepared to graph a polynomial function. Two additional topics related to polynomial functions—the multiplicity of a zero and the number of turning points—can provide valuable insight into the function's graph.

Multiplicity

When a polynomial function is factored, the number of times a factor occurs is the **multiplicity** of the zero. Consider the polynomial function $f(x) = (x - 3)(x - 1)^2(x + 2)$. The graph of the function is shown in Figure 4-21.

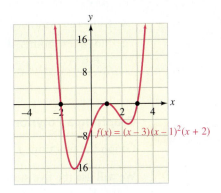

FIGURE 4-21

The polynomial function has the zeros $x = -2$, 1, and 3, and these zeros correspond to x-intercepts $(-2, 0)$, $(1, 0)$, and $(3, 0)$ on the graph of the function. Note that the factors $(x - 3)$ and $(x + 2)$ each occur once, so the zeros $x = 3$ and $x = -2$ have multiplicity one. The factor $(x - 1)^2$ occurs twice, so the zero $x = 1$ has multiplicity two. The multiplicity of a zero affects the behavior of the function at the zero.

Look at the graph of $f(x) = (x - 3)(x - 1)^2(x + 2)$. The graph crosses the x-axis where the multiplicity is one and touches the x-axis and turns where the multiplicity is two. The multiplicity of a zero determines whether the graph touches the x-axis and turns or crosses the x-axis.

Multiplicity and Zeros   If a is a zero with an **odd multiplicity**, the graph **crosses** the x-axis at $x = a$.

If a is a zero with an even multiplicity, the graph **touches** the x-axis and turns at $x = a$.

Number of Turning Points

There is also a relationship between the degree of the polynomial function and the number of turning points it has. Note in Figure 4-21 that the graph is of a fourth-degree polynomial function and that the graph has three turning points. This suggests the following result, which helps us understand the shape of many polynomial graphs.

Number of Turning Points   If $f(x)$ is a polynomial function of degree n, then the graph of $f(x)$ will have $n - 1$, or fewer, turning points.

The turning points are where local extrema occur. The graph of the polynomial function in Figure 4-21 has two local minima and one local maximum. Although we cannot find these values without using calculus, we can approximate them by plotting points or by using the [TRACE] feature on a graphing calculator. We can also use the CALC Minimum and CALC Maximum features on a graphing calculator to find these local extrema.

To graph polynomial functions, we can use the following steps.

Strategy for Graphing Polynomial Functions

1. Find the x- and y-intercepts of the graph.
2. Determine the end behavior.
3. Make a sign chart and determine where the graph is above and below the x-axis.
4. Find any symmetries of the graph.
5. Plot a few points, if necessary, and draw the graph as a smooth, continuous curve.

EXAMPLE 4   **Graphing a Polynomial Function of Degree 3**

Graph the function: $f(x) = x^3 - 4x$.

SOLUTION   We will use the five steps stated above to graph the polynomial function.

Step 1: Find the x- and y-intercepts of the graph.
To find the x-intercepts, we let $f(x) = 0$ and solve for x.

$$x^3 - 4x = 0$$
$$x(x^2 - 4) = 0 \qquad \text{Factor out } x.$$
$$x(x + 2)(x - 2) = 0 \qquad \text{Factor } x^2 - 4.$$
$$x = 0 \quad \text{or} \quad x + 2 = 0 \quad \text{or} \quad x - 2 = 0 \qquad \text{Set each factor equal to 0.}$$
$$x = -2 \qquad\qquad x = 2$$

The x-intercepts are $(0, 0)$, $(-2, 0)$, and $(2, 0)$. If we let $x = 0$ and solve for $f(x)$, we see that the y-intercept is also $(0, 0)$.

Step 2: Determine the end behavior.
The degree of the function $f(x) = x^3 - 4x$ is 3, which is odd, and the leading coefficient is 1, which is positive. The end behavior of the function is like that of the function $f(x) = x^3$. The graph falls on the left and rises on the right.

Step 3: Make a sign chart to determine where the graph is above and below the x-axis.
We will make a sign chart to determine the behavior of the function between the x-intercepts. The zeros of the polynomial function divide the x-axis into four intervals: $(-\infty, -2)$, $(-2, 0)$, $(0, 2)$, and $(2, \infty)$. We will test a value of x from

each interval and determine the sign of $f(x)$. We will then know if the function lies above or below the x-axis in that interval. See Figure 4-22.

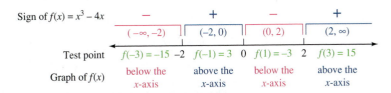

Sign of $f(x) = x^3 - 4x$	$-$	$+$	$-$	$+$
	$(-\infty, -2)$	$(-2, 0)$	$(0, 2)$	$(2, \infty)$
Test point	$f(-3) = -15$ -2	$f(-1) = 3$ 0	$f(1) = -3$ 2	$f(3) = 15$
Graph of $f(x)$	below the x-axis	above the x-axis	below the x-axis	above the x-axis

FIGURE 4-22

Step 4: Find any symmetries of the graph.
To test for symmetry about the y-axis, we check to see whether $f(x) = f(-x)$.
To test for symmetry about the origin, we check to see whether $f(-x) = -f(x)$.

$$f(x) = x^3 - 4x$$

$$f(-x) = (-x)^3 - 4(-x) \qquad \text{Substitute } -x \text{ for } x.$$

$$f(-x) = -x^3 + 4x \qquad \text{Simplify.}$$

Since $f(x) \neq f(-x)$, there is no symmetry about the y-axis. However, since $f(-x) = -f(x)$, there is symmetry about the origin.

Step 5: Plot a few points and draw the graph as a smooth continuous curve.
We now plot the intercepts and one additional point. In Step 3 we found that $f(1) = -3$. This will be the additional point we plot $(1, -3)$. Making use of our knowledge of symmetry and where the graph is above and below the x-axis we now draw the graph as shown in Figure 4-23(a). A calculator graph, using the standard viewing window, is shown in Figure 4-23(b).

$f(x) = x^3 - 4x$		
x	$f(x)$	$(x, f(x))$
-2	0	$(-2, 0)$
0	0	$(0, 0)$
1	-3	$(1, -3)$
2	0	$(2, 0)$

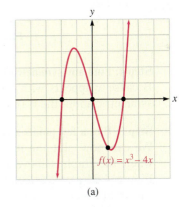

(a)

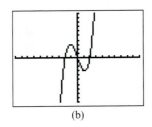

(b)

FIGURE 4-23

Self Check 4   Graph: $f(x) = x^3 - 9x$.

Now Try Exercise 51.

Graphing a Polynomial Function of Degree 4

EXAMPLE 5   Graph the function: $f(x) = x^4 - 5x^2 + 4$.

SOLUTION   We will use the five steps for graphing a polynomial function.

Step 1: Find the x- and y-intercepts of the graph.
To find the x-intercepts, we let $f(x) = 0$ and use factoring to solve for x.

$$x^4 - 5x^2 + 4 = 0$$
$$(x^2 - 4)(x^2 - 1) = 0$$
$$(x + 2)(x - 2)(x + 1)(x - 1) = 0$$

$$x + 2 = 0 \quad \text{or} \quad x - 2 = 0 \quad \text{or} \quad x + 1 = 0 \quad \text{or} \quad x - 1 = 0$$
$$x = -2 \qquad\qquad x = 2 \qquad\qquad x = -1 \qquad\qquad x = 1$$

The x-intercepts are $(-2, 0)$, $(2, 0)$, $(-1, 0)$, and $(1, 0)$. To find the y-intercept, we let $x = 0$ and see that the y-intercept is $(0, 4)$.

Step 2: Determine the end behavior.

The degree of the function $f(x) = x^4 - 5x^2 + 4$ is 4, which is even, and the leading coefficient is 1, which is positive. The end behavior of the function is like that of the function $f(x) = x^2$. The graph rises on the left and rises on the right.

Step 3: Make a sign chart to determine where the graph is above and below the x-axis.

We will make a sign chart to determine the behavior of the function between the x-intercepts. The zeros of the polynomial function divide the x-axis into five intervals: $(-\infty, -2)$, $(-2, -1)$, $(-1, 1)$, $(1, 2)$, and $(2, \infty)$. We will test a value of x from each interval and determine the sign of $f(x)$. We will then know if the function lies above or below the x-axis in that interval. See Figure 4-24.

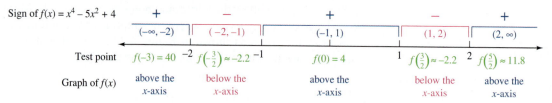

FIGURE 4-24

Step 4: Find any symmetries of the graph.

Because x appears with only even exponents, $f(x) = f(-x)$, and the graph is symmetric about the y-axis. The graph is not symmetric about the origin.

Step 5: Plot a few points and draw the graph as a smooth continuous curve.

We can now plot the intercepts and use our knowledge of symmetry and where the graph is above and below the x-axis to draw the graph, as shown in Figure 4-25(a). A calculator graph, using the standard viewing window, is shown in Figure 4-25(b).

$f(x) = x^4 - 5x^2 + 4$		
x	$f(x)$	$(x, f(x))$
-2	0	$(-2, 0)$
-1	0	$(-1, 0)$
0	4	$(0, 4)$
1	0	$(1, 0)$
$\dfrac{3}{2}$	$-\dfrac{35}{16}$	$\left(\dfrac{3}{2}, -\dfrac{35}{16}\right)$
2	0	$(2, 0)$

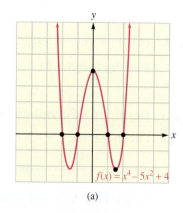

(a)

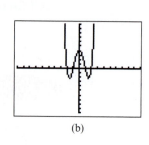

(b)

FIGURE 4-25

Self Check 5   Graph: $f(x) = x^4 - 10x^2 + 9$.

Now Try Exercise 61.

6. Use the Intermediate Value Theorem

The following theorem is called the **Intermediate Value Theorem**. It can be used to locate an interval that contains a zero. Sometimes the notation $P(x)$ is used to represent a polynomial function.

Intermediate Value Theorem   Let $P(x)$ be a polynomial function with real coefficients. If $P(a) \neq P(b)$ for $a < b$, then $P(x)$ takes on all values between $P(a)$ and $P(b)$ on the closed interval $[a, b]$.

This theorem becomes clear when we consider the graph of the polynomial function $y = P(x)$, shown in Figure 4-26. We have seen that graphs of polynomials are *continuous* curves, a technical term that means, roughly, that they can be drawn without lifting the pencil from the paper. If $P(a) \neq P(b)$, the continuous curve joining the points $A(a, P(a))$ and $B(b, P(b))$ must take on all values between $P(a)$ and $P(b)$ in the interval $[a, b]$, because the curve has no gaps in it.

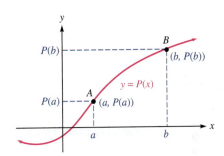

FIGURE 4-26

If $P(a)$ and $P(b)$ have opposite signs, there is at least one number r in the interval (a, b) for which $P(r) = 0$. See Figure 4-27. By the Intermediate Value Theorem, $P(x)$ takes on all values between $P(a)$ and $P(b)$. Since $P(a)$ and $P(b)$ have opposite signs, the number 0 lies between them. Thus, there is a number r between a and b for which $P(r) = 0$. This number r is a zero of $P(x)$ and a root of the equation $P(x) = 0$.

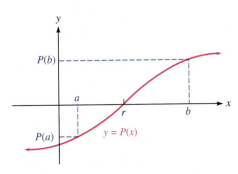

FIGURE 4-27

EXAMPLE 6   **Using the Intermediate Value Theorem**

Show that $P(x) = 2x^3 - x^2 - 8x + 4$ has at least one real zero between 0 and 1.

SOLUTION   Let $P(x) = 2x^3 - x^2 - 8x + 4$. We will evaluate the polynomial function at $x = 0$ and $x = 1$. If the resulting values have opposite signs, we know that the equation has a zero that lies between 0 and 1.

Evaluate $P(0)$ and $P(1)$ as follows:

$$P(0) = 2(0)^3 - 0^2 - 8(0) + 4 = 4$$

$$P(1) = 2(1)^3 - 1^2 - 8(1) + 4 = -3$$

Because $P(0)$ and $P(1)$ have opposite signs, we know that there is at least one real zero between 0 and 1, as shown in Figure 4-28. The zero shown is $\frac{1}{2}$.

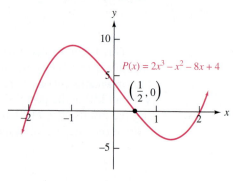

FIGURE 4-28

Self Check 6   Show that $P(x) = 2x^3 - 9x^2 + 7x + 6$ has at least one real zero between -1 and 0.

Now Try Exercise 73.

Self Check Answers

1. **a.** yes; degree 4
   b. not a polynomial function
2. **a.** $0, 2, -1$
   b. $0, 1, -1$
3. **a.** falls on the left and rises on the right
   b. falls on the left and falls on the right

4.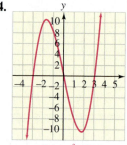

   $f(x) = x^3 - 9x$

5.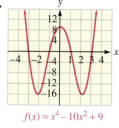

   $f(x) = x^4 - 10x^2 + 9$

6. $P(-1) = -12$; $P(0) = 6$ The results have opposite signs.

Exercises 4.2

Getting Ready

You should be able to complete these vocabulary and concept statements before you proceed to the practice exercises.

Fill in the blanks.

1. The degree of the function $f(x) = x^4 - 3$ is ___.

2. Peaks and valleys on a polynomial graph are called _____ points.

3. The roots of the polynomial equation $f(x) = 0$ are known as _____ of the polynomial function.

4. The zeros of a polynomial function appear as _____ on the graph of the polynomial function.

5. If the degree of the polynomial is odd and the leading coefficient is positive, then the graph of the polynomial function _____ on the left and _____ on the right.

6. If the degree of the polynomial is odd and the leading coefficient is negative, then the graph of the polynomial function _____ on the left and _____ on the right.

7. If the degree of the polynomial is even and the leading coefficient is positive, then the graph of the polynomial function _____ on the left and _____ on the right.

8. If the degree of the polynomial is even and the leading coefficient is negative, then the graph of the polynomial function _____ on the left and _____ on the right.

9. If $(x + 5)^3$ occurs as a factor of a polynomial function, then the _____ of the zero $x = -5$ is 3.

10. The graph of a nth degree polynomial function can have at most _____ turning points.

11. If $P(x)$ is a polynomial with real coefficients and $P(a) \neq P(b)$ for $a < b$, then $P(x)$ takes on all values between _____ in the interval $[a, b]$.

12. If $P(x)$ has real coefficients and $P(a)$ and $P(b)$ have opposite signs, there is at least one number r in (a, b) for which _____.

Practice

Determine whether or not the functions are polynomial functions. For those that are, state the degree.

13. $f(x) = \dfrac{1}{2}x^5 - 5x^3 + 3x - 10$

14. $f(x) = 0.8x^6 - 5x^3 - 2x + \sqrt{5}$

15. $f(x) = -11x^7 - 3x^2 + \sqrt{11}x - 1$

16. $f(x) = 2x^8 - \sqrt[3]{6}x^4 - 2x^3 + 2x - 9$

17. $f(x) = x^4 + \sqrt{x} + 7$

18. $f(x) = 3x^{-2} + 5x^{-1} - 7$

19. $f(x) = -6x^2 + 13 - \dfrac{1}{x}$

20. $f(x) = \sqrt[3]{7}x^3 - 9x^2 + |x|$

Determine whether or not the graph of the functions shown are polynomial functions.

21.
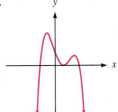

22.
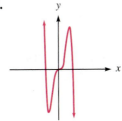

23.
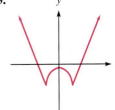

24.
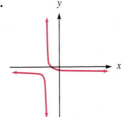

Find the zeros of each polynomial function and state the multiplicity of each. State whether the graph touches the x-axis and turns or crosses the x-axis at each zero.

25. $f(x) = 4x^2 - 25$

26. $f(x) = 64 - 9x^2$

27. $f(x) = 2x^2 + 7x - 15$

28. $f(x) = 6x^2 - x - 2$

29. $g(x) = 2x^3 - 7x^2 - 15x$

30. $g(x) = x^3 - 8x^2 + 16x$

31. $g(x) = x^3 + 6x^2 - 4x - 24$

32. $g(x) = x^3 - 2x^2 - 9x + 18$

33. $f(x) = x^4 + 2x^3 - 3x^2$

34. $f(x) = x^4 - 3x^3 + 2x^2$

35. $f(x) = x^4 - 15x^2 + 44$

36. $f(x) = x^4 - 19x^2 + 48$

37. $h(x) = 3x^2(x + 4)^2(x - 5)$

38. $h(x) = 2x(x - 3)^2(x + 1)^2$

39. $h(x) = (2x - 5)(x + 3)(x - 1)^2$

40. $h(x) = (3x + 1)^3(x + 1)^2$

Use the Leading Coefficient Test to determine the end behavior of each polynomial.

41. $f(x) = \sqrt{5}x^7 + 10x^3 - 2x$

42. $f(x) = 4x^9 - 7x^2 + 5x - 12$

43. $g(x) = -\frac{1}{2}x^5 + 3x^4 + 2x^2 - 4$

44. $g(x) = -3x^7 + 2x^4 - 5x + 2$

45. $f(x) = 7x^4 - 2x^2 + 1$

46. $f(x) = \frac{2}{3}x^6 + 3x^3 + 2x$

47. $h(x) = -3x^4 - 5x - 1$

48. $h(x) = -x^6 + 3x^2 + 2$

Graph each polynomial function.

49. $f(x) = x^3 - 9x$

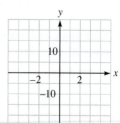

50. $f(x) = x^3 - 16x$

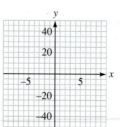

51. $f(x) = -x^3 - 4x^2$

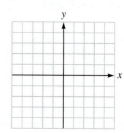

52. $f(x) = -x^3 + 2x$

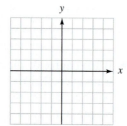

53. $f(x) = x^3 + x^2$

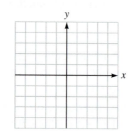

54. $f(x) = x^3 - x$

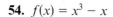

55. $f(x) = x^3 - 9x^2 + 18x$

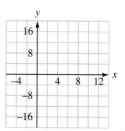

56. $f(x) = -x^3 - 9x^2 - 18x$

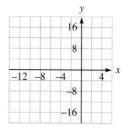

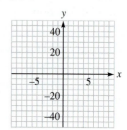
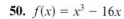
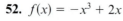

57. $f(x) = x^3 - x^2 - 4x + 4$

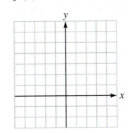

58. $f(x) = 4x^3 - 4x^2 - x + 1$

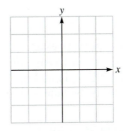

59. $f(x) = x^4 - 2x^2 + 1$ **60.** $f(x) = x^4 - 5x^2 + 4$

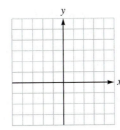   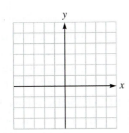

61. $f(x) = -x^4 + 5x^2 - 4$ **62.** $f(x) = -x^4 + 11x^2 - 18$

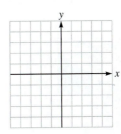   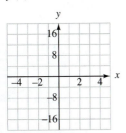

63. $f(x) = -x^4 + 6x^3 - 8x^2$ **64.** $f(x) = -x^4 + 2x^3 + 8x^2$

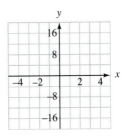   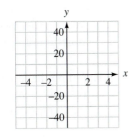

65. $f(x) = \dfrac{1}{2}x^4 - \dfrac{9}{2}x^2$ **66.** $f(x) = -\dfrac{1}{2}x^4 + 8x^2$

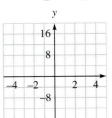   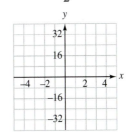

67. $f(x) = x(x - 3)(x - 2)(x + 1)$

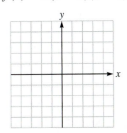

68. $f(x) = -(x - 4)(x - 2)(x + 2)(x + 4)$

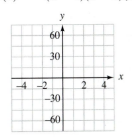

69. $f(x) = x^5 - 4x^3$ **70.** $f(x) = -x^5 + 8x^3$

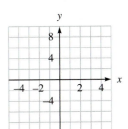   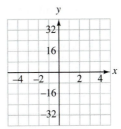

Use the Intermediate Value Theorem to show that each equation has at least one real zero between the specified numbers.

71. $P(x) = 2x^2 + x - 3$; -2 and -1

72. $P(x) = 2x^3 + 17x^2 + 31x - 20$; -1 and 2

73. $P(x) = 3x^3 - 11x^2 - 14x$; 4 and 5

74. $P(x) = 2x^3 - 3x^2 + 2x - 3$; 1 and 2

75. $P(x) = x^4 - 8x^2 + 15$; 1 and 2

76. $P(x) = x^4 - 8x^2 + 15$; 2 and 3

77. $P(x) = 30x^3 - 61x^2 + 39x + 10$; 2 and 3

78. $P(x) = 30x^3 - 61x^2 + 39x + 10$; −1 and 0

79. $P(x) = 30x^3 - 61x^2 + 39x + 10$; 0 and 1

80. $P(x) = 5x^3 - 9x^2 - 4x + 9$; −1 and 0

Applications

81. Maximize volume An open box is to be constructed from a piece of cardboard 20 inches by 24 inches by cutting a square of length x from each corner and folding up the sides as shown in the figure.

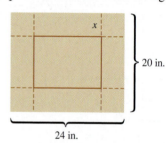

a. Write a polynomial function $V(x)$ that expresses the volume of the constructed box as a function of x.

b. Use the theory learned about graphing polynomial functions and graph $V(x)$.

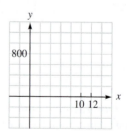

c. What is the domain of the function as it relates to the application problem?

d. Use a graphing calculator to graph the function, and estimate the value of x that gives the maximum volume and then estimate the maximum volume. Round to one decimal place.

82. Maximize volume Repeat Exercise 81 using a piece of cardboard dimensions 30 inches by 36 inches.

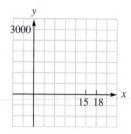

83. Maximize production If 270 apples trees are planted per acre, the production per tree is 840 pounds. For every tree, x, over 270 planted per acre, the production per tree decreases by $(840 - 0.1x^2)$ pounds per tree.

a. Write a production function, $P(x)$, for the number of pounds per acre as a function of x.

b. Use a graphing calculator to graph the function $P(x)$ and determine the number of trees that should be planted per acre to produce the maximum number of pounds of apples per acre. Round to the nearest unit.

84. Maximize volume for luggage Most airlines restrict the size of carry-on luggage to a total of 45 inches (length plus width plus height). The height of a piece of luggage is to be 4 inches less than the width.

a. Write a function for the volume V as a function of x, the width of the luggage.

b. Graph the function and determine the dimensions that will create a piece of luggage with the maximum volume. Round to the nearest inch.

$$\begin{array}{r|rrrr} -2 & 5 & 3 & -21 & -1 \\ & & -10 & & \\ \hline & 5 & -7 & & \end{array}$$ $-2(5) = -10;\ 3 + (-10) = -7$

$$\begin{array}{r|rrrr} -2 & 5 & 3 & -21 & -1 \\ & & -10 & 14 & \\ \hline & 5 & -7 & -7 & \end{array}$$ $-2(-7) = 14;\ -21 + 14 = -7$

$$\begin{array}{r|rrrr} -2 & 5 & 3 & -21 & -1 \\ & & -10 & 14 & 14 \\ \hline & 5 & -7 & -7 & 13 \end{array}$$ $-2(-7) = 14;\ -1 + 14 = 13$

Because the remainder is 13, $P(-2) = 13$.

Self Check 6   Find $P(3)$.

Now Try Exercise 49.

Using Synthetic Division to Evaluate a Polynomial Function at i

EXAMPLE 7

If $P(x) = x^3 - x^2 + x - 1$, find $P(i)$, where $i = \sqrt{-1}$.

SOLUTION   We will use synthetic division and find the remainder.

Take Note

Example 7 illustrates that synthetic division can be used with complex numbers.

$$\begin{array}{r|rrrr} i & 1 & -1 & 1 & -1 \\ & & i & -1-i & 1 \\ \hline & 1 & i-1 & -i & 0 \end{array}$$

Since the remainder is 0, $P(i) = 0$ and i is a zero of $P(x)$.

Self Check 7   Find $P(-i)$.

Now Try Exercise 53.

6. Use Synthetic Division to Solve Polynomial Equations

Solving a Polynomial Equation when Given One Solution

EXAMPLE 8

Let $P(x) = 3x^3 - 5x^2 + 3x - 10$. Completely solve the polynomial equation $P(x) = 0$ given that 2 is one solution.

SOLUTION   Since 2 is a solution of the equation $P(x) = 0$, we know that 2 is a zero of $P(x)$. We will use synthetic division and divide $P(x)$ by $x - 2$, obtaining a remainder of 0. Then, we will use the result of the synthetic division to help factor the polynomial and solve the equation.

1. We use synthetic division to divide $P(x)$ by $x - 2$.

$$\begin{array}{r|rrrr} 2 & 3 & -5 & 3 & -10 \\ & & 6 & 2 & 10 \\ \hline & 3 & 1 & 5 & 0 \end{array}$$

2. We use the result of the synthetic division and factor the polynomial.

$$3x^3 - 5x^2 + 3x - 10 = (x - 2)(3x^2 + x + 5)$$

3. Finally, we solve the polynomial equation $P(x) = 0$.

$$3x^3 - 5x^2 + 3x - 10 = 0$$

$$(x - 2)(3x^2 + x + 5) = 0$$

To solve for x, we set each factor equal to 0 and apply the Quadratic Formula to the equation $3x^2 + x + 5 = 0$.

$$x - 2 = 0 \quad \text{or} \quad 3x^2 + x + 5 = 0$$

$$x = 2 \qquad\qquad x = \frac{-1 \pm \sqrt{1^2 - 4(3)(5)}}{2(3)}$$

$$x = \frac{-1 \pm i\sqrt{59}}{6}$$

The solution set is $\left\{ 2, \ -\dfrac{1}{6} \pm \dfrac{\sqrt{59}}{6}i \right\}$.

Self Check 8   Solve: $x^3 - 1 = 0$ given that 1 is one solution.

Now Try Exercise 87.

EXAMPLE 9

Finding a Polynomial Function when Given the Zeros

Find a third-degree polynomial function $P(x)$ with three zeros of 3, 3, and -5.

SOLUTION   We will use the Factor Theorem and write the three factors that correspond to the three zeros of 3, 3, and -5. We will then multiply the resulting binomials.

If 3, 3, and -5 are the three zeros of $P(x)$, then $x - 3$, $x - 3$, and $x - (-5)$ are the three factors of $P(x)$.

$$P(x) = (x - 3)(x - 3)(x + 5)$$

$$P(x) = (x^2 - 6x + 9)(x + 5) \qquad \text{Multiply } x - 3 \text{ and } x - 3.$$

$$P(x) = x^3 - x^2 - 21x + 45 \qquad \text{Multiply using the Distributive Property.}$$

The polynomial function $P(x) = x^3 - x^2 - 21x + 45$ has zeros of 3, 3, and -5. Because 3 occurs twice as a zero, we say that 3 is a **zero of multiplicity 2**.

Self Check 9   Find a third-degree polynomial function $P(x)$ with zeros of -2, 2, and 3.

Now Try Exercise 97.

Self Check Answers

1. $P(2)$ is the remainder.   **2.** $P(-3)$ is the remainder.   **3.** -11 **4.** no

5. $2x^2 + 2x + 11 + \dfrac{26}{x - 3}$ **6.** 98   **7.** 0   **8.** $1, -\dfrac{1}{2} \pm \dfrac{\sqrt{3}}{2}i$

9. $P(x) = x^3 - 3x^2 - 4x + 12$

Exercises 4.3

Getting Ready

You should be able to complete these vocabulary and concept statements before you proceed to the practice exercises.

Fill in the blanks.

1. The variables in a polynomial have _____-number exponents.
2. A zero of $P(x)$ is any number c for which _____.
3. The Remainder Theorem holds when c is ____ number.
4. If $P(x)$ is a polynomial function and $P(x)$ is divided by _____, the remainder will be $P(c)$.
5. If $P(x)$ is a polynomial function, then $P(c) = 0$ if and only if $x - c$ is a _____ of $P(x)$.
6. A shortcut method for dividing a polynomial by a binomial of the form $x - c$ is called _____ division.

Practice

Use long division to perform each division.

7. $\dfrac{4x^3 - 2x^2 - x + 1}{x - 1}$

8. $\dfrac{2x^3 + 3x^2 - 5x + 1}{x + 3}$

9. $\dfrac{2x^4 + x^3 + 2x^2 + 15x - 5}{x + 2}$

10. $\dfrac{x^4 + 6x^3 - 2x^2 + x - 1}{x - 1}$

Find each value by substituting the given value of x into the polynomial and simplifying. Then find the value by performing long division and finding the remainder.

11. $P(x) = 3x^3 - 2x^2 - 5x - 7$; $P(2)$
12. $P(x) = 5x^3 + 4x^2 + x - 1$; $P(-2)$
13. $P(x) = 7x^4 + 2x^3 + 5x^2 - 1$; $P(-1)$
14. $P(x) = 2x^4 - 2x^3 + 5x^2 - 1$; $P(2)$
15. $P(x) = 2x^5 + x^4 - x^3 - 2x + 3$; $P(1)$
16. $P(x) = 3x^5 + x^4 - 3x^2 + 5x + 7$; $P(-2)$

Use the Remainder Theorem to find the remainder that occurs when $P(x) = 3x^4 + 5x^3 - 4x^2 - 2x + 1$ is divided by each binomial.

17. $x + 2$
18. $x - 1$
19. $x - 2$
20. $x + 1$
21. $x + 3$
22. $x - 3$
23. $x - 4$
24. $x + 4$

Use the Factor Theorem to determine whether each statement is true. If the statement is not true, so indicate.

25. $x - 1$ is a factor of $P(x) = x^7 - 1$.
26. $x - 2$ is a factor of $P(x) = x^3 - x^2 + 2x - 8$.
27. $x - 1$ is a factor of $P(x) = 3x^5 + 4x^2 - 7$.
28. $x + 1$ is a factor of $P(x) = 3x^5 + 4x^2 - 7$.
29. $x + 3$ is a factor of $P(x) = 2x^3 - 2x^2 + 1$.
30. $x - 3$ is a factor of
    $P(x) = 3x^5 - 3x^4 + 5x^2 - 13x - 6$.
31. $x - 1$ is a factor of
    $P(x) = x^{1984} - x^{1776} + x^{1492} - x^{1066}$.
32. $x + 1$ is a factor of
    $P(x) = x^{1984} + x^{1776} - x^{1492} - x^{1066}$.

Use the Division Algorithm and synthetic division to express the polynomial function $P(x) = 3x^3 - 2x^2 - 6x - 4$ in the form (divisor) (quotient) + remainder for each divisor.

33. $x - 1$
34. $x - 2$
35. $x - 3$
36. $x - 4$
37. $x + 1$
38. $x + 2$
39. $x + 3$
40. $x + 4$

Use synthetic division to perform each division.

41. $\dfrac{x^3 + x^2 + x - 3}{x - 1}$

42. $\dfrac{x^3 - x^2 - 5x + 6}{x - 2}$

43. $\dfrac{7x^3 - 3x^2 - 5x + 1}{x + 1}$

44. $\dfrac{2x^3 + 4x^2 - 3x + 8}{x - 3}$

45. $\dfrac{4x^4 - 3x^3 - x + 5}{x - 3}$

46. $\dfrac{x^4 + 5x^3 - 2x^2 + x - 1}{x + 1}$

47. $\dfrac{3x^5 - 768x}{x - 4}$

48. $\dfrac{x^5 - 4x^2 + 4x + 4}{x + 3}$

Let $P(x) = 5x^3 + 2x^2 - x + 1$. Use synthetic division to find each value.

49. $P(2)$

50. $P(-2)$

51. $P(-5)$

52. $P(3)$

53. $P(i)$

54. $P(-i)$

Let $P(x) = 2x^4 - x^2 + 2$. Use synthetic division to find each value.

55. $P\left(\dfrac{1}{2}\right)$

56. $P\left(\dfrac{1}{3}\right)$

57. $P(i)$

58. $P(-i)$

Let $P(x) = x^4 - 8x^3 + 8x + 14x^2 - 15$. Write the terms of $P(x)$ in descending powers of x and use synthetic division to find each value.

59. $P(1)$

60. $P(0)$

61. $P(-3)$

62. $P(-1)$

63. $P(-i)$

64. $P(i)$

Let $P(x) = 8 - 8x^2 + x^5 - x^3$. Write the terms of $P(x)$ in descending powers of x and use synthetic division to find each value.

65. $P(i)$

66. $P(-i)$

67. $P(-2i)$

68. $P(2i)$

Use the Factor Theorem and synthetic division to determine whether the given polynomial is a factor of the polynomial function $P(x)$.

69. $P(x) = 3x^3 - 13x^2 - 10x + 56; x + 2$

70. $P(x) = 2x^3 + 3x^2 - 32x + 15; x + 5$

71. $P(x) = x^4 - 3x^3 + 4x^2 - 2x + 4; x - 1$

72. $P(x) = 2x^4 - x^3 - 2x^2 + x + 1; x - 2$

73. $P(x) = 3x^5 - 22x^3 + 15x^2 + 3x + 9; x + 3$

74. $P(x) = x^5 + 5x^4 - 17x^2 - 2x + 8; x + 4$

Determine whether the given number is a zero of the polynomial function $P(x)$.

75. $P(x) = -3x^3 + 13x^2 + 10x - 56; 4$

76. $P(x) = -2x^3 - 3x^2 + 32x - 15; 3$

77. $P(x) = 4x^4 + x^3 + 20x^2 - 4; -2$

78. $P(x) = 2x^4 - x^3 - 64x^2 - 2; -6$

79. $P(x) = 4x^5 - 2x^4 + 6x^3 + 5x^2 - 6x + 1; \dfrac{1}{2}$

80. $P(x) = 6x^5 + 7x^4 - 3x^3 + 6x^2 + 13x - 5; \dfrac{1}{3}$

A partial solution set is given for each polynomial equation. Find the complete solution set.

81. $x^3 + 3x^2 - 13x - 15 = 0; \{-1\}$

82. $x^3 + 6x^2 + 5x - 12 = 0; \{1\}$

83. $2x^3 + x^2 - 18x - 9 = 0; \left\{-\dfrac{1}{2}\right\}$

84. $2x^3 - 3x^2 - 11x + 6 = 0; \left\{\dfrac{1}{2}\right\}$

85. $x^3 - 6x^2 + 7x + 2 = 0; \{2\}$

86. $x^3 + x^2 - 8x - 6 = 0; \{-3\}$

87. $x^3 - 3x^2 + x + 57 = 0; \{-3\}$

88. $2x^3 - x^2 + x - 2 = 0; \{1\}$

89. $x^4 - 2x^3 - 2x^2 + 6x - 3 = 0; \{1, 1\}$

90. $x^5 + 4x^4 + 4x^3 - x^2 - 4x - 4 = 0; \{1, -2, -2\}$

91. $x^4 - 5x^3 + 7x^2 - 5x + 6 = 0; \{2, 3\}$

92. $x^4 + 2x^3 - 3x^2 - 4x + 4 = 0; \{1, -2\}$

Find a polynomial function $P(x)$ with the given zeros. There is no unique answer for $P(x)$.

93. $4, 5$

94. $-3, 5$

95. $1, 1, 1$

96. $1, 0, -1$

97. $2, 4, 5$

98. $7, 6, 3$

99. $1, -1, \sqrt{2}, -\sqrt{2}$

100. $0, 0, 0, \sqrt{3}, -\sqrt{3}$

101. $\sqrt{2}, i, -i$

102. $i, i, 2$

103. $0, 1 + i, 1 - i$

104. $i, 2 + i, 2 - i$

Discovery and Writing

105. State the Division Algorithm and explain how it can be used to verify the results of long division.

106. State the Remainder Theorem and describe why it is used in algebra.

107. State the Factor Theorem and describe why it is used in algebra.

108. Describe the steps used to perform synthetic division.

109. If 0 is a zero of
$P(x) = a_n x^n + a_{n-1} x^{n-1} + \cdots + a_1 x + a_0$,
find a_0.

110. If 0 occurs twice as a zero of
$P(x) = a_n x^n + a_{n-1} x^{n-1} + \cdots + a_1 x + a_0$,
find a_1.

111. If $P(2) = 0$ and $P(-2) = 0$, explain why $x^2 - 4$ is a factor of $P(x)$.

112. If $P(x) = x^4 - 3x^3 + kx^2 + 4x - 1$ and $P(2) = 11$, find k.

Critical Thinking

Determine if the statement is true or false. If the statement is false, then correct it and make it true.

113. Synthetic division can be used to perform the division $\dfrac{55x^{44} + 44x^{33} + 33x^{22} + 11}{x - 66}$.

114. Synthetic division can be used to perform the division $\dfrac{55x^{44} + 44x^{33} + 33x^{22} + 11}{x^2 - 66}$.

115. -1 is a zero of $P(x) = x^{13,579} + x^{2468}$.

116. $(x - 1)$ is a factor of $P(x) = x^{13,579} - x^{2468}$.

117. If $P(x) = x^{444} + x^{44} + x^4$, then $P(i) = 3i$.

118. i is a zero of $P(x) = 222x^2 + 222$.

119. If $(135x - 246)$ is a factor of $P(x)$, then $x = \dfrac{135}{246}$ is a zero of $P(x)$.

120. If 0 is a zero of $P(x)$, then the constant term of $P(x)$ is 0.

4.4 Fundamental Theorem of Algebra and Descartes' Rule of Signs

In this section, we will learn to

1. Understand the Fundamental Theorem of Algebra.

2. Use the Conjugate Pairs Theorem.

3. Use Descartes' Rule of Signs.

4. Find integer bounds on roots.

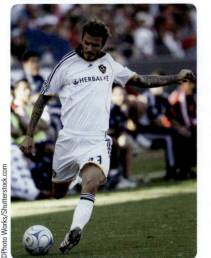

©Photo Works/Shutterstock.com

David Beckham is one of the best soccer players of all time. He's an English former professional soccer player who played for Manchester United, Preston North End, Real Madrid, Milan, LA Galaxy, Paris Saint-Germain, and the England national team. He was the first English player to win league titles in four countries: England, Spain, the United States, and France. He was twice runner-up for the International Federation of Association Football's Player of the Year award. He played the final game of his 20-year career in 2013.

In soccer, it is important for players to know the boundaries of the playing field. They need to know when it is advantageous to kick the ball out of bounds and when it is advantageous to keep the ball in bounds.

Knowing where the boundaries are can also be helpful in mathematics. Establishing integer bounds on roots can be an important aid in locating and finding the roots of polynomial equations. In this section, we will learn how to find integer bounds for the roots.

The Remainder Theorem and synthetic division provide a way of verifying that a particular number is a zero of a polynomial function, but they do not provide the zeros. We need some guidelines that indicate how many zeros to expect, what kind of zeros to expect, and where they are located. This section develops several theorems that provide such guidelines.

1. Understand the Fundamental Theorem of Algebra

Before attempting to find the zeros of a polynomial function, it would be useful to know whether any zeros exist. This question was answered by Carl Friedrich Gauss (1777–1855) when he proved the **Fundamental Theorem of Algebra**.

The Fundamental Theorem of Algebra	If $P(x)$ is a polynomial function with positive degree, then $P(x)$ has at least one complex zero.

The Fundamental Theorem of Algebra guarantees that polynomials such as

$$P(x) = 2x^3 - 2x^2 + 7x - 3 \text{ and } P(x) = 32x^{11} + x^3 - 2x - 5$$

all have zeros. Since all polynomials with positive degree have zeros, their corresponding polynomial equations $P(x) = 0$ all have roots. They are the zeros of the polynomial function.

The next theorem will help us show that every nth-degree polynomial function $P(x)$ has exactly n zeros.

The Polynomial Factorization Theorem	If $n > 0$ and $P(x)$ is an nth-degree polynomial function, then $P(x)$ has exactly n linear factors: $$P(x) = a_n(x - c_1)(x - c_2)(x - c_3) \cdot \cdots \cdot (x - c_n)$$ where $c_1, c_2, c_3, \ldots, c_n$ are numbers and a_n is the leading coefficient of $P(x)$.

PROOF Let $P(x)$ be a polynomial function of degree n ($n > 0$). Because of the Fundamental Theorem of Algebra, we know that $P(x)$ has a zero c_1 and that the equation $P(x) = 0$ has c_1 for a root. By the Factor Theorem, we know that $x - c_1$ is a factor of $P(x)$. Thus,

$$P(x) = (x - c_1)q_1(x)$$

If the leading coefficient of the nth-degree polynomial function $P(x)$ is a_n, then $q_1(x)$ is a polynomial function of degree $n - 1$ whose leading coefficient is also a_n.

By the Fundamental Theorem of Algebra, we know that $q_1(x)$ also has a zero, c_2. By the Factor Theorem, $x - c_2$ is a factor of $q_1(x)$, and

$$P(x) = (x - c_1)(x - c_2)q_2(x)$$

where $q_2(x)$ is a polynomial function of degree $n - 2$ with leading coefficient a_n.

This process can continue only to n factors of the form $x - c_i$ until the final quotient $q_n(x)$ is a polynomial function of degree $n - n$, or degree 0. Thus, the polynomial function $P(x)$ factors completely as

$$(1) \qquad P(x) = a_n(x - c_1)(x - c_2)(x - c_3) \cdot \cdots \cdot (x - c_n)$$

We can use the Polynomial Factorization Theorem and make the following conclusions:

1. If we substitute any one of the numbers $c_1, c_2, c_3, \ldots, c_n$ for x in Equation 1, $P(x)$ will equal 0. Thus, each value of c is a zero of $P(x)$ and a root of the equation $P(x) = 0$.
2. There can be no other zeros, because no single factor in Equation 1 is 0 for any value of x not included in the list $c_1, c_2, c_3, \ldots, c_n$.
3. The values of c in the previous list need not be distinct. Any number c_i that occurs k times as a zero of a polynomial function is called a **zero of multiplicity k**.

The following theorem summarizes the previous discussion.

Number of Zeros Theorem	If multiple zeros are counted individually, the polynomial function $P(x)$ with degree n ($n > 0$) has exactly n zeros among the complex numbers.

The previous theorems are illustrated by the examples shown below.

Polynomial Function	Properties
$P(x) = x - 6$	**Degree:** First **Number of Linear Factors:** 1 **Linear Factor:** $x - 6$ **Number of Zeros:** 1 **Zero:** 6 **Multiplicity of Each Zero:** • 6 has a multiplicity of one.
$P(x) = x^2 - 8x + 16$ $\quad = (x - 4)(x - 4)$	**Degree:** Second **Number of Linear Factors:** 2 **Linear Factors:** $x - 4$, $x - 4$ **Number of Zeros:** 2 **Zeros:** 4 and 4 **Multiplicity of Each Zero:** • 4 has a multiplicity of two.
$P(x) = x^3 + 100x$ $\quad = x(x^2 + 100)$ $\quad = x(x - 10i)(x + 10i)$	**Degree:** Third **Number of Linear Factors:** 3 **Linear Factors:** $x, x - 10i,$ and $x + 10i$ **Number of Zeros:** 3 **Zeros:** 0, $10i$, and $-10i$ **Multiplicity of Each Zero:** • 0 has a multiplicity of one. • $10i$ has a multiplicity of one. • $-10i$ has a multiplicity of one.
$P(x) = x^4 - 7x^2$ $\quad = x^2(x^2 - 7)$ $\quad = x \cdot x(x + \sqrt{7})(x - \sqrt{7})$	**Degree:** Fourth **Number of Linear Factors:** 4 **Linear Factors:** $x, x, x + \sqrt{7}, x - \sqrt{7}$ **Number of Zeros:** 4 **Zeros:** 0, 0, $-\sqrt{7}$, and $\sqrt{7}$ **Multiplicity of Each Zero:** • 0 has a multiplicity of two. • $-\sqrt{7}$ has a multiplicity of one. • $\sqrt{7}$ has a multiplicity of one.

Take Note

Remember that every nth degree polynomial function has exactly n linear factors and exactly n zeros. The n zeros are also the roots or solutions of the polynomial equation $P(x) = 0$.

2. Use the Conjugate Pairs Theorem

Recall that the complex numbers $a + bi$ and $a - bi$ are called **complex conjugates** of each other. The next theorem points out that *complex zeros of polynomial functions with real coefficients occur in complex conjugate pairs.*

The Conjugate Pairs Theorem   If a polynomial function $P(x)$ with real-number coefficients has a complex zero $a + bi$ with $b \neq 0$, then its conjugate $a - bi$ is also a zero.

A strategy is given next for finding a polynomial function.

Strategy for Finding a Polynomial Function with Real Coefficients	To find a polynomial function with real coefficients, apply the following two steps. **1.** Use the Conjugate Pairs Theorem to determine the zeros of the polynomial function. **2.** Write the linear factorization of the polynomial function. Then multiply.

Finding a Polynomial Function

EXAMPLE 1

Find a second-degree polynomial function with real coefficients that has a zero of $2 + i$.

SOLUTION   We can use the Conjugate Pairs Theorem to determine the second zero. Then, we will use the two zeros to write the linear factorization and polynomial function.

1. Determine the zeros.
Because $2 + i$ is a zero, its complex conjugate $2 - i$ is also a zero. The zeros are $2 + i$ and $2 - i$.

2. Write the linear factorization of the polynomial function. Then multiply.
Because $2 + i$ is a zero, $x - (2 + i)$ is a factor. Because $2 - i$ is a zero, $x - (2 - i)$ is a factor. The polynomial function is

$$P(x) = [x - (2 + i)][x - (2 - i)]$$

$$P(x) = (x - 2 - i)(x - 2 + i)$$

$$P(x) = x^2 - 4x + 5 \qquad \text{Multiply.}$$

The polynomial function $P(x)\ x^2 - 4x + 5$ will have zeros of $2 + i$ and $2 - i$. ∎

Self Check 1   Find a second-degree polynomial function $P(x)$ with real coefficients that has a zero of $1 - i$.

Now Try Exercise 23.

Finding a Polynomial Function

EXAMPLE 2

Find a fourth-degree polynomial function with real coefficients and i as a zero of multiplicity 2.

SOLUTION   We can use the Conjugate Pairs Theorem to determine the other two zeros. Then, we will use the four zeros to write the linear factorization and polynomial function.

1. Determine the zeros.
Because i is a zero twice and a fourth-degree polynomial function has four zeros, we must find the other two zeros. According to the Conjugate Pairs Theorem, the missing zeros are the conjugates of the given zeros. Thus, the four zeros are i, i, $-i$, and $-i$.

2. Write the linear factorization of the polynomial function. Then multiply.
Because i is a zero of multiplicity 2, the factor $x - i$ occurs twice. Because $-i$ is a zero of multiplicity 2, the factor $x - (-i)$ occurs twice. The function is

> **Take Note**
>
> The Conjugate Pairs Theorem applies only to polynomial function with real-number coefficients. In Examples 1 and 2, the theorem does apply and the results were polynomial functions with real-number coefficients.

$$P(x) = (x - i)(x - i)[x - (-i)][x - (-i)]$$

$$P(x) = (x - i)(x + i)(x - i)(x + i)$$

$$P(x) = (x^2 + 1)(x^2 + 1) \qquad \text{Multiply.}$$

$$P(x) = x^4 + 2x^2 + 1 \qquad \text{Multiply.}$$

Self Check 2   Find a fourth-degree polynomial function with real coefficients and $-i$ as a zero of multiplicity 2.

Now Try Exercise 29.

EXAMPLE 3   **Finding a Quadratic Function with a Repeated Zero**

Find a quadratic function with a repeated zero of i.

SOLUTION   We will use the two given zeros of i to write the linear factorization of the quadratic function.

Because i is a zero twice, the linear factorization of the quadratic function are

$$P(x) = (x - i)(x - i)$$

$$P(x) = x^2 - 2ix - 1 \qquad \text{Multiply.}$$

In this function, the coefficient of x is $-2i$, which is not a real number. Therefore, it is not surprising that the zeros are not in complex conjugate pairs.

Take Note

In Example 3, the Conjugate Pairs Theorem does **not** apply because the resulting polynomial function does **not** have real-number coefficients.

Self Check 3   Find a quadratic function with a repeated zero of $-i$.

Now Try Exercise 33.

©Brendan Howard/Shutterstock.com

I THINK THEREFORE "I" AM!

©Bruce Stanfield/Shutterstock.com

3. Use Descartes' Rule of Signs

René Descartes (1596–1650) was a French philosopher, mathematician, and writer who is known for the famous statement "I think, therefore I am." In mathematics, he is credited with a theorem known as **Descartes' Rule of Signs**, which enables us to estimate the number of positive, negative, and nonreal zeros of a polynomial function.

If a polynomial function is written in descending powers of x and we scan it from left to right, we say that a variation in sign occurs whenever successive terms have opposite signs. For example, the polynomial function

$$\overset{+ \text{ to } -}{P(x) = 3x^5} - 2x^4 \overset{- \text{ to } +}{-5x^3} \overset{+ \text{ to } -}{+ x^2} - x - 9$$

has three variations in sign, and the polynomial function

$$P(-x) = 3(-x)^5 - 2(-x)^4 - 5(-x)^3 + (-x)^2 - (-x) - 9$$

$$= -3x^5 \overset{- \text{ to } +}{- 2x^4} + 5x^3 + x^2 \overset{+ \text{ to } -}{+ x - 9}$$

has two variations in sign.

Descartes' Rule of Signs Let $P(x)$ be a polynomial function with real coefficients.

- The number of positive real zeros of $P(x)$ is either equal to the number of variations in sign of $P(x)$ or less than that by an even number.

- The number of negative real zeros of $P(x)$ is either equal to the number of variations in sign of $P(-x)$ or less than that by an even number.

Using Descartes' Rule of Signs

EXAMPLE 4

Find the number of possible positive, negative, and nonreal zeros of $P(x) = 3x^3 - 2x^2 + x - 5$.

SOLUTION We can use Descartes' Rule of Signs to determine the number of possible positive, negative, and nonreal zeros.

Take Note

Variation in signs means sign changes.

©DeiMosz/Shutterstock.com

Tip

To obtain $P(-x)$ quickly, you can simply multiply each odd degree term of $P(x)$ by negative one.

1. **Determine the number of possible positive zeros.**
   Let $P(x) = 3x^3 - 2x^2 + x - 5$. Since there are three variations of sign in $P(x) = 3x^3 - 2x^2 + x - 5$, there can be either 3 positive zeros or only 1 (1 is less than 3 by the even number 2).

2. **Determine the number of possible negative zeros.**
   Because

$$P(-x) = 3(-x)^3 - 2(-x)^2 + (-x) - 5$$
$$= -3x^3 - 2x^2 - x - 5$$

   has no variations in sign, there are 0 negative zeros. Furthermore, 0 is not a zero, because the terms of the polynomial function do not have a common factor of x.

3. **Determine the number of nonreal zeros.**
   If there are 3 positive zeros, then all of the zeros are accounted for. If there is 1 positive zero, the 2 remaining zeros must be nonreal complex numbers. The following chart shows these possibilities.

Number of Positive Zeros	Number of Negative Zeros	Number of Nonreal Zeros
3	0	0
1	0	2

The number of nonreal complex zeros is the number needed to bring the total number of zeros up to 3.

Self Check 4 Discuss the possibilities for the zeros of $P(x) = 5x^3 + 2x^2 - x + 3$.

Now Try Exercise 37.

Using Descartes' Rule of Signs

EXAMPLE 5

Find the number of possible positive, negative, and nonreal zeros of $P(x) = 5x^5 - 3x^3 - 2x^2 + x - 1$.

SOLUTION We can use Descartes' Rule of Signs to determine the number of possible positive, negative, and nonreal zeros.

Tip

Always make sure the polynomial function is written in descending powers of x before you apply Descartes' Rule of Signs. Then count the number of variation in signs very carefully.

1. **Determine the number of possible positive zeros.**

   Let $P(x) = 5x^5 - 3x^3 - 2x^2 + x - 1$. Since there are three variations of sign in $P(x)$, there are either 3 or 1 positive zeros.

2. **Determine the number of possible negative zeros.**

   Because $P(-x) = -5x^5 + 3x^3 - 2x^2 - x - 1$ has two variations in sign, there are 2 or 0 negative zeros.

3. **Determine the number of nonreal zeros.**

   The possibilities are shown as follows:

Number of Positive Zeros	Number of Negative Zeros	Number of Nonreal Zeros
1	0	4
3	0	2
1	2	2
3	2	0

   In each case, the number of nonreal complex zeros is an even number. This is expected, because this polynomial function has real coefficients, and its nonreal complex zeros will occur in conjugate pairs.

Self Check 5   Discuss the possibilities for the zeros of $P(x) = 5x^5 - 2x^2 - x - 1$.

Now Try Exercise 41.

4. Find Integer Bounds on Zeros

A final theorem provides a way to find **bounds** on the zeros of a polynomial function, enabling us to look for zeros where they can be found.

Upper and Lower Bounds on Zeros

Let $P(x)$ be a polynomial function with real coefficients and a positive leading coefficient.

1. If $P(x)$ is synthetically divided by a positive number c and each term in the last row of the division is nonnegative, then no number greater than c can be a zero of $P(x) = 0$. (c is called an **upper bound** of the real zeros.)

2. If $P(x)$ is synthetically divided by a negative number d and the signs in the last row alternate,* no value less than d can be a zero of $P(x) = 0$. (d is called a **lower bound** of the real zeros.)

*If 0 appears in the third row, that 0 can be assigned either a + or a − sign to help the signs alternate.

EXAMPLE 6   **Finding Integer Bounds on Zeros**

Establish integer bounds for the zeros of $P(x) = 2x^3 + 3x^2 - 5x - 7$.

SOLUTION   We will use the upper and lower bound rules previously stated to establish the integer bounds for the zeros.

1. We will perform several synthetic divisions by positive integers, looking for nonnegative values in the last row.

   Trying 1 first gives

$$
\begin{array}{r|rrrr}
\underline{1} & 2 & 3 & -5 & -7 \\
   &   & 2 & +5 & +0 \\
\hline
   & +2 & +5 & 0 & -7
\end{array}
$$

Because one of the signs in the last row is negative, we cannot claim that 1 is an upper bound. We now try 2.

$$
\begin{array}{r|rrrr}
\underline{2} & 2 & 3 & -5 & -7 \\
   &   & +4 & 14 & 18 \\
\hline
   & +2 & +7 & +9 & +11
\end{array}
$$

Because the last row is entirely nonnegative, we can claim that 2 is an upper bound. That is, no number greater than 2 can be a zero of the function.

2. We perform several synthetic divisions by negative divisors, looking for alternating signs in the last row.

   We begin with -3.

$$
\begin{array}{r|rrrr}
\underline{-3} & 2 & 3 & -5 & -7 \\
   &   & -6 & 9 & -12 \\
\hline
   & +2 & -3 & +4 & -19
\end{array}
$$

Since the signs in the last row alternate, -3 is a lower bound. That is, no number less than -3 can be a zero. To see whether there is a greater lower bound, we try -2.

$$
\begin{array}{r|rrrr}
\underline{-2} & 2 & 3 & -5 & -7 \\
   &   & -4 & 2 & 6 \\
\hline
   & +2 & -1 & -3 & -1
\end{array}
$$

Since the signs in the last row do not alternate, we cannot claim that -2 is a lower bound.

Since -3 is a lower bound and 2 is an upper bound, then all of the real zeros must be in the interval $(-3, 2)$.

Self Check 6 Establish integer bounds for the zeros of $P(x) = 2x^3 + 3x^2 - 11x - 7$.

Now Try Exercise 53.

It is important to understand what the theorem on the bounds of zeros says and what it doesn't say. The following two explanations will help clarify that for us.

1. If we divide synthetically by a positive number c and the last row of the synthetic division is entirely nonnegative, the theorem guarantees that c is an upper bound of the zeros. However, if the last row contains some negative values, c could still be an upper bound.

2. If we divide by a negative number d and the signs in the last row alternate, the theorem guarantees that d is a lower bound of the zeros. However, if the signs in the last row do not alternate, d could still be a lower bound. This is illustrated in Example 6. It can be shown that the smallest negative zero of the function is approximately -1.81. Thus, -2 is a lower bound for the zeros of the function. However, when we checked -2, the last row of the synthetic division did not have alternating signs. Unfortunately, the theorem does not always determine the best bounds for the zeros of the function.

Self Check Answers

1. $P(x) = x^2 - 2x + 2$ 2. $P(x) = x^4 + 2x^2 + 1$ 3. $P(x) = x^2 + 2ix - 1$

4.

Num. of Positive Zeros	Num. of Negative Zeros	Num. of Nonreal Zeros
2	1	0
0	1	2

5.

Num. of Positive Zeros	Num. of Negative Zeros	Num. of Nonreal Zeros
1	2	2
1	0	4

6. Zeros are in $(-4, 3)$.

Exercises 4.4

Getting Ready

You should be able to complete these vocabulary and concept statements before you proceed to the practice exercises.

Fill in the blanks.

1. If $P(x)$ is a polynomial function with positive degree, then $P(x)$ has at least one _____.

2. The statement in Exercise 1 is called the _____.

3. The _____ of $a + bi$ is $a - bi$.

4. The polynomial $6x^4 + 5x^3 - 2x^2 + 3$ has ___ variations in sign.

5. The polynomial $(-x)^3 - (-x)^2 - 4$ has ___ variations in sign.

6. The polynomial function $P(x) = 7x^4 + 5x^3 - 2x + 1$ can have at most ___ positive zeros.

7. The polynomial function $P(x) = 7x^4 + 5x^3 - 2x + 1$ can have at most ___ negative zeros.

8. Complex zeros occur in complex _____ pairs. (Assume that the equation has real coefficients.)

9. If no number less than d can be a zero of $P(x) = 0$, then d is called a(n) _____.

10. If no number greater than c can be a zero of $P(x) = 0$, then c is called a(n) _____.

Practice

Determine how many zeros each polynomial function has.

11. $P(x) = x^{10} - 1$ 12. $P(x) = x^{40} - 1$

13. $P(x) = 3x^4 - 4x^2 - 2x + 7$

14. $P(x) = -32x^{111} - x^5 - 1$

15. One zero of $P(x) = x(3x^4 - 2) - 12x$ is 0. How many other zeros are there?

16. Two zeros of $P(x) = 3x^2(x^7 - 14x + 3)$ are 0. How many other zeros are there?

Determine how many linear factors and zeros each polynomial function has.

17. $P(x) = x^4 - 81$ 18. $P(x) = x^{40} + x^{39}$

19. $P(x) = 4x^5 + 8x^3$ 20. $P(x) = x^3 + 144x$

Write a quadratic function with real coefficients and the given zero.

21. $2i$ 22. $-3i$

23. $3 - i$ 24. $4 + 2i$

Write a third-degree polynomial function with real coefficients and the given zeros.

25. $3, -i$ 26. $1, i$

27. $2, 2 + i$ 28. $-2, 3 - i$

Write a fourth-degree polynomial function with real coefficients and the given zeros.

29. $3, 2, i$

30. $1, 2, 1 + i$

31. $i, 1 - i$

32. $i, 2 - i$

Write a quadratic function with the given repeated zero.

33. $2i$

34. $-2i$

Use Descartes' Rule of Signs to find the number of possible positive, negative, and nonreal zeros of each function.

35. $P(x) = 3x^3 + 5x^2 - 4x + 3$

36. $P(x) = 3x^3 - 5x^2 - 4x - 3$

37. $P(x) = 2x^3 + 7x^2 + 5x + 5$

38. $P(x) = -2x^3 - 7x^2 - 5x - 4$

39. $P(x) = 8x^4 + 5$

40. $P(x) = -3x^3 + 5$

41. $P(x) = x^4 + 8x^2 - 5x - 10$

42. $P(x) = 5x^7 + 3x^6 - 2x^5 + 3x^4 + 9x^3 + x^2 + 1$

43. $P(x) = -x^{10} - x^8 - x^6 - x^4 - x^2 - 1$

44. $P(x) = x^{10} + x^8 + x^6 + x^4 + x^2 + 1$

45. $P(x) = x^9 + x^7 + x^5 + x^3 + x$ (Is 0 a zero?)

46. $P(x) = -x^9 - x^7 - x^5 - x^3 - x$ (Is 0 a zero?)

47. $P(x) = -2x^4 - 3x^2 + 2x + 3$

48. $P(x) = -7x^5 - 6x^4 + 3x^3 - 2x^2 + 7x - 4$

Find integer bounds for the zeros of each function. Answers can vary.

49. $P(x) = x^2 - 2x - 4$ **50.** $P(x) = 9x^2 - 6x - 1$

51. $P(x) = 18x^2 - 6x - 1$ **52.** $P(x) = 2x^2 - 10x - 9$

53. $P(x) = 6x^3 - 13x^2 - 110x$

54. $P(x) = 12x^3 + 20x^2 - x - 6$

55. $P(x) = x^5 + x^4 - 8x^3 - 8x^2 + 15x + 15$

56. $P(x) = 3x^4 - 5x^3 - 9x^2 + 15x$

57. $P(x) = 3x^5 - 11x^4 - 2x^3 + 38x^2 - 21x - 15$

58. $P(x) = 3x^6 - 4x^5 - 21x^4 + 4x^3 + 8x^2 + 8x + 32$

Discovery and Writing

59. Explain why the Fundamental Theorem of Algebra guarantees that every polynomial function of positive degree has at least one complex zero.

60. Explain why the Fundamental Theorem of Algebra and the Factor Theorem guarantee that an nth-degree polynomial function has n zeros.

61. State the Conjugate Pairs Theorem and explain why it is important in algebra.

62. What is Decartes' Rule of Signs? Explain how to apply the rule to a polynomial function.

63. Prove that any odd-degree polynomial function with real coefficients must have at least one real zero.

64. If a, b, c, and d are positive numbers, prove that $P(x) = ax^4 + bx^2 + cx - d$ has exactly two nonreal zeros.

Critical Thinking

Determine if the statement is true or false. If the statement is false, then correct it and make it true.

65. The Fundamental Theorem of Algebra states that every polynomial function of positive degree has at least one imaginary zero.

66. If a polynomial function is of degree n, then, counting multiple zeros separately, the function has n zeros.

67. If $55 - 77i$ is a zero of the polynomial function $P(x)$, then $55 + 77i$ is also a zero of $P(x)$.

68. If $22 + 44i$ is a zero of the polynomial function $P(x)$ with real coefficients, then $22 - 44i$ is also a zero of $P(x)$.

69. The polynomial function $P(x) = x^{123} + x^{456} + x^{789}$ has exactly 123 zeros.

70. The polynomial function $P(x) = x^4 - x^2 + 4x^7 - 3x^3 - 1$ has three variations in sign.

4.5   Zeros of Polynomial Functions

In this section, we will learn to

1. Find possible rational zeros of polynomial functions.
2. Find rational zeros of polynomial functions.
3. Find real and nonreal zeros of polynomial functions.
4. Solve applications.

©Daryl Lang/Shutterstock.com

©Restyler/Shutterstock.com

FedEx is a company that offers reliable shipping services across town and across the globe. It has a wide range of envelopes and boxes available to accommodate the shipping needs of most individuals and companies.

Suppose a FedEx box has the following characteristics:

- The length of the box is 6 inches more than its height.
- The width of the box is 3 inches more than its height.
- The volume of the box is 2080 cubic inches.

To find the dimensions of the box, we can let h represent the height (in inches). Then $h + 6$ will represent the length and $h + 3$ will represent the width. Since the volume of the box is given to be 2080, we have

$$V = l \cdot w \cdot h$$

$$2080 = (h + 6)(h + 3)h \qquad \text{Substitute.}$$

$$2080 = h^3 + 9h^2 + 18h \qquad \text{Multiply.}$$

$$0 = h^3 + 9h^2 + 18h - 2080 \qquad \text{Subtract 2080 from both sides.}$$

To find the height h, we must find the roots or solutions of the polynomial equation. The roots of the polynomial equation are zeros of the polynomial function $V(h) = h^3 + 9h^2 + 18h - 2080$. One zero of the function is $x = 10$, because $V(10) = 0$.

$$V(10) = 10^3 + 9(10)^2 + 18(10) - 2080 = 0$$

Therefore, the height of the box is $h = 10$ inches, the length is $h + 6 = 16$ inches, and the width is $h + 3 = 13$ inches.

In this section, we will develop a strategy for finding zeros of higher-degree polynomial functions.

1. Find Possible Rational Zeros of Polynomial Functions

Recall that a rational number is any number that can be written in the form $\frac{p}{q}$, where p and q are integers and $q \neq 0$. The following theorem enables us to list the possible rational zeros of such polynomial functions.

Rational Zero Theorem	Let the polynomial function

$$P(x) = a_n x^n + a_{n-1} x^{n-1} + a_{n-2} x^{n-2} + \cdots + a_1 x + a_0$$

have integer coefficients. If the rational number $\frac{p}{q}$ (written in lowest terms) is a zero of $P(x)$, then p is a factor of the constant a_0, and q is a factor of the leading coefficient a_n.

PROOF   Let $\frac{p}{q}$ (written in lowest terms) be a rational zero of $P(x)$. Then the equation $P(x) = 0$ is satisfied by $\frac{p}{q}$:

(1)
$$a_n\left(\frac{p}{q}\right)^n + a_{n-1}\left(\frac{p}{q}\right)^{n-1} + a_{n-2}\left(\frac{p}{q}\right)^{n-2} + \cdots + a_1\left(\frac{p}{q}\right) + a_0 = 0$$

We can clear Equation 1 of fractions by multiplying both sides by q^n.

(2)
$$a_np^n + a_{n-1}p^{n-1}q + a_{n-2}p^{n-2}q^2 + \cdots + a_1pq^{n-1} + a_0q^n = 0$$

We can factor p from all but the last term and subtract a_0q^n from both sides to get

$$p(a_np^{n-1} + a_{n-1}p^{n-2}q + a_{n-2}p^{n-3}q^2 + \cdots + a_1q^{n-1}) = -a_0q^n$$

Since p is a factor of the left side, it is also a factor of the right side. So p is a factor of $-a_0q^n$, but because $\frac{p}{q}$ is written in lowest terms, p cannot be a factor of q^n. Therefore, p is a factor of a_0.

We can factor q from all but the first term of Equation 2 and subtract a_np^n from both sides to get

$$q(a_{n-1}p^{n-1} + a_{n-2}p^{n-2}q + a_{n-3}p^{n-3}q^2 + \cdots + a_0q^{n-1}) = -a_np^n$$

Since q is a factor of the left side, it is also a factor of the right side. Because q is not a factor of p^n, it must be a factor of a_n.   ◾

To illustrate the Rational Zero Theorem, we consider the function

$$f(x) = \frac{1}{2}x^4 + \frac{2}{3}x^3 + 3x^2 - \frac{3}{2}x + 3$$

Because the theorem requires integer coefficients, we multiply by 6 to clear it of fractions and label the function $P(x)$.

$$P(x) = 3x^4 + 4x^3 + 18x^2 - 9x + 18$$

By the previous theorem, the only possible numerators for the rational zeros of the function are the factors of the constant term 18:

$$\pm 1, \pm 2, \pm 3, \pm 6, \pm 9, \text{ and } \pm 18$$

The only possible denominators are the factors of the leading coefficient 3:

$$\pm 1 \quad \text{and} \quad \pm 3$$

We can form a list of all possible rational zeros by listing the combinations of possible numerators and denominators:

> **Take Note**
>
> When we list possible rational zeros for a polynomial function, our list will not contain radicals or complex numbers.

$$\pm\frac{1}{1}, \pm\frac{2}{1}, \pm\frac{3}{1}, \pm\frac{6}{1}, \pm\frac{9}{1}, \pm\frac{18}{1}, \pm\frac{1}{3}, \pm\frac{2}{3}, \pm\frac{3}{3}, \pm\frac{6}{3}, \pm\frac{9}{3}, \pm\frac{18}{3}$$

Since several of these possibilities are duplicates, we can condense the list to get

Possible rational zeros

$$\pm 1, \quad \pm 2, \quad \pm 3, \quad \pm 6, \quad \pm 9, \quad \pm 18, \quad \pm\frac{1}{3}, \quad \pm\frac{2}{3}$$

2. Find Rational Zeros of Polynomial Functions

To find the rational zeros of a polynomial function, we will use the following steps.

Strategy for Finding Rational Zeros

1. Use Descartes' Rule of Signs to determine the number of possible positive, negative, and nonreal zeros.
2. Use the Rational Zero Theorem to list possible rational zeros.
3. Use synthetic division to find a zero.
4. If there are more zeros, repeat the previous steps.

Finding the Rational Zeros of a Polynomial Function

EXAMPLE 1

Find the rational zeros of $P(x) = 2x^3 + 3x^2 - 8x + 3$.

SOLUTION We will use the steps previously outlined to find the rational zeros. Since the function is of third degree, it has 3 zeros.

1. To determine the number of possible positive, negative, and nonreal zeros, we will use Descartes' Rule of Signs. We find that there are two possible combinations of positive, negative, and nonreal zeros. They are as follows:

Number of Positive Zeros	Number of Negative Zeros	Number of Nonreal Zeros
2	1	0
0	1	2

2. We then find the possible rational zeros that have the form $\frac{\text{factor of the constant 3}}{\text{factor of the leading coefficient 2}}$. They are

$$\pm\frac{3}{1}, \quad \pm\frac{1}{1}, \quad \pm\frac{3}{2}, \quad \pm\frac{1}{2}$$

or, written in order of increasing size,

$$-3, \quad -\frac{3}{2}, \quad -1, \quad -\frac{1}{2}, \quad \frac{1}{2}, \quad 1, \quad \frac{3}{2}, \quad 3$$

3. We then use synthetic division and check each possibility to see whether it is a zero. We can start with $\frac{3}{2}$.

$$
\begin{array}{r|rrrr}
\frac{3}{2} & 2 & 3 & -8 & 3 \\
  &   & 3 & 9 & \frac{3}{2} \\
\hline
  & 2 & 6 & 1 & \frac{9}{2}
\end{array}
$$

Since the remainder is not 0, $\frac{3}{2}$ is not a zero and we can cross it off the list. Since every number in the last row of the synthetic division is positive, $\frac{3}{2}$ is an upper bound. So 3 cannot be a zero either and we can cross it off the list.

$$-3, \quad -\frac{3}{2}, \quad -1, \quad -\frac{1}{2}, \quad \frac{1}{2}, \quad 1, \quad \cancel{\frac{3}{2}}, \quad \cancel{3}$$

We now try $\frac{1}{2}$:

$$
\begin{array}{r|rrrr}
\frac{1}{2} & 2 & 3 & -8 & 3 \\
  &   & 1 & 2 & -3 \\
\hline
  & 2 & 4 & -6 & 0
\end{array}
$$

This row represents the quotient $2x^2 + 4x - 6$.

Since the remainder is 0, $\frac{1}{2}$ is a zero and the binomial $x - \frac{1}{2}$ is a factor of $P(x)$.

4. The remaining zeros must be supplied by the remaining factor, which is the quotient $2x^2 + 4x - 6$. We can find the other zeros by solving the equation $2x^2 + 4x - 6 = 0$, called the **depressed equation**.

$$2x^2 + 4x - 6 = 0$$

$$x^2 + 2x - 3 = 0 \qquad \text{Divide both sides by 2.}$$

$$(x - 1)(x + 3) = 0 \qquad \text{Factor } x^2 + 2x - 3.$$

$$x - 1 = 0 \quad \text{or} \quad x + 3 = 0$$

$$x = 1 \qquad \qquad x = -3$$

Take Note

Example 1 illustrates that the upper and lower bound theorem is helpful when the list of possible rational zeros is long.

The rational zeros of the polynomial function are $\frac{1}{2}$, 1, and 3. Note that two positive zeros and one negative zero is a predicted possibility.

Self Check 1   Find the rational zeros of $P(x) = 3x^3 - 10x^2 + 9x - 2$.

Now Try Exercise 19.

Accent on Technology

Confirming Zeros of a Polynomial Function

We can confirm that the zeros found in Example 1 are correct by graphing the function $P(x) = 2x^3 + 3x^2 - 8x + 3$ and locating the resulting x-intercepts of the graph. If we use a graphing [WINDOW] as shown in Figure 4-29 (a) and [GRAPH] the function, we will obtain the graph shown in Figure 4-29 (b). From the graph, we can see that the graph crosses the x-axis at $x = -3$, $x = \frac{1}{2}$, and $x = 1$. These are the zeros of the polynomial equation.

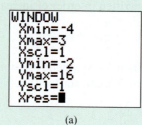

(a)   (b)

FIGURE 4-29

---

3. Find Real and Nonreal Zeros of Polynomial Functions

Finding All the Zeros of a Polynomial Function

EXAMPLE 2

Find the zeros of the polynomial function
$P(x) = x^7 - 2x^6 - 5x^5 + 6x^4 - x^3 + 2x^2 + 5x - 6$.

SOLUTION   We will use the steps outlined above to find the zeros of the polynomial function. Because the polynomial function is of seventh degree, it has 7 zeros.

1. To determine the number of possible positive, negative, and nonreal zeros, we will use Descartes' Rule of Signs. We find that there are several possible combinations of positive, negative, and nonreal zeros. They are as follows:

Number of Positive Zeros	Number of Negative Zeros	Number of Nonreal Zeros
5	2	0
3	2	2
1	2	4
5	0	2
3	0	4
1	0	6

Take Note

This example is designed to help you see the big picture; that is, to understand the overall process involved in finding the zeros of higher-degree polynomial functions. Along the way, you may feel like you do sometimes in your car, wondering how you are going to get to your final destination and if you ever will. Many cars are equipped with a navigation system that guides you. This example is just that—a navigational guide to help you find all zeros and to reach your algebra "destination."

©Michaeljung/Shutterstock.com

2. We then find the possible rational zeros that have the form $\frac{\text{factor of the constant } -6}{\text{factor of the leading coefficient } 1}$. They are as follows:

$$-6, \quad -3, \quad -2, \quad -1, \quad 1, \quad 2, \quad 3, \quad 6$$

3. We then use synthetic division and check each possibility to see whether it is a zero. We can start with -3.

$$
\begin{array}{r|rrrrrrr}
-3 & 1 & -2 & -5 & 6 & -1 & 2 & 5 & -6 \\
   &   & -3 & 15 & -30 & 72 & -213 & 633 & -1914 \\
\hline
   & 1 & -5 & 10 & -24 & 71 & -211 & 638 & -1920 \\
\end{array}
$$

Since the last number in the synthetic division is not 0, -3 is not a zero and can be crossed off the list. Since the signs in the last row alternate, -3 is a lower bound, and we can cross off -6 as well.

$$-\cancel{6}, \quad -\cancel{3}, \quad -2, \quad -1, \quad 1, \quad 2, \quad 3, \quad 6$$

We now try -2:

$$
\begin{array}{r|rrrrrrr}
-2 & 1 & -2 & -5 & 6 & -1 & 2 & 5 & -6 \\
   &   & -2 & 8 & -6 & 0 & 2 & -8 & 6 \\
\hline
   & 1 & -4 & 3 & 0 & -1 & 4 & -3 & 0 \\
\end{array}
$$

Since the remainder is 0, -2 is a zero.

4. Because the zero -2 is negative, we can revise the chart of possibilities to eliminate the possibility that there are 0 negative zeros.

Number of Positive Zeros	Number of Negative Zeros	Number of Nonreal Zeros
5	2	0
3	2	2
1	2	4

The remaining zeros must be supplied by the remaining factor, which is the quotient in the synthetic division shown above: $x^6 - 4x^5 + 3x^4 - x^2 + 4x - 3 = 0$. We can find the other zeros by repeating steps 1–3 and solving this equation, called the **depressed equation**.

Because the constant term of this depressed equation is different from the constant term of the original equation, we can cross off other possible rational zeros. For example, the number -2 cannot be a zero a second time, because it is not a factor of -3. The numbers 2 and 6 are no longer possible zeros, because neither is a factor of -3.

The list of possible zeros is now

$$-\cancel{6}, \quad -\cancel{3}, \quad -2, \quad -1, \quad 1, \quad \cancel{2}, \quad 3, \quad \cancel{6}$$

Since we know that there is one more negative zero, we will synthetically divide the coefficients of the depressed equation by -1.

$$
\begin{array}{r|rrrrrr}
-1 & 1 & -4 & 3 & 0 & -1 & 4 & -3 \\
   &   & -1 & 5 & -8 & 8 & -7 & 3 \\
\hline
   & 1 & -5 & 8 & -8 & 7 & -3 & 0
\end{array}
$$

Since the remainder is 0, -1 is a zero, and the zeros so far are -2 and -1. The number -1 cannot be a zero again, because we have found both negative zeros. When we cross off -1, we have only two possibilities left.

$$-\cancel{6}, \quad -\cancel{3}, \quad -2, \quad -\cancel{1}, \quad 1, \quad 2, \quad 3, \quad \cancel{6}$$

The depressed equation is now $x^5 - 5x^4 + 8x^3 - 8x^2 + 7x - 3 = 0$.

We can synthetically divide the coefficients of this equation by 1 to get

$$
\begin{array}{r|rrrrr}
1 & 1 & -5 & 8 & -8 & 7 & -3 \\
  &   & 1 & -4 & 4 & -4 & 3 \\
\hline
  & 1 & -4 & 4 & -4 & 3 & 0
\end{array}
$$

The depressed equation is now $x^4 - 4x^3 + 4x^2 - 4x + 3 = 0$.

Since the remainder is 0, 1 is a zero and our zeros so far are -2, -1, and 1. To see whether 1 is a zero a second time, we synthetically divide the coefficients of the new depressed equation by 1.

$$
\begin{array}{r|rrrrr}
1 & 1 & -4 & 4 & -4 & 3 \\
  &   & 1 & -3 & 1 & -3 \\
\hline
  & 1 & -3 & 1 & -3 & 0
\end{array}
$$

The depressed equation is now $x^3 - 3x^2 + x - 3 = 0$.

Again, 1 is a zero, and our zeros right now are -2, -1, 1, and 1

To see whether 1 is a zero a third time, we synthetically divide the coefficients of the new depressed equation by 1.

$$
\begin{array}{r|rrrr}
1 & 1 & -3 & 1 & -3 \\
  &   & 1 & -2 & -1 \\
\hline
  & 1 & -2 & -1 & -4
\end{array}
$$

Since the remainder is not 0, the number 1 is not a zero for a third time, and we can cross 1 off the list of possibilities, leaving only 3.

$$-\cancel{6}, \quad -\cancel{3}, \quad -2, \quad -\cancel{1}, \quad \cancel{1}, \quad 2, \quad 3, \quad \cancel{6}$$

To see whether 3 is a zero, we synthetically divide the coefficients of $x^3 - 3x^2 + x - 3 = 0$ by 3.

$$
\begin{array}{r|rrrr}
3 & 1 & -3 & 1 & -3 \\
  &   & 3 & 0 & 3 \\
\hline
  & 1 & 0 & 1 & 0
\end{array}
$$

Since the remainder is 0, 3 is a zero and our zeros currently are -2, -1, 1, 1, and 3.

The depressed equation is now $x^2 + 1 = 0$, which can be solved as a quadratic equation.

$$x^2 + 1 = 0$$
$$x^2 = -1$$
$$x = i \quad \text{or} \quad x = -i$$

Take Note

Example 2 illustrates that finding the seven zeros of a seventh-degree polynomial function can be a lengthy process. In such cases, the four-step process for finding rational zeros must be repeated.

The zeros of the polynomial function are -2, -1, 1, 1, 3, i, and $-i$. Note that there are 3 positive zeros, 2 negative zeros, and 2 nonreal zeros that are complex conjugates. This combination was one of the predicted possibilities.

Self Check 2   Find the zeros of the polynomial function $P(x) = x^5 - x^4 + 3x^3 - 3x^2 - 4x + 4$.

Now Try Exercise 51.

Accent on Technology

Confirming Zeros of a Polynomial Function

We can confirm that the zeros found in Example 2 are correct by graphing the function

$$P(x) = x^7 - 2x^6 - 5x^5 + 6x^4 - x^3 + 2x^2 + 5x - 6$$

and locating the resulting x-intercepts of the graph. If we use the graphing WINDOW shown in Figure 4-30(a) and GRAPH the function, we will obtain the graph shown in Figure 4-30(b). From the graph, we can see that the graph crosses or touches the x-axis at $x = -2$, $x = -1$, $x = 1$, and $x = 3$. We cannot detect the complex zeros from the graph.

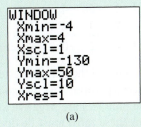

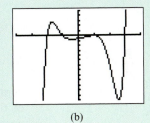

(a)   (b)

FIGURE 4-30

4. Solve Applications

Solving an Application

EXAMPLE 3

To protect cranberry crops from the damage of early freezes, growers flood the cranberry bogs. Three irrigation sources, used together, can flood a cranberry bog in one day. If the sources are used one at a time, the second source requires one day longer to flood the bog than the first, and the third requires four days longer than the first. If the bog must be flooded before a freeze that is predicted in three days, can the water in the last two sources be diverted to other bogs?

SOLUTION   We can use the given information to write an equation that models the situation. We will then solve the equation.

We can let x represent the number of days it would take the first irrigation source to flood the bog. Then $x + 1$ and $x + 4$ represent the number of days it would take the second and third sources, respectively, to flood the bog.

Because the first source, alone, requires x days to flood the bog, that source could fill $\frac{1}{x}$ of the bog in one day. In one day's time, the remaining sources could flood $\frac{1}{x+1}$ and $\frac{1}{x+4}$ of the bog. This gives the equation

©Olivier Le Queinec/Shutterstock.com

The part of the bog the first source can flood in one day	plus	the part the second source can flood in one day	plus	the part the third source can flood in one day	equals	one bog.
$\dfrac{1}{x}$	$+$	$\dfrac{1}{x+1}$	$+$	$\dfrac{1}{x+4}$	$=$	1

We multiply both sides of the equation by $x(x + 1)(x + 4)$ to clear it of fractions and then simplify to get

$$x(x + 1)(x + 4)\left(\frac{1}{x} + \frac{1}{x + 1} + \frac{1}{x + 4}\right) = 1 \cdot x(x + 1)(x + 4)$$

$$(x + 1)(x + 4) + x(x + 4) + x(x + 1) = x(x + 1)(x + 4)$$

$$x^2 + 5x + 4 + x^2 + 4x + x^2 + x = x^3 + 5x^2 + 4x$$

$$0 = x^3 + 2x^2 - 6x - 4$$

To solve the equation $x^3 + 2x^2 - 6x - 4 = 0$, we will find the zeros of the polynomial function $P(x) = x^3 + 2x^2 - 6x - 4$. We first list its possible rational zeros, which are the factors of the constant term, -4.

$-4, -2, -1, 1, 2,$ and 4

One zero of the polynomial function is $x = 2$, because when we synthetically divide by 2, the remainder is 0:

$$\begin{array}{r|rrrr} 2 & 1 & 2 & -6 & -4 \\ & & 2 & 8 & 4 \\ \hline & 1 & 4 & 2 & 0 \end{array}$$ The depressed equation is $x^2 + 4x + 2 = 0$.

We can find the remaining zeros by using the Quadratic Formula to solve the depressed equation. The two solutions are $-2 + \sqrt{2}$ and $-2 - \sqrt{2}$. Since both of these numbers are negative and the time it takes to flood the bog cannot be negative, these zeros must be discarded. The only meaningful zero is 2.

Since the first source, alone, can flood the bog in two days, and it is three days until the freeze, the other two water sources can be diverted to flood other bogs.

Self Check 3   If a freeze is predicted in one day, can the water in the last two sources be diverted?

Now Try Exercise 65.

Accent on Technology

Using the Table Feature on a Graphing Calculator to Find Zeros of a Polynomial Function

A graphing calculator can be used to reduce the amount of work involved in finding the zeros of a polynomial function. From a list of possible rational zeros, we can utilize the TABLE function and identify the rational zeros. Consider the following example.

- We will use the Rational Zero Theorem, a calculator, and synthetic division, if necessary, to find all the rational zeros of the polynomial function $P(x) = 2x^4 + x^3 - 9x^2 - 4x + 4$.
- First, we list all the possible rational zeros. For the polynomial, we see that

$$p = 4, q = 2 \rightarrow \frac{p}{q} = \frac{\text{all integer factors of 4}}{\text{all interger factors of 2}} = \frac{\pm 1, \pm 2, \pm 4}{\pm 1, \pm 2} = \pm 1, \pm 2, \pm 4, \pm \frac{1}{2}$$

- Next, we evaluate the function at these values. The work is shown in Figure 4-31.

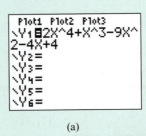
(a)

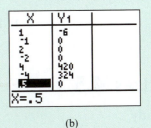
(b)

FIGURE 4-31

- From the table, we see that the rational zeros are $x = -1, 2, -2,$ and $\frac{1}{2}$. Because the polynomial function is of fourth degree and we have identified four zeros, our work is finished.

It wasn't necessary for us to use synthetic division to find the zeros. If, for example, we had only found two rational zeros using the table feature, we could be able to finish the problem using synthetic division. The table feature simply saves us time.

We can approximate the real zeros of an equation either by using the TRACE and ZOOM capabilities of a graphing calculator or by using the ZERO feature.

Accent on Technology

Using a Graphing Calculator to Approximate Real Zeros

We will use the ZERO feature on a graphing calculator to approximate the real zeros of the polynomial function $P(x) = x^4 - 6x^2 + 9$.

- We first enter the function, using the WINDOW shown, and GRAPH the function to obtain the graph shown in Figure 4-32.

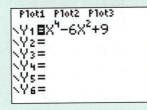

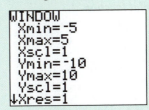

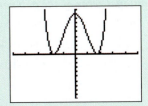

FIGURE 4-32

- We then select ZERO under the CALC menu. To find the positive zero, we guess a left bound and press ENTER, guess a right bound and press ENTER, and then press ENTER again. The zero of the equation is approximately 1.7320511. See Figure 4-33.

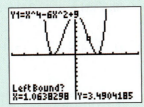

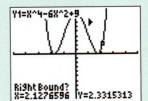

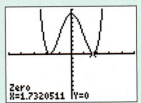

FIGURE 4-33

Because of symmetry, we know that the negative solution is -1.7320511. The real zeros are $x \approx \pm 1.7320511$.

Exercises 4.5

Getting Ready

You should be able to complete these vocabulary and concept statements before you proceed to the practice exercises.

Fill in the blanks.

1. The rational zeros of the function $P(x) = 3x^3 + 4x - 7$ will have the form $\frac{p}{q}$, where p is a factor of _____ and q is a factor of 3.

2. The rational zeros of the function $P(x) = 5x^3 + 3x^2 - 4$ will have the form $\frac{p}{q}$, where p is a factor of -4 and q is a factor of ___.

3. Consider the synthetic division of $P(x) = 5x^3 - 7x^2 - 3x - 63$ by $x - 3$.

$$\begin{array}{r|rrrr} 3 & 5 & -7 & -3 & -63 \\ & & 15 & 24 & 63 \\ \hline & 5 & 8 & 21 & 0 \end{array}$$

Since the remainder is 0, 3 is a _____ of the function.

4. In Exercise 3, the depressed equation is

_____.

Practice

Use the Rational Zero Theorem to list all possible rational zeros of the polynomial function.

5. $P(x) = x^3 + 10x^2 + 5x - 12$

6. $P(x) = -x^3 + 3x^2 - 4x - 8$

7. $P(x) = 2x^4 - x^3 + 10x^2 + 5x - 6$

8. $P(x) = 3x^4 - x^3 + 7x^2 - 5x - 8$

9. $P(x) = 4x^5 - x^4 - x^3 + x^2 + 5x - 10$

10. $P(x) = 6x^4 - 2x^3 + x^2 - x + 3$

Find all rational zeros of each polynomial function.

11. $P(x) = x^3 - 5x^2 - x + 5$

12. $P(x) = x^3 + 7x^2 - x - 7$

13. $P(x) = x^3 - 2x^2 - x + 2$

14. $P(x) = x^3 + x^2 - 4x - 4$

15. $P(x) = x^3 - x^2 - 4x + 4$

16. $P(x) = x^3 + 2x^2 - x - 2$

17. $P(x) = x^3 - 2x^2 - 9x + 18$

18. $P(x) = x^3 + 3x^2 - 4x - 12$

19. $P(x) = 2x^3 - x^2 - 2x + 1$

20. $P(x) = 3x^3 + x^2 - 3x - 1$

21. $P(x) = 3x^3 + 5x^2 + x - 1$

22. $P(x) = 2x^3 - 3x^2 + 1$

23. $P(x) = 30x^3 - 47x^2 - 9x + 18$

24. $P(x) = 20x^3 - 53x^2 - 27x + 18$

25. $P(x) = 15x^3 - 61x^2 - 2x + 24$

26. $P(x) = 20x^3 - 44x^2 + 9x + 18$

27. $P(x) = 24x^3 - 82x^2 + 89x - 30$

28. $P(x) = 3x^3 - 2x^2 + 12x - 8$

29. $P(x) = x^4 - 10x^3 + 35x^2 - 50x + 24$

30. $P(x) = x^4 + 4x^3 + 6x^2 + 4x + 1$

31. $P(x) = x^4 + 3x^3 - 13x^2 - 9x + 30$

32. $P(x) = x^4 - 8x^3 + 14x^2 + 8x - 15$

33. $P(x) = 4x^4 - 8x^3 - x^2 + 8x - 3$

34. $P(x) = 3x^4 - 14x^3 + 11x^2 + 16x - 12$

35. $P(x) = 2x^4 - x^3 - 2x^2 - 4x - 40$

36. $P(x) = 12x^4 + 20x^3 - 41x^2 + 20x - 3$

37. $P(x) = 36x^4 - x^2 + 2x - 1$

38. $P(x) = 12x^4 + x^3 + 42x^2 + 4x - 24$

39. $P(x) = x^5 + 3x^4 - 5x^3 - 15x^2 + 4x + 12$

40. $P(x) = x^5 - 3x^4 - 5x^3 + 15x^2 + 4x - 12$

41. $P(x) = 4x^5 - 12x^4 + 15x^3 - 45x^2 - 4x + 12$

42. $P(x) = 6x^5 - 7x^4 - 48x^3 + 81x^2 - 4x - 12$

43. $P(x) = x^7 - 12x^5 + 48x^3 - 64x$

44. $P(x) = x^7 + 7x^6 + 21x^5 + 35x^4 + 35x^3 + 21x^2 + 7x + 1$

Find all zeros of each polynomial function.

45. $P(x) = x^3 - 3x^2 - 2x + 6$

46. $P(x) = x^3 + 3x^2 - 3x - 9$

47. $P(x) = 2x^3 - x^2 + 2x - 1$

48. $P(x) = 3x^3 + x^2 + 3x + 1$

49. $P(x) = x^4 - 2x^3 - 8x^2 + 8x + 16$

50. $P(x) = x^4 - 2x^3 - 2x^2 + 2x + 1$

51. $P(x) = 2x^4 + x^3 + 17x^2 + 9x - 9$

52. $P(x) = 2x^4 - 4x^3 + 2x^2 + 4x - 4$

53. $P(x) = x^5 - 3x^4 + 28x^3 - 76x^2 + 75x - 25$

54. $P(x) = x^5 + 3x^4 - 2x^3 - 14x^2 - 15x - 5$

55. $P(x) = 2x^5 - 3x^4 + 6x^3 - 9x^2 - 8x + 12$

56. $P(x) = 3x^5 - x^4 + 36x^3 - 12x^2 - 192x + 64$

In Exercises 57–60, $1 + i$ is a zero of each polynomial function. Find the other zeros.

57. $P(x) = x^3 - 5x^2 + 8x - 6$

58. $P(x) = x^3 - 2x + 4$

59. $P(x) = x^4 - 2x^3 - 7x^2 + 18x - 18$

60. $P(x) = x^4 - 2x^3 - 2x^2 + 8x - 8$

Solve each equation.

61. $x^3 - \dfrac{4}{3}x^2 - \dfrac{13}{3}x - 2 = 0$

62. $x^3 - \dfrac{19}{6}x^2 + \dfrac{1}{6}x + 1 = 0$

63. $x^{-5} - 8x^{-4} + 25x^{-3} - 38x^{-2} + 28x^{-1} - 8 = 0$

64. $1 - x^{-1} - x^{-2} - 2x^{-3} = 0$

Applications

65. Parallel resistance If three resistors with resistances of R_1, R_2, and R_3 are wired in parallel, their combined resistance R is given by the following formula. The design of a voltmeter requires that the resistance R_2 be 10 ohms greater than the resistance R_1, that the resistance R_3 be 50 ohms greater than R_1, and that their combined resistance be 6 ohms. Find the value of each resistance.

$$\frac{1}{R} = \frac{1}{R_1} + \frac{1}{R_2} + \frac{1}{R_3}$$

66. Fabricating sheet metal The open tray shown in the illustration is to be manufactured from a 12-by-14-inch rectangular sheet of metal by cutting squares from each corner and folding up the sides. If the volume of the tray is to be 160 cubic inches and x is to be an integer, what size squares should be cut from each corner?

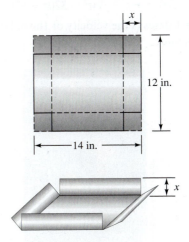

67. FedEx box The length of a FedEx 25-kg box is 7 inches more than its height. The width of the box is 4 inches more than its height. If the volume of the box is 4420 cubic inches, find the height of the box.

68. Dr. Pepper can A Dr. Pepper aluminum can is approximately the shape of a cylinder. If the height of the can is 9 centimeters more than its radius and the volume of the can is approximately 108π cubic centimeters, find the radius of the can. The formula for the volume of a cylinder is $V = \pi r^2 h$

©Andrey Armyagov/Shutterstock.com

69. Hilly terrain We are interested in the nature of some hilly terrain. Computer simulation has told us that for a cross-section from west to east, the height $h(x)$, in feet above sea level is related to the horizontal distance x (in miles) from a fixed point by the function,

$$h(x) = -x^4 + 5x^3 + 91x^2 - 545x + 550, \; x \in [0, 9].$$

At what distances from the fixed point is the height 100 feet above sea level?

70. Velocity of a hot-air balloon A hot-air balloon is tethered to the ground and only moves up and down. You and a friend take a ride on the ballon for approximately 25 minutes. On this particular ride the velocity of the balloon, $v(t)$ in feet per minute, as a function of time, t in minutes, is represented by the function

$$v(t) = -t^3 + 34t^2 - 320t + 850$$

At what times is the velocity of the balloon 50 feet per minute?

©Steve Bower/Shutterstock.com

Discovery and Writing

71. State the Rational Zero Theorem and explain its purpose in algebra.

72. Describe a strategy that can be used to determine all zeros of a polynomial function.

73. If n is an even integer and c is a positive constant, show that $P(x) = x^n + c$ has no real zeros.

74. If n is an even positive integer and c is a positive constant, show that $P(x) = x^n - c$ has two real zeros.

75. Precalculus A rectangle is inscribed in the parabola $y = 16 - x^2$, as shown in the illustration. Find the point (x, y) if the area of the rectangle is 42 square units.

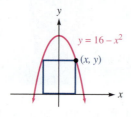

76. Precalculus One corner of the rectangle shown is at the origin, and the opposite corner (x, y) lies in the first quadrant on the curve $y = x^3 - 2x^2$. Find the point (x, y) if the area of the rectangle is 27 square units.

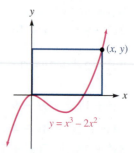

Critical Thinking

Determine if the statement is true or false. If the statement is false, then correct it and make it true.

77. $\dfrac{1}{3}$ is a possible rational zero of
$$P(x) = 3x^3 - 5x^4 - 2x^2 - x + 1$$

78. Every zero of a polynomial function has a corresponding x-intercept.

79. If we have identified one zero of a third-degree polynomial function, we can always find the remaining two zeros by factoring, using the Square Root Property or the quadratic formula.

80. If every coefficient of a polynomial function is positive, then the function has no positive real zeros.

81. A polynomial function with real coefficients and degree 3 will always have at least one rational zero.

82. A polynomial function can have possible rational zeros but no actual rational zeros.

4.6 Rational Functions

In this section, we will learn to

1. Find the domain of a rational function.
2. Understand the characteristics of rational functions and their graphs.
3. Find vertical asymptotes of rational functions.
4. Find horizontal asymptotes of rational functions.
5. Identify slant asymptotes of rational functions.
6. Graph rational functions.
7. Understand when a graph has a missing point.
8. Solve problems modeled by rational functions.

©Daniel Huerlimann-Beelde/Shutterstock.com

We have discussed polynomial functions and now focus on another class of functions called **rational functions**. Rational functions are defined by rational expressions that are quotients of polynomials.

For example, consider the time t it takes a Nascar driver such as Jeff Gordon to drive the 500 miles in the Daytona 500 race. The time t can be defined as a function of the average rate of the driver's speed r. That is

$$t = f(r) = \frac{500}{r}$$

If a driver averages a speed of 170 mph, we can evaluate $f(170)$ to determine the time in hours driven.

$$f(\mathbf{170}) = \frac{500}{\mathbf{170}} = \frac{50}{17} \qquad \text{This is approximately 2.94 hours.}$$

We will discuss this type of function in this section.

| **Rational Function** | A **rational function** is a function defined by an equation of the form $$f(x) = \frac{P(x)}{Q(x)}$$ where $P(x)$ and $Q(x)$ are polynomials and $Q(x) \neq 0$. |

1. Find the Domain of a Rational Function

Because rational functions are quotients of polynomials and $Q(x)$ is the denominator of a fraction, $Q(x)$ cannot equal 0. Thus, the domain of a rational function must exclude all values of x for which $Q(x) = 0$.

Here are some examples of rational functions and their domains:

Function	Domain
$f(x) = \dfrac{3}{x + 7}$	$(-\infty, -7) \cup (-7, \infty)$, x cannot equal -7
$f(x) = \dfrac{5x + 2}{x^2 - 4}$	$(-\infty, -2) \cup (-2, 2) \cup (2, \infty)$, x cannot equal -2 or 2
$f(x) = \dfrac{2x}{x^2 + 3}$	$(-\infty, \infty)$, x can equal any real number

Finding the Domain of a Rational Function

EXAMPLE 1

Find the domain of $f(x) = \dfrac{3x + 2}{x^2 - 7x + 12}$.

SOLUTION   We can factor the denominator to see what values of x will give 0's in the denominator. These values are not in the domain.

To find the numbers x that make the denominator 0, we set $x^2 - 7x - 12$ equal to 0 and solve for x.

> **Caution**
> Do not exclude values of the variable from the domain that make the numerator equal to zero, unless the denominator has the same factor.

$$x^2 - 7x + 12 = 0$$

$$(x - 4)(x - 3) = 0 \qquad \text{Factor } x^2 - 7x + 12.$$

$$x - 4 = 0 \quad \text{or} \quad x - 3 = 0 \qquad \text{Set each factor equal to 0.}$$

$$x = 4 \qquad\qquad x = 3 \qquad \text{Solve each linear equation.}$$

Since 4 and 3 make the denominator 0, the domain is the set of all real numbers except $x = 4$ and $x = 3$. In interval notation, we have $(-\infty, 3) \cup (3, 4) \cup (4, \infty)$.

Self Check 1   Find the domain of $f(x) = \dfrac{2x - 3}{x^2 - x - 2}$.

Now Try Exercise 23.

Accent on Technology

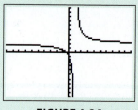

FIGURE 4-34

Domain and Range of a Rational Function

We can use a graphing calculator to find domains and ranges of rational functions. If we use settings of $[-10, 10]$ for x and $[-10, 10]$ for y and graph $f(x) = \frac{2x + 1}{x - 1}$, we will obtain Figure 4-34.

From the graph, we can see that every real number x except 1 gives a value of y. Thus, the domain of the function is $(-\infty, 1) \cup (1, \infty)$. We can also see that y can be any value except 2. The range of the function is $(-\infty, 2) \cup (2, \infty)$.

> **Tip**
> When using a graphing calculator to graph a rational function, make sure that the function is entered properly. To avoid making an error, place both the numerator and denominator in parentheses.

2. Understand the Characteristics of Rational Functions and Their Graphs

Consider the rational function $t = f(r) = \frac{500}{r}$ given at the beginning of the section. A graph of this rational function is shown in Figure 4-35.

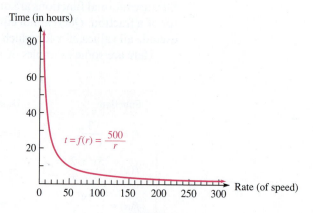

FIGURE 4-35

We see from the graph that as the rate of speed increases, the time it takes to complete the race decreases. In fact, if we drive at rocket speed, we will arrive in almost no time at all. We can express this by saying that "as the rate increases without bound (or approaches ∞), the time it takes to complete the race approaches 0 hours." When a graph approaches a line as shown in the figure, we call the line an **asymptote**. The horizontal line representing the rate axis shown in the graph is a **horizontal asymptote**.

We also see from the graph that as the rate of speed decreases, the time it takes to complete the race increases. In fact, if the car goes at turtle speed, it will take almost forever to finish the race. We can express this by saying that "as the rate gets slower and slower (or approaches 0 mph), the time approaches ∞." The vertical line representing the time axis shown on the graph is a **vertical asymptote**. A vertical asymptote is a vertical line that the graph approaches, but never touches.

One important characteristic of the graph of a rational function is that asymptotes often occur. We will first consider vertical and horizontal asymptotes. Later in the section, we will discuss slant asymptotes.

Vertical Asymptote	The line $x = a$ is a **vertical asymptote** of the graph of a function $y = f(x)$ if $f(x)$ either increases or decreases without bound (approaches ∞ or −∞) as x approaches a.

Horizontal Asymptote	The line $y = b$ is a **horizontal asymptote** of the graph of a function $y = f(x)$ if $f(x)$ approaches b as x increases or decreases without bound (approaches ∞ or −∞).

Figure 4-36 shows a typical vertical and typical horizontal asymptote.

Vertical asymptote at $x = a$

$f(x)$ approaches ∞ as x approaches a from left.

$f(x)$ approaches −∞ as x approaches a from right.

Horizontal asymptote at $y = b$

$f(x)$ approaches b as x approaches ∞ or as x approaches −∞.

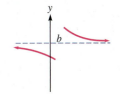

FIGURE 4-36

The graph of the rational function $f(x) = \frac{1}{x}$, called the **reciprocal function,** is shown in Figure 4-37. The domain of the function is $(-\infty, 0) \cup (0, \infty)$ and the range is $(-\infty, 0) \cup (0, \infty)$.

Take Note

The reciprocal function, $f(x) = \dfrac{1}{x}$, is a very important function. Please place it in your "library" of functions.

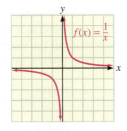

FIGURE 4-37

The reciprocal function has a vertical asymptote of $x = 0$ because

- We see from the graph that as x approaches 0 from the right that y or $f(x)$ approaches ∞.
- We see from the graph that as x approaches 0 from the left that y or $f(x)$ approaches $-\infty$.

The reciprocal function has a horizontal asymptote of $y = 0$ because

- We see from the graph that as x approaches ∞ that y or $f(x)$ approaches 0.
- We see from the graph that as x approaches $-\infty$ that y or $f(x)$ approaches 0.

3. Find Vertical Asymptotes of Rational Functions

To find the vertical asymptotes of a rational function written in simplest form, we must find the values of x for which the denominator of the rational function is 0 and the function is undefined. For example, since the denominator of $f(x) = \frac{2x - 1}{x + 2}$ is 0 when $x = -2$, there are no corresponding values of y and the line $x = -2$ is a vertical asymptote. We note that when x approaches -2 from the right or from the left, $f(x)$ approaches $-\infty$ and ∞ respectively. A graph of the function appears in Figure 4-38.

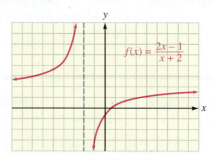

$$f(x) = \frac{2x - 1}{x + 2}$$

FIGURE 4-38

Strategy for Locating Vertical Asymptotes

To locate the vertical asymptotes of the rational function $f(x) = \frac{P(x)}{Q(x)}$, we follow these steps:

Step 1: Factor $P(x)$ and $Q(x)$ and remove any common factors.

Step 2: Set the denominator equal to 0 and solve the equation.

If a is a solution of the equation found in Step 2, $x = a$ is a vertical asymptote.

EXAMPLE 2

Finding Vertical Asymptotes of Rational Functions

Find the vertical asymptotes, if any, of each function:

a. $f(x) = \dfrac{2x}{x^2 - 16}$ **b.** $g(x) = \dfrac{x - 4}{x^2 - 16}$ **c.** $h(x) = \dfrac{5x}{x^2 + 16}$

SOLUTION

We will locate the vertical asymptotes of the graph of each function by factoring its numerator and/or denominator and removing any common factors. Then we will set the resulting denominator equal to zero, solve the equation, and identify the vertical asymptotes.

a. $f(x) = \dfrac{2x}{x^2 - 16}$

$f(x) = \dfrac{2x}{(x + 4)(x - 4)}$ Factor the denominator completely.

We then set the denominator equal to 0 and solve for x.

$$(x + 4)(x - 4) = 0$$

$$x + 4 = 0 \quad \text{or} \quad x - 4 = 0$$

$$x = -4 \quad | \quad x = 4$$

The vertical asymptotes are $x = 4$ and $x = -4$. See Figure 4-39.

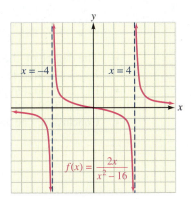

FIGURE 4-39

Tip

MC Hammer, a famous musician/rapper, had a huge hit song entitled "U Can't Touch This." The graph of a rational function will never touch a vertical asymptote.

©Everett Collection/Shutterstock.com

b. $g(x) = \dfrac{x - 4}{x^2 - 16}$

$$g(x) = \dfrac{x - 4}{(x + 4)(x - 4)} \qquad \text{Factor the denominator completely.}$$

$$g(x) = \dfrac{1}{x + 4} \qquad \text{Simplify: } \dfrac{x - 4}{x - 4} = 1.$$

We set the denominator equal to 0 and solve for x.

$$x + 4 = 0$$

$$x = -4$$

The vertical asymptote is $x = -4$. See Figure 4-40.

Take Note

A rational function can have zero, one, or several vertical asymptotes.

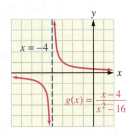

FIGURE 4-40

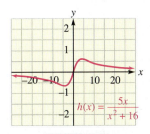

FIGURE 4-41

c. $h(x) = \dfrac{5x}{x^2 + 16}$

Because the denominator is the sum of two squared quantities, it cannot be factored over the set of real numbers. Thus, no values of x will make the denominator equal to 0. Thus, the rational function has no vertical asymptotes. See Figure 4-41.

Self Check 2   Find the vertical asymptotes, if any, of each function.

a. $g(x) = \dfrac{x + 5}{x^2 - 25}$ **b.** $h(x) = \dfrac{5}{x^2 + 25}$

Now Try Exercise 31.

4. Find Horizontal Asymptotes of Rational Functions

To find the horizontal asymptote of a rational function, we must find the value of y that the function approaches as x approaches ∞ or $-\infty$. To illustrate, we will consider three functions and some given function values.

First we consider a function where the degree of the numerator is less than the degree of the denominator.

$$f(x) = \frac{x + 2}{x^2 - 1} \qquad f(999) \approx 0.001 \qquad f(-999) \approx -0.001$$

- The degree of the numerator is 1 and the degree of the denominator is 2. Since $1 < 2$, the degree of the numerator is less than the degree of the denominator.
- If f is evaluated at large x-values (in both the positive and negative directions) such as 999 and -999, we obtain y-values that are close to 0.
- **Conclusion:** The line $y = 0$ (the x-axis) is a horizontal asymptote. See Figure 4-42.

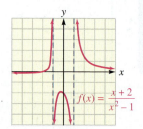

$$f(x) = \frac{x + 2}{x^2 - 1}$$

FIGURE 4-42

Second we consider a function where the degree of the numerator is equal to the degree of the denominator.

$$g(x) = \frac{2x}{x - 1} \qquad g(999) \approx 2.002 \qquad g(-999) \approx 1.998$$

- The degree of the numerator is 1 and the degree of the denominator is 1. Thus, the degrees of the numerator and denominator are equal.
- The leading coefficient of the numerator is 2 and the leading coefficient of the denominator is 1. Note that $2 \div 1$ is 2.
- If g is evaluated at large x-values (in both the positive and negative directions) such as 999 and -999, we obtain y-values that are close to 2.
- **Conclusion:** The line $y = 2$ is a horizontal asymptote. See Figure 4-43.

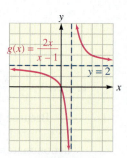

$$g(x) = \frac{2x}{x - 1}$$

$$y = 2$$

FIGURE 4-43

Finally we consider a function where the degree of the numerator is greater than the degree of the denominator.

$$h(x) = \frac{x^3}{x - 1} \qquad h(999) \approx 999{,}001 \qquad h(-999) \approx 997{,}003$$

- The degree of the numerator is 3 and the degree of the denominator is 1. Thus, the degree of the numerator is greater than the degree of the denominator.
- If h is evaluated at large x-values (in both the positive and negative directions) such as 999 and -999, we obtain y-values that are extremely large and do not seem to approach a finite number.
- **Conclusion:** There is no horizontal asymptote. See Figure 4-44.

$$h(x) = \frac{x^3}{x - 1}$$

FIGURE 4-44

We summarize the conclusions made in the previous examples.

Strategy for Locating Horizontal Asymptotes

To locate the horizontal asymptote of the rational function $f(x) = \frac{P(x)}{Q(x)}$, we consider three cases:

Case 1: If the degree of $P(x)$ *is less than* the degree of $Q(x)$, the line $y = 0$ is a horizontal asymptote.

Case 2: If the degree of $P(x)$ and $Q(x)$ are equal, the line $y = \frac{p}{q}$, where p and q are the leading coefficients of $P(x)$ and $Q(x)$, is a horizontal asymptote.

Case 3: If the degree of $P(x)$ is greater than the degree of $Q(x)$, there is no horizontal asymptote.

EXAMPLE 3

Finding Horizontal Asymptotes of Rational Functions

Find the horizontal asymptote, if any, of each function:

a. $f(x) = \dfrac{3x}{2x^2 - 1}$ **b.** $g(x) = \dfrac{3x^2 + 1}{x^2 - 2x + 1}$ **c.** $h(x) = \dfrac{x^3 + 2}{x - 5}$

SOLUTION

In each part, we will locate the horizontal asymptote by comparing the degree of the numerator to the degree of the denominator. Then we will decide which case applies and identify the horizontal asymptote.

a. Because the degree of the numerator is 1, the degree of the denominator is 2, and $1 < 2$, Case 1 applies. The horizontal asymptote is the line $y = 0$.

b. The degree of the numerator is 2 and the degree of the denominator is 2. Since the degrees are the same, Case 2 applies. Because the leading coefficient of the numerator is 3 and the leading coefficient of the denominator is 1, we divide 3 by 1 to obtain $\frac{3}{1} = 3$. The horizontal asymptote is the line $y = 3$.

c. Because the degree of the numerator is 3, the degree of the denominator is 1, and $3 > 1$, Case 3 applies. There is no horizontal asymptote.

Self Check 3   Find the horizontal asymptotes, if any, of each function.

a. $g(x) = \dfrac{4x - 5}{5 - x}$ **b.** $h(x) = \dfrac{3x^2}{x^3 - 5}$

Now Try Exercise 39.

5. Identify Slant Asymptotes of Rational Functions

A third type of asymptote is called a **slant asymptote**. These asymptotes occur when the degree of the numerator of a rational function is one more than the degree of the denominator. As the name implies, it is a slanted line, neither vertical nor horizontal.

To illustrate a slant asymptote, we consider the graph of $f(x) = \frac{x^2}{x - 2}$ shown in Figure 4-45.

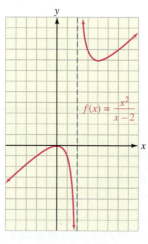

| FIGURE 4-45 | FIGURE 4-46 |

When x increases without bound to the right and to the left, the graph of the rational function approaches the slant asymptote shown in Figure 4-46. The equation of the slant asymptote is $y = x + 2$. To find this equation, we perform a long division, write the result in quotient $+ \frac{\text{remainder}}{\text{divisor}}$ form, and ignore the remainder.

$$
\begin{array}{r}
x + 2 \\
x - 2 \overline{\smash{\big)}\, x^2 + 0x + 0} \\
\underline{x^2 - 2x} \\
2x + 0 \\
\underline{2x - 4} \\
4
\end{array}
$$

Thus, $y = \frac{x^2}{x - 2} = \boldsymbol{x + 2} + \frac{4}{x - 2}$. Because the last fraction approaches 0 as x approaches ∞ and $-\infty$, the equation of the slant asymptote is $y = x + 2$.

Strategy for Locating Slant Asymptotes   If the degree of $P(x)$ is 1 greater than the degree of $Q(x)$ for the rational function $f(x) = \frac{P(x)}{Q(x)}$, there is a slant asymptote. To find it, divide $P(x)$ by $Q(x)$ and ignore the remainder.

Finding a Slant Asymptote of a Rational Function

EXAMPLE 4   Find the slant asymptote of $f(x) = \dfrac{3x^3 + 2x^2 + 2}{x^2 - 1}$.

SOLUTION   To find the slant asymptote, we divide the numerator by the denominator, write the result in quotient $+ \frac{\text{remainder}}{\text{divisor}}$ form, and ignore the remainder.

$$
\begin{array}{r}
3x + 2 \\
x^2 - 1 \overline{)3x^3 + 2x^2 + 2} \\
\underline{3x^3 - 3x } \\
2x^2 + 3x + 2 \\
\underline{2x^2 - 2} \\
3x + 4
\end{array}
$$

Thus,

$$y = \frac{3x^3 + 2x^2 + 2}{x^2 - 1} = \mathbf{3x + 2} + \frac{3x + 4}{x^2 - 1}$$

The last fraction approaches 0 as x approaches ∞ and $-\infty$. Thus the graph of the rational function approaches the slant asymptote with equation of $y = 3x + 2$. ∎

Self Check 4   Find the slant asymptote of $f(x) = \dfrac{2x^3 - 3x + 1}{x^2 - 4}$.

Now Try Exercise 45.

Tip
The graph of a rational function can cross horizontal and slant asymptotes but can never cross a vertical asymptote.

6. Graph Rational Functions

We will use the following strategy to graph the rational function $f(x) = \frac{P(x)}{Q(x)}$, where $P(x)$ and $Q(x)$ are polynomials written in descending powers of x and $\frac{P(x)}{Q(x)}$ is in simplest form (no common factors).

Strategy for Graphing Rational Functions of the Form
$$f(x) = \frac{P(x)}{Q(x)}$$

Step 1: Check for symmetry.
• If $f(-x) = f(x)$, the graph is symmetric about the y-axis.
• If $f(-x) = -f(x)$, the graph is symmetric about the origin.

Step 2: Find the vertical asymptotes.
The real roots of $Q(x) = 0$, if any, determine the vertical asymptotes of the graph.

Step 3: Find the y- and x-intercepts, if any.
• $f(0)$ is the y-coordinate of the y-intercept of the graph.
• Set $P(x) = 0$ and solve to find the x-intercepts of the graph.

Step 4: Find the horizontal asymptotes, if any.
• If the degree of $P(x)$ is less than the degree of $Q(x)$, the line $y = 0$ is a horizontal asymptote.
• If the degree of $P(x)$ is equal to the degree of $Q(x)$, the graph of $y = \frac{p}{q}$ is a horizontal asymptote, where p and q are the leading coefficients of $P(x)$ and $Q(x)$.

Step 5: Find the slant asymptotes, if any.
• If the degree of $P(x)$ is 1 greater than the degree of $Q(x)$, there is a slant asymptote. To find it, divide $P(x)$ by $Q(x)$ and ignore the remainder.

Step 6: Draw the graph.
Find additional points (if necessary) near the asymptotes.

Graphing a Rational Function

EXAMPLE 5

Graph: $f(x) = \dfrac{x^2 - 4}{x^2 - 1}$.

SOLUTION

We will use the steps outlined on the previous page to graph the rational function.

Step 1: Symmetry

Because x appears to even powers only, $f(-x) = f(x)$ and there is symmetry about the y-axis. There is no symmetry about the origin.

Step 2: Vertical asymptotes

To find the vertical asymptotes, we factor the numerator and denominator of $f(x)$ and simplify, if possible.

$$f(x) = \frac{x^2 - 4}{x^2 - 1} = \frac{(x + 2)(x - 2)}{(x + 1)(x - 1)} \qquad \textcolor{red}{\text{There are no common factors.}}$$

We then set the denominator equal to 0 and solve for x.

$$(x + 1)(x - 1) = 0$$

$$x + 1 = 0 \quad \text{or} \quad x - 1 = 0$$

$$x = -1 \qquad\qquad x = 1$$

There will be vertical asymptotes at $x = -1$ and $x = 1$.

Step 3: y- and x-intercepts

We can find the y-intercept by finding $f(0)$.

$$f(\textcolor{red}{0}) = \frac{\textcolor{red}{0}^2 - 4}{\textcolor{red}{0}^2 - 1} = \frac{-4}{-1} = 4$$

The y-intercept is $(0, 4)$.

We can find the x-intercepts by setting the numerator equal to 0 and solving for x:

$$x^2 - 4 = 0$$

$$(x + 2)(x - 2) = 0$$

$$x + 2 = 0 \quad \text{or} \quad x - 2 = 0$$

$$x = -2 \qquad\qquad x = 2$$

The x-intercepts are $(2, 0)$ and $(-2, 0)$.

Step 4: Horizontal asymptotes

Since the degrees of the numerator and denominator of the polynomials are the same, the line

$$y = \frac{1}{1} = 1 \qquad \textcolor{red}{\begin{array}{l}\text{The leading coefficient of the numerator is 1.} \\ \text{The leading coefficient of the denominator is 1.}\end{array}}$$

is a horizontal asymptote.

Step 5: Slant asymptotes

Since the degree of the numerator is not 1 greater than the degree of the denominator, there are no slant asymptotes.

Step 6: Graph

We plot the intercepts, draw the asymptotes, and make use of symmetry to graph the rational function. The graph is shown in Figure 4-47.

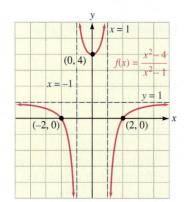

FIGURE 4-47

Self Check 5   Graph: $f(x) = \dfrac{x^2 - 9}{x^2 - 1}$.

Now Try Exercise 59.

Accent on Technology

Graphing a Rational Function

A graphing calculator can be very useful in helping us draw the graph of a rational function. However, certain precautions must be used or we will get a graph that is not going to give us a good idea of how the function looks. For example, in Figure 4-48(a) the graph of the function from Example 5 is shown. Using the ZOOM Standard window, a good representation of the graph is shown. In Figure 4-48(b) the ZOOM Decimal window is used, and part of the graph is not shown.

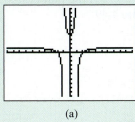
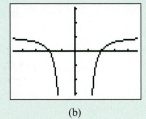

(a)   (b)

FIGURE 4-48

Graphing a Rational Function

EXAMPLE 6

Graph: $f(x) = \dfrac{3x}{x - 2}$.

SOLUTION   We will use the steps outlined earlier to graph the rational function.

Step 1: Symmetry
We find $f(-x)$.

$$f(-x) = \frac{3(-x)}{-x - 2} = \frac{-3x}{-x - 2} = \frac{3x}{x + 2}.$$

Because $f(-x) \neq f(x)$ and $f(-x) \neq -f(x)$, there is no symmetry about the y-axis or the origin.

Step 2: Vertical asymptotes
We first note that $f(x)$ is in simplest form. We then set the denominator equal to 0 and solve for x. Since the solution is 2, there will be a vertical asymptote at $x = 2$.

Step 3: y- and x-intercepts
We can find the y-intercept by finding $f(0)$.

$$f(0) = \frac{3(0)}{0 - 2} = \frac{0}{-2} = 0$$

The y-intercept is $(0, 0)$.
   We can find the x-intercepts by setting the numerator equal to 0 and solving for x:

$$3x = 0$$
$$x = 0$$

The x-intercept is $(0, 0)$.

Step 4: Horizontal asymptotes
Since the degrees of the numerator and denominator of the polynomials are the same, the line

$$y = \frac{3}{1} = 3 \qquad \text{The leading coefficient of the numerator is 3.}$$
$$\text{The leading coefficient of the denominator is 1.}$$

is a horizontal asymptote.

Step 5: Slant asymptotes

Since the degree of the numerator is not 1 greater than the degree of the denominator, there are no slant asymptotes.

Step 6: Graph

First, we plot the intercept $(0, 0)$ and draw the asymptotes. We then find one additional point on our graph to see what happens when x is greater than 2. To do so, we choose 3, a value of x that is greater than 2, and evaluate $f(3)$.

$$f(\mathbf{3}) = \frac{3(\mathbf{3})}{\mathbf{3} - 2} = \frac{9}{1} = 9$$

Since $f(3) = 9$, the point $(3, 9)$ lies on our graph. We sketch the graph as shown in Figure 4-49.

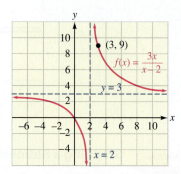

FIGURE 4-49

Self Check 6 Graph: $f(x) = \dfrac{3x}{x + 2}$.

Now Try Exercise 53.

Graphing a Rational Function

EXAMPLE 7

Graph the function: $f(x) = \dfrac{1}{x(x - 1)^2}$.

SOLUTION We will use the steps outlined earlier to graph the rational function.

Step 1: Symmetry

We find $f(-x)$.

$$f(x) = \frac{1}{x(x - 1)^2}$$

$$f(\mathbf{-x}) = \frac{1}{\mathbf{-x(-x} - 1)^2} \qquad \text{Replace } x \text{ with } -x.$$

$$= \frac{1}{-x[(-1)(x + 1)]^2} \qquad \text{Factor out } -1.$$

$$= \frac{1}{-x(-1)^2(x + 1)^2} \qquad \text{Square } -1 \text{ and } x + 1.$$

$$= \frac{1}{-x(x + 1)^2} \qquad \text{Simplify.}$$

$$= \frac{-1}{x(x + 1)^2} \qquad \text{Simplify.}$$

Because $f(-x) \neq f(x)$ and $f(-x) \neq -f(x)$, there is no symmetry about the y-axis or origin.

Step 2: Vertical asymptotes

We set the denominator equal to 0 and solve for x. Since 0 and 1 make the denominator 0, the vertical asymptotes are $x = 0$ and $x = 1$.

Step 3: y- and x-intercepts

Since x cannot be 0, the graph has no y-intercept. Since the numerator cannot be 0, y cannot be 0, and the graph has no x-intercepts.

Step 4: Horizontal asymptotes

Since the degree of the numerator is 0 and the degree of the denominator is 3, and $0 < 3$, we conclude that the horizontal asymptote is the line $y = 0$.

Step 5: Slant asymptotes

There are no slant asymptotes because the degree of the numerator is not 1 greater than the degree of the denominator.

Step 6: Graph

Because there are no intercepts to plot, we draw the asymptotes and find a few additional points to plot. The table gives three points lying in different intervals separated by the asymptotes. We sketch the graph as shown in Figure 4-50.

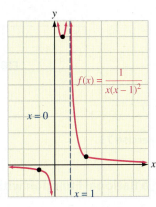

$f(x) = \dfrac{1}{x(x-1)^2}$		
x	y	(x, y)
-1	$-\dfrac{1}{4}$	$\left(-1, -\dfrac{1}{4}\right)$
$\dfrac{1}{2}$	8	$\left(\dfrac{1}{2}, 8\right)$
2	$\dfrac{1}{2}$	$\left(2, \dfrac{1}{2}\right)$

FIGURE 4-50

Self Check 7 Graph the function: $f(x) = \dfrac{1}{x(x-2)^2}$.

Now Try Exercise 69.

EXAMPLE 8

Graphing a Rational Function

Graph: $f(x) = \dfrac{1}{x^2 + 1}$.

SOLUTION We will use the steps outlined earlier to graph the rational function.

Step 1: Symmetry

Because x appears to an even power and $f(-x) = f(x)$, there is symmetry about the y-axis. There is no symmetry about the origin.

Step 2: Vertical asymptotes

Since no number x makes the denominator 0, the graph has no vertical asymptotes.

Step 3: y- and x-intercepts

We can find the y-intercept by finding $f(0)$.

$$f(\mathbf{0}) = \frac{1}{\mathbf{0}^2 + 1} = \frac{1}{1} = 1$$

The y-intercept is $(0, 1)$.

Because the denominator is always positive, the fraction is always positive and the graph lies entirely above the x-axis. There are no x-intercepts. We also see that if we set the numerator equal to 0 there are no solutions for x.

Step 4: Horizontal asymptotes
Since the degree of the numerator is less than the degree of the denominator, the line $y = 0$ is a horizontal asymptote.

Step 5: Slant asymptotes
There are no slant asymptotes because the degree of the numerator is not 1 greater than the degree of the denominator.

Step 6: Graph
We first plot the y-intercept $(0, 1)$. Then, we make use of the fact that the graph has y-axis symmetry and a horizontal asymptote of $y = 0$ and sketch the graph. The graph is shown in Figure 4-51.

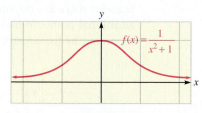

FIGURE 4-51

Self Check 8   Graph: $f(x) = \dfrac{4}{x^2 + 1}$.

Now Try Exercise 73.

EXAMPLE 9   **Graphing a Rational Function with a Slant Asymptote**

Graph: $f(x) = \dfrac{x^2 + x - 2}{x - 3}$.

SOLUTION   We first factor the numerator of the expression $f(x) = \dfrac{(x - 1)(x + 2)}{x - 3}$ and use the steps outlined earlier to graph the rational function.

Step 1: Symmetry
We find $f(-x)$.

$$f(x) = \frac{x^2 + x - 2}{x - 3}$$

$$f(-x) = \frac{(-x)^2 + (-x) - 2}{-x - 3} \qquad \text{Replace } x \text{ with } -x.$$

$$= \frac{x^2 - x - 2}{-x - 3} \qquad \text{Simplify.}$$

$$= \frac{x^2 - x - 2}{(-1)(x + 3)} \qquad \text{Factor } -1 \text{ out of the denominator.}$$

$$= \frac{-x^2 + x + 2}{x + 3} \qquad \text{Divide by } -1.$$

Because $f(-x) \neq f(x)$ and $f(x) \neq -f(x)$, there is no symmetry about the y-axis or origin.

Step 2: Vertical asymptotes
We set the denominator equal to 0 and solve for x. Since 3 makes the denominator 0, the vertical asymptote is $x = 3$.

Step 3: y- and x-intercepts

The y-intercept is $\left(0, \frac{2}{3}\right)$ because $f(0) = \frac{2}{3}$. The x-intercepts are $(1, 0)$ and $(-2, 0)$ because x values of 1 and -2 make the numerator 0.

Step 4: Horizontal asymptotes

Since the degree of the numerator is 2, the degree of the denominator is 1, and $2 > 1$, we conclude that there is no horizontal asymptote.

Step 5: Slant asymptotes

Because the degree of the numerator is 1 greater than the degree of the denominator, this graph will have a slant asymptote. To find it, we perform a long division.

$$
\begin{array}{r}
x + 4 \\
x - 3 \overline{\smash{)}\, x^2 + x - 2} \\
\underline{x^2 - 3x } \\
4x - 2 \\
\underline{4x - 12} \\
10
\end{array}
$$

We write the function as

$$
y = \frac{x^2 + x - 2}{x - 3} = x + 4 + \frac{10}{x - 3}
$$

The fraction $\frac{10}{x-3}$ approaches 0 as x approaches ∞ or $-\infty$, and the graph approaches a slant asymptote: the line $y = x + 4$.

Step 6: Graph

The graph appears in Figure 4-52.

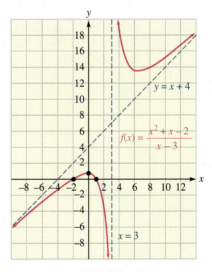

FIGURE 4-52

Self Check 9   Graph: $f(x) = \dfrac{x^2 + x + 2}{x + 3}$.

Now Try Exercise 75.

7. Understand when a Graph Has a Missing Point

We have discussed rational functions where the fraction is in simplified form. We now consider a rational function $f(x) = \frac{P(x)}{Q(x)}$, where $P(x)$ and $Q(x)$ have a common factor. Graphs of such functions have gaps or missing points that are not the result of vertical asymptotes.

Graphing a Rational Function that Has a Missing Point

EXAMPLE 10

Find the domain of the function $f(x) = \dfrac{x^2 - x - 12}{x - 4}$ and graph it.

SOLUTION

We find the values of x that make the denominator 0. These values of x will not be included in the domain. We will then write the rational function in simplest form and graph it using the methods of this section.

Since the denominator cannot be 0, $x \neq 4$. Therefore, the domain is the set of all real numbers except 4.

We now write the rational function in simplest form by factoring the numerator of the expression. We see that the numerator and denominator have a common factor of $x - 4$.

$$f(x) = \frac{x^2 - x - 12}{x - 4}$$

$$= \frac{(x + 3)(x - 4)}{x - 4} \qquad \text{Factor } x^2 - x - 12.$$

If $x \neq 4$, the common factor of $x - 4$ can be divided out. The resulting function is equivalent to the original function only when we keep the restriction that $x \neq 4$. Thus,

$$f(x) = \frac{(x + 3)(x - 4)}{x - 4} = x + 3 \qquad \text{(provided that } x \neq 4)$$

When $x = 4$, the function is not defined. The graph of the function appears in Figure 4-53. It is a line with the point with x-coordinate of 4 missing. The line has a hole in it.

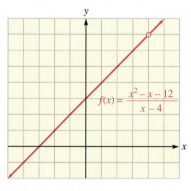

$$f(x) = \frac{x^2 - x - 12}{x - 4}$$

FIGURE 4-53

Caution

A calculator graph of a rational function does not indicate a missing point or hole.

Self Check 10 Graph: $f(x) = \dfrac{x^2 + x - 12}{x + 4}$.

Now Try Exercise 83.

8. Solve Problems Modeled by Rational Functions

Rational expressions often define functions that occur in the real world.

EXAMPLE 11

Solving an Application Using a Rational Function

Suppose the cost C of a no contract cell phone is \$20 per month plus \$0.15 per minute. Write a rational function that represents the average (mean) cost per minute, $\overline{C}$. Let x represent the number of minutes used per month. Find the mean cost per minute when the service is used for 180 minutes.

SOLUTION

©MJTH/Shutterstock.com

Because the mean cost is the total monthly cost divided by the number of minutes used, we can write a rational function that models the problem. We will then evaluate the mean cost function at 180 to find the mean cost for 180 minutes used.

We know that the mean cost $\overline{C}$ is total cost C divided by the number of access minutes x. That is,

$$\overline{C} = \frac{C}{x}$$

Since total cost C is \$20 per month plus \$0.15 for each minute x, we can write $C(x) = 0.15x + 20$.

Thus, the function

$$\overline{C}(x) = \frac{0.15x + 20}{x} \qquad (x > 0)$$

gives the mean cost per minute of using the service for x minutes per month.

To find the mean cost for 180 minutes of usage, we substitute 180 for x and simplify.

$$\overline{C}(\mathbf{180}) = \frac{0.15(\mathbf{180}) + 20}{\mathbf{180}} \approx 0.26$$

The mean cost per minute for 180 minutes of usage is approximately \$0.26 per minute. ∎

Self Check 11   Find the mean hourly cost when the cell phone described above is used for 240 minutes.

Now Try Exercise 89.

Accent on Technology

Graphs of Rational Functions

We can use a graphing calculator to graph the function $\overline{C}(x) = \frac{0.15x + 20}{x}$ in Example 11. We enter the function as shown in Figure 4-54(a) using the window settings as shown in Figure 4-54(b). The graph of the rational function appears in Figure 4-54(c). Note the following characteristics:

- The graph of the function passes the Vertical Line Test, as expected.
- From the graph in Figure 4-54(c), we see that the mean cost per minute decreases as the number of minutes of usage increases.

Since the cost of each extra minute of usage of time is \$0.15, the mean minute cost can approach \$0.15 but will never drop below it. The graph of the function approaches the line $y = 0.15$ as x increases without bound. The line $y = 0.15$ is a horizontal asymptote of the graph.

- As x gets smaller and approaches 0, the graph approaches the y-axis. The y-axis is a vertical asymptote of the graph.

(a)

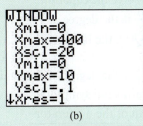
(b)

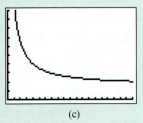
(c)

FIGURE 4-54

Self Check Answers

1. $(-\infty, -1) \cup (-1, 2) \cup (2, \infty)$ 2. **a.** $x = 5$ **b.** none   3. **a.** $y = -4$
b. $y = 0$ 4. $y = 2x$ 5.

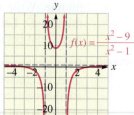

6.

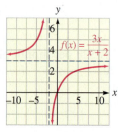

7.

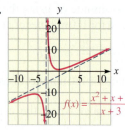

8.

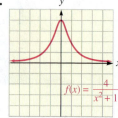

9.

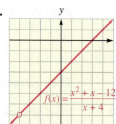

10.

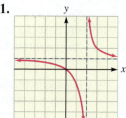

11. $0.23 per minute

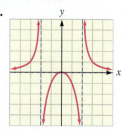

Exercises 4.6

Getting Ready
You should be able to complete these vocabulary and concept statements before you proceed to the practice exercises.

Fill in the blanks.

1. When a graph approaches a vertical line but never touches it, we call the line an _____.

2. A rational function is a function with a polynomial numerator and a _____ polynomial denominator.

3. To find a _____ asymptote of a rational function in simplest form, set the denominator polynomial equal to 0 and solve the equation.

4. To find the _____ of a rational function, let $x = 0$ and solve for y or find $f(0)$.

5. To find the _____ of a rational function, set the numerator equal to 0 and solve the equation.

6. In the function $f(x) = \frac{P(x)}{Q(x)}$, if the degree of $P(x)$ is less than the degree of $Q(x)$, the horizontal asymptote is _____.

7. In the function $f(x) = \frac{P(x)}{Q(x)}$, if the degree of $P(x)$ and $Q(x)$ are the _____, the horizontal symptote is
$$y = \frac{\text{the leading coefficient of the numerator}}{\text{the leading coefficient of the denominator}}.$$

8. In a rational function, if the degree of the numerator is 1 greater than the degree of the denominator, the graph will have a _____.

9. A graph can cross a _____ asymptote but can never cross a _____ asymptote.

10. The graph of $f(x) = \frac{x^2 - 4}{x + 2}$ will have a _____ point.

Find the equations of the vertical and horizontal asymptotes of each graph. Find the domain and range.

11.

12.

Practice

The time t it takes to travel 600 miles is a function of the mean rate of speed r:

$$t = f(r) = \frac{600}{r}$$

Find t for the given values of r.

13. 30 mph

14. 40 mph

15. 50 mph

16. 60 mph

Suppose the cost (in dollars) of removing p% of the pollution in a river is given by the function

$$C = f(p) = \frac{50{,}000p}{100 - p} \quad (0 \le p < 100)$$

Find the cost of removing each percent of pollution.

17. 10%

18. 30%

19. 50%

20. 80%

Find the domain of each rational function. Do not graph the function.

21. $f(x) = \dfrac{x^2}{x - 2}$

22. $f(x) = \dfrac{x^3 - 3x^2 + 1}{x + 3}$

23. $f(x) = \dfrac{2x^2 + 7x - 2}{x^2 - 25}$

24. $f(x) = \dfrac{5x^2 + 1}{x^2 + 5}$

25. $f(x) = \dfrac{x - 1}{x^3 - x}$

26. $f(x) = \dfrac{x + 2}{2x^2 - 9x + 9}$

27. $f(x) = \dfrac{3x^2 + 5}{x^2 + 1}$

28. $f(x) = \dfrac{7x^2 - x + 2}{x^4 + 4}$

Find the vertical asymptotes, if any, of each rational function. Do not graph the function.

29. $f(x) = \dfrac{x}{x - 3}$

30. $f(x) = \dfrac{2x}{2x + 5}$

31. $f(x) = \dfrac{x + 2}{x^2 - 1}$

32. $f(x) = \dfrac{x - 4}{x^2 - 16}$

33. $f(x) = \dfrac{1}{x^2 - x - 6}$

34. $f(x) = \dfrac{x + 2}{2x^2 - 6x - 8}$

35. $f(x) = \dfrac{x^2}{x^2 + 5}$

36. $f(x) = \dfrac{x^3 - 3x^2 + 1}{2x^2 + 3}$

Find the horizontal asymptotes, if any, of each rational function. Do not graph the function.

37. $f(x) = \dfrac{2x - 1}{x}$

38. $f(x) = \dfrac{x^2 + 1}{3x^2 - 5}$

39. $f(x) = \dfrac{x^2 + x - 2}{2x^2 - 4}$

40. $f(x) = \dfrac{5x^2 + 1}{5 - x^2}$

41. $f(x) = \dfrac{x + 1}{x^3 - 4x}$

42. $f(x) = \dfrac{x}{2x^2 - x + 11}$

43. $f(x) = \dfrac{x^2}{x - 2}$

44. $f(x) = \dfrac{x^4 + 1}{x - 3}$

Find the slant asymptote, if any, of each rational function. Do not graph the function.

45. $f(x) = \dfrac{x^2 - 5x - 6}{x - 2}$

46. $f(x) = \dfrac{x^2 - 2x + 11}{x + 3}$

47. $f(x) = \dfrac{2x^2 - 5x + 1}{x - 4}$

48. $f(x) = \dfrac{5x^3 + 1}{x + 5}$

49. $f(x) = \dfrac{x^3 + 2x^2 - x - 1}{x^2 - 1}$

50. $f(x) = \dfrac{-x^3 + 3x^2 - x + 1}{x^2 + 1}$

Graph each rational function. Check your work with a graphing calculator.

51. $y = \dfrac{1}{x - 2}$

52. $y = \dfrac{3}{x + 3}$

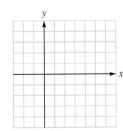

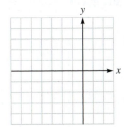

53. $y = \dfrac{x}{x - 1}$

54. $y = \dfrac{x}{x + 2}$

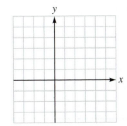

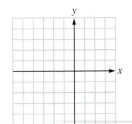

55. $f(x) = \dfrac{x + 1}{x + 2}$

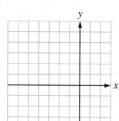

56. $f(x) = \dfrac{x - 1}{x - 2}$

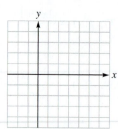

63. $y = \dfrac{x^2 + 2x - 3}{x^3 - 4x}$

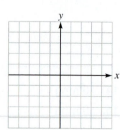

64. $y = \dfrac{3x^2 - 4x + 1}{2x^3 + 3x^2 + x}$

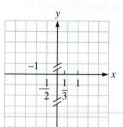

57. $f(x) = \dfrac{2x - 1}{x - 1}$

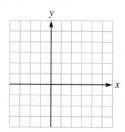

58. $f(x) = \dfrac{3x + 2}{x^2 - 4}$

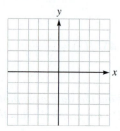

65. $y = \dfrac{x^2 - 9}{x^2}$

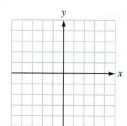

66. $y = \dfrac{3x^2 - 12}{x^2}$

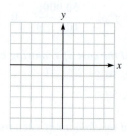

59. $g(x) = \dfrac{x^2 - 9}{x^2 - 4}$

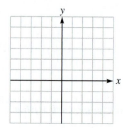

60. $g(x) = \dfrac{x^2 - 4}{x^2 - 9}$

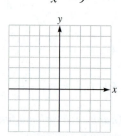

67. $f(x) = \dfrac{x}{(x + 3)^2}$

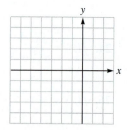

68. $f(x) = \dfrac{x}{(x - 1)^2}$

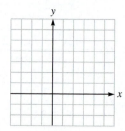

69. $f(x) = \dfrac{x + 1}{x^2(x - 2)}$

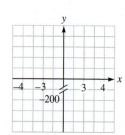

70. $f(x) = \dfrac{x - 1}{x^2(x + 2)^2}$

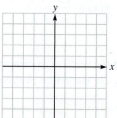

61. $g(x) = \dfrac{x^2 - x - 2}{x^2 - 4x + 3}$

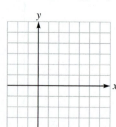

62. $g(x) = \dfrac{x^2 + 7x + 12}{x^2 - 7x + 12}$

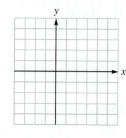

71. $y = \dfrac{x}{x^2 + 1}$

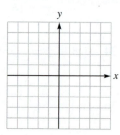

72. $y = \dfrac{x - 1}{x^2 + 2}$

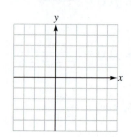

73. $y = \dfrac{3x^2}{x^2 + 1}$

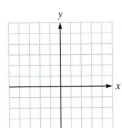

74. $y = \dfrac{x^2 - 9}{2x^2 + 1}$

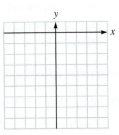

81. $f(x) = \dfrac{x^3 + x}{x}$

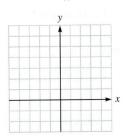

82. $f(x) = \dfrac{x^3 - x^2}{x - 1}$

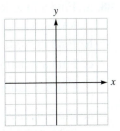

75. $h(x) = \dfrac{x^2 - 2x - 8}{x - 1}$

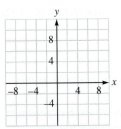

76. $h(x) = \dfrac{x^2 + x - 6}{x + 2}$

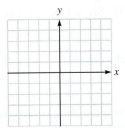

83. $f(x) = \dfrac{x^2 - 2x + 1}{x - 1}$

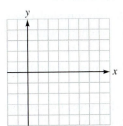

84. $f(x) = \dfrac{2x^2 + 3x - 2}{x + 2}$

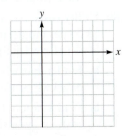

77. $f(x) = \dfrac{x^3 + x^2 + 6x}{x^2 - 1}$

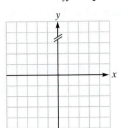

85. $f(x) = \dfrac{x^3 - 1}{x - 1}$

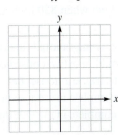

86. $f(x) = \dfrac{x^2 - x}{x^2}$

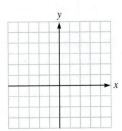

78. $f(x) = \dfrac{x^3 - 2x^2 + x}{x^2 - 4}$

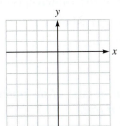

Graph each rational function. Note that the numerator and denominator of the fraction share a common factor.

79. $f(x) = \dfrac{x^2}{x}$

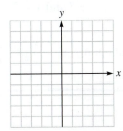

80. $f(x) = \dfrac{x^2 - 1}{x - 1}$

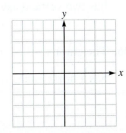

Applications

A service club wants to publish a directory of its members. Some investigation shows that the cost of typesetting and photography will be $700, and the cost of printing each directory will be $3.25.

87. a. Find a function that gives the total cost C of printing x directories.

b. Find the total cost of printing 500 directories.

c. Find a function that gives the mean cost per directory $\overline{C}$ of printing x directories.

d. Find the mean cost per directory if 500 directories are printed.

e. Find the mean cost per directory if 1000 directories are printed.

f. Find the mean cost per directory if 2000 directories are printed.

An electric company charges $10 per month plus 20¢ for each kilowatt-hour (kwh) of electricity used.

88. a. Find a function that gives the total cost C of x kwh of electricity.

  b. Find the total cost for using 775 kwh.

  c. Find a function that gives the mean cost per kwh, $\overline{C}$, when using x kwh.

  d. Find the mean cost per kwh when 775 kwh are used. Round to the nearest hundredth.

  e. Find the mean cost per kwh when 3200 kwh are used. Round to the nearest hundredth.

89. Utility costs An overseas electric company charges $8.50 per month plus 9.5¢ for each kilowatt-hour (kwh) of electricity used.

  a. Find a linear function that gives the total cost C of x kwh of electricity.

  b. Find a rational function that gives the average cost per kwh when using x kwh.

  c. Find the average cost per kwh when 850 kwh are used.

90. Scheduling work crews The following rational function gives the number of days it would take two construction crews, working together, to frame a house that crew 1 (working alone) could complete in x days and crew 2 (working alone) could complete in $(x + 3)$ days.

$$f(x) = \frac{x^2 + 3x}{2x + 3}$$

  a. If crew 1 could frame a certain house in 21 days, how long would it take both crews working together?

  b. If crew 2 could frame a certain house in 25 days, how long would it take both crews working together?

©Hans.slegers/Shutterstock.com

Discovery and Writing

91. What is a rational function?

92. If you are given the equation of a rational function, explain how you determine the vertical and horizontal asymptotes, if any.

93. How do you know when a rational function has a slant asymptote? If one exists, explain how to determine its equation.

94. Describe a strategy that can be used to graph a rational function.

In Exercises 95–98, a, b, c, and d are nonzero constants.

95. Show that $y = 0$ is a horizontal asymptote of the graph of $f(x) = \frac{ax + b}{cx^2 + d}$.

96. Show that $y = \frac{a}{c}x$ is a slant asymptote of the graph of $f(x) = \frac{ax^3 + b}{cx^2 + d}$.

97. Show that $y = \frac{a}{c}$ is a horizontal asymptote of the graph of $f(x) = \frac{ax^2 + b}{cx^2 + d}$.

98. Graph the rational function $f(x) = \frac{x^3 + 1}{x}$ and explain why the curve is said to have a *parabolic asymptote*.

Use a graphing calculator to perform each experiment. Write a brief paragraph describing your findings.

99. Investigate the positioning of the vertical asymptotes of a rational function by graphing $f(x) = \frac{x}{x - k}$ for several values of k. What do you observe?

100. Investigate the positioning of the vertical asymptotes of a rational function by graphing $f(x) = \frac{x}{x^2 - k}$ for $k = 4, 1, -1$, and 0. What do you observe?

101. Find the range of the rational function $f(x) = \frac{kx^2}{x^2 + 1}$ for several values of k. What do you observe?

102. Investigate the positioning of the x-intercepts of a rational function by graphing $f(x) = \frac{x^2 - k}{x}$ for $k = 1, -1$, and 0. What do you observe?

Critical Thinking

Determine if the statement is true or false. If the statement is false, then correct it and make it true.

103. A rational function can have two horizontal asymptotes.

104. All rational functions have vertical asymptotes.

105. The graph of the rational function $f(x) = \frac{x - 7}{x^2 - 49}$ has two vertical asymptotes, $x = -7$ and $x = 7$.

106. The graph of the rational function $f(x) = \frac{x^{100} + 100}{x^{101} - 101}$ has a horizontal asymptote at $y = 0$.

107. The graph of the rational function $f(x) = \dfrac{x^{101} + 101}{x^{100} - 100}$ has no horizontal asymptote.

108. The graph of a rational function will never cross a vertical asymptote.

109. The graph of a rational function will never cross a horizontal asymptote.

110. A rational function can have two slant asymptotes.

CHAPTER REVIEW

4.1 Quadratic Functions

Definitions and Concepts	Examples
A **quadratic function** is a second-degree polynomial function in one variable of the form $$f(x) = ax^2 + bx + c \quad \text{or} \quad y = ax^2 + bx + c$$ where a, b, and c are real numbers and $a \neq 0$. The graph of a quadratic function of the form $f(x) = ax^2 + bx + c$ $(a \neq 0)$ is a **parabola** with vertex at $\left(-\frac{b}{2a},\, c - \frac{b^2}{4a}\right)$ or $\left(-\frac{b}{2a},\, f\left(-\frac{b}{2a}\right)\right)$. • If $a > 0$, the parabola **opens upward**. • If $a < 0$, the parabola **opens downward**. The **standard form of an equation of a quadratic function** is: $$f(x) = a(x - h)^2 + k \quad (a \neq 0)$$ The vertex is at (h, k). • The parabola opens upward when $a > 0$ and downward when $a < 0$. • The axis of symmetry of the parabola is the vertical line graph of the equation $x = h$.	$f(x) = 3x^2 - 2x + 1$ • $a = 3$, the parabola opens upward. $y = -\dfrac{1}{2}x^2 - 4$ • $a = -\frac{1}{2}$, the parabola opens downward. $f(x) = 4(x - 2)^2 - 8$ • $a = 4$, the parabola opens upward. The vertex is $(2, -8)$. The axis of symmetry is $x = 2$. $y = -\dfrac{1}{3}(x + 2)^2 + 5$ • $a = -\frac{1}{3}$, the parabola opens downward. The vertex is $(-2, 5)$. The axis of symmetry is $x = -2$.

Graphing a quadratic function:
To graph a quadratic function:
 	
1. Determine whether the parabola opens upward or downward.
2. Find the vertex of the parabola.
3. Find the x-intercept(s).
4. Find the y-intercept.
5. Identify one additional point on the graph.
6. Draw a smooth curve through the points found in Steps 2–5.
	Graph the quadratic function $f(x) = 3(x + 2)^2 - 3$.
	Step 1: Determine whether the parabola opens upward or downward.
	
	   Standard Form: $f(x) = \boldsymbol{a}(x - h)^2 + k$
	   Given Form: $f(x) = \boldsymbol{3}(x + 2)^2 - 3$
	
	Since $a = 3$ and 3 is positive, the parabola opens upward.
	
	Step 2: Find the vertex of the parabola.
	
	   Standard Form: $f(x) = a(x - \boldsymbol{h})^2 + \boldsymbol{k}$
	   Given Form: $f(x) = 3(x + 2)^2 - 3$
	                   $f(x) = 3[x - (\boldsymbol{-2})]^2 + (\boldsymbol{-3})$
	
	Since $h = \boldsymbol{-2}$ and $k = \boldsymbol{-3}$ the vertex is the point $(-2, -3)$.

Definitions and Concepts	**Examples**
	Step 3: Find the x-intercept(s). To find the x-intercepts, we substitute 0 for $f(x)$ and solve for x.

$$f(x) = 3(x + 2)^2 - 3$$
$$0 = 3(x + 2)^2 - 3$$
$$3 = 3(x + 2)^2$$
$$1 = (x + 2)^2$$
$$x + 2 = \pm 1$$
$$x = -2 \pm 1$$
$$x = -1 \quad \text{or} \quad x = -3$$

The x-intercepts are the points $(-3, 0)$ and $(-1, 0)$.

Step 4: Find the y-intercept.
To find the y-intercept, we substitute 0 in for x and solve for y.

$$f(x) = 3(x + 2)^2 - 3$$
$$y = 3(x + 2)^2 - 3$$
$$y = 3(0 + 2)^2 - 3$$
$$y = 3(2)^2 - 3$$
$$y = 12 - 3$$
$$y = 9$$

The y-intercept is the point $(0, 9)$.

Step 5: Identify one additional point on the graph.
Because of symmetry, the point $(-4, -9)$ is on the graph.

Step 6: Draw a smooth curve through the points found in Steps 2–5.

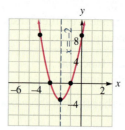

The axis of symmetry of the parabola is $x = -2$ because $h = -2$.

If the graph of the parabola opens downward, then the vertex is the **maximum point** on the graph of the parabola. If the graph of the parabola opens upward, then the vetex is the **minimum point** on the graph of the parabola.	**Minimum cost** A company has found that the total monthly cost C of producing x air-hockey tables is given by $C(x) = 1.5x^2 - 270x + 28{,}665$. Find the production level that minimizes the monthly cost and find that monthly minimum cost. The function $C(x)$ is a quadratic function whose graph is a parabola that opens upward. The minimum value of $C(x)$ occurs at the vertex of the parabola. We will use the vertex formula to find the vertex of the parabola.

Definitions and Concepts	**Examples**
	We compare the equations $C(x) = 1.5x^2 - 270x + 28{,}665$ and $f(x) = ax^2 + bx + c$ to see that $a = 1.5$, $b = -270$, and $c = 28{,}665$. Using the vertex formula, we see that the vertex of the parabola is the point with coordinates $$\left(-\frac{b}{2a}, c - \frac{b^2}{4a}\right) = \left(-\frac{-270}{2(1.5)}, 28{,}665 - \frac{(-270)^2}{4(1.5)}\right)$$ $$= (90,\ 16{,}515)$$ If the company makes 90 air-hockey tables each month, it will minimize its production cost. The minimum monthly cost will be $16,515.

Exercises

Determine whether the graph of each quadratic function opens upward or downward. State whether a maximum or minimum point occurs at the vertex of the parabola.

1. $f(x) = \dfrac{1}{2}x^2 + 4$

2. $f(x) = -4(x + 1)^2 + 5$

Find the vertex of each parabola.

3. $f(x) = 2(x - 1)^2 + 6$

4. $f(x) = -2(x + 4)^2 - 5$

5. $f(x) = x^2 + 6x - 4$

6. $f(x) = -4x^2 + 4x - 9$

Graph each quadratic function and find its vertex.

7. $f(x) = (x - 2)^2 - 3$

8. $f(x) = -(x - 4)^2 + 4$

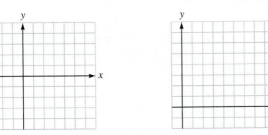

9. $y = x^2 - x$

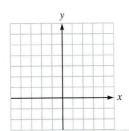

10. $y = x - x^2$

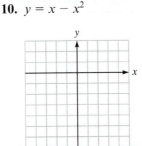

11. $y = x^2 - 3x - 4$

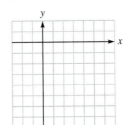

12. $y = 3x^2 - 8x - 3$

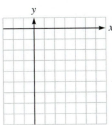

13. Architecture A parabolic arch has an equation of $3x^2 + y - 300 = 0$. Find the maximum height of the arch.

14. Puzzle problem The sum of two numbers is 1, and their product is as large as possible. Find the numbers.

15. Maximizing area A rancher wishes to enclose a rectangular corral with 1400 feet of fencing. What dimensions of the corral will maximize the area? Find the maximum area.

16. Digital cameras A company that produces and sells digital cameras has determined that the total weekly cost C of producing x digital cameras is given by the function $C(x) = 1.5x^2 - 150x + 4850$. Determine the production level that minimizes the weekly cost for producing the digital cameras and find that weekly minimum cost.

4.2 Polynomial Functions

Definitions and Concepts	Examples
A **polynomial function in one variable** (say, x) is a function of the form $$f(x) = a_n x^n + a_{n-1} x^{n-1} + \cdots + a_1 x + a_0$$ where $a_n, a_{n-1}, \ldots, a_1$, and a_0 are real numbers and n is a whole number. The **degree of a polynomial function** is the largest power of x that appears in the polynomial.	$f(x) = 3x^2 + 4x - 7$, degree of 2 $f(x) = -17x^4 + 3x^3 - 2x^2 + 13$, degree of 4
Zeros of polynomial functions: If $f(x)$ represents a polynomial function, then the solutions or roots of the polynomial equation $f(x) = 0$ are known as **zeros** of the polynomial function.	Use factoring to find the zeros of $$f(x) = x^5 - 16x^3$$ We set $f(x) = 0$ and use factoring to find the zeros. $x^5 - 16x^3 = 0$ $x^3(x^2 - 16) = 0$ Factor out x^3. $x^3(x + 4)(x - 4) = 0$ Factor $x^2 - 16$. $x^3 = 0$ or $x + 4 = 0$ or $x - 4 = 0$ $x = 0$ $x = -4$ $x = 4$ The zeros of the polynomial function are $x = 0$, $x = -4$, and $x = 4$.
Summary of the Leading Coefficient Test: **Case 1:** If the degree of the polynomial is odd and the leading coefficient is positive, then the graph of the polynomial function falls on the left and rises on the right.	The degree of $f(x) = 4x^3 - 6x + 3$ is 3, which is odd, and the leading coefficient is 4, which is positive. Case 1 applies. The graph falls on the left and rises on the right.
Case 2: If the degree of the polynomial is odd and the leading coefficient is negative, then the graph of the polynomial function rises on the left and falls on the right.	The degree of $f(x) = -2x^5 - 8x - 3$ is 5, which is odd, and the leading coefficient is -2, which is negative. Case 2 applies. The graph rises on the left and falls on the right.
Case 3: If the degree of the polynomial is even and the leading coefficient is positive, then the graph of the polynomial function rises on the left and rises on the right.	The degree of $f(x) = 5x^4 - 8x^3 + 4$ is 4, which is even, and the leading coefficient is 5, which is positive. Case 3 applies. The graph rises on the left and rises on the right.
Case 4: If the degree of the polynomial is even and the leading coefficient is negative, then the graph of the polynomial function falls on the left and falls on the right.	The degree of $f(x) = -7x^6 - x^3 + 2$ is 6, which is even, and the leading coefficient is -7, which is negative. Case 4 applies. The graph falls on the left and falls on the right.

Definitions and Concepts

Graphing polynomial functions:

1. Find the x- and y-intercepts of the graph.
2. Determine the end behavior.
3. Make a sign chart and determine where the graph is above and below the x-axis.
4. Find any symmetries of the graph.
5. Plot a few points, if necessary, and draw the graph as a smooth continuous curve.

If $f(-x) = f(x)$ for all x in the domain of f, the graph of the function is symmetric about the y-axis,

If $f(-x) = -f(x)$ for all x in the domain of f, the function is symmetric about the origin.

Examples

Graph the function $f(x) = -x^3 + 9x$.

Step 1: Find the x- and y-intercepts of the graph. To find the x-intercepts, we let $f(x) = 0$ and solve for x.

$$-x^3 + 9x = 0$$
$$-x(x^2 - 9) = 0$$
$$-x(x + 3)(x - 3) = 0$$

$-x = 0$ or $x + 3 = 0$ or $x - 3 = 0$

$x = 0$ | $x = -3$ | $x = 3$

The x-intercepts are $(0, 0)$, $(-3, 0)$, and $(3, 0)$.

If we let $x = 0$ and solve for $f(x)$, we see that the y-intercept is also $(0, 0)$.

Step 2: Determine the end behavior. The degree of $f(x) = -x^3 - 9x$ is 3, which is odd, and the leading coefficient is -1, which is negative. Case 2 applies. The graph of the polynomial function rises on the left and falls on the right.

Step 3: Determine where the graph is above or below the x-axis. To determine where the graph is above or below the x-axis, we plot the solutions of $-x^3 + 9x = 0$ on a number line and establish the four intervals shown in the figure. We then test a number from each interval to determine the sign of $f(x)$.

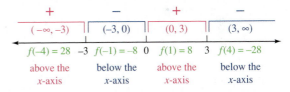

Step 4: Find any symmetries of the graph.
To test for symmetry about the y-axis, we check to see whether $f(x) = f(-x)$. To test for symmetry about the origin, we check to see whether $f(x) = -f(x)$.

$$f(x) = -x^3 + 9x$$
$$f(-x) = -(-x)^3 + 9(-x)$$
$$f(-x) = x^3 - 9x$$

Since $f(x) \neq f(-x)$, there is no symmetry about the y-axis. However, since $f(-x) = -f(x)$, there is symmetry about the origin.

Definitions and Concepts	**Examples**

Examples

Step 5: Plot a few points and draw the graph as a smooth continuous curve. We now plot the intercepts and one additional point. In the previous step we found that $f(1) = 8$. This will be the additional point we plot $(1, 8)$. Making use of our knowledge of symmetry about the origin and where the graph is above and below the x-axis, we now draw the graph as shown.

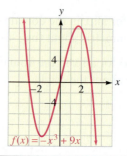

$$f(x) = -x^3 + 9x$$

The Intermediate Value Theorem:
Let $P(x)$ be a polynomial with real coefficients. If $P(a) \neq P(b)$ for $a < b$, then $P(x)$ takes on all values between $P(a)$ and $P(b)$ on the closed interval $[a, b]$.

If $P(a)$ and $P(b)$ have opposite signs, there is at least one number r in the interval (a, b) for which $P(r) = 0$.

Consider the polynomial function
$P(x) = 4x^3 + 2x^2 - 12x + 3$ and note that

$$P(0) = 4(0)^3 + 2(0)^2 - 12(0) + 3 = 3$$

$$P(1) = 4(1)^3 + 2(1)^2 - 12(1) + 3 = -3$$

Because $P(0) \neq P(1)$ and $0 < 1$, the Intermediate Value Theorem guarantees that $P(x)$ takes on all values between 3 and -3 on the closed interval $[0, 1]$.

Note that the polynomial function
$f(x) = 4x^3 + 2x^2 - 12x + 3$ has at least one real zero between 0 and 1. From above, we see that $P(0) = 3$ and $P(1) = -3$ have opposite signs. Therefore, we know that there is at least one real zero between 0 and 1, as shown in the figure.

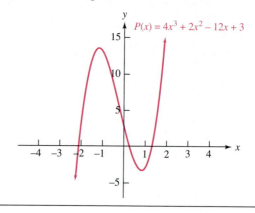

$$P(x) = 4x^3 + 2x^2 - 12x + 3$$

Exercises

Find the zeros of each polynomial function and state the multiplicity of each. State whether the graph touches the x-axis and turns or crosses the x-axis at each zero.

17. $g(x) = x^3 - 6x^2 + 9x$

18. $g(x) = x^3 + 7x^2 - 4x - 28$

19. $f(x) = x^4 - 4x^3 + 3x^2$

20. $f(x) = x^4 - 10x^2 + 24$

Use the Leading Coefficient Test to determine the end behavior of each polynomial function.

21. $f(x) = \sqrt{2}x^5 + 9x^3 - 7x$

22. $g(x) = -\frac{1}{2}x^7 + 5x^4 + 6x^2 - 7$

23. $f(x) = 7x^6 - 5x^2 + 4$

24. $h(x) = -2x^4 - 3x - 8$

Graph each polynomial function.

25. $y = x^3 - x$

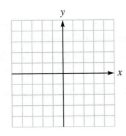

26. $y = x^3 - x^2$

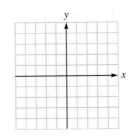

27. $f(x) = -x^3 - 7x^2 - 10x$

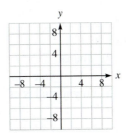

28. $f(x) = -x^4 + 18x^2 - 32$

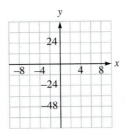

Use the Intermediate Value Theorem and show that each polynomial function has a zero between the two given numbers.

29. $f(x) = 5x^3 + 37x^2 + 59x + 18;\ -1$ and 0

30. $f(x) = 6x^3 - x^2 - 10x - 3;\ 1$ and 2

4.3 The Remainder and Factor Theorems; Synthetic Division

Definitions and Concepts	Examples
A **polynomial function** is a function that can be written in the form $P(x) = 0$, where $$P(x) = a_n x^n + a_{n-1} x^{n-1} + a_{n-2} x^{n-2} + \cdots + a_1 x + a_0$$ where n is a natural number and the polynomial is of degree n.	$P(x) = 3x - 10$; degree: 1 $P(x) = 7x^2 + 3x - 1$; degree: 2 $P(x) = 4x^3 + 7x^2 + 3x - 1$; degree: 3 $P(x) = x^4 - 4x^3 + 7x^2 + 3x - 1$; degree: 4
Zero of a polynomial function: A **zero of the polynomial** $P(x)$ is any number c for which $P(c) = 0$.	The real number 1 is a zero of the polynomial function $P(x) = x^3 - 4x^2 + 3x$ because $$P(1) = 1^3 - 4(1)^2 + 3(1)$$ $$= 1 - 4 + 3$$ $$= 0$$
The Remainder Theorem: If $P(x)$ is a polynomial function, c is any number, and $P(x)$ is divided by $x - c$, the remainder is $P(c)$.	Let $P(x) = 3x^3 - 5x^2 + 3x - 10$ and use the Remainder Theorem and long division to find $P(2)$. We will divide $P(x)$ by $x - 2$ as follows: $$\begin{array}{r} 3x^2 + x + 5 \\ x - 2 \overline{)\,3x^3 - 5x^2 + 3x - 10} \\ \underline{3x^3 - 6x^2} \\ x^2 + 3x \\ \underline{x^2 - 2x} \\ 5x - 10 \\ \underline{5x - 10} \\ \mathbf{0} \end{array}$$ The remainder is 0. $P(2) = 0$.
The Factor Theorem: If $P(x)$ is a polynomial function and c is any number, then if $P(c) = 0$, then $x - c$ is a factor of $P(x)$. If $x - c$ is a factor of $P(x)$, then $P(c) = 0$. **Alternate Form of the Factor Theorem:** If c is a zero of the polynomial function $P(x)$, then $x - c$ is a factor of $P(x)$. If $x - c$ is a factor of $P(x)$, then c is a zero of the polynomial function.	We see in the previous example that $P(2) = 0$. By the Factor Theorem, we know that $x - 2$ is a factor of the polynomial function $P(x) = 3x^3 - 5x^2 + 3x - 10$. We also know that 2 is a zero of the polynomial function.

Definitions and Concepts	Examples
Synthetic division is a fast way to divide higher-degree polynomials by binomials of the form $x - c$.	Divide $P(x) = 5x^3 + 3x^2 - 21x - 1$ by $x - 2$ using synthetic division. We first write the coefficients of the dividend, with its terms in descending powers of x, and the 2 from the divisor in the following form: $\underline{2}\,|\;\;5\quad 3\quad -21\quad -1$ $\underline{2}\,|\;\;5\quad 3\quad -21\quad -1$ Bring down the 5. $\quad\quad 5$ $\underline{2}\,|\;\;5\quad 3\quad -21\quad -1$ $2(5)=10;\ 3+10=13$ $\quad\quad\quad 10$ $\quad\;\; 5\quad 13\quad\quad\quad 9$ $\underline{2}\,|\;\;5\quad 3\quad -21\quad -1$ $2(13)=26;-21+26=5$ $\quad\quad\quad 10\quad 26$ $\quad\;\; 5\quad 13\quad 5\quad 9$ $\underline{2}\,|\;\;5\quad 3\quad -21\quad -1$ $2(5)=10;\ -1+10=9$ $\quad\quad\quad 10\quad 26\quad 10$ $\quad\;\; 5\quad 13\quad 5\quad 9$ The quotient is $5x^2 + 13x + 5$. The remainder is 9. Thus, $$\frac{5x^3 - 3x^2 - 21x - 1}{x - 2} = 5x^2 + 13x + 5 + \frac{9}{x - 2}$$ Because the remainder is 9, we know by the Remainder Theorem that $P(2) = 9$. Because $P(2) \neq 0$, we know by the Factor Theorem that $x - 2$ is not a factor of $5x^3 + 3x^2 - 21x - 1$.
Synthetic division can be used to solve polynomial equations.	Let $P(x) = x^3 + x^2 - 3x - 3$. Completely solve the polynomial equation $P(x) = 0$ given that -1 is a zero. 1. We use synthetic division to divide $P(x)$ by $x + 1$. $\underline{-1}\,|\;\;1\quad 1\quad -3\quad -3$ $\quad\quad\quad\;\, -1\quad 0\quad 3$ $\quad\;\; 1\quad 0\quad -3\quad 0$ 2. We use the result and factor $P(x)$ $$x^3 + x^2 - 3x - 3 = (x + 1)(x^2 - 3).$$

Definitions and Concepts	Examples
	3. Finally, we solve the polynomial equation $P(x) = 0$.
	$$x^3 + x^2 - 3x - 3 = 0$$
	$$(x + 1)(x^2 - 3) = 0$$
	We then set each factor equal to 0 and solve for x.
	$$x + 1 = 0 \quad \text{or} \quad x^2 - 3 = 0$$
	$$x = -1 \qquad\qquad x^2 = 3$$
	$$x = \pm\sqrt{3}$$
	The solution set is $\{-1, \sqrt{3}, -\sqrt{3}\}$.

Exercises

Let $P(x) = 4x^4 + 2x^3 - 3x^2 - 2$. Find the remainder when $P(x)$ is divided by each binomial.

31. $x - 1$ **32.** $x - 2$

33. $x + 3$ **34.** $x + 2$

Use the Factor Theorem to determine whether each statement is true.

35. $x - 2$ is a factor of $x^3 + 4x^2 - 2x + 4$.

36. $x + 3$ is a factor of $2x^4 + 10x^3 + 4x^2 + 7x + 21$.

37. $x - 5$ is a factor of $x^5 - 3125$.

38. $x - 6$ is a factor of $x^5 - 6x^4 - 4x + 24$.

Use synthetic division to divide the polynomial by the given polynomial.

39. $3x^4 + 2x^2 + 3x + 7; x - 3$

40. $2x^4 - 3x^2 + 3x - 1; x - 2$

41. $5x^5 - 4x^4 + 3x^3 - 2x^2 + x - 1; x + 2$

42. $4x^5 + 2x^4 - x^3 + 3x^2 + 2x + 1; x + 1$

Let $P(x) = 5x^3 + 2x^2 - x + 1$. Use synthetic division to find each value.

43. $P(3)$ **44.** $P(-3)$

45. $P\left(\dfrac{1}{2}\right)$ **46.** $P(i)$

A partial solution set is given for each polynomial equation. Find the complete solution set.

47. $2x^3 - 3x^2 - 11x + 6 = 0; \{3\}$

48. $x^4 + 4x^3 - x^2 - 20x - 20 = 0; \{-2, -2\}$

Find the polynomial function of lowest degree with integer coefficients and the given zeros.

49. $-1, 2$, and $\dfrac{3}{2}$ **50.** $1, -3$, and $\dfrac{1}{2}$

51. $2, -5, i$, and $-i$ **52.** $-3, 2, i$, and $-i$

4.4 Fundamental Theorem of Algebra and Descartes' Rule of Signs

Definitions and Concepts	Examples
The Fundamental Theorem of Algebra: If $P(x)$ is a polynomial function with positive degree, then $P(x)$ has at least one complex zero.	The Fundamental Theorem of Algebra guarantees that polynomial functions like $P(x) = 3x^3 - 5x^2 + 7x - 2$ and $P(x) = 4x^5 + x^3 - 6x - 15$ have zeros.

Definitions and Concepts	Examples
The Polynomial Factorization Theorem: If $n > 0$ and $P(x)$ is an nth-degree polynomial function, then $P(x)$ has exactly n linear factors: $$P(x) = a_n(x - c_1)(x - c_2)(x - c_3) \cdot \cdots \cdot (x - c_n)$$ where $c_1, c_2, c_3, \ldots, c_n$ are numbers and a_n is the leading coefficient of $P(x)$.	Consider the following polynomial function of degree $n = 3$. $$P(x) = x^3 + 144x$$ $$= x(x^2 + 144)$$ $$= x(x - 12i)(x + 12i)$$ The Polynomial Factorization Theorem guarantees that we have exactly 3 linear factors. They are $$x, x - 12i, \text{ and } x + 12i$$
If multiple zeros are counted individually, the polynomial function $P(x)$ with degree n ($n > 0$) has exactly n zeros among the complex numbers.	The polynomial function $P(x) = x^3 + 144x$ has exactly 3 zeros. They are $$0, 12i, \text{ and } -12i$$ Each zero occurs once and has a multiplicity of one.
The Conjugate Pairs Theorem: If a polynomial equation $P(x) = 0$ with real-number coefficients has a complex zero $a + bi$ with $b \neq 0$, then its conjugate $a - bi$ is also a zero.	The Conjugate Pairs Theorem applies to $P(x) = x^3 + 144x = 0$ because the polynomial has real coefficients. The complex zeros $12i$ and $-12i$ that occur are conjugate pairs.
Descartes' Rule of Signs: Let $P(x)$ be a polynomial function with real coefficients. • The number of positive zeros of $P(x)$ is either equal to the number of variations in sign of $P(x)$ or less than that by an even number. • The number of negative zeros of $P(x)$ is either equal to the number of variations in sign of $P(-x)$ or less than that by an even number.	Use Descartes' Rule of Signs to find the number of possible positive, negative, and nonreal zeros of $P(x) = 5x^3 - 7x^2 + x - 6$. 1. **Find the number of possible positive zeros.** Since there are three variations of sign in $P(x) = 5x^3 - 7x^2 + x - 6$, there can be either 3 positive zeros or only 1 (1 is less than 3 by the even number 2). 2. **Find the number of possible negative zeros.** Because $$P(-x) = 5(-x)^3 - 7(-x)^2 + (-x) - 6$$ $$= -5x^3 - 7x^2 - x - 6$$ has no variations in sign, there are 0 negative zeros. Furthermore, 0 is not a zero, because the terms of the polynomial do not have a common factor of x. 3. **Find the number of nonreal zeros.** If there are 3 positive zeros, then all of the zeros are accounted for. If there is 1 positive zero, the 2 remaining zeros must be nonreal complex numbers. The following chart shows these possibilities. | Number of Positive Zeros	Number of Negative Zeros	Number of Nonreal Zeros | |---|---|---| | 3 | 0 | 0 | | 1 | 0 | 2 | The number of nonreal complex zeros is the number needed to bring the total number of zeros up to 3.

Definitions and Concepts

Upper and lower bounds on zeros:

Let $P(x)$ be a polynomial function with real coefficients and a positive leading coefficient.

1. If $P(x)$ is synthetically divided by a positive number c and each term in the last row of the division is nonnegative, then no number greater than c can be a zero of $P(x)$. (c is called an **upper bound** of the real zeros.)

2. If $P(x)$ is synthetically divided by a negative number d and the signs in the last row alternate,* no value less than d can be a zero of $P(x)$. (d is called a **lower bound** of the real zeros.)

*If 0 appears in the third row, that 0 can be assigned either a + or a − sign to help the signs alternate.

Examples

Use the upper and lower bound rules to establish bounds for the zeros of $P(x) = 2x^3 + 2x^2 - 8x - 8$.

1. We will perform several synthetic divisions by positive integers, looking for nonnegative values in the last row.

   Trying 1 first gives

   $$\begin{array}{r|rrrr} 1 & 2 & 2 & -8 & -8 \\ & & 2 & 4 & -4 \\ \hline & +2 & +4 & -4 & -12 \end{array}$$

   Because at least one of the signs in the last row is negative, we cannot claim that 1 is an upper bound. We now try 2.

   $$\begin{array}{r|rrrr} 2 & 2 & 2 & -8 & -8 \\ & & 4 & 12 & 8 \\ \hline & +2 & +6 & +4 & 0 \end{array}$$

   Because the last row is entirely nonnegative, we can claim that 2 is an upper bound. That is, no number greater than 2 can be a zero of the equation.

2. We perform several synthetic divisions by negative divisors, looking for alternating signs in the last row. We begin with -3.

   $$\begin{array}{r|rrrr} -3 & 2 & 2 & -8 & -8 \\ & & -6 & 12 & -12 \\ \hline & +2 & -4 & +4 & -20 \end{array}$$

   Since the signs in the last row alternate, -3 is a lower bound. That is, no number less than -3 can be a zero. All of the real zeros must be in the interval $(-3, 2)$.

Exercises

How many zeros does each function have?

53. $P(x) = 3x^6 - 4x^5 + 3x + 2$

54. $P(x) = 2x^6 - 5x^4 + 5x^3 - 4x^2 + x - 12$

55. $P(x) = 3x^{65} - 4x^{50} + 3x^{17} + 2x$

56. $P(x) = x^{1984} - 12$

Determine how many linear factors and zeros each polynomial function has.

57. $P(x) = x^4 - 16$

58. $P(x) = x^{40} + x^{30}$

59. $P(x) = 4x^5 + 2x^3$

60. $P(x) = x^3 - 64x$

Find another zero of a polynomial function with real coefficients if the given quantity is one zero.

61. $2 + i$

62. $-i$

Write a third-degree polynomial function with real coefficients and the given zeros.

63. $4, -i$

64. $-5, i$

Find the number of possible positive, negative, and nonreal zeros for each polynomial function.

65. $P(x) = 3x^4 + 2x^3 - 4x + 2$

66. $P(x) = 2x^4 - 3x^3 + 5x^2 + x - 5$

67. $P(x) = 4x^5 + 3x^4 + 2x^3 + x^2 + x - 7$

68. $P(x) = 3x^7 - 4x^5 + 3x^3 + x - 4$

69. $P(x) = x^4 + x^2 + 24{,}567$

70. $P(x) = -x^7 - 5$

Find integer bounds for the zeros of each function. Answers can vary.

71. $P(x) = 5x^3 - 4x^2 - 2x + 4$

72. $P(x) = x^4 + 3x^3 - 5x^2 - 9x + 1$

4.5 Zeros of Polynomial Functions

Definitions and Concepts	Examples
Rational Zero Theorem: Let the polynomial function $P(x) = a_n x^n + a_{n-1}x^{n-1} + a_{n-2}x^{n-2} + \cdots + a_1 x + a_0$ have integer coefficients. If the rational number $\frac{p}{q}$ (written in lowest terms) is a zero of $P(x)$, then p is a factor of the constant a_0, and q is a factor of the leading coefficient a_n.	Consider $P(x) = 2x^3 - 7x^2 - 17x + 10$ By the Rational Zero Theorem, the only possible numerators for the rational zeros of the function are the factors of the constant term 10: $\pm 1, \pm 2, \pm 5,$ and ± 10 The only possible denominators are the factors of the leading coefficient 2: $\pm 1,$ and ± 2 We can form a list of all possible rational zeros by listing the combinations of possible numerators and denominators: $\pm\dfrac{1}{1}, \pm\dfrac{2}{1}, \pm\dfrac{5}{1}, \pm\dfrac{10}{1}, \pm\dfrac{1}{2}, \pm\dfrac{2}{2}, \pm\dfrac{5}{2}, \pm\dfrac{10}{2}$ Since several of these possibilities are duplicates, we can condense the list to get ***Possible Rational Zeros*** $\pm 1, \pm 2, \pm 5, \pm 10, \pm\dfrac{1}{2},$ and $\pm\dfrac{5}{2}$

Finding rational zeros:
1. Use Descartes' Rule of Signs to determine the number of possible positive, negative, and nonreal zeros.
2. Use the Rational Zero Theorem to list possible rational zeros.
3. Use synthetic division to find a zero.
4. If there are more zeros, repeat the previous steps.
 	
(Once the depressed equation is quadratic, synthetic division isn't required. It can be solved as a quadratic equation.)
	Find the zeros of the polynomial function
	$P(x) = 2x^3 - 7x^2 - 17x + 10$.
	
	1. To determine the number of possible positive, negative, and nonreal zeros, we will use Descartes' Rule of Signs.
	    Since there are two variations in sign of $P(x)$, the number of positive zeros is either 2 or 0.
	    Next, we consider $P(-x)$.
	
	$P(-x) = -2x^3 - 7x^2 + 17x + 10$.
	
	Since there is one variation in sign of $P(-x)$, the number of negative zeros is 1.

Definitions and Concepts	Examples

Examples

The following chart shows these possibilities:

Number of Positive Zeros	Number of Negative Zeros	Number of Nonreal Zeros
2	1	0
0	1	2

2. List the possible rational zeros.

The possible rational zeros were found in the previous example. They are:

Possible Rational Zeros

$$\pm 1, \ \pm 2, \ \pm 5, \ \pm 10, \ \pm \frac{1}{2}, \text{ and } \pm \frac{5}{2}$$

3. Use synthetic division to find a zero. We can start with -2.

$$
\begin{array}{r|rrrr}
-2 & 2 & -7 & -17 & 10 \\
   &   & -4 & 22 & -10 \\
\hline
   & 2 & -11 & 5 & 0
\end{array}
$$

Since the remainder is 0, -2 is a zero and the binomial $x + 2$ is a factor of $P(x)$.

4. The remaining zeros must be supplied by the remaining factor, which is the quotient $2x^2 - 11x + 5$. We can find the other zeros by solving the depressed equation $2x^2 - 11x + 5 = 0$.

$$2x^2 - 11x + 5 = 0$$

$$(2x - 1)(x - 5) = 0 \qquad \text{Factor.}$$

$$2x - 1 = 0 \quad \text{or} \quad x - 5 = 0$$

$$2x = 1 \qquad\qquad\quad x = 5$$

$$x = \frac{1}{2}$$

The zeros of the polynomial function are $-2, \frac{1}{2},$ and 5.

Note that two positive zeros and one negative zero is a predicted possibility.

Exercises

Use the Rational Zero Theorem to list all possible rational zeros of the polynomial function.

73. $P(x) = 2x^4 + x^3 - 3x^2 - 5x - 6$

74. $P(x) = 4x^5 - 2x^4 + 3x^3 - 5x - 10$

Find all rational zeros of each polynomial function.

75. $P(x) = x^3 - 10x^2 + 29x - 20$

76. $P(x) = x^3 - 8x^2 - x + 8$

77. $P(x) = 2x^3 + 17x^2 + 41x + 30$

78. $P(x) = 3x^3 + 2x^2 + 2x - 1$

79. $P(x) = 4x^4 - 25x^2 + 36$

80. $P(x) = 2x^4 - 11x^3 - 6x^2 + 64x + 32$

82. $P(x) = x^4 - 2x^3 - 9x^2 + 8x + 20$

Find all zeros of each function.

81. $P(x) = 3x^3 - x^2 + 48x - 16$

4.6 Rational Functions

Definitions and Concepts	**Examples**
A **rational function** is a function defined by an equation of the form $f(x) = \frac{P(x)}{Q(x)}$, where $P(x)$ and $Q(x)$ are polynomials and $Q(x) \neq 0$.	$f(x) = \dfrac{3}{x + 2} \qquad f(x) = \dfrac{3x + 4}{x^2 - 3x + 4}$

Locating vertical asymptotes:

To locate the vertical asymptotes of rational function $f(x) = \frac{P(x)}{Q(x)}$, we follow these steps:

Step 1: Factor $P(x)$ and $Q(x)$ and remove any common factors.

Step 2: Set the denominator equal to 0 and solve the equation.

If a is a solution of the equation found in Step 2, $x = a$ is a vertical asymptote.

Locating horizontal asymptotes:

To locate the horizontal asymptote of the rational function $f(x) = \frac{P(x)}{Q(x)}$, we consider three cases:

Case 1: If the degree of $P(x)$ is less than the degree of $Q(x)$, the line $y = 0$ is the horizontal asymptote.

Case 2: If the degree of $P(x)$ and $Q(x)$ are equal, the line $y = \frac{p}{q}$, where p and q are the leading coefficients of $P(x)$ and $Q(x)$, is the horizontal asymptote.

Case 3: If the degree of $P(x)$ is greater than the degree of $Q(x)$, there is no horizontal asymptote.

A third type of asymptote is called a **slant asymptote**. These asymptotes occur when the degree of the numerator of a rational function is 1 more than the degree of the denominator. As the name implies, it is a slanted line, neither vertical nor horizontal.

Locating slant asymptotes:

If the degree of $P(x)$ is 1 greater than the degree of $Q(x)$ for the rational function $f(x) = \frac{P(x)}{Q(x)}$, there is a slant asymptote. To find it, divide $P(x)$ by $Q(x)$ and ignore the remainder.

Graph: $y = f(x) = \dfrac{4x}{x - 2}$.

We will use the steps outlined to graph the function.

Step 1: Symmetry

We find $f(-x)$.

$$f(-x) = \frac{4(-x)}{(-x) - 2} = \frac{-4x}{-x - 2} = \frac{4x}{x + 2}$$

Because $f(-x) \neq f(x)$ and $f(-x) \neq -f(x)$, there is no symmetry about the y-axis or the origin.

Step 2: Vertical asymptotes

We first note that $f(x)$ is in simplest form. We then set the denominator equal to 0 and solve for x. Since the solution is 2, there will be a vertical asymptote at $x = 2$.

Step 3: y- and x-intercepts

We can find the y-intercept by finding $f(0)$.

$$f(0) = \frac{4(0)}{0 - 2} = \frac{0}{-2} = 0$$

The y-intercept is $(0, 0)$.

We can find the x-intercepts by setting the numerator equal to 0 and solving for x:

$$4x = 0$$
$$x = 0$$

The x-intercept is $(0, 0)$.

Step 4: Horizontal asymptotes

Since the degrees of the numerator and denominator of the polynomials are the same, the line

$$y = \frac{4}{1} = 4 \qquad \text{{\color{red}The leading coefficient of the numerator is 4.}}$$
$$\text{{\color{red}The leading coefficient of the denominator is 1.}}$$

is a horizontal asymptote.

Definitions and Concepts	Examples
Steps to graph a rational function: We will use the following steps to graph the rational function $f(x) = \frac{P(x)}{Q(x)}$, where $\frac{P(x)}{Q(x)}$ is in simplest form (no common factors). 1. Check symmetries. 2. Look for vertical asymptotes. 3. Look for the y- and x-intercepts. 4. Look for horizontal asymptotes. 5. Look for slant asymptotes. 6. Graph the function.	**Step 5: Slant asymptotes** Since the degree of the numerator is not 1 greater than the degree of the denominator, there are no slant asymptotes. **Step 6: Graph** First, we plot the intercept $(0, 0)$ and draw the asymptotes. We then find one additional point on our graph to see what happens when x is greater than 2. To do so, we choose 3, a value of x that is greater than 2, and evaluate $f(3)$. $$f(\mathbf{3}) = \frac{4(\mathbf{3})}{\mathbf{3} - 2} = \frac{12}{1} = 12$$ Since $f(3) = 12$, the point $(3, 12)$ lies on our graph. We sketch the graph as shown in the figure. 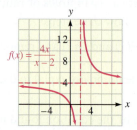

Exercises

Find the domain of each rational function.

83. $f(x) = \dfrac{3x^2 + x - 2}{x^2 - 25}$

84. $f(x) = \dfrac{2x^2 + 1}{x^2 + 7}$

Find the vertical asymptotes, if any, of each rational function.

85. $f(x) = \dfrac{x + 5}{x^2 - 1}$

86. $f(x) = \dfrac{x - 7}{x^2 - 49}$

87. $f(x) = \dfrac{x}{x^2 + x - 6}$

88. $f(x) = \dfrac{5x + 2}{2x^2 - 6x - 8}$

Find the horizontal asymptotes, if any, of each rational function.

89. $f(x) = \dfrac{2x^2 + x - 2}{4x^2 - 4}$

90. $f(x) = \dfrac{5x^2 + 4}{4 - x^2}$

91. $f(x) = \dfrac{x + 1}{x^3 - 4x}$

92. $f(x) = \dfrac{x^3}{2x^2 - x + 11}$

Find the slant asymptote, if any, for each rational function.

93. $f(x) = \dfrac{2x^2 - 5x + 1}{x - 4}$

94. $f(x) = \dfrac{5x^3 + 1}{x + 5}$

Graph each rational function.

95. $f(x) = \dfrac{2x}{x - 4}$

96. $f(x) = \dfrac{-4x}{x + 4}$

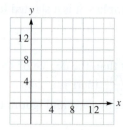

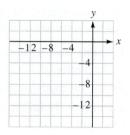

97. $f(x) = \dfrac{x}{(x-1)^2}$

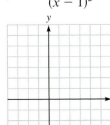

98. $f(x) = \dfrac{(x-1)^2}{x}$

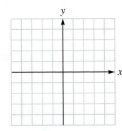

99. $f(x) = \dfrac{x^2 - x - 2}{x^2 + x - 2}$

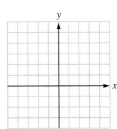

100. $f(x) = \dfrac{x^3 + x}{x^2 - 4}$

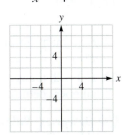

CHAPTER TEST

Find the vertex of each parabola.

1. $y = 3(x-7)^2 - 3$

2. $f(x) = 3x^2 - 24x + 38$

Graph the function.

3. $f(x) = (x-3)^2 + 1$

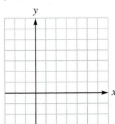

Assume that an object tossed vertically upward reaches a height of h feet after t seconds, where $h = 100t - 16t^2$.

4. In how many seconds does the object reach its maximum height?

5. What is that maximum height?

6. Suspension bridges The cable of a suspension bridge is in the shape of the parabola $x^2 - 2500y + 25,000 = 0$ in the coordinate system shown in the illustration. (Distances are in feet.) How far above the roadway is the cable's lowest point?

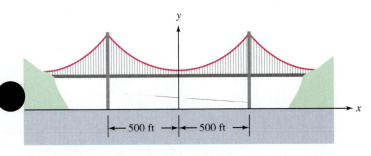

Graph each function.

7. $f(x) = x^4 - x^2$

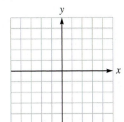

8. $f(x) = x^5 - x^3$

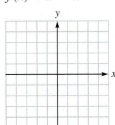

9. Is $x = -2$ a zero of $P(x) = x^2 + 5x + 6$?

Use long division and the Remainder Theorem to find each value.

10. $P(x) = x^5 + 2; \ P(-2)$

11. Use the Factor Theorem to determine whether $x - 3$ is a factor of $2x^4 - 10x^3 + 4x^2 + 7x + 21$.

Use synthetic division to express $P(x) = 2x^3 - 3x^2 - 4x - 1$ in the form (divisor)(quotient) + remainder for the divisor.

12. $x - 2$

Use synthetic division to perform each division.

13. $\dfrac{2x^2 - 7x - 15}{x - 5}$

14. $\dfrac{3x^3 + 7x^2 + 2x}{x + 2}$

Let $P(x) = 3x^3 - 2x^2 + 4$. Use synthetic division to find each value.

15. $P\left(-\dfrac{1}{3}\right)$

16. $P(i)$

Find a polynomial function with the given zeros.

17. $5, -1, 0$

18. $i, -i, \sqrt{3}, -\sqrt{3}$

19. How many linear factors and zeros does $P(x) = 3x^3 + 2x^2 - 4x + 1$ have?

20. If $3 - 2i$ is a zero of $P(x)$ where $P(x)$ has real-number coefficients, find another zero.

Use Descartes' Rule of Signs to find the number of possible positive, negative, and nonreal zeros of the polynomial function.

21. $P(x) = 3x^5 - 2x^4 + 2x^2 - x - 3$

Find integer bounds for the zeros of each polynomial function. Answers can vary.

22. $P(x) = x^5 - x^4 - 5x^3 + 5x^2 + 4x - 5$

23. Use the Rational Zero Theorem to list all possible rational zeros of $P(x) = 5x^3 + 4x^2 + 3x + 2$.

24. Find all zeros of the polynomial function.

$P(x) = 2x^3 + 3x^2 - 11x - 6$.

25. Find all zeros of the function
$P(x) = x^4 + x^3 + 3x^2 + 9x - 54$.

26. Does the polynomial $P(x) = 3x^3 + 2x^2 - 4x + 4$ have a zero between the values $x = 1$ and $x = 2$?

Find all asymptotes of the graph of each rational function. Do not graph the function.

27. $f(x) = \dfrac{x - 1}{x^2 - 9}$

28. $f(x) = \dfrac{x^2 - 5x - 14}{x - 3}$

Graph the rational function.

29. $f(x) = \dfrac{x^2}{x^2 - 9}$

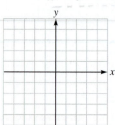

Graph the rational function. The numerator and denominator share a common factor.

30. $f(x) = \dfrac{x}{x^2 - x}$

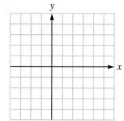

5

Exponential and Logarithmic Functions

In this chapter, we will discuss exponential functions, which are often used in banking, ecology, and science. We will also discuss logarithmic functions, which are applied in chemistry, geology, and environmental science.

Careers and Mathematics: Epidemiologists

©Kiselev Andrey Valerevich/Shutterstock.com

Epidemiologists investigate and describe the determinants and distribution of disease, disability, and other health outcomes. They also develop means for prevention and control. *Applied epidemiologists* typically work for state health agencies and are responsible for responding to disease outbreaks and determining the cause and method of containment. *Research epidemiologists* work in laboratories studying ways to prevent future outbreaks. This career can be quite rewarding, both mentally and financially. Epidemiologists spend a lot of time saving lives and finding solutions for better health.

Education and Mathematics Required
- Applied epidemiologists are generally required to have a master's degree from a school of public health. Research epidemiologists may need a Ph.D. or medical degree, depending on their area of work.
- College Algebra, Trigonometry, Calculus, Applied Data Analysis, Survey and Research Methods, Mathematical Statistics, and Biostatistics are required math courses.

How Epidemiologists Use Math and Who Employs Them
- Epidemiologists use mathematical models when they are tracking the progress of an infectious disease. The SIR model consists of three variables: S (for susceptible), I (for infectious), and R (for recovered). It is used for infectious diseases such as measles, mumps, and rubella.
- Government agencies employ 57%; hospitals employ 12%; colleges and universities employ 11%; and 9% are employed in areas of scientific research and developmental services like the American Cancer Society.

Career Outlook and Earnings
- Employment growth is projected to be 10% over the 2012–2022 decade, which is as fast as average. This is due to an increased threat of bioterrorism and rare but infectious diseases, such as West Nile Virus or Avian flu.
- The median annual income is $65,270, with the top 10% of salaries at $108,320.

For more information see: www.bls.gov/oco

5.1 Exponential Functions and Their Graphs

In this section, we will learn to

1. Approximate and simplify exponential expressions.
2. Graph exponential functions.
3. Solve compound interest problems.
4. Define e and graph base-e exponential functions.
5. Use transformations to graph exponential functions.

Extreme water slides, called "plunge" or "plummet" slides, are fearsome water slides because of their heights. With near vertical drops, the slides are designed to allow riders to reach the greatest possible speeds. Summit Plummet at Blizzard Beach, a part of Walt Disney World Resort in Florida, stands 120 ft tall. On this slide, riders can achieve speeds up to 55 mph.

The shapes of some extreme water slides can be modeled using *exponential functions*, the topic of this section.

Exponential functions are also important in business. Consider the graph shown in Figure 5-1. It shows the balance in a bank account in which $5000 was invested in 1990 at 8%, compounded monthly. The graph shows that in the year 2015, the value of the account will be approximately $38,000, and in the year 2030, the value will be approximately $121,000.

The curve in Figure 5-1 is the graph of an exponential function. From the graph, we can see that the longer the money is kept on deposit, the more rapidly it will grow.

Tip

Although money doesn't actually grow on trees, learning about compound interest can help you as a student make wise decisions with regard to your finances.

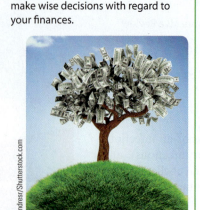

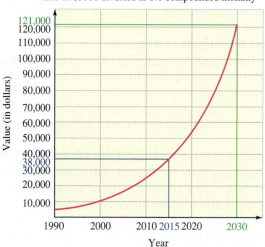

Value of $5000 invested at 8% compounded monthly

FIGURE 5-1

Before we can discuss exponential functions, we must define irrational exponents.

1. Approximate and Simplify Exponential Expressions

We have discussed expressions of the form b^x, where x is a rational number.

- 5^2 means "the square of 5."
- $4^{1/3}$ means "the cube root of 4."
- $6^{-2/5} = \frac{1}{6^{2/5}}$ means "the reciprocal of the fifth root of 6^2."

To understand exponential functions and their graphs requires that we give meaning to b^x when x is an irrational number. Consider the expression

$3^{\sqrt{2}}$ where $\sqrt{2}$ is the irrational number $1.414213562\ldots$

We can use closer and closer approximations as shown below. Since $\sqrt{2}$ is an irrational number, we will use a calculator to find the approximations.

$3^{\sqrt{2}} \approx 3^{1.4} \approx 4.655536722$

$3^{\sqrt{2}} \approx 3^{1.41} \approx 4.706965002$

$3^{\sqrt{2}} \approx 3^{1.414} \approx 4.727695035$

$3^{\sqrt{2}} \approx 3^{1.4142} \approx 4.72873393$

Since the exponents of the expressions in the list are getting closer to $\sqrt{2}$, the values of the expressions are getting closer to the value of $3^{\sqrt{2}}$.

On a scientific calculator, there is an exponential key, usually y^x. On a graphing calculator, we can use the $\boxed{\wedge}$ key.

Approximating Exponential Expressions

EXAMPLE 1

Approximate each expression correct to 4 decimals.

a. $4^{2/3}$ **b.** $5^{-\sqrt{3}}$ **c.** $\left(\dfrac{4}{7}\right)^{\pi}$

SOLUTION   We will use a calculator.

a. $4^{\wedge}(2/3) \approx 2.5198$ Enter 2/3 in parentheses.

b. $5^{\wedge}\left(-\sqrt{3}\right) \approx 0.0616$ Enter $-\sqrt{3}$ in parentheses.

c. $(4/7)^{\wedge}\pi \approx 0.1724$ Enter the base, 4/7, in parentheses.

Figure 5-2 shows the graphing calculator screens that were used to find the values.

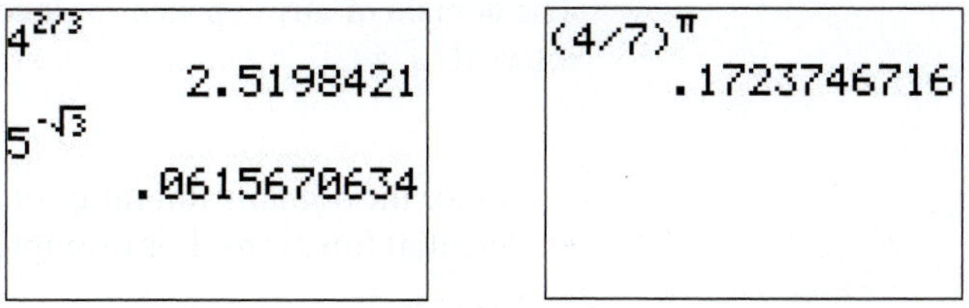

FIGURE 5-2

Self Check 1   Approximate each expression correct to 4 decimals.

a. $5^{3/5}$ **b.** $-3^{\sqrt{6}}$ **c.** $7^{-2.356}$

Now Try Exercise 15.

If b is a positive number and x is a real number, the expression b^x always represents a positive number. It is also true that the familiar properties of exponents hold for irrational exponents.

Simplifying Expressions with Irrational Exponents

EXAMPLE 2

Simplify each expression:

a. $\left(3^{\sqrt{2}}\right)^{\sqrt{2}}$

b. $a^{\sqrt{8}} \cdot a^{\sqrt{2}}$

SOLUTION   We will use properties of exponents to simplify each expression.

  a. $\left(3^{\sqrt{2}}\right)^{\sqrt{2}} = 3^{\sqrt{2}\sqrt{2}}$ Keep the base and multiply the exponents.

            $= 3^2$ $\sqrt{2}\sqrt{2} = \sqrt{4} = 2$

            $= 9$

  b. $a^{\sqrt{8}} \cdot a^{\sqrt{2}} = a^{\sqrt{8}+\sqrt{2}}$ Keep the base and add the exponents.

            $= a^{2\sqrt{2}+\sqrt{2}}$ $\sqrt{8} = \sqrt{4}\sqrt{2} = 2\sqrt{2}$

            $= a^{3\sqrt{2}}$ $2\sqrt{2} + \sqrt{2} = 3\sqrt{2}$

Self Check 2   Simplify: **a.** $\left(2^{\sqrt{3}}\right)^{\sqrt{12}}$ **b.** $x^{\sqrt{20}} \cdot x^{\sqrt{5}}$

Now Try Exercise 19.

2. Graph Exponential Functions

If $b > 0$ and $b \neq 1$, the function $y = b^x$ defines a function, because for each input x, there is exactly one output y. Since x can be any real number, the domain of the function is the set of real numbers. Since the base b of the expression b^x is positive, y is always positive, and the range is the set of positive numbers. Since b^x is an exponential expression, the function is called an **exponential function**.

    We make the restriction that $b > 0$ to exclude any imaginary numbers that might result from taking even roots of negative numbers. The restriction that $b \neq 1$ excludes the constant function $f(x) = 1^x$, in which $f(x) = 1$ for every real number x.

Exponential Function

An **exponential function with base b** is defined by the equation

$$f(x) = b^x \quad \text{or} \quad y = b^x \qquad (b > 0,\ b \neq 1,\ \text{and } x \text{ is a real number})$$

The **domain of any exponential function** is the interval $(-\infty, \infty)$. The **range** is the interval $(0, \infty)$.

Since the domain and range of $f(x) = b^x$ are sets of real numbers, we can graph exponential functions. For example, to graph

$$f(x) = 2^x$$

we find several points $(x, f(x))$ whose coordinates satisfy the equation, plot the points, and join them with a smooth curve, as in Figure 5-3(a). To graph the function

$$f(x) = \left(\frac{1}{2}\right)^x$$

we find several points $(x, f(x))$ whose coordinates satisfy the equation, plot the points, and join them with a smooth curve, as shown in Figure 5-3(b).

Caution

When we raise the real number 2 to the 0th power, we get 1 and not 0. Also, when we raise 2 to a negative power like -1, we get $\frac{1}{2}$ and not -2. Avoid making these two common errors when completing a table of value for exponential functions.

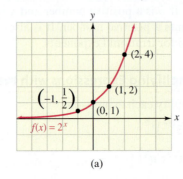

$f(x) = 2^x$		
x	$f(x)$	$(x, f(x))$
-1	$\dfrac{1}{2}$	$\left(-1, \dfrac{1}{2}\right)$
0	1	$(0, 1)$
1	2	$(1, 2)$
2	4	$(2, 4)$

(a)

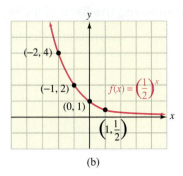

$f(x) = \left(\dfrac{1}{2}\right)^x$		
x	$f(x)$	$(x, f(x))$
-2	4	$(-2, 4)$
-1	2	$(-1, 2)$
0	1	$(0, 1)$
1	$\dfrac{1}{2}$	$\left(1, \dfrac{1}{2}\right)$

(b)

FIGURE 5-3

By looking at the graphs in Figure 5-3, we can see that the domain of each function is the interval $(-\infty, \infty)$ and that the range is the interval $(0, \infty)$.

Graphing an Exponential Function

EXAMPLE 3

Graph: $f(x) = 4^x$.

SOLUTION

We will find several points (x, y) that satisfy the equation, plot the points, and join them with a smooth curve, as in Figure 5-4.

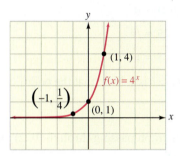

$f(x) = 4^x$		
x	$f(x)$	$(x, f(x))$
-1	$\dfrac{1}{4}$	$\left(-1, \dfrac{1}{4}\right)$
0	1	$(0, 1)$
1	4	$(1, 4)$

FIGURE 5-4

Self Check 3 Graph: $f(x) = \left(\dfrac{1}{4}\right)^x$.

Now Try Exercise 29.

The graph of $f(x) = 4^x$ in Example 3 has the following properties:

1. It passes through the point $(0, 1)$.
2. It passes through the point $(1, 4)$.
3. It approaches the x-axis. The x-axis is a horizontal asymptote.
4. The domain is the interval $(-\infty, \infty)$, and the range is the interval $(0, \infty)$.

This example illustrates the following properties of exponential functions.

Properties of Exponential Functions

- The **domain of the exponential function** $f(x) = b^x$ is $(-\infty, \infty)$, the set of real numbers.
- The **range** is $(0, \infty)$, the set of positive real numbers.
- The graph has a y-intercept at $(0, 1)$.
- The x-axis is a horizontal asymptote of the graph.
- The graph of $f(x) = b^x$ passes through the point $(1, b)$.

EXAMPLE 4

Determining the Base of an Exponential Function

The graph of an exponential function of the form $f(x) = b^x$ is shown in Figure 5-5. Find the value of b.

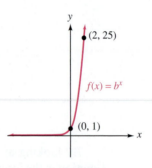

FIGURE 5-5

SOLUTION We note that the graph passes through the point $(0, 1)$, a property of exponential functions of this form. Since the graph also passes through the point $(2, 25)$, we can find the base b by substituting 2 for x and 25 for $f(2)$ in the equation $f(x) = b^x$ and solving for b.

$$f(x) = b^x$$

$$f(2) = b^2$$

$$25 = b^2$$

$$5 = b \qquad b \text{ must be positive.}$$

The base b is 5. Note that the points $(0, 1)$ and $(2, 25)$ satisfy the equation $f(x) = 5^x$. ∎

Self Check 4 Can a graph passing through $(0, 2)$ and $\left(1, \dfrac{3}{2}\right)$ be the graph of $f(x) = b^x$?

Now Try Exercise 45.

In Figure 5-3(a) (where $b = 2$ and $2 > 1$), the values of y increase as the values of x increase. Since the graph rises as we move to the right, the function is an increasing function. Such a function is said to model *exponential growth*.

In Figure 5-3(b), where $\left(b = \frac{1}{2} \text{ and } 0 < \frac{1}{2} < 1\right)$, the values of y decrease as the values of x increase. Since the graph drops as we move to the right, the function is a decreasing function. Such a function is said to model *exponential decay*. Two additional properties are stated here.

Additional Properties of Exponential Functions If $b > 1$, then $f(x) = b^x$ is an **increasing function**. This function models **exponential growth**.

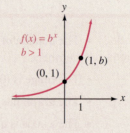

Increasing function

If $0 < b < 1$, then $f(x) = b^x$ is a **decreasing function**. This function models **exponential decay**.

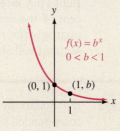

$f(x) = b^x$
$0 < b < 1$

$(0, 1)$ $(1, b)$
1

Decreasing function

Recall that $b^{-x} = \frac{1}{b^x} = \left(\frac{1}{b}\right)^x$. If $b > 1$, any function of the form $f(x) = b^{-x}$ models exponential decay, because $0 < \frac{1}{b} < 1$.

An exponential function $f(x) = b^x$ is either increasing (for $b > 1$) or decreasing (for $0 < b < 1$). Because different real numbers x always determine different values of b^x, an exponential function is one-to-one.

One-to-One Property of Exponential Functions

An exponential function defined by $f(x) = b^x$ or $y = b^x$, where $b > 0$ and $b \neq 1$, is one-to-one. This implies that

1. If $b^r = b^s$, then $r = s$.

2. If $r \neq s$, then $b^r \neq b^s$.

Accent on Technology

Graphing Exponential Functions

When using a graphing calculator to draw the graphs of exponential functions, use care entering the function. Proper use of parentheses is critical to obtaining the correct graph. Since we know the shape of an exponential function, it is usually easy to find a good window for the graph. For any function in the form $f(x) = b^x$, where $b > 1$, the graph will be steeper as the base increases. The opposite is true if the base is between 0 and 1. Several graphs are shown in Figure 5-6 along with a window that works well with these basic functions.

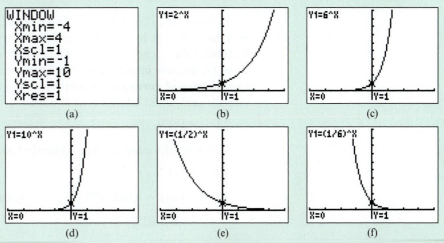

FIGURE 5-6

3. Solve Compound Interest Problems

Banks pay **interest** for using their customers' money. Interest is calculated as a percent of the amount on deposit in an account and is paid annually (once a year), quarterly (four times per year), monthly, or daily. Interest left on deposit in a bank account will also earn interest. Such accounts are said to earn **compound interest**.

Compound Interest Formula

If P dollars are deposited in an account earning interest at an annual rate r, compounded n times each year, the amount A in the account after t years is given by

$$A = P\left(1 + \frac{r}{n}\right)^{nt}$$

EXAMPLE 5

Solving a Compound Interest Problem

The parents of a newborn child invest $8000 in a plan that earns 9% interest, compounded quarterly. If the money is left untouched, how much will the child have in the account in 55 years?

SOLUTION

We will substitute 8000 for P, 0.09 for r, and 55 for t in the formula for compound interest. Because quarterly compounding means four times per year, we will substitute 4 for n.

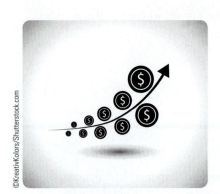

$$A = P\left(1 + \frac{r}{n}\right)^{nt}$$

$$A = 8000\left(1 + \frac{0.09}{4}\right)^{4 \cdot 55}$$

$$= 8000(1.0225)^{220}$$

$$\approx 1{,}069{,}103.266 \qquad \text{Use a calculator.}$$

In 55 years, the account will be worth $1,069,103.27.

Self Check 5

Would $20,000 invested at 7% interest, compounded monthly, have provided more income at age 55?

Now Try Exercise 81.

In financial calculations, the initial amount deposited is often called the **present value**, denoted by PV. The amount to which the account will grow is called the **future value**, denoted by FV. The interest rate for each compounding period is called the **periodic interest rate**, i, and the number of times interest is compounded is the **number of compounding periods**, n. Using these definitions, an alternate formula for compound interest is as follows.

$$FV = PV(1 + i)^n$$

To use this formula to solve Example 5, we proceed as follows:

$$FV = PV(1 + i)^n$$

$$FV = 8000(1 + 0.0225)^{220} \qquad i = \frac{0.09}{4} = 0.0225 \text{ and } n = 4(55) = 220.$$

$$= 8000(1.0225)^{220}$$

$$\approx 1{,}069{,}103.266 \qquad \text{Use a calculator.}$$

4. Define e and Graph Base-e Exponential Functions

In mathematical models of natural events, the number

$$e = 2.71828182845904\ldots$$

often appears as the base of an exponential function. e is known as Euler's constant. We can introduce this number by considering the compound interest formula

$$A = P\left(1 + \frac{r}{n}\right)^{nt} \qquad \text{\color{red}{\textit{A} is the amount, \textit{P} is the initial deposit, \textit{r} is the annual rate, \textit{n} is the number of compoundings per year, and \textit{t} is the time in years.}}$$

and allowing n to become very large. To see what happens, we let $n = rx$, where x is another variable.

$$A = P\left(1 + \frac{r}{\boldsymbol{n}}\right)^{\boldsymbol{n}t}$$

$$A = P\left(1 + \frac{r}{\boldsymbol{rx}}\right)^{\boldsymbol{rx}t} \qquad \text{\color{red}{Substitute \textit{rx} for \textit{n}.}}$$

$$A = P\left(1 + \frac{1}{x}\right)^{rxt} \qquad \text{\color{red}{Simplify $\frac{r}{rx}$.}}$$

$$A = P\left[\left(1 + \frac{1}{x}\right)^{x}\right]^{rt} \qquad \text{\color{red}{Remember that $(a^m)^n = a^{mn}$.}}$$

Since all variables in this formula are positive, r is a constant rate, and $n = rx$, it follows that as n becomes large, so does x. What happens to the value of A as n becomes large will depend on the value of $\left(1 + \frac{1}{x}\right)^x$ as x becomes large. Some results calculated for increasing values of x appear in the table shown.

x	$\left(1 + \dfrac{1}{x}\right)^x$
1	2
10	2.5937425
100	2.7048138
1000	2.7169239
1,000,000	2.7182805
1,000,000,000	2.7182818

From the table, we can see that as x increases, the value of $\left(1 + \frac{1}{x}\right)^x$ approaches the value of e, and the formula

$$A = P\left[\left(1 + \frac{1}{x}\right)^{x}\right]^{rt}$$

becomes

$$A = Pe^{rt} \qquad \text{\color{red}{Substitute \textit{e} for $\left(1 + \frac{1}{x}\right)^x$.}}$$

When the amount invested grows exponentially according to the formula $A = Pe^{rt}$, we say that interest is **compounded continuously**.

©Tschitscherin/Shutterstock.com

©Pixel 4 Images/Shutterstock.com

Continuous Compound Interest Formula	If P dollars are deposited in an account earning interest at an annual rate r, compounded continuously, the amount A after t years is given by the formula $$A = Pe^{rt}$$

EXAMPLE 6

Solving a Continuous Compound Interest Problem

If the parents of the newborn child in Example 5 had invested $8000 at an annual rate of 9%, compounded continuously, how much would the child have in the account in 55 years?

SOLUTION

We will substitute $8000 for P, 0.09 for r, and 55 for t in the continuous compound interest formula $A = Pe^{rt}$.

$$A = Pe^{rt}$$

$$A = 8000e^{(0.09)(55)}$$

$$= 8000e^{4.95}$$

$$\approx 1,129,399.711 \qquad \text{Use a calculator.}$$

In 55 years, the balance will be $1,129,399.71, which is $60,296.44 more than the amount earned with quarterly compounding.

Self Check 6

Find the balance in 60 years.

Now Try Exercise 89.

To graph the natural exponential function $f(x) = e^x$, we plot several points and join them with a smooth curve [as in Figure 5-7(a)] or use a graphing calculator [as in Figure 5-7(b)].

$f(x) = e^x$		
x	$f(x)$	$(x, f(x))$
-1	0.37	$(-1, 0.37)$
0	1	$(0, 1)$
1	2.72	$(1, 2.72)$
2	7.39	$(2, 7.39)$

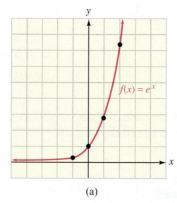

(a)

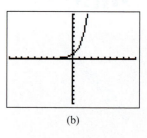

(b)

FIGURE 5-7

Take Note

The graph of $f(x) = e^x$ is very important in mathematics and should be memorized. The function is often referred to as the natural exponential function. Place it in your library of functions.

5. Use Transformations to Graph Exponential Functions

In Section 3.2, translations, reflections, and stretchings were applied to functions. These may also be applied to the graphs of exponential functions. A summary of these transformations, when $k > 0$, is shown in the table.

Function	Transformations of the Graph of $f(x) = b^x$
$g(x) = b^x + k$	Translates the graph of $f(x) = b^x$ upward k units
$g(x) = b^x - k$	Translates the graph of $f(x) = b^x$ downward k units
$g(x) = b^{x-k}$	Translates the graph of $f(x) = b^x$ to the right k units
$g(x) = b^{x+k}$	Translates the graph of $f(x) = b^x$ to the left k units
$g(x) = -b^x$	Reflects the graph of $f(x) = b^x$ about the x-axis
$g(x) = b^{-x}$	Reflects the graph of $f(x) = b^x$ about the y-axis
$g(x) = kb^x$	• Vertically stretches the graph of $f(x) = b^x$ if $k > 1$ • Vertically shrinks the graph of $f(x) = b^x$ if $0 < k < 1$
$g(x) = b^{kx}$	• Horizontally stretches the graph of $f(x) = b^x$ if $0 < k < 1$ • Horizontally shrinks the graph of $f(x) = b^x$ if $k > 1$

Using Translations to Graph an Exponential Function

EXAMPLE 7

On one set of axes, graph $f(x) = 2^x$ and $g(x) = 2^x + 3$.

SOLUTION   The graph of $g(x) = 2^x + 3$ is identical to the graph of $f(x) = 2^x$, except that it is translated 3 units upward. (See Figure 5-8.)

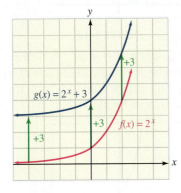

$f(x) = 2^x$		
x	$f(x)$	$(x, f(x))$
-4	$\dfrac{1}{16}$	$\left(-4, \dfrac{1}{16}\right)$
0	1	$(0, 1)$
2	4	$(2, 4)$

$g(x) = 2^x + 3$		
x	$g(x)$	$(x, g(x))$
-4	$3\dfrac{1}{16}$	$\left(-4, 3\dfrac{1}{16}\right)$
0	4	$(0, 4)$
2	7	$(2, 7)$

FIGURE 5-8

Self Check 7   On one set of axes, graph $f(x) = 2^x$ and $g(x) = 2^x - 2$.

Now Try Exercise 53.

Using Translations to Graph an Exponential Function with Base e

EXAMPLE 8

On one set of axes, graph $f(x) = e^x$ and $g(x) = e^{x-3}$.

SOLUTION   The graph of $g(x) = e^{x-3}$ is identical to the graph of $f(x) = e^x$, except that it is translated 3 units to the right. (See Figure 5-9.)

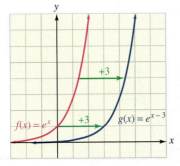

$f(x) = e^x$		
x	$f(x)$	$(x, f(x))$
-1	0.37	$(-1, 0.37)$
0	1	$(0, 1)$
1	2.72	$(1, 2.72)$
2	7.39	$(2, 7.39)$

$g(x) = e^{x-3}$		
x	$g(x)$	$(x, g(x))$
2	0.37	$(2, 0.37)$
3	1	$(3, 1)$
4	2.72	$(4, 2.72)$
5	7.39	$(5, 7.39)$

FIGURE 5-9

Self Check 8 On one set of axes, graph $f(x) = e^x$ and $g(x) = e^{x+2}$.

Now Try Exercise 61.

We can use a graphing calculator to graph exponential functions that are vertically or horizontally stretched or shrunk.

Accent on Technology

Graphing Exponential Functions

Some exponential functions are very difficult to graph by hand. A graphing calculator can be used.

- To graph the exponential function $f(x) = 2(3^{x/2})$, we enter the right side of the equation after $Y_1 =$. The display will show the equation $Y_1 = 2(3\wedge(X/2))$. If we use [WINDOW] settings of $[-10, 10]$ for x and $[-2, 18]$ for y and press [GRAPH], we will obtain the graph shown in Figure 5-10.

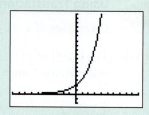

FIGURE 5-10

- To graph the exponential function $f(x) = 3e^{-x/2}$, we enter the right side of the equation $Y_1 =$. The display will show the equation $Y_1 = 3e\wedge(-x/2)$. If we use [WINDOW] settings of $[-10, 10]$ for x and $[-2, 18]$ for y and press [GRAPH], we will obtain the graph shown in Figure 5-11.

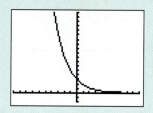

FIGURE 5-11

Self Check Answers

1. **a.** 2.6265   **b.** -14.7470 **c.** 0.0102   2. **a.** 64   **b.** $x^{3\sqrt{5}}$
3.   4. no   5. no   6. $1,771,251.33

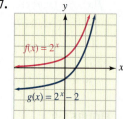

7.   8.

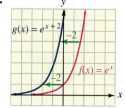

Exercises 5.1

Getting Ready

You should be able to complete these vocabulary and concept statements before you proceed to the practice exercises.

Fill in the blanks.

1. If $b > 0$ and $b \neq 1$, $f(x) = b^x$ represents an _____ function.

2. If $f(x) = b^x$ represents an increasing function, then $b > $ ___.

3. In interval notation, the domain of the exponential function $f(x) = b^x$ is _____.

4. The number b is called the _____ of the exponential function $f(x) = b^x$.

5. The range of the exponential function $f(x) = b^x$ is _____.

6. The graphs of all exponential functions $f(x) = b^x$ have the same ___-intercept, the point _____.

7. If $b > 0$ and $b \neq 1$, the graph of $f(x) = b^x$ approaches the x-axis, which is called a horizontal _____ of the curve.

8. If $f(x) = b^x$ represents a decreasing function, then ___ $< b <$ ___.

9. If $b > 1$, then $f(x) = b^x$ defines a (an) _____ function.

10. The graph of an exponential function $f(x) = b^x$ always passes through the points $(0, 1)$ and _____.

11. To two decimal places, the value of e is _____.

12. The continuous compound interest formula is $A = $ _____.

13. Since $e > 1$, the base-e exponential function is a (an) _____ function.

14. The graph of the exponential function $f(x) = e^x$ passes through the points $(0, 1)$ and _____.

Practice

 Use a calculator to find each value to four decimal places.

15. $4^{\sqrt{3}}$

16. $5^{\sqrt{2}}$

17. 7^{π}

18. $3^{-\pi}$

Simplify each expression.

19. $5^{\sqrt{2}}5^{\sqrt{2}}$

20. $\left(5^{\sqrt{2}}\right)^{\sqrt{2}}$

21. $\left(a^{\sqrt{8}}\right)^{\sqrt{2}}$

22. $a^{\sqrt{12}}a^{\sqrt{3}}$

Find $f(0)$ and $f(2)$ for each of the given exponential functions.

23. $f(x) = 5^x$

24. $f(x) = 4^{-x}$

25. $f(x) = \left(\dfrac{1}{3}\right)^{-x}$

26. $f(x) = \left(\dfrac{1}{4}\right)^{x}$

Graph each exponential function.

27. $f(x) = 3^x$

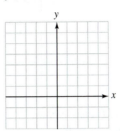

28. $f(x) = 5^x$

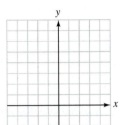

29. $f(x) = \left(\dfrac{1}{5}\right)^{x}$

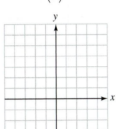

30. $f(x) = \left(\dfrac{1}{3}\right)^{x}$

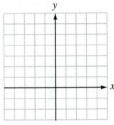

31. $f(x) = \left(\dfrac{3}{4}\right)^{x}$

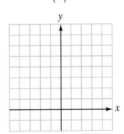

32. $f(x) = \left(\dfrac{4}{3}\right)^{x}$

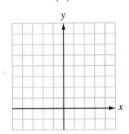

33. $f(x) = (1.5)^x$

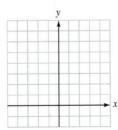

34. $f(x) = (0.3)^x$

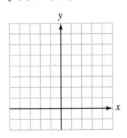

35. $f(x) = 3^{-x}$

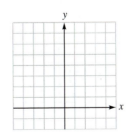

36. $f(x) = -5^x$

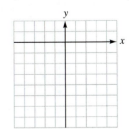

37. $f(x) = -\left(\dfrac{1}{5}\right)^x$ **38.** $f(x) = \left(\dfrac{1}{3}\right)^{-x}$

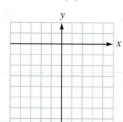   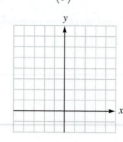

49.

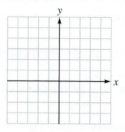

50.

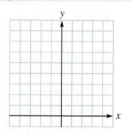

Graph each function by using transformations.

51. $f(x) = 3^x - 1$ **52.** $f(x) = 2^x + 3$

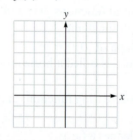   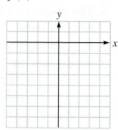

53. $f(x) = 2^x + 1$ **54.** $f(x) = 4^x - 4$

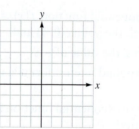   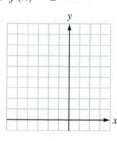

55. $f(x) = 3^{x-1}$ **56.** $f(x) = 2^{x+3}$

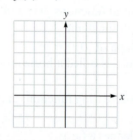   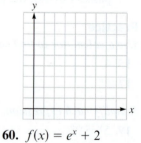

57. $f(x) = 3^{x+1}$ **58.** $f(x) = 2^{x-3}$

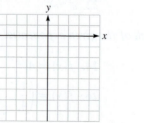   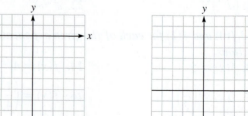

59. $f(x) = e^x - 4$ **60.** $f(x) = e^x + 2$

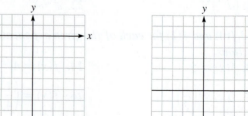

Determine whether the graph could represent an exponential function of the form $f(x) = b^x$.

39.

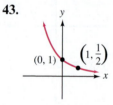

(0, 1)

40.

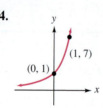

(1, 0)

41.

(0, 2)

42.

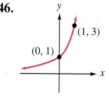

(0, 1)

Find the value of b, if any, that would cause the graph of $f(x) = b^x$ to look like the graph indicated.

43.

(0, 1) $\left(1, \dfrac{1}{2}\right)$

44.

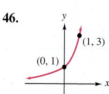

(0, 1)   (1, 7)

45.

(0, 2)   (1, 5)

46.

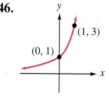

(0, 1)   (1, 3)

47.

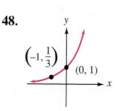

(0, 1)   (1, 2)

48.

$\left(-1, \dfrac{1}{3}\right)$ (0, 1)

61. $f(x) = e^{x-2}$

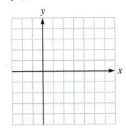

62. $f(x) = e^{x+3}$

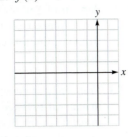

Use a graphing calculator to graph each function.

73. $f(x) = 5(2^x)$ **74.** $f(x) = 2(5^x)$

63. $f(x) = 2^{x+1} - 2$

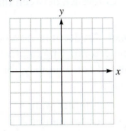

64. $f(x) = 3^{x-1} + 2$

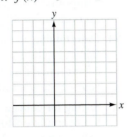

75. $f(x) = 3^{-x}$ **76.** $f(x) = 2^{-x}$

77. $f(x) = 2e^x$ **78.** $f(x) = 3e^{-x}$

79. $f(x) = 5e^{-0.5x}$ **80.** $f(x) = -3e^{2x}$

65. $y = 3^{x-2} + 1$

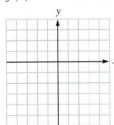

66. $y = 3^{x+2} - 1$

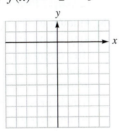

Applications

In Exercises 81–84, assume that there are no deposits or withdrawals.

81. Compound interest An initial deposit of $10,000 earns 8% interest, compounded quarterly. How much will be in the account in 10 years?

82. Compound interest An initial deposit of $1000 earns 9% interest, compounded monthly. How much will be in the account in $4\frac{1}{2}$ years?

83. Comparing interest rates How much more interest could $500 earn in 5 years, compounded semiannually (two times a year), if the annual interest rate were $5\frac{1}{2}\%$ instead of 5%?

84. Comparing savings plans Which institution in the ads provides the better investment?

67. $f(x) = -3^x + 1$

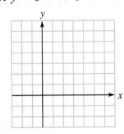

68. $f(x) = -2^x - 3$

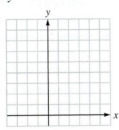

69. $f(x) = 2^{-x} - 3$

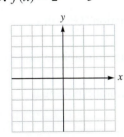

70. $f(x) = 4^{-x} + 4$

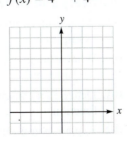

71. $f(x) = -e^x + 2$

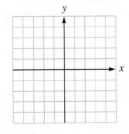

72. $f(x) = e^{-x} + 3$

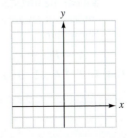

85. Compound interest If $1 had been invested on July 4, 1776, at 5% interest, compounded annually, what would it be worth on July 4, 2076?

86. 360/365 method Some financial institutions pay daily interest, compounded by the 360/365 method, using the following formula.

$$A = A_0\left(1 + \frac{r}{360}\right)^{365t} \quad (t \text{ is in years})$$

Using this method, what will an initial investment of $1000 be worth in 5 years, assuming a 7% annual interest rate?

87. Carrying charges A college student takes advantage of the ad shown and buys a bedroom set for $1100. He plans to pay the $1100 plus interest when his income tax refund comes in 8 months. At that time, what will he need to pay?

> BUY NOW,
> PAY LATER!
>
> Only $1\frac{3}{4}\%$ interest per month.

88. Credit card interest A bank credit card charges interest at the rate of 21% per year, compounded monthly. If a senior in college charges $1500 to pay for college expenses and intends to pay it in one year, what will he have to pay?

©Asier Romero/Shutterstock.com

89. Continuous compound interest An initial investment of $5000 earns 8.2% interest, compounded continuously. What will the investment be worth in 12 years?

90. Continuous compound interest An initial investment of $2000 earns 8% interest, compounded continuously. What will the investment be worth in 15 years?

91. Comparison of compounding methods An initial deposit of $5000 grows at an annual rate of 8.5% for 5 years. Compare the final balances resulting from continuous compounding and annual compounding.

92. Comparison of compounding methods An initial deposit of $30,000 grows at an annual rate of 8% for 20 years. Compare the final balances resulting from continuous compounding and annual compounding.

93. Frequency of compounding $10,000 is invested in each of two accounts, both paying 6% annual interest. In the first account, interest compounds quarterly, and in the second account, interest compounds daily. Find the difference between the accounts after 20 years.

94. Determining an initial deposit An account now contains $11,180 and has been accumulating interest at a 7% annual rate, compounded continuously, for 7 years. Find the initial deposit.

95. Saving for college In 20 years, a father wants to accumulate $40,000 to pay for his daughter's college expenses. If he can get 6% interest, compounded quarterly, how much must he invest now to achieve his goal?

96. Saving for college In Exercise 95, how much should he invest to achieve his goal if he can get 6% interest, compounded continuously?

97. Population of a city The population $P(t)$ of a small city can be approximated by the exponential function, $P(t) = 1200e^{0.2t}$, where t represents time in years. What will be the population of the city in 12 years? Round to a whole number.

98. Amount of drug present The amount of a drug $A(t)$, in mg, present in the bloodstream t hours after being intravenously administered can be approximated by the exponential function, $A(t) = -1000e^{-0.3t} + 1250$. How much of the drug is present in the bloodstream after 14 hours? Round to a whole number.

Discovery and Writing

99. What is an exponential function? Give three examples.

100. What strategy would you use to graph an exponential function?

101. Define e and describe the natural exponential function.

102. Explain compound interest.

103. Financial planning To have $P available in n years, $A can be invested now in an account paying interest at an annual rate r, compounded annually. Show that

$$A = P(1 + r)^{-n}$$

104. If $2^{t+4} = k2^t$, find k.

105. If $5^{3t} = k^t$, find k.

106. a. If $e^{t+3} = ke^t$, find k.

   b. If $e^{3t} = k^t$, find k.

Critical Thinking

Determine if the statement is true or false. If the statement is false, then correct it and make it true.

107. The domain of the exponential function $f(x) = 7^x$ is all real numbers.

108. The range of the exponential function is $f(x) = 7^x$ is $(7, \infty)$.

109. The graph of $f(x) = -7^{-x}$ has a y-intercept at $(0, -7)$.

110. The graph of $f(x) = -\left(\dfrac{1}{7}\right)^{-x}$ has a horizontal asymptote at $y = -7$.

111. The graph of $f(x) = 7^{x-7} + 7.7$ has a horizontal asymptote at $y = 7.7$.

112. The graphs of $f(x) = 7^x$ and $g(x) = \left(\dfrac{1}{7}\right)^{-x}$ are identical.

113. The graphs of $f(x) = 7^x$ and $g(x) = 7^{-x}$ intersect at the point $(0, 7)$.

114. To obtain the graph of $g(x) = e^{x+7} - 7$, we can use the graph of $f(x) = e^x$ and shift it 7 units to the left and 7 units down.

5.2 Applications of Exponential Functions

In this section, we will learn to

1. Solve radioactive decay problems.

2. Solve oceanography problems.

3. Solve Malthusian population growth problems.

4. Solve epidemiology problems.

Flu kills an estimated 36,000 Americans each year and results is a much larger number of hospitalizations. The influenza virus replicates quickly and can rapidly infect a population. The most effective method of preventing the virus infection and its severe complications is a flu vaccination.

An event that changes with time, such as the spread of the influenza virus, can by modeled by an exponential function. In this section, we will see several important applications of these functions: radioactive decay, oceanography, population growth, and epidemiology.

A mathematical description of an observed event is called a **model** of that event. Many real-world occurrences change with time and can be modeled by exponential functions of the form

$$f(t) = ab^{kt} \qquad \text{Remember that } ab^{kt} \text{ means } a(b^{kt}).$$

where a, b, and k are constants and t represents time. If f is an increasing function, we say that f *grows exponentially*. If f is a decreasing function, we say that f *decays exponentially*.

1. Solve Radioactive Decay Problems

The atomic structure of a radioactive material changes as the material emits radiation. Uranium, for example, changes (decays) into thorium, then into radium, and eventually into lead.

Experiments have determined the time it takes for one-half of a sample of a given radioactive element to decompose. That time is a constant, called the element's **half-life**. The amount present decays exponentially according to this formula.

Radioactive Decay Formula The amount A of radioactive material present at time t is given by

$$A = A_0 2^{-t/h}$$

where A_0 is the amount that was present initially (at $t = 0$) and h is the material's half-life.

Solving a Radioactive Decay Problem

EXAMPLE 1

The half-life of radium is approximately 1600 years. How much of a 1-gram sample will remain after 1000 years?

SOLUTION In this example, $A_0 = 1$, $h = 1600$, and $t = 1000$. We substitute these values into the formula for radioactive decay and simplify.

$$A = A_0 2^{-t/h}$$

$$A = 1 \cdot 2^{-1000/1600}$$

$$\approx 0.648419777 \qquad \text{Use a calculator.}$$

After 1000 years, approximately 0.65 gram of radium will remain.

Self Check 1 After 800 years, how much radium will remain?

Now Try Exercise 3.

2. Solve Oceanography Problems

Intensity of Light Formula The intensity I of light (in lumens) at a distance x meters below the surface of a body of water decreases exponentially according to the formula

$$I = I_0 k^x$$

where I_0 is the intensity of light above the water and k is a constant that depends on the clarity of the water.

Take Note

A lumen is a unit of standard measurement that describes how much light is contained in a certain area. The lumen is part of the photometry group which measures different aspects of light.

Solving an Intensity of Light Problem

EXAMPLE 2

At one location in the Atlantic Ocean, the intensity of light above water I_0 is 12 lumens and $k = 0.6$. Find the intensity of light at a depth of 5 meters.

SOLUTION We will substitute 12 for I_0, 0.6 for k, and 5 for x into the formula for light intensity and then simplify.

$$I = I_0 k^x$$

$$I = 12(0.6)^5$$

$$I = 0.93312$$

At a depth of 5 meters, the intensity of the light is slightly less than 1 lumen.

Self Check 2   Find the intensity at a depth of 10 meters.

Now Try Exercise 11.

3. Solve Malthusian Population Growth Problems

An equation based on the exponential function provides a model for **population growth**. One such model, called the **Malthusian model of population growth**, assumes a constant birth rate and a constant death rate. In this model, the population P grows exponentially according to the following formula.

Malthusian Model of Population Growth   If b is the annual birth rate, d is the annual death rate, t is the time (in years), P_0 is the initial population at $t = 0$, and P is the current population, then

$$P = P_0 e^{kt}$$

where $k = b - d$ is the **annual growth rate**, the difference between the annual birth rate and death rate.

Using the Malthusian Model to Predict Population Growth

EXAMPLE 3

The population of the United States is approximately 300 million people. Assuming that the annual birth rate is 19 per 1000 and the annual death rate is 7 per 1000, what does the Malthusian model predict the U.S. population will be in 50 years?

©Kurhan/Shutterstock.com

SOLUTION   We can use the stated information to write the Malthusian model for U.S. population. We will then substitute into the model to predict the population in 50 years.
   Since k is the difference between the birth and death rates, we have

$$k = b - d$$

$$k = \frac{19}{1000} - \frac{7}{1000} \qquad \text{Substitute } \frac{19}{1000} \text{ for } b \text{ and } \frac{7}{1000} \text{ for } d.$$

$$k = 0.019 - 0.007$$

$$= 0.012$$

We can now substitute 300,000,000 for P_0, 50 for t, and 0.012 for k in the formula for the Malthusian model of population growth and simplify.

$$P = P_0 e^{kt}$$

$$P = (300{,}000{,}000)e^{(0.012)(50)}$$

$$= (300{,}000{,}000)e^{0.6}$$

$$\approx 546{,}635{,}640.1 \qquad \text{Use a calculator.}$$

After 50 years, the U.S. population will exceed 546 million people.

Self Check 3   Find the population in 100 years.

Now Try Exercise 23.

   The English economist Thomas Robert Malthus (1766–1834) pioneered in population study. He believed that poverty and starvation were unavoidable, because the human population tends to grow exponentially, whereas the food supply tends to grow linearly.

EXAMPLE 4 **Using a Graphing Calculator to Solve a Population Problem**

Suppose that a country with a population of 1000 people is growing exponentially according to the population function

$$P(t) = 1000e^{0.02t}$$

where t is in years. Furthermore, assume that the food supply, measured in adequate food per day per person, is growing linearly according to the function

$$f(x) = 30.625x + 2000$$

In how many years will the population outstrip the food supply?

SOLUTION We can use a graphing calculator with [WINDOW] settings of $[0, 100]$ for x and $[0, 10{,}000]$ for y. After graphing the functions as shown in Figure 5-12, we find the point where the two graphs intersect. We can see that the food supply will be adequate for about 72 years. At that time, the population of approximately 4200 people will begin to have problems, assuming all conditions remain static.

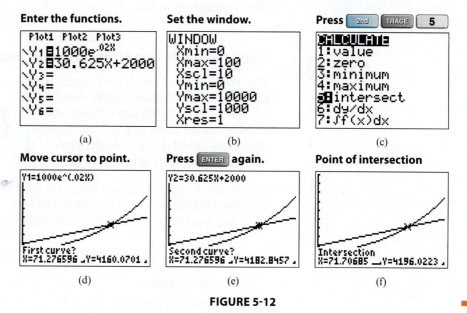

Enter the functions. (a)

Set the window. (b)

Press [2nd] [TRACE] [5] (c)

Move cursor to point. (d)

Press [ENTER] again. (e)

Point of intersection (f)

FIGURE 5-12

Self Check 4 In 80 years, what is the approximate number of people per day that will not have adequate food?

Now Try Exercise 43.

4. Solve Epidemiology Problems

Many infectious diseases, including some caused by viruses, spread most rapidly when they first infect a population, but then more slowly as the number of uninfected individuals decreases. These situations are often modeled by a function, called a *logistic function*.

Logistic Epidemiology Model The size P of an infected population at any time t in years is given by the logistic function

$$P = \frac{M}{1 + \left(\dfrac{M}{P_0} - 1\right)e^{-kt}}$$

where P_0 is the infected population size at $t = 0$, k is a constant determined by how contagious the virus is in a given environment, and M is the theoretical maximum size of the population P.

Solving an Epidemiology Problem

EXAMPLE 5

In a city with a population of 1,200,000, there are currently 1000 cases of infection with HIV. If the spread of the disease is projected by the function

$$P(t) = \frac{1{,}200{,}000}{1 + (1200 - 1)e^{-0.4t}}$$

how many people will be infected in 3 years?

SOLUTION   We can substitute 3 for t in the logistic function and calculate $P(t)$.

$$P(t) = \frac{1{,}200{,}000}{1 + (1200 - 1)e^{-0.4t}}$$

$$P(3) = \frac{1{,}200{,}000}{1 + (1199)e^{-0.4(3)}}$$

$$\approx 3{,}313.710094$$

In 3 years, approximately 3300 people are expected to be infected.

Self Check 5   How many will be infected in 10 years?

Now Try Exercise 35.

Accent on Technology

Exponential Regression

The table below shows the cooling temperatures of a hot cup of coffee after it is made.

Time in Minutes	Temperature °F
0	179.5
5	168.7
8	158.1
11	149.2
15	141.7
18	134.6
22	125.4
25	123.5
30	116.3
34	113.2
38	109.1
42	105.7
45	102.2
50	100.5

We can use a graphing calculator to determine an exponential regression model function that best fits or represents the given data.

- **Step 1:** Enter the data into lists in the calculator. Press STAT and then choose 1:Edit to access the list window and then enter the values. Enter the *x*-values (time in minutes) into L1 and the *y*-values (temperature) into L2. See Figure 5-13(a).
- **Step 2:** Set a window to fit the data. See Figure 5-13(b).
- **Step 3:** Create a scatterplot of the data. To do so, select STAT PLOT, located above the Y= button, and turn Plot 1 on. See Figure 5-14(c). Press GRAPH to see the scatterplot. See Figure 5-13(d).

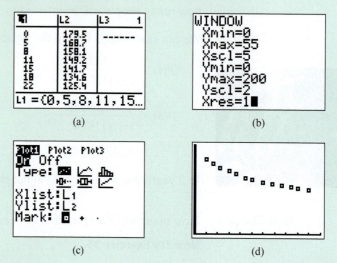

FIGURE 5-13

- **Step 4:** To determine the exponential regression function, we press STAT and scroll right to the CALC menu. We scroll downward to 0:ExpReg and press ENTER twice. This gives the equation. See Figure 5-14 (a) and (b). The function is $y = 171.4617283 \cdot (0.9882469577)^x$.
- **Step 5:** To graph the exponential function, we first press Y= , then we press VARS and scroll downward to 5: Statistics and press ENTER . Finally, we select EQ, then press ENTER and GRAPH . See Figure 5-14(c), (d), (e), and (f).

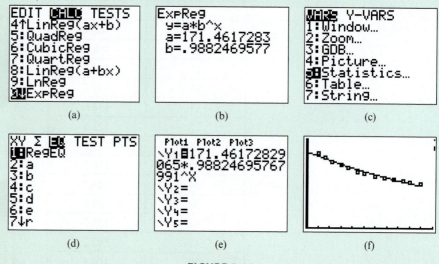

FIGURE 5-14

Exercises 5.2

Getting Ready

You should be able to complete these vocabulary and concept statements before you proceed to the practice exercises.

Fill in the blanks.

1. The Malthusian model assumes a constant _____ rate and a constant _____ rate.

2. The Malthusian prediction is pessimistic, because a _____ grows exponentially, but food supplies grow _____.

Applications

Use a calculator to help solve each problem.

3. **Tritium decay** Tritium, a radioactive isotope of hydrogen, has a half-life of 12.4 years. Of an initial sample of 50 grams, how much will remain after 100 years?

4. **Chernobyl** In April 1986, the world's worst nuclear power disaster occurred at Chernobyl in the former USSR. An explosion released about 1000 kilograms of radioactive cesium-137 (^{137}Cs) into the atmosphere. If the half-life of ^{137}Cs is 30.17 years, how much will remain in the atmosphere in 100 years?

5. **Chernobyl** Refer to Exercise 4. How much ^{137}Cs will remain in 200 years?

6. **Carbon-14 decay** The half-life of radioactive carbon-14 is 5700 years. How much of an initial sample will remain after 3000 years?

7. **Plutonium decay** One of the isotopes of plutonium, ^{237}Pu, decays with a half-life of 40 days. How much of an initial sample will remain after 60 days?

8. **Comparing radioactive decay** One isotope of holmium, ^{162}Ho, has a half-life of 22 minutes. The half-life of a second isotope, ^{164}Ho, is 37 minutes. Starting with a sample containing equal amounts, find the ratio of the amounts of ^{162}Ho to ^{164}Ho after one hour.

9. **Drug absorption in smokers** The biological half-life of the asthma medication theophylline is 4.5 hours for smokers. Find the amount of the drug retained in a smoker's system 12 hours after a dose of 1 unit is taken.

10. **Drug absorption in nonsmokers** For a nonsmoker, the biological half-life of theophylline is 8 hours. Find the amount of the drug retained in a nonsmoker's system 12 hours after taking a one-unit dose.

11. **Oceanography** The intensity I of light (in lumens) at a distance x meters below the surface is given by $I = I_0 k^x$, where I_0 is the intensity at the surface and k depends on the clarity of the water. At one location in the Arctic Ocean, $I_0 = 8$ lumens and $k = 0.5$. Find the intensity at a depth of 2 meters.

12. **Oceanography** At one location in the Atlantic Ocean, $I_0 = 14$ lumens and $k = 0.7$. Find the intensity of light at a depth of 12 meters. (See Exercise 11.)

13. **Oceanography** At a depth of 3 meters at one location in the Pacific Ocean, the intensity I of light is 1 lumen and $k = 0.5$. Find the intensity I_0 of light at the surface.

14. **Oceanography** At a depth of 2 meters at one location off the coast of Belize, the intensity I of light is 2 lumens and $k = 0.2$. Find the intensity I_0 of light at the surface.

15. **Bluegill population** A Wisconsin lake is stocked with 10,000 bluegill. The population is expected to grow exponentially according to the model $P(t) = P_0 2^{t/2}$. How many bluegill will be in the lake in 5 years?

DHNPhotos/iStockphoto.com

16. **Community growth** The population of Eagle River is growing exponentially according to the model $P(t) = 375(1.3)^t$, where t is measured in years from the present date. Find the population in 3 years.

17. **Newton's law of cooling** Some hot water, initially at 100°C, is placed in a room with a temperature of 40°C. The temperature T of the water after t hours is given by $T(t) = 40 + 60(0.75)^t$. Find the temperature in $3\frac{1}{2}$ hours.

18. **Bacterial cultures** A colony of 6 million bacteria is growing in a culture medium. The population P after t hours is given by the formula $P(t) = (6 \times 10^6)(2.3)^t$. Find the population after 4 hours.

©Macrovector/Shutterstock.com

19. **Population growth** The growth of a town's population is modeled by $P(t) = 173e^{0.03t}$. How large will the population be when $t = 20$?

20. **Population decline** The decline of a city's population is modeled by $P(t) = 1.2 \times 10^6 e^{-0.008t}$. How large will the population be when $t = 30$?

21. **Epidemics** The spread of hoof and mouth disease through a herd of cattle can be modeled by the formula $P(t) = P_0 e^{0.27t}$, where P is the size of the infected population, P_0 is the infected population size at $t = 0$, and t is in days. If a rancher does not act quickly to treat two cases, how many cattle will have the disease in one week?

22. **Alcohol absorption** In one individual, the percent of alcohol absorbed into the bloodstream after drinking two glasses of wine is given by the following formula. Find the percent of alcohol absorbed into the blood after $\frac{1}{2}$ hour.

$$P(t) = 0.3(1 - e^{-0.05t}) \text{ where } t \text{ is in minutes}$$

23. **World population growth** The population of the Earth is approximately 6 billion people and is growing at an annual rate of 1.9%. Assuming a Malthusian growth model, find the world population in 30 years.

24. **World population growth** See Exercise 23. Assuming a Malthusian growth model, find the world population in 40 years.

25. **World population growth** See Exercise 23. By what factor will the current population of the Earth increase in 50 years?

26. **World population growth** See Exercise 23. By what factor will the current population of the Earth increase in 100 years?

27. **Drug absorption** The percent P of the drug triazolam (a drug for treating insomnia) remaining in a person's bloodstream after t hours is given by $P(t) = e^{-0.3t}$. What percent will remain in the bloodstream after 24 hours?

28. **Medicine** The concentration x of a certain drug in an organ after t minutes is given by $y(t) = 0.08(1 - e^{-0.1t})$. Find the concentration of the drug in $\frac{1}{2}$ hour.

29. **Medicine** Refer to Exercise 28. Find the initial concentration of the drug (*Hint:* when $t = 0$).

30. **Spreading the news** Suppose the function

$$N(t) = P(1 - e^{-0.1t})$$

is used to model the length of time t (in hours) it takes for N people living in a town with population P to hear a news flash. How many people in a town of 50,000 will hear the news between 1 and 2 hours after it happened?

31. **Spreading the news** How many people in the town described in Problem 30 will not have heard the news after 10 hours?

32. **Epidemics** Refer to Example 5. How many people will be infected with HIV in 5 years?

33. **Epidemics** Refer to Example 5. How many people will be infected with HIV in 8 years?

34. **Epidemics** In a city with a population of 450,000, there are currently 1000 cases of hepatitis. If the spread of the disease is projected by the following logistic function, how many people will contract the hepatitis virus after 6 years?

$$P(t) = \frac{450,000}{1 + (450 - 1)e^{-0.2t}}$$

35. **Epidemics** In an Indonesian city with a population of 55,000, there are currently 100 cases of the avian bird flu. If the spread of the disease is projected by the following formula, how many people will contract the bird flu after 2 years?

$$P(t) = \frac{55,000}{1 + (550 - 1)e^{-0.8t}}$$

36. **Life expectancy** The life expectancy l of white females can be estimated by using the function $l(x) = 78.5(1.001)^x$, where x is the current age. Find the life expectancy of a white female who is currently 50 years old. Give the answer to the nearest tenth.

37. **Oceanography** The width w (in millimeters) of successive growth spirals of the sea shell *Catapulus voluto*, shown in the illustration, is given by the function $w(n) = 1.54e^{0.503n}$, where n is the spiral number. To the nearest tenth of a millimeter, find the width of the fifth spiral.

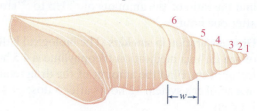

38. **Skydiving** Before the parachute opens, the velocity v (in meters per second) of a skydiver is given by $v(t) = 50(1 - e^{-0.2t})$. Find the initial velocity.

39. Skydiving Refer to Exercise 38 and find the velocity after 20 seconds.

40. Free-falling objects After t seconds, a certain falling object has a velocity v given by $v(t) = 50(1 - e^{-0.3t})$. Which is falling faster after 2 seconds, this object or the skydiver in Exercise 38?

41. Population growth In 1999, the male population of the United States was about 133 million, and the female population was about 139 million. Assuming a Malthusian growth model with a 1% annual growth rate, how many more females than males will there be in 20 years?

42. Population growth See Exercise 41. How many more females than males will there be in 50 years?

 Use a graphing calculator to solve each problem.

43. In Example 4, suppose that better farming methods change the formula for food growth to $f(x) = 31x + 2000$. How long will the food supply be adequate?

44. In Example 4, suppose that a birth control program changed the formula for population growth to $P(t) = 1000e^{0.01t}$. How long will the food supply be adequate?

Discovery and Writing

45. Exponential regression A population of *Escherichia coli* bacteria doubles every 20 minutes. Construct a table that shows the growth of a single *E. coli* bacterium for a 2-hour period. Then use a graphing calculator to plot the data and determine an exponential regression equation to model this growth.

46. Refer to Exercise 45. At what point will the population reach 200 cells?

 47. Graph the function defined by the equation $f(x) = \dfrac{e^x + e^{-x}}{2}$ from $x = -2$ to $x = 2$. The graph will look like a parabola, but it is not. The graph, called a **catenary**, is important in the design of power distribution networks, because it represents the shape of a uniform flexible cable whose ends are suspended from the same height. The function is called the **hyperbolic cosine function**. The hyperbolic cosine function was used in the design and construction of the Gateway Arch in St. Louis, Missouri.

 48. Graph the function defined by the equation $f(x) = \dfrac{e^x - e^{-x}}{2}$ from $x = -2$ to $x = 2$. The function is called the **hyperbolic sine function**.

 49. Graph the following logistic function, first discussed in Example 5. Use [WINDOW] settings of $[0, 20]$ for x and $[0, 1,500,000]$ for y.

$$P(t) = \frac{1,200,000}{1 + (1199)e^{-0.4t}}$$

 50. Use the [TRACE] capabilities of your graphing calculator to explore the logistic function of Example 5 and Exercise 49. As time passes, what value does P approach? How many years does it take for 20% of the population to become infected? For 80%?

 51. The value of e can be calculated to any degree of accuracy by adding the first several terms of the following list. The more terms that are added, the closer the sum will be to e. Add the first six numbers in the following list. To how many decimal places is the sum accurate?

$$1, 1, \frac{1}{2}, \frac{1}{2 \cdot 3}, \frac{1}{2 \cdot 3 \cdot 4}, \frac{1}{2 \cdot 3 \cdot 4 \cdot 5}, \dots$$

 52. Mixture problem The tank in the illustration initially contains 20 gallons of pure water. A brine solution containing 0.5 pounds of salt per gallon is pumped into the tank, and the well-stirred mixture leaves at the same rate. The amount A of salt in the tank after t minutes is given by $A(t) = 10(1 - e^{-0.03t})$.

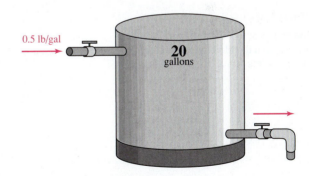

0.5 lb/gal

20 gallons

a. Graph this function.
b. What is A when $t = 0$? Explain why that value is expected.
c. What is A after 2 minutes? After 10 minutes?
d. What value does A approach after a long time (as t becomes large)? Explain why this is the value you would expect.

5.3 Logarithmic Functions and Their Graphs

In this section, we will learn to

1. Evaluate logarithms.
2. Evaluate common logarithms.
3. Evaluate natural logarithms.
4. Graph logarithmic functions.
5. Use transformations to graph logarithmic functions.

Guests aboard the Royal Caribbean's cruise ship *Freedom of the Seas* can now "hang ten" while out to sea. The flowrider surf simulator allows riders to body board surf against a wave-like water flow of 34,000 gallons per minute.

It is important that water in the flowrider has the proper pH value. For example, if it is too acidic, the water will make our eyes and nose burn, and it will make our skin get dry and itchy. The pH of the water is one of the most important factors in pool water balance and should be tested frequently. To calculate pH, we need to understand *logarithms*, the topic of this section. As we continue through this chapter, we will see several real-life applications of logarithms.

Since exponential functions are one-to-one functions, each one has an inverse. For example, to find the inverse of the function $y = 3^x$, we interchange the positions of x and y to obtain $x = 3^y$. The graphs of these two functions are shown in Figure 5-15(a).

To find the inverse of the function $y = \left(\frac{1}{3}\right)^x$, we again interchange the positions of x and y to obtain $x = \left(\frac{1}{3}\right)^y$. The graphs of these two functions are shown in Figure 5-15(b).

Take Note

The graphs of the functions are inverses and therefore symmetric about the line $y = x$.

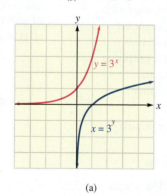
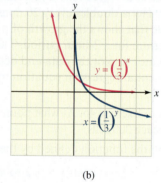

(a)   (b)

FIGURE 5-15

In general, the inverse of the function $y = b^x$ is $x = b^y$. When $b > 1$, their graphs appear as shown in Figure 5-16(a). When $0 < b < 1$, their graphs appear as shown in Figure 5-16(b).

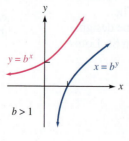
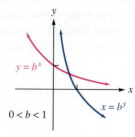

(a)   (b)

FIGURE 5-16

1. Evaluate Logarithms

Since an exponential function defined by $y = b^x$ is one-to-one, it has an inverse function that is defined by the equation $x = b^y$. To express this inverse function in the form $y = f^{-1}(x)$, we must solve the equation $x = b^y$ for y. To do this, we need the following definition.

Logarithmic Functions

If $b > 0$ and $b \neq 1$, the **logarithmic function with base b** is defined by

$$y = \log_b x \qquad \text{if and only if} \qquad x = b^y$$

The **domain of the logarithmic function** is the interval $(0, \infty)$. The **range** is the interval $(-\infty, \infty)$. The logarithmic function is also denoted as $f(x) = \log_b x$.

> **Caution**
>
> Since the domain of the logarithmic function is the set of positive numbers, the logarithm of 0 and the logarithm of a negative number are **undefined** in the set of real numbers. For example, both $\log_3 0$ and $\log_3(-27)$ are undefined.

The range of the logarithmic function is the set of real numbers, because the value of y in the equation $x = b^y$ can be any real number. The domain is the set of positive numbers, because the value of x in the equation $x = b^y$ $(b > 0)$ is always positive.

Since the function $y = \log_b x$ is the inverse of the one-to-one exponential function $y = b^x$, the logarithmic function is also one-to-one.

The expression $x = b^y$ is said to be written in *exponential form*. The equivalent expression $y = \log_b x$ is said to be written in *logarithmic form*. To translate from one form to the other, it is helpful to keep track of the base and the exponent.

> **Tip**
>
> The abbreviation "log" is used for the word "logarithm." A logarithm is an exponent.

Exponential Form

$$x = b^y$$

Base   Exponent

Logarithmic Form

$$y = \log_b x$$

Exponent   Base

> **Tip**
>
> If you prefer, you can write $b^y = x$ instead of writing $x = b^y$. Both are acceptable. Also, many students prefer writing $\log_b x = y$ instead of writing $y = \log_b x$. This too is acceptable.

Converting from Exponential Form to Logarithmic Form

EXAMPLE 1

Write each equation in logarithmic form.

a. $2^6 = 64$ **b.** $27^{\frac{1}{3}} = \sqrt[3]{27} = 3$ **c.** $6^{-2} = \dfrac{1}{36}$

SOLUTION   For each part, we can use the definition of the logarithm of a number.

> **Tip**
>
> Study closely the relationship between exponential and logarithmic statements. Understanding this definition can lead to success with the content in the rest of the chapter.

a. $2^6 = 64$ is equivalent to $\log_2 64 = 6$

b. $27^{\frac{1}{3}} = \sqrt[3]{27} = 3$ is equivalent to $\log_{27} 3 = \frac{1}{3}$

c. $6^{-2} = \frac{1}{36}$ is equivalent to $\log_6 \frac{1}{36} = -2$

Self Check 1   Write $13^{-2} = \dfrac{1}{169}$ in logarithmic form.

Now Try Exercise 21.

Converting from Logarithmic Form to Exponential Form

EXAMPLE 2

Write each equation in exponential form.

a. $\log_5 125 = 3$ **b.** $\log_{64} 8 = \dfrac{1}{2}$ **c.** $\log_{1/4} 16 = -2$

SOLUTION   For each part, we can use the definition of the logarithm of a number.

a. $\log_5 125 = 3$ is equivalent to $5^3 = 125$

b. $\log_{64} 8 = \frac{1}{2}$ is equivalent to $64^{\frac{1}{2}} = 8$

c. $\log_{1/4} 16 = -2$ is equivalent to $\left(\frac{1}{4}\right)^{-2} = 16$

Self Check 2   Write $\log_5 625 = 4$ in exponential form.

Now Try Exercise 29.

The definition of a logarithm can be used to find the values of many logarithms. Several examples are shown in the table.

> ### Examples of the Values of Logarithms
>
> $\log_5 25 = 2$ because **2** is the exponent to which 5 is raised to get 25: $5^2 = 25$
>
> $\log_7 1 = 0$ because **0** is the exponent to which 7 is raised to get 1: $7^0 = 1$
>
> $\log_{16} 4 = \dfrac{1}{2}$ because $\dfrac{1}{2}$ is the exponent to which 16 is raised to get 4: $16^{1/2} = \sqrt{16} = 4$
>
> $\log_2 \dfrac{1}{8} = -3$ because -3 is the exponent to which 2 is raised to get $\dfrac{1}{8}$: $2^{-3} = \dfrac{1}{8}$

Caution

$\log_5 25$ is **not** read as $\log_5$ times 25.

In each of these examples, the logarithm of a number is an exponent. In fact,

$\log_b x$ is the exponent to which b is raised to get x.

To express this as an equation, we write $b^{\log_b x} = x$

EXAMPLE 3   **Finding an Unknown Term in a Logarithmic Equation**

Find y in each equation: **a.** $\log_2 8 = y$ **b.** $\log_5 1 = y$ **c.** $\log_7 \dfrac{1}{49} = y$

SOLUTION   For each part, we can use the definition of the logarithm of a number to find y.

a. $\log_2 8 = y$ is equivalent to $2^y = 8$. Since $2^3 = 8$, we have $2^y = 2^3$ and $y = 3$.

b. $\log_5 1 = y$ is equivalent to $5^y = 1$. Since $5^0 = 1$, we have $5^y = 5^0$ and $y = 0$.

c. $\log_7 \frac{1}{49} = y$ is equivalent to $7^y = \frac{1}{49}$. Since $7^{-2} = \frac{1}{49}$, we have $7^y = 7^{-2}$ and $y = -2$.

Self Check 3   Find y in each equation:

a. $\log_3 9 = y$ b. $\log_2 16 = y$ c. $\log_5 \dfrac{1}{25} = y$

Now Try Exercise 35.

EXAMPLE 4   **Finding an Unknown Term in a Logarithmic Equation**

Find a in each equation: **a.** $\log_a 32 = 5$ **b.** $\log_9 a = -\dfrac{1}{2}$ **c.** $\log_9 3 = a$.

SOLUTION   For each part, we will use the definition of the logarithm of a number to find a.

a. $\log_a 32 = 5$ is equivalent to $a^5 = 32$. Since $2^5 = 32$, we have $a^5 = 2^5$ and $a = 2$.

b. $\log_9 a = -\frac{1}{2}$ is equivalent to $9^{-1/2} = a$. Since $9^{-1/2} = \frac{1}{3}$, it follows that $a = \frac{1}{3}$.

c. $\log_9 3 = a$ is equivalent to $9^a = 3$. Since $3 = 9^{1/2}$, we have $9^a = 9^{1/2}$ and $a = \frac{1}{2}$.

Self Check 4 Find d in each equation:

a. $\log_4 \dfrac{1}{16} = d$ **b.** $\log_d 36 = 2$ **c.** $\log_8 d = -\dfrac{1}{3}$

Now Try Exercise 49.

2. Evaluate Common Logarithms

Many applications use base-10 logarithms (also called **common logarithms**). When the base b is not indicated in the notation $\log x$, we assume that $b = 10$:

log x means $\log_{10} x$

Because base-10 logarithms appear so often, you should become familiar with the following base-10 logarithms:

Examples of Common Logarithms		
$\log_{10} \dfrac{1}{100} = -2$	because	$10^{-2} = \dfrac{1}{100}$
$\log_{10} \dfrac{1}{10} = -1$	because	$10^{-1} = \dfrac{1}{10}$
$\log_{10} 1 = 0$	because	$10^0 = 1$
$\log_{10} 10 = 1$	because	$10^1 = 10$
$\log_{10} 100 = 2$	because	$10^2 = 100$
$\log_{10} 1000 = 3$	because	$10^3 = 1000$

In general, we have

$\log_{10} 10^x = x$

Accent on Technology

Approximating Base-10 Logarithms

We can use the [LOG] key on a graphing calculator to evaluate base-10 logarithms. Figure 5-17 shows the evaluation of log 2.34 and demonstrates that **a logarithm is an exponent**. Notice that the logarithm is evaluated, and then the answer is used as the exponent of 10 to get 2.34. Remember: log 2.34 is the exponent of 10 that will yield 2.34.

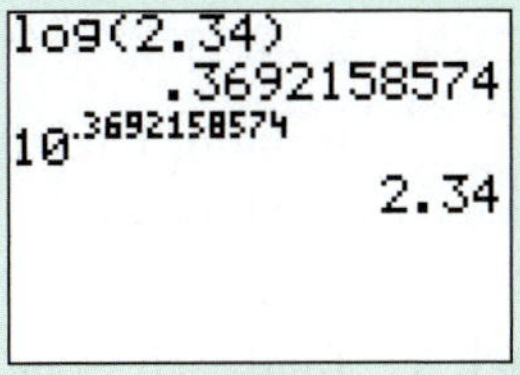

```
log(2.34)
        .3692158574
10^.3692158574
                2.34
```

FIGURE 5-17

Using a Calculator to Solve a Common Logarithmic Equation

EXAMPLE 5   Find x in the equation $\log x = 0.7482$ to four decimal places.

SOLUTION   We can use the definition of the common logarithm and a calculator to find x.

The equation $\log x = 0.7482$ is equivalent to $10^{0.7482} = x$.

Therefore we have $x = 10^{0.7482} \approx 5.6002$.

Self Check 5   Solve: $\log x = 1.87737$. Give the result to four decimal places.

Now Try Exercise 81.

3. Evaluate Natural Logarithms

We have seen the importance of the number e in mathematical models of events in nature. Base-e logarithms are just as important. They are called **natural logarithms** or **Napierian logarithms** after John Napier (1550–1617). They are usually written as $\ln x$, rather than $\log_e x$:

     ln x   **means**   **$\log_e x$**

Like all logarithmic functions, the domain of $f(x) = \ln x$ is the interval $(0, \infty)$, and the range is the interval $(-\infty, \infty)$.

To estimate the base-e logarithms of numbers, we can use a calculator. Scientific and graphing calculators have a natural logarithm key LN , which is used like the LOG key.

Accent on Technology

Approximating Natural Logarithms

We can use the LN key on a graphing calculator to evaluate base-e logarithms. Figure 5-18 shows the evaluation of $\ln 2.34$ and demonstrates that **a logarithm is an exponent**. Notice that the logarithm is evaluated, and then the answer is used as the exponent of e to get 2.34. Remember: $\ln 2.34$ is the exponent of e that will yield 2.34.

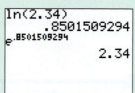

```
ln(2.34)
          .8501509294
e^.8501509294
               2.34
```

FIGURE 5-18

Using a Calculator to Approximate Natural Logarithms

EXAMPLE 6   Use a calculator to find   **a.** $\ln 17.32$ **b.** $\ln(\log 0.05)$

SOLUTION   **a.** $\ln 17.32 \approx 2.851861903$, which means $e^{2.851861903} \approx 17.32$

     b. $\ln(\log 0.05)$ is undefined because $\log(0.05) < 0$. Our calculator gives an error message.

Self Check 6   Find each value to four decimal places:

     a. $\ln \pi$ **b.** $\ln\left(\log \dfrac{1}{3}\right)$

Now Try Exercise 75.

Using a Calculator to Solve Natural Logarithmic Equations

EXAMPLE 7   Solve each equation and give the result to four decimal places:

     a. $\ln x = 1.335$ **b.** $\ln x = \log 5.5$

SOLUTION We will write the equation in exponential form and use our calculator to find x.

Tip

Every natural logarithmic equation has a corresponding exponential equation.

a. The equation $\ln x = 1.335$ is equivalent to $e^{1.335} = x$. We will use a calculator to evaluate the expression. $x = e^{1.335} \approx 3.8000$.

b. The equation $\ln x = \log 5.5$ is equivalent to $e^{\log 5.5} = x$. Using a calculator, we find $x = e^{\log 5.5} \approx 2.0967$.

Self Check 7 Solve each equation and give each result to four decimal places:

a. $\ln x = 1.9344$ **b.** $\ln x = \log 3.2$

Now Try Exercise 85.

4. Graph Logarithmic Functions

To draw the graph of a logarithmic function, we will use the fact that a logarithmic function is the inverse of an exponential function.

To graph the logarithmic function $f(x) = \log_2 x$, we calculate and plot several points with coordinates (x, y) that satisfy the equivalent equation $x = 2^y$. After joining these points with a smooth curve, we have the graph shown in Figure 5-19.

Tip

The base of the logarithmic function is 2. Choose values of x that are integer powers of 2.

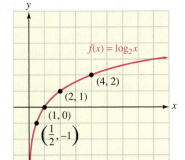

$f(x) = \log_2 x$		
x	$f(x)$	$(x, f(x))$
$\frac{1}{4}$	-2	$\left(\frac{1}{4}, -2\right)$
$\frac{1}{2}$	-1	$\left(\frac{1}{2}, -1\right)$
1	0	$(1, 0)$
2	1	$(2, 1)$
4	2	$(4, 2)$
8	3	$(8, 3)$

FIGURE 5-19

Graphing a Logarithmic Function

EXAMPLE 8 Graph: $f(x) = \log_{1/2} x$.

SOLUTION To graph $f(x) = \log_{1/2} x$. we calculate and plot several points with coordinates (x, y) that satisfy the equation $x = \left(\frac{1}{2}\right)^y$. After joining these points with a smooth curve, we have the graph shown in Figure 5-20.

Tip

The base of the logarithmic function is $\frac{1}{2}$. Choose values of x that are integer powers of $\frac{1}{2}$.

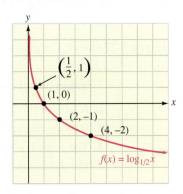

$f(x) = \log_{1/2} x$		
x	$f(x)$	$(x, f(x))$
$\frac{1}{4}$	2	$\left(\frac{1}{4}, 2\right)$
$\frac{1}{2}$	1	$\left(\frac{1}{2}, 1\right)$
1	0	$(1, 0)$
2	-1	$(2, -1)$
4	-2	$(4, -2)$
8	-3	$(8, -3)$

FIGURE 5-20

Self Check 8 Graph: $f(x) = \log_6 x$.

Now Try Exercise 105.

The graphs of all logarithmic functions are similar to those in Figure 5-21. If $b > 1$, the logarithmic function is an increasing function, as in Figure 5-21(a). If $0 < b < 1$, the logarithmic function is a decreasing function, as in Figure 5-21(b). We also know that the exponential function $f(x) = b^x$ and the logarithmic function $f(x) = \log_b x$ are inverse functions and therefore symmetric about the line $y = x$, as in Figure 5-21(c) and (d).

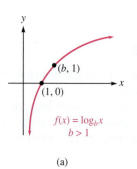
(a)

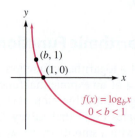
(b)

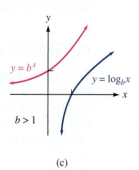
(c)

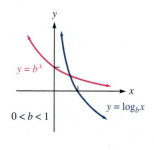
(d)

FIGURE 5-21

To graph, $f(x) = \log_{10} x$, we can plot points that satisfy the equation $x = 10^y$ and join them with a smooth curve, as shown in Figure 5-22.

> **Tip**
>
> The base of the common logarithmic function is 10. Choose x-values that are integer powers of 10.

The Graph of the Common Logarithm Function $f(x) = \log_{10} x$

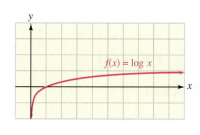

$f(x) = \log x$		
x	$f(x)$	$(x, f(x))$
$\dfrac{1}{100}$	-2	$\left(\dfrac{1}{100}, -2\right)$
$\dfrac{1}{10}$	-1	$\left(\dfrac{1}{10}, -1\right)$
1	0	$(1, 0)$
10	1	$(10, 1)$
100	2	$(100, 2)$

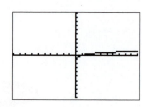

FIGURE 5-22

> **Take Note**
>
> The function $f(x) = \log_{10} x$ is very important and its graph should be memorized. Place in your library of functions.

As Figure 5-22 shows, the graph of $f(x) = \log_b x$ has these properties:

> **Properties of the Graph of $f(x) = \log_b x$**
>
> 1. It passes through the point $(1, 0)$.
> 2. It passes through the point $(b, 1)$.
> 3. The y-axis is a vertical asymptote.
> 4. The domain is $(0, \infty)$, and the range is $(-\infty, \infty)$.

To graph $f(x) = \ln x$, we can plot points that satisfy the equation $x = e^y$ and join them with a smooth curve, as shown in Figure 5-23.

The Graph of the Natural Logarithm Function $f(x) = \ln x$

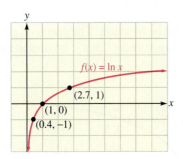

$f(x) = \ln x$		
x	$f(x)$	$(x, f(x))$
$\dfrac{1}{e} \approx 0.4$	-1	$(0.4, -1)$
1.0	0	$(1, 0)$
$e \approx 2.7$	1	$(2.7, 1)$
$e^2 \approx 7.4$	2	$(7.4, 2)$

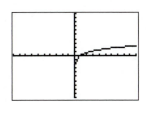

FIGURE 5-23

From the definition of logarithm, we know that

$$y = \ln x \quad \text{if and only if} \quad x = e^y \qquad \text{Remember that } \ln x = \log_e x.$$

Thus, $\ln e = 1$ because $e^1 = e$. This is an important property of the natural logarithm function. Two other important properties follow from the fact that $y = \ln x$ and $y = e^x$ are inverses:

$$\ln e^x = x \quad \text{and} \quad e^{\ln x} = x.$$

These two properties and examples are shown in the table:

Properties Involving e and ln	Examples
$\ln e^x = x$	$\ln e^8 = 8$
	$\ln e^{-5x} = -5x$
	$\ln e^{3x+2} = 3x + 2$
$e^{\ln x} = x$	$e^{\ln 4} = 4$
	$e^{\ln(7x)} = 7x$
	$e^{\ln(9x-2)} = 9x - 2$

5. Use Transformations to Graph Logarithmic Functions

The graphs of many functions involving logarithms are transformations of the basic logarithmic graphs. A summary of these is shown in the table for $k > 0$.

Function	Transformations of the Graph of $f(x) = \log_b x$
$g(x) = k + \log_b x$	Translates the graph of $f(x) = \log_b x$ upward k units
$g(x) = -k + \log_b x$	Translates the graph of $f(x) = \log_b x$ downward k units
$g(x) = \log_b(x - k)$	Translates the graph of $f(x) = \log_b x$ to the right k units
$g(x) = \log_b(x + k)$	Translates the graph of $f(x) = \log_b x$ to the left k units
$g(x) = -\log_b x$	Reflects the graph of $f(x) = \log_b x$ about the x-axis
$g(x) = \log_b(-x)$	Reflects the graph of $f(x) = \log_b x$ about the y-axis
$g(x) = k \log_b x$	• Vertically stretches the graph of $y = \log_b x$ if $k > 1$ • Vertically shrinks the graph of $y = \log_b x$ if $0 < k < 1$
$g(x) = \log_b(kx)$	• Horizontally stretches the graph of $y = \log_b x$ if $0 < k < 1$ • Horizontally shrinks the graph of $y = \log_b x$ if $k > 1$

EXAMPLE 9

Using a Transformation to Graph a Logarithmic Function

Graph: $g(x) = 3 + \log_2 x$.

SOLUTION We can use a translation to graph the function. The graph of $g(x) = 3 + \log_2 x$ is identical to the graph of $f(x) = \log_2 x$, except that it is translated 3 units upward. See Figure 5-24.

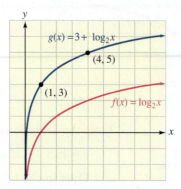

FIGURE 5-24

Self Check 9 Graph: $g(x) = -2 + \log_3 x$.

Now Try Exercise 111.

EXAMPLE 10

Using a Transformation to Graph a Logarithmic Function

Graph: $g(x) = \log_{1/2}(x - 1)$.

SOLUTION We can use a translation to graph the function. The graph of $g(x) = \log_{1/2}(x - 1)$ is identical to the graph of $f(x) = \log_{1/2} x$, except that it is translated 1 unit to the right. See Figure 5-25.

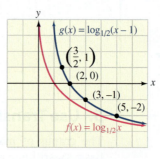

FIGURE 5-25

Self Check 10 Graph: $g(x) = \log_{1/3}(x + 2)$.

Now Try Exercise 113.

Many graphs of logarithmic functions involve translations of the graph of $f(x) = \ln x$. Consider the calculator graphs of the functions $f(x) = \ln x$, $g(x) = 2 + \ln x$, and $h(x) = -2 + \ln x$. See Figure 5-26.

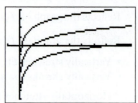

FIGURE 5-26

The graph of $f(x) = \ln x$ is the middle graph shown.

The graph of $g(x) = 2 + \ln x$ is 2 units above the graph of $f(x) = \ln x$.

The graph of $h(x) = -2 + \ln x$ is 2 units below the graph of $f(x) = \ln x$.

Figure 5-27 shows a calculator graph of the functions $f(x) = \ln x$, $g(x) = \ln(x + 2)$, and $h(x) = \ln(x - 2)$.

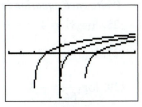

FIGURE 5-27

The graph of $f(x) = \ln x$ is the middle graph shown.

The graph of $g(x) = \ln(x + 2)$ is 2 units to the left of the graph of $f(x) = \ln x$.

The graph of $h(x) = \ln(x - 2)$ is 2 units to the right of the graph of $f(x) = \ln x$.

Accent on Technology

Graphing Logarithmic Functions

Graphing calculators can be used to graph logarithmic functions. However, the only bases that are built in are 10 and e. There is a way to graph logarithmic functions in other bases, but it involves changing the base. We will cover that in Section 5.5. In most applications either common logarithms or natural logarithms are used. When we are graphing logarithmic functions on a graphing calculator, we need to be aware of the domain so that we can set a proper window.

- Graph: $f(x) = -2 + \log_{10}\left(\frac{1}{2} x\right)$.

  To graph the function, we enter the right-hand side of the equation and use [WINDOW] settings of $[-1, 5]$ for x and $[-5, 1]$ for y. The graph is shown in Figure 5-28.

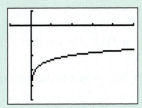

FIGURE 5-28

- Notice that the graph in Figure 5-28 appears to stop on the left and not continue toward negative infinity. That is a result of how the calculator plots points as it approaches the vertical asymptote.

Self Check Answers

1. $\log_{13} \dfrac{1}{169} = -2$ **2.** $5^4 = 625$ **3. a.** 2   **b.** 4   **c.** -2

4. a. -2 **b.** 6   **c.** $\dfrac{1}{2}$ **5.** 75.3998   **6. a.** 1.1447   **b.** undefined

7. a. 6.9199   **b.** 1.6572

8.
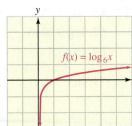

9.
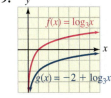

10.
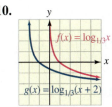

Exercises 5.3

Getting Ready

You should be able to complete these vocabulary and concept statements before you proceed to the practice exercises.

Fill in the blanks.

1. The equation $y = \log_b x$ is equivalent to _____.
2. The domain of a logarithmic function is the interval _____.
3. The _____ of a logarithmic function is the interval $(-\infty, \infty)$.
4. $b^{\log_b x} =$ ___.
5. Because the exponential function is one-to-one, it has an _____ function.
6. The inverse of an exponential function is called a _____ function.
7. $\log_b x$ is the _____ to which b is raised to get x.
8. The y-axis is an _____ of a graph of $f(x) = \log_b x$.
9. The graph of $f(x) = \log_b x$ passes through the points _____ and _____.
10. $\log_{10} 10^x =$ ___.
11. $\ln x$ means _____.
12. The domain of the function $f(x) = \ln x$ is the interval _____.
13. The range of the function $f(x) = \ln x$ is the interval _____.
14. The graph of $f(x) = \ln x$ has the _____ as an asymptote
15. In the expression $\log x$, the base is understood to be ___.
16. In the expression $\ln x$, the base is understood to be ___.

Practice

Write each equation in logarithmic form.

17. $8^2 = 64$
18. $10^3 = 1000$
19. $4^{-2} = \dfrac{1}{16}$
20. $3^{-4} = \dfrac{1}{81}$
21. $\left(\dfrac{1}{2}\right)^{-5} = 32$
22. $\left(\dfrac{1}{3}\right)^{-3} = 27$
23. $x^y = z$
24. $m^n = p$

Write each equation in exponential form.

25. $\log_3 81 = 4$
26. $\log_7 7 = 1$
27. $\log_{1/2} \dfrac{1}{8} = 3$
28. $\log_{1/5} 1 = 0$
29. $\log_4 \dfrac{1}{64} = -3$
30. $\log_6 \dfrac{1}{36} = -2$
31. $\log_\pi \pi = 1$
32. $\log_7 \dfrac{1}{49} = -2$

Find each value of x.

33. $\log_2 8 = x$
34. $\log_3 9 = x$
35. $\log_4 \dfrac{1}{64} = x$
36. $\log_6 216 = x$
37. $\log_{1/2} \dfrac{1}{8} = x$
38. $\log_{1/3} \dfrac{1}{81} = x$
39. $\log_9 3 = x$
40. $\log_{125} 5 = x$
41. $\log_{1/2} 8 = x$
42. $\log_{1/2} 16 = x$
43. $\log_8 x = 2$
44. $\log_7 x = 0$
45. $\log_7 x = 1$
46. $\log_2 x = 8$
47. $\log_{25} x = \dfrac{1}{2}$
48. $\log_4 x = \dfrac{1}{2}$
49. $\log_5 x = -2$
50. $\log_3 x = -4$
51. $\log_{36} x = -\dfrac{1}{2}$
52. $\log_{27} x = -\dfrac{1}{3}$
53. $\log_x 5^3 = 3$
54. $\log_x 5 = 1$
55. $\log_x \dfrac{9}{4} = 2$
56. $\log_x \dfrac{\sqrt{3}}{3} = \dfrac{1}{2}$
57. $\log_x \dfrac{1}{64} = -3$
58. $\log_x \dfrac{1}{100} = -2$
59. $\log_x \dfrac{9}{4} = -2$
60. $\log_x \dfrac{\sqrt{3}}{3} = -\dfrac{1}{2}$
61. $2^{\log_2 5} = x$
62. $3^{\log_3 4} = x$
63. $x^{\log_x 6} = 6$
64. $x^{\log_x 8} = 8$

Use a calculator to find each value to four decimal places.

65. $\log 3.25$
66. $\log 0.57$
67. $\log 0.00467$
68. $\log 375.876$
69. $\ln 45.7$
70. $\ln 0.005$
71. $\ln \dfrac{2}{3}$
72. $\ln \dfrac{12}{7}$

73. ln 35.15

74. ln 0.675

75. ln 7.896

76. ln 0.00465

77. log(ln 1.7)

78. ln(log 9.8)

79. ln(log 0.1)

80. log(ln 0.01)

 Use a calculator to find y to four decimal places, if possible.

81. log $y = 1.4023$

82. log $y = 0.926$

83. log $y = -3.71$

84. log $y = \log \pi$

85. ln $y = 1.4023$

86. ln $y = 2.6490$

87. ln $y = 4.24$

88. ln $y = 0.926$

89. ln $y = -3.71$

90. ln $y = -0.28$

91. log $y = \ln 8$

92. ln $y = \log 7$

Find each value without using a calculator.

93. log 10,000

94. log 1,000,000

95. log 0.001

96. $\log \dfrac{1}{100,000}$

97. $e^{\ln 7}$

98. $e^{\ln 9}$

99. $\ln(e^4)$

100. $\ln(e^{-6})$

Find the value of b, if any, that would cause the graph of $f(x) = \log_b x$ to look like the graph shown.

101.

102.

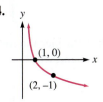

103.

104.

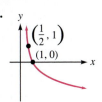

Graph each function.

105. $f(x) = \log_3 x$

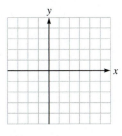

106. $f(x) = \log_4 x$

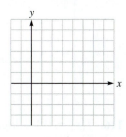

107. $f(x) = \log_{1/3} x$

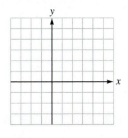

108. $f(x) = \log_{1/4} x$

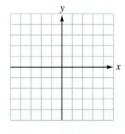

109. $f(x) = -\log_5 x$

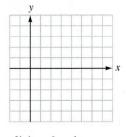

110. $f(x) = -\log_2 x$

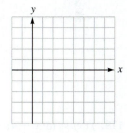

111. $f(x) = 2 + \log_2 x$

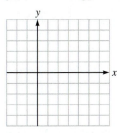

112. $f(x) = \log_2(x - 1)$

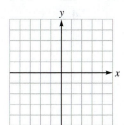

113. $f(x) = \log_3(x + 2)$

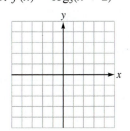

114. $f(x) = -3 + \log_3 x$

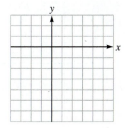

115. $f(x) = 3 + \log_3(x + 1)$

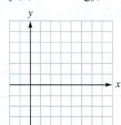

116. $f(x) = -3 + \log_3(x + 1)$

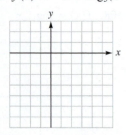

117. $f(x) = -3 + \ln x$ **118.** $f(x) = \ln(x + 1)$

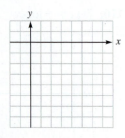 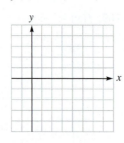

119. $f(x) = \ln(x - 4)$ **120.** $f(x) = 2 + \ln x$

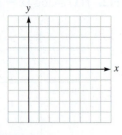 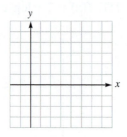

121. $f(x) = 1 - \ln x$ **122.** $f(x) = 2 - \ln x$

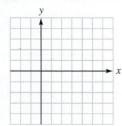 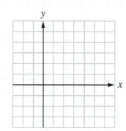

Use a graphing calculator to graph each function.

123. $f(x) = \log(3x)$ **124.** $f(x) = \log\left(\dfrac{x}{3}\right)$

125. $f(x) = \log(-x)$ **126.** $f(x) = -\log x$

127. $f(x) = \ln\left(\dfrac{1}{2}x\right)$ **128.** $f(x) = \ln x^2$

129. $f(x) = \ln(-x)$ **130.** $f(x) = \ln(3x)$

Discovery and Writing

131. Describe how to convert an equation in logarithmic form to an equation in exponential form.

132. Describe how to convert an equation in exponential form to an equation in logarithmic form.

133. Explain how you would evaluate the logarithmic expression $\log_2 256$.

134. What are the differences between common logarithms and natural logarithms?

135. How do you find the domain of a logarithmic function?

136. What strategy would you use to graph $f(x) = \log_5 x$?

137. Consider the following graphs. Which is larger, a or b, and why?

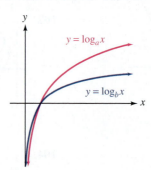

138. Consider the following graphs. Which is larger, a or b, and why?

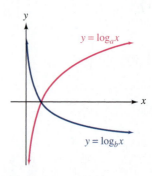

139. Choose two numbers and add their common logarithms. Then find the common logarithm of the product of those two numbers. What do you observe? Does it work for three numbers?

140. Choose two numbers and subtract their common logarithms. Then find the common logarithm of the quotient of those two numbers. What do you observe?

Critical Thinking

Determine if the statement is true or false. If the statement is false, then correct it and make it true.

141. $\log_{97,214} 1 = 0$

142. If $\log_a b = 88$, then $\log_b a = -88$.

143. $\log_{313} 0$ is undefined.

144. $\log_{10}(-1,000,000) = -6$

145. $\log_2 \dfrac{1}{1024} = 10$

146. $\log_{12} 12^{369} = 369$

147. $99^{\log_{99} 77} = 77$

148. $\log_5 \sqrt[19]{5} = 19$

149. $\log_{2468}(\ln e^{2468}) = 1$

150. The domain of $f(x) = \log_7(x^2 - 4)$ is $(0, \infty)$.

5.4 Applications of Logarithmic Functions

In this section, we will learn to

1. Use logarithms to solve electrical engineering problems.

2. Use logarithms to solve geology problems.

3. Use logarithms to solve charging battery problems.

4. Use logarithms to solve population growth problems.

5. Use logarithms to solve isothermal expansion problems.

Earthquakes are one of the deadliest natural disasters that can occur. In 2004, an earthquake in the Indian Ocean triggered a series of tsunamis killing more than 225,000 people in eleven countries. Indonesia, Sri Lanka, Thailand, and India were hit the hardest. The earthquake was the second largest ever recorded and had a magnitude between 9.1 and 9.3 on the Richter scale. In 2010, a 7.0 catastrophic earthquake hit 16 miles west of Port-au-Prince, Haiti's capital, affecting more than 3 million people.

In 2011, a massive earthquake of magnitude 9.0 hit off the coast of Japan, killing thousands and affecting Japan's nuclear reactors and triggering a nuclear crisis. A powerful earthquake rocked mountainous Nepal in 2015. It was a 7.8 magnitude earthquake that struck in Katmandu and killed thousands.

The intensity of an earthquake is based on a logarithmic function, the topic of this section. We will also use logarithmic functions to solve problems in engineering, geology, social science, and physics.

1. Use Logarithms to Solve Electrical Engineering Problems

Electronic engineers use common logarithms to measure the voltage gain of devices such as amplifiers or the length of a transmission line. The unit of gain, called the **decibel**, is defined by a logarithmic function.

Decibel Voltage Gain	If E_O is the output voltage of a device and E_I is the input voltage, the **decibel voltage gain** is given by

$$\text{dB gain} = 20 \log \frac{E_O}{E_I}$$

Finding the Decibel Gain of an Amplifier

EXAMPLE 1

Find the dB gain of an amplifier if its input is 0.5 volt and its output is 40 volts.

SOLUTION We can find the decibel voltage gain by substituting the given values into the dB gain formula.

$$\text{dB voltage gain} = 20 \log \frac{E_O}{E_I}$$

$$\text{dB voltage gain} = 20 \log \frac{40}{0.5} \qquad \text{Substitute 40 for } E_O \text{ and 0.5 for } E_I.$$

$$= 20 \log 80$$

$$\approx 38.06179974 \qquad \text{Use a calculator.}$$

To the nearest decibel, the db gain is 38 decibels.

©Zoran Milic/Shutterstock.com

Self Check 1 Find the dB gain if the input is 0.7 volt. Round to the nearest decibel.

Now Try Exercise 7.

2. Use Logarithms to Solve Geology Problems

Seismologists measure the intensity of earthquakes on the **Richter scale**, which is based on a logarithmic function.

Claudiad/iStockphoto.com

Haiti's Presidential Palace after 2010 earthquake

Richter Scale	If R is the intensity of an earthquake, A is the amplitude (measured in micrometers), and P is the period (the time of one oscillation of the Earth's surface, measured in seconds), then

$$R = \log \frac{A}{P}$$

Finding the Intensity of an Earthquake

EXAMPLE 2

Find the intensity of an earthquake with amplitude of 5000 micrometers $\left(\frac{1}{2} \text{ centimeter}\right)$ and a period of 0.07 second.

SOLUTION   We substitute 5000 for A and 0.07 for P in the Richter scale formula and simplify.

$$R = \log \frac{A}{P}$$

$$R = \log \frac{5000}{0.07}$$

$$\approx \log 71{,}428.57143 \qquad \text{Use a calculator.}$$

$$\approx 4.853871964$$

To the nearest tenth, the earthquake measures 4.9 on the Richter scale.

Self Check 2   Find the intensity of an aftershock with the same period but one-half of the amplitude. Round to the nearest tenth.

Now Try Exercise 13.

3. Use Logarithms to Solve Charging Battery Problems

A battery charges at a rate that depends on how close it is to being fully charged—it charges fastest when it is most discharged. The formula that determines the time required to charge a battery to a certain level is based on a natural logarithmic function.

Charging Batteries   If M is the theoretical maximum charge that a battery can hold and k is a positive constant that depends on the battery and the charger, the length of time t (in minutes) required to charge the battery to a given level C is given by

$$t = -\frac{1}{k} \ln\left(1 - \frac{C}{M}\right)$$

EXAMPLE 3   **Solving a Charging Battery Problem**

How long will it take to bring a fully discharged battery to 90% of full charge? Assume that $k = 0.025$ and that time is measured in minutes.

SOLUTION   90% of full charge means 90% of M. We can substitute $0.90M$ for C and 0.025 for k in the formula for charging batteries to find t.

$$t = -\frac{1}{k} \ln\left(1 - \frac{C}{M}\right)$$

$$t = -\frac{1}{0.025} \ln\left(1 - \frac{0.90M}{M}\right)$$

$$= -40 \ln(1 - 0.9)$$

$$= -40 \ln(0.1)$$

$$\approx 92.10340372 \qquad \text{Use a calculator.}$$

The battery will reach 90% charge in about 92 minutes.

©Joe Belanger/Shutterstock.Com

Self Check 3   To the nearest minute, how long will it take this battery to reach 80% of full charge?

Now Try Exercise 19.

4. Use Logarithms to Solve Population Growth Problems

If a population grows exponentially at a certain annual rate, the time required for the population to double is called the **doubling time** and is given by the following formula. You will be asked to prove this formula in Exercise 115 in Section 5.6.

©Hobbit/Shutterstock.com

| **Population Doubling Time** | If r is the annual growth rate and t is the time (in years) required for a population to double, then $$t = \frac{\ln 2}{r}$$ |

EXAMPLE 4   Finding the Doubling Time of a Population

The population of the Earth is growing at the approximate rate of 2% per year. If this rate continues, how long will it take the population to double?

SOLUTION   Because the population is growing at the rate of 2% per year, we can substitute 0.02 for r in the formula for doubling time and simplify.

$$t = \frac{\ln 2}{r}$$

$$t = \frac{\ln 2}{0.02}$$

$$\approx 34.65735903$$

It will take about 35 years for the Earth's population to double.

Self Check 4   If the world population's annual growth rate could be reduced to 1.5% per year, what would be the doubling time? Round to the nearest year.

Now Try Exercise 21.

5. Use Logarithms to Solve Isothermal Expansion Problems

When energy is added to a gas, its temperature and volume could increase. In **isothermal expansion**, the temperature remains constant—only the volume changes. The energy required is calculated as follows.

Isothermal Expansion If the temperature T is constant, the energy E required to increase the volume of 1 mole of gas from an initial volume V_i to a final volume V_f is given by

$$E = RT \ln\left(\frac{V_f}{V_i}\right)$$

E is measured in joules and T in Kelvins. R is the universal gas constant, which is 8.314 joules/mole/K.

EXAMPLE 5 Solving an Isothermal Expansion Problem

Find the amount of energy that must be supplied to triple the volume of 1 mole of gas at a constant temperature of 300 K.

SOLUTION We substitute 8.314 for R and 300 for T in the formula. Since the final volume is to be three times the initial volume, we also substitute $3V_i$ for V_f.

$$E = RT \ln\left(\frac{V_f}{V_i}\right)$$

$$E = (8.314)(300)\ln\left(\frac{3V_i}{V_i}\right)$$

$$= 2{,}494.2 \ln 3$$

$$\approx 2{,}740.15877$$

Approximately 2740 joules of energy must be added to triple the volume.

Self Check 5 To the nearest joule, what energy is required to double the volume?

Now Try Exercise 25.

Self Check Answers

1. 35 decibels **2.** 4.6 **3.** 64 min **4.** 46 years **5.** 1729 joules

Exercises 5.4

Getting Ready
You should be able to complete these vocabulary and concept statements before you proceed to the practice exercises.

Fill in the blanks.

1. dB gain = _____

2. The intensity of an earthquake is measured by the formula $R =$ _____.

3. The formula for charging batteries is

_____.

4. If a population grows exponentially at a rate r, the time it will take for the population to double is given by the formula $t =$ ____.

5. The formula for isothermal expansion is

_____.

6. The logarithm of a negative number is _____.

Applications
Use a calculator to solve each problem.

7. Gain of an amplifier An amplifier produces an output of 17 volts when the input signal is 0.03 volt. Find the decibel voltage gain. Round to the nearest decibel.

8. **Transmission lines** A 4.9-volt input to a long transmission line decreases to 4.7 volts at the other end. Find the decibel voltage loss. Round to two decimals.

9. **Gain of an amplifier** Find the dB gain of an amplifier whose input voltage is 0.71 volt and whose output voltage is 20 volts. Round to the nearest decibel.

10. **Gain of an amplifier** Find the dB gain of an amplifier whose output voltage is 2.8 volts and whose input voltage is 0.05 volt.

11. **dB gain** Find the dB gain of the amplifier shown below. Round to one decimal.

12. **dB gain** Find the dB gain of the amplifier shown below. Round to one decimal.

13. **Earthquakes** An earthquake has an amplitude of 5000 micrometers and a period of 0.2 second. Find its measure on the Richter scale. Round to one decimal.

14. **Earthquakes** An earthquake has an amplitude of 8000 micrometers and a period of 0.008 second. Find its measure on the Richter scale. Round to the nearest whole number.

15. **Earthquakes** An earthquake with a period of $\frac{1}{4}$ second has an amplitude of 2500 micrometers. Find its measure on the Richter scale. Round to the nearest whole number.

16. **Earthquakes** An earthquake has a period of $\frac{1}{2}$ second and an amplitude of 5 cm. Find its measure on the Richter scale. (*Hint:* 1 cm = 10,000 micrometers) Round to the nearest whole number.

17. **Earthquakes** An earthquake measuring between 3.5 and 5.4 on the Richter scale is often felt but rarely causes damage. Suppose an earthquake in Northern California has an amplitude of 6000 micrometers and a period of 0.3 second. Is it likely to cause damage?

18. **Earthquakes** An earthquake measuring between 7 and 7.9 on the Richter scale is a major earthquake and can cause serious damage over larger areas. Suppose an earthquake in Chile has an amplitude of 198.5 cm and a period of 0.1 second. Would it cause serious damage over large areas? (*Hint:* 1 cm = 10,000 micrometers)

19. **Battery charge** If $k = 0.116$, how long will it take a battery to reach a 90% charge? Assume that the battery was fully discharged when it began charging. Round to one decimal place.

20. **Battery charge** If $k = 0.201$, how long will it take a battery to reach a 40% charge? Assume that the battery was fully discharged when it began charging. Round to one decimal place.

21. **Population growth** A town's population grows at the rate of 12% per year. If this growth rate remains constant, how long will it take the population to double? Round to one decimal place.

22. **Fish population growth** One thousand bass were stocked in Catfish Lake in Eagle River, Wisconsin, a lake with no bass population. If the population of bass is expected to grow at a rate of 25% per year, how long will it take the population to double? Round to one decimal place.

23. **Population growth** A population growing at an annual rate r will triple in a time t given by the formula $t = \frac{\ln 3}{r}$. How long will it take the population of the town in Exercise 21 to triple? Round to one decimal place.

24. **Fish population growth** How long would it take the fish population in Exercise 22 to triple? Round to one decimal place.

25. **Isothermal expansion** One mole of gas expands isothermically to triple its volume. If the gas temperature is 400 K, what energy is absorbed? Round to the nearest joule.

26. **Isothermal expansion** One mole of gas expands isothermically to double its volume. If the gas temperature is 300K, what energy is absorbed? Round to the nearest joule.

If an investment is growing continuously for t years, its annual growth rate r is given by the following formula, where P is the current value and P₀ is the amount originally invested.

$$r = \frac{1}{t} \ln \frac{P}{P_0}$$

27. **Investing** Under Armour, Inc. grew continuously from 2005 to 2015. A $10,000 investment in the stock in 2005 would be worth $100,000 in 2015. Find Under Armour's average annual growth rate during this period. Round to the nearest percent.

28. Investing Nike, Inc. has grown continuously from 1980 to 2015. A $10,000 investment in the stock in 2005 would be worth $2,500,000 in 2015. Find Nike's average annual growth rate during this period. Round to the nearest percent.

29. Depreciation In business, equipment is often depreciated using the double declining-balance method. In this method, a piece of equipment with a life expectancy of N years, costing $\$C$, will depreciate to a value of $\$V$ in n years, where n is given by the following formula.

$$n = \frac{\log V - \log C}{\log\left(1 - \dfrac{2}{N}\right)}$$

If a computer that cost $37,000 has a life expectancy of 5 years and has depreciated to a value of $8000, how old is it?

30. Depreciation A word processor worth $470 when new had a life expectancy of 12 years. If it is now worth $189, how old is it? (See Exercise 29.) Round to the nearest year.

31. Annuities If $\$P$ is invested at the end of each year in an annuity earning interest at an annual rate r, the amount in the account will be $\$A$ after n years, where

$$n = \frac{\log\left[\dfrac{Ar}{P} + 1\right]}{\log(1 + r)}$$

If $1000 is invested each year in an annuity earning 12% annual interest, when will the account be worth $20,000? Round to one decimal place.

32. Annuities If $5000 is invested each year in an annuity earning 8% annual interest, when will the account be worth $50,000? (See Exercise 31.) Round to one decimal place.

33. Breakdown voltage The coaxial power cable shown has a central wire with radius $R_1 = 0.25$ centimeter. It is insulated from a surrounding shield with inside radius $R_2 = 2$ centimeters. The maximum voltage the cable can withstand is called the **breakdown voltage** V of the insulation. V is given by the formula

$$V = ER_1 \ln \frac{R_2}{R_1}$$

where E is the **dielectric strength** of the insulation. If $E = 400,000$ volts/centimeter, find V. Round to the nearest volt.

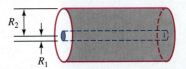

34. Breakdown voltage In Exercise 33, if the inside diameter of the shield were doubled, what voltage could the cable withstand? Round to the nearest volt.

35. Suppose you graph the function $f(x) = \ln x$ on a coordinate grid with a unit distance of 1 centimeter on the x- and y-axes. How far out must you go on the x-axis so that $f(x) = 12$? Give your result to the nearest mile.

36. Suppose you graph the function $f(x) = \log x$ on a coordinate grid with a unit distance of 1 centimeter on the x- and y-axes. How far out must you go on the x-axis so that $f(x) = 12$? Give the result to the nearest mile. Why is this result so much larger than the result in Exercise 35?

Discovery and Writing

37. Describe how to determine the intensity of an earthquake.

38. Explain how to determine the doubling time of a population.

39. One form of the logistic function is given by the following function. Explain how you would find the y-intercept of its graph.

$$f(x) = \frac{1}{1 + e^{-2x}}$$

40. Graph the function $f(x) = \ln |x|$. Explain why the graph looks the way it does.

5.5 Properties of Logarithms

In this section, we will learn to

1. Use properties of logarithms to simplify expressions.
2. Use the Change-of-Base Formula.
3. Use logarithms to solve pH problems.
4. Use logarithms to solve problems in electronics.
5. Use logarithms to solve physiology problems.

Rolling Stone magazine lists the rock band U2 at number 22 in their list of the top 100 artists of all time. The Irish band is noted for lead singer Bono's vocals and the band's anthem-like sound.

Attending a U2 rock concert or cranking up a car stereo and playing U2's song "With or Without You" is an enjoyable activity. Music is an important part of our pop culture. Since many students prefer their music loud, the loudness of sound and its intensity are interesting concepts. In this section, we will see that loudness of sound and the intensity of sound are related by a formula involving the natural logarithmic function.

1. Use Properties of Logarithms to Simplify Expressions

Since logarithms are exponents, the properties of exponents have counterparts in the theory of logarithms. We begin with four basic properties.

Properties of Logarithms   If b is a positive number and $b \neq 1$, then

    1. $\log_b 1 = 0$ **2.** $\log_b b = 1$

    3. $\log_b b^x = x$ **4.** $b^{\log_b x} = x$ $(x > 0)$

Properties 1 through 4 follow directly from the definition of logarithm.

1. $\log_b 1 = 0$, because $b^0 = 1$.
2. $\log_b b = 1$, because $b^1 = b$.
3. $\log_b b^x = x$, because $b^x = b^x$.
4. $b^{\log_b x} = x$, because $\log_b x$ is the exponent to which b is raised to get x.

Properties 3 and 4 also indicate that the composition of the exponential and logarithmic functions (in both directions) is the identity function. This is expected, because the exponential and logarithmic functions with the same base are inverse functions.

Using Properties of Logarithms to Simplify Expressions

EXAMPLE 1   Simplify each expression: **a.** $\log_3 1$ **b.** $\log_4 4$ **c.** $\log_7 7^3$ **d.** $b^{\log_b 3}$

SOLUTION   We can simplify each expression using the properties of logariathms.

a. By Property 1, $\log_3 1 = \mathbf{0}$, because $3^0 = 1$.

b. By Property 2, $\log_4 4 = \mathbf{1}$, because $4^1 = 4$.

c. By Property 3, $\log_7 7^3 = \mathbf{3}$, because $7^3 = 7^3$.

d. By Property 4, $b^{\log_b 3} = 3$, because $\log_b 3$ is the power to which b is raised to get 3.

Self Check 1   Simplify: **a.** $\log_4 1$ **b.** $\log_3 3$ **c.** $\log_2 2^4$ **d.** $5^{\log_5 2}$

Now Try Exercise 13.

The four properties also hold for natural logarithms.

1. $\ln 1 = 0$, because $e^0 = 1$.
2. $\ln e = 1$, because $e^1 = e$.
3. $\ln e^x = x$, because $e^x = e^x$.
4. $e^{\ln x} = x$, because $\ln x$ is the exponent to which e is raised to get x.

The next two properties state that

The logarithm of a product is the sum of the logarithms.

The logarithm of a quotient is the difference of the logarithms.

The Logarithm of a Product and a Difference

If M, N, and b are positive numbers and $b \neq 1$, then

5. $\log_b MN = \log_b M + \log_b N$ **6.** $\log_b \dfrac{M}{N} = \log_b M - \log_b N$

Property 5 is known as the **Product Rule**.
Property 6 is known as the **Quotient Rule**.

PROOF   To prove Property 5, we let $x = \log_b M$ and $y = \log_b N$ and use the definition of logarithm to write each equation in exponential form.

$M = b^x$ and $N = b^y$

Then $MN = b^x b^y$ and a property of exponents gives

$MN = b^{x+y}$ $b^x b^y = b^{x+y}$; keep the base and add the exponents.

We write this exponential equation in logarithmic form as

$\log_b MN = x + y$

Substituting the values of x and y completes the proof.

$\log_b MN = \log_b M + \log_b N$

The proof of Property 6 is similar. You will be asked to do it in an Exercise 115.

Caution

Avoid applying the properties incorrectly.
1. The log of a product **does not equal** the product of the logs.
2. The log of a quotient **does not equal** the quotient of the logs.
3. The log of a sum **does not equal** the sum of the logs.
4. The log of a difference **does not equal** the difference of the logs.

Properties 5 and 6 also hold for natural logarithms.

5. $\ln MN = \ln M + \ln N$ Product Rule

6. $\ln \dfrac{M}{N} = \ln M - \ln N$ Quotient Rule

Using the Product and Quotient Rules and Expanding a Logarithmic Expression

EXAMPLE 2

Assume that x, y, z, and b are positive numbers and $b \neq 1$. Write each expression in terms of the logarithms of x, y, and z: **a.** $\log_b xyz$ **b.** $\ln \frac{x}{yz}$

SOLUTION

We can use the properties of logarithms to write each logarithm as the sum or difference of several logarithms.

a. $\log_b xyz = \log_b(xy)z$

$\qquad\qquad = \log_b(xy) + \log_b z$ The log of a product is the sum of the logs.

$\qquad\qquad = \log_b x + \log_b y + \log_b z$ The log of a product is the sum of the logs.

> **Tip**
>
> Expanding a log expression means we begin with one log and use our log properties to rewrite or "expand" it. We end up with more than one log.

b. $\ln \dfrac{x}{yz} = \ln x - \ln(yz)$ The ln of a quotient is the difference of the natural logs.

$\qquad\qquad = \ln x - (\ln y + \ln z)$ The ln of a product is the sum of the natural logs.

$\qquad\qquad = \ln x - \ln y - \ln z$ Remove parentheses.

Self Check 2   Write the expression in terms of the logarithms of x, y, and z: $\log_b \dfrac{xy}{z}$

Now Try Exercise 27.

Accent on Technology

Product and Quotient Rules

A graphing calculator can be used to verify the Product and Quotient Rules.

- Using the Product Rule, we can write the expression

$$\ln(75 \cdot 20) = \ln 75 + \ln 20$$

Using a graphing calculator, we can verify that these two are equal as shown in Figure 5-29.

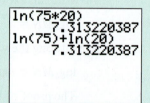

```
ln(75*20)
           7.313220387
ln(75)+ln(20)
           7.313220387
```

FIGURE 5-29

- Using the Quotient Rule, we can write the expression

$$\log \frac{240}{13} = \log 240 - \log 13.$$

Using a graphing calculator, we can verify that these two are equal as shown in Figure 5-30.

```
log(240/13)
           1.266267889
log(240)-log(13)
           1.266267889
```

FIGURE 5-30

Two more properties state that

The logarithm of a power is the power times the logarithm.

If the logarithms of two numbers are equal, the numbers are equal.

More Properties of Logarithms   If M and b are positive numbers and $b \neq 1$, then

7. $\log_b M^p = p \log_b M$ **8.** If $\log_b x = \log_b y$, then $x = y$.

Property 7 is known as the **Power Rule**.
Property 8 is known as the **One-to-One Property**.

PROOF   To prove Property 7, we let $x = \log_b M$, write the expression in exponential form, and raise both sides to the pth power:

$$M = b^x$$

$$(M)^p = (b^x)^p \qquad \text{Raise both sides to the } p\text{th power.}$$

$$M^p = b^{px} \qquad \text{Keep the base and multiply the exponents.}$$

Using the definition of logarithm gives

$$\log_b M^p = p\textcolor{red}{x}$$

Substituting the value for x completes the proof.

$$\log_b M^p = p \log_b M$$

Property 8 follows from the fact that the logarithmic function is a one-to-one function. Property 8 will be important in the next section when we solve logarithmic equations.

Properties 7 and 8 also hold for natural logarithms.

7. $\ln M^p = p \ln M$ Power Rule
8. If $\ln x = \ln y$, then $x = y$.

We can use the properties of logarithms to write a logarithm as the sum or difference of several logarithms.

EXAMPLE 3

Using the Power Rule and Expanding a Logarithmic Expression

Assume that x, y, z, and b are positive numbers and $b \neq 1$. Write each expression in terms of the logarithms of x, y, and z:

a. $\log_b(x^3 y^2 z)$ **b.** $\ln \dfrac{y^2 \sqrt{z}}{x}$

SOLUTION   We can use the properties of logarithms to write each logarithm as the sum or difference of several logarithms.

a. $\log_b(x^3 y^2 z) = \log_b x^3 + \log_b y^2 + \log_b z$ — The log of a product is the sum of the logs.

$\qquad\qquad\qquad = 3\log_b x + 2\log_b y + \log_b z$ — The log of a power is the power times the log.

> **Tip**
>
> Be sure to expand a log expression by working the problem in steps. Attempting to expand all at once can easily lead to an error.

b. $\ln \dfrac{y^2 \sqrt{z}}{x} = \ln(y^2 \sqrt{z}) - \ln x$ — The ln of a quotient is the difference of the natural logs.

$\qquad\qquad\quad = \ln y^2 + \ln z^{1/2} - \ln z$ — The ln of a product is the sum of the natural logs: $\sqrt{z} = z^{1/2}$.

$\qquad\qquad\quad = 2\ln y + \dfrac{1}{2}\ln z - \ln z$ — The ln of a power is the power times the ln.

Self Check 3 Write the expression in terms of the logarithms of x, y, and z: $\log_b \sqrt[3]{\dfrac{x^2 y}{z}}$.

Now Try Exercise 35.

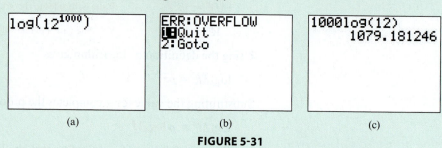

Accent on Technology

Applying the Power Rule

The development of calculators has made computations extremely easy and accurate, but hand-held calculators have a limit to the size of numbers they can display and evaluate.

- For example, if we were to try to calculate $\log 12^{1000}$, see Figure 5-31(a), we would most likely get an overflow error message as shown in Figure 5-31(b). However, by applying the Power Rule, the number can be evaluated with our calculator as shown in Figure 5-31(c).

`log(12^1000)`	`ERR:OVERFLOW`
	`1█Quit`
	`2:Goto`
		`1000log(12)`
		`        1079.181246`

(a)	(b)	(c)

FIGURE 5-31

We can use the properties of logarithms to combine several logarithms into one logarithm.

EXAMPLE 4 Combining Logarithmic Expressions

Assume that x, y, z, and b are positive numbers and $b \neq 1$. Write each expression as one logarithm:

a. $2 \log_b x + \dfrac{1}{3} \log_b y$ **b.** $\dfrac{1}{2} \log_b(x-2) - \log_b y + 3 \log_b z$

SOLUTION We can use the properties of logarithms to combine several logarithms into one logarithm.

a. $2 \log_b x + \dfrac{1}{3} \log_b y = \log_b x^2 + \log_b y^{1/3}$ A power times a log is the log of the power.

$\qquad\qquad = \log_b(x^2 y^{1/3})$ The sum of two logs is the log of the product.

$\qquad\qquad = \log_b\left(x^2 \sqrt[3]{y}\right)$

b. $\dfrac{1}{2} \log_b(x-2) - \log_b y + 3 \log_b z$

$\qquad = \log_b(x-2)^{1/2} - \log_b y + \log_b z^3$ A power times a log is the log of the power.

$\qquad = \log_b \dfrac{(x-2)^{1/2}}{y} + \log_b z^3$ The difference of two logs is the log of the quotient.

$\qquad = \log_b \dfrac{z^3 \sqrt{x-2}}{y}$ The sum of two logs is the log of the product.

> **Tip**
>
> Combining log expressions is referred to as condensing log expressions. Our final result is one logarithm. The first step in the process is to identify the coefficients of any log terms and then apply the Power Rule in reverse.

Self Check 4 Write as one logarithm: $2 \ln x + \dfrac{1}{2} \ln y - 3 \ln(x - y)$.

Now Try Exercise 45.

As we have seen in Examples 2, 3, and 4, understanding and applying logarithmic properties is important. Expanding and condensing logarithmic expressions are two important skills to master. The skills are much like playing with a Slinky™ toy. We can expand or condense as needed. We summarize the eight properties of logarithms which also apply to natural logarithms as follows.

Summary of the Properties of Logarithms

If b, M, and N are positive numbers and $b \neq 1$, then

1. $\log_b 1 = 0$
2. $\log_b b = 1$
3. $\log_b b^x = x$
4. $b^{\log_b x} = x$
5. $\log_b MN = \log_b M + \log_b N$
6. $\log_b \dfrac{M}{N} = \log_b M - \log_b N$
7. $\log_b M^p = p \log_b M$
8. If $\log_b x = \log_b y$, then $x = y$.

1. $\ln 1 = 0$
2. $\ln e = 1$
3. $\ln e^x = x$
4. $e^{\ln x} = x$
5. $\ln MN = \ln M + \ln N$
6. $\ln \dfrac{M}{N} = \ln M - \ln N$
7. $\ln M^p = p \ln M$
8. If $\ln x = \ln y$, then $x = y$.

EXAMPLE 5 **Using Properties of Logarithms to Find Approximations**

Given that $\log_{10} 2 \approx 0.3010$ and $\log_{10} 3 \approx 0.4771$, find approximations for

a. $\log_{10} 18$ **b.** $\log_{10} 2.5$

SOLUTION We will use the properties of logarithms to write each expression in terms of known logarithms. Then, we can substitute the values of the known logarithms and simplify.

a. $\log_{10} \mathbf{18} = \log_{10}(\mathbf{2 \cdot 3^2})$

$= \log_{10} 2 + \log_{10} 3^2$ The log of a product is the sum of the logs.

$= \log_{10} 2 + 2 \log_{10} 3$ The log of a power is the power times the log.

$\approx 0.3010 + 2(0.4771)$

≈ 1.2552

b. $\log_{10} \mathbf{2.5} = \log_{10}\left(\dfrac{\mathbf{5}}{\mathbf{2}}\right)$

$= \log_{10} 5 - \log_{10} 2$ The log of a quotient is the difference of the logs.

$= \log_{10} \dfrac{10}{2} - \log_{10} 2$ Write 5 as $\dfrac{10}{2}$.

$= \log_{10} 10 - \log_{10} 2 - \log_{10} 2$ The log of a quotient is the difference of the logs.

$= 1 - 2 \log_{10} 2$ $\log_{10} 10 = 1$

$\approx 1 - 2(0.3010)$

≈ 0.3980

Take Note

In Example 5, it is important to note that we can use logarithmic values that are not stated in the problem. For example, $\log 10 = 1$, $\log 100 = 2$, and $\log 1000 = 3$.

Self Check 5 Use the information given in Example 5 to find an approximation for $\log_{10} 0.75$.

Now Try Exercise 77.

2. Use the Change-of-Base Formula

We have seen how to use a calculator to find base-10 and base-e logarithms. To use a calculator to find logarithms with different bases, such as $\log_7 63$, we can divide the base-10 (or base-e) logarithm of 63 by the base-10 (or base-e) logarithm of 7.

$$\log_7 63 = \frac{\log 63}{\log 7} \qquad\qquad \log_7 63 = \frac{\ln 63}{\ln 7}$$

$$\approx 2.129150068 \qquad\qquad \approx 2.129150068$$

To check the result, we verify that $7^{2.129150063} \approx 63$. This example suggests that if we know the base-a logarithm of a number, we can find its logarithm to some other base b.

Change-of-Base Formula If a, b, and x are positive numbers and $a \neq 1$ and $b \neq 1$, then

$$\log_b x = \frac{\log_a x}{\log_a b}$$

To prove this formula, we begin with the equation $\log_b x = y$.

$$y = \log_b x$$

$$x = b^y \qquad\qquad \text{Change the equation from logarithmic to exponential form.}$$

$$\log_a x = \log_a b^y \qquad\qquad \text{Take the base-}a\text{ logarithm of both sides.}$$

$$\log_a x = y \log_a b \qquad\qquad \text{The log of a power is the power times the log.}$$

$$y = \frac{\log_a x}{\log_a b} \qquad\qquad \text{Divide both sides by } \log_a b.$$

$$\log_b x = \frac{\log_a x}{\log_a b} \qquad\qquad \text{Refer to the first equation and substitute } \log_b x \text{ for } y.$$

> **Caution**
>
> $\dfrac{\log_a x}{\log_a b}$ means that one logarithm is to be divided by the other. They are **not** to be subtracted.

If we know logarithms to base a (for example, $a = 10$), we can find the logarithm of x to a new base b by dividing the base-a logarithm of x by the base-a logarithm of b.

EXAMPLE 6 **Using the Change-of-Base Formula**

Use the Change-of-Base Formula to find $\log_3 5$.

SOLUTION We can substitute 3 for b, 10 for a, and 5 for x into the Change-of-Base Formula and simplify.

$$\log_b x = \frac{\log_a x}{\log_a b}$$

$$\log_3 5 = \frac{\log_{10} 5}{\log_{10} 3} \qquad\qquad \text{Divide the base-10 logarithm of 5 by the base-10 logarithm of 3.}$$

$$\approx 1.464973521$$

To four decimal places, $\log_3 5 = 1.4650$.

Self Check 6   Find $\log_5 3$ to four decimal places.

Now Try Exercise 89.

Accent on Technology

Using the Change-of-Base Formula to Graph a Logarithmic Function

We can use the Change-of-Base Formula to graph logarithmic functions that do not have a base of 10. Consider the function

$$f(x) = \log_4(x + 1)$$

- Since the base of the function $f(x) = \log_4(x + 1)$ is 4, to graph this function on a calculator, we use the Change-of-Base Formula and rewrite the function as

$$f(x) = \log_4(x + 1) = \frac{\log(x + 1)}{\log 4}$$

We enter the function as shown in Figure 5-32(a). We can also use the natural logarithm function if we prefer.

- The domain of this function is $(-1, \infty)$, so there will be a vertical asymptote of $x = -1$. This will help us set our WINDOW. We then GRAPH the function. See Figures 5-32(b) and (c).

- Notice that the graph appears to "stop" at $x = -1$, but that is not the case. We can have the calculator evaluate the function at a value very close to $x = -1$ to verify this. Notice in Figure 5-32(d) that the point on the graph is not shown, but the value is indicated.

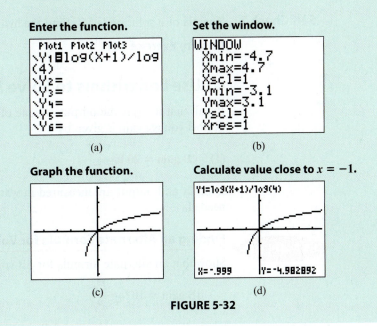

Enter the function.	Set the window.
(a)	(b)

Graph the function.	Calculate value close to $x = -1$.
(c)	(d)

FIGURE 5-32

We can use properties of logarithms to solve many problems.

3. Use Logarithms to Solve pH Problems

The more acidic a chemical solution, the greater the concentration of hydrogen ions. Chemists measure this concentration indirectly by the **pH scale**, or the **hydrogen ion index**.

pH of a Solution   If $[H^+]$ is the hydrogen ion concentration in gram-ions per liter, then

$$pH = -\log[H^+]$$

Since pure water has approximately 10^{-7} gram-ions per liter, its pH is

$$pH = -\log[\mathbf{H^+}]$$

$$pH = -\log \mathbf{10^{-7}}$$

$$= -(-7)\log 10 \qquad \text{The log of a power is the power times the log.}$$

$$= -(-7) \cdot 1 \qquad \text{Use Property 2 of logarithms: } \log_b b = 1.$$

$$= 7$$

EXAMPLE 7   **Solving an Application Involving pH**

Seawater has a pH of approximately 8.5. Find its hydrogen ion concentration.

SOLUTION   We can substitute 8.5 for pH and solve the equation $pH = -\log[H^+]$ for $[H^+]$.

$$8.5 = -\log[H^+]$$

$$-8.5 = \log[H^+]$$

$$[H^+] = 10^{-8.5} \qquad \text{Change the equation from logarithmic form to exponential form.}$$

We can then use a calculator to find that $[H^+] \approx 3.2 \times 10^{-9}$ gram-ions per liter.

Self Check 7   The pH of a solution is 5.7. Find the hydrogen ion concentration.

Now Try Exercise 101.

4. Use Logarithms to Solve Problems in Electronics

Recall that if E_O is the output voltage of a device and E_I is the input voltage, the decibel voltage gain is given by

(1)   **dB gain $= 20 \log \dfrac{E_O}{E_I}$**

If input and output are measured in watts instead of volts, a different formula is needed.

EXAMPLE 8   **Finding an Alternate Formula for Voltage Gain**

Show that an alternate formula for dB voltage gain is

$$\text{dB gain} = 10 \log \frac{P_O}{P_I}$$

where P_I is the power input and P_O is the power output.

SOLUTION   Power is directly proportional to the square of the voltage. So for some constant k,

$$P_I = k(E_I)^2 \qquad \text{and} \qquad P_O = k(E_O)^2$$

and

$$\frac{P_O}{P_I} = \frac{k(E_O)^2}{k(E_I)^2} = \left(\frac{E_O}{E_I}\right)^2$$

We raise both sides to the $\frac{1}{2}$ power to get

$$\frac{E_O}{E_I} = \left(\frac{P_O}{P_I}\right)^{1/2}$$

which we substitute into Equation (1) for dB gain.

$$\text{dB gain} = 20 \log \frac{E_O}{E_I}$$

$$= 20 \log \left(\frac{P_O}{P_I}\right)^{1/2}$$

$$= 20 \cdot \frac{1}{2} \log \frac{P_O}{P_I} \qquad \text{The log of a power is the power times the log.}$$

$$\text{dB gain} = 10 \log \frac{P_O}{P_I} \qquad \text{Simplify.}$$

Self Check 8   Find the dB gain of a device to the nearest hundredth when $P_O = 30$ watts and $P_I = 2$ watts. Round to two decimal places.

Now Try Exercise 103.

5. Use Logarithms to Solve Physiology Problems

In physiology, experiments suggest that the relationship between the loudness and the intensity of sound is a logarithmic one known as the Weber–Fechner Law.

Weber–Fechner Law   If L is the apparent loudness of a sound and I is the intensity, then

$$L = k \ln I$$

EXAMPLE 9

Using the Weber–Fechner Law

What increase in the intensity of a sound at a One Direction concert is necessary to cause a doubling of the apparent loudness?

SOLUTION   We use the formula $L = k \ln I$. To double the apparent loudness, we multiply both sides of the equation by 2 and use Property 7 of logarithms.

$$L = k \ln I$$

$$2L = 2k \ln I$$

$$= k \ln I^2$$

To double the apparent loudness, we must square the intensity.

Self Check 9   What increase is necessary to triple the apparent loudness?

Now Try Exercise 105.

Self Check Answers

1. a. 0   **b.** 1   **c.** 4   **d.** 2   **2.** $\log_b x + \log_b y - \log_b z$

3. $\frac{1}{3}(2 \log_b x + \log_b y - \log_b z)$ **4.** $\ln \dfrac{x^2\sqrt{y}}{(x-y)^3}$

5. -0.1249 **6.** 0.6826 **7.** 2×10^{-6} **8.** 11.76 decibels

9. Cube the intensity.

Exercises 5.5

Getting Ready

You should be able to complete these vocabulary and concept statements before you proceed to the practice exercises.

Fill in the blanks.

1. $\log_b 1 =$ ___

2. $\log_b b =$ ___

3. $\log_b MN = \log_b$ ___ $+ \log_b$ ___

4. $b^{\log_b x} =$ ___

5. If $\log_b x = \log_b y$, then ___ $=$ ___.

6. $\log_b \dfrac{M}{N} = \log_b M$ ___ $\log_b N$

7. $\log_b x^p = p \cdot \log_b$ ___

8. $\log_b b^x =$ ___

9. $\log_b(A + B)$ ___ $\log_b A + \log_b B$

10. $\log_b A + \log_b B$ ___ $\log_b AB$

Simplify each expression.

11. $\log_4 1 =$ ___

12. $\log_4 4 =$ ___

13. $\log_4 4^7 =$ ___

14. $4^{\log_4 8} =$ ___

15. $5^{\log_5 10} =$ ___

16. $\log_5 5^2 =$ ___

17. $\log_5 5 =$ ___

18. $\log_5 1 =$ ___

Practice

 Use a calculator to verify each equation.

19. $\log[(3.7)(2.9)] = \log 3.7 + \log 2.9$

20. $\ln \dfrac{9.3}{2.1} = \ln 9.3 - \ln 2.1$

21. $\ln(3.7)^3 = 3 \ln 3.7$

22. $\log\sqrt{14.1} = \dfrac{1}{2}\log 14.1$

23. $\log 3.2 = \dfrac{\ln 3.2}{\ln 10}$

24. $\ln 9.7 = \dfrac{\log 9.7}{\log e}$

Assume that x, y, z, and b are positive numbers and b ≠ 1. Use the properties of logarithms to write each expression in terms of the logarithms of x, y, and z.

25. $\log_b 2xy$

26. $\log_b 3xz$

27. $\log_b \dfrac{2x}{y}$

28. $\log_b \dfrac{x}{yz}$

29. $\log_b x^2 y^3$

30. $\log_b x^3 y^2 z$

31. $\log_b(xy)^{1/3}$

32. $\log_b x^{1/2} y^3$

33. $\log_b x\sqrt{z}$

34. $\log_b \sqrt{xy}$

35. $\log_b \dfrac{\sqrt[3]{x}}{\sqrt[3]{yz}}$

36. $\log_b \sqrt[4]{\dfrac{x^3 y^2}{z^4}}$

37. $\ln x^7 y^8$

38. $\ln \dfrac{4x}{y}$

39. $\ln \dfrac{x}{y^4 z}$

40. $\ln x\sqrt{y}$

Assume that x, y, z, and b are positive numbers and b ≠ 1. Use the properties of logarithms to write each expression as the logarithm of one quantity.

41. $\log_b(x + 1) - \log_b x$

42. $\log_b x + \log_b(x + 2) - \log_b 8$

43. $2\log_b x + \dfrac{1}{3}\log_b y$

44. $-2\log_b x - 3\log_b y + \log_b z$

45. $-3\log_b x - 2\log_b y + \dfrac{1}{2}\log_b z$

46. $3\log_b(x + 1) - 2\log_b(x + 2) + \log_b x$

47. $\log_b\left(\dfrac{x}{z} + x\right) - \log_b\left(\dfrac{y}{z} + y\right)$

48. $\log_b(xy + y^2) - \log_b(xz + yz) + \log_b z$

49. $\ln x + \ln(x + 5) - \ln 9$

50. $5\ln x + \dfrac{1}{5}\ln y$

51. $-6\ln x - 2\ln y + \ln z$

52. $-2\ln x - 3\ln y + \dfrac{1}{3}\ln z$

Determine whether each statement is true or false.

53. $\log_b ab = \log_b a + 1$

54. $\log_b \dfrac{1}{a} = -\log_b a$

55. $\log_b 0 = 1$

56. $\log_b 2 = \log_2 b$

57. $\log_b(x + y) \ne \log_b x + \log_b y$

58. $\log_b xy = (\log_b x)(\log_b y)$

59. If $\log_a b = c$, then $\log_b a = c$.

60. If $\log_a b = c$, then $\log_b a = \dfrac{1}{c}$.

61. $\log_7 7^7 = 7$

62. $7^{\log_7 7} = 7$

63. $\log_b(-x) = -\log_b x$

64. If $\log_b a = c$, then $\log_b a^p = pc$.

65. $\dfrac{\log_b A}{\log_b B} = \log_b A - \log_b B$

66. $\log_b(A - B) = \dfrac{\log_b A}{\log_b B}$

67. $\log_b \dfrac{1}{5} = -\log_b 5$

68. $3 \log_b \sqrt[3]{a} = \log_b a$

69. $\dfrac{1}{3} \log_b a^3 = \log_b a$

70. $\log_{4/3} y = -\log_{3/4} y$

71. $\log_b y + \log_{1/b} y = 0$

72. $\log_{10} 10^3 = 3(10^{\log_{10} 3})$

73. $\ln xy = (\ln x)(\ln y)$

74. $\dfrac{\ln A}{\ln B} = \ln A - \ln B$

75. $\dfrac{1}{5} \ln a^5 = \ln a$

76. $\ln y = -\ln \dfrac{1}{y}$

Given that $\log_{10} 4 \approx 0.6021$, $\log_{10} 7 \approx 0.8451$, and $\log_{10} 9 \approx 0.9542$, use these values and the properties of logarithms to approximate each value. Do not use a calculator.

77. $\log_{10} 28$

78. $\log_{10} \dfrac{7}{4}$

79. $\log_{10} 2.25$

80. $\log_{10} 36$

81. $\log_{10} \dfrac{63}{4}$

82. $\log_{10} \dfrac{4}{63}$

83. $\log_{10} 252$

84. $\log_{10} 49$

85. $\log_{10} 112$

86. $\log_{10} 324$

87. $\log_{10} \dfrac{144}{49}$

88. $\log_{10} \dfrac{324}{63}$

Use a calculator and the Change-of-Base Formula to find each logarithm. Round to four decimal places.

89. $\log_3 7$

90. $\log_7 3$

91. $\log_\pi 3$

92. $\log_3 \pi$

93. $\log_3 8$

94. $\log_5 10$

95. $\log_{\sqrt{2}} \sqrt{5}$

96. $\log_\pi e$

Applications

97. **pH of water slide** The water in the Abyss, a water slide at the Atlantis Resort in the Bahamas, has a hydrogen ion concentration of 6.3×10^{-8} gram-ions per liter. Find the pH. Round to two decimal places.

98. **pH of swimming pool** The ideal pH for a swimming pool is 7.2, the same pH as our eyes. The swimming pool at the local YMCA has a hydrogen ion concentration of 1.6×10^{-7} gram-ions per liter. Find the pH of the pool. Round to two decimal places. Is this ideal?

99. **pH of a solution** Find the pH of a solution with a hydrogen ion concentration of 1.7×10^{-5} gram-ions per liter. Round to two decimal places.

100. **pH of calcium hydroxide** Find the hydrogen ion concentration of a saturated solution of calcium hydroxide whose pH is 13.2.

101. **pH of apples** The pH of apples can range from 2.9 to 3.3. Find the range in the hydrogen ion concentration.

©Samuel Borges Photography/Shutterstock.com

102. **pH of sour pickles** The hydrogen ion concentration of sour pickles is 6.31×10^{-4}. Find the pH. Round to one decimal place.

103. **dB gain** An amplifier produces a 40-watt output with a $\frac{1}{2}$-watt input. Find the dB gain. Round to the nearest decibel.

104. dB loss Losses in a long telephone line reduce a 12-watt input signal to an output of 3 watts. Find the dB gain. (Because it is a loss, the "gain" will be negative.) Round to the nearest decibel.

105. Weber–Fechner Law What increase in intensity is necessary to quadruple the loudness?

106. Weber–Fechner Law What decrease in intensity is necessary to make a sound half as loud?

107. Isothermal expansion If a certain amount E of energy is added to one mole of a gas, it expands from an initial volume of 1 liter to a final volume V without changing its temperature according to the formula

$$E = 8300 \ln V$$

Find the volume if twice that energy is added to the gas.

108. Richter scale By what factor must the amplitude of an earthquake change to increase its severity by 1 point on the Richter scale? Assume that the period remains constant. The Richter scale is given by

$$R = \log \frac{A}{P}$$

where A is the amplitude and P the period of the tremor.

Discovery and Writing

109. Explain the Product Rule of logarithms and give an example.

110. Explain the Quotient Rule of logarithms and give an example.

111. Explain the Power Rule of logarithms and give an example.

112. Explain the Change-of-Base Formula and give an example.

113. Simplify: $3^{4\log_3 2} + 5^{\frac{1}{2}\log_5 25}$

114. Find the value of $a - b$:

$$5 \log x + \frac{1}{3} \log y - \frac{1}{2} \log x - \frac{5}{6} \log y = \log(x^a y^b)$$

115. Prove Property 6 of logarithms:

$$\log_b \frac{M}{N} = \log_b M - \log_b N$$

116. Show that $-\log_b x = \log_{1/b} x$.

117. Show that $e^{x \ln a} = a^x$.

118. Show that $e^{\ln x} = x$.

119. Show that $\ln(e^x) = x$.

120. If $\log_b 3x = 1 + \log_b x$, find b.

121. Explain why $\ln(\log 0.9)$ is undefined.

122. Explain why $\log_b(\ln 1)$ is undefined.

In Exercises 123–124, A and B both are negative. Thus, AB and $\frac{A}{B}$ are positive, and log AB and log $\frac{A}{B}$ are defined.

123. Is it still true that $\log AB = \log A + \log B$? Explain.

124. Is it still true that $\log \frac{A}{B} = \log A - \log B$? Explain.

Critical Thinking

In Exercises 125–132, match the logarithmic expression on the left with an equivalent logarithmic expression on the right. Assume x, y, z, and w are positive numbers.

125. $\log_{500} \dfrac{w^{100} x^{200} z^{300}}{y^{400}}$

a. $100 \log_{500} w + 200 \log_{500} x - 300 \log_{500} z - 400 \log_{500} y$

126. $\log_{500} \dfrac{w^{100} x^{200}}{y^{400} z^{300}}$

b. $100 \log_{500} w + 200 \log_{500} x + 300 \log_{500} z - 400 \log_{500} y$

127. $\log_{200} x \sqrt[100]{y}$

c. $\dfrac{1}{100} \log_{200} x + \dfrac{1}{100} \log_{200} y$

128. $\log_{200} \sqrt[100]{xy}$

d. $\log_{200} x + \dfrac{1}{100} \log_{200} y$

129. $\log_{300} x^{100} y^{200}$

e. $200 \log_{300} x + 100 \log_{300} y$

130. $\log_{300} x^{200} y^{100}$

f. $100 \log_{300} x + 200 \log_{300} y$

131. $-\log_{300} x^{100} y^{200}$

g. $\log_{300} x^{100} + \log_{300} y^{200}$

132. $-\log_{300} \dfrac{1}{x^{100} y^{200}}$

h. $\log_{300} \dfrac{1}{x^{100}} + \log_{300} \dfrac{1}{y^{200}}$

5.6 Exponential and Logarithmic Equations

In this section, we will learn to

1. Use like bases to solve exponential equations.
2. Use logarithms to solve exponential equations.
3. Solve logarithmic equations.
4. Solve carbon-14 dating problems.
5. Solve population growth problems.

Coffee is one of the most popular beverages worldwide. Suppose we visit a coffee shop and order a white chocolate mocha. After blending smooth white chocolate with rich espresso and steamed milk and topping it with whipped cream, the coffee is served to us at a temperature of 180°F.

Suppose the exponential function $T(t) = 70 + 110e^{-0.2t}$ models the temperature T of the mocha after t minutes. If we are interested in determining how long it will take for the temperature of the coffee to reach 80°F, we would substitute 80 for $T(t)$ and solve the resulting equation $80 = 70 + 110e^{-0.2t}$ for t. This equation is called an *exponential equation* because the variable t occurs as an exponent.

In this section, we will learn to solve exponential equations. When we complete Exercise 105, we will see that it takes the mocha approximately 12 minutes to reach a temperature of 80°F. We will also learn to solve *logarithmic equations* in this section.

An **exponential equation** is an equation with a variable in one of its exponents. Some examples of exponential equations are

$$3^x = 5 \qquad e^{2x} = 7 \qquad 6^{x-3} = 2^x \qquad 3^{2x+1} - 10(3^x) + 3 = 0$$

A **logarithmic equation** is an equation with logarithmic expressions that contain a variable. Some examples of logarithmic equations are

$$\log 2x = 25 \qquad \ln x - \ln(x - 12) = 24 \qquad \log x = \log \frac{1}{x} + 4$$

1. Use Like Bases to Solve Exponential Equations

Some exponential equations can be solved by using like bases.

One-to-One Property of Exponents	If $b^x = b^y$ and $b \neq 1$, $b \neq 0$, then $x = y$. That is, equal quantities with like bases have equal exponents.

EXAMPLE 1

Using Like Bases to Solve an Exponential Equation

Solve: $4^{x+3} = 8^{2x}$.

SOLUTION

We will use like bases to solve the exponential equation.

$$4^{x+3} = 8^{2x}$$
$$\left(2^2\right)^{x+3} = \left(2^3\right)^{2x} \qquad \text{Write 4 as } 2^2 \text{ and 8 as } 2^3.$$
$$2^{2(x+3)} = 2^{6x} \qquad \text{Multiply exponents.}$$

$$2(x + 3) = 6x \qquad \text{Equal quantities with like bases have equal exponents.}$$

$$2x + 6 = 6x \qquad \text{Use the Distributive Property.}$$

$$-4x = -6 \qquad \text{Subtract 6 and } 6x \text{ from both sides.}$$

$$x = \frac{3}{2} \qquad \text{Divide both sides by } -4 \text{ and simplify.}$$

Self Check 1   Solve: $3^{3x-5} = 81$.

Now Try Exercise 13.

EXAMPLE 2   **Using Like Bases to Solve an Exponential Equation**

Solve: $2^{x^2+2x} = \dfrac{1}{2}$.

SOLUTION   Since $\frac{1}{2} = 2^{-1}$, we can write the equation in the form $2^{x^2+2x} = 2^{-1}$ and use like bases to solve the equation.

Because equal quantities with like bases have equal exponents, we have

$$x^2 + 2x = -1$$

$$x^2 + 2x + 1 = 0 \qquad \text{Add 1 to both sides.}$$

$$(x + 1)(x + 1) = 0 \qquad \text{Factor the trinomial.}$$

$$x + 1 = 0 \quad \text{or} \quad x + 1 = 0 \qquad \text{Set each factor equal to 0.}$$

$$x = -1 \qquad \qquad x = -1$$

Verify that -1 satisfies the equation.

Self Check 2   Solve: $3^{x^2+2x} = 27$.

Now Try Exercise 19.

Accent on Technology

Verifying Solutions of an Exponential Equation

We can verify that -1 satisfies the equation shown in Example 2 by using the TABLE feature on a graphing calculator.

- Enter each side of the equation as shown in Figure 5-33.
- Go to TABLE and enter $x = -1$. Notice in Figure 5-33 that the y-values are exactly the same. This confirms that $x = -1$ is a solution.

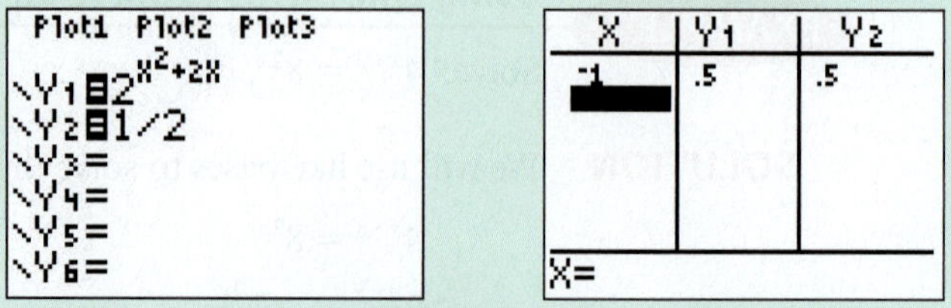

Plot1 Plot2 Plot3
\Y₁▪2^x²+2x
\Y₂▪1/2
\Y₃=
\Y₄=
\Y₅=
\Y₆=

X	Y₁	Y₂
-1	.5	.5

X=

FIGURE 5-33

EXAMPLE 3 **Using Like Bases to Solve an Exponential Equation**

Solve: $e^{6x^2} = e^{-x+1}$.

SOLUTION   We can use like bases to solve the exponential equation.

Because equal quantities with like bases have equal exponents, we have

$$6x^2 = -x + 1$$

$$6x^2 + x - 1 = 0 \qquad \text{Add } x - 1 \text{ to both sides.}$$

$$(3x - 1)(2x + 1) = 0 \qquad \text{Factor the trinomial.}$$

$$3x - 1 = 0 \quad \text{or} \quad 2x + 1 = 0 \qquad \text{Set each factor equal to 0.}$$

$$x = \frac{1}{3} \qquad\qquad x = -\frac{1}{2} \qquad \text{Solve each equation.}$$

Verify that $\frac{1}{3}$ and $-\frac{1}{2}$ satisfy the equation.

Self Check 3   Solve: $e^{4x^2} = e^{24}$.

Now Try Exercise 23.

2. Use Logarithms to Solve Exponential Equations

If the bases of the terms in an exponential equation are not equal, as is often the case, we use logarithms to solve the equation.

Since logarithmic functions are one-to-one, we can take the logarithm of both sides of an equation. This allows us to use the properties of logarithms to help solve the equation. A logarithm of any base may be used, but it is most efficient to use natural logarithms or common logarithms because a calculator is often needed to get an approximate answer. A strategy for solving exponential equations with different bases is given next.

Strategy for Solving Exponential Equations with Different Bases

To solve an exponential equation we can follow these steps:

Step 1: Isolate the exponential expression.

Step 2: Take the same logarithm of both sides.

Step 3: Simplify using the rules of logarithms.

Step 4: Solve for the variable.

When using the strategy given, please keep the following things in mind:

- If the equation contains base e, it is most efficient to use natural logarithms because

    $$\ln e^x = x$$

- If the equation contains base 10, it is most efficient to use common logarithms because

    $$\log 10^x = x$$

- If the equation contains exponential terms with bases other than base e or 10, use either natural logarithms or common logarithms.

EXAMPLE 4

Using Logarithms to Solve an Exponential Equation with Different Bases

Solve the exponential equation: $3^x = 5$.

SOLUTION

We will use the strategy outlined above to solve the exponential equation. First, we note that the exponential expression 3^x is isolated. Since logarithms of equal numbers are equal, we can take the common logarithm of each side of the equation. We can then use the Power Rule and move the variable x from its position as an exponent to a position as a coefficient and solve the equation.

$$3^x = 5$$

> **Tip**
>
> The Power Rule of logarithms provides a way of moving the variable x from its position as an exponent to a factor or $x \log 3$. We can then solve for x.

$$\log 3^x = \log 5 \qquad \text{Take the common logarithm of each side.}$$

$$x \log 3 = \log 5 \qquad \text{The log of a power is the power times the log.}$$

$$(1) \qquad x = \frac{\log 5}{\log 3} \qquad \text{Divide both sides by } \log 3.$$

$$\approx 1.464973521 \qquad \text{Use a calculator.}$$

To four decimal places, $x = 1.4650$.

Self Check 4   Solve $5^x = 3$ to four decimal places.

Now Try Exercise 25.

> **Caution**
>
> A careless reading of Equation 4 leads to a common error. The right side of the equation calls for a division, **not** a subtraction.
>
> $$\frac{\log 5}{\log 3} \quad \text{means} \quad (\log 5) \div (\log 3)$$
>
> It is the expression $\log \frac{5}{3}$ that means $\log 5 - \log 3$.

EXAMPLE 5

Using Logarithms to Solve an Exponential Equation with Different Bases

Solve the exponential equation: $6^{x-3} = 2^x$.

SOLUTION

We will use logarithms to solve the exponential equation and apply the strategy stated earlier.

$$6^{x-3} = 2^x$$

$$\log 6^{x-3} = \log 2^x \qquad \text{Take the common logarithm of each side.}$$

$$(x - 3)\log 6 = x \log 2 \qquad \text{The log of a power is the power times the log.}$$

$$x \log 6 - 3 \log 6 = x \log 2 \qquad \text{Use the Distributive Property.}$$

$$x \log 6 - x \log 2 = 3 \log 6 \qquad \text{Add } 3 \log 6 \text{ and subtract } x \log 2 \text{ from both sides.}$$

$$x(\log 6 - \log 2) = 3 \log 6 \qquad \text{Factor out } x \text{ on the left-hand side.}$$

$$x = \frac{3 \log 6}{\log 6 - \log 2} \qquad \text{Divide both sides by } \log 6 - \log 2.$$

$$x \approx 4.892789261 \qquad \text{Use a calculator.}$$

To four decimal places, $x = 4.8928$.

Self Check 5   Solve: $5^{x+3} = 3^x$.

Now Try Exercise 29.

In the next example, we will take the natural logarithm of both sides of an exponential equation. However, before we do, it is a good idea to review the following properties of natural logarithms.

Natural Logarithm Properties

$\ln 1 = 0 \qquad \ln e = 1 \qquad \ln e^x = x$

Product Rule: $\ln MN = \ln M + \ln N$

Quotient Rule: $\ln \dfrac{M}{N} = \ln M - \ln N$

Power Rule: $\ln M^p = p \ln M$

Using Natural Logarithms to Solve Exponential Equations

EXAMPLE 6

Use natural logarithms to solve: **a.** $e^x - 7 = 0$ **b.** $4^{x+3} = 8^{2x}$

SOLUTION   For each problem, we will use the strategy stated earlier. We will isolate the exponential expression, take the natural logarithm of both sides, simplify using the natural logarithms properties, and solve for x.

a. $e^x - 7 = 0$

$e^x - 7 + 7 = 0 + 7$	Isolate the exponential expression. Add 7 to both sides.
$e^x = 7$	
$\ln e^x = \ln 7$	Take the natural logarithm of both sides.
$x = \ln 7$	Substitute x for $\ln e^x$: $\ln e^x = x$.
$x \approx 1.95$	Use a calculator and round to two decimal places.

b. $4^{x+3} = 8^{2x}$

$\ln 4^{x+3} = \ln 8^{2x}$	Take the natural logarithm of both sides.
$(x + 3)\ln 4 = (2x)\ln 8$	The log of a power is the power times the log.
$x \ln 4 + 3 \ln 4 = 2x \ln 8$	Use the Distributive Property on the left side.
$x \ln 4 - 2x \ln 8 = -3 \ln 4$	Subtract $2x \ln 8$ and $3 \ln 4$ from both sides.
$x(\ln 4 - 2 \ln 8) = -3 \ln 4$	Factor out x on the left-hand side.
$x = \dfrac{-3 \ln 4}{\ln 4 - 2 \ln 8}$	Divide both sides by $\ln 4 - 2 \ln 8$.
$x = 1.5$	Use a calculator.

Self Check 6   Use natural logarithms to solve: **a.** $3e^x = 12$ **b.** $8^{x+1} = 4^{2x}$

Now Try Exercise 39.

Accent on Technology

Finding Approximate Solutions to an Exponential Equation

If only a decimal approximation of a solution is needed, a graphing calculator can be used to solve exponential equations. Our strategy is to graph each side of the equation and then find the point of intersection.

We will use a graphing calculator to solve the exponential equation $\sqrt{3}^x = 2^{\pi x - 3}$.

- Enter each side of the equation in the graph menu and find the point of intersection. We see in Figure 5-34 that the solution is $x \approx 1.2770786$.

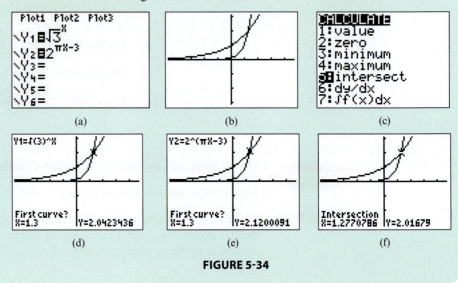

(a)   (b)   (c)

(d)   (e)   (f)

FIGURE 5-34

3. Solve Logarithmic Equations

We can use the One-to-One Property of logarithms to change some logarithmic equations into algebraic equations that we can solve. Here's a reminder of that property.

One-to-One Property of Logarithms	If $\log_b x = \log_b y$, then $x = y$. That is, logarithms of equal numbers are equal.

Using the One-to-One Property of Logarithms to Solve a Logarithmic Equation

EXAMPLE 7

Solve: $\log_5(3x + 2) = \log_5(2x - 3)$.

SOLUTION

We will use the property of logarithms stated above to solve the equation. Then we will check our solutions.

$$\log_5(3x + 2) = \log_5(2x - 3)$$

$$3x + 2 = 2x - 3 \qquad \text{If the logs of two numbers are equal, the numbers are equal.}$$

$$x = -5 \qquad \text{Subtract } 2x \text{ and 2 from both sides.}$$

Check: $\log_5(3x + 2) = \log_5(2x - 3)$

$$\log_5[3(-5) + 2] \stackrel{?}{=} \log_5[2(-5) - 3]$$

$$\log_5(-13) \stackrel{?}{=} \log_5(-13)$$

Since the logarithm of a negative number does not exist, -5 is extraneous and must be discarded. The equation has no roots.

Self Check 7   Solve: $\log_3(5x + 2) = \log_3(6x + 1)$.

Now Try Exercise 57.

Take Note

It is incorrect to state that the logarithms in Example 7 are canceled. The logarithms aren't canceled; the One-to-One Property is used.

Caution

Example 7 illustrates that we **must** check the solutions of a logarithmic equation. If we fail to do so, we risk having incorrect answers.

EXAMPLE 8

Solving a Logarithmic Equation Using the Properties of Logarithms

Solve: $\log x + \log(x - 3) = 1$.

SOLUTION   We will first combine the two logarithms as a single logarithm. Next, we will use the definition of logarithm to write the equation in exponential form. Finally, we will solve the equation and check the solutions.

Tip

Use the Product Rule to "condense" the left side of the equation. We desire to write an equation in which x appears in a single logarithmic expression.

$$\log x + \log(x - 3) = 1$$

$$\log x(x - 3) = 1 \qquad \text{The sum of two logs is the log of a product.}$$

$$x(x - 3) = 10^1 \qquad \text{Use the definition of logarithms to change the equation to exponential form.}$$

$$x^2 - 3x - 10 = 0 \qquad \text{Remove parentheses and subtract 10 from both sides.}$$

$$(x + 2)(x - 5) = 0 \qquad \text{Factor the trinomial.}$$

$$x + 2 = 0 \quad \text{or} \quad x - 5 = 0$$

$$x = -2 \quad | \quad x = 5$$

Caution

Don't automatically discard proposed solutions that are negative. Only discard proposed solutions that produce undefined logarithms in the original equation.

Check: The number -2 is not a solution, because it does not satisfy the equation (a negative number does not have a logarithm). We check the remaining number, 5.

$$\log x + \log(x - 3) = 1$$

$$\log 5 + \log(5 - 3) \stackrel{?}{=} 1 \qquad \text{Substitute 5 for } x.$$

$$\log 5 + \log 2 \stackrel{?}{=} 1$$

$$\log 10 \stackrel{?}{=} 1 \qquad \text{The sum of two logs is the log of a product.}$$

$$1 = 1 \qquad \log_b b = 1$$

Since 5 does check, it is a solution.

Self Check 8   Solve: $\log x + \log(x - 15) = 2$.

Now Try Exercise 59.

EXAMPLE 9

Solving Logarithmic Equations

Solve: $\ln(2x + 5) - \ln(x + 5) = \ln\dfrac{5}{3}$.

SOLUTION First, we combine the two logarithms on the left. Then we will use the One-to-One Property and solve for x.

$$\ln(2x + 5) - \ln(x + 5) = \ln\frac{5}{3}$$

$$\ln\frac{2x + 5}{x + 5} = \ln\frac{5}{3} \qquad \text{Use the Quotient Rule and combine logs.}$$

$$\frac{2x + 5}{x + 5} = \frac{5}{3} \qquad \text{Use the One-to-One Property of logs.}$$

$$3(2x + 5) = 5(x + 5) \qquad \text{Multiply both sides by } 3(x + 5).$$

$$6x + 15 = 5x + 25$$

$$x = 10$$

Check: The solution $x = 10$ checks as we see here.

$$\ln(2x + 5) - \ln(x + 5) \stackrel{?}{=} \ln\frac{5}{3}$$

$$\ln[2(\mathbf{10}) + 5] - \ln[(\mathbf{10}) + 5] \stackrel{?}{=} \ln\frac{5}{3} \qquad \text{Substitute 10 in for } x.$$

$$\ln 25 - \ln 15 \stackrel{?}{=} \ln\frac{5}{3}$$

$$\ln\frac{25}{15} \stackrel{?}{=} \ln\frac{5}{3} \qquad \text{Use the Quotient Rule.}$$

$$\ln\frac{5}{3} = \ln\frac{5}{3} \qquad \text{This is a true statement.}$$

Self Check 9 Solve: $\log_7(2x + 2) - \log_7(x + 6) = \log_7\dfrac{4}{3}$.

Now Try Exercise 69.

EXAMPLE 10

Solving Logarithmic Equations

Solve **a.** $\ln x = 5$ **b.** $\dfrac{\ln(5x - 6)}{\ln x} = 2$

SOLUTION For part (a), we will write the natural logarithm as $\log_e x = 5$ and use the definition of logarithm to solve the equation. For part (b), we will multiply both sides by $\ln x$ and then use properties of logarithms to write an algebraic equation. We will then solve the equation. We must verify the solutions to both parts.

a. $\ln x = 5$

$$\log_e x = 5 \qquad \text{Rewrite using the definition of natural logarithm.}$$

$$x = e^5 \qquad \text{Write in exponential form.}$$

Verify that e^5 satisfies the solution.

Tip

We can also solve Example 10a by introducing e on both sides of the equation. This conveniently produces $e^{\ln x}$, which equals x, on the left side of the equation.

$$\ln x = 5$$

$$e^{\ln x} = e^5$$

$$x = e^5$$

b. We multiply both sides of $\dfrac{\ln(5x - 6)}{\ln x} = 2$ by $\ln x$ to get

$$\ln(5x - 6) = 2 \ln x$$

and apply the Power Rule of logarithms to get

$$\ln(5x - 6) = \ln x^2$$

By the property of logarithms stated above $5x - 6 = x^2$, because they have equal logarithms. So

$$x^2 = 5x - 6$$

$$x^2 - 5x + 6 = 0$$

$$(x - 3)(x - 2) = 0$$

$$x - 3 = 0 \quad \text{or} \quad x - 2 = 0$$

$$x = 3 \quad \Big| \quad x = 2$$

Verify that both 2 and 3 satisfy the equation.

Self Check 10   Solve: **a.** $6 \ln x = 12$ **b.** $\dfrac{\ln(8x - 15)}{\ln x} = 2$

Now Try Exercise 79.

The strategy we used to solve a logarithmic equation was dependent upon the nature of the equation. A summary of these strategies appears here.

Summary of Strategies Used to Solve Logarithmic Equations

1. For logarithmic equations containing one logarithm, we can use the definition of the logarithm. We write the logarithm in exponential form and then solve for the variable.

   Note: This strategy can be used to solve the following logarithmic equations.

   $$\log_3(x + 5) = 2, \log(x^2 - 4) = 5, \ln(4x) = 7$$

2. For more complicated logarithmic equations, we can combine and isolate the logarithmic expressions. We can then use the definition of the logarithm and write the logarithm in exponential form to solve for the variable.

   Note: This strategy can be used to solve the following logarithmic equations.

   $$\log(x) + \log(x - 1) = 3, \log_5(3x + 2) - \log_5(2x - 3) = 2,$$
   $$\ln x - \ln(x + 3) = 4$$

3. We can use the One-to-One Property of logarithms for logarithmic equations of the form $\log_b x = \log_b y$.

   Note: This strategy can be used to solve the following logarithmic equations.

   $$\log(5x - 6) = \log x^2, \ln(2x - 7) = \ln(3x + 5)$$

4. For other types of logarithmic equations, we sometimes combine logarithms on one or both sides and then use the One-to-One Property of logarithms to solve the equations.

   Note: This strategy can be used to solve the following logarithmic equations.

   $$\ln(x^2) + \ln(x - 3) = \ln(x + 4), \log_4 x - \log_4(x + 2) = \log_4 5 + \log_4(3x)$$

Finding Approximate Solutions to a Logarithmic Equation

We can use a graphing calculator to approximate the solutions of a logarithmic equation.

Consider the log equation $\ln x^2 + \ln(x - 3) = \ln(x + 4)$.

- Enter each side of the equation in the graph menu, and then find the point of intersection. We see that in Figure 5-35 that $x \approx 3.5891327$. Use **ZOOM** and select ZDecimal to graph.

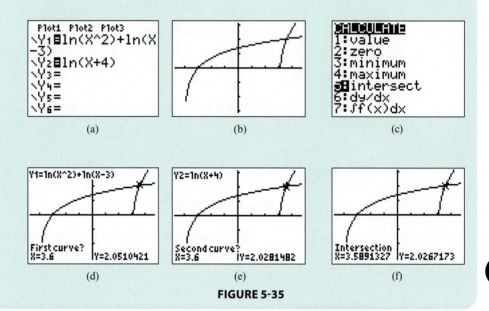

(a) (b) (c)

(d) (e) (f)

FIGURE 5-35

4. Solve Carbon-14 Dating Problems

When a living organism dies, the oxygen/carbon dioxide cycle common to all living things ceases; then carbon-14, a radioactive isotope with a half-life of 5700 years, is no longer absorbed. By measuring the amount of carbon-14 present in ancient objects, archaeologists can estimate the object's age.

The Shroud of Turin, the piece of linen long believed to have been wrapped around Jesus's body after the crucifixion, is much older than the date suggested by radiocarbon tests, according to new microchemical research. Published in the current issue of *Thermochimica Acta*, a chemistry peer-reviewed scientific journal, the study dismisses the results of the 1988 carbon-14 dating. At that time, three reputable laboratories in Oxford, Zurich, and Tucson, Ariz., concluded that the cloth on which the smudged outline of the body of a man is indelibly impressed was a medieval fake dating from 1260 to 1390 and not the burial cloth wrapped around the body of Christ.

The amount A of radioactive material present at time t is given by the model

$$A = A_0 2^{-t/h}$$

where A_0 is the amount present initially and h is the half-life of the material.

Solving a Carbon-14 Dating Problem

EXAMPLE 11

An archeologist finds a wooden statue in the tomb of an ancient Egyptian ruler. If the statue contains two-thirds of its original carbon-14 content, how old is it?

SOLUTION

To find the time t when $A = \frac{2}{3}A_0$, we substitute $\frac{2A_0}{3}$ for A and 5700 for h in the radioactive decay formula and solve for t:

$$A = A_0 2^{-t/h}$$

$$\frac{2A_0}{3} = A_0 2^{-t/5700}$$

$$1 = \frac{3}{2}\left(2^{-t/5700}\right) \qquad \text{Divide both sides by } A_0 \text{ and multiply both sides by } \frac{3}{2}.$$

$$\log 1 = \log \frac{3}{2}\left(2^{-t/5700}\right) \qquad \text{Take the common logarithm of each side.}$$

$$0 = \log \frac{3}{2} + \log 2^{-t/5700} \qquad \text{The log of a product is the sum of the logs.}$$

$$-\log \frac{3}{2} = -\frac{t}{5700}\log 2 \qquad \text{Subtract } \log \frac{3}{2} \text{ from both sides and use the Power Rule of logarithms.}$$

$$5700\left(\frac{\log \frac{3}{2}}{\log 2}\right) = t \qquad \text{Multiply both sides by } -\frac{5700}{\log 2}.$$

$$t \approx 3{,}334.286254 \qquad \text{Use a calculator.}$$

The wooden statue is approximately 3300 years old.

©Milosk50/Shutterstock.com

Self Check 11

How old is an artifact that has 60% of its original carbon-14 content? Round to the nearest hundred years.

Now Try Exercise 95.

5. Solve Population Growth Problems

When there is sufficient food and space, populations of living organisms tend to increase exponentially according to the Malthusian growth model

$$P = P_0 e^{kt}$$

where P_0 is the initial population at $t = 0$ and k depends on the rate of growth.

Solving a Population Growth Problem

EXAMPLE 12

Streptococcus bacteria taken from the lungs of a patient is placed in a laboratory culture increased from an initial population of 500 to 1500 in 3 hours. Find the time it will take for the population to reach 10,000.

SOLUTION

We will apply the following two steps to solve the problem:

Step 1: Substitute 1500 for P, 500 for P_0, and 3 for t into the Malthusian growth model and find k.

Step 2: Substitute 10,000 for P, 500 for P_0, and the value of k found in Step 1 into the model and use logarithms to solve for t.

1. Substitute 1500 for P, 500 for P_0, and 3 for t into the Malthusian growth model and find k:

$$P = P_0 e^{kt}$$

$$1500 = 500\left(e^{k3}\right) \qquad \text{Substitute 1500 for } P, 500 \text{ for } P_0, \text{ and 3 for } t.$$

$$3 = e^{3k} \qquad \text{Divide both sides by 500.}$$

$$3k = \ln 3 \qquad \text{Take the natural log of both sides.}$$

$$k = \frac{\ln 3}{3} \qquad \text{Divide both sides by 3.}$$

2. To find out when the population will reach 10,000, we substitute 10,000 for P, 500 for P_0, and $\frac{\ln 3}{3}$ for k in the equation $P = P_0 e^{kt}$, and solve for t:

$$P = P_0 e^{kt}$$

$$10{,}000 = 500 e^{\left(\frac{\ln 3}{3}\right)t}$$

$$20 = e^{\left(\frac{\ln 3}{3}\right)t} \qquad \text{Divide both sides by 500.}$$

$$\ln 20 = \ln\left[e^{\left(\frac{\ln 3}{3}\right)t}\right] \qquad \text{Take the natural log of both sides.}$$

$$\ln 20 = \frac{\ln 3}{3} t \qquad \text{Simplify the right side using the natural log property } \ln(e^x) = x.$$

$$t = \frac{3 \ln 20}{\ln 3} \qquad \text{Multiply both sides by } \frac{3}{\ln 3}.$$

$$\approx 8.180499084 \qquad \text{Use a calculator.}$$

The culture will reach 10,000 bacteria in a little more than 8 hours.

Self Check 12 If the population increases from 1000 to 3000 in 3 hours, how long will it take to reach 20,000? Round to the nearest hour.

Now Try Exercise 103.

Self Check Answers

1. 3 **2.** 1, −3 **3.** $\pm\sqrt{6}$ **4.** $x = 0.6826$ **5.** −9.4520
6. a. $\ln 4$ **b.** 3 **7.** 1 **8.** 20; −5 is extraneous **9.** 9
10. a. e^2 **b.** 3, 5 **11.** 4200 years **12.** 8 hr

Exercises 5.6

Getting Ready

You should be able to complete these vocabulary and concept statements before you proceed to the practice exercises.

Fill in the blanks.

1. An equation with a variable in its exponent is called a(n) _____ equation.
2. An equation with a logarithmic expression that contains a variable is a(n) _____ equation.
3. The formula for carbon dating is $A =$ _____.
4. The formula for population growth is $P =$ _____.

Practice

Solve each exponential equation using like bases.

5. $2^{3x+2} = 16^x$
6. $32^{x+2} = 2^{7x+12}$

7. $27^{x+1} = 3^{2x+1}$
8. $3^{x-1} = 9^{2x}$

9. $5^{4x+1} = 25^{-x-2}$
10. $5^{2x+1} = 125^x$

11. $4^{x-2} = 8^x$
12. $16^{x+1} = 8^{2x+1}$

13. $81^{2x} = 27^{2x-5}$
14. $625^{x-9} = 125^{x-12}$

15. $2^{x^2-2x} = 8$
16. $5^{x^2+3x} = 625$

17. $36^{x^2} = 216^{x^2-3}$
18. $25^{x^2-5x} = 3125^{4x}$

19. $7^{x^2+3x} = \dfrac{1}{49}$
20. $3^{x^2+4x} = \dfrac{1}{81}$

21. $e^{-x+6} = e^x$
22. $e^{2x+1} = e^{3x-11}$

23. $e^{x^2-1} = e^{24}$
24. $e^{x^2+7x} = \dfrac{1}{e^{12}}$

Solve each exponential equation using logarithms. Give the answer in decimal form, rounding to four decimal places.

25. $4^x = 5$
26. $7^x = 12$

27. $13^{x-1} = 2$
28. $5^{x+1} = 3$

29. $2^{x+1} = 3^x$
30. $5^{x-3} = 3^{2x}$

31. $2^x = 3^x$
32. $3^{2x} = 4^x$

33. $7^{x^2} = 10$
34. $8^{x^2} = 11$

35. $8^{x^2} = 9^x$
36. $5^{x^2} = 2^{5x}$

Find the exact solution to each exponential equation.

37. $e^x = 10$
38. $8e^x = 16$

39. $4e^{2x} = 24$
40. $2e^{5x} = 18$

Solve each equation. If an answer is not exact, give the answer in decimal form. Round to four decimal places.

41. $4^{x+2} - 4^x = 15$ (*Hint:* $4^{x+2} = 4^x 4^2$.)
42. $3^{x+3} + 3^x = 84$ (*Hint:* $3^{x+3} = 3^x 3^3$.)
43. $2(3^x) = 6^{2x}$
44. $2(3^{x+1}) = 3(2^{x-1})$
45. $2^{2x} - 10(2^x) + 16 = 0$ (*Hint:* Let $y = 2^x$.)
46. $3^{2x} - 10(3^x) + 9 = 0$ (*Hint:* Let $y = 3^x$.)
47. $2^{2x+1} - 2^x = 1$ (*Hint:* $2^{a+b} = 2^a 2^b$.)
48. $3^{2x+1} - 10(3^x) + 3 = 0$ (*Hint:* $3^{a+b} = 3^a 3^b$.)

Solve each logarithmic equation. Use the definition of logarithm or the definition of natural logarithm.

49. $\log x^2 = 2$
50. $\log x^3 = 3$

51. $\log \dfrac{4x + 1}{2x + 9} = 0$
52. $\log \dfrac{5x + 2}{2(x + 7)} = 0$

53. $\ln x = 6$
54. $\ln x = 3$

55. $\ln(2x - 7) = 4$
56. $\ln(3x - 5) = 7$

Solve each logarithmic equation.

57. $\log_2(2x - 3) = \log_2(x + 4)$
58. $\log_3(3x + 5) - \log_3(2x + 6) = 0$
59. $\log x + \log(x - 48) = 2$
60. $\log x + \log(x + 9) = 1$
61. $\log x + \log(x - 15) = 2$
62. $\log x + \log(x + 21) = 2$
63. $\log(x + 90) = 3 - \log x$
64. $\log(x - 3) - \log 6 = 2$
65. $\log(5000) - \log(x - 2) = 3$
66. $\log_4(2x - 3) - \log_4(x - 1) = 0$
67. $\log_7 x + \log_7(x - 5) = \log_7 6$
68. $\ln x + \ln(x - 2) = \ln 120$
69. $\ln 15 - \ln(x - 2) = \ln x$
70. $\ln 10 - \ln(x - 3) = \ln x$
71. $\log_6 8 - \log_6 x = \log_6(x - 2)$
72. $\log(x - 6) - \log(x - 2) = \log \dfrac{5}{x}$
73. $\log_8(x - 1) - \log_8 6 = \log_8(x - 2) - \log_8 x$
74. $\log x^2 = (\log x)^2$

75. $\log(\log x) = 1$

76. $\log_3(\log_3 x) = 1$

77. $\dfrac{\log(3x - 4)}{\log x} = 2$

78. $\dfrac{\ln(8x - 7)}{\ln x} = 2$

79. $\dfrac{\ln(5x + 6)}{2} = \ln x$

80. $\dfrac{1}{2}\log(4x + 5) = \log x$

81. $\log_3 x = \log_3\left(\dfrac{1}{x}\right) + 4$

82. $\log_5(7 + x) + \log_5(8 - x) - \log_5 2 = 2$

83. $2 \log_2 x = 3 + \log_2(x - 2)$

84. $2 \log_3 x - \log_3(x - 4) = 2 + \log_3 2$

85. $\ln(7y + 1) = 2 \ln(y + 3) - \ln 2$

86. $2 \log(y + 2) = \log(y + 2) - \log 12$

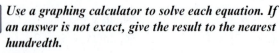 *Use a graphing calculator to solve each equation. If an answer is not exact, give the result to the nearest hundredth.*

87. $\log x + \log(x - 15) = 2$

88. $\log x + \log(x + 3) = 1$

89. $2^{x+1} = 7$

90. $\ln(2x + 5) - \ln 3 = \ln(x - 1)$

Applications

Use a calculator to help solve each problem.

91. **Tritium decay** The half-life of tritium is 12.4 years. How long will it take for 25% of a sample of tritium to decompose? Round to one decimal place.

92. **Radioactive decay** In 2 years, 20% of a radioactive element decays. Find its half-life. Round to one decimal place.

93. **Thorium decay** An isotope of thorium, ^{227}Th, has a half-life of 18.4 days. How long will it take 80% of the sample to decompose? Round to one decimal place.

94. **Lead decay** An isotope of lead, ^{201}Pb, has a half-life of 8.4 hours. How many hours ago was there 30% more of the substance? Round to one decimal place.

95. **Carbon-14 dating** A cloth fragment is found in an ancient tomb. It contains 70% of the carbon-14 that it is assumed to have had initially. How old is the cloth? Round to the nearest hundred.

96. **Carbon-14 dating** Only 25% of the carbon-14 in a wooden bowl remains. How old is the bowl? Round to the nearest thousand.

97. **Compound interest** If $500 is deposited in an account paying 8.5% annual interest, compounded semiannually, how long will it take for the account to increase to $800? Round to the nearest tenth.

98. **Continuous compound interest** In Exercise 97, how long will it take if the interest is compounded continuously? Round to the nearest tenth.

99. **Compound interest** If $1300 is deposited in a savings account paying 9% interest, compounded quarterly, how long will it take the account to increase to $2100? Round to the nearest tenth.

100. **Compound interest** A sum of $5000 deposited in an account grows to $7000 in 5 years. Assuming annual compounding, what interest rate is being paid? Round to two decimal places.

101. **Rule of Seventy** A rule of thumb for finding how long it takes an investment to double is called the **Rule of Seventy.** To apply the rule, divide 70 by the interest rate (expressed as a percent). At 5%, it takes $\frac{70}{5} = 14$ years to double the investment. At 7%, it takes $\frac{70}{7} = 10$ years. Explain why this formula works.

102. **Oceanography** The intensity I of a light a distance x meters beneath the surface of a lake decreases exponentially. If the light intensity at 6 meters is 70% of the intensity at the surface, at what depth will the intensity be 20%? Round to the nearest meter.

103. **Bacterial growth** A staphylococcus bacterial culture grows according to the formula $P = P_0 a^t$. If it takes 5 days for the culture to triple in size, how long will it take to double in size? Round to one decimal place.

104. **Rodent control** The rodent population in a city is currently estimated at 30,000. If it is expected to double every 5 years, when will the population reach 1 million? Round to one decimal place.

105. **Temperature of coffee** Refer to the section opener and find the time it takes for the white chocolate mocha to reach a temperature of 80°F. Round to the nearest minute.

106. **Time of death** The exponential function $T(t) = 17e^{-0.0626t} + 20$ models the temperature T in °C of a person's body t hours after death. If a dead body is discovered at 8:30 a.m. and the body's temperature is 30°C, what was the person's approximate time of death?

107. **Newton's Law of Cooling** Water whose temperature is at 100°C is left to cool in a room where the temperature is 60°C. After 3 minutes, the water temperature is 90°. If the water temperature T is a function of time t given by $T = 60 + 40e^{kt}$, find k.

108. Newton's Law of Cooling Refer to Exercise 107 and find the time for the water temperature to reach 70°C. Round to one decimal place.

109. Newton's Law of Cooling A block of steel, initially at 0°C, is placed in an oven heated to 300°C. After 5 minutes, the temperature of the steel is 100°C. If the steel temperature T is a function of time t given by $T = 300 - 300e^{kt}$, find the value of k.

110. Newton's Law of Cooling Refer to Exercise 109 and find the time for the steel temperature to reach 200°C. Round to one decimal place.

Discovery and Writing

111. Explain how to solve the exponential equation $5^{x+1} = 125$.

112. Explain how to solve the exponential equation $5^{x+1} = 126$.

113. Explain why it is necessary to check the solutions of a logarithmic equation.

114. What is meant by the term "half-life"?

115. Use the population growth formula to show that the doubling time for population growth is given by $t = \frac{\ln 2}{r}$.

116. Use the population growth formula to show that the tripling time for population growth is given by $t = \frac{\ln 3}{r}$.

117. Can you solve $x = \log x$ algebraically? Can you find an approximate solution?

118. Can you solve $x = \ln x$ algebraically? Can you find an approximate solution?

Find x.

119. $\log_2(\log_5(\log_7 x)) = 2$

120. $\log_8\left[16\sqrt[3]{4096}\right]^{\frac{1}{6}} = x$

Critical Thinking

Determine if the statement is true or false. If the statement is false, then correct it and make it true.

121. The exponential equation $8^x = 0$ has no solution.

122. The exponential equation $8^x = 1$ has no solution.

123. The equations $\left(\frac{3}{2}\right)^x = \frac{27}{8}$ and $x^{\frac{3}{2}} = \frac{27}{8}$ are exponential equations.

124. The exponential equation $7^{x+5} = \left(\frac{1}{49}\right)^{3x-8}$ can be solved by writing each side of the equation as a power of the same base.

125. The exponential equation $e^{x+5} = \frac{1}{e^{3x-8}}$ can be solved by writing each side of the equation as a power of the same base.

126. The exponential equation $6^{x+5} = 12^{3x-8}$ can be solved by writing each side of the equation as a power of the same base.

127. To solve the exponential equation, $e^x = 15$, take the common logarithm of both sides.

128. The logarithmic equations $\log_3(2x - 7) = 4$ and $\log_3(2x - 7) = \log_3 4$ can be solved using the same method.

129. To solve the logarithmic equation $\log_2 8 - \log_2 x = 5$, we combine the two logarithms on the left side of the equation and then write the equation in the exponential form $2^5 = 8 - x$.

130. Proposed solutions of logarithmic equations that are negative must be discarded.

CHAPTER REVIEW

5.1   Exponential Functions and Their Graphs

Definitions and Concepts	Examples
An **exponential function** with base b is defined by the equation $$f(x) = b^x \qquad (b > 0, b \neq 1)$$	**Graph:** $f(x) = 6^x$. We will find several points $(x, f(x))$ that satisfy the equation, plot the points, and join them with a smooth curve, as shown in the figure.

<table>
<tr><th colspan="3">$f(x) = 6^x$</th></tr>
<tr><th>x</th><th>$f(x)$</th><th>$(x, f(x))$</th></tr>
<tr><td>-1</td><td>$\dfrac{1}{6}$</td><td>$\left(-1, \dfrac{1}{6}\right)$</td></tr>
<tr><td>0</td><td>1</td><td>$(0, 1)$</td></tr>
<tr><td>1</td><td>6</td><td>$(1, 6)$</td></tr>
</table>

Compound interest formula:

If P dollars are deposited in an account earning interest at an annual rate r, compounded n times each year, the amount A in the account after t years is given by

$$A = P\left(1 + \frac{r}{n}\right)^{nt}$$

The grandparents of a newborn child invest $10,000 in an educational savings plan that earns 8% interest, compounded quarterly. If the money is left untouched, how much will the child have in the account in 18 years?

We will substitute 10,000 for P, 0.08 for r, and 18 for t in the formula for compound interest. Because quarterly compounding means four times per year, we will substitute 4 for n.

$$A = P\left(1 + \frac{r}{n}\right)^{nt}$$

$$A = 10{,}000\left(1 + \frac{0.08}{4}\right)^{4 \cdot 18}$$

$$= 10{,}000(1.02)^{72}$$

$$\approx 41{,}611.40 \qquad \text{Use a calculator and round to two decimals.}$$

In 18 years, the account will be worth $41,611.40.

The number $e \approx 2.718281828$.

The graph of $f(x) = e^x$ is:

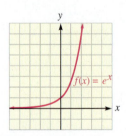

$f(x) = e^x$

Continuous compound interest formula:

If P dollars are deposited in an account earning interest at an annual rate r, compounded continuously, the amount A after t years is given by the formula

$$A = Pe^{rt}$$

If the grandparents of the newborn child in the previous example had invested $10,000 at an annual rate of 8%, compounded continuously, how much would the child have in the account in 18 years?

We will substitute $10,000 for P, 0.08 for r, and 18 in for t in the continuous compound interest formula $A = Pe^{rt}$.

$$A = Pe^{rt}$$

$$A = 10{,}000e^{(0.08)(18)}$$

$$= 10{,}000e^{1.44}$$

$$\approx 42{,}206.96 \qquad \text{Use a calculator and round to two decimal places.}$$

In 18 years, the balance will be $42,206.96.

Exercises

Use properties of exponents to simplify.

1. $5^{\sqrt{2}} \cdot 5^{\sqrt{2}}$

2. $\left(2^{\sqrt{5}}\right)^{\sqrt{2}}$

Graph the function defined by each equation.

3. $f(x) = 3^x$

4. $f(x) = \left(\dfrac{1}{3}\right)^x$

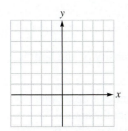
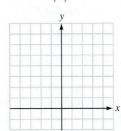

5. The graph of $f(x) = 7^x$ will pass through the points $(0, p)$ and $(1, q)$. Find p and q.

6. Give the domain and range of the function $f(x) = b^x$, with $b > 0$ and $b \neq 1$.

Use translations to help graph each function.

7. $g(x) = \left(\dfrac{1}{2}\right)^x - 2$

8. $g(x) = \left(\dfrac{1}{2}\right)^{x+2}$

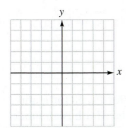
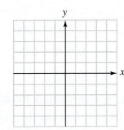

Graph each function.

9. $f(x) = -5^x$

10. $f(x) = -5^x + 4$

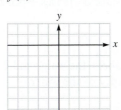
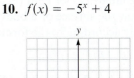

11. $f(x) = e^x + 1$

12. $f(x) = e^{x-3}$

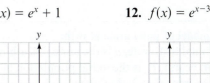
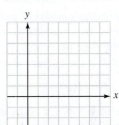

13. Compound interest How much will $10,500 become if it earns 9% per year for 60 years, compounded quarterly?

14. Continuous compound interest If $10,500 accumulates interest at an annual rate of 9%, compounded continuously, how much will be in the account in 60 years?

5.2 Applications of Exponential Functions

Definitions and Concepts

Radioactive decay formula:

The amount A of radioactive material present at time t is given by

$$A = A_0 2^{-t/h}$$

where A_0 is the amount that was present initially (at $t = 0$) and h is the material's half-life.

Examples

The half-life of radium is approximately 1600 years. To find how much of a 1-gram sample will remain after 500 years, we will substitute $A_0 = 1$, $h = 1600$, and $t = 500$ into the formula for radioactive decay and simplify.

$$A = A_0 2^{-t/h}$$

$$A = 1 \cdot 2^{-500/1600}$$

$$\approx 0.81 \quad \text{Use a calculator. Round to two decimal places.}$$

After 500 years, approximately 0.81 gram of radium will remain.

Definitions and Concepts	Examples
Intensity of light formula: The intensity I of light (in lumens) at a distance x meters below the surface of a body of water decreases exponentially according to the formula $\qquad I = I_0 k^x$ where I_0 is the intensity of light above the water and k is a constant that depends on the clarity of the water.	At one location in the Atlantic Ocean, the intensity of light above water I_0 is 10 lumens and $k = 0.4$. To find the intensity of light at a depth of 3 meters, we will substitute 10 for I_0, 0.4 for k, and 3 for x into the formula for light intensity and simplify. $\qquad I = I_0 k^x$ $\qquad I = 10(0.4)^3$ $\qquad I = 0.64$ At a depth of 3 meters, the intensity of the light is 0.64 lumen.
Malthusian model of population growth: If b is the annual birth rate, d is the annual death rate, t is the time (in years), P_0 is the initial population at $t = 0$, and P is the current population, then $\qquad P = P_0 e^{kt}$ where $k = b - d$ is the **annual growth rate**, the difference between the annual birth rate and death rate.	The population of the United States is approximately 300 million people. Assuming that the annual birth rate is 19 per 1000 and the annual death rate is 7 per 1000, what does the Malthusian model predict the U.S. population will be in 30 years? We can use the stated information to write the Malthusian model for U.S. population. We will then substitute into the model to predict the population in 30 years. Since k is the difference between the birth and death rates, we have $\qquad k = b - d$ $\qquad k = \dfrac{19}{1000} - \dfrac{7}{1000}$ $\quad$ Substitute $\frac{19}{1000}$ for b and $\frac{7}{1000}$ for d. $\qquad k = 0.019 - 0.007$ $\qquad\quad\, = 0.012$ We can substitute 300,000,000 for P_0, 30 for t, and 0.012 for k in the formula for the Malthusian model of population growth and simplify. $\qquad P = P_0 e^{kt}$ $\qquad P = (300,000,000)e^{(0.012)(30)}$ $\qquad\;\; = (300,000,000)e^{0.36}$ $\qquad\;\; \approx 429,998,824.4$ $\qquad$ Use a calculator. After 30 years, the U.S. population will be approximately 430 million people.

Exercises

15. The half-life of a radioactive material is about 34.2 years. How much of the material is left after 20 years?

16. Find the intensity of light at a depth of 12 meters if $I_0 = 14$ and $k = 0.7$. Round to two decimals.

17. The population of the United States is approximately 300,000,000 people. Find the population in 50 years if $k = 0.015$. Round to the nearest million.

18. Spread of hepatitis In a city with a population of 450,000, there are currently 1000 cases of hepatitis. If the spread of the disease is projected by the following logistic function, how many people will contract the hepatitis virus after 5 years? Round to the nearest whole number.

$$P(t) = \frac{450,000}{1 + (450 - 1)e^{-0.2t}}$$

5.3 Logarithmic Functions and Their Graphs

Definitions and Concepts	Examples

Logarithmic functions:

If $b > 0$ and $b \neq 1$, the **logarithmic function with base b** is defined by $y = \log_b x$ if and only if $x = b^y$.

The **domain of the logarithmic function** is the interval $(0, \infty)$. The **range** is the interval $(-\infty, \infty)$.

Base-10 logarithms are called **common logarithms**. The notation $\log x$ represents $\log_{10} x$.

$\log_5 125 = \mathbf{3}$ because $5^3 = 125$

$\log_8 1 = \mathbf{0}$ because $8^0 = 1$

$\log_9 3 = \dfrac{\mathbf{1}}{\mathbf{2}}$ because $9^{1/2} = \sqrt{9} = 3$

$\log_2 \dfrac{1}{16} = \mathbf{-4}$ because $2^{-4} = \dfrac{1}{16}$

$\log 10{,}000 = \mathbf{4}$ because $10^4 = 10{,}000$

To find x in the equation $\log_8 \frac{1}{64} = x$, we note that $\log_8 \frac{1}{64} = x$ is equivalent to $8^x = \frac{1}{64}$. Since $8^{-2} = \frac{1}{64}$, we have $8^x = 8^{-2}$ and $x = -2$.

To graph $f(x) = \log_5 x$. We will find several points $(x, f(x))$ that satisfy the equation, plot the points, and join them with a smooth curve, as shown in the figure.

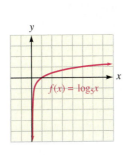

$f(x) = \log_5 x$		
x	$f(x)$	$(x, f(x))$
$\dfrac{1}{25}$	-2	$\left(\dfrac{1}{25}, -2\right)$
$\dfrac{1}{5}$	-1	$\left(\dfrac{1}{5}, -1\right)$
1	0	$(1, 0)$
5	1	$(5, 1)$
25	2	$(25, 2)$

Natural logarithms:

Base-e logarithmms are called **natural logarithms**. The notation $\ln x$ means $\log_e x$.

The graph of $f(x) = \ln x$ is:

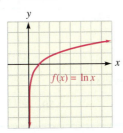

Since the functions $y = \ln x$ and $y = e^x$ are inverses
$$\ln e^x = x \quad \text{and} \quad e^{\ln x} = x$$

To graph $g(x) = \ln(x + 4)$, we will translate the graph of $f(x) = \ln x$ four units to the left as shown in the figure.

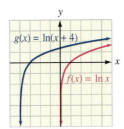

$\ln e^{-5x} = -5x$ $e^{\ln(7x)} = 7x$

Exercises

19. Give the domain and range of the logarithmic function $f(x) = \log_3 x$.

20. Give the domain and range of the natural logarithm function, $f(x) = \ln x$.

Find each value.

21. $\log_3 9$

22. $\log_9 \dfrac{1}{3}$

23. $\log_x 1$

24. $\log_5 0.04$

25. $\log_a \sqrt{a}$

26. $\log_a \sqrt[3]{a}$

Find x.

27. $\log_2 x = 5$

28. $\log_{\sqrt{3}} x = 4$

29. $\log_{\sqrt{2}} x = 6$

30. $\log_{0.1} 10 = x$

31. $\log_x 2 = -\dfrac{1}{3}$

32. $\log_x 32 = 5$

33. $\log_{0.25} x = -1$

34. $\log_{0.125} x = -\dfrac{1}{3}$

35. $\log_{\sqrt{2}} 32 = x$

36. $\log_{\sqrt{5}} x = -4$

37. $\log_{\sqrt{3}} 9\sqrt{3} = x$

38. $\log_{\sqrt{5}} 5\sqrt{5} = x$

Graph each function.

39. $f(x) = \log(x - 2)$

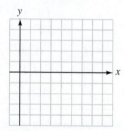

40. $f(x) = 3 + \log x$

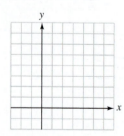

Graph each pair of equations on one set of coordinate axes.

41. $y = 4^x$ and $y = \log_4 x$

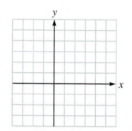

42. $y = \left(\dfrac{1}{3}\right)^x$ and $y = \log_{1/3} x$

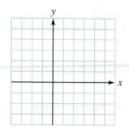

Use a calculator to find each value to four decimal places.

43. $\ln 452$

44. $\ln(\log 7.85)$

Use a calculator to solve each equation. Round each answer to four decimal places.

45. $\ln x = 2.336$

46. $\ln x = \log 8.8$

Graph each function.

47. $f(x) = 1 + \ln x$

48. $f(x) = \ln(x + 1)$

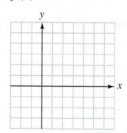

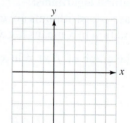

Simplify each expression.

49. $\ln(e^{12})$

50. $e^{\ln 14x}$

5.4 Applications of Logarithmic Functions

Definitions and Concepts	Examples
Decibel voltage gain: If E_O is the output voltage of a device and E_I is the input voltage, the **decibel voltage gain** is given by $$\text{dB gain} = 20 \log \frac{E_O}{E_I}$$	To find the dB gain of an amplifier with an input of 0.4 volt and an output of 50 volts, we can substitute the given values into the dB gain formula. $$\text{dB gain} = 20 \log \frac{E_O}{E_I}$$ $$\text{dB gain} = 20 \log \frac{50}{0.4} \qquad \text{Substitute 50 for } E_O \text{ and } 0.4 \text{ for } E_I.$$ $$= 20 \log 125$$ $$\approx 41.93820026 \qquad \text{Use a calculator.}$$ To the nearest decibel, the dB gain is 42 decibels.
Richter scale: If R is the intensity of an earthquake, A is the amplitude (measured in micrometers), and P is the period (the time of one oscillation of the Earth's surface, measured in seconds), then $$R = \log \frac{A}{P}$$	To find the intensity of an earthquake with amplitude of 5000 micrometers $\left(\frac{1}{2}\text{ cm}\right)$ and a period of 0.08 second, we substitute 5000 for A and 0.08 for P in the Richter scale formula and simplify. $$R = \log \frac{A}{P}$$ $$R = \log \frac{5000}{0.08}$$ $$= \log 62{,}500$$ $$\approx 4.795880017 \qquad \text{Use a calculator}$$ To the nearest tenth, the earthquake measures 4.8 on the Richter scale.
Charging batteries: If M is the theoretical maximum charge that a battery can hold and k is a positive constant that depends on the battery and the charger, the length of time (in minutes) required to charge the battery to a given level C is given by $$t = -\frac{1}{k} \ln\left(1 - \frac{C}{M}\right)$$	To find how long will it take to bring a fully discharged battery to 80% of full charge, we will assume that $k = 0.02$ and that time is measured in minutes. Since 80% of full charge means 80% of M, we can substitute $0.80M$ for C and 0.02 for k in the formula for charging batteries and find t. $$t = -\frac{1}{k} \ln\left(1 - \frac{C}{M}\right)$$ $$t = -\frac{1}{0.02} \ln\left(1 - \frac{0.80M}{M}\right)$$ $$= -50 \ln(1 - 0.8)$$ $$= -50 \ln(0.2)$$ $$\approx 80.47189562 \qquad \text{Use a calculator.}$$ The battery will reach 80% charge in about 80 minutes.

Definitions and Concepts	Examples
Population doubling time: If r is the annual growth rate and t is the time (in years) required for a population to double, then $$t = \frac{\ln 2}{r}$$	The population of the Earth is growing at the approximate rate of 2.5% per year. If this rate continues, how long will it take the population to double? Because the population is growing at the rate of 2.5% per year, we can substitute 0.025 for r in the formula for doubling time and simplify. $$t = \frac{\ln 2}{r}$$ $$t = \frac{\ln 2}{0.025}$$ $$\approx 27.72588722$$ It will take about 28 years for the Earth's population to double.
Isothermal expansion: If the temperature T is constant, the energy E required to increase the volume of 1 mole of gas from an initial volume V_i to a final volume V_f is given by $$E = RT \ln\left(\frac{V_f}{V_i}\right)$$ E is measured in joules and T in Kelvins. R is the universal gas constant, which is 8.314 joules/mole/K.	To find the amount of energy that must be supplied to double the volume of 1 mole of gas at a constant temperature of 300 K, we substitute 8.314 for R and 300 for T in the formula. Since the final volume is to be two times the initial volume, we also substitute $2V_i$ for V_f. $$E = RT \ln\left(\frac{V_f}{V_i}\right)$$ $$E = 8.314(300)\ln\left(\frac{2V_i}{V_i}\right)$$ $$= 2494.2 \ln 2$$ $$\approx 1728.847698$$ Approximately 1729 joules of energy must be added to double the volume.

Exercises

51. Decibel gain An amplifier has an output of 18 volts when the input is 0.04 volt. Find the dB gain. Round to the nearest decibel.

52. Intensity of an earthquake An earthquake had a period of 0.3 second and an amplitude of 7500 micrometers. Find its measure on the Richter scale. Round to the nearest tenth.

53. Charging batteries How long will it take a dead battery to reach an 80% charge? (Assume $k = 0.17$.) Round to one decimal place.

54. Doubling time How long will it take the population of the United States to double if the growth rate is 3% per year? Round to the nearest year.

55. Isothermal energy Find the amount of energy that must be supplied to double the volume of 1 mole of gas at a constant temperature of 350K. (*Hint:* $R = 8.314$.) Round to the nearest joule.

5.5 Properties of Logarithms

Definitions and Concepts	Examples
Properties of logarithms: If b is a positive number and $b \neq 1$,	
1. $\log_b 1 = 0$	By Property 1, $\log_9 1 = \mathbf{0}$, because $9^0 = 1$.
2. $\log_b b = 1$	By Property 2, $\log_{11} 11 = \mathbf{1}$, because $11^1 = 11$.
3. $\log_b b^x = x$	By Property 3, $\log_4 4^3 = \mathbf{3}$, because $4^3 = 4^3$.
4. $b^{\log_b x} = x$	By Property 4, $6^{\log_6 3} = 3$, because $\log_6 3$ is the power to which 6 is raised to get 3.
5. Product Rule: $\log_b MN = \log_b M + \log_b N$	By Property 5, $\log(2 \cdot 3) = \log 2 + \log 3$.
6. Quotient Rule: $\log_b \dfrac{M}{N} = \log_b M - \log_b N$	By Property 6, $\log \dfrac{17}{5} = \log 17 - \log 5$.
7. Power Rule: $\log_b M^p = p \log_b M$	By Property 7, $\log 3^2 = 2 \log 3$.
8. One-to-One Property: If $\log_b x = \log_b y$, then $x = y$.	By Property 8, if $\log 8 = \log y$, then $8 = y$.
The properties of logarithms also hold for natural logarithms. **Properties of natural logarithms:**	
1. $\ln 1 = 0$	By Property 1, $\ln 1 = \mathbf{0}$, because $e^0 = 1$.
2. $\ln e = 1$	By Property 2, $\ln e = \mathbf{1}$, because $e^1 = e$.
3. $\ln e^x = x$	By Property 3, $\ln e^{13} = 13$.
4. $e^{\ln x} = x$	By Property 4, $e^{\ln 23} = 23$.
5. Product Rule: $\ln MN = \ln M + \ln N$	By Property 5, $\ln(2 \cdot 3) = \ln 2 + \ln 3$.
6. Quotient Rule: $\ln \dfrac{M}{N} = \ln M - \ln N$	By Property 6, $\ln \dfrac{17}{5} = \ln 17 - \ln 5$.
7. Power Rule: $\ln M^p = p \ln M$	By Property 7, $\ln 3^2 = 2 \ln 3$.
8. One-to-One Property: If $\ln x = \ln y$, then $x = y$.	By Property 8, if $\ln y = \ln 8$, then $y = 8$.

Definitions and Concepts	Examples
Expanding or condensing logarithmic expressions: The properties of logarithms can be used to expand or condense logarithmic expressions.	Write the expression $\log_2 \dfrac{x^2}{y}$ in terms of logarithms of x and y. $\log_2 \dfrac{x^2}{y} = \log_2 x^2 - \log_2 y$ Use the Quotient Rule. $\qquad\quad = 2 \log_2 x - \log_2 y$ Use the Power Rule.
	Write $5 \ln x + \dfrac{1}{2} \ln y$ as a single natural logarithm. $5 \ln x + \dfrac{1}{2} \ln y = \ln x^5 + \ln y^{1/2}$ Use the Power Rule. $\qquad\qquad\quad = \ln(x^5 y^{1/2})$ Use the Product Rule. $\qquad\qquad\quad = \ln(x^5 \sqrt{y})$
Change-of-Base Formula: $\log_b y = \dfrac{\log_a y}{\log_a b}$	Use the Change-of-Base Formula to find $\log_3 10$. We will substitute 3 for b and 10 for y in the Change-of-Base Formula. $\log_3 10 = \dfrac{\log_{10} 10}{\log_{10} 3} = \dfrac{1}{0.4771212547} = 2.095903274$
pH of a solution: If $[H^+]$ is the hydrogen ion concentration in gram-ions per liter, then $pH = -\log[H^+]$	Lemon juice in a bottle has a pH of approximately 4. To find its hydrogen ion concentration, we can substitute 4 for pH in the pH formula and solve it for $[H^+]$. $4 = -\log[H^+]$ $-4 = \log[H^+]$ $[H^+] = 10^{-4}$ Change the equation from logarithmic form to exponential form. We can then use a calculator to find that $[H^+] = 1 \times 10^{-4}$ gram-ions per liter.
Weber–Fechner Law: If L is the apparent loudness of a sound and I is the intensity, then $L = k \ln I$	To find what increase in the intensity of a sound is necessary to cause a quadrupling of the apparent loudness, we can use the formula $L = k \ln I$. To quadruple the apparent loudness, we multiply both sides of the equation by 4 and use Property 7 of natural logarithms. $L = k \ln I$ $4L = 4k \ln I$ $\quad = k \ln I^4$ To quadruple the apparent loudness, we must raise the intensity to the fourth power.

Exercises

Simplify each expression.

56. $\log_7 1$

57. $\log_7 7$

58. $\log_7 7^3$

59. $7^{\log_7 4}$

60. $\ln e^4$

61. $\ln 1$

62. $10^{\log_{10} 7}$

63. $e^{\ln 3}$

64. $\log_b b^4$

65. $\ln e^9$

Assume x, y, z, and b are positive numbers and b ≠ 1. Write each expression in terms of the logarithms of x, y, and z.

66. $\log_b \dfrac{x^2 y^3}{z^4}$

67. $\log_8 \sqrt{\dfrac{x}{yz^2}}$

68. $\ln \dfrac{x^4}{y^5 z^6}$

69. $\ln \sqrt[3]{xyz}$

Assume x, y, z, and b are positive numbers and b ≠ 1. Write each expression as the logarithm of one quantity.

70. $3\log_b x - 5\log_b y + 7\log_b z$

71. $\dfrac{1}{2}(\log_b x + 3\log_b y) - 7\log_b z$

72. $4\ln x - 5\ln y - 6\ln z$

73. $\dfrac{1}{2}\ln x + 3\ln y - \dfrac{1}{3}\ln z$

Given that $\log a \approx 0.6$, $\log b \approx 0.36$, *and* $\log c \approx 2.4$, *approximate the value of each expression.*

74. $\log abc$

75. $\log a^2 b$

76. $\log \dfrac{ac}{b}$

77. $\log \dfrac{a^2}{c^3 b^2}$

78. To four decimal places, find $\log_5 17$.

79. **pH of grapefruit** The pH of grapefruit juice is about 3.1. Find its hydrogen ion concentration. Write the answer using scientific notation and round to two decimal places.

80. **Loudness of sound** Find the decrease in loudness if the intensity is cut in half.

5.6 Exponential and Logarithmic Equations

Definitions and Concepts	Examples
One-to-One Property of exponents: If $b^x = b^y$ then $x = y$. That is, equal quantities with like bases have equal exponents. The One-to-One Property of exponents can be used to solve exponential equations with like bases.	Solve using like bases: $5^{x^2 - 2x} = \dfrac{1}{5}$. Since $\frac{1}{5} = 5^{-1}$, we can write the equation in the form $5^{x^2-2x} = 5^{-1}$ and use like bases to solve the equation. Because equal quantities with like bases have equal exponents, we have $x^2 - 2x = -1$ $x^2 - 2x + 1 = 0$ Add 1 to both sides. $(x-1)(x-1) = 0$ Factor the trinomial. $x - 1 = 0$ or $x - 1 = 0$ Set each factor equal to 0. $x = 1$ $x = 1$

Definitions and Concepts	Examples
Solving exponential equations with different bases: Exponential equations with different bases can be solved using logarithms.	Solve the exponential equation: $3^x = 15$. Since logarithms of equal numbers are equal, we can take the common logarithm of each side of the equation. We can then use the Power Rule and move the variable x from its position as an exponent to a position as a coefficient and solve the equation. $$3^x = 5$$ $\log 3^x = \log 15$ Take the common logarithm of each side. $x \log 3 = \log 15$ Use the Power Rule. $x = \dfrac{\log 15}{\log 3}$ Divide both sides by $\log 3$. ≈ 2.464973521 Use a calculator. To four decimal places, $x = 2.4650$.
Solving exponential equations with e: Use natural logarithms to solve exponential equations with a base of e.	Solve $e^x = 19$. We will take the natural logarithm of both sides, simplify using the natural logarithms properties, and solve for x. $e^x = 19$ $\ln e^x = \ln 19$ Take the natural logarithm of both sides. $x = \ln 19$ Use the property $\ln(e^x) = x$. $x \approx 2.94$ Use a calculator and round to two decimal places.
Solving logarithmic equations: The One-to-One Property of logarithms can be used to solve some logarithmic equations. **One-to-One Property:** If $\log_b x = \log_b y$, then $x = y$. The solution(s) to a logarithmic equation must be checked.	Solve: $\log_4(5x + 3) = \log_4(-5x + 23)$. We will use the One-to-One Property to solve the equation. $\log_4(5x + 3) = \log_4(-5x + 23)$ $5x + 3 = -5x + 23$ Use the log property stated above. $10x = 20$ Add $5x$ and subtract 3 from both sides. $x = 2$ Divide both sides by 10. **Check:** $\log_4(5x + 3) = \log_4(-5x + 23)$ $\log_4[5(2) + 3] \overset{?}{=} \log_4[-5(2) + 23]$ $\log_4(13) = \log_4(13)$ Since $x = 2$ checks, it is the solution.

Definitions and Concepts	Examples
Solving logarithmic equations: Logarithmic properties can be used to solve many logarithmic equations.	Solve: $\log x + \log(x - 9) = 1$. $\log x + \log(x - 9) = 1$

$\log x(x - 9) = 1$ Use the Product Rule.

$x(x - 9) = 10^1$ Use the definition of logarithm and write in exponential form.

$x^2 - 9x - 10 = 0$ Remove parentheses and subtract 10 from both sides.

$(x - 10)(x + 1) = 0$ Factor the trinomial.

$x - 10 = 0$ or $x + 1 = 0$

$x = 10$ | $x = -1$

Check: The number -1 is not a solution, because it does not satisfy the equation (a negative number does not have a logarithm). We check the remaining number, 10.

$$\log x + \log(x - 9) = 1$$

$$\log 10 + \log(10 - 9) \overset{?}{=} 1 \quad \text{Substitute 10 for } x.$$

$$\log 10 + \log 1 \overset{?}{=} 1$$

$$1 + 0 \overset{?}{=} 1 \quad \log_{10} 10 = 1; \log_{10} 1 = 0$$

$$1 = 1$$

Since 10 does check, it is a solution.

| **Carbon dating:**
The amount A of radioactive material present at time t is given by the model

$A = A_0 2^{-t/h}$

where A_0 is the amount present initially and h is the half-life of the material. | To find the age of an artifact that contains three-fourths of its original carbon-14 content, we substitute $\frac{3}{4}A_0$ for A and 5700 for h in the formula for carbon dating and solve for t. |

$$A = A_0 2^{-t/h}$$

$$\frac{3}{4}A_0 = A_0 2^{-t/5700}$$

$$\frac{3}{4} = 2^{-t/5700} \qquad \text{Divide both sides by } A_0.$$

$$\log \frac{3}{4} = \log 2^{-t/5700} \qquad \text{Take the base-10 log of both sides.}$$

$$-0.1249387366 = -\frac{t}{5700} \log 2 \qquad \text{Use Property 7 of logarithms.}$$

$$t \approx 2366 \qquad \text{Solve for } t.$$

The artifact is about 2300 years old.

Exercises

Solve each equation for x.

81. $81^{x+2} = 27$

82. $2^{x^2+4x} = \dfrac{1}{8}$

83. $e^x = e^{-6x+14}$

84. $e^{2x^2} = e^{18}$

85. $3^x = 7$

86. $2^x = 3^{x-1}$

87. $2e^x = 16$

88. $-5e^x = -35$

Solve each equation for x.

89. $\log_7(-7x + 2) = \log_7(3x + 32)$

90. $\ln(x + 3) = \ln(-5x + 51)$

91. $\log x + \log(29 - x) = 2$

92. $\log_2 x + \log_2(x - 2) = 3$

93. $\log_2(x + 2) + \log_2(x - 1) = 2$

94. $\dfrac{\log(7x - 12)}{\log x} = 2$

95. $\ln x + \ln(x - 5) = \ln 6$

96. $\log 3 - \log(x - 1) = -1$

97. $e^{x \ln 2} = 9$

98. $\ln x = \ln(x - 1)$

99. $\ln x - 3 = 4$

100. $\ln x = \ln(x - 1) + 1$

101. $\ln x = \log_{10} x$ (*Hint:* Use the Change-of-Base Formula.)

102. Carbon-14 dating A wooden statue found in Egypt has a carbon-14 content that is two-thirds of that found in living wood. If the half-life of carbon-14 is 5700 years, how old is the statue? Round to the nearest hundred year.

CHAPTER TEST

Graph each function.

1. $f(x) = 2^x + 1$

2. $f(x) = e^{x-2}$

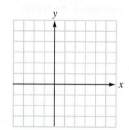

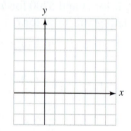

Solve each problem.

3. Radioactive decay A radioactive material decays according to the formula $A = A_0(2)^{-t}$. How much of a 3-gram sample will be left in 6 years?

4. Compound interest An initial deposit of $1000 earns 6% interest, compounded twice a year. How much will be in the account in one year?

5. Continuous compound interest An account contains $2000 and has been earning 8% interest, compounded continuously. How much will be in the account in 10 years? Round to two decimal places.

Find each value.

6. $\log_7 343$

7. $\log_3 \dfrac{1}{27}$

8. $\log_{10} 10^{12} + 10^{\log_{10} 5}$

9. $\log_{3/2} \dfrac{9}{4}$

10. $\log_{2/3} \dfrac{27}{8}$

Graph each function.

11. $f(x) = \log(x - 1)$

12. $f(x) = 2 + \ln x$

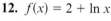

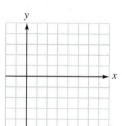

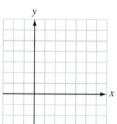

Write each expression in terms of the logarithms of a, b, and c. Assume a, b, and c are positive numbers.

13. $\log a^2 bc^3$

14. $\ln \sqrt{\dfrac{a}{b^2 c}}$

Write each expression as a logarithm of a single quantity. Assume a, b, and c are positive numbers.

15. $\frac{1}{2}\log(a + 2) + \log b - 2\log c$

16. $\frac{1}{3}(\ln a - 2\ln b) - \ln c$

Given that $\log 2 \approx 0.3010$ and $\log 3 \approx 0.4771$, approximate each value. Do not use a calculator.

17. $\log 24$ **18.** $\log\dfrac{8}{3}$

Use the Change-of-Base Formula to find each logarithm. Do not attempt to simplify the answer.

19. $\log_7 3$ **20.** $\log_\pi e$

Determine whether each statement is true or false.

21. $\log_a ab = 1 + \log_a b$

22. $\dfrac{\log a}{\log b} = \log a - \log b$

Find the solution.

23. pH of a solution Find the pH of a solution with a hydrogen ion concentration of 3.7×10^{-7}. (*Hint:* pH $= -\log[H^+]$.) Round to one decimal place.

24. Decibel gain Find the dB gain of an amplifier when $E_O = 60$ volts and $E_I = 0.3$ volt. (*Hint:* dB gain $= 20\log(E_O/E_I)$.) Round to the nearest decibel.

Solve each equation.

25. $3^{x^2 - 2x} = 27$

26. $3^{x-1} = 100^x$

27. $5e^x = 45$

28. $\ln(5x + 2) = \ln(2x + 5)$

29. $\log x + \log(x - 9) = 1$

30. $\log_6 18 - \log_6(x - 3) = \log_6 3$

CUMULATIVE REVIEW EXERCISES

Graph each function.

1. $f(x) = 2(x + 5)^2 - 8$

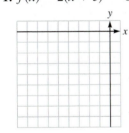

2. $f(x) = -x^2 - 6x - 5$

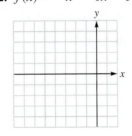

3. $f(x) = x^3 + x$

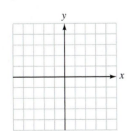

4. $f(x) = -x^4 + 2x^2 + 1$

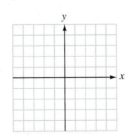

Let $P(x) = 4x^3 + 3x + 2$. Use synthetic division to find each value.

5. $P(1)$ **6.** $P(-2)$

7. $P\left(\dfrac{1}{2}\right)$ **8.** $P(i)$

Determine whether each binomial is a factor of $P(x) = x^3 + 2x^2 - x - 2$. Use synthetic division.

9. $x + 1$ **10.** $x - 2$

11. $x - 1$ **12.** $x + 2$

Determine how many zeros each function has.

13. $P(x) = x^{12} - 4x^8 + 2x^4 + 12$

14. $P(x) = x^{2000} - 1$

Determine the number of possible positive, negative, and nonreal zeros of each function.

15. $P(x) = x^4 + 2x^3 - 3x^2 + x + 2$

16. $P(x) = x^4 - 3x^3 - 2x^2 - 3x - 5$

Find the zeros of each polynomial function.

17. $P(x) = x^3 + x^2 - 9x - 9$

18. $P(x) = x^3 - 2x^2 - x + 2$

Graph each function. Show all asymptotes.

19. $f(x) = \dfrac{x}{x - 3}$

20. $f(x) = \dfrac{x^2 - 1}{x^2 - 9}$

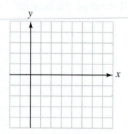 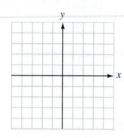

Graph the function defined by each equation.

21. $f(x) = 3^x - 2$

22. $f(x) = 2e^x$

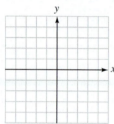 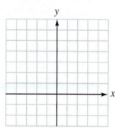

23. $f(x) = \log_3 x$

24. $f(x) = \ln(x - 2)$

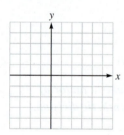 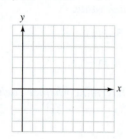

Find each value.

25. $\log_2 64$

26. $\log_{1/2} 8$

27. $\ln e^3$

28. $2^{\log_2 2}$

Solve for x.

29. $\log_5 x = -3$

30. $\log_x 72 = 2$

Write each expression in terms of the logarithms of a, b, and c. Assume a, b, and c are positive numbers.

31. $\log abc$

32. $\log \dfrac{a^2 b}{c}$

33. $\log \sqrt{\dfrac{ab}{c^3}}$

34. $\ln \dfrac{\sqrt{ab^2}}{c}$

Write each expression as the logarithm of a single quantity. Assume a, b, and c are positive numbers.

35. $3 \ln a - 3 \ln b$

36. $\dfrac{1}{2}\log a + 3 \log b - \dfrac{2}{3}\log c$

Solve each equation.

37. $3^{x+1} = 8$

38. $3^{x-1} = 3^{2x}$

39. $\log x + \log 2 = 3$

40. $\log(x + 1) + \log(x - 1) = 1$

Linear Systems

In this chapter, we will learn to solve systems of equations—sets of several equations, most with more than one variable. One method of solution uses matrices, an important tool in mathematics and its applications.

Careers and Mathematics: Cryptanalyst

©Andrey_Popov/Shutterstock.com

Cryptanalysts are essential to the implementation of protection systems employed by corporations and private citizens to keep hackers out of important data systems. They also play an important part in the safety of cities, states, and countries by analyzing and deciphering secret coding systems used by terrorists and other enemies. With increased reliance on the Internet to conduct business, store data, and pass information between groups, the job of a cryptanalyst is becoming increasingly more important to ensure both private and governmental security of data.

Education and Mathematics Required

- Most are required to have at least an undergraduate degree in mathematics or computer science. Many will also have a graduate degree in mathematics, with a Ph.D. being required for those wishing to be employed in a research facility or by a university.
- College Algebra, Trigonometry, Calculus I, II, and III, Linear Algebra, Differential and Partial Differential Equations, Elementary Number Theory, Introduction to Real Analysis, Analysis I and II, Methods of Complex Analysis, and Mathematical Cryptography are math courses required.

How Cyptanalysts Use Math and Who Employs Them

- Among the many ways that cryptanalysts use mathematics are the encoding and encrypting of systems and databases, devising systems used to prevent hackers from stealing consumer data, and consulting with business and industry to solve problems related to security.
- Cryptanalysts are employed in all levels of government, including intelligence agencies and special services, universities, financial institutions, insurance companies, telecommunications companies, computer design firms, consulting firms, and science and engineering firms.

Career Outlook and Earnings

- Employment of cryptanalysts is expected to grow by 23% through 2022, which is much faster than the average for all occupations.
- The median annual wages of cryptanalysts is approximately $101,360.

For more information see: www.bls.gov/oco

6.1 Systems of Linear Equations

In this section, we will learn to

1. Solve systems using the graphing method.
2. Solve systems using the substitution method.
3. Solve systems using the elimination method.
4. Solve systems with infinitely many solutions.
5. Solve inconsistent systems.
6. Solve systems involving three equations in three variables.
7. Solve applications involving systems of equations.

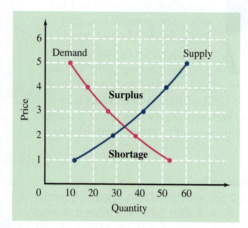

In economics we often talk about the law of supply and demand.

- As the illustration shows, few units of a product will be in demand when the price is high. However, more units will be in demand when the price is low.
- As the illustration also shows, few units of product will be supplied when the price is low. However, more product will be supplied when the price is high.

In a free economy, the intersection of the graph of the supply function and the demand function will be the market price of the product.

To see that this is true, suppose that the price of a small bag of Lindt milk chocolate Lindor truffles is $3. At this price, less than 30 units of product would be in demand, but over 40 units would be supplied. Since supply is greater than demand, the price would come down.

At a price of $2, a little less than 40 units would be in demand, but less than 30 units would be supplied. Since demand is greater than supply, the price would go up.

Only when the graphs cross would the demand and supply be in equilibrium.

In this section, we will begin to discuss how to find the common solution of two equations that occur simultaneously.

1. Solve Systems Using the Graphing Method

Equations with two variables have infinitely many solutions. For example, the tables shown give a few of the solutions of $x + y = 5$ and $x - y = 1$.

$x + y = 5$	
x	y
1	4
2	3
3	2
5	0

$x - y = 1$	
x	y
8	7
3	2
1	0
0	-1

Only the pair $x = 3$ and $y = 2$ satisfies both equations. The pair of equations

$$\begin{cases} x + y = 5 \\ x - y = 1 \end{cases}$$

is called a **system of equations**, and the solution $x = 3$, $y = 2$ is called its **simultaneous solution**, or just its **solution**. The process of finding the solution of a system of equations is called **solving the system**.

The graph of an equation in two variables displays the equation's infinitely many solutions. For example, the graph of one of the lines in Figure 6-1 represents the infinitely many solutions of the equation $5x - 2y = 1$. The other line in the figure is the graph of the infinitely many solutions of $2x + 3y = 8$.

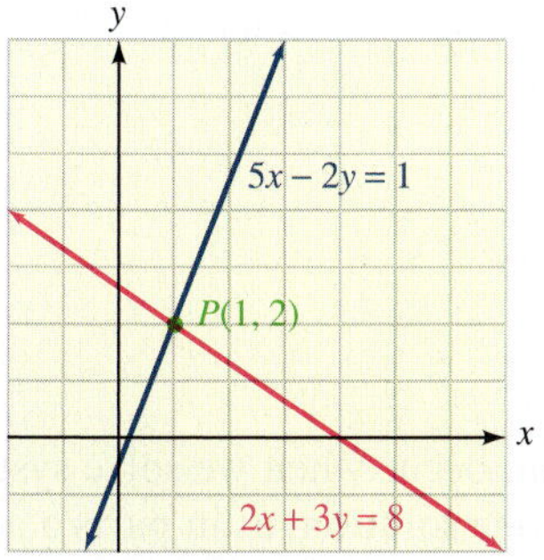

Because only point $P(1, 2)$ lies on both lines, only its coordinates satisfy both equations. Thus, the simultaneous solution of the system of equations

$$\begin{cases} 5x - 2y = 1 \\ 2x + 3y = 8 \end{cases}$$

is the pair of numbers $x = 1$ and $y = 2$ or simply the pair $(1, 2)$.

This discussion suggests a graphical method of solving systems of equations in two variables.

We can use the following steps to solve a system of two equations in two variables.

5x − 2y = 1

P(1, 2)

2x + 3y = 8

FIGURE 6-1

Strategy for Using the Graphing Method

1. On one coordinate grid, graph each equation.

2. Find the coordinates of the point or points where all of the graphs intersect. These coordinates give the solutions of the system.

3. If the graphs have no point in common, the system has no solution.

Using the Graphing Method to Solve a System of Equations

EXAMPLE 1

Use the graphing method to solve each system:

a. $\begin{cases} 3x + y = 1 \\ -x + 2y = 9 \end{cases}$　　b. $\begin{cases} 2x - 3y = 4 \\ 4x = -4 + 6y \end{cases}$　　c. $\begin{cases} y = 4 - x \\ 2x + 2y = 8 \end{cases}$

SOLUTION

a. The graphs of the equations are the lines shown in Figure 6-2(a). The solution of this system is given by the coordinates of the point $(-1, 4)$, where the lines intersect. By checking both values in both equations, we can verify that the solution is $x = -1$ and $y = 4$.

b. The graphs of the equations are the parallel lines shown in Figure 6-2(b). Since parallel lines do not intersect, the system has no solution.

c. The graphs of the equations are the lines shown in Figure 6-2(c). Since the lines are the same, they have infinitely many points in common, and the system has infinitely many solutions. All ordered pairs whose coordinates satisfy one of the equations satisfy the other also.

Tip

To ensure accuracy when using the graphing method, use graph paper, a sharp pencil, and a straightedge.

To find some solutions, we substitute numbers for x in the first equation and solve for y. If $x = 3$ for example, then $y = 1$. One solution is the pair $(3, 1)$. Other solutions are $(0, 4)$ and $(5, -1)$.

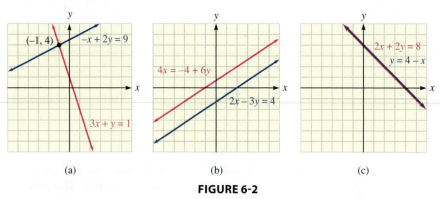

(a)   (b)   (c)

FIGURE 6-2

Self Check 1   Use the graphing method to solve $\begin{cases} y = 2x \\ x + y = 3 \end{cases}$.

Now Try Exercise 13.

Example 1 illustrates three possibilities that can occur when we solve systems of equations. If a system of equations has at least one solution, as in parts a and c of Example 1, the system is called **consistent**. If it has no solution, as in part b, it is called **inconsistent**.

If a system of two equations in two variables has exactly one solution as in part a, or no solution as in part b, the equations in the system are called **independent**. If a system of linear equations has infinitely many solutions, as in part c, the equations of the system are called **dependent**.

There are three possibilities that can occur when two equations, each with two variables, are graphed.

Possibilities		Conclusion
	The lines are distinct and intersect.	The equations are independent, and the system is consistent. The system has one solution.
	The lines are distinct and parallel.	The equations are independent, and the system is inconsistent. The system has no solution.
	The lines are identical and coincide.	The equations are dependent, and the system is consistent. The system has infinitely many solutions.

Accent on Technology

Solving a System of Linear Equations

In order to use a graphing calculator to solve a system of equations, each equation must be written with y as a function of x. This will allow us to enter each equation into the graph editor. Let's solve the system $\begin{cases} x - 4y = 7 \\ x + 2y = 4 \end{cases}$.

Solving each equation for y gives us $\begin{cases} y = \frac{x-7}{4} \\ y = 4 - \frac{x}{2} \end{cases}$. Now we are ready to solve the system using the graphing calculator.

The point where the lines intersect is the solution for the system. We will use the INTERSECT function to solve this system as shown in Figure 6-3.

1. Enter the equations. Make sure you put parentheses around the numerators.

2. Set a proper WINDOW.

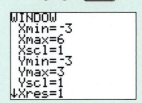

3. GRAPH the lines. Select the intersect function by pressing 2nd TRACE : 5 . Move the cursor near the intersection. It doesn't have to be really close, just near.

4. Now press the ENTER key three times, and the intersection will be shown.

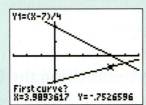

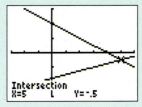

FIGURE 6-3

We now consider two algebraic methods to find solutions of systems containing any number of variables.

2. Solve Systems Using the Substitution Method

We use the following steps to solve a system of two equations in two variables by substitution.

Strategy for Using the Substitution Method	1. Solve one equation for a variable—say, y.
	2. Substitute the expression obtained for y for every y in the second equation.
	3. Solve the equation that results.
	4. Substitute the solution found in Step 3 into the equation found in Step 1 and solve for y.

Using the Substitution Method to Solve a System of Equations

EXAMPLE 2

Use the substitution method to solve $\begin{cases} 3x + y = 1 \\ -x + 2y = 9 \end{cases}$.

SOLUTION   We can solve the first equation for y and then substitute that result for y in the second equation.

Tip
If possible, choose to solve for a variable with a coefficient of 1 or −1 and avoid fractions.

$$\begin{cases} 3x + y = 1 \rightarrow y = \boxed{1 - 3x} \\ -x + 2y = 9 \end{cases}$$

The substitution gives one linear equation with one variable, which we can solve for x:

$$-x + 2(\mathbf{1 - 3x}) = 9$$

$$-x + 2 - 6x = 9$$

$$-7x = 7 \qquad \text{Combine like terms and subtract 2 from both sides.}$$

$$x = -1 \qquad \text{Divide both sides by } -7.$$

To find y, we substitute -1 for x in the equation $y = 1 - 3x$ and simplify:

$$y = 1 - 3\mathbf{x}$$

$$= 1 - 3(\mathbf{-1})$$

$$= 1 + 3$$

$$= 4$$

The solution is the pair $(-1, 4)$.

Self Check 2   Use the substitution method to solve $\begin{cases} 2x + 3y = 1 \\ x - y = 3 \end{cases}$.

Now Try Exercise 29.

3. Solve Systems Using the Elimination Method

The substitution method for solving a system of equations can be difficult to use if none of the variables has a coefficient of 1 or −1. Fortunately, we can solve systems like this one using an easier method called the **elimination method**. As with substitution, the elimination method combines the equations of a system to eliminate terms involving one of the variables. The elimination method is also known as the **addition method**.

Strategy for Using the Elimination Method

1. Write the equations of the system in general form so that terms with the same variable are aligned vertically.

2. Multiply all the terms of one or both of the equations by constants chosen to make the coefficients of x (or y) differ only in sign.

3. Add the equations. Solve the equation that results, if possible.

4. Substitute the value obtained in Step 3 into either of the original equations, and solve for the remaining variable.

5. The results obtained in Steps 3 and 4 are the solution of the system.

EXAMPLE 3   **Using the Elimination Method to Solve a System of Equations**

Use the elimination method to solve $\begin{cases} 3x + 2y = 8 \\ 2x - 5y = 18 \end{cases}$.

SOLUTION   To eliminate y, we multiply the terms of the first equation by 5 and the terms of the second equation by 2, to obtain an equivalent system in which the coefficients of y differ only in sign. Then we add the equations to eliminate y.

Take Note

Take Note
To make the coefficients of y differ only in sign and be the smallest numbers possible, we multiply each equation by a number that makes the new coefficients of y the least common multiple of the original coefficients.

$$\begin{cases} 3x + 2y = 8 & \rightarrow & 15x + 10y = 40 \\ 2x - 5y = 18 & \rightarrow & \underline{4x - 10y = 36} \\ & & 19x \qquad = 76 \\ & & x = 4 \end{cases}$$

Multiply by 5.
Multiply by 2.
Add the equations.
Divide both sides by 19.

We now substitute 4 for x into either of the original equations and solve for y. If we use the first equation,

$$3x + 2y = 8$$
$$3(4) + 2y = 8 \qquad \text{Substitute 4 for } x.$$
$$12 + 2y = 8 \qquad \text{Simplify.}$$
$$2y = -4 \qquad \text{Subtract 12 from both sides.}$$
$$y = -2 \qquad \text{Divide both sides by 2.}$$

The solution is $(4, -2)$.

Self Check 3   Use the elimination method to solve $\begin{cases} 2x + 5y = 8 \\ 2x - 4y = -10 \end{cases}$.

Now Try Exercise 37.

EXAMPLE 4   **Using the Elimination Method to Solve a System of Equations**

Use the elimination method to solve $\begin{cases} \dfrac{x + 2y}{4} + \dfrac{x - y}{5} = \dfrac{6}{5} \\ \dfrac{x + y}{7} - \dfrac{x - y}{3} = -\dfrac{12}{7} \end{cases}$.

SOLUTION   To clear the first equation of fractions, we multiply both sides by the lowest common denominator, 20. To clear the second equation of fractions, we multiply both sides of the second equation by the LCD, which is 21.

$$\begin{cases} 20\left(\dfrac{x + 2y}{4} + \dfrac{x - y}{5}\right) = 20\left(\dfrac{6}{5}\right) \\ 21\left(\dfrac{x + y}{7} - \dfrac{x - y}{3}\right) = 21\left(-\dfrac{12}{7}\right) \end{cases}$$

$$\begin{cases} 5(x + 2y) + 4(x - y) = 24 \\ 3(x + y) - 7(x - y) = -36 \end{cases}$$

$$\begin{cases} 9x + 6y = 24 \\ -4x + 10y = -36 \end{cases}$$

Remove parentheses and combine like terms.

$$\begin{cases} 3x + 2y = 8 \\ 2x - 5y = 18 \end{cases}$$

Divide both sides by 3.
Divide both sides by −2.

In Example 3, we saw that the solution to this final system is $(4, -2)$.

Self Check 4   Use the elimination method to solve $\begin{cases} \dfrac{x + y}{3} + \dfrac{x - y}{4} = \dfrac{3}{4} \\ \dfrac{x - y}{3} - \dfrac{2x - y}{2} = -\dfrac{1}{3} \end{cases}$.

Now Try Exercise 51.

4. Solve Systems with Infinitely Many Solutions

EXAMPLE 5

Solving a System with Infinitely Many Solutions

Use the elimination method to solve $\begin{cases} x + 2y = 3 \\ 2x + 4y = 6 \end{cases}$.

SOLUTION

To eliminate the variable y, we can multiply both sides of the first equation by -2 and add the result to the second equation to get

$$\begin{array}{r} -2x - 4y = -6 \\ \underline{2x + 4y = 6} \\ 0 = 0 \end{array}$$

Although the result $0 = 0$ is true, it does not give the value of y.

Since the second equation in the original system is twice the first, the equations are equivalent. If we were to solve this system by graphing, the two lines would coincide. The (x, y) coordinates of points on that one line form the infinite set of solutions to the given system. The system is consistent, but the equations are dependent.

To find a general solution, we can solve either equation of the system for y. If we solve the first equation for y, we obtain

$$x + 2y = 3$$
$$2y = 3 - x$$
$$y = \frac{3 - x}{2}$$

All ordered pairs that are solutions will have the form $\left(x, \frac{3-x}{2}\right)$. For example:

- If $x = 3$, then $y = 0$ and $(3, 0)$ is a solution.
- If $x = 0$, then $y = \frac{3}{2}$ and $\left(0, \frac{3}{2}\right)$ is a solution.
- If $x = -7$, then $y = 5$ and $(-7, 5)$ is a solution.

Self Check 5 Use the elimination method to solve $\begin{cases} 3x - y = 2 \\ 6x - 2y = 4 \end{cases}$.

Now Try Exercise 43.

5. Solve Inconsistent Systems

EXAMPLE 6

Solving an Inconsistent System

Use the elimination method to solve $\begin{cases} x + y = 3 \\ x + y = 2 \end{cases}$.

SOLUTION

We can multiply both sides of the second equation by -1 and add the results to the first equation to get

$$\begin{array}{r} x + y = 3 \\ \underline{-x - y = -2} \\ 0 = 1 \end{array}$$

Because $0 \neq 1$ the system has no solution. If we graph each equation in this system, the graphs will be parallel lines. This system is inconsistent, and the equations are independent.

Tip

If you obtain a false statement at any time during the solution process, the system has no solution.

©TotallyPic.com/
Shutterstock.com

Self Check 6   Use the elimination method to solve $\begin{cases} 3x + 2y = 2 \\ 3x + 2y = 3 \end{cases}$.

Now Try Exercise 45.

6. Solve Systems Involving Three Equations in Three Variables

To solve three equations in three variables, we use addition to eliminate one variable. This will produce a system of two equations in two variables, which we can solve using the methods previously discussed.

Solving a System Involving Three Equations in Three Variables

EXAMPLE 7

Solve the system:

$$\begin{matrix}(1) \\ (2) \\ (3)\end{matrix} \begin{cases} x + 2y + z = 8 \\ 2x + y - z = 1 \\ x + y - 2z = -3 \end{cases}$$

SOLUTION   We can add Equations 1 and 2 to eliminate z

$$\begin{aligned} &(1) \quad x + 2y + z = 8 \\ &(2) \quad \underline{2x + y - z = 1} \\ &(4) \quad 3x + 3y \; = 9 \end{aligned}$$

> **Take Note**
> Numbering the equations helps keep track of the equations and describe how the system is solved.

and divide both sides of Equation 4 by 3 to get

$(5) \quad x + y = 3$

We now choose a different pair of equations—say, Equations 1 and 3—and eliminate z again. If we multiply both sides of Equation 1 by 2 and add the result to Equation 3, we get

$$\begin{aligned} &\quad 2x + 4y + 2z = 16 \\ &(3) \quad \underline{x + y - 2z = -3} \\ &(6) \quad 3x + 5y \; = 13 \end{aligned}$$

> **Take Note**
> We used elimination to reduce the system of three equations in three variables to two equations in two variables. We can finish solving the system by using substitution, elimination, or graphing.

Equations 5 and 6 form the system $\begin{cases} x + y = 3 \\ 3x + 5y = 13 \end{cases}$, which we can solve by substitution.

$$\begin{aligned} (5) &\qquad x + y = 3 \rightarrow y = \boxed{3 - x} \\ (6) &\qquad 3x + 5y = 13 \\ &\qquad 3x + 5(\mathbf{3 - x}) = 13 \qquad \text{Substitute } 3 - x \text{ for } y. \\ &\qquad 3x + 15 - 5x = 13 \qquad \text{Remove parentheses.} \\ &\qquad -2x = -2 \qquad \text{Combine like terms and subtract 15 from both sides.} \\ &\qquad x = 1 \qquad \text{Divide both sides by } -2. \end{aligned}$$

To find y, we substitute 1 for x in the equation $y = 3 - x$.

$y = 3 - \mathbf{1}$

$y = 2$

To find z, we substitute 1 for x and 2 for y in any one of the original equations that includes z, and we find that $z = 3$. The solution is the triple

$(x, y, z) = (1, 2, 3)$

Because there is one solution, the system is consistent, and its equations are independent.

Self Check 7 Solve: $\begin{cases} x - y + 2z = 2 \\ 2x + y + z = 4 \\ -x - 2y + 3z = 0 \end{cases}$.

Now Try Exercise 53.

EXAMPLE 8 Solving a System Involving Three Equations in Three Variables

Solve the system:

(1) $\quad\begin{cases} x + 2y + z = 8 \\ 2x + y - z = 1 \\ x - y - 2z = -7 \end{cases}$
(2)
(3)

SOLUTION We can add Equations 1 and 2 to eliminate z.

$$\begin{array}{r} x + 2y + z = 8 \\ \underline{2x + y - z = 1} \\ \text{(4)} \quad 3x + 3y = 9 \end{array}$$

We can now multiply Equation 1 by 2 and add it to Equation 3 to eliminate z again.

$$\begin{array}{r} 2x + 4y + 2z = 16 \\ \underline{x - y - 2z = -7} \\ \text{(5)} \quad 3x + 3y = 9 \end{array}$$

Since Equations 4 and 5 are the same, the system is consistent, but the equations are dependent. There will be infinitely many solutions. To find a general solution, we can solve either Equation 4 or 5 for y to get

(6) $\quad y = 3 - x$

We can find the value of z in terms of x by substituting the right side of Equation 6 into any of the first three equations—say Equation 1.

$$x + 2y + z = 8$$

$$x + 2(3 - x) + z = 8 \qquad \text{Substitute } 3 - x \text{ for } y.$$

$$x + 6 - 2x + z = 8 \qquad \text{Use the Distributive Property to remove parentheses.}$$

$$-x + 6 + z = 8 \qquad \text{Combine terms.}$$

$$z = x + 2 \qquad \text{Solve for } z.$$

A general solution to this system is $(x, y, z) = (x, 3 - x, x + 2)$. To find some specific solutions, we can substitute numbers for x and compute y and z. For example,

If $x = 1$, then $y = 2$ and $z = 3$. One possible solution is $(1, 2, 3)$.

If $x = 0$, then $y = 3$ and $z = 2$. Another possible solution is $(0, 3, 2)$.

Self Check 8 Solve: $\begin{cases} x + y + z = 3 \\ x - y + 3z = 4 \\ 2x + 2y + 2z = 5 \end{cases}$.

Now Try Exercise 61.

7. Solve Applications Involving Systems of Equations

Solving an Application Problem

EXAMPLE 9

An airplane flies 600 miles with the wind for 2 hours and returns against the wind in 3 hours. Find the speed of the wind and the air speed of the plane.

SOLUTION

If a represents the air speed and w represents wind speed, the ground speed of the plane with the wind is the combined speed $a + w$. On the return trip, against the wind, the ground speed is $a - w$. The information in this problem is organized in Figure 6-4 and can be used to give a system of two equations in the variables a and w.

	d	$=$	r	$\cdot$	t
Outbound trip	600		$a + w$		2
Return trip	600		$a - w$		3

FIGURE 6-4

Since $d = rt$ we have $\begin{cases} 600 = 2(a + w) \\ 600 = 3(a - w) \end{cases}$, which can be written as

$$(7) \quad \begin{cases} 300 = a + w \\ 200 = a - w \end{cases}$$

We can add these equations to get

$$500 = 2a$$

$$a = 250 \qquad \text{Divide both sides by 2.}$$

To find w, we substitute 250 for a into either of the previous equations (we'll use Equation 7) and solve for w:

$$(7) \quad 300 = \boldsymbol{a} + w$$

$$300 = \boldsymbol{250} + w$$

$$w = 50 \qquad \text{Subtract 250 from both sides.}$$

The air speed of the plane is 250 mph. With a 50-mph tailwind, the ground speed is $250 + 50$, or 300 mph. At 300 mph, the 600-mile trip will take 2 hours.

With a 50-mph headwind, the ground speed is $250 - 50$, or 200 mph. At 200 mph, the 600-mile trip will take 3 hours. The answers check.

Self Check 9   An airplane flies 500 miles with the wind for 2 hours and returns against the wind in 3 hours. Find the speed of the wind and the air speed of the plane. Round to tenths.

Now Try Exercise 75.

Self Check Answers

1. $(1, 2)$ **2.** $(2, -1)$ **3.** $(-1, 2)$ **4.** $(1, 2)$ **5.** A general solution is $(x, 3x - 2)$.   **6.** no solution; inconsistent system   **7.** $(1, 1, 1)$
8. no solution; inconsistent system   **9.** speed of wind: 41.7 mph; air speed of the plane: 208.3 mph.

Exercises 6.1

Getting Ready

You should be able to complete these vocabulary and concept statements before you proceed to the practice exercises.

Fill in the blanks.

1. A set of several equations with several variables is called a _____ of equations.

2. Any set of numbers that satisfies each equation of a system is called a _____ of the system.

3. If a system of equations has a solution, the system is _____.

4. If a system of equations has no solution, the system is _____.

5. If a system of equations has only one solution, the equations of the system are _____.

6. If a system of equations has infinitely many solutions, the equations of the system are _____.

7. The system $\begin{cases} x + y = 5 \\ x - y = 1 \end{cases}$ is _____ (consistent, inconsistent).

8. The system $\begin{cases} x + y = 5 \\ x + y = 1 \end{cases}$ is _____ (consistent, inconsistent).

9. The equations of the system $\begin{cases} x + y = 5 \\ 2x + 2y = 10 \end{cases}$ are _____ (dependent, independent).

10. The equations of the system $\begin{cases} x + y = 5 \\ x - y = 1 \end{cases}$ are _____ (dependent, independent).

11. The pair $(1, 3)$ ___ (is, is not) a solution of the system $\begin{cases} x + 2y = 7 \\ 2x - y = -1 \end{cases}$.

12. The pair $(1, 3)$ ___ (is, is not) a solution of the system $\begin{cases} 3x + y = 6 \\ x - 3y = -8 \end{cases}$.

Practice

Solve each system of equations by graphing.

13. $\begin{cases} y = -3x + 5 \\ x - 2y = -3 \end{cases}$

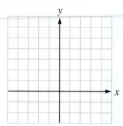

14. $\begin{cases} x - 2y = -3 \\ 3x + y = -9 \end{cases}$

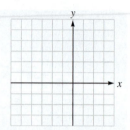

15. $\begin{cases} 3x + 2y = 2 \\ -2x + 3y = 16 \end{cases}$

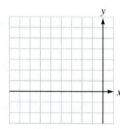

16. $\begin{cases} x + y = 0 \\ 5x - 2y = 14 \end{cases}$

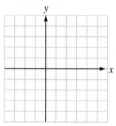

17. $\begin{cases} y = -x + 5 \\ 3x + 3y = 30 \end{cases}$

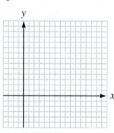

18. $\begin{cases} x - 3y = -3 \\ 2x - 6y = 12 \end{cases}$

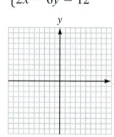

19. $\begin{cases} y = -x + 6 \\ 5x + 5y = 30 \end{cases}$

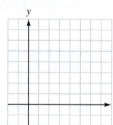

20. $\begin{cases} 2x - y = -3 \\ 8x - 4y = -12 \end{cases}$

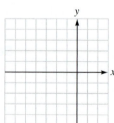

Use a graphing calculator to approximate the solutions of each system. Give answers to the nearest tenth.

21. $\begin{cases} y = -5.7x + 7.8 \\ y = 37.2 - 19.1x \end{cases}$ **22.** $\begin{cases} y = 3.4x - 1 \\ y = -7.1x + 3.1 \end{cases}$

23. $\begin{cases} y = \dfrac{5.5 - 2.7x}{3.5} \\ 5.3x - 9.2y = 6.0 \end{cases}$ **24.** $\begin{cases} 29x + 17y = 7 \\ -17x + 23y = 19 \end{cases}$

Solve each system by substitution, if possible.

25. $\begin{cases} y = x - 1 \\ y = 2x \end{cases}$ **26.** $\begin{cases} y = 2x - 1 \\ x + y = 5 \end{cases}$

27. $\begin{cases} 2x + 3y = 0 \\ y = 3x - 11 \end{cases}$ **28.** $\begin{cases} 2x + y = 3 \\ y = 5x - 11 \end{cases}$

29. $\begin{cases} 4x + 3y = 3 \\ 2x - 6y = -1 \end{cases}$ **30.** $\begin{cases} 4x + 5y = 4 \\ 8x - 15y = 3 \end{cases}$

31. $\begin{cases} x + 3y = 1 \\ 2x + 6y = 3 \end{cases}$ **32.** $\begin{cases} x - 3y = 14 \\ 3(x - 12) = 9y \end{cases}$

33. $\begin{cases} y = 3x - 6 \\ x = \dfrac{1}{3}y + 2 \end{cases}$ **34.** $\begin{cases} 3x - y = 12 \\ y = 3x - 12 \end{cases}$

Solve each system by the elimination method, if possible.

35. $\begin{cases} 5x - 3y = 12 \\ 2x - 3y = 3 \end{cases}$ **36.** $\begin{cases} 2x + 3y = 8 \\ -5x + y = -3 \end{cases}$

37. $\begin{cases} x - 7y = -11 \\ 8x + 2y = 28 \end{cases}$ **38.** $\begin{cases} 3x + 9y = 9 \\ -x + 5y = -3 \end{cases}$

39. $\begin{cases} 3(x - y) = y - 9 \\ 5(x + y) = -15 \end{cases}$ **40.** $\begin{cases} 2(x + y) = y + 1 \\ 3(x + 1) = y - 3 \end{cases}$

41. $\begin{cases} 2 = \dfrac{1}{x + y} \\ 2 = \dfrac{3}{x - y} \end{cases}$ **42.** $\begin{cases} \dfrac{1}{x + y} = 12 \\ \dfrac{3x}{y} = -4 \end{cases}$

43. $\begin{cases} y + 2x = 5 \\ 0.5y = 2.5 - x \end{cases}$

44. $\begin{cases} -0.3x + 0.1y = -0.1 \\ 6x - 2y = 2 \end{cases}$

45. $\begin{cases} x + 2(x - y) = 2 \\ 3(y - x) - y = 5 \end{cases}$ **46.** $\begin{cases} 3x = 4(2 - y) \\ 3(x - 2) + 4y = 0 \end{cases}$

47. $\begin{cases} x + \dfrac{y}{3} = \dfrac{5}{3} \\ \dfrac{x + y}{3} = 3 - x \end{cases}$ **48.** $\begin{cases} 3x - y = 0.25 \\ x + \dfrac{3}{2}y = 2.375 \end{cases}$

49. $\begin{cases} \dfrac{3}{2}x + \dfrac{1}{3}y = 2 \\ \dfrac{2}{3}x + \dfrac{1}{9}y = 1 \end{cases}$ **50.** $\begin{cases} \dfrac{x + y}{2} + \dfrac{x - y}{5} = 2 \\ x = \dfrac{y}{2} + 1 \end{cases}$

51. $\begin{cases} \dfrac{x - y}{5} + \dfrac{x + y}{2} = 6 \\ \dfrac{x - y}{2} - \dfrac{x + y}{4} = 3 \end{cases}$ **52.** $\begin{cases} \dfrac{x - 2}{5} + \dfrac{y + 3}{2} = 5 \\ \dfrac{x + 3}{2} + \dfrac{y - 2}{3} = 6 \end{cases}$

Solve each system by any method.

53. $\begin{cases} x + y + z = 3 \\ 2x + y + z = 4 \\ 3x + y - z = 5 \end{cases}$ **54.** $\begin{cases} x - y - z = 0 \\ x + y - z = 0 \\ x - y + z = 2 \end{cases}$

55. $\begin{cases} x - y + z = 0 \\ x + y + 2z = -1 \\ -x - y + z = 0 \end{cases}$ **56.** $\begin{cases} 2x + y - z = 7 \\ x - y + z = 2 \\ x + y - 3z = 2 \end{cases}$

57. $\begin{cases} 2x + y = 4 \\ x - z = 2 \\ y + z = 1 \end{cases}$ **58.** $\begin{cases} 3x + y + z = 0 \\ 2x - y + z = 0 \\ 2x + y + z = 0 \end{cases}$

59. $\begin{cases} x + y + z = 6 \\ 2x + y + 3z = 17 \\ x + y + 2z = 11 \end{cases}$ **60.** $\begin{cases} x + y + z = 3 \\ 2x + y + z = 6 \\ x + 2y + 3z = 2 \end{cases}$

61. $\begin{cases} x + y + z = 3 \\ x + z = 2 \\ 2x + 2y + 2z = 3 \end{cases}$ **62.** $\begin{cases} x + y + z = 3 \\ x + z = 2 \\ 2x + y + 2z = 5 \end{cases}$

63. $\begin{cases} x + 2y - z = 2 \\ 2x - y = -1 \\ 3x + y + z = 1 \end{cases}$ **64.** $\begin{cases} x + y = 2 \\ y + z = 2 \\ 3x + 3y = 2 \end{cases}$

65. $\begin{cases} 3x + 4y + 2z = 4 \\ 6x - 2y + z = 4 \\ 3x - 8y - 6z = -3 \end{cases}$ **66.** $\begin{cases} x + y = 2 \\ y + z = 2 \\ x - z = 0 \end{cases}$

67. $\begin{cases} 2x - y - z = 0 \\ x - 2y - z = -1 \\ x - y - 2z = -1 \end{cases}$ **68.** $\begin{cases} x + 3y - z = 5 \\ 3x - y + z = 2 \\ 2x + y = 1 \end{cases}$

69. $\begin{cases} (x + y) + (y + z) + (z + x) = 6 \\ (x - y) + (y - z) + (z - x) = 0 \\ x + y + 2z = 4 \end{cases}$

70. $\begin{cases} (x + y) + (y + z) = 1 \\ (x + z) + (x + z) = 3 \\ (x - y) - (x - z) = -1 \end{cases}$

Applications

Use systems of equations to solve each problem.

71. Price of food items If Jonathan purchases two hamburgers and four orders of french fries for $8 and Hannah purchases three hamburgers and two orders of fries for $8, what is the price of each item?

72. Price of tennis equipment Hunter purchases two tennis rackets and four cans of tennis balls for $102. Jana purchases three tennis rackets and two cans of tennis balls for $141. What is the price of each item?

73. Planning for harvest A farmer raises corn and soybeans on 350 acres of land. Because of expected prices at harvest time, he thinks it would be wise to plant 100 more acres of corn than of soybeans. How many acres of each does he plant?

74. Club memberships There is an initiation fee to join the Pine River Country Club, as well as monthly dues. The total cost after 7 months' membership will be $3025, and after $1\frac{1}{2}$ years, $3850. Find both the initiation fee and the monthly dues.

75. Boating A General Jackson riverboat can travel 30 kilometers downstream in 3 hours and can make the return trip in 5 hours. Find the speed of the boat in still water.

76. Framing pictures A rectangular picture frame has a perimeter of 1900 centimeters and a width that is 250 centimeters less than its length. Find the area of the picture.

77. Making an alloy A metallurgist wants to make 60 grams of an alloy that is to be 34% copper. She has samples that are 9% copper and 84% copper. How many grams of each must she use?

78. Archimedes' law of the lever The two weights shown will be in balance if the product of one weight and its distance from the fulcrum is equal to the product of the other weight and its distance from the fulcrum. Two weights are in balance when one is 2 meters and the other 3 meters from the fulcrum. If the fulcrum remained in the same spot and the weights were interchanged, the closer weight would need to be increased by 5 pounds to maintain balance. Find the weights.

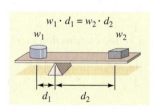

79. Lifting weights A 112-pound force can lift the 448-pound load shown. If the fulcrum is moved 1 additional foot away from the load, a 192-pound force is required. Find the length of the lever.

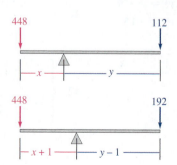

80. Writing test questions For a test question, a mathematics teacher wants to find two constants a and b such that the test item "Simplify $a(x + 2y) - b(2x - y)$" will have an answer of $-3x + 9y$. What constants a and b should the teacher use?

81. Break-even point Rollowheel, Inc., can manufacture a pair of in-line skates for $43.53. Daily fixed costs of manufacturing in-line skates amount to $742.72. A pair of in-line skates can be sold for $89.95. Find equations expressing the expenses E and the revenue R as functions of x, the number of pairs manufactured and sold. At what production level will expenses equal revenues?

82. Choosing salary options For its sales staff, a company offers two salary options. One is $326 per week plus a commission of $3\frac{1}{2}\%$ of sales. The other is $200 per week plus $4\frac{1}{4}\%$ of sales. Find equations that express incomes I_1 and I_2 as functions of sales x, and find the weekly sales level that produces equal salaries.

Use systems of three equations in three variables to solve each problem.

83. Work schedules A college student earns $198.50 per week working three part-time jobs. Half of his 30-hour work week is spent cooking hamburgers at a fast-food chain, earning $5.70 per hour. In addition, the student earns $6.30 per hour working at a gas station and $10 per hour doing janitorial work. How many hours per week does the student work at each job?

84. Investment income A woman invested a $22,000 rollover IRA account in three banks paying 5%, 6%, and 7% annual interest. She invested $2000 more at 6% than at 5%. The total annual interest she earned was $1370. How much did she invest at each rate?

85. Age distribution Approximately 3 million people live in Costa Rica. 2.61 million are younger than 50 years, and 1.95 million are older than 14 years. How many people are in each of the categories 0–14 years, 15–49 years, and 50 years and older?

86. Designing arches The engineer designing a parabolic arch knows that its equation has the form $y = ax^2 + bx + c$. Use the information in the illustration to find a, b, and c. Assume that the distances are given in feet. (*Hint:* The coordinates of points on the parabola satisfy its equation.)

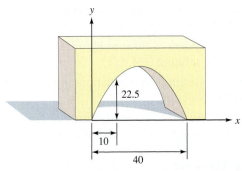

87. Geometry The sum of the angles of a triangle is 180°. In a certain triangle, the largest angle is 20° greater than the sum of the other two and is 10° greater than 3 times the smallest. How large is each angle?

88. Ballistics The path of a thrown object is a parabola with the equation $f(x) = ax^2 + bx + c$. Use the information in the illustration to find a, b, and c. (Distances are in feet.)

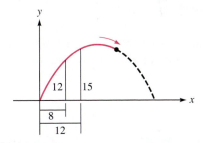

Discovery and Writing

89. If no method is stated, describe how you would determine the most efficient method to use to solve a linear system.

90. Describe how a system of three equations in three variables can be reduced to a system of two equations and two variables.

91. When using the elimination method, how can you tell whether the system has no solution?

92. When using the elimination method, how can you tell whether the system has infinitely many solutions?

 93. Use a graphing calculator to attempt to find the solution of the system $\begin{cases} x - 8y = -51 \\ 3x - 25y = -160 \end{cases}$.

94. Solve the system of Exercise 93 algebraically. Which method is easier, and why?

 95. Use a graphing calculator to attempt to find the solution of the system $\begin{cases} 17x - 23y = -76 \\ 29x + 19y = -278 \end{cases}$.

96. Solve the system of Exercise 95 algebraically. Which method is easier, and why?

97. Write a system of two equations in two variables with the solution $(-2, 5)$.

98. Write a system of three equations in three variables with the solution $(-4, 5, 1)$.

99. Write a system of two equations in two variables with no solution.

100. Write a system of three equations in three variables with an infinite number of solutions.

Critical Thinking

Determine if the statement is true or false. If the statement is false, then correct it and make it true.

101. If a system of two equations in two variables is represented by two lines with the same slope and different y-intercepts, then the system has an infinite number of solutions.

102. If a system of two equations in two variables is represented by two lines with negative reciprocal slopes, then the system has an infinite number of solutions.

103. If a linear system of three equations in three variables has infinitely many solutions, then any ordered triple is a solution of the system.

104. When using the graphing method, a system of two equations in two variables can appear to have no solution and yet have a unique one.

105. A linear system of two equations in three variables cannot have a unique solution.

106. If a linear system of two equations in two variables has a solution set involving fractions, then use the graphing method to ensure accuracy.

107. To solve a linear system of three equations in three variables, we use the graphing method.

108. The system of equations $999x - 999y = 999$ and $-999x + 999y = -999$ has an infinite number of solutions.

109. The system of equations $-777x + 777y = -777$ and $777x - 777y = -777$ has no solution.

110. The system of equations $555x + 555y = 555$ and $555x - 555y = -555$ has no solution.

6.2 Gaussian Elimination and Matrix Methods

In this section, we will learn to

1. Write a system matrix in row-echelon form and solve the system.
2. Solve an inconsistent system.
3. Solve a system of dependent equations.
4. Solve systems using Gauss–Jordan elimination.

Since the invention of modern day electronics, dot matrix screens with rectangular arrays of dots (or pixels) have become important. For example, we all are familiar with signs like the football scoreboard shown, a sign in front of a bank that gives the date and temperature, or a sign that shows which lanes on the toll road are open.

In this section, we will show how rectangular arrays of numbers can be used to solve systems of equations. We begin by introducing a method for solving systems, called **Gaussian elimination**, after the German mathematician Carl Friedrich Gauss (1777–1855). In this method, we transform a system of equations into an equivalent system that can be solved by a process called **back substitution**.

Using Gaussian elimination, we solve systems of equations by working only with the coefficients of the variables. From the system of equations

$$\begin{cases} x + 2y + z = 8 \\ 2x + y - z = 1 \\ x + y - 2z = -3 \end{cases}$$

for example, we can form three rectangular arrays of numbers. Each is called a **matrix**. The first is the **coefficient matrix**, which contains the coefficients of the variables of the system. The second matrix contains the constants from the right side of the system. We can write the third matrix, called the **augmented matrix** or the **system matrix**, by joining the two. The coefficient matrix appears to the left of the dashed line, and the matrix of constants appears to the right.

Coefficient matrix   *Constants*   *Augmented matrix*

$$\begin{bmatrix} 1 & 2 & 1 \\ 2 & 1 & -1 \\ 1 & 1 & -2 \end{bmatrix} \qquad \begin{bmatrix} 8 \\ 1 \\ -3 \end{bmatrix} \qquad \left[\begin{array}{ccc|c} 1 & 2 & 1 & 8 \\ 2 & 1 & -1 & 1 \\ 1 & 1 & -2 & -3 \end{array}\right]$$

> **Tip**
>
> The rows and columns of a classroom setup can serve as an example of a matrix.

Each row of the augmented matrix represents one equation of the system.

- The first row represents the equation $x + 2y + z = 8$.
- The second row represents the equation $2x + y - z = 1$.
- The third row represents the equation $x + y - 2z = -3$.

Because the system matrix has three rows and four columns, it is called a 3×4 matrix (read as "3 by 4"). The coefficient matrix is 3×3, and the constants form a 3×1 matrix.

To illustrate Gaussian elimination, we solve this system by the elimination method of the previous section. In a second column, we keep track of the changes to the augmented or system matrix.

Using Gaussian Elimination to Solve a System of Equations

EXAMPLE 1

Use Gaussian elimination to solve
$$\begin{array}{ll} (1) \\ (2) \\ (3) \end{array} \quad \begin{cases} x + 2y + z = 8 \\ 2x + y - z = 1 \\ x + y - 2z = -3 \end{cases}$$

SOLUTION   We multiply each term in Equation 1 by -2 and add the result to Equation 2 to obtain Equation 4. Then we multiply each term of Equation 1 by -1 and add the result to Equation 3 to obtain Equation 5. This gives the equivalent system

> **Tip**
>
> Always align the variables in the equations vertically. Use 0 to indicate coefficients of zero in the matrix. The last column represents the constant terms.

$$\begin{array}{l} (1) \\ (4) \\ (5) \end{array} \quad \begin{cases} x + 2y + z = 8 \\ -3y - 3z = -15 \\ -y - 3z = -11 \end{cases} \qquad \left[\begin{array}{ccc|c} 1 & 2 & 1 & 8 \\ 0 & -3 & -3 & -15 \\ 0 & -1 & -3 & -11 \end{array}\right]$$

Now we divide both sides of Equation 4 by -3 to obtain Equation 6

$$\begin{array}{l} (1) \\ (6) \\ (5) \end{array} \quad \begin{cases} x + 2y + z = 8 \\ y + z = 5 \\ -y - 3z = -11 \end{cases} \qquad \left[\begin{array}{ccc|c} 1 & 2 & 1 & 8 \\ 0 & 1 & 1 & 5 \\ 0 & -1 & -3 & -11 \end{array}\right]$$

and add Equation 6 to Equation 5 to obtain Equation 7.

$$\begin{array}{l} (1) \\ (6) \\ (7) \end{array} \quad \begin{cases} x + 2y + z = 8 \\ y + z = 5 \\ -2z = -6 \end{cases} \qquad \left[\begin{array}{ccc|c} 1 & 2 & 1 & 8 \\ 0 & 1 & 1 & 5 \\ 0 & 0 & -2 & -6 \end{array}\right]$$

Finally, we divide both sides of Equation 7 by -2 to obtain the system

$$\begin{array}{l} (1) \\ (6) \end{array} \quad \begin{cases} x + 2y + z = 8 \\ y + z = 5 \\ z = 3 \end{cases} \qquad \left[\begin{array}{ccc|c} 1 & 2 & 1 & 8 \\ 0 & 1 & 1 & 5 \\ 0 & 0 & 1 & 3 \end{array}\right]$$

The system can now be solved by back substitution. Because $z = 3$, we can substitute 3 for z in Equation 6 and solve for y:

(6) $y + z = 5$

$\quad\quad y + 3 = 5$

$\quad\quad\quad y = 2$

We can now substitute 2 for y and 3 for z in Equation 1 and solve for x:

(1) $x + 2y + z = 8$

$\quad\quad x + 2(2) + 3 = 8$

$\quad\quad\quad x + 7 = 8$

$\quad\quad\quad\quad x = 1$

The solution is the ordered triple $(1, 2, 3)$.

Self Check 1   Use Gaussian elimination to solve $\begin{cases} x - 3y + z = 2 \\ 2x + y + z = 10 \\ 3x - y - 2z = 9 \end{cases}$

Now Try Exercise 17.

1. Write an Augmented Matrix in Row-Echelon Form and Solve the System

In Example 1, we performed operations on the equations as well as on the corresponding matrices. These matrix operations are called **elementary row operations**.

Elementary Row Operations   If any of the following operations are performed on the rows of a system matrix, the matrix of an equivalent system results:

Type 1 row operation: Two rows of a matrix can be interchanged.

Type 2 row operation: The elements of a row of a matrix can be multiplied by a nonzero constant.

Type 3 row operation: Any row can be changed by adding a multiple of another row to it.

- A type 1 row operation is equivalent to writing the equations of a system in a different order.
- A type 2 row operation is equivalent to multiplying both sides of an equation by a nonzero constant.
- A type 3 row operation is equivalent to adding a multiple of one equation to another.

Any matrix that can be obtained from another matrix by a sequence of elementary row operations is **row equivalent** to the original matrix. This means that row equivalent matrices represent equivalent systems of equations.

The process of Example 1 changed a system matrix into a special row equivalent form, called **row-echelon form**.

| Row-Echelon Form of a Matrix | A matrix is in **row-echelon form** if it has these three properties: |

1. The first nonzero entry in each row (called the **leading entry**) is 1.
2. Leading entries appear farther to the right as we move down the rows of the matrix.
3. Any rows containing only 0's are at the bottom of the matrix.

Tip

Echelon is a term which means "a steplike for mation" as in the way geese fly.

©Myotis/Shutterstock.com

Examples of Matrices in Row-Echelon Form

The matrices

$$\begin{bmatrix} 1 & 2 & 3 & 6 & 5 \\ 0 & 1 & -2 & -7 & 4 \\ 0 & 0 & 1 & 2 & -2 \\ 0 & 0 & 0 & 1 & 2 \end{bmatrix}, \begin{bmatrix} 1 & 3 & 0 & 7 \\ 0 & 0 & 1 & 8 \\ 0 & 0 & 0 & 0 \end{bmatrix}, \begin{bmatrix} 0 & 1 & 2 \\ 0 & 0 & 1 \end{bmatrix}, \text{ and } \begin{bmatrix} 1 & 1 \\ 0 & 0 \\ 0 & 0 \\ 0 & 0 \end{bmatrix}$$

are in row-echelon form, because the first nonzero entry in each row is 1, the leading entries appear farther to the right as we move down the rows, and any rows that consist entirely of 0's are at the bottom of the matrix.

Examples of Matrices That Are Not in Row-Echelon Form

The next three matrices are not in row-echelon form.

$$\begin{bmatrix} 1 & 3 & 0 & 7 \\ 0 & 0 & 1 & 8 \\ 0 & 0 & 1 & 0 \end{bmatrix}$$
The leading entry in the last row is not to the right of the leading entry of the middle row.

$$\begin{bmatrix} 0 & 3 & 0 \\ 0 & 0 & 1 \end{bmatrix}$$
The leading entry of the first row is not 1.

$$\begin{bmatrix} 0 & 0 \\ 1 & 0 \\ 0 & 1 \end{bmatrix}$$
The row of 0's is not last.

Writing an Augmented Matrix in Row-Echelon Form and Using Back Substitution to Solve the System

EXAMPLE 2

Solve the following system by transforming its augmented matrix into row-echelon form and back substituting.

$$\begin{cases} x + 2y + 3z = 4 \\ 2x - y - 2z = 0 \\ x - 3y - 3z = -2 \end{cases}$$

SOLUTION   This system is represented by the augmented matrix

$$\begin{bmatrix} 1 & 2 & 3 & | & 4 \\ 2 & -1 & -2 & | & 0 \\ 1 & -3 & -3 & | & -2 \end{bmatrix}$$

We will reduce the augmented matrix to row-echelon form. We can use the 1 in the upper left corner to zero out the rest of the first column. To do so, we multiply the first row by -2 and add the result to the second row. We indicate this operation with the notation $(-2)R1 + R2$.

$$\begin{bmatrix} 1 & 2 & 3 & | & 4 \\ 2 & -1 & -2 & | & 0 \\ 1 & -3 & -3 & | & -2 \end{bmatrix} \begin{matrix} \\ (-2)R1 + R2 \rightarrow \\ \\ \end{matrix} \begin{bmatrix} 1 & 2 & 3 & | & 4 \\ 0 & -5 & -8 & | & -8 \\ 1 & -3 & -3 & | & -2 \end{bmatrix}$$

Next, we multiply row one by -1 and add the result to row three to get a new row three. This fills the rest of the first column with 0's.

$(-1)R1 + R3 \rightarrow \begin{bmatrix} 1 & 2 & 3 & \vdots & 4 \\ 0 & -5 & -8 & \vdots & -8 \\ 0 & -5 & -6 & \vdots & -6 \end{bmatrix}$

We multiply row two by –1 and add the result to row three to get

$(-1)R2 + R3 \rightarrow \begin{bmatrix} 1 & 2 & 3 & \vdots & 4 \\ 0 & -5 & -8 & \vdots & -8 \\ 0 & 0 & 2 & \vdots & 2 \end{bmatrix}$

Finally, we multiply row two by $-\frac{1}{5}$ and row three by $\frac{1}{2}$:

$\begin{matrix} (-\frac{1}{5})R_2 \rightarrow \\ (\frac{1}{2})R3 \rightarrow \end{matrix} \begin{bmatrix} 1 & 2 & 3 & \vdots & 4 \\ 0 & 1 & \frac{8}{5} & \vdots & \frac{8}{5} \\ 0 & 0 & 1 & \vdots & 1 \end{bmatrix}$

The final matrix, now in row-echelon form, represents the equivalent system

$\begin{cases} x + 2y + 3z = 4 \\ y + \frac{8}{5}z = \frac{8}{5} \\ z = 1 \end{cases}$

We can now use back substitution to solve the system. To find y, we substitute 1 for z in the second equation.

$y + \frac{8}{5}z = \frac{8}{5}$

$y + \frac{8}{5}(1) = \frac{8}{5}$

$y = 0$

To solve for x, we substitute 0 for y and 1 for z in the first equation.

$x + 2y + 3z = 4$

$x + 2(0) + 3(1) = 4$

$x + 3 = 4$

$x = 1$

The solution of the original system is the triple $(x, y, z) = (1, 0, 1)$.

Self Check 2 Solve: $\begin{cases} x + y + 2z = 5 \\ y - 2z = 0 \\ x - z = 0 \end{cases}$.

Now Try Exercise 35.

Accent on Technology

Reduce a Matrix to Row-Echelon Form

A matrix can be reduced to row-echelon form with a graphing calculator. As an example, we will reduce the matrix $\begin{bmatrix} 1 & 2 & 3 & 4 \\ 2 & -1 & -2 & 0 \\ 1 & -3 & -3 & -2 \end{bmatrix}$ to row-echelon form by following the steps shown in Figure 6-5.

1. The MATRIX menu is accessed by pressing `2nd` `x⁻¹`.

2. The window we will see will list the matrices that are available for edit.

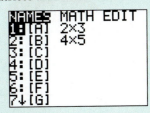

3. Move to EDIT, select matrix [A], and press `ENTER`. You will see this screen.

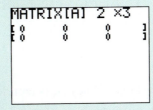

4. Change the dimensions of the matrix to 3X4, and use the cursor to enter the values of the matrix.

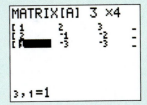

5. To change the matrix to row-echelon form, go to the home screen; access the MATRIX MATH menu; select A:ref(.

6. Now select the MATRIX menu again and select [A]. Press `ENTER` and the row-echelon form is shown.

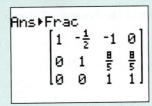

7. If we want the answers in fraction form, we can add that command by selecting `MATH` 1: Frac.

$$\begin{bmatrix} 1 & -\frac{1}{2} & -1 & 0 \\ 0 & 1 & \frac{8}{5} & \frac{8}{5} \\ 0 & 0 & 1 & 1 \end{bmatrix}$$

FIGURE 6-5

Caution

The row-echelon form we obtained using the graphing calculator as shown in Figure 6-5 is not the same as the row-echelon form found in Example 2. However, if we multiply row two of the matrix shown in Figure 6-5 by $\frac{5}{2}$ and add the results to row one, we obtain the equivalent row-echelon form found in Example 2.

$$\begin{bmatrix} 1 & -\frac{1}{2} & -1 & 0 \\ 0 & 1 & \frac{8}{5} & \frac{8}{5} \\ 0 & 0 & 1 & 1 \end{bmatrix} \quad \left(\tfrac{5}{2}\right)R2 + R1 \rightarrow \begin{bmatrix} 1 & 2 & 3 & 4 \\ 0 & 1 & \frac{8}{5} & \frac{8}{5} \\ 0 & 0 & 1 & 1 \end{bmatrix}$$

If we back substitute into the equations represented by either row-echelon form, we will obtain the same values for x, y, and z.

Since the systems of Examples 1 and 2 have solutions, each system is consistent. The next two examples illustrate a system that is inconsistent and a system whose equations are dependent.

2. Solve an Inconsistent System

Using Matrix Methods to Solve an Inconsistent System of Equations

EXAMPLE 3

Use matrix methods to solve $\begin{cases} x + y + z = 3 \\ 2x - y + z = 2 \\ 3y + z = 1 \end{cases}$.

SOLUTION

We form the augmented matrix and use row operations to write it in row-echelon form. We use the 1 in the top left position to zero out the rest of the first column.

$$\begin{bmatrix} 1 & 1 & 1 & 3 \\ 2 & -1 & 1 & 2 \\ 0 & 3 & 1 & 1 \end{bmatrix} \xrightarrow{(-2)R1 + R2} \begin{bmatrix} 1 & 1 & 1 & 3 \\ 0 & -3 & -1 & -4 \\ 0 & 3 & 1 & 1 \end{bmatrix}$$

$$\xrightarrow{R2 + R3} \begin{bmatrix} 1 & 1 & 1 & 3 \\ 0 & -3 & -1 & -4 \\ 0 & 0 & 0 & -3 \end{bmatrix}$$

Since the last row of the matrix represents the equation

$$0x + 0y + 0z = -3$$

and no values of x, y, and z could make $0 = -3$, there is no point in continuing. The given system has no solution and is inconsistent.

Self Check 3

Solve: $\begin{cases} x + y + z = 5 \\ y - z = 1 \\ x + 2y = 3 \end{cases}$.

Now Try Exercise 43.

3. Solve a System of Dependent Equations

Using Matrix Methods to Solve a System of Dependent Equations

EXAMPLE 4

Use matrices to solve $\begin{cases} x + 2y + z = 8 \\ 2x + y - z = 1 \\ x - y - 2z = -7 \end{cases}$.

SOLUTION

We can set up the augmented matrix and use row operations to reduce it to row-echelon form:

$$\begin{bmatrix} 1 & 2 & 1 & 8 \\ 2 & 1 & -1 & 1 \\ 1 & -1 & -2 & -7 \end{bmatrix} \begin{matrix} \\ \xrightarrow{(-2)R1 + R2} \\ \xrightarrow{(-1)R1 + R3} \end{matrix} \begin{bmatrix} 1 & 2 & 1 & 8 \\ 0 & -3 & -3 & -15 \\ 0 & -3 & -3 & -15 \end{bmatrix}$$

$$\begin{matrix} \xrightarrow{(-\frac{1}{3})R2} \\ \xrightarrow{(-\frac{1}{3})R3} \end{matrix} \begin{bmatrix} 1 & 2 & 1 & 8 \\ 0 & 1 & 1 & 5 \\ 0 & 1 & 1 & 5 \end{bmatrix}$$

$$\xrightarrow{(-1)R2 + R3} \begin{bmatrix} 1 & 2 & 1 & 8 \\ 0 & 1 & 1 & 5 \\ 0 & 0 & 0 & 0 \end{bmatrix}$$

The final matrix is in row-echelon form and represents the system

$$\begin{array}{ll}(1) & \\ (2) & \\ (3) & \end{array} \left\{ \begin{array}{l} x + 2y + z = 8 \\ y + z = 5 \\ 0x + 0y + 0z = 0 \end{array} \right.$$

Since all coefficients in Equation 3 are 0, it can be ignored. To solve this system by back substitution, we solve Equation 2 for y in terms of z:

$$y = 5 - z$$

and then substitute $5 - z$ for y in Equation 1 and solve for x in terms of z.

$$(1) \qquad x + 2\textcolor{red}{y} + z = 8$$

$$x + 2(\textcolor{red}{5 - z}) + z = 8$$

$$x + 10 - 2z + z = 8 \qquad \text{\textcolor{red}{Remove parentheses.}}$$

$$x + 10 - z = 8 \qquad \text{\textcolor{red}{Combine like terms.}}$$

$$x = -2 + z \qquad \text{\textcolor{red}{Solve for } x.}$$

The solution of this system is $(x, y, z) = (-2 + z, 5 - z, z)$.

There are infinitely many solutions to this system. We can choose any real number for z, but once it is chosen, x and y are determined. For example,

- If $z = 3$, then $x = 1$ and $y = 2$. One possible solution of this system is $x = 1$, $y = 2$, and $z = 3$.
- If $z = 2$, then $x = 0$ and $y = 3$. Another solution is $x = 0$, $y = 3$, and $z = 2$.

Because this system has solutions, it is consistent. Because there are infinitely many solutions, the equations are dependent.

Self Check 4   Solve: $\left\{ \begin{array}{l} x + 2y + z = 2 \\ x - z = 2 \\ 2x + 2y = 4 \end{array} \right.$.

Now Try Exercise 41.

4. Solve Systems Using Gauss–Jordan Elimination

In Gaussian elimination, we perform row operations until the system matrix is in row-echelon form. Then we convert the matrix back into equation form and solve by back substitution. A modification of this method, called **Gauss–Jordan elimination**, uses row operations to produce a matrix in *reduced* **row-echelon** form. With this method, we can obtain the solution of the system from that matrix directly, and back substitution is not needed.

Reduced Row-Echelon Form of a Matrix   A matrix is in **reduced row-echelon form** if

1. It is in row-echelon form.
2. Entries above each leading entry are also zero.

Examples of Matrices in Reduced Row-Echelon Form

The matrices

$$\begin{bmatrix} 1 & 0 & 0 \\ 0 & 1 & 0 \end{bmatrix}, \quad \begin{bmatrix} 1 & 0 & 0 & 7 \\ 0 & 1 & 0 & 8 \\ 0 & 0 & 1 & 3 \end{bmatrix}$$

are in reduced row-echelon form, because each matrix is in row-echelon form and the entries above each leading entry are 0's.

Examples of Matrices That Are Not in Reduced Row-Echelon Form

$$\begin{bmatrix} 1 & 3 & 0 & 7 \\ 0 & 0 & 1 & 8 \\ 0 & 0 & 1 & 0 \end{bmatrix} \quad \text{The matrix is not in row-echelon form.}$$

$$\begin{bmatrix} 0 & 3 & 5 \\ 0 & 1 & 1 \end{bmatrix} \quad \text{The number above the leading entry of 1 is not 0.}$$

In the next example, we solve a system by Gauss–Jordan elimination.

EXAMPLE 5

Using Gauss–Jordan Elimination to Solve a System of Equations

Use Gauss–Jordan elimination to solve $\begin{cases} w + 2x + 3y + z = 4 \\ x + 4y - z = 0 \\ w - x - y + 2z = 2 \end{cases}$.

SOLUTION We will use row operations to reduce the augmented matrix as follows. We first use the 1 in the upper left corner to zero out the rest of the first column:

$$\begin{bmatrix} 1 & 2 & 3 & 1 & 4 \\ 0 & 1 & 4 & -1 & 0 \\ 1 & -1 & -1 & 2 & 2 \end{bmatrix} \xrightarrow{(-1)R1 + R3} \begin{bmatrix} 1 & 2 & 3 & 1 & 4 \\ 0 & 1 & 4 & -1 & 0 \\ 0 & -3 & -4 & 1 & -2 \end{bmatrix}$$

The leading entry in the second row is already 1. We use it to zero out the rest of the second column.

$$\begin{matrix} (-2)R2 + R1 \to \\ \\ (3)R2 + R3 \to \end{matrix} \begin{bmatrix} 1 & 0 & -5 & 3 & 4 \\ 0 & 1 & 4 & -1 & 0 \\ 0 & 0 & 8 & -2 & -2 \end{bmatrix}$$

To make the leading entry in the third row equal to 1, we multiply the third row by $\frac{1}{8}$ and use it to zero out the rest of the third column.

$$\left(\tfrac{1}{8}\right)R3 \to \begin{bmatrix} 1 & 0 & -5 & 3 & 4 \\ 0 & 1 & 4 & -1 & 0 \\ 0 & 0 & 1 & -\frac{1}{4} & -\frac{1}{4} \end{bmatrix}$$

$$\begin{matrix} (5)R3 + R1 \to \\ (-4)R3 + R2 \to \\ \\ \end{matrix} \begin{bmatrix} 1 & 0 & 0 & \frac{7}{4} & \frac{11}{4} \\ 0 & 1 & 0 & 0 & 1 \\ 0 & 0 & 1 & -\frac{1}{4} & -\frac{1}{4} \end{bmatrix}$$

The final matrix is in reduced row-echelon form. Note that each leading entry is 1, and each is alone in its column. The matrix represents the system of equations

$$\begin{cases} w + \dfrac{7}{4}z = \dfrac{11}{4} \\ x = 1 \\ y - \dfrac{1}{4}z = -\dfrac{1}{4} \end{cases} \quad \text{or} \quad \begin{cases} w = \dfrac{11}{4} - \dfrac{7}{4}z \\ x = 1 \\ y = -\dfrac{1}{4} + \dfrac{1}{4}z \end{cases}$$

A general solution is $(w, x, y, z) = \left(\dfrac{11}{4} - \dfrac{7}{4}z, 1, -\dfrac{1}{4} + \dfrac{1}{4}z, z\right)$. To find some specific solutions, we choose values for z, and the corresponding values of w, x, and y will be determined. For example, if $z = 1$, then $w = 1$, $x = 1$, and $y = 0$. Thus, $(w, x, y, z) = (1, 1, 0, 1)$ is a solution. If $z = -1$, another solution is $(w, x, y, z) = \left(\dfrac{9}{2}, 1, -\dfrac{1}{2}, -1\right)$.

Self Check 5   Solve: $\begin{cases} w + x + y - z = 2 \\ 2w + 2x + 2y - 2z = 3 \\ w + x - y + 3z = 4 \\ w + 3x + 2y - 2z = -1 \end{cases}$.

Now Try Exercise 59.

In Example 5, there were more variables than equations. In the next example, there are more equations than variables.

EXAMPLE 6   **Using Gauss–Jordan Elimination to Solve a System of Equations**

Use Gauss–Jordan elimination to solve $\begin{cases} 2x + y = 4 \\ x - 3y = 9 \\ x + 4y = -5 \end{cases}$.

SOLUTION   To get a 1 in the top left corner, we could multiply the first row by $\frac{1}{2}$, but that would introduce fractions. Instead, we can exchange the first two rows and proceed as follows:

$$\begin{bmatrix} 2 & 1 & | & 4 \\ 1 & -3 & | & 9 \\ 1 & 4 & | & -5 \end{bmatrix} \begin{array}{c} R1 \leftrightarrow R2 \end{array} \begin{bmatrix} 1 & -3 & | & 9 \\ 2 & 1 & | & 4 \\ 1 & 4 & | & -5 \end{bmatrix}$$

$$\begin{array}{c} (-2)R1 + R2 \rightarrow \\ (-1)R1 + R3 \rightarrow \end{array} \begin{bmatrix} 1 & -3 & | & 9 \\ 0 & 7 & | & -14 \\ 0 & 7 & | & -14 \end{bmatrix}$$

$$\begin{array}{c} \left(\frac{1}{7}\right)R2 \rightarrow \\ \left(\frac{1}{7}\right)R3 \rightarrow \end{array} \begin{bmatrix} 1 & -3 & | & 9 \\ 0 & 1 & | & -2 \\ 0 & 1 & | & -2 \end{bmatrix}$$

$$\begin{array}{c} (3)R2 + R1 \rightarrow \\ \\ (-1)R2 + R3 \rightarrow \end{array} \begin{bmatrix} 1 & 0 & | & 3 \\ 0 & 1 & | & -2 \\ 0 & 0 & | & 0 \end{bmatrix}$$

This final matrix is in reduced row-echelon form. It represents the system

$$\begin{cases} x = 3 \\ y = -2 \end{cases}$$

Verify that $x = 3$ and $y = -2$ satisfy the equations of the original system.

Self Check 6   Solve: $\begin{cases} x + y = 5 \\ x - y = 1 \\ x + 2y = 7 \end{cases}$.

Now Try Exercise 67.

Accent on Technology

Reduce a Matrix to Reduced Row-Echelon Form

A graphing calculator can be used to place a matrix in reduced row-echelon form. Enter the augmented matrix found in Example 6 and then go to the home screen and select the function shown.

 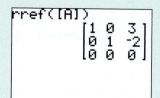

FIGURE 6-6

Self Check Answers

1. $(4, 1, 1)$ **2.** $(1, 2, 1)$ **3.** no solution; inconsistent system
4. dependent equations; a general solution is $(2 + z, -z, z)$
5. no solution; inconsistent system   **6.** $(3, 2)$

Exercises 6.2

Getting Ready

You should be able to complete these vocabulary and concept statements before you proceed to the practice exercises.

Fill in the blanks.

1. A rectangular array of numbers is called a _____.

2. A 3×5 matrix has ___ rows and ___ columns.

3. The matrix containing the coefficients of the variables is called the _____ matrix.

4. The coefficient matrix joined to the column of constants is called the _____ matrix or the _____ matrix.

5. Each row of a system matrix represents one _____.

6. The rows of the system matrix are changed using elementary _____.

7. If one augmented matrix is changed to another using row operations, the matrices are _____.

8. If two augmented matrices are row equivalent, then the systems have the _____.

9. In a type 1 row operation, two rows of a matrix can be _____.

10. In a type 2 row operation, one entire row can be _____ by a nonzero constant.

11. In a type 3 row operation, any row can be changed by _____ to it any _____ of another row.

12. The first nonzero entry in a row is called that row's _____.

Practice

Use Gaussian elimination to solve each system.

13. $\begin{cases} x + y = 5 \\ x - 2y = -4 \end{cases}$

14. $\begin{cases} x + 3y = 8 \\ 2x - 5y = 5 \end{cases}$

15. $\begin{cases} x - y = 1 \\ 2x - y = 8 \end{cases}$

16. $\begin{cases} x - 5y = 4 \\ 2x + 3y = 21 \end{cases}$

17. $\begin{cases} x + 2y - z = 2 \\ x - 3y + 2z = 1 \\ x + y - 3z = -6 \end{cases}$

18. $\begin{cases} x + 5y - z = 2 \\ x + 2y + z = 3 \\ x + y + z = 2 \end{cases}$

19. $\begin{cases} x - y - z = -3 \\ 5x + y = 6 \\ y + z = 4 \end{cases}$ **20.** $\begin{cases} x + y = 1 \\ x + z = 3 \\ y + z = 2 \end{cases}$

41. $\begin{cases} x + y - z = 5 \\ x + y + z = 2 \\ 3x + 3y - z = 12 \end{cases}$ **42.** $\begin{cases} x + y + 2z = 4 \\ -x - y - 3z = -5 \\ 2x + y + z = 2 \end{cases}$

Determine whether each matrix is in row-echelon form, reduced row-echelon form, or neither.

21. $\begin{bmatrix} 1 & 3 & 0 & 5 \\ 0 & 1 & 2 & 7 \\ 0 & 0 & 1 & 0 \end{bmatrix}$ **22.** $\begin{bmatrix} 1 & 3 & 0 & 5 \\ 0 & 1 & 2 & 7 \\ 0 & 0 & 0 & 0 \end{bmatrix}$

43. $\begin{cases} 2x - y + z = 6 \\ 3x + y - z = 2 \\ -x + 3y - 3z = 8 \end{cases}$ **44.** $\begin{cases} -x + 3y + 2z = -10 \\ 3x - 2y - 2z = 7 \\ -2x + y - z = -10 \end{cases}$

23. $\begin{bmatrix} 1 & 0 & 1 \\ 0 & 1 & 5 \\ 0 & 0 & 0 \\ 0 & 0 & 0 \end{bmatrix}$ **24.** $\begin{bmatrix} 1 & 0 & 1 \\ 0 & 1 & 5 \\ 0 & 0 & 1 \\ 0 & 0 & 0 \end{bmatrix}$

Write each system as a matrix and solve it by Gauss–Jordan elimination. If a system has infinitely many solutions, show a general solution.

45. $\begin{cases} x - 2y = 7 \\ y = 3 \end{cases}$ **46.** $\begin{cases} x - 2y = 7 \\ y = 8 \end{cases}$

Write each system as a matrix and solve it by Gaussian elimination. If a system has infinitely many solutions, show a general solution.

47. $\begin{cases} x + 2y - z = 3 \\ y + 3z = 1 \\ z = -2 \end{cases}$ **48.** $\begin{cases} x - 3y + 2z = -1 \\ y - 2z = 3 \\ z = 5 \end{cases}$

25. $\begin{cases} 2x + y = 3 \\ x - 3y = 5 \end{cases}$ **26.** $\begin{cases} x + 2y = -1 \\ 3x - 5y = 19 \end{cases}$

49. $\begin{cases} x - y = 7 \\ x + y = 13 \end{cases}$ **50.** $\begin{cases} x + 2y = 7 \\ 2x - y = -1 \end{cases}$

27. $\begin{cases} x - 7y = -2 \\ 5x - 2y = -10 \end{cases}$ **28.** $\begin{cases} 3x - y = 3 \\ 2x + y = -3 \end{cases}$

51. $\begin{cases} x - \dfrac{1}{2}y = 0 \\ x + 2y = 0 \end{cases}$ **52.** $\begin{cases} x - y = 5 \\ -x + \dfrac{1}{5}y = -9 \end{cases}$

29. $\begin{cases} 2x - y = 5 \\ x + 3y = 6 \end{cases}$ **30.** $\begin{cases} 3x - 5y = -25 \\ 2x + y = 5 \end{cases}$

31. $\begin{cases} x - 2y = 3 \\ -2x + 4y = 6 \end{cases}$ **32.** $\begin{cases} 2(2y - x) = 6 \\ 4y = 2(x + 3) \end{cases}$

53. $\begin{cases} x + y + 2z = 0 \\ x + y + z = 2 \\ x + z = 1 \end{cases}$ **54.** $\begin{cases} x + 2y = -3 \\ x + 4y = -2 \\ 2x + z = -8 \end{cases}$

55. $\begin{cases} 2x + y - 2z = 1 \\ -x + y - 3z = 0 \\ 4x + 3y = 4 \end{cases}$ **56.** $\begin{cases} 3x + y = 3 \\ 3x + y - z = 2 \\ 6x + z = 5 \end{cases}$

33. $\begin{cases} 2x - y = 7 \\ -x + \dfrac{1}{3}y = -\dfrac{7}{3} \end{cases}$ **34.** $\begin{cases} 45x - 6y = 60 \\ 30x + 15y = 63.75 \end{cases}$

35. $\begin{cases} x - y + z = 3 \\ 2x - y + z = 4 \\ x + 2y - z = -1 \end{cases}$ **36.** $\begin{cases} 2x + y - z = 1 \\ x + y - z = 0 \\ 3x + y + 2z = 2 \end{cases}$

57. $\begin{cases} 2x - 2y + 3z + t = 2 \\ x + y + z + t = 5 \\ -x + 2y - 3z + 2t = 2 \\ x + y + 2z - t = 4 \end{cases}$ **58.** $\begin{cases} x + y + 2z + t = 1 \\ x + 2y + z + t = 2 \\ 2x + y + z + t = 4 \\ x + y + z + 2t = 3 \end{cases}$

37. $\begin{cases} x + y - z = -1 \\ 3x + y = 4 \\ y - 2z = -4 \end{cases}$ **38.** $\begin{cases} 3x + y = 7 \\ x - z = 0 \\ y - 2z = -8 \end{cases}$

59. $\begin{cases} x + y + t = 4 \\ x + z + t = 2 \\ 2x + 2y + z + 2t = 8 \\ x - y + z - t = -2 \end{cases}$ **60.** $\begin{cases} x - y + 2z + t = 3 \\ 3x - 2y - z - t = 4 \\ 2x + y + 2z - t = 10 \\ x + 2y + z - 3t = 8 \end{cases}$

39. $\begin{cases} x - y + z = 2 \\ 2x + y + z = 5 \\ 3x - 4z = -5 \end{cases}$ **40.** $\begin{cases} x + z = -1 \\ 3x + y = 2 \\ 2x + y + 5z = 3 \end{cases}$

61. $\begin{cases} \dfrac{1}{3}x + \dfrac{3}{4}y - \dfrac{2}{3}z = -2 \\ x + \dfrac{1}{2}y + \dfrac{1}{3}z = 1 \\ \dfrac{1}{6}x - \dfrac{1}{8}y - z = 0 \end{cases}$ **62.** $\begin{cases} \dfrac{1}{4}x + y + 3z = 1 \\ \dfrac{1}{2}x - 4y + 6z = -1 \\ \dfrac{1}{3}x - 2y - 2z = -1 \end{cases}$

63. $\begin{cases} \dfrac{1}{2}x + \dfrac{1}{4}y - z = 2 \\ \dfrac{2}{3}x + \dfrac{1}{4}y + \dfrac{1}{2}z = \dfrac{3}{2} \\ \dfrac{2}{3}x + z = -\dfrac{1}{3} \end{cases}$ **64.** $\begin{cases} \dfrac{5}{7}x - \dfrac{1}{3}y + z = 0 \\ \dfrac{2}{7}x + y + \dfrac{1}{8}z = 9 \\ 6x + 4y - \dfrac{27}{4}z = 20 \end{cases}$

65. $\begin{cases} 3x - 6y + 9z = 18 \\ 2x - 4y + 3z = 12 \\ x - 2y + 3z = 6 \end{cases}$ **66.** $\begin{cases} x + 2y - z = 7 \\ 2x - y + z = 2 \\ 3x - 4y + 3z = -3 \end{cases}$

Each system contains a different number of equations than variables. Solve each system using Gauss–Jordan elimination. If a system has infinitely many solutions, show a general solution.

67. $\begin{cases} x + y = -2 \\ 3x - y = 6 \\ 2x + 2y = -4 \\ x - y = 4 \end{cases}$ **68.** $\begin{cases} x - y = -3 \\ 2x + y = -3 \\ 3x - y = -7 \\ 4x + y = -7 \end{cases}$

69. $\begin{cases} x + 2y + z = 4 \\ 3x - y - z = 2 \end{cases}$ **70.** $\begin{cases} x + 2y - 3z = 5 \\ 5x + y - z = -11 \end{cases}$

71. $\begin{cases} w + x = 1 \\ w + y = 0 \\ x + z = 0 \end{cases}$ **72.** $\begin{cases} w + x - y + z = 2 \\ 2w - x - 2y + z = 0 \\ w - 2x - y + z = -1 \end{cases}$

73. $\begin{cases} x + y = 3 \\ 2x + y = 1 \\ 3x + 2y = 2 \end{cases}$ **74.** $\begin{cases} x + 2y + z = 4 \\ x - y + z = 1 \\ 2x + y + 2z = 2 \\ 3x + 3z = 6 \end{cases}$

Applications

Use matrix methods to solve each problem.

75. Flight range The speed of an airplane with a tailwind is 300 miles per hour and with a headwind is 220 miles per hour. On a day with no wind, how far could the plane travel on a 5-hour fuel supply?

76. Resource allocation 120,000 gallons of fuel are to be divided between two airlines. Triple A Airways requires twice as much as UnityAir. How much fuel should be allocated to Triple A?

77. Library shelving To use space effectively, librarians like to fill shelves completely. One 35-inch shelf can hold 3 dictionaries, 5 atlases, and 1 thesaurus; or 6 dictionaries and 2 thesauruses; or 2 dictionaries, 4 atlases, and 3 thesauruses. How wide is one copy of each book?

78. Copying machine productivity When both copying machines A and B are working, an office assistant can make 100 copies in one minute. In one minute's time, copiers A and C together produce 140 copies, and all three working together produce 180 copies. How many copies per minute can each machine produce separately?

©lightpoet/Shutterstock.com

79. Nutritional planning One ounce of each of three foods has the vitamin and mineral content shown in the table. How many ounces of each must be used to provide exactly 22 milligrams (mg) of niacin, 12 mg of zinc, and 20 mg of vitamin C?

Food	Niacin	Zinc	Vitamin C
A	1 mg	1 mg	2 mg
B	2 mg	1 mg	1 mg
C	2 mg	1 mg	2 mg

80. Chainsaw sculpting A wood sculptor carves three types of statues with a chainsaw. The number of hours required for carving, sanding, and painting a totem pole, a bear, and a deer are shown in the table. How many of each should be produced to use all available labor hours?

	Totem Pole	Bear	Deer	Time Available
Carving	2 hr	2 hr	1 hr	14 hr
Sanding	1 hr	2 hr	2 hr	15 hr
Painting	3 hr	2 hr	2 hr	21 hr

Discovery and Writing

81. What is a matrix?

82. Describe the three elementary row operations that can be used to produce an equivalent system.

83. Explain the difference between the row-echelon form and the reduced row-echelon form of a matrix.

84. Explain the differences between Gaussian elimination and Gauss–Jordan elimination.

85. Describe the steps you would use to solve a system of linear equations using Gaussian elimination.

86. Describe the steps you would use to solve a system of linear equations using Gauss–Jordan elimination.

87. If the upper-left corner entry of a matrix is zero, what row operation might you do first?

88. What characteristics of a row-reduced matrix would let you conclude that the system is inconsistent?

Use matrix methods to solve each system.

89. $\begin{cases} x^2 + y^2 + z^2 = 14 \\ 2x^2 + 3y^2 - 2z^2 = -7 \\ x^2 - 5y^2 + z^2 = 8 \end{cases}$

(*Hint:* Solve first as a system in x^2, y^2, and z^2.)

90. $\begin{cases} 5\sqrt{x} + 2\sqrt{x} + \sqrt{z} = 22 \\ \sqrt{x} + \sqrt{y} - \sqrt{z} = 5 \\ 3\sqrt{x} - 2\sqrt{y} - 3\sqrt{z} = 10 \end{cases}$

Critical Thinking

Determine if the statement is true or false. If the statement is false, then correct it and make it true.

91. A 7×5 matrix has 5 rows.

92. A 2×9 matrix has 2 columns.

93. Adding a multiple of a row to another row produces a new augmented matrix corresponding to an equivalent system of linear equations.

94. Multiplying a row by any constant produces a new augmented matrix corresponding to an equivalent system of linear equations.

95. Every matrix has a unique reduced row-echelon form.

96. If the augmented matrix is $\begin{bmatrix} 1 & 0 & 6 & | & 10 \\ 0 & 1 & 8 & | & 2 \\ 0 & 0 & 4 & | & 0 \end{bmatrix}$, then the system has a unique solution.

97. If the augmented matrix is $\begin{bmatrix} 1 & 0 & 6 & | & 10 \\ 0 & 1 & 8 & | & 2 \\ 0 & 0 & 0 & | & 4 \end{bmatrix}$, then the system has an infinite number of solutions.

98. If the augmented matrix is $\begin{bmatrix} 1 & 0 & 6 & | & 10 \\ 0 & 1 & 8 & | & 2 \\ 0 & 0 & 0 & | & 0 \end{bmatrix}$, then the system has an infinite number of solutions.

99. If the row-echelon form of the augmented matrix contains the row $\begin{bmatrix} 1 & 0 & 0 & : & 0 \end{bmatrix}$, then the original system is consistent.

100. If an augmented matrix is square and contains one row of zeros, then the linear system has infinitely many solutions.

6.3 Matrix Algebra

In this section, we will learn to

1. Add and subtract matrices.
2. Multiply a matrix by a constant.
3. Solve matrix equations.
4. Multiply matrices.
5. Solve applications using matrices.
6. Recognize the identity matrix.

©John Roman Images/Shutterstock.com

With certain restrictions, we can add, subtract, and multiply matrices. In fact, many problems can be solved using the arithmetic of matrices.

To illustrate the addition of matrices, suppose there are 108 police officers employed at two different locations:

Downtown Station	Male	Female
Day Shift	21	18
Night Shift	12	6

Suburban Station	Male	Female
Day Shift	14	12
Night Shift	15	10

The employment information about the police officers is contained in two matrices.

$$D = \begin{bmatrix} 21 & 18 \\ 12 & 6 \end{bmatrix} \quad \text{and} \quad S = \begin{bmatrix} 14 & 12 \\ 15 & 10 \end{bmatrix}$$

The entry 21 in matrix D indicates that 21 male officers work the day shift at the downtown station. The entry 10 in matrix S indicates that 10 female officers work the night shift at the suburban station.

To find the city-wide totals, we can add the corresponding entries of matrices D and S:

$$D + S = \begin{bmatrix} 21 & 18 \\ 12 & 6 \end{bmatrix} + \begin{bmatrix} 14 & 12 \\ 15 & 10 \end{bmatrix} = \begin{bmatrix} 35 & 30 \\ 27 & 16 \end{bmatrix}$$

We interpret the total to mean:

$$\begin{array}{c} \text{Male} \quad \text{Female} \\ \begin{array}{c} \text{Day Shift} \\ \text{Night Shift} \end{array} \begin{bmatrix} 35 & 30 \\ 27 & 16 \end{bmatrix} \end{array}$$

To illustrate how to multiply a matrix by a number, suppose that one-third of the officers at the downtown station retire. The downtown staff would then consist of $\frac{2}{3}D$ officers. We can compute $\frac{2}{3}D$ by multiplying each entry of matrix D by $\frac{2}{3}$.

After retirements, the downtown staff will be

$$\begin{array}{c} \text{Male} \quad \text{Female} \\ \begin{array}{c} \text{Day Shift} \\ \text{Night Shift} \end{array} \begin{bmatrix} 14 & 12 \\ 8 & 4 \end{bmatrix} \end{array}$$

$$\frac{2}{3}D = \frac{2}{3}\begin{bmatrix} 21 & 18 \\ 12 & 6 \end{bmatrix} = \begin{bmatrix} 14 & 12 \\ 8 & 4 \end{bmatrix}$$

These examples illustrate two calculations used in the algebra of matrices, which is the topic of this section. We begin by giving a formal definition of a matrix and defining when two matrices are equal.

Matrix   An $m \times n$ **matrix** is a rectangular array of mn numbers arranged in m rows and n columns. We say that the matrix is of **size** (or **order**) $m \times n$.

Matrices are often denoted by letters such as A, B, and C. To denote the entries in an $m \times n$ matrix A, we use double-subscript notation: The entry in the first row, third column is a_{13}, and the entry in the ith row, jth column is a_{ij}. We can use any of the following notations to denote the $m \times n$ matrix A:

$$A,\ [a_{ij}],\ \begin{bmatrix} a_{11} & a_{12} & a_{13} & \cdots & a_{1n} \\ a_{21} & a_{22} & a_{23} & \cdots & a_{2n} \\ \vdots & \vdots & \vdots & \ddots & \vdots \\ a_{m1} & a_{m2} & a_{m3} & \cdots & a_{mn} \end{bmatrix} \Big\}\ m \text{ rows}$$

n columns

Two matrices are equal if they are the same size, with the same entries in corresponding positions.

Equality of Matrices   If $A = [a_{ij}]$ and $B = [b_{ij}]$ are both $m \times n$ matrices, then

$$A = B$$

provided that each entry a_{ij} in matrix A is equal to the corresponding entry b_{ij} in matrix B.

The following matrices are equal, because they are the same size and corresponding entries are equal.

$$\begin{bmatrix} \sqrt{9} & 0.5 \\ 1 & 4 \end{bmatrix} = \begin{bmatrix} 3 & \frac{1}{2} \\ 1 & 2^2 \end{bmatrix}$$

The following matrices are not equal, because they are not the same size.

$$\begin{bmatrix} 1 & 2 & 3 \\ 1 & 2 & 3 \end{bmatrix} \neq \begin{bmatrix} 1 & 2 & 3 \\ 1 & 2 & 3 \\ 1 & 2 & 3 \end{bmatrix}$$ The first matrix is 2×3, and the second is 3×3.

1. Add and Subtract Matrices

We can add matrices of the same size by adding the entries in corresponding positions.

Sum of Two Matrices   Let A and B be two $m \times n$ matrices. The **sum**, $A + B$, is the $m \times n$ matrix C found by adding the corresponding entries of matrices A and B:

$$A + B = C$$

where each entry c_{ij} in C is equal to the sum of a_{ij} in A and b_{ij} in B.

Adding Matrices

EXAMPLE 1

Add: $\begin{bmatrix} 2 & 1 & 3 \\ 1 & -1 & 0 \end{bmatrix} + \begin{bmatrix} 1 & -1 & 2 \\ -1 & 1 & 5 \end{bmatrix}$.

SOLUTION   Since each matrix is 2×3, we can find their sum by adding their corresponding entries.

$$\begin{bmatrix} 2 & 1 & 3 \\ 1 & -1 & 0 \end{bmatrix} + \begin{bmatrix} 1 & -1 & 2 \\ -1 & 1 & 5 \end{bmatrix} = \begin{bmatrix} 2+1 & 1-1 & 3+2 \\ 1-1 & -1+1 & 0+5 \end{bmatrix}$$

$$= \begin{bmatrix} 3 & 0 & 5 \\ 0 & 0 & 5 \end{bmatrix}$$

Caution
Matrices that are not the same size cannot be added.

Self Check 1 Add: $\begin{bmatrix} 3 & -5 \\ 2 & 0 \\ -6 & 5 \end{bmatrix} + \begin{bmatrix} -3 & 4 \\ -1 & -3 \\ 7 & 0 \end{bmatrix}$.

Now Try Exercise 13.

In arithmetic, 0 is the **additive identity**, because $a + 0 = 0 + a = a$ for any real number a. In matrix algebra, the matrix $\mathbf{0} = \begin{bmatrix} 0 & 0 \\ 0 & 0 \end{bmatrix}$ is called the **additive identity** for 2×2 matrices, because $A + \mathbf{0} = \mathbf{0} + A$. For example,

$$\begin{bmatrix} 1 & 2 \\ 3 & 4 \end{bmatrix} + \begin{bmatrix} \mathbf{0} & \mathbf{0} \\ \mathbf{0} & \mathbf{0} \end{bmatrix} = \begin{bmatrix} \mathbf{0} & \mathbf{0} \\ \mathbf{0} & \mathbf{0} \end{bmatrix} + \begin{bmatrix} 1 & 2 \\ 3 & 4 \end{bmatrix} = \begin{bmatrix} 1 & 2 \\ 3 & 4 \end{bmatrix}$$

Zero Matrix or Additive Identity Matrix

Let A be any $m \times n$ matrix. There is an $m \times n$ matrix $\mathbf{0}$, called the **zero matrix** or the **additive identity matrix**, for which

$$A + \mathbf{0} = \mathbf{0} + A = A$$

The matrix $\mathbf{0}$ consists of m rows and n columns of 0's.

Every matrix also has an additive inverse.

Additive Inverse of a Matrix

Any $m \times n$ matrix A has an **additive inverse**, an $m \times n$ matrix $-A$, with the property that the sum of A and $-A$ is the zero matrix:

$$A + (-A) = (-A) + A = \mathbf{0}$$

The entries of $-A$ are the negatives of the corresponding entries of A.

The additive inverse of $A = \begin{bmatrix} 1 & -3 & 2 \\ 0 & 1 & -5 \end{bmatrix}$ is the matrix

$$-A = \begin{bmatrix} -1 & 3 & -2 \\ 0 & -1 & 5 \end{bmatrix} \qquad \text{Each entry of } -A \text{ is the negative of the corresponding entry of } A.$$

because their sum is the zero matrix:

$$A + (-A) = \begin{bmatrix} 1 & -3 & 2 \\ 0 & 1 & -5 \end{bmatrix} + \begin{bmatrix} -1 & 3 & -2 \\ 0 & -1 & 5 \end{bmatrix}$$

$$= \begin{bmatrix} 1-1 & -3+3 & 2-2 \\ 0+0 & 1-1 & -5+5 \end{bmatrix}$$

$$= \begin{bmatrix} 0 & 0 & 0 \\ 0 & 0 & 0 \end{bmatrix}$$

Subtraction of matrices is similar to the subtraction of real numbers.

Difference of Two Matrices

If A and B are $m \times n$ matrices, their **difference** $A - B$ is the sum of A and the additive inverse of B:

$$A - B = A + (-B)$$

For example,

$$\begin{bmatrix} 3 & 7 \\ -4 & 0 \end{bmatrix} - \begin{bmatrix} -1 & 4 \\ -5 & 1 \end{bmatrix} = \begin{bmatrix} 3 & 7 \\ -4 & 0 \end{bmatrix} + \begin{bmatrix} 1 & -4 \\ 5 & -1 \end{bmatrix} = \begin{bmatrix} 4 & 3 \\ 1 & -1 \end{bmatrix}$$

2. Multiply a Matrix by a Constant

As we have seen in the police officers example at the beginning of the section, we can multiply a matrix by a constant by multiplying each of its entries by that constant. For example, if $A = \begin{bmatrix} 1 & -2 \\ 3 & 4 \end{bmatrix}$, then

$$5A = 5\begin{bmatrix} 1 & -2 \\ 3 & 4 \end{bmatrix} = \begin{bmatrix} 5(1) & 5(-2) \\ 5(3) & 5(4) \end{bmatrix} = \begin{bmatrix} 5 & -10 \\ 15 & 20 \end{bmatrix}$$

$$-A = -1A = -1\begin{bmatrix} 1 & -2 \\ 3 & 4 \end{bmatrix} = \begin{bmatrix} -1(1) & -1(-2) \\ -1(3) & -1(4) \end{bmatrix} = \begin{bmatrix} -1 & 2 \\ -3 & -4 \end{bmatrix}$$

The second example above illustrates that we can find the additive inverse of a matrix by multiplying the matrix by -1.

If A is a matrix, the real number k in the product kA is called a **scalar**.

Multiplying a Matrix by a Scalar	If A and B are two $m \times n$ matrices and k is a scalar, then kA, the **scalar multiple of k times A**, is B where each entry b_{ij} in B is equal to k times the corresponding entry a_{ij} in A.

Performing Scalar Multiplication

EXAMPLE 2

If $A = \begin{bmatrix} -2 & 5 \\ 6 & 1 \end{bmatrix}$ and $B = \begin{bmatrix} 3 & -4 \\ 5 & -2 \end{bmatrix}$, find $4A + 5B$.

SOLUTION   We will use scalar multiplication to find $4A$ and $5B$. Then we will use matrix addition to determine $4A + 5B$.

$$4A + 5B = 4\begin{bmatrix} -2 & 5 \\ 6 & 1 \end{bmatrix} + 5\begin{bmatrix} 3 & -4 \\ 5 & -2 \end{bmatrix}$$

$$= \begin{bmatrix} 4(-2) & 4(5) \\ 4(6) & 4(1) \end{bmatrix} + \begin{bmatrix} 5(3) & 5(-4) \\ 5(5) & 5(-2) \end{bmatrix}$$ Perform both scalar multiplications.

$$= \begin{bmatrix} -8 & 20 \\ 24 & 4 \end{bmatrix} + \begin{bmatrix} 15 & -20 \\ 25 & -10 \end{bmatrix}$$ Add the two matrices.

$$= \begin{bmatrix} 7 & 0 \\ 49 & -6 \end{bmatrix}$$

Self Check 2   Use the matrices given in Example 2 and find $2A - 5B$.

Now Try Exercise 23.

Using Scalar Multiplication and Finding Values of Variables

EXAMPLE 3

Let $\begin{bmatrix} 5 & y \\ 15 & z \end{bmatrix} = 5\begin{bmatrix} x & 3 \\ 3 & y \end{bmatrix}$. Find y and z.

SOLUTION We simplify the right side of the expression by multiplying each entry of the matrix by 5.

$$\begin{bmatrix} 5 & y \\ 15 & z \end{bmatrix} = \begin{bmatrix} 5x & 15 \\ 15 & 5y \end{bmatrix}$$

Because the matrices are equal, their corresponding entries are equal. So $y = 15$, and $z = 5y$. We conclude that $y = 15$ and $z = 5 \cdot 15 = 75$.

Self Check 3 Find x.

Now Try Exercise 11.

3. Solving Matrix Equations

There are many similarities between addition of real numbers and addition of matrices, subtraction of real numbers and subtraction of matrices, and multiplication of real numbers and multiplication of a matrix by a scalar. We can solve matrix equations using these similarities.

Solving a Matrix Equation

EXAMPLE 4

Let $A = \begin{bmatrix} 1 & -2 \\ 4 & 3 \\ -1 & 0 \end{bmatrix}$ and $B = \begin{bmatrix} 4 & -2 \\ 3 & 9 \\ 5 & 12 \end{bmatrix}$. Solve $3X + A = B$ for X.

SOLUTION We will solve $3X + A = B$ for X. Then we will use matrices A and B to determine X.

$3X + A = B$ This is the given matrix equation.

$3X = B - A$ Subtract matrix A from both sides.

$X = \dfrac{1}{3}(B - A)$ Multiply both sides by $\dfrac{1}{3}$ and solve for X.

Now, we use matrices A and B and determine X.

$X = \dfrac{1}{3}(B - A)$

$= \dfrac{1}{3}\left(\begin{bmatrix} 4 & -2 \\ 3 & 9 \\ 5 & 12 \end{bmatrix} - \begin{bmatrix} 1 & -2 \\ 4 & 3 \\ -1 & 0 \end{bmatrix}\right)$ Substitute matrices A and B into the equation.

$= \dfrac{1}{3}\begin{bmatrix} 3 & 0 \\ -1 & 6 \\ 6 & 12 \end{bmatrix}$ Subtract the matrices.

$= \begin{bmatrix} 1 & 0 \\ -\dfrac{1}{3} & 2 \\ 2 & 4 \end{bmatrix}$ Perform the scalar multiplication.

Self Check 4   Use the matrices given in Example 4 and solve the matrix equation $2X - A = B$ for X.

Now Try Exercise 25.

4. Multiply Matrices

To introduce multiplication of matrices, we consider the grading policy at a school where a student's final grade is based on a test score average and a homework average.

©Monkey Business Images/Shutterstock

The grades for three students are shown below, along with how the test scores and the homework scores are weighted.

	Test Score Average	Homework Average
Ann	83	95
Tonya	72	80
Carlos	85	92

	Weighting System 1	System 2
Test Average	0.5	0.3
Homework Average	0.5	0.7

To calculate the students' final grades, we must use information from both tables. For example, if their teacher used weighting system 1, their final grades would be

- Ann's grade = $83(0.5) + 95(0.5) = 41.5 + 47.5 = 89$
- Tonya's grade = $72(0.5) + 80(0.5) = 36 + 40 = 76$
- Carlos's grade = $85(0.5) + 92(0.5) = 42.5 + 46 = 88.5$

If we calculate their final grades using weighting system 2, we get

- Ann's grade = $83(0.3) + 95(0.7) = 24.9 + 66.5 = 91.4$
- Tonya's grade = $72(0.3) + 80(0.7) = 21.6 + 56 = 77.6$
- Carlos's grade = $85(0.3) + 92(0.7) = 25.5 + 64.4 = 89.9$

The result of each grade calculation is the sum of the products of the elements in one row of the first matrix and one column of the second matrix. Each calculation is a part of the product of the matrices

$$A = \begin{bmatrix} 83 & 95 \\ 72 & 80 \\ 85 & 92 \end{bmatrix} \quad \text{and} \quad B = \begin{bmatrix} 0.5 & 0.3 \\ 0.5 & 0.7 \end{bmatrix}$$

To further illustrate how to find the product of two matrices, we will find the product of the $3 \times \mathbf{2}$ matrix A and the $\mathbf{2} \times 2$ matrix B. This product will exist because the number of columns in A is equal to the number of rows in B. The result will be a 3×2 matrix C.

$$AB = \begin{bmatrix} 83 & 95 \\ 72 & 80 \\ 85 & 92 \end{bmatrix} \begin{bmatrix} 0.5 & 0.3 \\ 0.5 & 0.7 \end{bmatrix} = \begin{bmatrix} ? & ? & ? \\ ? & ? & ? \\ ? & ? & ? \end{bmatrix} = C$$

Each entry of matrix C will be the result of multiplying the entries in one row of A and the entries in one column of B and adding the results. As shown below, the first-row, first-column entry of matrix C is the sum of the products of entries of the first row of A and the first column of B. The second-row, second-column entry of matrix C is the sum of the products of entries of the second row of A and the second column of B.

$$AB = \begin{bmatrix} 83 & 95 \\ 72 & 80 \\ 85 & 92 \end{bmatrix} \begin{bmatrix} 0.5 & 0.3 \\ 0.5 & 0.7 \end{bmatrix} = \begin{bmatrix} 83(0.5) + 95(0.5) & ? \\ ? & 72(0.3) + 80(0.7) \\ ? & ? \end{bmatrix} = \begin{bmatrix} 89 & ? \\ ? & 77.6 \\ ? & ? \end{bmatrix} = C$$

As shown below, the first-row, second-column entry of matrix C is the sum of the products of entries of the first row of A and the second column of B. The second-row, first-column entry of matrix C is the sum of the products of entries of the second row of A and the first column of B.

$$AB = \begin{bmatrix} 83 & 95 \\ 72 & 80 \\ 85 & 92 \end{bmatrix} \begin{bmatrix} 0.5 & 0.3 \\ 0.5 & 0.7 \end{bmatrix} = \begin{bmatrix} 83(0.5) + 95(0.5) & 83(0.3) + 95(0.7) \\ 72(0.5) + 80(0.5) & 72(0.3) + 80(0.7) \\ ? & ? \end{bmatrix} = \begin{bmatrix} 89 & 91.4 \\ 76 & 77.6 \\ ? & ? \end{bmatrix} = C$$

As shown below, the third-row, first-column entry of matrix C is the sum of the products of entries of the third row of A and the first column of B. The third-row, second-column entry of matrix C is the sum of the products of entries of the third row of A and the second column of B.

$$AB = \begin{bmatrix} 83 & 95 \\ 72 & 80 \\ 85 & 92 \end{bmatrix} \begin{bmatrix} 0.5 & 0.3 \\ 0.5 & 0.7 \end{bmatrix} = \begin{bmatrix} 83(0.5) + 95(0.5) & 83(0.3) + 95(0.7) \\ 72(0.5) + 80(0.5) & 72(0.3) + 80(0.7) \\ 85(0.5) + 92(0.5) & 85(0.3) + 92(0.7) \end{bmatrix} = \begin{bmatrix} 89 & 91.4 \\ 76 & 77.6 \\ 88.5 & 89.9 \end{bmatrix} = C$$

The resulting 3×2 matrix C is the product of matrix A and B, and gives the final grades of each student with respect to each weighting system.

	Weighting System 1	Weighting System 2
Ann	89	91.4
Tonya	76	77.6
Carlos	88.5	89.9

For the product AB of two matrices to exist, the number of columns of A must equal the number of rows of B. If the product exists, it will have as many rows as A and as many columns as B:

$$\begin{array}{ccccc} A & \cdot & B & = & C \\ m \times n & & n \times p & & m \times p \end{array}$$

These must agree.

The product is of size $m \times p$.

We now formally define the product of two matrices.

Product of Two Matrices

Let A be an $m \times n$ matrix and B be an $n \times p$ matrix. The **product**, AB, is the $m \times p$ matrix C

$$AB = C$$

where each entry c_{ij} in C is the sum of the products of the corresponding entries in the ith row of A and the jth column of B, where $i = 1, 2, 3, \ldots, m$ and $j = 1, 2, 3, \ldots, p$.

Multiplying Matrices

EXAMPLE 5

Find $C = AB$ if $A = \begin{bmatrix} 1 & 2 & 4 \\ -2 & 1 & -1 \end{bmatrix}$ and $B = \begin{bmatrix} 1 & 5 \\ -2 & 4 \\ 1 & -3 \end{bmatrix}$.

SOLUTION Because matrix A is **2 × 3** and B is **3 × 2**, the number of columns of A is the same as the number of rows of B. Therefore, the product C exists and it will be a 2×2 matrix. To find entry c_{11} of C, we find the sum of the products of the entries in the first row of A and the first column of B:

$$c_{11} = 1(1) + 2(-2) + 4(1) = 1$$

$$\begin{bmatrix} \mathbf{1} & \mathbf{2} & \mathbf{4} \\ -2 & 1 & -1 \end{bmatrix} \begin{bmatrix} \mathbf{1} & 5 \\ \mathbf{-2} & 4 \\ \mathbf{1} & -3 \end{bmatrix} = \begin{bmatrix} \mathbf{1} & ? \\ ? & ? \end{bmatrix}$$

To find entry c_{12}, we move across the first row of A and down the second column of B:

$$c_{12} = 1(5) + 2(4) + 4(-3) = 1$$

$$\begin{bmatrix} \mathbf{1} & \mathbf{2} & \mathbf{4} \\ -2 & 1 & -1 \end{bmatrix} \begin{bmatrix} 1 & \mathbf{5} \\ -2 & \mathbf{4} \\ 1 & \mathbf{-3} \end{bmatrix} = \begin{bmatrix} 1 & \mathbf{1} \\ ? & ? \end{bmatrix}$$

To find entry c_{21}, we move across the second row of A and down the first column of B:

$$c_{21} = (-2)(1) + 1(-2) + (-1)(1) = -5$$

$$\begin{bmatrix} 1 & 2 & 4 \\ \mathbf{-2} & \mathbf{1} & \mathbf{-1} \end{bmatrix} \begin{bmatrix} \mathbf{1} & 5 \\ \mathbf{-2} & 4 \\ \mathbf{1} & -3 \end{bmatrix} = \begin{bmatrix} 1 & 1 \\ \mathbf{-5} & ? \end{bmatrix}$$

To find entry c_{22}, we move across the second row of A and down the second column of B:

$$c_{22} = (-2)(5) + 1(4) + (-1)(-3) = -3$$

$$\begin{bmatrix} 1 & 2 & 4 \\ \mathbf{-2} & \mathbf{1} & \mathbf{-1} \end{bmatrix} \begin{bmatrix} 1 & \mathbf{5} \\ -2 & \mathbf{4} \\ 1 & \mathbf{-3} \end{bmatrix} = \begin{bmatrix} 1 & 1 \\ -5 & \mathbf{-3} \end{bmatrix}$$

Tip

To obtain the element in the ith row and the jth column of the product AB, place your left index finger at the beginning of the ith row of A and your right index finger at the top of the jth column of B. Then move the left finger across and the right finger down, multiplying and summing as you proceed.

Caution

When we multiply matrices of the same size, keep in mind that we do not multiply corresponding entries.

Self Check 5 Find $D = EF$ if $E = \begin{bmatrix} 1 & -2 \\ 2 & 0 \end{bmatrix}$ and $F = \begin{bmatrix} 2 & 1 \\ 3 & 5 \end{bmatrix}$.

Now Try Exercise 37.

EXAMPLE 6

Multiplying Matrices

Find the product: $\begin{bmatrix} 1 & -1 & 2 \\ 1 & 3 & 0 \\ 0 & 1 & 1 \end{bmatrix} \begin{bmatrix} 2 & 1 \\ 1 & 3 \\ 0 & 1 \end{bmatrix}$.

SOLUTION Because the matrices are 3×3 and 3×2, the number of columns of the first matrix is the same as the number of rows of the second matrix. The product will be a 3×2 matrix.

$$\begin{bmatrix} 1 & -1 & 2 \\ 1 & 3 & 0 \\ 0 & 1 & 1 \end{bmatrix} \begin{bmatrix} 2 & 1 \\ 1 & 3 \\ 0 & 1 \end{bmatrix} = \begin{bmatrix} 1(2) + (-1)(1) + 2(0) & 1(1) + (-1)(3) + 2(1) \\ 1(2) + 3(1) + 0(0) & 1(1) + 3(3) + 0(1) \\ 0(2) + 1(1) + 1(0) & 0(1) + 1(3) + 1(1) \end{bmatrix}$$

$$= \begin{bmatrix} 1 & 0 \\ 5 & 10 \\ 1 & 4 \end{bmatrix}$$

Self Check 6 Find the product: $\begin{bmatrix} 1 & -2 \\ 3 & 1 \\ -4 & 0 \end{bmatrix} \begin{bmatrix} 4 & -3 \\ 1 & -1 \end{bmatrix}$.

Now Try Exercise 39.

EXAMPLE 7

Multiplying Matrices

Find each product: **a.** $\begin{bmatrix} 1 & 2 & 3 \end{bmatrix} \begin{bmatrix} 4 \\ 5 \\ 6 \end{bmatrix}$ **b.** $\begin{bmatrix} 1 \\ 2 \\ 3 \end{bmatrix} \begin{bmatrix} 4 & 5 & 6 \end{bmatrix}$

SOLUTION **a.** Since the first matrix is 1×3 and the second matrix is 3×1 the product is a 1×1 matrix:

$$\begin{bmatrix} 1 & 2 & 3 \end{bmatrix} \begin{bmatrix} 4 \\ 5 \\ 6 \end{bmatrix} = [1(4) + 2(5) + 3(6)] = [32]$$

b. Since the first matrix is 3×1 and the second matrix is 1×3 the product is a 3×3 matrix:

$$\begin{bmatrix} 1 \\ 2 \\ 3 \end{bmatrix} \begin{bmatrix} 4 & 5 & 6 \end{bmatrix} = \begin{bmatrix} 1(4) & 1(5) & 1(6) \\ 2(4) & 2(5) & 2(6) \\ 3(4) & 3(5) & 3(6) \end{bmatrix} = \begin{bmatrix} 4 & 5 & 6 \\ 8 & 10 & 12 \\ 12 & 15 & 18 \end{bmatrix}$$

Self Check 7 Find each product:

a. $\begin{bmatrix} 1 & 3 & 5 \end{bmatrix} \begin{bmatrix} 1 \\ 0 \\ 1 \end{bmatrix}$ **b.** $\begin{bmatrix} 1 \\ 2 \end{bmatrix} \begin{bmatrix} 3 & 4 \end{bmatrix}$

Now Try Exercise 31.

Take Note

There are cases where the product of two matrices is commutative, $AB = BA$, but, in general, the multiplication of matrices is not commutative.

The multiplication of matrices is not commutative. To show this, we compute AB and BA, where $A = \begin{bmatrix} 1 & 1 \\ 0 & 0 \end{bmatrix}$ and $B = \begin{bmatrix} 0 & 1 \\ 0 & 1 \end{bmatrix}$.

$$AB = \begin{bmatrix} 1 & 1 \\ 0 & 0 \end{bmatrix}\begin{bmatrix} 0 & 1 \\ 0 & 1 \end{bmatrix} = \begin{bmatrix} 0 & 2 \\ 0 & 0 \end{bmatrix}$$

$$BA = \begin{bmatrix} 0 & 1 \\ 0 & 1 \end{bmatrix}\begin{bmatrix} 1 & 1 \\ 0 & 0 \end{bmatrix} = \begin{bmatrix} 0 & 0 \\ 0 & 0 \end{bmatrix}$$

Since the products are not equal, matrix multiplication is not commutative.

Accent on Technology

Add and Multiply Matrices

Most graphing calculators can perform numerous operations on matrices. We must always be mindful that the sizes of the matrices allow these operations. This example will demonstrate the operations of addition and multiplication.

Given the two matrices, $A = \begin{bmatrix} 2 & 3.7 \\ -2.1 & 3 \end{bmatrix}$ and $B = \begin{bmatrix} 2 & -1 \\ 0 & 0.3 \end{bmatrix}$, add and multiply the matrices using a graphing calculator as shown in Figure 6-7.

1. Enter the values for the matrix by pressing [2nd] [x⁻¹] and using the MATRIX menu. Then select EDIT. We can enter the values from this menu.

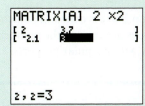

2. On the home screen, enter the names of the matrices by using the MATRIX menu. Use the operations of addition and multiplication on the home screen.

[A]+[B]
$$\begin{bmatrix} 4 & 2.7 \\ -2.1 & 3.3 \end{bmatrix}$$

[A]*[B]
$$\begin{bmatrix} 4 & -.89 \\ -4.2 & 3 \end{bmatrix}$$

FIGURE 6-7

5. Solve Applications Using Matrices

Solving an Application Problem

EXAMPLE 8

Suppose that supplies must be purchased for the police officers discussed at the beginning of the section. The quantities and prices of each item required for each shift are as follows:

	Quantities		
	Uniforms	Badges	Whistles
Day Shift	17	13	19
Night Shift	14	24	27

Unit Prices (in $)	
Uniforms	47
Badges	7
Whistles	5

Find the cost of supplies of each shift.

©Mike McDonald/Shutterstock.com

SOLUTION We can write the quantities Q and prices P in matrix form and multiply them to get a cost matrix.

$$C = QP$$

$$= \begin{bmatrix} 17 & 13 & 19 \\ 14 & 24 & 27 \end{bmatrix} \begin{bmatrix} 47 \\ 7 \\ 5 \end{bmatrix}$$

$$= \begin{bmatrix} 17(47) + 13(7) + 19(5) \\ 14(47) + 24(7) + 27(5) \end{bmatrix}$$

(17 uniforms)(\$47) + (13 badges)(\$7) + (19 whistles)(\$5)
(14 uniforms)(\$47) + (24 badges)(\$7) + (27 whistles)(\$5)

$$= \begin{bmatrix} 985 \\ 961 \end{bmatrix}$$

It will cost \$985 to buy supplies for the day shift and \$961 to buy supplies for the night shift.

Self Check 8 If the unit prices (in \$) in Example 8 increase and are \$50 for uniforms, \$10 for badges, and \$6 for whistles, how much will it cost to buy supplies for the day shift and the night shift?

Now Try Exercise 65.

6. Recognize the Identity Matrix

The number 1 is called the *identity for multiplication* because multiplying a number by 1 does not change the number: $a \cdot 1 = 1 \cdot a = a$. There is a **multiplicative identity matrix** with a similar property.

Identity Matrix Let A be an $n \times n$ matrix. There is an $n \times n$ **identity matrix** I for which

$$AI = IA = A$$

It is the matrix I consisting of 1's on its diagonal and 0's elsewhere.

$$I = \begin{bmatrix} 1 & 0 & 0 & \cdots & 0 \\ 0 & 1 & 0 & \cdots & 0 \\ 0 & 0 & 1 & \cdots & 0 \\ \vdots & \vdots & \vdots & \ddots & \vdots \\ 0 & 0 & 0 & \cdots & 1 \end{bmatrix}$$

Take Note

An identity matrix is always a square matrix—a matrix that has the same number of rows and columns.

Tip

The 3×3 identity matrix can be denoted by writing I_3. Similarly, I_n denotes the identity matrix of order n.

We illustrate the previous definition for the 3×3 identity matrix.

$$\begin{bmatrix} 1 & 0 & 0 \\ 0 & 1 & 0 \\ 0 & 0 & 1 \end{bmatrix} \begin{bmatrix} 1 & 2 & 3 \\ 4 & 5 & 6 \\ 7 & 8 & 9 \end{bmatrix} = \begin{bmatrix} 1+0+0 & 2+0+0 & 3+0+0 \\ 0+4+0 & 0+5+0 & 0+6+0 \\ 0+0+7 & 0+0+8 & 0+0+9 \end{bmatrix}$$

$$= \begin{bmatrix} 1 & 2 & 3 \\ 4 & 5 & 6 \\ 7 & 8 & 9 \end{bmatrix}$$

$$\begin{bmatrix} 1 & 2 & 3 \\ 4 & 5 & 6 \\ 7 & 8 & 9 \end{bmatrix} \begin{bmatrix} 1 & 0 & 0 \\ 0 & 1 & 0 \\ 0 & 0 & 1 \end{bmatrix} = \begin{bmatrix} 1 & 2 & 3 \\ 4 & 5 & 6 \\ 7 & 8 & 9 \end{bmatrix}$$

The following properties of real numbers carry over to matrices. In the exercises, you will be asked to illustrate many of these properties.

Properties of Matrices   Let A, B, and C be matrices and a and b be scalars.

Commutative Property of Addition: $A + B = B + A$

Associative Property of Addition: $A + (B + C) = (A + B) + C$

Associative Properties of Scalar Multiplication: $\begin{cases} a(bA) = (ab)A \\ a(AB) = (aA)B \end{cases}$

Distributive Properties of Scalar Multiplication: $\begin{cases} (a + b)A = aA + bA \\ a(A + b) = aA + aB \end{cases}$

Associative Property of Multiplication: $A(BC) = (AB)C$

Distributive Properties of Matrix Multiplication: $\begin{cases} A(B + C) = AB + AC \\ (A + B)C = AC + BC \end{cases}$

Verifying the Distributive Property of Matrix Multiplication

EXAMPLE 9

Verify the Distributive Property $A(B + C) = AB + AC$ using the matrices

$$A = \begin{bmatrix} 2 & 1 \\ 3 & 1 \end{bmatrix}, \quad B = \begin{bmatrix} 3 & -4 \\ 1 & -1 \end{bmatrix}, \quad \text{and} \quad C = \begin{bmatrix} -1 & 3 \\ 0 & 1 \end{bmatrix}$$

SOLUTION   We do the operations on the left side and the right side separately and compare the results.

$$\begin{bmatrix} 2 & 1 \\ 3 & 1 \end{bmatrix}\left(\begin{bmatrix} 3 & -4 \\ 1 & -1 \end{bmatrix} + \begin{bmatrix} -1 & 3 \\ 0 & 1 \end{bmatrix}\right) = \begin{bmatrix} 2 & 1 \\ 3 & 1 \end{bmatrix}\left(\begin{bmatrix} 2 & -1 \\ 1 & 0 \end{bmatrix}\right)$$ Do the addition within the parentheses.

$$= \begin{bmatrix} 5 & -2 \\ 7 & -3 \end{bmatrix}$$ Multiply.

$$\begin{bmatrix} 2 & 1 \\ 3 & 1 \end{bmatrix}\begin{bmatrix} 3 & -4 \\ 1 & -1 \end{bmatrix} + \begin{bmatrix} 2 & 1 \\ 3 & 1 \end{bmatrix}\begin{bmatrix} -1 & 3 \\ 0 & 1 \end{bmatrix} = \begin{bmatrix} 7 & -9 \\ 10 & -13 \end{bmatrix} + \begin{bmatrix} -2 & 7 \\ -3 & 10 \end{bmatrix}$$ Do the multiplications.

$$= \begin{bmatrix} 5 & -2 \\ 7 & -3 \end{bmatrix}$$ Add.

Because the left and right sides agree, this example illustrates the Distributive Property.

Self Check 9   Verify the Distributive Property for Multiplication

$$(A + B)C = AC + BC$$

for the matrices given in Example 9.

Now Try Exercise 53.

1. $\begin{bmatrix} 0 & -1 \\ 1 & -3 \\ 1 & 5 \end{bmatrix}$ 2. $\begin{bmatrix} -19 & 30 \\ -13 & 12 \end{bmatrix}$ 3. 1   4. $\begin{bmatrix} \frac{5}{2} & -2 \\ \frac{7}{2} & 6 \\ 2 & 6 \end{bmatrix}$

5. $\begin{bmatrix} -4 & -9 \\ 4 & 2 \end{bmatrix}$ 6. $\begin{bmatrix} 2 & -1 \\ 13 & -10 \\ -16 & 12 \end{bmatrix}$ 7. **a.** $[6]$ **b.** $\begin{bmatrix} 3 & 4 \\ 6 & 8 \end{bmatrix}$

8. day shift $1094 and night shift $1102

Exercises 6.3

Getting Ready
You should be able to complete these vocabulary and concept statements before you proceed to the practice exercises.

Fill in the blanks.

1. In a matrix A, the symbol a_{ij} is the entry in row ___ and column ___.
2. For matrices A and B to be equal, they must be the same ____, and corresponding entries must be _____.
3. To find the sum of matrices A and B, we add the _____ entries.
4. To multiply a matrix by a scalar, we multiply _____ by that scalar.
5. The product of a 3×2 matrix A and a 2×4 matrix B will exist because the number of _____ of A is equal to the number of _____ of B.
6. The product of the matrices A and B in Exercise 5 will be a _____ matrix.
7. Among 2×2 matrices, $\begin{bmatrix} 0 & 0 \\ 0 & 0 \end{bmatrix}$ is the _____ or zero matrix.
8. Among 2×2 matrices, $\begin{bmatrix} 1 & 0 \\ 0 & 1 \end{bmatrix}$ is the _____ matrix.

Practice
Find values of x and y, if any, that will make the matrices equal.

9. $\begin{bmatrix} x & y \\ 1 & 3 \end{bmatrix} = \begin{bmatrix} 2 & 5 \\ 1 & 3 \end{bmatrix}$

10. $\begin{bmatrix} x & 5 \\ 3 & y \end{bmatrix} = \begin{bmatrix} 0 & 5 \\ 3 & 2 \end{bmatrix}$

11. $\begin{bmatrix} x+y & 3+x \\ -2 & 5y \end{bmatrix} = \begin{bmatrix} 3 & 4 \\ -2 & 10 \end{bmatrix}$

12. $\begin{bmatrix} x+y & x-y \\ 2x & 3y \end{bmatrix} = \begin{bmatrix} -x & x-2 \\ -y & 8-y \end{bmatrix}$

Find A + B.

13. $A = \begin{bmatrix} 2 & 1 & -1 \\ -3 & 2 & 5 \end{bmatrix}, B = \begin{bmatrix} -3 & 1 & 2 \\ -3 & -2 & -5 \end{bmatrix}$

14. $A = \begin{bmatrix} 3 & 2 & 1 \\ -2 & 3 & -3 \\ -4 & -2 & -1 \end{bmatrix}, B = \begin{bmatrix} -2 & 6 & -2 \\ 5 & 7 & -1 \\ -4 & -6 & 7 \end{bmatrix}$

Find the additive inverse of each matrix.

15. $A = \begin{bmatrix} 5 & -2 & 7 \\ -5 & 0 & 3 \\ -2 & 3 & -5 \end{bmatrix}$

16. $A = \begin{bmatrix} 3 & -\frac{2}{3} & -5 & \frac{1}{2} \end{bmatrix}$

Find A − B.

17. $A = \begin{bmatrix} -3 & 2 & -2 \\ -1 & 4 & -5 \end{bmatrix}, B = \begin{bmatrix} 3 & -3 & -2 \\ -2 & 5 & -5 \end{bmatrix}$

18. $A = \begin{bmatrix} 2 & 2 & 0 \\ -2 & 8 & 1 \\ 3 & -3 & -8 \end{bmatrix}, B = \begin{bmatrix} -4 & 3 & 7 \\ -1 & 2 & 0 \\ 1 & 4 & -1 \end{bmatrix}$

Find 5A.

19. $A = \begin{bmatrix} 3 & -3 \\ 0 & -2 \end{bmatrix}$

20. $A = \begin{bmatrix} 3 & \frac{3}{5} \\ 0 & -1 \end{bmatrix}$

21. $A = \begin{bmatrix} 5 & 15 & -2 \\ -2 & -5 & 1 \end{bmatrix}$

22. $A = \begin{bmatrix} -3 & 1 & 2 \\ -8 & -2 & -5 \end{bmatrix}$

Find 5A + 3B.

23. $A = \begin{bmatrix} 3 & 1 & -2 \\ -4 & 3 & -2 \end{bmatrix}, B = \begin{bmatrix} 1 & -2 & 2 \\ -5 & -5 & 3 \end{bmatrix}$

24. $A = \begin{bmatrix} 2 & -5 \\ -5 & 2 \end{bmatrix}, B = \begin{bmatrix} 5 & -2 \\ 2 & -5 \end{bmatrix}$

Let $A = \begin{bmatrix} 1 & -3 & 4 \\ 2 & -1 & 2 \end{bmatrix}$ *and* $B = \begin{bmatrix} 5 & 2 & -1 \\ 4 & 1 & -2 \end{bmatrix}$. *Solve each matrix equation for X.*

25. $X + A = B$

26. $X + B = A$

27. $2B - X = A$

28. $2A - X = B$

29. $X + 2A = 3B$

30. $X + 3B = 2A$

31. $2X - 3A = B$

32. $2X - 3B = A$

33. $3A + 5B = -3X$

34. $5A + 3B = -3X$

Find each product, if possible.

35. $\begin{bmatrix} 2 & 3 \\ 3 & -2 \end{bmatrix}\begin{bmatrix} 1 & 2 \\ 0 & -2 \end{bmatrix}$

36. $\begin{bmatrix} -2 & 3 \\ 3 & -2 \end{bmatrix}\begin{bmatrix} 2 & 4 \\ -5 & 7 \end{bmatrix}$

37. $\begin{bmatrix} -4 & -2 \\ 21 & 0 \end{bmatrix}\begin{bmatrix} -5 & 6 \\ 21 & -1 \end{bmatrix}$

38. $\begin{bmatrix} -5 & 4 \\ 4 & -5 \end{bmatrix}\begin{bmatrix} 6 & -2 \\ 1 & 3 \end{bmatrix}$

39. $\begin{bmatrix} 2 & 1 & 3 \\ 1 & 2 & -1 \\ 0 & 1 & 0 \end{bmatrix}\begin{bmatrix} 1 & 2 & 3 \\ 2 & -2 & 1 \\ 0 & 0 & 1 \end{bmatrix}$

40. $\begin{bmatrix} 2 & 1 & 1 \\ 1 & 1 & 2 \\ 1 & -2 & -1 \end{bmatrix}\begin{bmatrix} 1 & 2 & 3 \\ 1 & 2 & -3 \\ -1 & -1 & 3 \end{bmatrix}$

41. $\begin{bmatrix} 1 \\ -2 \\ -3 \end{bmatrix}\begin{bmatrix} 4 & -5 & -6 \end{bmatrix}$

42. $\begin{bmatrix} 1 & -2 & -3 \\ 2 & 0 & 1 \end{bmatrix}\begin{bmatrix} 4 \\ -5 \\ -6 \end{bmatrix}$

43. $\begin{bmatrix} 1 & 2 & 3 \end{bmatrix}\begin{bmatrix} 4 & 5 & 6 \\ 7 & 8 & 9 \end{bmatrix}$

44. $\begin{bmatrix} 2 & 5 \\ -1 & 7 \end{bmatrix}\begin{bmatrix} 3 & 5 & -8 \\ -2 & 7 & 5 \\ 3 & -6 & 2 \end{bmatrix}$

45. $\begin{bmatrix} 2 & 3 & 4 \\ 1 & 2 & 3 \\ -2 & 2 & 2 \end{bmatrix}\begin{bmatrix} -1 \\ 2 \\ 3 \end{bmatrix}$

46. $\begin{bmatrix} 2 & 5 \\ -3 & 1 \\ 0 & -2 \\ 1 & -5 \end{bmatrix}\begin{bmatrix} 3 & -2 & 4 \\ -2 & -3 & 1 \end{bmatrix}$

47. $\begin{bmatrix} 1 & 2 & 3 \\ 4 & 5 & 6 \\ 7 & 8 & 9 \end{bmatrix}\begin{bmatrix} 1 & 2 \\ 3 & 4 \end{bmatrix}$

48. $\begin{bmatrix} 1 & 4 & 0 & 0 \\ -4 & 1 & 0 & -2 \\ 0 & 0 & 1 & 0 \\ 0 & 2 & 0 & 1 \end{bmatrix}\begin{bmatrix} 1 \\ 2 \\ -2 \\ -1 \end{bmatrix}$

Let $A = \begin{bmatrix} 2.3 & -1.7 & 3.1 \\ -2 & 3.5 & 1 \\ -8 & 4.7 & 9.1 \end{bmatrix}, B = \begin{bmatrix} -2.5 \\ 5.2 \\ -7 \end{bmatrix}$, *and*

$C = \begin{bmatrix} -5.8 \\ 2.9 \\ 4.1 \end{bmatrix}$. *Use a graphing calculator to find each result.*

49. AB

50. $B + C$

51. A^2

52. $AB + C$

Let $A = \begin{bmatrix} 2 & 3 \\ 1 & 3 \end{bmatrix}$, $B = \begin{bmatrix} 2 & 1 & -5 \\ 1 & 1 & 2 \end{bmatrix}$,

$C = \begin{bmatrix} -2 & -1 & 6 \\ 0 & -1 & -1 \end{bmatrix}$, $D = \begin{bmatrix} 1 & 2 \\ 1 & 3 \end{bmatrix}$, and $E = \begin{bmatrix} 1 & -2 \\ 2 & 3 \end{bmatrix}$.

Verify each property by doing the operations on each side of the equation and comparing the results.

53. Distributive Property: $A(B + C) = AB + AC$

54. Associative Property of Scalar Multiplication:
$5(6A) = (5 \cdot 6)A$

55. Associative Property of Scalar Multiplication:
$3(AB) = (3A)B$

56. Associative Property of Multiplication:
$A(DE) = (AD)E$

Let $A = \begin{bmatrix} 1 & 3 \\ 2 & 5 \end{bmatrix}$, $B = \begin{bmatrix} -1 \\ 3 \end{bmatrix}$, and $C = \begin{bmatrix} 3 & 2 \end{bmatrix}$. *Perform the operations, if possible.*

57. $A - BC$

58. $AB + B$

59. $CB - AB$

60. CAB

61. ABC

62. $CA + C$

63. A^2B

64. $(BC)^2$

Applications

Use a graphing calculator to help solve each problem.

65. Sporting goods Two suppliers manufactured footballs, baseballs, and basketballs in the quantities and costs given in the tables. Find matrices Q and C that represent the quantities and costs, find the product QC, and interpret the result.

	Quantities		
	Footballs	Baseballs	Basketballs
Supplier 1	200	300	100
Supplier 2	100	200	200

Unit Costs	(in $)
Footballs	5
Baseballs	2
Basketballs	4

66. Retailing Three ice cream stores sold cones, sundaes, and milkshakes in the quantities and prices given in the tables. Find matrices Q and P that represent the quantities and prices, find the product QP, and interpret the results.

©Mandy Godbhear/Shutterstock

	Quantities		
	Cones	Sundaes	Shakes
Store 1	75	75	32
Store 2	80	69	27
Store 3	62	40	30

Unit Price	
Cones	$1.50
Sundaes	$1.75
Shakes	$3.00

67. Beverage sales Beverages were sold to parents and children at a school basketball game in the quantities and prices given in the tables. Find matrices Q and P that represent the quantities and prices, find the product QP, and interpret the result.

	Quantities		
	Coffee	Milk	Cola
Adult Males	217	23	319
Adult Females	347	24	340
Children	3	97	750

Price	
Coffee	$0.75
Milk	$1.00
Cola	$1.25

68. Production costs Each of four factories manufactures three products in the daily quantities and unit costs given in the tables. Find a suitable matrix product to represent production costs.

Production Quantities			
Factory	Product A	Product B	Product C
Ashtabula	19	23	27
Boston	17	21	22
Chicago	21	18	20
Denver	27	25	22

Unit Production Costs		
	Day Shift	Night Shift
Product A	$1.20	$1.35
Product B	$0.75	$0.85
Product C	$3.50	$3.70

72. Communication on one-way channels Three communication centers are linked as indicated in the illustration, with communication only in the direction of the arrows. Thus, location 1 can send a message directly to location 2 along two paths, but location 2 can return a message directly on only one path. Entry c_{ij} of matrix C indicates the number of channels from i to j. Find and interpret C^2.

$$C = \begin{bmatrix} 0 & 2 & 2 \\ 1 & 0 & 1 \\ 1 & 0 & 0 \end{bmatrix}$$

Discovery and Writing

73. Explain how to add two matrices. Give an example.

74. Explain how to subtract two matrices. Give an example.

75. What is scalar multiplication and how is it performed? Give an example.

76. Explain the steps you would use to multiply two matrices. Give an example.

77. If A and B are 2×2 matrices, is $(AB)^2$ equal to A^2B^2? Support your answer.

69. Connectivity matrix An entry of 1 in the following **connectivity matrix** A indicates that the person associated with that row knows the address of the person associated with that column. For example, the 1 in Bill's row and Alan's column indicates that Bill can write to Alan. The 0 in Bill's row and Carl's column indicates that Bill cannot write to Carl. However, Bill could ask Alan to forward his letter to Carl. The matrix A^2 indicates the number of ways that one person can write to another with a letter that is forwarded exactly once. Find A^2.

$$\begin{array}{c} \\ \text{Alan} \\ \text{Bill} \\ \text{Carl} \end{array} \begin{array}{ccc} \text{Alan} & \text{Bill} & \text{Carl} \\ \begin{bmatrix} 0 & 1 & 1 \\ 1 & 0 & 0 \\ 0 & 1 & 0 \end{bmatrix} \end{array} = A$$

78. Let a, b, and c be real numbers. If $ab = ac$ and $a \neq 0$, then $b = c$. Find 2×2 matrices A, B, and C, where $A \neq 0$ to show that such a law does not hold for all matrices.

70. Communication routing Refer to Exercise 69. Find the matrix $A + A^2$. Can everyone receive a letter from everyone else with at most one forwarding?

79. Another property of the real numbers is that if $ab = 0$, then either $a = 0$ or $b = 0$. To show that this property is not true for matrices, find two nonzero 2×2 matrices A and B, such that $AB = 0$.

71. Routing telephone calls A long-distance telephone carrier has established several direct microwave links among four cities. In the following connectivity matrix, entries a_{ij} and a_{ji} indicate the number of direct links between cities i and j. For example, cities 2 and 4 are not linked directly but could be connected through city 3. Find and interpret matrix A^2.

$$A = \begin{bmatrix} 0 & 2 & 1 & 0 \\ 2 & 0 & 1 & 0 \\ 1 & 1 & 0 & 2 \\ 0 & 0 & 2 & 0 \end{bmatrix}$$

80. Find 2×2 matrices to show that $(A + B)(A - B) \neq A^2 - B^2$.

Critical Thinking

Determine if the statement is true or false. If the statement is false, then correct it and make it true.

81. If $A = \begin{bmatrix} 2 & 2 \\ 2 & 2 \end{bmatrix}$ and $B = \begin{bmatrix} 5 & 5 \\ 5 & 5 \end{bmatrix}$, then

$$AB = \begin{bmatrix} 10 & 10 \\ 10 & 10 \end{bmatrix}.$$

82. If $A = \begin{bmatrix} 2 & 4 \\ 6 & 8 \end{bmatrix}$, then $A^2 = \begin{bmatrix} 4 & 16 \\ 36 & 64 \end{bmatrix}$.

83. Matrix addition is commutative.

84. Matrix multiplication is commutative.

85. For the product of two matrices to be defined, the number of rows of the first matrix must equal the number of columns of the second matrix.

86. If A is a 20×30 matrix and B is a 30×20 matrix, then AB is a 600×600 matrix.

87. The additive inverse of $\begin{bmatrix} -1 & -2 & -3 & -4 & -5 \\ -6 & -7 & -8 & -9 & -10 \end{bmatrix}$

is $\begin{bmatrix} 1 & 2 & 3 & 4 & 5 \\ 6 & 7 & 8 & 9 & 10 \end{bmatrix}$.

88. $\begin{bmatrix} 1 & 0 & 0 \\ 0 & 1 & 0 \end{bmatrix}$ is an example of an identity matrix.

89. The associative property of multiplication holds for matrices.

90. If $AC = BC$, then $A = B$.

6.4 Matrix Inversion

In this section, we will learn to

1. Find the inverse of a square matrix using row operations.
2. Solve a system of equations by matrix inversion.
3. Solve applications using matrix inversion.

©Everett Historical/Shutterstock.com

Since the beginning of civilized society, governments and businesses have been interested in communicating in secret. This required the development of secret codes. Today, credit card information, bank account information, and even many e-mail messages are *encrypted*. Encryption is a process of converting ordinary text into unreadable text called *cipher text*.

Codes and code breaking have a long history in warfare. A very important event that helped the United States win WWII was the capture of German U-boat 110, which carried an Enigma encryption machine like the one shown. By breaking the code, the British and United States Navies knew the movements of German submarines.

Some of the most sophisticated codes we have today involve matrices. In Exercises 39 and 40, we must find the inverse of a matrix to decode a message.

Two real numbers are called **multiplicative inverses** if their product is the multiplicative identity 1. Some matrices have multiplicative inverses also.

Inverse of a Matrix   If A and B are $n \times n$ matrices, I is the $n \times n$ identity matrix, and

$$AB = BA = I,$$

then A and B are called **multiplicative inverses**. Matrix A is the **inverse** of B, and B is the **inverse** of A.

It can be shown that if a matrix A has an inverse, it only has one inverse. The inverse of A is written as A^{-1}.

$$AA^{-1} = A^{-1}A = I$$

Showing that Two Matrices Are Inverses

EXAMPLE 1

Show that A and B are inverses.

$$A = \begin{bmatrix} 1 & 1 & 0 \\ 4 & 3 & 0 \\ 2 & 1 & -1 \end{bmatrix} \qquad B = \begin{bmatrix} -3 & 1 & 0 \\ 4 & -1 & 0 \\ -2 & 1 & -1 \end{bmatrix}$$

SOLUTION

We must show that both AB and BA are equal to I.

$$AB = \begin{bmatrix} 1 & 1 & 0 \\ 4 & 3 & 0 \\ 2 & 1 & -1 \end{bmatrix}\begin{bmatrix} -3 & 1 & 0 \\ 4 & -1 & 0 \\ -2 & 1 & -1 \end{bmatrix}$$

$$= \begin{bmatrix} -3+\ 4+0 & 1-1+0 & 0+0+0 \\ -12+12+0 & 4-3+0 & 0+0+0 \\ -6+\ 4+2 & 2-1-1 & 0+0+1 \end{bmatrix}$$

$$= \begin{bmatrix} 1 & 0 & 0 \\ 0 & 1 & 0 \\ 0 & 0 & 1 \end{bmatrix}$$

$$BA = \begin{bmatrix} -3 & 1 & 0 \\ 4 & -1 & 0 \\ -2 & 1 & -1 \end{bmatrix}\begin{bmatrix} 1 & 1 & 0 \\ 4 & 3 & 0 \\ 2 & 1 & -1 \end{bmatrix} = \begin{bmatrix} 1 & 0 & 0 \\ 0 & 1 & 0 \\ 0 & 0 & 1 \end{bmatrix}$$

Self Check 1 Are C and D inverses? $C = \begin{bmatrix} 2 & 5 \\ 1 & 3 \end{bmatrix}$ and $D = \begin{bmatrix} 3 & -5 \\ -1 & 2 \end{bmatrix}$

Now Try Exercise 1.

1. Find the Inverse of a Square Matrix Using Row Operations

> **Tip**
>
> *Invertible* (has an inverse) and *singular* (does not have an inverse) are the preferred terms.

A matrix that has an inverse is called a **nonsingular matrix** and is said to be **invertible**. If it does not have an inverse, it is a **singular matrix** and is **noninvertible**. The following method provides a way to find the inverse of an invertible matrix.

Strategy for Finding the Inverse of a Matrix

If a sequence of row operations performed on the $n \times n$ matrix A reduces A to the $n \times n$ identity matrix I, then those same row operations, performed in the same order on I, will transform I into A^{-1}.

If no sequence of row operations will reduce A to I, then A is not invertible.

> **Take Note**
>
> The row operations we use to change the matrix into the identity matrix are the same row operations we use in Gauss–Jordan elimination.

To use this method to find the inverse of an invertible matrix, we perform row operations on matrix A to change it to the identity matrix I. At the same time, we perform the same row operations on I. This changes I into A^{-1}.

A notation for this process uses an n-row-by-$2n$-column matrix, with matrix A as the left half and matrix I as the right half. If A is invertible, the proper row operations performed on $[A \,|\, I]$ will transform it into $[I \,|\, A^{-1}]$.

Finding the Inverse of a 2 × 2 Matrix

EXAMPLE 2

Find the inverse of matrix A if $A = \begin{bmatrix} 2 & -4 \\ 4 & -7 \end{bmatrix}$.

SOLUTION We can set up a 2×4 matrix with A on the left and I on the right of the dashed line:

$$[A \mid I] = \begin{bmatrix} 2 & -4 & 1 & 0 \\ 4 & -7 & 0 & 1 \end{bmatrix}$$

Tip

Here's an alternate (fast) method for finding the inverse of a 2×2 invertible matrix.

If $A = \begin{bmatrix} a & b \\ c & d \end{bmatrix}$, then

$$A^{-1} = \frac{1}{ad - bc}\begin{bmatrix} d & -b \\ -c & a \end{bmatrix}$$

Consider the matrix given in Example 2.

If $A = \begin{bmatrix} 2 & -4 \\ 4 & -7 \end{bmatrix}$, then

$$A^{-1} = \frac{1}{-14 - (-16)}\begin{bmatrix} -7 & 4 \\ -4 & 2 \end{bmatrix}$$

$$= \frac{1}{2}\begin{bmatrix} -7 & 4 \\ -4 & 2 \end{bmatrix}$$

$$= \begin{bmatrix} -\frac{7}{2} & 2 \\ -2 & 1 \end{bmatrix}$$

We perform row operations on the entire matrix to transform the left half into I:

$$\begin{bmatrix} 2 & -4 & 1 & 0 \\ 4 & -7 & 0 & 1 \end{bmatrix} \begin{array}{c} (\frac{1}{2})R1 \to \\ (-2)R1 + R2 \to \end{array} \begin{bmatrix} 1 & -2 & \frac{1}{2} & 0 \\ 0 & 1 & -2 & 1 \end{bmatrix}$$

$$(2)R2 + R1 \to \begin{bmatrix} 1 & 0 & -\frac{7}{2} & 2 \\ 0 & 1 & -2 & 1 \end{bmatrix}$$

Since matrix A has been transformed into I, the right side of the previous matrix is A^{-1}. That is, $A^{-1} = \begin{bmatrix} -\frac{7}{2} & 2 \\ -2 & 1 \end{bmatrix}$. We can verify this by finding AA^{-1} and $A^{-1}A$ and showing that each product is I:

$$AA^{-1} = \begin{bmatrix} 2 & -4 \\ 4 & -7 \end{bmatrix}\begin{bmatrix} -\frac{7}{2} & 2 \\ -2 & 1 \end{bmatrix} = \begin{bmatrix} 1 & 0 \\ 0 & 1 \end{bmatrix}$$

$$A^{-1}A = \begin{bmatrix} -\frac{7}{2} & 2 \\ -2 & 1 \end{bmatrix}\begin{bmatrix} 2 & -4 \\ 4 & -7 \end{bmatrix} = \begin{bmatrix} 1 & 0 \\ 0 & 1 \end{bmatrix}$$

Self Check 2 Find the inverse of $A = \begin{bmatrix} 3 & 2 \\ 4 & 3 \end{bmatrix}$.

Now Try Exercise 5.

EXAMPLE 3

Finding the Inverse of a 3×3 Matrix

Find the inverse of matrix A if $A = \begin{bmatrix} 1 & 1 & 0 \\ 1 & 2 & 1 \\ 2 & 3 & 2 \end{bmatrix}$.

SOLUTION We set up a 3×6 matrix with A on the left and I on the right of the dashed line.

$$[A \mid I] = \begin{bmatrix} 1 & 1 & 0 & 1 & 0 & 0 \\ 1 & 2 & 1 & 0 & 1 & 0 \\ 2 & 3 & 2 & 0 & 0 & 1 \end{bmatrix}$$

Caution

We do not find the inverse of a square matrix by finding the multiplicative inverse of each of its entries.

We then perform row operations on the matrix to transform the left half into I.

$$\begin{bmatrix} 1 & 1 & 0 & 1 & 0 & 0 \\ 1 & 2 & 1 & 0 & 1 & 0 \\ 2 & 3 & 2 & 0 & 0 & 1 \end{bmatrix} \begin{array}{c} \\ (-1)R1 + R2 \to \\ (-2)R1 + R3 \to \end{array} \begin{bmatrix} 1 & 1 & 0 & 1 & 0 & 0 \\ 0 & 1 & 1 & -1 & 1 & 0 \\ 0 & 1 & 2 & -2 & 0 & 1 \end{bmatrix}$$

$$\begin{array}{c} (-1)R2 + R1 \to \\ \\ (-1)R2 + R3 \to \end{array} \begin{bmatrix} 1 & 0 & -1 & 2 & -1 & 0 \\ 0 & 1 & 1 & -1 & 1 & 0 \\ 0 & 0 & 1 & -1 & -1 & 1 \end{bmatrix}$$

$$\begin{array}{c} R3 + R1 \to \\ (-1)R3 + R2 \to \\ \end{array} \begin{bmatrix} 1 & 0 & 0 & 1 & -2 & 1 \\ 0 & 1 & 0 & 0 & 2 & -1 \\ 0 & 0 & 1 & -1 & -1 & 1 \end{bmatrix}$$

Since the left half of the matrix has been transformed into the identity matrix, the right half has become A^{-1}, and

$$A^{-1} = \begin{bmatrix} 1 & -2 & 1 \\ 0 & 2 & -1 \\ -1 & -1 & 1 \end{bmatrix}$$

Self Check 3   Find the inverse of $B = \begin{bmatrix} 1 & 1 & 1 \\ 2 & 1 & 4 \\ 2 & 2 & 3 \end{bmatrix}$.

Now Try Exercise 9.

EXAMPLE 4   **Finding the Inverse (If Possible) of a Square Matrix**

If possible, find the inverse of $A = \begin{bmatrix} 1 & 2 \\ 2 & 4 \end{bmatrix}$.

SOLUTION   We form the 2×4 matrix

$$[A \mid I] = \begin{bmatrix} 1 & 2 & \vdots & 1 & 0 \\ 2 & 4 & \vdots & 0 & 1 \end{bmatrix}$$

and begin to transform the left side of the matrix into the identity matrix I:

$$\begin{bmatrix} 1 & 2 & \vdots & 1 & 0 \\ 2 & 4 & \vdots & 0 & 1 \end{bmatrix} (-2)R1 + R2 \rightarrow \begin{bmatrix} 1 & 2 & \vdots & 1 & 0 \\ \mathbf{0} & \mathbf{0} & \vdots & -2 & 1 \end{bmatrix}$$

In obtaining the second-row, first-column position of A, the entire second row of A is zeroed out. Since we cannot transform matrix A to the identity matrix, A is singular.

Self Check 4   If possible, find the inverse of $B = \begin{bmatrix} 1 & -2 \\ -3 & 6 \end{bmatrix}$.

Now Try Exercise 13.

> **Caution**
>
> Do not attempt to find the inverse of a matrix that isn't square. It is singular.

The previous examples suggest the following facts:

- Matrices that are not square matrices do not have inverses.
- Some square matrices have inverses and others do not.

Accent on Technology

Finding the Inverse of a Matrix

If a matrix is invertible, the inverse can be found using a graphing calculator by following the procedure in Figure 6-8.

1. Access the MATRIX menu; select EDIT, and enter the values of the matrix.

2. Go to the HOME screen, and press the key $\boxed{x^{-1}}$.

```
MATRIX[A] 3 ×3
[ 2    2    3  ]
[ 1    2    3  ]
[ 1    0    ▮  ]



3,3=1
```

```
[A] ⁻¹
    [ 1   -1    0  ]
    [ 1  -.5 -1.5  ]
    [-1    1    1  ]
```

3. If we are going to use the inverse matrix, then we store (use [STO>] key) it in matrix B.

4. To verify that $AA^{-1} = I$, multiply matrix A by the inverse stored in B.

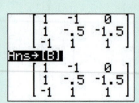

FIGURE 6-8

2. Solve a System of Equations by Matrix Inversion

If we multiply the matrices on the left side of the equation

$$\begin{bmatrix} 1 & 1 & 0 \\ 1 & 2 & 1 \\ 2 & 3 & 2 \end{bmatrix} \begin{bmatrix} x \\ y \\ z \end{bmatrix} = \begin{bmatrix} 20 \\ 30 \\ 55 \end{bmatrix}$$

and set the corresponding entries equal, we get the following system of equations:

$$\begin{cases} x + y = 20 \\ x + 2y + z = 30 \\ 2x + 3y + 2z = 55 \end{cases}$$

A system of equations can always be written as a matrix equation $AX = B$, where A is the coefficient matrix of the system, X is a column matrix of variables, and B is the column matrix of constants. If matrix A is invertible, the matrix equation $AX = B$ is easy to solve, as we will see in the next example.

EXAMPLE 5

Using the Inverse of a Matrix to Solve a System of Equations

Solve: $\begin{bmatrix} 1 & 1 & 0 \\ 1 & 2 & 1 \\ 2 & 3 & 2 \end{bmatrix} \begin{bmatrix} x \\ y \\ z \end{bmatrix} = \begin{bmatrix} 20 \\ 30 \\ 55 \end{bmatrix}$.

SOLUTION

The 3×3 matrix on the left is the matrix whose inverse was found in Example 3. We multiply both sides of the equation on the left by this inverse to obtain an equivalent system of equations.

$$\begin{bmatrix} 1 & -2 & 1 \\ 0 & 2 & -1 \\ -1 & -1 & 1 \end{bmatrix} \begin{bmatrix} 1 & 1 & 0 \\ 1 & 2 & 1 \\ 2 & 3 & 2 \end{bmatrix} \begin{bmatrix} x \\ y \\ z \end{bmatrix} = \begin{bmatrix} 1 & -2 & 1 \\ 0 & 2 & -1 \\ -1 & -1 & 1 \end{bmatrix} \begin{bmatrix} 20 \\ 30 \\ 55 \end{bmatrix}$$

$$\begin{bmatrix} 1 & 0 & 0 \\ 0 & 1 & 0 \\ 0 & 0 & 1 \end{bmatrix} \begin{bmatrix} x \\ y \\ z \end{bmatrix} = \begin{bmatrix} 15 \\ 5 \\ 5 \end{bmatrix} \qquad \text{Multiply the matrices. On the left, remember that } A^{-1}A = I.$$

$$\begin{bmatrix} x \\ y \\ z \end{bmatrix} = \begin{bmatrix} 15 \\ 5 \\ 5 \end{bmatrix} \qquad IX = X$$

The solution of this system can be read directly from the matrix on the right side. Verify that the values $x = 15$, $y = 5$, $z = 5$ satisfy the original equations.

Self Check 5   Solve: $\begin{bmatrix} 2 & -4 \\ 4 & -7 \end{bmatrix}\begin{bmatrix} x \\ y \end{bmatrix} = \begin{bmatrix} -4 \\ 2 \end{bmatrix}$. (The inverse was found in Example 2.)

Now Try Exercise 29.

Example 5 suggests the following result.

Solving Systems of Equations   If A is invertible, the solution of the matrix equation $AX = B$ is

$$X = A^{-1}B$$

Accent on Technology

Solving a System of Equations Using Matrix Inversion

To solve the system $\begin{cases} x - 2y + 3z = 9 \\ -x + 3y + z = -2 \\ 2x - 5y + 5z = 17 \end{cases}$ with a graphing calculator, follow

the procedure shown in Figure 6-9.

1. Enter the coefficient matrix in A.

2. Enter the column matrix in B.

3. Multiply $A^{-1}B$ and obtain the solution.

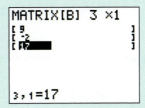

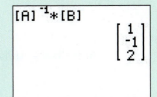

$$[A]^{-1}*[B]$$

$$\begin{bmatrix} 1 \\ -1 \\ 2 \end{bmatrix}$$

FIGURE 6-9

The solution is $x = 1$, $y = -1$, and $z = 2$. Always check the results in the original equation.

This method is especially useful for finding solutions of several systems of equations that differ from each other only in the column matrix B. If the coefficient matrix A remains unchanged from one system of equations to the next, then A^{-1} needs to be found only once. The solution of each system is found by a single matrix multiplication, $A^{-1}B$.

3. Solve Applications Involving Matrix Inversion

Solving an Application Problem Using the Inverse of a Matrix

EXAMPLE 6

A company that manufactures medical equipment spends time on paperwork, manufacture, and testing for each of three versions of a circuit board. The times spent on each and the total time available are given in the following tables.

	Hours Required per Unit		
	Product A	**Product B**	**Product C**
Paperwork	1	1	0
Manufacture	1	2	1
Testing	2	3	2

Hours Available	
Paperwork	20
Manufacture	30
Testing	55

SOLUTION We can let x, y, and z represent the number of units of products A, B, and C to be manufactured, respectively. We can then set up the following system of equations:

> *Paperwork:* $x + y = 20$ — One hour is needed for every A, and one hour for every B.
>
> *Manufacture:* $x + 2y + z = 30$ — One hour is needed for every A and C, and two hours for every B.
>
> *Testing:* $2x + 3y + 2z = 55$ — Two hours are needed for every A and C, and three hours for every B.

In matrix form, the system becomes

$$\begin{bmatrix} 1 & 1 & 0 \\ 1 & 2 & 1 \\ 2 & 3 & 2 \end{bmatrix} \begin{bmatrix} x \\ y \\ z \end{bmatrix} = \begin{bmatrix} 20 \\ 30 \\ 55 \end{bmatrix}$$

We solved this equation in Example 5 to get $x = 15$, $y = 5$, and $z = 5$. To use all of the available time, the company should manufacture 15 units of product A and 5 units each of products B and C.

Self Check 6 In Example 6 if 18 hours are available for paperwork, 32 for manufacture, and 57 for testing, how many of each product should the company manufacture?

Now Try Exercise 37.

Self Check Answers

1. yes **2.** $\begin{bmatrix} 3 & -2 \\ -4 & 3 \end{bmatrix}$ **3.** $\begin{bmatrix} 5 & 1 & -3 \\ -2 & -1 & 2 \\ -2 & 0 & 1 \end{bmatrix}$ **4.** no inverse

5. $\begin{bmatrix} x \\ y \end{bmatrix} = \begin{bmatrix} 18 \\ 10 \end{bmatrix}$ **6.** The company should manufacture 11 units of product A and 7 units each of products B and C.

Exercises 6.4

Getting Ready
You should be able to complete these vocabulary and concept statements before you proceed to the practice exercises.

Fill in the blanks.

1. Matrices A and B are multiplicative inverses if _____.

2. A nonsingular matrix ___ (is, is not) invertible.

3. If A is invertible, elementary row operations can change $[A\,|\,I]$ into _____.

4. If A is invertible, the solution of $AX = B$ is _____.

Practice
Find the inverse of each matrix, if possible.

5. $\begin{bmatrix} 3 & -4 \\ -2 & 3 \end{bmatrix}$

6. $\begin{bmatrix} 2 & 3 \\ 3 & 5 \end{bmatrix}$

7. $\begin{bmatrix} 3 & 7 \\ 2 & 5 \end{bmatrix}$

8. $\begin{bmatrix} 1 & -2 \\ 2 & -5 \end{bmatrix}$

9. $\begin{bmatrix} 1 & 0 & 3 \\ -1 & 1 & 3 \\ -2 & 1 & 1 \end{bmatrix}$

10. $\begin{bmatrix} 2 & 1 & -1 \\ 2 & 2 & -1 \\ -1 & -1 & 1 \end{bmatrix}$

11. $\begin{bmatrix} 3 & 2 & 1 \\ 1 & 1 & -1 \\ 4 & 3 & 1 \end{bmatrix}$

12. $\begin{bmatrix} -2 & 1 & -3 \\ 2 & 3 & 0 \\ 1 & 0 & 1 \end{bmatrix}$

13. $\begin{bmatrix} 1 & 3 & 5 \\ 0 & 1 & 6 \\ 1 & 4 & 11 \end{bmatrix}$

14. $\begin{bmatrix} 1 & 2 & 3 \\ 4 & 5 & 6 \\ 7 & 8 & 9 \end{bmatrix}$

15. $\begin{bmatrix} 1 & 2 & 3 \\ 0 & 1 & 2 \\ 0 & 0 & 1 \end{bmatrix}$

16. $\begin{bmatrix} 1 & 2 & 3 \\ 0 & 1 & 1 \\ 0 & -1 & 0 \end{bmatrix}$

17. $\begin{bmatrix} 1 & 6 & 4 \\ 1 & -2 & -5 \\ 2 & 4 & -1 \end{bmatrix}$

18. $\begin{bmatrix} 1 & 1 & 1 \\ 1 & 0 & -1 \\ 1 & 2 & 3 \end{bmatrix}$

Use a graphing calculator to find the inverse of each matrix.

19. $\begin{bmatrix} 1 & 2 & 3 & 4 \\ 0 & 1 & 2 & 3 \\ 0 & 0 & 1 & 2 \\ 0 & 0 & 0 & 1 \end{bmatrix}$

20. $\begin{bmatrix} 1 & 0 & 0 & 0 \\ 1 & 1 & 0 & 0 \\ 1 & 1 & 1 & 0 \\ 1 & 2 & 2 & 1 \end{bmatrix}$

21. $\begin{bmatrix} 1 & 1 & -1 \\ 0.5 & 1 & 0.5 \\ 1 & 1 & -1.5 \end{bmatrix}$

22. $\begin{bmatrix} -2 & -1 & 1 \\ 0.5 & -1.5 & -0.5 \\ 0 & 1 & 0.5 \end{bmatrix}$

23. $\begin{bmatrix} 3 & 3 & -3 & 2 \\ 1 & -4 & 3 & -5 \\ 3 & 0 & -2 & -1 \\ -1 & 5 & -3 & 6 \end{bmatrix}$

24. $\begin{bmatrix} 1 & 0 & 0 & 0 \\ 2 & 1 & 0 & 0 \\ 3 & 2 & 1 & 0 \\ 4 & 3 & 2 & 1 \end{bmatrix}$

Use matrix inversion to solve each system of equations. Note that several systems have the same coefficient matrix.

25. $\begin{cases} 3x - 4y = 1 \\ -2x + 3y = 5 \end{cases}$

26. $\begin{cases} 3x - 4y = -1 \\ -2x + 3y = 3 \end{cases}$

27. $\begin{cases} 3x - 4y = 0 \\ -2x + 3y = 0 \end{cases}$

28. $\begin{cases} 3x - 4y = -3 \\ -2x + 3y = -2 \end{cases}$

29. $\begin{cases} 2x + y - z = 2 \\ 2x + 2y - z = 4 \\ -x - y + z = -1 \end{cases}$

30. $\begin{cases} 2x + y - z = 3 \\ 2x + 2y - z = -1 \\ -x - y + z = 4 \end{cases}$

31. $\begin{cases} -2x + y - 3z = 2 \\ 2x + 3y = -3 \\ x + z = 5 \end{cases}$

32. $\begin{cases} -2x + y - 3z = 5 \\ 2x + 3y = 1 \\ x + z = -2 \end{cases}$

Use a graphing calculator to solve each system of equations. Use matrix inversion.

33. $\begin{cases} 5x + 3y = 13 \\ -7x + 5y = -9 \end{cases}$

34. $\begin{cases} 8x - 3y = 7 \\ -3x + 2y = 0 \end{cases}$

35. $\begin{cases} 5x + 2y + 3z = 12 \\ 2x + 5z = 7 \\ 3x + z = 4 \end{cases}$

36. $\begin{cases} 3x + 2y - z = 0 \\ 5x - 2y = 5 \\ 3x + y + z = 6 \end{cases}$

Applications

37. **Manufacturing and testing** The numbers of hours required to manufacture and test each of two models of heart monitor are given in the first table, and the numbers of hours available each week for manufacturing and testing are given in the second table.

	Hours Required per Unit	
	Model A	Model B
Manufacturing	23	27
Testing	21	22

Hours Available	
Manufacturing	127
Testing	108

How many of each model can be manufactured each week?

38. Making clothes A clothing manufacturer makes coats, shirts, and slacks. The times required for cutting, sewing, and packaging each item are shown in the table. How many of each should be made to use all available labor hours?

©Wrangler/Shutterstock.com

	Coats	Shirts	Slacks
Cutting	20 min	15 min	10 min
Sewing	60 min	30 min	24 min
Packaging	5 min	12 min	6 min

Time Available	
Cutting	115 hr
Sewing	280 hr
Packaging	65 hr

39. Cryptography The letters of a message, called **plain text**, are assigned values $1-26$ (for a−z) and are written in groups of 2 as 2×1 matrices. To write the message in **cipher text**, each 2×1 matrix B is multiplied by a matrix A, where

$$A = \begin{bmatrix} 1 & 1 \\ 2 & 3 \end{bmatrix}$$

Find the plain text if the cipher text of one message is

$$AB = \begin{bmatrix} 17 \\ 43 \end{bmatrix}$$

40. Cryptography The letters of a message, called **plain text**, are assigned values $1-26$ (for a−z) and are written in groups of 3 as 3×1 matrices. To write the message in **cipher text**, each 3×1 matrix Y is multiplied by matrix A, where

$$A = \begin{bmatrix} 1 & 1 & 0 \\ 2 & 3 & 3 \\ 1 & 1 & 1 \end{bmatrix}$$

Find the plain text if the cipher text of one message is

$$AY = \begin{bmatrix} 30 \\ 122 \\ 49 \end{bmatrix}$$

Discovery and Writing

41. Explain what is meant by "multiplicative inverse of a matrix."

42. Describe a strategy to use to determine the inverse of an invertible matrix.

43. Once you have applied the steps to find the inverse of a matrix, how can you check your answer?

44. Describe a strategy you would use to solve the matrix equation $AX = B$.

Let $A = \begin{bmatrix} -1 & -1 \\ 1 & 1 \end{bmatrix}$.

45. Show that $A^2 = \mathbf{0}$.

46. Show that the inverse of $I - A$ is $I + A$.

Let $A = \begin{bmatrix} 3 & 0 & 0 \\ -2 & -1 & -2 \\ 3 & 6 & 3 \end{bmatrix}$ *and* $X = \begin{bmatrix} x \\ y \\ z \end{bmatrix}$. *Solve each equation. Each solution is called an eigenvector of the matrix A.*

47. $(A - 2I)X = 0$

48. $(A - 3I)X = 0$

49. Suppose that A, B, and C are $n \times n$ matrices and A is invertible. If $AB = AC$, prove that $B = C$.

50. Prove that $\begin{bmatrix} a & b \\ c & d \end{bmatrix}$ has an inverse if and only if $ad - bc \neq 0$. (*Hint:* Try to find the inverse and see what happens.)

51. Suppose that B is any matrix for which $B^2 = 0$. Show that $I - B$ is invertible by showing that the inverse of $I - B$ is $I + B$.

52. Suppose that C is any matrix for which $C^3 = 0$. Show that $I - C$ is invertible by showing that the inverse of $I - C$ is $I + C + C^2$.

Critical Thinking

Determine if the statement is true or false. If the statement is false, then correct it and make it true.

53. All matrices have inverses.

54. If $A = \begin{bmatrix} 1 & 2 \\ 3 & 4 \end{bmatrix}$, then $A^{-1} = \begin{bmatrix} -1 & -2 \\ -3 & -4 \end{bmatrix}$.

55. If $A = \begin{bmatrix} 2 & 4 \\ 6 & 8 \end{bmatrix}$, then $A^{-1} = \begin{bmatrix} \frac{1}{2} & \frac{1}{4} \\ \frac{1}{6} & \frac{1}{8} \end{bmatrix}$.

56. $(A^{-1})^{-1} = A$

57. The inverse of a matrix is unique.

58. $(5A)^{-1} = 5A^{-1}$

59. $(AB)^{-1} = B^{-1}A^{-1}$

60. If C is invertible, then $CA = CB$ implies that $A = B$.

6.5   Determinants

In this section, we will learn to

1. Evaluate determinants of higher-order matrices.

2. Understand and use properties of determinants.

3. Use determinants to solve systems of equations.

4. Write equations of lines.

5. Find areas of triangles.

There is a function, called the **determinant function**, that associates a number with every square matrix. The domain of this function is the set of all square matrices, and the range is the set of numbers. Historically, determinants were considered before matrices. They first appeared in the 3rd century BC when they were considered in a Chinese textbook, called *The Nine Chapters on the Mathematical Art.* Gerolamo Cardano, an Italian mathematician, considered two-by-two determinants toward the end of the 16th century. However, it was the great Swiss mathematician, Gabriel Cramer, who made them popular.

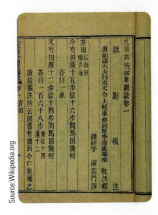

Source: Wikipedia.org

In this section, we will use determinants to solve systems of linear equations, and to find an equation of a line and the area of a triangle.

The determinant function is written as $\det(A)$ or as $|A|$.

Determinant of a 2×2 Matrix	If $a, b, c,$ and d are numbers, the **determinant** of $A = \begin{bmatrix} a & b \\ c & d \end{bmatrix}$ is

$$\det(A) = \begin{vmatrix} a & b \\ c & d \end{vmatrix} = ad - bc$$

Finding the Determinant of a 2 × 2 Matrix

EXAMPLE 1

a. $\begin{vmatrix} 1 & 2 \\ 3 & 4 \end{vmatrix} = 1(4) - 2(3)$ **b.** $\begin{vmatrix} -2 & 3 \\ -\pi & \frac{1}{2} \end{vmatrix} = (-2)\left(\frac{1}{2}\right) - 3(-\pi)$

$= 4 - 6$ $= -1 + 3\pi$

$= -2$

Caution

Do not confuse the notation $|A|$ with absolute value symbols.

Self Check 1 Evaluate: $\begin{vmatrix} 3 & -2 \\ 5 & -4 \end{vmatrix}$.

Now Try Exercise 7.

1. Evaluate Determinants of Higher-Order Matrices

To evaluate determinants of higher-order matrices, we must define the **minor** and the **cofactor** of an element in a matrix.

Minor and Cofactor of a Matrix

Let $A = [a_{ij}]$ be a square matrix of size (or order) $n \geq 2$.

1. The **minor** of a_{ij}, denoted as M_{ij}, is the determinant of the $n - 1 \times n - 1$ matrix formed by deleting the ith row and the jth column of A.

2. The **cofactor** of a_{ij}, denoted as C_{ij}, is $\begin{cases} M_{ij} \text{ when } i + j \text{ is even} \\ -M_{ij} \text{ when } i + j \text{ is odd} \end{cases}$.

Finding Minors and Cofactors of a Matrix

EXAMPLE 2

In $A = \begin{bmatrix} 1 & 2 & 3 \\ 4 & 5 & 6 \\ 7 & 8 & 9 \end{bmatrix}$ find the minor and cofactor of **a.** a_{31} **b.** a_{12}

SOLUTION **a.** The minor M_{31} is the minor of $a_{31} = 7$ appearing in row 3, column 1. It is found by deleting row 3 and column 1:

$$M_{31} = \begin{vmatrix} 1 & 2 & 3 \\ 4 & 5 & 6 \\ 7 & 8 & 9 \end{vmatrix} = \begin{vmatrix} 2 & 3 \\ 5 & 6 \end{vmatrix} = 2(6) - 3(5) = -3$$

Tip

A 2 × 2 matrix will have four cofactors. The sign pattern is

$\begin{bmatrix} + & - \\ - & + \end{bmatrix}$

A 3 × 3 matrix will have nine cofactors. The sign pattern is

$\begin{bmatrix} + & - & + \\ - & + & - \\ + & - & + \end{bmatrix}$

Because $i + j$ is even ($3 + 1 = 4$) the cofactor of the minor M_{31} is M_{31}:

$$C_{31} = M_{31} = \begin{vmatrix} 2 & 3 \\ 5 & 6 \end{vmatrix} = 2 \cdot 6 - 3 \cdot 5 = 12 - 15 = -3$$

b. The minor M_{12} is the minor of $a_{12} = 2$ appearing in row 1, column 2. It is found by deleting row 1 and column 2.

$$M_{12} = \begin{vmatrix} 1 & 2 & 3 \\ 4 & 5 & 6 \\ 7 & 8 & 9 \end{vmatrix} = \begin{vmatrix} 4 & 6 \\ 7 & 9 \end{vmatrix} = 4(9) - 6(7) = 36 - 42 = -6$$

Because $i + j$ is odd ($1 + 2 = 3$) the cofactor of the minor M_{12} is $-M_{12}$:

$$C_{12} = -M_{12} = -(-6) = 6$$

Self Check 2   Find the cofactor of a_{23}.

Now Try Exercise 13.

We are now ready to evaluate determinants of higher-order matrices.

Determinant of a Higher-Order Matrix	If A is a square matrix of order $n \geq 2$, the **determinant** of A, det (A) or $|A|$, is the sum of the products of the elements in any row (or column) and the cofactors of those elements.

In Example 3, we will evaluate a 3×3 determinant by expanding the determinant in three ways: along two different rows and along a column. This method is called **expanding a determinant by cofactors**.

EXAMPLE 3   **Finding the Determinant of a 3 × 3 Matrix**

Evaluate $\begin{vmatrix} 1 & 2 & -3 \\ -1 & 0 & 1 \\ -2 & 2 & 1 \end{vmatrix}$ by expanding by cofactors along the designated row or column.

Expanding on row 1:

$$\begin{vmatrix} \mathbf{1} & \mathbf{2} & \mathbf{-3} \\ -1 & 0 & 1 \\ -2 & 2 & 1 \end{vmatrix} = a_{11}C_{11} + a_{12}C_{12} + a_{13}C_{13}$$

$$= 1\begin{vmatrix} 0 & 1 \\ 2 & 1 \end{vmatrix} + 2\left(-\begin{vmatrix} -1 & 1 \\ -2 & 1 \end{vmatrix}\right) + (-3)\begin{vmatrix} -1 & 0 \\ -2 & 2 \end{vmatrix}$$

$$= 1(-2) + 2(-1) - 3(-2)$$

$$= -2 - 2 + 6$$

$$= 2$$

Expanding on row 3:

$$\begin{vmatrix} 1 & 2 & -3 \\ -1 & 0 & 1 \\ \mathbf{-2} & \mathbf{2} & \mathbf{1} \end{vmatrix} = a_{31}C_{31} + a_{32}C_{32} + a_{33}C_{33}$$

$$= -2\begin{vmatrix} 2 & -3 \\ 0 & 1 \end{vmatrix} + 2\left(-\begin{vmatrix} 1 & -3 \\ -1 & 1 \end{vmatrix}\right) + 1\begin{vmatrix} 1 & 2 \\ -1 & 0 \end{vmatrix}$$

$$= -2(2) + 2(2) + 1(2)$$

$$= -4 + 4 + 2$$

$$= 2$$

Expanding on column 2:

$$\begin{vmatrix} 1 & 2 & -3 \\ -1 & 0 & 1 \\ -2 & 2 & 1 \end{vmatrix} = a_{12}C_{12} + a_{22}C_{22} + a_{32}C_{32}$$

$$= 2\left(-\begin{vmatrix} -1 & 1 \\ -2 & 1 \end{vmatrix}\right) + 0\begin{vmatrix} 1 & -3 \\ -2 & 1 \end{vmatrix} + 2\left(-\begin{vmatrix} 1 & -3 \\ -1 & 1 \end{vmatrix}\right)$$

$$= 2(-1) + 0(-5) + 2(2)$$

$$= -2 + 4$$

$$= 2$$

In each case, the result is 2.

> **Take Note**
>
> Note that we don't need to evaluate the determinant $\begin{vmatrix} 1 & -3 \\ -2 & 1 \end{vmatrix}$ because it has a coefficient of 0. When we expand by cofactors along a row or column containing some 0's, the work is easier because we have fewer determinants to evaluate.

Self Check 3 Evaluate $\begin{vmatrix} 1 & 0 & 1 \\ 2 & -1 & 0 \\ 3 & 1 & -1 \end{vmatrix}$ by expanding on its first row and second columns.

Now Try Exercise 19.

EXAMPLE 4 **Finding the Determinant of a 4 × 4 Matrix**

Evaluate: $\begin{vmatrix} 0 & 0 & 2 & 0 \\ 1 & 2 & 17 & -3 \\ -1 & 0 & 28 & 1 \\ -2 & 2 & -37 & 1 \end{vmatrix}$.

SOLUTION Because row 1 contains three 0's, we expand the determinant along row 1. Then only one cofactor needs to be evaluated.

> **Tip**
>
> The row or column containing the most zeros is usually the best choice for expansion by cofactors. So examine the matrix and look for zeros.
>
> ©In Tune/Shutterstock.com

$$\begin{vmatrix} 0 & 0 & 2 & 0 \\ 1 & 2 & 17 & -3 \\ -1 & 0 & 28 & 1 \\ -2 & 2 & -37 & 1 \end{vmatrix} = 0 + 0 + 2\begin{vmatrix} 1 & 2 & -3 \\ -1 & 0 & 1 \\ -2 & 2 & 1 \end{vmatrix} + 0$$

$$= 2(2) \qquad \text{See Example 3.}$$

$$= 4$$

Self Check 4 Evaluate: $\begin{vmatrix} 1 & 0 & 0 & 0 \\ 2 & 0 & 1 & 0 \\ -1 & 2 & 11 & -2 \\ -2 & 1 & 13 & 1 \end{vmatrix}$.

Now Try Exercise 27.

Example 4 suggests the following theorem.

Zero Row or Column Theorem If every entry in a row or column of a square matrix A is 0, then $|A| = 0$.

Accent on Technology

Evaluating Determinants

The MATRIX: MATH menu can be used to evaluate a determinant. For example, as shown in Figure 6-10, we will evaluate the determinant

$$\begin{vmatrix} 2 & 3 & 0.5 & 6 \\ 1 & 4 & -2 & -3 \\ 3 & 4 & -3 & -2 \\ -0.7 & 6 & 2 & 1 \end{vmatrix}.$$

1. Enter the values of the determinant into MATRIX A.

2. Go to the home screen; use the MATRIX menu, and select MATH 1:det(.

3. Enter the matrix name and press ENTER. The value of the determinant is 99.3.

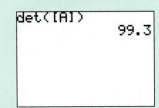

FIGURE 6-10

2. Understand and Use Properties of Determinants

We have seen that there are row operations for transforming matrices. There are similar row and column operations for transforming determinants.

Row and Column Operations and Determinants

Let A be a square matrix and k be a real number.

1. If a matrix B is obtained from matrix A by interchanging two rows (or columns), then $|B| = -|A|$.

2. If B is obtained from A by multiplying every element in a row (or column) of A by k, then $|B| = k|A|$.

3. If B is obtained from A by adding k times any row (or column) of A to another row (or column) of A, then $|B| = |A|$.

We will illustrate each row operation by showing that the operation holds true for the 2×2 determinant.

$$(1) \qquad |A| = \begin{vmatrix} 3 & 4 \\ 1 & 2 \end{vmatrix} = 3(2) - 4(1) = 6 - 4 = 2$$

- **Interchanging two rows**
  Let $|B|$ be the determinant obtained from $|A|$ by interchanging its two rows. Then

$$|B| = \begin{vmatrix} 1 & 2 \\ 3 & 4 \end{vmatrix} = 1(4) - 2(3) = 4 - 6 = -2$$

Since $-2 = -1(2)$, we have $|B| = -|A|$.

In general, if $|A|$ and $|B|$ are as follows, we have

$$|A| = \begin{vmatrix} a & b \\ c & d \end{vmatrix} = ad - bc$$

$$|B| = \begin{vmatrix} c & d \\ a & b \end{vmatrix} = cb - da = -(ad - bc) = -|A|$$

- **Multiplying every element in a row by k**

  Let $|B|$ be the determinant obtained from $|A|$ in Equation 1 by multiplying its second row by 3. Then

$$|B| = \begin{vmatrix} 3 & 4 \\ 3(1) & 3(2) \end{vmatrix} = \begin{vmatrix} 3 & 4 \\ 3 & 6 \end{vmatrix} = 3(6) - 4(3) = 18 - 12 = 6$$

  Since $6 = 3(2)$, we have $|B| = 3|A|$.

  In general, if $|A|$ and $|B|$ are as follows, we have

(2) $\quad |A| = \begin{vmatrix} a & b \\ c & d \end{vmatrix} = ad - bc$

$$|B| = \begin{vmatrix} a & b \\ kc & kd \end{vmatrix} = akd - bkc = k(ad - bc) = k|A|$$

- **Adding k times any row of A to another row of A**

  Let $|B|$ be the determinant obtained from $|A|$ in Equation 1 by adding 3 times its first row to its second row. Then

$$|B| = \begin{vmatrix} 3 & 4 \\ 3(3) + 1 & 3(4) + 2 \end{vmatrix} = 3(14) - 4(10) = 42 - 40 = 2$$

  Since $2 = 2$, we have $|A| = |B|$.

  In general, if $|B|$ is obtained from $|A|$ in Equation 2 by adding k times its first row to its second row, then

$$|B| = \begin{vmatrix} a & b \\ ka + c & kb + d \end{vmatrix} = a(kb + d) - b(ka + c)$$

$$= akb + ad - bka - bc$$

$$= ad - bc$$

$$= |A|$$

Evaluating higher-order determinants can be very time consuming. However, as we have seen, if we use row and column operations to introduce as many 0's as possible in a row (or column) and then expand the determinant by cofactors along that row (or column), the work is much easier. In the next example, we will use several row and column operations to introduce 0's into the determinant.

EXAMPLE 5

Using Row and Column Operations to Help Evaluate the Determinant

Use row and column operations to evaluate: $\begin{vmatrix} 10 & 20 & -10 & 20 \\ 2 & 1 & 1 & 1 \\ 1 & 2 & -3 & 2 \\ 2 & -1 & -1 & 1 \end{vmatrix}$.

SOLUTION

To get smaller numbers in the first row, we can use a type 2 row operation and multiply each entry in the first row by $\frac{1}{10}$. However, we must then multiply the resulting determinant by 10 to retain its original value.

$$(1)\quad \begin{vmatrix} 10 & 20 & -10 & 20 \\ 2 & 1 & 1 & 1 \\ 1 & 2 & -3 & 2 \\ 2 & -1 & -1 & 1 \end{vmatrix} \begin{array}{c} \left(\frac{1}{10}\right)R1 \to \\ = 10 \end{array} \begin{vmatrix} 1 & 2 & -1 & 2 \\ 2 & 1 & 1 & 1 \\ 1 & 2 & -3 & 2 \\ 2 & -1 & -1 & 1 \end{vmatrix}$$

To get three 0's in the first row of the second determinant in Equation 1, we perform a type 3 row operation and expand the new determinant along its first row.

$$10\begin{vmatrix} 1 & 2 & -1 & 2 \\ 2 & 1 & 1 & 1 \\ 1 & 2 & -3 & 2 \\ 2 & -1 & -1 & 1 \end{vmatrix} \begin{array}{c} (-1)R3 + R1 \to \\ = 10 \end{array} \begin{vmatrix} 0 & 0 & \mathbf{2} & 0 \\ 2 & 1 & 1 & 1 \\ 1 & 2 & -3 & 2 \\ 2 & -1 & -1 & 1 \end{vmatrix}$$

$$(2)\qquad\qquad\qquad\qquad = 10(\mathbf{2})\begin{vmatrix} 2 & 1 & 1 \\ 1 & 2 & 2 \\ 2 & -1 & 1 \end{vmatrix}$$

To introduce 0's into the 3×3 determinant in Equation 2, we perform a column operation on the 3×3 determinant and expand the result on its first column.

$$\overset{\displaystyle (-2)C3 + C1}{\underset{\downarrow}{}}$$

$$20\begin{vmatrix} 2 & 1 & 1 \\ 1 & 2 & 2 \\ 2 & -1 & 1 \end{vmatrix} = 20\begin{vmatrix} 0 & 1 & 1 \\ \mathbf{-3} & 2 & 2 \\ 0 & -1 & 1 \end{vmatrix}$$

$$= 20\left[(\mathbf{-3})(-1)\begin{vmatrix} 1 & 1 \\ -1 & 1 \end{vmatrix} \right]$$

$$= 20(3)[1 - (-1)]$$

$$= 60(2)$$

$$= 120$$

Self Check 5   Evaluate: $\begin{vmatrix} 15 & 20 & 5 & 10 \\ 1 & 2 & 2 & 3 \\ 0 & 3 & -1 & 1 \\ 1 & 0 & 0 & -1 \end{vmatrix}$.

Now Try Exercise 29.

We will consider one final theorem.

Zero Row or Column Theorem   If A is a square matrix with two identical rows (or columns), then $|A| = 0$.

PROOF   If the square matrix A has two identical rows (or columns), we can apply a type 3 row (or column) operation to zero out one of those rows (or columns). Since the matrix would then have an all-zero row (or column), its determinant would be 0 when we expand by cofactors along the all-zero row (or column).

3. Use Determinants to Solve Systems of Equations

We can solve the system $\begin{cases} ax + by = e \\ cx + dy = f \end{cases}$ by multiplying the first equation by d, multiplying the second equation by $-b$, and adding to get

$$adx + bdy = ed$$
$$\underline{-bcx - bdy = -bf}$$
$$adx - bcx = ed - bf$$

If $ad \neq bc$ we can solve the resulting equation for x:

$$adx - bcx = ed - bf$$

$$(ad - bc)x = ed - bf \qquad \textcolor{red}{\text{Factor out } x.}$$

(3) $$\qquad x = \frac{ed - bf}{ad - bc} \qquad \textcolor{red}{\text{Divide both sides by } ad - bc.}$$

If $ad \neq bc$, we can also solve the system for y to get

(4) $$\quad y = \frac{af - ec}{ad - bc}$$

We can write the values of x and y in Equations 3 and 4 using determinants.

$$x = \frac{\begin{vmatrix} e & b \\ f & d \end{vmatrix}}{\begin{vmatrix} a & b \\ c & d \end{vmatrix}} = \frac{ed - bf}{ad - bc} \qquad\qquad y = \frac{\begin{vmatrix} a & e \\ c & f \end{vmatrix}}{\begin{vmatrix} a & b \\ c & d \end{vmatrix}} = \frac{af - ec}{ad - bc}$$

If we compare these formulas with the original system,

$$\begin{cases} \textcolor{red}{a}x + \textcolor{red}{b}y = e \\ \textcolor{red}{c}x + \textcolor{red}{d}y = f \end{cases}$$

we see that the denominators are the determinant of the coefficient matrix:

$$\text{Denominator determinant } = \begin{vmatrix} a & b \\ c & d \end{vmatrix}$$

To find the numerator determinant for x, we replace the a and c in the first column of the denominator determinant with the constants e and f.

To find the numerator determinant for y, we replace the b and d in the second column of the denominator determinant with the constants e and f.

$$x = \frac{\begin{vmatrix} e & b \\ f & d \end{vmatrix}}{\begin{vmatrix} a & b \\ c & d \end{vmatrix}} \qquad\qquad y = \frac{\begin{vmatrix} a & e \\ c & f \end{vmatrix}}{\begin{vmatrix} a & b \\ c & d \end{vmatrix}}$$

This method of using determinants to solve systems of equations is called **Cramer's Rule**.

Cramer's Rule for Two Equations in Two Variables   If the system $\begin{cases} ax + by = e \\ cx + dy = f \end{cases}$ has a single solution, it is given by

$$x = \frac{D_x}{D} \quad \text{and} \quad y = \frac{D_y}{D}$$

where $D = \begin{vmatrix} a & b \\ c & d \end{vmatrix}$, $D_x = \begin{vmatrix} e & b \\ f & d \end{vmatrix}$, and $D_y = \begin{vmatrix} a & e \\ c & f \end{vmatrix}$.

If D, D_x, and D_y are all 0, the system is consistent, but the equations are dependent. If $D = 0$ and $D_x \neq 0$ or $D_y \neq 0$, the system is inconsistent.

Using Cramer's Rule to Solve a System of Linear Equations

EXAMPLE 6

Use Cramer's Rule to solve $\begin{cases} 3x + 2y = 7 \\ -x + 5y = 9 \end{cases}$.

SOLUTION

$$x = \frac{\begin{vmatrix} 7 & 2 \\ 9 & 5 \end{vmatrix}}{\begin{vmatrix} 3 & 2 \\ -1 & 5 \end{vmatrix}} = \frac{7(5) - 2(9)}{3(5) - 2(-1)} = \frac{35 - 18}{15 + 2} = \frac{17}{17} = 1$$

$$y = \frac{\begin{vmatrix} 3 & 7 \\ -1 & 9 \end{vmatrix}}{\begin{vmatrix} 3 & 2 \\ -1 & 5 \end{vmatrix}} = \frac{3(9) - 7(-1)}{3(5) - 2(-1)} = \frac{27 + 7}{15 + 2} = \frac{34}{17} = 2$$

Verify that the ordered pair $(1, 2)$ satisfies both of the equations in the system.

Self Check 6   Solve: $\begin{cases} 2x + 5y = 9 \\ 3x + 7y = 13 \end{cases}$.

Now Try Exercise 39.

We can use Cramer's Rule to solve systems of n equations in n variables where each equation has the form

$$a_1 x_1 + a_2 x_2 + \cdots + a_n x_n = c$$

To do so, we let D be the determinant of the coefficient matrix of the system and let D_x be the determinant formed by replacing the ith column of D by the column of constants from the right of the equal signs. If $D \neq 0$, Cramer's Rule provides the following solution:

$$x_1 = \frac{D_{x_1}}{D}, x_2 = \frac{D_{x_2}}{D}, \ldots, x_n = \frac{D_{x_n}}{D}$$

Using Cramer's Rule to Solve a System of Linear Equations

EXAMPLE 7

Use Cramer's Rule to solve the system $\begin{cases} 2x - y + 2z = 3 \\ x - y + z = 2 \\ x + y + 2z = 3 \end{cases}$.

SOLUTION   Each of the values x, y, and z is the quotient of two 3×3 determinants. The denominator of each quotient is the determinant consisting of the nine coefficients of the variables. The numerators for x, y, and z are modified copies of this denominator determinant. We substitute the column of constants for the coefficients of the variable for which we are solving.

Tip

To solve a system of three equations with three variables using Cramer's Rule, you will need to find the determinant of four 3×3 matrices.

$$\begin{cases} 2x - y + 2z = 3 \\ x - y + z = 2 \\ x + y + 2z = 3 \end{cases}$$

$$x = \frac{\begin{vmatrix} 3 & -1 & 2 \\ 2 & -1 & 1 \\ 3 & 1 & 2 \end{vmatrix}}{\begin{vmatrix} 2 & -1 & 2 \\ 1 & -1 & 1 \\ 1 & 1 & 2 \end{vmatrix}} = \frac{3\begin{vmatrix} -1 & 1 \\ 1 & 2 \end{vmatrix} + (-1)(-1)\begin{vmatrix} 2 & 1 \\ 3 & 2 \end{vmatrix} + 2\begin{vmatrix} 2 & -1 \\ 3 & 1 \end{vmatrix}}{2\begin{vmatrix} -1 & 1 \\ 1 & 2 \end{vmatrix} + (-1)(-1)\begin{vmatrix} 1 & 1 \\ 1 & 2 \end{vmatrix} + 2\begin{vmatrix} 1 & -1 \\ 1 & 1 \end{vmatrix}} = \frac{2}{-1} = -2$$

$$y = \frac{\begin{vmatrix} 2 & 3 & 2 \\ 1 & 2 & 1 \\ 1 & 3 & 2 \end{vmatrix}}{\begin{vmatrix} 2 & -1 & 2 \\ 1 & -1 & 1 \\ 1 & 1 & 2 \end{vmatrix}} = \frac{2\begin{vmatrix} 2 & 1 \\ 3 & 2 \end{vmatrix} + 3(-1)\begin{vmatrix} 1 & 1 \\ 1 & 2 \end{vmatrix} + 2\begin{vmatrix} 1 & 2 \\ 1 & 3 \end{vmatrix}}{-1} = \frac{1}{-1} = -1$$

$$z = \frac{\begin{vmatrix} 2 & -1 & 3 \\ 1 & -1 & 2 \\ 1 & 1 & 3 \end{vmatrix}}{\begin{vmatrix} 2 & -1 & 2 \\ 1 & -1 & 1 \\ 1 & 1 & 2 \end{vmatrix}} = \frac{2\begin{vmatrix} -1 & 2 \\ 1 & 3 \end{vmatrix} + (-1)(-1)\begin{vmatrix} 1 & 2 \\ 1 & 3 \end{vmatrix} + 3\begin{vmatrix} 1 & -1 \\ 1 & 1 \end{vmatrix}}{-1} = \frac{-3}{-1} = 3$$

Verify that the ordered triple $(-2, -1, 3)$ satisfies each equation in the system.

Self Check 7 Use Cramer's rule to solve $\begin{cases} 2x - y + 2z = 6 \\ x - y + z = 2 \\ x + y + 2z = 9 \end{cases}$.

Now Try Exercise 45.

4. Write Equations of Lines

If we are given the coordinates of two points in the xy-plane, we can use determinants to write an equation of the line passing through those points.

Two-Point Form of an Equation of a Line

An equation of the line passing through points $P(x_1, y_1)$ and $Q(x_2, y_2)$ is given by

$$\begin{vmatrix} x & y & 1 \\ x_1 & y_1 & 1 \\ x_2 & y_2 & 1 \end{vmatrix} = 0$$

EXAMPLE 8

Using the Determinant to Write the Equation of a Line

Write an equation of the line in standard form passing through $P(-2, 3)$ and $Q(4, -5)$.

SOLUTION   We set up the equation $\begin{vmatrix} x & y & 1 \\ -2 & 3 & 1 \\ 4 & -5 & 1 \end{vmatrix} = 0$ and expand along the first row to get

$$[3(1) - 1(-5)]x - [-2(1) - 1(4)]y + [(-2)(-5) - 3(4)]1 = 0$$

$$8x + 6y - 2 = 0$$

$$4x + 3y = 1 \qquad \text{Add 2 to both sides and divide both sides by 2.}$$

An equation of the line in standard form is $4x + 3y = 1$.

Self Check 8   Find an equation in standard form of the line passing through $(1, 3)$ and $(3, 5)$.

Now Try Exercise 53.

5. Find Areas of Triangles

Area of a Triangle   If points $P(x_1, y_1)$, $Q(x_2, y_2)$, and $R(x_3, y_3)$ are the vertices of a triangle, then the area of the triangle is given by

$$A = \pm\frac{1}{2}\begin{vmatrix} x_1 & y_1 & 1 \\ x_2 & y_2 & 1 \\ x_3 & y_3 & 1 \end{vmatrix} \qquad \text{Pick either + or − to make the area positive.}$$

Using the Determinant to Find the Area of a Triangle

EXAMPLE 9   Find the area of the triangle shown in Figure 6-11.

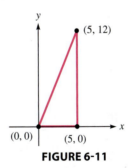

FIGURE 6-11

SOLUTION   We set up the equation

$$A = \pm\frac{1}{2}\begin{vmatrix} 0 & 0 & 1 \\ 5 & 0 & 1 \\ 5 & 12 & 1 \end{vmatrix}$$

and expand the determinant along the first row to get

$$= \pm\frac{1}{2}\left[1\begin{vmatrix} 5 & 0 \\ 5 & 12 \end{vmatrix} \right]$$

$$= \pm\frac{1}{2}(60 - 0)$$

$$= 30$$

The area of the triangle is 30 square units.

Self Check 9 Find the area of the triangle with vertices at $(1, 2)$, $(2, 3)$, and $(-1, 4)$.

Now Try Exercise 57.

Self Check Answers

1. -2 **2.** 6 **3.** first row 6; second column 6   **4.** -4 **5.** 30
6. $(2, 1)$ **7.** $(1, 2, 3)$ **8.** $x - y = -2$ **9.** 2 sq. units

Exercises 6.5

Getting Ready

You should be able to complete these vocabulary and concept statements before you proceed to the practice exercises.

Fill in the blanks.

1. The determinant of a square matrix A is written as ____ or _____.

2. $\begin{vmatrix} a & b \\ c & d \end{vmatrix} =$ _____

3. If every entry in one row or one column of A is zero, then $|A| =$ ___.

4. If a matrix B is obtained from matrix A by adding one row to another, then $|B| =$ ____.

5. If two columns of A are identical, then $|A| =$ ___.

6. In Cramer's Rule, the denominator is the determinant of the _____.

Practice

Evaluate each determinant.

7. $\begin{vmatrix} 2 & 1 \\ -2 & 3 \end{vmatrix}$

8. $\begin{vmatrix} -3 & -6 \\ 2 & -5 \end{vmatrix}$

9. $\begin{vmatrix} 2 & -3 \\ -3 & 5 \end{vmatrix}$

10. $\begin{vmatrix} 5 & 8 \\ -6 & -2 \end{vmatrix}$

In Exercises 11–18, $A = \begin{vmatrix} 1 & -2 & 3 \\ 4 & 5 & -6 \\ -7 & 8 & 9 \end{vmatrix}$. *Find each minor or cofactor.*

11. M_{21}

12. M_{13}

13. M_{33}

14. M_{32}

15. C_{21}

16. C_{13}

17. C_{33}

18. C_{32}

Evaluate each determinant by expanding by cofactors.

19. $\begin{vmatrix} 2 & -3 & 5 \\ -2 & 1 & 3 \\ 1 & 3 & -2 \end{vmatrix}$

20. $\begin{vmatrix} 1 & 3 & 1 \\ -2 & 5 & 3 \\ 3 & -2 & -2 \end{vmatrix}$

21. $\begin{vmatrix} 1 & -1 & 2 \\ 2 & 1 & 3 \\ 1 & 1 & -1 \end{vmatrix}$

22. $\begin{vmatrix} 1 & 3 & 1 \\ 2 & 1 & -1 \\ 2 & -1 & 1 \end{vmatrix}$

23. $\begin{vmatrix} 2 & 1 & -1 \\ 1 & 3 & 5 \\ 2 & -5 & 3 \end{vmatrix}$

24. $\begin{vmatrix} 3 & 1 & -2 \\ -3 & 2 & 1 \\ 1 & 3 & 0 \end{vmatrix}$

25. $\begin{vmatrix} 0 & 1 & -3 \\ -3 & 5 & 2 \\ 2 & -5 & 3 \end{vmatrix}$

26. $\begin{vmatrix} 1 & -7 & -2 \\ -2 & 0 & 3 \\ -1 & 7 & 1 \end{vmatrix}$

27. $\begin{vmatrix} 0 & 0 & 1 & 0 \\ -2 & 1 & 0 & 1 \\ 1 & 0 & 1 & 2 \\ 2 & 0 & 1 & 2 \end{vmatrix}$

28. $\begin{vmatrix} 1 & 0 & -2 & 1 \\ 0 & 1 & 0 & 1 \\ 0 & 3 & -1 & 2 \\ 0 & -1 & 0 & 1 \end{vmatrix}$

29. $\begin{vmatrix} 10 & 20 & 10 & 30 \\ -2 & 1 & -3 & 1 \\ -1 & 0 & 1 & -2 \\ 2 & -1 & -1 & 3 \end{vmatrix}$

30. $\begin{vmatrix} -1 & 3 & -2 & 5 \\ 2 & 1 & 0 & 1 \\ 1 & 3 & -2 & 5 \\ 2 & -1 & 0 & -1 \end{vmatrix}$

Determine whether each statement is true. Do not evaluate the determinants.

31. $\begin{vmatrix} 1 & 3 & -4 \\ -2 & 1 & 3 \\ 1 & 3 & 2 \end{vmatrix} = -\begin{vmatrix} -2 & 1 & 3 \\ 1 & 3 & -4 \\ 1 & 3 & 2 \end{vmatrix}$

32. $\begin{vmatrix} 4 & 6 & 8 \\ 10 & 5 & 15 \\ 20 & 5 & 10 \end{vmatrix} = \begin{vmatrix} 2 & 3 & 4 \\ 10 & 5 & 15 \\ 20 & 5 & 10 \end{vmatrix}$

33. $\begin{vmatrix} -2 & -3 & -4 \\ 5 & -1 & 2 \\ 1 & 2 & 3 \end{vmatrix} = -\begin{vmatrix} 2 & 3 & 4 \\ -5 & 1 & -2 \\ 1 & 2 & 3 \end{vmatrix}$

34. $\begin{vmatrix} 1 & 2 & 3 \\ 4 & 5 & 6 \\ 7 & 8 & 9 \end{vmatrix} = \begin{vmatrix} 5 & 7 & 9 \\ 4 & 5 & 6 \\ 7 & 8 & 9 \end{vmatrix}$

If $\begin{vmatrix} a & b & c \\ d & e & f \\ g & h & i \end{vmatrix} = 3$, **find the value of each determinant.**

35. $\begin{vmatrix} d & e & f \\ a & b & c \\ -g & -h & -i \end{vmatrix}$

36. $\begin{vmatrix} 5a & 5b & 5c \\ -d & -e & -f \\ 3g & 3h & 3i \end{vmatrix}$

37. $\begin{vmatrix} a+g & b+h & c+i \\ d & e & f \\ g & h & i \end{vmatrix}$

38. $\begin{vmatrix} g & h & i \\ a & b & c \\ d & e & f \end{vmatrix}$

Use Cramer's Rule to find the solution of each system, if possible.

39. $\begin{cases} 3x + 2y = 7 \\ 2x - 3y = -4 \end{cases}$

40. $\begin{cases} x - 5y = -6 \\ 3x + 2y = -1 \end{cases}$

41. $\begin{cases} x - y = 3 \\ 3x - 7y = 9 \end{cases}$

42. $\begin{cases} 2x - y = -6 \\ x + y = 0 \end{cases}$

43. $\begin{cases} x + 2y + z = 2 \\ x - y + z = 2 \\ x + y + 3z = 4 \end{cases}$

44. $\begin{cases} x + 2y - z = -1 \\ 2x + y - z = 1 \\ x - 3y - 5z = 17 \end{cases}$

45. $\begin{cases} 2x - y + z = 5 \\ 3x - 3y + 2z = 10 \\ x + 3y + z = 0 \end{cases}$

46. $\begin{cases} x - y - z = 2 \\ x + y + z = 2 \\ -x - y + z = -4 \end{cases}$

47. $\begin{cases} \dfrac{x}{2} + \dfrac{y}{3} + \dfrac{z}{2} = 11 \\ \dfrac{x}{3} + y - \dfrac{z}{6} = 6 \\ \dfrac{x}{2} + \dfrac{y}{6} + z = 16 \end{cases}$

48. $\begin{cases} \dfrac{x}{2} + \dfrac{y}{5} + \dfrac{z}{3} = 17 \\ \dfrac{x}{5} + \dfrac{y}{2} + \dfrac{z}{5} = 32 \\ x + \dfrac{y}{3} + \dfrac{z}{2} = 30 \end{cases}$

49. $\begin{cases} 2p - q + 3r - s = 0 \\ p + q - s = -1 \\ 3p - r = 2 \\ p - 2q + 3s = 7 \end{cases}$

50. $\begin{cases} a + b + c + d = 8 \\ a + b + c + 2d = 7 \\ a + b + 2c + 3d = 3 \\ a + 2b + 3c + 4d = 4 \end{cases}$

Use determinants to write an equation of the line in standard form that passes through the given points.

51. $P(0, 0)$, $Q(4, 6)$

52. $P(2, 3)$, $Q(6, 8)$

53. $P(-2, 3)$, $Q(5, -3)$

54. $P(1, -2)$, $Q(-4, 3)$

Use determinants to find the area of each triangle with vertices at the given points.

55. $P(0, 0)$, $Q(12, 0)$, $R(12, 5)$

56. $P(0, 0)$, $Q(0, 5)$, $R(12, 5)$

57. $P(2, 3)$, $Q(10, 8)$, $R(0, 20)$

58. $P(1, 1)$, $Q(6, 6)$, $R(2, 10)$

In Exercises 59–61, illustrate each column operation by showing that it is true for the determinant $\begin{vmatrix} a & b \\ c & d \end{vmatrix}$.

59. Interchanging two columns

60. Multiplying each element in a column by k

61. Adding k times any column to another column

62. Use the method of addition to solve $\begin{cases} ax + by = e \\ cx + dy = f \end{cases}$

for y, and thereby show that $y = \dfrac{af - ec}{ad - bc}$.

Expand the determinants and solve for x.

63. $\begin{vmatrix} 3 & x \\ 1 & 2 \end{vmatrix} = \begin{vmatrix} 2 & -1 \\ x & -5 \end{vmatrix}$

64. $\begin{vmatrix} 4 & x^2 \\ 1 & -1 \end{vmatrix} = \begin{vmatrix} x & 4 \\ 2 & 3 \end{vmatrix}$

65. $\begin{vmatrix} 3 & x & 1 \\ x & 0 & -2 \\ 4 & 0 & 1 \end{vmatrix} = \begin{vmatrix} 2 & x \\ x & 4 \end{vmatrix}$

66. $\begin{vmatrix} x & -1 & 2 \\ -2 & x & 3 \\ 4 & -3 & -1 \end{vmatrix} = \begin{vmatrix} 2 & 2 \\ 5 & x \end{vmatrix}$

Use a graphing calculator to evaluate each determinant.

67. $\begin{vmatrix} 2.3 & 5.7 & 6.1 \\ 3.4 & 6.2 & 8.3 \\ 5.8 & 8.2 & 9.2 \end{vmatrix}$

68. $\begin{vmatrix} 0.32 & -7.4 & -6.7 \\ 3.3 & 5.5 & -0.27 \\ -8 & -0.13 & 5.47 \end{vmatrix}$

Applications

69. Investing A student wants to average a 6.6% return by investing \$20,000 in the three stocks listed in the table. Because HiTech is a high-risk investment, he wants to invest three times as much in SaveTel and OilCo combined as he invests in HiTech. How much should he invest in each stock?

Stock	Rate of Return
HiTech	10%
SaveTel	5%
OilCo	6%

70. Ice skating The illustration shows three circles traced out by a figure skater during her performance. If the centers of the circles are the given distances apart, find the radius of each circle.

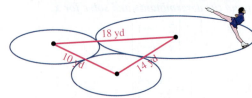

Discovery and Writing

71. Explain how to find the determinant of a 2 × 2 matrix.

72. Explain how to find the determinant of a 3 × 3 matrix using cofactor expansion.

73. Explain why applying row or column operations can help evaluate the determinant.

74. What is Cramer's Rule? Describe how it can be used to solve systems of equations.

In Exercises 75–78, evaluate each determinant. What do you discover?

75. $\begin{vmatrix} 1 & 3 & 4 \\ 0 & 5 & 2 \\ 0 & 0 & 2 \end{vmatrix}$

76. $\begin{vmatrix} 2 & 1 & -2 \\ 0 & 3 & 4 \\ 0 & 0 & -1 \end{vmatrix}$

77. $\begin{vmatrix} 1 & 2 & 4 & 3 \\ 0 & 2 & 2 & 1 \\ 0 & 0 & 3 & 2 \\ 0 & 0 & 0 & 4 \end{vmatrix}$

78. $\begin{vmatrix} 2 & 1 & -2 & 1 \\ 0 & 2 & 2 & -1 \\ 0 & 0 & 3 & 1 \\ 0 & 0 & 0 & 2 \end{vmatrix}$

79. Another way to evaluate a 3 × 3 determinant is to copy its first two columns to the right of the determinant as shown. Then find the product of the numbers on each red diagonal and find their sum. Then find the product of the numbers on each blue diagonal and find

their sum. Then subtract the sum of the products on the blue diagonals from the sum of the products on the red diagonals. Find the value of the determinant.

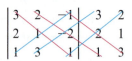

80. Use the method of Exercise 79 to evaluate the determinant $\begin{vmatrix} 0 & 1 & -3 \\ -3 & 5 & 2 \\ 2 & -5 & 3 \end{vmatrix}$.

81. A determinant is a function that associates a number with every square matrix. Give the domain and the range of that function.

82. Use an example chosen from 2 × 2 matrices to show that for $n \times n$ matrices A and B, $AB \neq BA$ but $|AB| = |BA|$.

83. If A and B are matrices and $|AB| = 0$, must $|A| = 0$ or $|B| = 0$? Explain.

84. If A and B are matrices and $|AB| = 0$, must $A = 0$ or $B = 0$? Explain.

Critical Thinking

Determine if the statement is true or false. If the statement is false, then correct it and make it true.

85. If A and B are square matrices of the same order, then $|A + B| = |A| + |B|$.

86. If A and B are square matrices of the same order, then $|AB| = |A||B|$.

87. $\begin{vmatrix} 999 & 888 \\ 777 & 666 \end{vmatrix} = \begin{vmatrix} 777 & 666 \\ 999 & 888 \end{vmatrix}$

88. $\begin{vmatrix} 111 & 222 & 0 \\ 333 & 444 & 0 \\ 555 & 666 & 0 \end{vmatrix} = 0$

89. $\begin{vmatrix} 111 & 222 & 333 & 444 \\ 555 & 666 & 777 & 888 \\ 111 & 222 & 333 & 444 \\ 555 & 666 & 777 & 888 \end{vmatrix} = 0$

90. The **transpose** of a matrix is formed by writing its columns as rows. If A is a square matrix and A^T denotes its transpose, then $|A| = |A^T|$.

6.6 Partial Fractions

In this section, we will learn to

1. Decompose a fraction when the denominator has distinct linear factors.
2. Decompose a fraction when the denominator has distinct quadratic factors.
3. Decompose a fraction when the denominator has repeated linear factors.
4. Decompose a fraction when the denominator has repeated quadratic factors.
5. Decompose a fraction when the degree of the numerator is equal to or greater than the degree of the denominator.

©Jurgen Vogt/Shutterstock.com

In this section, we will discuss how to write a complicated fraction as a sum of simpler fractions. This skill is used in calculus to solve problems such as modeling a population where there is a maximum population (called the **carrying capacity**) that an environment can sustain.

Suppose there are P gorillas in a specific region in Africa. If there is a carrying capacity of 300 gorillas in that region, the fraction $\frac{1}{P^2 - 300P}$ will occur in the model of population growth of gorillas in that region. To solve this modeling problem, we need to rewrite this fraction as a sum of two simpler fractions. In this section, we will learn to do that. The process is called **partial fraction decomposition**.

You will be asked to decompose this fraction in Problem 10 in the exercises.

We begin the discussion by reviewing how to add fractions. For example, to find the sum

$$\frac{2}{x} + \frac{6}{x + 1} + \frac{-1}{(x + 1)^2}$$

we write each fraction with an LCD of $x(x + 1)^2$, add the fractions by adding their numerators and keeping the common denominator, and simplify.

$$\frac{2}{x} + \frac{6}{x + 1} + \frac{-1}{(x + 1)^2} = \frac{2(x + 1)^2}{x(x + 1)^2} + \frac{6x(x + 1)}{(x + 1)x(x + 1)} + \frac{-1x}{(x + 1)^2 x}$$

$$= \frac{2x^2 + 4x + 2 + 6x^2 + 6x - x}{x(x + 1)^2}$$

$$= \frac{8x^2 + 9x + 2}{x(x + 1)^2}$$

To reverse the addition process and write a fraction as the sum of simpler fractions with denominators of smallest possible degree, we must **decompose a fraction into partial fractions**. To decompose the fraction

$$\frac{8x^2 + 9x + 2}{x(x + 1)^2}$$

into partial fractions, we will assume that there are constants A, B, and C such that

$$\frac{8x^2 + 9x + 2}{x(x + 1)^2} = \frac{A}{x} + \frac{B}{x + 1} + \frac{C}{(x + 1)^2}$$

After writing the terms on the right side as fractions with an LCD of $x(x + 1)^2$, we add the fractions to get

$$\frac{8x^2 + 9x + 2}{x(x+1)^2} = \frac{A(x+1)^2}{x(x+1)^2} + \frac{Bx(x+1)}{x(x+1)(x+1)} + \frac{Cx}{(x+1)^2 x}$$

$$= \frac{Ax^2 + 2Ax + A + Bx^2 + Bx + Cx}{x(x+1)^2}$$

$$(1) \quad \frac{8x^2 + 9x + 2}{x(x+1)^2} = \frac{(A+B)x^2 + (2A+B+C)x + A}{x(x+1)^2}$$

Factor x^2 from $Ax^2 + Bx^2$.
Factor x from $2Ax + Bx + Cx$.

Since the fractions on the left and right sides of Equation 1 are equal and their denominators are equal, the coefficients of their polynomial numerators are equal.

$$\begin{cases} A + B = 8 \\ 2A + B + C = 9 \\ A = 2 \end{cases}$$

These are the coefficients of x^2.
These are the coefficients of x.
These are the constants.

We can solve this system of equations to find that $A = 2$, $B = 6$, and $C = -1$. Then we know that

$$\frac{8x^2 + 9x + 2}{x(x+1)^2} = \frac{2}{x} + \frac{6}{x+1} + \frac{-1}{(x+1)^2}$$

To check the result, we can add the previous fractions and show that the sum is the original fraction.

To find the partial fraction decomposition of a fraction, we will follow these steps:

Strategy for Decomposing a Fraction with Distinct Linear Factors into Partial Fractions

1. Set up the decomposition with unknown constants A, B, C, ... in the numerator of the composition.
2. Write each fraction with a common denominator and add the fractions and simplify.
3. Set the coefficients of the corresponding terms of each numerator equal to each other.
4. Solve the resulting system of linear equations.
5. Use the values of A, B, C, ... to write the decomposition.

1. Decompose a Fraction when the Denominator Has Distinct Linear Factors

Decomposing a Fraction with a Denominator with Distinct Linear Factors

EXAMPLE 1

Decompose $\dfrac{9x + 2}{(x+2)(3x-2)}$ into partial fractions.

SOLUTION **Step 1:** Since each factor in the denominator is linear, there are constants A and B such that

$$\frac{9x+2}{(x+2)(3x-2)} = \frac{A}{x+2} + \frac{B}{3x-2}$$

Step 2: After writing the terms on the right side as fractions with a LCD of $(x+2)(3x-2)$, we add the fractions to get

$$(2) \quad \frac{9x + 2}{(x + 2)(3x - 2)} = \frac{A(3x - 2)}{(x + 2)(3x - 2)} + \frac{B(x + 2)}{(x + 2)(3x - 2)}$$

$$\frac{9x + 2}{(x + 2)(3x - 2)} = \frac{3Ax - 2A + Bx + 2B}{(x + 2)(3x - 2)}$$

$$(3) \quad \frac{9x + 2}{(x + 2)(3x - 2)} = \frac{(3A + B)x - 2A + 2B}{(x + 2)(3x - 2)} \qquad \text{Factor } x \text{ from } 3Ax + Bx.$$

Step 3: Since the fractions in Equation 3 are equal, the coefficients of their polynomial numerators are equal, and we have

$$\begin{cases} 3A + B = 9 & \text{These are the coefficients of } x. \\ -2A + 2B = 2 & \text{These are the constants.} \end{cases}$$

Step 4: The solution of the system of linear equations in step 3 is $A = 2$ and $B = 3$.

Step 5: Finally, we will substitute the values of A and B and write the decomposition.

$$\frac{9x + 2}{(x + 2)(3x - 2)} = \frac{2}{x + 2} + \frac{3}{3x - 2}$$

Self Check 1   Decompose $\dfrac{2x + 2}{x(x + 2)}$ into partial fractions.

Now Try Exercise 5.

> **Tip**
>
> The values of A and B in Example 1 can also be found in the following way. From Equation 2, we know that $9x + 2 = A(3x - 2) + B(x + 2)$. In this equation, we can then let $x = -2$ to eliminate the term involving B and solve for A. Then we can let $x = \frac{2}{3}$ to eliminate the term involving A and solve for B. As before, we will find that $A = 2$ and $B = 3$.

2. Decompose a Fraction when the Denominator Has Distinct Quadratic Factors

We can use the following theorem to decompose a fraction with linear and non-linear factors in the denominator. We begin with an algebraic fraction—like $\frac{P(x)}{Q(x)}$, the quotient of two polynomials—and write the fraction as the sum of two or more fractions with simpler denominators. By the theorem, we know that they will be either first-degree or irreducible second-degree polynomials, or powers of those.

Polynomial Factorization Theorem   The factorization of any polynomial $Q(x)$ with real coefficients is the product of polynomials of the forms

$$(ax + b)^n \quad \text{and} \quad (ax^2 + bx + c)^n$$

where n is a positive integer and $ax^2 + bx + c$ is irreducible over the real numbers.

To decompose a fraction when one of the factors in the denominator is a prime quadratic factor, keep the following in mind.

Partial Fraction Decomposition of $\dfrac{P(x)}{Q(x)}$: $Q(x)$ Has a Prime Quadratic Factor	If $Q(x)$ has a prime quadratic factor of the form $ax^2 + bx + c$, the partial fraction decomposition of $\dfrac{P(x)}{Q(x)}$ will have a corresponding term of the form $$\dfrac{Ax + B}{ax^2 + bx + c}$$

Decomposing a Fraction when One of the Factors in the Denominator Is a Prime Quadratic Factor

EXAMPLE 2   Decompose $\dfrac{2x^2 + x + 1}{x^3 + x}$ into partial fractions.

SOLUTION   **Step 1:** Since the denominator can be written as a product of a linear factor and a prime quadratic factor, the partial fractions have the form

$$\frac{2x^2 + x + 1}{x(x^2 + 1)} = \frac{A}{x} + \frac{Bx + C}{x^2 + 1}$$

> **Tip**
> If the given denominator of a rational expression is not factored, begin by factoring the denominator completely.

Step 2: We add the fractions and simplify.

$$\frac{2x^2 + x + 1}{x(x^2 + 1)} = \frac{A(x^2 + 1)}{x(x^2 + 1)} + \frac{(Bx + C)x}{x(x^2 + 1)}$$

$$= \frac{Ax^2 + A + Bx^2 + Cx}{x(x^2 + 1)}$$

$$= \frac{(A + B)x^2 + Cx + A}{x(x^2 + 1)} \qquad \text{Factor } x^2 \text{ from } Ax^2 + Bx^2.$$

Step 3: We can equate the corresponding coefficients of the numerators $2x^2 + 1x + 1$ and $(A + B)x^2 + Cx + A$ to get the system

$$\begin{cases} A + B = 2 & \text{These are the coefficients of } x^2. \\ C = 1 & \text{These are the coefficients of } x. \\ A = 1 & \text{These are the constants.} \end{cases}$$

Step 4: This system has solutions of $A = 1$, $B = 1$, and $C = 1$.

Step 5: Substituting the values of A, B, and C, we find that the partial fraction decomposition is

$$\frac{2x^2 + x + 1}{x(x^2 + 1)} = \frac{A}{x} + \frac{Bx + C}{x^2 + 1}$$

$$= \frac{1}{x} + \frac{1x + 1}{x^2 + 1}$$

$$= \frac{1}{x} + \frac{x + 1}{x^2 + 1}$$

Self Check 2   Decompose $\dfrac{2x^2 + 1}{x^3 + x}$ into partial fractions.

Now Try Exercise 19.

3. Decompose a Fraction when the Denominator Has Repeated Linear Factors

If $Q(x)$ has n linear factors of $ax + b$, then $(ax + b)^n$ is a factor of $Q(x)$. When this occurs, the partial fraction decomposition will contain a sum of n fractions for this term. We will include one fraction with a constant numerator for each power of $ax + b$.

Partial Fraction Decomposition of $\dfrac{P(x)}{Q(x)}$: $Q(x)$ Has Repeated Linear Factors

If $Q(x)$ has n linear factors of $ax + b$, then $(ax + b)^n$ is a factor of $Q(x)$. Each factor of the form $(ax + b)^n$ generates the following sum of n partial fractions:

$$\frac{A}{ax + b} + \frac{B}{(ax + b)^2} + \frac{C}{(ax + b)^3} + \cdots + \frac{D}{(ax + b)^n}$$

Decomposing a Fraction when the Denominator Has Repeated Linear Factors

EXAMPLE 3

Decompose $\dfrac{3x^2 - x + 1}{x(x - 1)^2}$ into partial fractions.

SOLUTION **Step 1:** Here, each factor in the denominator is linear. The linear factor x appears once, and the linear factor $x - 1$ appears twice. Thus, there are constants A, B, and C such that

$$\frac{3x^2 - x + 1}{x(x - 1)^2} = \frac{A}{x} + \frac{B}{x - 1} + \frac{C}{(x - 1)^2}$$

> **Caution**
>
> Note that $x - 1$ is a repeated factor. It would be **incorrect** to write
>
> $$\frac{3x^2 - x + 1}{x(x - 1)^2} = \frac{A}{x} + \frac{B}{x - 1} + \frac{C}{x - 1}$$

Step 2: After writing the terms on the right side as fractions with a LCD of $x(x - 1)^2$, we combine them to get

$$\frac{3x^2 - x + 1}{x(x - 1)^2} = \frac{A(x - 1)^2}{x(x - 1)^2} + \frac{Bx(x - 1)}{x(x - 1)(x - 1)} + \frac{Cx}{(x - 1)^2 x}$$

$$= \frac{Ax^2 - 2Ax + A + Bx^2 - Bx + Cx}{x(x - 1)^2}$$

$$= \frac{(A + B)x^2 + (-2A - B + C)x + A}{x(x - 1)^2} \qquad \begin{array}{l}\text{Factor } x^2 \text{ from } Ax^2 + Bx^2.\\ \text{Factor } x \text{ from } -2Ax - Bx + Cx.\end{array}$$

Step 3: Since the fractions are equal, the coefficients of the polynomial numerators are equal, and we have

$$\begin{cases} A + B = 3 & \text{These are the coefficients of } x^2.\\ -2A - B + C = -1 & \text{These are the coefficients of } x.\\ A = 1 & \text{These are the constants.}\end{cases}$$

Step 4: We can solve this system to find that $A = 1$, $B = 2$, and $C = 3$.

Step 5: Substituting the values of A, B, and C, we find that the partial fraction decomposition is

$$\frac{3x^2 - x + 1}{x(x - 1)^2} = \frac{A}{x} + \frac{B}{x - 1} + \frac{C}{(x - 1)^2}$$

$$= \frac{1}{x} + \frac{2}{x - 1} + \frac{3}{(x - 1)^2}$$

Self Check 3 Decompose $\dfrac{3x^2 + 7x + 1}{x(x + 1)^2}$ into partial fractions.

Now Try Exercise 23.

4. Decompose a Fraction when the Denominator Has Repeated Quadratic Factors

If $Q(x)$ has n prime factors of $ax^2 + bx + c$, then $(ax^2 + bx + c)^n$ is a factor. Each factor of the form $(ax^2 + bx + c)^n$ generates a sum of n partial fractions for this term. We will include one fraction with a linear numerator for each power of $ax^2 + bx + c$.

Partial Fraction Decomposition of $\dfrac{P(x)}{Q(x)}$: $Q(x)$ Has Repeated Quadratic Factors

If $Q(x)$ has n prime factors of $ax^2 + bx + c$, then $(ax^2 + bx + c)^n$ is a factor. Each factor of the form $(ax^2 + bx + c)^n$ generates a sum of n partial fractions of the form

$$\frac{Ax + B}{ax^2 + bx + c} + \frac{Cx + D}{(ax^2 + bx + c)^2} + \cdots + \frac{Ex + F}{(ax^2 + bx + c)^n}$$

EXAMPLE 4

Decomposing a Fraction when the Denominator Has Repeated Quadratic Factors

Decompose $\dfrac{3x^2 + 5x + 5}{(x^2 + 1)^2}$ into partial fractions.

SOLUTION **Step 1:** Since the quadratic factor $x^2 + 1$ is used twice, we must find constants A, B, C, and D such that

$$\frac{3x^2 + 5x + 5}{(x^2 + 1)^2} = \frac{Ax + B}{x^2 + 1} + \frac{Cx + D}{(x^2 + 1)^2}$$

Tip

When the denominator of a rational expression contains a repeated prime quadratic expression like $x^2 + 1$, set up the partial fraction decomposition with linear numerators, $Ax + B$, $Cx + D$, and so forth.

Step 2: We add the fractions on the right side to get

$$\frac{3x^2 + 5x + 5}{(x^2 + 1)^2} = \frac{(Ax + B)(x^2 + 1)}{(x^2 + 1)(x^2 + 1)} + \frac{Cx + D}{(x^2 + 1)^2}$$

$$= \frac{Ax^3 + Ax + Bx^2 + B + Cx + D}{(x^2 + 1)^2}$$

$$= \frac{Ax^3 + Bx^2 + (A + C)x + B + D}{(x^2 + 1)^2} \qquad \textcolor{red}{\text{Factor } x \text{ from } Ax + Cx.}$$

Step 3: If we add the term $0x^3$ to the numerator on the left side, we can equate the corresponding coefficients in the numerators to get

$A = 0$ These are the coefficients of x^3.

$B = 3$ These are the coefficients of x^2.

$A + C = 5$ These are the coefficients of x.

$B + D = 5$ These are the constants.

Step 4: The solution to this system is $A = 0$, $B = 3$, $C = 5$, and $D = 2$.

Step 5: Substituting the values of A, B, C, and D, we find that the partial fraction decomposition is

$$\frac{3x^2 + 5x + 5}{(x^2 + 1)^2} = \frac{Ax + B}{x^2 + 1} + \frac{Cx + D}{(x^2 + 1)^2}$$

$$= \frac{0x + 3}{x^2 + 1} + \frac{5x + 2}{(x^2 + 1)^2}$$

$$= \frac{3}{x^2 + 1} + \frac{5x + 2}{(x^2 + 1)^2}$$

Self Check 4   Decompose $\dfrac{x^3 + 2x + 3}{(x^2 + 2)^2}$ into partial fractions.

Now Try Exercise 39.

5. Decompose a Fraction when the Degree of the Numerator Is Equal to or Greater Than the Degree of the Denominator

When the degree of $P(x)$ is equal to or greater than the degree of $Q(x)$ in the fraction $\dfrac{P(x)}{Q(x)}$, we do a long division before decomposing the fraction into partial fractions.

Decomposing a Fraction when the Degree of the Numerator Is Greater Than or Equal to the Degree of the Denominator

EXAMPLE 5   Decompose $\dfrac{x^2 + 4x + 2}{x^2 + x}$ into partial fractions.

SOLUTION   Because the degree of the numerator and denominator are the same, we must do a long division and express the fraction in quotient $+ \frac{\text{remainder}}{\text{divisor}}$ form:

$$
\begin{array}{r}
1 \\
x^2 + x \overline{)\, x^2 + 4x + 2} \\
\underline{x^2 + x } \\
3x + 2
\end{array}
$$

So we can write

(4) $\dfrac{x^2 + 4x + 2}{x^2 + x} = 1 + \dfrac{3x + 2}{x^2 + x}$

Because the degree of the numerator of the fraction on the right side of Equation 3 is less than the degree of the denominator, we can find its partial fraction decomposition:

$$
\begin{aligned}
\frac{3x + 2}{x^2 + x} &= \frac{3x + 2}{x(x + 1)} \\
&= \frac{A}{x} + \frac{B}{x + 1} \\
&= \frac{A(x + 1) + Bx}{x(x + 1)} \\
&= \frac{(A + B)x + A}{x(x + 1)}
\end{aligned}
$$

We equate the corresponding coefficients in the numerator and solve the resulting system of equations to find that the solution is $A = 2$, $B = 1$. So we have

$$
\frac{x^2 + 4x + 2}{x^2 + x} = 1 + \frac{2}{x} + \frac{1}{x + 1}
$$

Self Check 5   Decompose $\dfrac{x^2 + x - 1}{x^2 - x}$ into partial fractions.

Now Try Exercise 41.

Tip

Always note the degree of the numerator and the degree of the denominator. If the numerator's degree is the same as or greater than the denominator, always use long division first.

©Archiwiz/Shutterstock.com

Self Check Answers

1. $\dfrac{1}{x} + \dfrac{1}{x + 2}$ **2.** $\dfrac{1}{x} + \dfrac{x}{x^2 + 1}$ **3.** $\dfrac{1}{x} + \dfrac{2}{x + 1} + \dfrac{3}{(x + 1)^2}$

4. $\dfrac{x}{x^2 + 2} + \dfrac{3}{(x^2 + 2)^2}$ **5.** $1 + \dfrac{1}{x} + \dfrac{1}{x - 1}$

Exercises 6.6

Getting Ready
You should be able to complete these vocabulary and concept statements before you proceed to the practice exercises.

Fill in the blanks.

1. A polynomial with real coefficients factors as the product of _____ and _____ factors or powers of those.

2. The second-degree factors of a polynomial with real coefficients are _____, which means they don't factor further over the real numbers.

Practice
Decompose each fraction into partial fractions.

3. $\dfrac{3x - 1}{x(x - 1)}$

4. $\dfrac{4x + 6}{x(x + 2)}$

5. $\dfrac{2x - 15}{x(x - 3)}$

6. $\dfrac{5x + 21}{x(x + 7)}$

7. $\dfrac{3x + 1}{(x + 1)(x - 1)}$

8. $\dfrac{9x - 3}{(x + 1)(x - 2)}$

9. $\dfrac{-4}{x^2 - 2x}$

10. $\dfrac{1}{P^2 - 300P}$

11. $\dfrac{-2x + 11}{x^2 - x - 6}$

12. $\dfrac{7x + 2}{x^2 + x - 2}$

13. $\dfrac{3x - 23}{x^2 + 2x - 3}$

14. $\dfrac{-x - 17}{x^2 - x - 6}$

15. $\dfrac{9x - 31}{2x^2 - 13x + 15}$

16. $\dfrac{-2x - 6}{3x^2 - 7x + 2}$

17. $\dfrac{4x^2 + 4x - 2}{x(x^2 - 1)}$

18. $\dfrac{x^2 - 6x - 13}{(x + 2)(x^2 - 1)}$

19. $\dfrac{x^2 + x + 3}{x(x^2 + 3)}$

20. $\dfrac{5x^2 + 2x + 2}{x^3 + x}$

21. $\dfrac{3x^2 + 8x + 11}{(x + 1)(x^2 + 2x + 3)}$

22. $\dfrac{-3x^2 + x - 5}{(x + 1)(x^2 + 2)}$

23. $\dfrac{5x^2 + 9x + 3}{x(x + 1)^2}$

24. $\dfrac{2x^2 - 7x + 2}{x(x - 1)^2}$

25. $\dfrac{-2x^2 + x - 2}{x^2(x - 1)}$

26. $\dfrac{x^2 + x + 1}{x^3}$

27. $\dfrac{3x^2 - 13x + 18}{x^3 - 6x^2 + 9x}$

28. $\dfrac{3x^2 + 13x + 20}{x^3 + 4x^2 + 4x}$

29. $\dfrac{x^2 - 2x - 3}{(x - 1)^3}$

30. $\dfrac{x^2 + 8x + 18}{(x + 3)^3}$

31. $\dfrac{x^3 + 4x^2 + 2x + 1}{x^4 + x^3 + x^2}$

32. $\dfrac{3x^3 + 5x^2 + 3x + 1}{x^2(x^2 + x + 1)}$

33. $\dfrac{4x^3 + 5x^2 + 3x + 4}{x^2(x^2 + 1)}$

34. $\dfrac{2x^2 + 1}{x^4 + x^2}$

35. $\dfrac{-x^2 - 3x - 5}{x^3 + x^2 + 2x + 2}$

36. $\dfrac{-2x^3 + 7x^2 + 6}{x^2(x^2 + 2)}$

37. $\dfrac{x^3 + 4x^2 + 3x + 6}{(x^2 + 2)(x^2 + x + 2)}$

38. $\dfrac{x^3 + 3x^2 + 2x + 4}{(x^2 + 1)(x^2 + x + 2)}$

39. $\dfrac{2x^4 + 6x^3 + 20x^2 + 22x + 25}{x(x^2 + 2x + 5)^2}$

40. $\dfrac{x^3 + 3x^2 + 6x + 6}{(x^2 + x + 5)(x^2 + 1)}$

41. $\dfrac{x^3}{x^2 + 3x + 2}$

42. $\dfrac{2x^3 + 6x^2 + 3x + 2}{x^3 + x^2}$

43. $\dfrac{3x^3 + 3x^2 + 6x + 4}{3x^3 + x^2 + 3x + 1}$

44. $\dfrac{x^4 + x^3 + 3x^2 + x + 4}{(x^2 + 1)^2}$

45. $\dfrac{x^3 + 3x^2 + 2x + 1}{x^3 + x^2 + x}$

46. $\dfrac{x^4 + x^3 + 3x^2 + x + 1}{(x^2 + 1)^2}$

47. $\dfrac{2x^4 + 2x^3 + 3x^2 - 1}{(x^2 - x)(x^2 + 1)}$

48. $\dfrac{x^4 - x^3 + 5x^2 + x + 6}{(x^2 + 3)(x^2 + 1)}$

Discovery and Writing

49. Describe what is meant by partial fraction decomposition.

50. How can you check your result of partial fraction decomposition?

51. Explain how to use partial fraction decomposition when the denominator of a rational expression has distinct linear factors.

52. Explain how to use partial fraction decomposition when the denominator of a rational expression has repeated linear factors.

53. Explain how to use partial fraction decomposition when the denominator of a rational expression has a prime quadratic factor.

54. Explain how to use partial fraction decomposition when the denominator of a rational expression has a repeated prime quadratic factor.

55. Is the polynomial $x^3 + 1$ prime?

56. Decompose $\dfrac{1}{x^3 + 1}$ into partial fractions.

Critical Thinking

Match the rational expression on the left with the correct partial fraction decomposition form on the right.

57. $\dfrac{x^2 - 2x + 3}{x(x - 4)(x^2 + 5)}$

58. $\dfrac{x^2 - 2x + 3}{x^2(x - 4)(x^2 + 5)}$

59. $\dfrac{x^2 - 2x + 3}{x^3(x - 4)^2(x^2 + 5)}$

60. $\dfrac{x^2 - 2x + 3}{x(x - 4)^3(x^2 + 5)^2}$

61. $\dfrac{x^2 - 2x + 3}{x^4(x - 4)(x^2 + 5)}$

62. $\dfrac{x^2 - 2x + 3}{x^2(x - 4)(x^3 + 5)}$

a. $\dfrac{A}{x} + \dfrac{B}{x^2} + \dfrac{C}{x^3} + \dfrac{D}{x^4}$
$+ \dfrac{E}{x - 4} + \dfrac{Fx + G}{x^2 + 5}$

b. $\dfrac{A}{x} + \dfrac{B}{x - 4} + \dfrac{C}{(x - 4)^2}$
$+ \dfrac{D}{(x - 4)^3} + \dfrac{Fx + G}{x^2 + 5}$
$+ \dfrac{Hx + I}{(x^2 + 5)^2}$

c. $\dfrac{A}{x} + \dfrac{B}{x^2} + \dfrac{C}{x - 4}$
$+ \dfrac{Dx^2 + Ex + F}{x^3 + 5}$

d. $\dfrac{A}{x} + \dfrac{B}{x^2} + \dfrac{C}{x^3}$
$+ \dfrac{D}{x - 4} + \dfrac{E}{(x - 4)^2}$
$+ \dfrac{Fx + G}{x^2 + 5}$

e. $\dfrac{A}{x} + \dfrac{B}{x - 4} + \dfrac{Cx + D}{x^2 + 5}$

f. $\dfrac{A}{x} + \dfrac{B}{x^2} + \dfrac{C}{x - 4}$
$+ \dfrac{Dx + E}{x^2 + 5}$

6.7 Graphs of Inequalities

In this section, we will learn to

1. Graph inequalities.

2. Graph systems of inequalities.

A company builds 84-inch and 120-inch motorized-drive projection screens. If the company needs 2 hours to build an 84-inch screen, it will take $2x$ hours to build x of them. If the company needs 3 hours to build a 120-inch screen, it will take $3y$ hours to build y of them. If the company has 300 hours of labor available each week and cannot build a negative number of either screen, the following system of inequalities provides restrictions on x and y, the number of screens it can build.

$$\begin{cases} 2x + 3y \le 300 \\ x \ge 0 \\ y \ge 0 \end{cases}$$

The time it takes to build x small screens plus the time it takes to build y large screens must be less than or equal to 300. Both x and y must be greater than or equal to 0.

In Example 6, we will solve this system of inequalities and determine the number of possible 84-inch and 120-inch projections screens the company can build each week. We will show how to find the solution sets of many types of systems of inequalities in this section.

1. Graph Inequalities

The **graph of an inequality** in x and y is the graph of all ordered pairs (x, y) that satisfy the inequality. We will start by considering graphs of **linear inequalities**—inequalities that can written in one of the following forms:

- $Ax + By < C$
- $Ax + By > C$
- $Ax + By \le C$ or
- $Ax + By \ge C$

To graph the inequality $y > 3x + 2$, we note that one of the following statements is true:

$$y = 3x + 2, \quad y < 3x + 2, \quad \text{or} \quad y > 3x + 2$$

The graph of $y = 3x + 2$ is a line, as shown in Figure 6-12(a). The graphs of the inequalities are half-planes, one on each side of that line. We can think of the graph of $y = 3x + 2$ as a boundary separating the two half-planes. The graph of $y = 3x + 2$ is drawn with a dashed line to show that it is not part of the graph of $y > 3x + 2$.

To find which half-plane is the graph of $y > 3x + 2$ we can substitute the coordinates of any point on one side of the line—say, the origin $(0, 0)$—into the inequality and simplify:

$$y > 3x + 2$$

$$0 > 3(0) + 2$$

$$0 > 2 \qquad \text{False}$$

Since $0 > 2$ is false, the coordinates $(0, 0)$ do not satisfy the inequality, and the origin is not in the half-plane that is the graph of $y > 3x + 2$ Thus, the graph is the half-plane located on the other side of the dashed line. The graph of the inequality $y > 3x + 2$ is shown in Figure 6-12(b).

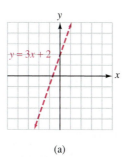
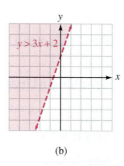
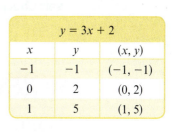

(a)   (b)

FIGURE 6-12

We can use the following steps to graph linear inequalities.

Strategy for Graphing Linear Inequalities in Two Variables	1. Write the inequality as an equation by replacing the inequality symbol with an equal sign.
	2. Graph the resulting equation to establish a boundary line. If the inequality is $<$ or $>$, draw a dashed line. If the inequality is $\leq$ or $\geq$, draw a solid line.
	3. Select a test point that is not on the boundary line.
	4. Substitute the coordinates of the test point into the inequality.
	5. Shade the region that contains the test point if the coordinates of the test point satisfy the inequality. If they don't, shade the region that does not contain the test point.

EXAMPLE 1

Graphing a Linear Inequality

Graph the inequality: $2x - 3y \leq 6$.

SOLUTION **Step 1:** This inequality is the combination of $2x - 3y < 6$ and $2x - 3y = 6$.

Step 2: We start by graphing $2x - 3y = 6$ to establish the boundary line that separates the plane into two half-planes. We draw a solid line, because equality is permitted. See Figure 6-13(a).

Step 3: Because the computations will be easy, we select the origin as a test point.

Step 4: We will substitute the coordinates of the origin into the inequality and see whether the coordinates satisfy the inequality:

$$2x - 3y \leq 6$$
$$2(0) - 3(0) \leq 6$$
$$0 \leq 6 \quad \text{True}$$

> **Tip**
>
> The easiest point to test is the origin $(0, 0)$. Always make sure $(0, 0)$ is not a point on the line. If it is, then choose another point to test.

Step 5: Because $0 \leq 6$ is true, we shade the half-plane that contains the test point $(0, 0)$. The graph is shown in Figure 6-13(b).

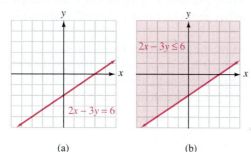

(a)   (b)

FIGURE 6-13

Self Check 1   Graph: $3x + 2y \leq 6$.

Now Try Exercise 5.

EXAMPLE 2

Graphing a Linear Inequality

Graph the inequality: $y < 2x$.

SOLUTION   **Step 1:** We write the equation $y = 2x$.

> **Take Note**
>
> $(0, 0)$ is a point on the line and therefore isn't used as our test point.

Step 2: Since the graph of $y = 2x$ is not part of the graph of the inequality, we graph the boundary with a dashed line, as in Figure 6-14(a).

Step 3: We cannot use the origin as a test point, because the boundary line passes through the origin. So we choose some other point—say, $(3, 1)$.

Step 4: To determine which half-plane represents the graph $y < 2x$ we check whether the coordinates of the test point $(3, 1)$ satisfy the inequality.

$$y < 2x$$

$$1 < 2(3)$$

$$1 < 6 \qquad \text{True}$$

Step 5: Since $1 < 6$ the point $(3, 1)$ lies in the graph. The graph is shown in Figure 6-14(b).

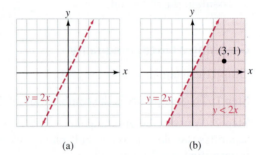

\(y = 2x\)		
x	y	(x, y)
0	0	$(0, 0)$
1	2	$(1, 2)$

(a)   (b)

FIGURE 6-14

Self Check 2   Graph: $y > 3x$.

Now Try Exercise 11.

Accent on Technology

Graphing Inequalities

Many calculators have the capability of graphing inequalities by shading areas either above or below a curve (Figure 6-15).

Move the cursor over to the left of the equation name, Y1, Y2, ..., and press [ENTER]. Each time the [ENTER] key is used, the type of curve that will be graphed is changed.

To graph the inequality $2x - 3y \geq 6$ in Example 1, solve for y in terms of x, $y \leq \frac{2}{3}x - 2$. Enter the expression in the graph editor and select the style that shades the area below the line.

Press [ZOOM] and scroll down to 6:ZStandard for the WINDOW.

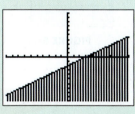

Press [ENTER], and the inequality is graphed.

FIGURE 6-15

We can graph many inequalities that are not linear inequalities.

EXAMPLE 3

Graphing Inequalities That Are Not Linear

Graph the inequality: $x^2 + y^2 > 25$.

SOLUTION   **Step 1:** We form the equation $x^2 + y^2 = 25$.

Step 2: The graph of $x^2 + y^2 = 25$ is a circle. Since the inequality is $>$, we draw the circle as a dashed circle, as in Figure 6-16(a).

Step 3: Because the work is easy, we select the origin as the test point.

Step 4: We substitute the coordinates of the origin into the inequality.

$$x^2 + y^2 > 25$$
$$0^2 + 0^2 > 25$$
$$0 > 25 \quad \text{False}$$

Step 5: Since the coordinates of the origin do not satisfy the inequality, we shade the area outside the circle, as shown in Figure 6-16(b).

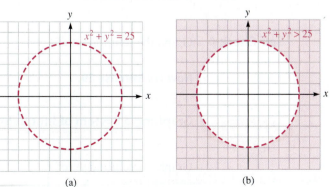

(a)   (b)

FIGURE 6-16

Self Check 3 Graph the inequality: $x^2 + y^2 \leq 36$.

Now Try Exercise 19.

2. Graph Systems of Inequalities

We now consider systems of inequalities. To graph the solution set of the system

$$\begin{cases} y < 5 \\ x \leq 6 \end{cases}$$

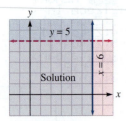

FIGURE 6-17

we graph each inequality on the same set of coordinate axes, as in Figure 6-17. The graph of the inequality $y < 5$ is the half-plane that lies below the line $y = 5$. The graph of the inequality $x \leq 6$ includes the half-plane that lies to the left of the line $x = 6$ together with the line $x = 6$.

The portion of the xy-plane where the two graphs intersect is the graph of the system. Any point that lies in the doubly shaded region has coordinates that satisfy both inequalities in the system.

Graphing a System of Linear Inequalities

EXAMPLE 4

Graph the solution set of $\begin{cases} x + y \leq 1 \\ 2x - y > 2 \end{cases}$.

SOLUTION On the same set of coordinate axes, we graph each inequality, as in Figure 6-18. The graph of $x + y \leq 1$ includes the graph of $x + y = 1$ and all points below it. Because the boundary line is included, we draw it as a solid line.

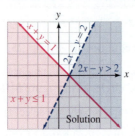

$x + y = 1$		
x	y	(x, y)
0	1	$(0, 1)$
1	0	$(1, 0)$

$2x - y = 2$		
x	y	(x, y)
0	-2	$(0, -2)$
1	0	$(1, 0)$

FIGURE 6-18

The graph of $2x - y > 2$ contains only those points below the line graph of $2x - y = 2$. Because the boundary line is not included, we draw it as a dashed line.

The area that is shaded twice represents the solution of the system of inequalities. Any point in the doubly-shaded region has coordinates that satisfy both inequalities.

Tip

To be sure you graph the correct side of the boundary line, always use a test point.

Self Check 4 Graph the solution set of $\begin{cases} x + y < 2 \\ x - 2y \geq 2 \end{cases}$.

Now Try Exercise 29.

Graphing a System of Inequalities

EXAMPLE 5

Graph the solution set of $\begin{cases} y < x^2 \\ y > \dfrac{x^2}{4} - 2 \end{cases}$.

SOLUTION   The graph of $y = x^2$ is a parabola opening upward with vertex at the origin, as shown in Figure 6-19. The points with coordinates that satisfy the inequality are the points below the parabola.

   The graph of $y = \frac{x^2}{4} - 2$ is also a parabola opening upward. The points that satisfy the inequality are the points above the parabola. The graph of the solution set is the shaded area between the two parabolas.

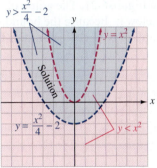

$y = x^2$		
x	y	(x, y)
0	0	$(0, 0)$
1	1	$(1, 1)$
−1	1	$(−1, 1)$
2	4	$(2, 4)$
−2	4	$(−2, 4)$

$y = \frac{x^2}{4} - 2$		
x	y	(x, y)
0	−2	$(0, −2)$
2	−1	$(2, −1)$
−2	−1	$(−2, −1)$
4	2	$(4, 2)$
−4	2	$(−4, 2)$

FIGURE 6-19

Self Check 5   Graph the solution set of $\begin{cases} y \le 4 - x^2 \\ y > \dfrac{x^2}{4} \end{cases}$.

Now Try Exercise 35.

Solving an Application Problem

EXAMPLE 6   Graph the solution set of the system that began the section:

$$\begin{cases} 2x + 3y \le 300 \\ x \ge 0 \\ y \ge 0 \end{cases}$$

SOLUTION   We graph each line, as shown in Figure 6-20. The graph of $2x + 3y \le 300$ includes the line $2x + 3y = 300$ and all points below it. Because the boundary line is included, we will draw it as a solid line.

   The graph of $x \ge 0$ includes the y-axis and all points to the right of it. The graph of $y \ge 0$ includes the x-axis and all points above it.

   The solution set of the system is the set of points in the shaded region.

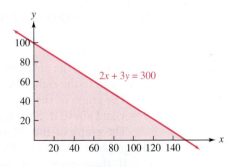

FIGURE 6-20

Self Check 6   Graph the solution set of the system $\begin{cases} x + 2y \le 50 \\ x \ge 0 \\ y \ge 0 \end{cases}$.

Now Try Exercise 47.

EXAMPLE 7

Graphing a System of Inequalities

Graph the solution set of $\begin{cases} x + y \le 4 \\ x - y \le 6. \\ x \ge 0 \end{cases}$

SOLUTION

We graph each inequality, as in Figure 6-21. The graph of $x + y \le 4$ includes the line $x + y = 4$ and all points below it. Because the boundary line is included, we draw it as a solid line. The graph of $x - y \le 6$ contains the line $x - y = 6$ and all points above it.

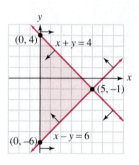

FIGURE 6-21

The graph of the inequality $x \ge 0$ contains the y-axis and all points to the right of the y-axis. The solution of the system of inequalities is the shaded area in the figure.

The coordinates of the corner points of the shaded area are $(0, 4)$, $(0, -6)$, and $(5, -1)$.

Self Check 7 Graph the solution set of $\begin{cases} x + y \le 5 \\ x - 3y \le -3. \\ x \ge 0 \end{cases}$

Now Try Exercise 37.

EXAMPLE 8

Graphing a System of Inequalities

Graph the solution set of the system $\begin{cases} x \ge 1 \\ y \ge x \\ 4x + 5y < 20 \end{cases}$.

SOLUTION

The graph of the solution set of $x \ge 1$ includes those points on the graph of $x = 1$ and to the right. See Figure 6-22(a).

The graph of the solution set of $y \ge x$ includes those points on the graph of $y = x$ and above it. See Figure 6-22(b).

The graph of the solution set of $4x + 5y < 20$ includes those points below the graph of $4x + 5y = 20$. See Figure 6-22(c).

If these graphs are merged onto a single set of coordinate axes, as in Figure 6-22(d), the graph of the original system of inequalities includes those points within the shaded triangle together with the points on the sides of the triangle drawn as solid lines. The coordinates of the corner points are $\left(1, \frac{16}{5}\right)$, $(1, 1)$, and $\left(\frac{20}{9}, \frac{20}{9}\right)$.

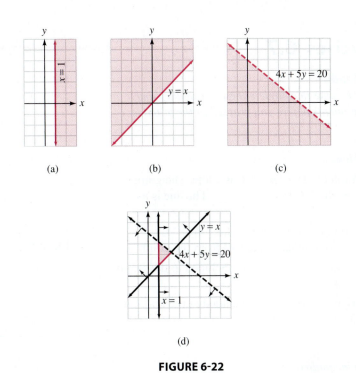

(a)   (b)   (c)

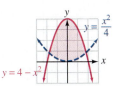

(d)

FIGURE 6-22

Self Check 8   Is the point $(2, 2)$ in the solution set of Example 8?

Now Try Exercise 41.

Self Check Answers

1. 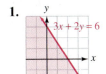   2.   3.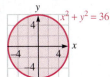

4. 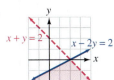   5.

6. 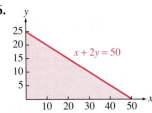   7.   8. yes

Exercises 6.7

Getting Ready

You should be able to complete these vocabulary and concept statements before you proceed to the practice exercises.

Fill in the blanks.

1. The graph of $Ax + By = C$ is a line. The graph of $Ax + By \leq C$ is a _____. The line is its _____.

2. The boundary of the graph $Ax + By < C$ is _____ (included, excluded) from the graph.

3. The origin _____ (is, is not) included in the graph of $3x - 4y > 4$.

4. The origin ___ (is, is not) included in the graph of $4x + 3y \leq 5$.

Practice

Graph each inequality.

5. $2x + 3y < 12$ **6.** $4x - 3y > 6$

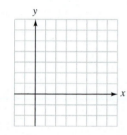   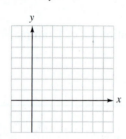

7. $x < 3$ **8.** $y > -1$

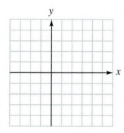   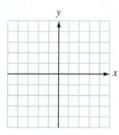

9. $4x - y > 4$ **10.** $x - 2y < 5$

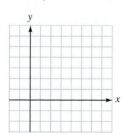   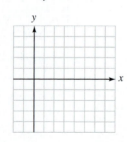

11. $y > 2x$ **12.** $y < 3x$

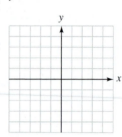   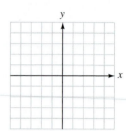

13. $y \leq \dfrac{1}{2}x + 1$ **14.** $y \geq \dfrac{1}{3}x - 1$

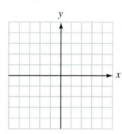   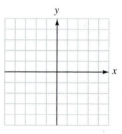

15. $2y \geq 3x - 2$ **16.** $3y \leq 2x + 3$

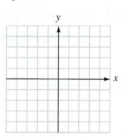   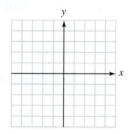

17. $y < x^2$ **18.** $y \geq |x|$

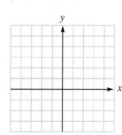   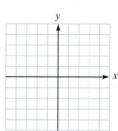

19. $x^2 + y^2 \leq 4$ **20.** $x^2 + y^2 > 4$

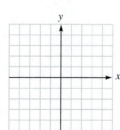   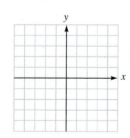

Graph the solution set of each system of inequalities.

21. $\begin{cases} y < 3 \\ x \geq 2 \end{cases}$

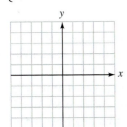

22. $\begin{cases} y \geq -2 \\ x < 0 \end{cases}$

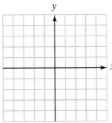

31. $\begin{cases} 2x - 3y \geq 6 \\ 3x + 2y < 6 \end{cases}$

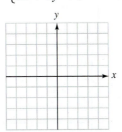

32. $\begin{cases} 4x + 2y \leq 6 \\ 2x - 4y \geq 10 \end{cases}$

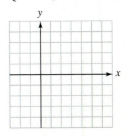

23. $\begin{cases} y \geq 1 \\ x < 2 \end{cases}$

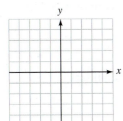

24. $\begin{cases} y \leq -1 \\ x > -1 \end{cases}$

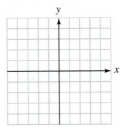

33. $\begin{cases} y \geq x^2 - 4 \\ y \leq \dfrac{1}{2}x \end{cases}$

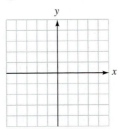

34. $\begin{cases} y \leq -x^2 + 4 \\ y > -x - 1 \end{cases}$

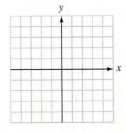

25. $\begin{cases} y \leq x - 2 \\ y \geq 2x + 1 \end{cases}$

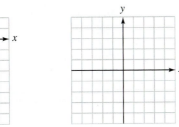

26. $\begin{cases} y < 3x + 2 \\ y < -2x + 3 \end{cases}$

35. $\begin{cases} y \geq x^2 \\ y < 4 - x^2 \end{cases}$

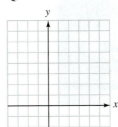

36. $\begin{cases} x^2 + y \leq 1 \\ y - x^2 \geq -1 \end{cases}$

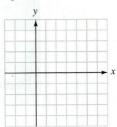

27. $\begin{cases} x + y < 2 \\ x + y \leq 1 \end{cases}$

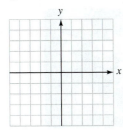

28. $\begin{cases} 3x + 2y \geq 6 \\ x + 3y \leq 2 \end{cases}$

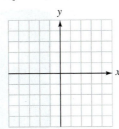

37. $\begin{cases} 2x - y \leq 0 \\ x + 2y \leq 10 \\ y \geq 0 \end{cases}$

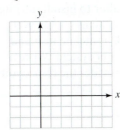

38. $\begin{cases} x - 2y \geq 0 \\ x - y \leq 2 \\ x \geq 0 \end{cases}$

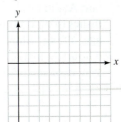

29. $\begin{cases} x + 2y < 3 \\ 2x - 4y < 8 \end{cases}$

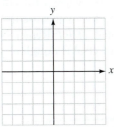

30. $\begin{cases} 3x + y \leq 1 \\ -x + 2y \geq 9 \end{cases}$

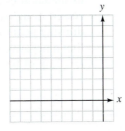

39. $\begin{cases} 3x - 2y \geq 5 \\ 2x + y \geq 8 \\ x \leq 5 \end{cases}$

40. $\begin{cases} 2x + 3y \leq 6 \\ x - y \geq 4 \\ y \geq -4 \end{cases}$

41. $\begin{cases} x + y \le 4 \\ x - y \le 4 \\ x \ge 0 \\ y \ge 0 \end{cases}$

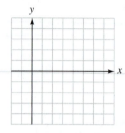

42. $\begin{cases} 2x + 3y \ge 12 \\ 2x - 3y \le 6 \\ x \ge 0 \\ y \le 4 \end{cases}$

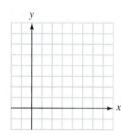

43. $\begin{cases} 3x - 2y \le 6 \\ x + 2y \le 10 \\ x \ge 0 \\ y \ge 0 \end{cases}$

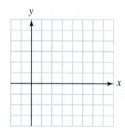

44. $\begin{cases} 3x + 2y \ge 12 \\ 5x - y \le 15 \\ x \ge 0 \\ y \le 4 \end{cases}$

Applications

45. Building furniture A furniture maker has 60 hours of labor to make sofas (s) and loveseats (l). It takes 6 hours to make a sofa and 4 hours to make a loveseat. Write a system of inequalities that provides the restrictions on the variables. (*Hint*: Remember that a negative number of pieces of furniture cannot be made.)

46. Installing video Each week, Prime Time Video and Audio has 90 hours of labor to install satellite dishes (d) and home theater systems (t). On average, it takes 5 hours to install a satellite dish and 6 hours to install a home theater system. Write a system of inequalities that provides the restrictions

on the variables. (*Hint*: Remember that a negative number of units cannot be installed.)

47. Fundraising A college club is selling baskets of fruit and blocks of cheese to raise at least $600 for a local children's hospital.

a. If the profit for selling a basket of fruit is $5 and for selling a block of cheese is $6, write a system of inequalities that describes when x boxes of fruit and y blocks of cheese will cause the fundraising goal to be reached. (*Hint*: Remember that a negative number of baskets of fruit or blocks of cheese cannot be sold.)

b. Graph the system of inequalities.

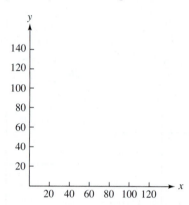

48. Fundraising A cheerleading team is selling cookie dough and pizza kits to raise at least $3600 for their summer camp expenses.

a. If the profit for selling a tub of cookie dough is $6 and for selling a pizza kit is $8, write a system of inequalities that describes when x tubs of cookie dough and y pizza kits will cause the fundraising goal to be reached. (*Hint*: Remember that a negative number of cookie dough or pizza kits cannot be sold.)

b. Graph the system of inequalities.

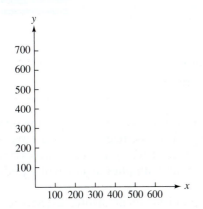

Discovery and Writing

49. Explain how to graph an inequality in two variables.

50. When graphing an inequality in two variables, explain how you decide which side to shade.

51. Explain how you determine whether the graph of a linear inequality in two variables is a dashed or solid line.

52. What is a system of inequalities?

53. Explain how to graph the solution set of a system of inequalities.

54. Explain why it is possible for a system of inequalities to have no solution.

Critical Thinking

Determine if the statement is true or false. If the statement is false, then correct it and make it true.

55. The origin (0, 0) is always used as a test point when graphing a linear inequality in two variables.

56. The solution of a linear inequality in two variables is always a half-plane.

57. A system of inequalities always has a solution.

58. If the inequality in two variables contains a $>$ or $<$ symbol, a dashed curve is drawn.

59. If the inequality in two variables contains a $\leq$ or $\geq$ symbol, a solid curve is drawn.

60. An inequality representing the graph shown is $3x + 5y \geq 15$.

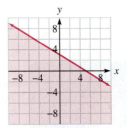

61. A system of inequalities that represents the graph shown is $\begin{cases} x \geq -5 \\ y > 2 \end{cases}$.

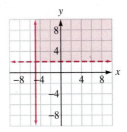

62. A system of inequalities that represents the graph shown is $\begin{cases} x^2 + y^2 \leq 36 \\ x^2 + y^2 \leq 25 \end{cases}$.

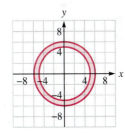

6.8 Linear Programming

In this section, we will learn to

1. Solve linear programming problems.
2. Solve applications of linear programming.

Linear programming is a mathematical technique used to find the optimal allocation of resources in the military, business, telecommunications, and other fields. It got its start during World War II when it became necessary to move huge quantities of people, materials, and supplies as efficiently and economically as possible.

A simple linear programming problem might involve a television program director who wants to schedule comedy skits and musical acts such as Alicia Keys for a prime-time variety show. Of course, the director wants to do this in a way that earns the maximum possible income for her network. Such an example is provided in Example 5 in this section.

1. Solve Linear Programming Problems

To solve linear programming problems, we must maximize (or minimize) a function (called the **objective function**) subject to given restrictions on its variables. These restrictions (called **constraints**) are usually given as a system of linear inequalities. For example, suppose that the annual profit (in millions of dollars) earned by a business is given by the equation $P = 2x + y$ and that x and y are subject to the following constraints:

$$\begin{cases} 3x + y \le 120 \\ x + y \le 60 \\ x \ge 0 \\ y \ge 0 \end{cases}$$

To find the maximum profit P that can be earned by the business, we solve the system of inequalities as shown in Figure 6-23(a) and find the coordinates of each corner point of the region R. This region is often called a **feasibility region**. We can then write the profit equation

$$P = y + 2x \quad \text{in the form} \quad y = -2x + P$$

The equation $y = -2x + P$ is the equation of a set of parallel lines, each with a slope of -2 and a y-intercept of P. The graph of $y = -2x + P$ for three values of P is shown as red lines in Figure 6-23(b). To find the red line that passes through region R and provides the maximum value of P, we locate the red line with the greatest y-intercept. Since line l has the greatest y-intercept and intersects region R at the corner point (30, 30), the maximum value of P (subject to the given constraints) is

$$P = 2x + y$$

$$= 2(30) + 30$$

$$= 90$$

Thus, the maximum profit P that can be earned is $90 million. This profit occurs when $x = 30$ and $y = 30$.

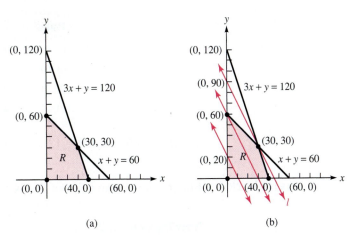

FIGURE 6-23

The preceding discussion illustrates the following important fact.

Maximum or Minimum of an Objective Function	If a linear function, subject to the constraints of a system of linear inequalities in two variables, attains a maximum or a minimum value, that value will occur at a corner point or along an entire edge of the region R that represents the solution of the system.

Finding the Maximum Value of an Objective Function Given Certain Constraints

EXAMPLE 1

If $P = 2x + 3y$, find the maximum value of P subject to the following constraints:

$$\begin{cases} x + y \le 4 \\ 2x + y \le 6 \\ x \ge 0 \\ y \ge 0 \end{cases}$$

SOLUTION   We solve the system of inequalities to find the feasibility region R shown in Figure 6-24. The coordinates of its corner points are $(0, 0)$, $(3, 0)$, $(0, 4)$, and $(2, 2)$.

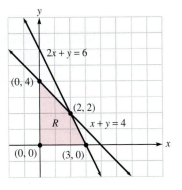

FIGURE 6-24

Since the maximum value of P will occur at a corner of R, we substitute the coordinates of each corner point into the objective function $P = 2x + 3y$ and find the one that gives the maximum value of P.

Point	$P = 2x + 3y$
$(0, 0)$	$P = 2(0) + 3(0) = 0$
$(3, 0)$	$P = 2(3) + 3(0) = 6$
$(2, 2)$	$P = 2(2) + 3(2) = 10$
$(0, 4)$	$P = 2(0) + 3(4) = 12$

The maximum value $P = 12$ occurs when $x = 0$ and $y = 4$.

Self Check 1   Find the maximum value of $P = 4x + 3y$ subject to the constraints of Example 1.

Now Try Exercise 7.

Finding the Minimum Value of an Objective Function Given Certain Constraints

EXAMPLE 2   If $P = 3x + 2y$, find the minimum value of P subject to the following constraints:

$$\begin{cases} x + y \geq 1 \\ x - y \leq 1 \\ x - y \geq 0 \\ x \leq 2 \end{cases}$$

SOLUTION   We refer to the feasibility region shown in Figure 6-25 with corner points at $\left(\frac{1}{2}, \frac{1}{2}\right)$, $(2, 2)$, $(2, 1)$, and $(1, 0)$.

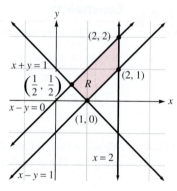

FIGURE 6-25

Since the minimum value of P occurs at a corner point of region R, we substitute the coordinates of each corner point into the objective function $P = 3x + 2y$ and find the one that gives the minimum value of P.

Point	$P = 3x + 2y$
$\left(\frac{1}{2}, \frac{1}{2}\right)$	$P = 3\left(\frac{1}{2}\right) + 2\left(\frac{1}{2}\right) = \frac{5}{2}$
$(2, 2)$	$P = 3(2) + 2(2) = 10$
$(2, 1)$	$P = 3(2) + 2(1) = 8$
$(1, 0)$	$P = 3(1) + 2(0) = 3$

The minimum value $P = \frac{5}{2}$ occurs when $x = \frac{1}{2}$ and $y = \frac{1}{2}$.

Self Check 2   Find the minimum value of $P = 2x + y$ subject to the constraints of Example 2.

Now Try Exercise 15.

2. Solve Applications of Linear Programming

Linear programming problems can be very complex and involve hundreds of variables. In this section, we will consider only a few simple problems. Since they involve only two variables, we can solve them using graphical methods.

To solve a linear programming problem, we will follow these steps.

Strategy for Solving Linear Programming Problems

1. Find the objective function and constraints.
2. Find the feasibility region by graphing the system of inequalities and identifying the coordinates of its corner points.
3. Find the maximum (or minimum) value by substituting the coordinates of the corner points into the objective function.

EXAMPLE 3

Solving an Application Problem

An accountant prepares tax returns for individuals and for small businesses. On average, each individual return requires 3 hours of her time and 1 hour of computer time. Each business return requires 4 hours of her time and 2 hours of computer time. Because of other business considerations, her time is limited to 240 hours, and the computer time is limited to 100 hours. If she earns a profit of $80 on each individual return and a profit of $150 on each business return, how many returns of each type should she prepare to maximize her profit?

SOLUTION   First, we organize the given information into a table.

	Individual Tax Return	Business Tax Return	Time Available
Accountant's Time	3	4	240 hours
Computer Time	1	2	100 hours
Profit	$80	$150	

Then we solve the problem using the following steps.

Find the objective function and constraints   Suppose that x represents the number of individual returns to be completed and y represents the number of business returns to be completed. Since each of the x individual returns will earn an $80 profit and each of the y business returns will earn a $150 profit, the total profit is given by the equation

$$P = 80x + 150y$$

Since the number of individual returns and business returns cannot be negative, we know that $x \geq 0$ and $y \geq 0$.

Since each of the x individual returns will take 3 hours of her time and each of the y business returns will take 4 hours of her time, the total number of hours she will work will be $(3x + 4y)$ hours. This amount must be less than or equal to her available time, which is 240 hours. Thus, the inequality $3x + 4y \leq 240$ is a constraint on the accountant's time.

Since each of the x individual returns will take 1 hour of computer time and each of the y business returns will take 2 hours of computer time, the total number of hours of computer time will be $(x + 2y)$ hours. This amount must be less than or equal to the available computer time, which is 100 hours. Thus, the inequality $x + 2y \leq 100$ is a constraint on the computer time.

We have the following constraints on the values of x and y.

$$\begin{cases} x \geq 0 & \text{The number of individual returns is nonnegative.} \\ y \geq 0 & \text{The number of business returns is nonnegative.} \\ 3x + 4y \leq 240 & \text{The accountant's time must be less than or equal to 240 hours.} \\ x + 2y \leq 100 & \text{The computer time must be less than or equal to 100 hours.} \end{cases}$$

Find the feasibility region To find the feasibility region, we graph each of the constraints to find region R, as in Figure 6-26. The four corner points of this region have coordinates of $(0, 0)$, $(80, 0)$, $(40, 30)$, and $(0, 50)$.

FIGURE 6-26

> **Tip**
>
> The maximum and minimum values of an objective function will occur at the vertices of the shaded region. The vertices are referred to as our corner points.

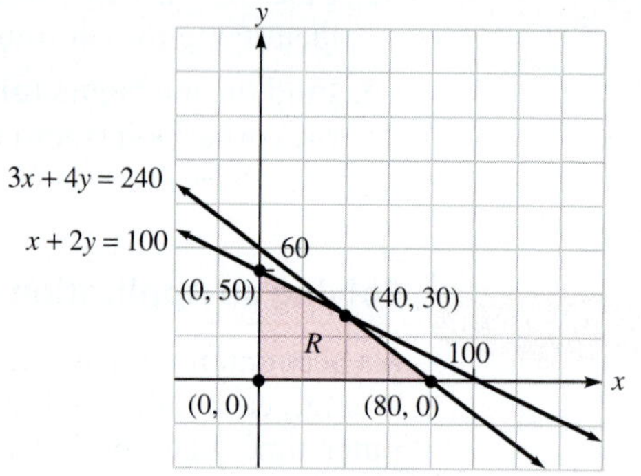

Find the maximum profit To find the maximum profit, we substitute the coordinates of each corner point into the objective function $P = 80x + 150y$.

Point	$P = 80x + 150y$
$(0, 0)$	$P = 80(0) + 150(0) = 0$
$(80, 0)$	$P = 80(80) + 150(0) = 6400$
$(40, 30)$	$P = 80(40) + 150(30) = 7700$
$(0, 50)$	$P = 80(0) + 150(50) = 7500$

From the table, we can see that the accountant will earn a maximum profit of $7700 if she prepares 40 individual returns and 30 business returns.

Self Check 3 In Example 3, if the accountant earns a profit of $100 on each individual return and a profit of $175 on each business return, find the maximum profit.

Now Try Exercise 21.

Solving an Application Problem

EXAMPLE 4

Vigortab and Robust are two diet supplements. Each Vigortab tablet costs 50¢ and contains 3 units of calcium, 20 units of vitamin C, and 40 units of iron. Each Robust tablet costs 60¢ and contains 4 units of calcium, 40 units of vitamin C, and 30 units of iron. At least 24 units of calcium, 200 units of vitamin C, and 120 units of iron are required for the daily needs of one patient. How many tablets of each supplement should be taken daily for a minimum cost? Find the daily minimum cost.

SOLUTION   First, we organize the given information into a table.

	Vigortab	Robust	Amount Required
Calcium	3	4	24
Vitamin C	20	40	200
Iron	40	30	120
Cost	50¢	60¢	

Find the objective function and constraints   We can let x represent the number of Vigortab tablets to be taken daily and y the corresponding number of Robust tablets. Because each of the x Vigortab tablets will cost 50¢ and each of the y Robust tablets will cost 60¢, the total cost will be given by the equation

$$C = 0.50x + 0.60y \qquad \text{50¢ = \$0.50 and 60¢ = \$0.60.}$$

Since there are requirements for calcium, vitamin C, and iron, there is a constraint for each. Note that neither x nor y can be negative.

$$\begin{cases} 3x + 4y \geq 24 & \text{The amount of calcium must be greater than or equal to 24 units.} \\ 20x + 40y \geq 200 & \text{The amount of vitamin C must be greater than or equal to 200 units.} \\ 40x + 30y \geq 120 & \text{The amount of iron must be greater than or equal to 120 units.} \\ x \geq 0, y \geq 0 & \text{The number of tablets taken must be greater than or equal to 0.} \end{cases}$$

Find the feasibility region   We graph the inequalities to find the feasibility region and the coordinates of its corner points, as in Figure 6-27.

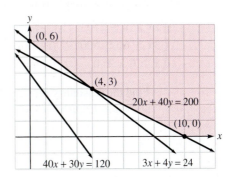

FIGURE 6-27

Find the minimum cost   In this case, the feasibility region is not bounded on all sides. The coordinates of the corner points are (0, 6), (4, 3), and (10, 0). To find the minimum cost, we substitute each pair of coordinates into the objective function.

Point	$C = 0.50x + 0.60y$
(0, 6)	$C = 0.50(0) + 0.60(6) = 3.60$
(4, 3)	$C = 0.50(4) + 0.60(3) = 3.80$
(10, 0)	$C = 0.50(10) + 0.60(0) = 5.00$

A minimum cost will occur if no Vigortab and 6 Robust tablets are taken daily. The minimum daily cost is $3.60.

Self Check 4   If the cost of each Robust tablet increases to 75¢ and the cost of each Vigortab increases to 80¢, find the minimum cost.

Now Try Exercise 24.

EXAMPLE 5 Solving an Application Problem

A television program director must schedule comedy skits and musical numbers for prime-time variety shows. Each comedy skit requires 2 hours of rehearsal time, costs $3000, and brings in $20,000 from the show's sponsors. Each musical number requires 1 hour of rehearsal time, costs $6000, and generates $12,000. If 250 hours are available for rehearsal and $600,000 is budgeted for comedy and music, how many segments of each type should be produced to maximize income? Find the maximum income.

SOLUTION First, we organize the given information into a table.

	Comedy	Musical	Available
Rehearsal Time (hours)	2	1	250
Cost (in $1000s)	3	6	600
Generated Income (in $1000s)	20	12	

Find the objective function and constraints We can let x represent the number of comedy skits and y the number of musical numbers to be scheduled. Since each of the x comedy skits generates $20 thousand, the income generated by the comedy skits is $20x$ thousand. The musical numbers produce $12y$ thousand. The objective function to be maximized is

$$I = 20x + 12y$$

Since there are limits on rehearsal time and budget, there is a constraint for each. Note that neither x nor y can be negative.

$$\begin{cases} 2x + y \le 250 & \text{The total rehearsal time must be less than or equal to 250 hours.} \\ 3x + 6y \le 600 & \text{The total cost must be less than or equal to \$600 thousand.} \\ x \ge 0, y \ge 0 & \text{The numbers of skits and musical numbers must be greater than or equal to 0.} \end{cases}$$

Find the feasibility region We graph the inequalities to find the feasibility region shown in Figure 6-28 and find the coordinates of each corner point.

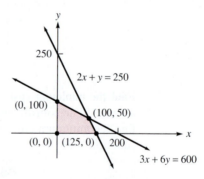

FIGURE 6-28

Find the maximum income The coordinates of the corner points of the feasible region are (0, 0), (0, 100), (100, 50), and (125, 0). To find the maximum income, we substitute each pair of coordinates into the objective function.

Point	$I = 20x + 12y$
(0, 0)	$I = 20(0) + 12(0) = 0$
(0, 100)	$I = 20(0) + 12(100) = 1200$
(100, 50)	$I = 20(100) + 12(50) = 2600$
(125, 0)	$I = 20(125) + 12(0) = 2500$

Maximum income will occur if 100 comedy skits and 50 musical numbers are scheduled. The maximum income will be 2600 thousand dollars, or $2,600,000. ∎

Self Check 5 If during the following year it is predicted that each comedy skit will generate $30 thousand and each musical number $20 thousand, find the maximum income for the year.

Now Try Exercise 25.

Self Check Answers

1. 14   **2.** $\dfrac{3}{2}$ **3.** $9250   **4.** $4.80   **5.** $4,000,000

Exercises 6.8

Getting Ready

You should be able to complete these vocabulary and concept statements before you proceed to the practice exercises.

Fill in the blanks.

1. In a linear program, the inequalities are called _____.

2. Ordered pairs that satisfy the constraints of a linear program are called _____ solutions.

3. The function to be maximized (or minimized) in a linear program is called the _____ function.

4. The objective function of a linear program attains a maximum (or minimum), subject to the constraints, at a _____ or along an _____ of the feasibility region.

Practice

Maximize P subject to the following constraints.

5. $P = 2x + 3y$
$$\begin{cases} x \geq 0 \\ y \geq 0 \\ x + y \leq 4 \end{cases}$$

6. $P = 3x + 2y$
$$\begin{cases} x \geq 0 \\ y \geq 0 \\ x + y \leq 4 \end{cases}$$

7. $P = y + \dfrac{1}{2}x$
$$\begin{cases} x \geq 0 \\ y \geq 0 \\ 2y - x \leq 1 \\ y - 2x \geq -2 \end{cases}$$

8. $P = 4y - x$
$$\begin{cases} x \geq 2 \\ y \geq 0 \\ x + y \geq 1 \\ 2y - x \leq 1 \end{cases}$$

9. $P = 2x + y$
$$\begin{cases} y \geq 0 \\ y - x \leq 2 \\ 2x + 3y \leq 6 \\ 3x + y \leq 3 \end{cases}$$

10. $P = x - 2y$
$$\begin{cases} x + y \leq 5 \\ y \leq 3 \\ x \leq 2 \\ x \geq 0 \\ y \geq 0 \end{cases}$$

11. $P = 3x - 2y$
$$\begin{cases} x \leq 1 \\ x \geq -1 \\ y - x \leq 1 \\ x - y \leq 1 \end{cases}$$

12. $P = x - y$
$$\begin{cases} 5x + 4y \leq 20 \\ y \leq 5 \\ x \geq 0 \\ y \geq 0 \end{cases}$$

Minimize P subject to the following constraints.

13. $P = 5x + 12y$
$$\begin{cases} x \geq 0 \\ y \geq 0 \\ x + y \leq 4 \end{cases}$$

14. $P = 3x + 6y$
$$\begin{cases} x \geq 0 \\ y \geq 0 \\ x + y \leq 4 \end{cases}$$

15. $P = 3y + x$
$$\begin{cases} x \geq 0 \\ y \geq 0 \\ 2y - x \leq 1 \\ y - 2x \geq -2 \end{cases}$$

16. $P = 5y + x$
$$\begin{cases} x \geq 0 \\ y \geq 0 \\ x + y \geq 1 \\ 2y - x \leq 1 \end{cases}$$

17. $P = 6x + 2y$
$$\begin{cases} y \geq 0 \\ y - x \leq 2 \\ 2x + 3y \leq 6 \\ 3x + y \leq 3 \end{cases}$$

18. $P = 2y - x$
$$\begin{cases} x \geq 0 \\ y \geq 0 \\ x + y \leq 5 \\ x + 2y \geq 2 \end{cases}$$

19. $P = 2x - 2y$

$$\begin{cases} x \le 1 \\ x \ge -1 \\ y - x \le 1 \\ x - y \le 1 \end{cases}$$

20. $P = y - 2x$

$$\begin{cases} x + 2y \le 4 \\ 2x + y \le 4 \\ x + 2y \ge 2 \\ 2x + y \ge 2 \end{cases}$$

	Snowman	Santa Claus	Time Available
Rob's Time (hr)	2	4	20
Nina's Time (hr)	4	3	20
Income ($)	80	64	

Applications

Write the objective function and the inequalities that describe the constraints in each problem. Graph the feasibility region, showing the corner points. Then find the maximum or minimum value of the objective function.

21. Making furniture Two woodworkers, Chase and Devin, get $100 for making a table and $80 for making a chair. On average, Chase must work 3 hours and Devin 2 hours to make a chair. Chase must work 2 hours and Devin 6 hours to make a table. If neither wishes to work more than 42 hours per week, how many tables and how many chairs should they make each week to maximize their income? Find the maximum income.

©WorldWide/Shutterstock.com

	Table	Chair	Time Available
Devin's Time (hr)	6	2	42
Chase's Time (hr)	2	3	42
Income ($)	100	80	

22. Making crafts Two artists, Nina and Rob, make yard ornaments. They get $80 for each wooden snowman they make and $64 for each wooden Santa Claus. On average, Nina must work 4 hours and Rob 2 hours to make a snowman. Nina must work 3 hours and Rob 4 hours to make a Santa Claus. If neither wishes to work more than 20 hours per week, how many of each ornament should they make each week to maximize their income? Find the maximum income.

23. Inventories An electronics store manager stocks from 20 to 30 IBM-compatible computers and from 30 to 50 Apple computers. There is room in the store to stock up to 60 computers. The manager receives a commission of $50 on the sale of each IBM-compatible computer and $40 on the sale of each Apple computer. If the manager can sell all of the computers, how many should she stock to maximize her commissions? Find the maximum commission.

Inventory	IBM	Apple
Minimum	20	30
Maximum	30	50
Commission	$50	$40

24. Diet problems A diet requires at least 16 units of vitamin C and at least 34 units of vitamin B complex. Two food supplements are available that provide these nutrients in the amounts and costs shown in the table. How much of each should be used to minimize the cost?

Supplement	Vitamin C	Vitamin B	Cost
A	3 units/g	2 units/g	3¢/g
B	2 units/g	6 units/g	4¢/g

25. Production Manufacturing DVRs and TVs requires the use of the electronics, assembly, and finishing departments of a factory, according to the following schedule:

	Hours for DVR	Hours for TV	Hours Available per Week
Electronics	3	4	180
Assembly	2	3	120
Finishing	2	1	60

Each DVR has a profit of $40, and each TV has a profit of $32. How many DVRs and TVs should be manufactured weekly to maximize profit? Find the maximum profit.

26. Production problems A company manufactures one type of computer chip that runs at 2.0 GHz and another that runs at 2.8 GHz. The company can make a maximum of 50 fast chips per day and a maximum of 100 slow chips per day. It takes 6 hours to make a fast chip and 3 hours to make a slow chip, and the company's employees can provide up to 360 hours of labor per day. If the company makes a profit of $20 on each 2.8-GHz chip and $27 on each 2.0-GHz chip, how many of each type should be manufactured to earn the maximum profit?

27. Financial planning A stockbroker has $200,000 to invest in stocks and bonds. She wants to invest at least $100,000 in stocks and at least $50,000 in bonds. If stocks have an annual yield of 9% and bonds have an annual yield of 7%, how much should she invest in each to maximize her income? Find the maximum return.

28. Production A small country exports soybeans and flowers. Soybeans require 8 workers per acre, flowers require 12 workers per acre, and 100,000 workers are available. Government contracts require that there be at least 3 times as many acres of soybeans as flowers planted. It costs $250 per acre to plant soybeans and $300 per acre to plant flowers, and there is a budget of $3 million. If the profit from soybeans is $1600 per acre and the profit from flowers is $2000 per acre, how many acres of each crop should be planted to maximize profit? Find the maximum profit.

29. Band trips A college band trip will require renting buses and trucks to transport no fewer than 100 students and 18 or more large instruments. Each bus can accommodate 40 students plus three large instruments; it costs $350 to rent. Each truck can accommodate 10 students plus 6 large instruments and costs $200 to rent. How many of each type of vehicle should be rented for the cost to be minimum? Find the minimum cost.

30. Making ice cream An ice cream store sells two new flavors: Fantasy and Excess. Each barrel of Fantasy requires 4 pounds of nuts and 3 pounds of chocolate

and has a profit of $500. Each barrel of Excess requires 4 pounds of nuts and 2 pounds of chocolate and has a profit of $400. There are 16 pounds of nuts and 18 pounds of chocolate in stock, and the owner does not want to buy more for this batch. How many barrels of each should be made for a maximum profit? Find the maximum profit.

Discovery and Writing

31. Describe what linear programming is and some of the types of problems it can be used to solve.

32. Explain what an objective function is in a linear programming problem.

33. In a linear programming problem, describe what constraints are and how they are represented.

34. Describe a strategy that can be used to solve a linear programming problem.

35. Does the objective function attain a maximum at the corners of a region defined by following nonlinear inequalities? Attempt to maximize $P(x) = x + y$ on the region and write a paragraph on your findings.

$$\begin{cases} x \geq 0 \\ y \geq 0 \\ y \leq 4 - x^2 \end{cases}$$

36. Attempt to minimize the objective function of Exercise 35.

Critical Thinking

Determine if the statement is true or false. If the statement is false, then correct it and make it true.

37. An objective function always has a maximum or minimum.

38. A system of linear equations is used to write constraints.

39. The minimum value of objective function occurs at exactly one point.

40. If the feasibility region is unbounded, then it is possible that no maximum value of the objective function exists.

CHAPTER REVIEW

6.1 Systems of Linear Equations

Definitions and Concepts	Examples
Solving a system by graphing: To solve a system of equations by graphing, graph each equation in the system and find the coordinates of the point where all of the graphs intersect. If the graphs have no point in common, the system has no solutions. If the graphs coincide, the system has infinitely many solutions.	Solve $\begin{cases} x - 2y = -9 \\ 3x + y = 1 \end{cases}$ by graphing. We graph each equation as shown in the illustration and find the coordinates of the point where the graphs intersect. From the illustration, we see that the solution is $(-1, 4)$. 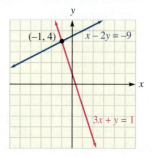

Solving a system by substitution: To solve a system by substitution, solve one equation for one variable, and substitute that result into the other equation. Then solve for the other variable using back substitution.	Solve $\begin{cases} x - 2y = -9 \\ 3x + y = 1 \end{cases}$ by substitution. We can solve the first equation for x and substitute the result for x into the other equation. $x - 2y = -9 \rightarrow x = \mathbf{2y - 9}$ $3(\mathbf{2y - 9}) + y = 1$ Substitute $2y - 9$ for x. $6y - 27 + y = 1$ $7y = 28$ $y = 4$ To find x, we can substitute 4 for y in the equation $x = 2y - 9$ to get $x = -1$. The solution is $(-1, 4)$.

Solving a system by elimination: To solve a system by elimination, multiply one or both equations by suitable constants so that when the results are added, one variable will be eliminated. Then back substitute to find the other variable.	Solve $\begin{cases} x - 2y = -9 \\ 3x + y = 1 \end{cases}$ by elimination. We can multiply the second equation by 2 and add the equations to eliminate y. $\begin{cases} x - 2y = -9 \\ 3x + y = 1 \end{cases} \rightarrow \begin{cases} x - 2y = -9 \\ 6x + 2y = 2 \end{cases}$ $7x \quad\quad = -7$ $x = -1$ We can substitute -1 for x in any equation and find that $y = 4$. The solution is $(-1, 4)$.

Exercises

Solve each system by graphing.

1. $\begin{cases} 2x - y = -1 \\ x + y = 7 \end{cases}$ 2. $\begin{cases} 5x + 2y = 1 \\ 2x - y = -5 \end{cases}$

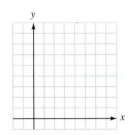   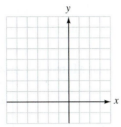

3. $\begin{cases} y = 5x + 7 \\ x = y - 7 \end{cases}$ 4. $\begin{cases} 3x + 2y = 6 \\ y = -\dfrac{3}{2}x + 3 \end{cases}$

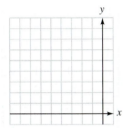   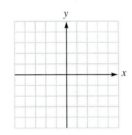

5. $\begin{cases} 4x - y = 4 \\ y = 4(x - 2) \end{cases}$

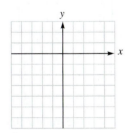

Solve each system by substitution.

6. $\begin{cases} 2y + x = 0 \\ x = y + 3 \end{cases}$ 7. $\begin{cases} 2x + y = -3 \\ x - y = 3 \end{cases}$

8. $\begin{cases} \dfrac{x + y}{2} + \dfrac{x - y}{3} = 1 \\ y = 3x - 2 \end{cases}$ 9. $\begin{cases} y = 3x - 4 \\ 9x - 3y = 12 \end{cases}$

10. $\begin{cases} x = -\dfrac{3}{2}y + 3 \\ 2x + 3y = 4 \end{cases}$

Solve each system by elimination.

11. $\begin{cases} x + 5y = 7 \\ 3x + y = -7 \end{cases}$ 12. $\begin{cases} 2x + 3y = 11 \\ 3x - 7y = -41 \end{cases}$

13. $\begin{cases} 2(x + y) - x = 0 \\ 3(x + y) + 2y = 1 \end{cases}$ 14. $\begin{cases} 8x + 12y = 24 \\ 2x + 3y = 4 \end{cases}$

15. $\begin{cases} 3x - y = 4 \\ 9x - 3y = 12 \end{cases}$

Solve each system by any method.

16. $\begin{cases} 3x + 2y - z = 2 \\ x + y - z = 0 \\ 2x + 3y - z = 1 \end{cases}$ 17. $\begin{cases} 5x - y + z = 3 \\ 3x + y + 2z = 2 \\ x + y = 2 \end{cases}$

18. $\begin{cases} 2x - y + z = 1 \\ x - y + 2z = 3 \\ x - y + z = 1 \end{cases}$

19. **Department store order** The buyer for a large department store must order 40 coats, some fake fur and some leather. He is unsure of the expected sales. He can buy 25 fur coats and the rest leather for $9300, or 10 fur coats and the rest leather for $12,600. How much does he pay if he decides to split the order evenly?

20. **Ticket sales** Adult tickets for the championship game are usually $5, but on Seniors' Day, seniors paid $4. Children's tickets were $2.50. Sales of 1800 tickets totaled $7425, and children and seniors accounted for one-half of the tickets sold. How many of each were sold?

6.2 Gaussian Elimination and Matrix Methods

Definitions and Concepts	**Examples**

Three matrices associated with a system:
There are three matrices associated with a system of equations, a *coefficient matrix*, a *matrix of constants*, and an *augmented* or *system matrix*.

There are three matrices associated with the following system of equations: $\begin{cases} x + 3y + z = 2 \\ 3x - 2y - z = 5. \\ 4x + 2y + z = 9 \end{cases}$

Coefficient Matrix
$$\begin{bmatrix} 1 & 3 & 1 \\ 3 & -2 & -1 \\ 4 & 2 & 1 \end{bmatrix}$$

Constants
$$\begin{bmatrix} 2 \\ 5 \\ 9 \end{bmatrix}$$

Augmented Matrix
$$\left[\begin{array}{ccc|c} 1 & 3 & 1 & 2 \\ 3 & -2 & -1 & 5 \\ 4 & 2 & 1 & 9 \end{array}\right]$$

Elementary row operations:
1. In a type 1 row operation, any two rows of a matrix can be interchanged.
2. In a type 2 row operation, the elements of any row of a matrix can be multiplied by any non-zero constant.
3. In a type 3 row operation, any row of a matrix can be changed by adding a multiple of another row to it.

Row-echelon form of a matrix:
1. The first nonzero entry in each row is 1.
2. Leading entries appear farther to the right as you move down the rows of the matrix.
3. Any rows containing only 0's are at the bottom of the matrix.

Elementary row operations can be used to write an augmented or system matrix in row-echelon form and solve a system of equations.

To solve the previous system, we can use row operations to write the augmented or system matrix in row-echelon form.

$$\left[\begin{array}{ccc|c} 1 & 3 & 1 & 2 \\ 3 & -2 & -1 & 5 \\ 4 & 2 & 1 & 9 \end{array}\right] \begin{array}{l} \\ (-3)R_1 + R_2 \to \\ (-4)R_1 + R_3 \to \end{array} \left[\begin{array}{ccc|c} 1 & 3 & 1 & 2 \\ 0 & -11 & -4 & -1 \\ 0 & -10 & -3 & 1 \end{array}\right]$$

$$(-1)R_3 + R_2 \to \left[\begin{array}{ccc|c} 1 & 3 & 1 & 2 \\ 0 & -1 & -1 & -2 \\ 0 & -10 & -3 & 1 \end{array}\right]$$

$$(-1)R_2 \to \left[\begin{array}{ccc|c} 1 & 3 & 1 & 2 \\ 0 & 1 & 1 & 2 \\ 0 & -10 & -3 & 1 \end{array}\right]$$

$$(10)R_2 + R_3 \to \left[\begin{array}{ccc|c} 1 & 3 & 1 & 2 \\ 0 & 1 & 1 & 2 \\ 0 & 0 & 7 & 21 \end{array}\right]$$

$$\left(\tfrac{1}{7}\right)R_3 \to \left[\begin{array}{ccc|c} 1 & 3 & 1 & 2 \\ 0 & 1 & 1 & 2 \\ 0 & 0 & 1 & 3 \end{array}\right]$$

From the last line, we see that $z = 3$. From the middle line, we see that $y + z = 2$. Since $z = 3$, we know that $y = -1$. From the top equation, we know that $x + 3y + z = 2$. Substituting the values for y and z, we can determine that $x = 2$. The solution is $(2, -1, 3)$.

Exercises

Solve each system by matrix methods, if possible.

21. $\begin{cases} 2x + 5y = 7 \\ 3x - y = 2 \end{cases}$

22. $\begin{cases} 3x - y = -4 \\ -6x + 2y = 8 \end{cases}$

23. $\begin{cases} x + 3y - z = 8 \\ 2x + y - 2z = 11 \\ x - y + 5z = -8 \end{cases}$

24. $\begin{cases} x + 3y + z = 3 \\ 2x - y + z = -11 \\ 3x + 2y + 3z = 2 \end{cases}$

25. $\begin{cases} x + y + z = 4 \\ 3x - 2y - 2z = -3 \\ 4x - y - z = 0 \end{cases}$

26. $\begin{cases} x + y + z - t = 4 \\ 2x - y + 2z + 3t = -8 \\ -x + 2y - 3z + t = 4 \\ 3x + y + 2z - 3t = 9 \end{cases}$

6.3 Matrix Algebra

Definitions and Concepts	Examples
Equality of matrices: Matrices are equal if and only if they are the same size and have the same corresponding entries.	The following matrices are equal because they are the same size and their corresponding entries are equal. $$\begin{bmatrix} 2 & \sqrt{9} \\ \frac{8}{4} & 7 \end{bmatrix} = \begin{bmatrix} \sqrt{4} & 3 \\ 2 & \sqrt{49} \end{bmatrix}$$
Matrix operations: • Two $m \times n$ matrices are added by adding the corresponding elements of those matrices.	$$\begin{bmatrix} 2 & -3 \\ -7 & 5 \end{bmatrix} + \begin{bmatrix} -3 & 4 \\ -2 & 2 \end{bmatrix} = \begin{bmatrix} -1 & 1 \\ -9 & 7 \end{bmatrix}$$ Add the corresponding entries.
• Two $m \times n$ matrices are subtracted by the rule $A - B = A + (-B)$.	$$\begin{bmatrix} 2 & -3 \\ -7 & 5 \end{bmatrix} - \begin{bmatrix} -3 & 4 \\ -2 & 2 \end{bmatrix} = \begin{bmatrix} 5 & -7 \\ -5 & 3 \end{bmatrix}$$ Subtract the corresponding entries
• The product AB of the $m \times n$ matrix A and the $n \times p$ matrix B is the $m \times p$ matrix C. The ith-row, jth-column entry of C is found by keeping a running total of the products of the elements in the ith row of A with the corresponding elements in the jth column of B.	$$\begin{bmatrix} 2 & -3 \\ -7 & 5 \end{bmatrix}\begin{bmatrix} -3 & 4 \\ -2 & 2 \end{bmatrix} = \begin{bmatrix} 2(-3) + (-3)(-2) & 2(4) + (-3)(2) \\ (-7)(-3) + 5(-2) & -7(4) + (5)(2) \end{bmatrix}$$ $$= \begin{bmatrix} 0 & 2 \\ 11 & -18 \end{bmatrix}$$

Exercises

Solve for x and y.

27. $\begin{bmatrix} 1 & -4 \\ x & 2 \\ 0 & x+7 \end{bmatrix} = \begin{bmatrix} 1 & x \\ -4 & 2 \\ x+4 & y \end{bmatrix}$

Perform the matrix operations, if possible.

28. $\begin{bmatrix} 3 & 2 & 1 \\ 3 & 2 & 1 \end{bmatrix} + \begin{bmatrix} -2 & 1 & 3 \\ 1 & -2 & 1 \end{bmatrix}$

29. $\begin{bmatrix} 2 & 3 & 5 \\ 1 & -2 & 4 \\ 2 & 1 & -2 \end{bmatrix} - \begin{bmatrix} 0 & -2 & 1 \\ 3 & 4 & -2 \\ 6 & -4 & 1 \end{bmatrix}$

30. $\begin{bmatrix} 1 & -2 \\ -3 & 1 \end{bmatrix}\begin{bmatrix} 2 & 3 \\ -1 & 2 \end{bmatrix}$

31. $\begin{bmatrix} -2 & 3 & 5 \\ 1 & -2 & -3 \end{bmatrix}\begin{bmatrix} 2 & 1 \\ -1 & 2 \\ -2 & 3 \end{bmatrix}$

32. $\begin{bmatrix} 1 & -3 & 2 \end{bmatrix}\begin{bmatrix} 2 \\ 1 \\ 3 \end{bmatrix}$

33. $\begin{bmatrix} 1 \\ 2 \\ 1 \\ 5 \end{bmatrix}\begin{bmatrix} 2 & -1 & 1 & 3 \end{bmatrix}$

34. $\begin{bmatrix} 1 & -5 & 3 \\ 2 & 1 & -1 \end{bmatrix}\begin{bmatrix} 2 \\ -2 \\ 3 \end{bmatrix}\begin{bmatrix} 1 & -1 \\ -1 & 3 \end{bmatrix}\begin{bmatrix} 1 \\ -2 \end{bmatrix}$

35. $\begin{bmatrix} 1 & -3 & 2 \end{bmatrix}\begin{bmatrix} 2 \\ 1 \\ -5 \end{bmatrix} + \begin{bmatrix} 1 & -3 \end{bmatrix}\begin{bmatrix} 2 \\ 5 \end{bmatrix}$

36. $\left(\begin{bmatrix} 1 & -3 \\ 3 & 1 \end{bmatrix} + \begin{bmatrix} -1 & 3 \\ 1 & 1 \end{bmatrix}\right)\begin{bmatrix} 1 \\ -5 \end{bmatrix}$

For Exercise 37 and 38, let $A = \begin{bmatrix} 0 & -2 \\ -3 & 3 \\ -1 & 0 \end{bmatrix}$ and $B = \begin{bmatrix} 1 & -2 \\ 3 & 9 \\ 5 & 1 \end{bmatrix}$.

37. Solve $X + A = -B$ for X.

38. Solve $4X - A = B$ for X.

6.4 Matrix Inversion

Definitions and Concepts	Examples

Definitions and Concepts

The inverse of a matrix:

The **inverse** of an $n \times n$ matrix A is A^{-1}, where

$$AA^{-1} = A^{-1}A = I$$

Use elementary row operations to transform $[A\,|\,I]$ into $[I\,|\,A^{-1}]$, where I is an identity matrix. If A cannot be transformed into I, then A is singular.

If A is invertible, then the solution of $AX = B$ is $X = A^{-1}B$.

Examples

To find the inverse of matrix $A = \begin{bmatrix} 1 & 4 \\ 3 & 2 \end{bmatrix}$, we form the matrix $\begin{bmatrix} 1 & 4 & | & 1 & 0 \\ 3 & 2 & | & 0 & 1 \end{bmatrix}$ and use row operations to transform matrix A to the left of the dashed line into the identity matrix. These row operations performed on the identity matrix to the right of the dashed line will transform it into the inverse.

$$\begin{bmatrix} 1 & 4 & | & 1 & 0 \\ 3 & 2 & | & 0 & 1 \end{bmatrix} (-3)R_1 + R_2 \rightarrow \begin{bmatrix} 1 & 4 & | & 1 & 0 \\ 0 & -10 & | & -3 & 1 \end{bmatrix}$$

$$\left(-\tfrac{1}{10}\right)R_2 \rightarrow \begin{bmatrix} 1 & 4 & | & 1 & 0 \\ 0 & 1 & | & \dfrac{3}{10} & -\dfrac{1}{10} \end{bmatrix}$$

$$-4R_2 + R_1 \rightarrow \begin{bmatrix} 1 & 0 & | & -\dfrac{1}{5} & \dfrac{2}{5} \\ 0 & 1 & | & \dfrac{3}{10} & -\dfrac{1}{10} \end{bmatrix}$$

The inverse is $\begin{bmatrix} -\dfrac{1}{5} & \dfrac{2}{5} \\ \dfrac{3}{10} & -\dfrac{1}{10} \end{bmatrix}$. Check by showing that the product of matrix A and the inverse is the identity matrix.

Exercises

Find the inverse of each matrix, if possible.

39. $\begin{bmatrix} 2 & 3 \\ 3 & 5 \end{bmatrix}$

40. $\begin{bmatrix} 2 & -1 \\ -6 & 4 \end{bmatrix}$

41. $\begin{bmatrix} -6 & 4 \\ -3 & 2 \end{bmatrix}$

42. $\begin{bmatrix} 1 & 0 & 0 \\ 2 & 0 & -2 \\ 1 & 2 & 2 \end{bmatrix}$

43. $\begin{bmatrix} 1 & 0 & 8 \\ 3 & 7 & 6 \\ 1 & 2 & 3 \end{bmatrix}$

44. $\begin{bmatrix} 4 & 4 & 1 \\ 1 & 1 & 1 \\ -1 & -1 & 0 \end{bmatrix}$

Use the inverse of the coefficient matrix to solve each system of equations.

45. $\begin{cases} 3x - y = 8 \\ x + 2y = 5 \end{cases}$

46. $\begin{cases} 4x - y + 2z = 0 \\ x + y + 2z = 1 \\ x + z = 0 \end{cases}$

47. $\begin{cases} w + 3x + y + 3z = 1 \\ w + 4x + y + 3z = 2 \\ x + y = 1 \\ w + 2x - y + 2z = 1 \end{cases}$

6.5 Determinants

Definitions and Concepts	**Examples**

The determinants:

- The **determinant** of $A = \begin{bmatrix} a & b \\ c & d \end{bmatrix}$ is

$$\det(A) = |A| = \begin{vmatrix} a & b \\ c & d \end{vmatrix} = ad - bc.$$

$$\begin{vmatrix} 2 & -3 \\ -4 & 5 \end{vmatrix} = 2(5) - (-3)(-4) = 10 - 12 = -2$$

- The **determinant** of an $n \times n$ matrix A is the sum of the products of the elements of any row (or column) and the cofactors of those elements.

Evaluate the following determinant by expanding along the first row.

$$\begin{vmatrix} 2 & 3 & -4 \\ 3 & 2 & 1 \\ -2 & 5 & -1 \end{vmatrix} = 2\begin{vmatrix} 2 & 1 \\ 5 & -1 \end{vmatrix} - 3\begin{vmatrix} 3 & 1 \\ -2 & -1 \end{vmatrix} - 4\begin{vmatrix} 3 & 2 \\ -2 & 5 \end{vmatrix}$$

$$= 2(-2 - 5) - 3[-3 - (-2)] - 4[15 - (-4)]$$
$$= 2(-7) - 3(-1) - 4(19)$$
$$= -14 + 3 - 76$$
$$= -87$$

Properties of determinants:

- If two rows (or columns) of a matrix are interchanged, the sign of its determinant is reversed.

$$\begin{vmatrix} 2 & 3 \\ 4 & 5 \end{vmatrix} = -\begin{vmatrix} 4 & 5 \\ 2 & 3 \end{vmatrix} \qquad \begin{vmatrix} 2 & 3 \\ 4 & 5 \end{vmatrix} = -\begin{vmatrix} 3 & 2 \\ 5 & 4 \end{vmatrix}$$

- If a row (or column) of a matrix is multiplied by a constant k, the value of its determinant is multiplied by k.

$$\begin{vmatrix} 3 & 4 \\ -2 & 1 \end{vmatrix} = 11 \text{ If we multiply row 1 by 5 and obtain}$$

$$\begin{vmatrix} 15 & 20 \\ -2 & 1 \end{vmatrix}, \text{ its determinant is 55. Note that } 55 = 5(11).$$

- If a row (or column) of a matrix is altered by adding to it a multiple of another row (or column), the value of its determinant is unchanged.

$$\begin{vmatrix} 3 & 4 \\ -2 & 1 \end{vmatrix} = 11 \quad \text{If we multiply row 2 by 5 and add it to row 1, the}$$

value of the determinant is still 11.

$$\begin{vmatrix} 3 & 4 \\ -2 & 1 \end{vmatrix} = \begin{vmatrix} 3 + 5(-2) & 4 + 5(1) \\ -2 & 1 \end{vmatrix} = \begin{vmatrix} -7 & 9 \\ -2 & 1 \end{vmatrix} = 11$$

Cramer's Rule:

Form quotients of two determinants. The denominator is the determinant of the coefficient matrix, A. The numerator is the determinant of a modified coefficient matrix. When solving for the ith variable, replace the ith column of A with a column of constants, B.

Use Cramer's Rule to solve the following system for x.

$$\begin{cases} 2x + 3y - 4z = 1 \\ 3x + 2y + z = 6 \\ -2x + 5y - z = 2 \end{cases}$$

$$x = \frac{\begin{vmatrix} 1 & 3 & -4 \\ 6 & 2 & 1 \\ 2 & 5 & -1 \end{vmatrix}}{\begin{vmatrix} 2 & 3 & -4 \\ 3 & 2 & 1 \\ -2 & 5 & -1 \end{vmatrix}} = \frac{1\begin{vmatrix} 2 & 1 \\ 5 & -1 \end{vmatrix} - 3\begin{vmatrix} 6 & 1 \\ 2 & -1 \end{vmatrix} - 4\begin{vmatrix} 6 & 2 \\ 2 & 5 \end{vmatrix}}{-87 \text{ (See above.)}}$$

$$= \frac{1(-7) - 3(-8) - 4(26)}{-87}$$

$$= \frac{-87}{-87}$$

$$= 1$$

Exercises

Evaluate each determinant.

48. $\begin{vmatrix} 3 & -2 \\ 1 & -3 \end{vmatrix}$ **49.** $\begin{vmatrix} 1 & -2 & 3 \\ 2 & -1 & 3 \\ 1 & -1 & 0 \end{vmatrix}$

50. $\begin{vmatrix} 1 & 3 & -1 \\ 1 & 2 & 1 \\ 1 & 0 & 2 \end{vmatrix}$ **51.** $\begin{vmatrix} 1 & 2 & 3 & 4 \\ -1 & 3 & -3 & 2 \\ 0 & 0 & 0 & -1 \\ 3 & 3 & 4 & 3 \end{vmatrix}$

Use Cramer's Rule to solve each system.

52. $\begin{cases} x + 3y = -5 \\ -2x + y = -4 \end{cases}$

53. $\begin{cases} x - y + z = -1 \\ 2x - y + 3z = -4 \\ x - 3y + z = -1 \end{cases}$

54. $\begin{cases} x - 3y + z = 7 \\ x + y - 3z = -9 \\ x + y + z = 3 \end{cases}$

55. $\begin{cases} w + x - y + z = 4 \\ 2w + x + z = 4 \\ x + 2y + z = 0 \\ w + y + z = 2 \end{cases}$

If $\begin{vmatrix} a & b & c \\ d & e & f \\ g & h & i \end{vmatrix} = 7$, *evaluate each determinant.*

56. $\begin{vmatrix} 3a & 3b & 3c \\ d & e & f \\ g & h & i \end{vmatrix}$

57. $\begin{vmatrix} a & b & c \\ d+g & e+h & f+i \\ g & h & i \end{vmatrix}$

6.6 Partial Fractions

Definitions and Concepts

Partial fraction decomposition:

The fraction $\frac{P(x)}{Q(x)}$ can be written as the sum of simpler fractions with denominators determined by the prime factors of $Q(x)$. For a review of all cases, see Section 6.6.

Examples

Decompose $\dfrac{4x - 1}{2x^2 + 3x + 1}$ into partial fractions.

We factor the denominator to get $\dfrac{4x - 1}{(2x + 1)(x + 1)}$.

Since each denominator is linear, there are constants A and B such that

$$\frac{4x - 1}{(2x + 1)(x + 1)} = \frac{A}{2x + 1} + \frac{B}{x + 1}$$

Adding the fractions on the right, we get

$$\frac{4x - 1}{(2x + 1)(x + 1)} = \frac{A(x + 1)}{(2x + 1)(x + 1)} + \frac{B(2x + 1)}{(x + 1)(2x + 1)}$$

$$= \frac{Ax + A + 2Bx + B}{(2x + 1)(x + 1)}$$

$$= \frac{(A + 2B)x + A + B}{(2x + 1)(x + 1)}$$

Setting the coefficients of the polynomial numerators equal, we have the following system of equations:

$$\begin{cases} A + 2B = 4 \\ A + B = -1 \end{cases}$$

whose solution is $A = -6$ and $B = 5$. Thus,

$$\frac{4x - 1}{(2x + 1)(x + 1)} = \frac{-6}{2x + 1} + \frac{5}{x + 1}$$

Exercises

Decompose into partial fractions.

58. $\dfrac{7x + 3}{x^2 + x}$

59. $\dfrac{4x^3 + 3x + x^2 + 2}{x^4 + x^2}$

60. $\dfrac{x^2 + 5}{x^3 + x^2 + 5x}$

61. $\dfrac{x^2 + 1}{(x + 1)^3}$

6.7 Graphs of Linear Inequalities

Definitions and Concepts	Examples
Solving a system of inequalities: Systems of inequalities in two variables can be solved by graphing. On the same set of coordinate axes, we graph each inequality. The area that is shaded twice represents the solution of the system of inequalities. The solution is represented by a plane region with boundaries determined by graphing the inequalities as if they were equations.	To use graphing to solve the system $\begin{cases} y \le -x + 1 \\ 2x > y + 2 \end{cases}$, we graph each inequality on the same coordinate axes as shown below. 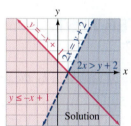 The solution is the overlapping region shown in purple.

Exercises

62. Graph: $y \ge -2x - 1$.

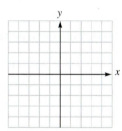

63. Graph: $x^2 + y^2 > 4$.

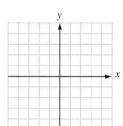

Solve each system by graphing.

64. $\begin{cases} 3x + 2y \le 6 \\ x - y > 3 \end{cases}$

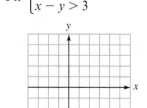

65. $\begin{cases} y \le x^2 + 1 \\ y \ge x^2 - 1 \end{cases}$

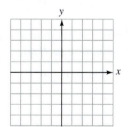

6.8 Linear Programming

Definitions and Concepts	Examples
Solving linear programming problems: The maximum and the minimum values of a linear function in two variables, subject to the constraints of a system of linear inequalities, are attained at a corner or along an entire edge of the region determined by the system of inequalities.	See Example 1 on page 677 in the text.

Exercises

Maximize P subject to the given conditions.

66. $P = 2x + y$

$$\begin{cases} x \geq 0 \\ y \geq 0 \\ x + y \leq 3 \end{cases}$$

67. $P = 3x - y$

$$\begin{cases} y \geq 1 \\ y \leq 2 \\ y \leq 3x + 1 \\ x \leq 1 \end{cases}$$

68. $P = 2x - 3y$

$$\begin{cases} x \geq 0 \\ y \leq 3 \\ x - y \leq 4 \end{cases}$$

69. $P = y - 2x$

$$\begin{cases} x + y \geq 1 \\ x \leq 1 \\ y \leq \dfrac{x}{2} + 2 \\ x + y \leq 2 \end{cases}$$

70. A company manufactures two fertilizers, x and y. Each 50-pound bag of fertilizer requires three ingredients, which are available in the limited quantities shown in the table. The profit on each bag of fertilizer x is \$6 and on each bag of y is \$5. How many bags of each product should be produced to maximize the profit?

Ingredient	Number of Pounds in Fertilizer x	Number of Pounds in Fertilizer y	Total Number of Pounds Available
Nitrogen	6	10	20,000
Phosphorus	8	6	16,400
Potash	6	4	12,000

CHAPTER TEST

Solve each system of equations by the graphing method.

1. $\begin{cases} x - 3y = -5 \\ 2x - y = 0 \end{cases}$

2. $\begin{cases} x = 2y + 5 \\ y = 2x - 4 \end{cases}$

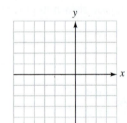

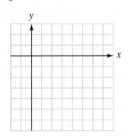

Solve each system of equations by the substitution or elimination method.

3. $\begin{cases} 3x + y = 0 \\ 2x - 5y = 17 \end{cases}$

4. $\begin{cases} \dfrac{x + y}{2} + x = 7 \\ \dfrac{x - y}{2} - y = -6 \end{cases}$

5. Mixing solutions A chemist has two solutions; one has a 20% concentration and the other a 45% concentration. How many liters of each must she mix to obtain 10 liters of 30% concentration?

6. Wholesale distribution Ace Electronics, Hi-Fi Stereo, and CD World buy a total of 175 DVD players from the same distributor each month. Because CD World buys 25 more units than the other two stores combined, CD World's cost is only \$160 per unit. The players cost Hi-Fi \$165 each and Ace \$170 each. How many players does each retailer buy each month if the distributor receives \$28,500 each month from the sale of the players to the three stores?

Write each system of equations as a matrix and solve it by Gaussian elimination.

7. $\begin{cases} 3x - 2y = 4 \\ 2x + 3y = 7 \end{cases}$

8. $\begin{cases} x + 3y - z = 6 \\ 2x - y - 2z = -2 \\ x + 2y + z = 6 \end{cases}$

Write each system of equations as a matrix and solve it by Gauss–Jordan elimination. If the system has infinitely many solutions, write a general solution.

9. $\begin{cases} x + 2y + 3z = -5 \\ 3x + y - 2z = 7 \\ y - z = 2 \end{cases}$

10. $\begin{cases} x + 2y + z = 0 \\ 3x - 2y - 2z = 7 \\ 4x - z = 7 \end{cases}$

Perform the operations.

11. $3\begin{bmatrix} 2 & -3 & 5 \\ 0 & 3 & -1 \end{bmatrix} - 5\begin{bmatrix} -2 & 1 & -1 \\ 0 & 3 & 2 \end{bmatrix}$

12. $\begin{bmatrix} 1 & 2 & 3 \end{bmatrix}\begin{bmatrix} 2 & -2 \\ -2 & 2 \\ 1 & 0 \end{bmatrix}\begin{bmatrix} 3 \\ -2 \end{bmatrix}$

Find the inverse of each matrix, if possible.

13. $\begin{bmatrix} 5 & 19 \\ 2 & 7 \end{bmatrix}$

14. $\begin{bmatrix} -1 & 3 & -2 \\ 4 & 1 & 4 \\ 0 & 3 & -1 \end{bmatrix}$

Use the inverses found in Questions 13 and 14 to solve each system.

15. $\begin{cases} 5x + 19y = 3 \\ 2x + 7y = 2 \end{cases}$

16. $\begin{cases} -x + 3y - 2z = 1 \\ 4x + y + 4z = 3 \\ 3y - z = -1 \end{cases}$

Evaluate each determinant.

17. $\begin{vmatrix} 3 & -5 \\ -3 & 1 \end{vmatrix}$

18. $\begin{vmatrix} 3 & 5 & -1 \\ -2 & 3 & -2 \\ 1 & 5 & -3 \end{vmatrix}$

Use Cramer's Rule to solve each system for y.

19. $\begin{cases} 3x - 5y = 3 \\ -3x + y = 2 \end{cases}$

20. $\begin{cases} 3x + 5y - z = 2 \\ -2x + 3y - 2z = 1 \\ x + 5y - 3z = 0 \end{cases}$

Decompose each fraction into partial fractions.

21. $\dfrac{5x}{2x^2 - x - 3}$

22. $\dfrac{3x^2 + x + 2}{x^3 + 2x}$

Graph the solution set of each system.

23. $\begin{cases} x - 3y \geq 3 \\ x + 3y \leq 3 \end{cases}$

24. $\begin{cases} 3x + 4y \leq 12 \\ 3x + 4y \geq 6 \\ x \geq 0 \\ y \geq 0 \end{cases}$

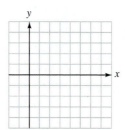

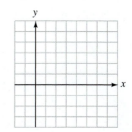

25. Maximize $P = 3x + 2y$ subject to

$$\begin{cases} y \geq 0 \\ x \geq 0 \\ 2x + y \leq 4 \\ y \leq 2 \end{cases}$$

26. Minimize $P = y - x$ subject to

$$\begin{cases} x \geq 0 \\ y \geq 0 \\ x + y \leq 8 \\ 2x + y \geq 2 \end{cases}$$

Conic Sections and Quadratic Systems

7

Careers and Mathematics: Engineer

©Shots Studio/Shutterstock.com

Engineers are the link between a discovery in science and the practical application of that discovery to meet consumer demands. Engineers will often design and develop new products. They are also supervisors in manufacturing facilities. Most engineers have a specialty. Aerospace engineers, biological engineers, chemical engineers, electrical engineers, mechanical engineers, civil engineers, and petroleum engineers are a few of the specialists in engineering.

Education and Mathematics Required

- Engineers typically have a bachelor's degree in an engineering specialty. Research positions will often require a graduate degree.
- College Algebra, Geometry, Trigonometry, the Calculus sequence, Linear Algebra, Differential Equations, and Statistics are math courses required.

How Engineers Use Math and Who Employs Them

- Mathematics is important for all types of engineers. It is the language of engineering and physical science. Mathematics can develop the intellectual maturity to solve difficult problems. An understanding of mathematical analysis is essential to validate the work of computer programs and to write code that can make programs even more powerful.
- About 37% of engineers are employed in manufacturing industries, and 28% in professional, scientific, and technical services. Federal, state, and local government agencies employ engineers in highway and public works departments. Potential employers are determined by the engineering specialty.

Career Outlook and Earnings

- Employment of engineers is expected to grow about as fast as the average for all occupations, although growth will vary by specialty. Job opportunities are expected to be good.
- Earnings vary significantly among the specialties. Median salaries run from $130,280 for petroleum engineers to $74,000 for agricultural engineers. Highest salaries can be over $165,000.

For more information see: www.bls.gov/oco

7.1 The Circle and the Parabola

In this section, we will learn to

1. Write an equation of a circle.
2. Graph circles.
3. Write an equation of a parabola in standard form.
4. Graph parabolas.
5. Solve applied problems involving parabolas.

The suspension bridge is one of the oldest types of bridges. An example is the Golden Gate Bridge that spans the opening of San Francisco Bay and the Pacific Ocean. Completed in 1937, it is currently the second longest suspension bridge in the United States. As the photo shows, the main cables that hold up the bridge hang in the shape of a parabola. A parabola is an example of a *conic section*.

Second-degree equations in x and y have the general form

$$Ax^2 + Bxy + Cy^2 + Dx + Ey + F = 0$$

where at least one of the coefficients A, B, and C is not zero. The graphs of these equations fall into one of several categories: a point, a pair of lines, a circle, a parabola, an ellipse, a hyperbola, or no graph at all. These graphs are called **conic sections**, because each one is the intersection of a plane and a right-circular cone, as shown in Figure 7-1.

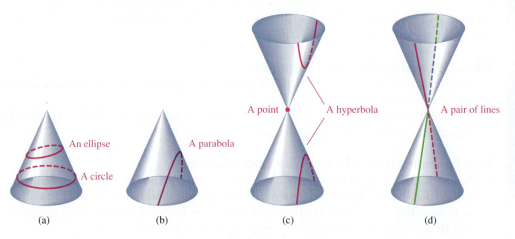

FIGURE 7-1

Although these shapes have been known since the time of the ancient Greeks, it wasn't until the 17th century that René Descartes (1596–1650) and Blaise Pascal (1623–1662) developed the mathematics needed to study them in detail.

In this section, we will consider two conic sections, the circle and the parabola.

1. Write an Equation of a Circle

In Section 2.5, we developed the following standard equations for circles:

The Standard Form of an Equation of a Circle with Center at (h, k)

The graph of any equation that can be written in the form

$$(x - h)^2 + (y - k)^2 = r^2$$

is a circle with radius r and center at point (h, k).

If $h = 0$ and $k = 0$, we have this result.

The Standard Form of an Equation of a Circle with Center at $(0, 0)$	The graph of any equation that can be written in the form

$$x^2 + y^2 = r^2$$

is a circle with radius r and center at the origin.

The graphs of two circles, one with center at (h, k) and one with center at $(0, 0)$ are shown in Figure 7-2.

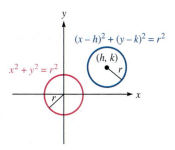

FIGURE 7-2

EXAMPLE 1

Writing an Equation of a Circle

Find an equation of the circle shown in Figure 7-3. Write both the standard and general forms of the equation.

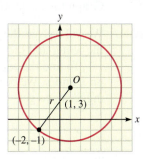

FIGURE 7-3

SOLUTION   We will use distance formula to find the radius of the circle and then substitute the coordinates of the center and radius into the standard equation of a circle.

To find the radius of the circle, we substitute the coordinates of the points $(1, 3)$ and $(-2, -1)$ into the distance formula and simplify:

$$r = \sqrt{(x_2 - x_1)^2 + (y_2 - y_1)^2}$$

$$r = \sqrt{(-2 - 1)^2 + (-1 - 3)^2} \qquad \text{\color{red}Substitute 1 for } x_1\text{, 3 for } y_1\text{, } -2 \text{ for } x_2\text{, and } -1 \text{ for } y_2.$$

$$= \sqrt{(-3)^2 + (-4)^2}$$

$$= \sqrt{9 + 16}$$

$$= \sqrt{25}$$

$$= 5$$

To find an equation of a circle with radius 5 and center at (1, 3), we substitute 1 for h, 3 for k, and 5 for r in the standard form of the circle.

$$(x - h)^2 + (y - k)^2 = r^2$$
$$(x - 1)^2 + (y - 3)^2 = 5^2$$

To write the equation in general form, we square the binomials and simplify.

$$x^2 - 2x + 1 + y^2 - 6y + 9 = 25$$
$$x^2 + y^2 - 2x - 6y - 15 = 0 \qquad \text{Subtract 25 from both sides and simplify.}$$

Self Check 1   Find the general equation of a circle with center at $(-2, 1)$ and radius of 4.

Now Try Exercise 17.

The final equation in Example 1 can be written as

$$1x^2 + 0xy + 1y^2 - 2x - 6y - 15 = 0$$

which illustrates that the graph of

$$Ax^2 + Bxy + Cy^2 + Dx + Ey + F = 0$$

is a circle whenever $B = 0$ and $A = C$.

2. Graph Circles

Graphing a Circle

EXAMPLE 2   Graph the circle whose equation is $2x^2 + 2y^2 + 4x + 3y = 3$.

SOLUTION   We will convert the general form of the circle into standard form so we can identify the center and radius. Then we will graph the circle.

To find the coordinates of the center and the radius, we complete the square on x and y. We begin by dividing both sides of the equation by 2 and rearranging terms to get

$$x^2 + 2x + y^2 + \frac{3}{2}y = \frac{3}{2}$$

To complete the square on x and y, we add 1 and $\frac{9}{16}$ to both sides.

$$x^2 + 2x + \mathbf{1} + y^2 + \frac{3}{2}y + \frac{\mathbf{9}}{\mathbf{16}} = \frac{3}{2} + \mathbf{1} + \frac{\mathbf{9}}{\mathbf{16}}$$

We can factor $x^2 + 2x + 1$ and $y^2 + \frac{3}{2}y + \frac{9}{16}$ and simplify on the right side to get

$$(x + 1)^2 + \left(y + \frac{3}{4}\right)^2 = \frac{49}{16}$$

$$[x - (\mathbf{-1})]^2 + \left[y - \left(-\frac{\mathbf{3}}{\mathbf{4}}\right)\right]^2 = \left(\frac{\mathbf{7}}{\mathbf{4}}\right)^2$$

From the equation, we see that the coordinates of the center of the circle are $h = -1$ and $k = -\frac{3}{4}$ and that the radius of the circle is $\frac{7}{4}$. The graph is shown in Figure 7-4.

Take Note

Since the graphs of circles fail the vertical line test, their equations do not represent functions. It is somewhat more difficult to use a graphing calculator to graph equations that are not functions. Please see the Accent on Technology, Graphing a Circle, in Section 2.5. In many cases, it is easier to graph the circle by hand.

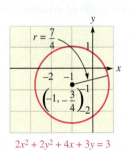

$$2x^2 + 2y^2 + 4x + 3y = 3$$

FIGURE 7-4

Self Check 2   Graph the circle whose equation is $2x^2 + 2y^2 - 12x - 4y = -12$.

Now Try Exercise 23.

EXAMPLE 3

Solving an Application Problem

The effective broadcast area of a radio station is bounded by the circle

$$x^2 + y^2 = 2500$$

where x and y are measured in miles. Another station's broadcast area is bounded by the circle

$$(x - 100)^2 + (y - 100)^2 = 900$$

Can any location receive both stations?

SOLUTION It is possible to receive both stations only if their circular broadcast areas overlap. (See Figure 7-5.) This happens when the sum of the radii, r and r', of the two circles is greater than the distance, d, between their centers. That is, $r + r' > d$.

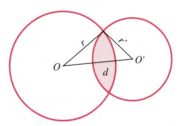

FIGURE 7-5

The center of the circle $x^2 + y^2 = 2500$ is $(x_1, y_1) = (0, 0)$, and its radius r is 50 miles. The center of the circle $(x - 100)^2 + (y - 100)^2 = 900$ is $(x_2, y_2) = (100, 100)$, and its radius r' is 30 miles.

We can use the distance formula to find the distance d between the centers.

$$d = \sqrt{(x_2 - x_1)^2 + (y_2 - y_1)^2}$$

$$d = \sqrt{(100 - 0)^2 + (100 - 0)^2}$$

$$= \sqrt{100^2 + 100^2}$$

$$= \sqrt{100^2 \cdot 2}$$

$$= 100\sqrt{2}$$

$$\approx 141 \text{ miles}$$

The sum of the radii, $r + r'$, of the two circles is $(50 + 30)$ miles, or 80 miles. Since this is less than the distance d between their centers (141 miles), there is no location where both stations can be received.

Self Check 3 In Example 3, if the station at the origin boosted its power to cover the area within $x^2 + y^2 = 9000$, would the coverage overlap?

Now Try Exercise 67.

3. Write an Equation of a Parabola in Standard Form

In Chapter 4, we saw that the graphs of quadratic functions of the forms

$$f(x) = ax^2 + bx + c \quad \text{or} \quad f(x) = a(x - h)^2 + k \quad (a \neq 0)$$

were parabolas that open upward or downward. We now discuss parabolas that open to the left or to the right and examine the properties of all parabolas in greater detail.

Parabola A **parabola** is the set of all points in a plane equidistant from a line *l* (called the **directrix**) and a fixed point *F* (called the **focus**) that is not on line *l*. See Figure 7-6.

The point on the parabola that is closest to the directrix is called the **vertex**, and the line passing through the vertex and the focus is called the **axis of the parabola**.

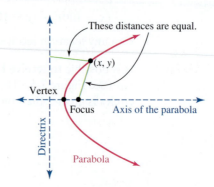

FIGURE 7-6

A basic parabola is one whose vertex occurs at the origin and opens upward, as shown in Figure 7-7. Consider such a parabola with the characteristics shown in Figure 7-7.

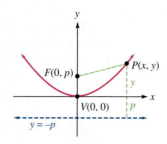

FIGURE 7-7

To find the equation of the parabola, we will find the distance from point $P(x, y)$ on the parabola to the focus and find the distance from point $P(x, y)$ to the directrix and set the two distances equal. Using the distance formula, we can find that the distance from $P(x, y)$ to $F(0, p)$.

$$d(PF) = \sqrt{x^2 + (y - p)^2}$$

From the figure, we see that the distance from $P(x, y)$ to the directrix $y = -p$ is:

$$|y - (-p)| = |y + p|$$

We can equate these distances and simplify.

$$\sqrt{x^2 + (y - p)^2} = |y + p|$$

$$x^2 + (y - p)^2 = |y + p|^2 \qquad \text{Square both sides.}$$

$$x^2 + (y - p)^2 = (y + p)^2 \qquad |y + p|^2 = (y + p)^2$$

$$x^2 + y^2 - 2yp + p^2 = y^2 + 2py + p^2 \qquad \text{Expand the binomials.}$$

$$x^2 - 2py = 2py \qquad \text{Subtract } y^2 + p^2 \text{ from both sides.}$$

$$(1) \qquad x^2 = 4py \qquad \text{Add } 2py \text{ to both sides.}$$

Equation 1 is one of the **standard equations of a parabola** with vertex at the origin. If $p > 0$ as in Equation 1, the graph of the equation will be a parabola that opens upward. If $p < 0$, the graph of the equation will be a parabola that opens downward.

It can be shown that a parabola whose vertex is at the origin and opens to the left or to the right has an equation of the form $y^2 = 4px$.

The standard equations of a parabola with vertex $V(0, 0)$ and some of the characteristics of the parabola are summarized below.

Standard Form of an Equation of a Parabola with Vertex (0, 0)		
Equation	$x^2 = 4py$	$y^2 = 4px$
Focus	$(0, p)$	$(p, 0)$
Directrix	$y = -p$	$x = -p$
Axis of Symmetry	Vertical y-axis	Horizontal x-axis
$p > 0$	Opens upward	Opens right
$p < 0$	Opens downward	Opens left

> **Tip**
>
> The equation of a parabola has either an x^2 term or a y^2 term, but not both.

> **Tip**
>
> Consider p as the directed distance from the vertex to the focus.
>
> 1. If p is positive, then the focus lies p units to the right or p units above the vertex.
>
> 2. If p is negative, then the focus lies $|p|$ units to the left of the vertex or $|p|$ units below the vertex.

©Vladimir Jankovic/Shutterstock.com

Consider the parabola with equation $x^2 = 8y$, which is shown in Figure 7-8. Since the equation of the parabola is written in standard form, $x^2 = 4py$, we know that $4p = 8$ and $p = 2$. Thus, the focus of the parabola is $F(0, p) = F(0, 2)$ and the directrix is $y = -p$ or $y = -2$. Because $p = 2$ and $2 > 0$, the parabola opens upward. The axis of symmetry is a vertical line, which is the y-axis.

FIGURE 7-8

Finding an Equation of a Parabola in Standard Form

EXAMPLE 4

Find an equation of the parabola in standard form with vertex at the origin and focus at $(3, 0)$.

SOLUTION   A sketch of the parabola is shown in Figure 7-9. Because the focus is to the right of the vertex, the parabola opens to the right, and because the vertex is the origin, the standard equation is $y^2 = 4px$. The distance between the focus and the vertex is $p = 3$. We can substitute 3 for p in the standard equation to get

$$y^2 = 4px$$

$$y^2 = 4(3)x$$

$$y^2 = 12x$$

An equation of the parabola in standard form is $y^2 = 12x$.

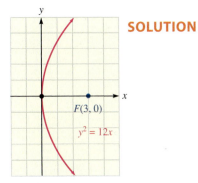

FIGURE 7-9

Self Check 4   Find an equation of the parabola in standard form with vertex at the origin and focus at $(-3, 0)$.

Now Try Exercise 31.

In Section 3.2 we used translations to sketch graphs of functions. Translations can also be applied to graphs that are parabolas. The following table summarizes the characteristics of parabolas with vertex at $V(h, k)$.

Standard Form of an Equation of a Parabola with Vertex (*h, k*)		
Equation	$(x - h)^2 = 4p(y - k)$	$(y - k)^2 = 4p(x - h)$
Focus	$(h, k + p)$	$(h + p, k)$
Directrix	$y = -p + k$	$x = -p + h$
Axis of Symmetry	Vertical $x = h$	Horizontal $y = k$
$p > 0$	Opens upward	Opens right
$p < 0$	Opens downward	Opens left

Take Note

Horizontal and vertical translations are accomplished by replacing x with $x - h$ and y with $y - k$ in the standard form of the parabola's equation.

EXAMPLE 5

Finding an Equation of a Parabola in Standard Form

Find an equation of the parabola in standard form that opens upward, has vertex at the point $(4, 5)$, and passes through the point $(0, 7)$.

SOLUTION Because the parabola opens upward, we will substitute the coordinates of the given vertex and the point into the equation $(x - h)^2 = 4p(y - k)$ and solve for p. Then we will substitute the coordinates of the vertex and the value of p into the previous equation.

Since $(h, k) = (4, 5)$ and the point $(0, 7)$ is on the curve, we can substitute 4 for h, 5 for k, 0 for x, and 7 for y in the standard equation and solve for p.

$$(x - h)^2 = 4p(y - k)$$

$$(0 - 4)^2 = 4p(7 - 5)$$

$$16 = 8p$$

$$2 = p$$

To find an equation of the parabola, we substitute 4 for h, 5 for k, and 2 for p in the standard equation and simplify:

$$(x - h)^2 = 4p(y - k)$$

$$(x - 4)^2 = 4(2)(y - 5)$$

$$(x - 4)^2 = 8(y - 5)$$

The graph of the equation appears in Figure 7-10.

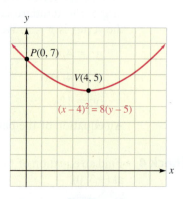

FIGURE 7-10

Self Check 5   Find an equation of the parabola in standard form that opens upward, has vertex at (4, 5), and passes through (0, 9).

Now Try Exercise 43.

EXAMPLE 6   **Finding Equations of Two Parabolas in Standard Form**

Find equations of two parabolas in standard form with a vertex at (2, 4) that pass through (0, 0).

SOLUTION   The two parabolas are shown in Figure 7-11.

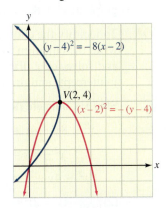

FIGURE 7-11

Part 1: To find the parabola that opens to the left, we use the equation $(y - k)^2 = 4p(x - h)$. Since the curve passes through the point $(x, y) = (0, 0)$ and the vertex $(h, k) = (2, 4)$, we substitute 0 for x, 0 for y, 2 for h, and 4 for k in the standard equation and solve for p:

$$(y - k)^2 = 4p(x - h)$$
$$(0 - 4)^2 = 4p(0 - 2)$$
$$16 = -8p$$
$$-2 = p$$

Since $h = 2$, $k = 4$, $p = -2$, and the parabola opens to the left, its equation is

$$(y - k)^2 = 4p(x - h)$$
$$(y - 4)^2 = 4(-2)(x - 2)$$
$$(y - 4)^2 = -8(x - 2)$$

Part 2: To find an equation of the parabola that opens downward, we use the equation $(x - h)^2 = 4p(y - k)$ and substitute 2 for h, 4 for k, 0 for x, and 0 for y and solve for p:

$$(x - h)^2 = 4p(y - k)$$
$$(0 - 2)^2 = 4p(0 - 4)$$
$$4 = -16p$$
$$p = -\frac{1}{4}$$

Since $h = 2$, $k = 4$, $p = -\frac{1}{4}$, the equation is

$$(x - h)^2 = 4p(y - k)$$
$$(x - 2)^2 = 4\left(-\frac{1}{4}\right)(y - 4) \qquad \text{Substitute 2 for } h, -\frac{1}{4} \text{ for } p, \text{ and 4 for } k.$$
$$(x - 2)^2 = -(y - 4)$$

Self Check 6   Find equations of two parabolas in standard form with vertex at $(2, 4)$ and that pass through $(0, 8)$.

Now Try Exercise 45.

4. Graph Parabolas

EXAMPLE 7

Graphing a Parabola

Find the vertex and y-intercepts of the parabola with the following equation and graph it: $y^2 + 8x - 4y = 28$.

SOLUTION   To identify the coordinates of the vertex, we will complete the square on y and write the equation in standard form. To find the y-intercepts, we will let $x = 0$ and solve for y. Finally, we will graph the parabola by plotting the vertex and y-intercepts and drawing a smooth curve through the points.

Step 1: Complete the square and write the equation in standard form.

$$y^2 + 8x - 4y = 28$$

$$y^2 - 4y = -8x + 28 \qquad \text{Subtract } 8x \text{ from both sides.}$$

$$y^2 - 4y + 4 = -8x + 28 + 4 \qquad \text{Add 4 to both sides.}$$

$$(2) \qquad (y - 2)^2 = -8(x - 4) \qquad \text{Factor both sides.}$$

Equation 2 represents a parabola opening to the left with vertex at $(4, 2)$.

Step 2: Find the y-intercept.
To find the y-intercepts, we substitute 0 for x in Equation 2 and solve for y.

$$(y - 2)^2 = -8(x - 4)$$

$$(y - 2)^2 = -8(0 - 4) \qquad \text{Substitute 0 for } x.$$

$$y^2 - 4y + 4 = 32 \qquad \text{Remove parentheses.}$$

$$(3) \quad y^2 - 4y - 28 = 0$$

We can use the quadratic formula to find that the solutions of Equation 3 are $y = 2 \pm 4\sqrt{2}$ or $y \approx 7.7$ and $y \approx -3.7$. So the points with coordinates of approximately $(0, 7.7)$ and $(0, -3.7)$ lie on the graph of the parabola and are the y-intercepts.

Step 3: Graph the parabola.
We can use the information above and the knowledge that the graph opens to the left and has vertex at $(4, 2)$ to draw the graph, as shown in Figure 7-12.

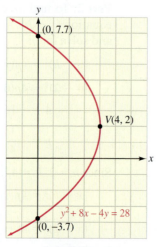

FIGURE 7-12

Self Check 7   Find the vertex and y-intercepts of the parabola with an equation of $y^2 - x + 2y = 3$. Then graph it.

Now Try Exercise 53.

Graphing Parabolas

Parabolas with a vertical axis, such as $y = x^2 - 3x + 5$, can be entered into a graphing calculator as it is written. However, a parabola with a horizontal axis, such as $y^2 = 4x - 4$, cannot be entered into a calculator in this form. To graph this parabola using a graphing calculator, we must first solve for y by taking the square root of both sides, getting the equations $y = \sqrt{4x - 4}$ and $y = -\sqrt{4x - 4}$. We enter these as shown in Figure 7-13.

Enter the equations.   Set an appropriate WINDOW.   GRAPH the equations.

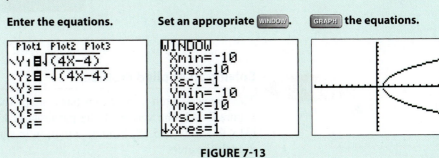

FIGURE 7-13

5. Solve Applied Problems Involving Parabolas

Parabolas arise in many real world settings. A few examples are:

- As shown in the illustration, a bouncing ball travels in a series of parabolic paths. Air resistance causes the ball to deviate slightly from a perfect parabola shape.

- A satellite dish is a type of parabolic antenna designed with the specific purpose of transmitting signals to and/or receiving signals from satellites.

- Objects extended in space often follow parabolic paths, such as a diver jumping from a diving board or cliff. The diver may follow a complex motion, but his center of mass forms a parabola.

EXAMPLE 8 Solving an Applied Problem

The Gateway Arch in St. Louis has a shape that approximates a parabola. (See Figure 7-14.) The vertex of the parabola is $V(0, 630)$ and the x-intercepts are $(315, 0)$ and $(-315, 0)$. Find an equation in standard form of the parabola that models the arch.

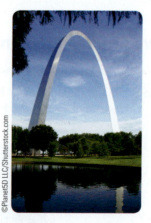

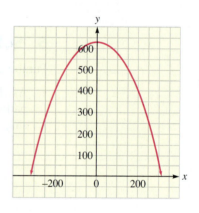

FIGURE 7-14

SOLUTION To find an equation in standard form of the parabola that opens downward, we use the equation $(x - h)^2 = 4p(y - k)$ and substitute the vertex $V(0, 630)$ and the x-intercept $(315, 0)$ and solve for p. Then, we write an equation of the parabola.

We substitute 0 for h, 630 for k, 315 for x, and 0 for y and solve for p:

$$(x - h)^2 = 4p(y - k)$$

$$(315 - 0)^2 = 4p(0 - 630)$$

$$99{,}225 = -2520p$$

$$p = -\frac{99{,}225}{2520}$$

$$= -\frac{315}{8}$$

Since $h = 0$, $k = 630$, $p = -\frac{315}{8}$, an equation is

$$(x - h)^2 = 4p(y - k)$$

$$(x - 0)^2 = 4\left(-\frac{315}{8}\right)(y - 630) \qquad \text{Substitute 0 for } h, -\frac{315}{8} \text{ for } p, \text{ and } 630 \text{ for } k.$$

$$x^2 = -\frac{315}{2}(y - 630) \qquad \text{Simplify.}$$

Self Check 8   What is the height of the arch 115 feet from the base? Round to the nearest foot.

Now Try Exercise 83.

Solving an Applied Problem

EXAMPLE 9   If water is propelled straight up by a "super nozzle" during a water fountain show in Las Vegas, the equation $s = 128 - 16t^2$ expresses the water's height s (in feet), t seconds after it is propelled. Find the maximum height reached by the water.

SOLUTION   The graph of $s = 128t - 16t^2$, which expresses the height s of the water t seconds after it is propelled, is the parabola shown in Figure 7-15. To find the maximum height reached by the water, we find the s-coordinate k of the vertex of the parabola. To find k, we write the equation of the parabola in standard form by completing the square on t.

©Littleny/Shutterstock.com

$$s = 128t - 16t^2$$

$$16t^2 - 128t = -s \qquad \text{Multiply both sides by } -1.$$

$$t^2 - 8t = -\frac{s}{16} \qquad \text{Divide both sides by 16.}$$

$$t^2 - 8t + 16 = -\frac{s}{16} + 16 \qquad \text{Add 16 to both sides to complete the square.}$$

$$(t - 4)^2 = -\frac{s - 256}{16} \qquad \text{Factor } t^2 - 8t + 16 \text{ and combine terms.}$$

$$(t - 4)^2 = -\frac{1}{16}(s - 256) \qquad \text{Factor out } \frac{1}{16}.$$

This equation indicates that the maximum height is 256 feet.

Caution
The parabola shown in Figure 7-15 is not the path of the water. The water goes straight up and straight down.

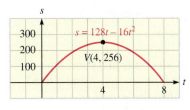

FIGURE 7-15

Self Check 9   At what time will the water strike the ground? (*Hint:* Find the t-intercept.)

Now Try Exercise 79.

Self Check Answers

1. $x^2 + y^2 + 4x - 2y - 11 = 0$ 2.

   3. no

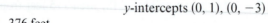

4. $y^2 = -12x$ 5. $(x - 4)^2 = 4(y - 5)$ 6. $(y - 4)^2 = -8(x - 2)$ or
   $(x - 2)^2 = y - 4$ 7. vertex $(-4, -1)$;
   y-intercepts $(0, 1)$, $(0, -3)$

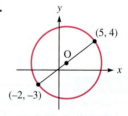

8. 376 feet
9. $t = 8$ seconds

Exercises 7.1

Getting Ready

You should be able to complete these vocabulary and concept statements before you proceed to the practice exercises.

Fill in the blanks.

1. $(x - 2)^2 + (y + 5)^2 = 9$: center (__, __); radius __

2. $x^2 + y^2 - 36 = 0$: center (__, __); radius __

3. $x^2 + y^2 = 5$: center (__, __); radius ____

4. $2(x - 9)^2 + 2y^2 = 7$: center (__, __); radius ____

Determine whether the graph of the parabola opens upward, downward, to the left, or to the right.

5. $y^2 = -4x$: opens ____

6. $y^2 = 10x$: opens ____

7. $x^2 = -8(y - 3)$: opens _____

8. $(x - 2)^2 = (y + 3)$: opens _____

Fill in the blanks.

9. A parabola is the set of all points in a plane equidistant from a line, called the _____, and a fixed point not on the line, called the _____.

10. The general form of a second-degree equation in the variables x and y is
   $Ax^2 +$ _____ $= 0$.

Identify the conic as a circle or parabola.

11. $x^2 - 5x + y^2 = 12$

12. $3x^2 + 3y^2 + 18x + 6y = 24$

13. $x^2 - 8y = 6x - 1$ 14. $2y^2 + 4y - 6x = 4$

Practice

Write an equation of each circle shown or described. Write your answer in standard form and then general form.

15.
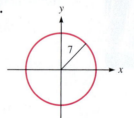

16.
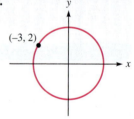

17.
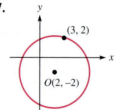

18.

19. Radius of 6; center at the intersection of $3x + y = 1$ and $-2x - 3y = 4$

20. Radius of 8; center at the intersection of $x + 2y = 8$ and $2x - 3y = -5$

Graph each circle.

21. $x^2 + y^2 = 4$

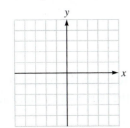

22. $x^2 - 2x + y^2 = 15$

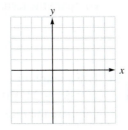

23. $3x^2 + 3y^2 - 12x - 6y = 12$

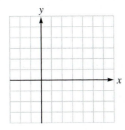

24. $2x^2 + 2y^2 + 4x - 8y + 2 = 0$

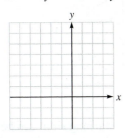

Find the vertex, focus, and directrix of each parabola.

25. $x^2 = 12y$

26. $y^2 = -12y$

27. $(y - 3)^2 = 20x$

28. $x^2 = -\dfrac{1}{2}(y + 5)$

29. $(x + 2)^2 = -24(y - 1)$

30. $(y + 1)^2 = 28(x - 2)$

Find an equation in standard form of each parabola described.

31. Vertex at $(0, 0)$; focus at $(0, 3)$

32. Vertex at $(0, 0)$; focus at $(0, -3)$

33. Vertex at $(0, 0)$; focus at $(-3, 0)$

34. Vertex at $(0, 0)$; focus at $(3, 0)$

35. Vertex at $(3, 5)$; focus at $(3, 2)$

36. Vertex at $(3, 5)$; focus at $(-3, 5)$

37. Vertex at $(3, 5)$; focus at $(3, -2)$

38. Vertex at $(3, 5)$; focus at $(6, 5)$

39. Vertex at $(0, 2)$; directrix at $y = 3$

40. Vertex at $(-3, 4)$; directrix at $y = 2$

41. Vertex at $(1, -5)$; directrix at $x = -1$

42. Vertex at $(3, 5)$; directrix at $x = 6$

43. Vertex at $(2, 2)$; passes through $(0, 0)$

44. Vertex at $(-2, -2)$; passes through $(0, 0)$

45. Vertex at $(-4, 6)$; passes through $(0, 3)$

46. Vertex at $(-2, 3)$; passes through $(0, -3)$

47. Vertex at $(6, 8)$; passes through $(5, 10)$ and $(5, 6)$

48. Vertex at $(2, 3)$; passes through $\left(1, \dfrac{13}{4}\right)$ and $\left(-1, \dfrac{21}{4}\right)$

49. Vertex at $(3, 1)$; passes through $(4, 3)$ and $(2, 3)$

50. Vertex at $(-4, -2)$; passes through $(-3, 0)$ and $\left(\dfrac{9}{4}, 3\right)$

Write each parabola in standard form and graph it.

51. $y = x^2 + 4x + 5$

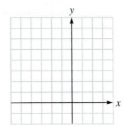

52. $2x^2 - 12x - 7y = 10$

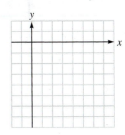

53. $y^2 + 4x - 6y = -1$

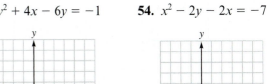

54. $x^2 - 2y - 2x = -7$

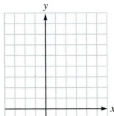

55. $y^2 - 4y = 4x - 8$

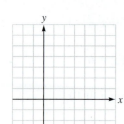

56. $y^2 + 2x - 2y = 5$

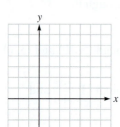

63. Use a graphing calculator to graph the parabola $y^2 = 4x - 12$. Sketch the parabola by hand and compare the results.

64. Use a graphing calculator to graph the parabola $y^2 + 8x - 24 = 0$. Sketch the parabola by hand and compare the results.

57. $y^2 - 4y = -8x + 20$

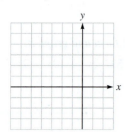

58. $y^2 - 2y = 9x + 17$

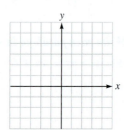

Applications

65. Broadcast range A television tower broadcasts a signal with a circular range, as shown in the illustration. Can a city 50 miles east and 70 miles north of the tower receive the signal?

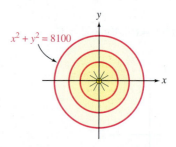

$x^2 + y^2 = 8100$

59. $x^2 - 6y + 22 = -4x$

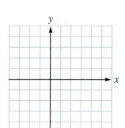

60. $4y^2 - 4y + 16x = 7$

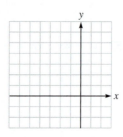

66. Warning sirens A tornado warning siren can be heard in the circular range shown in the illustration. Can a person 4 miles west and 5 miles south of the siren hear its sound?

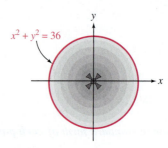

$x^2 + y^2 = 36$

61. $4x^2 - 4x + 32y = 47$

62. $4y^2 - 16x + 17 = 20y$

67. Radio translators Some radio stations extend their broadcast range by installing a translator—a remote device that receives the signal and retransmits it. A station with a broadcast range given by $x^2 + y^2 = 1600$, where x and y are in miles, installs a translator with a broadcast area bounded by $x^2 + y^2 - 70y + 600 = 0$. Find the greatest distance from the main transmitter that the signal can be received.

68. Ripples in a pond When a stone is thrown into the center of a pond, the ripples spread out in a circular pattern, moving at a rate of 3 feet per second. If the stone is dropped at the point (0, 0) in the illustration, when will the ripple reach the seagull floating at the point (15, 36)?

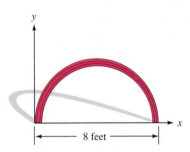

69. Writing equations of circles Find an equation in standard form of the circle whose outer rim is the circular arch shown in the illustration.

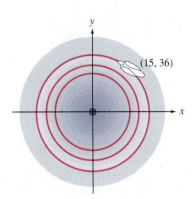

8 feet

70. Writing equations of circles The shape of the window shown is a combination of a rectangle and a semicircle. Find an equation in standard form of the circle of which the semicircle is a part.

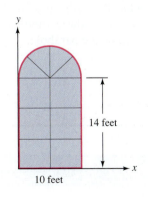

14 feet

10 feet

71. Meshing gears For design purposes, the large gear is described by the circle $x^2 + y^2 = 16$. The smaller gear is a circle centered at (7, 0) and tangent to the larger circle. Find an equation in standard form of the smaller gear.

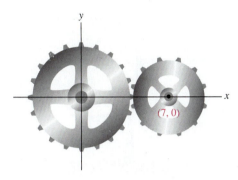

(7, 0)

72. Walkways The walkway shown is bounded by the two circles $x^2 + y^2 = 2500$ and $(x - 10)^2 + y^2 = 900$, measured in feet. Find the largest and the smallest width of the walkway.

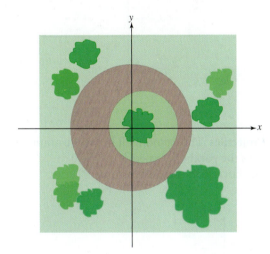

73. Solar furnaces A parabolic mirror collects rays of the sun and concentrates them at its focus. In the illustration, how far from the vertex of the parabolic mirror will it get the hottest? (All measurements are in feet.)

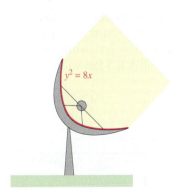

$y^2 = 8x$

74. Searchlight reflectors A parabolic mirror reflects light in a beam when the light source is placed at its focus. In the illustration, how far from the vertex of the parabolic reflector should the light source be placed? (All measurements are in feet.)

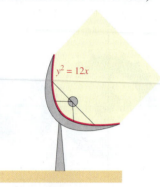

$y^2 = 12x$

75. Writing equations of parabolas Derive an equation of the parabolic arch shown.

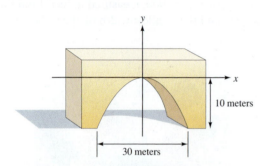

10 meters

30 meters

76. Projectiles The cannonball in the illustration follows the parabolic trajectory $y = 30x - x^2$. How far short of the castle does it land?

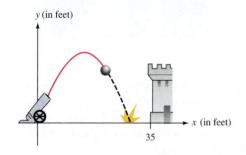

y (in feet)

x (in feet)

35

77. Satellite antennas The cross section of the satellite antenna in the illustration is a parabola given by the equation $y = \frac{1}{16}x^2$, with distances measured in feet. If the dish is 8 feet wide, how deep is it?

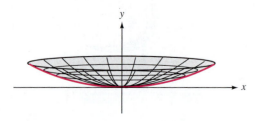

y

x

78. Design of a satellite antenna The cross section of the satellite antenna shown is a parabola with the pickup at its focus. Find the distance d from the pickup to the center of the dish.

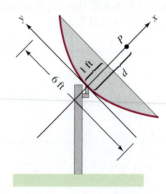

P

1 ft

6 ft

d

79. Toy rockets A toy rocket is s feet above the Earth at the end of t seconds, where $s = -16t^2 + 80\sqrt{3}t$. Find the maximum height of the rocket.

80. Operating a resort A resort owner plans to build and rent n cabins for d dollars per week. The price d that she can charge for each cabin depends on the number of cabins she builds, where $d = -45\left(\frac{n}{32} - \frac{1}{2}\right)$. Find the number of cabins she should build to maximize her weekly income.

81. Design of a parabolic reflector Find the outer diameter (the length $\overline{AB}$) of the parabolic reflector shown.

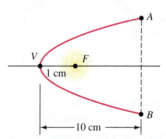

A

V F

1 cm

B

10 cm

82. Design of a suspension bridge The cable between the towers of the suspension bridge shown in the illustration has the shape of a parabola with vertex 15 feet above the roadway. Find an equation in standard form of the parabola.

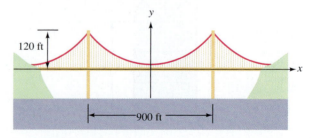

y

120 ft

900 ft

x

83. Gateway Arch The Gateway Arch in St. Louis has a shape that approximates a parabola. (See the illustration.) Find the width w of the arch 200 feet above the ground. Round to the nearest foot.

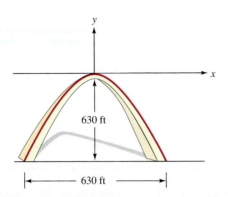

630 ft

630 ft

84. Building tunnels A construction firm plans to build a tunnel whose arch is in the shape of a parabola. (See the illustration.) The tunnel will span a two-lane highway 8 meters wide. To allow safe passage for vehicles, the tunnel must be 5 meters high at a distance of 1 meter from the tunnel's edge. Find the maximum height of the tunnel.

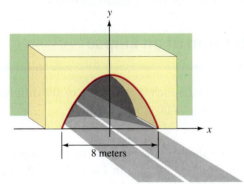

8 meters

Discovery and Writing

85. Describe a parabola.

86. How can you recognize the equation of a parabola when compared to other equations?

87. Show that the standard form of the equation of a parabola $(y - 2)^2 = 8(x - 1)$ is a special case of the general form of a second-degree equation in two variables.

88. Show that the standard form of the equation of a circle $(x + 2)^2 + (y - 5)^2 = 36$ is a special case of the general form of a second-degree equation in two variables.

Find an equation, in the form $(x - h)^2 + (y - k)^2 = r^2$, of the circle passing through the given points.

89. $(0, 8)$, $(5, 3)$, and $(4, 6)$

90. $(-2, 0)$, $(2, 8)$, and $(5, -1)$

Find an equation of the parabola passing through the given points. Give the equation in the form $y = ax^2 + bx + c$.

91. $(1, 8)$, $(-2, -1)$, and $(2, 15)$

92. $(1, -3)$, $(-2, 12)$, and $(-1, 3)$

93. Projectile motion A stone tossed upward is s feet above the Earth after t seconds, where $s = -16t^2 + 128t$. Show that the stone's height x seconds after it is thrown is equal to its height x seconds before it hits the ground.

94. Ballistics Show that the stone in Exercise 93 reaches its greatest height in one-half of the time it takes until it strikes the ground.

Critical Thinking

In Exercises 95–98, match the equation of the parabola with its graph.

95. $x^2 = 9y$

96. $x^2 = -9y$

97. $y^2 = 9x$

98. $y^2 = -9x$

a.

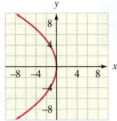

b.

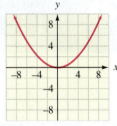

c.

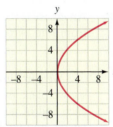

d.

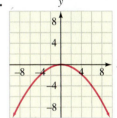

In Exercises 99–102, match the equation of the parabola with its graph.

99. $(y + 2)^2 = 8(x - 2)$

100. $(y + 2)^2 = -8(x - 2)$

101. $(x - 2)^2 = 8(y + 2)$

102. $(x - 2)^2 = -8(y + 2)$

a.

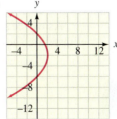

b.

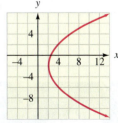

c.

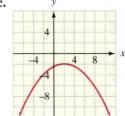

d.

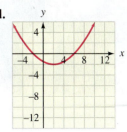

7.2 The Ellipse

In this section, we will learn to

1. Understand the definition of an ellipse.
2. Write an equation of an ellipse.
3. Graph ellipses.
4. Solve applied problems using ellipses.

Washington, D.C., has many historical sites. Many people touring the United States Capitol visit the National Statuary Hall, originally the chamber of the House of Representatives. It is said that because of the oval shape of the ceiling, politicians on one side of the hall could eavesdrop on politicians talking on the opposite side of the hall. Such rooms are often called *whispering galleries*. Problem 54 in the Exercises discusses some properties of whispering galleries.

Today, Statuary Hall houses a collection of statues donated by individual states. Sam Houston, Daniel Webster, and Will Rogers are among those honored with a statue.

A third important conic is an oval-shaped curve, called an *ellipse*.

1. Understand the Definition of an Ellipse

Ellipse   An **ellipse** is the set of all points P in a plane such that the sum of the distances from P to two other fixed points F and F' is a positive constant.

We can illustrate this definition by showing how to construct an ellipse. To do so, we place two thumbtacks fairly close together, as in Figure 7-16. We then tie each end of a piece of string to a thumbtack, catch the loop with the point of a pencil, and (while keeping the string stretched tightly) draw the ellipse. Note that $d(F_1P) + d(PF_2)$ will be a positive constant.

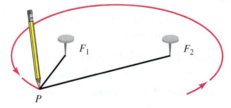

FIGURE 7-16

As we see in the figure, the graph of an ellipse is egg-shaped. The graphs of two ellipses are shown in Figure 7-17.

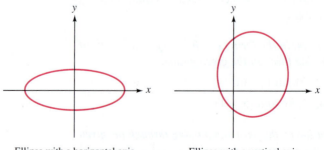

Ellipse with a horizontal axis   Ellipse with a vertical axis

(a)   (b)

FIGURE 7-17

Ellipses occur in many real-life settings.

- Rotating an ellipse about its axis produces a football.

- Planets such as Mars revolve around the sun in an elliptical path.

- A workout machine, called an *elliptic trainer*, uses elliptical movements for exercise without causing damage to the joints.

- An ellipse has an interesting property. Any light or sound that starts at one focus will be reflected through the other. This property is the basis of a medical procedure for treating kidney stones, called **lithotripsy**. The patient is placed in an elliptical tank of water with the kidney stone at one focus. Shock waves from a controlled explosion at the other focus are concentrated on the stone, pulverizing it. See Figure 7-18. Though some machines are still in use today, most are being replaced and patients lie down on a soft water-filled cushion.

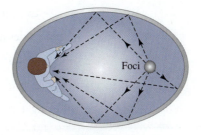

FIGURE 7-18

2. Write an Equation of an Ellipse

In the ellipse shown in Figure 7-19(a), the fixed points F and F' are called **foci** (each is a **focus**), the midpoint of the chord FF' is called the **center**, and the chord VV' is called the **major axis**. Each of the endpoints V and V' of the major axis is called a **vertex**. The chord BB', perpendicular to the major axis and passing through the center C of the ellipse, is called the **minor axis**.

Deriving the Standard Form for an Equation of an Ellipse

To derive an equation of the ellipse shown in Figure 7-19(b), we note that point O is the midpoint of chord FF' and let $d(OF) = d(OF') = c$, where $c > 0$. Then the coordinates of point F are $(c, 0)$, and the coordinates of F' are $(-c, 0)$. We also let $P(x, y)$ be any point on the ellipse.

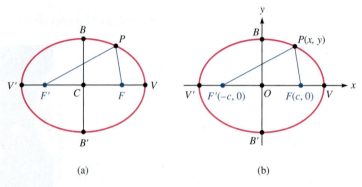

(a)   (b)

FIGURE 7-19

By the definition of an ellipse, $d(F'P) + d(PF)$ must be a positive constant, which we will call $2a$. Thus,

(1) $d(F'P) + d(PF) = 2a$

We can use the distance formula to compute the lengths of $F'P$ and PF,

$$d(F'P) = \sqrt{[x - (-c)]^2 + y^2} \quad \text{and} \quad d(PF) = \sqrt{(x - c)^2 + y^2}$$

and substitute these values into Equation 1 to obtain

$$\sqrt{[x - (-c)]^2 + y^2} + \sqrt{(x - c)^2 + y^2} = 2a$$

or

(2) $\sqrt{[x + c)]^2 + y^2} = 2a - \sqrt{(x - c)^2 + y^2}$ Subtract $\sqrt{(x - c)^2 + y^2}$ from both sides.

We can square both sides of Equation 2 and simplify to get

$$(x + c)^2 + y^2 = 4a^2 - 4a\sqrt{(x - c)^2 + y^2} + [(x - c)^2 + y^2]$$

$$x^2 + 2cx + c^2 + y^2 = 4a^2 - 4a\sqrt{(x - c)^2 + y^2} + x^2 - 2cx + c^2 + y^2$$

$$4cx = 4a^2 - 4a\sqrt{(x - c)^2 + y^2}$$

$$cx = a^2 - a\sqrt{(x - c)^2 + y^2}$$

$$cx - a^2 = -a\sqrt{(x - c)^2 + y^2}$$

We square both sides again and simplify to get

$$c^2x^2 - 2a^2cx + a^4 = a^2[(x - c)^2 + y^2]$$

$$c^2x^2 - 2a^2cx + a^4 = a^2(x^2 - 2cx + c^2 + y^2)$$

$$c^2x^2 - 2a^2cx + a^4 = a^2x^2 - 2a^2cx + a^2c^2 + a^2y^2$$

$$c^2x^2 + a^4 = a^2x^2 + a^2c^2 + a^2y^2$$

$$a^4 - a^2c^2 = a^2x^2 - c^2x^2 + a^2y^2$$

(3) $\qquad a^2(a^2 - c^2) = (a^2 - c^2)x^2 + a^2y^2$

Because the shortest distance between two points is a line segment, $d(F'P) + d(PF) > d(F'F)$. Therefore, $2a > 2c$. Thus, $a > c$, and $a^2 - c^2$ is a positive number, which we will call b^2. Letting $b^2 = a^2 - c^2$ and substituting into Equation 3, we have

(4) $a^2b^2 = b^2x^2 + a^2y^2$

Dividing both sides of Equation 4 by a^2b^2 gives the equation

$$\frac{x^2}{a^2} + \frac{y^2}{b^2} = 1 \qquad \text{where } a > b > 0$$

This is the standard form of an equation of an ellipse centered at the origin and major axis on the x-axis.

To find the coordinates of the vertices V and V', we substitute 0 for y and solve for x:

$$\frac{x^2}{a^2} + \frac{y^2}{b^2} = 1$$

$$\frac{x^2}{a^2} + \frac{0^2}{b^2} = 1$$

$$\frac{x^2}{a^2} = 1$$

$$x^2 = a^2$$

$$x = a \quad \text{or} \quad x = -a$$

Since the coordinates of V are $(a, 0)$ and the coordinates of V' are $(-a, 0)$, a is the distance between the center of the ellipse and either of its vertices. Thus, the center of the ellipse is the midpoint of the major axis.

To find the coordinates of B and B', we substitute 0 for x and solve for y:

$$\frac{x^2}{a^2} + \frac{y^2}{b^2} = 1$$

$$\frac{0^2}{a^2} + \frac{y^2}{b^2} = 1$$

$$y^2 = b^2$$

$$y = b \quad \text{or} \quad y = -b$$

Since the coordinates of B are $(0, b)$ and the coordinates of B' are $(0, -b)$, the distance between the center of the ellipse and either endpoint of the minor axis is b. We have the following results.

The Ellipse: Major Axis on x-Axis, Center at $(0, 0)$

The standard form of an equation of an ellipse with center at the origin and the major axis (horizontal) on the x-axis is

$$\frac{x^2}{a^2} + \frac{y^2}{b^2} = 1 \text{ where } \quad a > b > 0$$

Vertices (ends of the major axis): $V(a, 0)$ and $V'(-a, 0)$

Length of major axis: $2a$

Ends of the minor axis: $B(0, b)$ and $B'(0, -b)$

Length of minor axis: $2b$

Foci: $F(c, 0)$ and $F'(-c, 0)$ where $c^2 = a^2 - b^2$

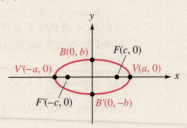

A similar equation results if the ellipse has a major axis on the y-axis and center at the origin.

The Ellipse: Major Axis on y-Axis, Center at $(0, 0)$

The standard form of an equation of an ellipse with center at the origin and the major axis (vertical) on the y-axis is

$$\frac{x^2}{b^2} + \frac{y^2}{a^2} = 1 \text{ where } a > b > 0$$

Vertices (ends of the major axis): $V(0, a)$ and $V'(0, -a)$

Length of major axis: $2a$

Ends of the minor axis: $B(b, 0)$ and $B'(-b, 0)$

Length of minor axis: $2b$

Foci: $F(0, c)$ and $F'(0, -c)$ where $c^2 = a^2 - b^2$

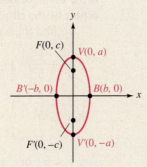

Writing an Equation of an Ellipse in Standard Form

EXAMPLE 1

Find an equation of the ellipse in standard form with center at the origin, major axis of length 6 units located on the x-axis, and minor axis of length 4 units.

SOLUTION

To find an equation, we determine a and b and substitute into the standard equation of an ellipse with center at the origin and major axis on the x-axis:

$$\frac{x^2}{a^2} + \frac{y^2}{b^2} = 1$$

The center of the ellipse is given to be the origin and the length of the major axis is 6. Since the length of the major axis of an ellipse centered at the origin is $2a$, we have $2a = 6$ or $a = 3$. The coordinates of the vertices are $(3, 0)$ and $(-3, 0)$, as shown in Figure 7-20.

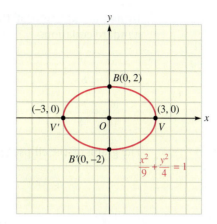

FIGURE 7-20

The length of the minor axis is given to be 4. Since the length of the minor axis of an ellipse centered at the origin is $2b$, we have $2b = 4$ or $b = 2$. The coordinates of B and B' are $(0, 2)$ and $(0, -2)$.

To find an equation of the ellipse, we substitute 3 for a and 2 for b in the standard equation of an ellipse with center at the origin and major axis on the x-axis.

$$\frac{x^2}{a^2} + \frac{y^2}{b^2} = 1$$

$$\frac{x^2}{3^2} + \frac{y^2}{2^2} = 1$$

$$\frac{x^2}{9} + \frac{y^2}{4} = 1$$

Self Check 1   Find an equation in standard form of the ellipse with center at the origin, major axis of length 10 on the y-axis, and minor axis of length 8.

Now Try Exercise 15.

EXAMPLE 2   **Writing an Equation of an Ellipse in Standard Form**

Find an equation in standard form of the ellipse with center at the origin, focus at $\left(2\sqrt{3}, 0\right)$, and vertex at $(4, 0)$.

SOLUTION   Because the vertex and focus are on the x-axis, the major axis of the ellipse is on the x-axis. (See Figure 7-21.) To find an equation of the ellipse, we will determine a^2 and b^2 and substitute into the standard equation

$$\frac{x^2}{a^2} + \frac{y^2}{b^2} = 1$$

The distance between the center of the ellipse and the vertex is $a = 4$. The distance between the focus and the center is $c = 2\sqrt{3}$.

Since $b^2 = a^2 - c^2$, we can substitute 4 for a and $2\sqrt{3}$ for c and solve for b^2:

$$b^2 = a^2 - c^2$$

$$b^2 = 4^2 - \left(2\sqrt{3}\right)^2$$

$$= 16 - 12$$

$$= 4$$

To find an equation of the ellipse, we substitute 16 for a^2, and 4 for b^2 in the standard equation.

$$\frac{x^2}{a^2} + \frac{y^2}{b^2} = 1$$

$$\frac{x^2}{16} + \frac{y^2}{4} = 1$$

The graph of the ellipse $\frac{x^2}{16} + \frac{y^2}{4} = 1$ is shown in the figure.

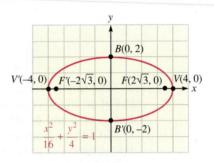

FIGURE 7-21

Self Check 2 Find an equation of the ellipse with center at the origin, focus at $\left(0, 2\sqrt{3}\right)$, and vertex at $(0, 4)$.

Now Try Exercise 17.

To translate an ellipse to a new position centered at the point (h, k) instead of the origin, we replace x and y in the standard equations with $x - h$ and $y - k$, respectively, to get the following results.

The Ellipse: Major Axis Horizontal, Center at (h, k)

The standard form of an equation of an ellipse with center at (h, k) and major axis horizontal is

$$\frac{(x - h)^2}{a^2} + \frac{(y - k)^2}{b^2} = 1 \quad \text{where } a > b > 0$$

Vertices (ends of the major axis): $V(a + h, k)$ and $V'(-a + h, k)$

Ends of the minor axis: $B(h, b + k)$ and $B'(h, -b + k)$

Foci: $F(h + c, k)$ and $F'(h - c, k)$ where $c^2 = a^2 - b^2$

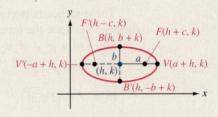

There is a similar result when the major axis is vertical.

The Ellipse: Major Axis Vertical, Center at (h, k)

The standard form of an equation of an ellipse with center at (h, k) and major axis vertical is

$$\frac{(x - h)^2}{b^2} + \frac{(y - k)^2}{a^2} = 1 \quad \text{where } a > b > 0$$

Vertices (ends of the major axis): $V(h, a + k)$ and $V'(h, -a + k)$

Ends of the minor axis: $B(b + h, k)$ and $B'(-b + h, k)$

Foci: $F(h, k + c)$ and $F'(h, k - c)$ where $c^2 = a^2 - b^2$

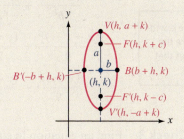

In each case, the length of the major axis is $2a$, and the length of the minor axis is $2b$.

Writing an Equation of an Ellipse in Standard Form

EXAMPLE 3

Find an equation in standard form of the ellipse with focus at $(-1, 7)$ and vertices at $V(-1, 8)$ and $V'(-1, -2)$.

SOLUTION

We will use the coordinates of the focus and the vertices to determine the center of the ellipse and a and b. We will then substitute these coordinates into the appropriate standard equation of an ellipse. Because the major axis passes through points V and V', we see that the major axis is a vertical line that is parallel to the y-axis. So the standard equation to use is

$$\frac{(x - h)^2}{b^2} + \frac{(y - k)^2}{a^2} = 1 \quad \text{where } a > b > 0$$

Since the midpoint of the major axis is the center of the ellipse, the coordinates of the center are $(-1, 3)$, as in Figure 7-22.

From Figure 7-22, we see that the distance between the center of the ellipse and either vertex is $a = 5$. We also see that the distance between the focus and the center is $c = 4$.

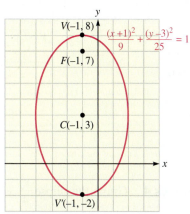

FIGURE 7-22

Since $b^2 = a^2 - c^2$ is an ellipse, we can substitute 5 for a and 4 for c and solve for b^2:

$$b^2 = a^2 - c^2$$
$$= 5^2 - 4^2$$
$$= 9$$

To find an equation of the ellipse, we substitute -1 for h, 3 for k, 25 for a^2, and 9 for b^2 in the standard equation and simplify:

$$\frac{(x - h)^2}{b^2} + \frac{(y - k)^2}{a^2} = 1$$

$$\frac{[x - (-1)]^2}{9} + \frac{(y - 3)^2}{25} = 1$$

$$\frac{(x + 1)^2}{9} + \frac{(y - 3)^2}{25} = 1$$

> **Tip**
>
> When the equation of an ellipse is written in standard form, h and k are the numbers within the parentheses that follow the subtraction symbols.

Self Check 3   Find an equation in standard form of the ellipse with focus at $(3, 1)$ and vertices at $V(5, 1)$ and $V'(-5, 1)$.

Now Try Exercise 29.

3. Graph Ellipses

Graphing an Ellipse

EXAMPLE 4   Graph the ellipse $\dfrac{x^2}{49} + \dfrac{y^2}{4} = 1$.

SOLUTION   To graph the ellipse, we will first determine the coordinates of the vertices and endpoints of the minor axis. We will then plot these points and draw an ellipse through them.

Because the equation is the standard form of an ellipse centered at the origin with major axis on the x-axis, we know that the center of the ellipse is at $(0, 0)$ and that the vertices lie on the x-axis. Because $a = 7$, the vertices are 7 units to the right and left of the origin at points $(7, 0)$ and $(-7, 0)$. Because $b^2 = 4$, $b = 2$, and the endpoints of the minor axis are 2 units above and below the origin at points $(0, 2)$ and $(0, -2)$. Using these points, we can sketch the ellipse shown in Figure 7-23.

> **Take Note**
>
> In Example 4, note that $49 > 4$. Because 49 is the denominator of the x^2 term, the axis of the ellipse is horizontal and is the x-axis.

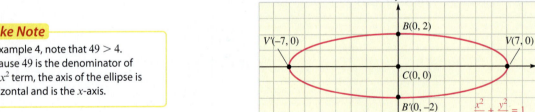

FIGURE 7-23

Self Check 4   Graph the ellipse $\dfrac{x^2}{4} + \dfrac{y^2}{9} = 1$.

Now Try Exercise 35.

EXAMPLE 5   **Graphing an Ellipse**

Graph the ellipse $\dfrac{(x+2)^2}{4} + \dfrac{(y-2)^2}{9} = 1$.

SOLUTION   The equation is the standard form of the equation of an ellipse whose major axis is vertical and whose center is at $(-2, 2)$. We will use translations to graph the ellipse.

Because $a^2 = 9$, we have $a = 3$ and the vertices are 3 units above and 3 units below the center at points $(-2, 5)$ and $(-2, -1)$. Because $b^2 = 4$, we have $b = 2$ and the endpoints of the minor axis are 2 units to the right and 2 units to the left of the center at points $(0, 2)$ and $(-4, 2)$. Using these points, we can sketch the ellipse as shown in Figure 7-24.

> **Take Note**
>
> In Example 5, note that $9 > 4$. Because 9 is the denominator of the y^2 term, the axis of the ellipse is vertical.

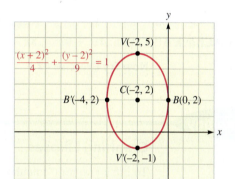

FIGURE 7-24

Self Check 5   Graph the ellipse $\dfrac{(x-2)^2}{9} + \dfrac{(y+1)^2}{25} = 1$.

Now Try Exercise 37.

EXAMPLE 6   **Graphing an Ellipse**

Graph: $4x^2 + 9y^2 - 16x - 18y = 11$.

SOLUTION   We write the equation in standard form by completing the square on x and y and then use translations to graph the ellipse.

$$4x^2 + 9y^2 - 16x - 18y = 11$$

$$4x^2 - 16x + 9y^2 - 18y = 11$$

$$4(x^2 - 4x) + 9(y^2 - 2y) = 11$$

$$4(x^2 - 4x + 4) + 9(y^2 - 2y + 1) = 11 + 16 + 9$$

$$4(x-2)^2 + 9(y-1)^2 = 36$$

$$\frac{(x-2)^2}{9} + \frac{(y-1)^2}{4} = 1$$

We can now see that the graph is an ellipse with center at $(2, 1)$ and major axis parallel to the x-axis. Because $a^2 = 9$, we have $a = 3$ and the vertices are 3 units to the left and right of the center at $(-1, 1)$ and $(5, 1)$. Because $b^2 = 4$, we have $b = 2$ and the endpoints of the minor axis are 2 units above and below the center at $(2, -1)$ and $(2, 3)$. Using these points, we can sketch the ellipse as shown in Figure 7-25.

> **Caution**
>
> Circles are ellipses, but only some ellipses are circles. If the coefficients of x^2 and y^2 in an equation of an ellipse are equal, the ellipse is a circle. If the coefficients of x^2 and y^2 are not equal, the ellipse is not a circle.

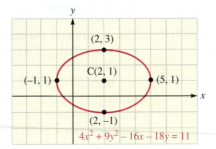

FIGURE 7-25

Self Check 6 Graph: $4x^2 + 9y^2 - 8x + 36y = -4$.

Now Try Exercise 47.

Accent on Technology

Graphing Ellipses

To use a graphing calculator to graph $\dfrac{(x + 2)^2}{4} + \dfrac{(y - 2)^2}{25} = 1$, we must solve for y in terms of x.

$$25(x + 2)^2 + 4(y - 1)^2 = 100 \qquad \text{Multiply both sides by 100.}$$

$$4(y - 1)^2 = 100 - 25(x + 2)^2 \qquad \text{Subtract } 25(x + 2)^2 \text{ from both sides.}$$

$$(y - 1)^2 = \frac{100 - 25(x + 2)^2}{4} \qquad \text{Divide both sides by 4.}$$

$$y - 1 = \pm\frac{\sqrt{100 - 25(x + 2)^2}}{2} \qquad \text{Take the square root of both sides.}$$

$$y = 1 \pm \frac{\sqrt{100 - 25(x + 2)^2}}{2} \qquad \text{Add 1 to both sides.}$$

$$y = 1 + \frac{\sqrt{100 - 25(x + 2)^2}}{2} \quad \text{or} \quad y = 1 - \frac{\sqrt{100 - 25(x + 2)^2}}{2}$$

The equation is then entered into the calculator as shown in Figure 7-26.

Enter the equations.   **Set a** WINDOW.   GRAPH **the equations.**

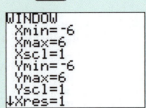 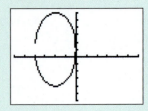

FIGURE 7-26

The eccentricity of an ellipse provides a measure of how much the curve resembles a true circle. Specifically, the eccentricity of a true circle equals 0.

4. Solve Applied Problems Using Ellipses

Solving an Application Problem

EXAMPLE 7

The orbit of the Earth is approximately an ellipse, with the sun at one focus. The ratio of c to a (called the **eccentricity** of the ellipse) is about $\frac{1}{62}$, and the length of the major axis is approximately 186,000,000 miles. How close does the Earth get to the sun?

SOLUTION   We will assume that the ellipse has its center at the origin and vertices V' and V at $(-93{,}000{,}000, 0)$ and $(93{,}000{,}000, 0)$, as shown in Figure 7-27.

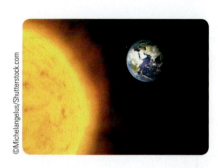

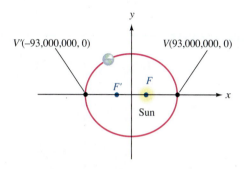

FIGURE 7-27

Because the eccentricity $\frac{c}{a}$ is given to be $\frac{1}{62}$ and $a = 93{,}000{,}000$, we have

$$\frac{c}{a} = \frac{1}{62}$$

$$c = \frac{1}{62}a$$

$$c = \frac{1}{62}(\mathbf{93{,}000{,}000})$$

$$= 1{,}500{,}000$$

The distance $d(FV)$ is the shortest distance between the Earth and the sun. (You'll be asked to prove this in Exercise 65.) Thus,

$$d(FV) = a - c = 93{,}000{,}000 - 1{,}500{,}000 = 91{,}500{,}000 \text{ mi}$$

The Earth's point of closest approach to the sun (called the **perihelion**) is approximately 91.5 million miles.

Self Check 7   Find the eccentricity of the ellipse $\dfrac{(x + 2)^2}{9} + \dfrac{(y - 5)^2}{25} = 1$.

Now Try Exercise 57.

The eccentricity of an ellipse provides a measure of how much the curve resembles a true circle. Specifically, the eccentricity of a true circle equals 0. We can use the eccentricity of an ellipse to judge its shape. If the eccentricity is close to 1, the ellipse is relatively flat, as in the ellipse shown.

If the eccentricity is close to 0, the ellipse is more circular, as shown.

Since the eccentricity of the Earth's orbit is $\frac{1}{62}$, the Earth's orbit is almost a circle.

Self Check Answers

1. $\dfrac{x^2}{16} + \dfrac{y^2}{25} = 1$ 2. $\dfrac{x^2}{4} + \dfrac{y^2}{16} = 1$ 3. $\dfrac{x^2}{25} + \dfrac{(y-1)^2}{16} = 1$

4.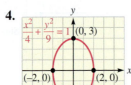

5.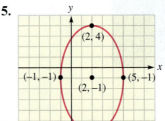

$$\dfrac{(x-2)^2}{9} + \dfrac{(y+1)^2}{25} = 1$$

6.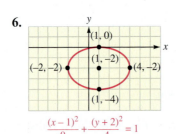

$$\dfrac{(x-1)^2}{9} + \dfrac{(y+2)^2}{4} = 1$$

7. $\dfrac{4}{5}$

Exercises 7.2

Getting Ready
You should be able to complete these vocabulary and concept statements before you proceed to the practice exercises.

Fill in the blanks.

1. An ellipse is the set of all points in the plane such that the _____ of the distances from two fixed points is a positive _____.

2. Each of the two fixed points in the definition of an ellipse is called a _____ of the ellipse.

3. The chord that joins the _____ is called the major axis of the ellipse.

4. The chord through the center of an ellipse and perpendicular to the major axis is called the _____ axis.

5. In the ellipse $\frac{x^2}{a^2} + \frac{y^2}{b^2} = 1$ $(a > b > 0)$, the vertices are $V(\_\_, \_\_)$ and $V'(\_\_, \_\_)$.

6. In an ellipse, the relationship between a, b, and c is _____.

7. To draw an ellipse that is 26 inches wide and 10 inches tall, how long should the piece of string be, and how far apart should the two thumbtacks be?

8. To draw an ellipse that is 20 centimeters wide and 12 centimeters tall, how long should the piece of string be, and how far apart should the two thumbtacks be?

Identify the conic as a circle, parabola, or an ellipse.

9. $x^2 - 6x + y^2 = 7$

10. $5x^2 + 5y^2 + 10y - \dfrac{24}{5} = 0$

11. $x^2 - 4y + 5 = -2x$

12. $y^2 - 8y = -2x - 20$

13. $7x^2 + 5y^2 - 35 = 0$

14. $5x^2 + 2y^2 - 10x + 4y = 13$

Practice

Write an equation in standard form of the ellipse described. The center of each ellipse is the origin.

15. Major axis of length 8 units located on the x-axis and minor axis of length 6 units

16. Major axis of length 14 units located on the y-axis and minor axis of length 10 units

17. Focus at $(3, 0)$; vertex at $(5, 0)$

18. Focus at $(0, 4)$; vertex at $(0, 7)$

19. Focus at $(0, 1)$; $\dfrac{4}{3}$ is one-half the length of the minor axis

20. Focus at $(1, 0)$; $\dfrac{4}{3}$ is one-half the length of the minor axis

21. Focus at $(0, 3)$; major axis equal to 8

22. Focus at $(5, 0)$; major axis equal to 12

Write an equation in standard form of each ellipse described.

23. Center at $(3, 4)$; $a = 3$, $b = 2$; major axis parallel to the y-axis

24. Center at $(3, 4)$; passes through $(3, 10)$ and $(3, -2)$; $b = 2$

25. Center at $(3, 4)$; $a = 3$, $b = 2$; major axis parallel to the x-axis

26. Center at $(3, 4)$; passes through $(8, 4)$ and $(-2, 4)$; $b = 2$

27. Foci at $(-2, 4)$ and $(8, 4)$; $b = 4$

28. Foci at $(8, 5)$ and $(4, 5)$; $b = 3$

29. Vertex at $(6, 4)$; foci at $(-4, 4)$ and $(4, 4)$

30. Center at $(-4, 5)$; $\dfrac{c}{a} = \dfrac{1}{3}$; vertex at $(-4, -1)$

31. Foci at $(6, 0)$ and $(-6, 0)$; $\dfrac{c}{a} = \dfrac{3}{5}$

32. Vertices at $(2, 0)$ and $(-2, 0)$; $\dfrac{2b^2}{a} = 2$

Graph each ellipse.

33. $\dfrac{x^2}{25} + \dfrac{y^2}{9} = 1$

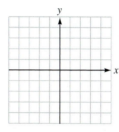

34. $\dfrac{x^2}{36} + \dfrac{y^2}{25} = 1$

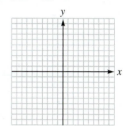

35. $\dfrac{x^2}{25} + \dfrac{y^2}{49} = 1$

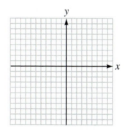

36. $4x^2 + y^2 = 4$

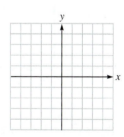

37. $\dfrac{x^2}{16} + \dfrac{(y + 2)^2}{36} = 1$

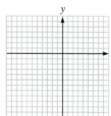

38. $(x - 1)^2 + \dfrac{4y^2}{25} = 4$

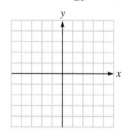

39. $\dfrac{(x - 4)^2}{49} + \dfrac{(y - 2)^2}{9} = 1$

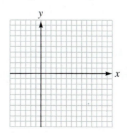

40. $\dfrac{(x - 1)^2}{25} + \dfrac{y^2}{4} = 1$

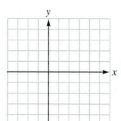

Write each ellipse in standard form.

41. $4x^2 + y^2 - 2y = 15$

42. $4x^2 + 25y^2 + 8x - 96 = 0$

43. $9x^2 + 4y^2 + 18x + 16y - 11 = 0$

44. $x^2 + 4y^2 - 10x - 8y = -13$

Graph each ellipse.

45. $x^2 + 4y^2 - 4x + 8y + 4 = 0$

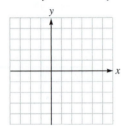

46. $x^2 + 4y^2 - 2x - 16y = -13$

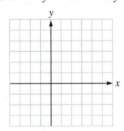

47. $16x^2 + 25y^2 - 160x - 200y + 400 = 0$

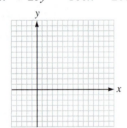

48. $3x^2 + 2y^2 + 7x - 6y = -1$

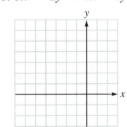

49. Use a graphing calculator to graph the ellipse $\frac{x^2}{4} + \frac{y^2}{36} = 1$. Sketch the ellipse by hand and compare the results.

50. Use a graphing calculator to graph the ellipse $\frac{(x+3)^2}{4} + \frac{(y-2)^2}{25} = 1$. Sketch the ellipse by hand and compare the results.

Applications

51. Pool tables Find an equation in standard form of the outer edge of the elliptical pool table shown below.

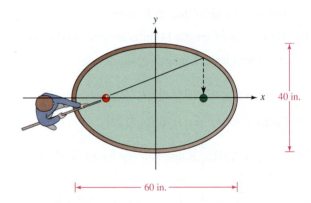

52. Equation of an arch An arch is a semiellipse 12 meters wide and 5 meters high. Write an equation in standard form of the ellipse if the ellipse is centered at the origin.

53. Design of a track A track is built in the shape of an ellipse with a maximum length of 100 meters and a maximum width of 60 meters. Write an equation in standard form of the ellipse and find its **focal width**. That is, find the length of a chord that is perpendicular to the major axis and passes through either focus of the ellipse.

54. Whispering galleries Any sound from one focus of an ellipse reflects off the ellipse directly back to the other focus. This property explains whispering galleries such as Statuary Hall in Washington, D.C. The ceiling of the whispering gallery shown has the shape of a semiellipse. Find the distance sound travels as it leaves focus F and returns to focus F'.

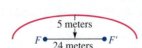

55. Finding the width of a mirror Many mirrors are oval shaped. The dimensions of a mirror are shown, and the mirror is in the shape of an ellipse. Find the width of the mirror 12 inches above its base.

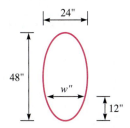

56. Finding the height of a window The window shown has the shape of an ellipse. Find the height of the window 20 inches from one end.

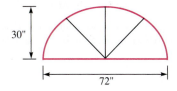

57. Astronomy The moon has an orbit that is an ellipse, with the Earth at one focus. If the major axis of the orbit is 378,000 miles and the ratio of c to a is approximately $\frac{11}{200}$, how far does the moon get from the Earth? (This farthest point in an orbit is called the **apogee**.)

58. Area of an ellipse The area A of the ellipse

$$\frac{x^2}{a^2} + \frac{y^2}{b^2} = 1$$

is given by $A = \pi ab$. Find the area of the ellipse $9x^2 + 16y^2 = 144$.

Discovery and Writing

59. Describe an ellipse.

60. Explain the difference between the foci and vertices of an ellipse.

61. How do you distinguish among the equations of circles, parabolas, and ellipses?

62. If F is a focus of the ellipse shown and B is an endpoint of the minor axis, use the distance formula to prove that the length of segment FB is a. (*Hint:* In an ellipse, $b^2 = a^2 - c^2$.)

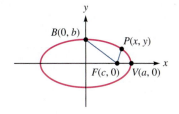

63. If F is a focus of the ellipse shown and P is any point on the ellipse, use the distance formula to show that the length of FP is $a - \frac{c}{a}x$. (*Hint:* In an ellipse, $c^2 = a^2 - b^2$.)

64. Finding the focal width In the ellipse shown, chord AA' passes through the focus F and is perpendicular to the major axis. Show that the length of AA' (called the **focal width**) is $\frac{2b^2}{a}$.

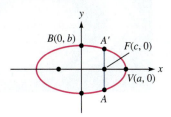

65. Prove that segment FV in Example 7 is the shortest distance between the Earth and the sun. (*Hint:* Refer to Exercise 63.)

66. Constructing an ellipse The ends of a piece of string 6 meters long are attached to two thumbtacks that are 2 meters apart. A pencil catches the loop and draws it tight. As the pencil is moved about the thumbtacks (always keeping the tension), an ellipse is produced, with the thumbtacks as foci. Write an equation in standard form of the ellipse. (*Hint:* You'll have to establish a coordinate system.)

67. The distance between point $P(x, y)$ and the point $(0, 2)$ is $\frac{1}{3}$ of the distance of point P from the line $y = 18$. Find an equation in standard form of the ellipse on which point P lies.

68. Prove that $a > b$ in the development of the standard equation of an ellipse.

69. Show that the expansion of the standard equation of an ellipse is a special case of the general second-degree equation in two variables.

70. The eccentricity of an ellipse provides a measure of how much the curve resembles a true circle. Specifically, the eccentricity of a true circle equals 0. Note that the semimajor axis is perpendicular to the axis containing the foci of an ellipse. When analyzing planetary orbits, astronomers plot the relationship between the length of the orbit's semimajor axis (measured in Astronomical Units, where 1 AU = 149,598,000 km) and the eccentricity of the orbit. Use the given data plot to estimate how many of the 75 planets shown follow orbits that are true circles.

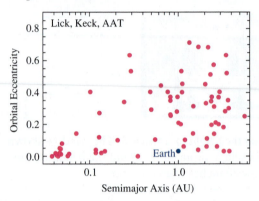

Source for right panel: Eccentricity vs. semimajor axis for extrasolar planets. The 75 planets shown were found in a Doppler survey of 1300 FGKM main sequence stars using the Lick, Keck, and AAT telescopes. The survey was carried out by the California-Carnegie planet search team. http://exoplanets.org/newsframe.html

Critical Thinking

In Exercise 71–74, match the equation of the ellipse with its graph.

71. $\dfrac{x^2}{64} + \dfrac{y^2}{9} = 1$

72. $\dfrac{x^2}{9} + \dfrac{y^2}{64} = 1$

73. $\dfrac{4x^2}{49} + \dfrac{y^2}{81} = 1$

74. $\dfrac{x^2}{81} + \dfrac{4y^2}{49} = 1$

In Exercise 75–78, match the equation of the ellipse with its graph.

75. $\dfrac{(x-2)^2}{16} + \dfrac{(y+2)^2}{49} = 1$

76. $\dfrac{(x+2)^2}{16} + \dfrac{(y-2)^2}{49} = 1$

77. $\dfrac{(x-2)^2}{49} + \dfrac{(y+2)^2}{16} = 1$

78. $\dfrac{(x+2)^2}{49} + \dfrac{(y-2)^2}{16} = 1$

a.

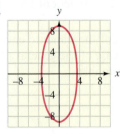

b.

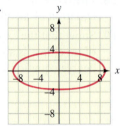

a.

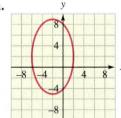

b.

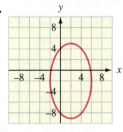

c.

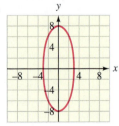

d.

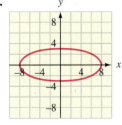

c.

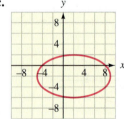

d.

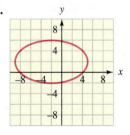

7.3 The Hyperbola

In this section, we will learn to

1. Write equations of hyperbolas.
2. Graph hyperbolas.

When a military plane exceeds the speed of sound, it is said to break the sound barrier. When this happens, a visible vapor cloud appears and an explosion-like sound results that we call a *sonic boom*.

As the plane moves faster beyond the speed of sound, a cone shape intersects the ground forming one branch of a curve called a *hyperbola*, a fourth type of conic section. The sonic boom is heard at points inside this hyperbola.

Hyperbolas appear in many real-life applications.

- Hyperbolas are the basis of a navigational system known as LORAN (LOng RAnge Navigation). It is a radio navigation system that uses time intervals between radio signals to determine the location of a ship or aircraft.
- Hyperbolas are the basis for the design of hypoid and spiral bevel gears that are used in motor vehicles.
- Hyperbolas describe the orbits of some comets.

The definition of a hyperbola is similar to the definition of an ellipse, except that we require a constant difference of $2a$ instead of a constant sum.

The Hyperbola   A **hyperbola** is the set of all points P in a plane such that the absolute value of the difference of the distances from point P to two other points in the plane is a positive constant.

Although the definition of an ellipse and a hyperbola are quite similar, their graphs are very different. The graphs of two hyperbolas are shown in Figure 7-28.

> **Tip**
> Cylindrical lampshades cast shadows that can help us remember the shape of a hyperbola's graph. The shadow contains two disjoint parts. These look like parabolas, but they are very different.
>
>

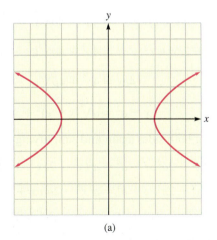

(a)

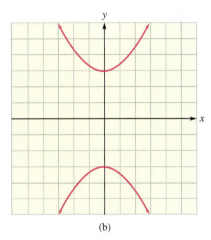

(b)

FIGURE 7-28

1. Write Equations of Hyperbolas

In the graph of the hyperbola shown in Figure 7-29, points F and F' are called the **foci** of the hyperbola and the midpoint of chord FF' is called the **center**. The points

V and V', where the hyperbola intersects FF', are called **vertices**. The segment VV' is called the **transverse axis**.

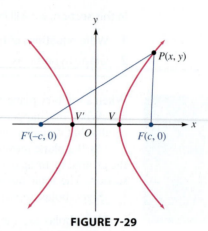

FIGURE 7-29

Deriving the Standard Form of an Equation of a Hyperbola

To develop the equation of the hyperbola centered at the origin, we note that the origin is the midpoint of chord FF', and we let $d(F'O) = d(OF) = c$, where $c > 0$. Then F is at $(c, 0)$, and F' is at $(-c, 0)$. The definition of a hyperbola requires that $d(F'P) - d(PF) = 2a$, where $2a$ is a positive constant. We use the distance formula to compute the lengths of $F'P$ and PF:

$$d(F'P) = \sqrt{[x - (-c)]^2 + y^2} \qquad d(PF) = \sqrt{(x - c)^2 + y^2}$$

Substituting these values into the equation $|d(F'P) - d(PF)| = 2a$ gives

$$\sqrt{(x + c)^2 + y^2} - \sqrt{(x - c)^2 + y^2} = 2a$$

or

$$\sqrt{(x + c)^2 + y^2} = 2a + \sqrt{(x - c)^2 + y^2}$$

After squaring, we have

$$(x + c)^2 + y^2 = 4a^2 + 4a\sqrt{(x - c)^2 + y^2} + (x - c)^2 + y^2$$

$$x^2 + 2cx + c^2 + y^2 = 4a^2 + 4a\sqrt{(x - c)^2 + y^2} + x^2 - 2cx + c^2 + y^2$$

$$4cx = 4a^2 + 4a\sqrt{(x - c)^2 + y^2}$$

$$cx - a^2 = a\sqrt{(x - c)^2 + y^2}$$

Squaring both sides again and simplifying gives

$$c^2x^2 - 2a^2cx + a^4 = a^2(x^2 - 2cx + c^2 + y^2)$$

$$c^2x^2 - 2a^2cx + a^4 = a^2x^2 - 2a^2cx + a^2c^2 + a^2y^2$$

$$c^2x^2 + a^4 = a^2x^2 + a^2c^2 + a^2y^2$$

$$(1) \quad (c^2 - a^2)x^2 - a^2y^2 = a^2(c^2 - a^2)$$

Because $c > a$ (you will be asked to prove this in Exercise 69), $c^2 - a^2$ is a positive number. So we can let $b^2 = c^2 - a^2$ and substitute b^2 for $c^2 - a^2$ in Equation 1 to get

$$b^2x^2 - a^2y^2 = a^2b^2 \qquad \text{Substitute } b^2 \text{ for } c^2 - a^2.$$

If we divide both sides of the previous equation by a^2b^2, we will obtain the following equation.

$$\frac{x^2}{a^2} - \frac{y^2}{b^2} = 1$$

This equation is standard form of the equation of a hyperbola with center at the origin and horizontal axis the x-axis.

To find the x-intercepts of the graph, we let $y = 0$ and solve for x. We get

$$\frac{x^2}{a^2} = 1$$

$$x^2 = a^2$$

$$x = a \quad \text{or} \quad x = -a$$

We now know that the x-intercepts are the vertices $V(a, 0)$ and $V'(-a, 0)$. The distance between the center of the hyperbola and either vertex is a, and the center of the hyperbola is the midpoint of the segment $V'V$ as well as that of the segment FF'.

We attempt to find the y-intercepts by letting $x = 0$. Then the equation becomes

$$\frac{-y^2}{b^2} = 1 \quad \text{or} \quad y^2 = -b^2$$

Because $-b^2$ represents a negative number, and y^2 cannot be negative, the equation has no real solutions. Since there are no y-values corresponding to $x = 0$, the hyperbola does not intersect the y-axis.

This discussion suggests the following results.

Hyperbola: Foci on x-Axis, Center at $(0, 0)$ The standard form of an equation of a hyperbola with center at the origin and foci on the x-axis is

$$\frac{x^2}{a^2} - \frac{y^2}{b^2} = 1$$

where $a^2 + b^2 = c^2$.

Vertices: $V(a, 0)$ and $V'(-a, 0)$

Foci: $F(c, 0)$ and $F'(-c, 0)$

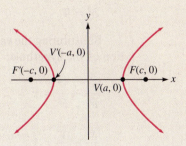

If the foci are on the y-axis, a similar equation results.

Hyperbola: Foci on y-Axis, Center at $(0, 0)$

The standard form of an equation of a hyperbola with center at the origin and foci on the y-axis is

$$\frac{y^2}{a^2} - \frac{x^2}{b^2} = 1$$

where $a^2 + b^2 = c^2$.

Vertices: $V(0, a)$ and $V'(0, -a)$

Foci: $F(0, c)$ and $F'(0, -c)$

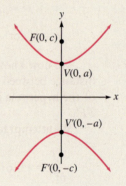

EXAMPLE 1

Writing an Equation of a Hyperbola in Standard Form

Write an equation in standard form of the hyperbola with vertices $V(4, 0)$ and $V'(-4, 0)$ and a focus at $F(5, 0)$.

SOLUTION   Because the foci lie on the x-axis, we use the standard equation $\frac{x^2}{a^2} - \frac{y^2}{b^2} = 1$. We will find a^2 and b^2 and substitute the results into the standard equation.

The center of the hyperbola is midway between the vertices V and V'. Thus, the center is the origin $(0, 0)$. The distance between the vertex and the center is $a = 4$, and the distance between the focus and the center is $c = 5$. We can find b^2 by substituting 4 for a and 5 for c in the following equation to get

$$b^2 = c^2 - a^2 \quad \text{In a hyperbola, } a^2 + b^2 = c^2 \text{ or } b^2 = c^2 - a^2.$$

$$b^2 = 5^2 - 4^2$$

$$b^2 = 9$$

Substituting the values for a^2 and b^2 in the standard equation gives an equation of the hyperbola:

$$\frac{x^2}{a^2} - \frac{y^2}{b^2} = 1$$

$$\frac{x^2}{16} - \frac{y^2}{9} = 1$$

The graph of $\frac{x^2}{16} - \frac{y^2}{9} = 1$ is shown in Figure 7-30.

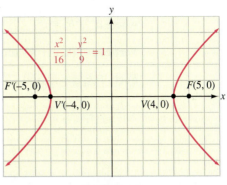

FIGURE 7-30

Self Check 1   Write an equation in standard form of the hyperbola with vertices $(0, 4)$ and $(0, -4)$ and a focus at $(0, 5)$.

Now Try Exercise 15.

To translate the hyperbola to a new position centered at the point (h, k) instead of the origin, we replace x and y with $x - h$ and $y - k$, respectively. We get the following results.

Hyperbola: Transverse Axis Horizontal, Center at (h, k)

The standard form of an equation of a hyperbola with center at the point (h, k) and foci on a line parallel to the x-axis is

$$\frac{(x - h)^2}{a^2} - \frac{(y - k)^2}{b^2} = 1$$

where $a^2 + b^2 = c^2$.

Vertices: $V(a + h, k)$ and $V'(-a + h, k)$

Foci: $F(c + h, k)$ and $F'(-c + h, k)$

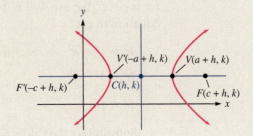

Hyperbola: Transverse Axis Vertical, Center at (h, k)

The standard form of an equation of a hyperbola with center at (h, k) and foci on a line parallel to the y-axis is

$$\frac{(y - k)^2}{a^2} - \frac{(x - h)^2}{b^2} = 1$$

where $a^2 + b^2 = c^2$.

Vertices: $V(h, a + k)$ and $V'(h, -a + k)$

Foci: $F(h, c + k)$ and $F'(h, -c + k)$

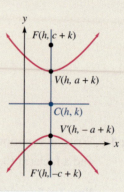

EXAMPLE 2

Writing an Equation of a Hyperbola in Standard Form

Write an equation in standard form of the hyperbola with vertices $(3, -3)$ and $(3, 3)$ and a focus at $(3, 5)$.

SOLUTION Because the foci lie on a vertical line, as shown in Figure 7-31, we will use the standard form of the equation

$$\frac{(y - k)^2}{a^2} - \frac{(x - h)^2}{b^2} = 1$$

and determine h, k, a^2, and b^2, and substitute the results into the standard equation.

The center of the hyperbola is midway between the vertices V and V'. Thus, the center is point $(3, 0)$; $h = 3$; and $k = 0$. The distance between the vertex and the center is $a = 3$, and the distance between the focus and the center is $c = 5$. We can find b^2 by substituting 3 for a and 5 for c in the following equation to get

$$b^2 = c^2 - a^2 \quad \text{In a hyperbola, } a^2 + b^2 = c^2 \text{ or } b^2 = c^2 - a^2.$$

$$b^2 = 5^2 - 3^2$$

$$b^2 = 16$$

Substituting the values for h, k, a^2, and b^2 in the standard equation gives an equation of the hyperbola:

$$\frac{(y - k)^2}{a^2} - \frac{(x - h)^2}{b^2} = 1$$

$$\frac{(y - 0)^2}{9} - \frac{(x - 3)^2}{16} = 1$$

$$\frac{y^2}{9} - \frac{(x - 3)^2}{16} = 1$$

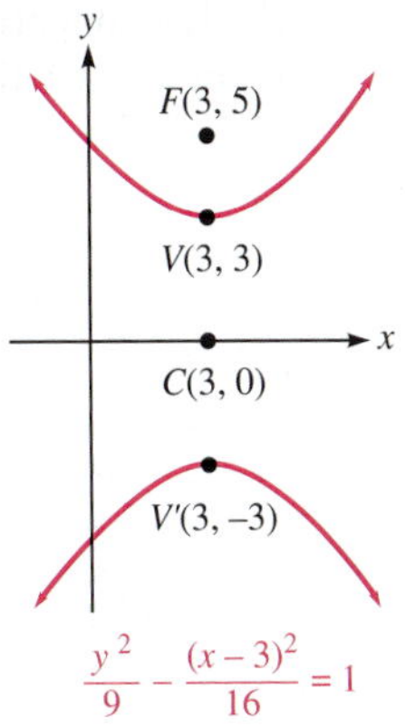

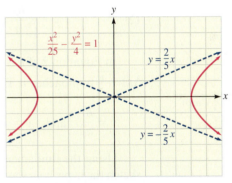

$F(3, 5)$

$V(3, 3)$

$C(3, 0)$

$V'(3, -3)$

$$\frac{y^2}{9} - \frac{(x-3)^2}{16} = 1$$

FIGURE 7-31

Self Check 2   Write an equation of the hyperbola with vertices $(3, 1)$ and $(-3, 1)$ and a focus at $(5, 1)$.

Now Try Exercise 23.

2. Graph Hyperbolas

Consider the graph of the hyperbola with equation $\frac{x^2}{25} - \frac{y^2}{4} = 1$ shown in Figure 7-32. From the graph, we can see that

- The values of a and b are $a = 5$ and $b = 2$.
- The equations of the asymptotes of the graph are $y = \frac{2}{5}x$ and $y = -\frac{2}{5}x$.

$$\frac{x^2}{25} - \frac{y^2}{4} = 1$$

$$y = \frac{2}{5}x$$

$$y = -\frac{2}{5}x$$

FIGURE 7-32

The values of a and b are important aids in finding the equations of the asymptotes of the hyperbola. In fact, the equations of the asymptotes are

$$y = \frac{b}{a}x \quad \text{and} \quad y = -\frac{b}{a}x$$

$$= \frac{2}{5}x \qquad\qquad = -\frac{2}{5}x$$

The values of a and b are also aids in drawing a special rectangle that can be used to help graph hyperbolas.

To show that this is true, we consider the hyperbola with equation of $\frac{x^2}{a^2} - \frac{y^2}{b^2} = 1$.

This hyperbola has its center at the origin and vertices at $V(a, 0)$ and $V'(-a, 0)$. We can plot points V, V', $B(0, b)$, and $B'(0, -b)$ and form rectangle RSQP, called the **fundamental rectangle**, as shown in Figure 7-33. The extended diagonals of this rectangle are the asymptotes of the hyperbola.

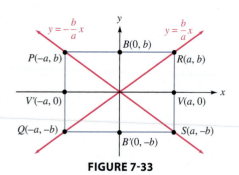

FIGURE 7-33

Verifying the Extended Diagonals Are Asymptotes

To show that the extended diagonals are asymptotes of the hyperbola, we solve the equation $\frac{x^2}{a^2} - \frac{y^2}{b^2} = 1$ for y and modify its form as follows:

$$\frac{x^2}{a^2} - \frac{y^2}{b^2} = 1$$

$$b^2x^2 - a^2y^2 = a^2b^2 \qquad \text{Multiply both sides by } a^2b^2.$$

$$y^2 = \frac{b^2x^2 - a^2b^2}{a^2} \qquad \begin{array}{l}\text{Subtract } b^2x^2 \text{ from both sides}\\ \text{and divide both sides by } -a^2.\end{array}$$

$$y^2 = \frac{b^2x^2}{a^2}\left(1 - \frac{a^2}{x^2}\right) \qquad \text{Factor out the fraction } \frac{b^2x^2}{a^2}.$$

(2) $$y = \pm\frac{bx}{a}\sqrt{1 - \frac{a^2}{x^2}} \qquad \text{Take the square root of both sides.}$$

In Equation 2, if a is constant and $|x|$ approaches ∞, then $\frac{a^2}{x^2}$ approaches 0, and $\sqrt{1 - \frac{a^2}{x^2}}$ approaches 1. Thus, the hyperbola approaches the lines

$$y = \frac{b}{a}x \quad \text{and} \quad y = -\frac{b}{a}x$$

Using the fundamental rectangle, asymptotes, and vertices as guides, we can sketch the hyperbola shown in Figure 7-34. The segment BB' is called the **conjugate axis** of the hyperbola.

If the hyperbola is not in standard form, we simply convert it to standard form by completing the square and making use of translations to graph the hyperbola. This information is summarized as follows:

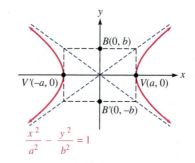

FIGURE 7-34

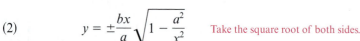

Strategy for Graphing a Hyperbola	1. Write the hyperbola in standard form.
	2. Identify the center and a and b.
	3. Draw the fundamental rectangle and asymptotes.
	4. Sketch the hyperbola beginning with the vertices and using the asymptotes as guides.

Graphing a Hyperbola

EXAMPLE 3

Graph the hyperbola: $\dfrac{y^2}{9} - \dfrac{x^2}{25} = 1$.

SOLUTION   We will apply the four steps listed above to graph the hyperbola.

Step 1: Write the hyperbola in standard form. The equation of the hyperbola $\dfrac{y^2}{9} - \dfrac{x^2}{25} = 1$ is already written in the standard form $\dfrac{y^2}{a^2} - \dfrac{x^2}{b^2} = 1$.

Step 2: Identify the center and *a* and *b*. From the standard equation of a hyperbola, we see that the center is $(0, 0)$, that $a = 3$ and $b = 5$, and that the vertices are on the *y*-axis.

Step 3: Draw the fundamental rectangle and asymptotes. Vertices V and V' are 3 units above and below the origin and have coordinates of $(0, 3)$ and $(0, -3)$. Points B and B' are 5 units right and left of the origin and have coordinates of $(5, 0)$ and $(-5, 0)$. We use the points V, V', B, and B' to construct the fundamental rectangle and extend the diagonals to graph the asymptotes as shown in Figure 7-35.

Step 4: Sketch the hyperbola. We begin with the vertices and sketch the graph of the hyperbola, as shown in Figure 7-35, using the asymptotes as guides.

> **Caution**
> The graphs of hyperbolas look very similar to the graphs of parabolas, but hyperbolas are not parabolas.

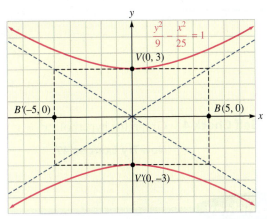

FIGURE 7-35

Self Check 3   Graph the hyperbola: $\dfrac{y^2}{9} - \dfrac{(x + 1)^2}{25} = 1$.

Now Try Exercise 41.

Graphing a Hyperbola

EXAMPLE 4

Graph the hyperbola $x^2 - y^2 - 2x + 4y = 12$.

SOLUTION   We will use the four steps listed above to graph the hyperbola.

Step 1: Write the hyperbola in standard form. We complete the square on *x* and *y* to write the equation in standard form:

$$x^2 - y^2 - 2x + 4y = 12$$

$$x^2 - 2x - (y^2 - 4y) = 12$$

$$x^2 - 2x + 1 - (y^2 - 4y + 4) = 12 + 1 - 4$$

$$(x - 1)^2 - (y - 2)^2 = 9$$

$$\frac{(x - 1)^2}{9} - \frac{(y - 2)^2}{9} = 1$$

> **Tip**
>
> When the equation of a hyperbola is written in standard form, h and k are the numbers within parentheses that follow the subtraction symbol.

Step 2: Identify the center and a and b. From the standard equation of a hyperbola, we see that the center is $(1, 2)$, that $a = 3$ and $b = 3$, and that the vertices are on a line segment parallel to the x-axis, as shown in Figure 7-36.

Step 3: Draw the fundamental rectangle and asymptotes. The vertices V and V' are 3 units to the right and left of the center and have coordinates of $(4, 2)$ and $(-2, 2)$. Points B and B', 3 units above and below the center, have coordinates of $(1, 5)$ and $(1, -1)$. We use the points V, V', B, and B' to construct the fundamental rectangle and extend the diagonals to draw the asymptotes as shown in Figure 7-36.

Step 4: Sketch the hyperbola. We begin with the vertices and sketch the graph of the hyperbola, as shown in Figure 7-36, using the asymptotes as guides.

> **Tip**
>
> When sketching the graph of a hyperbola, the branches should not touch the asymptotes.

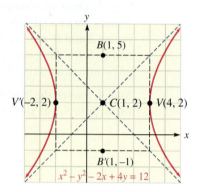

FIGURE 7-36

Self Check 4 Graph the hyperbola $x^2 - y^2 + 6x + 2y = 8$.

Now Try Exercise 47.

Accent on Technology

Graphing Hyperbolas

To graph the equation $x^2 - y^2 - 2x + 4y = 12$ with a graphing calculator, we first solve the equation for y.

$$x^2 - y^2 - 2x + 4y = 12$$

$$-x^2 + y^2 + 2x - 4y = -12 \qquad \text{Multiply both sides by } -1.$$

$$y^2 - 4y = x^2 - 2x - 12 \qquad \text{Add } x^2 - 2x \text{ to both sides.}$$

$$y^2 - 4y + 4 = x^2 - 2x - 12 + 4 \qquad \text{Complete the square by adding 4 to both sides.}$$

$$(y - 2)^2 = x^2 - 2x - 8 \qquad \text{Factor the left side.}$$

$$y - 2 = \pm\sqrt{x^2 - 2x - 8} \qquad \text{Use the Square Root Property.}$$

$$y = 2 \pm \sqrt{x^2 - 2x - 8} \qquad \text{Solve for } y.$$

If we use window settings of $[-4, 6]$ for x and $[-3, 7]$ for y and graph the functions

$$y = 2 + \sqrt{x^2 - 2x - 8} \quad \text{and} \quad y = 2 - \sqrt{x^2 - 2x - 8}$$

we will get a graph as shown in Figure 7-37.

Enter the equations.

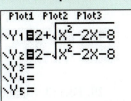

Set a WINDOW.

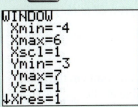

GRAPH **the equations.**

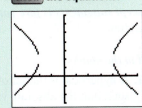

FIGURE 7-37

We have considered only those hyperbolas with a major axis that is horizontal or vertical. However, some hyperbolas have nonhorizontal or nonvertical major axes. For example, the graph of the equation $xy = 4$ is a hyperbola with vertices at $(2, 2)$ and $(-2, -2)$, as shown in Figure 7-38.

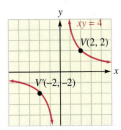

FIGURE 7-38

Self Check Answers

1. $\dfrac{x^2}{9} - \dfrac{y^2}{16} = 1$ **2.** $\dfrac{x^2}{9} - \dfrac{(y-1)^2}{16} = 1$

3.

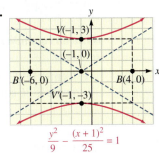

4.

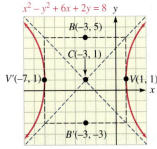

Exercises 7.3

Getting Ready

You should be able to complete these vocabulary and concept statements before you proceed to the practice exercises.

Fill in the blanks.

1. A hyperbola is the set of all points in the plane such that the absolute value of the _____ of the distances from two fixed points is a positive _____.

2. Each of the two fixed points in the definition of a hyperbola is called a _____ of the hyperbola.

3. The vertices of the hyperbola $\dfrac{x^2}{a^2} - \dfrac{y^2}{b^2} = 1$ are $V(\_\_, \_\_)$ and $V'(\_\_, \_\_)$.

4. The vertices of the hyperbola $\dfrac{y^2}{a^2} - \dfrac{x^2}{b^2} = 1$ are $V(\_\_, \_\_)$ and $V'(\_\_, \_\_)$

5. The chord that joins the vertices is called the _____ of the hyperbola.

6. In a hyperbola, the relationship between a, b, and c is _____.

Identify the conic as a circle, parabola, ellipse, or hyperbola.

7. $x^2 + (y - 4)^2 = 12$

8. $7x^2 + 7y^2 + 70x - 14y = 119$

9. $y^2 - 2x + 23 = 5y$ 10. $(x - 8)^2 = 5(y + 6)$

11. $\dfrac{x^2}{35} + \dfrac{y^2}{9} = 1$

12. $2x^2 + 32y^2 + 4x - 30 = 0$

13. $x^2 - 2y^2 - 4y - 6 = 0$

14. $\dfrac{(x - 3)^2}{4} - \dfrac{y^2}{9} = 1$

Practice

Write an equation in standard form of each hyperbola described.

15. Vertices (5, 0) and (−5, 0); focus (7, 0)

16. Focus (3, 0); vertex (2, 0); center (0, 0)

17. Center (2, 4); $a = 2$, $b = 3$; transverse axis is horizontal

18. Center (−1, 3); vertex (1, 3); focus (2, 3)

19. Center (5, 3); vertex (5, 6); passes through (1, 8)

20. Foci (0, 10) and (0, −10); $\dfrac{c}{a} = \dfrac{5}{4}$

21. Vertices (0, 3) and (0, −3); $\dfrac{c}{a} = \dfrac{5}{3}$

22. Focus (4, 0); vertex (2, 0); center (0, 0)

23. Center (1, 4); focus (7, 4); vertex (3, 4)

24. Center (1, −3); $a^2 = 4$; $b^2 = 16$

25. Center at the origin; passes through (4, 2) and (8, −6)

26. Center (3, −1); y-intercept (0, −1); x-intercept $\left(3 + \dfrac{3\sqrt{5}}{2}, 0\right)$

Find the area of the fundamental rectangle of each hyperbola.

27. $4(x - 1)^2 - 9(y + 2)^2 = 36$

28. $x^2 - y^2 - 4x - 6y = 6$

29. $x^2 + 6x - y^2 + 2y = -11$

30. $9x^2 - 4y^2 = 18x + 24y + 63$

Write an equation in standard form of each hyperbola described.

31. Center (−2, −4); $a = 2$; area of fundamental rectangle is 36 square units

32. Center (3, −5); $b = 6$; area of fundamental rectangle is 24 square units

33. Vertex (6, 0); one end of conjugate axis at $\left(0, \dfrac{5}{4}\right)$

34. Vertex (3, 0); focus (−5, 0); center (0, 0)

Graph each hyperbola.

35. $\dfrac{x^2}{9} - \dfrac{y^2}{4} = 1$

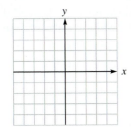

36. $\dfrac{y^2}{4} - \dfrac{x^2}{9} = 1$

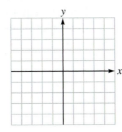

37. $4x^2 - 3y^2 = 36$

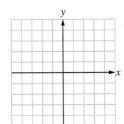

38. $3x^2 - 4y^2 = 36$

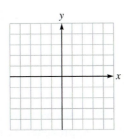

39. $y^2 - x^2 = 1$

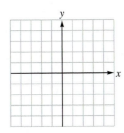

40. $x^2 - \dfrac{y^2}{4} = 1$

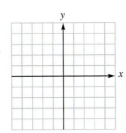

41. $\dfrac{(x + 2)^2}{9} - \dfrac{y^2}{4} = 1$

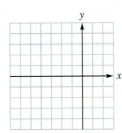

42. $\dfrac{y^2}{9} - \dfrac{(x - 2)^2}{36} = 1$

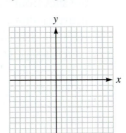

43. $4(y - 2)^2 - 9(x + 1)^2 = 36$

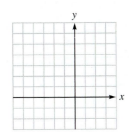

44. $9(y + 2)^2 - 4(x - 1)^2 = 36$

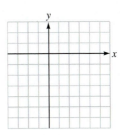

45. $4x^2 - 2y^2 + 8x - 8y = 8$

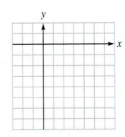

46. $x^2 - y^2 - 4x - 6y = 6$

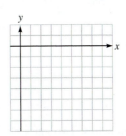

47. $y^2 - 4x^2 + 6y + 32x = 59$

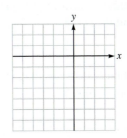

48. $x^2 + 6x - y^2 + 2y = -11$

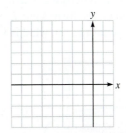

49. $-xy = 6$

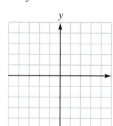

50. $xy = 20$

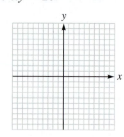

58. Astronomy Some comets have a hyperbolic orbit, with the sun as one focus. When the comet shown in the illustration is far away from Earth, it appears to be approaching Earth along the line $y = 2x$. Find the equation in standard form of its orbit if the comet comes within 100 million miles of Earth.

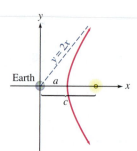

Find an equation in standard form of the hyperbola on which point P lies.

51. The difference of the distances between $P(x, y)$ and the points $(-2, 1)$ and $(8, 1)$ is 6.

52. The difference of the distances between $P(x, y)$ and the points $(3, -1)$ and $(3, 5)$ is 5.

53. The distance between point $P(x, y)$ and the point $(0, 3)$ is $\frac{3}{2}$ of the distance between P and the line $y = -2$.

54. The distance between point $P(x, y)$ and the point $(5, 4)$ is $\frac{5}{3}$ of the distance between P and the line $x = -3$.

59. Alpha particles The particle in the illustration approaches the nucleus at the origin along the path $9y^2 - x^2 = 81$ in the coordinate system shown. How close does the particle come to the nucleus?

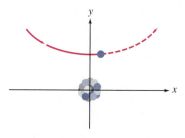

 Use a graphing calculator to graph each hyperbola. Then sketch the hyperbola by hand and compare the results.

55. $x^2 - \dfrac{y^2}{4} = 1$

56. $\dfrac{(x + 3)^2}{4} - \dfrac{(y - 2)^2}{25} = 1$

60. Physics Parallel beams of similarly charged particles are shot from two atomic accelerators 20 meters apart, as shown in the illustration. If the particles were not deflected, the beams would be 2.0×10^{-4} meters apart. However, because the charged particles repel each other, the beams follow the hyperbolic path $y = \frac{k}{x}$, for some k. Find k.

Applications

57. Fluids See the illustration below. Two glass plates in contact at the left, and separated by about 5 millimeters on the right, are dipped in beet juice, which rises by capillary action to form a hyperbola. The hyperbola is modeled by an equation of the form $xy = k$. If the curve passes through the point $(12, 2)$, what is k?

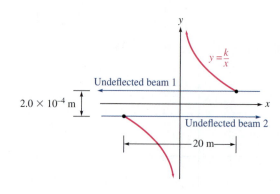

61. Navigation The LORAN (LOng RAnge Navigation) system in the illustration uses two radio transmitters 26 miles apart to send simultaneous signals. The navigator on a ship at $P(x, y)$ receives the closer signal first and determines that the difference of the distances between the ship and each transmitter is 24 miles. That places the ship on a certain curve. Identify the curve and find an equation in standard form.

62. Navigation By determining the difference of the distances between the ship in the illustration and two radio transmitters, the LORAN navigation system places the ship on the hyperbola $x^2 - 4y^2 = 576$ in the coordinate system shown. If the ship is 5 miles out to sea, find its coordinates.

63. Wave propagation Stones dropped into a calm pond at points A and B create ripples that propagate in widening circles. In the illustration, points A and B are 20 feet apart, and the radii of the circles differ by 12 feet. The point $P(x, y)$ where the circles intersect moves along a curve.

Identify the curve and find an equation in standard form.

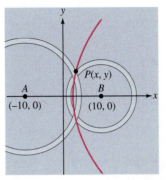

64. Sonic boom The position of a sonic boom caused by the faster-than-sound aircraft is one branch of the hyperbola $y^2 - x^2 = 25$ in the coordinate system shown. How wide is the hyperbola 5 miles from its vertex?

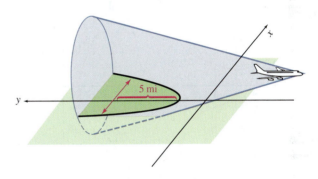

Discovery and Writing

65. Describe a hyperbola.

66. Explain a strategy you would use to graph a hyperbola.

67. Explain how to determine the dimensions of the fundamental rectangle.

68. How do you distinguish among the equations of circles, parabolas, ellipses, and hyperbolas?

69. Prove that $c > a$ for a hyperbola with center at $(0, 0)$ and line segment FF' on the x-axis.

70. Show that the extended diagonals of the fundamental rectangle of the hyperbola $\frac{x^2}{a^2} - \frac{y^2}{b^2} = 1$ are $y = \frac{b}{a}x$ and $y = -\frac{b}{a}x$.

71. Show that the expansion of the standard equation of a hyperbola is a special case of the general equation of second degree with $B = 0$.

72. Write a paragraph describing how you can tell from the equation of a hyperbola whether the transverse axis is vertical or horizontal.

Critical Thinking

In Exercises 73–76, match the equation of the hyperbola with its graph.

73. $\dfrac{x^2}{25} - \dfrac{y^2}{9} = 1$

74. $\dfrac{x^2}{9} - \dfrac{y^2}{25} = 1$

75. $\dfrac{y^2}{9} - \dfrac{x^2}{25} = 1$

76. $\dfrac{y^2}{25} - \dfrac{x^2}{9} = 1$

In Exercises 77–80, match the equation of the hyperbola with its graph.

77. $\dfrac{(y-3)^2}{9} - \dfrac{(x-3)^2}{16} = 1$

78. $\dfrac{(y-3)^2}{9} - \dfrac{(x+3)^2}{16} = 1$

79. $\dfrac{(x+3)^2}{16} - \dfrac{(y+3)^2}{9} = 1$

80. $\dfrac{(x-3)^2}{16} - \dfrac{(y+3)^2}{9} = 1$

a.

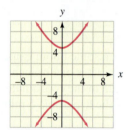

b.

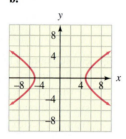

a.

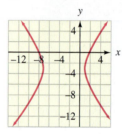

b.

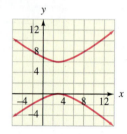

c.

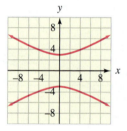

d.

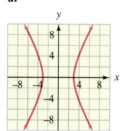

c.

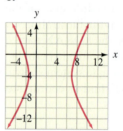

d.

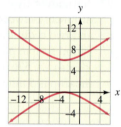

7.4 Solving Nonlinear Systems of Equations

In this section, we will learn to

1. Solve systems by graphing.
2. Solve systems by substitution.
3. Solve systems by elimination.
4. Solve problems using systems of nonlinear equations.

Air-traffic controllers maintain an orderly flow of air traffic. Since air-traffic controllers keep planes a safe distance apart, the paths of two planes will not intersect. This guarantees that there will be no mid-air collisions. Knowing where the graphs of two or more equations intersect can be very important. For example, the intersection point of the graphs of the flight paths of two planes will be a potential point of collision.

We will now discuss techniques for solving systems of two equations in two variables, where at least one of the equations is nonlinear. These systems of equations are called **nonlinear systems**. We will use three methods to solve such systems: graphing, substitution, and elimination.

1. Solve Systems by Graphing

To solve systems by the graphing method, we can graph each equation and identify the point(s) of intersection of the two graphs.

Solving a System of Equations by Graphing

EXAMPLE 1

Solve $\begin{cases} x^2 + y^2 = 25 \\ 2x + y = 10 \end{cases}$ by graphing.

SOLUTION   To solve this system, we will graph each equation and identify the point(s) of intersection of the two graphs.

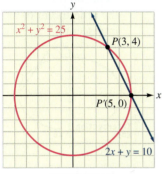

FIGURE 7-39

The graph of $x^2 + y^2 = 25$ is a circle with center at the origin and radius of 5. The equation of $2x + y = 10$ is a line with x-intercept at $(5, 0)$ and y-intercept at $(0, 10)$.

After graphing the circle and the line, as shown in Figure 7-39, we see that there are two intersection points, $P(3, 4)$ and $P'(5, 0)$. The two solutions to the system are $x = 3$ and $y = 4$, which we write as $(3, 4)$, and $x = 5$ and $y = 0$, which we write as $(5, 0)$.

The line $2x + y = 10$ is a called a **secant line** because it intersects the graph of the circle at two points. In this case, the system has two solutions. If a line intersects the graph of a circle at one point, the line is called a **tangent line**. In this case, there is one solution. If the line does not intersect the circle, there is no solution.

Self Check 1   Solve $\begin{cases} x^2 + y^2 = 25 \\ 2x - y = 5 \end{cases}$ by graphing.

Now Try Exercise 9.

Solving Systems of Equations

To solve the system of equations in Example 1 using a graphing calculator, we must solve each of the equations for y as shown in Figure 7-40. From the equation $x^2 + y^2 = 25$, we get two equations to graph:

$$y_1 = \sqrt{25 - x^2} \quad \text{and} \quad y_2 = -\sqrt{25 - x^2}$$

From the equation $2x + y = 10$, we get a third equation to graph:

$$y_3 = 10 - 2x$$

1. Enter each equation.   2. Set a WINDOW.   3. GRAPH the equations.

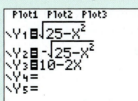

4. Press 2nd TRACE to access the CALCULATE menu.

5. Move the cursor close to the point of intersection.

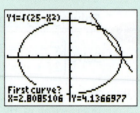

6. Press ENTER and the cursor will move to Y2. This is not the curve that you need.

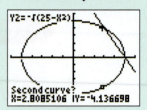

7. Press the DOWN CURSOR key to move the point to Y3.

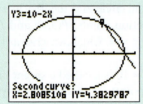

8. Press ENTER twice to get to the point of intersection.

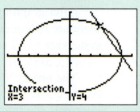

9. Repeat to find the other point of intersection.

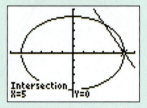

FIGURE 7-40

2. Solve Systems by Substitution

The substitution method can be used to find exact solutions. By making an appropriate substitution, we can convert a nonlinear system of two equations in two variables into one equation in one variable. We can then solve the resulting equation for one variable and back substitute to find the other variable.

EXAMPLE 2

Solving a System of Equations Using Substitution

Solve $\begin{cases} x^2 + y^2 = 25 \\ 2x + y = 10 \end{cases}$ by substitution.

SOLUTION We will solve the linear equation for y to obtain an equation called the **substitution equation**. We will then substitute the expression that represents y into the second-degree equation to obtain an equation in the one variable x. After solving the equation for x, we can find y by back substituting the resulting values of x into the substitution equation and simplifying.

We begin by solving the linear equation for y.

$$2x + y = 10 \qquad \text{This equation is linear.}$$

$$y = -2x + 10 \qquad \text{This is the substitution equation.}$$

We can now substitute $-2x + 10$ for y in the second-degree equation and solve the resulting quadratic equation for x:

$$x^2 + y^2 = 25 \qquad \text{This equation is nonlinear.}$$

$$x^2 + (-2x + 10)^2 = 25 \qquad \text{Substitute } -2x + 10 \text{ for } y.$$

$$x^2 + 4x^2 - 40x + 100 = 25 \qquad \text{Square the binomial.}$$

> **Tip**
>
> Our objective is to make an appropriate substitution to obtain one equation in one variable.

$$5x^2 - 40x + 75 = 0 \qquad \text{\textcolor{red}{Subtract 25 from both sides and combine like terms.}}$$

$$x^2 - 8x + 15 = 0 \qquad \text{\textcolor{red}{Divide both sides by 5.}}$$

$$(x - 5)(x - 3) = 0 \qquad \text{\textcolor{red}{Factor }} x^2 - 8x + 15.$$

$$x - 5 = 0 \quad \text{or} \quad x - 3 = 0$$

$$x = 5 \qquad \qquad x = 3$$

Take Note

The solutions we obtained by graphing and by substitution are the same.

Because $y = -2x + 10$, if $x = 5$, then $y = 0$; and if $x = 3$, then $y = 4$. The two solutions are $(5, 0)$ and $(3, 4)$.

Self Check 2   Solve $\begin{cases} x^2 + y^2 = 25 \\ 2x - y = 5 \end{cases}$ by substitution.

Now Try Exercise 21.

EXAMPLE 3

Solving a System of Equations by Substitution

Solve $\begin{cases} 4x^2 + 9y^2 = 5 \\ y = x^2 \end{cases}$ by substitution.

SOLUTION   In this example, both equations are nonlinear. To solve it, we will substitute y for x^2 in the top equation to obtain one equation in the variable y. Then, we will solve the resulting equation for y and use back substitution to find x.

$$4\textcolor{red}{x^2} + 9y^2 = 5$$

$$4\textcolor{red}{y} + 9y^2 = 5 \qquad \text{\textcolor{red}{Substitute }} y \text{ \textcolor{red}{for }} x^2.$$

$$9y^2 + 4y - 5 = 0 \qquad \text{\textcolor{red}{Add }} -5 \text{ \textcolor{red}{to both sides.}}$$

$$(9y - 5)(y + 1) = 0 \qquad \text{\textcolor{red}{Factor }} 9y^2 + 4y - 5.$$

$$9y - 5 = 0 \quad \text{or} \quad y + 1 = 0$$

$$y = \frac{5}{9} \qquad \qquad y = -1$$

Tip

The substitution method is a good choice when one variable in the system is raised to the first power.

Because $y = x^2$, we can find x by solving the equations

$$x^2 = \frac{\textcolor{red}{5}}{\textcolor{red}{9}} \quad \text{and} \quad x^2 = \textcolor{red}{-1}$$

The solutions of $x^2 = \frac{5}{9}$ are

$$x = \frac{\sqrt{5}}{3} \quad \text{or} \quad x = -\frac{\sqrt{5}}{3}$$

Since the equation $x^2 = -1$ has no real solutions, the only solutions of the system are

$$\left(\frac{\sqrt{5}}{3}, \frac{5}{9}\right) \quad \text{and} \quad \left(-\frac{\sqrt{5}}{3}, \frac{5}{9}\right)$$

Self Check 3 Solve $\begin{cases} x^2 + 3y^2 = 13 \\ x = y^2 - 1 \end{cases}$ by substitution.

Now Try Exercise 25.

3. Solve Systems by Elimination

When we have two second-degree equations of the form $ax^2 + by^2 = c$, we can solve the system by using the elimination method and eliminating one of the variables.

Solving a System of Equations by Elimination

EXAMPLE 4

Solve $\begin{cases} 3x^2 + 2y^2 = 36 \\ 4x^2 - y^2 = 4 \end{cases}$ by elimination.

SOLUTION In this example, we have two second-degree equations of the form $ax^2 + by^2 = c$. In such cases, we can solve the system by eliminating one of the variables by addition, solving the resulting equation, and using back substitution to solve for the second variable.

To eliminate the terms involving y^2, we copy the first equation and multiply the second equation by 2 to obtain the following equivalent system.

> **Tip**
>
> The elimination method is also known as the addition method. The method is a good choice when both equations are of the form $Ax^2 + By^2 = C$.

$$\begin{cases} 3x^2 + 2y^2 = 36 \\ 8x^2 - 2y^2 = 8 \end{cases}$$

We can then add the equations and solve the resulting equation for x:

$$11x^2 = 44$$
$$x^2 = 4$$
$$x = 2 \quad \text{or} \quad x = -2$$

To find y, we substitute 2 for x and then -2 for x in the first equation.

For $x = 2$	For $x = -2$
$3x^2 + 2y^2 = 36$	$3x^2 + 2y^2 = 36$
$3(2)^2 + 2y^2 = 36$	$3(-2)^2 + 2y^2 = 36$
$12 + 2y^2 = 36$	$12 + 2y^2 = 36$
$2y^2 = 24$	$2y^2 = 24$
$y^2 = 12$	$y^2 = 12$
$y = \sqrt{12} \quad \text{or} \quad y = -\sqrt{12}$	$y = \sqrt{12} \quad \text{or} \quad y = -\sqrt{12}$
$y = 2\sqrt{3} \quad \mid \quad y = -2\sqrt{3}$	$y = 2\sqrt{3} \quad \mid \quad y = -2\sqrt{3}$

> **Take Note**
>
> $3x^2 + 2y^2 = 36$ is an equation of an ellipse. $8x^2 - 2y^2 = 8$ is an equation of a hyperbola. It is possible to have four points of intersection. In Example 4, we *do* have four solutions.

The four solutions of this system are

$$\left(2, 2\sqrt{3}\right), \quad \left(2, -2\sqrt{3}\right), \quad \left(-2, 2\sqrt{3}\right), \quad \text{and} \quad \left(-2, -2\sqrt{3}\right)$$

Self Check 4 Solve: $\begin{cases} 2x^2 + y^2 = 23 \\ 3x^2 - 2y^2 = 17 \end{cases}$ by addition.

Now Try Exercise 29.

4. Solve Problems Using Systems of Nonlinear Equations

EXAMPLE 5

Solving an Application Problem

The area of a tennis court for singles matches is 2106 square feet, and the perimeter is 210 feet. Find the dimensions of the court.

SOLUTION

We will set up a system of two equations that models the problem and then use the substitution method to solve it.

Step 1: Write a system of two equations that models the problem. We let x represent the length of the tennis court and y represent its width as shown in Figure 7-41.

FIGURE 7-41

Arthur Ashe Tennis Stadium

Because the area of the tennis court (2106 square feet) is the product of its length and width, we can form the equation $xy = 2106$. Because the perimeter of the court (210 feet) is the sum of twice its length and twice its width, we can write the equation $2x + 2y = 210$, or $x + y = 105$. This gives the following system of equations that models the problem.

$$\begin{cases} xy = 2106 & \text{This equation is nonlinear.} \\ x + y = 105 & \text{This equation is linear.} \end{cases}$$

Step 2: Use substitution to solve the system. First, we solve the linear equation for y.

$$x + y = 105$$

$$y = -x + 105 \qquad \text{This is the substitution equation.}$$

We can now substitute $-x + 105$ for y in the nonlinear equation and solve the resulting equation for x:

$$xy = 2106$$

$$x(-x + 105) = 2106 \qquad \text{Substitute } -x + 105 \text{ for } y.$$

$$-x^2 + 105x = 2106 \qquad \text{Remove parentheses.}$$

$$-x^2 + 105x - 2106 = 0 \qquad \text{Subtract 2106 from both sides.}$$

$$x^2 - 105x + 2106 = 0 \qquad \text{Multiply both sides by } -1.$$

$$(x - 78)(x - 27) = 0 \qquad \text{Factor } x^2 - 105x + 2106.$$

$$x - 78 = 0 \quad \text{or} \quad x - 27 = 0$$

$$x = 78 \qquad \qquad x = 27$$

Because $y = -x + 105$, if $x = 78$, then $y = 27$; and if $x = 27$, then $y = 78$. The two solutions are (78, 27) and (27, 78).

The dimensions of the tennis court are 78 feet by 27 feet.

Self Check 5 If the area of a tennis court for doubles matches is 2808 square feet and the perimeter is 228 feet, find the dimensions of the court.

Now Try Exercise 45.

Self Check Answers

1. $(4, 3)$ and $(0, -5)$ **2.** $(4, 3)$ and $(0, -5)$ **3.** $\left(2, \sqrt{3}\right)$ and $\left(2, -\sqrt{3}\right)$

4. $\left(3, \sqrt{5}\right), \left(3, -\sqrt{5}\right), \left(-3, \sqrt{5}\right),$ and $\left(-3, -\sqrt{5}\right)$

5. 78 ft by 36 ft

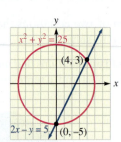

Exercises 7.4

Getting Ready

You should be able to complete these vocabulary and concept statements before you proceed to the practice exercises.

Fill in the blanks.

1. Solutions of nonlinear systems of equations are the points of intersection of the _____ of conic sections.

2. Approximate solutions of nonlinear systems can be found _____, and exact solutions can be found algebraically using the methods of _____ or _____.

Practice
Solve each system of equations by graphing.

3. $\begin{cases} 8x^2 + 32y^2 = 256 \\ x = 2y \end{cases}$

4. $\begin{cases} x^2 + y^2 = 2 \\ x + y = 2 \end{cases}$

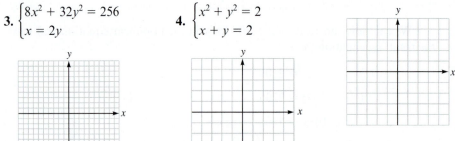

5. $\begin{cases} x^2 + y^2 = 90 \\ y = x^2 \end{cases}$

6. $\begin{cases} x^2 + y^2 = 5 \\ x + y = 3 \end{cases}$

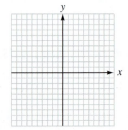

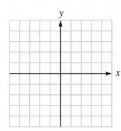

7. $\begin{cases} x^2 + y^2 = 25 \\ 12x^2 + 64y^2 = 768 \end{cases}$

8. $\begin{cases} x^2 + y^2 = 13 \\ y = x^2 - 1 \end{cases}$

9. $\begin{cases} x^2 - 13 = -y^2 \\ y = 2x - 4 \end{cases}$

10. $\begin{cases} x^2 + y^2 = 20 \\ y = x^2 \end{cases}$

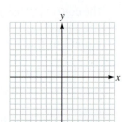

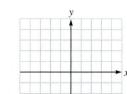

11. $\begin{cases} x^2 - 6x - y = -5 \\ x^2 - 6x + y = -5 \end{cases}$

12. $\begin{cases} x^2 - y^2 = -5 \\ 3x^2 + 2y^2 = 30 \end{cases}$

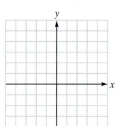

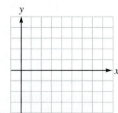

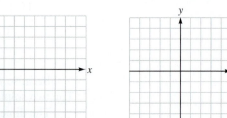

Use a graphing calculator to solve each system of equations.

13. $\begin{cases} y = x + 1 \\ y = x^2 + x \end{cases}$

14. $\begin{cases} y = 6 - x^2 \\ y = x^2 - x \end{cases}$

15. $\begin{cases} 6x^2 + 9y^2 = 10 \\ 3y - 2x = 0 \end{cases}$

16. $\begin{cases} x^2 + y^2 = 68 \\ y^2 - 3x^2 = 4 \end{cases}$

Solve each system of equations using substitution or elimination for real values of x and y.

17. $\begin{cases} 25x^2 + 9y^2 = 225 \\ 5x + 3y = 15 \end{cases}$

18. $\begin{cases} x^2 + y^2 = 20 \\ y = x^2 \end{cases}$

19. $\begin{cases} x^2 + y^2 = 2 \\ x + y = 2 \end{cases}$

20. $\begin{cases} x^2 + y^2 = 36 \\ 49x^2 + 36y^2 = 1764 \end{cases}$

21. $\begin{cases} x^2 + y^2 = 5 \\ x + y = 3 \end{cases}$

22. $\begin{cases} x^2 - x - y = 2 \\ 4x - 3y = 0 \end{cases}$

23. $\begin{cases} x^2 + y^2 = 13 \\ y = x^2 - 1 \end{cases}$

24. $\begin{cases} x^2 + y^2 = 25 \\ 2x^2 - 3y^2 = 5 \end{cases}$

25. $\begin{cases} x^2 + y^2 = 30 \\ y = x^2 \end{cases}$

26. $\begin{cases} 9x^2 - 7y^2 = 81 \\ x^2 + y^2 = 9 \end{cases}$

27. $\begin{cases} 2x^2 + y^2 = 6 \\ x^2 - y^2 = 3 \end{cases}$

28. $\begin{cases} x^2 + y^2 = 13 \\ x^2 - y^2 = 5 \end{cases}$

29. $\begin{cases} x^2 + y^2 = 20 \\ x^2 - y^2 = -12 \end{cases}$

30. $\begin{cases} xy = -\dfrac{9}{2} \\ 3x + 2y = 6 \end{cases}$

31. $\begin{cases} y^2 = 40 - x^2 \\ y = x^2 - 10 \end{cases}$

32. $\begin{cases} x^2 - 6x - y = -5 \\ x^2 - 6x + y = -5 \end{cases}$

33. $\begin{cases} y = x^2 - 4 \\ x^2 - y^2 = -16 \end{cases}$

34. $\begin{cases} 6x^2 + 8y^2 = 182 \\ 8x^2 - 3y^2 = 24 \end{cases}$

35. $\begin{cases} x^2 - y^2 = -5 \\ 3x^2 + 2y^2 = 30 \end{cases}$

36. $\begin{cases} \dfrac{1}{x} + \dfrac{1}{y} = 5 \\ \dfrac{1}{x} - \dfrac{1}{y} = -3 \end{cases}$

37. $\begin{cases} \dfrac{1}{x} + \dfrac{2}{y} = 1 \\ \dfrac{2}{x} - \dfrac{1}{y} = \dfrac{1}{3} \end{cases}$

38. $\begin{cases} \dfrac{1}{x} + \dfrac{3}{y} = 4 \\ \dfrac{2}{x} - \dfrac{1}{y} = 7 \end{cases}$

39. $\begin{cases} 3y^2 = xy \\ 2x^2 + xy - 84 = 0 \end{cases}$

40. $\begin{cases} x^2 + y^2 = 10 \\ 2x^2 - 3y^2 = 5 \end{cases}$

41. $\begin{cases} xy = \dfrac{1}{6} \\ y + x = 5xy \end{cases}$

42. $\begin{cases} xy = \dfrac{1}{12} \\ y + x = 7xy \end{cases}$

Applications

43. Geometry The area of a rectangle is 63 square centimeters, and its perimeter is 32 centimeters. Find the dimensions of the rectangle.

44. Dimensions of a whiteboard The area of a SMART Board interactive whiteboard is 2880 square inches, and its perimeter is 216 inches. Find the dimensions of the whiteboard.

45. Fencing pastures The rectangular pasture shown below is to be fenced in along a riverbank. If 260 feet of fencing is to enclose an area of 8000 square feet, find the dimensions of the pasture.

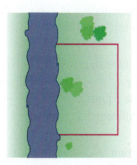

46. Investments Grant receives $225 annual income from one investment. Jeff invested $500 more than Grant, but at an annual rate of 1% less. Jeff's annual income is $240. Find the amount and rate of Grant's investment.

47. Investments Carol receives $67.50 annual income from one investment. John invested $150 more than Carol at an annual rate of $1\frac{1}{2}\%$ more. John's annual income is $94.50. Find the amount and rate of Carol's investment. (*Hint:* There are two answers.)

48. Finding the rate and time Jim drove 306 miles. Jim's brother made the same trip at a speed 17 miles per hour slower than Jim did and required an extra $1\frac{1}{2}$ hours. Find Jim's rate and time.

49. Paintball See the illustration. A liquid-filled paint-ball is shot from the base of an incline and follows the parabolic path $y = -\frac{1}{300}x^2 + \frac{1}{5}x$, with distances measured in feet. The incline has a slope of $\frac{1}{10}$. Find the coordinates of the point of intersection of the paintball and the ground at impact.

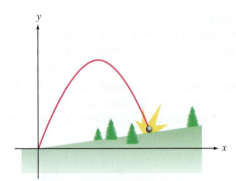

50. Artillery See the illustration for Exercise 49. A shell fired from the base of a hill follows the parabolic path $y = -\frac{1}{6}x^2 + 2x$, with distances measured in miles. The hill has a slope of $\frac{1}{3}$. How far from the base of the hill is the point of impact? (*Hint:* Find the coordinates of the point and then the distance.)

51. Air-traffic control A plane is flying over an airport on a path whose equation is $y = x^2$. If a second plane, flying at the same altitude, is traveling on a path whose equation is $x + y = 2$, is there any danger of a mid-air collision?

52. Ship traffic One ship is steaming on a path whose equation is $y = x^2 + 1$ and another is steaming on a path whose equation is $x + y = -4$. Is there any danger of collision?

53. Radio reception A radio station located 120 miles due east of Collinsville has a listening radius of 100 miles. A straight road joins Collinsville with Harmony, a town 200 miles to the east and 100 miles north. See the illustration. If a driver leaves Collinsville and heads toward Harmony, how far from Collinsville will the driver pick up the station?

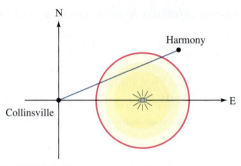

54. Listening ranges For how many miles will a driver in Exercise 53 continue to receive the signal?

Discovery and Writing

55. What is a system of nonlinear equations? Give an example to support your answer.

56. Describe three methods that can be used to solve a nonlinear system of equations.

57. Describe any disadvantages in using the graphing method to solve nonlinear systems of equations.

58. Explain why the elimination method, not the substitution method, is the better method to solve the system $\begin{cases} x^2 + 4y^2 = 16 \\ x^2 - y^2 = 1 \end{cases}$.

Critical Thinking

Determine if the statement is true or false. If the statement is false, then correct it and make it true.

59. The graphing method is the most precise method for solving nonlinear systems of equations

60. The substitution method is the best method for solving systems consisting of a first-degree equation and a second-degree equation.

61. If both equations of the system are of the form $Ax^2 + By^2 = C$, the substitution method is the best method to use to solve the system.

62. It is possible for a system of two equations and two variables whose graphs are a line and a parabola to have no solution.

63. It is possible for a system of two equations and two variables whose graphs are a parabola and a circle to have exactly one real ordered-pair solution.

64. It is possible for a system of two equations and two variables whose graphs are a hyperbola and an ellipse to have exactly five real ordered-pair solutions.

65. The product of two integers is 32 and their sum is 12. If you find the two numbers and subtract the smaller one from the larger one, you will obtain 2.

66. The sum of the squares of two numbers is 221, and the sum of the numbers is 9. If you find the two numbers and subtract the smaller one from the larger one, you will obtain 19.

CHAPTER REVIEW

7.1 The Circle and the Parabola

Definitions and Concepts	Examples
Standard form of an equation of a circle with center at $(0, 0)$ and radius r: $$x^2 + y^2 = r^2$$	The standard equation of the circle with center $(0, 0)$ and radius 5 is: $x^2 + y^2 = 5^2$ or $x^2 + y^2 = 25$. The graph of the circle is: 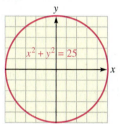
Standard form of an equation of a circle with center at (h, k) and radius r: $$(x - h)^2 + (y - k)^2 = r^2$$	The standard equation of the circle with center $(4, 6)$ and radius 3 is: $$(x - 4)^2 + (y - 6)^2 = 9$$
General form of a second-degree equation in x and y: $$Ax^2 + Bxy + Cy^2 + Dx + Ey + F = 0$$ The second degree equation is a circle if $A = C$ and $B = 0$.	The general form of an equation of a circle is $x^2 + y^2 + 4x - 10y + 25 = 0$. To write the equation in standard form, we first subtract 25 from both sides to obtain $$x^2 + y^2 + 4x - 10y = -25$$ and then complete the square on x and y. $$x^2 + 4x + y^2 - 10y = -25 \quad \text{Rearrange terms.}$$ To complete the square, we add 4 and 25 to both sides. $$x^2 + 4x + 4 + y^2 - 10y + 25 = -25 + 4 + 25$$ We can factor and simplify on the right side to get $$(x + 2)^2 + (y - 5)^2 = 4$$ The circle is now written in standard form. From the equation, we can see that the coordinates of the center of the circle are $(-2, 5)$ and that the radius of the circle is 2.
Parabola: A **parabola** is the set of all points in a plane equidistant from a line l (called the **directrix**) and fixed point F (called the **focus**) that is not on line l. 	Find an equation of the parabola with vertex at the origin and focus at $(4, 0)$. Because the focus is to the right of the vertex, the parabola opens to the right, and because the vertex is the origin, the standard equation is $y^2 = 4px$. Since the distance between the focus and the vertex is $p = 4$, we can substitute 4 for p in the standard equation to get $$y^2 = 4px$$ $$y^2 = 4(4)x$$ $$y^2 = 16x$$ We can easily graph the parabola by plotting the vertex $(0, 0)$ and two points such as $(1, 4)$ and $(1, -4)$ which satisfy the equations.

Parabola Opening	Vertex at Origin
Right	$y^2 = 4px \quad (p > 0)$
Left	$y^2 = 4px \quad (p < 0)$
Upward	$x^2 = 4py \quad (p > 0)$
Downward	$x^2 = 4py \quad (p < 0)$

For a parabola that opens right or left with vertex at the origin, the directrix is $x = -p$ and the focus is $(p, 0)$.

Definitions and Concepts

For a parabola that opens upward or downward with vertex at the origin, the directrix is $y = -p$ and the focus is $(0, p)$.

Parabola Opening	Vertex at $V(h, k)$	
Right	$(y - k)^2 = 4p(x - h)$	$(p > 0)$
Left	$(y - k)^2 = 4p(x - h)$	$(p < 0)$
Upward	$(x - h)^2 = 4p(y - k)$	$(p > 0)$
Downward	$(x - h)^2 = 4p(y - k)$	$(p < 0)$

For a parabola that opens right or left with vertex at (h, k), the directrix is $x = -p + h$ and the focus is $(h + p, k)$.

For a parabola that opens upward or downward with vertex at (h, k), the directrix is $y = -p + k$ and the focus is $(h, k + p)$.

Examples

The graph of the parabola is shown in the figure.

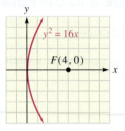

Exercises

Write an equation in standard form of each circle described.

1. Center $(0, 0)$; radius 4

2. Center $(0, 0)$; passes through $(6, 8)$

3. Center $(3, -2)$; radius 5

4. Center $(-2, 4)$; passes through $(1, 0)$

5. Endpoints of diameter $(-2, 4)$ and $(12, 16)$

6. Endpoints of diameter $(-3, -6)$ and $(7, 10)$

Write an equation of each circle in standard form and graph the circle.

7. $x^2 + y^2 - 6x + 4y = 3$

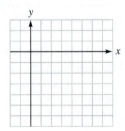

8. $x^2 + 4x + y^2 - 10y = -13$

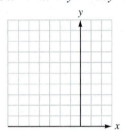

Write an equation in standard form of each parabola described.

9. Vertex $(0, 0)$; passes through $(-8, 4)$ and $(-8, -4)$

10. Vertex $(0, 0)$; passes through $(-8, 4)$ and $(8, 4)$

11. Find an equation in standard form of the parabola with vertex at $(-2, 3)$, curve passing through point $(-4, -8)$, and opening down.

Graph each parabola.

12. $x^2 - 4y - 2x + 9 = 0$

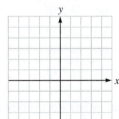

13. $y^2 - 6y = 4x - 13$

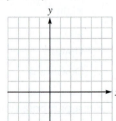

7.2 The Ellipse

Definitions and Concepts	**Examples**

Ellipse:

An **ellipse** is the set of all points P in a plane such that the sum of the distances from P to two other fixed points F and F' is a positive constant.

Ellipse with the major axis on the x-axis and center at $(0, 0)$:

The standard form of an equation of an ellipse with center at the origin and the major axis (horizontal) on the x-axis is

$$\frac{x^2}{a^2} + \frac{y^2}{b^2} = 1 \text{ where } a > b > 0$$

Vertices (ends of the major axis):

$$V(a, 0) \text{ and } V'(-a, 0)$$

Length of major axis: $2a$

Ends of the minor axis: $B(0, b)$ and $B'(0, -b)$

Length of minor axis: $2b$

Foci: $(c, 0)$ and $(-c, 0)$ where $c^2 = a^2 - b^2$

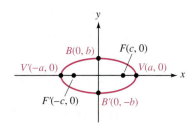

Find an equation in standard form of the ellipse with center at the origin, major axis of length 10 units located on the x-axis, and minor axis of length 4 units.

To find an equation, we determine a and b and substitute into the standard equation of an ellipse

$$\frac{x^2}{a^2} + \frac{y^2}{b^2} = 1$$

with center at the origin and major axis on the x-axis.

The center of the ellipse is given to be the origin and the length of the major axis is 10. Since the length of the major axis of an ellipse centered at the origin is $2a$, we have $2a = 10$ or $a = 5$ The coordinates of the vertices are $(5, 0)$ and $(-5, 0)$.

The length of the minor axis is given to be 4. Since the length of the minor axis of an ellipse centered at the origin is $2b$, we have $2b = 4$ or $b = 2$. The coordinates of B and B' are $(0, 2)$ and $(0, -2)$.

To find an equation of the ellipse, we substitute 5 for a and 2 for b in the standard equation of an ellipse with center at the origin and major axis on the x-axis.

$$\frac{x^2}{a^2} + \frac{y^2}{b^2} = 1$$

$$\frac{x^2}{5^2} + \frac{y^2}{2^2} = 1$$

$$\frac{x^2}{25} + \frac{y^2}{4} = 1$$

The graph of the ellipse is shown below.

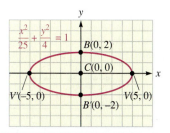

Ellipse with the major axis on the y-axis and center at $(0, 0)$:

The standard form of an equation of an ellipse with center at the origin and the major axis (vertical) on the y-axis is

$$\frac{x^2}{b^2} + \frac{y^2}{a^2} = 1 \text{ where } a > b > 0$$

An example of an ellipse in standard form $\frac{x^2}{b^2} + \frac{y^2}{a^2} = 1$ with center at the origin and major axis (vertical) and on the y-axis is $\frac{x^2}{4} + \frac{y^2}{36} = 1$.

From the equation, we can see that $a = 6$ and the vertices of the ellipse are $V(0, 6)$ and $V'(0, -6)$. We can also see that $b = 2$ and the ends of the minor axis are $B(2, 0)$ and $B'(-2, 0)$.

Definitions and Concepts	**Examples**

Vertices (ends of the major axis):

$V(0, a)$ and $V'(0, -a)$

Length of major axis: $2a$

Ends of the minor axis: $B(b, 0)$ and $B'(-b, 0)$

Length of minor axis: $2b$

Foci: $F(0, c)$ and $F'(0, -c)$ where $c^2 = a^2 - b^2$

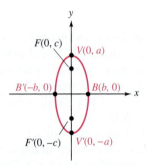

Ellipse with the major axis horizontal and center at (h, k):
The standard form of an equation of an ellipse with center at (h, k) and major axis horizontal is

$$\frac{(x - h)^2}{a^2} + \frac{(y - k)^2}{b^2} = 1 \text{ where } a > b > 0$$

Vertices (ends of the major axis):

$V(a + h, k)$ and $V'(-a + h, k)$

Ends of the minor axis:

$B(h, b + k)$ and $B'(h, -b + k)$

Foci: $F(h + c, k)$ and $F'(h - c, k)$ where
$c^2 = a^2 - b^2$

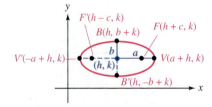

Ellipse with the major axis vertical and center at (h, k):
The standard form of an equation of an ellipse with center at (h, k) and major axis vertical is

$$\frac{(x - h)^2}{b^2} + \frac{(y - k)^2}{a^2} = 1 \text{ where } a > b > 0$$

Graph the ellipse $9x^2 + 4y^2 - 18x + 16y = 11$.

We first write the equation in standard form by completing the square on x and y.

$$9x^2 + 4y^2 - 18x + 16y = 11$$

$$9x^2 - 18x + 4y^2 + 16y = 11$$

$$9(x - 2x) + 4(y + 4y) = 11$$

$$9(x^2 - 2x + \mathbf{1}) + 4(y^2 + 4y + \mathbf{4}) = 11 + \mathbf{9} + \mathbf{16}$$

$$9(x - 1)^2 + 4(y + 2)^2 = 36$$

$$\frac{(x - 1)^2}{4} + \frac{(y + 2)^2}{9} = 1$$

The major axis of the ellipse is vertical and the center of the ellipse is $(1, -2)$. The graph of the ellipse is shown below.

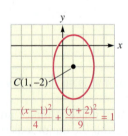

Definitions and Concepts	**Examples**
Vertices (ends of the major axis): $V(h, a + k)$ and $V'(h, -a + k)$ **Ends of the minor axis:** $B(b + h, k)$ and $B'(-b + h, k)$ **Foci:** $F(h, k + c)$ and $F'(h, k - c)$ where $c^2 = a^2 - b^2$ 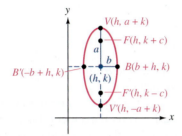	

Exercises

14. Write an equation of the ellipse in standard form with center at the origin, major axis that is horizontal and 12 units long, and minor axis 8 units long.

15. Write an equation of the ellipse in standard form with center at the origin, major axis that is vertical and 10 units long, and minor axis 4 units long.

16. Write an equation of the ellipse in standard form with center at point $(-2, 3)$ and curve passing through points $(-2, 0)$ and $(2, 3)$.

17. Write the equation of the ellipse in standard form and graph it. $4x^2 + y^2 - 16x + 2y = -13$

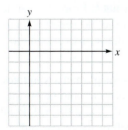

7.3 The Hyperbola

Definitions and Concepts

Hyperbola:
A **hyperbola** is the set of all points P in a plane such that the absolute value of the difference of the distances from point P to two other points in the plane is a positive constant.

Hyperbola with the foci on the x-axis and center at (0, 0):

The standard form of an equation of a hyperbola with center at the origin and foci on the x-axis is

$$\frac{x^2}{a^2} - \frac{y^2}{b^2} = 1 \text{ where } a^2 + b^2 = c^2$$

Vertices: $V(a, 0)$ and $V'(-a, 0)$

Foci: $F(c, 0)$ and $F'(-c, 0)$

Asymptotes: $y = \pm\dfrac{b}{a}x$

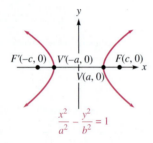

Hyperbola with the foci on the y-axis and center at (0, 0):
The standard form of an equation of a hyperbola with center at the origin and foci on the y-axis is

$$\frac{y^2}{a^2} - \frac{x^2}{b^2} = 1 \text{ where } a^2 + b^2 = c^2$$

Vertices: $V(0, a)$ and $V'(0, -a)$

Foci: $F(0, c)$ and $F'(0, -c)$

Asymptotes: $y = \pm\dfrac{a}{b}x$

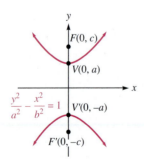

Examples

Write an equation in standard form of the hyperbola with vertices $V(3, 0)$ and $V'(-3, 0)$ and a focus at $F(5, 0)$.

Because the foci lie on the x-axis, we use the standard equation

$$\frac{x^2}{a^2} - \frac{y^2}{b^2} = 1.$$

We will find a^2 and b^2 and substitute the results into the standard equation.

The center of the hyperbola is midway between the vertices V and V'. Thus, the center is the origin $(0, 0)$. The distance between the vertex and the center is $a = 3$, and the distance between the focus and the center is $c = 5$. We can find b^2 by substituting 3 for a and 5 for c in the following equation to get

$$b^2 = c^2 - a^2 \quad \text{In a hyperbola, } a^2 + b^2 = c^2 \text{ or } b^2 = c^2 - a^2.$$

$$b^2 = 5^2 - 3^2$$

$$b^2 = 16$$

Substituting the values for a^2 and b^2 in the standard equation gives an equation of the hyperbola:

$$\frac{x^2}{a^2} - \frac{y^2}{b^2} = 1$$

$$\frac{x^2}{9} - \frac{y^2}{16} = 1$$

Graph the hyperbola: $\dfrac{y^2}{4} - \dfrac{x^2}{25} = 1$.

Step 1: Write the hyperbola in standard form. The equation of the hyperbola $\dfrac{y^2}{4} - \dfrac{x^2}{25} = 1$ is already written in the standard form $\dfrac{y^2}{a^2} - \dfrac{x^2}{b^2} = 1$.

Step 2: Identify the center and a and b. From the standard equation of a hyperbola, we see that the center is $(0, 0)$, that $a = 2$ and $b = 5$, and that the vertices are on the y-axis.

Step 3: Draw the fundamental rectangle and asymptotes. The vertices V and V' are 2 units above and below the origin and have coordinates of $(0, 2)$ and $(0, -2)$. Points B and B' are 5 units right and left of the origin and have coordinates of $(5, 0)$ and $(-5, 0)$. We use the points V, V', B, and B' to construct the fundamental rectangle and extend the diagonals to graph the asymptotes, as shown in the following figure.

Definitions and Concepts	**Examples**

Hyperbola with the transverse axis horizontal and center at (h, k):

The standard form of an equation of a hyperbola with center at the (h, k) and foci on a line parallel to the x-axis is

$$\frac{(x - h)^2}{a^2} - \frac{(y - k)^2}{b^2} = 1 \text{ where } a^2 + b^2 = c^2$$

Vertices: $V(a + h, k)$ and $V'(-a + h, k)$

Foci: $F(c + h, k)$ and $F'(-c + h, k)$

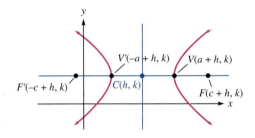

Hyperbola with the transverse axis vertical and center at (h, k):

The standard equation of a hyperbola with center at (h, k) and foci on a line parallel to the y-axis is

$$\frac{(y - k)^2}{a^2} - \frac{(x - h)^2}{b^2} = 1 \text{ where } a^2 + b^2 = c^2$$

Vertices: $V(h, a + k)$ and $V'(h, -a + k)$

Foci: $F(h, c + k)$ and $F'(h, -c + k)$

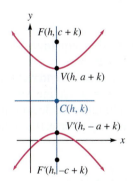

Step 4: Sketch the hyperbola. We begin with the vertices and sketch the graph of the hyperbola, as shown in the figure below using the asymptotes as guides.

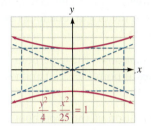

Graphing hyperbolas:

1. Write the hyperbola in standard form.
2. Identify the center and a and b.
3. Draw the fundamental rectangle and asymptotes.
4. Sketch the hyperbola beginning with the vertices and using the asymptotes as guides.

Exercises

18. Write an equation of the hyperbola in standard form with center at the origin, passing through points $(-2, 0)$ and $(2, 0)$, and having a focus at $(4, 0)$.

19. Write an equation of the hyperbola in standard form with center at the origin, one focus at $(0, 5)$, and one vertex at $(0, 3)$.

20. Write an equation of the hyperbola in standard form with vertices at points $(-3, 3)$ and $(3, 3)$ and a focus at point $(5, 3)$.

21. Write an equation of the hyperbola in standard form with vertices at points $(3, -3)$ and $(3, 3)$ and a focus at point $(3, 5)$.

22. Write equations of the asymptotes of the hyperbola $\frac{x^2}{25} - \frac{y^2}{16} = 1$.

23. Write the equation in standard form and graph it.

$$9x^2 - 4y^2 - 16y - 18x = 43$$

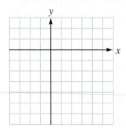

24. Graph: $4xy = 1$.

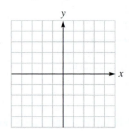

7.4 Solving Nonlinear Systems of Equations

Definitions and Concepts	Examples

There are three methods that can be used to solve
nonlinear systems of equations: graphing, substitution, and elimination.

Graphing method:
Graph each equation and identify the point(s) of intersection of the graphs.

Solve by graphing: $\begin{cases} x^2 + y^2 = 16 \\ x + y = 4 \end{cases}$.

To solve the system, we will graph each equation and identify the point(s) of intersection of the two graphs.
The graph of $x^2 + y^2 = 16$ is a circle with center at the origin and radius of 4. The equation of $x + y = 4$ is a line with x-intercept $(4, 0)$ and y-intercept $(0, 4)$. After graphing the circle and the line, as shown in the figure, we see that there are two intersection points or solutions, $P(0, 4)$ and $P'(4, 0)$.

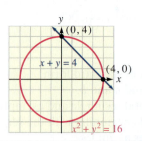

Definitions and Concepts	Examples
Substitution method: We convert the two-variable, two-equation system into one equation and one variable using substitution. To do so, solve one equation for one variable and then substitute into the second equation. Next, solve for the one variable. Finally, use back substitution to find the second variable. This method is recommended when one of the equations is second degree and one equation is linear.	**Solve by substitution:** $\begin{cases} x^2 + y^2 = 2 \\ y = x \end{cases}$. We will substitute y for x into the first equation and convert the system into one equation and one variable y. Then, we will solve for x and use back substitution to solve for the second variable. $x^2 + y^2 = 2$ $y^2 + y^2 = 2$ Substitute y for x. $2y^2 = 2$ $y^2 = 1$ $y = 1$ or $y = -1$ Solve by Square Root Property. Because $y = x$, the solutions are $(1, 1)$ and $(-1, -1)$.
Elimination method: We eliminate one of the variables by addition and then solve the system. Use this method when a system consists of two second-degree equations of the form $ax^2 + by^2 = c$.	**Solve by elimination:** $\begin{cases} x^2 + y^2 = 4 \\ 4x^2 + y^2 = 16 \end{cases}$. To eliminate the terms involving y^2, we copy the first equation and multiply the second equation by -1 to obtain the following equivalent system. $\begin{cases} x^2 + y^2 = 4 \\ -4x^2 - y^2 = -16 \end{cases}$ We can then add the equations and solve the resulting equation for x: $-3x^2 = -12$ $x^2 = 4$ $x = 2$ or $x = -2$ To find y, we substitute 2 for x and then -2 for x in the first equation. For $x = 2$ For $x = -2$ $x^2 + y^2 = 4$ $x^2 + y^2 = 4$ $2^2 + y^2 = 4$ $(-2)^2 + y^2 = 4$ $4 + y^2 = 4$ $4 + y^2 = 4$ $y^2 = 0$ $y^2 = 0$ $y = 0$ $y = 0$ The two solutions are $(2, 0)$ and $(-2, 0)$.

Exercises

25. Solve by graphing: $\begin{cases} x^2 + y^2 = 16 \\ y = x + 4 \end{cases}$.

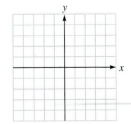

26. Solve by graphing: $\begin{cases} 3x^2 + y^2 = 52 \\ x^2 - y^2 = 12 \end{cases}$.

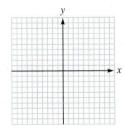

27. Solve by graphing: $\begin{cases} \dfrac{x^2}{16} + \dfrac{y^2}{12} = 1 \\ x^2 - \dfrac{y^2}{3} = 1 \end{cases}$.

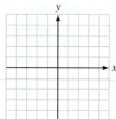

28. Solve by substitution or elimination: $\begin{cases} 3x^2 + y^2 = 52 \\ x^2 - y^2 = 12 \end{cases}$.

29. Solve by substitution or elimination:

$$\begin{cases} x^2 + y^2 = 16 \\ -\sqrt{3}y + 4\sqrt{3} = 3x \end{cases}$$

30. Solve by substitution or elimination: $\begin{cases} \dfrac{x^2}{16} + \dfrac{y^2}{12} = 1 \\ x^2 - \dfrac{y^2}{3} = 1 \end{cases}$.

CHAPTER TEST

Write an equation in standard form of each circle described.

1. Center $(2, 3)$; $r = 3$

2. Ends of diameter at $(-2, -2)$ and $(6, 8)$

3. Center $(2, -5)$, passes through $(7, 7)$

4. Change the equation of the circle
$x^2 + y^2 - 4x + 6y + 4 = 0$ to standard form and graph it.

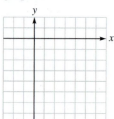

Find an equation in standard form of each parabola described.

5. Vertex $(3, 2)$; focus at $(3, 6)$

6. Vertex $(4, -6)$; passes through $(3, -8)$ and $(3, -4)$

7. Vertex $(2, -3)$; passes through $(0, 0)$

8. Change the equation of the parabola
$x^2 - 6x - 8y = 7$ into standard form and graph it.

Find an equation in standard form of each ellipse described.

9. Vertex $(10, 0)$, center at the origin, focus at $(6, 0)$

10. Minor axis 24, center at the origin, focus at $(5, 0)$

11. Center $(2, 3)$; passes through $(2, 9)$ and $(0, 3)$

12. Change the equation of the ellipse
$9x^2 + 4y^2 - 18x - 16y - 11 = 0$ into standard form and graph it.

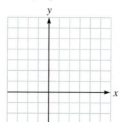

Find an equation in standard form of each hyperbola described.

13. Center at the origin, focus at $(13, 0)$, vertex at $(5, 0)$

14. Vertices $(6, 0)$ and $(-6, 0)$; $\dfrac{c}{a} = \dfrac{13}{12}$

15. Center $(2, -1)$, major axis horizontal and of length 16, distance of 20 between foci

16. Change the equation of the hyperbola $x^2 - 4y^2 + 16y = 8$ into standard form and graph it.

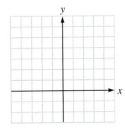

Solve each system using substitution or elimination.

17. $\begin{cases} x^2 + y^2 = 23 \\ y = x^2 - 3 \end{cases}$

18. $\begin{cases} 2x^2 - 3y^2 = 9 \\ x^2 + y^2 = 27 \end{cases}$

Complete the square to write each equation in standard form, and identify the curve.

19. $y^2 - 4y - 6x - 14 = 0$

20. $2x^2 + 3y^2 - 4x + 12y + 8 = 0$

CUMULATIVE REVIEW EXERCISES

Simplify each expression. Assume that all variables represent positive numbers, and write answers without using negative exponents.

1. $64^{2/3}$

2. $8^{-1/3}$

3. $\dfrac{y^{2/3}y^{5/3}}{y^{1/3}}$

4. $\dfrac{(x^{5/3})(x^{1/2})}{x^{3/4}}$

5. $(x^{2/3} - x^{1/3})(x^{2/3} + x^{1/3})$

6. $(x^{-1/2} + x^{1/2})^2$

7. $\sqrt[3]{-27x^3}$

8. $\sqrt{48t^3}$

9. $\sqrt[3]{\dfrac{128x^4}{2x}}$

10. $\sqrt{x^2 + 6x + 9}$

11. $\sqrt{50} - \sqrt{8} + \sqrt{32}$

12. $-3\sqrt[4]{32} - 2\sqrt[4]{162} + 5\sqrt[4]{48}$

13. $3\sqrt{2}(2\sqrt{3} - 4\sqrt{12})$

14. $\dfrac{5}{\sqrt[3]{x}}$

15. $\dfrac{\sqrt{x} + 2}{\sqrt{x} - 1}$

16. $\sqrt[6]{x^3 y^3}$

Solve each equation.

17. $5\sqrt{x + 2} = x + 8$

18. $\sqrt{x} + \sqrt{x + 2} = 2$

19. Use the method of completing the square to solve the equation $2x^2 + x - 3 = 0$.

20. Use the Quadratic Formula to solve the equation $3x^2 + 4x - 1 = 0$.

Perform each operation and write each complex number in a + bi form.

21. $(3 + 5i) + (4 - 3i)$

22. $(7 - 4i) - (12 + 3i)$

23. $(2 - 3i)(2 + 3i)$

24. $(3 + i)(3 - 3i)$

25. $(3 - 2i) - (4 + i)^2$

26. $\dfrac{5}{3 - i}$

Find each value.

27. $|3 + 2i|$

28. $|5 - 6i|$

29. For what values of k will the solutions of $2x^2 + 4x = k$ be equal?

30. Find the coordinates of the vertex of the graph of the equation $y = \frac{1}{2}x^2 - x + 1$.

Solve each inequality and give the result in interval notation.

31. Solve: $x^2 - x - 6 > 0$.

32. Solve: $x^2 - x - 6 \leq 0$.

Let $f(x) = 3x^2 + 2$ and $g(x) = 2x - 1$.

Find each value or function.

33. $f(-1)$

34. $(g \circ f)(2)$

35. $(f \circ g)(x)$

36. $(g \circ f)(x)$

37. Write $y = \log_2 x$ in exponential notation.

38. Write $3^b = a$ in logarithmic notation.

Find x.

39. $\log_x 25 = 2$

40. $\log_5 125 = x$

41. $\log_3 x = -3$

42. $\log_5 x = 0$

43. Find the inverse of $y = \log_2 x$.

44. If $\log_{10} 10^x = y$, then y equals what quantity?

log 7 ≈ 0.8451 *and* log 14 ≈ 1.1461

Approximate each expression without using a calculator or tables.

45. $\log 98$

46. $\log 2$

47. $\log 49$

48. $\log \dfrac{7}{5}$ (*Hint:* log 10 = 1.)

49. Solve: $2^{x+2} = 3^x$.

50. Solve: $2 \log 5 + \log x - \log 4 = 2$.

Use a calculator for Exercises 51 and 52.

51. Boat depreciation How much will a $9000 boat be worth after 9 years if it depreciates 12% per year?

52. Find $\log_6 8$.

53. Use graphing to solve $\begin{cases} 2x + y = 5 \\ x - 2y = 0 \end{cases}$.

54. Use substitution to solve $\begin{cases} 3x + y = 4 \\ 2x - 3y = -1 \end{cases}$.

55. Use elimination to solve $\begin{cases} x + 2y = -2 \\ 2x - y = 6 \end{cases}$.

56. Use any method to solve $\begin{cases} \dfrac{x}{10} + \dfrac{y}{5} = \dfrac{1}{2} \\ \dfrac{x}{2} - \dfrac{y}{5} = \dfrac{13}{10} \end{cases}$.

57. Evaluate: $\begin{vmatrix} 3 & -2 \\ 1 & -1 \end{vmatrix}$.

58. Use Cramer's Rule and solve for y only: $\begin{cases} 4x - 3y = -1 \\ 3x + 4y = -7 \end{cases}$

59. Solve: $\begin{cases} x + y + z = 1 \\ 2x - y - z = -4 \\ x - 2y + z = 4 \end{cases}$.

60. Solve: $\begin{cases} x + 2y + 3z = 6 \\ 3x + 2y + z = 6 \\ 2x + 3y + z = 6 \end{cases}$.

61. Identify the vertex of the parabola $(y - 3)^2 = 8(x + 3)$.

62. Graph $x^2 = -8y$.

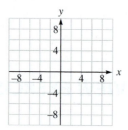

63. Graph $\dfrac{(x - 1)^2}{9} + \dfrac{(y - 3)^2}{25} = 1$.

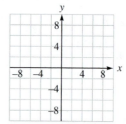

64. Graph $\dfrac{(y - 3)^2}{9} - \dfrac{(x - 2)^2}{16} = 1$.

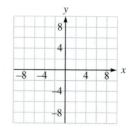

Sequences, Series, and Probability

8

Careers and Mathematics: Actuary

©Andrey_Popov/Shutterstock.com

Actuaries deal with the financial impact of risk and uncertainty. Actuaries are in high demand and are well paid for their services. They work for insurance companies and estimate the probability and likely cost of events such as death, sickness, injury, disability, or loss of property. They use this information to help design insurance policies. In the finance sector, they aid companies in how to invest their resources to maximize the return on investment and help ensure that pension plans are maintained on a sound financial basis.

Education and Mathematics Required

- Actuaries generally have a bachelor's degree and are required to pass a series of exams in order to become a certified actuary.
- College Algebra, Trigonometry, Calculus I and II, Linear Algebra, Probability and Mathematical Statistics, Applied Statistics, and Actuarial Mathematics are math courses required. Optional courses in numerical analysis, training in operations research, and substantial training in computer science are also necessary.

How Actuaries Use Math and Who Employs Them

- Actuaries estimate the probability and likely cost of an event, such as death, sickness, injury, or loss of property. To do this, they must gather and correctly analyze data. Actuaries also have a broad knowledge of statistics, finance, and business.
- About 69% of all actuaries are employed by insurance carriers. Another 17% work for professional, scientific, and technical consulting services. A small number are employed by government agencies.

Career Outlook and Earnings

- Employment is expected to grow much faster than the average for all occupations. Employment of actuaries is expected to increase by 26% over the 2012–2022 period. Greater job growth will occur in the financial and consulting areas than in the insurance sector.
- The median annual salary in 2012 was $93,680, with the top 10% earning more than $175,330.

For more information see: www.bls.gov/oco

8.1 The Binomial Theorem

In this section, we will learn to

1. Use Pascal's Triangle to expand a binomial.
2. Define and use factorial notation.
3. Use the Binomial Theorem to expand binomials.
4. Find a specific term of a binomial expansion.

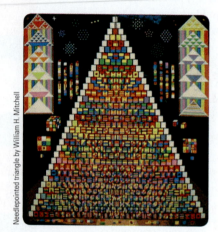

Needlepointed triangle by William H. Mitchell

We begin this chapter by discussing a triangle, called **Pascal's Triangle**, which is named after the French mathematician Blaise Pascal. Although the Western world gives credit for the triangle to Pascal, other mathematicians in other parts of the world studied it centuries before him. Around 1068, the Indian mathematician Bhattotpala presented the first seventeen rows of the triangle. Around the same time, the great Persian mathematician Omar Khayyám discussed the triangle. Today, in Iran, the triangle is referred to as the Khayyám Triangle. Around 1250, the Chinese mathematician Yang Hui developed the triangle, and in China, the triangle is known as Yang Hui's Triangle.

Over the years many people have searched the triangle for the many number patterns it contains. One such person, William H. Mitchell, was so impressed with the triangle that he created a needlepoint design of the triangle.

We introduce the triangle by discussing a way to expand binomials of the form $(x + y)^n$, where n is a natural number. When we expand the power of a binomial, the result is called a **binomial expansion**. Consider the following binomial expansions.

$$(a + b)^0 = 1$$
$$(a + b)^1 = a + b$$
$$(a + b)^2 = a^2 + 2ab + b^2$$
$$(a + b)^3 = a^3 + 3a^2b + 3ab^2 + b^3$$
$$(a + b)^4 = a^4 + 4a^3b + 6a^2b^2 + 4ab^3 + b^4$$
$$(a + b)^5 = a^5 + 5a^4b + 10a^3b^2 + 10a^2b^3 + 5ab^4 + b^5$$
$$(a + b)^6 = a^6 + 6a^5b + 15a^4b^2 + 20a^3b^3 + 15a^2b^4 + 6ab^5 + b^6$$

Four patterns are apparent in the above expansions:

1. Each expansion has one more term than the power of the binomial.
2. The degree of each term in each expansion equals the exponent of the binomial.
3. The first term in each expansion is a raised to the power of the binomial.
4. The exponents of a decrease by 1 in each successive term, and the exponents on b, beginning with b^0 in the first term, increase by 1 in each successive term.

1. Use Pascal's Triangle to Expand a Binomial

To see another pattern, we write the coefficients of the above expansions in a triangular array:

$(a + b)^0 =$						1					Row 0
$(a + b)^1 =$					1		1				Row 1
$(a + b)^2 =$				1		2		1			Row 2
$(a + b)^3 =$			1		3		3		1		Row 3
$(a + b)^4 =$		1		4		6		4		1	Row 4
$(a + b)^5 =$	1		5		10		10		5		1	Row 5
$(a + b)^6 =$	1	6		15		20		15		6	1	Row 6

Blaise Pascal statue in Saint-Jacques Tower, Paris, France

In this array, each entry other than the 1's is the sum of the closest pair of numbers in the line above it. For example, the first 3 in the third row is the sum of the 1 and 2 above it. The 20 in the sixth row is the sum of the 10's above it.

This array, called **Pascal's Triangle** after Blaise Pascal (1623–1662), continues with the same pattern forever. The next two lines are

$(a + b)^7 =$		1		7		21		35		35		21		7		1	**Row 7**
$(a + b)^8 =$	1		8		28		56		70		56		28		8		1	**Row 8**

EXAMPLE 1

Using Pascal's Triangle to Expand a Binomial

Expand: $(x + y)^6$.

SOLUTION The first term is x^6, and the exponents on x will decrease by 1 in each successive term. A y will appear in the second term, and the exponents on y will increase by 1 in each successive term, concluding when the term y^6 is reached. The variables in the expansion are

$$x^6 \qquad x^5y \qquad x^4y^2 \qquad x^3y^3 \qquad x^2y^4 \qquad xy^5 \qquad y^6$$

We can use Pascal's Triangle to find the coefficients of the variables. Because the binomial is raised to the sixth power, we choose Row 6 of Pascal's Triangle. The coefficients of the variables are the numbers in that row.

$$1 \qquad 6 \qquad 15 \qquad 20 \qquad 15 \qquad 6 \qquad 1$$

Putting this information together gives the binomial expansion:

$$(x + y)^6 = x^6 + 6x^5y + 15x^4y^2 + 20x^3y^3 + 15x^2y^4 + 6xy^5 + y^6$$

Self Check 1   Expand: $(p + q)^3$.

Now Try Exercise 21.

EXAMPLE 2

Expanding a Binomial

Expand: $(x - y)^6$.

SOLUTION We write the binomial as $[x + (-y)]^6$ and substitute $-y$ for y in the result of Example 1.

$$[x + (-y)]^6$$
$$= x^6 + 6x^5(-y) + 15x^4(-y)^2 + 20x^3(-y)^3 + 15x^2(-y)^4 + 6x(-y)^5 + (-y)^6$$
$$= x^6 - 6x^5y + 15x^4y^2 - 20x^3y^3 + 15x^2y^4 - 6xy^5 + y^6$$

> **Tip**
>
> In general, the signs in the expansion of $(x - y)^n$ alternate. The sign of the first term is $+$, the sign of the second term is $-$, and so on.

Self Check 2   Expand: $(p - q)^3$.

Now Try Exercise 23.

2. Define and Use Factorial Notation

To use the Binomial Theorem to expand a binomial, we will need **factorial notation**.

Factorial Notation   If n is a natural number, the symbol $n!$ (read either as "**n factorial**" or as "**factorial n**") is defined as

$$n! = n(n - 1)(n - 2)(n - 3) \cdot \cdots \cdot (3)(2)(1)$$

EXAMPLE 3

Evaluating Factorials

Evaluate: **a.** 3!   **b.** 6!   **c.** 10!

SOLUTION   We will apply the definition of factorial notation.

a. $3! = 3 \cdot 2 \cdot 1 = 6$

b. $6! = 6 \cdot 5 \cdot 4 \cdot 3 \cdot 2 \cdot 1 = 720$

c. $10! = 10 \cdot 9 \cdot 8 \cdot 7 \cdot 6 \cdot 5 \cdot 4 \cdot 3 \cdot 2 \cdot 1 = 3,628,800$

Self Check 3   Evaluate: **a.** 4!   **b.** 7!

Now Try Exercise 9.

Accent on Technology

Factorials

A graphing calculator can be used to evaluate factorials. Consider 10!. The calculator operation that will be used is !, as shown in Figure 8-1.

To find 10!, first enter 10 in your calculator.

Press MATH **and move the cursor to PRB. Scroll down to 4:!.**

Press ENTER **twice and obtain a value of 3,628,800.**

FIGURE 8-1

There are two fundamental properties of factorials.

Properties of Factorials   **1.** By definition, $0! = 1$.

2. If n is a natural number, $n(n - 1)! = n!$.

EXAMPLE 4

Using Factorials

Show that **a.** $6 \cdot 5! = 6!$ and   **b.** $8 \cdot 7! = 8!$.

SOLUTION   We will apply the definition of factorial notation.

a. $6 \cdot 5! = 6(5 \cdot 4 \cdot 3 \cdot 2 \cdot 1)$

$\quad\quad\quad = 6 \cdot 5 \cdot 4 \cdot 3 \cdot 2 \cdot 1$

$\quad\quad\quad = 6!$

b. $8 \cdot 7! = 8(7 \cdot 6 \cdot 5 \cdot 4 \cdot 3 \cdot 2 \cdot 1)$

$\quad\quad\quad = 8 \cdot 7 \cdot 6 \cdot 5 \cdot 4 \cdot 3 \cdot 2 \cdot 1$

$\quad\quad\quad = 8!$

Self Check 4   Show that $4 \cdot 3! = 4!$.

Now Try Exercise 7.

3. Use the Binomial Theorem to Expand Binomials

We can now state the Binomial Theorem.

The Binomial Theorem   If n and r are positive numbers, then

$$(a + b)^n = a^n + \frac{n!}{1!(n-1)!} a^{n-1}b + \frac{n!}{2!(n-2)!} a^{n-2}b^2$$

$$+ \frac{n!}{3!(n-3)!} a^{n-3}b^3 + \cdots + \frac{n!}{r!(n-r)!} a^{n-r}b^r + \cdots + b^n$$

Take Note

In the expansion of $(a + b)^n$, the term containing b^r is given by

$$\frac{n!}{r!(n-r)!} a^{n-r}b^r$$

A proof of the Binomial Theorem appears in Appendix I.

In the Binomial Theorem, the exponents on the variables in each term on the right side follow the familiar patterns:

1. The sum of the exponents on a and b in each term is n.
2. The exponents on a decrease by 1 in each successive term.
3. The exponents on b increase by 1 in each successive term.

However, the method of finding the coefficients is different. Except for the first and last terms, $n!$ is the numerator of each fractional coefficient. When the exponent on b is 2, the factors in the denominator of the fractional coefficient are 2! and $(n-2)!$. When the exponent on b is 3, the factors in the denominator are 3! and $(n-3)!$. When the exponent on b is r, the factors in the denominator are $r!$ and $(n-r)!$.

Using the Binomial Theorem to Expand a Binomial

EXAMPLE 5   Use the Binomial Theorem to expand $(a + b)^5$.

SOLUTION   We substitute directly into the Binomial Theorem.

$$(a + b)^5 = a^5 + \frac{5!}{1!(5-1)!} a^4b + \frac{5!}{2!(5-2)!} a^3b^2 + \frac{5!}{3!(5-3)!} a^2b^3 + \frac{5!}{4!(5-4)!} ab^4 + b^5$$

$$= a^5 + \frac{5 \cdot 4!}{1 \cdot 4!} a^4b + \frac{5 \cdot 4 \cdot 3!}{2 \cdot 1 \cdot 3!} a^3b^2 + \frac{5 \cdot 4 \cdot 3!}{3! \cdot 2 \cdot 1} a^2b^3 + \frac{5 \cdot 4!}{4! \cdot 1} ab^4 + b^5$$

$$= a^5 + 5a^4b + 10a^3b^2 + 10a^2b^3 + 5ab^4 + b^5$$

Tip

For powers greater than or equal to 2, an expansion has more terms than the original binomial.

We note that the coefficients are the same numbers as in the fifth row of Pascal's Triangle.

Self Check 5   Use the Binomial Theorem to expand $(p + q)^4$.

Now Try Exercise 25.

Using the Binomial Theorem to Expand a Binomial

EXAMPLE 6   Use the Binomial Theorem to expand $(2x - 3y)^4$.

SOLUTION   We first find the expansion of $(a + b)^4$.

$$(a + b)^4 = a^4 + \frac{4!}{1!(4-1)!} a^3b + \frac{4!}{2!(4-2)!} a^2b^2 + \frac{4!}{3!(4-3)!} ab^3 + b^4$$

$$= a^4 + \frac{4 \cdot 3!}{1 \cdot 3!} a^3b + \frac{4 \cdot 3 \cdot 2!}{2 \cdot 1 \cdot 2!} a^2b^2 + \frac{4 \cdot 3!}{3! \cdot 1} ab^3 + b^4$$

$$= a^4 + 4a^3b + 6a^2b^2 + 4ab^3 + b^4$$

We then substitute $2x$ for a and $-3y$ for b in the result.

$$(a + b)^4 = a^4 + 4a^3b + 6a^2b^2 + 4ab^3 + b^4$$

$$[2x + (-3y)]^4 = (2x)^4 + 4(2x)^3(-3y) + 6(2x)^2(-3y)^2 + 4(2x)(-3y)^3 + (-3y)^4$$

$$(2x - 3y)^4 = 16x^4 - 96x^3y + 216x^2y^2 - 216xy^3 + 81y^4$$

Self Check 6   Use the Binomial Theorem to expand $(3x - 2y)^4$.

Now Try Exercise 37.

4. Find a Specific Term of a Binomial Expansion

Suppose that we wish to find the fifth term of the expansion of $(a + b)^{11}$. It would be tedious to raise the binomial to the 11th power and then look at the fifth term. The Binomial Theorem provides an easier way.

EXAMPLE 7   **Finding a Specific Term of a Binomial Expansion**

Find the fifth term of the expansion of $(a + b)^{11}$.

SOLUTION   In the fifth term, the exponent on b is 4 (the exponent on b is always 1 less than the number of the term). Since the exponent on b added to the exponent on a equals 11, the exponent on a is 7. The variables of the fifth term are a^7b^4.

The number in the numerator of the fractional coefficient is $n!$, which in this case is 11!. The factors in the denominator are 4! and $(11 - 4)!$. The complete fifth term is

$$\frac{11!}{4!(11 - 4)!}\, a^7b^4 = \frac{11!}{4!7!}\, a^7b^4$$

$$= \frac{11 \cdot 10 \cdot 9 \cdot 8 \cdot 7!}{4 \cdot 3 \cdot 2 \cdot 1 \cdot 7!}\, a^7b^4$$

$$= 330a^7b^4$$

> **Take Note**
>
> We can also use the formula for the term containing b^r, where $r = 4$ and $n = 11$.
>
> $$\frac{n!}{r!(n - r)!}\, a^{n-r}b^r = \frac{11!}{4!(11 - 4)!}\, a^{11-4}b^4$$
>
> $$= \frac{11!}{4!7!}a^7b^4$$
>
> $$= 330a^7b^4$$

Self Check 7   Find the sixth term of the expansion in Example 7.

Now Try Exercise 43.

EXAMPLE 8   **Finding a Specific Term of a Binomial Expansion**

Find the sixth term of the expansion of $(a + b)^9$.

SOLUTION   In the sixth term, the exponent on b is 5, and the exponent on a is $9 - 5$, or 4. The numerator of the fractional coefficient is 9!, and the factors in the denominator are 5! and $(9 - 5)!$. The sixth term of the expansion is

$$\frac{9!}{5!(9 - 5)!}\, a^4b^5 = \frac{9 \cdot 8 \cdot 7 \cdot 6 \cdot 5!}{5! \cdot 4!}\, a^4b^5$$

$$= \frac{9 \cdot 8 \cdot 7 \cdot 6}{4 \cdot 3 \cdot 2 \cdot 1}\, a^4b^5$$

$$= 126a^4b^5$$

> **Take Note**
>
> We can also use the formula for the term containing b^r, where $r = 5$ and $n = 9$.
>
> $$\frac{n!}{r!(n - r)!}\, a^{n-r}b^r = \frac{9!}{5!(9 - 5)!}\, a^{9-5}b^5$$
>
> $$= \frac{9!}{5!4!}a^4b^5$$
>
> $$= 126a^4b^5$$

Self Check 8   Find the fifth term of the expansion in Example 8.

Now Try Exercise 47.

Finding a Specific Term of a Binomial Expansion

EXAMPLE 9 Find the third term of the expansion of $(3x - 2y)^6$.

SOLUTION We begin by finding the third term of the expansion of $(a + b)^6$.

(1) $\dfrac{6!}{2!(6-2)!} a^4 b^2 = \dfrac{6 \cdot 5 \cdot 4!}{2 \cdot 1 \cdot 4!} a^4 b^2 = 15a^4 b^2$

We can then substitute $3x$ for a and $-2y$ for b in Equation 1 to obtain the third term of the expansion of $(3x - 2y)^6$.

$$15a^4 b^2 = 15(3x)^4(-2y)^2$$
$$= 15(3^4)(-2)^2 x^4 y^2$$
$$= 4860x^4 y^2$$

Self Check 9 Find the fourth term of the expansion in Example 9.

Now Try Exercise 53.

Self Check Answers

1. $p^3 + 3p^2 q + 3pq^2 + q^3$ 2. $p^3 - 3p^2 q + 3pq^2 - q^3$ 3. **a.** 24 **b.** 5040
4. $4(3 \cdot 2 \cdot 1) = 4 \cdot 3 \cdot 2 \cdot 1 = 4!$ 5. $p^4 + 4p^3 q + 6p^2 q^2 + 4pq^3 + q^4$
6. $81x^4 - 216x^3 y + 216x^2 y^2 - 96xy^3 + 16y^4$ 7. $462a^6 b^5$ 8. $126a^5 b^4$
9. $-4320x^3 y^3$

Exercises 8.1

Getting Ready

You should be able to complete these vocabulary and concept statements before you proceed to the practice exercises.

Fill in the blanks.

1. In the expansion of a binomial, there will be one more term than the _____ of the binomial.

2. The _____ of each term in a binomial expansion is the same as the exponent of the binomial.

3. The ____ term in a binomial expansion is the first term raised to the power of the binomial.

4. In the expansion of $(a + b)^n$, the _____ on a decrease by 1 in each successive term.

5. Expand 7!: _____

6. $0! = $ ___

7. $n \cdot$ _____ $= n!$

8. In the seventh term of $(a + b)^{11}$, the exponent on a is ___.

Practice

Evaluate each expression.

9. $5!$

10. $-5!$

11. $3! \cdot 6!$

12. $0! \cdot 7!$

13. $6! + 6!$

14. $5! - 2!$

15. $\dfrac{9!}{12!}$

16. $\dfrac{8!}{5!}$

17. $\dfrac{5! \cdot 7!}{9!}$

18. $\dfrac{3! \cdot 5! \cdot 7!}{1!8!}$

19. $\dfrac{18!}{6!(18-6)!}$

20. $\dfrac{15!}{9!(15-9)!}$

Use Pascal's Triangle to expand each binomial.

21. $(a + b)^5$

22. $(a + b)^7$

23. $(x - y)^3$

24. $(x - y)^7$

Use the Binomial Theorem to expand each binomial.

25. $(a + b)^3$

26. $(a + b)^4$

27. $(a - b)^5$

28. $(x - y)^4$

29. $(2x + y)^3$

30. $(x + 2y)^3$

31. $(x - 2y)^3$

32. $(2x - y)^3$

33. $(2x + 3y)^4$

34. $(2x - 3y)^4$

35. $(x - 2y)^4$

36. $(x + 2y)^4$

37. $(x - 3y)^5$

38. $(3x - y)^5$

39. $\left(\dfrac{x}{2} + y\right)^4$

40. $\left(x + \dfrac{y}{2}\right)^4$

Find the required term in each binomial expansion.

41. $(a + b)^4$; 3rd term

42. $(a - b)^4$; 2nd term

43. $(a + b)^7$; 5th term

44. $(a + b)^5$; 4th term

45. $(a - b)^5$; 6th term

46. $(a - b)^8$; 7th term

47. $(a + b)^{17}$; 5th term

48. $(a - b)^{12}$; 3rd term

49. $\left(a - \sqrt{2}\right)^4$; 2nd term

50. $\left(a - \sqrt{3}\right)^8$; 3rd term

51. $\left(a + \sqrt{3}b\right)^9$; 5th term

52. $\left(\sqrt{2}a - b\right)^7$; 4th term

53. $\left(\dfrac{x}{2} + y\right)^4$; 3rd term

54. $\left(m + \dfrac{n}{2}\right)^8$; 3rd term

55. $\left(\dfrac{r}{2} - \dfrac{s}{2}\right)^{11}$; 10th term

56. $\left(\dfrac{p}{2} - \dfrac{q}{2}\right)^9$; 6th term

57. $(a + b)^n$; 4th term

58. $(a - b)^n$; 5th term

59. $(a + b)^n$; rth term

60. $(a + b)^n$; $(r + 1)$th term

Discovery and Writing

61. Describe how to construct Pascal's Triangle.

62. What is binomial expansion?

63. Explain why the terms alternate in the binomial expansion of $(x - y)^8$.

64. Define factorial notation and explain how to evaluate 10!.

65. With a calculator, evaluate 69!. Explain why we cannot find 70! with a calculator.

66. Find the sum of the numbers in each row of the first ten rows of Pascal's Triangle. Do you see a pattern?

67. Show that the sum of the coefficients in the binomial expansion of $(x + y)^n$ is 2^n. (*Hint:* Let $x = y = 1$.)

68. Explain how the rth term of a binomial expansion is constructed.

69. If we applied the pattern of coefficients to the coefficient of the first term in the Binomial Theorem, it would be $\dfrac{n!}{0!(n - 0)!}$. Show that this expression equals 1.

70. If we applied the pattern of coefficients to the coefficient of the last term in the Binomial Theorem, it would be $\dfrac{n!}{n!(n - n)!}$. Show that this expression equals 1.

Critical Thinking

Determine if the statement is true or false. If the statement is false, then correct it and make it true.

71. $0! = 0$

72. The first term in the expansion of $(a + b)^{999}$ is a^{999}.

73. The last term in the expansion of $(a - b)^{888}$ is b^{888}.

74. For the expansion of $(a + b)^{777}$, the exponents on a increase by 1 in each successive term.

75. For the expansion of $(a - b)^{666}$, the exponents on b decrease by 1 in each successive term.

76. The sum of the exponents on the variables in any term in the expansion of $(a + b)^{555}$ is 555.

77. The number of terms in the binomial expansion of $(a - b)^{444}$ is 444.

78. To find the binomial expansion of $(x^{333} - y^{222})^{111}$, it is helpful to rewrite the expression inside the parentheses as $x^{333} + (-y^{222})$.

79. The constant term in the expansion of $\left(a - \dfrac{1}{a}\right)^{10}$ is 252.

80. The coefficient of x^5 in the expansion of $\left(x - \dfrac{1}{x}\right)^9$ is 36.

8.2 Sequences, Series, and Summation Notation

In this section, we will learn to

1. Define and use sequences.
2. Find terms of a sequence given the general term.
3. Define series.
4. Define and use summation notation.

In this section, we will introduce a function whose domain is the set of natural numbers. This function, called a **sequence**, is a list of numbers in a specific order. Sequences are seen all around us. For example, we can unlock a digital lock by entering a correct sequence of numbers; and a pilot performs the same sequence of checks before each takeoff. Our lives are often changed by a sequence of events, as simulated in the board game Life® by Milton Bradley. Your life will undergo many changes, some of which may include job changes, marriage, and parenthood. We even use sequences for entertainment when we play the Sequence Game.

1. Define and Use Sequences

The meaning of the word *sequence* in mathematics is very similar to the meaning of the word in everyday English. It refers to a list of events in a particular order.

Sequence	An **infinite sequence** is a function whose domain is the set of natural numbers.
	A **finite sequence** is a function whose domain is the set of the first n natural numbers.

Since a sequence is a function whose domain is the set of natural numbers, we can write its terms as a list of numbers. For example, if n is a natural number, the function defined by $f(n) = 2n - 1$ generates the following infinite sequence

$$f(\mathbf{1}) = 2(\mathbf{1}) - 1 = 1$$
$$f(\mathbf{2}) = 2(\mathbf{2}) - 1 = 3$$
$$f(\mathbf{3}) = 2(\mathbf{3}) - 1 = 5$$
$$f(\mathbf{n}) = 2(\mathbf{n}) - 1 = 2n - 1$$

Writing the results horizontally, we have

$$1, 3, 5, \ldots, 2n - 1, \ldots$$

The number 1 is the first term, 3 is the second term, 5 is the third term, and $2n - 1$ is the **general**, or **nth term**.

If n is a natural number, the function $f(n) = 3n^2 + 1$ generates the infinite sequence

$$4, 13, 28, \ldots, 3n^2 + 1, \ldots$$

The number 4 is the first term, 13 is the second term, 28 is the third term, and $3n^2 + 1$ is the general term.

If n is one of the first four natural numbers, the function defined by $f(n) = n^2 + 1$ generates the following finite sequence

$$2, 5, 10, 17$$

A constant function such as $g(n) = 1$ is a sequence, because it generates the infinite sequence

$$1, 1, 1, \ldots$$

> **Take Note**
>
> We seldom use function notation to denote a sequence, because it is often difficult or even impossible to write the general term. In such cases, if there is a pattern that is assumed to be continued, we simply list several terms of the sequence.

Some additional examples of infinite sequences are:

- $1^2, 2^2, 3^2, \ldots, n^2, \ldots$
- $3, 9, 19, 33, \ldots, 2n^2 + 1, \ldots$
- $1, 3, 6, 10, 15, 21, \ldots, \dfrac{n(n + 1)}{2}, \ldots$
- $2, 3, 5, 7, 11, 13, 17, 19, 23, \ldots$ **(prime numbers)**
- $1, 1, 2, 3, 5, 8, 13, 21, \ldots$ **(Fibonacci Sequence)**

The **Fibonacci Sequence** is named after the 12th-century mathematician Leonardo of Pisa, also known as Fibonacci. After the two 1's in the Fibonacci Sequence, each term is the sum of the two terms that immediately precede it. The Fibonacci Sequence occurs in many fields, such as music, the growth patterns of plants, and the reproductive habits of bees. For example, in one octave of a piano keyboard, there are 2 black keys followed by 3 black keys for a total of 5 black keys. In one octave, there are 8 white keys and a total of 13 black and white keys. These are numbers in the Fibonacci Sequence.

C D E F G A B C D E F G A B C D E F G A B

The fruitlets of a pineapple are arranged in interlocking spirals, 8 spirals in one direction and 13 in the other.

©Igor Borodin/Shutterstock.com

2. Find Terms of a Sequence Given the General Term

Instead of using $f(n)$ notation, we use a_n notation, to write the value of a sequence at a number n. For the infinite sequence introduced earlier, $f(n) = 3n^2 + 1$, we write $a_n = 3n^2 + 1$. The variable n represents the position of the term in the sequence, while the notation a_n represents the value of the term in that position. To find the 5th term of the sequence, we substitute 5 in for n.

$$a_5 = 3(5)^2 + 1$$
$$= 3(25) + 1$$
$$= 76$$

Writing the Terms of a Sequence and Finding a Specific Term

EXAMPLE 1

Given $a_n = 3n - 5$, find each of the following:

a. the first four terms of the sequence   **b.** a_{50}

SOLUTION   **a.** To find the first four terms of the sequence, we will substitute 1, 2, 3, and 4 in the formula that defines the sequence.

$$a_1 = 3(\mathbf{1}) - 5 = -2 \qquad \text{Substitute 1 for } n.$$

$$a_2 = 3(\mathbf{2}) - 5 = 1 \qquad \text{Substitute 2 for } n.$$

$$a_3 = 3(\mathbf{3}) - 5 = 4 \qquad \text{Substitute 3 for } n.$$

$$a_4 = 3(\mathbf{4}) - 5 = 7 \qquad \text{Substitute 4 for } n.$$

The first four terms are $-2, 1, 4,$ and 7.

b. To find a_{50}, the 50th term of the sequence, we will substitute 50 in the formula for the nth term.

$$a_{50} = 3(\mathbf{50}) - 5 = 145 \qquad \text{Substitute 50 for } n.$$

Self Check 1   Given an infinite sequence $a_n = 4n + 7$, find each of the following:

a. the first five terms of the sequence   **b.** a_{100}

Now Try Exercise 21.

Writing the Terms of a Sequence and Finding a Specific Term

EXAMPLE 2

Given $a_n = \dfrac{(-1)^n}{2^n}$, find each of the following:

a. the first four terms of the sequence   **b.** a_{10}

SOLUTION   **a.** To find the first four terms of the sequence, we will substitute 1, 2, 3, and 4 in the formula that defines the sequence.

Tip

$(-1)^n$ causes the signs of the terms to alternate between negative (when n is odd) and positive (when n is even).

©Style-photography/Shutterstock.com

$$a_1 = \frac{(-1)^{\mathbf{1}}}{2^{\mathbf{1}}} = -\frac{1}{2} \qquad \text{Substitute 1 for } n.$$

$$a_2 = \frac{(-1)^{\mathbf{2}}}{2^{\mathbf{2}}} = \frac{1}{4} \qquad \text{Substitute 2 for } n.$$

$$a_3 = \frac{(-1)^{\mathbf{3}}}{2^{\mathbf{3}}} = -\frac{1}{8} \qquad \text{Substitute 3 for } n.$$

$$a_4 = \frac{(-1)^{\mathbf{4}}}{2^{\mathbf{4}}} = \frac{1}{16} \qquad \text{Substitute 4 for } n.$$

The first four terms are $-\dfrac{1}{2}, \dfrac{1}{4}, -\dfrac{1}{8},$ and $\dfrac{1}{16}$.

b. To find a_{10}, the 10th term of the sequence, we will substitute 10 in the formula for the nth term.

$$a_{10} = \frac{(-1)^{\mathbf{10}}}{2^{\mathbf{10}}} = \frac{1}{1024} \qquad \text{Substitute 10 for } n.$$

Self Check 2   Given an infinite sequence $a_n = \dfrac{(-1)^n}{3^n}$, find each of the following:

a. the first three terms of the sequence   **b.** a_6

Now Try Exercise 29.

A sequence can be defined **recursively** by giving its first term and a rule showing how to obtain the $(n + 1)$th term from the nth term. For example, the information

$$a_1 = 5 \quad \text{(the first term)} \quad \text{and} \quad a_{n+1} = 3a_n - 2 \quad \text{(the rule showing how to get the}$$
$$(n + 1)\text{th term from the } n\text{th term)}$$

defines a sequence recursively. To find the first five terms of this sequence, we proceed as follows:

$$a_1 = 5$$
$$a_2 = 3(a_1) - 2 = 3(5) - 2 = 13 \qquad \text{Substitute 5 for } a_1 \text{ and simplify to get } a_2.$$
$$a_3 = 3(a_2) - 2 = 3(13) - 2 = 37 \qquad \text{Substitute 13 for } a_2 \text{ and simplify to get } a_3.$$
$$a_4 = 3(a_3) - 2 = 3(37) - 2 = 109 \qquad \text{Substitute 37 for } a_3 \text{ and simplify to get } a_4.$$
$$a_5 = 3(a_4) - 2 = 3(109) - 2 = 325 \qquad \text{Substitute 109 for } a_4 \text{ and simplify to get } a_5.$$

3. Define Series

To add the terms of a sequence, we replace each comma between its terms with a + sign to form a **series**. If a sequence is infinite, the number of terms in the series associated with it is infinite also. Two examples of infinite series are

$$1^2 + 2^2 + 3^2 + \cdots + n^2 + \cdots$$

and

$$1 + 2 + 3 + 5 + 8 + 13 + 21 + \cdots$$

If a sequence is finite, the number of terms in the series associated with it is finite also. An example of a finite series is

$$3 + 7 + 11 + 15$$

If the signs between successive terms of a series alternate, the series is called an **alternating series**. Two examples of alternating infinite series are

$$-3 + 6 - 9 + 12 - 15 + 18 - \cdots + (-1)^n 3n + \cdots$$

and

$$2 - 4 + 8 - 16 + \cdots + (-1)^{n+1} 2^n + \cdots$$

> **Take Note**
>
> The series $1 + \frac{1}{2} + \frac{1}{3} + \cdots$ is called a **harmonic series** and is important in calculus. The series $1 - \frac{1}{2} + \frac{1}{3} - \cdots$ is an alternating harmonic series.

4. Define and Use Summation Notation

Summation notation is a shorthand way to indicate the sum of the first n terms, or the **nth partial sum**, of a sequence. For example, the expression

$$\sum_{n=1}^{4} n \qquad \text{The symbol } \Sigma \text{ is the capital letter sigma in the Greek alphabet.}$$

indicates the sum of the four terms obtained when we successively substitute 1, 2, 3, and 4 for n, called the **index of summation**.

$$\sum_{n=1}^{4} n = 1 + 2 + 3 + 4 = 10$$

The expression

$$\sum_{n=2}^{4} n^2$$

indicates the sum of the three terms obtained when we successively substitute 2, 3, and 4 for n.

$$\sum_{n=2}^{4} n^2 = \mathbf{2}^2 + \mathbf{3}^2 + \mathbf{4}^2$$

$$= 4 + 9 + 16$$

$$= 29$$

Any letter can be used for the index of summation. The letters i, j, k, and n are commonly used.

The expression

$$\sum_{k=1}^{3} (2k^2 + 1)$$

indicates the sum of the three terms obtained if we successively substitute 1, 2, and 3 for k in the expression $2k^2 + 1$.

$$\sum_{k=1}^{3} (2k^2 + 1) = [2(\mathbf{1}^2) + 1] + [2(\mathbf{2}^2) + 1] + [2(\mathbf{3}^2) + 1]$$

$$= 3 + 9 + 19$$

$$= 31$$

Evaluating a Sum

EXAMPLE 3

Evaluate: $\displaystyle\sum_{k=1}^{4} (k^2 - 1)$.

SOLUTION Since k runs from 1 to 4, we substitute 1, 2, 3, and 4 for k in the expression $k^2 - 1$ and find the sum of the resulting terms:

$$\sum_{k=1}^{4} (k^2 - 1) = (\mathbf{1}^2 - 1) + (\mathbf{2}^2 - 1) + (\mathbf{3}^2 - 1) + (\mathbf{4}^2 - 1)$$

$$= 0 + 3 + 8 + 15$$

$$= 26$$

Self Check 3 Evaluate: $\displaystyle\sum_{k=1}^{5} (k^2 - 1)$.

Now Try Exercise 53.

Evaluating a Sum

EXAMPLE 4

Evaluate: $\displaystyle\sum_{k=3}^{5} (3k + 2)$.

SOLUTION Since k runs from 3 to 5, we substitute 3, 4, and 5 for k in the expression $3k + 2$ and find the sum of the resulting terms:

$$\sum_{k=3}^{5} (3k + 2) = [3(\mathbf{3}) + 2] + [3(\mathbf{4}) + 2] + [3(\mathbf{5}) + 2]$$

$$= 11 + 14 + 17$$

$$= 42$$

Self Check 4 Evaluate: $\displaystyle\sum_{k=2}^{5} (3k + 2)$.

Now Try Exercise 61.

We can use summation notation to state the Binomial Theorem concisely. In Exercise 72, you will be asked to explain why the Binomial Theorem can be stated as

$$\sum_{r=0}^{n} \frac{n!}{r!(n-r)!} a^{n-r}b^r$$

There are three basic properties of summations. The first states that *the summation of a constant as k runs from 1 to n is n times the constant.*

Summation of a Constant   If c is a constant, then $\displaystyle\sum_{k=1}^{n} c = nc$.

PROOF   Because c is a constant, each term is c for each value of k as k runs from 1 to n.

$$\overbrace{\sum_{k=1}^{n} c = c + c + c + c + \cdots + c}^{n \text{ number of } c\text{'s}} = nc$$

EXAMPLE 5   **Using the Summation of a Constant Property**

Evaluate: $\displaystyle\sum_{k=1}^{5} 13$.

SOLUTION $\displaystyle\sum_{k=1}^{5} 13 = 13 + 13 + 13 + 13 + 13$

$$= 5(13)$$

$$= 65$$

Self Check 5   Evaluate using the Constant Property: $\displaystyle\sum_{k=1}^{6} 12$.

Now Try Exercise 59.

A second property states that *a constant factor can be brought outside a summation sign.*

Summation of a Product   If c is a constant, then $\displaystyle\sum_{k=1}^{n} cf(k) = c\sum_{k=1}^{n} f(k)$.

PROOF $\displaystyle\sum_{k=1}^{n} cf(k) = cf(1) + cf(2) + cf(3) + \cdots + cf(n)$

$$= c[f(1) + f(2) + f(3) + \cdots + f(n)] \qquad \text{Factor out } c.$$

$$= c\sum_{k=1}^{n} f(k)$$

Using the Summation of a Product Property

EXAMPLE 6

Show that $\displaystyle\sum_{k=1}^{3} 5k^2 = 5\sum_{k=1}^{3} k^2$.

SOLUTION

$$\sum_{k=1}^{3} 5k^2 = \mathbf{5}(1)^2 + \mathbf{5}(2)^2 + \mathbf{5}(3)^2 \qquad\qquad 5\sum_{k=1}^{3} k^2 = \mathbf{5}[(1)^2 + (2)^2 + (3)^2]$$

$$= 5 + 20 + 45 \qquad\qquad\qquad\qquad\qquad = 5[1 + 4 + 9]$$

$$= 70 \qquad\qquad\qquad\qquad\qquad\qquad\qquad = 5(14)$$

$$\qquad\qquad\qquad\qquad\qquad\qquad\qquad\qquad\qquad = 70$$

The quantities are equal.

Self Check 6   Evaluate using the Product Property: $\displaystyle\sum_{k=1}^{4} 3k$.

Now Try Exercise 55.

The third property states that *the summation of a sum is equal to the sum of the summations*.

Summation of a Sum $\displaystyle\sum_{k=1}^{n} [f(k) + g(k)] = \sum_{k=1}^{n} f(k) + \sum_{k=1}^{n} g(k)$

PROOF

$$\sum_{k=1}^{n} [f(k) + g(k)] = [f(1) + g(1)] + [f(2) + g(2)] + [f(3) + g(3)] + \cdots + [f(n) + g(n)]$$

$$= [f(1) + f(2) + f(3) + \cdots + f(n)] + [g(1) + g(2) + g(3) + \cdots + g(n)]$$

$$= \sum_{k=1}^{n} f(k) + \sum_{k=1}^{n} g(k)$$

Using the Summation of a Sum Property

EXAMPLE 7

Show that $\displaystyle\sum_{k=1}^{3} (k + k^2) = \sum_{k=1}^{3} k + \sum_{k=1}^{3} k^2$.

SOLUTION

$$\sum_{k=1}^{3} (k + k^2) = (1 + 1^2) + (2 + 2^2) + (3 + 3^2)$$

$$= 2 + 6 + 12$$

$$= 20$$

$$\sum_{k=1}^{3} k + \sum_{k=1}^{3} k^2 = (1 + 2 + 3) + (1^2 + 2^2 + 3^2)$$

$$= 6 + 14$$

$$= 20$$

The quantities are equal.

Self Check 7   Evaluate using the summation properties: $\displaystyle\sum_{k=1}^{3} (k^2 + 2k)$.

Now Try Exercise 57.

EXAMPLE 8

Evaluating a Sum Directly and by Using Properties

Evaluate $\displaystyle\sum_{k=1}^{5}(2k-1)^2$ directly. Then expand the binomial, apply the previous properties, and evaluate the expression again.

SOLUTION **Part 1:** $\displaystyle\sum_{k=1}^{5}(2k-1)^2 = 1 + 9 + 25 + 49 + 81 = 165$

Part 2: $\displaystyle\sum_{k=1}^{5}(2k-1)^2 = \sum_{k=1}^{5}(4k^2 - 4k + 1)$

$$= \sum_{k=1}^{5}4k^2 + \sum_{k=1}^{5}(-4k) + \sum_{k=1}^{5}1$$ The summation of a sum is the sum of the summations.

$$= 4\sum_{k=1}^{5}k^2 - 4\sum_{k=1}^{5}k + \sum_{k=1}^{5}1$$ Bring the constant factors outside the summation signs.

$$= 4\sum_{k=1}^{5}k^2 - 4\sum_{k=1}^{5}k + 5$$ The summation of a constant as k runs from 1 to 5 is 5 times that constant.

$$= 4(1 + 4 + 9 + 16 + 25) - 4(1 + 2 + 3 + 4 + 5) + 5$$

$$= 4(55) - 4(15) + 5$$

$$= 220 - 60 + 5$$

$$= 165$$

Either way, the sum is 165.

Self Check 8   Evaluate directly. Then expand the binomial, apply properties, and evaluate the expression: $\displaystyle\sum_{k=1}^{4}(2k-1)^2$.

Now Try Exercise 63.

Accent on Technology

Sequences, Series, and Summation

A graphing calculator will list a designated number of terms in a sequence as well as add a designated number of terms in a series. For example, let's find the first six terms of the sequence defined by the function

$$f(n) = 3n(n + 2) \text{ and then find } \sum_{n=1}^{6}3n(n + 2).$$

The calculator operation that will be used is **seq(**, and the order for this operation is *seq(function, variable, start, end, increment)*. The default increment is 1, and it is not necessary to enter the 1. If we want the numbers listed in an increment other than 1, then we will have to put that value in the operation. This process is shown in Figure 8-2.

Step 1

- Press [2nd] [STAT] to access LIST.
- Move the cursor to OPS and then scroll down to 5:seq(.

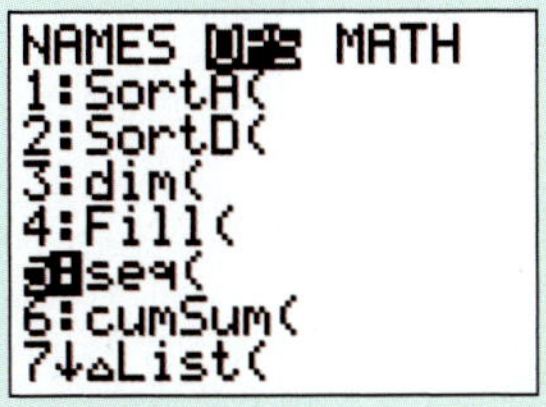

Step 2

- Press [ENTER], and type the function, variable, starting value, and ending value.
- Press [ENTER] again to see the values.

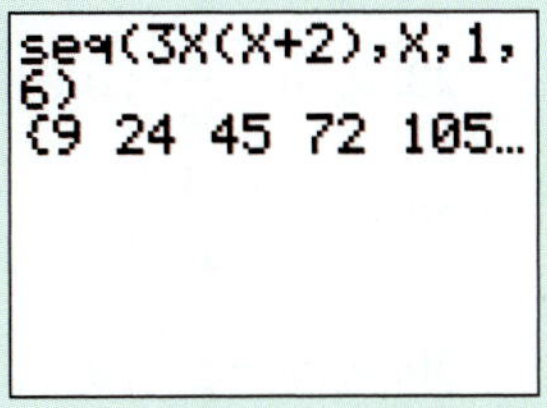

Step 3

- Scroll to the right to see all the values.

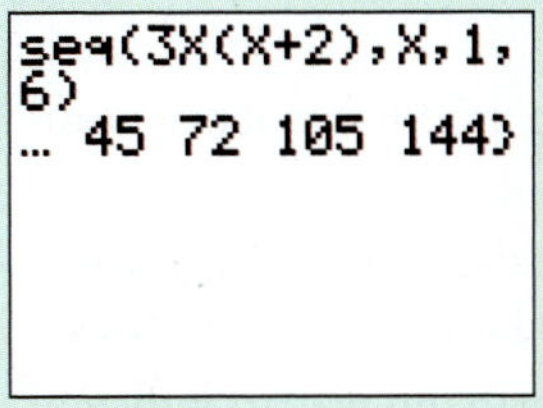

Step 4

To evaluate

$$\sum_{n=1}^{6} 3n(n + 2),$$

we need to store the list of values in the sequence to a list.

- Press [STO▸], then [2nd] [STAT] to access the LIST menu.
- Press [ENTER] twice to store the values in List 1.

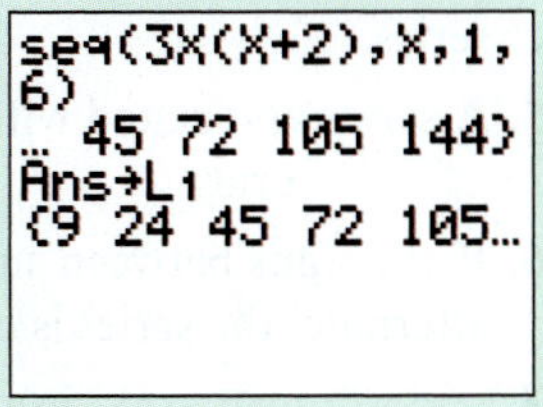

```
NAMES OPS MATH
1:SortA(
2:SortD(
3:dim(
4:Fill(
5:seq(
6:cumSum(
7↓△List(
```

```
seq(3X(X+2),X,1,
6)
{9 24 45 72 105…
```

```
seq(3X(X+2),X,1,
6)
… 45 72 105 144}
```

```
seq(3X(X+2),X,1,
6)
… 45 72 105 144}
Ans→L₁
{9 24 45 72 105…
```

Step 5

Now we will access the operation cumSum(, which is located directly below the seq(operation.

- Press [2nd] [STAT].
- Scroll right to OPS.
- Scroll down to 6:cumSum(.

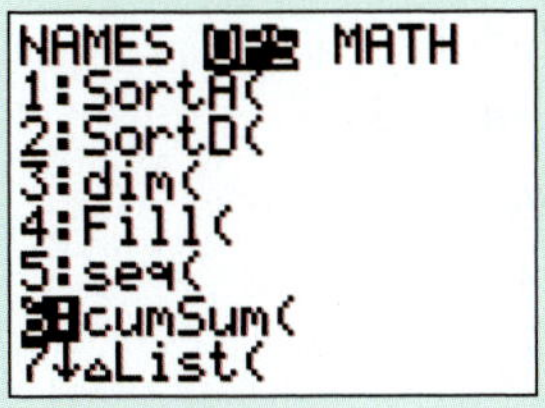

Step 6

- Put L₁ into the operation by pressing [2nd] [STAT] [1].
- Close the parentheses and press [ENTER].

The second number is the sum of the first two numbers in the sequence; the third number is the sum of the first 3 terms in the sequence, etc.

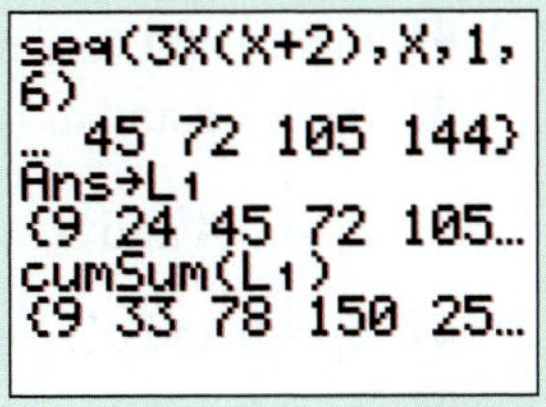

Step 7

- Move the cursor to the right until the last number is shown and you will see 399.

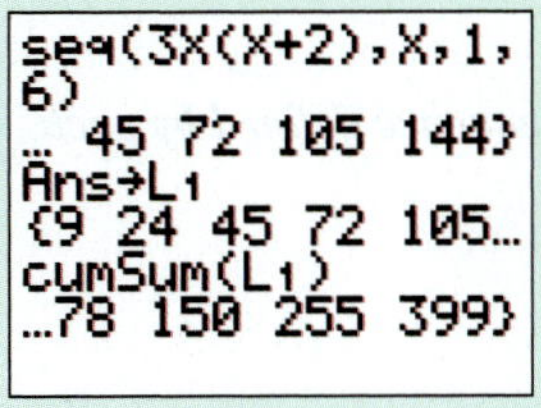

Conclusion

The sum of the first 6 terms in the sequence is 399.

$$\sum_{n=1}^{6} 3n(n + 2) = 399$$

```
NAMES OPS MATH
1:SortA(
2:SortD(
3:dim(
4:Fill(
5:seq(
6:cumSum(
7↓△List(
```

```
seq(3X(X+2),X,1,
6)
… 45 72 105 144}
Ans→L₁
{9 24 45 72 105…
cumSum(L₁)
{9 33 78 150 25…
```

```
seq(3X(X+2),X,1,
6)
… 45 72 105 144}
Ans→L₁
{9 24 45 72 105…
cumSum(L₁)
…78 150 255 399}
```

FIGURE 8-2

Self Check Answers

1. **a** 11, 15, 19, 23, 27   **b.** 407   2. **a.** $-\dfrac{1}{3}, \dfrac{1}{9}, -\dfrac{1}{27}$ **b.** $\dfrac{1}{729}$

3. 50   4. 50   5. 72   6. 30   7. 26   8. 84

Exercises 8.2

Getting Ready

You should be able to complete these vocabulary and concept statements before you proceed to the practice exercises.

Fill in the blanks.

1. An infinite sequence is a function whose _____ is the set of natural numbers.

2. A finite sequence is a function whose _____ is the set of the first n natural numbers.

3. A _____ is formed when we add the terms of a sequence.

4. A series associated with a finite sequence is a _____ series.

5. A series associated with an infinite sequence is an _____ series.

6. If the signs between successive terms of a series alternate, the series is called an _____ series.

7. _____ is a shorthand way to indicate the sum of the first n terms of a sequence.

8. The symbol $\sum_{k=1}^{5} (k^2 - 3)$ indicates the _____ of the five terms obtained when we successively substitute 1, 2, 3, 4, and 5 for k.

9. $\sum_{k=1}^{5} 6k^2 = \underline{\quad} \sum_{k=1}^{5} k^2$

10. $\sum_{k=1}^{5} (k^2 + 3k) = \sum_{k=1}^{5} k^2 + \underline{\quad}$

11. $\sum_{k=1}^{5} c$, where c is a constant, equals ___.

12. The summation of a sum is equal to the _____ of the summations.

Practice

Write the first six terms of the sequence defined by each function.

13. $f(n) = 5n(n - 1)$

14. $f(n) = n\left(\dfrac{n-1}{2}\right)\left(\dfrac{n-2}{3}\right)$

Find the next term of each sequence.

15. $1, 6, 11, 16, \ldots$

16. $1, 8, 27, 64, \ldots$

17. $a, a + d, a + 2d, a + 3d, \ldots$

18. $a, ar, ar^2, ar^3, \ldots$

19. $1, 3, 6, 10, \ldots$

20. $20, 17, 13, 8, \ldots$

Write the first five terms of each sequence and then find the specified term.

21. $a_n = 9n - 1; a_{30}$

22. $a_n = 7n + 3; a_{14}$

23. $a_n = -n^2 + 5; a_{20}$

24. $a_n = -2n^2 + n; a_{13}$

25. $a_n = n^3 + 6; a_{10}$

26. $a_n = -n^3 - 7; a_{15}$

27. $a_n = \dfrac{n - 1}{n}; a_{30}$

28. $a_n = \dfrac{n + 1}{3n}; a_{15}$

29. $a_n = \dfrac{(-1)^n}{4^n}; a_8$

30. $a_n = \dfrac{(-1)^n}{5^n}; a_9$

31. $a_n = \dfrac{(-1)^{n+1}}{n^2}; a_{16}$

32. $a_n = \dfrac{(-1)^{n+1}}{n^3}; a_{11}$

Find the sum of the first five terms of the sequence with the given general term.

33. $a_n = n$

34. $a_n = 2n$

35. $a_n = 3$

36. $a_n = 4n^2$

37. $a_n = 2\left(\dfrac{1}{3}\right)^n$

38. $a_n = (-1)^n$

39. $a_n = 3n - 2$

40. $a_n = 2n + 1$

Assume that each sequence is defined recursively. Find the first four terms of each sequence.

41. $a_1 = 3$ and $a_{n+1} = 2a_n + 1$

42. $a_1 = -5$ and $a_{n+1} = -a_n - 3$

43. $a_1 = -4$ and $a_{n+1} = \dfrac{a_n}{2}$

44. $a_1 = 0$ and $a_{n+1} = 2a_n^2$

45. $a_1 = k$ and $a_{n+1} = a_n^2$

46. $a_1 = 3$ and $a_{n+1} = ka_n$

47. $a_1 = 8$ and $a_{n+1} = \dfrac{2a_n}{k}$

48. $a_1 = m$ and $a_{n+1} = \dfrac{a_n^2}{m}$

Determine whether each series is an alternating infinite series.

49. $-1 + 2 - 3 + \cdots + (-1)^n n + \cdots$

50. $1 + 4 - 9 + \ldots + (-1)^{n+1} n^2 + \ldots$

51. $a + a^2 + a^3 + \cdots + a^n + \cdots; a = 3$

52. $a + a^2 + a^3 + \cdots + a^n + \cdots; a = -2$

Evaluate each sum.

53. $\displaystyle\sum_{k=1}^{5} 2k$

54. $\displaystyle\sum_{k=3}^{6} 3k$

55. $\displaystyle\sum_{k=3}^{4} (-2k^2)$

56. $\displaystyle\sum_{k=1}^{100} 5$

57. $\displaystyle\sum_{k=1}^{5} (3k - 1)$

58. $\displaystyle\sum_{k=2}^{5} (k^2 + 3k)$

59. $\displaystyle\sum_{k=1}^{1000} \frac{1}{2}$

60. $\displaystyle\sum_{k=4}^{5} \frac{2}{k}$

61. $\displaystyle\sum_{k=3}^{4} \frac{1}{k}$

62. $\displaystyle\sum_{k=2}^{6} (3k^2 + 2k) - 3\sum_{k=2}^{6} k^2$

63. $\displaystyle\sum_{k=1}^{4} (4k + 1)^2 - \sum_{k=1}^{4} (4k - 1)^2$

64. $\displaystyle\sum_{k=0}^{10} (2k - 1)^2 + 4\sum_{k=0}^{10} k(1 - k)$

65. $\displaystyle\sum_{k=6}^{8} (5k - 1)^2 + \sum_{k=6}^{8} (10k - 1)$

66. $\displaystyle\sum_{k=2}^{7} (3k + 1)^2 - 3\sum_{k=2}^{7} k(3k + 2)$

Discovery and Writing

67. What is the difference between a sequence and a series?

68. What is the symbol $\sum$ and how is it used in this section?

69. Find a counterexample to disprove the proposition that the summation of a product is the product of the summations. In other words, prove that

$$\sum_{k=1}^{n} f(k)g(k) \neq \sum_{k=1}^{n} f(k) \sum_{k=1}^{n} g(k)$$

70. Find a counterexample to disprove the proposition that the summation of a quotient is the quotient of the summations. In other words, prove that

$$\sum_{k=1}^{n} \frac{f(k)}{g(k)} \neq \frac{\displaystyle\sum_{k=1}^{n} f(k)}{\displaystyle\sum_{k=1}^{n} g(k)}$$

71. Explain what it means to define a sequence recursively.

72. Explain why the Binomial Theorem can be stated as

$$\sum_{r=0}^{n} \frac{n!}{r!(n - r)!} a^{n-r} b^r$$

Critical Thinking

Determine if the statement is true or false. If the statement is false, then correct it and make it true.

73. The next term in the sequence $1, -8, 27, -64, 125, \ldots$ is 216.

74. The next term in the sequence $1, 1, 2, 3, 5, 8, 13, 21, 34, 55, 89, 144, \ldots$ is 243.

75. If $a_n = \dfrac{(-1)^{n+1}}{\sqrt{n}}$, then $a_{324} = \dfrac{1}{-18}$.

76. As n increases without bound, the terms of the sequence $a_n = \dfrac{2n}{n + 1}$ approach 2.

77. $\displaystyle\sum_{2}^{1000} 5 = 5000$

78. The graph of a sequence is a set of discrete points.

79. $\displaystyle\sum_{k=1}^{999} 9k^{99} = \left(\sum_{k=1}^{999} 9\right)\left(\sum_{k=1}^{999} k^{99}\right)$

80. $\displaystyle\sum_{k=8}^{8888} (k^{88} - k^{888}) = \sum_{k=8}^{8888} k^{888} - \sum_{k=8}^{8888} k^{88}$

8.3 Arithmetic Sequences and Series

In this section, we will learn to

1. Define and use arithmetic sequences.
2. Find arithmetic means.
3. Find the sum of the first n terms of an arithmetic series.
4. Solve problems involving sequences and series.

Critical thinking is one of the major aims of mathematics education. The German mathematician Carl Friedrich Gauss (1777–1855) was once a student in the class of a strict teacher. One day, the teacher asked the students to add together all of the natural numbers from 1 through 100. Gauss, being a critical thinker, recognized that in the sum

$$1 + 2 + 3 + \cdots + 98 + 99 + 100$$

the first number (1) added to the last number (100) is 101, the second number (2) added to the second from the last number (99) is 101, and the third number (3) added to the third from the last number (98) is 101. He reasoned that there would be fifty pairs of such numbers, and that there would be fifty sums of 101. He multiplied 101 by 50 to get the correct answer of 5050.

This story illustrates a problem involving the sum of the terms of a sequence, called an **arithmetic sequence**, in which each term except the first is found by adding a constant to the preceding term.

1. Define and Use Arithmetic Sequences

Arithmetic Sequence

An **arithmetic sequence** is a sequence of the form

$$a, \quad a + d, \quad a + 2d, \quad a + 3d, \quad \ldots, \quad a + (n - 1)d, \quad \ldots$$

where a is the **first term**, $a + (n - 1)d$ is the **nth term**, and d is the **common difference**.

In this definition, the second term has an addend of d, the third term has an addend of $2d$, the fourth term has an addend of $3d$, and so on. This is why the nth term has an addend of $(n - 1)d$.

If an arithmetic sequence has infinitely many terms, it is called an **infinite arithmetic sequence**. If it has a finite number of terms, it is called a **finite arithmetic sequence**.

> **Tip**
>
> For an arithmetic sequence, you can use subscript notation and write the first term as a_1 if you prefer.

Writing Terms of an Arithmetic Sequence

EXAMPLE 1

Write the first six terms and the 21st term of an arithmetic sequence with a first term of 7 and a common difference of 5.

SOLUTION

Since the first term a is 7 and the common difference d is 5, the first six terms are

$$7, \quad 7 + 5, \quad 7 + 2(5), \quad 7 + 3(5), \quad 7 + 4(5), \quad 7 + 5(5)$$

or

$$7, \quad 12, \quad 17, \quad 22, \quad 27, \quad 32$$

To find the 21st term, we substitute 21 for n in the formula for the nth term:

$$n\text{th term} = a + (n - 1)d$$

$$21\text{st term} = 7 + (21 - 1)5$$

$$= 7 + (20)5$$

$$= 107$$

The 21st term is 107.

> **Tip**
>
> There are three important characteristics of an arithmetic sequence:
> a: the first term
> d: the constant difference
> n: the number of terms

Self Check 1 Write the first five terms and the 18th term of an arithmetic sequence with a first term of 3 and a common difference of 6.

Now Try Exercise 9.

Determining a Specific Term of an Arithmetic Sequence

EXAMPLE 2

Find the 98th term of an arithmetic sequence whose first three terms are 2, 6, and 10.

SOLUTION

Here $a = 2$, $n = 98$, and $d = 6 - 2 = 10 - 6 = 4$. Because we want to find the 98th term, we substitute these numbers into the formula for the nth term:

$$n\text{th term} = a + (n - 1)d$$

$$98\text{th term} = 2 + (98 - 1)4$$

$$= 2 + (97)4$$

$$= 390$$

Self Check 2 Write the 50th term of the arithmetic sequence whose first three terms are 3, 8, and 13.

Now Try Exercise 19.

2. Find Arithmetic Means

Numbers inserted between a first and last term to form a segment of an arithmetic sequence are called **arithmetic means**. When finding arithmetic means, we consider the last term, a_n, to be the nth term:

$$a_n = a + (n - 1)d$$

Inserting Arithmetic Means between Real Numbers

EXAMPLE 3

Insert three arithmetic means between -3 and 12.

SOLUTION Since we are inserting three arithmetic means between -3 and 12, the total number of terms is five. Thus, $a = -3$, $a_n = 12$, and $n = 5$. To find the common difference, we substitute -3 for a, 12 for a_n, and 5 for n in the formula for the last term and solve for d:

$$a_n = a + (n - 1)d$$

$$12 = -3 + (5 - 1)d$$

$$15 = 4d \qquad \text{Add 3 to both sides and simplify.}$$

$$\frac{15}{4} = d \qquad \text{Divide both sides by 4.}$$

Once we know d, we can find the other terms of the sequence:

$$a + d = -3 + \frac{15}{4} = \frac{3}{4}$$

$$a + 2d = -3 + 2\left(\frac{15}{4}\right) = -3 + \frac{30}{4} = 4\frac{1}{2}$$

$$a + 3d = -3 + 3\left(\frac{15}{4}\right) = -3 + \frac{45}{4} = 8\frac{1}{4}$$

The three arithmetic means are $\frac{3}{4}$, $4\frac{1}{2}$, and $8\frac{1}{4}$.

Self Check 3 Find three arithmetic means between -5 and 23.

Now Try Exercise 23.

3. Find the Sum of the First n Terms of an Arithmetic Series

If we replace the commas in an infinite arithmetic sequence with $+$ signs, we form an **infinite arithmetic series**.

$$a + (a + d) + (a + 2d) + (a + 3d) + \cdots + [a + (n - 1)d] + \cdots$$

If we replace the commas in a finite arithmetic sequence with $+$ signs, we form a **finite arithmetic series**.

$$a + (a + d) + (a + 2d) + (a + 3d) + \cdots + [a + (n - 1)d]$$

To find the sum of the first n terms of an arithmetic series, we use the following formula.

Sum of the First n Terms of an Arithmetic Series

The formula

$$S_n = \frac{n(a + a_n)}{2}$$

gives the sum of the first n terms of an arithmetic series. In this formula, a is the first term, a_n is the last (or nth) term, and n is the number of terms.

PROOF We write the first n terms of an arithmetic series (letting S_n represent their sum). Then we write the same sum in reverse order and add the equations term by term:

$$S_n = \qquad a \qquad + \qquad (a + d) \qquad + \cdots + [a + (n - 2)d] + [a + (n - 1)d]$$

$$S_n = [a + (n - 1)d] + [a + (n - 2)d] + \cdots + \qquad (a + d) \qquad + \qquad a$$

$$\overline{2S_n = [2a + (n - 1)d] + [2a + (n - 1)d] + \cdots + [2a + (n - 1)d] + [2a + (n - 1)d]}$$

Because there are n equal terms on the right side of the previous equation,

$$2S_n = n[2a + (n - 1)d]$$

or

(1) $2S_n = n\{a + [a + (n - 1)d]\}$ Write $2a$ as $a + a$.

We can substitute a_n for $a + (n - 1)d$ on the right side of Equation 1 and divide both sides by 2 to get

$$S_n = \frac{n(a + a_n)}{2}$$

EXAMPLE 4

Finding the Sum of the Terms of an Arithmetic Series

Find the sum of the first 30 terms of the arithmetic series $5 + 8 + 11 + \cdots$.

SOLUTION   Here $a = 5$, $n = 30$, $d = 3$, and $a_n = a_{30} = 5 + 29(3) = 92$.
Substituting these values into the formula for the sum of the first n terms of an arithmetic series gives

$$S_n = \frac{n(a + a_n)}{2}$$

$$S_{30} = \frac{30(5 + 92)}{2}$$

$$= 15(97)$$

$$= 1455$$

The sum of the first 30 terms is 1455.

Caution

Finding the sum by adding up all of the numbers isn't recommended. Use the formula because it is faster. Don't risk running out of time while testing or completing timed assignments.

©PathDoc/Shutterstock.com

Self Check 4   Find the sum of the first 50 terms of the arithmetic series $-2 + 5 + 12 + \cdots$.

Now Try Exercise 27.

4. Solve Problems Involving Sequences and Series

EXAMPLE 5

Solving an Applied Problem

A student deposits \$50 in a non-interest-bearing account and plans to add \$7 a week. How much will he have in the account one year after his first deposit?

SOLUTION   His weekly balances form an arithmetic sequence:

$$50, 57, 64, 71, 78, \ldots$$

with a first term of 50 and a common difference of 7. To find his balance in one year (52 weeks), we substitute 50 for a, 7 for d, and 52 for n in the formula for the last term.

$$a_n = a + (n - 1)d$$

$$a_{52} = 50 + (52 - 1)7$$

$$= 50 + (51)7$$

$$= 407$$

After one year, the balance will be \$407.

©PathDoc/Shutterstock.com

Self Check 5 How much will she have in the account after 60 weeks?

Now Try Exercise 37.

Solving an Applied Problem

EXAMPLE 6

The equation $s = 16t^2$ represents the distance s (in feet) that a raindrop will fall in t seconds.

- In 1 second, the raindrop will fall 16 feet.
- In 2 seconds, the raindrop will fall 64 feet.
- In 3 seconds, the raindrop will fall 144 feet.

The raindrop fell 16 feet during the first second, 48 feet during the next second, and 80 feet during the third second. Find the distance the raindrop will fall during the 12th second.

SOLUTION The sequence 16, 48, 80, . . . is an arithmetic sequence with $a = 16$ and $d = 32$. To find the 12th term, we substitute these values into the formula for the last term.

$$a_n = a + (n - 1)d$$

$$a_{12} = 16 + (12 - 1)32$$

$$= 16 + 11(32)$$

$$= 368$$

During the 12th second, the raindrop falls 368 feet.

Self Check 6 How far will the raindrop fall during the 20th second?

Now Try Exercise 41.

Self Check Answers

1. 3, 9, 15, 21, 27; 105 **2.** 248 **3.** 2, 9, 16 **4.** 8475 **5.** $463
6. 624 ft

Exercises 8.3

Getting Ready
You should be able to complete these vocabulary and concept statements before you proceed to the practice exercises.

Fill in the blanks.

1. An arithmetic sequence is a sequence of the form
   $a, a + d, a + 2d, a + 3d, . . . , a +$ _____ $d, . . .$

2. An arithmetic series is a series of the form
   $a + (a + d) + (a + 2d) + (a + 3d) + \cdots$
   $+ [a + (n - 1)$__$] + \cdots$

3. If an arithmetic series has infinitely many terms, it is called an _____ arithmetic series.

4. In an arithmetic sequence, a is the _____ term, d is the common _____, and n is the _____ of terms.

5. The last term of an arithmetic sequence is given by the formula _____.

6. The formula for the sum of the first n terms of an arithmetic series is given by the formula

   _____.

7. _____ are numbers inserted between a first and last term of a sequence to form an arithmetic sequence.

8. The formula _____ gives the distance (in feet) that an object will fall in t seconds.

Practice
Write the first six terms of the arithmetic sequences with the given properties.

9. $a = 1; d = 2$

10. $a = -12; d = -5$

11. $a = 5$; 3rd term is 2

12. $a = 4$; 5th term is 12

13. 7th term is 24; common difference is $\frac{5}{2}$

14. 20th term is -49; common difference is -3

Find the missing term in each arithmetic sequence.

15. Find the 40th term of an arithmetic sequence with a first term of 6 and a common difference of 8.

16. Find the 35th term of an arithmetic sequence with a first term of 50 and a common difference of -6.

17. The 6th term of an arithmetic sequence is 28, and the first term is -2. Find the common difference.

18. The 7th term of an arithmetic sequence is -42, and the common difference is -6. Find the first term.

19. Find the 55th term of an arithmetic sequence whose first three terms are -8, -1, and 6.

20. Find the 37th term of an arithmetic sequence whose second and third terms are -4 and 6.

21. If the fifth term of an arithmetic sequence is 14 and the second term is 5, find the 15th term.

22. If the fourth term of an arithmetic sequence is 13 and the second term is 3, find the 24th term.

Find the required means.

23. Insert three arithmetic means between 10 and 20.

24. Insert five arithmetic means between 5 and 15.

25. Insert four arithmetic means between -7 and $\frac{2}{3}$.

26. Insert three arithmetic means between -11 and -2.

Find the sum of the first n terms of each arithmetic series.

27. $5 + 7 + 9 + \cdots$ (to 15 terms)

28. $-3 + (-4) + (-5) + \cdots$ (to 10 terms)

29. $\sum_{n=1}^{20} \left(\frac{3}{2} n + 12 \right)$

30. $\sum_{n=1}^{10} \left(\frac{2}{3} n + \frac{1}{3} \right)$

Solve each problem.

31. Find the sum of the first 30 terms of an arithmetic sequence with 25th term of 10 and a common difference of $\frac{1}{2}$.

32. Find the sum of the first 100 terms of an arithmetic sequence with 15th term of 86 and first term of 2.

33. Find the sum of the first 200 natural numbers.

34. Find the sum of the first 1,000 natural numbers.

Applications

35. Interior angles The sums of the angles of several polygons are given in the table. Assuming that the pattern continues, complete the table.

Figure	Number of Sides	Sum of Angles
Triangle	3	180°
Quadrilateral	4	360°
Pentagon	5	540°
Hexagon	6	720°
Octagon	8	
Dodecagon	12	

36. Borrowing money To pay for college, a student borrows $5000 interest-free from his father. If he pays his father back at the rate of $200 per month, how much will he still owe after 12 months?

37. Borrowing money If Ellie borrows $5500 interest-free from her mother to buy a new car and agrees to pay her mother back at the rate of $105 per month, how much will she still owe after 4 years?

38. Jogging One day, some students jogged $\frac{1}{2}$ mile. Because it was fun, they decided to increase the jogging distance each day by a certain amount. If they jogged $6\frac{3}{4}$ miles on the 51st day, how much was the distance increased each day?

©Pressmaster/Shutterstock.com

39. Sales The year it incorporated, a company had sales of $237,500. Its sales were expected to increase by $150,000 annually for the next several years. If the forecast was correct, what will sales be in 10 years?

40. Falling objects Find how many feet a brick will travel during the 10th second of its fall.

41. Falling objects If a rock is dropped from the Golden Gate Bridge, how far will it fall in the third second? Us the formula $s = t^2$.

42. Designing patios Each row of bricks in the following triangular patio is to have one more brick than the previous row, ending with the longest row of 150 bricks. How many bricks will be needed?

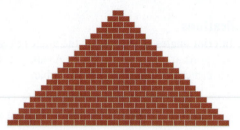

43. Pile of logs Several logs are stored in a pile with 20 logs on the bottom layer, 19 on the second layer, 18 on the third layer, and so on. If the top layer has one log, how many logs are in the pile?

44. Theater seating The first row in a movie theater contains 24 seats. As you move toward the back, each row has 1 additional seat. If there are 30 rows, what is the capacity of the theater?

Discovery and Writing

45. Define *arithmetic sequence* and provide two examples.

46. Explain what the common distance d is in an arithmetic sequence.

47. Describe how to determine a specific term of an arithmetic sequence.

48. What formula is used to determine the sum of the first n terms of an arithmetic series? Explain how it is used.

49. In an arithmetic sequence, can a and d be negative, but a_n positive?

50. Can an arithmetic sequence be an alternating sequence? Explain.

Critical Thinking

Determine if the statement is true or false. If the statement is false, then correct it and make it true.

51. $1, 4, 8, 13, 19, 26, \ldots$ is not an arithmetic sequence.

52. The common difference for the arithmetic sequence $14, 9, 4, -1, -6, \ldots$ is $d = 5$.

53. An arithmetic sequence can have a first term of 4, a 25th term of 106, and a common difference of $4\frac{1}{4}$.

54. Between 5 and $10\frac{1}{3}$ are three arithmetic means. One of them is 9 and the other two are $6\frac{1}{3}$ and $7\frac{2}{3}$.

55. If we know the first term, last term, and number of terms of an arithmetic series, then we can find the sum of the terms of the series.

56. The formula $S_n = \dfrac{n(a - a_n)}{2}$ gives the sum of the first n terms of an arithmetic sequence.

57. The sum of the first 200 terms of the arithmetic sequence $1, 3, 5, 7, 9, \ldots$ is 40,000.

58. The sum of the first 200 terms of the arithmetic sequence $2, 4, 6, 8, 20, \ldots$ is 40,000.

59. Each row of a formation of the members of a college marching band has one more person in it than the previous row. If 5 people are in the front row and 24 are in the 20th (and last) row, then there are 300 band members.

60. The discrete points on the graph of an arithmetic sequence are collinear.

8.4 Geometric Sequences and Series

In this section, we will learn to

1. Define and use geometric sequences.
2. Find geometric means.
3. Find the sum of the first n terms of a geometric series.
4. Define and find the sum of infinite geometric series.
5. Solve problems involving geometric sequences and series.

Bungee jumping is a sport invented by daredevil A. J. Hackett. His first jump was off the 43-meter high Kawarau Bridge in New Zealand. Since he was attached to a long rubber cord, he bounced up and down for a considerable time giving him a thrilling ride. Today, this sport is gaining in popularity. But it is not for the faint of heart.

©Marko Marcello/Shutterstock.com

Suppose a jumper is attached to a cord that stretches to a length of 100 feet. Also suppose that on each bounce, he rebounds to a height that is 60% of the distance from which he fell. If we list the distances that he falls, we will get the following sequence:

100, 60, 36, . . .   Each number in the list is 60% of the number before it.

If we list the distances that he rebounds, we will get the following sequence:

60, 36, 21.6, . . .   Each number in the list is 60% of the number before it.

Each of these sequences is an example of a common sequence called a *geometric sequence*.

1. Define and Use Geometric Sequences

A **geometric sequence** is a sequence in which each term, except the first, is found by multiplying the preceding term by a constant.

Geometric Sequence   A **geometric sequence** is a sequence of the form

$$a, \quad ar, \quad ar^2, \quad ar^3, \quad \ldots, \quad ar^{n-1}, \quad \ldots$$

where a is the **first term**, ar^{n-1} is the **nth term**, and r is the **common ratio**.

> **Caution**
>
> Please note that the nth term of a geometric sequence is not $a_n = ar^n$. The nth term is $a_n = ar^{n-1}$.

In this definition, the second term of the sequence has a factor of r^1, the third term has a factor of r^2, the fourth term has a factor of r^3, and so on. This explains why the nth term has a factor of r^{n-1}.

If a geometric sequence has infinitely many terms, it is called an **infinite geometric sequence**. If it has a finite number of terms, it is called a **finite geometric sequence**.

EXAMPLE 1   **Writing the Terms of a Geometric Sequence**

Write the first six terms and the 15th term of the geometric sequence whose first term is 3 and whose common ratio is 2.

SOLUTION   We begin by writing the first six terms of the geometric sequence:

$$3, \quad 3(2), \quad 3(2^2), \quad 3(2^3), \quad 3(2^4), \quad 3(2^5)$$

> **Caution**
>
> A common error is to multiply first. Always evaluate exponents first.

or

$$3, \quad 6, \quad 12, \quad 24, \quad 48, \quad 96$$

To find the 15th term, we substitute 15 for n, 3 for a, and 2 for r in the formula for the nth term:

$$n\text{th term} = ar^{n-1}$$

$$15\text{th term} = 3(2^{15-1})$$

$$= 3(2^{14})$$

$$= 3(16{,}384)$$

$$= 49{,}152$$

Self Check 1   Write the first five terms of the geometric sequence whose first term is 2 and whose common ratio is 3. Find the 10th term.

Now Try Exercise 11.

EXAMPLE 2

Finding a Specific Term of a Geometric Sequence

Find the eighth term of a geometric sequence whose first three terms are 9, 3, and 1.

SOLUTION Here $a = 9$, $r = \frac{1}{3}$, and $n = 8$. To find the eighth term, we substitute these values into the formula for the nth term.

$$n\text{th term} = ar^{n-1}$$

$$8\text{th term} = 9\left(\frac{1}{3}\right)^{8-1}$$

$$= 9\left(\frac{1}{3}\right)^{7}$$

$$= \frac{1}{243}$$

> **Tip**
>
> The common ratio r of a geometric sequence is the ratio between any two successive terms.

Self Check 2 Find the eighth term of a geometric sequence whose first three terms are $\frac{1}{3}$, 1, and 3.

Now Try Exercise 17.

2. Find Geometric Means

Numbers inserted between a first and last term to form a segment of a geometric sequence are called **geometric means**. When finding geometric means, we consider the last term, a_n, to be the nth term.

EXAMPLE 3

Inserting Geometric Means between Two Integers

Insert two geometric means between 4 and 256.

SOLUTION The first term is $a = 4$, and because 256 is the fourth term we know that, $n = 4$ and $a_n = a_4 = 256$. To find the common ratio, we substitute these values into the formula for the nth term and solve for r:

$$ar^{n-1} = a_n$$

$$4r^{4-1} = 256$$

$$r^3 = 64$$

$$r = 4$$

The common ratio is 4. The two geometric means are the second and third terms of the geometric sequence:

$$ar = 4 \cdot 4 = 16$$

$$ar^2 = 4 \cdot 4^2 = 4 \cdot 16 = 64$$

The first four terms of the geometric sequence are 4, 16, 64, and 256. The two geometric means between 4 and 256 are 16 and 64.

Self Check 3 Insert two geometric means between -3 and 192.

Now Try Exercise 23.

3. Find the Sum of the First n Terms of a Geometric Series

If we replace the commas in an infinite geometric sequence with $+$ signs, we form an **infinite geometric series**.

$$a + ar + ar^2 + ar^3 + \cdots + ar^{n-1} + \cdots$$

If we replace the commas in a finite geometric sequence with $+$ signs, we form a **finite geometric series**.

$$a + ar + ar^2 + ar^3 + \cdots + ar^{n-1}$$

To find the sum of the first n terms of a geometric series, we can use the following formula.

Sum of the First n Terms of a Geometric Series

The formula

$$S_n = \frac{a - ar^n}{1 - r} \qquad (r \neq 1)$$

gives the sum of the first n terms of a geometric series. In the formula, S_n is the sum, a is the first term, r is the common ratio, and n is the number of terms.

PROOF We write the sum of the first n terms of the geometric series:

$$(1) \quad S_n = a + ar + ar^2 + \cdots + ar^{n-3} + ar^{n-2} + ar^{n-1}$$

Multiplying both sides of this equation by r gives

$$(2) \quad S_n r = ar + ar^2 + \cdots + ar^{n-2} + ar^{n-1} + ar^n$$

We now subtract Equation 2 from Equation 1 and solve for S_n:

$$S_n - S_n r = a - ar^n$$

$$S_n(1 - r) = a - ar^n \qquad \text{Factor out } S_n.$$

$$S_n = \frac{a - ar^n}{1 - r} \qquad \text{Divide both sides by } 1 - r.$$

Finding the Sum of the Terms of a Geometric Series

EXAMPLE 4

Find the sum of the first six terms of the geometric series $8 + 4 + 2 + \cdots$.

SOLUTION Here $a = 8$, $n = 6$, and $r = \frac{1}{2}$. Substituting these values in the formula for the sum of the first n terms of a geometric series gives

$$S_n = \frac{a - ar^n}{1 - r}$$

$$S_6 = \frac{8 - 8\left(\frac{1}{2}\right)^6}{1 - \frac{1}{2}}$$

$$= 2\left(\frac{63}{8}\right)$$

$$= \frac{63}{4}$$

The sum of the first six terms is $\frac{63}{4}$.

Self Check 4 Find the sum of the first eight terms of the geometric series $81, 27, 9, \ldots$.

Now Try Exercise 25.

4. Define and Find the Sum of Infinite Geometric Series

Under certain conditions, we can find the sum of all of the terms in an **infinite geometric series**. To define this sum, we consider the infinite geometric series

$$a + ar + ar^2 + \cdots$$

- The first partial sum, S_1, of the series is $S_1 = a$.
- The second partial sum, S_2, of the series is $S_2 = a + ar$.
- The nth partial sum, S_n, of the series is $S_n = a + ar + ar^2 + \cdots + ar^{n-1}$.

If the nth partial sum, S_n, of an infinite geometric series approaches some number S as n approaches ∞, then S is called the **sum of the infinite geometric series**. The following symbol denotes the sum, S, of an infinite geometric series, provided the sum exists.

$$S = \sum_{n=1}^{\infty} ar^{n-1}$$

n	$1\left(\dfrac{1}{2}\right)^n$
1	$\dfrac{1}{2}$
2	$\dfrac{1}{4}$
3	$\dfrac{1}{8}$
4	$\dfrac{1}{16}$
10	$\dfrac{1}{1024}$

To develop a formula for finding the sum of all the terms in an infinite geometric series, we consider the formula

$$(3) \quad S_n = \frac{a - ar^n}{1 - r} \quad (r \neq 1)$$

If $|r| < 1$ and a is a constant, then as n approaches ∞, ar^n approaches 0, and the term ar^n in Equation 3 can be dropped. As an illustration, suppose that $a = 1$ and $r = \frac{1}{2}$. We can see from the table on the left that as n increases without bound, $ar^n = 1\left(\frac{1}{2}\right)^n$ approaches 0.

This argument gives the following formula.

Sum of the Terms of an Infinite Geometric Series

If $|r| < 1$, the sums of the terms of an infinite geometric series is given by

$$S = \frac{a}{1 - r}$$

where a is the first term and r is the common ratio.

Caution

If $|r| \geq 1$, the terms get larger and larger, and the sum does not approach a number. In this case, the previous theorem does not apply.

EXAMPLE 5 **Changing a Repeating Decimal to a Fraction**

Change $0.\overline{4}$ to a common fraction.

SOLUTION We write the decimal as an infinite geometric series and find its sum:

$$S = \frac{4}{10} + \frac{4}{100} + \frac{4}{1000} + \frac{4}{10,000} + \cdots$$

$$= \frac{4}{10} + \frac{4}{10}\left(\frac{1}{10}\right) + \frac{4}{10}\left(\frac{1}{10}\right)^2 + \frac{4}{10}\left(\frac{1}{10}\right)^3 + \cdots$$

Since the common ratio r equals $\frac{1}{10}$ and $\left|\frac{1}{10}\right| < 1$, we can use the formula for the sum of an infinite geometric series:

$$S = \frac{a}{1 - r} = \frac{\frac{4}{10}}{1 - \frac{1}{10}} = \frac{\frac{4}{10}}{\frac{9}{10}} = \frac{4}{9}$$

Take Note

Using a generalization of the method used in Example 5, we can write any repeating decimal as a fraction.

Long division will verify that $\frac{4}{9} = 0.\overline{4}$.

Self Check 5   Change $0.\overline{7}$ to a common fraction.

Now Try Exercise 35.

5. Solve Problems Involving Geometric Sequences and Series

Many of the exponential growth problems discussed in Chapter 4 can be solved using the concepts of geometric sequences.

Solving an Application Problem

EXAMPLE 6

A statistician knows that a town with a population of 3500 people has a predicted growth rate of 6% per year for the next 20 years. What should she predict the population to be 20 years from now?

SOLUTION   Let P_0 be the initial population of the town. After 1 year, the population P_1 will be the initial population (P_0) plus the growth (the product of P_0 and the rate of growth, r).

$$P_1 = P_0 + P_0 r$$
$$= P_0(1 + r) \qquad \text{Factor out } P_0.$$

The population P_2 at the end of 2 years will be

$$P_2 = P_1 + P_1 r$$
$$= P_1(1 + r) \qquad\qquad \text{Factor out } P_1.$$
$$= P_0(1 + r)(1 + r) \qquad \text{Substitute } P_0(1 + r) \text{ for } P_1.$$
$$= P_0(1 + r)^2$$

The population at the end of the third year will be $P_3 = P_0(1 + r)^3$. Writing the terms as a sequence gives

$$P_0, \qquad P_0(1 + r), \qquad P_0(1 + r)^2, \qquad P_0(1 + r)^3, \qquad P_0(1 + r)^4, \qquad \ldots$$

This is a geometric sequence with P_0 as the first term and $1 + r$ as the common ratio. In this example, $P_0 = 3500$, $1 + r = 1.06$, and (since the population after 20 years will be the value of the 21st term of the geometric sequence) $n = 21$. We can substitute these values into the formula for the last term of a geometric sequence to get

$$P_n = P_0 r^{n-1}$$

$$P_{21} = 3500(1.06)^{21-1}$$

$$= 3500(1.06)^{20}$$

$$\approx 11{,}224.97415 \qquad \text{Use a calculator.}$$

The population after 20 years will be approximately 11,225.

Self Check 6 After 30 years, what will be the population? Round to the nearest whole number.

Now Try Exercise 39.

EXAMPLE 7

Solving an Application Problem

A student deposits $2500 in a bank at 7% annual interest, compounded daily. If the investment is left untouched for 60 years, how much money will be in the account?

SOLUTION We let the initial amount in the account be A_0 and r be the rate. At the end of the first day, the account is worth

$$A_1 = A_0 + A_0\left(\frac{r}{365}\right) = A_0\left(1 + \frac{r}{365}\right)$$

After the second day, the account is worth

$$A_2 = A_1 + A_1\left(\frac{r}{365}\right) = A_1\left(1 + \frac{r}{365}\right) = A_0\left(1 + \frac{r}{365}\right)^2$$

The daily amounts form the following geometric sequence

$$A_0, \qquad A_0\left(1 + \frac{r}{365}\right), \qquad A_0\left(1 + \frac{r}{365}\right)^2, \qquad A_0\left(1 + \frac{r}{365}\right)^3, \qquad \ldots$$

where A_0 is the initial deposit and r is the annual rate of interest.

Because interest is compounded daily for 60 years (21,900 days), the amount at the end of 60 years will be the 21,901st term of the sequence.

$$A_{21{,}901} = 2500\left(1 + \frac{0.07}{365}\right)^{21{,}900}$$

We can use a calculator to find that $A_{21{,}901} \approx \$166{,}648.71$.

Self Check 7 After 40 years, how much money was in the account? Round to two decimal places.

Now Try Exercise 45.

EXAMPLE 8

Solving an Application Problem

A pump can remove 20% of the gas in a container with each stroke. Find the percentage of gas that remains in the container after six strokes.

SOLUTION We let V represent the volume of the container. Because each stroke of the pump removes 20% of the gas, 80% remains after each stroke, and we have the geometric sequence

$$V, \qquad 0.80V, \qquad 0.80(0.80V), \qquad 0.80[0.80(0.80V)], \qquad \ldots$$

or

$$V, \qquad 0.8V, \qquad (0.8)^2V, \qquad (0.8)^3V, \qquad \ldots$$

The amount of gas remaining after six strokes is the seventh term of the sequence:

$$a_n = ar^{n-1}$$
$$a_7 = V(0.8)^{7-1}$$
$$= V(0.8)^6$$

We can use a calculator to find that approximately 26% of the gas remains after six strokes.

Self Check 8   What percentage would remain after seven strokes? Round to the nearest percent.

Now Try Exercise 47.

Self Check Answers

1. 2, 6, 18, 54, 162; 39,366   2. 729   3. 12, −48   4. $\dfrac{3280}{27}$ 5. $\dfrac{7}{9}$

6. 20,102   7. $41,100.58   8. 21%

Exercises 8.4

Getting Ready
You should be able to complete these vocabulary and concept statements before you proceed to the practice exercises.

Fill in the blanks.

1. A geometric sequence is a sequence of the form a, ar, ar^2, ar^3, ... The nth term is $a(\underline{\quad})$.

2. In a geometric sequence, a is the _____ term, r is the common _____, and n is the _____ of terms.

3. The last term of a geometric sequence is given by the formula $a_n = $ _____.

4. A geometric _____ is the sum of the terms of a geometric sequence.

5. A geometric series with infinitely many terms is called an _____ geometric series.

6. The formula for the sum of the first n terms of a geometric series is given by _____.

7. _____ are numbers inserted between a first and a last term to form a geometric sequence.

8. If $|r| < 1$, the formula _____ gives the sum of the terms of an infinite geometric series.

Practice
Write the first four terms of each geometric sequence with the given properties.

9. $a = 10$; $r = 2$

10. $a = -3$; $r = 2$

11. $a = -2$ and $r = 3$

12. $a = 64$; $r = \frac{1}{2}$

13. $a = 3$; $r = \sqrt{2}$

14. $a = 2$; $r = \sqrt{3}$

15. $a = 2$; 4th term is 54

16. 3rd term is 4; $r = \frac{1}{2}$

Find the requested term of each geometric sequence.

17. Find the sixth term of the geometric sequence whose first three terms are $\frac{1}{4}$, 1, and 4.

18. Find the eighth term of the geometric sequence whose second and fourth terms are 0.2 and 5.

19. Find the fifth term of a geometric sequence whose second term is 6 and whose third term is -18.

20. Find the sixth term of a geometric sequence whose second term is 3 and whose fourth term is $\frac{1}{3}$.

Solve each problem.

21. Insert three positive geometric means between 10 and 20.

22. Insert five geometric means between -5 and 5, if possible.

23. Insert four geometric means between 2 and 2048.

24. Insert three geometric means between 162 and 2. (There are two possibilities.)

Find the sum of the indicated terms of each geometric series.

25. $4 + 8 + 16 + \cdots$ (to 5 terms)

26. $9 + 27 + 81 + \cdots$ (to 6 terms)

27. $2 + (-6) + 18 + \cdots$ (to 10 terms)

28. $\frac{1}{8} + \frac{1}{4} + \frac{1}{2} + \cdots$ (to 12 terms)

29. $\sum_{n=1}^{6} 3\left(\frac{3}{2}\right)^{n-1}$ **30.** $\sum_{n=1}^{6} 12\left(-\frac{1}{2}\right)^{n-1}$

Find the sum of each infinite geometric series.

31. $6 + 4 + \frac{8}{3} + \cdots$ **32.** $8 + 4 + 2 + 1 + \cdots$

33. $\sum_{n=1}^{\infty} 12\left(-\frac{1}{2}\right)^{n-1}$ **34.** $\sum_{n=1}^{\infty} \left(\frac{1}{3}\right)^{n-1}$

Change each repeating decimal to a common fraction.

35. $0.\overline{5}$ **36.** $0.\overline{6}$

37. $0.\overline{25}$ **38.** $0.\overline{37}$

Applications

  Use a calculator to help solve each problem.

39. Staffing a department The number of students studying algebra at State College is 623. The department chair expects enrollment to increase 10% each year. How many professors will be needed in 8 years to teach algebra if one professor can handle 60 students?

40. Bouncing balls On each bounce, the rubber ball in the illustration rebounds to a height one-half of that from which it fell. Find the total vertical distance the ball travels.

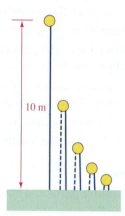

10 m

41. Bungee jumping A bungee jumper is attached to a cord that stretches to a length of 100 feet. If he rebounds to 60% of the height jumped, how far will he fall on his fifth descent? How far will he travel when he comes to rest?

42. Bungee jumping A bungee jumper is attached to a cord that stretches to a length of 100 feet. If he rebounds to 70% of the height jumped, how far will he travel upward on the fifth rebound? How far will he have traveled when he comes to rest?

43. Bouncing balls A SuperBall rebounds to approximately 95% of the height from which it is dropped. If the ball is dropped from a height of 10 meters, how high will it rebound after the 13th bounce? Round to two decimal places.

©Keith Bell/Shutterstock.com

44. Genealogy The following family tree spans 3 generations and lists 7 people. How many names would be listed in a family tree that spans 10 generations?

45. Investing money If a married couple invests $1000 in a 1-year certificate of deposit at $6\frac{3}{4}\%$ annual interest, compounded daily, how much interest will be earned during the year? Round to two decimal places.

46. Biology If a single cell divides into two cells every 30 minutes, how many cells will there be at the end of 10 hours?

47. Depreciation A lawn tractor, costing C dollars when new, depreciates 20% of the previous year's value each year. How much is the lawn tractor worth after 5 years? Round to two decimal places.

48. Financial planning Lily can invest $1000 at $7\frac{1}{2}\%$, compounded annually, or at $7\frac{1}{4}\%$, compounded daily. If she invests the money for a year, which is the better investment?

49. Population study If the population of the Earth were to double every 30 years, approximately how many people would there be in the year 3020? (Consider the population in 2000 to be 5 billion and use 2000 as the base year.)

50. Investing money If Emma deposits $1300 in a bank at 7% interest, compounded annually, how much will be in the bank 17 years later? Round to two decimal places. (Assume that there are no other transactions on the account.)

51. Real estate appreciation If a house purchased for $50,000 in 1998 appreciates in value by 6% each year, how much will the house be worth in the year 2020? Round to two decimal places.

52. Compound interest Find the value of $1000 left on deposit for 10 years at an annual rate of 7%, compounded annually. Round to two decimal places.

53. Compound interest Find the value of $1000 left on deposit for 10 years at an annual rate of 7%, compounded quarterly. Round to two decimal places.

54. Compound interest Find the value of $1000 left on deposit for 10 years at an annual rate of 7%, compounded monthly. Round to two decimal places.

55. Compound interest Find the value of $1000 left on deposit for 10 years at an annual rate of 7%, compounded daily. Round to two decimal places.

56. Compound interest Find the value of $1000 left on deposit for 10 years at an annual rate of 7%, compounded hourly. Round to two decimal places.

57. Saving for retirement When Grayson was 20 years old, he opened an individual retirement account by investing $2000 at 11% interest, compounded quarterly. How much will his investment be worth when he is 65 years old? Round to two decimal places.

58. Biology One bacterium divides into two bacteria every 5 minutes. If two bacteria multiply enough to completely fill a petri dish in 2 hours, how long will it take one bacterium to fill the dish?

59. Pest control To reduce the population of a destructive moth, biologists release 1000 sterilized male moths each day into the environment. If 80% of these moths alive one day survive until the next, then after a long time the population of sterile males is the sum of the infinite geometric series

$$1000 + 1000(0.8) + 1000(0.8)^2 + 1000(0.8)^3 + \cdots$$

Find the long-term population.

60. Pest control If mild weather increases the day-to-day survival rate of the sterile male moths in Exercise 59 to 90%, find the long-term population.

61. Mathematical myth A legend tells of a king who offered to grant the inventor of the game of chess any request. The inventor said, "Simply place one grain of wheat on the first square of a chessboard, two grains on the second, four on the third, and so on, until the board is full. Then give me the wheat." The king agreed. How many grains did the king need to fill the chessboard?

©Ilike/Shutterstock.com

62. Mathematical myth Estimate the size of the wheat pile in Exercise 61. (*Hint:* There are about one-half million grains of wheat in a bushel.)

Discovery and Writing

63. What is a geometric sequence? Give an example to support your description.

64. What is an infinite geometric series? Give an example to support your description.

65. What is the common ratio r in a geometric sequence or series?

66. How do you determine the sum of the first n terms of a geometric series?

67. How do you know whether or not an infinite geometric series has a sum?

68. If an infinite geometric series has a sum, how do you determine the sum?

69. Does $0.999999 = 1$? Explain.

70. Does $0.999\ldots = 1$? Explain.

Critical Thinking

Determine if the statement is true or false. If the statement is false, then correct it and make it true.

71. The sequence $5, 10, 20, 40, 80, \ldots$ is a geometric sequence.

72. The sequence $5, 10, 15, 20, 25, \ldots$ is a geometric sequence.

73. The nth term of a geometric sequence is ar^n.

74. The common ratio of a geometric series is always positive.

75. The sum of an infinite geometric series can always be found.

76. The nth term a_n of the geometric sequence

$$-1, \frac{1}{5}, -\frac{1}{25}, \frac{1}{125}, \ldots \text{ is } a_n = (-1)\left(\frac{1}{5}\right)^{n-1} = \frac{-1}{5^{n-1}}.$$

77. The sum of the first n terms of the geometric series $2 + 4 + 8 + 16 + \ldots$ is $2(2^n - 1)$.

78. $\displaystyle\sum_{n=1}^{\infty} 2(5)^{n-1} = -\frac{1}{2}$

8.5 Mathematical Induction

In this section, we will learn to

1. State the Principle of Mathematical Induction.

2. Use mathematical induction to prove formulas.

Suppose you are waiting in line to get into a Broadway show and are wondering if you will be admitted. Even if the first person in line gets in, your worries will still be justified. Perhaps there is room in the theater for only a few more people.

It would be good news to hear the manager say, "If anyone gets in, the next person in line will also get in." However, this promise does not guarantee that anyone will be admitted. Perhaps the theater is full, and no one will get in.

However, when you see the first person in line walk in, you know that everyone will be admitted because you know two things:

- The first person was admitted.
- Because of the promise, if the first person is admitted, then so is the second, and when the second person is admitted, then so is the third, and so on until everyone gets in.

This situation is similar to a game played with dominoes. Suppose that some dominoes are placed on end, as in Figure 8-3. When the first domino is knocked over, it knocks over the second. The second domino, in turn, knocks over the third, which knocks over the fourth and so on until all of the dominoes fall. Two things must happen to guarantee that all of the dominoes will fall:

- The first domino must be knocked over.
- Every domino that falls must knock over the next one.

When both conditions are met, it is certain that all of the dominoes will fall.

These two examples illustrate the basic Principle of Mathematical Induction.

FIGURE 8-3

1. State the Principle of Mathematical Induction

In Section 8.3, we developed the following formula for the sum of the first n terms of an arithmetic series:

$$S_n = \frac{n(a + a_n)}{2}$$

If we apply this formula to the arithmetic series $1 + 2 + 3 + \cdots + n$, we have

$$S_n = 1 + 2 + 3 + \cdots + n = \frac{n(1 + n)}{2}$$

To see that this formula is true, we can check it for some positive numbers n:

For $n = 1$: $\quad 1 = \dfrac{1(1 + 1)}{2}$ is a true statement, because $1 = 1$.

For $n = 2$: $\quad 1 + 2 = \dfrac{2(1 + 2)}{2}$ is a true statement, because $3 = 3$.

For $n = 3$: $\quad 1 + 2 + 3 = \dfrac{3(1 + 3)}{2}$ is a true statement, because $6 = 6$.

For $n = 6$: $\quad 1 + 2 + 3 + 4 + 5 + 6 = \dfrac{6(1 + 6)}{2}$ is a true statement, because $21 = 21$.

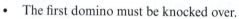

Because the set of positive numbers is infinite, it is impossible to prove the formula by verifying it for all positive numbers. To verify this sequence formula for all positive numbers n, we must have a method of proof called **mathematical induction**, a method first used extensively by Giuseppe Peano (1858–1932).

Principle of Mathematical Induction	If a statement involving the natural number n has the following two properties **1.** The statement is true for $n = 1$, and **2.** If the statement is true for $n = k$, then it is true for $n = k + 1$, then the statement is true for all natural numbers.

Mathematical induction provides a way to prove many formulas. Any proof by induction involves two parts. First, we must show that the formula is true for the number 1. Second, we must show that if the formula is true for any natural number k, then it also is true for the natural number $k + 1$. A proof by induction is complete only when both of these properties are established.

2. Use Mathematical Induction to Prove Formulas

Using Mathematical Induction to Prove a Formula

EXAMPLE 1

Use induction to prove that the following formula is true for every natural number n.

$$1 + 2 + 3 + \cdots + n = \frac{n(n + 1)}{2}$$

SOLUTION   **Proof by induction:**

Part 1: Verify that the formula is true for $n = 1$. When $n = 1$, there is a single term, the number 1, on the left side of the equation. Substituting 1 for n on the right side, we have

$$1 = \frac{n(n + 1)}{2}$$

$$1 = \frac{1(1 + 1)}{2}$$

$$1 = 1$$

The formula is true when $n = 1$. Part 1 of the proof is complete.

Part 2: We assume that the given formula is true when $n = k$. By this assumption, called the **induction hypothesis**, we accept that

$$(1) \quad 1 + 2 + 3 + \cdots + k = \frac{k(k + 1)}{2}$$

is a true statement. We must show that the induction hypothesis forces the given formula to be true when $n = k + 1$. We can show this by verifying the statement

$$(2) \quad 1 + 2 + 3 + \cdots + k + (k + 1) = \frac{(k + 1)[(k + 1) + 1]}{2}$$

which is obtained from the given formula by replacing n with $k + 1$.

Tip

Proofs are very important in mathematics, but writing a proof can initially be challenging.

©Filipe Frazao/Shutterstock.com

To use induction, simply memorize and apply the two steps or parts stated.

Comparing the left sides of Equations 1 and 2 shows that the left side of Equation 2 contains an extra term of $k + 1$. Thus, we add $k + 1$ to both sides of Equation 1 (which was assumed to be true) to obtain the equation

$$1 + 2 + 3 + \cdots + k + (k + 1) = \frac{k(k + 1)}{2} + (k + 1)$$

Because both terms on the right side of this equation have a common factor of $k + 1$, the right side factors, and the equation can be written as follows:

$$1 + 2 + 3 + \cdots + k + (k + 1) = (k + 1)\left(\frac{k}{2} + 1\right)$$

$$= (k + 1)\left(\frac{k + 2}{2}\right)$$

$$= \frac{(k + 1)(k + 2)}{2}$$

$$= \frac{(k + 1)[(k + 1) + 1]}{2}$$

This final result is Equation 2. Because the truth of Equation 1 implies the truth of Equation 2, part 2 of the proof is complete. Parts 1 and 2 together establish that the formula is true for any natural number n.

Self Check 1 Verify that the formula holds for $n = 4$.

Now Try Exercise 9.

EXAMPLE 2 **Using Mathematical Induction to Prove a Formula**

Use induction to prove the following formula for all natural numbers n.

$$1 + 5 + 9 + \cdots + (4n - 3) = n(2n - 1)$$

SOLUTION **Proof by induction:**

Part 1: First we verify the formula for $n = 1$. When $n = 1$, there is a single term, the number 1, on the left side of the equation. Substituting 1 for n on the right side, we have

$$1 = 1[2(1) - 1]$$

$$1 = 1$$

The formula is true for $n = 1$. Part 1 of the proof is complete.

Part 2: We assume that the formula is true for $n = k$. Hence,

(3) $1 + 5 + 9 + \cdots + (4k - 3) = k(2k - 1)$

is a true statement. To show that the induction hypothesis guarantees the truth of the formula for $k + 1$ terms, we add the $(k + 1)$th term to both sides of Equation 3. Because the terms on the left side increase by 4, the $(k + 1)$th term is $(4k - 3) + 4$, or $4k + 1$. Adding $4k + 1$ to both sides of Equation 3 gives

$$1 + 5 + 9 + \cdots + (4k - 3) + (4k + 1) = k(2k - 1) + (4k + 1)$$

We can simplify the right side and write the previous equation as follows:

$$1 + 5 + 9 + \cdots + (4k - 3) + [4(k + 1) - 3] = 2k^2 + 3k + 1$$

$$= (k + 1)(2k + 1)$$

$$= (k + 1)[2(k + 1) - 1]$$

Since this result has the same form as the given formula, except that $k + 1$ replaces n, the truth of the formula for $n = k$ implies the truth of the formula for $n = k + 1$. Part 2 of the proof is complete.

Because both of the induction requirements are true, the formula is true for all natural numbers n.

Self Check 2   Verify that the formula holds for $n = 6$.

Now Try Exercise 11.

Using Mathematical Induction

EXAMPLE 3

Prove that $\dfrac{1}{2} + \dfrac{1}{4} + \dfrac{1}{8} + \cdots + \dfrac{1}{2^n} < 1.$

SOLUTION   **Proof by induction:**

Part 1: We verify the formula for $n = 1$. When $n = 1$, there is a single term, the fraction $\frac{1}{2}$, on the left side of the equation. On the right side, there is the single term of 1. We have the following true statement:

$$\frac{1}{2} < 1$$

The formula is therefore true for $n = 1$. Part 1 of the proof is complete.

Part 2: We assume that the inequality is true for $n = k$. Thus,

$$\frac{1}{2} + \frac{1}{4} + \frac{1}{8} + \cdots + \frac{1}{2^k} < 1$$

We can multiply both sides of the above inequality by $\frac{1}{2}$ to get

$$\frac{1}{2}\left(\frac{1}{2} + \frac{1}{4} + \frac{1}{8} + \cdots + \frac{1}{2^k}\right) < 1\left(\frac{1}{2}\right)$$

or

$$\frac{1}{4} + \frac{1}{8} + \frac{1}{16} + \cdots + \frac{1}{2^{k+1}} < \frac{1}{2}$$

We now add $\frac{1}{2}$ to both sides of this inequality to get

$$\frac{1}{2} + \frac{1}{4} + \frac{1}{8} + \frac{1}{16} + \cdots + \frac{1}{2^{k+1}} < \frac{1}{2} + \frac{1}{2}$$

or

$$\frac{1}{2} + \frac{1}{4} + \frac{1}{8} + \frac{1}{16} + \cdots + \frac{1}{2^{k+1}} < 1$$

The resulting inequality is the same as the original inequality, except that $k + 1$ appears in place of n. Thus, the truth of the inequality for $n = k$ implies the truth of the inequality for $n = k + 1$. Part 2 of the proof is complete.

Because both of the induction requirements have been verified, this inequality is true for all natural numbers.

Self Check 3   Verify that the formula holds for $n = 8$.

Now Try Exercise 21.

Take Note

Mathematical induction can also be used to show that a statement is true for all values of n greater than or equal to a fixed natural number.

Some statements are not true when $n = 1$ but are true for all natural numbers equal to or greater than some given natural number (say, q). In these cases, we verify the given statements for $n = q$ in part 1 of the induction proof. After establishing part 2 of the induction proof, the given statement is proved for all natural numbers that are greater than q.

Self Check Answers

1. $10 = 10$ **2.** $66 = 66$ **3.** $\dfrac{255}{256} < 1$

Exercises 8.5

Getting Ready
You should be able to complete these vocabulary and concept statements before you proceed to the practice exercises.

Fill in the blanks.

1. Any proof by induction requires ____ parts.

2. Part 1 is to show that the statement is true for ____.

3. Part 2 is to show that the statement is true for _____ whenever it is true for $n = k$.

4. When we assume that a formula is true for $n = k$, we call the assumption the induction _____.

Practice
Verify each formula for $n = 1, 2, 3,$ and 4.

5. $5 + 10 + 15 + \cdots + 5n = \dfrac{5n(n + 1)}{2}$

6. $1^2 + 2^2 + 3^2 + \cdots + n^2 = \dfrac{n(n + 1)(2n + 1)}{6}$

7. $7 + 10 + 13 + \cdots + (3n + 4) = \dfrac{n(3n + 11)}{2}$

8. $1(3) + 2(4) + 3(5) + \cdots + n(n + 2) = \dfrac{n}{6}(n + 1)(2n + 7)$

Prove each formula by mathematical induction, if possible.

9. $2 + 4 + 6 + \cdots + 2n = n(n + 1)$

10. $1 + 3 + 5 + \cdots + (2n - 1) = n^2$

11. $3 + 7 + 11 + \cdots + (4n - 1) = n(2n + 1)$

12. $4 + 8 + 12 + \cdots + 4n = 2n(n + 1)$

13. $10 + 6 + 2 + \cdots + (14 - 4n) = 12n - 2n^2$

14. $8 + 6 + 4 + \cdots + (10 - 2n) = 9n - n^2$

15. $2 + 5 + 8 + \cdots + (3n - 1) = \dfrac{n(3n + 1)}{2}$

16. $3 + 6 + 9 + \cdots + 3n = \dfrac{3n(n + 1)}{2}$

17. $1^2 + 2^2 + 3^2 + \cdots + n^2 = \dfrac{n(n + 1)(2n + 1)}{6}$

18. $1 + 2 + 3 + \cdots + (n - 1) + n + (n - 1) + \cdots + 3 + 2 + 1 = n^2$

19. $\dfrac{1}{3} + 2 + \dfrac{11}{3} + \cdots + \left(\dfrac{5}{3}n - \dfrac{4}{3}\right) = n\left(\dfrac{5}{6}n - \dfrac{1}{2}\right)$

20. $\dfrac{1}{1 \cdot 2} + \dfrac{1}{2 \cdot 3} + \dfrac{1}{3 \cdot 4} + \cdots + \dfrac{1}{n(n + 1)} = \dfrac{n}{n + 1}$

21. $\dfrac{1}{2} + \dfrac{1}{4} + \dfrac{1}{8} + \cdots + \left(\dfrac{1}{2}\right)^n = 1 - \left(\dfrac{1}{2}\right)^n$

22. $\dfrac{1}{3} + \dfrac{2}{9} + \dfrac{4}{27} + \cdots + \dfrac{1}{3}\left(\dfrac{2}{3}\right)^{n-1} = 1 - \left(\dfrac{2}{3}\right)^n$

23. $2^0 + 2^1 + 2^2 + 2^3 + \cdots + 2^{n-1} = 2^n - 1$

24. $1^3 + 2^3 + 3^3 + \cdots + n^3 = \left[\dfrac{n(n + 1)}{2}\right]^2$

25. Prove by induction that $x - y$ is a factor of $x^n - y^n$. (*Hint:* Consider subtracting and adding xy^k to the binomial $x^{k+1} - y^{k+1}$.)

26. Prove by induction that $n < 2^n$.

27. There are $180°$ in the sum of the angles of any triangle. Prove by induction that $(n - 2)180°$ is the sum of the angles of any simple polygon when n is its number of sides. (*Hint:* If a polygon has $k + 1$ sides, it has $k - 2$ sides plus three more sides.)

28. Consider the equation

$$1 + 3 + 5 + \cdots + 2n - 1 = 3n - 2$$

a. Is the equation true for $n = 1$?

b. Is the equation true for $n = 2$?

c. Is the equation true for all natural numbers n?

29. If $1 + 2 + 3 + \cdots + n = \frac{n}{2}(n + 1) + 1$ were true for $n = k$, show that it would be true for $n = k + 1$. Is it true for $n = 1$?

30. Prove by induction that $n + 1 = 1 + n$ for each natural number n.

31. If n is any natural number, prove by induction that $7^n - 1$ is divisible by 6.

32. Prove by induction that $1 + 2n < 3^n$ for $n > 1$.

33. Prove by induction that, if r is a real number where $r \neq 1$, then

$$1 + r + r^2 + \cdots + r^n = \frac{1 - r^{n+1}}{1 - r}.$$

34. Prove the formula for the sum of the first n terms of an arithmetic series:

$$a + [a + d] + [a + 2d] + [a + (n - 1)d] = \frac{n(a + a_n)}{2}$$

where $a_n = a + (n - 1)d$.

Discovery and Writing

35. Describe how to apply the Principle of Mathematical Induction to prove that a statement is true for every natural number n.

36. Explain why proofs in mathematics are very important.

37. The expression a^m, where m is a natural number, was defined in Section 0.2. An alternative definition of a^m is (part 1) $a^1 = a$ and (part 2) $a^{m+1} = a^m \cdot a$. Use induction on n to prove the Product Rule for Exponents, $a^m a^n = a^{m+n}$.

38. Use induction on n to prove the Power Rule for Exponents, $(a^m)^n = a^{mn}$. (See Exercise 37.)

39. Tower of Hanoi A well-known problem in mathematics is "The Tower of Hanoi," first attributed to Edouard Lucas in 1883. In this problem, several disks, each of a different size and with a hole in the center, are placed on a board, with progressively smaller disks going up the stack. The object is to transfer the stack of disks to another peg by moving only one disk at a time and never placing a disk over a smaller one.

a. Find the minimum number of moves required if there is only one disk.

b. Find the minimum number of moves required if there are two disks.

c. Find the minimum number of moves required if there are three disks.

d. Find the minimum number of moves required if there are four disks.

40. Tower of Hanoi The results in Exercise 39 suggest that the minimum number of moves required to transfer n disks from one peg to another is given by the formula $2^n - 1$. Use the following outline to prove that this result is correct using mathematical induction.

a. Verify the formula for $n = 1$.

b. Write the induction hypothesis.

c. How many moves are needed to transfer all but the largest of $k + 1$ disks to another peg?

d. How many moves are needed to transfer the largest disk to an empty peg?

e. How many moves are needed to transfer the first k disks back onto the largest one?

f. How many moves are needed to accomplish steps **c, d,** and **e**?

g. Show that part **f** can be written in the form $2^{(k+1)} - 1$.

h. Write the conclusion of the proof.

Critical Thinking

Determine if the statement is true or false. If the statement is false, then correct it and make it true.

41. Mathematical induction is used to prove that statements are true for all real numbers.

42. To prove that if $x > 1$, then $x^n > 1$ for all natural numbers n, we would use mathematical induction and begin by showing that the statement is true for $n = 1$.

43. When mathematical induction is used, we assume that the statement is true for $n = k + 1$.

44. $3^n > 3n + 1$ is true for all natural numbers n greater than 2.

8.6 Permutations and Combinations

In this section, we will learn to

1. Use the Fundamental Counting Principle.
2. Solve permutation problems.
3. Solve combination problems.
4. Use combination notation to write the Binomial Theorem.
5. Find the number of permutations of like things.

1. Use the Fundamental Counting Principle

Ava plans to go to dinner and come home and watch a Netflix movie. If she has a choice of four restaurants and three movies in mind, in how many ways can she spend her evening? There are four choices of restaurants and, for any one of these choices, there are three choices of movies, as shown in the tree diagram in Figure 8-4.

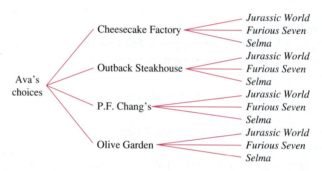

FIGURE 8-4

The diagram shows that Ava has 12 ways to spend her evening. One possibility is to eat at the Cheesecake Factory and watch *Jurassic World*. Another is to eat at the Olive Garden and watch *Selma*.

Any situation that has several outcomes is called an **event**. Ava's first event (choosing a restaurant) can occur in 4 ways. Her second event (choosing a movie) can occur in 3 ways. Thus, she has $4 \cdot 3$, or 12, ways to spend her evening. This example illustrates the **Fundamental Counting Principle**.

Fundamental Counting Principle	Let E_1 and E_2 be two events. If E_1 can be done in a_1 ways, and if—after E_1 has occurred—E_2 can be done in a_2 ways, then the event "E_1 followed by E_2" can be done in $a_1 \cdot a_2$ ways.

The Fundamental Counting Principle can be extended to n events.

Using the Fundamental Counting Principle

EXAMPLE 1

If a traveler has 4 ways to go from New York to Chicago, 3 ways to go from Chicago to Denver, and 6 ways to go from Denver to San Francisco, in how many ways can she go from New York to San Francisco?

SOLUTION We can let E_1 be the event "going from New York to Chicago," E_2 the event "going from Chicago to Denver," and E_3 the event "going from Denver to San Francisco." Since there are 4 ways to accomplish E_1, 3 ways to accomplish E_2, and 6 ways to accomplish E_3, the number of routes available is

$$4 \cdot 3 \cdot 6 = 72$$

Self Check 1 If a man has 4 sweaters and 5 pairs of slacks, how many different outfits can he wear?

Now Try Exercise 29.

2. Solve Permutation Problems

Suppose we want to arrange 7 books in order on a shelf. We can fill the first space with any one of the 7 books, the second space with any of the remaining 6 books, the third space with any of the remaining 5 books, and so on, until there is only one space left to fill with the last book. According to the Fundamental Counting Principle, the number of ordered arrangements of the books is

$$7 \cdot 6 \cdot 5 \cdot 4 \cdot 3 \cdot 2 \cdot 1 = 5040$$

When finding the number of possible ordered arrangements of books on a shelf, we are finding the number of **permutations**. The number of permutations of 7 books, using all the books, is 7! or 5040. The symbol $P(n, r)$ is read as "the number of permutations of n things r at a time." Thus, $P(7, 7) = 5040$.

EXAMPLE 2

Finding the Number of Permutations

Assume that there are 9 signal flags of different designs available to hang on a mast. How many different signals can be sent when 3 flags are used?

©Maigi/Shutterstock.com

SOLUTION We are asked to find $P(9, 3)$, the number of permutations (ordered arrangements) of 9 things using 3 of them. Any one of the 9 flags can hang in the top position on the mast. Any one of the 8 remaining flags can hang in the middle position, and any one of the remaining 7 flags can hang in the bottom position. By the Fundamental Counting Principle, we have

$$P(9, 3) = 9 \cdot 8 \cdot 7 = 504$$

It is possible to send 504 different signals.

Self Check 2 How many different signals can be sent, when three flags are used, if two of the 9 flags are missing?

Now Try Exercise 37.

Although it is correct to write $P(9, 3) = 9 \cdot 8 \cdot 7$, we will change the form of the answer to obtain a convenient formula. To derive this formula, we proceed as follows:

$$P(9, 3) = 9 \cdot 8 \cdot 7$$

$$= \frac{9 \cdot 8 \cdot 7 \cdot \mathbf{6 \cdot 5 \cdot 4 \cdot 3 \cdot 2 \cdot 1}}{\mathbf{6 \cdot 5 \cdot 4 \cdot 3 \cdot 2 \cdot 1}} \qquad \text{Multiply numerator and denominator by } 6 \cdot 5 \cdot 4 \cdot 3 \cdot 2 \cdot 1.$$

$$= \frac{9!}{6!}$$

$$= \frac{9!}{(9-3)!}$$

The generalization of this idea gives the following formula.

Formula for $P(n, r)$ The number of permutations of n things r at a time is given by

$$P(n, r) = \frac{n!}{(n - r)!}$$

Using the Permutation Formula

EXAMPLE 3   Find: **a.** $P(8, 4)$ **b.** $P(n, n)$ **c.** $P(n, 0)$

SOLUTION   We will substitute into the formula for $P(n, r)$ and simplify.

a. $P(8, 4) = \dfrac{8!}{(8 - 4)!} = \dfrac{8 \cdot 7 \cdot 6 \cdot 5 \cdot 4!}{4!} = 1680$

b. $P(n, n) = \dfrac{n!}{(n - n)!} = \dfrac{n!}{0!} = \dfrac{n!}{1} = n!$

c. $P(n, 0) = \dfrac{n!}{(n - 0)!} = \dfrac{n!}{n!} = 1$

Self Check 3   Find: **a.** $P(7, 5)$ **b.** $P(6, 0)$

Now Try Exercise 13.

Parts (b) and (c) of Example 3 establish the following formulas.

Formulas for $P(n, n)$ and $P(n, 0)$ The number of permutations of n things n at a time and n things 0 at a time are given by the formulas

$$P(n, n) = n! \quad \text{and} \quad P(n, 0) = 1$$

Solving a Permutation Problem

EXAMPLE 4

In how many ways can a baseball manager arrange a batting order of 9 players if there are 25 players on the team?

SOLUTION   To find the number of permutations of 25 things 9 at a time, we substitute 25 for n and 9 for r in the formula for finding $P(n, r)$.

$$P(n, r) = \frac{n!}{(n - r)!}$$

$$P(25, 9) = \frac{25!}{(25 - 9)!}$$

$$= \frac{25!}{16!}$$

$$= \frac{25 \cdot 24 \cdot 23 \cdot 22 \cdot 21 \cdot 20 \cdot 19 \cdot 18 \cdot 17 \cdot 16!}{16!}$$

$$= 741{,}354{,}768{,}000$$

The number of permutations is 741,354,768,000.

Self Check 4   In how many ways can the manager arrange a batting order if 2 players can't play?

Now Try Exercise 41.

Accent on Technology

Permutations

A graphing calculator can be used to find $P(n, r)$. For example, let's look at Figure 8-5 and consider the permutation in Example 4 and determine $P(25, 9)$. The calculator operation that will be used is nPr.

To find $P(25, 9)$, first enter 25 in your calculator.

```
25
```

Press [MATH] and move the cursor to PRB. Scroll down to 2:nPr and press [ENTER].

```
MATH NUM CPX PRB
1:rand
2:nPr
3:nCr
4:!
5:randInt(
6:randNorm(
7:randBin(
```

Enter 9 on your calculator and press [ENTER]. We see that $P(25, 9)$ is equal to 741,354,768,000.

```
25 nPr 9
        7.41354768E11
```

FIGURE 8-5

EXAMPLE 5

Solving a Permutation Problem

In how many ways can 5 people stand in a line if 2 people refuse to stand next to each other?

SOLUTION

The total number of ways that 5 people can stand in line is

$$P(5, 5) = 5! = 5 \cdot 4 \cdot 3 \cdot 2 \cdot 1 = 120$$

To find the number of ways that 5 people can stand in line if 2 people insist on standing together, we consider the two people as one person. Then there are 4 people to stand in line, and this can be done in $P(4, 4) = 4! = 24$ ways. However, because either of the two who are paired could be first, there are two arrangements for the pair who insist on standing together. Thus, there are $2 \cdot 4!$, or 48 ways that 5 people can stand in line if 2 people insist on standing together.

The number of ways that 5 people can stand in line if 2 people refuse to stand together is $5! = 120$ (the total number of ways to line up 5 people) minus $2 \cdot 4! = 48$ (the number of ways to line up the 5 people if 2 do stand together):

$$120 - 48 = 72$$

There are 72 ways to line up the people.

Self Check 5 In how many ways can 5 people stand in a line if one person demands to be first?

Now Try Exercise 39.

EXAMPLE 6

Solving a Permutation Problem

In how many ways can 5 people be seated at a round table?

SOLUTION

If we were to seat 5 people in a row, there would be 5! possible arrangements. However, at a round table, each person has a neighbor to the left and to the right. If each person moves one, two, three, four, or five places to the left, everyone has the same neighbors and the arrangement has not changed. Thus, we must divide 5! by 5 to get rid of these duplications. The number of ways that 5 people can be seated at a round table is

$$\frac{5!}{5} = \frac{5 \cdot 4!}{5} = 4! = 4 \cdot 3 \cdot 2 \cdot 1 = 24$$

Self Check 6   In how many ways can 6 people be seated at a round table?

Now Try Exercise 43.

The results of Example 6 suggest the following fact.

Circular Arrangements   There are $(n - 1)!$ ways to arrange n things in a circle.

---

3. Solve Combination Problems

Suppose that a class of 12 students selects a committee of 3 persons to plan a party. With committees, order is not important. A committee of Caden, Logan, and Cooper is the same as a committee of Logan, Cooper, and Caden. However, if we assume for the moment that order is important, we can find the number of permutations of 12 things 3 at a time.

$$P(12, 3) = \frac{12!}{(12 - 3)!} = \frac{12 \cdot 11 \cdot 10 \cdot 9!}{9!} = 1320$$

However, since we do not care about order, this result of 1320 ways is too large. Because there are 6 ways ($3! = 6$) of ordering every committee of 3 students, the result of $P(12, 3) = 1320$ is exactly 6 times too big. To get the correct number of committees, we must divide $P(12, 3)$ by 6:

$$\frac{P(12, 3)}{6} = \frac{1320}{6} = 220$$

> **Tip**
>
> When discussing permutations, order counts. When discussing combinations, order doesn't count.

In cases of selection where order is not important, we are interested in **combinations**, not permutations. The symbols $C(n, r)$ and $\binom{n}{r}$ both mean the number of combinations of n things r at a time.

If a committee of r people is chosen from a total of n people, the number of possible committees is $C(n, r)$, and there will be $r!$ arrangements of each committee. If we consider the committee as an ordered grouping, the number of orderings is $P(n, r)$. Thus, we have

(1) $r!\,C(n, r) = P(n, r)$

We can divide both sides of Equation 1 by $r!$ to obtain the formula for finding $C(n, r)$.

$$C(n, r) = \binom{n}{r} = \frac{P(n, r)}{r!} = \frac{n!}{r!(n - r)!}$$

Formula for $C(n, r)$ The number of combinations of n things r at a time is given by

$$C(n, r) = \binom{n}{r} = \frac{n!}{r!(n - r)!}$$

In Exercises 81 and 82, you will be asked to prove the following formulas.

Formulas for $C(n, n)$ and $C(n, 0)$ If n is a whole number, then

$$C(n, n) = 1 \quad \text{and} \quad C(n, 0) = 1$$

EXAMPLE 7   Solving a Combination Problem

If Christian must read 4 books from a reading list of 10 books, how many choices does he have?

SOLUTION   Because the order in which the books are read is not important, we find the number of combinations of 10 things 4 at a time:

$$C(10, 4) = \frac{10!}{4!(10-4)!} = \frac{10 \cdot 9 \cdot 8 \cdot 7 \cdot 6!}{4 \cdot 3 \cdot 2 \cdot 1 \cdot 6!}$$

$$= \frac{10 \cdot 9 \cdot 8 \cdot 7}{4 \cdot 3 \cdot 2}$$

$$= 210$$

Christian has 210 choices.

Self Check 7   How many choices would Christian have if he had to read 5 books?

Now Try Exercise 49.

Accent on Technology

Combinations

A graphing calculator can be used to find $C(n, r)$. For example, let's consider the combination in Example 7 and determine $C(10, 4)$. The calculator operation that will be used is nCr, as shown in Figure 8-6.

To find $C(10, 4)$, first enter 10 in your calculator.	Press **MATH** and move the cursor to PRB. Scroll down to 3:nCr and press **ENTER**.	Enter 4 on your calculator and press **ENTER**. We see that $C(10, 4)$ is equal to 210.

`10`	`MATH NUM CPX PRB`
	`1:rand`
	`2:nPr`
	`3:nCr`
	`4:!`
	`5:randInt(`
	`6:randNorm(`
	`7:randBin(`
		`10 nCr 4`
		`          210`

FIGURE 8-6

EXAMPLE 8   Solving a Combination Problem

A class consists of 15 men and 8 women. In how many ways can a debate team be chosen with 3 men and 3 women?

SOLUTION   There are $C(15, 3)$ ways of choosing 3 men and $C(8, 3)$ ways of choosing 3 women. By the Fundamental Counting Principle, there are $C(15, 3) \cdot C(8, 3)$ ways of choosing members of the debate team:

$$C(15, 3) \cdot C(8, 3) = \frac{15!}{3!(15-3)!} \cdot \frac{8!}{3!(8-3)!}$$

$$= \frac{15 \cdot 14 \cdot 13}{6} \cdot \frac{8 \cdot 7 \cdot 6}{6}$$

$$= 25{,}480$$

There are 25,480 ways to choose the debate team.

Self Check 8 In how many ways can the debate team be chosen if it is to have 4 men and 2 women?

Now Try Exercise 59.

4. Use Combination Notation to Write the Binomial Theorem

The formula

$$C(n, r) = \frac{n!}{r!(n-r)!}$$

gives the coefficient of the $(r + 1)$th term of the binomial expansion of $(a + b)^n$. This implies that the coefficients of a binomial expansion can be used to solve problems involving combinations. The Binomial Theorem is restated below—this time listing the $(r + 1)$th term and using combination notation.

Binomial Theorem If n is any positive integer, then

$$(a + b)^n = \binom{n}{0}a^n + \binom{n}{1}a^{n-1}b + \binom{n}{2}a^{n-2}b^2 + \cdots + \binom{n}{r}a^{n-r}b^r + \cdots + \binom{n}{n}b^n$$

Take Note

In the expansion of $(a + b)^n$, the term containing b^r is given by

$$\binom{n}{r}a^{n-r}b^r$$

EXAMPLE 9 **Using Pascal's Triangle to Compute a Combination**

Use Pascal's Triangle to compute $C(7, 5)$.

SOLUTION Consider the eighth row of Pascal's Triangle and the corresponding combinations:

1	7	21	35	35	21	7	1
$\binom{7}{0}$	$\binom{7}{1}$	$\binom{7}{2}$	$\binom{7}{3}$	$\binom{7}{4}$	$\binom{7}{5}$	$\binom{7}{6}$	$\binom{7}{7}$

$$C(7, 5) = \binom{7}{5} = 21.$$

Self Check 9 Use Pascal's Triangle to compute $C(6, 5)$.

Now Try Exercise 75.

5. Find the Number of Permutations of Like Things

A *word* is a distinguishable arrangement of letters. For example, six words can be formed with the letters a, b, and c if each letter is used exactly once. The six words are

abc,   acb,   bac,   bca,   cab,   and   cba

If there are n distinct letters and each letter is used once, the number of distinct words that can be formed is $n! = P(n, n)$. It is more complicated to compute the number of distinguishable words that can be formed with n letters when some of the letters are duplicates.

EXAMPLE 10   Finding the Number of Permutations of Like Things

There's an inspirational saying "enjoy the little things" in life. Find the number of "words" that can be formed if each of the 6 letters of the word *little* is used once.

SOLUTION   For the moment, we assume that the letters of the word *little* are distinguishable: "LitTle." The number of words that can be formed using each letter once is $6! = P(6, 6)$. However, in reality we cannot tell the *l*'s or the *t*'s apart. Therefore, we must divide by a number to get rid of these duplications. Because there are 2! orderings of the two *l*'s and 2! orderings of the two *t*'s, we divide by $2! \cdot 2!$. The number of words that can be formed using each letter of the word *little* is

©Claudia Balasoiu/Shutterstock.com

$$\frac{P(6, 6)}{2! \cdot 2!} = \frac{6!}{2! \cdot 2!} = \frac{6 \cdot 5 \cdot 4 \cdot 3 \cdot 2 \cdot 1}{2 \cdot 1 \cdot 2 \cdot 1} = 180$$

Self Check 10   How many words can be formed if each letter of the word *balloon* is used once?

Now Try Exercise 55.

Example 10 illustrates the following general principle.

Permutations of Like Things   The number of permutations of n things with a things alike, b things alike, and so on, is

$$\frac{n!}{a! b! \cdots}$$

Self Check Answers

1. 20   **2.** 210   **3. a.** 2520   **b.** 1   **4.** 296,541,907,200   **5.** 24
6. 120   **7.** 252   **8.** 38,220   **9.** 6   **10.** 1260

Exercises 8.6

Getting Ready

You should be able to complete these vocabulary and concept statements before you proceed to the practice exercises.

Fill in the blanks.

1. If E_1 and E_2 are two events and E_1 can be done in 4 ways and E_2 can be done in 6 ways, then the event E_1 followed by E_2 can be done in ___ ways.

2. An arrangement of n objects is called a _____.

3. $P(n, r) =$ _____

4. $P(n, n) =$ ___

5. $P(n, 0) =$ ___

6. There are _____ ways to arrange n things in a circle.

7. $C(n, r) =$ _____

8. Using combination notation, $C(n, r) =$ ____.

9. $C(n, n) =$ ___

10. $C(n, 0) =$ ___

11. If a word with n letters has a of one letter, b of another letter, and so on, the number of different words that can be formed is _____.

12. Where the order of selection is not important, we are interested in _____, not _____.

Practice

Evaluate each expression.

13. $P(7, 4)$ **14.** $P(8, 3)$

15. $C(7, 4)$ **16.** $C(8, 3)$

17. $P(5, 5)$ **18.** $P(5, 0)$

19. $\binom{5}{4}$ **20.** $\binom{8}{4}$

21. $\binom{5}{0}$ **22.** $\binom{5}{5}$

23. $P(5, 4) \cdot C(5, 3)$ **24.** $P(3, 2) \cdot C(4, 3)$

25. $\binom{5}{3}\binom{4}{3}\binom{3}{3}$

26. $\binom{5}{5}\binom{6}{6}\binom{7}{7}\binom{8}{8}$

27. $\binom{68}{66}$

28. $\binom{100}{99}$

Applications

29. **Choosing lunch** A lunchroom has a machine with eight kinds of sandwiches, a machine with four kinds of soda, a machine with both white and chocolate milk, and a machine with three kinds of ice cream. How many different lunches can be chosen? (Consider a lunch to be one sandwich, one drink, and one ice cream.)

30. **Manufacturing license plates** How many six-digit license plates can be manufactured if no license plate number begins with 0?

31. **Available phone numbers** How many different seven-digit phone numbers can be used in one area code if no phone number begins with 0 or 1?

32. **Arranging letters** In how many ways can the letters of the word *number* be arranged?

33. **Arranging letters with restrictions** In how many ways can the letters of the word *number* be arranged if the e and r must remain next to each other?

34. **Arranging letters with restrictions** In how many ways can the letters of the word *number* be arranged if the e and r cannot be side by side?

35. **Arranging letters with repetitions** How many ways can five Scrabble tiles bearing the letters, F, F, F, L, and U be arranged to spell the word *fluff*?

36. **Arranging letters with repetitions** How many ways can six Scrabble tiles bearing the letters B, E, E, E, F, and L be arranged to spell the word *feeble*?

37. **Placing people in line** In how many arrangements can 8 women be placed in a line?

38. **Placing people in line** In how many arrangements can 5 women and 5 men be placed in a line if the women and men alternate?

39. **Placing people in line** In how many arrangements can 5 women and 5 men be placed in a line if all the men line up first?

40. **Placing people in line** In how many arrangements can 5 women and 5 men be placed in a line if all the women line up first?

41. **Combination locks** How many permutations does a combination lock have if each combination has 3 numbers, no two numbers of the combination are the same, and the lock dial has 30 notches?

42. **Combination locks** How many permutations does a combination lock have if each combination has 3 numbers, no two numbers of the combination are the same, and the lock dial has 100 notches?

43. **Seating at a table** In how many ways can 8 people be seated at a round table?

44. **Seating at a table** In how many ways can 7 people be seated at a round table?

45. **Seating at a table** In how many ways can 6 people be seated at a round table if 2 of the people insist on sitting together?

46. **Seating arrangements with conditions** In how many ways can 6 people be seated at a round table if 2 of the people refuse to sit together?

47. **Arrangements in a circle** In how many ways can 7 children be arranged in a circle if Ella and Eli want to sit together and Jayden and Jackson want to sit together?

48. **Arrangements in a circle** In how many ways can 8 children be arranged in a circle if Laura, Scott, and Grace want to sit together?

49. **Selecting candy bars** In how many ways can 4 candy bars be selected from 10 different candy bars?

50. **Selecting surfboards** In how many ways can 6 surfboards be selected from 24 different surfboards?

51. **Circuit wiring** A wiring harness containing a red, a green, a white, and a black wire must be attached to a control panel. In how many different orders can the wires be attached?

52. **Grading homework** A professor grades homework by randomly checking 7 of the 20 problems assigned. In how many different ways can this be done?

53. Forming words with distinct letters How many words can be formed from the letters of the word *plastic* if each letter is to be used once?

54. Forming words with distinct letters How many words can be formed from the letters of the word *computer* if each letter is to be used once?

55. Forming words with repeated letters How many words can be formed from the letters of the word *banana* if each letter is to be used once?

56. Forming words with repeated letters How many words can be formed from the letters of the word *laptop* if each letter is to be used once?

57. Manufacturing license plates How many license plates can be made using two different letters followed by four different digits if the first digit cannot be 0 and the letter O is not used?

58. Planning class schedules If there are 7 class periods in a school day, and a typical student takes 5 classes, how many different time patterns are possible for the student?

59. Selecting golf balls From a bucket containing 6 red and 8 white golf balls, in how many ways can we draw 6 golf balls of which 3 are red and 3 are white?

60. Selecting a committee In how many ways can you select a committee of 3 Republicans and 3 Democrats from a group containing 18 Democrats and 11 Republicans?

61. Selecting a committee In how many ways can you select a committee of 4 Democrats and 3 Republicans from a group containing 12 Democrats and 10 Republicans?

62. Drawing cards In how many ways can you select a group of 5 red cards and 2 black cards from a deck containing 10 red cards and 8 black cards?

63. Planning dinner In how many ways can a husband and wife choose 2 different dinners from a menu of 17 dinners?

64. Placing people in line In how many ways can 7 people stand in a row if 2 of the people refuse to stand together?

65. Geometry How many lines are determined by 8 points if no 3 points lie on a straight line?

66. Geometry How many lines are determined by 10 points if no 3 points lie on a straight line?

67. Coaching basketball How many different teams can a basketball coach start if the entire squad consists

of 10 players? (Assume that a starting team has 5 players and each player can play all positions.)

68. Managing baseball How many different teams can a manager start if the entire squad consists of 25 players? (Assume that a starting team has 9 players and each player can play all positions.)

69. Selecting job applicants There are 30 qualified applicants for 5 openings in the sales department. In how many different ways can the group of 5 be selected?

70. Sales promotions If a customer purchases a new stereo system during the spring sale, he may choose any 6 CDs from 20 classical and 30 jazz selections. In how many ways can the customer choose 3 of each?

71. Guessing on matching questions Ten words are to be paired with the correct 10 out of 12 possible definitions. How many ways are there of guessing?

72. Guessing on true-false exams How many possible ways are there of guessing on a 10-question true-false exam, if it is known that the instructor will have 5 true and 5 false responses?

73. Number of Wendy's® hamburgers Wendy's® Old Fashioned Hamburgers offers eight toppings for their single hamburger. How many different single hamburgers can be ordered?

74. Number of ice cream sundaes A restaurant offers ten toppings for their ice cream sundaes. How many different sundaes can be ordered?

Practice
Use Pascal's Triangle to compute each combination.

75. $C(8, 3)$ **76.** $C(7, 4)$

Discovery and Writing

77. Describe the Fundamental Counting Principle.

78. What is a permutation?

79. What is a combination?

80. Explain the difference between a permutation and a combination.

81. Prove that $C(n, n) = 1$.

82. Prove that $C(n, 0) = 1$.

83. Prove that $\binom{n}{r} = \binom{n}{n-r}$.

84. Show that the Binomial Theorem can be expressed in the form

$$(a + b)^n = \sum_{k=0}^{n} \binom{n}{k} a^{n-k} b^k$$

85. Explain how to use Pascal's Triangle to find $C(8, 5)$.

86. Explain how to use Pascal's Triangle to find $C(10, 8)$.

Critical Thinking

Determine if the statement is true or false. If the statement is false, then correct it and make it true.

87. Permutation problems involve situations in which the order of the items makes no difference.

88. Combination problems involve situations in which order matters.

89. The number of permutations of *n* distinct objects is greater than the number of combinations of those *n* objects.

90. The number of permutations of *n* things taken *r* at a time can be found using the Fundamental Counting Principle.

91. The number of combinations of *n* things taken *r* at a time cannot be found using the Fundamental Counting Principle.

92. The number of ways to choose 11 people out of 25 to form a soccer team is $_{25}P_{11}$.

93. The number of ways to choose 3 company employees out of 25, one to work in Italy, one to work in Aruba, and one to work in Hawaii, is $_{25}P_3$.

94. The digits 1–9 are used to create a four-digit ATM personal identification number (PIN) for your debit bank card. If no digits are repeated, the number of possible numbers is $_{10}C_4$.

8.7 Probability

In this section, we will learn to

1. Compute probabilities.

2. Use the Multiplication Property of Probabilities.

In some parts of the United States, casinos are a big part of the entertainment industry. Casinos design their games using the laws of probability to ensure they make huge amounts of money at their customers' expense.

Gerolamo Cardano, an Italian doctor and mathematician, was an accomplished gambler who developed much of the theory of probability in the casino. His book *Liber de Ludo Aleae* (published after his death in 1663) was the first book to develop the topics of probability. Cardano predicted the day of his death. On that day, when it looked like he would live, it is said that he committed suicide to make his prediction come true.

1. Compute Probabilities

The probability that an event will occur is a measure of the likelihood of that event. A tossed coin, for example, can land in two ways, either heads or tails. Because one of these two equally likely outcomes is heads, we expect that out of several tosses, about half will be heads. We say that the probability of obtaining heads in a single toss of the coin is $\frac{1}{2}$.

If records show that out of 100 days with weather conditions like today's, 30 have received rain, the weather service will report, "There is a $\frac{30}{100}$ or 30% probability of rain today."

An **experiment** is a process for which the outcome is uncertain. Tossing a coin, rolling a die, drawing a card, and predicting rain are examples of experiments. For any experiment, the set of all possible outcomes is called a **sample space**.

The sample space, *S*, for the experiment of tossing two coins is the set

$$S = \{(H, H), (H, T), (T, H), (T, T)\}$$

where the ordered pair (H, T) represents the outcome "heads on the first coin and tails on the second coin." Because there are two possible outcomes for the first coin and two for the second coin, we know (by the Fundamental Counting Principle) that there are $2 \cdot 2 = 4$ possible outcomes. Since there are 4 elements in the sample space S, we write

$n(S) = 4$ Read as "The number of elements in set S is 4."

An **event** associated with an experiment is any subset of the sample space of that experiment. For example, if E is the event "getting at least one heads" in the experiment of tossing two coins, then

$E = \{(H, H), (H, T), (T, H)\}$

and $n(E) = 3$. Because the outcome of getting at least one heads can occur in 3 out of 4 possible ways, we say that the **probability** of a favorable outcome is $\frac{3}{4}$.

$$P(E) = P(\text{at least one heads}) = \frac{3}{4}$$

We define the probability of an event as follows.

Probability of an Event   If S is the sample space of an experiment with n distinct and equally likely outcomes, and E is an event that occurs in s of those ways, then the **probability of E** is

$$P(E) = \frac{n(E)}{n(S)} = \frac{s}{n}$$

> **Take Note**
>
> Probabilities are expressed as numbers ranging from 0 to 1.

Because $0 \le s \le n$, it follows that $0 \le \frac{s}{n} \le 1$. This implies that all probabilities have values from 0 to 1. An event that cannot happen has probability 0. An event that is certain to happen has probability 1.

To say that the probability of tossing heads on one toss of a coin is $\frac{1}{2}$ means that if a fair coin is tossed a large number of times, the ratio of the number of heads to the total number of tosses is nearly $\frac{1}{2}$.

To say that the probability of rolling 5 on one roll of a die is $\frac{1}{6}$ means that as the number of rolls approaches infinity, the ratio of the number of favorable outcomes (rolling a 5) to the total number of outcomes (rolling a 1, 2, 3, 4, 5, or 6) approaches $\frac{1}{6}$.

Showing the Sample Space of an Experiment

EXAMPLE 1

Show the sample space of the experiment "rolling two dice a single time."

©Flat Design/Shutterstock.com

SOLUTION   We can list ordered pairs and let the first number be the result on the first die and the second number the result on the second die. The sample space, S, is the set with the following elements:

(1, 1)	(1, 2)	(1, 3)	(1, 4)	(1, 5)	(1, 6)
(2, 1)	(2, 2)	(2, 3)	(2, 4)	(2, 5)	(2, 6)
(3, 1)	(3, 2)	(3, 3)	(3, 4)	(3, 5)	(3, 6)
(4, 1)	(4, 2)	(4, 3)	(4, 4)	(4, 5)	(4, 6)
(5, 1)	(5, 2)	(5, 3)	(5, 4)	(5, 5)	(5, 6)
(6, 1)	(6, 2)	(6, 3)	(6, 4)	(6, 5)	(6, 6)

> **Tip**
>
> It is important that you become very familiar with the sample space for rolling a pair of dice. Note that there are 36 outcomes.

Since there are 6 possible outcomes with the first die and 6 possible outcomes with the second die, we expect $6 \cdot 6 = 36$ equally likely possible outcomes, and we have $n(S) = 36$.

Self Check 1 How many pairs in the above sample space have a sum of 7?

Now Try Exercise 5.

EXAMPLE 2   Finding the Probability of an Event

Find the probability of the event "rolling a sum of 7 on one roll of two dice."

SOLUTION The sample space is listed in Example 1. We let E be the set of favorable outcomes, those that give a sum of 7:

$$E = \{(1, 6), (2, 5), (3, 4), (4, 3), (5, 2), (6, 1)\}$$

Since there are 6 favorable outcomes among the 36 equally likely outcomes, $n(E) = 6$, and

$$P(E) = P(\text{rolling a 7}) = \frac{n(E)}{n(S)} = \frac{6}{36} = \frac{1}{6}$$

Self Check 2 Find the probability of rolling a sum of 10.

Now Try Exercise 21.

A standard playing deck of 52 cards has two red suits, hearts and diamonds, and two black suits, clubs and spades. Each suit has 13 cards, including an ace, a king, a queen, a jack, and cards numbered from 2 to 10. We will refer to a standard deck of cards in many examples and exercises.

EXAMPLE 3   Finding the Probability of an Event

Find the probability of drawing 5 cards, all hearts, from a standard deck of cards.

SOLUTION Since the number of ways to draw 5 hearts from the 13 hearts is $C(13, 5)$, we have $n(E) = C(13, 5)$. Since the number of ways to draw 5 cards from the deck is $C(52, 5)$, we have $n(S) = C(52, 5)$. The probability of drawing 5 hearts is the ratio of the number of favorable outcomes to the number of possible outcomes.

$$P(5 \text{ hearts}) = \frac{C(13, 5)}{C(52, 5)}$$

$$P(5 \text{ hearts}) = \frac{\dfrac{13!}{5!8!}}{\dfrac{52!}{5!47!}}$$

$$= \frac{13!}{5!8!} \cdot \frac{5!47!}{52!}$$

$$= \frac{13 \cdot 12 \cdot 11 \cdot 10 \cdot 9 \cdot 8!}{8!} \cdot \frac{47!}{52 \cdot 51 \cdot 50 \cdot 49 \cdot 48 \cdot 47!}$$

$$= \frac{13 \cdot 12 \cdot 11 \cdot 10 \cdot 9}{52 \cdot 51 \cdot 50 \cdot 49 \cdot 48}$$

$$= \frac{33}{66{,}640}$$

©Photobar/Shutterstock.com

The probability of drawing 5 hearts is $\frac{33}{66{,}640}$.

Self Check 3   Find the probability of drawing 6 cards, all hearts, from the deck.

Now Try Exercise 33.

2. Use the Multiplication Property of Probabilities

There is a property of probabilities that is similar to the Multiplication Principle for Events. In the following theorem, we read $P(A \cap B)$ as "the probability of A and B" and $P(B|A)$ as "the probability of B given A." If A and B are events, the set $A \cap B$ contains the outcomes that are in both A and B.

Multiplication Property of Probabilities	If $P(A)$ represents the probability of event A, and $P(B\|A)$ represents the probability that event B will occur after event A, then $$P(A \cap B) = P(A) \cdot P(B\|A)$$

EXAMPLE 4

Finding the Probability of an Event

A box contains 40 wooden blocks of the same size. Of these blocks, 17 are red, 13 are blue, and the rest are yellow. If 2 blocks are drawn at random, without replacement, find the probability that 2 yellow blocks will be drawn.

SOLUTION   Of the 40 blocks in the box, 10 are yellow. The probability of getting a yellow block on the first draw is

$$P(\text{yellow block on the first draw}) = \frac{10}{40} = \frac{1}{4}$$

Because there is no replacement after the first draw, 39 blocks remain in the box, and 9 of these are yellow. The probability of drawing a yellow block on the second draw is

$$P(\text{yellow block on the second draw}) = \frac{9}{39} = \frac{3}{13}$$

The probability of drawing 2 yellow blocks in succession is the product of the probability of drawing a yellow block on the first draw and the probability of drawing a yellow block on the second draw.

$$P(\text{drawing two yellow blocks}) = \frac{1}{4} \cdot \frac{3}{13} = \frac{3}{52}$$

Self Check 4   Find the probability that 2 blue blocks will be drawn.

Now Try Exercise 23.

EXAMPLE 5

Using the Multiplication Property to Determine the Probability of an Event

Repeat Example 3 using the Multiplication Property of Probabilities.

SOLUTION   The probability of drawing a heart on the first draw is $\frac{13}{52}$. The probability of drawing a heart on the second draw *given that we got a heart on the first draw* is $\frac{12}{51}$. The probability is $\frac{11}{50}$ on the third draw, $\frac{10}{49}$ on the fourth draw, and $\frac{9}{48}$ on the fifth draw. By the Multiplication Property of Probabilities,

$$P(\text{5 hearts in a row}) = \frac{13}{52} \cdot \frac{12}{51} \cdot \frac{11}{50} \cdot \frac{10}{49} \cdot \frac{9}{48}$$

$$= \frac{33}{66,640}$$

Self Check 5 Use the Multiplication Property of Probabilities to find the probability of drawing 6 cards, all hearts, from the deck.

Now Try Exercise 27.

EXAMPLE 6

Finding the Probability of an Event

In a school, 30% of the students are gifted in mathematics and 10% are gifted in art and mathematics. If a student is gifted in mathematics, find the probability that the student is also gifted in art.

SOLUTION Let $P(M)$ be the probability that a randomly chosen student is gifted in mathematics, and let $P(M \cap A)$ be the probability that the student is gifted in both art and mathematics. We must find $P(A|M)$, the probability that the student is gifted in art, given that he or she is gifted in mathematics. To do so, we substitute the given values

$$P(M) = 0.3 \quad \text{and} \quad P(M \cap A) = 0.1$$

in the formula for multiplication of probabilities and solve for $P(A|M)$:

$$P(M \cap A) = P(M) \cdot P(A|M)$$

$$0.1 = (0.3)P(A|M)$$

$$P(A|M) = \frac{0.1}{0.3}$$

$$= \frac{1}{3}$$

©Serg Zastavkin/Shutterstock.com

If a student is gifted in mathematics, there is a probability of $\frac{1}{3}$ that he or she is also gifted in art.

Self Check 6 If 40% of the students are gifted in art, find the probability that a student gifted in art is also gifted in mathematics.

Now Try Exercise 55.

Self Check Answers

1. 6 2. $\frac{1}{12}$ 3. $\frac{33}{391,510}$ 4. $\frac{1}{10}$ 5. $\frac{33}{391,510}$ 6. $\frac{1}{4}$

Exercises 8.7

Getting Ready
You should be able to complete these vocabulary and concept statements before you proceed to the practice exercises.

Fill in the blanks.

1. An _____ is any process for which the outcome is uncertain.

2. A list of all possible outcomes for an experiment is called a _____.

3. The probability of an event E is defined as

$$P(E) = \underline{\hspace{1cm}} = \frac{s}{n}$$

4. $P(A \cap B) = $ _____

Practice
List the sample space of each experiment.

5. Rolling one die and tossing one coin

6. Tossing three coins

7. Selecting a letter of the alphabet

8. Picking a one-digit number

An ordinary die is rolled. Find the probability of each event.

Holly Kuchera/Shutterstock.com

9. Rolling a 2

10. Rolling a number greater than 4

11. Rolling a number greater than 1 but less than 6

12. Rolling an odd number

Balls numbered from 1 to 42 are placed in a container. If one is drawn at random, find the probability of each result.

13. The number is less than 20.

14. The number is less than 50.

15. The number is a prime number.

16. The number is less than 10 or greater than 40.

If the spinner shown below is spun, find the probability of each event. Assume that the spinner never stops on a line.

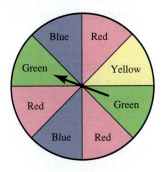

17. The spinner stops on red.

18. The spinner stops on green.

19. The spinner stops on orange.

20. The spinner stops on yellow.

Find the probability of each event.

21. Rolling a sum of 4 on one roll of two dice

22. Drawing a diamond on one draw from a card deck

23. Drawing two aces in succession from a card deck if the card is replaced and the deck is shuffled after the first draw

24. Drawing two aces from a card deck without replacing the card after the first draw

25. Drawing a red egg from a basket containing 5 red eggs and 7 blue eggs

26. Getting 2 red eggs in a single scoop from a bucket containing 5 red eggs and 7 yellow eggs

27. Drawing a bridge hand of 13 cards, all of one suit

28. Drawing 6 diamonds from a card deck without replacing the cards after each draw

29. Drawing 5 aces from a card deck without replacing the cards after each draw

30. Drawing 5 clubs from the black cards in a card deck

31. Drawing a face card (king, queen, or jack) from a card deck

32. Drawing 6 face cards in a row from a card deck without replacing the cards after each draw

33. Drawing 5 orange cubes from a bowl containing 5 orange cubes and 1 beige cube

34. Rolling a sum of 4 with one roll of three dice

35. Rolling a sum of 11 with one roll of three dice

36. Picking, at random, 5 Republicans from a group containing 8 Republicans and 10 Democrats

37. Tossing 3 heads in 5 tosses of a fair coin

38. Tossing 5 heads in 5 tosses of a fair coin

Assume that the probability that an airplane engine will fail during a torture test is $\frac{1}{2}$ and that the aircraft in question has 4 engines.

39. Construct a sample space for the torture test. Use S for survive and F for fail.

40. Find the probability that all engines will survive the test.

41. Find the probability that exactly 1 engine will survive.

42. Find the probability that exactly 2 engines will survive.

43. Find the probability that exactly 3 engines will survive.

44. Find the probability that no engines will survive.

45. Find the sum of the probabilities in Exercises 40 through 44.

Assume that a survey of 282 people is taken to determine the opinions of doctors, teachers, and lawyers on a proposed piece of legislation, with the results as shown in the table. A person is chosen at random from those surveyed. Refer to the table to find each probability.

	Number that Favor	Number that Oppose	Number with No Opinion	Total
Doctors	70	32	17	119
Teachers	83	24	10	117
Lawyers	23	15	8	46
Total	176	71	35	282

46. The person favors the legislation.

47. A doctor opposes the legislation.

48. A person who opposes the legislation is a lawyer.

49. Quality control In a batch of 10 tires, 2 are known to be defective. If 4 tires are chosen at random, find the probability that all 4 tires are good.

50. Medicine Out of a group of 9 patients treated with a new drug, 4 suffered a relapse. Find the probability that 3 patients of this group, chosen at random, will remain disease-free.

Use the Multiplication Property of Probabilities.

51. If $P(A) = 0.3$ and $P(B|A) = 0.6$, find $P(A \cap B)$.

52. If $P(A \cap B) = 0.3$ and $P(B|A) = 0.6$, find $P(A)$.

53. Conditional probability The probability that a person owns a luxury car is 0.2, and the probability that the owner of such a car also owns a personal computer is 0.7. Find the probability that a person, chosen at random, owns both a luxury car and a computer.

54. Conditional probability If 40% of the population have completed college, and 85% of college graduates are registered to vote, what percent of the population are both college graduates and registered voters?

55. Conditional probability About 25% of the population watches the evening television news coverage as well as the soap operas. If 75% of the population watches the news, what percent of those who watch the news also watch the soaps?

56. Conditional probability The probability of rain today is 0.40. If it rains, the probability that Cameron will forget his raincoat is 0.70. Find the probability that Cameron will get wet.

Discovery and Writing

57. What is an experiment? Give two examples.

58. What is meant by the sample space of an experiment?

59. Describe how to determine the probability of an event.

60. Explain the Multiplication Property of Probabilities.

61. If $P(A \cap B) = 0.7$, is it possible that $P(B|A) = 0.6$? Explain.

62. Is it possible that $P(A \cap B) = P(A)$? Explain.

Critical Thinking

Determine if the statement is true or false. If the statement is false, then correct it and make it true.

63. The probability that an event occurs can be a negative number.

64. The probability of a certain event is 0.

65. The probability of an impossible event is 1.

66. If the probability that you will graduate is 0.63, then the probability that you will not graduate is -0.63.

67. Events that cannot occur simultaneously are called **mutually exclusive events**. If one card is randomly selected from a deck of cards, drawing a jack or a queen would be mutually exclusive events.

68. Events that can occur simultaneously are called **non–mutually exclusive events**. If one card is randomly selected from a deck of cards, drawing a jack or a heart would be non–mutually exclusive events.

69. The probability of a couple having four boys is $\frac{1}{2}$ because the probability each time of having a boy is $\frac{1}{2}$.

70. If the probability of a couple of five girls is $\frac{1}{32}$, then the probability of having five boys is $\frac{31}{32}$.

CHAPTER REVIEW

8.1 The Binomial Theorem

Definitions and Concepts	Examples

Pascal's Triangle:

Pascal's Triangle is a triangular array of numbers showing the coefficients of the expansion of $(a + b)^n$. The example shows the first several rows of Pascal's triangle. Except for the 1's, each number in the triangle is the sum of the two numbers above it.

					1					Row 0
				1		1				Row 1
			1		2		1			Row 2
		1		3		3		1		Row 3
	1		4		6		4		1	Row 4
1		5		10		10		5	1	Row 5

Factorial notation:

$$n! = n(n - 1)(n - 2) \cdot \cdots \cdot 3 \cdot 2 \cdot 1$$

$$0! = 1 \qquad n(n - 1)! = n!$$

$7! = 7 \cdot 6 \cdot 5 \cdot 4 \cdot 3 \cdot 2 \cdot 1 = 5040$

$7 \cdot 6! = 7 \cdot 6 \cdot 5 \cdot 4 \cdot 3 \cdot 2 \cdot 1 = 7!$

The Binomial Theorem:

If n is any positive integer, then

$$(a + b)^n = a^n + \frac{n!}{1!(n - 1)!} a^{n-1}b$$

$$+ \frac{n!}{2!(n - 2)!} a^{n-2}b^2 + \frac{n!}{3!(n - 3)!} a^{n-3}b^3 + \cdots$$

$$+ \frac{n!}{r!(n - r)!} a^{n-r}b^r + \cdots + b^n$$

$$(a + b)^3 = a^3 + \frac{3!}{1!(3 - 1)!} a^{3-1}b + \frac{3!}{2!(3 - 2)!} a^{3-2}b^2 + b^3$$

$$= a^3 + \frac{3 \cdot 2!}{1 \cdot 2!} a^2b + \frac{3 \cdot 2 \cdot 1}{2 \cdot 1 \cdot 1} ab^2 + b^3$$

$$= a^3 + 3a^2b + 3ab^2 + b^3$$

Note that the coefficients 1, 3, 3, and 1 are the numbers in the third row of Pascal's Triangle.

Find the fourth term of $(a + b)^9$.

In the 4th term, the exponent on b is 3, and the exponent on a is $9 - 3$ or 6. The numerator of the fractional coefficient is 9! and the factors of the denominator are 3! and $(9 - 3)!$. Thus the 4th term is

$$\frac{9!}{3!(9 - 3)!} a^6b^3 = 84a^6b^3$$

Exercises

Find each value.

1. $6!$

2. $7! \cdot 0! \cdot 1! \cdot 3!$

3. $\dfrac{8!}{7!}$

4. $\dfrac{5! \cdot 7! \cdot 8!}{6! \cdot 9!}$

Expand each expression.

5. $(x + y)^3$

6. $(p + q)^4$

7. $(a - b)^5$

8. $(2a - b)^3$

Find the required term of each expansion.

9. $(a + b)^8$; 4th term

10. $(2x - y)^5$; 3rd term

11. $(x - y)^9$; 7th term

12. $(4x + 7)^6$; 4th term

8.2 Sequences, Series, and Summation Notation

Definitions and Concepts	Examples
Sequences and series: An **infinite sequence** is a function whose domain is the set of natural numbers.	**Infintie sequence:** $2, 5, 8, 11, 14, 17, 20, \ldots$
A **finite sequence** is a function whose domain is the set of the first n natural numbers.	**Finite sequence:** $2, 5, 8, 11, 14, 17, 20$
If the commas in a sequence are replaced with $+$ signs, the result is called a **series**.	**Infinite series:** $2 + 5 + 8 + 11 + 14 + 17 + 20 + \ldots$
	Finite series: $2 + 5 + 8 + 11 + 14 + 17 + 20$

nth term or general term of a sequence notation:
We use a_n notation to write the value of a sequence at a number n. The variable n represents the position of the term in the sequence, while the notation a_n represents the value of the term in that position.

Given $a_n = \dfrac{(-1)^{n+1}}{2^n}$, find the first four terms of the sequence and a_{10}.

To find the first four terms of the sequence, we will substitute 1, 2, 3, and 4 in the formula that defines the sequence.

$a_1 = \dfrac{(-1)^{1+1}}{2^1} = \dfrac{1}{2}$ Substitute 1 for n.

$a_2 = \dfrac{(-1)^{2+1}}{2^2} = -\dfrac{1}{4}$ Substitute 2 for n.

$a_3 = \dfrac{(-1)^{3+1}}{2^3} = \dfrac{1}{8}$ Substitute 3 for n.

$a_4 = \dfrac{(-1)^{4+1}}{2^4} = -\dfrac{1}{16}$ Substitute 4 for n.

The first four terms are $\dfrac{1}{2}, -\dfrac{1}{4}, \dfrac{1}{8},$ and $-\dfrac{1}{16}$.

To find a_{10}, the 10th term of the sequence, we will substitute 10 in the formula for the nth term.

$a_{10} = \dfrac{(-1)^{10+1}}{2^{10}} = -\dfrac{1}{1024}$ Substitute 10 for n.

Define a sequence recursively:
A sequence can be defined **recursively** by giving its first term and a rule showing how to obtain the $(n + 1)$th term from the nth term.

Write the first 4 terms of the sequence with a first term of 3 and a general term of $a_{n+1} = 2a_n + 1$.

$a_1 = 3$

$a_2 = 2(\textbf{3}) + 1 = 7$

$a_3 = 2(\textbf{7}) + 1 = 15$

$a_4 = 2(\textbf{15}) + 1 = 31$

The first four terms are 3, 7, 15, and 31.

Definitions and Concepts	Examples

Summation notation:

$\sum_{k=1}^{4} n^3$ indicates the sum of the four terms obtained when 1, 2, 3, and 4 are substituted for n.

$$\sum_{k=1}^{4} n^3 = 1^3 + 2^3 + 3^3 + 4^3 = 1 + 8 + 27 + 64 = 100$$

- If c is a constant, then

$$\sum_{k=1}^{n} c = nc$$

$$\sum_{k=1}^{5} 9 = 5 \cdot 9 = 45$$

- If c is a constant, then

$$\sum_{k=1}^{n} cf(k) = c\sum_{k=1}^{n} f(k)$$

$$\sum_{k=1}^{8} 4k^2 = 4\sum_{k=1}^{8} k^2$$

- $\sum_{k=1}^{n} [f(k) + g(k)] = \sum_{k=1}^{n} f(k) + \sum_{k=1}^{n} g(k)$

$$\sum_{k=1}^{5} (k + k^2) = \sum_{k=1}^{5} k + \sum_{k=1}^{5} k^2$$

Exercises

Write the fourth term in each sequence.

13. $0, 7, 26, \ldots, n^3 - 1, \ldots$

14. $\dfrac{3}{2}, 3, \dfrac{11}{2}, \ldots, \dfrac{n^2 + 2}{2}, \ldots$

Find the first five terms of the sequence and then find a_{10}.

15. $a_n = 2n^2 - 2$

Find the first four terms of the sequence.

16. $a_1 = -2$ and $a_{n+1} = 2a_n^2$

Evaluate each expression.

17. $\sum_{k=1}^{4} 3k^2$

18. $\sum_{k=1}^{10} 6$

19. $\sum_{k=5}^{8} (k^3 + 3k^2)$

20. $\sum_{k=1}^{30} \left(\dfrac{3}{2}k - 12\right) - \dfrac{3}{2}\sum_{k=1}^{30} k$

8.3 Arithmetic Sequences

Definitions and Concepts	Examples

Arithmetic sequence:

An **arithmetic sequence** is of the form

$$a, a + d, a + 2d, \ldots, a + (n - 1)d$$

where a is the first term, d is the common difference, and $a_n = a + (n - 1)d$ is the n term.

In the arithmetic sequence $2, 5, 8, 11, 14, 17, \ldots,$ 2 is the first term, 3 is the common difference, and the sixth term is 17.

The twentieth term of the sequence above is given by

$$a_n = a + (n - 1)d$$

$$a_{20} = 2 + (20 - 1)3$$

$$= 2 + (19)3$$

$$= 59$$

Definitions and Concepts	Examples
Arithmetic means: Numbers inserted between a first and last term to form an arithmetic sequence are called **arithmetic means**.	To insert three arithmetic means between -4 and 16, we first determine the common difference d by substituting -4 for a, 16 for a_n, and 5 for n in the formula $a_n = a + (n-1)d$ and solving for d. $a_n = a + (n-1)d$ $16 = -4 + (5-1)d$ $20 = 4d$ $5 = d$ Since the common difference is 5, the three arithmetic means are $-4 + 5,\ -4 + 2(5),\ -4 + 3(5)$ or $1, 6, 11$
Sum of the first n terms of an arithmetic series: The formula $S_n = \dfrac{n(a + a_n)}{2}$ gives the sum of the first n terms of an arithmetic series, where S_n is the sum, a is the first term, a_n is the last (or nth) term, and n is the number of terms.	The sum of the first 20 terms of the arithmetic series $2 + 5 + 8 + 11 + 14 + 17 + 20 + \cdots$ is given by $S_n = \dfrac{n(a + a_n)}{2}$ $S_{20} = \dfrac{20(2 + 59)}{2}$ $a_n = 59$ was found earlier. $= 610$

Exercises

Find the required term of each arithmetic sequence.

21. $5, 9, 13, \ldots$; 29th term

22. $8, 15, 22, \ldots$; 40th term

23. $6, -1, -8, \ldots$; 15th term

24. $\dfrac{1}{2}, -\dfrac{3}{2}, -\dfrac{7}{2}, \ldots$; 35th term

25. Find three arithmetic means between 2 and 8.

26. Find five arithmetic means between 10 and 100.

Find the sum of the first 40 terms in each sequence.

27. $5, 9, 13, \ldots$

28. $8, 15, 22$

29. $6, -1, -8, \ldots$

30. $\dfrac{1}{2}, -\dfrac{3}{2}, -\dfrac{7}{2}, \ldots$

8.4 Geometric Sequences

Definitions and Concepts	Examples
Geometric sequence: A **geometric sequence** is of the form $a, ar, ar^2, ar^3, \ldots, ar^{n-1}, \ldots$ where a is the first term and r is the common ratio. The formula $a_n = ar^{n-1}$ gives the nth term of the sequence.	In the geometric sequence $2, 6, 18, 54, 162, \ldots$, 2 is the first term, 3 is the common ratio, and the fifth term is 162. The tenth term of the sequence above is given by $a_n = ar^{n-1}$ $a_{10} = 2 \cdot 3^{10-1}$ $= 2 \cdot 3^9$ $= 39{,}366$

Definitions and Concepts	Examples
Geometric means: Numbers inserted between a first and last term to form a geometric sequence are called **geometric means**.	To insert a positive geometric mean between 1 and 16, we first determine the common ratio r by substituting 16 for a_n, 1 for a and 3 for n in the formula $a_n = ar^{n-1}$ and solving for r. $$a_n = ar^{n-1}$$ $$16 = 1r^{3-1}$$ $$16 = r^2$$ $$\pm 4 = r$$ The positive geometric mean is 4.
Sum of the first n terms of a geometric series: The formula $$S_n = \frac{a - ar^n}{1 - r} \quad (r \neq 1)$$ gives the sum of the first n terms of a geometric series, where S_n is the sum, a is the first term, r is the common ratio, and n is the number of terms.	Find the sum of the first 6 terms of the geometric series $2 + 6 + 18 + 54 + 162 + \cdots$. In this series, the first term is 2, the common ratio is 3, and the sixth term is $a_n = ar^{n-1} = 2(3^5) = 486$. So the sum of the first 6 terms is $$S_n = \frac{a - ar^n}{1 - r}$$ $$S_6 = \frac{2 - 2(3^6)}{1 - 3}$$ $$= \frac{2 - 2(727)}{-2}$$ $$= 728$$
Sum of the terms of an infinite geometric series: If $|r| < 1$, the formula $$S = \frac{a}{1 - r}$$ gives the sum of the terms of an infinite geometric series, where S_∞ is the sum, a is the first term, and r is the common ratio.	Find the sum of the infinite series $162 + 54 + 18 + \cdots$. In this infinite series, the first term is 162 and the common ratio is $\frac{1}{3}$. So the sum of the series is $$S = \frac{a}{1 - r}$$ $$= \frac{162}{1 - \frac{1}{3}}$$ $$= 243$$

Exercises

Find the required term of each geometric sequence.

31. $81, 27, 9, \ldots$; 11th term

32. $2, 6, 18, \ldots$; 9th term

33. $9, \frac{9}{2}, \frac{9}{4}, \ldots$; 15th term

34. $8, -\frac{8}{5}, \frac{8}{25}, \ldots$; 7th term

35. Find three positive geometric means between 2 and 8.

36. Find four geometric means between -2 and 64.

37. Find the positive geometric mean between 4 and 64.

Find the sum of the first 8 terms in each geometric sequence.

38. $81, 27, 9, \ldots$

39. $2, 16, 18, \ldots$

40. $9, \frac{9}{2}, \frac{9}{4}, \ldots$

41. $8, -\frac{8}{5}, \frac{8}{25}$

42. Find the sum of the first eight terms of the geometric sequence $\frac{1}{3}, 1, 3, \ldots$.

43. Find the seventh term of the geometric sequence $2\sqrt{2}, 4, 4\sqrt{2}, \ldots$.

Find the sum of each infinite geometric sequence, if possible.

44. $\frac{1}{3}, \frac{1}{6}, \frac{1}{12}, \cdots$

45. $\frac{1}{5}, -\frac{2}{15}, \frac{4}{45}, \cdots$

46. $1, \frac{3}{2}, \frac{9}{4}, \cdots$

47. $0.5, 0.25, 0.125, \ldots$

Change each decimal into a common fraction.

48. $0.\overline{3}$

49. $0.\overline{9}$

50. $0.\overline{17}$

51. $0.\overline{45}$

52. Investment problem If Landon invests $3000 in a 6-year certificate of deposit at the annual rate of 7.75%, compounded daily, how much money will be in the account when it matures?

53. College enrollments The enrollment at Hometown College is growing at the rate of 5% over each previous year's enrollment. If the enrollment is currently 4000 students, what will it be 10 years from now? What was it 5 years ago?

54. Mobile home depreciation A mobile home that originally cost $10,000 depreciates in value at the rate of 10% per year. How much will the mobile home be worth after 10 years?

8.5 Mathematical Induction

Definitions and Concepts	Examples
Principle of Mathematical Induction: If a statement involving the natural number n has the two properties **1.** The statement is true for $n = 1$, and **2.** If the statement is true for $n = k$, then it is true for $n = k + 1$, then the statement is true for all natural numbers.	See Examples 1, 2, and 3 in the section.

Exercises

55. Verify the following formula for $n = 1, n = 2, n = 3$, and $n = 4$:

$$1^3 + 2^3 + 3^3 + \cdots + n^3 = \frac{n^2(n + 1)^2}{4}$$

56. Prove the formula given in Exercise 55 by mathematical induction.

8.6 Permutations and Combinations

Definitions and Concepts	Examples
Fundamental Counting Principle If event E_1 can be done in a_1 ways and event E_2 can be done in a_2 ways, then event E_1 followed by E_2 can be done in $a_1 \cdot a_2$ ways.	If you choose 1 drink from **3** choices and 1 sandwich from **6** choices, in how many ways can you choose 1 drink and 1 sandwich? By the Fundamental Counting Principle, the number of ways is $3 \cdot 6 = 18$.

Definitions and Concepts	Examples
Formulas for permutations and combinations: $$P(n, r) = \frac{n!}{(n-r)!}$$ $$P(n, n) = n!$$ $$P(n, 0) = 1$$ $$C(n, r) = \binom{n}{r} = \frac{n!}{r!(n-r)!}$$ $$C(n, n) = 1$$ $$C(n, 0) = 1$$	$$P(7, 2) = \frac{7!}{(7-2)!} = \frac{7 \cdot 6 \cdot 5!}{5!} = 42$$ $$P(7, 7) = \frac{7!}{(7-7)!} = \frac{7!}{0!} = \frac{7!}{1} = 7!$$ $$P(7, 0) = \frac{7!}{(7-0)!} = \frac{7!}{7!} = 1$$ $$C(7, 2) = \frac{7!}{2!(7-2)!} = \frac{7 \cdot 6 \cdot 5!}{2 \cdot 5!} = 7 \cdot 3 = 21$$ $$C(7, 7) = \frac{7!}{7!(7-7)!} = \frac{7!}{7! \cdot 0!} = \frac{7!}{7! \cdot 1} = 1$$ $$C(7, 0) = \frac{7!}{0!(7-0)!} = \frac{7!}{1 \cdot 7!} = \frac{7!}{1 \cdot 7!} = 1$$
Circular arrangements: There are $(n - 1)!$ ways to place n things in a circle.	There are $(6 - 1)! = 5! = 120$ ways to seat 6 people at a round table.
Permutations of like objects: The number of permutations of n things with a things alike, b things alike, and so on, is $$\frac{n!}{a!b! \cdots}$$	The number of distinguishable words that can be formed from the letters of the word *happy* is $$\frac{5!}{1! \cdot 1! \cdot 2! \cdot 1!} \qquad \text{There are 5 letters. } h, a, \text{ and } y \text{ occur once; } p \text{ occurs twice.}$$ $$= \frac{5 \cdot 4 \cdot 3 \cdot 2!}{2!} = 60$$

Exercises

Evaluate each expression.

57. $P(8, 5)$

58. $C(7, 4)$

59. $0! \cdot 1!$

60. $P(10, 2) \cdot C(10, 2)$

61. $P(8, 6) \cdot C(8, 6)$

62. $C(8, 5) \cdot C(6, 2)$

63. $C(7, 5) \cdot P(4, 0)$

64. $C(12, 10) \cdot C(11, 0)$

65. $\dfrac{P(8, 5)}{C(8, 5)}$

66. $\dfrac{C(8, 5)}{C(13, 5)}$

67. $\dfrac{C(6, 3)}{C(10, 3)}$

68. $\dfrac{C(13, 5)}{C(52, 5)}$

69. In how many ways can 10 teenagers be seated at a round table if 2 girls wish to sit with their boyfriends?

70. How many distinguishable words can be formed from the letters of the word *casserole* if each letter is used exactly once?

8.7 Probability

Definitions and Concepts	Examples
Probability of an event: An event that cannot happen has a probability of 0. An event that is certain to happen has a probability of 1. All other events have probabilities between 0 and 1.	The probability that pigs fly is 0. The probability that every pig will die is 1.

Definitions and Concepts	Examples
If S is the sample space of an experiment with n distinct and equally likely outcomes, and if E is an event that occurs in s of those ways, then the **probability of E** is $$P(E) = \frac{n(E)}{n(S)} = \frac{s}{n}$$	Since there are 28 days and 4 Fridays in February, the probability that a day chosen at random in February will be a Friday is $$P(\text{Friday}) = \frac{4}{28} \qquad \begin{array}{l}\text{the number of Fridays}\\ \text{the number of days}\end{array}$$ $$= \frac{1}{7}$$
Multiplication Property of Probabilities: $$P(A \cap B) = P(A) \cdot P(B \mid A)$$	A jar contains 25 marbles of the same size. Of these marbles, 10 are red, 10 are blue, and 5 are yellow. If 2 marbles are drawn (without replacement), find the probability that both marbles will be red. Of the 25 marbles in the jar, 10 are red. The probability of getting a red marble on the first draw is $$P(\text{red marble on the first draw}) = \frac{10}{25} = \frac{2}{5}$$ Because there is no replacement after the first draw, 24 marbles remain in the jar and 9 of these are red. The probability of drawing a red marble on the second draw is $$P(\text{red marble on the second draw}) = \frac{9}{24} = \frac{3}{8}$$ The probability of drawing 2 red marbles in a row is the product of the probabilities: $$P(\text{drawing 2 red marbles}) = \frac{2}{5} \cdot \frac{3}{8} = \frac{3}{20}$$

Exercises

71. Make a tree diagram (like Figure 8-4 on page 810) to illustrate the possible results of tossing a coin four times.

72. In how many ways can you draw a 5-card poker hand of 3 aces and 2 kings?

73. Find the probability of drawing the hand described in Exercise 72.

74. Find the probability of not drawing the hand described in Exercise 72.

75. Find the probability of having a 13-card bridge hand consisting of 4 aces, 4 kings, 4 queens, and 1 jack.

76. Find the probability of choosing a committee of 3 men and 2 women from a group of 8 men and 6 women.

77. Find the probability of drawing a club or a spade on one draw from a card deck.

78. Find the probability of drawing a black card or a king on one draw from a card deck.

79. Find the probability of getting an ace-high royal flush in hearts (ace, king, queen, jack, and ten of hearts) in poker.

80. Find the probability of being dealt 5 cards of one suit in a poker hand.

CHAPTER TEST

Find each value.

1. $3! \cdot 0! \cdot 4! \cdot 1!$

2. $\dfrac{2! \cdot 4! \cdot 6! \cdot 8!}{3! \cdot 5! \cdot 7!}$

Find the required term in each binomial expansion.

3. $(x + 2y)^5$; 2nd term

4. $(2a - b)^8$; 7th term

Find each sum.

5. $\displaystyle\sum_{k=1}^{3}(4k + 1)$

6. $\displaystyle\sum_{k=2}^{4}(3k - 21)$

Find the sum of the first ten terms of each sequence.

7. $2, 5, 8, \ldots$

8. $5, 1, -3, \ldots$

9. Find three arithmetic means between 4 and 24.

10. Find two geometric means between -2 and -54.

Find the sum of the first ten terms of each sequence.

11. $\dfrac{1}{4}, \dfrac{1}{2}, 1, \ldots$

12. $6, 2, \dfrac{2}{3}, \ldots$

13. A car costing $\$C$ when new depreciates 25% of the previous year's value each year. How much is the car worth after 3 years?

14. A house costing $\$C$ when new appreciates 10% of the previous year's value each year. How much will the house be worth after 4 years?

15. Prove by mathematical induction:

$$3 + 4 + 5 + \cdots + (n + 2) = \frac{1}{2}n(n + 5)$$

16. How many six-digit license plates can be made if no plate begins with 0 or 1?

Find each value.

17. $P(7, 2)$

18. $P(4, 4)$

19. $C(8, 2)$

20. $C(12, 0)$

21. How many ways can 4 men and 4 women stand in line if all the women are first?

22. How many different ways can 6 people be seated at a round table?

23. How many different words can be formed from the letters of the word *bluff* if each letter is used once?

24. Show the sample space of the experiment: toss a fair coin three times.

Find each probability.

25. Rolling a 5 on one roll of a die

26. Drawing a jack or a queen from a standard card deck

27. Receiving 5 hearts for a 5-card poker hand

28. Rolling a sum of 9 on one roll of two dice.

29. A box contains 50 cubes of the same size. Of these cubes, 20 are red, and 30 are blue. If 2 cubes are drawn at random, without replacement, find the probability that 2 blue cubes are drawn.

30. In a batch of 20 tires, 2 are known to be defective. If 4 tires are chosen at random, find the probability that all 4 tires are good.

APPENDIX I

A PROOF OF THE BINOMIAL THEOREM

The Binomial Theorem can be proved for positive-integer exponents using mathematical induction.

The Binomial Theorem   If n is a positive integer, then

$$(a + b)^n = a^n + \frac{n!}{1!(n-1)!}a^{n-1}b^1 + \frac{n!}{2!(n-2)!}a^{n-2}b^2 + \cdots$$

$$+ \frac{n!}{r!(n-r)!}a^{n-r}b^r + \cdots + b^n$$

PROOF   As in all induction proofs, there are two parts.

Part 1: Substituting the number 1 for n on both sides of the equation, we have

$$(a + b)^1 = a^1 + \frac{1!}{1!(1-1)!}a^{1-1}b^1$$

$$a + b = a + a^0 b$$

$$a + b = a + b$$

and the theorem is true when $n = 1$. Part 1 is complete.

Part 2: We write expressions for two general terms in the statement of the induction hypothesis. We assume that the theorem is true for $n = k$:

$$(a + b)^k = a^k + \frac{k!}{1!(k-1)!}a^{k-1}b + \frac{k!}{2!(k-2)!}a^{k-2}b^2 + \cdots$$

$$+ \frac{k!}{(r-1)!(k-r+1)!}a^{k-r+1}b^{r-1}$$

$$+ \frac{k!}{r!(k-r)!}a^{k-r}b^r + \cdots + b^k$$

We multiply both sides of this equation by $a + b$ and hope to obtain a similar equation in which the quantity $k + 1$ replaces all of the n values in the Binomial Theorem:

$$(a + b)^k(a + b)$$

$$= (a + b)\left[a^k + \frac{k!}{1!(k-1)!}a^{k-1}b + \frac{k!}{2!(k-2)!}a^{k-2}b^2 + \cdots \right.$$

$$\left. + \frac{k!}{(r-1)!(k-r+1)!}a^{k-r+1}b^{r-1} + \frac{k!}{r!(k-r)!}a^{k-r}b^r + \cdots + b^k \right]$$

We distribute the multiplication first by a and then by b:

$$(a + b)^{k+1} = \left[a^{k+1} + \frac{k!}{1!(k-1)!}a^k b + \frac{k!}{2!(k-2)!}a^{k-1}b^2 + \cdots \right.$$

$$\left. + \frac{k!}{(r-1)!(k-r+1)!}a^{k-r+2}b^{r-1} + \frac{k!}{r!(k-r)!}a^{k-r+1}b^r + \cdots + ab^k \right]$$

$$+ \left[a^k b + \frac{k!}{1!(k-1)!}a^{k-1}b^2 + \frac{k!}{2!(k-2)!}a^{k-2}b^3 + \cdots \right.$$

$$\left. + \frac{k!}{(r-1)!(k-r+1)!}a^{k-r+1}b^r + \frac{k!}{r!(k-r)!}a^{k-r}b^{r+1} + \cdots + b^{k+1} \right]$$

Combining like terms, we have

$$(a + b)^{k+1} = a^{k+1} + \left[\frac{k!}{1!(k-1)!} + 1 \right]a^k b$$

$$+ \left[\frac{k!}{2!(k-2)!} + \frac{k!}{1!(k-1)!} \right]a^{k-1}b^2 + \cdots$$

$$+ \left[\frac{k!}{r!(k-r)!} + \frac{k!}{(r-1)!(k-r+1)!} \right]a^{k-r+1}b^r + \cdots + b^{k+1}$$

These results may be written as

$$(a + b)^{k+1} = a^{k+1} + \frac{(k+1)!}{1!(k+1-1)!}a^{(k+1)-1}b + \frac{(k+1)!}{2!(k+1-2)!}a^{(k+1)-2}b^2$$

$$+ \cdots + \frac{(k+1)!}{r!(k+1-r)!}a^{(k+1)-r}b^r + \cdots + b^{k+1}$$

This formula has precisely the same form as the Binomial Theorem, with the quantity $k + 1$ replacing all of the original n values. Therefore, the truth of the theorem for $n = k$ implies the truth of the theorem for $n = k + 1$. Because both parts of the axiom of mathematical induction are verified, the theorem is proved.

Answers to Selected Exercises

Section 0.1
1. set **3.** union **5.** decimal **7.** 2 **9.** composite
11. decimals **13.** negative **15.** $x + (y + z)$
17. $5m + 5 \cdot 2$ **19.** interval **21.** two **23.** positive
25. true **27.** false **29.** true **31.** {a, b, c, d, e, f, g}
33. {a, c, e} **35.** terminates **37.** repeats
39. 1, 2, 6, 7 **41.** $-5, -4, 0, 1, 2, 6, 7$ **43.** $\sqrt{2}$
45. 6 **47.** $-5, 1, 7$
49.
51.
53.
55.
57. $(2, \infty)$
59. $(0, 5)$
61. $(-4, \infty)$
63. $[-2, 2)$
65. $(-\infty, 5]$
67. $(-5, 0]$
69. $[-2, 3]$
71. $[2, 6]$
73. $(-5, \infty) \cap (-\infty, 4)$
75. $[-8, \infty) \cap (-\infty, -3]$
77. $(-\infty, -2) \cup (2, \infty)$
79. $(-\infty, -1] \cup [3, \infty)$
81. 13 **83.** 0 **85.** -8 **87.** -32 **89.** $5 - \pi$
91. 0 **93.** $x + 1$ **95.** $-(x - 4)$ **97.** 5 **99.** 5
101. natural numbers **103.** integers
105. 22°F **113.** false There are five integers between -3
and 3. **115.** false ∞ is not a real number. **117.** true
119. false There are eight subsets of {11, 22, 33}.

Section 0.2
1. factor **3.** 3, $2x$ **5.** scientific, integer **7.** x^{m+n}
9. $x^n y^n$ **11.** 1 **13.** 169 **15.** -25
17. $4 \cdot x \cdot x \cdot x$ **19.** $(-5x)(-5x)(-5x)(-5x)$
21. $-8 \cdot x \cdot x \cdot x \cdot x$ **23.** $7x^3$ **25.** x^2 **27.** $-27t^3$
29. $x^3 y^2$ **31.** 10.648 **33.** -0.0625 **35.** x^5 **37.** z^6
39. y^{21} **41.** z^{26} **43.** a^{14} **45.** $27x^3$ **47.** $x^6 y^3$

49. $\dfrac{a^6}{b^3}$ **51.** 1 **53.** 1 **55.** $\dfrac{1}{z^4}$ **57.** $\dfrac{1}{y^5}$ **59.** x^2

61. x^4 **63.** a^4 **65.** x **67.** $\dfrac{m^9}{n^6}$ **69.** $\dfrac{1}{a^9}$ **71.** $\dfrac{a^{12}}{b^4}$

73. $\dfrac{1}{r^4}$ **75.** $\dfrac{x^{32}}{y^{16}}$ **77.** $\dfrac{9x^{10}}{25y^2}$ **79.** $\dfrac{4y^{12}}{x^{20}}$ **81.** $\dfrac{64z^7}{25y^6}$

83. -50 **85.** 4 **87.** -8 **89.** 216 **91.** -12

93. 20 **95.** $\dfrac{3}{64}$ **97.** 3.72×10^5 **99.** -1.77×10^8

101. 7×10^{-3} **103.** -6.93×10^{-7} **105.** 1×10^{12}
107. 937,000 **109.** 0.0000221 **111.** 3.2
113. -0.0032 **115.** 1.17×10^4 **117.** 7×10^4
119. 5.3×10^{19} **121.** 1.986×10^4 meters per min
123. 1.67248×10^{-15} g **125.** $10^3 \cdot 26^3$; 1.7576×10^7
127. polar radius: 6.356750×10^3 km;
equatorial radius: 6.378135×10^3 km
129. x^{n+2} **131.** x^{m-1} **133.** x^{m+4}

143. false 0^0 is undefined. **145.** false $x^{-n} = \dfrac{1}{x^n}$ **147.** true

149. 1.264725×10^{12}

Section 0.3
1. 0 **3.** not **5.** $a^{1/n}$ **7.** $\sqrt[n]{ab}$ **9.** $\neq$ **11.** 3

13. $\dfrac{1}{5}$ **15.** -3 **17.** 10 **19.** $-\dfrac{3}{2}$

21. not a real number **23.** $4|a|$ **25.** $2|a|$

27. $-2a$ **29.** $-6b^2$ **31.** $\dfrac{4a^2}{5|b|}$ **33.** $-\dfrac{10x^2}{3y}$

35. 8 **37.** -64 **39.** -100 **41.** $\dfrac{1}{8}$ **43.** $\dfrac{1}{512}$

45. $-\dfrac{1}{27}$ **47.** $\dfrac{32}{243}$ **49.** $\dfrac{16}{9}$ **51.** $10s^2$ **53.** $\dfrac{1}{2y^2 z}$

55. $x^6 y^3$ **57.** $\dfrac{1}{r^6 s^{12}}$ **59.** $\dfrac{4a^4}{25b^6}$ **61.** $\dfrac{100s^8}{9r^4}$ **63.** a

65. 7 **67.** 5 **69.** -5 **71.** $-\dfrac{1}{5}$ **73.** $6|x|$

75. $3y^2$ **77.** $2y$ **79.** $\dfrac{|x|y^2}{|z^3|}$ **81.** $\sqrt{2}$ **83.** $17x\sqrt{2}$

85. $2y^2\sqrt{3y}$ **87.** $12\sqrt[3]{3}$ **89.** $6z\sqrt[4]{3z}$ **91.** $6x\sqrt{2y}$

93. 0 **95.** $\sqrt{3}$ **97.** $\dfrac{2\sqrt{x}}{x}$ **99.** $\sqrt[3]{4}$ **101.** $\sqrt[3]{5a^2}$

103. $\dfrac{2b\sqrt[4]{27a^2}}{3a}$ **105.** $\dfrac{u\sqrt[3]{6uv^2}}{3v}$ **107.** $\dfrac{1}{2\sqrt{5}}$ **109.** $\dfrac{1}{\sqrt[3]{3}}$

111. $\dfrac{b}{32a\sqrt[5]{2b^2}}$ **113.** $\dfrac{2\sqrt{3}}{9}$ **115.** $-\dfrac{\sqrt{2x}}{8}$ **117.** $\sqrt{3}$

119. $\sqrt[5]{4x^3}$ **121.** $2\sqrt{30}$ inches **123.** $\sqrt[6]{32}$ **125.** $\dfrac{\sqrt[4]{12}}{2}$

133. g **135.** d **137.** h **139.** e

Section 0.4

1. monomial, variables **3.** trinomial **5.** one **7.** like
9. coefficients, variables **11.** yes, trinomial, 2nd degree
13. no **15.** yes, binomial, 3rd degree **17.** yes, monomial,
0th degree **19.** yes, monomial, no defined degree
21. $6x^3 - 3x^2 - 8x$ **23.** $4y^3 + 14$ **25.** $-x^2 + 14$
27. $-28t + 96$ **29.** $-4y^2 + y$ **31.** $2x^2y - 4xy^2$

33. $8x^3y^7$ **35.** $\dfrac{m^4n^4}{2}$ **37.** $-4r^3s - 4rs^3$

39. $12a^2b^2c^2 + 18ab^3c^3 - 24a^2b^4c^2$ **41.** $a^2 + 4a + 4$
43. $a^2 - 12a + 36$ **45.** $x^2 - 16$ **47.** $x^2 + 2x - 15$
49. $3u^2 + 4u - 4$ **51.** $10x^2 + 13x - 3$
53. $9a^2 - 12ab + 4b^2$ **55.** $9m^2 - 16n^2$
57. $6y^2 - 16xy + 8x^2$ **59.** $9x^3 - x^2y - 27xy + 3y^2$
61. $5z^3 - 5tz + 2tz^2 - 2t^2$ **63.** $-3\sqrt{5}x^2 + x + 2\sqrt{5}$
65. $27x^3 - 27x^2 + 9x - 1$ **67.** $6x^3 + 14x^2 - 5x - 3$
69. $6x^3 - 5x^2y + 6xy^2 + 8y^3$ **71.** $6y^{2n} + 2$
73. $-10x^{4n} - 15y^{2n}$ **75.** $x^{2n} - x^n - 12$
77. $6r^{2n} - 25r^n + 14$ **79.** $xy + x^{3/2}y^{1/2}$
81. $a - b$ **83.** $\sqrt{3} + 1$ **85.** $x(\sqrt{7} - 2)$

87. $\dfrac{x(x + \sqrt{3})}{x^2 - 3}$ **89.** $\dfrac{y^2 + 2y\sqrt{2} + 2}{y^2 - 2}$

91. $\dfrac{\sqrt{3} + 3 - \sqrt{2} - \sqrt{6}}{2}$ **93.** $\dfrac{x - 2\sqrt{xy} + y}{x - y}$

95. $\dfrac{1}{2(\sqrt{2} - 1)}$ **97.** $\dfrac{y^2 - 3}{y^2 + 2y\sqrt{3} + 3}$

99. $\dfrac{1}{\sqrt{x + 3} + \sqrt{x}}$ **101.** $\dfrac{2a}{b^3}$ **103.** $-\dfrac{2z^9}{3x^3y^2}$

105. $\dfrac{x}{2y} + \dfrac{3xy}{2}$ **107.** $\dfrac{2y^3}{5} - \dfrac{3y}{5x^3} + \dfrac{1}{5x^4y^3}$ **109.** $3x + 2$

111. $x - 7 + \dfrac{2}{2x - 5}$ **113.** $2x^2 + 2x + 2 + \dfrac{3}{x - 1}$

115. $x - 3$ **117.** $x^2 - 2 + \dfrac{-x^2 + 5}{x^3 - 2}$

119. $x^4 + 2x^3 + 4x^2 + 8x + 16$ **121.** $6x^2 + x - 12$
123. $(x^2 + 3x - 10)$ ft^2 **125.** $(4x^3 - 48x^2 + 144x)$ in.^{3}
133. false Some polynomials are trinomials. **135.** true
137. false $(x^{1/3} - 6)(4x^{1/3} + 7) = 4x^{2/3} - 17x^{1/3} - 42$
139. $x^2 + 400x - 500$

Section 0.5

1. factor **3.** $x(a + b)$ **5.** $(x + y)^2$
7. $(x + y)(x^2 - xy + y^2)$ **9.** $3(x - 2)$ **11.** $4x^2(2 + x)$
13. $7x^2y^2(1 + 2x)$ **15.** $(x + y)(a + b)$
17. $(4a + b)(1 - 3a)$ **19.** $(2x + 3)(2x - 3)$
21. $(2 + 3r)(2 - 3r)$ **23.** $(9x^2 + 1)(3x + 1)(3x - 1)$
25. $(x + z + 5)(x + z - 5)$ **27.** $(x + 4)^2$
29. $(b - 5)^2$ **31.** $(m + 2n)^2$
33. $(4x - 3y)(3x + 2y)$ **35.** $(x + 7)(x + 3)$
37. $(x - 6)(x + 2)$ **39.** $(2p + 3)(3p - 1)$
41. $(t + 7)(t^2 - 7t + 49)$ **43.** $(5y + 6z)(25y^2 - 30yz + 36z^2)$
45. $(2z - 3)(4z^2 + 6z + 9)$ **47.** $(7y - z)(49y^2 + 7yz + z^2)$
49. $3abc(a + 2b + 3c)$ **51.** $(x + 1)(3x^2 - 1)$
53. $t(y + c)(2x - 3)$ **55.** $(a + b)(x + y + z)$
57. $(x + y - z)(x - y + z)$ **59.** $-4xy$
61. $(x^2 + y^2)(x + y)(x - y)$ **63.** $3(x + 2)(x - 2)$
65. $2x(3y + 2)(3y - 2)$ **67.** prime

69. $(6a + 5)(4a - 3)$ **71.** $(3x + 7y)(2x + 5y)$
73. $2(6p - 35q)(p + q)$ **75.** $-(6m - 5n)(m - 7n)$
77. $-x(6x + 7)(x - 5)$ **79.** $x^2(2x - 7)(3x + 5)$
81. $(x^2 + 5)(x^2 - 3)$ **83.** $(a^n - 3)(a^n + 1)$
85. $(3x^n - 2)(2x^n - 1)$ **87.** $(2x^n + 3y^n)(2x^n - 3y^n)$
89. $(5y^n + 2)(2y^n - 3)$ **91.** $2(x + 10)(x^2 - 10x + 100)$
93. $(x + y - 4)(x^2 + 2xy + y^2 + 4x + 4y + 16)$
95. $(2a + y)(2a - y)(4a^2 - 2ay + y^2)(4a^2 + 2ay + y^2)$
97. $(a - b)(a^2 + ab + b^2 + 1)$
99. $(4x^2 + y^2)(16x^4 - 4x^2y^2 + y^4)$
101. $(x - 3 + 12y)(x - 3 - 12y)$
103. $(a + b - 5)(a + b + 2)$
105. $(x + 2)(x^2 - 2x + 4)(x - 1)(x^2 + x + 1)$
107. $(x^2 + 1 + x)(x^2 + 1 - x)$
109. $(x^2 + x + 4)(x^2 - x + 4)$
111. $(2a^2 + a + 1)(2a^2 - a + 1)$

113. $\dfrac{4}{3}\pi(r_1 - r_2)(r_1^2 + r_1r_2 + r_2^2)$ **119.** $2\left(\dfrac{3}{2}x + 1\right)$

121. $2\left(\dfrac{1}{2}x^2 + x + 2\right)$ **123.** $a\left(1 + \dfrac{b}{a}\right)$

125. $x^{1/2}(x^{1/2} + 1)$ **127.** $\sqrt{2}(\sqrt{2}x + y)$
129. $ab(b^{1/2} - a^{1/2})$ **131.** $(x - 2)(x + 3 + y)$
133. $(a + 1)(a^3 + a^2 + 1)$ **135.** true
137. false $p^3q^3r^3 + 64$ factors as $(pqr + 4)(p^2q^2r^2 - 4pqr + 16)$
139. true

Section 0.6

1. numerator **3.** $ad = bc$ **5.** $\dfrac{ac}{bd}$ **7.** $\dfrac{a + c}{b}$

9. equal **11.** not equal **13.** $\dfrac{a}{3b}$ **15.** $\dfrac{8x}{35a}$

17. $\dfrac{16}{3}$ **19.** $\dfrac{z}{c}$ **21.** $\dfrac{2x^2y}{a^2b^3}$ **23.** $\dfrac{2}{x + 2}$

25. $-\dfrac{x + 2}{x - 3}$ **27.** $\dfrac{3x - 4}{2x - 1}$ **29.** $\dfrac{x^2 + 2x + 4}{x + a}$

31. $\dfrac{x(x - 1)}{x + 1}$ **33.** $\dfrac{x - 1}{x}$ **35.** $\dfrac{x(x + 1)^2}{x + 2}$

37. $\dfrac{1}{2}$ **39.** $\dfrac{z - 4}{(z + 2)(z - 2)}$ **41.** 1 **43.** 1

45. $\dfrac{x - 5}{x + 5}$ **47.** $\dfrac{x + 5}{x + 3}$ **49.** 4 **51.** $\dfrac{-1}{x - 5}$

53. $\dfrac{5x - 1}{(x + 1)(x - 1)}$ **55.** $\dfrac{2(a - 2)}{(a + 4)(a - 4)}$

57. $\dfrac{2}{(x + 2)(x - 2)}$ **59.** $\dfrac{2(x^2 - 3x + 1)}{(x + 1)^2(x - 1)}$

61. $\dfrac{3y - 2}{y - 1}$ **63.** $\dfrac{1}{x + 2}$ **65.** $\dfrac{2x - 5}{2x(x - 2)}$

67. $\dfrac{2x^2 + 19x + 1}{(x + 4)(x - 4)}$ **69.** 0

71. $\dfrac{-x^4 + 3x^3 - 43x^2 - 58x + 697}{(x + 5)(x - 5)(x + 4)(x - 4)}$ **73.** $\dfrac{b}{2c}$

75. $81a$ **77.** -1 **79.** $\dfrac{y + x}{x^2y^2}$ **81.** $\dfrac{y + x}{y - x}$

83. $\dfrac{a^2(3x - 4ab)}{ax + b}$ **85.** $\dfrac{x - 2}{x + 2}$ **87.** $\dfrac{3x^2y^2}{xy - 1}$

89. $\dfrac{3x^2}{x^2 + 1}$ **91.** $\dfrac{x^2 - 3x - 4}{x^2 + 5x - 3}$ **93.** $\dfrac{x}{x + 1}$

95. $\dfrac{5x+1}{x-1}$ **97.** $\dfrac{k_1k_2}{k_2+k_1}$ **107.** $\dfrac{3x}{3+x}$

109. $\dfrac{x+1}{2x+1}$ **111.** false The numerator can be 0.

113. false $\dfrac{x+7}{x+7}=1$ for all values of x except for $x=-7$.

115. true

117. false The sum is $\dfrac{5y+5x}{xy}$. **119.** All real numbers

except $x=6$

Chapter Review
1. $3,6,8$ **2.** $0,3,6,8$ **3.** $-6,-3,0,3,6,8$

4. $-6,-3,0,\dfrac{1}{2},3,6,8$ **5.** $\pi,\sqrt{5}$

6. $-6,-3,0,\dfrac{1}{2},3,\pi,\sqrt{5},6,8$ **7.** 3 **8.** $6,8$

9. $-6,0,6,8$ **10.** $-3,3$

11. Associative Property of Addition

12. Commutative Property of Addition

13. Associative Property of Multiplication

14. Distributive Property

15. Commutative Property of Multiplication

16. Commutative Property of Addition

17. Double Negative Rule

18.

19.

20. **21.**

22. **23.**

24.

25. 6 **26.** 25 **27.** $\sqrt{2}-1$ **28.** $\sqrt{3}-1$

29. 12 **30.** $-5\cdot a\cdot a\cdot a$ **31.** $(-5a)(-5a)$

32. $3t^3$ **33.** $-6b^2$ **34.** n^6 **35.** p^6 **36.** $x^{12}y^8$

37. $\dfrac{a^{12}}{b^6}$ **38.** $\dfrac{1}{m^6}$ **39.** $\dfrac{q^6}{8p^6}$ **40.** $\dfrac{1}{a^3}$ **41.** $\dfrac{b^6}{a^4}$

42. $\dfrac{y^8}{9}$ **43.** $\dfrac{a^8}{b^{10}}$ **44.** $\dfrac{y^4}{9x^4}$ **45.** $-\dfrac{8m^{12}}{n^3}$

46. 18 **47.** 6.75×10^3 **48.** 2.3×10^{-4}

49. 480 **50.** 0.00025 **51.** 1.5×10^{14} **52.** 11

53. $\dfrac{3}{5}$ **54.** $2x$ **55.** $3|a|$ **56.** $-10x^2$

57. not a real number **58.** $x^6|y|$ **59.** $\dfrac{y^2}{x^6}$

60. $-c$ **61.** a^2 **62.** 16 **63.** $\dfrac{1}{8}$ **64.** $\dfrac{8}{27}$ **65.** $\dfrac{4}{9}$

66. $\dfrac{9}{4}$ **67.** $\dfrac{125}{8}$ **68.** $36x^2$ **69.** $p^{2a/3}$ **70.** 6

71. -7 **72.** $\dfrac{3}{5}$ **73.** $\dfrac{3}{5}$ **74.** $|x|y^2$ **75.** x

76. $\dfrac{m^2|n|}{p^4}$ **77.** $\dfrac{a^3b^2}{c}$ **78.** $7\sqrt{2}$ **79.** 0 **80.** $x\sqrt[3]{3x}$

81. $\dfrac{\sqrt{35}}{5}$ **82.** $2\sqrt{2}$ **83.** $\dfrac{\sqrt[3]{4}}{2}$ **84.** $\dfrac{2\sqrt[3]{5}}{5}$

85. $\dfrac{2}{5\sqrt{2}}$ **86.** $\dfrac{1}{\sqrt{5}}$ **87.** $\dfrac{2x}{3\sqrt{2x}}$ **88.** $\dfrac{21x}{2\sqrt[3]{49x^2}}$

89. 3rd degree, binomial **90.** 2nd degree, trinomial

91. 2nd degree, monomial **92.** 4th degree, trinomial

93. $5x-6$ **94.** $2x^3-7x^2-6x$ **95.** $9x^2+12x+4$

96. $6x^2-7xy-3y^2$ **97.** $8a^2-8ab-6b^2$

98. $3z^3+10z^2+2z-3$ **99.** $a^{2n}+a^n-2$

100. $2+2x\sqrt{2}+x^2$ **101.** $\sqrt{6}+\sqrt{2}+\sqrt{3}+1$

102. -5 **103.** $\sqrt{3}+1$ **104.** $-2(\sqrt{3}+\sqrt{2})$

105. $\dfrac{2x(\sqrt{x}+2)}{x-4}$ **106.** $\dfrac{x-2\sqrt{xy}+y}{x-y}$

107. $\dfrac{x-4}{5(\sqrt{x}-2)}$ **108.** $\dfrac{1-a}{a(1+\sqrt{a})}$ **109.** $\dfrac{y}{2x}$

110. $2a^2b+3ab^2$ **111.** x^2+2x+1

112. $x^3+2x-3-\dfrac{6}{x^2-1}$ **113.** $3t(t+1)(t-1)$

114. $5(r-1)(r^2+r+1)$ **115.** $(3x+8)(2x-3)$

116. $(3a+x)(a-1)$ **117.** $(2x-5)(4x^2+10x+25)$

118. $2(3x+2)(x-4)$ **119.** $(x+3+t)(x+3-t)$

120. prime **121.** $(2z+7)(4z^2-14z+49)$

122. $(7b+1)^2$ **123.** $(11z-2)^2$

124. $8(2y-5)(4y^2+10y+25)$ **125.** $(y-2z)(2x-w)$

126. $(x^2+1+x)(x^2+1-x)(x^4+1-x^2)$

127. $\dfrac{-1}{x-2}$ **128.** $\dfrac{a+3}{a-3}$ **129.** $(x-2)(x+3)$

130. $\dfrac{2y-5}{y-2}$ **131.** $\dfrac{t+1}{5}$ **132.** $\dfrac{p(p+4)}{(p^2+8p+4)(p-3)}$

133. $\dfrac{(x-2)(x+3)(x-3)}{(x-1)(x+2)^2}$ **134.** 1

135. $\dfrac{3x^2-10x+10}{(x-4)(x+5)}$ **136.** $\dfrac{2(2x^2+3x+6)}{(x+2)(x-2)}$

137. $\dfrac{3x^3-12x^2+11x}{(x-1)(x-2)(x-3)}$ **138.** $\dfrac{-5x-6}{(x+1)(x+2)}$

139. $\dfrac{-x^3+x^2-12x-15}{x^2(x+1)}$ **140.** $\dfrac{3x}{x+1}$

141. $\dfrac{20}{3x}$ **142.** $\dfrac{y}{2}$ **143.** $\dfrac{y+x}{xy(x-y)}$ **144.** $\dfrac{y+x}{x-y}$

Chapter Test
1. $-7,1,3$ **2.** 3 **3.** Commutative Property of

Addition **4.** Distributive Property

5. **6.**

7. 17 **8.** $-(x-7)$ **9.** 16 **10.** 8 **11.** x^{11}

12. $\dfrac{r}{s}$ **13.** a **14.** x^{24} **15.** 4.5×10^5

16. 3.45×10^{-4} **17.** 3700 **18.** 0.0012

19. $5a^2$ **20.** $\dfrac{216}{729}$ **21.** $\dfrac{9s^6}{4t^4}$ **22.** $3a^2$

23. $5\sqrt{3}$ **24.** $-4x\sqrt[3]{3x}$ **25.** $\dfrac{x(\sqrt{x}+2)}{x-4}$

26. $\dfrac{x-y}{x+2\sqrt{xy}+y}$ **27.** $-a^2+7$ **28.** $-6a^6b^6$

29. $6x^2+13x-28$ **30.** $a^{2n}-a^n-6$ **31.** x^4-16

32. $2x^3-5x^2+7x-6$ **33.** $6x+19+\dfrac{34}{x-3}$

34. x^2+2x+1 **35.** $3(x+2y)$ **36.** $(x+10)(x-10)$

37. $5(3x+2y)(3x-2y)$ **38.** $(5t-2w)(2t-3w)$

39. $(4m + 5n)(16m^2 - 20mn + 25n^2)$
40. $3(a - 6)(a^2 + 6a + 36)$ **41.** $(x + 2)(x - 2)(x^2 + 3)$
42. $(3x^2 - 2)(2x^2 + 5)$ **43.** $\dfrac{4q^4}{3p}$ **44.** $-\dfrac{x - 7}{x + 7}$
45. 1 **46.** $\dfrac{-2x}{(x + 1)(x - 1)}$ **47.** $\dfrac{(x + 5)^2}{x + 4}$
48. $\dfrac{1}{(x + 1)(x - 2)}$ **49.** $\dfrac{b + a}{a}$ **50.** $\dfrac{y}{y + x}$

Section 1.1

1. root; solution   **3.** no   **5.** linear   **7.** one   **9.** no restrictions   **11.** $x \neq 0$ **13.** $x \neq 6$ or -2
15. $x \neq 3, 4,$ or -4 **17.** 5; conditional equation
19. no solution; contradiction   **21.** 7; conditional equation
23. no solution; contradiction   **25.** all real numbers; identity
27. 6; conditional equation   **29.** all real numbers; identity
31. 1 **33.** 6 **35.** $\dfrac{5}{2}$ **37.** 9 **39.** 10 **41.** -3
43. 3 **45.** $\dfrac{21}{19}$ **47.** $-\dfrac{14}{11}$ **49.** 2 **51.** all real numbers
53. -5 **55.** -4 **57.** no solution   **59.** 17
61. $-\dfrac{2}{5}$ **63.** $\dfrac{2}{3}$ **65.** 3 **67.** 5 **69.** no solution
71. 2 **73.** $m = \dfrac{f}{a}$ **75.** $w = \dfrac{P - 2l}{2}$ **77.** $r^2 = \dfrac{3V}{\pi h}$
79. $s = \dfrac{f(P_n - L)}{i}$ **81.** $m = \dfrac{r^2 F}{Mg}$ **83.** $y = b\left(1 - \dfrac{x}{a}\right)$
85. $r = \dfrac{r_1 r_2}{r_1 + r_2}$ **87.** $n = \dfrac{l - a + d}{d}$ **89.** $n = \dfrac{360}{180 - a}$
91. $r_1 = \dfrac{-Rr_2 r_3}{Rr_3 + Rr_2 - r_2 r_3}$ **97.** false The equation is an identity.
99. true   **101.** true

Section 1.2

1. add   **3.** amount   **5.** rate; time   **7.** 82   **9.** 84
11. 94   **13.** 7   **15.** 10 ft by 36 ft   **17.** $2\frac{1}{2}$ ft   **19.** 10 m
21. 20 ft by 8 ft   **23.** 20 ft   **25.** $7250 at 4% and $8750 at 6%
27. $45,714.29   **29.** $29,100   **31.** 327   **33.** $449
35. $215   **37.** 200 units   **39.** 21   **41.** $2\frac{6}{11}$ days
43. $1\frac{1}{3}$ hr   **45.** $21\frac{1}{9}$ hr   **47.** 10 oz   **49.** 1 liter
51. $\frac{1}{15}$ liter   **53.** about 4.5 gal   **55.** 4 liters   **57.** 50 lb
59. 30 gal   **61.** 600 lb barley; 1637 lb oats; 163 lb soybean meal
63. 39 mph going; 65 mph returning   **65.** $2\frac{1}{2}$ hr   **67.** 50 sec
69. 12 mph   **71.** about 11.2 mm

Section 1.3

1. imaginary   **3.** imaginary   **5.** $2 - 5i$ **7.** real
9. $12i$ **11.** $-8i\sqrt{2}$ **13.** $-4i\sqrt{6}$ **15.** $\dfrac{5\sqrt{2}}{3}i$
17. $-\dfrac{7\sqrt{6}}{4}i$ **19.** $x = 3; y = 5$ **21.** $x = \dfrac{2}{3}; y = -\dfrac{2}{9}$
23. $5 - 6i$ **25.** $-2 - 10i$ **27.** $4 + 10i$ **29.** $1 - i$
31. $7 + i$ **33.** $4 - 2i$ **35.** $-15 - 25i$ **37.** $56 + 28i$
39. $-9 + 19i$ **41.** $-5 + 12i$ **43.** $52 - 56i$ **45.** $-6 + 17i$

47. $0 + i$ **49.** $0 + \dfrac{4}{3}i$ **51.** $\dfrac{2}{5} - \dfrac{1}{5}i$ **53.** $\dfrac{1}{25} + \dfrac{7}{25}i$
55. $\dfrac{1}{2} + \dfrac{1}{2}i$ **57.** $-\dfrac{7}{13} - \dfrac{22}{13}i$ **59.** $-\dfrac{12}{17} + \dfrac{11}{34}i$
61. $\dfrac{6 + \sqrt{3}}{10} + \dfrac{3\sqrt{3} - 2}{10}i$ **63.** i **65.** -1 **67.** $-i$
69. 1 **71.** -1 **73.** -1 **75.** i **77.** 4
79. 5 **81.** $\sqrt{13}$ **83.** $7\sqrt{2}$ **85.** $\dfrac{\sqrt{2}}{2}$ **87.** 6
89. $\sqrt{2}$ **91.** $\dfrac{3\sqrt{5}}{5}$ **93.** 1 **95.** $(x + 2i)(x - 2i)$
97. $(5p + 6qi)(5p - 6qi)$ **99.** $2(y + 2zi)(y - 2zi)$
101. $2(5m + ni)(5m - ni)$ **103.** $21 + 12i$ **105.** $6.2 - 0.7i$
113. false $\sqrt{-300} = 10i\sqrt{3}$ **115.** true   **117.** true
119. true

Section 1.4

1. $ax^2 + bx + c = 0$ **3.** $\sqrt{c}; -\sqrt{c}$
5. rational numbers   **7.** $3, -2$ **9.** $12, -12$
11. $2, -\dfrac{5}{2}$ **13.** $2, \dfrac{3}{5}$ **15.** $\dfrac{3}{5}, -\dfrac{5}{3}$ **17.** $\dfrac{3}{2}, \dfrac{1}{2}$
19. ± 3 **21.** $\pm 13i$ **23.** $\pm 5\sqrt{2}$ **25.** $\pm 3i\sqrt{6}$
27. $\pm 2\sqrt{5}$ **29.** $\pm 3i\sqrt{5}$ **31.** $\pm\dfrac{\sqrt{7}}{2}$ **33.** $\pm\dfrac{\sqrt{7}}{3}i$
35. $\pm\dfrac{\sqrt{26}}{2}$ **37.** $\pm\dfrac{\sqrt{30}}{2}i$ **39.** $-1 \pm 2\sqrt{2}$
41. $-1 \pm 2i\sqrt{3}$ **43.** $\dfrac{-1 \pm 3\sqrt{3}}{2}$ **45.** $-\dfrac{1}{5} \pm \dfrac{2\sqrt{2}}{5}i$
47. $x^2 + 6x + 9$ **49.** $x^2 - 4x + 4$ **51.** $a^2 + 5a + \dfrac{25}{4}$
53. $r^2 - 11r + \dfrac{121}{4}$ **55.** $y^2 + \dfrac{3}{4}y + \dfrac{9}{64}$
57. $q^2 - \dfrac{1}{5}q + \dfrac{1}{100}$ **59.** $-6 \pm 2\sqrt{7}$ **61.** $5 \pm 2i\sqrt{3}$
63. $\dfrac{-5 \pm \sqrt{5}}{2}$ **65.** $-\dfrac{11}{2} \pm \dfrac{5\sqrt{3}}{2}i$ **67.** $\dfrac{10 \pm \sqrt{2}}{2}$
69. $\dfrac{-2 \pm \sqrt{7}}{3}$ **71.** $\dfrac{3 \pm \sqrt{17}}{4}$ **73.** $1 \pm \dfrac{\sqrt{5}}{3}i$
75. $\dfrac{7}{2} \pm \dfrac{\sqrt{11}}{2}i$ **77.** $\dfrac{-5 \pm \sqrt{13}}{6}$ **79.** $\dfrac{-7 \pm 3\sqrt{5}}{2}$
81. $\dfrac{-3 \pm \sqrt{6}}{3}$ **83.** $\dfrac{1 \pm \sqrt{15}}{7}$ **85.** $-1 \pm i$
87. $-2 \pm i$ **89.** $1 \pm 2i$ **91.** $\dfrac{1}{3} \pm \dfrac{1}{3}i$
93. $t = \pm\dfrac{\sqrt{2hg}}{g}$ **95.** $t = \dfrac{8 \pm \sqrt{64 - h}}{4}$
97. $y = \pm\dfrac{b\sqrt{a^2 - x^2}}{a}$ **99.** $a = \pm\dfrac{bx\sqrt{b^2 + y^2}}{b^2 + y^2}$
101. $x = \dfrac{-y \pm y\sqrt{5}}{2}$ **103.** one repeated rational number
105. two different nonreal complex numbers
107. two different rational numbers
109. two different irrational numbers   **111.** $2, 10$
113. $3, -4$ **115.** $\dfrac{3}{2}, -\dfrac{1}{4}$ **117.** $\dfrac{1}{2}, -\dfrac{4}{3}$ **119.** $\dfrac{5}{6}, -\dfrac{2}{5}$
121. $1, -1$ **123.** $-\dfrac{1}{2}, 5$ **125.** -2 **127.** $\dfrac{7 \pm \sqrt{145}}{4}$
129. $\pm 6i$ **137.** b   **139.** c   **141.** true

Section 1.5

1. $A = lw$ **3.** 4 ft by 8 ft   **5.** 72 ft by 160 ft   **7.** 9 cm

9. 12 cm by 15 cm   **11.** 2 in.   **13.** 4 m

15. 10 m and 24 m   **17.** 40.1 in. by 22.6 in.
19. 20 mph going and 10 mph returning   **21.** 7 hr
23. 25 sec   **25.** about 9.5 sec   **27.** 1.6 sec   **29.** 10¢
31. 1440   **33.** Morgan at 7%; Chloe at 8%   **35.** 10
37. 4 hr   **39.** about 9.5 hr   **41.** 221   **43.** 24.3 ft
45. 1.70 in.

Section 1.6

1. equal   **3.** extraneous   **5.** $0, -5, -4$ **7.** $0, \dfrac{4}{3}, -\dfrac{1}{2}$

9. $\pm 1, \pm 5$ **11.** $\pm 3\sqrt{2}, \pm 3\sqrt{5}$ **13.** $\pm 3, \pm i$

15. $\pm\dfrac{3}{2}, \pm 2i$ **17.** $\pm 1, \pm 6$ **19.** $\dfrac{1}{8}, -8$ **21.** 1, 144

23. $\dfrac{1}{9}$ **25.** $\dfrac{1}{64}$ **27.** $\dfrac{1}{8}, \dfrac{1}{2}$ **29.** 0, 1   **31.** 27

33. $-\dfrac{1}{3}$ **35.** 2   **37.** $-\dfrac{3}{2}, -\dfrac{5}{2}$ **39.** 2   **41.** 12

43. 5   **45.** 3, 4   **47.** $\dfrac{1}{5}, -1$ **49.** 3, 5   **51.** 0, 4
53. -2 **55.** no solution   **57.** 3   **59.** 1   **61.** 9
63. $-2, 1$ **65.** $2, -\dfrac{5}{2}$ **67.** 20   **69.** $\dfrac{25}{2}$ **71.** 400 ft

73. 8 ft   **75.** about \$3109
83. false $x^4 + 6x^2 + 5 = 0$ can be solved by factoring.
85. true   **87.** false $x = 256$
89. false   We isolate one radical and then square both sides.

Section 1.7

1. right   **3.** $a < c$ **5.** $b - c$ **7.** $>$ **9.** linear
11. equivalent

13. $(-\infty, 1)$
1

15. $[1, \infty)$
1

17. $(-\infty, 1)$
1

19. $[1, \infty)$
1

21. $(-\infty, 3]$
3

23. $(-10/3, \infty)$
$-10/3$

25. $(5, \infty)$
5

27. $[14, \infty)$
14

29. $(-\infty, 15/4]$
$15/4$

31. $(-44/41, \infty)$
$-44/41$

33. $(6, 9]$
6   9

35. $(8, 22]$
8   22

37. $[-11, 4]$
-11 4

39. $[5, 21]$
5   21

41. $(-\infty, 0)$
0

43. $(0, \infty)$
0

45. $(3, 14/3)$
3   $14/3$

47. $(2, \infty)$
2

49. $(-4, 5/6)$
-4 $5/6$

51. $[-2, \infty)$
-2

53. $(-4, -3)$
-4 -3

55. $(-\infty, 2] \cup [3, \infty)$
2   3

57. $(-3, -2)$
-3 -2

59. $(-\infty, -1/2] \cup [-1/3, \infty)$
$-1/2$ $-1/3$

61. $(1/3, 1/2)$
$1/3$ $1/2$

63. $(-\infty, -3/2] \cup [1, \infty)$
$-3/2$ 1

65. $(-\infty, -\sqrt{3}] \cup [\sqrt{3}, \infty)$
$-\sqrt{3}$ $\sqrt{3}$

67. $(-\sqrt{11}, \sqrt{11})$
$-\sqrt{11}$ $\sqrt{11}$

69. $(-3, 2)$
-3 2

71. $(-\infty, -1) \cup (-1, 0) \cup (1, \infty)$
-1 0   1

73. $(-\infty, -3) \cup (-3, -2] \cup (2, \infty)$
-3 -2 2

75. $(-\infty, -2) \cup (-2, -1/3) \cup (1/2, \infty)$
-2 $-1/3$ $1/2$

77. $(0, 3/2)$
0   $3/2$

79. $(-\infty, 0) \cup (3/2, \infty)$
0   $3/2$

81. $(-\infty, 2) \cup [13/5, \infty)$
2   $13/5$

83. $(-\infty, -\sqrt{7}) \cup (-1, 1) \cup (\sqrt{7}, \infty)$
$-\sqrt{7}$ -1 1 $\sqrt{7}$

85. 12   **87.** 19 min

89. 12   **91.** $p \le \$1124.12$ **93.** $x > 183$
95. $a > \$50,000$ **97.** anything over \$1800
99. $16\dfrac{2}{3}$ cm $< s <$ 20 cm   **101.** between 1 and 9 seconds
107. false $[-10, 10]$ **109.** false   The first step is to subtract 2 from both sides.

Section 1.8

1. x **3.** $x = k$ or $x = -k$ **5.** $-k < x < k$
7. $x \le -k$ or $x \ge k$ **9.** 7   **11.** 0   **13.** 2
15. $\pi - 2$ **17.** $x - 5$ **19.** x^3 if $x \ge 0$; $-x^3$ if $x < 0$
21. $0, -4$ **23.** $2, -\dfrac{4}{3}$ **25.** $\dfrac{14}{3}, -2$ **27.** $7, -3$

29. no solution   **31.** 5   **33.** $\dfrac{2}{7}, 2$ **35.** $x \ge 0$

37. $-\dfrac{3}{2}$ **39.** $0, -6$ **41.** 0   **43.** $\dfrac{3}{5}, 3$

45. $-\dfrac{3}{13}, \dfrac{9}{5}$

47. $(-3, 9)$
-3 9

49. $(-\infty, -9) \cup (3, \infty)$
-9 3

51. $(-\infty, -7] \cup [3, \infty)$
-7 3

53. $[-13/3, 1]$
$-13/3$ 1

55. $(-\infty, -3) \cup (-3, \infty)$
-3

57. $(-1, 1/5)$
-1 $1/5$

59. $(-\infty, -7/9) \cup (13/9, \infty)$
$-7/9$ $13/9$

61. (−5, 7)

63. (−2, −1/2) ∪ (−1/2, 1)

65. (−7/3, −2/3) ∪ (4/3, 3)

67. (−7, −1) ∪ (11, 17)

69. (−18, −6) ∪ (10, 22)

71. (−10, −7] ∪ [5, 8)

73. $\left[-\dfrac{1}{2}, \infty\right)$ **75.** (−∞, 0) **77.** $\left(-\infty, -\dfrac{1}{2}\right)$

79. [0, ∞) **81.** 70° ≤ t ≤ 86° **83.** |c − 0.6°| ≤ 0.5°
85. |h − 55| < 17 **87. a.** 26.45%, 24.76%
b. It is less than or equal to 1%. **95.** false Absolute value equations can have zero, one, or two solutions.
97. false The solution set is (−∞, −5] ∪ [5, ∞). **99.** true

Chapter Review

1. no restrictions **2.** $x \neq 0$ **3.** $x \neq 1$
4. $x \neq 2, x \neq 3$ **5.** $\dfrac{16}{27}$; conditional equation
6. −14; conditional equation **7.** $\dfrac{16}{5}$; conditional equation
8. no solution; contradiction **9.** 7; conditional equation
10. all real numbers except −9; identity
11. 7; conditional equation **12.** $\dfrac{1}{3}$; conditional equation
13. −2; conditional equation **14.** 0; conditional equation
15. $F = \dfrac{9}{5}C + 32$ **16.** $f = \dfrac{is}{P_n - l}$ **17.** $f_1 = \dfrac{ff_2}{f_2 - f}$
18. $l = \dfrac{a - S + Sr}{r}$ **19.** 60% **20.** 22.5 ft by 27.5 ft
21. 3 hr **22.** 0.5 hr **23.** 1.5 liters **24.** about 3.9 hr
25. $5\dfrac{1}{7}$ hr **26.** $3\dfrac{1}{3}$ oz **27.** $4500 at 11%; $5500 at 14%
28. 10 **29.** $30i\sqrt{3}$ **30.** $-\dfrac{3\sqrt{5}}{2}i$ **31.** $-2 - i$
32. $5 - 2i$ **33.** $-2 - 5i$ **34.** $146 + 0i$ **35.** $55 - 48i$
36. $21 - 9i$ **37.** $0 - 3i$ **38.** $0 + \dfrac{5}{6}i$ **39.** $\dfrac{3}{2} - \dfrac{3}{2}i$
40. $-\dfrac{2}{5} + \dfrac{4}{5}i$ **41.** $\dfrac{4}{5} + \dfrac{3}{5}i$ **42.** $\dfrac{1}{2} - \dfrac{5}{2}i$ **43.** $0 + i$
44. $0 - i$ **45.** $0 - 2i$ **46.** $\sqrt{10} + 0i$ **47.** $1 + 0i$
48. $(8r + 3si)(8r - 3si)$ **49.** $2, -\dfrac{3}{2}$ **50.** $\dfrac{1}{4}, -\dfrac{4}{3}$
51. $0, \dfrac{8}{5}$ **52.** $\dfrac{2}{3}, \dfrac{4}{9}$ **53.** $\pm 2\sqrt{2}$ **54.** $\pm i\sqrt{5}$
55. $\dfrac{5 \pm 4\sqrt{2}}{4}$ **56.** $\dfrac{7}{5} \pm \dfrac{3\sqrt{5}}{5}i$ **57.** 3, 5 **58.** −4, −2
59. $\dfrac{1 \pm \sqrt{21}}{10}$ **60.** $0, \dfrac{1}{5}$ **61.** $\dfrac{1}{3} \pm \dfrac{\sqrt{2}}{3}i$ **62.** 2, −7
63. $9, -\dfrac{2}{3}$ **64.** $\dfrac{-1 \pm \sqrt{21}}{10}$ **65.** $-1 \pm \sqrt{6}$
66. $\dfrac{1}{3} \pm \dfrac{\sqrt{11}}{3}i$ **67.** 1 **68.** two different rational numbers
69. $\dfrac{1}{3}$ **70.** 10, 2 **71.** 2, −3 **72.** $\dfrac{8}{5}, 5$
73. either 95 by 110 yd or 55 by 190 yd **74.** 320 mph for
prop plane; 440 mph for jet plane **75.** 1 sec **76.** $1\dfrac{1}{2}$ ft

77. 0, −6, 2 **78.** $0, \dfrac{2}{3}, -2$ **79.** ±1 **80.** ±1, ±6i
81. 9 **82.** 8, −27 **83.** $3, -\dfrac{2}{5}$ **84.** 3 **85.** 5
86. 0 **87.** ±4 **88.** no solution **89.** 2 **90.** 18
91. (−∞, 7) **92.** [−1/5, ∞)
93. (−∞, 5/3) **94.** (−∞, −12/7)
95. [−3, 5) **96.** (2, ∞)
97. (−∞, −2) ∪ (4, ∞) **98.** (−4, 1)
99. [−1, 3] **100.** (−∞, −3/2) ∪ (1, ∞)
101. (−∞, −2] ∪ (3, ∞) **102.** (−4, 1]
103. [−2, 1] ∪ (3, ∞) **104.** (−∞, 0) ∪ (5/2, ∞)
105. 5, −7 **106.** $-\dfrac{4}{3}, -6$ **107.** $-\dfrac{3}{8}, \dfrac{3}{10}$
108. 0 **109.** no solution
110. (−6, 0) **111.** (−∞, 2] ∪ [8/3, ∞)
112. (−5, 1) **113.** (−∞, −29) ∪ (35, ∞)
114. (−7/2, −2) ∪ (−1, 1/2) **115.** (−1, 4/3) ∪ (4/3, 11/3)

Chapter Test

1. $x \neq 0, x \neq 1$ **2.** $x \neq \dfrac{2}{3}$ **3.** $\dfrac{5}{2}$ **4.** $\dfrac{34}{7}$
5. $x = \mu + z\sigma$ **6.** $a = \dfrac{bc}{c + b}$ **7.** 87.5 **8.** $14,000
9. $12i\sqrt{6}$ **10.** $\dfrac{3\sqrt{10}}{5}i$ **11.** $7 - 12i$ **12.** $-23 - 43i$
13. $\dfrac{4}{5} + \dfrac{2}{5}i$ **14.** $0 + i$ **15.** i **16.** 7 **17.** $\dfrac{1}{2}, \dfrac{3}{2}$
18. $\dfrac{3}{2}, -4$ **19.** $\pm 3i\sqrt{3}$ **20.** $7 \pm 6\sqrt{2}$ **21.** $\dfrac{5 \pm \sqrt{133}}{6}$
22. 37 **23.** 5, −3 **24.** 8 sec **25.** 13 **26.** $\dfrac{\sqrt{10}}{10}$
27. ±2, ±3 **28.** $1, -\dfrac{1}{32}$ **29.** 139 **30.** −1
31. (−∞, 2] **32.** (−∞, 5)
33. [3, 4) **34.** (2, ∞)

35. $(-\infty, -1] \cup [8, \infty)$

36. $[-2, 1)$

37. $2, -\dfrac{10}{3}$ **38.** 0

39. $(-\infty, 3/2) \cup (7/2, \infty)$

40. $[-9, 6]$

Cumulative Review

1. $-2, 0, 2, 6$ **2.** $2, 5, 11$

3. $[-4, 7)$

4. $(-\infty, 0) \cup [2, \infty)$

5. Comm. Prop. of Addition   **6.** Transitive Prop.   **7.** $9a^2$

8. $81a^2$ **9.** $a^6 b^4$ **10.** $\dfrac{9y^2}{x^4}$ **11.** $\dfrac{x^4}{16y^2}$ **12.** $\dfrac{4y^{10}}{9x^6}$

13. $a^3 b^3$ **14.** abc^2 **15.** $\sqrt{3}$ **16.** $\dfrac{\sqrt[3]{2x^2}}{x}$

17. $\dfrac{3(y + \sqrt{3})}{y^2 - 3}$ **18.** $\dfrac{3x(\sqrt{x} + 1)}{x - 1}$ **19.** $5\sqrt{3} - 3\sqrt{5}$

20. $3\sqrt{2}$ **21.** $5 - 2\sqrt{6}$ **22.** 4 **23.** $-8x + 8$

24. $9x^4 - 4x^3$ **25.** $6x^2 + 11x - 35$ **26.** $z^3 + z^2 + 4$

27. $2x^2 - x + 1$ **28.** $3x^2 + 1 - \dfrac{x}{x^2 + 2}$ **29.** $3t(t - 2)$

30. $(3x + 2)(x - 4)$ **31.** $(x + 1)^2(x - 1)^2(x^2 + 1)^2$

32. $(x + 1)(x^2 - x + 1)(x - 1)(x^2 + x + 1)$ **33.** $\dfrac{x - 5}{x + 5}$

34. $(2x + 1)(x + 2)$ **35.** $\dfrac{5x^2 + 17x - 6}{(x + 3)(x - 3)}$

36. $\dfrac{-x^2 + x + 7}{(x + 2)(x + 3)}$ **37.** $b + a$ **38.** $-\dfrac{1}{xy}$

39. $0, 10$ **40.** 34 **41.** $R = \dfrac{R_1 R_2}{R_1 + R_2}$

42. $r = \dfrac{a - S}{l - S}$ or $r = \dfrac{S - a}{S - l}$

43. either 8 ft by 24 ft or 12 ft by 16 ft   **44.** $8000

45. $\dfrac{3}{5} + \dfrac{4}{5}i$ **46.** $\dfrac{3}{2} - \dfrac{1}{2}i$ **47.** 5 **48.** $0 + 10i$

49. $\dfrac{7}{5}, -\dfrac{1}{3}$ **50.** $\dfrac{4}{7} \pm \dfrac{2\sqrt{2}}{7}i$ **51.** 3 **52.** $\pm 2, \pm 3$

53. $2, 7$ **54.** $64, 729$

55. $\left(-\infty, \dfrac{11}{5}\right]$

56. $(-\infty, 3) \cup (5, \infty)$

57. $[-3, -1] \cup (2, \infty)$

58. $(-\infty, -3) \cup (0, 3)$

59. $(-\infty, -1] \cup [4, \infty)$

60. $\left(\dfrac{1}{3}, 3\right)$

Section 2.1

1. function   **3.** domain   **5.** $y = f(x)$ **7.** dependent

9. **(a)** The domain is $\{2, 3, 4, 5\}$.
      The range is $\{3, 4, 5, 6\}$.
   (b) function

11. **(a)** The domain is $\{1, 2, -5\}$.
      The range is $\{3, 4, 5, 2\}$.
   (b) not a function

13. **(a)** {(LSU, Tigers), (Georgia, Bulldogs), (MSU, Bulldogs), (Auburn Tigers)}
   (b) The domain is {LSU, Georgia, MSU, Auburn}.
      The range is {Tigers, Bulldogs}.
   (c) function

15. **(a)** {(76, September 9), (76, October 12), (78, May 10), (80, June 1)}
   (b) The domain is $\{76, 78, 80\}$.
      The range is {September 9, October 12, May 10, June 1}.
   (c) not a function

17. function   **19.** not a function   **21.** function
23. not a function   **25.** function   **27.** not a function
29. function   **31.** function   **33.** not a function
35. domain: $(-\infty, \infty)$ **37.** domain: $(-\infty, \infty)$
39. domain: $[2, \infty)$ **41.** domain: $(-\infty, 4]$
43. domain: $(-\infty, -1] \cup [1, \infty)$ **45.** domain: $(-\infty, \infty)$
47. domain: $(-\infty, -1) \cup (-1, \infty)$
49. domain: $(-\infty, 3) \cup (3, \infty)$
51. domain: $(-\infty, -2) \cup (-2, 2) \cup (2, \infty)$
53. domain: $(-\infty, -1) \cup (-1, 5) \cup (5, \infty)$ **55.** $(-\infty, \infty)$

57. $4; -11; 3k - 2; 3k^2 - 5$ **59.** $4; \dfrac{3}{2}; \dfrac{1}{2}k + 3; \dfrac{1}{2}k^2 + \dfrac{5}{2}$

61. $4; 9; k^2; k^4 - 2k^2 + 1$
63. $9; -1; k^2 + 3k - 1; k^4 + k^2 - 3$
65. $6; -29; k^3 - 2; k^6 - 3k^4 + 3k^2 - 3$
67. $5; 10; k^2 + 1; k^4 - 2k^2 + 2$

69. $\dfrac{1}{3}; 2; \dfrac{2}{k + 4}; \dfrac{2}{k^2 + 3}$ **71.** $\dfrac{1}{3}; \dfrac{1}{8}; \dfrac{1}{k^2 - 1}; \dfrac{1}{k^4 - 2k^2}$

73. $\sqrt{5}; \sqrt{10}; \sqrt{k^2 + 1}; \sqrt{k^4 - 2k^2 + 2}$
75. $\sqrt[3]{2} - 1; -\sqrt[3]{3} - 1; \sqrt[3]{k} - 1; \sqrt[3]{k^2 - 1} - 1$

77. 3 **79.** -7 **81.** $2x + h$ **83.** $8x + 4h$
85. $2x + h + 3$ **87.** $4x + 2h - 4$ **89.** $-2x - h + 1$

91. $3x^2 + 3xh + h^2$ **93.** $-\dfrac{1}{x(x + h)}$ **95.** 117

97. $\dfrac{15}{32}$ seconds   **99.** 109,500 gallons   **101.** $A(x) = x^2 + 5x$

103. **a.** $C(x) = 8x + 75$ **b.** $755
105. **a.** $C(x) = 0.07x + 9.99$ **b.** $11.39
117. e   **119.** b   **121.** f   **123.** e

Section 2.2

1. quadrants   **3.** to the right   **5.** first   **7.** linear
9. x-intercept   **11.** horizontal   **13.** midpoint
15. $A(2, 3)$ **17.** $C(-2, -3)$ **19.** $E(0, 0)$
21. $G(-5, -5)$ **23.** QI   **25.** QIII   **27.** QI   **29.** x-axis
31.

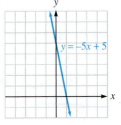

33.

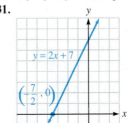

35.
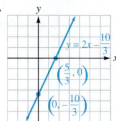

37.
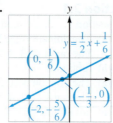

39.
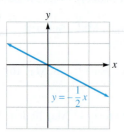

41.
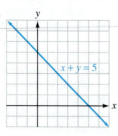

43.
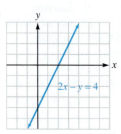

45.
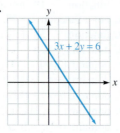

47.
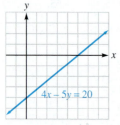

49.
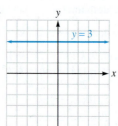

51.
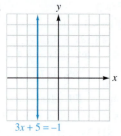

53.
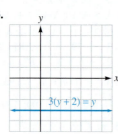

55.
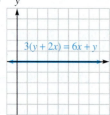

57. 1.22   **59.** 4.67   **61.** 5   **63.** $\sqrt{13}$ **65.** $\sqrt{2}$
67. 2   **69.** 5   **71.** 13   **73.** $2\sqrt{13}$ **75.** $8\sqrt{2}$
77. 7   **79.** (4, 6)   **81.** (0, 1)   **83.** $\left(-1, \frac{1}{2}\right)$

85. $\left(\frac{\sqrt{5}}{2}, \frac{\sqrt{5}}{2}\right)$ **87.** (5, 6)   **89.** (5, 15)   **93.** $\sqrt{2}$ units
97. $412,500   **99.** 200   **101.** 100 rpm
103. $15\sqrt{5}$ yards   **105.** (15, 17.5)   **107.** approx. 171 mi
113. true   **115.** false   Equations of vertical lines are
not functions.   **117.** false   The vertical line $x = 0$ has an
infinite number of y-intercepts.   **119.** true

Section 2.3

1. divided   **3.** run   **5.** the change in   **7.** vertical
9. perpendicular   **11.** 1   **13.** $-\frac{5}{12}$ **15.** $-\frac{7}{4}$
17. -2 **19.** 0   **21.** undefined   **23.** $\frac{5}{3}$
25. -1 **27.** 3   **29.** 4   **31.** $\frac{1}{2}$ **33.** $\frac{2}{3}$
35. 0   **37.** 0   **39.** 0   **41.** undefined   **43.** negative
45. positive   **47.** undefined   **49.** perpendicular
51. parallel   **53.** perpendicular   **55.** perpendicular
57. perpendicular   **59.** parallel   **61.** perpendicular
63. neither   **65.** 5   **67.** 6   **69.** not on same line
71. on same line   **73.** No two are perpendicular.
75. PQ and PR are perpendicular.   **77.** PQ and PR are
perpendicular.   **87.** 7 students per yr
89. $642.86 per year   **91.** $\frac{\Delta T}{\Delta t}$ is the hourly rate of change in
temperature.

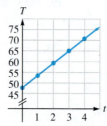

93. The slope is the speed of the plane.

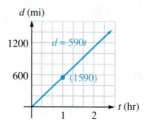

99. false $m = \dfrac{y_2 - y_1}{x_2 - x_1}$ **101.** true
103. false   The slope is 0.   **105.** true

Section 2.4

1. slope-intercept   **3.** y-intercept   **5.** $Ax + By = C$
7. $y = 3x - 2$ **9.** $y = 5x - \frac{1}{5}$ **11.** $y = ax + \frac{1}{a}$
13. $y = ax + a$ **15.** $3x - 2y = 0$ **17.** $3x + y = -4$
19. $\sqrt{2}x - y = -\sqrt{2}$ **21.** $\frac{3}{2}$, (0, −4)
23. $-\frac{1}{3}$, $\left(0, -\frac{5}{6}\right)$ **25.** $\frac{7}{2}$, (0, 2)

27. 1, $(0, -1)$

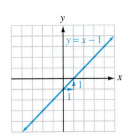

29. $\frac{2}{3}$, $(0, 2)$

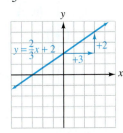

31. $-\frac{2}{3}$, $(0, 6)$

33. parallel

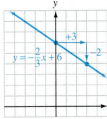

35. perpendicular   **37.** parallel   **39.** neither
41. perpendicular   **43.** perpendicular   **45.** $2x - y = 0$
47. $4x - 2y = -7$ **49.** $2x - 5y = -7$ **51.** $y = -3$
53. $x = -6$ **55.** $\pi x - y = \pi^2$ **57.** $2x - 3y = -11$
59. $y = x$ **61.** $y = \frac{7}{3}x - 3$ **63.** $y = -\frac{9}{5}x + \frac{2}{5}$
65. $y = 4x$ **67.** $y = 4x - 3$ **69.** $y = \frac{4}{5}x - \frac{26}{5}$
71. $y = -\frac{1}{4}x$ **73.** $y = -\frac{1}{4}x + \frac{11}{2}$ **75.** $y = -\frac{5}{4}x + 3$
77. $x = -2$ **79.** $x = 5$ **81.** $y = -3200x + 24,300$
83. $y = 47,500x + 475,000$ **85.** $y = -\frac{710}{3}x + 1900$
87. \$90   **89.** \$890   **91.** \$37,200   **93.** \$230
95. about 838   **97.** $C = \frac{5}{9}(F - 32)$
99. $y = -\frac{9}{10}x + 47$; 2%   **101.** $y = 3.75x + 37.5$; \$75
103. 1655 barrels per day
105. a. Chirps/min

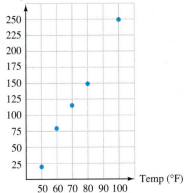

b. $y = \frac{23}{5}x - 210$ (answers may vary)
c. 204 (answers may vary)   **107.** $y = 4.44x - 196.62$
119. false $-\frac{A}{B}$ **121.** true   **123.** false $x = -99$
125. true

Section 2.5
1. x-intercept   **3.** axis of symmetry   **5.** x-axis
7. circle; center   **9.** $x^2 + y^2 = r^2$
11. $(-2, 0), (2, 0); (0, -4)$ **13.** $(0, 0), \left(\frac{1}{2}, 0\right); (0, 0)$
15. $(-1, 0), (5, 0); (0, -5)$ **17.** $(1, 0), (-2, 0); (0, -2)$
19. $(-3, 0), (0, 0), (3, 0); (0, 0)$ **21.** $(-1, 0), (1, 0); (0, -1)$
23.

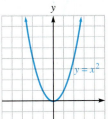

25.

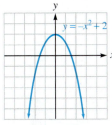

27.

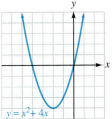

29.

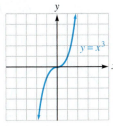

31. about the y-axis   **33.** about the x-axis
35. about the x-axis, the y-axis, and the origin
37. about the y-axis   **39.** none   **41.** about the x-axis
43. about the y-axis   **45.** about the x-axis
47.

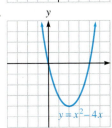

49.

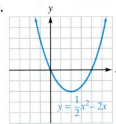

51.

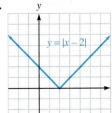

53.

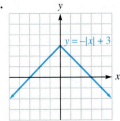

55.

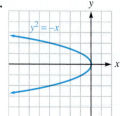
$y^2 = -x$

57.

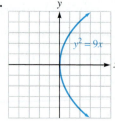
$y^2 = 9x$

107.

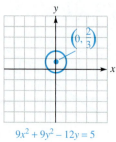
$\left(0, \frac{2}{3}\right)$
$9x^2 + 9y^2 - 12y = 5$

109.

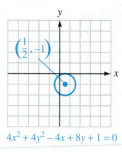
$\left(\frac{1}{2}, -1\right)$
$4x^2 + 4y^2 - 4x + 8y + 1 = 0$

59.

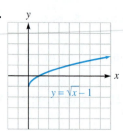
$y = \sqrt{x} - 1$

61.

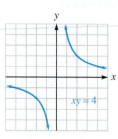
$xy = 4$

111. (0.25, 0.88)

113. (0.50, 7.25)

115. ±2.65

117. 1.44

119. 4 sec   **121.**

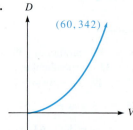
D
(60, 342)
V

63.

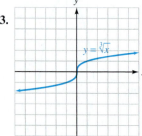
$y = \sqrt[3]{x}$

65. (0, 0); 10   **67.** (0, 5); 7   **69.** $(-6, 0); \frac{1}{2}$

71. (4, 1); 3   **73.** $\left(\frac{1}{4}, -2\right); 3\sqrt{5}$ **75.** $x^2 + y^2 = 25$

77. $x^2 + (y + 6)^2 = 36$ **79.** $(x - 8)^2 + y^2 = \frac{1}{25}$

81. $(x + 2)^2 + (y - 12)^2 = 169$ **83.** $x^2 + y^2 - 1 = 0$

85. $x^2 + y^2 - 12x - 16y + 84 = 0$

87. $x^2 + y^2 - 6x + 8y + 23 = 0$

89. $x^2 + y^2 - 6x - 6y - 7 = 0$

91. $x^2 + y^2 + 6x - 8y = 0$ **93.** $(x - 3)^2 + (y + 2)^2 = 9$

95. $(x - 5)^2 + (y - 6)^2 = 4$ **97.** $(x - 2)^2 + (y - 4)^2 = 9$

99.

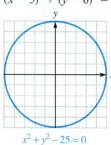
$x^2 + y^2 - 25 = 0$

101.

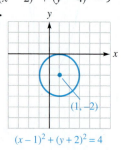
$(1, -2)$
$(x - 1)^2 + (y + 2)^2 = 4$

103.

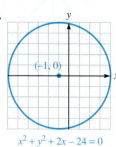
$(-1, 0)$
$x^2 + y^2 + 2x - 24 = 0$

105.

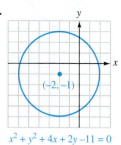
$(-2, -1)$
$x^2 + y^2 + 4x + 2y - 11 = 0$

123. $x^2 + y^2 = 36$ **125.** $x^2 + (y - 35)^2 = 900$
127. $x^2 + y^2 - 14x - 8y + 40 = 0$
133. a single point
135. false The graphs are symmetric with respect to the y-axis.
137. true   **139.** true
141. false The graph is the single point $\left(4, -\frac{1}{7}\right)$.

Section 2.6
1. quotient   **3.** means   **5.** extremes; means
7. inverse   **9.** joint   **11.** 14   **13.** 2, −3   **15.** 18
17. $\frac{1}{2}$ **19.** 1000   **21.** 1   **23.** $\frac{21}{4}$ **25.** 6   **27.** −8
29. direct variation   **31.** neither   **33.** 247,520
35. about 2 gal   **37.** $14\frac{6}{11}$ ft^3 **39.** 3 sec   **41.** $\sqrt{2}$ m
43. 15 lumens   **45.** It is multiplied by 18.
47. The force is multiplied by $\frac{9}{4}$.   **49.** $\frac{\sqrt{3}}{4}$ **59.** d
61. b   **63.** c   **65.** a

Chapter Review
1. (a) The domain is {3, 4, 5, 6}.
   The range is {4, 5, 6, 7}.
   (b) function
2. (a) The domain is {2, 3, −4}.
   The range is {4, 5, 6, 3}.
   (b) not a function
3. function   **4.** function   **5.** not a function
6. function   **7.** domain: $(-\infty, \infty)$
8. domain: $(-\infty, 5) \cup (5, \infty)$
9. $(-\infty, -2) \cup (-2, 2) \cup (2, \infty)$ **10.** domain: $[1, \infty)$

11. $(-\infty, 5)$ **12.** domain: $(-\infty, \infty)$ **13.** $8; -17; 5k - 2$

14. $-2; -\dfrac{3}{4}; \dfrac{6}{k-5}$ **15.** $0; 5; |k - 2|$ **16.** $\dfrac{1}{7}; \dfrac{1}{2}; \dfrac{k^2 - 3}{k^2 + 3}$

17. 5   **18.** $4x + 2h - 7$ **19.** 105 beats per minute

20. a. $I(x) = 3.5x - 50$ **b.** \$650   **21.** $(2, 0)$ **22.** $(-2, 1)$

23. $(0, -1)$ **24.** $(3, -1)$

25-28.

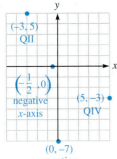

29.

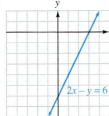

30.

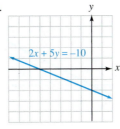

31.

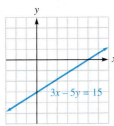

32.

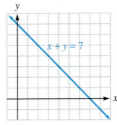

33.

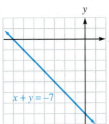

34.

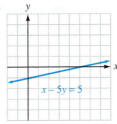

35.

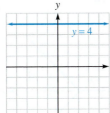

36.

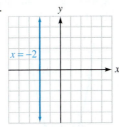

37. \$12,150   **38.** \$332,500   **39.** 10   **40.** $4\sqrt{2}$

41. 2   **42.** $2\sqrt{2}|a|$ **43.** $(0, 3)$ **44.** $\left(-6, \dfrac{15}{2}\right)$

45. $\left(\sqrt{3}, 8\right)$ **46.** $(0, 0)$ **47.** -6 **48.** 2

49. undefined   **50.** 0   **51.** -1 **52.** 1   **53.** 3

54. $-\dfrac{1}{5}$ **55.** 0   **56.** undefined   **57.** negative

58. positive   **59.** perpendicular   **60.** neither

61. $y = 7$ **62.** $x = 3$ **63.** 200 ft/min

64. \$48,750 per year   **65.** $y = \dfrac{2}{3}x + 3$

66. $y = -\dfrac{3}{2}x - 5$ **67.** $\dfrac{3}{2}, (0, -5)$ **68.** $-\dfrac{1}{2}, (0, -2)$

69. $\dfrac{3}{2}, (0, -5)$ **70.** $-\dfrac{1}{2}, (0, -2)$ **71.** $-\dfrac{5}{2}, \left(0, \dfrac{7}{2}\right)$

72. $\dfrac{3}{4}, \left(0, -\dfrac{7}{2}\right)$

73.

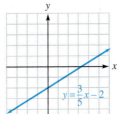

74.

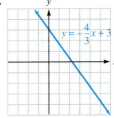

75. parallel   **76.** perpendicular   **77.** $7x + 5y = 0$

78. $4x + y = -7$ **79.** $x + 5y = -3$ **80.** $2x + y = 9$

81. $y = 17$ **82.** $x = -5$ **83.** $y = \dfrac{3}{4}x - \dfrac{3}{2}$

84. $y = -7x + 47$ **85.** $y = 3x + 5$ **86.** $y = \dfrac{1}{7}x - 3$

87. \$385   **88.** 140 hr

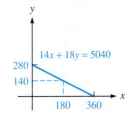

89. $(0, 0), \left(\dfrac{1}{2}, 0\right); (0, 0)$ **90.** $(12, 0), (-2, 0); (0, -24)$

91. about the x-axis   **92.** about the y-axis

93. about the y-axis   **94.** none

95.

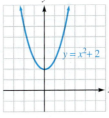

96.

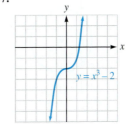

97.

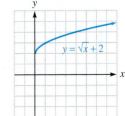

98.

99.

100.
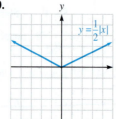

101.
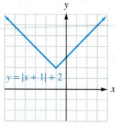

102.
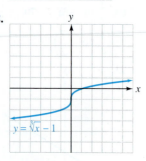

103.

104.

105.

106.

107. $(0, 0)$; 8   **108.** $(0, 6)$; 10   **109.** $(-7, 0)$; $\dfrac{1}{2}$

110. $(5, -1)$; 3   **111.** $x^2 + y^2 = 49$

112. $(x - 3)^2 + y^2 = \dfrac{1}{25}$ **113.** $(x + 2)^2 + (y - 12)^2 = 25$

114. $\left(x - \dfrac{2}{7}\right)^2 + (y - 5)^2 = 81$

115. $(x + 3)^2 + (y - 4)^2 = 144$ or $x^2 + y^2 + 6x - 8y - 119 = 0$

116. $\left(x + \dfrac{1}{2}\right)^2 + \left(y - \dfrac{5}{2}\right)^2 = \dfrac{121}{2}$ or $x^2 + y^2 + x - 5y - 54 = 0$

117. $(x + 3)^2 + (y - 2)^2 = 9$
118. $(x - 2)^2 + (y - 4)^2 = 25$

119.
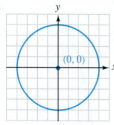

120.
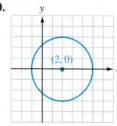

121.
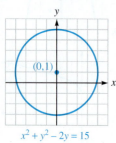

122.
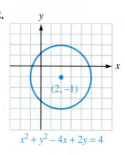

123. 3.32, -3.32 **124.** $-1, 0, 1$ **125.** $\pm 1.73, -1.73,$ $\pm 1, -1$ **126.** 4.19, -1.19 **127.** $x = 2, x = 5$
128. $x = -5, x = 5$ **129.** 400 mg   **130.** $\dfrac{9}{5}$ lb
131. $\dfrac{25}{9}$ **132.** $333\dfrac{1}{3}$ cc   **133.** 1   **134.** about 117 ohms

Chapter Test

1. $\left(-\infty, \dfrac{5}{2}\right) \cup \left(\dfrac{5}{2}, \infty\right)$ **2.** $[-3, \infty)$ **3.** $\dfrac{1}{2}, 2$

4. $\sqrt{6}, 3$ **5.** $2x + h - 1$ **6.** QII   **7.** y-axis

8.
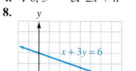

9.
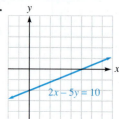

10.
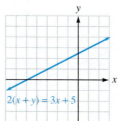

11.
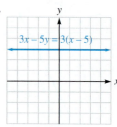

12.
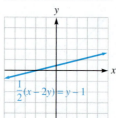

13.
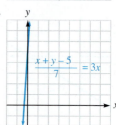

14. $\sqrt{41}$ **15.** approximately 4.44   **16.** $(0, 0)$

17. $\left(\sqrt{2}, 2\sqrt{2}\right)$ **18.** $-\dfrac{5}{4}$ **19.** $\dfrac{\sqrt{3}}{3}$ **20.** neither

21. perpendicular   **22.** $y = 2x - 11$ **23.** $y = 3x + \dfrac{1}{2}$

24. $y = 2x + 5$ **25.** $y = -\dfrac{1}{2}x + 5$ **26.** $y = 2x - \dfrac{11}{2}$

27. $x = 3$ **28.** $(-4, 0), (0, 0), (4, 0); (0, 0)$
29. $(4, 0); (0, 4)$ **30.** about the x-axis
31. about the y-axis

32.
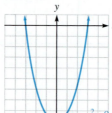

33.
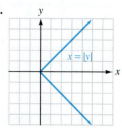

34.

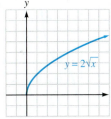

$y = 2\sqrt{x}$

35.

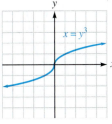

$x = y^3$

36. $(x - 5)^2 + (y - 7)^2 = 64$ **37.** $(x - 2)^2 + (y - 4)^2 = 32$

38.

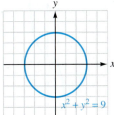

$x^2 + y^2 = 9$

39.

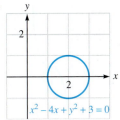

$x^2 - 4x + y^2 + 3 = 0$

40. $y = kz^2$ **41.** $w = krs^2$ **42.** $P = \dfrac{35}{2}$ **43.** $x = \dfrac{27}{32}$

44. $x = 2.65$ **45.** $x = 5.85$

Section 3.1

1. (x, y); domain; range **3.** identity **5.** cubing
7. square root

9.

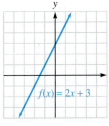

$f(x) = 2x + 3$

domain: $(-\infty, \infty)$
range: $(-\infty, \infty)$

11.

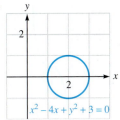

$f(x) = -\dfrac{3}{4}x + 4$

domain: $(-\infty, \infty)$
range: $(-\infty, \infty)$

13.

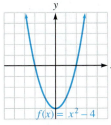

$f(x) = x^2 - 4$

domain: $(-\infty, \infty)$
range: $[-4, \infty]$

15.

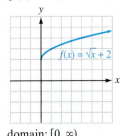

$f(x) = -\dfrac{1}{2}x^2 + 5$

domain: $(-\infty, \infty)$
range: $(-\infty, 5]$

17.

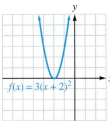

$f(x) = 3(x + 2)^2$

domain: $(-\infty, \infty)$
range: $[0, \infty)$

19.

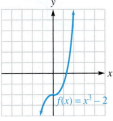

$f(x) = x^3 - 2$

domain: $(-\infty, \infty)$
range: $(-\infty, \infty)$

21. $f(x) = -x^3 + 1$

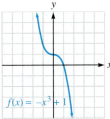

$f(x) = -x^3 + 1$

domain: $(-\infty, \infty)$
range: $(-\infty, \infty)$

23.

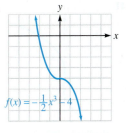

$f(x) = -\dfrac{1}{2}x^3 - 4$

domain: $(-\infty, \infty)$
range: $(-\infty, \infty)$

25.

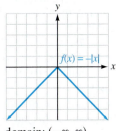

$f(x) = -|x|$

domain: $(-\infty, \infty)$
range: $(-\infty, 0]$

27.

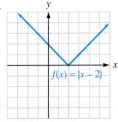

$f(x) = |x - 2|$

domain: $(-\infty, \infty)$
range: $[0, \infty)$

29.

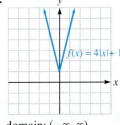

$f(x) = \left|\dfrac{1}{2}x + 3\right|$

domain: $(-\infty, \infty)$
range: $[0, \infty)$

31.

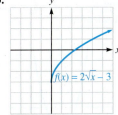

$f(x) = 4|x| + 1$

domain: $(-\infty, \infty)$
range: $[1, \infty)$

33. $f(x) = \sqrt{x} + 2$

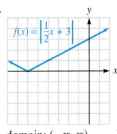

$f(x) = \sqrt{x} + 2$

domain: $[0, \infty)$
range: $[2, \infty)$

35.

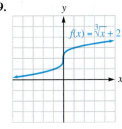

$f(x) = 2\sqrt{x} - 3$

domain: $[0, \infty)$
range: $[-3, \infty)$

37.

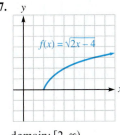

$f(x) = \sqrt{2x - 4}$

domain: $[2, \infty)$
range: $[0, \infty)$

39.

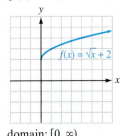

$f(x) = \sqrt[3]{x} + 2$

domain: $(-\infty, \infty)$
range: $(-\infty, \infty)$

41.

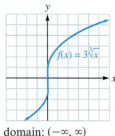

domain: $(-\infty, \infty)$
range: $(-\infty, \infty)$

43.

domain: $(-\infty, \infty)$
range: $(-\infty, \infty)$

45. function   **47.** not a function   **49.** function
51. function   **53.** 2   **55.** 0   **57.** $(2, 0)$ and $(8, 0)$
59. -6 **61.** -25 **63.** $(-2, 0)$ and $(2, 0)$ **65.** 0
67. 2   **69.** -3 **71.** $(-4, 0)$ **73.** -5
75.   **77.**   **79.**
domain: $(-\infty, \infty)$ domain: $(-\infty, \infty)$ domain: $(-\infty, \infty)$
range: $(-\infty, \infty)$ range: $[-4, \infty)$ range: $(-\infty, \infty)$

81.   **83.**   **85.**
domain: $(-\infty, \infty)$ domain: $[-2, 1)$ domain: $(-\infty, 0) \cup (0, \infty)$
range: $(-\infty, 1]$ range: $(-3, 3]$ range: $(-\infty, 0) \cup (0, \infty)$
87.   **89.**
domain: $(-\infty, 0]$ domain: $(-\infty, \infty)$
range: $[2, \infty)$ range: $\{-2, 2\}$

91.

93.

domain: $(-\infty, \infty)$;   domain: $(-\infty, \infty)$;
range: $[0, \infty)$ range: $(-\infty, \infty)$

95. a. domain: $[1, 6]$, range: $[2, 5]$ **b.** 5   **c.** 4   **d.** August
103. false Functions always pass the Vertical Line Test.
105. false The range could be any set of real numbers.
107. b   **109.** a

Section 3.2
1. upward   **3.** to the right   **5.** 2; downward   **7.** y-axis
9. horizontally
11.

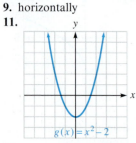

13.

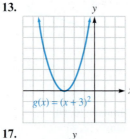

15.

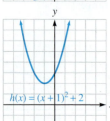

17.

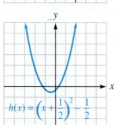

19.

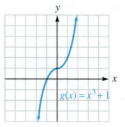

21.

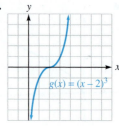

23.

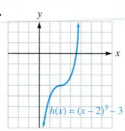

25.

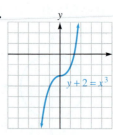

27.

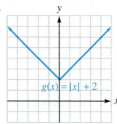

29.

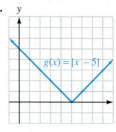

31.

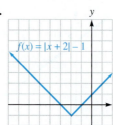

33.

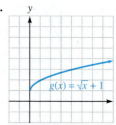

35.

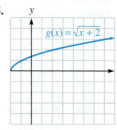

37.

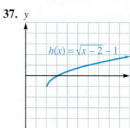

39.

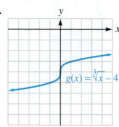

41.

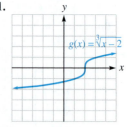

43.

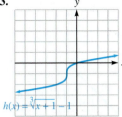

$h(x) = \sqrt[3]{x+1} - 1$

45.

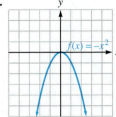

$f(x) = -x^2$

67.

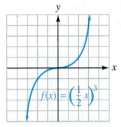

$f(x) = \left(\frac{1}{2}x\right)^3$

69.

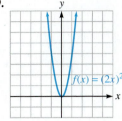

$f(x) = (2x)^2$

47.

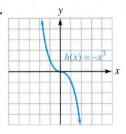

$h(x) = -x^3$

49. $f(x) = -|x|$

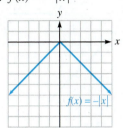

$f(x) = -|x|$

71.

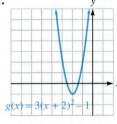

$g(x) = 3(x+2)^2 - 1$

73.

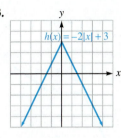

$h(x) = -2|x| + 3$

51.

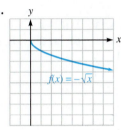

$f(x) = -\sqrt{x}$

53. $f(x) = -\sqrt[3]{x}$

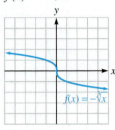

$f(x) = -\sqrt[3]{x}$

75.

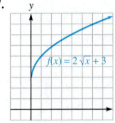

$f(x) = 2|x-2| + 1$

77.

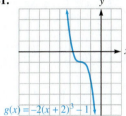

$f(x) = 2\sqrt{x} + 3$

55.

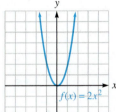

$f(x) = 2x^2$

57.

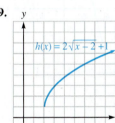

$h(x) = -3x^2$

79.

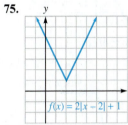

$h(x) = 2\sqrt{x-2} + 1$

81.

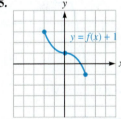

$g(x) = -2(x+2)^3 - 1$

59.

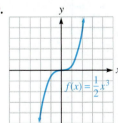

$f(x) = \frac{1}{2}x^3$

61.

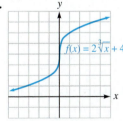

$h(x) = -3|x|$

83.

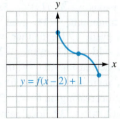

$f(x) = 2\sqrt[3]{x} + 4$

85.

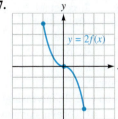

$y = f(x) + 1$

63. $f(x) = 3\sqrt{x}$

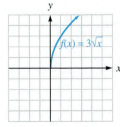

$f(x) = 3\sqrt{x}$

65. $f(x) = \frac{1}{2}\sqrt[3]{x}$

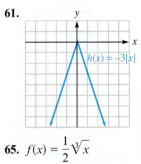
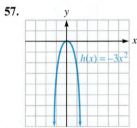

$f(x) = \frac{1}{2}\sqrt[3]{x}$

87.

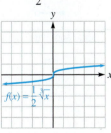

$y = 2f(x)$

89.

$y = f(x-2) + 1$

91.

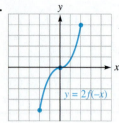

93. $y = f(x) + 2$

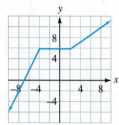

95. $y = f(x + 2)$

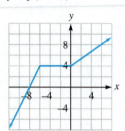

97. $y = f(x - 4) - 2$

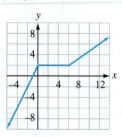

99. $y = f(-x)$

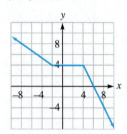

101. $y = 4f(x)$

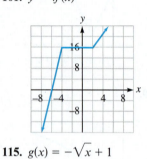

103. $y = f(4x)$

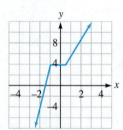

115. $g(x) = -\sqrt{x} + 1$

117. $g(x) = -2\sqrt[3]{x}$

119. $g(x) = -(x - 2)^3 + 2$

Section 3.3

1. y-axis **3.** $f(x)$ **5.** increasing **7.** constant
9. minimum **11.** even **13.** odd **15.** neither
17. even **19.** neither **21.** odd **23.** odd
25. odd **27.** even **29.** neither **31.** odd
33. decreasing on $(-\infty, 0)$; increasing on $(0, \infty)$
35. increasing on $(-\infty, 0)$; decreasing on $(0, \infty)$
37. decreasing on $(-\infty, -2)$; constant on $(-2, 2)$; increasing on $(2, \infty)$ **39.** local minimum is 2 **41.** local maximum is 5 **43.** local maximum is 2; local minimum is 1
45. local maximum is 1; local minimum is 0
47. local maximum is 3; local minimum is -3
49. a. -2 **b.** 3 **51. a.** 2 **b.** 1 **c.** 3

53.

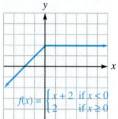

$f(x) = \begin{cases} x + 2 & \text{if } x < 0 \\ 2 & \text{if } x \geq 0 \end{cases}$

55.

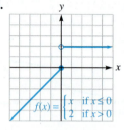

$f(x) = \begin{cases} x & \text{if } x \leq 0 \\ 2 & \text{if } x > 0 \end{cases}$

57.

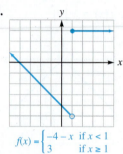

$f(x) = \begin{cases} -4 - x & \text{if } x < 1 \\ 3 & \text{if } x \geq 1 \end{cases}$

59.

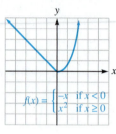

$f(x) = \begin{cases} -x & \text{if } x < 0 \\ x^2 & \text{if } x \geq 0 \end{cases}$

61.

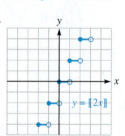

$f(x) = \begin{cases} 0 & \text{if } x < 0 \\ x^2 & \text{if } 0 \leq x \leq 2 \\ 4 - 2x & \text{if } x > 2 \end{cases}$

63. a. 3 **b.** -4
c. -3 **65. a.** 2 **b.** 3
c. 4

67.

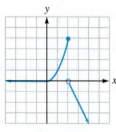

$y = [\![2x]\!]$

69.

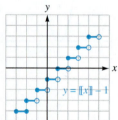

$y = [\![x]\!] - 1$

71. B

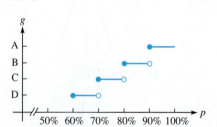

73. $32

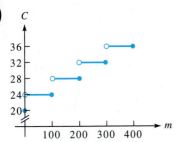

75. $1.60

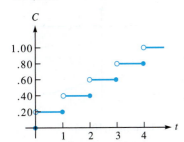

77.

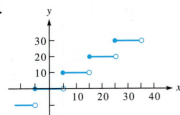

79. no; this is not defined at $x = 0$

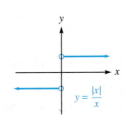

91. true **93.** true **95.** true **97.** true
99. false $f(x_1)$ is a local maximum value.

Section 3.4

1. $f(x) + g(x)$ **3.** $f(x)g(x)$ **5.** intersection
7. $g(f(x))$ **9.** commutative
11. $(f + g)(x) = 5x - 1; (-\infty, \infty)$
13. $(f \cdot g)(x) = 6x^2 - x - 2; (-\infty, \infty)$
15. $(f - g)(x) = x + 1; (-\infty, \infty)$
17. $(f/g)(x) = \dfrac{x^2 + x}{x^2 - 1} = \dfrac{x}{x - 1}; (-\infty, -1) \cup (-1, 1) \cup (1, \infty)$
19. $(f + g)(x) = 2x^2 - 12x + 9; (-\infty, \infty)$
21. $(f \cdot g)(x) = x^4 - 12x^3 + 44x^2 - 57x + 18; (-\infty, \infty)$
23. $(f + g)(x) = x^2 + \sqrt{x} - 7; [0, \infty)$
25. $(f/g)(x) = \dfrac{x^2 - 7}{\sqrt{x}}; (0, \infty)$ **27.** 7 **29.** 1 **31.** 12
33. undefined **35.** 13 **37.** -56 **39.** 7 **41.** $-\dfrac{19}{2}$

43. $f(x) = 3x^2; g(x) = 2x$ **45.** $f(x) = 3x^2; g(x) = x^2 - 1$
47. $f(x) = 3x^3; g(x) = -x$ **49.** $f(x) = x + 9; g(x) = x - 2$
51. $(-\infty, \infty); (f \circ g)(x) = 3x + 3$ **53.** $(-\infty, \infty); (f \circ f)(x) = 9x$
55. $(-\infty, \infty); (g \circ f)(x) = 2x^2$ **57.** $(-\infty, \infty); (g \circ g)(x) = 4x$
59. $(-\infty, \infty); (f \circ g)(x) = 32x^2 - 28x + 12$
61. $(-\infty, \infty); (f \circ f)(x) = 8x^4 - 24x^3 + 68x^2 - 75x + 84$
63. $[-1, \infty); (f \circ g)(x) = \sqrt{x + 1}$ **65.** $[0, \infty); (f \circ f)(x) = \sqrt[4]{x}$
67. $[-1, \infty); (g \circ f)(x) = x$
69. $(-\infty, \infty); (g \circ g)(x) = x^4 - 2x^2$
71. $(-\infty, 2) \cup (2, 3) \cup (3, \infty); (f \circ g)(x) = \dfrac{x - 2}{3 - x}$
73. $(-\infty, 1) \cup (1, 2) \cup (2, \infty); (f \circ f)(x) = \dfrac{x - 1}{2 - x}$ **75.** 11
77. -57 **79.** -17 **81.** 190 **83.** 104 **85.** 145
87. $\dfrac{1}{5}$ **89.** 8 **91.** $\dfrac{3}{4}$ **93.** $f(x) = x - 2; g(x) = 3x$
95. $f(x) = x - 2; g(x) = x^2$ **97.** $f(x) = x^2; g(x) = x - 2$
99. $f(x) = \sqrt{x}; g(x) = x + 2$
101. $f(x) = x + 2; g(x) = \sqrt{x}$
103. $f(x) = x; g(x) = x$ **105.** 0 **107.** 0 **109.** 1
111. 1 **113.** 8 **115.** 9
117. a. $(R - C)(x) = 260x - 60,000$ **b.** $70,000
119. $A(t) = \dfrac{9}{4}\pi t^2; 101,787.6$ square inches **121.** $P = 4\sqrt{A}$
133. false $(f - g)(x) = -(g - f)(x)$ **135.** true
137. false $(g \circ g \circ g)(x) = -x^{27}$ **139.** true

Section 3.5

1. one-to-one **3.** identity **5.** not one-to-one
7. one-to-one **9.** not one-to-one **11.** one-to-one
13. not one-to-one **15.** not one-to-one **17.** one-to-one
19. one-to-one **21.** one-to-one **23.** one-to-one
25. not a function **27.** one-to-one
37. $f^{-1}(x) = \dfrac{1}{3}x$ **39.** $f^{-1}(x) = \dfrac{x - 2}{3}$
41. $f^{-1}(x) = \sqrt[3]{x - 2}$ **43.** $f^{-1}(x) = x^5$
45. $f^{-1}(x) = \dfrac{1}{x} - 3$ **47.** $f^{-1}(x) = \dfrac{1}{2x}$

49.
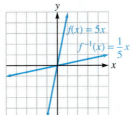

51.
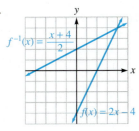

53.
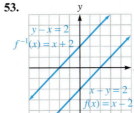

55.
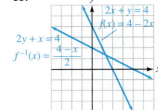

57.

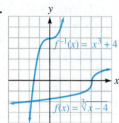

59.

5.

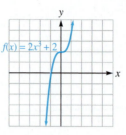

6.

61.

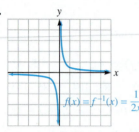

63.

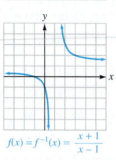

domain: $(-\infty, \infty)$
range: $(-\infty, \infty)$

domain: $(-\infty, \infty)$
range: $(-\infty, \infty)$

7.

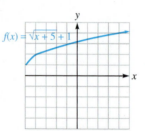

8.

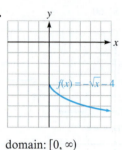

65. $f^{-1}(x) = \sqrt{x - 5}$ $(x \geq 5)$

67. $f^{-1}(x) = \frac{1}{2}\sqrt{x}$ $(x \geq 0)$

69. $f^{-1}(x) = -\sqrt{x + 3}$ $(x \geq -3)$

71. $f^{-1}(x) = \sqrt[4]{x + 8}$ $(x \geq -8)$

73. $f^{-1}(x) = \sqrt{4 - x^2}$ $(0 \leq x \leq 2)$

75. domain: $(-\infty, 2) \cup (2, \infty)$; range: $(-\infty, 1) \cup (1, \infty)$

77. domain: $(-\infty, 0) \cup (0, \infty)$; range: $(-\infty, -2) \cup (-2, \infty)$

79. a. $f(x) = 0.75x + 8.50$ **b.** $11.50

c. $f^{-1}(x) = \dfrac{x - 8.50}{0.75}$ **d.** 2 **85.** 0 **87.** $a \geq 0$

89. false Only one-to-one functions have inverses.
91. false The squaring function is not one-to-one and doesn't have an inverse. **93.** true
95. false The graph of a function and its inverse are symmetric about the line $y = x$.

domain: $[-5, \infty)$
range: $[1, \infty)$

domain: $[0, \infty)$
range: $(-\infty, -4]$

9.

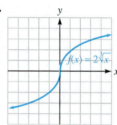

10.

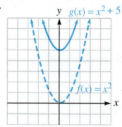

domain: $(-\infty, \infty)$
range: $(-\infty, \infty)$

domain: $(-\infty, \infty)$
range: $(-\infty, \infty)$

11. function **12.** not a function
13. domain is $(-\infty, \infty)$; range is $(-\infty, 4]$ **14.** 0 **15.** 3
16. $-2, 0, 2$ **17.** domain is $(-\infty, \infty)$; range is $(-\infty, -2]$
18. -6 **19.** -4 **20.** -6 and 0
21. 121 beats per minute **22.** 225 feet

Chapter Review

1. domain: $(-\infty, \infty)$
range: $[0, \infty)$

2.

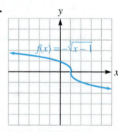

domain: $(-\infty, \infty)$
range: $(-\infty, 4]$

23.

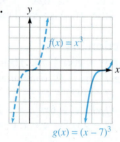

24.

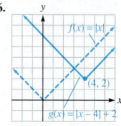

3.

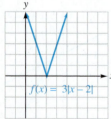

domain: $(-\infty, \infty)$
range: $[0, \infty)$

4.

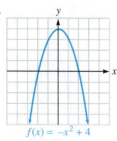

domain: $(-\infty, \infty)$
range: $(-\infty, 3]$

25.

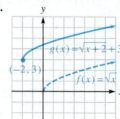

26.

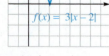

27.
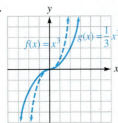

28.
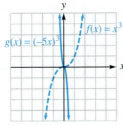

51.
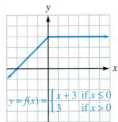

52.
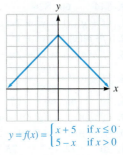

29.
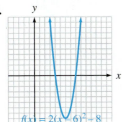

30.
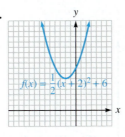

increasing on $(-\infty, 0)$;
constant on $(0, \infty)$

increasing on $(-\infty, 0)$;
decreasing on $(0, \infty)$

53. $f(x) = \begin{cases} -3x + 1 & \text{if } x < 0 \\ \dfrac{1}{3}x^2 - 4 & \text{if } x \geq 0 \end{cases}$

31.
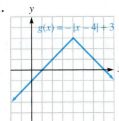

32.
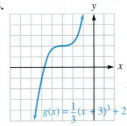

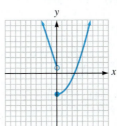

Decreasing on $(-\infty, 0)$; increasing on $(0, \infty)$

33.
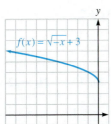

34.
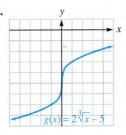

54. $f(x) = \begin{cases} \dfrac{1}{2}(x + 1)^2 + 2 & \text{if } x \leq -1 \\ -2 & \text{if } -1 < x < 1 \\ 2x + \dfrac{1}{2} & \text{if } x \geq 1 \end{cases}$

35.
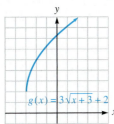

36.
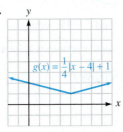

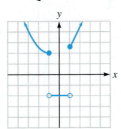

37. even **38.** odd **39.** neither **40.** odd
41. odd **42.** neither **43.** neither **44.** even
45. increasing on $(-\infty, 4)$; decreasing on $(4, \infty)$
46. increasing on $(-\infty, -2) \cup (2, \infty)$; decreasing on $(-2, 2)$
47. local maximum 2; local minimum 0
48. local maximum 2; local minimum -2
49. a. -4 **b.** 9 **50. a.** $\dfrac{1}{2}$ **b.** 3

increasing on $(1, \infty)$; decreasing on $(-\infty, -1)$;
constant on $(-1, 1)$

55. 3 **56.** -1
57.
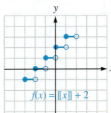

58.
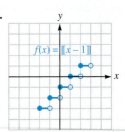

59. $44   **60.** $26

61. $(f + g)(x) = x^2 + 2x$; domain: $(-\infty, \infty)$

62. $(f \cdot g)(x) = 2x^3 + x^2 - 2x - 1$; domain: $(-\infty, \infty)$

63. $(f - g)(x) = x^2 - 2x - 2$; domain: $(-\infty, \infty)$

64. $(f/g)(x) = \dfrac{f(x)}{g(x)} = \dfrac{x^2 - 1}{2x + 1}$; domain: $\left(-\infty, -\dfrac{1}{2}\right) \cup \left(-\dfrac{1}{2}, \infty\right)$

65. 10   **66.** 60   **67.** 21   **68.** undefined

69. $(f \circ g)(x) = f(g(x)) = 4x^2 + 4x$; domain: $(-\infty, \infty)$

70. $(g \circ f)(x) = g(f(x)) = 2x^2 - 1$; domain: $(-\infty, \infty)$

71. 20   **72.** -2 **73.** $f(x) = \sqrt{x}$; $g(x) = x - 5$

74. $f(x) = x^3$; $g(x) = x + 6$ **75.** not one-to-one

76. one-to-one   **77.** one-to-one   **78.** not one-to-one

81. $f^{-1}(x) = \dfrac{x + 1}{7}$ **82.** $f^{-1}(x) = \dfrac{1}{5}x + \dfrac{8}{5}$

83. $f^{-1}(x) = \sqrt[3]{x + 10}$ **84.** $f^{-1}(x) = (x - 5)^3$

85. $f^{-1}(x) = \dfrac{5}{x}$ **86.** $f^{-1}(x) = 2 - \dfrac{1}{x}$

87. $f^{-1}(x) = \dfrac{x}{x + 1}$ **88.** $f^{-1}(x) = \sqrt[3]{\dfrac{3}{x}} = \dfrac{\sqrt[3]{3x^2}}{x}$

89. $f^{-1}(x) = \dfrac{x + 5}{2}$ **90.** $\left(-\infty, \dfrac{2}{5}\right) \cup \left(\dfrac{2}{5}, \infty\right)$

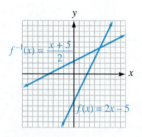

Chapter Test

1.

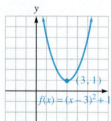

2.

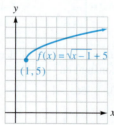

3. domain: $(-\infty, \infty)$; range $(-\infty, 5]$ **4.** 4

5.

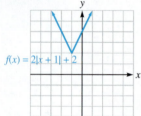

6.

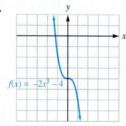

7.

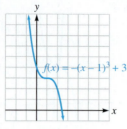

8.

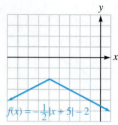

9.

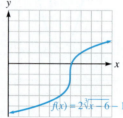

10. odd   **11.** even

12. local maxima is 5; local minima is 4   **13.** $\dfrac{3}{2}$ **14.** 5

15. Graph $f(x) = \begin{cases} -x - 1 & \text{if } x < 1 \\ 4 & \text{if } x \geq 1 \end{cases}$.

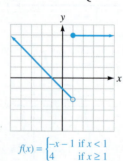

16. $(f + g)(x) = f(x) + g(x) = x^2 + 3x + 2$

17. $(f/g)(x) = \dfrac{f(x)}{g(x)} = \dfrac{3x}{x^2 + 2}$

18. $(g \circ f)(x) = g(f(x)) = 9x^2 + 2$

19. $(f \circ g)(x) = f(g(x)) = 3x^2 + 6$ **20.** 10   **21.** -12

22. -32 **23.** 1   **24.** 53   **25.** 171

26. $f^{-1}(x) = \dfrac{1}{5}x + \dfrac{2}{5}$ **27.** $f^{-1}(x) = \dfrac{x + 1}{x - 1}$

28. $f^{-1}(x) = \sqrt[3]{x + 3}$ **29.** range: $(-\infty, -2) \cup (-2, \infty)$

30. range: $(-\infty, 3) \cup (3, \infty)$

Cumulative Review

1.

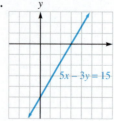

2.

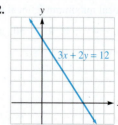

3. $\sqrt{41}$; $\left(\dfrac{1}{2}, \dfrac{3}{2}\right)$; $-\dfrac{4}{5}$ **4.** $2\sqrt{29}$; $(-2, 5)$; $\dfrac{2}{5}$ **5.** 5

6. 0   **7.** $y = -2x - 1$ **8.** $y = \dfrac{7}{2}x - \dfrac{11}{4}$

9. $y = \dfrac{3}{5}x + 6$ **10.** $y = -4x$

11. 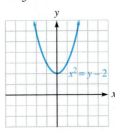   **12.**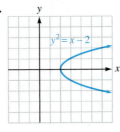

13. center is $(0, 7)$; radius is $\dfrac{1}{2}$

14. center is $(5, -4)$; radius is 12

15. 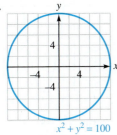   **16.**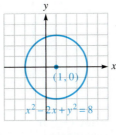

17. $x = 1$, $x = 10$ **18.** $x = 2$, $x = 8$ **19.** $62.50

20. $\dfrac{25}{4}$ **21.**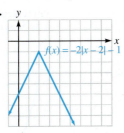

22. 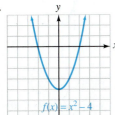   **23.**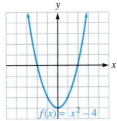

24. 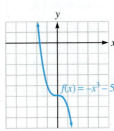   **25.**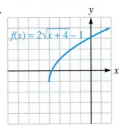

26.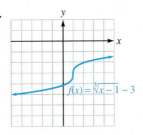

27. domain is $(-\infty, \infty)$; range is $[0, \infty)$

28. $(f - g)(x) = f(x) - g(x) = -x^2 + 3x - 5$; domain: $(-\infty, \infty)$

29. $(f \cdot g)(x) = f(x) \cdot g(x) = 3x^3 - 4x^2 + 3x - 4$; domain: $(-\infty, \infty)$

30. $(f/g)(x) = \dfrac{f(x)}{g(x)} = \dfrac{3x - 4}{x^2 + 1}$; domain: $(-\infty, \infty)$

31. $(f \circ g)(2) = 11$ **32.** $(g \circ f)(2) = 5$

33. $(f \circ g)(x) = 3x^2 - 1$ **34.** $(g \circ f)(x) = 9x^2 - 24x + 17$

35. $f^{-1}(x) = \dfrac{x - 2}{3}$ **36.** $f^{-1}(x) = \dfrac{1}{x} + 3$

37. $f^{-1}(x) = \sqrt{x - 5}$ **38.** $f^{-1}(x) = \dfrac{x + 1}{3}$

39. $y = kwz$ **40.** $y = \dfrac{kx}{t^2}$

Section 4.1

1. $f(x) = ax^2 + bx + c$ **3.** $(3, 5)$ **5.** upward

7. $-\dfrac{b}{2a}$ **9.** upward; minimum

11. downward; maximum   **13.** downward; maximum

15. $(0, -1)$ **17.** $(3, 5)$ **19.** $(-6, -4)$ **21.** $(3, 0)$

23. $(2, 0)$ **25.** $(-3, -12)$ **27.** $(3, 1)$

29. $\left(\dfrac{2}{3}, \dfrac{11}{3}\right)$ **31.** $(-4, -11)$

33.

vertex: $(0, -4)$
x-intercepts: $(-2, 0)$
and $(2, 0)$
y-intercept: $(0, -4)$
axis of symmetry: $x = 0$

35.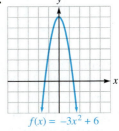

vertex: $(0, 6)$
x-intercepts: $\left(-\sqrt{2}, 0\right)$
and $\left(\sqrt{2}, 0\right)$
y-intercept: $(0, 6)$
axis of symmetry: $x = 0$

37.

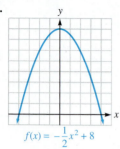

$$f(x) = -\frac{1}{2}x^2 + 8$$

vertex: $(0, 8)$
x-intercepts: $(-4, 0)$
and $(4, 0)$
y-intercept: $(0, 8)$
axis of symmetry: $x = 0$

39.

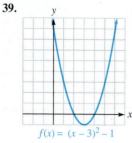

$$f(x) = (x - 3)^2 - 1$$

vertex: $(3, -1)$
x-intercepts: $(2, 0)$
and $(4, 0)$
y-intercept: $(0, 8)$
axis of symmetry: $x = 3$

49.

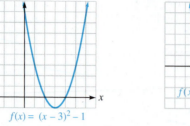

$f(x) = x^2 + 2x$

vertex: $(-1, -1)$
x-intercepts: $(-2, 0)$
and $(0, 0)$
y-intercept: $(0, 0)$
axis of symmetry: $x = -1$

51.

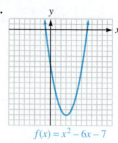

$$f(x) = x^2 - 6x - 7$$

vertex: $(3, -16)$
x-intercepts: $(-1, 0)$
and $(7, 0)$
y-intercept: $(0, -7)$
axis of symmetry: $x = 3$

41.

$$f(x) = 2(x + 1)^2 - 2$$

vertex: $(-1, -2)$
x-intercepts: $(-2, 0)$
and $(0, 0)$
y-intercept: $(0, 0)$
axis of symmetry: $x = -1$

43.

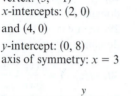

$$f(x) = -(x + 4)^2 + 1$$

vertex: $(-4, 1)$
x-intercepts: $(-5, 0)$
and $(-3, 0)$
y-intercept: $(0, -15)$
axis of symmetry: $x = -4$

53.

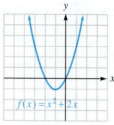

$f(x) = -x^2 - 4x + 1$

vertex: $(-2, 5)$
x-intercepts: $\left(-2 - \sqrt{5}, 0\right)$
and $\left(-2 + \sqrt{5}, 0\right)$
y-intercept: $(0, 1)$
axis of symmetry: $x = -2$

55.

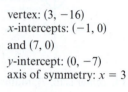

$$f(x) = 2x^2 - 12x + 10$$

vertex: $(3, -8)$
x-intercepts: $(1, 0)$
and $(5, 0)$
y-intercept: $(0, 10)$
axis of symmetry: $x = 3$

45.

$$f(x) = -3(x - 2)^2 + 6$$

vertex: $(2, 6)$
x-intercepts: $\left(2 - \sqrt{2}, 0\right)$
and $\left(2 + \sqrt{2}, 0\right)$
y-intercept: $(0, -6)$
axis of symmetry: $x = 2$

47.

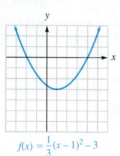

$$f(x) = \frac{1}{3}(x - 1)^2 - 3$$

vertex: $(1, -3)$
x-intercepts: $(-2, 0)$
and $(4, 0)$
y-intercept: $\left(0, -\frac{8}{3}\right)$
axis of symmetry: $x = 1$

57.

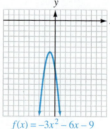

$f(x) = -3x^2 - 6x - 9$

vertex: $(-1, -6)$

x-intercepts: none

y-intercept: $(0, -9)$

axis of symmetry: $x = -1$

59.

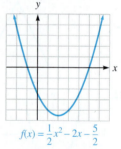

$$f(x) = \frac{1}{2}x^2 - 2x - \frac{5}{2}$$

vertex: $\left(2, -\frac{9}{2}\right)$

x-intercepts: $(-1, 0)$
and $(5, 0)$

y-intercept: $\left(0, -\frac{5}{2}\right)$

axis of symmetry: $x = 2$

61. 75 ft by 75 ft, 5625 ft^2 **63.** 200 ft by 400 ft; 80,000 ft^2
65. 25 ft by 25 ft   **67.** $w = 12$ in.; $d = 6$ in.   **69.** 20 ft
71. 15.4 ft   **73.** 26 ft   **75.** $600
77. 48 digital cameras; minimum cost $2400   **79.** $1.65

81. $95 **83.** $\frac{5}{2}$ sec **85.** 100 ft **87.** $(-2.25, -66.13)$
89. $(3.3, -68.5)$ **91.** $f(x) = 1.679x^2 - 3.907x - 0.229$

93. $f(x) = 0.086616x^2 - 11.317553x + 410.484123$; 403 lb

99. 6 by $4\frac{1}{2}$ units **101.** Both numbers are 3. **103.** true

105. false The graph of a quadratic function always has exactly one y-intercept. **107.** false The range is $[k, \infty]$.
109. true

Section 4.2

1. 4 **3.** zeros **5.** falls; rises **7.** rises; rises
9. multiplicity **11.** $P(a)$ and $P(b)$ **13.** polynomial function; degree 5 **15.** polynomial function; degree 7
17. not a polynomial function **19.** not a polynomial function **21.** polynomial function **23.** not a
polynomial function **25.** $x = -\frac{5}{2}$, multiplicity 1, crosses;
$x = \frac{5}{2}$, multiplicity 1, crosses **27.** $x = -5$, multiplicity
1, crosses; $x = \frac{3}{2}$, multiplicity 1, crosses **29.** $x = -\frac{3}{2}$,
multiplicity 1, crosses; $x = 0$, multiplicity 1, crosses; $x = 5$,
multiplicity 1, crosses **31.** $x = -6$, multiplicity 1, crosses;
$x = -2$, multiplicity 1, crosses; $x = 2$, multiplicity 1, crosses
33. $x = -3$, multiplicity 1, crosses; $x = 0$, multiplicity 2,
touches and turns; $x = 1$, multiplicity 1, crosses
35. $x = -\sqrt{11}$, multiplicity 1, crosses; $x = -2$, multiplicity 1,
crosses; $x = 2$, multiplicity 1, crosses; $x = \sqrt{11}$, multiplicity
1, crosses **37.** $x = -4$, multiplicity 2, touches and turns;
$x = 0$, multiplicity 2, touches; $x = 5$, multiplicity 1, crosses
39. $x = -3$, multiplicity 1, crosses; $x = 1$, multiplicity
2, touches; $x = \frac{5}{2}$, multiplicity 1, crosses **41.** falls on
the left and rises on the right **43.** rises on the left and
falls on the right **45.** rises on the left and rises on the
right **47.** falls on the left and falls on the right

49.

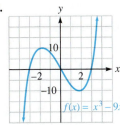

$f(x) = x^3 - 9x$

51.

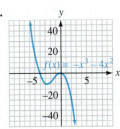

$f(x) = -x^3 - 4x^2$

53.

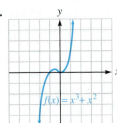

$f(x) = x^3 + x^2$

55. $f(x) = x^3 - 9x^2 + 18x$

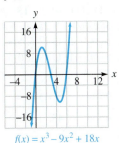

$f(x) = x^3 - 9x^2 + 18x$

57.

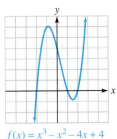

$f(x) = x^3 - x^2 - 4x + 4$

59.

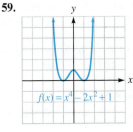

$f(x) = x^4 - 2x^2 + 1$

61.

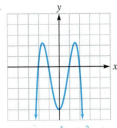

$f(x) = -x^4 + 5x^2 - 4$

63.

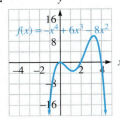

$f(x) = -x^4 + 6x^3 - 8x^2$

65.

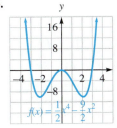

$f(x) = \frac{1}{2}x^4 - \frac{9}{2}x^2$

67.

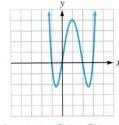

$f(x) = x(x - 3)(x - 2)(x + 1)$

69.

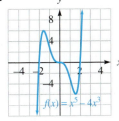

$f(x) = x^5 - 4x^3$

71. $P(-2) = 3$; $P(-1) = -2$ The signs of the results are opposites.
73. $P(4) = -40$; $P(5) = 30$ The signs of the results are opposites.
75. $P(1) = 8$; $P(2) = -1$ The signs of the results are opposites.
77. $P(2) = -72$; $P(3) = 154$ The signs of the results are opposites.
79. $P(0) = 10$; $P(1) = -60$ The signs of the results are opposites.
81. a. $V(x) = x(20 - 2x)(24 - 2x)$ or $V(x) = 4x^3 - 88x^2 + 480x$

b.

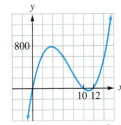

$V(x) = x(20 - 2x)(24 - 2x) = 4x^3 - 88x^2 + 480x$

c. $[0, 10]$
d. 3.6 in; 774.2 in^3

83. a. $P(x) = (270 + x)(840 - 0.1x^2)$ or
$\quad\quad P(x) = -0.1x^3 - 27x^2 + 840x + 226{,}800$
  b. 14 trees
85. a. \$5400   **b.** March and July   **87.** yes
101. b   **103.** d   **105.** true

Section 4.3

1. whole   **3.** any   **5.** factor

7. $4x^2 + 2x + 1 + \dfrac{2}{x - 1}$

9. $2x^3 - 3x^2 + 8x - 1 + \dfrac{-3}{x + 2}$

11. -1 **13.** 9   **15.** 3   **17.** -3 **19.** 69
21. 79   **23.** 1017   **25.** true   **27.** true
29. false   **31.** true   **33.** $(x - 1)(3x^2 + x - 5) - 9$
35. $(x - 3)(3x^2 + 7x + 15) + 41$
37. $(x + 1)(3x^2 - 5x - 1) - 3$
39. $(x + 3)(3x^2 - 11x + 27) - 85$

41. $x^2 + 2x + 3$ **43.** $7x^2 - 10x + 5 + \dfrac{-4}{x + 1}$

45. $4x^3 + 9x^2 + 27x + 80 + \dfrac{245}{x - 3}$

47. $3x^4 + 12x^3 + 48x^2 + 192x$ **49.** 47   **51.** -569

53. $-1 - 6i$ **55.** $\dfrac{15}{8}$ **57.** 5   **59.** 0   **61.** 384

63. $-28 - 16i$ **65.** $16 + 2i$ **67.** $40 - 40i$ **69.** yes
71. no   **73.** yes   **75.** yes   **77.** no   **79.** yes

81. $\{-1, -5, 3\}$ **83.** $\left\{-\dfrac{1}{2}, \pm 3\right\}$

85. $\left\{2, 2 \pm \sqrt{5}\right\}$

87. $\left\{-3, 3 \pm i\sqrt{10}\right\}$

89. $\left\{1, 1, \pm \sqrt{3}\right\}$ **91.** $\{2, 3, \pm i\}$
93. $P(x) = x^2 - 9x + 20$ **95.** $P(x) = x^3 - 3x^2 + 3x - 1$
97. $P(x) = x^3 - 11x^2 + 38x - 40$
99. $P(x) = x^4 - 3x^2 + 2$
101. $P(x) = x^3 - \sqrt{2}x^2 + x - \sqrt{2}$
103. $P(x) = x^3 - 2x^2 + 2x$ **109.** 0   **113.** true
115. true   **117.** false $P(i) = 3$

119. false $x = \dfrac{246}{135}$ would be a zero

Section 4.4

1. complex zero   **3.** conjugate   **5.** 0   **7.** 2
9. lower bound   **11.** 10   **13.** 4   **15.** 4   **17.** 4, 4
19. 5, 5   **21.** $P(x) = x^2 + 4$ **23.** $P(x) = x^2 - 6x + 10$
25. $P(x) = x^3 - 3x^2 + x - 3$
27. $P(x) = x^3 - 6x^2 + 13x - 10$
29. $P(x) = x^4 - 5x^3 + 7x^2 - 5x + 6$
31. $P(x) = x^4 - 2x^3 + 3x^2 - 2x + 2$
33. $P(x) = x^2 - 4ix - 4$
35. 0 or 2 positive; 1 negative; 0 or 2 nonreal
37. 0 positive; 1 or 3 negative; 0 or 2 nonreal
39. 0 positive; 0 negative; 4 nonreal

41. 1 positive; 1 negative; 2 nonreal
43. 0 positive; 0 negative; 10 nonreal
45. 0 positive; 0 negative; 8 nonreal; yes
47. 1 positive; 1 negative; 2 nonreal   **49.** $(-2, 4)$
51. $(-1, 1)$ **53.** $(-4, 6)$ **55.** $(-4, 3)$ **57.** $(-2, 4)$
65. false   The theorem states that every polynomial function has at least one complex zero.
67. false $P(x)$ would have to have real coefficients for the Conjugate Pairs Theorem to apply.
69. false   789 zeros

Section 4.5

1. -7 **3.** zero   **5.** $\pm 1, \pm 2, \pm 3, \pm 4, \pm 6, \pm 12$

7. $\pm 1, \pm 2, \pm 3, \pm 6, \pm\dfrac{1}{2}, \pm\dfrac{3}{2}$

9. $\pm 1, \pm 2, \pm 5, \pm 10, \pm\dfrac{1}{2}, \pm\dfrac{5}{2}, \pm\dfrac{1}{4}, \pm\dfrac{5}{4}$

11. $1, -1, 5$ **13.** $1, 2, -1$ **15.** $1, 2, -2$ **17.** $3, -3, 2$

19. $1, -1, \dfrac{1}{2}$ **21.** $-1, -1, \dfrac{1}{3}$ **23.** $\dfrac{3}{2}, \dfrac{2}{3}, -\dfrac{3}{5}$

25. $\dfrac{2}{3}, -\dfrac{3}{5}, 4$ **27.** $\dfrac{3}{2}, \dfrac{2}{3}, \dfrac{5}{4}$ **29.** $1, 2, 3, 4$ **31.** $2, -5$

33. $1, -1, \dfrac{1}{2}, \dfrac{3}{2}$ **35.** $-2, \dfrac{5}{2}$ **37.** $\dfrac{1}{3}, -\dfrac{1}{2}$

39. $1, -1, 2, -2, -3$ **41.** $3, \dfrac{1}{2}, -\dfrac{1}{2}$

43. $0, 2, 2, 2, -2, -2, -2$ **45.** $3, \sqrt{2}, -\sqrt{2}$

47. $\dfrac{1}{2}, i, -i$ **49.** $2, -2, 1 + \sqrt{5}, 1 - \sqrt{5}$

51. $\dfrac{1}{2}, -1, 3i, -3i$ **53.** $1, 1, 1, 5i, -5i$

55. $\dfrac{3}{2}, 1, -1, 2i, -2i$ **57.** $3, 1 - i$ **59.** $3, -3, 1 - i$

61. $-\dfrac{2}{3}, 3, -1$ **63.** $\dfrac{1}{2}, \dfrac{1}{2}, \dfrac{1}{2}, 1, 1$ **65.** 10, 20, 60 ohms

67. 13 in.   **69.** 1, 5, and 9 miles
75. $(3, 7)$ or approx. $(1.54, 13.63)$

77. false   The possible rational zeros are ± 1 and $\pm\dfrac{1}{5}$.

79. true
81. false   It will always have at least one real zero.

Section 4.6

1. asymptote   **3.** vertical   **5.** x-intercept   **7.** same
9. horizontal; vertical   **11.** vertical asymptote: $x = 2$;
horizontal asymptote: $y = 1$; domain: $(-\infty, 2) \cup (2, \infty)$; range:
$(-\infty, 1) \cup (1, \infty)$ **13.** 20 hr   **15.** 12 hr   **17.** \$5,555.56
19. \$50,000   **21.** $(-\infty, 2) \cup (2, \infty)$
23. $(-\infty, -5) \cup (-5, 5) \cup (5, \infty)$
25. $(-\infty, -1) \cup (-1, 0) \cup (0, 1) \cup (1, \infty)$ **27.** $(-\infty, \infty)$
29. $x = 3$ **31.** $x = 1, x = -1$ **33.** $x = -2, x = 3$

35. none   **37.** $y = 2$ **39.** $y = \dfrac{1}{2}$ **41.** $y = 0$

43. none   **45.** $y = x - 3$ **47.** $y = 2x + 3$
49. $y = x + 2$

51.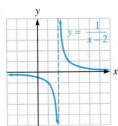

53.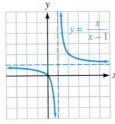

75.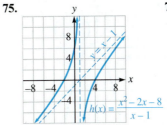

77.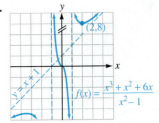

55.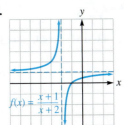

57.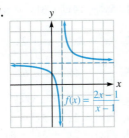

79.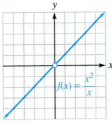

81.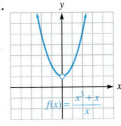

59.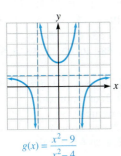

61.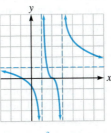

83.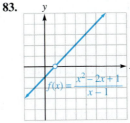

85.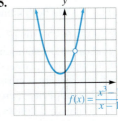

87. a. $C(x) = 3.25x + 700$ **b.** \$2325

c. $\overline{C}(x) = \dfrac{3.25x + 700}{x}$ **d.** \$4.65   **e.** \$3.95   **f.** \$3.60

89. a. $C(x) = 0.095x + 8.50$ **b.** $\overline{C}(x) = \dfrac{0.095x + 8.50}{x}$

c. 10.5¢   **103.** true

105. false The rational function has one vertical asymptote, $x = -7$.

107. true

109. false A rational function can cross a horizontal asymptote.

63.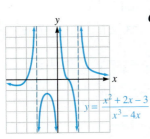

65.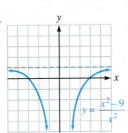

Chapter Review

1. upward; minimum   **2.** downward; maximum

3. $(1, 6)$ **4.** $(-4, -5)$ **5.** $(-3, -13)$ **6.** $\left(\dfrac{1}{2}, -8\right)$

67.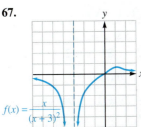

69.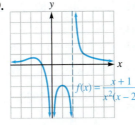

7.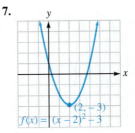

8.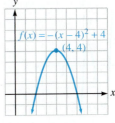

71.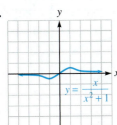

73.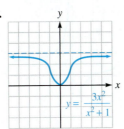

9.

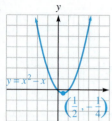

$y = x^2 - x$, $\left(\frac{1}{2}, -\frac{1}{4}\right)$

10.

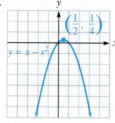

$\left(\frac{1}{2}, \frac{1}{4}\right)$, $y = x - x^2$

11.

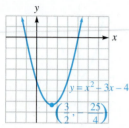

$y = x^2 - 3x - 4$, $\left(\frac{3}{2}, -\frac{25}{4}\right)$

12.

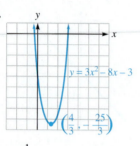

$y = 3x^2 - 8x - 3$, $\left(\frac{4}{3}, -\frac{25}{3}\right)$

13. 300 units   **14.** Both numbers are $\frac{1}{2}$.

15. 350 ft by 350 ft; 122,500 ft^2

16. 50 digital cameras; minimum cost $1100

17. $x = 0$, multiplicity 1, crosses; $x = 3$, multiplicity 2, touches

18. $x = -7$, multiplicity 1, crosses; $x = -2$, multiplicity 1, crosses; $x = 2$, multiplicity 1, crosses

19. $x = 0$, multiplicity 2, touches; $x = 1$, multiplicity 1, crosses; $x = 3$, multiplicity 1, crosses

20. $x = -\sqrt{6}$, multiplicity 1, crosses; $x = -2$, multiplicity 1, crosses; $x = 2$, multiplicity 1, crosses; $x = \sqrt{6}$, multiplicity 1, crosses

21. falls on the left and rises on the right

22. rises on the left and falls on the right

23. rises on the left and rises on the right

24. falls on the left and falls on the right

25.

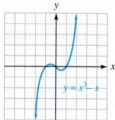

$y = x^3 - x$

26.

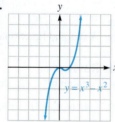

$y = x^3 - x^2$

27.

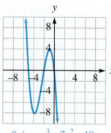

$f(x) = -x^3 - 7x^2 - 10x$

28.

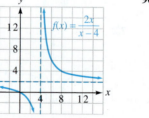

$f(x) = -x^4 + 18x^2 - 32$

29. $f(-1) = -9$; $f(0) = 18$

30. $f(1) = -8$; $f(2) = 21$

31. 1   **32.** 66   **33.** 241   **34.** 34

35. false   **36.** false   **37.** true   **38.** true

39. $3x^3 + 9x^2 + 29x + 90 + \dfrac{277}{x - 3}$

40. $2x^3 + 4x^2 + 5x + 13 + \dfrac{25}{x - 2}$

41. $5x^4 - 14x^3 + 31x^2 - 64x + 129 + \dfrac{-259}{x + 2}$

42. $4x^4 - 2x^3 + x^2 + 2x + \dfrac{1}{x + 1}$ **43.** 151   **44.** -113

45. $\dfrac{13}{8}$ **46.** $-1 - 6i$ **47.** $\left\{3, \dfrac{1}{2}, -2\right\}$

48. $\left\{-2, -2, \sqrt{5}, -\sqrt{5}\right\}$

49. $P(x) = 2x^3 - 5x^2 - x + 6$

50. $P(x) = 2x^3 + 3x^2 - 8x + 3$

51. $P(x) = x^4 + 3x^3 - 9x^2 + 3x - 10$

52. $P(x) = x^4 + x^3 - 5x^2 + x - 6$ **53.** 6   **54.** 6

55. 65   **56.** 1984   **57.** 4, 4   **58.** 40, 40   **59.** 5, 5

60. 3, 3   **61.** $2 - i$ **62.** i

63. $P(x) = x^3 - 4x^2 + x - 4$ **64.** $P(x) = x^3 + 5x^2 + x + 5$

65. 0 or 2 positive; 0 or 2 negative; 0, 2, or 4 nonreal

66. 1 or 3 positive; 1 negative; 0 or 2 nonreal

67. 1 positive; 0, 2, or 4 negative; 0, 2, or 4 nonreal

68. 1 or 3 positive; 0 or 2 negative; 2, 4, or 6 nonreal

69. 0 positive; 0 negative; 4 nonreal

70. 0 positive; 1 negative; 6 nonreal

71. $(-1, 2)$ **72.** $(-5, 2)$ **73.** $\pm 1, \pm 2, \pm 3, \pm 6, \pm\dfrac{1}{2}, \pm\dfrac{3}{2}$

74. $\pm 1, \pm 2, \pm 5, \pm 10, \pm\dfrac{1}{2}, \pm\dfrac{5}{2}, \pm\dfrac{1}{4}, \pm\dfrac{5}{4}$ **75.** 1, 4, 5

76. $1, -1, 8$ **77.** $-5, -\dfrac{3}{2}, -2$ **78.** $\dfrac{1}{3}$

79. $2, -2, \dfrac{3}{2}, -\dfrac{3}{2}$ **80.** $4, 4, -2, -\dfrac{1}{2}$

81. $\dfrac{1}{3}, 4i, -4i$ **82.** $2, -2, 1 + \sqrt{6}, 1 - \sqrt{6}$

83. $(-\infty, -5) \cup (-5, 5) \cup (5, \infty)$ **84.** $(-\infty, \infty)$

85. $x = 1, x = -1$ **86.** $x = -7$ **87.** $x = 2, x = -3$

88. $x = 4, x = -1$ **89.** $y = \dfrac{1}{2}$ **90.** $y = -5$

91. $y = 0$ **92.** none   **93.** $y = 2x + 3$ **94.** none

95.

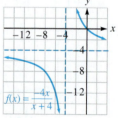

$f(x) = \dfrac{2x}{x - 4}$

96.

$f(x) = \dfrac{-4x}{x + 4}$

97.

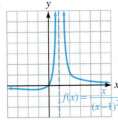

98.

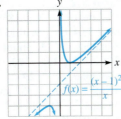

99.

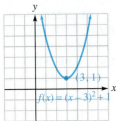

100.

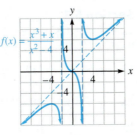

Chapter Test

1. $(7, -3)$ **2.** $(4, -10)$

3.

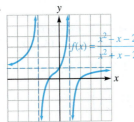

4. $\dfrac{25}{8}$ sec   **5.** $\dfrac{625}{4}$ ft   **6.** 10 ft

7.

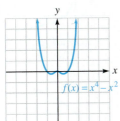

8.

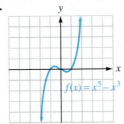

9. yes   **10.** -30 **11.** no

12. $(x - 2)(2x^2 + x - 2) - 5$

13. $2x + 3$ **14.** $3x^2 + x$

15. $\dfrac{11}{3}$ **16.** $6 - 3i$

17. $P(x) = x^3 - 4x^2 - 5x$ **18.** $P(x) = x^4 - 2x^2 - 3$

19. $3, 3$ **20.** $3 + 2i$

21. 1 or 3 positive; 0 or 2 negative; 0, 2, or 4 nonreal

22. $(-3, 3)$

23. $\pm 1, \pm 2, \pm \dfrac{1}{5}, \pm \dfrac{2}{5}$

24. $2, -3, -\dfrac{1}{2}$ **25.** $2, -3, 3i, -3i$

26. no

27. vertical asymptotes: $x = -3$ and $x = 3$; horizontal asymptote: $y = 0$

28. vertical asymptote: $x = 3$; horizontal asymptote: none; slant asymptote: $y = x - 2$

29.

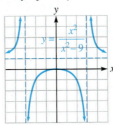

30.

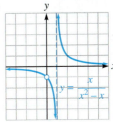

Section 5.1

1. exponential   **3.** $(-\infty, \infty)$ **5.** $(0, \infty)$ **7.** asymptote

9. increasing   **11.** 2.72   **13.** increasing   **15.** 11.0357

17. 451.8079   **19.** $5^{2\sqrt{2}} = 25^{\sqrt{2}}$ **21.** a^4 **23.** 1, 25

25. 1, 9

27.

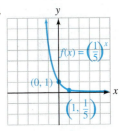

29.

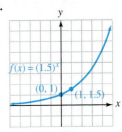

31.

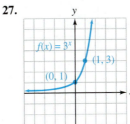

33.

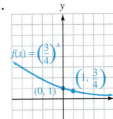

35.

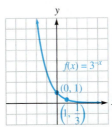

37.

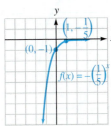

39. yes   **41.** no   **43.** $b = \dfrac{1}{2}$ **45.** no value of b

47. $b = 2$ **49.** $b = e$

51.

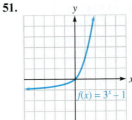

53.

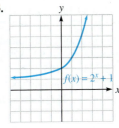

55.
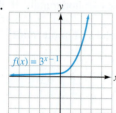
$f(x) = 3^{x-1}$

57.
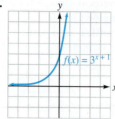
$f(x) = 3^{x+1}$

59.
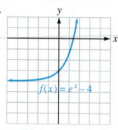
$f(x) = e^x - 4$

61.
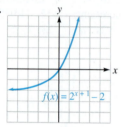
$f(x) = e^{x-2}$

63.
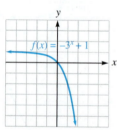
$f(x) = 2^{x+1} - 2$

65.
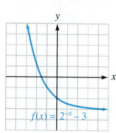
$y = 3^{x-2} + 1$

67.
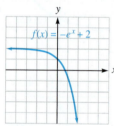
$f(x) = -3^x + 1$

69.

$f(x) = 2^{-x} - 3$

71.

$f(x) = -e^x + 2$

73.

75.

77.
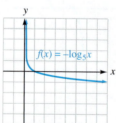

79.
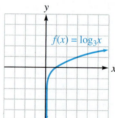

81. \$22,080.40   **83.** \$15.79   **85.** \$2,273,996.13
87. \$1263.77   **89.** \$13,375.68   **91.** \$7647.95
from continuous compounding, \$7518.28 from annual
compounding   **93.** \$291.27   **95.** \$12,155.61
97. 13,228   **105.** 125   **107.** true
109. false The graph has a y-intercept at $(0, -1)$.
111. true   **113.** false The graphs intercept at $(0, 1)$.

Section 5.2

1. birth; death   **3.** 0.1868 g   **5.** about 10 kg
7. about 35.4%   **9.** 0.1575 unit   **11.** 2 lumens
13. 8 lumens   **15.** about 56,570   **17.** 61.9°C
19. 315   **21.** 13   **23.** 10.6 billion   **25.** 2.6
27. about 0.07%   **29.** 0   **31.** 18,394
33. about 24,060   **35.** about 492   **37.** 19.0 mm
39. 49 mps   **41.** about 7 million   **43.** about 72.2 years
45. $y = 1.035264924^x$

Section 5.3

1. $x = b^y$ **3.** range   **5.** inverse   **7.** exponent
9. $(b, 1); (1, 0)$ **11.** $\log_e x$ **13.** $(-\infty, \infty)$ **15.** 10
17. $\log_8 64 = 2$ **19.** $\log_4 \dfrac{1}{16} = -2$ **21.** $\log_{1/2} 32 = -5$
23. $\log_x z = y$ **25.** $3^4 = 81$ **27.** $\left(\dfrac{1}{2}\right)^3 = \dfrac{1}{8}$
29. $4^{-3} = \dfrac{1}{64}$ **31.** $\pi^1 = \pi$ **33.** 3   **35.** -3 **37.** 3
39. $\dfrac{1}{2}$ **41.** -3 **43.** 64   **45.** 7   **47.** 5   **49.** $\dfrac{1}{25}$
51. $\dfrac{1}{6}$ **53.** 5   **55.** $\dfrac{3}{2}$ **57.** 4   **59.** $\dfrac{2}{3}$ **61.** 5
63. 4   **65.** 0.5119   **67.** -2.3307 **69.** 3.8221
71. -0.4055 **73.** 3.5596   **75.** 2.0664   **77.** -0.2752
79. undefined   **81.** 25.2522   **83.** 1.9498×10^{-4}
85. 4.0645   **87.** 69.4079   **89.** 0.0245   **91.** 120.0719
93. 4   **95.** -3 **97.** 7   **99.** 4   **101.** $b = 2$
103. $b = 2$

105.
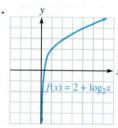
$f(x) = \log_3 x$

107.
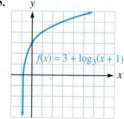
$f(x) = \log_{1/3} x$

109.
$f(x) = -\log_5 x$

111.
$f(x) = 2 + \log_2 x$

113.
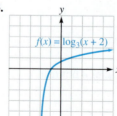
$f(x) = \log_3(x + 2)$

115.
$f(x) = 3 + \log_3(x + 1)$

117.

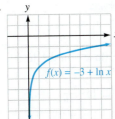

$f(x) = -3 + \ln x$

119.

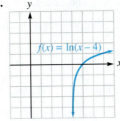

$f(x) = \ln(x - 4)$

121.

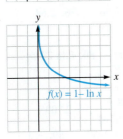

$f(x) = 1 - \ln x$

123.

125.

127.

129.

137. b is larger.   **141.** true   **143.** true

145. false $\log_2 \dfrac{1}{1024} = -10$ **147.** true   **149.** true

Section 5.4

1. $20 \log \dfrac{E_O}{E_I}$ **3.** $t = -\dfrac{1}{k} \ln\left(1 - \dfrac{C}{M}\right)$

5. $E = RT \ln\left(\dfrac{V_f}{V_i}\right)$ **7.** 55 dB   **9.** 29 dB   **11.** 49.5 dB

13. 4.4   **15.** 4   **17.** no   **19.** 19.8 min
21. 5.8 yr   **23.** 9.2 yr   **25.** 3654 joules   **27.** 23%
29. 3 yr old   **31.** 10.8 yr   **33.** 208,000 V
35. 1 mi

Section 5.5

1. 0   **3.** $M; N$ **5.** $x; y$ **7.** x **9.** $\neq$ **11.** 0
13. 7   **15.** 10   **17.** 1   **25.** $\log_b 2 + \log_b x + \log_b y$
27. $\log_b 2 + \log_b x - \log_b y$ **29.** $2 \log_b x + 3 \log_b y$

31. $\dfrac{1}{3}(\log_b x + \log_b y)$ **33.** $\log_b x + \dfrac{1}{2} \log_b z$

35. $\dfrac{1}{3} \log_b x - \dfrac{1}{3} \log_b y - \dfrac{1}{3} \log_b z$ **37.** $7 \ln x + 8 \ln y$

39. $\ln x - 4 \ln y - \ln z$ **41.** $\log_b \dfrac{x + 1}{x}$

43. $\log_b x^2 \sqrt[3]{y}$ **45.** $\log_b \dfrac{\sqrt{z}}{x^3 y^2}$ **47.** $\log_b \dfrac{\dfrac{x}{z} + x}{\dfrac{y}{z} + y} = \log_b \dfrac{x}{y}$

49. $\ln \dfrac{x(x + 5)}{9}$ **51.** $\ln \dfrac{z}{x^6 y^2}$ **53.** true   **55.** false
57. true   **59.** false   **61.** true   **63.** false   **65.** false

67. true   **69.** true   **71.** true   **73.** false   **75.** true
77. 1.4472   **79.** 0.3521   **81.** 1.1972   **83.** 2.4014
85. 2.0493   **87.** 0.4682   **89.** 1.7712   **91.** 0.9597
93. 1.8928   **95.** 2.3219   **97.** 7.20   **99.** 4.77
101. from 5.01×10^{-4} to 1.26×10^{-3} **103.** 19 dB
105. The original intensity must be raised to the 4th power.
107. The volume V is squared.   **125.** b   **127.** d
129. f   **131.** h

Section 5.6

1. exponential   **3.** $A_0 2^{-t/h}$ **5.** 2   **7.** -2 **9.** $-\dfrac{5}{6}$

11. -4 **13.** $-\dfrac{15}{2}$ **15.** $3, -1$ **17.** ± 3

19. $-2, -1$ **21.** 3   **23.** ± 5 **25.** 1.1610
27. 1.2702   **29.** 1.7095   **31.** 0   **33.** ± 1.0878

35. $0, 1.0566$ **37.** $\ln 10$ **39.** $\dfrac{1}{2} \ln 6$ **41.** 0

43. 0.2789   **45.** $1, 3$ **47.** 0   **49.** $10, -10$ **51.** 4

53. e^6 **55.** $\dfrac{1}{2}(e^4 + 7)$ **57.** 7   **59.** 50   **61.** 20

63. 10   **65.** 7   **67.** 6   **69.** 5   **71.** 4   **73.** $3, 4$
75. 10^{10} **77.** no solution   **79.** 6   **81.** 9   **83.** 4
85. $7, 1$ **87.** 20   **89.** 1.81   **91.** 5.1 yr
93. 42.7 days   **95.** 2900 yr   **97.** 5.6 yr   **99.** 5.4 yr
101. because $\ln 2 \approx 0.70$ **103.** 3.2 days

105. 12 min   **107.** $k = \dfrac{\ln 0.75}{3}$ **109.** $\dfrac{\ln\left(\dfrac{2}{3}\right)}{5}$

121. true

123. false $x^{\frac{3}{2}} = \dfrac{27}{8}$ is not an exponential equation.

125. true
127. false Take the natural logarithm of both sides.
129. false Combine the two logarithms on the left side of the equation and then write the equation in the exponential form $2^5 = \dfrac{8}{x}$.

Chapter Review

1. $5^{2\sqrt{2}}$ **2.** $2^{\sqrt{10}}$

3.

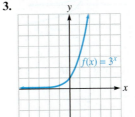

$f(x) = 3^x$

4.

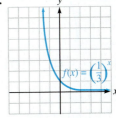

$f(x) = \left(\dfrac{1}{3}\right)^x$

5. $p = 1, q = 7$ **6.** domain: $(-\infty, \infty)$; range: $(0, \infty)$

7.

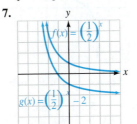

$f(x) = \left(\dfrac{1}{2}\right)^x$

$g(x) = \left(\dfrac{1}{2}\right)^x - 2$

8.

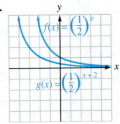

$f(x) = \left(\dfrac{1}{2}\right)^x$

$g(x) = \left(\dfrac{1}{2}\right)^{x+2}$

9.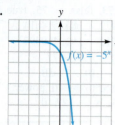

10.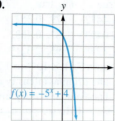

11.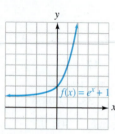

12.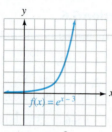

13. $2,189,703.45   **14.** $2,324,767.37   **15.** $\dfrac{2}{3}$

16. 0.19 lumen   **17.** 635,000,000   **18.** 2708

19. $(0, \infty); (-\infty, \infty)$ **20.** $(0, \infty); (-\infty, \infty)$ **21.** 2

22. $-\dfrac{1}{2}$ **23.** 0   **24.** -2 **25.** $\dfrac{1}{2}$ **26.** $\dfrac{1}{3}$ **27.** 32

28. 9   **29.** 8   **30.** -1 **31.** $\dfrac{1}{8}$ **32.** 2   **33.** 4

34. 2   **35.** 10   **36.** $\dfrac{1}{25}$ **37.** 5   **38.** 3

39.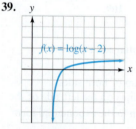

40.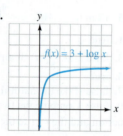

41.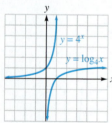

42.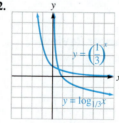

43. 6.1137   **44.** -0.1111 **45.** 10.3398   **46.** 2.5715

47.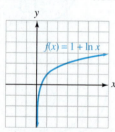

48.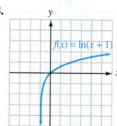

49. 12   **50.** $14x$ **51.** 53 dB   **52.** 4.4   **53.** 9.5 min

54. 23 yr   **55.** 2017 joules   **56.** 0   **57.** 1   **58.** 3

59. 4   **60.** 4   **61.** 0   **62.** 7   **63.** 3   **64.** 4

65. 9   **66.** $2 \log_b x + 3 \log_b y - 4 \log_b z$

67. $\dfrac{1}{2}(\log_8 x - \log_8 y - 2 \log_8 z)$

68. $4 \ln x - 5 \ln y - 6 \ln z$ **69.** $\dfrac{1}{3}(\ln x + \ln y + \ln z)$

70. $\log_b \dfrac{x^3 z^7}{y^5}$ **71.** $\log_b \dfrac{\sqrt{xy^3}}{z^7}$ **72.** $\ln \dfrac{x^4}{y^5 z^6}$

73. $\ln \dfrac{y^3 \sqrt{x}}{\sqrt[9]{z}}$ **74.** 3.36   **75.** 1.56   **76.** 2.64

77. -6.72 **78.** 1.7604

79. 7.94×10^{-4} gram-ions per liter   **80.** $k \ln 2$ less

81. $-\dfrac{5}{4}$ **82.** $-1, -3$ **83.** 2   **84.** 3, -3

85. $\dfrac{\log 7}{\log 3} \approx 1.7712$ **86.** $\dfrac{\log 3}{\log 3 - \log 2} \approx 2.7095$

87. $\ln 8 \approx 2.0794$ **88.** $\ln 7 \approx 1.9459$ **89.** -3

90. 8   **91.** 25, 4   **92.** 4   **93.** 2   **94.** 4, 3   **95.** 6

96. 31   **97.** $\dfrac{\ln 9}{\ln 2} \approx 3.1699$ **98.** no solution

99. $e^7 \approx 1096.6332$ **100.** $\dfrac{e}{e - 1} \approx 1.5820$ **101.** 1

102. 3300 yr

Chapter Test

1.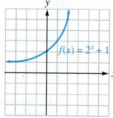

2.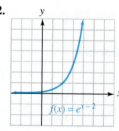

3. $\dfrac{3}{64}$ g   **4.** $1060.90   **5.** $4451.08   **6.** 3

7. -3 **8.** 17   **9.** 2   **10.** -3

11.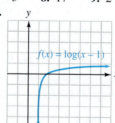

12.

13. $2 \log a + \log b + 3 \log c$ **14.** $\dfrac{1}{2}(\ln a - 2 \ln b - \ln c)$

15. $\log \dfrac{b\sqrt{a+2}}{c^2}$ **16.** $\ln \dfrac{\sqrt[3]{\dfrac{a}{b^2}}}{c}$ **17.** 1.3801

18. 0.4259   **19.** $\dfrac{\log 3}{\log 7}$ or $\dfrac{\ln 3}{\ln 7}$ **20.** $\dfrac{\log e}{\log \pi}$ or $\dfrac{1}{\ln \pi}$

21. true   **22.** false   **23.** 6.4   **24.** 46 dB   **25.** $-1, 3$

26. $\dfrac{\log 3}{\log 3 - 2} \approx -0.3133$ **27.** $\ln 9 \approx 2.1972$ **28.** 1

29. 10   **30.** 9

Cumulative Review

1.

$f(x) = 2(x + 5)^2 - 8$

2.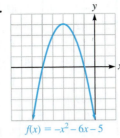

$f(x) = -x^2 - 6x - 5$

3.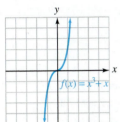

$f(x) = x^3 + x$

4.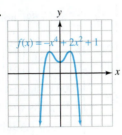

$f(x) = -x^4 + 2x^2 + 1$

5. 9   **6.** -36 **7.** 4   **8.** $2 - i$ **9.** a factor
10. not a factor   **11.** a factor   **12.** a factor   **13.** 12
14. 2000   **15.** 2 or 0 positive; 2 or 0 negative; 4, 2, or 0
nonreal   **16.** 1 positive; 3 or 1 negative; 2 or 0 nonreal
17. $-1, -3, 3$ **18.** $-1, 1, 2$

19.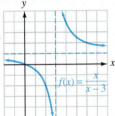

$f(x) = \dfrac{x}{x - 3}$

20.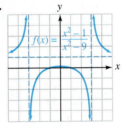

$f(x) = \dfrac{x^2 - 1}{x^2 - 9}$

21.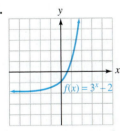

$f(x) = 3^x - 2$

22.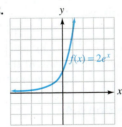

$f(x) = 2e^x$

23.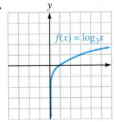

$f(x) = \log_3 x$

24.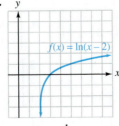

$f(x) = \ln(x - 2)$

25. 6   **26.** -3 **27.** 3   **28.** 2   **29.** $\dfrac{1}{125}$

30. $6\sqrt{2}$ **31.** $\log a + \log b + \log c$

32. $2 \log a + \log b - \log c$ **33.** $\dfrac{1}{2}(\log a + \log b - 3 \log c)$

34. $\dfrac{1}{2} \ln a + \ln b - \ln c$ **35.** $\ln \dfrac{a^3}{b^3}$ **36.** $\log \dfrac{\sqrt{ab^3}}{\sqrt[3]{c^2}}$

37. $x = \dfrac{\log 8}{\log 3} - 1$ **38.** $x = -1$ **39.** $x = 500$

40. $x = \sqrt{11}$

Section 6.1

1. system   **3.** consistent   **5.** independent
7. consistent   **9.** dependent   **11.** is

13.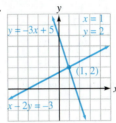

15.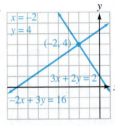

17.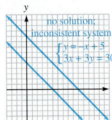

19.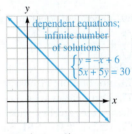

21. $(2.2, -4.7)$ **23.** $(1.7, 0.3)$ **25.** $(-1, -2)$

27. $(3, -2)$ **29.** $\left(\dfrac{1}{2}, \dfrac{1}{3}\right)$ **31.** no solution; inconsistent

system   **33.** dependent equations; a general solution is
$(x, 3x - 6)$ **35.** $(3, 1)$ **37.** $(3, 2)$ **39.** $(-3, 0)$

41. $\left(1, -\dfrac{1}{2}\right)$ **43.** dependent equations; a general solution
is $(x, 5 - 2x)$ **45.** no solution; inconsistent system
47. $(4, -7)$ **49.** $(2, -3)$ **51.** $(9, -1)$ **53.** $(1, 2, 0)$

55. $\left(0, -\dfrac{1}{3}, -\dfrac{1}{3}\right)$ **57.** $(1, 2, -1)$ **59.** $(1, 0, 5)$

61. no solution; inconsistent system   **63.** $(0, 1, 0)$

65. $\left(\dfrac{2}{3}, \dfrac{1}{4}, \dfrac{1}{2}\right)$ **67.** $\left(\dfrac{1}{2}, \dfrac{1}{2}, \dfrac{1}{2}\right)$

69. dependent equations; a general solution is $(x, 2 - x, 1)$
71. price of a hamburger is \$2; price of one order of fries is \$1
73. 225 acres of corn; 125 acres of soybeans   **75.** 8 km/h
77. 40 g and 20 g   **79.** 10 ft
81. $E(x) = 43.53x + 742.72$, $R(x) = 89.95x$; 16 pairs per day
83. 15 hr cooking hamburgers, 10 hr pumping gas, 5 hr
janitorial   **85.** 1.05 million in 0−14 group, 1.56 million in
15−49 group, 0.39 million in 50-and-older group
87. 30°, 50°, 100°   **101.** false The lines are parallel and
the system has no solution.   **103.** false The system would
consists of only the ordered triples that satisfy the system.
105. true   **107.** false Use the substitution or elimination
methods.   **109.** true

Section 6.2

1. matrix **3.** coefficient **5.** equation
7. row equivalent **9.** interchanged
11. adding, multiple **13.** (2, 3) **15.** (7, 6)
17. (1, 2, 3) **19.** (1, 1, 3) **21.** row-echelon form
23. reduced row-echelon form **25.** (2, −1)
27. (−2, 0) **29.** (3, 1)
31. no solution; inconsistent system **33.** (0, −7)
35. (1, 0, 2) **37.** (2, −2, 1) **39.** (1, 1, 2)
41. dependent equations; a general solution is
$\left(x, -x + \dfrac{7}{2}, -\dfrac{3}{2}\right)$ **43.** no solution; inconsistent system

45. (13, 3) **47.** (−13, 7, −2) **49.** (10, 3)

51. (0, 0) **53.** (3, 1, −2) **55.** $\left(\dfrac{1}{4}, 1, \dfrac{1}{4}\right)$

57. (1, 2, 1, 1) **59.** (1, 2, 0, 1) **61.** $\left(\dfrac{9}{4}, -3, \dfrac{3}{4}\right)$

63. $\left(0, \dfrac{20}{3}, -\dfrac{1}{3}\right)$ **65.** dependent equations; a general

solution is $\left(x, \dfrac{1}{2}x - 3, 0\right)$ **67.** (1, −3)

69. dependent equations; a general solution is
$\left(\dfrac{8}{7} + \dfrac{1}{7}z, \dfrac{10}{7} - \dfrac{4}{7}z, z\right)$ **71.** dependent equations; a general

solution is $(1 + z, -z, -1 - z, z)$
73. no solution; inconsistent system **75.** 1300 mi
77. Dictionaries are 4.5 in. wide; atlases are 3.5 in. wide;
thesauruses are 4 in. wide. **79.** 2 niacin, 4 zinc, 6 vitamin C
89. (±2, 1, ±3) **91.** false The matrix has
7 rows. **93.** true **95.** true **97.** false The system has
no solution. **99.** true

Section 6.3

1. i, j **3.** corresponding **5.** columns, rows
7. additive identity **9.** $x = 2, y = 5$ **11.** $x = 1, y = 2$

13. $\begin{bmatrix} -1 & 2 & 1 \\ -6 & 0 & 0 \end{bmatrix}$ **15.** $\begin{bmatrix} -5 & 2 & -7 \\ 5 & 0 & -3 \\ 2 & -3 & 5 \end{bmatrix}$

17. $\begin{bmatrix} -6 & 5 & 0 \\ 1 & -1 & 0 \end{bmatrix}$ **19.** $\begin{bmatrix} 15 & -15 \\ 0 & -10 \end{bmatrix}$

21. $\begin{bmatrix} 25 & 75 & -10 \\ -10 & -25 & 5 \end{bmatrix}$ **23.** $\begin{bmatrix} 18 & -1 & -4 \\ -35 & 0 & -1 \end{bmatrix}$

25. $\begin{bmatrix} 4 & 5 & -5 \\ 2 & 2 & -4 \end{bmatrix}$ **27** $\begin{bmatrix} 9 & 7 & -6 \\ 6 & 3 & -6 \end{bmatrix}$

29. $\begin{bmatrix} 13 & 12 & -11 \\ 8 & 5 & -10 \end{bmatrix}$

31. $\begin{bmatrix} 4 & -\dfrac{7}{2} & \dfrac{11}{2} \\ 5 & -1 & 2 \end{bmatrix}$ **33.** $\begin{bmatrix} -\dfrac{28}{3} & -\dfrac{1}{3} & -\dfrac{7}{3} \\ -\dfrac{26}{3} & -\dfrac{2}{3} & \dfrac{4}{3} \end{bmatrix}$

35. $\begin{bmatrix} 2 & -2 \\ 3 & 10 \end{bmatrix}$ **37.** $\begin{bmatrix} -22 & -22 \\ -105 & 126 \end{bmatrix}$

39. $\begin{bmatrix} 4 & 2 & 10 \\ 5 & -2 & 4 \\ 2 & -2 & 1 \end{bmatrix}$ **41.** $\begin{bmatrix} 4 & -5 & -6 \\ -8 & 10 & 12 \\ -12 & 15 & 18 \end{bmatrix}$

43. not possible **45.** $\begin{bmatrix} 16 \\ 12 \\ 12 \end{bmatrix}$ **47.** not possible

49. $\begin{bmatrix} -36.29 \\ 16.2 \\ -19.26 \end{bmatrix}$ **51.** $\begin{bmatrix} -16.11 & 4.71 & 33.64 \\ -19.6 & 20.35 & 6.4 \\ -100.6 & 72.82 & 62.71 \end{bmatrix}$

57. $\begin{bmatrix} 4 & 5 \\ -7 & -1 \end{bmatrix}$ **59.** not possible

61. $\begin{bmatrix} 24 & 16 \\ 39 & 26 \end{bmatrix}$ **63.** $\begin{bmatrix} 47 \\ 81 \end{bmatrix}$

65. $QC = \begin{bmatrix} 2000 \\ 1700 \end{bmatrix}$ The cost to Supplier 1 is $2000.
The cost to Supplier 2 is $1700.

67. $QP = \begin{bmatrix} 584.50 \\ 709.25 \\ 1036.75 \end{bmatrix}$ Adult males spent $584.50.
Adult females spent $709.25.
Children spent $1036.75.

69. $\begin{bmatrix} 1 & 1 & 0 \\ 0 & 1 & 1 \\ 1 & 0 & 0 \end{bmatrix}$

71. $A^2 = \begin{bmatrix} 5 & 1 & 2 & 2 \\ 1 & 5 & 2 & 2 \\ 2 & 2 & 6 & 0 \\ 2 & 2 & 0 & 4 \end{bmatrix}$ indicates the number of ways two cities can be linked with exactly one intermediate city to relay messages.

77. No; if $A = \begin{bmatrix} 1 & 1 \\ 1 & 1 \end{bmatrix}$ and $B = \begin{bmatrix} 1 & 0 \\ 0 & 0 \end{bmatrix}$, then $(AB)^2 \neq A^2B^2$.

79. Let $A = \begin{bmatrix} 1 & 2 \\ 1 & 2 \end{bmatrix}$ and $B = \begin{bmatrix} 2 & 2 \\ -1 & -1 \end{bmatrix}$. Neither is the zero

matrix, yet $AB = 0$.

81. false $AB = \begin{bmatrix} 20 & 20 \\ 20 & 20 \end{bmatrix}$. **83.** true **85.** false The

number of columns of the first matrix must equal the number
of rows of the second matrix. **87.** true **89.** true

Section 6.4

1. $AB = BA = I$ **3.** $[I \mid A^{-1}]$ **5.** $\begin{bmatrix} 3 & 4 \\ 2 & 3 \end{bmatrix}$

7. $\begin{bmatrix} 5 & -7 \\ -2 & 3 \end{bmatrix}$ **9.** $\begin{bmatrix} -2 & 3 & -3 \\ -5 & 7 & -6 \\ 1 & -1 & 1 \end{bmatrix}$

11. $\begin{bmatrix} 4 & 1 & -3 \\ -5 & -1 & 4 \\ -1 & -1 & 1 \end{bmatrix}$ **13.** no inverse

15. $\begin{bmatrix} 1 & -2 & 1 \\ 0 & 1 & -2 \\ 0 & 0 & 1 \end{bmatrix}$ **17.** no inverse

19. $\begin{bmatrix} 1 & -2 & 1 & 0 \\ 0 & 1 & -2 & 1 \\ 0 & 0 & 1 & -2 \\ 0 & 0 & 0 & 1 \end{bmatrix}$ **21.** $\begin{bmatrix} 8 & -2 & -6 \\ -5 & 2 & 4 \\ 2 & 0 & -2 \end{bmatrix}$

23. $\begin{bmatrix} -2.5 & 5 & 3 & 5.5 \\ 5.5 & -8 & -6 & -9.5 \\ -1 & 3 & 1 & 3 \\ -5.5 & 9 & 6 & 10.5 \end{bmatrix}$

25. $x = 23, y = 17$ **27.** $x = 0, y = 0$
29. $x = 1, y = 2, z = 2$ **31.** $x = 54, y = -37, z = -49$
33. $x = 2, y = 1$ **35.** $x = 1, y = 2, z = 1$
37. 2 of model A, 3 of model B **39.** hi

47. $X = \begin{bmatrix} 0 \\ 0 \\ 0 \end{bmatrix}$ **53.** false Some square matrices have inverses.

55. false $A^{-1} = \begin{bmatrix} -1 & \dfrac{1}{2} \\ \dfrac{3}{4} & -\dfrac{1}{4} \end{bmatrix}$ **57.** true **59.** true

Section 6.5

1. $|A|$, $\det(A)$ **3.** 0 **5.** 0 **7.** 8 **9.** 1
11. -42 **13.** 13 **15.** 42 **17.** 13 **19.** -54
21. -7 **23.** 86 **25.** -2 **27.** 2 **29.** 120
31. true **33.** false **35.** 3 **37.** 3 **39.** $(1, 2)$
41. $(3, 0)$ **43.** $(1, 0, 1)$ **45.** $(1, -1, 2)$ **47.** $(6, 6, 12)$
49. $\left(\dfrac{5}{6}, \dfrac{2}{3}, \dfrac{1}{2}, \dfrac{5}{2}\right)$ **51.** $3x - 2y = 0$ **53.** $6x + 7y = 9$
55. 30 sq. units **57.** 73 sq. units **63.** 8 **65.** -1
67. 21.468 **69.** $5000 in HiTech, $8000 in SaveTel, $7000
in OilCo **75.** 10 **77.** 24 **79.** 8
81. domain: $n \times n$ matrices; range: all real numbers
83. yes **85.** false In general, $|A + B| \neq |A| + |B|$.

87. false $\begin{vmatrix} 999 & 888 \\ 777 & 666 \end{vmatrix} = -\begin{vmatrix} 777 & 666 \\ 999 & 888 \end{vmatrix}$ **89.** true

Section 6.6

1. first-degree; second-degree **3.** $\dfrac{1}{x} + \dfrac{2}{x - 1}$

5. $\dfrac{5}{x} - \dfrac{3}{x - 3}$ **7.** $\dfrac{1}{x + 1} + \dfrac{2}{x - 1}$ **9.** $\dfrac{2}{x} - \dfrac{2}{x - 2}$

11. $\dfrac{1}{x - 3} - \dfrac{3}{x + 2}$ **13.** $\dfrac{8}{x + 3} - \dfrac{5}{x - 1}$

15. $\dfrac{5}{2x - 3} + \dfrac{2}{x - 5}$ **17.** $\dfrac{2}{x} + \dfrac{3}{x - 1} - \dfrac{1}{x + 1}$

19. $\dfrac{1}{x} + \dfrac{1}{x^2 + 3}$ **21.** $\dfrac{3}{x + 1} + \dfrac{2}{x^2 + 2x + 3}$

23. $\dfrac{3}{x} + \dfrac{2}{x + 1} + \dfrac{1}{(x + 1)^2}$ **25.** $\dfrac{1}{x} + \dfrac{2}{x^2} - \dfrac{3}{x - 1}$

27. $\dfrac{2}{x} + \dfrac{1}{x - 3} + \dfrac{2}{(x - 3)^2}$ **29.** $\dfrac{1}{x - 1} - \dfrac{4}{(x - 1)^3}$

31. $\dfrac{1}{x} + \dfrac{1}{x^2} + \dfrac{2}{x^2 + x + 1}$ **33.** $\dfrac{3}{x} + \dfrac{4}{x^2} + \dfrac{x + 1}{x^2 + 1}$

35. $-\dfrac{1}{x + 1} - \dfrac{3}{x^2 + 2}$ **37.** $\dfrac{x + 1}{x^2 + 2} + \dfrac{2}{x^2 + x + 2}$

39. $\dfrac{1}{x} + \dfrac{x}{x^2 + 2x + 5} + \dfrac{x + 2}{(x^2 + 2x + 5)^2}$

41. $x - 3 - \dfrac{1}{x + 1} + \dfrac{8}{x + 2}$ **43.** $1 + \dfrac{2}{3x + 1} + \dfrac{1}{x^2 + 1}$

45. $1 + \dfrac{1}{x} + \dfrac{x}{x^2 + x + 1}$ **47.** $2 + \dfrac{1}{x} + \dfrac{3}{x - 1} + \dfrac{2}{x^2 + 1}$

55. No, it's the sum of two cubes. **57.** e **59.** d **61.** a

Section 6.7

1. half-plane; boundary **3.** is not

5. 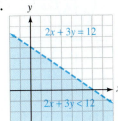 **7.**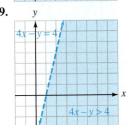

9. 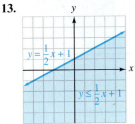 **11.**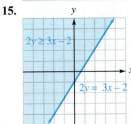

13. 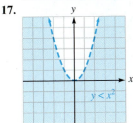 **15.**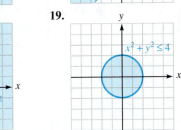

17. 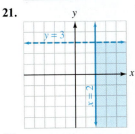 **19.**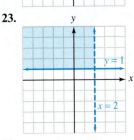

21. 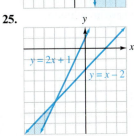 **23.**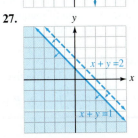

25. 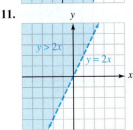 **27.**

29.
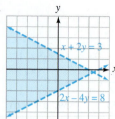

31.
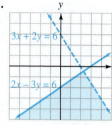

33.
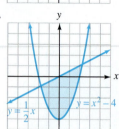

35.
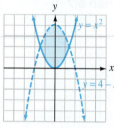

37.
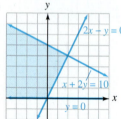

39.
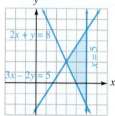

41.
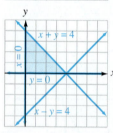

43.
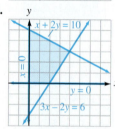

45. $\begin{cases} 6s + 4l \le 60 \\ s \ge 0 \\ l \ge 0 \end{cases}$ **47. a.** $\begin{cases} 5x + 6y \ge 600 \\ x \ge 0 \\ y \ge 0 \end{cases}$

b.
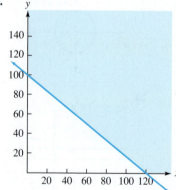

55. false If $(0, 0)$ is a point on the line, it cannot be a test point.
57. false A system of inequalities can have no solution.
59. true   **61.** true

Section 6.8
1. constraints   **3.** objective   **5.** $P = 12$ at $(0, 4)$
7. $P = \dfrac{13}{6}$ at $\left(\dfrac{5}{3}, \dfrac{4}{3}\right)$ **9.** $P = \dfrac{18}{7}$ at $\left(\dfrac{3}{7}, \dfrac{12}{7}\right)$

11. $P = 3$ at $(1, 0)$ **13.** $P = 0$ at $(0, 0)$
15. $P = 0$ at $(0, 0)$ **17.** $P = -12$ at $(-2, 0)$
19. $P = -2$ at the edge joining $(1, 2)$ and $(-1, 0)$
21. 3 tables, 12 chairs; $1260
23. 30 IBMs, 30 Apple; $2700
25. 15 DVRs, 30 TVs; $1560
27. $150,000 in stocks, $50,000 in bonds; $17,000
29. 2 buses, 2 trucks; $1100
37. false For an objective function to have a maximum and minimum value, it must be subjected to constraints that correspond to a bounded region.
39. false The minimum value of an objective function can occur at more than one point.

Chapter Review
1.
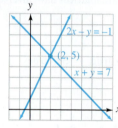

2.
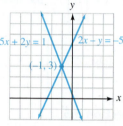

3.
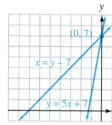

4.
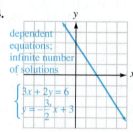

a general solution is
$\left(x, -\dfrac{3}{2}x + 3\right)$

5.
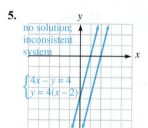

6. $(2, -1)$ **7.** $(0, -3)$ **8.** $(1, 1)$
9. dependent equations; a general solution is $(x, 3x - 4)$
10. no solution; inconsistent system   **11.** $(-3, 2)$
12. $(2, 5)$ **13.** $(2, -1)$
14. no solution; inconsistent system   **15.** dependent equations; a general solution is $(x, 3x - 4)$
16. $(1, 0, 1)$ **17.** $(1, 1, -1)$ **18.** $(0, 1, 2)$
19. $10,400
20. 900 adult tickets, 450 senior tickets, 450 children's tickets
21. $(1, 1)$ **22.** dependent equations; a general solution is $(x, 3x + 4)$ **23.** $(3, 1, -2)$ **24.** $(-10, 1, 10)$
25. no solution; inconsistent system   **26.** $(1, 2, -1, -2)$
27. $x = -4, y = 3$ **28.** $\begin{bmatrix} 1 & 3 & 4 \\ 4 & 0 & 2 \end{bmatrix}$

29. $\begin{bmatrix} 2 & 5 & 4 \\ -2 & -6 & 6 \\ -4 & 5 & -3 \end{bmatrix}$ **30.** $\begin{bmatrix} 4 & -1 \\ -7 & -7 \end{bmatrix}$

31. $\begin{bmatrix} -17 & 19 \\ 10 & -12 \end{bmatrix}$ **32.** $[5]$ **33.** $\begin{bmatrix} 2 & -1 & 1 & 3 \\ 4 & -2 & 2 & 6 \\ 2 & -1 & 1 & 3 \\ 10 & -5 & 5 & 15 \end{bmatrix}$

34. not possible **35.** $[-24]$ **36.** $\begin{bmatrix} 0 \\ -6 \end{bmatrix}$

37. $\begin{bmatrix} -1 & 4 \\ 0 & -12 \\ -4 & -1 \end{bmatrix}$ **38.** $\begin{bmatrix} \frac{1}{4} & -1 \\ 0 & 3 \\ 1 & \frac{1}{4} \end{bmatrix}$

39. $\begin{bmatrix} 5 & -3 \\ -3 & 2 \end{bmatrix}$ **40.** $\begin{bmatrix} 2 & \frac{1}{2} \\ 3 & 1 \end{bmatrix}$ **41.** No inverse exists.

42. $\begin{bmatrix} 1 & 0 & 0 \\ -\frac{3}{2} & \frac{1}{2} & \frac{1}{2} \\ 1 & -\frac{1}{2} & 0 \end{bmatrix}$ **43.** $\begin{bmatrix} -9 & 16 & -56 \\ -3 & -5 & 18 \\ -1 & -2 & 7 \end{bmatrix}$

44. No inverse exists. **45.** $(3, 1)$
46. $(1, 2, -1)$
47. $(1, 1, 0, -1)$ **48.** -7 **49.** -6
50. 3 **51.** -25 **52.** $(1, -2)$
53. $(1, 0, -2)$ **54.** $(1, -1, 3)$
55. $(1, 0, -1, 2)$ **56.** 21 **57.** 7
58. $\dfrac{3}{x} + \dfrac{4}{x+1}$ **59.** $\dfrac{3}{x} + \dfrac{2}{x^2} + \dfrac{x-1}{x^2+1}$
60. $\dfrac{1}{x} - \dfrac{1}{x^2+x+5}$ **61.** $\dfrac{1}{x+1} - \dfrac{2}{(x+1)^2} + \dfrac{2}{(x+1)^3}$

62.
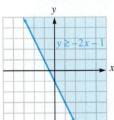

63.
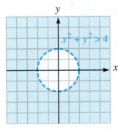

64.
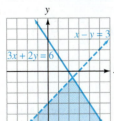

65.
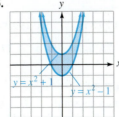

66. $P = 6$ at $(3, 0)$ **67.** $P = 2$ at $(1, 1)$
68. $P = 12$ at $(0, -4)$ **69.** $P = 3$ at $\left(-\dfrac{2}{3}, \dfrac{5}{3}\right)$
70. 1000 bags of x, 1400 bags of y

Chapter Test

1.
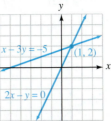

2.
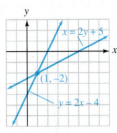

3. $(1, -3)$ **4.** $(3, 5)$
5. 6 liters of 20% solution, 4 liters of 45% solution
6. CD World 100 units, Ace 25 units, Hi-Fi 50 units
7. $(2, 1)$ **8.** $(1, 2, 1)$ **9.** $(1, 0, -2)$
10. A general solution is $\left(-\dfrac{2}{5}y + \dfrac{7}{5},\, y,\, -\dfrac{8}{5}y - \dfrac{7}{5}\right)$

11. $\begin{bmatrix} 16 & -14 & 20 \\ 0 & -6 & -13 \end{bmatrix}$ **12.** $[-1]$ **13.** $\begin{bmatrix} -\frac{7}{3} & \frac{9}{3} \\ \frac{2}{3} & -\frac{5}{3} \end{bmatrix}$

14. $\begin{bmatrix} -13 & -3 & 14 \\ 4 & 1 & -4 \\ 12 & 3 & -13 \end{bmatrix}$ **15.** $\left(\dfrac{17}{3}, -\dfrac{4}{3}\right)$

16. $(-36, 11, 34)$ **17.** -12 **18.** -24 **19.** $-\dfrac{5}{4}$
20. 1 **21.** $\dfrac{3}{2x-3} + \dfrac{1}{x+1}$ **22.** $\dfrac{1}{x} + \dfrac{2x+1}{x^2+2}$

23.
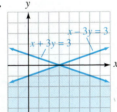

24.
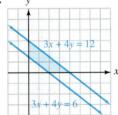

25. $P = 7$ at $(1, 2)$ **26.** $P = -8$ at $(8, 0)$

Section 7.1
1. $2, -5, 3$ **3.** $0, 0, \sqrt{5}$ **5.** to the left
7. downward **9.** directrix, focus **11.** circle
13. parabola **15.** $x^2 + y^2 = 49$, $x^2 + y^2 - 49 = 0$
17. $(x - 2)^2 + (y + 2)^2 = 17$, $x^2 + y^2 - 4x + 4y - 9 = 0$
19. $(x - 1)^2 + (y + 2)^2 = 36$, $x^2 + y^2 - 2x + 4y - 31 = 0$
21.
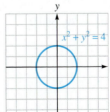

23.
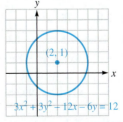

25. $(0, 0), (0, 3), y = -3$ **27.** $(0, 3), (5, 3), x = -5$
29. $(-2, 1), (-2, -5), y = 7$ **31.** $x^2 = 12y$
33. $y^2 = -12x$ **35.** $(x - 3)^2 = -12(y - 5)$
37. $(x - 3)^2 = -28(y - 5)$ **39.** $x^2 = -4(y - 2)$

41. $(y + 5)^2 = 8(x - 1)$

43. $(y - 2)^2 = -2(x - 2)$ or $(x - 2)^2 = -2(y - 2)$

45. $(x + 4)^2 = -\dfrac{16}{3}(y - 6)$ or $(y - 6)^2 = \dfrac{9}{4}(x + 4)$

47. $(y - 8)^2 = -4(x - 6)$ **49.** $(x - 3)^2 = \dfrac{1}{2}(y - 1)$

51.
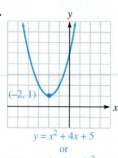
$y = x^2 + 4x + 5$
or
$y - 1 = (x + 2)^2$

53.
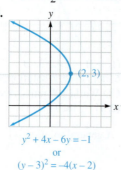
$y^2 + 4x - 6y = -1$
or
$(y - 3)^2 = -4(x - 2)$

55.
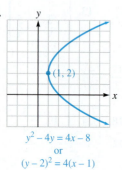
$y^2 - 4y = 4x - 8$
or
$(y - 2)^2 = 4(x - 1)$

57.
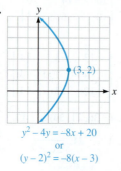
$y^2 - 4y = -8x + 20$
or
$(y - 2)^2 = -8(x - 3)$

59.
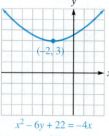
$x^2 - 6y + 22 = -4x$
or
$(x + 2)^2 = 6(y - 3)$

61.
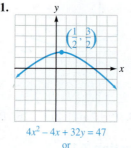
$4x^2 - 4x + 32y = 47$
or
$\left(x - \dfrac{1}{2}\right)^2 = -8\left(y - \dfrac{3}{2}\right)$

63.   **65.** yes   **67.** 60 mi

69. $(x - 4)^2 + y^2 = 16$ **71.** $(x - 7)^2 + y^2 = 9$

73. 2 ft   **75.** $x^2 = -\dfrac{45}{2}y$ **77.** 1 ft   **79.** 300 ft

81. about 12.6 cm   **83.** 520 ft

87. $0x^2 + 0xy + y^2 - 8x - 4y + 12 = 0$

89. $x^2 + (y - 3)^2 = 25$ **91.** $y = x^2 + 4x + 3$

95. b   **97.** c   **99.** b   **101.** d

Section 7.2

1. sum, constant   **3.** vertices   **5.** $a, 0; -a, 0$

7. 26-in. string; thumbtacks 24 in. apart   **9.** circle

11. parabola   **13.** ellipse   **15.** $\dfrac{x^2}{16} + \dfrac{y^2}{9} = 1$

17. $\dfrac{x^2}{25} + \dfrac{y^2}{16} = 1$ **19.** $\dfrac{9x^2}{16} + \dfrac{9y^2}{25} = 1$

21. $\dfrac{x^2}{7} + \dfrac{y^2}{16} = 1$ **23.** $\dfrac{(x - 3)^2}{4} + \dfrac{(y - 4)^2}{9} = 1$

25. $\dfrac{(x - 3)^2}{9} + \dfrac{(y - 4)^2}{4} = 1$

27. $\dfrac{(x - 3)^2}{41} + \dfrac{(y - 4)^2}{16} = 1$ **29.** $\dfrac{x^2}{36} + \dfrac{(y - 4)^2}{20} = 1$

31. $\dfrac{x^2}{100} + \dfrac{y^2}{64} = 1$

33.
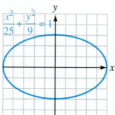
$\dfrac{x^2}{25} + \dfrac{y^2}{9} = 1$

35.
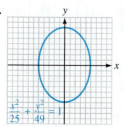
$\dfrac{x^2}{25} + \dfrac{y^2}{49} = 1$

37.
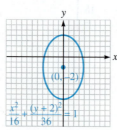
$\dfrac{x^2}{16} + \dfrac{(y + 2)^2}{36} = 1$

39.
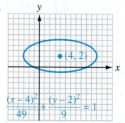
$\dfrac{(x - 4)^2}{49} + \dfrac{(y - 2)^2}{9} = 1$

41. $\dfrac{x^2}{4} + \dfrac{(y - 1)^2}{16} = 1$ **43.** $\dfrac{(x + 1)^2}{4} + \dfrac{(y + 2)^2}{9} = 1$

45.
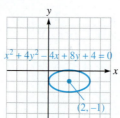
$x^2 + 4y^2 - 4x + 8y + 4 = 0$

47.
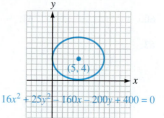
$16x^2 + 25y^2 - 160x - 200y + 400 = 0$

49.   **51.** $\dfrac{x^2}{900} + \dfrac{y^2}{400} = 1$ **53.** $\dfrac{x^2}{2500} + \dfrac{y^2}{900} = 1$; 36 m

55. about 20.8 in.   **57.** 199,395 mi   **67.** $\dfrac{x^2}{32} + \dfrac{y^2}{36} = 1$

71. d   **73.** a   **75.** b   **77.** c

Section 7.3

1. difference, constant   **3.** $a, 0; -a, 0$ **5.** transverse axis

7. circle   **9.** parabola   **11.** ellipse   **13.** hyperbola

15. $\dfrac{x^2}{25} - \dfrac{y^2}{24} = 1$ **17.** $\dfrac{(x - 2)^2}{4} - \dfrac{(y - 4)^2}{9} = 1$

19. $\dfrac{(y - 3)^2}{9} - \dfrac{(x - 5)^2}{9} = 1$ **21.** $\dfrac{y^2}{9} - \dfrac{x^2}{16} = 1$

23. $\dfrac{(x-1)^2}{4} - \dfrac{(y-4)^2}{32} = 1$ **25.** $\dfrac{x^2}{10} - \dfrac{3y^2}{20} = 1$

27. 24 sq. units   **29.** 12 sq. units

31. $\dfrac{(x+2)^2}{4} - \dfrac{4(y+4)^2}{81} = 1$ or $\dfrac{(y+4)^2}{4} - \dfrac{4(x+2)^2}{81} = 1$

33. $\dfrac{x^2}{36} - \dfrac{16y^2}{25} = 1$

35.
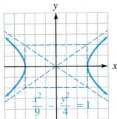

37.
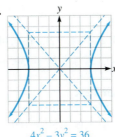
$4x^2 - 3y^2 = 36$

39.
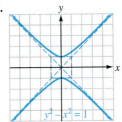
$y^2 - x^2 = 1$

41.
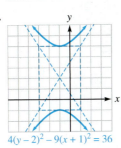
$\dfrac{(x+2)^2}{9} - \dfrac{y^2}{4} = 1$

43.
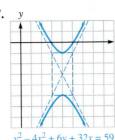
$4(y-2)^2 - 9(x+1)^2 = 36$

45.
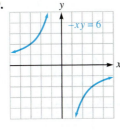
$4x^2 - 2y^2 + 8x - 8y = 8$

47.

$y^2 - 4x^2 + 6y + 32x = 59$

49.
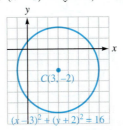
$-xy = 6$

51. $\dfrac{(x-3)^2}{9} - \dfrac{(y-1)^2}{16} = 1$ **53.** $\dfrac{(y+6)^2}{36} - \dfrac{x^2}{45} = 1$

55.   **57.** 24   **59.** 3 units

61. hyperbola; $\dfrac{x^2}{144} - \dfrac{y^2}{25} = 1$

63. hyperbola; $\dfrac{x^2}{36} - \dfrac{y^2}{64} = 1$

73. b   **75.** c   **77.** b   **79.** a

Section 7.4

1. graphs

3.
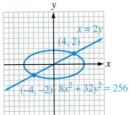
$x = 2y$
$(4, 2)$
$(-4, -2)$ $8x^2 + 32y^2 = 256$

5.
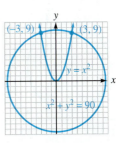
$(-3, 9)$ $(3, 9)$
$y = x^2$
$x^2 + y^2 = 90$

7.
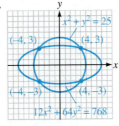
$x^2 + y^2 = 25$
$(-4, 3)$ $(4, 3)$
$(-4, -3)$ $(4, -3)$
$12x^2 + 64y^2 = 768$

9.
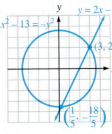
$y = 2x - 4$
$x^2 - 13 = -y^2$
$(3, 2)$
$\left(\dfrac{1}{5}, -\dfrac{18}{5}\right)$

11.
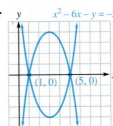
$x^2 - 6x - y = -5$
$(1, 0)$ $(5, 0)$
$x^2 - 6x + y = -5$

13. $(1, 2), (-1, 0)$
15. $(1, 0.67), (-1, -0.67)$
17. $(3, 0), (0, 5)$
19. $(1, 1)$
21. $(1, 2), (2, 1)$
23. $(-2, 3), (2, 3)$
25. $\left(\sqrt{5}, 5\right), \left(-\sqrt{5}, 5\right)$
27. $\left(\sqrt{3}, 0\right), \left(-\sqrt{3}, 0\right)$

29. $(2, 4), (2, -4), (-2, 4), (-2, -4)$
31. $\left(-\sqrt{15}, 5\right), \left(\sqrt{15}, 5\right), (-2, -6), (2, -6)$
33. $(0, -4), (-3, 5), (3, 5)$
35. $(-2, 3), (2, 3), (-2, -3), (2, -3)$
37. $(3, 3)$ **39.** $(6, 2), (-6, -2), \left(\sqrt{42}, 0\right), \left(-\sqrt{42}, 0\right)$
41. $\left(\dfrac{1}{2}, \dfrac{1}{3}\right), \left(\dfrac{1}{3}, \dfrac{1}{2}\right)$ **43.** 7 cm by 9 cm
45. 80 ft by 100 ft or 50 ft by 160 ft
47. either $750 at 9% or $900 at 7.5%
49. $(30, 3)$ **51.** yes there are potential collision points at $(-2, 4)$ and $(1, 1)$ **53.** about 23 mi   **59.** false The substitution and elimination methods are more precise.
61. false In general, the elimination method is best.   **63.** true
65. false You will obtain 4.

Chapter Review

1. $x^2 + y^2 = 16$ **2.** $x^2 + y^2 = 100$
3. $(x-3)^2 + (y+2)^2 = 25$ **4.** $(x+2)^2 + (y-4)^2 = 25$
5. $(x-5)^2 + (y-10)^2 = 85$ **6.** $(x-2)^2 + (y-2)^2 = 89$
7. $(x-3)^2 + (y+2)^2 = 16$

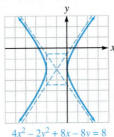
$C(3, -2)$
$(x-3)^2 + (y+2)^2 = 16$

8. $(x + 2)^2 + (y - 5)^2 = 16$ **9.** $y^2 = -2x$

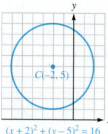

$(x + 2)^2 + (y - 5)^2 = 16$

10. $x^2 = 16y$ **11.** $(x + 2)^2 = -\dfrac{4}{11}(y - 3)$

12.   **13.**

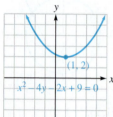

$x^2 - 4y - 2x + 9 = 0$

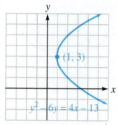

$y^2 - 6y = 4x - 13$

14. $\dfrac{x^2}{36} + \dfrac{y^2}{16} = 1$ **15.** $\dfrac{x^2}{4} + \dfrac{y^2}{25} = 1$

16. $\dfrac{(x + 2)^2}{16} + \dfrac{(y - 3)^2}{9} = 1$

17. $(x - 2)^2 + \dfrac{(y + 1)^2}{4} = 1$ **18.** $\dfrac{x^2}{4} - \dfrac{y^2}{12} = 1$

19. $\dfrac{y^2}{9} - \dfrac{x^2}{16} = 1$

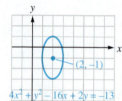

$4x^2 + y^2 - 16x + 2y = -13$

20. $\dfrac{x^2}{9} - \dfrac{(y - 3)^2}{16} = 1$

21. $\dfrac{y^2}{9} - \dfrac{(x - 3)^2}{16} = 1$ **22.** $y = \pm\dfrac{4}{5}x$

23. $\dfrac{(x - 1)^2}{4} - \dfrac{(y + 2)^2}{9} = 1$

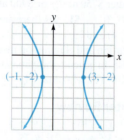

$9x^2 - 4y^2 - 16y - 18x = 43$

24.

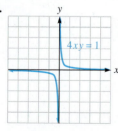

$4xy = 1$

25.

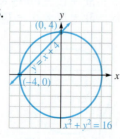

$(0, 4)$
$y = x$
$(-4, 0)$
$x^2 + y^2 = 16$

26.

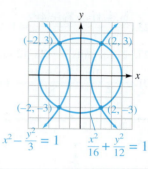

$(-4, 2)$ $(4, 2)$
$(-4, -2)$ $(4, -2)$
$x^2 - y^2 = 12$ $3x^2 + y^2 = 52$

27.

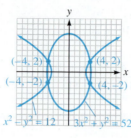

$(-2, 3)$ $(2, 3)$
$(-2, -3)$ $(2, -3)$
$x^2 - \dfrac{y^2}{3} = 1$ $\dfrac{x^2}{16} + \dfrac{y^2}{12} = 1$

28. $(-4, 2), (-4, -2), (4, 2), (4, -2)$

29. $(0, 4), \left(2\sqrt{3}, -2\right)$ **30.** $(-2, 3), (-2, -3), (2, 3), (2, -3)$

Chapter Test

1. $(x - 2)^2 + (y - 3)^2 = 9$ **2.** $(x - 2)^2 + (y - 3)^2 = 41$

3. $(x - 2)^2 + (y + 5)^2 = 169$

4.

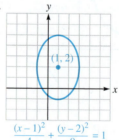

$(2, -3)$
$(x - 2)^2 + (y + 3)^2 = 9$

5. $(x - 3)^2 = 16(y - 2)$

6. $(y + 6)^2 = -4(x - 4)$

7. $(x - 2)^2 = \dfrac{4}{3}(y + 3)$ or $(y + 3)^2 = -\dfrac{9}{2}(x - 2)$

8.

$(3, -2)$
$(x - 3)^2 = 8(y + 2)$

9. $\dfrac{x^2}{100} + \dfrac{y^2}{64} = 1$

10. $\dfrac{x^2}{169} + \dfrac{y^2}{144} = 1$

11. $\dfrac{(x - 2)^2}{4} + \dfrac{(y - 3)^2}{36} = 1$

12.

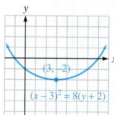

$(1, 2)$
$\dfrac{(x - 1)^2}{4} + \dfrac{(y - 2)^2}{9} = 1$

13. $\dfrac{x^2}{25} - \dfrac{y^2}{144} = 1$

14. $\dfrac{x^2}{36} - \dfrac{y^2}{\frac{25}{4}} = 1$

15. $\dfrac{(x - 2)^2}{64} - \dfrac{(y + 1)^2}{36} = 1$

16.

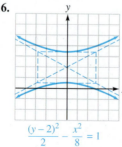

$\dfrac{(y - 2)^2}{2} - \dfrac{x^2}{8} = 1$

17. $\left(\sqrt{7}, 4\right), \left(-\sqrt{7}, 4\right)$

18. $\left(3\sqrt{2}, 3\right), \left(-3\sqrt{2}, 3\right),$
$\left(3\sqrt{2}, -3\right), \left(-3\sqrt{2}, -3\right)$

19. $(y - 2)^2 = 6(x + 3)$; parabola

20. $\dfrac{(x - 1)^2}{3} + \dfrac{(y + 2)^2}{2} = 1$; ellipse

Cumulative Review

1. 16 **2.** $\dfrac{1}{2}$ **3.** y^2 **4.** $x^{17/12}$ **5.** $x^{4/3} - x^{2/3}$

6. $\dfrac{1}{x} + 2 + x$ **7.** $-3x$ **8.** $4t\sqrt{3t}$ **9.** $4x$

10. $x + 3$ **11.** $7\sqrt{2}$ **12.** $-12\sqrt[4]{2} + 10\sqrt[4]{3}$

13. $-18\sqrt{6}$ **14.** $\dfrac{5\sqrt[3]{x^2}}{x}$ **15.** $\dfrac{x + 3\sqrt{x} + 2}{x - 1}$

16. $\sqrt{xy}$ **17.** 2, 7 **18.** $\dfrac{1}{4}$ **19.** $1, -\dfrac{3}{2}$

20. $\dfrac{-2 \pm \sqrt{7}}{3}$ **21.** $7 + 2i$ **22.** $-5 - 7i$

23. $13 + 0i$ **24.** $12 - 6i$ **25.** $-12 - 10i$

26. $\dfrac{3}{2} + \dfrac{1}{2}i$ **27.** $\sqrt{13}$ **28.** $\sqrt{61}$ **29.** -2

30. $\left(1, \dfrac{1}{2}\right)$ **31.** $(-\infty, -2) \cup (3, \infty)$ **32.** $[-2, 3]$

33. 5 **34.** 27 **35.** $12x^2 - 12x + 5$ **36.** $6x^2 + 3$

37. $2^y = x$ **38.** $\log_3 a = b$ **39.** 5 **40.** 3

41. $\dfrac{1}{27}$ **42.** 1 **43.** $y = 2^x$ **44.** x

45. 1.9912 **46.** 0.301 **47.** 1.6902 **48.** 0.1461

49. $\dfrac{2 \log 2}{\log 3 - \log 2}$ **50.** 16 **51.** \$2848.31

52. 1.16056 **53.** (2, 1) **54.** (1, 1) **55.** (2, −2)
56. (3, 1) **57.** −1 **58.** −1 **59.** (−1, −1, 3)
60. (1, 1, 1) **61.** (−3, 3)

62.
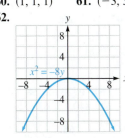

63.
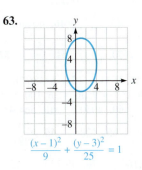
$\dfrac{(x-1)^2}{9} + \dfrac{(y-3)^2}{25} = 1$

64.
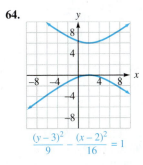
$\dfrac{(y-3)^2}{9} - \dfrac{(x-2)^2}{16} = 1$

Section 8.1

1. power **3.** first **5.** $7 \cdot 6 \cdot 5 \cdot 4 \cdot 3 \cdot 2 \cdot 1$
7. $(n - 1)!$ **9.** 120 **11.** 4320 **13.** 1440

15. $\dfrac{1}{1320}$ **17.** $\dfrac{5}{3}$ **19.** 18,564

21. $a^5 + 5a^4b + 10a^3b^2 + 10a^2b^3 + 5ab^4 + b^5$
23. $x^3 - 3x^2y + 3xy^2 - y^3$ **25.** $a^3 + 3a^2b + 3ab^2 + b^3$
27. $a^5 - 5a^4b + 10a^3b^2 - 10a^2b^3 + 5ab^4 - b^5$

29. $8x^3 + 12x^2y + 6xy^2 + y^3$
31. $x^3 - 6x^2y + 12xy^2 - 8y^3$
33. $16x^4 + 96x^3y + 216x^2y^2 + 216xy^3 + 81y^4$
35. $x^4 - 8x^3y + 24x^2y^2 - 32xy^3 + 16y^4$
37. $x^5 - 15x^4y + 90x^3y^2 - 270x^2y^3 + 405xy^4 - 243y^5$

39. $\dfrac{x^4}{16} + \dfrac{x^3y}{2} + \dfrac{3x^2y^2}{2} + 2xy^3 + y^4$ **41.** $6a^2b^2$

43. $35a^3b^4$ **45.** $-b^5$ **47.** $2380a^{13}b^4$ **49.** $-4\sqrt{2}a^3$

51. $1134a^5b^4$ **53.** $\dfrac{3x^2y^2}{2}$ **55.** $-\dfrac{55r^2s^9}{2048}$

57. $\dfrac{n!}{3!(n-3)!}a^{n-3}b^3$ **59.** $\dfrac{n!}{(r-1)!(n-r+1)!}a^{n-r+1}b^{r-1}$

71. false $0! = 1$ **73.** true
75. false The exponents on b increase by 1 in each successive term. **77.** false 445 **79.** false -252

Section 8.2

1. domain **3.** series **5.** infinite
7. Summation notation **9.** 6 **11.** $5c$
13. 0, 10, 30, 60, 100, 150 **15.** 21 **17.** $a + 4d$
19. 15 **21.** 8, 17, 26, 35, 44; 269
23. 4, 1, −4, −11, −20; −395 **25.** 7, 14, 33, 70, 131; 1006

27. $0, \dfrac{1}{2}, \dfrac{2}{3}, \dfrac{3}{4}, \dfrac{4}{5}; \dfrac{29}{30}$ **29.** $-\dfrac{1}{4}, \dfrac{1}{16}, -\dfrac{1}{64}, \dfrac{1}{256}, -\dfrac{1}{1024}; \dfrac{1}{65,536}$

31. $1, -\dfrac{1}{4}, \dfrac{1}{9}, -\dfrac{1}{16}, \dfrac{1}{25}; -\dfrac{1}{256}$ **33.** 15 **35.** 15

37. $\dfrac{242}{243}$ **39.** 35 **41.** 3, 7, 15, 31

43. $-4, -2, -1, -\dfrac{1}{2}$ **45.** k, k^2, k^4, k^8

47. $8, \dfrac{16}{k}, \dfrac{32}{k^2}, \dfrac{64}{k^3}$ **49.** an alternating infinite series

51. not an alternating infinite series **53.** 30 **55.** −50

57. 40 **59.** 500 **61.** $\dfrac{7}{12}$ **63.** 160 **65.** 3725

73. false −216 **75.** true **77.** false $\displaystyle\sum_{2}^{1000} 5 = 4995$

79. false $\displaystyle\sum_{k=1}^{999} 9k^{99} = 9\sum_{k=1}^{999} k^{99}$

Section 8.3

1. $(n - 1)$ **3.** infinite **5.** $a_n = a + (n - 1)d$
7. Arithmetic means **9.** 1, 3, 5, 7, 9, 11

11. $5, \dfrac{7}{2}, 2, \dfrac{1}{2}, -1, -\dfrac{5}{2}$ **13.** $9, \dfrac{23}{2}, 14, \dfrac{33}{2}, 19, \dfrac{43}{2}$

15. 318 **17.** 6 **19.** 370 **21.** 44 **23.** $\dfrac{25}{2}, 15, \dfrac{35}{2}$

25. $-\dfrac{82}{15}, -\dfrac{59}{15}, -\dfrac{12}{5}, -\dfrac{13}{15}$ **27.** 285 **29.** 555

31. $157\dfrac{1}{2}$ **33.** 20,100 **35.** 1080°; 1800° **37.** \$460
39. \$1,587,500 **41.** 80 ft **43.** 210 **51.** true
53. true **55.** true **57.** true
59. false There are 290 band members.

Section 8.4

1. r^{n-1} **3.** ar^{n-1} **5.** infinite **7.** Geometric means
9. 10, 20, 40, 80 **11.** −2, −6, −18, −54

13. $3, 3\sqrt{2}, 6, 6\sqrt{2}$ **15.** 2, 6, 18, 54 **17.** 256
19. -162 **21.** $10\sqrt[4]{2}, 10\sqrt[4]{4}$ or $10\sqrt{2}, 10\sqrt[4]{8}$
23. 8, 32, 128, 512 **25.** 124 **27.** $-29,524$
29. $\dfrac{1995}{32}$ **31.** 18 **33.** 8 **35.** $\dfrac{5}{9}$ **37.** $\dfrac{25}{99}$
39. 23 **41.** 12.96 ft, 400 ft **43.** 5.13 m
45. $69.82 **47.** $0.33C$ **49.** about 1.03×10^{20}
51. $180,176.87 **53.** $2001.60 **55.** $2013.62
57. $264,094.58 **59.** 5000 **61.** 1.8447×10^{19} grains
69. no **71.** true **73.** false The nth term is ar^{n-1}.
75. false The sum can be found if the common ratio r is between -1 and 1. That is, $|r| < 1$. **77.** true

Section 8.5

1. two **3.** $n = k + 1$
5. $5 = 5; 15 = 15; 30 = 30; 50 = 50$
7. $7 = 7; 17 = 17; 30 = 30; 46 = 46$
29. no **39. a.** 1 **b.** 3 **c.** 7 **d.** 15
41. false Mathematical induction is used to prove that statements are true for all natural numbers.
43. false We assume the statement is true for $n = k$.

Section 8.6

1. 24 **3.** $\dfrac{n!}{(n-r)!}$ **5.** 1 **7.** $\dfrac{n!}{r!(n-r)!}$ **9.** 1
11. $\dfrac{n!}{a!b!\cdots}$ **13.** 840 **15.** 35 **17.** 120 **19.** 5
21. 1 **23.** 1200 **25.** 40 **27.** 2278 **29.** 144
31. 8,000,000 **33.** 240 **35.** 6 **37.** 40,320
39. 14,400 **41.** 24,360 **43.** 5040 **45.** 48
47. 96 **49.** 210 **51.** 24 **53.** 5040 **55.** 60
57. 2,721,600 **59.** 1120 **61.** 59,400 **63.** 272
65. 28 **67.** 252 **69.** 142,506 **71.** 66 **73.** 256
75. 56
87. false Permutations involve situations in which order matters.
89. true **91.** true **93.** true

Section 8.7

1. experiment **3.** $\dfrac{n(E)}{n(S)}$
5. {(1, H), (2, H), (3, H), (4, H), (5, H), (6, H), (1, T), (2, T), (3, T), (4, T), (5, T), (6, T)} **7.** {a, b, c, d, e, f, g, h, i, j, k, l, m, n, o, p, q, r, s, t, u, v, w, x, y, z} **9.** $\dfrac{1}{6}$ **11.** $\dfrac{2}{3}$
13. $\dfrac{19}{42}$ **15.** $\dfrac{13}{42}$ **17.** $\dfrac{3}{8}$ **19.** 0 **21.** $\dfrac{1}{12}$
23. $\dfrac{1}{169}$ **25.** $\dfrac{5}{12}$ **27.** about 6.3×10^{-12} **29.** 0
31. $\dfrac{3}{13}$ **33.** $\dfrac{1}{6}$ **35.** $\dfrac{1}{8}$ **37.** $\dfrac{5}{16}$
39. {SSSS, SSSF, SSFS, SFSS, FSSS, SSFF, SFSF, FSSF, SFFS, FSFS, FFSS, SFFF, FSFF, FFSF, FFFS, FFFF}
41. $\dfrac{1}{4}$ **43.** $\dfrac{1}{4}$ **45.** 1 **47.** $\dfrac{32}{119}$ **49.** $\dfrac{1}{3}$
51. 0.18 **53.** 0.14 **55.** about 33% **61.** no

63. false Probabilities are expressed as numbers ranging from 0 to 1.
65. false The probability of an impossible event is 0.
67. true
69. false The probability of having four boys is
$\dfrac{1}{2} \cdot \dfrac{1}{2} \cdot \dfrac{1}{2} \cdot \dfrac{1}{2} = \dfrac{1}{16}$.

Chapter Review

1. 720 **2.** 30,240 **3.** 8 **4.** $\dfrac{280}{3}$
5. $x^3 + 3x^2y + 3xy^2 + y^3$
6. $p^4 + 4p^3q + 6p^2q^2 + 4pq^3 + q^4$
7. $a^5 - 5a^4b + 10a^3b^2 - 10a^2b^3 + 5ab^4 - b^5$
8. $8a^3 - 12a^2b + 6ab^2 - b^3$ **9.** $56a^5b^3$ **10.** $80x^3y^2$
11. $84x^3y^6$ **12.** $439,040x^3$ **13.** 63 **14.** 9
15. 0, 6, 16, 30, 48; $a_{10} = 198$ **16.** $-2, 8, 128, 32,768$
17. 90 **18.** 60 **19.** 1718 **20.** -360
21. 117 **22.** 281 **23.** -92 **24.** $-\dfrac{135}{2}$
25. $\dfrac{7}{2}, 5, \dfrac{13}{2}$ **26.** 25, 40, 55, 70, 85 **27.** 3320
28. 5780 **29.** -5220 **30.** -1540 **31.** $\dfrac{1}{729}$
32. 13,122 **33.** $\dfrac{9}{16,384}$ **34.** $\dfrac{8}{15,625}$ **35.** $2\sqrt{2}, 4, 4\sqrt{2}$
36. $4, -8, 16, -32$ **37.** 16 **38.** $\dfrac{3280}{27}$ **39.** 6560
40. $\dfrac{2295}{128}$ **41.** $\dfrac{520,832}{78,125}$ **42.** $\dfrac{3280}{3}$ **43.** $16\sqrt{2}$
44. $\dfrac{2}{3}$ **45.** $\dfrac{3}{25}$ **46.** no sum **47.** 1 **48.** $\dfrac{1}{3}$
49. 1 **50.** $\dfrac{17}{99}$ **51.** $\dfrac{5}{11}$ **52.** $4775.81
53. 6516; 3134 **54.** $3486.78
55. $1 = 1; 9 = 9; 36 = 36; 100 = 100$ **57.** 6720 **58.** 35
59. 1 **60.** 4050 **61.** 564,480 **62.** 840 **63.** 21
64. 66 **65.** 120 **66.** $\dfrac{56}{1287}$ **67.** $\dfrac{1}{6}$ **68.** $\dfrac{33}{66,640}$
69. 20,160 **70.** 90,720 **72.** 24 **73.** $\dfrac{1}{108,290}$
74. $\dfrac{108,289}{108,290}$ **75.** about 6.3×10^{-12} **76.** $\dfrac{60}{143}$ **77.** $\dfrac{1}{2}$
78. $\dfrac{7}{13}$ **79.** $\dfrac{1}{2,598,960}$ **80.** $\dfrac{33}{16,660}$

Chapter Test

1. 144 **2.** 384 **3.** $10x^4y$ **4.** $112a^2b^6$ **5.** 27
6. -36 **7.** 155 **8.** -130 **9.** 9, 14, 19
10. $-6, -18$ **11.** 255.75 **12.** about 9
13. about $0.42C **14.** about $1.46C **16.** 800,000
17. 42 **18.** 24 **19.** 28 **20.** 1 **21.** 576
22. 120 **23.** 60 **24.** {(H, H, H), (H, H, T), (H, T, H), (H, T, T), (T, H, H), (T, H, T), (T, T, H), (T, T, T)}
25. $\dfrac{1}{6}$ **26.** $\dfrac{2}{13}$ **27.** $\dfrac{33}{66,640}$ **28.** $\dfrac{1}{9}$ **29.** $\dfrac{87}{245}$
30. $\dfrac{12}{19}$

Index

Properties of Logarithms

If b is a positive number and $b \neq 1$,

1. $\log_b 1 = 0$
2. $\log_b b = 1$
3. $\log_b b^x = x$
4. $b^{\log_b x} = x$
5. **Product Rule:**
   $\log_b MN = \log_b M + \log_b N$
6. **Quotient Rule:**
   $\log_b \dfrac{M}{N} = \log_b M - \log_b N$
7. **Power Rule:**
   $\log_b M^p = p \log_b M$
8. **One-to-One Property:**

If $\log_b x = \log_b y$, then $x = y$.

Graphs of $f(x) = e^x$ and $f(x) = \ln x$

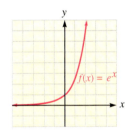 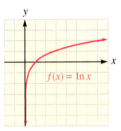

Natural Logarithm Properties

1. $\ln 1 = 0$
2. $\ln e = 1$
3. $\ln e^x = x$
4. $e^{\ln x} = x$

Theorems

Remainder Theorem:
If $P(x)$ is a polynomial, r is any number, and $P(x)$ is divided by $x - r$, the remainder is $P(r)$.

Factor Theorem:
If $P(x)$ is a polynomial and r is any number, then

If $P(r) = 0$, then $x - r$ is a factor of $P(x)$.

If $x - r$ is a factor of $P(x)$, then $P(r) = 0$.

Binomial Theorem:
If n is any positive integer, then

$$(a + b)^n = a^n + \frac{n!}{1!(n - 1)!}a^{n-1}b$$

$$+ \frac{n!}{2!(n - 2)!}a^{n-2}b^2 + \frac{n!}{3!(n - 3)!}a^{n-3}b^3$$

$$+ \frac{n!}{r!(n - r)!}a^{n-r}b^r + \cdots + b^n$$

Parabolas

A **parabola** is the set of all points in a plane equidistant from a line l (called the **directrix**) and fixed point F (called the **focus**) that is not on line l.

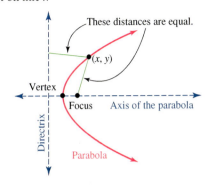

Parabola opening	Vertex at origin
Right	$y^2 = 4px$ $(p > 0)$
Left	$y^2 = 4px$ $(p < 0)$
Upward	$x^2 = 4py$ $(p > 0)$
Downward	$x^2 = 4py$ $(p < 0)$

- For a parabola that opens right or left with vertex at the origin, the directrix is $x = -p$ and the focus is $(p, 0)$.
- For a parabola that opens upward or downward with vertex at the origin, the directrix is $y = -p$ and the focus is $(0, p)$.

Parabola opening	Vertex at $V(h, k)$
Right	$(y - k)^2 = 4p(x - h)$ $(p > 0)$
Left	$(y - k)^2 = 4p(x - h)$ $(p < 0)$
Upward	$(x - h)^2 = 4p(y - k)$ $(p > 0)$
Downward	$(x - h)^2 = 4p(y - k)$ $(p < 0)$

- For a parabola that opens right or left with vertex at (h, k), the directrix is $x = -p + h$ and the focus is $(h + p, k)$.
- For a parabola that opens upward or downward with vertex at (h, k), the directrix is $y = -p + k$ and the focus is $(h, k + p)$.

Ellipses

An **ellipse** is the set of all points P in a plane such that the sum of the distances from P to two other fixed points F and F' is a positive constant.

The standard equations of an ellipse with center (h, k) and major axis horizontal is

$$\frac{(x - h)^2}{a^2} + \frac{(y - k)^2}{b^2} = 1,$$

where $a > b > 0$.

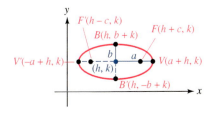

The standard equations of an ellipse with center (h, k) and major axis vertical is

$$\frac{(x - h)^2}{b^2} + \frac{(y - k)^2}{a^2} = 1,$$

where $a > b > 0$.

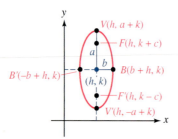

Hyberbolas

A **hyperbola** is the set of all points in a plane such that the absolute value of the difference of the distances from point P to two other points in the plane is a positive constant.

The standard equation of a hyperbola with center at (h, k) and foci on a line parallel to the x-axis is

$$\frac{(x - h)^2}{a^2} - \frac{(y - k)^2}{b^2} = 1,$$

where $a^2 + b^2 = c^2$.

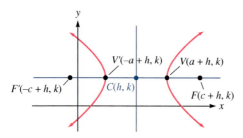

The standard equation of a hyperbola with center at (h, k) and foci on a line parallel to the y-axis is

$$\frac{(y - h)^2}{a^2} - \frac{(x - k)^2}{b^2} = 1,$$

where $a^2 + b^2 = c^2$.

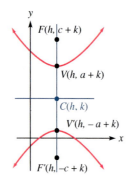

Permutations and Combinations

$$P(n, r) = \frac{n!}{(n - r)!}; \; P(n, n) = n!; \; P(n, 0) = 1$$

$$C(n, r) = \frac{n!}{r!(n - r)!}; \; C(n, n) = 1; \; C(n, 0) = 1$$

Geometry Formulas

Rectangle: $A = l \cdot w; \; P = 2l + 2w$

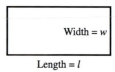

Width = w

Length = l

Triangle: $A = \frac{1}{2}b \cdot h$

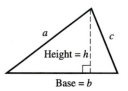

Height = h

Base = b

Circle: $C = 2\pi r; \; A = \pi r^2$

Rectangular solid: $V = l \cdot w \cdot h$

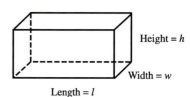

Height = h

Width = w

Length = l

Sphere: $V = \frac{4}{3}\pi r^3$

Right circular cone: $V = \frac{1}{3}\pi r^2 h$

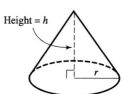

Height = h

Pythagorean Theorem

$$a^2 + b^2 = c^2$$